KIRK-OTHMER

ENCYCLOPEDIA OF CHEMICAL TECHNOLOGY

Third Edition

SUPPLEMENT VOLUME

Alcohol Fuels
to
Toxicology

KIRK-OTHMER

ENCYCLOPEDIA OF CHEMICAL TECHNOLOGY

THIRD EDITION

SUPPLEMENT VOLUME

ALCOHOL FUELS
TO
TOXICOLOGY

A WILEY-INTERSCIENCE PUBLICATION

John Wiley & Sons

NEW YORK • CHICHESTER • BRISBANE • TORONTO • SINGAPORE

Library of Congress Cataloging in Publication Data:

Main entry under title:
Encyclopedia of chemical technology.

At head of title: Kirk-Othmer.
"A Wiley-Interscience publication."
Includes bibliographies.
1. Chemistry, Technical—Dictionaries. I. Kirk, Raymond
Eller, 1890–1957. II. Othmer, Donald Frederick, 1904–
III. Grayson, Martin. IV. Eckroth, David. V. Title:
Kirk-Othmer encyclopedia of chemical technology.

TP9.E685 1978 660'.03 77-15820
ISBN 0-471-89214-9

Printed in the United States of America
10 9 8 7 6 5 4 3 2

CONTENTS

EDITORIAL STAFF FOR SUPPLEMENT VOLUME

Executive Editor: **Martin Grayson**
Editor: **David Eckroth**
Production Supervisor: **Michalina Bickford**
Editors: **Joyce Brown** **Caroline L. Eastman** **Carolyn Golojuch** **Anna Klingsberg** **Mimi Wainwright**

CONTRIBUTORS TO SUPPLEMENT VOLUME

E. J. Anderson, *Central States Consulting, Louisville, Kentucky,* Conveying
Bryan Ballantyne, *Union Carbide Corporation, South Charleston, West Virginia,* Toxicology
Alan Bleier, *Massachusetts Institute of Technology, Cambridge, Massachusetts,* Colloids
Allan Brown, *Beecham Research Laboratories, Betchworth, Surrey, UK,* Antibiotics, β-lactams, clavulanic acid, thienamycin, and others
Thomas H. Burgess, *Fischer-Porter Company, Warminster, Pennsylvania,* Flow measurement
D. A. Carlson, *University of Florida, Gainesville, Florida,* Repellents
Peter Cervoni, *Lederle Laboratories, American Cyanamid Company, Pearl River, New York,* Cardiovascular agents
Christopher M. Cimarusti, *Squibb Institute for Medical Research, Princeton, New Jersey,* Antibiotics, β-lactams, monobactams
David A. Clark, *Pfizer Inc., Groton, Connecticut,* Opiods, endogenous

G. Collin, *Rütgerswerke AG, Duisberg, FRG,* Coal chemicals and feedstocks, carbonization and coking

B. Cornils, *Ruhrchemie AG, Oberhausen, FRG,* Coal chemicals and feedstocks, gasification

John Davis, *NASA, Hampton, Virginia,* Composites, high performance

J. Elks, *Glaxo Pharmaceuticals Ltd., Glaxo Group Research Ltd., Greenford, Middlesex, UK,* Antibiotics, β-lactams, cephalosporins and penems

J. Falbe, *Ruhrchemie AG, Oberhausen, FRG,* Coal chemicals and feedstocks, gasification

C. D. Frohning, *Ruhrchemie AG, Oberhausen, FRG,* Coal chemicals and feedstocks, gasification

Robert Geyer, *Harvard University, Cambridge, Massachusetts,* Blood-replacement preparations

Edward Hohmann, *California State Polytechnic Institute, Pomona, California,* Heat exchange technology, network synthesis

Alfons Jankowski, *Ruhrkohle Oel und Gas GmbH, Bottrop, FRG,* Coal chemicals and feedstocks from hydrogenation

Ernest Jaworski *Monsanto Company, St. Louis, Missouri,* Genetic engineering

M. Ross Johnson, *Pfizer Inc., Groton, Connecticut,* Opiods, endogenous

Keith Johnston, *University of Texas, Austin, Austin, Texas,* Supercritical fluids

Donald L. Klass, *Institute of Gas Technology, Chicago, Illinois,* Alcohol fuels

Georg Kölling, *Bergbau-Forschung GmbH, Essen, FRG,* Coal chemicals and feedstocks, introduction and from hydrogenation

Fong M. Lai, *Lederle Laboratories, American Cyanamid Company, Pearl River, New York,* Cardiovascular agents

Josef Langhoff, *Ruhrkohle Oel und Gas GmbH, Bottrop, FRG,* Coal chemicals and feedstocks from hydrogenation

G. Löhnert, *Ruetgers-Nease Chemical Company, State College, Pennsylvania,* Coal chemicals and feedstocks, carbonization and coking

Peter Luckie, *The Pennsylvania State University, University Park, Pennsylvania,* Size separation

Robert D. Lundberg, *Exxon Research and Engineering Company, Linden, New Jersey,* Ionomers

Tom Mix, *Merix Corporation, Babson Park, Massachusetts,* Separations, low energy

Booker Morey, *Telic Technical Services, Bedford, Massachusetts,* Dewatering

Douglas R. Morton, Jr., *The Upjohn Company, Kalamazoo, Michigan,* Prostaglandins

D. J. Nagel, *Naval Research Laboratory, Washington, D.C.,* Plasma technology

Richard Neerken, *The Ralph M. Parsons Company, Pasadena, California,* Pumps

Cynthia O'Callaghan, *Glaxo Pharmaceuticals Ltd., Glaxo Group Research Ltd., Greenford, Middlesex, UK,* Antibiotics, β-lactams, cephalosporins and penems

G. D. Parfitt, *Carnegie-Mellon University, Pittsburgh, Pennsylvania,* Dispersion of powders in liquids

Paul Patterson, *Cyanamid Canada Inc., Toronto, Canada,* Guanidine and guanidine salts

K. Porter, *ICI Fibres, Harrogate, North Yorkshire, UK,* Fibers, multicomponent

W. Portz, *Hoechst AG, Knapsack, FRG,* Coal chemicals and feedstocks, carbide production

James E. Potts, *Union Carbide Corporation, Bound Brook, New York,* Plastics, environmentally degradable

George Prokopakis, *Columbia University, New York, New York,* Azeotropic and extractive distillation

Curt W. Reimann, *National Bureau of Standards, Washington, D.C.,* Analytical methods
E. Eugene Stauffer, *The Ansul Company, Marinette, Wisconsin,* Fire-extinguishing agents
D. Steinmeyer, *Monsanto Inc., St. Louis, Missouri,* Process energy conservation
Richard B. Sykes, *Squibb Institute for Medical Research, Princeton, New Jersey,* Antibiotics, β-lactams, monobactams
David Tiemeier, *Monsanto Company, St. Louis, Missouri,* Genetic engineering
Aaron Twerski, *Hofstra Law School, Hempstead, New York,* Product liability
L. van Zelst, *Museum of Fine Arts, Boston, Massachusetts,* Fine art examination and conservation
Rance A. Velapoldi, *National Bureau of Standards, Washington, D.C.,* Analytical methods
Masaaki Yamabe, *Asahi Glass Company, Ltd., Yokohama, Japan,* Perfluorinated ionomer membranes

PREFACE

The third edition of the *Encyclopedia of Chemical Technology* spans the alphabet from Abherents to Zone refining with 1293 articles in the period 1978–1983. The purposes of the Supplement volume are to update some of the articles that appeared in volumes 1 to 24, add some topics on recent developments, add a few articles that the editors forgot to include, and broaden the horizons of chemical technology. The primary purpose of the Supplement volume is to describe subjects that in recent years have undergone changes of a fundamental character rather than to make minor additions to all of the articles that have appeared in the 24 volumes. The 39 articles in the Supplement volume are in alphabetical order and contain cross references to the original articles as well as to those in this volume. The full cumulative index to be published in a few months will cover all of the articles in the Encyclopedia including those in this Supplement volume.

NOTE ON CHEMICAL ABSTRACTS SERVICE REGISTRY NUMBERS AND NOMENCLATURE

Chemical Abstracts Service (CAS) Registry Numbers are unique numerical identifiers assigned to substances recorded in the CAS Registry System. They appear in brackets in the *Chemical Abstracts* (CA) substance and formula indexes following the names of compounds. A single compound may have many synonyms in the chemical literature. A simple compound like phenethylamine can be named β-phenylethylamine or, as in *Chemical Abstracts*, benzeneethanamine. The usefulness of the *Encyclopedia* depends on accessibility through the most common correct name of a substance. Because of this diversity in nomenclature careful attention has been given the problem in order to assist the reader as much as possible, especially in locating the systematic CA index name by means of the Registry Number. For this purpose, the reader may refer to the CAS Registry Handbook-Number Section which lists in numerical order the Registry Number with the *Chemical Abstracts* index name and the molecular formula; eg, **458-88-8,** Piperidine, 2-propyl-, (*S*)-, $C_8H_{17}N$; in the *Encyclopedia* this compound would be found under its common name, coniine [*458-88-8*]. The Registry Number is a valuable link for the reader in retrieving additional published information on substances and also as a point of access for such on-line data bases as Chemline, Medline, and Toxline.

In all cases, the CAS Registry Numbers have been given for title compounds in articles and for all compounds in the index. All specific substances indexed in *Chemical Abstracts* since 1965 are included in the CAS Registry System as are a large number of substances derived from a variety of reference works. The CAS Registry System identifies a substance on the basis of an unambiguous computer-language description of its molecular structure including stereochemical detail. The Registry Number is a machine-checkable number (like a Social Security number) assigned in sequential order to each substance as it enters the registry system. The value of the number lies in the fact that it is a concise and unique means of substance identification, which is independent of, and therefore bridges, many systems of chemical nomenclature. For polymers, one Registry Number is used for the entire family; eg, polyoxyethylene (20) sorbitan monolaurate has the same number as all of its polyoxyethylene homologues.

Registry numbers for each substance are provided in the third edition cumulative index and appear as well in the annual indexes (eg, Alkaloids shows the Registry Number of all alkaloids (title compounds) in a table in the article as well, but the intermediates have their Registry Numbers shown only in the index). Articles such as Analytical methods, Batteries and electric cells, Chemurgy, Distillation, Economic evaluation, and Fluid mechanics have no Registry Numbers in the text.

Cross-references are inserted in the index for many common names and for some systematic names. Trademark names appear in the index. Names that are incorrect, misleading or ambiguous are avoided. Formulas are given very frequently in the text to help in identifying compounds. The spelling and form used, even for industrial names, follow American chemical usage, but not always the usage of *Chemical Abstracts* (eg, *coniine* is used instead of *(S)-2-propylpiperidine*, *aniline* instead of *benzenamine*, and *acrylic acid* instead of *2-propenoic acid*).

There are variations in representation of rings in different disciplines. The dye industry does not designate aromaticity or double bonds in rings. All double bonds and aromaticity are shown in the *Encyclopedia* as a matter of course. For example, tetralin has an aromatic ring and a saturated ring and its structure appears in the

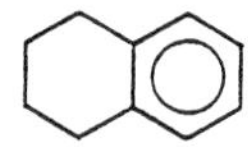

Encyclopedia with its common name, Registry Number enclosed in brackets, and parenthetical CA index name, ie, tetralin, [*119-64-2*] (1,2,3,4-tetrahydronaphthalene). With names and structural formulas, and especially with CAS Registry Numbers the aim is to help the reader have a concise means of substance identification.

CONVERSION FACTORS, ABBREVIATIONS, AND UNIT SYMBOLS

SI Units (Adopted 1960)

A new system of measurement, the International System of Units (abbreviated SI), is being implemented throughout the world. This system is a modernized version of the MKSA (meter, kilogram, second, ampere) system, and its details are published and controlled by an international treaty organization (The International Bureau of Weights and Measures) (1).

SI units are divided into three classes:

BASE UNITS

length	meter† (m)
mass‡	kilogram (kg)
time	second (s)
electric current	ampere (A)
thermodynamic temperature§	kelvin (K)
amount of substance	mole (mol)
luminous intensity	candela (cd)

† The spellings "metre" and "litre" are preferred by ASTM; however "-er" are used in the Encyclopedia.

‡ "Weight" is the commonly used term for "mass."

§ Wide use is made of "Celsius temperature" (t) defined by

$$t = T - T_0$$

where T is the thermodynamic temperature, expressed in kelvins, and $T_0 = 273.15$ K by definition. A temperature interval may be expressed in degrees Celsius as well as in kelvins.

SUPPLEMENTARY UNITS

plane angle	radian (rad)
solid angle	steradian (sr)

DERIVED UNITS AND OTHER ACCEPTABLE UNITS

These units are formed by combining base units, supplementary units, and other derived units (2–4). Those derived units having special names and symbols are marked with an asterisk in the list below:

Quantity	*Unit*	*Symbol*	*Acceptable equivalent*
*absorbed dose	gray	Gy	J/kg
acceleration	meter per second squared	m/s^2	
*activity (of ionizing radiation source)	becquerel	Bq	1/s
area	square kilometer	km^2	
	square hectometer	hm^2	ha (hectare)
	square meter	m^2	
*capacitance	farad	F	C/V
concentration (of amount of substance)	mole per cubic meter	mol/m^3	
*conductance	siemens	S	A/V
current density	ampere per square meter	A/m^2	
density, mass density	kilogram per cubic meter	kg/m^3	g/L; mg/cm^3
dipole moment (quantity)	coulomb meter	C·m	
*electric charge, quantity of electricity	coulomb	C	A·s
electric charge density	coulomb per cubic meter	C/m^3	
electric field strength	volt per meter	V/m	
electric flux density	coulomb per square meter	C/m^2	
*electric potential, potential difference, electromotive force	volt	V	W/A
*electric resistance	ohm	Ω	V/A
*energy, work, quantity of heat	megajoule	MJ	
	kilojoule	kJ	
	joule	J	N·m
	electron volt†	eV†	
	kilowatt hour†	kW·h†	
energy density	joule per cubic meter	J/m^3	

† This non-SI unit is recognized by the CIPM as having to be retained because of practical importance or use in specialized fields (1).

Quantity	Unit	Symbol	Acceptable equivalent
*force	kilonewton	kN	
	newton	N	kg·m/s^2
*frequency	megahertz	MHz	
	hertz	Hz	1/s
heat capacity, entropy	joule per kelvin	J/K	
heat capacity (specific), specific entropy	joule per kilogram kelvin	J/(kg·K)	
heat transfer coefficient	watt per square meter kelvin	W/(m^2·K)	
*illuminance	lux	lx	lm/m^2
*inductance	henry	H	Wb/A
linear density	kilogram per meter	kg/m	
luminance	candela per square meter	cd/m^2	
*luminous flux	lumen	lm	cd·sr
magnetic field strength	ampere per meter	A/m	
*magnetic flux	weber	Wb	V·s
*magnetic flux density	tesla	T	Wb/m^2
molar energy	joule per mole	J/mol	
molar entropy, molar heat capacity	joule per mole kelvin	J/(mol·K)	
moment of force, torque	newton meter	N·m	
momentum	kilogram meter per second	kg·m/s	
permeability	henry per meter	H/m	
permittivity	farad per meter	F/m	
*power, heat flow rate, radiant flux	kilowatt	kW	
	watt	W	J/s
power density, heat flux density, irradiance	watt per square meter	W/m^2	
*pressure, stress	megapascal	MPa	
	kilopascal	kPa	
	pascal	Pa	N/m^2
sound level	decibel	dB	
specific energy	joule per kilogram	J/kg	
specific volume	cubic meter per kilogram	m^3/kg	
surface tension	newton per meter	N/m	
thermal conductivity	watt per meter kelvin	W/(m·K)	
velocity	meter per second	m/s	
	kilometer per hour	km/h	
viscosity, dynamic	pascal second	Pa·s	
	millipascal second	mPa·s	
viscosity, kinematic	square meter per second	m^2/s	
	square millimeter per second	mm^2/s	

Quantity	*Unit*	*Symbol*	*Acceptable equivalent*
volume	cubic meter	m^3	
	cubic decimeter	dm^3	L(liter) (5)
	cubic centimeter	cm^3	mL
wave number	1 per meter	m^{-1}	
	1 per centimeter	cm^{-1}	

In addition, there are 16 prefixes used to indicate order of magnitude, as follows:

Multiplication factor	*Prefix*	*Symbol*	*Note*
10^{18}	exa	E	
10^{15}	peta	P	
10^{12}	tera	T	
10^{9}	giga	G	
10^{6}	mega	M	
10^{3}	kilo	k	
10^{2}	hecto	h[a]	[a] Although hecto, deka, deci, and centi are SI prefixes, their use should be avoided except for SI unit-multiples for area and volume and nontechnical use of centimeter, as for body and clothing measurement.
10	deka	da[a]	
10^{-1}	deci	d[a]	
10^{-2}	centi	c[a]	
10^{-3}	milli	m	
10^{-6}	micro	μ	
10^{-9}	nano	n	
10^{-12}	pico	p	
10^{-15}	femto	f	
10^{-18}	atto	a	

For a complete description of SI and its use the reader is referred to ASTM E 380 (4) and the article Units and Conversion Factors in Vol. 23.

A representative list of conversion factors from non-SI to SI units is presented herewith. Factors are given to four significant figures. Exact relationships are followed by a dagger. A more complete list is given in ASTM E 380-79(4) and ANSI Z210.1-1976 (6).

Conversion Factors to SI Units

To convert from	*To*	*Multiply by*
acre	square meter (m^2)	4.047×10^3
angstrom	meter (m)	1.0×10^{-10}†
are	square meter (m^2)	1.0×10^{2}†
astronomical unit	meter (m)	1.496×10^{11}
atmosphere	pascal (Pa)	1.013×10^{5}
bar	pascal (Pa)	1.0×10^{5}†
barn	square meter (m^2)	1.0×10^{-28}†
barrel (42 U.S. liquid gallons)	cubic meter (m^3)	0.1590

† Exact.

To convert from	*To*	*Multiply by*
Bohr magneton (μ_β)	J/T	9.274×10^{-24}
Btu (International Table)	joule (J)	1.055×10^{3}
Btu (mean)	joule (J)	1.056×10^{3}
Btu (thermochemical)	joule (J)	1.054×10^{3}
bushel	cubic meter (m^3)	3.524×10^{-2}
calorie (International Table)	joule (J)	4.187
calorie (mean)	joule (J)	4.190
calorie (thermochemical)	joule (J)	4.184†
centipoise	pascal second (Pa·s)	1.0×10^{-3}†
centistokes	square millimeter per second (mm^2/s)	1.0†
cfm (cubic foot per minute)	cubic meter per second (m^3/s)	4.72×10^{-4}
cubic inch	cubic meter (m^3)	1.639×10^{-5}
cubic foot	cubic meter (m^3)	2.832×10^{-2}
cubic yard	cubic meter (m^3)	0.7646
curie	becquerel (Bq)	3.70×10^{10}†
debye	coulomb·meter (C·m)	3.336×10^{-30}
degree (angle)	radian (rad)	1.745×10^{-2}
denier (international)	kilogram per meter (kg/m)	1.111×10^{-7}
	tex‡	0.1111
dram (apothecaries')	kilogram (kg)	3.888×10^{-3}
dram (avoirdupois)	kilogram (kg)	1.772×10^{-3}
dram (U.S. fluid)	cubic meter (m^3)	3.697×10^{-6}
dyne	newton (N)	1.0×10^{-5}†
dyne/cm	newton per meter (N/m)	1.0×10^{-3}†
electron volt	joule (J)	1.602×10^{-19}
erg	joule (J)	1.0×10^{-7}†
fathom	meter (m)	1.829
fluid ounce (U.S.)	cubic meter (m^3)	2.957×10^{-5}
foot	meter (m)	0.3048†
footcandle	lux (lx)	10.76
furlong	meter (m)	2.012×10^{-2}
gal	meter per second squared (m/s^2)	1.0×10^{-2}†
gallon (U.S. dry)	cubic meter (m^3)	4.405×10^{-3}
gallon (U.S. liquid)	cubic meter (m^3)	3.785×10^{-3}
gallon per minute (gpm)	cubic meter per second (m^3/s)	6.308×10^{-5}
	cubic meter per hour (m^3/h)	0.2271
gauss	tesla (T)	1.0×10^{-4}
gilbert	ampere (A)	0.7958
gill (U.S.)	cubic meter (m^3)	1.183×10^{-4}
grad	radian	1.571×10^{-2}
grain	kilogram (kg)	6.480×10^{-5}
gram-force per denier	newton per tex (N/tex)	8.826×10^{-2}
hectare	square meter (m^2)	1.0×10^{4}†

† Exact.

‡ See footnote on p. xvi.

To convert from	*To*	*Multiply by*
horsepower (550 ft·lbf/s)	watt (W)	7.457×10^{2}
horsepower (boiler)	watt (W)	9.810×10^{3}
horsepower (electric)	watt (W)	$7.46 \times 10^{2\dagger}$
hundredweight (long)	kilogram (kg)	50.80
hundredweight (short)	kilogram (kg)	45.36
inch	meter (m)	$2.54 \times 10^{-2\dagger}$
inch of mercury (32°F)	pascal (Pa)	3.386×10^{3}
inch of water (39.2°F)	pascal (Pa)	2.491×10^{2}
kilogram-force	newton (N)	9.807
kilowatt hour	megajoule (MJ)	$3.6^{\dagger}$
kip	newton (N)	4.48×10^{3}
knot (international)	meter per second (m/s)	0.5144
lambert	candela per square meter (cd/m^2)	3.183×10^{3}
league (British nautical)	meter (m)	5.559×10^{3}
league (statute)	meter (m)	4.828×10^{3}
light year	meter (m)	9.461×10^{15}
liter (for fluids only)	cubic meter (m^3)	$1.0 \times 10^{-3\dagger}$
maxwell	weber (Wb)	$1.0 \times 10^{-8\dagger}$
micron	meter (m)	$1.0 \times 10^{-6\dagger}$
mil	meter (m)	$2.54 \times 10^{-5\dagger}$
mile (statute)	meter (m)	1.609×10^{3}
mile (U.S. nautical)	meter (m)	$1.852 \times 10^{3\dagger}$
mile per hour	meter per second (m/s)	0.4470
millibar	pascal (Pa)	1.0×10^{2}
millimeter of mercury (0°C)	pascal (Pa)	$1.333 \times 10^{2\dagger}$
minute (angular)	radian	2.909×10^{-4}
myriagram	kilogram (kg)	10
myriameter	kilometer (km)	10
oersted	ampere per meter (A/m)	79.58
ounce (avoirdupois)	kilogram (kg)	2.835×10^{-2}
ounce (troy)	kilogram (kg)	3.110×10^{-2}
ounce (U.S. fluid)	cubic meter (m^3)	2.957×10^{-5}
ounce-force	newton (N)	0.2780
peck (U.S.)	cubic meter (m^3)	8.810×10^{-3}
pennyweight	kilogram (kg)	1.555×10^{-3}
pint (U.S. dry)	cubic meter (m^3)	5.506×10^{-4}
pint (U.S. liquid)	cubic meter (m^3)	4.732×10^{-4}
poise (absolute viscosity)	pascal second (Pa·s)	$0.10^{\dagger}$
pound (avoirdupois)	kilogram (kg)	0.4536
pound (troy)	kilogram (kg)	0.3732
poundal	newton (N)	0.1383
pound-force	newton (N)	4.448
pound force per square inch (psi)	pascal (Pa)	6.895×10^{3}
quart (U.S. dry)	cubic meter (m^3)	1.101×10^{-3}
quart (U.S. liquid)	cubic meter (m^3)	9.464×10^{-4}
quintal	kilogram (kg)	$1.0 \times 10^{2\dagger}$

† Exact.

To convert from	*To*	*Multiply by*
rad	gray (Gy)	$1.0 \times 10^{-2\dagger}$
rod	meter (m)	5.029
roentgen	coulomb per kilogram (C/kg)	2.58×10^{-4}
second (angle)	radian (rad)	4.848×10^{-6}
section	square meter (m^2)	2.590×10^{6}
slug	kilogram (kg)	14.59
spherical candle power	lumen (lm)	12.57
square inch	square meter (m^2)	6.452×10^{-4}
square foot	square meter (m^2)	9.290×10^{-2}
square mile	square meter (m^2)	2.590×10^{6}
square yard	square meter (m^2)	0.8361
stere	cubic meter (m^3)	$1.0^{\dagger}$
stokes (kinematic viscosity)	square meter per second (m^2/s)	$1.0 \times 10^{-4\dagger}$
tex	kilogram per meter (kg/m)	$1.0 \times 10^{-6\dagger}$
ton (long, 2240 pounds)	kilogram (kg)	1.016×10^{3}
ton (metric)	kilogram (kg)	$1.0 \times 10^{3\dagger}$
ton (short, 2000 pounds)	kilogram (kg)	9.072×10^{2}
torr	pascal (Pa)	1.333×10^{2}
unit pole	weber (Wb)	1.257×10^{-7}
yard	meter (m)	$0.9144^{\dagger}$

Abbreviations and Unit Symbols

Following is a list of commonly used abbreviations and unit symbols appropriate for use in the *Encyclopedia.* In general they agree with those listed in *American National Standard Abbreviations for Use on Drawings and in Text* (*ANSI Y1.1*) (6) and *American National Standard Letter Symbols for Units in Science and Technology* (*ANSI Y10*) (6). Also included is a list of acronyms for a number of private and government organizations as well as common industrial solvents, polymers, and other chemicals.

Rules for Writing Unit Symbols (4):

1. Unit symbols should be printed in upright letters (roman) regardless of the type style used in the surrounding text.
2. Unit symbols are unaltered in the plural.
3. Unit symbols are not followed by a period except when used as the end of a sentence.
4. Letter unit symbols are generally written in lower-case (eg, cd for candela) unless the unit name has been derived from a proper name, in which case the first letter of the symbol is capitalized (W,Pa). Prefix and unit symbols retain their prescribed form regardless of the surrounding typography.
5. In the complete expression for a quantity, a space should be left between the numerical value and the unit symbol. For example, write 2.37 lm, *not* 2.37lm, and 35 mm, *not* 35mm. When the quantity is used in an adjectival sense, a hyphen is often used, for example, 35-mm film. *Exception:* No space is left between the numerical value and the symbols for degree, minute, and second of plane angle, and degree Celsius.
6. No space is used between the prefix and unit symbols (eg, kg).
7. Symbols, not abbreviations, should be used for units. For example, use "A," not "amp," for ampere.

8. When multiplying unit symbols, use a raised dot:

N·m for newton meter

In the case of W·h, the dot may be omitted, thus:

Wh

An exception to this practice is made for computer printouts, automatic typewriter work, etc, where the raised dot is not possible, and a dot on the line may be used.

9. When dividing unit symbols use one of the following forms:

$$\mathrm{m/s} \ \textit{or} \ \mathrm{m{\cdot}s^{-1}} \ \textit{or} \ \frac{\mathrm{m}}{\mathrm{s}}$$

In no case should more than one slash be used in the same expression unless parentheses are inserted to avoid ambiguity. For example, write:

$$\mathrm{J/(mol{\cdot}K)} \ \textit{or} \ \mathrm{J{\cdot}mol^{-1}\cdot K^{-1}} \ \textit{or} \ \mathrm{(J/mol)/K}$$

but *not*

J/mol/K

10. Do not mix symbols and unit names in the same expression. Write:

joules per kilogram *or* J/kg *or* $\mathrm{J{\cdot}kg^{-1}}$

but *not*

joules/kilogram *nor* joules/kg *nor* $\mathrm{joules{\cdot}kg^{-1}}$

ABBREVIATIONS AND UNITS

A	ampere
A	anion (eg, H*A*); mass number
a	atto (prefix for 10^{-18})
AATCC	American Association of Textile Chemists and Colorists
ABS	acrylonitrile–butadiene–styrene
abs	absolute
ac	alternating current, *n.*
a-c	alternating current, *adj.*
ac-	alicyclic
acac	acetylacetonate
ACGIH	American Conference of Governmental Industrial Hygienists
ACS	American Chemical Society
AGA	American Gas Association
Ah	ampere hour
AIChE	American Institute of Chemical Engineers
AIME	American Institute of Mining, Metallurgical, and Petroleum Engineers
AIP	American Institute of Physics
AISI	American Iron and Steel Institute
alc	alcohol(ic)
Alk	alkyl
alk	alkaline (not alkali)
amt	amount
amu	atomic mass unit
ANSI	American National Standards Institute
AO	atomic orbital
AOAC	Association of Official Analytical Chemists
AOCS	American Oil Chemists' Society
APHA	American Public Health Association
API	American Petroleum Institute
aq	aqueous
Ar	aryl
ar-	aromatic
as-	asymmetric(al)

ASHRAE	American Society of Heating, Refrigerating, and Air Conditioning Engineers
ASM	American Society for Metals
ASME	American Society of Mechanical Engineers
ASTM	American Society for Testing and Materials
at no.	atomic number
at wt	atomic weight
av(g)	average
AWS	American Welding Society
b	bonding orbital
bbl	barrel
bcc	body-centered cubic
BCT	body-centered tetragonal
Bé	Baumé
BET	Brunauer-Emmett-Teller (adsorption equation)
bid	twice daily
Boc	*t*-butyloxycarbonyl
BOD	biochemical (biological) oxygen demand
bp	boiling point
Bq	becquerel
C	coulomb
°C	degree Celsius
C-	denoting attachment to carbon
c	centi (prefix for 10^{-2})
c	critical
ca	circa (approximately)
cd	candela;current density; circular dichroism
CFR	Code of Federal Regulations
cgs	centimeter–gram–second
CI	Color Index
cis-	isomer in which substituted groups are on same side of double bond between C atoms
cl	carload
cm	centimeter
cmil	circular mil
cmpd	compound
CNS	central nervous system
CoA	coenzyme A
COD	chemical oxygen demand
coml	commercial(ly)
cp	chemically pure
cph	close-packed hexagonal
CPSC	Consumer Product Safety Commission
cryst	crystalline
cub	cubic
D	debye
D-	denoting configurational relationship
d	differential operator
d-	*dextro*-, dextrorotatory
da	deka (prefix for 10^1)
dB	decibel
dc	direct current, *n.*
d-c	direct current, *adj.*
dec	decompose
detd	determined
detn	determination
Di	didymium, a mixture of all lanthanons
dia	diameter
dil	dilute
DIN	Deutsche Industrie Normen
dl-; DL-	racemic
DMA	dimethylacetamide
DMF	dimethylformamide
DMG	dimethyl glyoxime
DMSO	dimethyl sulfoxide
DOD	Department of Defense
DOE	Department of Energy
DOT	Department of Transportation
DP	degree of polymerization
dp	dew point
DPH	diamond pyramid hardness
dstl(d)	distill(ed)
dta	differential thermal analysis
(*E*)-	entgegen; opposed
ϵ	dielectric constant (unitless number)
e	electron
ECU	electrochemical unit
ed.	edited, edition, editor
ED	effective dose
EDTA	ethylenediaminetetraacetic acid
emf	electromotive force

emu	electromagnetic unit
en	ethylene diamine
eng	engineering
EPA	Environmental Protection Agency
epr	electron paramagnetic resonance
eq.	equation
esca	electron-spectroscopy for chemical analysis
esp	especially
esr	electron-spin resonance
est(d)	estimate(d)
estn	estimation
esu	electrostatic unit
exp	experiment, experimental
ext(d)	extract(ed)
F	farad (capacitance)
F	faraday (96,487 C)
f	femto (prefix for 10^{-15})
FAO	Food and Agriculture Organization (United Nations)
fcc	face-centered cubic
FDA	Food and Drug Administration
FEA	Federal Energy Administration
FHSA	Federal Hazardous Substances Act
fob	free on board
fp	freezing point
FPC	Federal Power Commission
FRB	Federal Reserve Board
frz	freezing
G	giga (prefix for 10^9)
G	gravitational constant = 6.67×10^{11} N·m²/kg²
g	gram
(g)	gas, only as in $H_2O(g)$
g	gravitational acceleration
gc	gas chromatography
gem-	geminal
glc	gas-liquid chromatography
g-mol wt; gmw	gram-molecular weight
GNP	gross national product
gpc	gel-permeation chromatography
GRAS	Generally Recognized as Safe
grd	ground
Gy	gray
H	henry
h	hour; hecto (prefix for 10^2)
ha	hectare
HB	Brinell hardness number
Hb	hemoglobin
hcp	hexagonal close-packed
hex	hexagonal
HK	Knoop hardness number
hplc	high pressure liquid chromatography
HRC	Rockwell hardness (C scale)
HV	Vickers hardness number
hyd	hydrated, hydrous
hyg	hygroscopic
Hz	hertz
^{i}(eg, Pr^i)	iso (eg, isopropyl)
i-	inactive (eg, *i*-methionine)
IACS	International Annealed Copper Standard
ibp	initial boiling point
IC	inhibitory concentration
ICC	Interstate Commerce Commission
ICT	International Critical Table
ID	inside diameter; infective dose
ip	intraperitoneal
IPS	iron pipe size
IPTS	International Practical Temperature Scale (NBS)
ir	infrared
IRLG	Interagency Regulatory Liaison Group
ISO	International Organization for Standardization
IU	International Unit
IUPAC	International Union of Pure and Applied Chemistry
IV	iodine value
iv	intravenous
J	joule
K	kelvin
k	kilo (prefix for 10^3)
kg	kilogram
L	denoting configurational relationship
L	liter (for fluids only)(5)
l-	*levo*-, levorotatory

(l)	liquid, only as in NH_3(l)
LC_{50}	conc lethal to 50% of the animals tested
LCAO	linear combination of atomic orbitals
LCD	liquid crystal display
lcl	less than carload lots
LD_{50}	dose lethal to 50% of the animals tested
LED	light-emitting diode
liq	liquid
lm	lumen
ln	logarithm (natural)
LNG	liquefied natural gas
log	logarithm (common)
LPG	liquefied petroleum gas
ltl	less than truckload lots
lx	lux
M	mega (prefix for 10^6); metal (as in M*A*)
M	molar; actual mass
$\overline{M}_w$	weight-average mol wt
$\overline{M}_n$	number-average mol wt
m	meter; milli (prefix for 10^{-3})
m	molal
m-	meta
max	maximum
MCA	Chemical Manufacturers' Association (was Manufacturing Chemists Association)
MEK	methyl ethyl ketone
meq	milliequivalent
mfd	manufactured
mfg	manufacturing
mfr	manufacturer
MIBC	methyl isobutyl carbinol
MIBK	methyl isobutyl ketone
MIC	minimum inhibiting concentration
min	minute; minimum
mL	milliliter
MLD	minimum lethal dose
MO	molecular orbital
mo	month
mol	mole
mol wt	molecular weight
mp	melting point
MR	molar refraction
ms	mass spectrum
mxt	mixture
μ	micro (prefix for 10^{-6})
N	newton (force)
N	normal (concentration); neutron number
N-	denoting attachment to nitrogen
n (as n_D^{20})	index of refraction (for 20°C and sodium light)
^{n} (as Bu^n), *n*-	normal (straight-chain structure)
n	neutron
n	nano (prefix for 10^9)
na	not available
NAS	National Academy of Sciences
NASA	National Aeronautics and Space Administration
nat	natural
NBS	National Bureau of Standards
neg	negative
NF	*National Formulary*
NIH	National Institutes of Health
NIOSH	National Institute of Occupational Safety and Health
nmr	nuclear magnetic resonance
NND	New and Nonofficial Drugs (AMA)
no.	number
NOI-(BN)	not otherwise indexed (by name)
NOS	not otherwise specified
nqr	nuclear quadruple resonance
NRC	Nuclear Regulatory Commission; National Research Council
NRI	New Ring Index
NSF	National Science Foundation
NTA	nitrilotriacetic acid
NTP	normal temperature and pressure (25°C and 101.3 kPa or 1 atm)
NTSB	National Transportation Safety Board
O-	denoting attachment to

	oxygen
o-	ortho
OD	outside diameter
OPEC	Organization of Petroleum Exporting Countries
o-phen	*o*-phenanthridine
OSHA	Occupational Safety and Health Administration
owf	on weight of fiber
Ω	ohm
P	peta (prefix for 10^{15})
p	pico (prefix for 10^{-12})
p-	para
p	proton
p.	page
Pa	pascal (pressure)
pd	potential difference
pH	negative logarithm of the effective hydrogen ion concentration
phr	parts per hundred of resin (rubber)
p-i-n	positive-intrinsic-negative
pmr	proton magnetic resonance
p-n	positive-negative
po	per os (oral)
POP	polyoxypropylene
pos	positive
pp.	pages
ppb	parts per billion (10^9)
ppm	parts per million (10^6)
ppmv	parts per million by volume
ppmwt	parts per million by weight
PPO	poly(phenyl oxide)
ppt(d)	precipitate(d)
pptn	precipitation
Pr (no.)	foreign prototype (number)
pt	point; part
PVC	poly(vinyl chloride)
pwd	powder
py	pyridine
qv	quod vide (which see)
R	univalent hydrocarbon radical
(*R*)-	rectus (clockwise configuration)
r	precision of data
rad	radian; radius
rds	rate determining step
ref.	reference
rf	radio frequency, *n.*
r-f	radio frequency, *adj.*
rh	relative humidity
RI	Ring Index
rms	root-mean square
rpm	rotations per minute
rps	revolutions per second
RT	room temperature
^{s} (eg, Bu^s); *sec*-	secondary (eg, secondary butyl)
S	siemens
(*S*)-	sinister (counterclockwise configuration)
S-	denoting attachment to sulfur
s-	symmetric(al)
s	second
(s)	solid, only as in $H_2O(s)$
SAE	Society of Automotive Engineers
SAN	styrene–acrylonitrile
sat(d)	saturate(d)
satn	saturation
SBS	styrene–butadiene–styrene
sc	subcutaneous
SCF	self-consistent field; standard cubic feet
Sch	Schultz number
SFs	Saybolt Furol seconds
SI	Le Système International d'Unités (International System of Units)
sl sol	slightly soluble
sol	soluble
soln	solution
soly	solubility
sp	specific; species
sp gr	specific gravity
sr	steradian
std	standard
STP	standard temperature and pressure (0°C and 101.3 kPa)
sub	sublime(s)
SUs	Saybolt Universal seconds

syn	synthetic
t (eg, But), *t*-, *tert*-	tertiary (eg, tertiary butyl)
T	tera (prefix for 10^{12}); tesla (magnetic flux density)
t	metric ton (tonne); temperature
TAPPI	Technical Association of the Pulp and Paper Industry
tex	tex (linear density)
T_g	glass-transition temperature
tga	thermogravimetric analysis
THF	tetrahydrofuran
tlc	thin layer chromatography
TLV	threshold limit value
trans-	isomer in which substituted groups are on opposite sides of double bond between C atoms
TSCA	Toxic Substance Control Act
TWA	time-weighted average
Twad	Twaddell
UL	Underwriters' Laboratory
USDA	United States Department of Agriculture
USP	*United States Pharmacopeia*
uv	ultraviolet
V	volt (emf)
var	variable
vic-	vicinal
vol	volume (not volatile)
vs	versus
v sol	very soluble
W	watt
Wb	weber
Wh	watt hour
WHO	World Health Organization (United Nations)
wk	week
yr	year
(*Z*)-	zusammen; together; atomic number

Non-SI (Unacceptable and Obsolete) Units		*Use*
Å	angstrom	nm
at	atmosphere, technical	Pa
atm	atmosphere, standard	Pa
b	barn	cm^2
bar†	bar	Pa
bbl	barrel	m^3
bhp	brake horsepower	W
Btu	British thermal unit	J
bu	bushel	m^3; L
cal	calorie	J
cfm	cubic foot per minute	m^3/s
Ci	curie	Bq
cSt	centistokes	mm^2/s
c/s	cycle per second	Hz
cu	cubic	exponential form
D	debye	C·m
den	denier	tex
dr	dram	kg
dyn	dyne	N
dyn/cm	dyne per centimeter	mN/m
erg	erg	J
eu	entropy unit	J/K
°F	degree Fahrenheit	°C; K
fc	footcandle	lx
fl	footlambert	lx
fl oz	fluid ounce	m^3; L
ft	foot	

† Do not use bar (10^5Pa) or millibar (10^2Pa) because they are not SI units, and are accepted internationally only for a limited time in special fields because of existing usage.

Non-SI (Unacceptable and Obsolete) Units		*Use*
ft·lbf	foot pound-force	J
gf/den	gram-force per denier	N/tex
G	gauss	T
Gal	gal	m/s^2
gal	gallon	m^3; L
Gb	gilbert	A
gpm	gallon per minute	(m^3/s); (m^3/h)
gr	grain	kg
hp	horsepower	W
ihp	indicated horsepower	W
in.	inch	m
in. Hg	inch of mercury	Pa
in. H_2O	inch of water	Pa
in.-lbf	inch pound-force	J
kcal	kilogram-calorie	J
kgf	kilogram-force	N
kilo	for kilogram	kg
L	lambert	lx
lb	pound	kg
lbf	pound-force	N
mho	mho	S
mi	mile	m
MM	million	M
mm Hg	millimeter of mercury	Pa
mμ	millimicron	nm
mph	mile per hour	km/h
μ	micron	μm
Oe	oersted	A/m
oz	ounce	kg
ozf	ounce-force	N
η	poise	Pa·s
P	poise	Pa·s
ph	phot	lx
psi	pound-force per square inch	Pa
psia	pound-force per square inch absolute	Pa
psig	pound-force per square inch gauge	Pa
qt	quart	m^3; L
°R	degree Rankine	K
rd	rad	Gy
sb	stilb	lx
SCF	standard cubic foot	m^3
sq	square	exponential form
thm	therm	J
yd	yard	m

BIBLIOGRAPHY

1. The International Bureau of Weights and Measures, BIPM (Parc de Saint-Cloud, France) is described on page 22 of Ref. 4. This bureau operates under the exclusive supervision of the International Committee of Weights and Measures (CIPM).
2. *Metric Editorial Guide* (*ANMC-78-1*) 3rd ed., American National Metric Council, 5410 Grosvenor Lane, Bethesda, Md. 20814, 1981.
3. *SI Units and Recommendations for the Use of Their Multiples and of Certain Other Units* (*ISO 1000-1981*), American National Standards Institute, 1430 Broadway, New York, N. Y. 10018, 1981.
4. Based on *ASTM E 380-82* (*Standard for Metric Practice*), American Society for Testing and Materials, 1916 Race Street, Philadelphia, Pa. 19103, 1982.
5. *Fed. Regist.*, Dec. 10, 1976 (41 FR 36414).
6. For ANSI address, see Ref. 3.

R. P. LUKENS
American Society for Testing and Materials

A

ALCOHOL FUELS

The use of alcohols as fuels for motor vehicles generated more interest and controversy in the 1970s and early 1980s than almost any other renewable energy source (1–2). Although modern commercial marketing of gasohol, a blend of 90 vol % unleaded gasoline and 10 vol % ethanol [*64-17-5*] began in the United States in 1979, many of the problems associated with alcohol costs and performance have not been solved (3–7). Controversial aspects involve net energy production efficiencies of the alcohol manufacturing processes, fuel usage, and tax subsidies. Generally, the popular press has given the consumer the impression that motor-fuel shortages could be alleviated or even eliminated if alcohol fuels were marketed by the oil industry. A careful review of the facts and the technology of alcohol-fuel production and usage, particularly of methanol and ethanol, establishes the technical feasibility of neat alcohols and alcohol–gasoline blends as motor fuels and the areas where improvements must be made to make it economically practical to use alcohols as motor fuels on a large scale (see Gasoline and other motor fuels).

By far, most alcohol usage as motor fuel has occurred with ethanol, either as a fuel extender in ethanol–gasoline blends or as neat ethanol. Its ready availability by fermentation of grains and its relatively good combustion characteristics in internal-combustion engines probably accounted for the periodic resurgence of ethanol as a motor fuel during the first half of this century. The invention of the four-cycle, internal-combustion engine in 1877 by Otto and the two-cycle automobile engine in 1879 by Benz involved the testing of ethanol, other alcohols, and many other organic liquids as potential fuels. Another factor that played an important role in the development of ethanol fuel was the passage of laws that permitted the production of tax-free ethanol for industrial use: in the UK in 1855, the Netherlands in 1865, France in 1872, Germany in 1879, and later in the United States (8–9).

During World War I, various alcohol blends were used by the European military forces as fuels because of gasoline shortages. After the war, numerous countries other than the United States began a serious effort to extend their motor-fuel supplies by blending ethanol in gasoline (10). However, in the late 1930s, many of the countries that had enacted laws requiring the blending of ethanol in motor fuels suspended legal requirements because of the instability of alcohol supplies (11).

Synthetic methanol manufactured from coal-based hydrogen and carbon monoxide was introduced into the U.S. markets at about half the price of wood alcohol (12) (see Coal chemicals and feedstocks). By the early 1930s, when synthetic methanol process technology was well-established and comprised ca 75% of the methanol market in the United States, methanol was considered a possible alternative fuel for gasoline (12). With few exceptions, it was only used as an anti-icing additive for aircraft and as a racing fuel where advantage could be taken of the increased power obtainable, however costly (13). During World War II, thousands of on-board gasifiers were used in Europe to fuel vehicles by thermal gasification of wood (qv), and the resulting fuel gases undoubtedly contained small amounts of wood alcohol (12,14).

Little information has been reported on the use of C_3 and higher alcohols as motor fuels. Many oil companies have blended alcohols, eg, 2-propanol [*67-63-0*] and 2-methyl-1-propanol [*78-83-1*], in gasolines as anti-icing additives but not as primary fuel components. Also, neat 2-methyl-2-propanol [*75-65-0*], the corresponding methyl ether [*1634-04-4*] (MTBE, or methyl *t*-butyl ether), and blends of 2-methyl-2-propanol with methanol (Arco Chemical Company's Oxinol and American Methyl Corporation's Petrocoal) are marketed as octane boosters in the United States.

After World War I, car racing teams began to use alcohol-fuel blends formulated with aviation gasoline. Ethanol–benzene [*71-43-2*]–gasoline blends generally ranged from volumetric ratios of ca 20:20:60 to 80:10:10 (15). Ethanol, because of its high latent heat of vaporization and low air:fuel ratio requirements, compared to those of gasoline, can be used at higher inducted fuel energy densities than gasoline alone and delivers increased power outputs. Some typical alcohol racing fuel blends are described in ref. 15. After World War II, engine compression ratios were increased, and the racing community shifted to neat methanol and methanol–nitroparaffin blends. With the exception of some racing events, neat methanol is used almost exclusively.

Properties

Comparison with Gasoline. Table 1 presents selected key properties of methanol, ethanol, 2-methyl-2-propanol, methyl 2-methyl-2-propyl ether (methyl *t*-butyl ether, MTBE), typical unleaded regular gasoline, typical diesel fuel, and isooctane (2,2,4-trimethylpentane). Methanol and ethanol have ca 50% and 66% of the volumetric heating value of gasoline, whereas 2-methyl-2-propanol and its methyl ether have ca 80% of the volumetric heating value. All other factors being equal, one might expect that the distance per unit volume of these compounds as neat fuels would be correspondingly less than gasoline. This is not necessarily the case because many complex factors influence the performance of an engine–fuel combination.

The alcohols are pure compounds and exhibit specific boiling points, whereas commercial hydrocarbon fuels consist of many paraffinic, aromatic, and naphthenic compounds and, therefore, exhibit boiling ranges. This difference coupled with the higher latent heats of vaporization of the alcohols suggest that there may be significant

Table 1. Comparison of Some Properties for Several Liquid Fuels and Gasoline Additives[a]

Property	Methanol	Ethanol	2-Methyl-2-propanol	Methyl 2-methyl-2-propyl ether	Unleaded regular gasoline	Diesel fuel	Isooctane
formula	CH_3OH	CH_3CH_2OH	$(CH_3)_3COH$	$(CH_3)_3COCH_3$	C_4–C_{12}	C_{14}–C_{19}	C_8H_{18}
molecular weight	32.04	46.08	74.12	88.15	100–115	233 avg	114.23
element, wt %							
C	37.48	52.13	64.82	68.13	85–88	85–88	84.12
H	12.58	13.12	13.60	13.72	12–15	12–15	15.88
O	49.94	34.72	21.59	18.14	0	0	
density (at 20°C), g/cm^3	0.7914	0.7893	0.7887	0.7405	0.69–0.80	0.82–0.86	0.6919
boiling point, °C	65.0	78.5	82.2	55.2	27–225	240–360	99.238
latent heat of vaporization (at 20°C)							
MJ/kg[b]	1.177	0.839	0.600	0.321	0.349	0.256	0.314
MJ/L[c]	0.931	0.662	0.474	0.238	0.251	0.237	0.217
flashpoint, °C	11.1	12.8	11.1		−43	38	
autoignition temperature, °C	464	423	478		495	260	447
flammability limits, vol % in air	6.7–36.0	4.3–19.0	2.4–8.0		1.4–7.6		1.1–6.0
higher heating value (at 20°C), MJ/kg[b]	22.3	29.8	35.5	38.1	47.2		47.8
lower heating value (at 20°C), MJ/L[c]	15.76	21.09	25.92	26.02	32.16	35.4	33.07
stoichiometric air:fuel mass ratio	6.45	8.97	11.15	12.50	14.73	15.00	15.07
stoichiometric air:fuel volumetric ratio	7.16	14.32	28.65	38.18	57.28		59.68
water solubility, %	infinite	infinite	infinite	1.4	0.009	nil	0
research octane number (RON)	112	111	113	117	91–93		100
motor octane number (MON)	91	92	110	101	82–84		100
cetane number	3	8			8–14	40–60	10
Reid vapor pressure, kPa[d]	32	17		54	48–103	nil	70

[a] Refs. 5–6, 16–24.
[b] To convert MJ/kg to Btu/lb, multiply by 430.2.
[c] To convert MJ/L to Btu/gal, multiply by 3590.
[d] To convert kPa to psi, multiply by 0.145.

differences in the carburetion of a given engine with neat methanol and ethanol. Also, fuel volatility differences might be expected with alcohol–gasoline blends as compared to neat gasoline, particularly because of the interactions of associated liquids, eg, alcohols, in nonassociated liquids, eg, hydrocarbons.

Another significant difference in properties is that methanol and ethanol have much lower stoichiometric air:fuel ratios for combustion than gasoline. Despite the greater flammability range of alcohols in air, this is the reason why alcohol–air mixtures supplied by a gasoline-set carburetor to an engine are too lean for combustion to occur. Alcohol–gasoline blends containing up to ca 20 vol % alcohol generally operate satisfactorily, although there is a blend-leaning effect on delivery of the fuel mixture to the engine with a conventionally adjusted gasoline carburetor. The leaning effect of the blend can often improve fuel mileage if the carburetor is set too rich for gasoline alone, even though the alcohols have a lower energy content. If the carburetor is set properly for gasoline alone, the fuel mileage should be lower with the blend. In many of the late-model cars, the air:fuel ratio is automatically controlled by feedback from an oxygen sensor in the exhaust, and the blend-leaning effect should not occur.

The higher octane numbers of the alcohols compared to those of unleaded regular gasoline suggest greater volumetric efficiencies of neat alcohols than gasoline, provided the compression ratios of the engine are high enough to take advantage of the higher octane values. In addition to compression ratio, volumetric efficiency depends on engine-design factors, eg, timing, type of fuel-induction system, the breathing capacity, and on fuel parameters, eg, mixture strength, latent heat of vaporization of the fuel, and the heat received by the ingoing charge during its passage through the induction system (16). For ethanol and especially for methanol, the latent heats of vaporization are sufficiently large that the fuel does not completely evaporate during the suction stroke and continues to evaporate during the compression stroke (16). For the same amount of fuel evaporated before the inlet valves close, air cooling and the quantity of air drawn into the cylinder are greater. Hence, the temperature of the whole cycle is lower, and the density of the fuel–air charge is higher, resulting in increased efficiency. It should be emphasized that the heating value of a fuel only determines fuel consumption for a specific amount of work, not efficiency or power output. Thus, the higher octane values of methanol and ethanol and the higher operational engine compression ratios imply higher latent heats of vaporization and the resulting increases in fuel–air cooling, density, and mass flow as well as the more favorable combustion-products-to-charge molar ratio. These factors suggest greater efficiency of alcohols than hydrocarbon fuels in terms of mileage per unit energy expended. Note that this is unrelated to mileage per unit liquid volume of fuel. The greater energy efficiency means that, for the same power outputs as gasoline engines, smaller alcohol-fueled engines could be used. For 10-vol % blends, the lower volumetric energy content reduces fuel economy in properly adjusted induction systems but not necessarily in proportion to the amount of alcohol used. Losses are usually ca 2–5% (25–27). This undoubtedly results from the higher thermal efficiency of alcohol fuels.

Several of the key properties of methyl 2-methyl-2-propyl ether, which is used as an octane booster and gasoline extender, are more similar to those of unleaded regular gasoline than those of methanol and ethanol. This is expected because the polar hydroxyl group is substituted by a less-polar ether linkage, and the molecule is more hydrocarbonlike than the lower alcohols. Indeed, compared to the lower alcohols, this ether exerts less blend-leaning effect, less effect on evaporative emissions, and less water-induced phase-separation problems in gasoline blends (22).

Comparison with Diesel Fuel. As for applications of methanol and ethanol as diesel-fuel extenders or substitutes, the properties listed in Table 1 suggest several problems. The cetane numbers, water-solubility characteristics, stoichiometric combustion ratios, heating values, ignition temperatures, vaporization characteristics, and boiling points are vastly different. Only anhydrous ethanol forms solutions with diesel fuel. Up to ca 30 vol % anhydrous ethanol can be added to diesel fuel without a need for engine modification. As the percentage of ethanol is increased, power is reduced, fuel consumption and engine noise increase, and the delay period (the time needed for ignition after the fuel enters the combustion chamber) is extended (28–29). Performance with 10 vol % ethanol is about equal to that of diesel fuel, and no significant performance characteristic is enhanced by ethanol (28). Nevertheless, severe problems do occur. For example, only 0.5 vol % water contamination causes phase separation of ethanol–diesel fuel blends at 0°C (29). Severe knocking caused by ethanol's low cetane value can also occur. The use of additives, eg, vegetable oils and organic nitrates, to improve cetane numbers of ethanol–diesel fuel blends and neat ethanol is being studied (30–31).

Other techniques that are being evaluated to permit use of ethanol and methanol as diesel fuels include separate injection of the alcohol and diesel fuel into the combustion chamber, formulation of stable alcohol–diesel-fuel emulsions, and introduction of the alcohol fuel by direct carburetion in the intake air stream (fumigation) (31–37). About 50% of the energy requirements of a diesel engine are met with ethanol containing up to 30 vol % water by means of fumigation, and thermal efficiency is higher than with diesel fuel alone when anhydrous or 50 vol % ethanol is delivered through a separate injector. Although the effects on emissions and engine life have not been fully assessed, the data compiled from the work done to date indicate that alcohols can serve as diesel-fuel extenders without requiring significant engine and fuel-system modifications or loss of power and efficiency (38). Diesel fuel cannot be completely replaced by methanol and ethanol because of the latters' low cetane numbers and high autoignition temperatures; these properties would require extensive engine modifications or use of cetane-improving additives.

Combustion. The stoichiometric equations for complete combustion of the lower molecular weight alcohols, methyl 2-methyl-2-propyl ether, isooctane, and gasoline are shown in Table 2. The stoichiometric air:fuel ratio of neat methanol and ethanol is low compared to that of gasoline. As the molecular weight increases, this ratio increases, but it is still less than the ratio for gasoline up to C_7 alcohols. The mole percent of air needed for stoichiometric combustion decreases with increasing molecular weight and is the same as for gasoline (1.72 mol %) with C_7 alcohols. The relatively high volume of air needed for methanol and ethanol to combust stoichiometrically results in the blend-leaning effect. If a carburetor that is set for delivery of gasoline–air mixtures to a spark-ignition engine is converted to methanol–gasoline mixtures without adjustment, the same volume of fuel and air is supplied, but the methanol mixture is significantly deficient in oxygen. The stoichiometric relationships presented in Table 2 also show that, with the exception of benzyl alcohol, the molar ratios of products-to-charge for the alcohols are higher than for gasoline and isooctane. This suggests that one of the reasons for better thermal efficiency of the lower molecular weight alcohols is higher pressure in the combustion chambers and more power output. The product:charge molar ratio for methyl 2-methyl-2-propyl ether is about the same as that of gasoline (1.051). Thus, neat ether may not exhibit proportionately increased

Table 2. Stoichiometric Combustion Air Requirements for Pure Liquid Alcohols, Methyl-*t*-Butyl Ether, Isooctane, and Gasoline[a]

Fuel	Stoichiometry	Product:charge molar ratio	Air:fuel ratio		Wt %	Fuel in air, mol
			wt	mol		
methanol	$CH_3OH + 1.5\ O_2 + 5.66\ N_2 = CO_2 + 2\ H_2O + 5.66\ N_2$	1.061	6.45	7.16	13.43	12.25
ethanol	$C_2H_5OH + 3\ O_2 + 11.32\ N_2 = 2\ CO_2 + 3\ H_2O + 11.32\ N_2$	1.065	8.97	14.32	10.03	6.53
1-propanol 2-propanol	$C_3H_7OH + 4.5\ O_2 + 16.98\ N_2 = 3\ CO_2 + 4\ H_2O + 16.98\ N_2$	1.067	10.31	21.48	8.84	4.45
all butanols	$C_4H_9OH + 6\ O_2 + 22.65\ N_2 = 4\ CO_2 + 5\ H_2O + 22.65\ N_2$	1.067	11.15	28.65	8.23	3.37
all pentanols	$C_5H_{11}OH + 7.5\ O_2 + 28.31\ N_2 = 5\ CO_2 + 6\ H_2O + 28.31\ N_2$	1.068	11.72	35.81	7.86	2.72
cyclohexanol	$C_6H_{11}OH + 8.5\ O_2 + 32.08\ N_2 = 6\ CO_2 + 6\ H_2O + 32.08\ N_2$	1.060	11.69	40.58	7.88	2.41
all hexanols	$C_6H_{13}OH + 9\ O_2 + 33.97\ N_2 = 6\ CO_2 + 7\ H_2O + 33.97\ N_2$	1.068	12.13	42.97	7.62	2.27
benzyl alcohol	$C_7H_7OH + 8.5\ O_2 + 32.08\ N_2 = 7\ CO_2 + 4\ H_2O + 32.08\ N_2$	1.036	10.83	40.58	8.46	2.41
all heptanols	$C_7H_{15}OH + 10.5\ O_2 + 39.63\ N_2 = 7\ CO_2 + 8\ H_2O + 39.63\ N_2$	1.068	12.45	50.13	7.44	1.96
all octanols	$C_8H_{17}OH + 12\ O_2 + 45.29\ N_2 = 8\ CO_2 + 9\ H_2O + 45.29\ N_2$	1.069	12.69	57.29	7.30	1.72
methyl 2-methyl-2-propyl ether	$(CH_3)_3COCH_3 + 7.5\ O_2 + 28.31\ N_2 = 5\ CO_2 + 6\ H_2O + 28.31\ N_2$	1.068	11.72	35.81	7.86	2.72
isooctane	$C_8H_{18} + 12.5\ O_2 + 47.18\ N_2 = 8\ CO_2 + 9\ H_2O + 47.18\ N_2$	1.058	15.07	59.68	7.96	1.65
gasoline[b]	$C_nH_{2n} + 1.5\ n\ O_2 + 5.66\ n\ N_2 = n\ CO_2 + n\ H_2O + 5.66\ n\ N_2$	1.051	14.73	57.28	8.13	1.72

[a] Air is assumed to contain 20.946 mol % O_2.

[b] Gasoline is assumed to have an average molecular formula of C_8H_{16}.

power output compared to gasoline. A 30-vol % ether blend with gasoline shows no measurable effect on wide-open throttle performance (39). For a 20-vol % ether blend, the combustion rate characteristics are equivalent to gasoline (40).

The calculated higher heating values and the heating values per unit volume of stoichiometric air:fuel mixtures of several alcohols and other fuels are listed in Table 3. Despite the many differences in density, heating value, and fuel–air requirement for complete combustion, the heating values of the stoichiometric mixtures for all of the alcohols, methyl 2-methyl-2-propyl ether, isooctane, and gasoline are in a very narrow range, ie, ca 3.727 ± 0.07 MJ/m^3 (100.1 ± 1.9 Btu/ft^3) at 20°C. In theory, a properly aspirated and timed spark-ignition engine would thus be expected to deliver the same power outputs, independent of which fuel was used (41). At suitable air:fuel ratios, the specific energy contents of the fuels vary only slightly and the differences in maximum power output are insignificant. Furthermore, this argument assumes that complete combustion occurs, and that there are no gross differences between fuels. However, as pointed out, there are many differences that cause deviation from idealized behavior.

Table 3. Higher Heating Values of Pure Fuels and Heating Values of Stoichiometric Air–Fuel Mixtures

Fuel	CAS Registry No.	Mol wt	Higher heating values for pure fuels[a]		Heating values for air–fuel mixtures[b], MJ/m^3 [c]
			MJ/kg[d]	MJ/L[e]	
methanol	*[67-56-1]*	32.04	22.33	17.70	3.643
ethanol		46.07	29.77	23.50	3.724
1-propanol	*[71-23-8]*	60.10	33.48	26.93	3.722
2-propanol	*[67-63-0]*	60.10	33.08	25.98	3.678
1-butanol	*[71-36-3]*	74.12	36.07	29.21	3.746
2-methyl-1-propanol		74.12	36.05	28.88	3.743
2-methyl-2-propanol		74.12	35.55	28.04	3.692
1-pentanol	*[71-41-0]*	88.15	37.70	30.70	3.758
2-methyl-2-butanol	*[75-85-4]*	88.15	37.27	30.15	3.715
cyclohexanol	*[108-93-0]*	100.16	37.23	35.83	3.736
3-methyl-3-pentanol	*[77-74-7]*	102.18	37.99	31.48	3.663
benzyl alcohol	*[100-51-6]*	108.14	34.62	36.07	3.751
1-heptanol	*[111-70-6]*	116.20	39.81	32.72	3.769
3-ethyl-3-pentanol	*[597-49-9]*	116.20	38.91	32.64	3.684
methyl 2-methyl-2-propyl ether		88.15	38.12	28.23	3.800
isooctane	*[540-84-1]*	114.23	47.79	33.07	3.745
gasoline[f]		112.21	47.20	34.0–36.8	3.787

[a] Conditions for combustion: atmospheric pressure, temperature = 20°C, and product water in liquid state.

[b] Calculated from heats of combustion at 20°C and mole percent of fuel in stoichiometric air–fuel mixture in Table 2 by (MJ/kg) (mol wt) (mol % fuel) (4.1572×10^{-4}).

[c] To convert MJ/m^3 to Btu/ft^3, multiply by 26.9.

[d] Calculated from heats of combustion 4.1868 (kcal/mol × mol/kg); to convert MJ/kg to Btu/lb, multiply by 430.2.

[d] To convert MJ/L to Btu/gal, multiply by 3590.

[f] Gasoline is assumed to have an average molecular formula of C_8H_{16}.

Table 4. Octane Numbers of Alcohol–Gasoline Blends[a]

Alcohol	Concentration, wt %	Gasoline 1[b] RON[c]	Gasoline 1[b] MON[c]	Gasoline 2[b] RON[c]	Gasoline 2[b] MON[c]	Gasoline 1[b] BRON[c]	Gasoline 1[b] BMON[c]	Gasoline 2[b] BRON[c]	Gasoline 2[b] BMON[c]
none (base gasoline only)	0	78.5	72.0	90.1	83.5				
methanol	5	80.9	74.2	92.8	84.0				
	10	84.1	76.4	95.1	84.5	137	111	135	94
	15	87.1	78.0	97.2	85.0				
ethanol	5	81.6	74.4	92.8	84.2				
	10	83.9	76.7	94.7	84.8	135	115	130	96
	15	86.7	78.7	96.5	85.3				
1-propanol	5	80.5	73.8	91.9	83.9				
	10	82.8	75.1	93.4	84.2	121	101	119	90
	15	84.5	76.7	94.8	84.5				
2-methyl-1-propanol	5	80.4	73.5	91.4	83.7				
	10	81.7	74.7	92.9	84.0	113	96	113	88
	15	83.6	75.8	94.0	84.2				

[a] Ref. 42.
[b] Base gasolines 1 and 2 are CRC-RMFD (Coordinating Research Council Reference Motor-Fuel Detonation) 286 and CRC-RMFD 287, respectively.
[c] RON, research octane number; MON, motor octane number; BRON, blending research octane number; BMON, blending motor octane number.

Octane Numbers. The octane numbers of neat alcohols are high, as shown in Table 1, and would be expected to increase the octane value of alcohol–gasoline blends (see Table 4) (42). The addition of the lower molecular weight alcohols increases the research octane number (RON) proportionately more than the motor octane number (MON). Also, the blending octane values are higher for the lower octane number base stock, and octane improvement decreases with increasing molecular weight of the alcohol at equivalent weight percentage additions. Blending value is an octane number calculated from the experimentally determined values by a simple linear equation and corresponds to a hypothetical rating at 100% concentration of the additive. It is a measure of the synergistic octane improvement capability of the additive and is often much higher than the octane number determined with neat additive. The octane-enhancing properties of methanol and ethanol appear to be about the same. Many petroleum refiners are marketing regular and premium grades of gasoline that contain alcohols primarily as octane-improving substitutes for tetraethyllead and not as gasoline extenders. Some refiners have even discarded the name gasohol.

Manufacture

Ethanol. Ethanol is manufactured from various natural feedstocks by fermentation (qv) and from ethylene by hydration, as illustrated in the following equations (see Ethanol; Fuels from biomass):

Anaerobic fermentation of biomass

(*1*) $(C_6H_{10}O_5)_x + x\ H_2O \rightarrow x\ C_6H_{12}O_6$

(*2*) $x\ C_6H_{12}O_6 \rightarrow 2\ x\ C_2H_5OH + 2\ x\ CO_2$

Direct ethylene hydration

$CH_2{=}CH_2 + H_2O \rightarrow C_2H_5OH$

Indirect ethylene hydration

$$CH_2{=}CH_2 + H_2SO_4 \rightarrow C_2H_5OSO_3H$$
$$2\ C_2H_5OSO_3H \rightarrow (C_2H_5O)_2SO_2 + H_2SO_4$$
$$C_2H_5OSO_3H + CH_2{=}CH_2 \rightarrow (C_2H_5O)_2SO_2$$
$$C_2H_5OSO_3H + H_2O \rightarrow C_2H_5OH + H_2SO_4$$
$$(C_2H_5O)_2SO_2 + H_2O \rightarrow C_2H_5OH + C_2H_5OSO_3H$$

Alcohol fermentation under anaerobic conditions can be conducted with three basic kinds of feedstocks: sugar crops and sugar-containing by-products, eg, sugarcane, molasses, sugar beets, sweet sorghum, Jerusalem artichoke, fodder beets, fruit crops, and cheese whey; starchy crops, eg, corn, sorghum, wheat, barley, cassava, potatoes, and sweet potatoes; and lignocellulosics, eg, woody materials, sulfite waste liquors from paper pulping, and crop residues. Most of the sugars in the sugar crops are individual hexoses or disaccharides and only need to be extracted for direct fermentation with specific yeasts, eg, *Saccharomyces cerevisiae*, as shown in step *2*. The starch fraction of starchy crops contains sugar polymers of water-soluble amylose and water-insoluble amylopectin. Since the ethanol-forming yeasts act only on the simple sugars and not on these polymers, the polymers must be broken down first. Starchy crops used for ethanol manufacture are usually sequentially reduced in particle size by milling or grinding to break the starch walls, slurried in water, liquefied by heating the slurry in the presence of α-amylase to break the cell walls of the starch, treated with the saccharifying enzyme glucoamylase or dilute acid to liberate the sugars in step *1*, and then fermented to produce ethanol in step *2*. With the exception of the sulfite waste liquors from sulfite pulping processes, the third group, the lignocellulosics, contains the glucose polymer cellulose in complexed form. The complexed cellulose is more resistant to hydrolysis than the starch polymers and more difficult to separate from the other components, ie, lignin and hemicellulose, in the lignocellulose complex. More severe treatment is required to liberate the sugars in step *1*, but the basic processes of cellulose separation and hydrolysis are needed to prepare the feedstock for fermentation. By-product sulfite waste liquors from paper pulping are derived from wood and contain 2–3 wt % free sugars after pulping. About 65% of these sugars are fermentable to ethanol after neutralization of the formic, acetic, and sulfurous acids in the liquid.

Synthetic ethanol refers to all ethanol made by nonfermentative chemical routes. Direct ethylene hydration is the primary production method for synthetic ethanol. It is carried out in the vapor phase with liquid or solid catalysts, steam, and ethylene under temperatures and pressures that minimize ethylene polymerization, ethanol dehydration, and by-product formation. Some of the catalysts used for direct ethylene hydration are tungsten oxides promoted with zinc oxide, phosphoric acid on special supports, and certain ion-exchange (qv) resins. Until the early 1970s, indirect hydration was also used for ethanol production; it involves the absorption of ethylene in concentrated sulfuric acid to ethylsulfuric acid and diethyl sulfate. Dilution of the reaction mixture with water and hydrolysis of the sulfates yield ethanol, by-product diethyl ether, and dilute sulfuric acid, which is concentrated and recycled.

Because of the sensitivity of gasohol-type formulations to phase separation on contamination with small amounts of water, anhydrous ethanol should be used for gasoline-blending applications. A fermentation plant for the manufacture of ethanol fuel includes distillation units to produce anhydrous ethanol from the fermentation beer, which contains ca 8–10 vol % ethanol, by successive distillations. The first distillation yields an overhead containing ca 50–55 vol % ethanol, the second distillation yields 95 vol % ethanol, and the last is usually a distillation with an azeotropic agent, eg, benzene, cyclohexane, gasoline, or diethyl ether, to remove the remaining water by formation of a ternary (alcohol–water–benzene, –cyclohexane, or –gasoline) or binary (water–ether) azeotrope. These energy-intensive purification steps must be carefully performed to maximize the thermal efficiency of the plant and to minimize process steam needs. Some of the older corn fermentation plants use up to 42.4 MJ for fermenting and distilling one liter of ethanol (152×10^3 Btu/gal), which has a higher heating value of 23.6 MJ/L (84.7×10^3 Btu/gal) (43). Net energy production would obviously be very poor for this kind of plant, especially if fossil fuels were used as the process fuel.

The potential fermentation ethanol yields of several biomass species are reported in refs. 44–46. Of the biomass considered as feedstocks, corn and grains offer the highest potential yields. In commercial plants, yields of ca 355 L/t (85 gal/short ton, 2.5 gal/bu) are realized from corn. The potential ethanol yield from sugarcane is only about 20% of that from corn. The theoretical yield of ethanol from a pure hexose, based on the stoichiometry of two moles of ethanol per mole of hexose is 645.7 L/t or 51.14 wt % of the sugar.

Almost all synthetic ethanol is manufactured by direct vapor-phase catalytic hydration of ethylene rather than by ethylene sulfation-hydrolysis. In a commercial process used by U.S. Industrial Chemicals at its Tuscola, Illinois plant, steam and high purity ethylene are passed over a fixed bed of a solid-supported phosphoric acid catalyst at elevated pressure and temperature, and the dilute crude ethanol is separated from unreacted ethylene and is concentrated in a series of several distillation towers (8). The first distillation yields concentrated crude ethanol as overhead; the second distillation removes diethyl ether by-product as overhead after the crude concentrate is hydrogenated to reduce unsaturates; the third distillation removes oils and impurities as overhead and ethanol as bottoms; the fourth distillation yields 190-proof ethanol as overhead after the bottoms are stripped to remove more impurities; and the last column involves azeotropic distillation with benzene to yield anhydrous ethanol. If the ethanol were used for fuel purposes, such an extensive treatment would not be required. This process is designed to remove impurities in the parts-per-million (10^6) range to yield high purity 190- and 200-proof ethanol.

Although they are not used commercially at present, several other routes to synthetic ethanol are also feasible, eg,

Direct ethane oxidation

$$CH_3CH_3 + \tfrac{1}{2}\,O_2 \rightarrow C_2H_5OH$$

Higher hydrocarbon oxidation

$$C_nH_{2n+2} + O_2 \rightarrow C_2H_5OH \text{ (by-product)}$$

Ethyl ester hydrolysis

$$RC(=O)OC_2H_5 + H_2O \longrightarrow RC(=O)OH + C_2H_5OH$$

Diethyl ether reduction

$$C_2H_5OC_2H_5 + H_2 \rightarrow C_2H_5OH + CH_3CH_3$$

Acetaldehyde reduction

$$CH_3CHO + H_2 \rightarrow C_2H_5OH$$

Synthesis gas conversion

$$2\,CO + 4\,H_2 \rightarrow C_2H_5OH + H_2O$$

Homologation

$$2\,CH_3OH \rightarrow C_2H_5OH + H_2O$$

$$CH_3OH + 2\,H_2 + CO \rightarrow C_2H_5OH + H_2O$$

Direct hydrocarbon oxidation suffers from poor selectivity and the formation of many other products. Natural ethyl esters are not sufficiently abundant to make them practical raw materials for ethanol manufacture, whereas the synthetic ethyl esters are usually made from ethanol. Similarly, diethyl ether reduction is not a preferred route to ethanol, because most diethyl ether is made from ethanol. Acetaldehyde reduction is not a practical route because over 80% of the acetaldehyde marketed today is made directly from ethylene, as is almost all synthetic ethanol. Acetaldehyde would simply be an unnecessary intermediate. Some ethanol has been produced as a byproduct in Fischer-Tropsch plants for the conversion of synthesis gas to hydrocarbons, but the selectivities are poor.

Total U.S. nameplate capacity for ethanol fuel produced by fermentation is projected to be 1.58×10^6 metric tons per year or 1.99 GL/yr (5.26×10^8 gal/yr) in January 1983. Estimated world synthetic and fermentation ethanol production in 1977 is shown in Table 5. At that time, fermentation ethanol production was only slightly greater than synthetic ethanol production but, since considerable quantities of fermentation ethanol are not reported in international statistics, it is likely that fermentation ethanol production is much higher than that indicated in Table 5 (45). Large increases in fermentation ethanol production have occurred in the United States and Brazil since 1977 because of increased gasohol usage.

Table 5. World Ethanol Production, 1977[a]

	Synthetic		Fermentation	
Country or Region	ML[b]	1000 t	ML[b]	1000 t
United States	673	533	48.1	38.1
Canada	192	152	nil	nil
UK	246	195	44.7	35.4
France	109	86	259	205
FRG	116	91.6	93.9	74.4
other European nations	na	na	245	194
Japan	91.7	72.6	126	99.8
India	4.5	3.6	389	308
Brazil	nil	nil	601	476
Eastern Europe	97.3	77.1	na	na
others	115	90.7	na	na
Total	*1644.5*	*1301.6*	*1806.7*	*1430.7*

[a] Ref. 45.

[b] To convert L to gal, divide by 3.785.

Methanol. Methanol has been produced by wood pyrolysis, noncatalytic oxidation of hydrocarbons, as a by-product of Fischer-Tropsch synthesis, and by reduction of carbon monoxide (see Methanol; Hydrocarbon oxidation). Now, synthetic methanol is manufactured almost entirely from synthesis gas.

$$CO + 2\,H_2 \rightarrow CH_3OH$$

$$CO_2 + 3\,H_2 \rightarrow CH_3OH + H_2O$$

Several improvements have been made since the process was commercialized, particularly in terms of catalysts and operating pressures. The so-called low pressure process originally developed by Imperial Chemical Industries in the 1960s is exemplary. Subsequent research to develop new methanol processes has usually been patterned after the ICI method which is characterized by heterogeneous copper-based catalysts, pressures ca 5–10 MPa (50–100 atm), and temperatures in the 220°–270°C range. Natural gas is by far the largest source of synthesis gas for synthetic methanol plants (70–75%); petroleum residues, naphtha, and coal provide about 15%, 5%, and 2%, respectively, of the remainder.

Commercial catalysts for the conversion of synthesis gas to methanol generally have the empirical composition $Cu/ZnO/Al_2O_3$ or $Cu/ZnO/Cr_2O_3$ (47). Specific catalyst compositions and their performance are shown in Table 6. Work on the function of the various catalyst components indicates that the activity for methanol formation resides in the CuO/ZnO system and that the Cr_2O_3 and Al_2O_3 act as structural promoters (47). In the most efficient process designs, removal of the low boiling components and water is achieved with two distillation columns. Modern plants almost attain theoretical thermal efficiencies; additional improvements are generally marginal and can require high capital costs. Process energy consumption in modern plants may be as high as ca 24.9 MJ/L (89,389 Btu/gal) of methanol product, which has a higher heating value of 17.7 MJ/L (63,541 Btu/gal).

In nature, small amounts of methanol can form from the breakdown of natural methyl esters and ethers in plant material but, as a commercial supply, this source of methanol is insignificant. Also, even though fermentation is well established as a source of ethanol, similar processes for methanol have not been developed. A fermentation methanol process with methane as the feedstock is theoretically possible because methanol is an intermediate in the microbial oxidation of methane (48). Agents, such as iodoacetate, that block the oxidation of methanol in the fermentation broth cause methanol to accumulate in small concentrations. Natural gas could be converted directly to methanol by aerobic fermentation in a practical process if the yields could be increased to a suitable level. This would permit methanol to be manufactured at low temperatures and pressures and eliminate the necessity of synthesis gas intermediates. Another direct route to methanol that has been considered is the direct thermal oxidation of methane with oxygen. This route would also eliminate the need for synthesis gas production. Slow oxidation of methane at ca 10 MPa (100 atm) and ca 340°C yields 22 mol % methanol in the product gas with a molar feed ratio of CH_4 to O of 8.1:1 (49). Other main products are formaldehyde and carbon oxides; methanol selectivity is low.

Methane has been converted to methanol at selectivities up to ca 60% by direct oxidation with nitrous oxide in the presence of water over silica-supported molybdenum oxide at 0.1 MPa (1 atm) and 560–600°C in a continuous-flow reactor (50). Methane conversion was 8–16% with 3.8:1:0.96 CH_4:N_2O:H_2O molar feed ratios. The

Table 6. Methanol Catalyst Composition and Performance[a]

Catalyst composition, wt %				Reactants	Temperature, °C	Pressure, MPa[b]	Space velocity, h^{-1}	Yield, kg/(L·h)	Company
CuO	ZnO	Al_2O_3	Cr_2O_3						
12	62	25		H_2, CO, CO_2, CH_4	230	20	10,000	3.290	BASF
12	62	25		H_2, CO, CO_2, CH_4	230	10	10,000	2.086	BASF
23	46	30		H_2, CO	240		20,000	2.5	CCI
24	38	38		H_2, CO, CO_2, CH_4	226	5	12,000	0.7	ICI
60	22	8		H_2, CO, CO_2	250	5	40,000	0.5	ICI
60	22	8		H_2, CO, CO_2, CH_4	226	10	9,600	0.5	ICI
66	17	17		H_2, CO, CO_2	275	7	200 mol/h	4.75	DuPont
15	48		37	H_2, CO	270	14.7	10,000	1.95 kg/(kg·h)	Japanese Gas-Chemical Co.
31	38		5	H_2, CO	230	5	10,000	0.755	BASF
40	40		20	H_2, CO, CO_2, CH_4	250	4	6,000	0.26	ICI
40	40		20	H_2, CO, CO_2, CH_4	250	8	10,000	0.77	ICI
60	30		10	H_2, CO, CO_2	250	10	9,800	2.28	Metall-Gesellshaft

[a] Ref. 47.

[b] To convert MPa to psi, multiply by 145.

other products were formaldehyde and carbon oxides. Trace amounts of ethane and ethylene were detected in all products.

Methanol is a principal heavy organic chemical and is produced in over 30 countries and approximately 82 plants with a total capacity of ca 17.8 GL/yr (4.7×10^9 gal/yr) (51). The U.S. nameplate capacity in 1982 was 5.68 GL/yr (1.5×10^9 gal/yr) and production has increased by ca 60% since 1975 (52–53).

The relationship of U.S. methanol and ethanol plant capacities and production to U.S. gasoline consumption is illustrated in Table 7. Only a few percent of total gasoline consumption could be displaced if all methanol and ethanol plants were devoted exclusively to motor-fuel production. There appears to be no possibility of replacing gasoline with alcohol fuels within the next 20 yr unless there is a nationwide effort to expand production.

Research

Fermentation Ethanol. Much research has been conducted in the last decade to improve the ethanol fermentation process. This work has concentrated mainly on minimizing the energy inputs, particularly of the distillation steps needed to prepare 95% and anhydrous ethanol, on the fermentation process itself to increase ethanol yields and reduce fermentation times, and on the pretreatment processes needed to make cellulosics and their hexose derivatives suitable feedstocks for fermentation. Since the hemicelluloses comprise about one quarter of the organic components in biomass, considerable research has also been done on the fermentation of pentoses to ethanol.

The basic problem with conventional yeast fermentation of carbohydrates and conventional purification to yield fuel-grade ethanol is that more energy can be utilized

Table 7. Comparison of U.S. Gasoline Consumption with Methanol and Ethanol Plant Capacities and Production, 10^9 L[a]

	1981	1985	1990
gasoline[b]			
regular unleaded	147	161	154
premium unleaded	49.2	86.7	103
regular leaded	175	92.4	32.6
premium leaded	3.79		
Total	*374.99*	*340.1*	*289.6*
methanol nameplate capacity[c]	5.7		
methanol production[d]	4.8		
synthetic ethanol nameplate capacity[d]	1.0		
synthetic ethanol production[d]	0.7		
fermentation ethanol nameplate capacity[e]	0.7		
fermentation ethanol production[f]	0.28		

[a] To convert L to gal, divide by 3.785.
[b] Ref. 51.
[c] Ref. 53.
[d] Ref. 52.
[e] For fuel use; calculated from data in ref. 54; 1.99 GL (5.26×10^8 gal) capacity by January 1, 1983 (ref. 55).
[f] Ref. 56.

to manufacture a unit of ethanol than the energy contained in that unit. The perception by many is that the net energy production is negative, that is, if the process energy inputs are fossil fuels and the ethanol is used to replace petroleum-derived fuels, the entire system may be operating at a net energy loss. This argument fails if the process energy inputs are renewable energy sources, even if excessive energy is used to manufacture a unit of ethanol. In any case, many net energy analyses have been conducted on ethanol production from carbohydrates. Table 8 is a summary of a few energy consumption/production figures for the growth and conversion of corn to ethanol (1). The figures were converted to the net energy production ratio N, as given by:

$$N = \frac{E_p - E_x}{E_x}$$

where E_p is the energy content of salable fuels and E_x is the sum of nonsolar energy inputs into the system. An N greater than zero indicates that an amount of energy equivalent to the sum of the external nonfeed energy inputs and an additional energy increment of salable fuel are produced. If N is equal to zero, only E_x is replaced; whereas if N is less than zero, the system operates at a net loss in energy. Table 8 shows the variations in net energy production ratios that can be encountered in a net energy analysis. A net energy analysis based on reported energy consumption data for an integrated corn production–ethanol fermentation system was performed to help clarify this apparent anomaly (57). The boundary of the system depicted in Figure 1 circumscribes all the operations necessary to grow and harvest the corn, collect the residual cobs and stalks if they are used as process fuel, operate the fermentation plant for the production of anhydrous ethanol and by-product chemicals, and dry the stillage to produce distillers' dried grains plus solubles for sale as cattle feed. The capital energy investment is not included within the boundary. Various net energy production ratios were calculated, as shown in Table 9. The ratio can be either positive or negative, depending upon whether credit is taken for the by-product chemicals and cattle feed, whether the energy for drying the stillage is included as an input to the system, and whether a portion of the residual corn cobs and stalks is collected and used as fuel within the system to replace fossil fuel inputs. Comparisons require that net energy analyses be clearly specified as to all details, including the boundary of the system. On the whole, net energy production in a modern corn-to-ethanol plant would appear to be borderline if petroleum fuels comprise most of the nonfeed energy inputs. Net energy production can be improved by replacing these fuels with feeds generated within the system or with renewable fuels, and credit can be taken for the by-products. Using more of the corn plant or the stillage as feedstock to the fermentation plant also would

Table 8. Net Energy Production Ratios for Ethanol Production from Corn[a]

Source	Energy consumed	Energy as alcohol	N
Office of Technology Assessment	320 MJ[b]	128 MJ[b]	−0.60
Amoco Oil Company	46.29 MJ/L[c]	21.19 MJ/L[c]	−0.54
U.S. Department of Energy	0.612 MJ/h[b]	0.643 MJ/h[b]	0.05
Archer Daniels Midland Company	6.44×10^6 MJ[b]	8.97×10^6 MJ[b]	0.39

[a] Ref. 57.

[b] To convert MJ to Btu, divide by 1.054×10^{-3}.

[c] To convert MJ/L to Btu/gal, multiply by 3590.

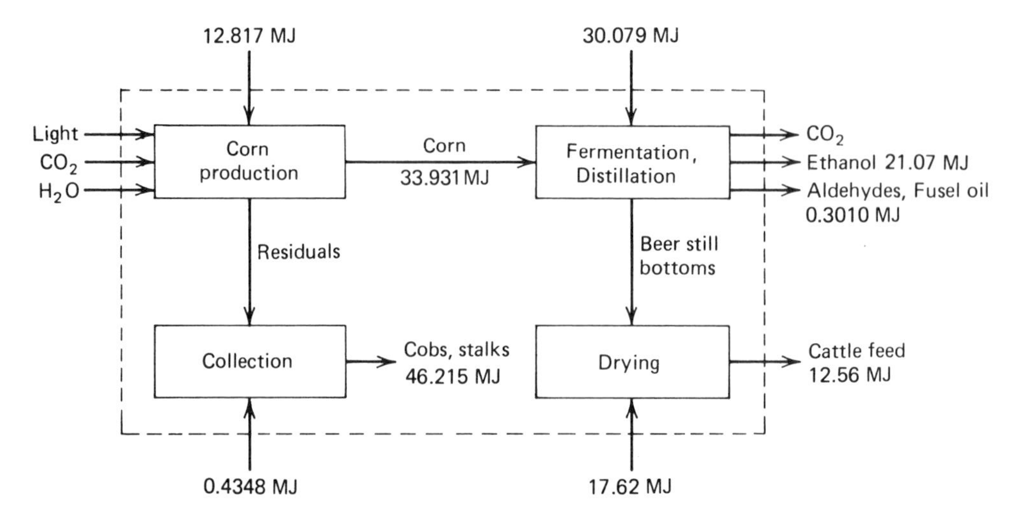

Figure 1. Energy inputs and outputs to manufacture one liter of anhydrous ethanol from corn (57). Ethanol yield: 279 L/m^3 (2.6 gal/bu corn). - - - denotes system boundary; all figures are lower heating values. To convert MJ to Btu, divide by 1.054×10^{-3}.

Table 9. Net Energy Production Ratios for Ethanol Production from Corn in an Integrated System[a]

Saleable energy products, E_p	Nonfeed energy inputs, E_x, for	N
ethanol	corn production, fermentation, bottoms drying	−0.65
ethanol	corn production, fermentation	−0.51
ethanol, chemicals	corn production, fermentation	−0.50
ethanol, chemicals, cattle feed	corn production, fermentation, bottoms drying	−0.44
ethanol, chemicals, cattle feed	corn production, fermentation, bottoms drying, 50% residuals[b]	−0.10
ethanol, chemicals, cattle feed	corn production, fermentation, bottoms drying, 75% residuals[b]	0.29
ethanol	corn production, fermentation, 75% residuals[b]	1.43
ethanol, chemicals	corn production, fermentation, 75% residuals[b]	1.47

[a] Ref. 57.
[b] Percent of cobs and stalks collected and used as fuel within system to replace fossil fuel inputs.

increase net energy production. Some of the information available on the designs of corn wet-milling plants, in which corn oil and other products are produced, indicate that integration with an alcohol plant may be more efficient than a conventional corn alcohol plant. Also, several reports recommend that the net energy production efficiency of ethanol plants be adjusted upward because ethanol effectively displaces more than an equivalent volume of gasoline in gasohol. Arguments are usually based on slight volumetric expansion upon mixing of ethanol and gasoline, slightly improved mileage with gasohol, and less crude oil consumption at the refinery because of the lower reforming severity since ethanol increases the octane of the blend.

Energy consumption in the ethanol recovery section of an ethanol fermentation plant can be excessive and can significantly reduce net energy production. A typical distillation train, consisting of successive units for beer distillation, removal of low boiling impurities such as aldehydes, refining crude ethanol to 190-proof (95 vol %), dehydrating to 200-proof (100 vol %) ethanol, and recovery of drying agent, might exhibit the energy consumption characteristics shown in Table 10. The most efficient configuration in this table for recovery of 96 vol % ethanol uses about one third the energy equivalent of the product, whereas conventional azeotropic distillation with benzene to produce anhydrous ethanol uses substantially more energy. The diethyl ether azeotropic system of Vulcan Cincinnati, Inc., apparently uses much less energy than the benzene system. Many opportunities exist for reducing the expenditures of energy in the distillation of ethanol by waste-heat recovery and reuse as well as use of innovative methods of dehydration. In addition, fuel-grade anhydrous ethanol need not be of high purity. For example, by energy reuse, pressure cascading, and waste-heat recovery, the steam consumption at one of the newly developed distillation systems for high grade industrial ethanol has been reduced to 3.0–4.2 kg/L (25–35 lb/gal) of 200-proof ethanol (59). For fuel-grade ethanol, steam consumption is 1.8–2.4 kg/L (15–20 lb/gal) of 199-proof ethanol, and 1.2–1.4 kg/L (10–12 lb/gal) of 192-proof ethanol (58).

Less energy-consuming methods for separating ethanol from water are also being developed. Drying of the partially concentrated ethanol solution with solid dehydrating agents, including cracked corn, starch, cellulose, and carboxymethyl cellulose, is reported to be effective for producing nearly anhydrous ethanol; the energy content of the product ethanol is ten times that needed for dehydration (60–62). In a variation of this technique, in which 80–90 vol % aqueous ethanol is dried with the saponified graft polymer of starch and acrylonitrile (HSPAN) in the present of gasoline, HSPAN selectively absorbs the water, thereby allowing the ethanol to dissolve in the gasoline phase (63). This suggests that gasohol blends might be made directly from aqueous ethanol. Other drying techniques involving solid absorbents, eg, molecular sieves (qv), solvent extraction, and ethanol-selective permeable membranes, are being studied (65). Commercial liquid-phase dehydration of aqueous ethanol with molecular sieves has been reported by Pall Pneumatic Products Corp. to produce 199.8-proof fuel-grade ethanol for less than one fifth the energy required for benzene distillation (2.65 MJ/L (9500 Btu/gal)).

Recent approaches to improving the ethanol fermentation process include the use of reduced pressure to remove ethanol as fast as it forms in the fermentation broth; use of bacteria instead of yeasts to shorten fermentation times; continuous fermentation to shorten fermentation times; simultaneous saccharification and fermentation of low grade cellulosics with enzymes and yeasts; use of thermophilic anaerobes for the one-step hydrolysis and fermentation of cellulosics; use of packed columns containing live, immobilized yeast cells or both enzymes and yeast cells through which glucose solutions are passed; and recombinant-DNA techniques to develop new yeast strains for rapid conversion of starch to sugar (60,65–68) (see Genetic engineering). For example, packed columns of live *Saccharomyces* yeast cells entrapped in carragenan gel convert 20 wt % aqueous glucose solutions containing nutrients to 12.8 vol % ethanol in 2.5 h (66). Biomass not normally used for alcohol production, eg, pineapple, has also been evaluated for ethanol fermentation (69). This biomass species, which requires much less water than sugarcane or cassava for growth, was projected to yield ethanol in greater quantities per unit biomass growth area than sugarcane or cassava. One of the most interesting research reports on the production of ethanol by fermentation involves the use of immobilized *Zymomonas mobilis* on fibers, eg, cotton, acrylics, and polyester, in a plug-flow, inclined, rotating fermentor (70). Passing sugar solutions through the unit produces 12 wt % ethanol solutions in 80% yields at

Table 10. Energy Requirements for Processing Fermentation Beer to Produce Ethanol[a]

Process	Ethanol feed, vol %	Columns used: Beer	Puri-fying	Recti-fying	Dehy-drating	Water stripping	Supple-mental rectifying and dehydrating	Ethanol product, vol %	Steam consumption: kg/L product[b,c]	MJ/L[d] product	Ethanol energy equivalent, %
continuous beer distillation, rectification	6.25	+		+[e]				96	2.40	5.11	25.2[f]
continuous beer distillation, purification, rectification	6.25	+	+	+[g]				96	4.07	8.67	42.8
continuous beer distillation, purification, rectification, with vapor reuse	6.25	+	+	+[g]				96	2.40	5.11	25.2
continuous dehydration of 96% ethanol with benzene at atmospheric pressure	96.0				+	+	+[g]	100	1.02	2.17	10.3
continuous beer distillation, purification, and benzene dehydration	6.25	+	+	+[g]	+	+	+[g,h]	100	5.26	11.21	53.2
continuous beer distillation, purification, and diethyl ether dehydration at 862 kPa (125 psi) with vapor reuse	6.25	+	+	+[g]	+	+		100	1.68	3.42	16.2

[a] Ref. 58.
[b] Assumes 27.2-kg (60-lb) saturated steam used with a latent heat of 2.131 MJ/kg (916 Btu/lb) except for the last process (Vulcan Cincinnati, Inc.) which uses 56.7-kg (125-lb) saturated steam with a latent heat of 2.038 MJ/kg (877 Btu/lb).
[c] To convert kg/L to kg/gal, multiply by 3.785.
[d] To convert MJ/L to Btu/gal, multiply by 3590.
[e] Enriching section only.
[f] Ethanol lower heating value: 21.09 MJ/L (7.571×10^4 Btu/gal) for anhydrous ethanol, and 20.25 MJ/L (7.27×10^4 Btu/gal) for 96 vol % ethanol.
[g] Enriching and exhausting sections.
[h] The main rectifying column may be used for concentrating aqueous ethanol from the stripping column, providing that there is sufficient excess capacity.

18-min residence times; 24–48-h residence times are usually required in conventional systems. If these results are valid and reproducible over long periods of time, this technique has the potential of greatly advancing ethanol fermentation technology.

Much of the current research on ethanol-fuel production by fermentation is directed at the development of pretreatment processes involving the conversion of low-value cellulosics to fermentable materials that can be used as ethanol feedstocks. The starchy components of corn biodegrade fairly easily, but the cellulose polymers in wood, for example, must first be converted to lower molecular weight sugar fragments before fermentation occurs with standard organisms. The use of cellulosic feedstocks offers large cost reductions. Table 11 illustrates that relatively low cost cellulosics, eg, wood and wheat straw, contribute much less on a percentage basis to product price than other established feedstocks. Table 12 is a listing of some of the processes being developed for converting low cost cellulosics to ethanol. The acid hydrolysis processes for wood, which were developed in the 1930s and 1940s, are included for comparison. Basically, these processes involve either acid- or enzyme-catalyzed hydrolysis of cellulose to sugars and separate or simultaneous fermentation. It appears that several of these processes will eventually be commercialized.

Thermochemical Ethanol. Considerable effort has been spent on the catalytic conversion of synthesis gas to mixed alcohols, in which ethanol is an important product. The Institut Français du Pétrole has patented a catalyst that contains a mixture of Cu and Co oxides, alkali-metal salts, and oxides of either Cr, Fe, V, or Mn, and that converts synthesis gas to mixed alcohols (45–46). At 250°C and 6.08 MPa (60 atm), conversion is 35% per pass; selectivity to alcohols is over 95%; and selectivity to two-carbon or normal higher alcohols is greater than 71%. A typical product distribution is methanol 20 wt %, ethanol 35 wt %, 1-propanol 21 wt %, 2-propanol 3 wt %, and 1-butanol 17 wt %. Other recent work done at the Sagami Chemical Research Center in Japan shows that ethanol is catalytically produced from synthesis gas in reasonably high selectivities at 0.1–5 MPa (1–50 atm) and 150–290°C over Rh, Pt, and Ir carbonyl clusters, eg, $Rh_4(CO)_{12}$, $[Pt_3(CO)_6]_{2-5}.2N(C_2H_5)_4$, and $Ir_4(CO)_{12}$ impregnated on basic oxides, such as MgO, CaO, La_2O_3, ZrO_2, and TiO_2 (72). For example, $Rh_4(CO)_{12}$ on La_2O_3 at 220°C and atmospheric pressure gave a product slate containing 19% methanol (carbon efficiency), 49% ethanol, 14% methane, 8% carbon dioxide, and 8% other compounds at 23% conversion of synthesis gas.

These types of processes would appear to have considerable merit for production of ethanol or mixed-alcohol fuels because synthesis gas can be manufactured from a broad range of fossil and nonfossil raw materials, including coal, peat, biomass, and organic wastes. Ethanol can also be obtained by homologation of methanol or its re-

Table 11. Capital, Operating, and Feedstock Costs as Average Percentages of Total Ethanol Costs[a]

Feedstock used	Capital cost	Operating cost	Feedstock cost
corn	11.6	26.5	61.9
sugarcane and juice	13.8	17.4	68.8
molasses	8.5	20.1	71.4
grain sorghum and milo	11.0	29.3	59.8
wheat straw	12.4	73.8	13.8
wood	28.9	55.1	16.1

[a] Ref. 71.

Table 12. Selected Cellulose-to-Ethanol Processes

Feedstock	Developer	Description
waste cellulosics (refuse-derived fuel (RDF), straw), wood	U.S. Army Natick Development Center	cellulase from mutant strains of *Trichoderma* catalyzes feed hydrolysis to sugars; small particles needed; sugars fermented
waste cellulosics (newsprint, corn stover)	SRI	shredded feed heated with dilute acid to remove hemicellulose; enzyme-catalyzed hydrolysis; fermentation of sugars under vacuum
RDF and agricultural wastes	Purdue University	dilute acid treatment to remove hemicellulose; strong acid treatment to solubilize cellulose which is separated and hydrolyzed with enzymes; sugars fermented
poplar wood	University of Pennsylvania–General Electric Co.	chips delignified with aqueous butanol; cellulose solids hydrolyzed with cellulases from *Thermomonospora*; sugars fermented
cellulosics	MIT	*Clostridium thermocellum* simultaneously produces cellulase, hydrolyzes cellulose, and ferments sugars
cellulosics	Auburn University	hemicellulose fraction hydrolyzed with dilute acid
waste cellulosics	Gulf Oil–University of Arkansas	cellulose hydrolyzed with enzymes from *Trichoderma reesei* and simultaneously fermented with yeasts
cellulosics	Iotech	steam-exploded cellulose hydrolyzed with enzymes from *Trichoderma reesei;* sugars fermented separately
cellulosics	Stake Technology–Vulcan Cincinnati	autohydrolysis; solubilized hemicellulose removed in water; lignins in alkali solution; cellulose saccharified and fermented
cellulosics (wood chips, sawdust, agricultural residues)	New York University	short-residence acid hydrolysis in an extruder; glucose syrup extracted and fermented
wood	Scholler (1930s)	batch sulfuric acid hydrolysis; sugars separated and fermented; the USSR has 25 commercial plants; Brazil is installing process for Eucalyptus
wood wastes	U.S. Forest Products Laboratory (1940s)	rapid acid percolation; sugars fermented; plant built in Oregon and operated for a short time

action with synthesis gas in the presence of cobalt octacarbonyl catalysts. Recently, a new catalyst system based on iron pentacarbonyl in the presence of a tertiary amine was reported to promote the reaction of methanol with synthesis gas to produce ethanol in high yield without formation of water (73).

$$CH_3OH + H_2 + 2\ CO \rightarrow C_2H_5OH + CO_2$$

A methanol conversion rate of 14%/h at 220°C, with ethanol accounting for 72 wt %

and methane 28 wt % of the product, was obtained for an iron carbonyl–manganese carbonyl catalyst. Although this work is still being carried out on a laboratory scale, the ability to synthesize ethanol without coproduct water could lead to lower-cost processes that do not require extensive energy-consuming distillations to manufacture fuel-grade ethanol.

Another nonfermentation route to ethanol that may provide leads to new commercial processes involves the conversion of furfural from rice hulls, corn cobs, and material from southern pines. Furfural undergoes ring cleavage and reduction in the presence of lithium metal and alkyl amine solvent to form ethanol (74). Bombardment with gamma rays of less expensive lithium salts in amine solvents may also promote the same reaction (75).

Fermentation Methanol. Fermentation methanol processes have not been developed, although an aerobic fermentation in which methanol forms directly from methane or natural gas by means of agents that capture methanol before it is oxidized to other products has been reported (49). Conversely, the biological reduction of carbon oxides produces methane, so anaerobic processes in which methanol intermediate is trapped and removed from the system are possible. However, there has been little research involving these concepts.

Thermochemical Methanol. Virtually all methanol is manufactured from synthesis gas. An advancement over the ICI vapor-phase process for conversion of synthesis gas to methanol is the Chem System, Inc., liquid-phase catalytic process (76). In this process, which is still being developed, an inert hydrocarbon liquid in the presence of a heterogeneous catalyst effects high conversions of synthesis gas. The liquid controls the reaction temperatures more efficiently than vapor and allows closer approach to equilibrium and more efficient recovery of the reaction heat as steam. The methanol stream produced is claimed to be suitable for direct use as fuel without distillation. A C_{14}–C_{21} mineral oil (72 wt % paraffinic and 28 wt % cycloparaffinic) with an ASTM boiling range of 270–349°C is preferred over aromatic liquids as the liquid phase. Product compositions obtained with synthesis gases that simulate those from a Koppers-Totzek coal gasifier after acid gas removal and those obtained with a Lurgi coal gasifier after complete hydrogen sulfide removal and partial carbon dioxide removal are shown in Table 13 (see Coal chemicals and feedstocks, gasification). The

Table 13. Typical Crude Product Compositions from Liquid-Phase Methanol Synthesis[a], wt %

Component	Lurgi synthesis gas feed	Koppers-Totzek synthesis gas feed
methanol	96.16	91.44
methyl formate	0.17	0.24
ethanol	0.32	2.55
2-propanol	trace	trace
methyl acetate	0.07	0.78
1-propanol	0.14	1.23
C_4 alcohols	0.23	1.43
C_5 alcohols	0.33	1.40
$C_{\geq 6}$ alcohols	0.06	0.56
water	2.71	0.51

[a] Ref. 77.
[b] Compositions are given on a mineral oil-free basis.

water would have to be removed from these products for gasoline-blending applications, because methanol–gasoline blends are even more sensitive to phase separation upon contamination with water than gasohol.

Another advancement being developed in synthetic methanol production involves the use of three different catalysts instead of one in successive beds that are intercooled for removal of the heat of reaction (78). The catalysts are chosen for their optimum activities at several successively higher temperatures, at which they are used to promote the successive reactions that occur on conversion of carbon oxides to methanol. Better catalysts, reactor design, and control of temperatures and space velocities afford maximum methanol yields. The raw reaction product contains ca 97 wt % methanol, 2 wt % higher alcohols, and not more than 1 wt % water and can be used as fuel without distillation. A main advantage is the small amount of water that forms from hydrogen in the synthesis gas in contrast to 10–20% yields produced by other processes. Overall thermal efficiency from coal-derived synthesis gas is 58%.

Processes are being developed for the coproduction of fuel-grade methanol containing higher alcohols; C_2–C_4 alcohols form by catalytic conversion of synthesis gas with modified zinc chromite catalysts in yields up to 30 wt % of the methanol (79–80). An increase in temperature, a decrease in space velocity, or a decrease in pressure results in an increase in higher alcohol concentration, and the presence of carbon dioxide adversely affects the yields of higher alcohols and the total product yield with increasing space velocity and pressure. Typical gas compositions of 70 vol % hydrogen, 17 vol % carbon monoxide, and 13 vol % nitrogen at 13–28 MPa (130–280 atm) and 300–425°C and gaseous hourly space velocities of 7000–20,000 h^{-1} give products containing 70–80 wt % methanol, 3–7 wt % ethanol, 2.5–5.5 wt % 1-propanol, 7–23 wt % 2-methyl-1-propanol, trace quantities of 1-butanol and 2-methyl-2-propanol, and small amounts of C_5 alcohols and ethers. This type of product has been termed methyl fuel and costs ca $0.183–$0.199/L ($0.69–$0.75/gal) when the higher alcohols are 20–30 wt % of the methanol. The higher alcohols are expected to reduce or prevent phase separation when methyl fuel–gasoline blends are contaminated with water. A similar process has been developed on a pilot-plant scale and evaluated in gasoline blends (81).

Consumption

United States. In the U.S. Midwest, Amoco Oil Company began marketing gasohol as an unleaded premium fuel on July 1, 1979, becoming the first large refiner in recent times to market an alcohol–unleaded gasoline blend in the United States (82). Other oil companies began to market gasohol soon thereafter. In 1979, ca 378,000 L/d (100,000 gal/d, 869,000 bbl/yr) of fermentation ethanol was produced for gasohol by several firms and sold through 2000 retail outlets out of a total of 175,000 outlets (1,83). Excluding synthetic and beverage-ethanol production capacity, because synthetic ethanol is derived from petroleum and is not affected by all of the fuel-ethanol tax benefits and because beverage ethanol is heavily taxed, total U.S. fermentation ethanol capacity in 1978 was about 189×10^6 L (50×10^6 gal) (1). About 242×10^6 L (64×10^6 gal) additional capacity existed in the form of idle plants and older plants that could be overhauled to produce fuel-grade ethanol (1). In July 1979, the Federal government set a 1990 production goal for ethanol fuel of $\geq 15.9 \times 10^9$ L/d (4.19×10^6 gal/d, 1.53×10^9 bbl/yr) or more than an order of magnitude increase in fermentation ethanol capacity over that of 1978 (1). This projected amount corresponds to only ca 1.4% of total U.S. gasoline consumption in 1978 (1).

Since 1979, fuel-ethanol sales have increased rapidly from 3.78×10^5 L/d (10^5 gal/d) to ca 1.76×10^6 L/d (4.65×10^5 gal/d) in July 1982 in the 21 states that produce most of the fuel ethanol in the United States (84). A list of U.S. anhydrous fermentation ethanol-fuel plants and their capacities is given in Table 14 (55–56,85). If the fuel-ethanol plants being constructed are added, the total capacity would be ca 2.118 GL/yr (5.6×10^8 gal/yr). This is ca 37% of the Federal goal for 1990.

Methanol is also marketed in the United States for blending with gasoline as an octane-enhancing agent, but it is not legally blended as a principal fuel component, eg, ethanol (see Legal Blending Limitations). Methanol-fuel usage was 340–568 ML [$(90–150) \times 10^6$ gal] in 1982 (86). This includes 189–378 ML [$(50–100) \times 10^6$ gal] for blending into gasoline, some of it illegally but mostly in admixture with 2-methyl-2-propanol; 151–189 ML [$(40–50) \times 10^6$ gal] as the additive methyl 2-methyl-2-propyl ether; and ca 3.8 ML (10^6 gal) as neat fuel. Current synthetic methanol capacity is far in excess of this level of methanol-fuel usage (53). Projections indicate that by the 1990s, methanol-fuel markets could include 1.13–2.27 GL/yr [$(2.3–6) \times 10^8$ gal/yr] for octane enhancement, ca 3.8 GL/yr (10^9 gal/yr) for blending with gasoline, and 15–38 GL/yr [$(4.0–10) \times 10^9$ gal/yr] as neat fuel if methanol-fueled cars are sold (86). Considerably more manufacturing capacity would be needed if methanol demand for the latter application occurs.

Brazil. Large-scale use of ethanol fuel was started in Brazil in October 1975 at a concentration of 10 vol % in gasoline (87). This was rapidly increased to 20 vol %, and all gasoline used in the country was converted to this blend (87). Brazilian ethanol production, primarily from sugarcane, reached 3.41 GL (0.9×10^9 gal) in 1980, out of which 2.68 GL (0.708×10^9 gal) was consumed as vehicle fuel (87). This program, which is called Proalcool by the Brazilians and which was started by government decree, continues to expand. The goal of Proalcool is to produce 10.7 GL (2.83×10^9 gal) ethanol by 1985 or ca 23% of all transportation fuels (88–89). Approximately 6.1 GL (1.6×10^9 gal) of 95 vol % ethanol is allocated for use as neat fuel, 3.1 GL (0.82×10^9 gal) anhydrous ethanol is allocated for blending with gasoline, and the remaining 1.5 GL (0.40×10^9 gal) is for the chemical industry (88). By the year 2000, Brazil's Otto-cycle-powered vehicles will require 40–45 GL [$(10.5–11.9) \times 10^9$ gal] of fuel, ca 20–25 GL [$(5.3–6.6) \times 10^9$ gal] of which will be supplied by ethanol and the balance gasoline (88).

In addition to Proalcool's goal of producing 10.7 GL (2.83×10^9 gal) ethanol by 1985, Brazil's National Energy Council approved a plan for large-scale production of neat-ethanol-fueled vehicles beginning in 1980 (1). This plan authorized manufacture of 900,000 vehicles during 1980–1982 and the adaptation of existing engines to ethanol. This represents an important step toward complete displacement of petroleum-based automobile fuels in Brazil. The automobile manufacturers involved in this program in Brazil are Volkswagen, Ford, General Motors, Fiat, and Chrysler. Although over 450,000 automobiles operated on neat ethanol near the end of 1981, problems arose regarding consumer acceptance of neat-alcohol-fueled cars, illegal conversions of gasoline-fueled cars to neat ethanol, and ethanol availability (90). The Brazilian Government will probably scale down the goals of Proalcool (90).

Table 14. Anhydrous Fermentation Ethanol-Fuel Plants in the United States[a]

Company	Plant location	Type of facility	Feedstock[b]	Anhydrous ethanol capacity, ML/yr[c]	Coproducts[d]
White Flame Fuels, Inc.	Van Buren, Ariz.	dry milling	corn (milo)	15.1	DDG
National Biofuels	Fayetteville, Ariz.	dry milling	starch waste, whey (corn)	1.9	DDG
Raven Alcohol Distillery	Selma, Calif.	conventional fermentation	molasses (cull fruit)	34.1	none
Parallel Products, Inc.	Dixon, Calif.	mobile	agricultural waste (sugar waste)	2.8	none
Baca Food & Fuel Corp.	Campo, Colo.	dry milling	milo (corn)	4.9	DDG
Eugene L. Schroder	Campo, Colo.	dry milling	milo (corn)	2.8	DDG
Colorado Argo-Energy, Inc.	Monte Vista, Colo.	dry milling	barley (potatoes)	8.3	DDG
Colorado Gasohol, Inc.	Walsh, Colo.	dry milling	corn (milo)	13.2	stillage
Biochemical Energy, Ltd.	Brooksville, Fla.	conventional fermentation	molasses (brewery waste, corn)	2.3	none
Southeast Georgia Agricultural Production & Development Corp.	Douglas, Ga.	dry milling	corn (wheat)	8.3	DDG or DWG
Syncorp, Inc.	Roberta, Ga.	dry milling	corn	10.6	stillage
Snake River Ethanol	Bliss, Ind.	dry milling	wheat (corn)	6.1	DDG
Beard Alcohol Farm	Grant, Ind.	dry and wet milling	potatoes (barley)	3.0	stillage
Spudcohol	Pingree, Ind.	dry milling	wheat (molasses)	3.8	stillage
Archer Daniels Midland	Decatur, Ill.; Peoria, Ill.; Cedar Rapids, Iowa	wet milling	corn	833 (total of 3 plants)[e]	na
Midstate Energy Resources	Lanark, Ill.	dry milling	corn	2.3	stillage
Pekin Energy Co.	Pekin, Ill.	wet milling	corn	227	traditional corn wet milling
Van Buren County Alcohol, Inc.	Bonaparte, Iowa	dry milling	corn	9.5	DDG
Appropriate Community Technology	Sheldon, Iowa	dry milling	corn	2.6	DDG
Farm Fuel Products Corp.	Storm Lake, Iowa	dry milling	corn	12.5	DDG
Midwest Solvents Co., Inc.	Atchison, Kan.	dry milling	milo (corn, wheat, starch)	22.7	DDG
Reeves Agri Energy	Garden City, Kan.	dry milling	milo (corn)	7.6	stillage
Ese Alcohol, Inc.	Leoti, Kan.	dry milling	corn (milo)	2.5	DDG
Fuel Alcohol	Bardstown, Ky.	dry milling	low proof ethanol (corn, barley)	17.0	stillage
National Distillery Products Co.	Louisville, Ky.	dry milling	corn	22.7	DDG
American Synfuels	Federalsburg, Md.	dry milling	corn (barley)	13.2	DDG
Jeritsma Farms	Bronson, Mich.	dry milling	corn	2.3	stillage
Alcon Industries, Inc.	Houston, Minn.	dry milling	corn	2.7	DDG
Zumbro Valley Alcohol & Feed	Kasson, Minn.	dry milling	corn (cellulose)	3.0	DDG
G&S Gasohol, Inc.	Mankato, Minn.	molecular sieves	190-proof ethanol	5.7	none

Kraft, Inc.	Melrose, Minn.	whey processing	cheese whey	3.0	none
Southeast Missouri Distillers, Inc.	Morehouse, Mo.	dry milling	corn (milo)	3.8	DDG
A. E. Montana, Inc.	Amsterdam, Mont.	dry milling	barley (wheat)	2.5	stillage
Alco Tech	Ringling, Mont.	dry milling	barley (wheat)	5.7	DDG
Ecological Energy, Inc.	Roca, Neb.	dry milling	corn (milo)	3.8	stillage
Deboer Brothers Renewable Energy, Inc.	Smithfield, Neb.	dry milling	corn (milo)	2.5	stillage
N-M Energy, Inc.	Clovis, N.M.	dry milling	corn (milo)	2.5	stillage
Cochran Oil	Jefferson, Ohio	dry milling	corn	1.3	stillage
Southpoint Ethanol	Southpoint, Ohio	dry milling	corn	227	DDG
Oklahoma Gasohol, Inc.	Hydro, Okla.	dry milling	milo (corn)	11.4	DDG
A. E. Staley Mfg. Co.	Loudon, Tenn.	wet milling	corn	151	traditional corn wet milling
Jonton Alcohol, TX	Edinburg, Texas	dry milling	grain sorghum	2.5	stillage
Charmec Energy Co.	Muleshoe, Texas	dry milling	corn	2.8	stillage
Agri-Ethanol Chemicals Corp.	Doswell, Va.	dry milling	corn	6.8	DDG
A. Smith Bowman Distillery, Inc.	Reston, Va.	ACR Process Corp. process	190-proof ethanol	10.9	none
Georgia Pacific Corp.	Bellingham, Wash.	pulp mill	lignin liquor	7.6	low sugar lignin
Olympia Brewing Co.	Olympia, Wash.	conventional fermentation	brewery waste	2.6	na
Matrix Energy Co.	Warren, Wash.	dry milling	corn (barley)	7.6	stillage
TCV Alcohol	Greenleaf, Wisc.	dry milling	corn	4.5	stillage
Summary[c]					
operating plants with capacities greater than 1.89×10^6 L/yr as of September 30, 1982				1763.3 (51 plants)	
operating plants with capacities of $(0.379\text{–}1.89) \times 10^6$ L/yr as of September 1981				56.0 (23 plants)	
stand-by plants with capacities greater than 1.89×10^6 L/yr as of September 30, 1982				62.3 (11 plants)	
stand-by plants with capacities of $(0.379\text{–}1.89) \times 10^6$ L/yr as of September 1981				1.9 (2 plants)	
Total				*1883.5*	
plants with capacities greater than 1.89×10^6 L/yr as of January 1, 1983				226.4 (14 plants)	
plants with capacities of $(0.379\text{–}1.89) \times 10^6$ L/yr as of September 1981				7.9 (9 plants)	
projected capacity as of January 1, 1983				2117.8[g]	

[a] Refs. 55–56, 85.
[b] Secondary feedstock indicated in parentheses.
[c] To convert L/yr to gal/yr, divide by 3.785.
[d] DDG is distillers' dried grains; DWG, distillers' wet grains.
[e] Ref. 55 states that anhydrous ethanol capacities are 276×10^6 L/yr (72.9×10^6 gal/yr) for the Decatur plant and 170×10^6 L/yr (45×10^6 gal/yr) for the Cedar Rapids plant.
[f] Includes 188–200-proof ethanol plants.
[g] Assumes that plants with capacities of $(0.379\text{–}1.89) \times 10^6$ L/yr $[(0.1\text{–}0.5) \times 10^6$ gal/yr$]$ are operational in January 1, 1983.

Economic Aspects

After wood alcohol was displaced by petrochemical methanol, the cost of methanol has almost always been less than that of fermentation and synthetic ethanol. Although the processes for manufacturing the alcohols by synthesis gas reduction (methanol), ethylene hydration (ethanol), and fermentation (ethanol) are very dissimilar, the basic embedded raw-material cost per unit volume of alcohol has been a principal factor in cost differentials among these products (as indicated in Table 15). The selling prices of the alcohols correlate with the embedded feedstock costs. This simple analysis ignores the value of by-products and processing differences, but it points out one of the main reasons why methanol has been the least expensive. The data in Table 15 also indicate that fermentation ethanol was competitive with synthetic ethanol in November 1982, in contrast to several years ago when synthetic ethanol was less expensive than fermentation ethanol (1). This resulted from the depressed prices of corn and the increasing price of ethylene in 1982. Forecasts have been made which suggest that, by 1990, fermentation ethanol will have a 6.6¢/L ($0.25/gal) price advantage over synthetic ethanol (91). This is based on 1990 prices of $1.10–1.32/kg for ethylene and $113–128/m^3 ($4.00–4.50/bu) of corn. In the early 1980s, fermentation ethanol is expected to maintain a slight price advantage over synthetic ethanol as long as corn prices remain low; this could reverse if the expected economic recovery in the middle 1980s disproportionately boosts corn prices compared to ethylene prices. Ethylene prices, however, generally follow those of oil and natural gas. These arguments are predicated on the continued use of existing processes and established feedstocks. The use of lower-cost feedstocks can change the alcohol pricing structure by 1990 through the commercialization of new processes based on coal, wood, wood wastes, and other feedstocks.

The Federal government and several state governments provide economic in-

Table 15. Estimated Embedded Feedstock Cost in Methanol and Ethanol

Alcohol	Feedstock	Feedstock price[a]	Embedded feedstock cost, $/L[b] 100% yield	Embedded feedstock cost, $/L[b] 60% yield	Posted price[c] $/L[b]	Posted price[c] $/GJ[d]	Feedstock as percent of posted price
methanol	natural gas	$3.32/GJ ($3.50/10^6 Btu)	0.07	0.12	0.20/L	12.69/GJ	35–60
200-proof synthetic ethanol	ethylene	$0.485/kg (0.22/lb)	0.235	0.39	0.48/L	22.76/GJ	49–81
200-proof fermentation ethanol	corn	$0.098/kg[e] ($2.50/bu)	0.14	0.23	0.45/L	21.34/GJ	31–51

[a] Estimated average price paid by alcohol producers for November 1982.
[b] To convert $/L to $/gal, multiply by 3.785.
[c] Average posted price in November 1982.
[d] To convert $/GL to $/Btu, multiply by 1.054×10^{12}.
[e] One bushel of corn is assumed to weigh 25.4 kg (56 lb).

centives to stimulate the production of alcohols for fuel applications (1,92). Of a total Federal excise tax of $0.09/gal ($0.0232/L), $0.05/gal ($0.132/L) has been forgiven until 1992 for fuel containing 10 vol % ethanol or methanol not made from natural gas, petroleum, or coal. This corresponds to a tax exemption of $0.13/L ($0.50/gal) ethanol. Several states have also exempted these blends from all or a portion of state motor-fuel or sales taxes, equivalent to a tax savings up to $0.29/L ($1.10/gal) ethanol. In addition, there is a 20% investment tax credit provided by Title III of the Federal Energy Tax Act of 1978 that helps stimulate new plant construction. With all the credits and subsidies, the USDA believes that a 150 mL/yr ethanol plant, based on corn wet milling, might be expected to have a 20–25% after-tax return on equity. Most of the state tax forgiveness subsidies are limited to fermentation ethanol. The Federal excise tax subsidies are effectively limited to fermentation ethanol, because all methanol is currently manufactured from synthesis gas made from natural gas or other fossil fuels. Some reports indicate that peat-derived methanol qualifies for the Federal excise tax subsidy. Peat, however, is regarded by many to be a young coal and may not meet the Federal requirements for the exemption. Several Federal loan guarantees have also been awarded for new fermentation ethanol plant construction (57).

Table 16 presents the results of an engineering cost analysis of methanol and ethanol production from several feedstocks at two plant sizes for each alcohol and two costs believed to be limiting for each feedstock (93). With the assumption that the market prices for methanol and ethanol in 1981 were $0.20/L ($0.75/gal) and $0.53/L ($2.00/gal), it was concluded that methanol from wood and coal are potentially competitive and that both methanol and ethanol could be made from biomass at prices competitive with petroleum-based products by the late 1980s.

Table 16. Estimated Plant Investment and Selling Price Ranges for Methanol and Ethanol, 1981 Dollars[a]

	Investment for different capacity plant, $\$10^6$ [b]					Alcohol prices from plants with different capacities, \$/L[c]			
Alcohol	94.6 $\times 10^6$ L/yr[c]	189 $\times 10^6$ L/yr[c]	379 $\times 10^6$ L/yr[c]	757 $\times 10^6$ L/yr[c]	Feedstock cost[c]	9.46 $\times 10^6$ L/yr[c]	189 $\times 10^6$ L/yr[c]	379 $\times 10^6$ L/yr[c]	757 $\times 10^6$ L/yr[c]
methanol									
from natural gas		40.1		108	\$0.071/m^3		0.20		0.13
					\$0.176/m^3		0.27		0.21
from coal		132		315	\$33.07/t		0.42		0.26
					\$66.14/t		0.49		0.33
from wood		113		299	\$16.53/dry t		0.33		0.25
					\$55.12/dry t		0.46		0.36
190-proof ethanol									
from ethylene	35		94		\$0.35/kg	0.41		0.32	
					\$0.70/kg	0.61		0.53	
from corn	44		117		\$56.76/m^3	0.46		0.36	
					\$169.68/m^3	0.85		0.74	
from wood by acid hydrolysis	124		327		\$16.53/dry t	0.74		0.46	
					\$55.12/dry t	0.90		0.63	

[a] Ref. 93.
[b] Plant investments based on 1981 costs, no escalation included, includes a 25% contingency.
[c] To convert L to gal, divide by 3.785.

Legal Blending Limitations

Fermentation ethanol has a *de facto* advantage over methanol, even though methanol has a substantially lower price than fermentation ethanol on both a volumetric and energy cost basis as shown in Table 15, because of the tax forgiveness requirements and the Federal controls on blending. Any company wishing to blend specific alcohols with unleaded gasoline in the United States must first obtain permission from the EPA if not already granted to another applicant. Up to 10 vol % ethanol can be blended in unleaded gasoline. As of early 1983, however, neat methanol in unleaded gasoline is limited to a maximum concentration of only 0.3 vol %, whereas neat 2-methyl-2-propanol and the corresponding methyl ether are limited to 7 vol %, and mixtures of methanol and 2-methyl-2-propanol in approximately equal amounts are limited to 9.5 vol %. The rationale for limiting the use of neat methanol to 0.3 vol % in unleaded gasolines relates to the existence of questions concerning the effects of methanol–gasoline mixtures upon fuel-system components as well as on water separation and evaporative emission characteristics. The 0.3 vol % limitation does not apply to blends with leaded gasolines; there are no Federal concentration limitations for alcohols in leaded gasolines. This rationale seems weak at best because the same basic problems that exist with neat methanol-unleaded gasoline blends would be expected to be common to neat methanol-leaded gasoline blends, since the lead additives are present at very small concentrations. DuPont's waiver request to increase the concentration maximum of neat methanol in unleaded gasoline blends to 3.0 vol % was denied by the EPA in February 1983.

Neat methanol at concentrations up to 3 vol % has been a common component of gasoline in the FRG for several years (94). Fuel marketeers in the FRG use methanol freely, with or without other alcohol cosolvents. As a result, the independent marketers sell methanol blends to ca 20% of the FRG market. Also, the Permanent Bureau of European Automobile Manufacturers Association accepts gasoline blends with 3 vol % neat methanol and has approved the proprietary methanol formulation Petrocoal for use in unleaded gasolines at methanol concentrations up to 12 vol %. This waiver has been contested by the Motor Vehicle Manufacturers' Association of the United States and other groups (95).

Some of the methanol–gasoline blends sold in the United States are in violation of the legal requirements, but little has been done to control distribution of these fuels. One of the reasons that methanol is apparently marketed illegally is that the alcohol was ca \$0.07/L (\$0.26/gal) lower in cost than unleaded gasoline in November 1982 and, thus, offered a larger profit margin to the retailer than gasoline.

Storage and Handling

The material-compatibility problems associated with vehicle fuel-storage and delivery systems also apply to the commercial storage of large quantities of alcohol fuels (19–20). The storage stability of alcohol–gasoline blends over long periods of time increases the rate of gum formation and loosens existing deposits of rust and other sediments (21). Some of the properties of alcohol fuels, eg, their evaporative emissions, phase-separation properties, and wide flammability range, and their low odor intensity, flame luminosity, and ability to form explosive mixtures in closed containers, in contrast to gasoline, also present storage and handling problems. None of these problems, however, appears to be insurmountable.

Operating Problems

Water Tolerance and Phase Separation. Water contamination can cause separation of alcohol–gasoline blends into a lower water-rich layer and an upper hydrocarbon-rich layer. The amount of water that the blend can tolerate before separation occurs depends on the alcohol structure and concentration, gasoline composition, and the effect of cosolvents, eg, higher alcohols, in the blend. Figure 2 illustrates the tolerance of methanol and ethanol blends with benzene upon addition of water. The curves show that ethanol is much less sensitive to phase separation in benzene than methanol, and that relatively small amounts of water would cause separation in the alcohol concentration ranges permitted in gasoline blends by the EPA. Table 17 shows the effects in unleaded gasolines blended with methanol and ethanol of alcohol concentration, temperature, cosolvent, and aromatic content of the gasoline. Again, ethanol has a much higher water tolerance than methanol. Water tolerance increases with both methanol– and ethanol–gasoline blends as the temperature and aromatic content of the gasoline increase and with addition of a cosolvent.

With methanol– and ethanol–gasoline blends, very small amounts of water contamination results in phase separation, which can cause engine-operating difficulties, corrosion, and plugging of small orifices with gums and deposits caused by the solvency of the alcohols. The reduction of alcohol concentration in the fuel blend, if the upper hydrocarbon-rich layer were withdrawn from storage tanks for distribution and use, can also adversely effect engine operation. Addition of cosolvents, eg, higher alcohols, or use of mixed alcohols containing methanol or ethanol as the dominant alcohols appears to be the most practical solution. Long-term storage of alcohol–gasoline blends should be avoided to preclude absorption of moisture from air, especially in humid climates.

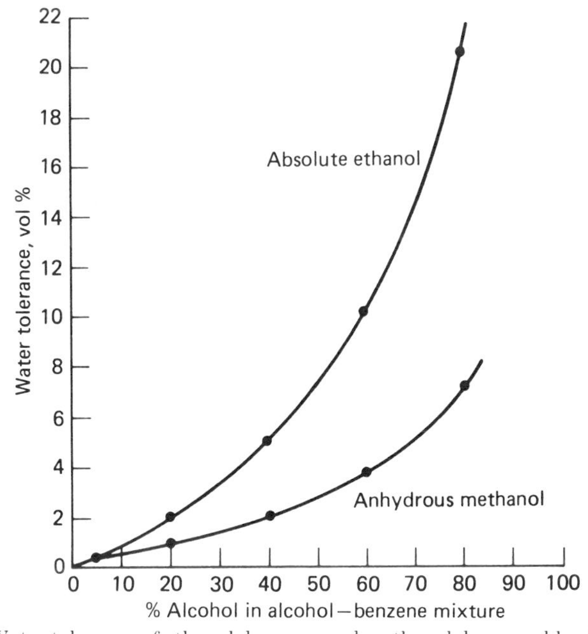

Figure 2. Water tolerance of ethanol–benzene and methanol–benzene blends at 0°C (16).

Table 17. Water Tolerance of Alcohol-Gasoline Blends[a]

Alcohol	Alcohol concentration, vol %	Temperature, °C						
		−30	−20	−10	0	10	15	20
methanol	10[b]					0.02	0.04	0.06
	10[c]			0.03	0.07	0.11	0.14	0.16
	10[d]		0.04	0.08	0.12	0.17	0.19	0.23
	10[e]		0.17	0.22	0.27	0.34	0.37	0.43
	15[c]				0.06	0.13	0.17	0.20
	20[c]					0.05	0.20	0.26
ethanol	5[c]	0.16	0.18	0.19	0.22	0.26	0.28	0.30
	10[b]	0.30	0.34	0.40	0.47	0.53	0.57	0.60
	10[c]	0.37	0.42	0.47	0.54	0.60	0.64	0.67
	10[d]	0.42	0.47	0.52	0.58	0.65	0.68	0.73
	10[f]	0.68	0.74	0.82	0.90	0.98	1.3	1.7
	15[c]	0.61	0.50	0.80	0.85	1.04	1.09	1.16

[a] Ref. 21.
[b] In unleaded gasoline containing 14 vol % aromatics.
[c] In unleaded gasoline containing 26 vol % aromatics.
[d] In unleaded gasoline containing 38 vol % aromatics.
[e] In unleaded gasoline containing 26 vol % aromatics; 3.2 vol % 2-methyl-1-propanol cosolvent added.
[f] In unleaded gasoline containing 26 vol % aromatics; 3.2 vol % 1-butanol cosolvent added.

Neat-alcohol fuels would not present the problem of phase separation, since water is soluble in methanol and ethanol in essentially all proportions, except at very low temperatures. It is not necessary to use anhydrous neat-alcohol fuels in spark-ignition engines. The neat-ethanol-fueled cars in Brazil operate with 95–96 vol % ethanol, which precludes the energy-consuming step of azeotropic distillation (88). The addition of water to alcohols may even improve their fuel characteristics. The addition of 10 and 20 wt % water to methanol raises its octane value from 102 (RON + MON/2) to ca 107 and 112, and the corresponding effect with ethanol raises its octane number from 100 to ca 108 and 115 (96). Higher thermal efficiencies would be expected for the aqueous alcohols as engine fuels, because water has a higher latent heat of vaporization than the alcohols. The resulting performance should be analogous to the effect of methanol on the performance of methanol–gasoline blends.

Emissions. Methanol and ethanol form azeotropes with many of the hydrocarbons in gasolines. The boiling points and compositions of a few of the constant-boiling mixtures are listed in Table 18. Thus, in alcohol–gasoline blends, the boiling points of many of the components are reduced and, at lower temperatures, the vapor pressure of the blend is higher than that of the gasoline alone. This is illustrated by the distillation curves shown in Figure 3. The depression of the initial portion of the curves is apparent; the relatively flat portion of the curves for the alcohol blends is sometimes referred to as the alcohol flat (16). As might be expected from the structures of methanol and ethanol and the data in Table 18, the alcohol flat is less depressed with ethanol. The inclusion of mixed C_2–C_4 alcohols in the methanol blend also provides less depression. The increased front-end (lower left) volatility from the alcohol flats increases evaporative emissions. At concentrations of 10 wt % in gasoline blends, methanol and ethanol increase evaporative emissions by as much as 130–220% and 49–62%, respectively (98–99). This can have an adverse effect on the volatility balance

Table 18. Constant-Boiling Mixtures of Methanol and Ethanol with Selected Hydrocarbons[a]

		Constant-boiling mixture			
Component	Boiling point, °C	Boiling point with methanol, °C	Methanol in mixture, wt %	Boiling point with ethanol, °C	Ethanol in mixture, wt %
2-methylbutane	27.95	24.5	4.0		
2-methyl-2-butene	37.15	31.75	7.0		
1,5-hexadiene	60.2	47.05	22.5		
n-hexane	68.95	50.0	26.4		
benzene	80.2	58.34	39.55	68.25	32.41
1,3-cyclohexadiene	80.6	56.38	38.8	66.7	34.0
cyclohexane	80.75	54.2	37.2	64.9	30.5
cyclohexene	82.75	55.9	40.0	66.7	35.0
n-heptane	98.45	60.5	62.0		
methylcyclohexane	101.8	60.0	70.0	73.0	53.0
toluene	110.6			76.7	68.0

[a] Ref. 16; all boiling points at 101.3 kPa (1 atm). The boiling points of methanol and ethanol are 65.0°C and 78.5°C, respectively.

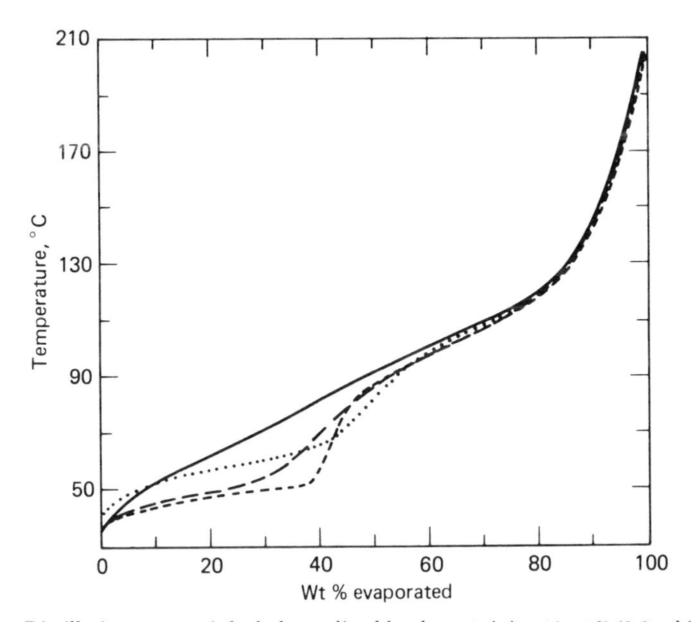

Figure 3. Distillation curves of alcohol–gasoline blends containing 10 wt % (9.2 vol %) alcohol (97). — base gasoline; --- + methanol; – – + 70% methanol, 30 wt % C_2–C_4 alcohols; ····· + ethanol.

of the fuel and can promote vapor lock. The latter is caused by fuel vaporization at some point in the fuel system where temperature exceeds the boiling points of the light fractions. It can stop the flow of fuel to the engine and, thus, interrupt normal engine operation. To overcome these problems, refiners adjust the lower boiling fractions of gasolines, particularly the butanes and pentanes, that are to be used for blending with alcohols.

Exhaust emissions with neat alcohols and alcohol–gasoline blends are essentially the same as those with gasolines: NO_x, CO, and unburned fuel and derivatives (UBFs).

For the same equivalence ratios (The equivalence ratio is the actual fuel:air ratio divided by the stoichiometric fuel:air ratio. It is 1.0 for stoichiometric air and fuel and less than 1.0 at lean-burn conditions), the NO_x emissions are lower than those formed with gasoline, because alcohol combustion temperatures are lower. At equivalence ratios slightly less than 1.0, the NO_x emissions peak, so the blend-leaning effect of alcohol–gasoline blends can either yield higher or lower NO_x emissions than gasoline, depending on whether the degree of leaning is away from or closer to the NO_x maximum. Under lean-burn conditions when excess combustion air is present, CO and UBF emissions usually decrease, so the leaning effect of alcohol–gasoline blends typically reduces these pollutants. Modern automobiles equipped with three-way conversion catalysts, oxygen sensors, and feedback loops to control NO_x, CO, and UBFs generally compensate for any changes that might occur with alternate fuels (100). The aldehyde emissions, which are greater from alcohol fuels than gasolines, are low enough so that they are not regulated by the EPA. A large displacement of hydrocarbon fuels by alcohols could lead to control of aldehyde emissions.

Driveability. The term driveability refers to the response of the vehicle to control manipulations by the driver. Modern automobiles powered with gasohol-type fuels generally respond satisfactorily to all throttle inputs. Some of the older vehicles that are not equipped with exhaust-sensing feedback controls may suffer from excessive blend-leaning effects, particularly when operating on fuels containing methanol (23). Manual adjustments to the carburetion system may be necessary if the fuel cannot be changed. Conventional cars should operate satisfactorily on alcohol–gasoline blends containing 5–20 vol % alcohol, but the driveability of the engine when it is warm deteriorates in proportion to the alcohol content (21). This is accounted for by the blend-leaning effect and closely follows the amount of oxygen added to the fuel. For example, 10 vol % methanol (50 wt % oxygen) or ethanol (35 wt % oxygen) in the fuel results in increased driveability demerits of about 90% and 60% with standard carburetion (21). Carburetor modifications to give the same equivalence ratios for the alcohol–gasoline blends as the unmodified carburetor gives with gasoline provides about the same driveability demerits for all fuels (21).

Vapor lock and difficult starting at high ambient or underhood temperatures with alcohol–gasoline blends would seem to be an inherent problem because of the alcohol flat, but there has been considerable variability in the results of fleet tests. The phenomenon seems to be highly vehicle-dependent with ethanol blends but common with methanol blends when no volatility adjustments are made (101–102). Methanol blends would be expected to exhibit greater tendency to vapor lock because of the greater depression of the alcohol flat. The increased volatility of alcohol–gasoline blends should facilitate cold starting and, indeed, there does not appear to be a problem. As already mentioned, air-neat alcohol mixtures delivered to an engine by a carburetor adjusted for gasoline are too lean to burn. Even if the carburetion system is adjusted to supply the proper air–fuel mixture, cold starting may still not be possible unless special measures have been taken. Sufficient vaporization of neat methanol and neat ethanol to form a flammable air mixture does not occur below 9–10°C and 15–17°C, respectively (21,23). The cold-starting problem with neat-alcohol fuels can be solved by several techniques, including addition of more volatile fuels to the alcohol, use of an auxiliary starting system that operates on another fuel such as gasoline, use of a small electric heater to warm a small amount of fuel, and use of a small catalytic alcohol reformer to produce enough hydrogen to start the engine. In Brazil, neat-ethanol-fueled cars are equipped with a small, two-liter (0.5-gal) auxiliary tank of gasoline (103). During start-up, the gasoline is fed by means of an electric pump and solenoid injector into the ethanol-fuel stream to give mixtures containing 5–10 wt % gasoline. This system has proven satisfactory. In normal operations, the neat-ethanol-fuel-intake system is preheated with hot exhaust gas.

Materials Compatibility. Neat-ethanol-fueled cars require special materials of construction in certain areas of the fuel system because of potential corrosion problems. The Ford Motor Company has reported that necessary materials changes include electroless nickel plating on the fuel tank instead of terneplate, ie, lead and tin, as well as on the zinc carburetor (104). The metal fuel-pump parts are protected with dichromate and apparently are resistant to deterioration by alcohols. The steel fuel lines must also be replaced with nylon 7 lines, but relatively few changes of the plastic and rubber parts are needed (104). Other sources indicate that polyethylene fuel lines are satisfactory and that the carburetor and upper part of the fuel pump should be chromium-plated (105). Methanol is more corrosive than ethanol towards metals, eg, magnesium and aluminum, especially if water is present, so the selection of suitable materials is more difficult. However, none of the material problems encountered with neat-methanol and ethanol fuels are insurmountable (104). The fuel systems of modern vehicles are constructed of materials compatible with gasohol and thus present little or no problem.

Additional potential compatibility problems are lubricant compatability, engine wear, and rust (19–20). Excessive cylinder-bore wear, apparently caused by use of neat-alcohol fuels, has been identified as a potentially severe problem and is being studied (23).

Fuel Economy. Brazilian automobiles equipped with engines designed specifically for neat-ethanol fuel consume only ca 15–20 vol % more fuel than the corresponding gasoline-fueled cars (103). Based on the heating values of 95% neat ethanol used in Brazilian cars and typical gasolines and assuming the same distance per energy unit consumed, these vehicles would be expected to consume about 40 vol % more ethanol than gasoline. Volkswagen (VW) in Brazil has also reported that VW engines designed for operation in Brazil on neat ethanol consume about 22% more fuel than comparable VW gasoline engines (106). The Ford Motor Company reported that neat-methanol- and ethanol-fueled Escorts equipped with alcohol engines operate at ca 64 vol % and 85 vol % of the fuel economy of a gasoline-fueled Escort (104). This means that, on a distance basis, 1.56 L (0.41 gal) methanol and 1.18 L (0.31 gal) ethanol are equivalent to 1.0 L (0.26 gal) gasoline. At the same thermal efficiencies, the equivalency should be 2.04 L (0.54 gal) methanol and 1.53 L (0.40 gal) ethanol per liter of gasoline, so the thermal efficiencies and the distance traveled for each energy unit expended in the neat-alcohol-fueled cars are considerably higher than in gasoline-fueled cars. The Brazilian automobile engines have 10–11:1 compression ratios and operate under lean-burn conditions, and the Ford Escorts have 11.4:1 compression ratios. This accounts in part for the performance of the neat-alcohol fuels, but the fuel economy is still higher than might be expected. In general, gasohol fuels would be expected to have about the same or slightly less fuel economy than gasoline in the same vehicle.

Toxicology

A summary of the toxicological characteristics of methanol, ethanol, and 2-methyl-2-propanol is given in Table 19. The data in this table were produced by many different sources and, in most cases, cannot be quantitatively compared, but a number of generalizations can be made. Although it has been known for many years that human ingestion of small amounts of methanol can cause blindness and death, the toxic doses listed in Table 19 for small animals indicate that ethanol can be as toxic as methanol. However, these materials are moderate hazards and should be handled accordingly. The organic components in petroleum fuels have a broad range of lethal toxic doses, so the compositional variations of a petroleum fuel would be expected to result in corresponding variations of toxicity.

It is apparent that tetraethyl lead presents a serious hazard in the pure state. At the small concentrations used in leaded gasolines [0.79 mL/L (3 mL/gal)], the toxicological effect of this additive is substantially diluted. Based on available data, the toxicity of methyl 2-methyl-2-propyl ether, which can be used at concentrations up to 7 vol % (70 mL/L) in unleaded gasoline, is much less toxic than tetraethyl lead.

Additional toxicological data for alcohol–gasoline blends should be researched and reported, because little or no data are available on the toxic effects of gasohol-type mixtures. At least it should be established that there are no unexpected trends in toxicity ratings for the blends and that toxicities can be estimated reasonably accurately from fuel compositions and from data for the individual components.

Uses

Conversion to Other Fuels. Uses of alcohols as intermediates in the production of other fuels and for nonvehicular fuel applications are being developed. An indirect route to gasoline that has received much attention is the Mobil Oil Company process, in which methanol is converted to gasoline over zeolite catalysts. Basic studies have revealed a possible mechanism for the process (110). Based on product-distribution analysis, the reaction path for methanol conversion to hydrocarbons appears to be reasonably represented by the following scheme:

$$2\,CH_3OH \underset{+H_2O}{\overset{-H_2O}{\rightleftharpoons}} CH_3OCH_3 \xrightarrow{-H_2O} C_2\text{–}C_5 \text{ olefins} \rightarrow \text{paraffins} + \text{olefins}$$

The initial dehydration reaction is sufficiently fast that an equilibrium among methanol, dimethyl ether, and water is established. Analysis of the kinetic data suggests that autocatalytic reactions between methanol/dimethyl ether and olefins are important steps. The liquid hydrocarbon yields are high and no coke or residuals form. The patented ZSM-5 zeolite catalysts are crystalline aluminosilicates, and some have been used for several years for ethylbenzene production, xylene isomerization, and toluene disproportionation. The unusual shape-selective properties of these catalysts are apparently responsible for the high liquid-hydrocarbon yields. For example, at 450°C and a weight hourly space velocity of 0.67 h^{-1}, methanol is converted to 48 wt % gasoline, 42 wt % C_3–C_4 hydrocarbons, 8 wt % C_1–C_2 hydrocarbons, and 2 wt % distillate; no coke forms (57,111). A detailed analysis and comparison of the Mobil and Fischer-Tropsch processes has led New Zealand to the decision to build a large natural-gas-to-gasoline plant based on methanol (112). A summary of their analysis

Table 19. Toxicity of Selected Fuels and Fuel Components

Substance	Oral toxicity[a,b]	Action on skin[a,c]	Irritation dose[d]				Toxic dose[d]					Aquatic toxicity, ppm (10^6)[d]	Threshold limit, ppm (10^6)[d]
			Tissue	Species	Dose, mg	Time, h	Effect	Entry	Species	Type	Dose		
methanol	3	D	eye	rabbit	40	72	moderate	oral	rat	LD_{50}	13 g/kg	>1000	200
								intraperitoneal	rat	LD_{50}	9.54 g/kg		
								skin	rabbit	LD_{50}	20 g/kg		
								inhalation	rat	LC_{50}	64,000 ppm/4 h		
ethanol	2	D	eye	rabbit	100	24	severe	oral	rat	LD_{50}	14 g/kg	>1000	1000
			skin	rabbit	500	24	severe	intraperitoneal	rat	LD_{50}	6.06 g/kg		
								intravenous	rat	LD_{50}	1.44 g/kg		
								oral	rabbit	LD_{50}	6.3 g/kg		
								inhalation	rat	LC_{50}	20,000 ppm/10 h		
2-methyl-2-propanol								oral	rat	LD_{50}	3.5 g/kg	>1000	100
								intraperitoneal	mouse	LD_{50}	0.933 g/kg		
methyl 2-methyl-2-propyl ether			eye	rabbit			mild	oral	rat	LD_{50}	4.0 mL/kg		
								skin	rabbit	LD_{50}	>10 mL/kg		
								inhalation	rat	LC_{50}	85 mL/(L·4 h)		
tetraethyl lead	6							parenteral	rat	LD_{50}	0.015 g/kg	<1	0.1 mg/m^3
								inhalation	rat	LC_{50}	0.85 g/(m^3·h)		

[a] Ref. 107.

[b] Rating is based on acute toxicity of a single dose taken by mouth to produce death when swallowed by an average (70-kg) man; 1, practically nontoxic [takes more than 1 L (ca 1 qt)]; 2, slightly toxic (0.5–1 L); 3, moderately toxic (30 mL–0.5 L); 4, very toxic (1.3–80 mL); 5, extremely toxic (7 drops to 1.3 mL); 6, super toxic (less than 7 drops).

[c] D, can cause solvent irritation-type dermatitis.

[d] Refs. 108–109.

Table 20. Comparison of Yields and Economics of Mobil Oil's Methanol-to-Gasoline Process and Fischer-Tropsch Processes[a]

Property	Mobil process[b]	Arge process[c]	Synthol process[d]
gasoline, 1000 L/PJ[e] feed gas	15.4	6.5	8.0
diesel fuel, 1000 L/PJ[e] feed gas		6.0	4.9
fuel oil 1000 L/PJ[e] feed gas			0.6
Total liquids, 1000 L/PJ[e]	*15.4*	*12.5*	*13.5*
Total fuel gas, 1000 L/PJ[e]		*0.6*	
capital investment, $\$10^6$ [f]			
at gas feed rate of 50 PJ/yr[g]	280	410	430
at gas feed rate of 75 PJ/yr[g]	390	540	570
at gas feed rate of 100 PJ/yr[g]	480	680	700
annual operating costs, $\$10^6$ [f]			
at gas feed rate of 50 PJ/yr[g]	19	27	28
at gas feed rate of 75 PJ/yr[g]	26	36	38
at gas feed rate of 100 PJ/yr[g]	33	46	47
average product cost, $\$/m^3$ [f,h]			
at gas feed rate of 50 PJ/yr[g]	145	226	214
at gas feed rate of 75 PJ/yr[g]	138	208	201
at gas feed rate of 100 PJ/yr[g]	132	201	126

[a] Ref. 112; natural gas feedstock; costs in \$ New Zealand, mid-1979.
[b] Including recycling of liquid petroleum gas.
[c] Including gas recycle.
[d] Net yield after gas recycle.
[e] To convert L/PJ to bbl/10^{15} Btu, multiply by 6,631.
[f] Product costs are based on the assumption of a 10% DCF return, a 15-yr project life, and a gas price of \$1/GJ (\$1054/10^9 Btu); costs are for synthetic fuels considered individually, ie, refinery costs for other products are excluded; investment figures are ±25%, but all costs have been calculated on a consistent basis.
[g] To convert PJ/yr to 10^{15} Btu/yr, divide by 1054.
[h] To convert $\$/m^3$ to \$/bbl, multiply by 0.1590.

is shown in Table 20 (112). The New Zealand plant will be the first commercial plant to convert natural gas to gasoline. It will use Mobil's fixed-bed reactor configuration and is scheduled for completion in mid-1985 (113). Two methanol plants, each with 2000 t/d capacities, will be used (113).

The use of shape-selective catalysts for the direct conversion of ethanol to gasoline has also been reported (114). Methanol conversion takes place with anhydrous methanol, but with ethanol, 28 wt % water is necessary. The conversion of aqueous ethanol to hydrocarbon fuels is potentially competitive with the distillation route to ethanol fuel because of the energy consumed in the removal of water. Cost estimates of \$0.33/L (\$1.26/gal) gasoline have been made.

An indirect method of utilizing alcohols as motor fuels is to first reform them in catalytic reactors with exhaust heat from the engine, and then operate the engine on the hydrogen and carbon monoxide in the reformer gas. The advantages of this approach are that the strongly endothermic dissociation effectively increases the energy content of the alcohol in the fuel tank through energy capture and reuse. It is also well established that gas-fueled engines are reliable, more efficient, and do not suffer from some of the operational problems of liquid-fueled engines. For methanol, catalytic reforming increases the equivalent lower heating value by:

$$CH_3OH \rightarrow 2\ H_2 + CO \qquad +22\%$$

$$CH_3OH + H_2O \rightarrow 3\ H_2 + CO_2 \qquad +15\%$$

At 300°C and 150 kPa (1.5 atm), equilibrium favors the first reaction going to about 99.9% completion. In recent work on this concept, operation of a dissociated methanol-fuel system gave improvements in brake thermal efficiency of up to 100% over gasoline in engine dynamometer tests, even though the fuel did not completely dissociate (115).

Nonmotor Fuel Uses. Methanol is being evaluated as a substitute for conventional fossil fuels in stationary applications, eg, for boilers and turbines. Boiler fuel uses would not appear to be economically feasible at this time, because the advantage of methanol could not be realized as in spark-ignition engines and the fuel costs would be strictly related to energy content. If the price of petroleum escalates, large-scale methanol plants based on coal or biomass might provide incentives for use of methanol boiler fuels.

Turbine fuel uses of methanol by electric utilities appear to offer several long-range benefits. In studies conducted by the Electric Power Research Institute, the coproduction of methanol from surplus fuel gas produced for electric power generation in a coal-gasification plant would provide a means to effectively store the fuel gas during periods of low electric power demand (116–118). Economic projections indicate that methanol coproduced by a utility company by means of the best available coal-gasification and methanol-synthesis technology would cost about one-third less than that purchased from a nonregulated company (116). It is believed that by coproducing methanol in large baseload gasification-combined-cycle power plants or in gasification-based fuel-gas plants, the electric utility industry could achieve declining constant dollar costs for liquid fuels at a considerable reduction in cost while ensuring a secure supply (118). Methanol has never been used in the United States as a turbine fuel, but a 532-h test in 1979 showed that it would be an excellent turbine fuel from both an operational and an emissions viewpoint (116,118).

Other stationary applications of alcohol fuels that are being studied and developed include their use to power fuel cells and in coal slurries for transport and direct combustion, eg, of Methacoal or Ethacoal. Several years ago, the concept of converting natural gas in some of the gas-producing areas to methanol, transporting the methanol by ship, and gasifying the methanol at its destination for distribution as fuel gas or conversion to substitute natural gas (SNG) was considered to be an alternative to the liquefaction of natural gas, its shipment in cryogenic tankers as liquefied natural gas (LNG), and vaporization at its destination. Wasted flare gas was also considered to be a logical source of methanol for the same application. The concept has apparently not been field-tested.

Higher Alcohols. The emphasis on alcohol fuels has been almost entirely devoted to methanol and ethanol. There is commercial use of 2-methyl-2-propanol as a co-solvent with methanol in gasoline blends, and some work is in progress on the coproduction of mixed alcohol fuels containing higher alcohols, but methanol or ethanol usually predominates in these mixtures. One of the notable exceptions is the development of mixtures of 1-butanol and acetone as vehicular fuels (119). These mixtures are prepared by the Weizman process, in which pentoses and hexoses are converted by anaerobic fermentation. So-called power-grade fuels are produced and contain ca 17 wt % water and 83 wt % 1-butanol, acetone, and ethanol in ratios of ca 7.75:4:1. These mixtures as such or in gasoline blends provide good performance in spark-ignition engines (120).

Outlook

Technology. Processes for the utilization of cellulosic feedstocks for ethanol manufacture are expected to be commercialized in the 1980s. This should reduce the cost of fuel-grade ethanol and help to displace grains as the source of ethanol in gasohol. It is conceivable that the development of selective thermochemical processes for the manufacture of ethanol from synthesis gas derived from the less-expensive feedstocks, eg, coal and biomass wastes, will make the use of fermentation ethanol in gasohol obsolete. The availability of thermochemical ethanol from coal and biomass could profoundly affect methanol fuels. The price differential between thermochemical methanol and ethanol made from synthesis gas would be expected to be much smaller than the current differential. This could result in total displacement of fermentation ethanol and methanol motor fuels. At similar prices, the energy cost of ethanol would be substantially less than that of methanol. The price reduction of thermochemical ethanol plus ethanol's overall characteristics and performance in spark-ignition engines would tend to make it the preferred fuel.

Engine-design advances that take advantage of methanol and ethanol fuels are expected to be developed to extract the most brake horsepower possible. The dissociation technique that permits engine operation at thermal efficiencies higher than provided by methanol alone marks the beginning of this trend.

Gasohol will continue to be marketed in the 1980s and should have no competition from other alcohol fuels until practical methods are developed that permit the use of methanol–gasoline blends. In the 1990s, neat-alcohol-fueled cars should be available to the general public. Barring the development of high selectivity processes for thermochemical ethanol made from synthesis gas, neat methanol is expected to be the fuel of choice.

Impact on Petroleum Use. If it is presumed that fermentation-ethanol continues to be the primary alcohol fuel, little or no impact should be felt by petroleum refiners and marketeers in terms of gasoline sales. Fermentation-ethanol capacity for fuels is capacity-limited and is not expected to displace more than a small percentage of petroleum usage by 1990. Even if Federal targets for 1990 are met by fermentation ethanol, the effect on gasoline sales will be minimal because of their enormous size. The higher price of fermentation ethanol relative to gasoline should persist if ethanol fuel continues to be manufactured from grains, and this is another reason why petroleum usage should not deteriorate.

In the United States, the most obvious way to replace petroleum gasoline is to convert coal to a gasoline substitute, ie, methanol, ethanol, or synthetic gasoline. For example, the largest U.S. methanol plant today has a capacity of ca 2.6–2.9 ML/d (16,000–18,000 bbl/d). Fourteen or fifteen 8-ML/d (50,000-bbl/d) methanol plants based on coal would be able to supply enough methanol to displace ca 10% of the country's daily gasoline consumption. Since the lead time to build such plants is about five years, some analysts have projected that by the 1990s, sufficient methanol capacity could be developed for fuel applications to begin a transition from gasoline. Technology also exists for efficient conversion of methanol (and ethanol) to high octane gasoline if it is decided that hydrocarbons should be retained as the primary fuel for motor vehicles.

Impact on Food and Feed Sources. Although expanded grain-based ethanol-fuel manufacture is not expected to have much impact on gasoline markets, it can have serious adverse effects on food and feed supplies. Corn is projected to be in short supply in the 1990s because of diversion of corn resources to meet ethanol-fuel requirements. Shortfalls of 10–20%, termination of grain exports, and no expansion of livestock feeding have been predicted. Beef producers are concerned about the effects on meat prices of increased feeding of distillers' grains; some fear that a subsidized alcohol market would sharply raise feed prices. Some feed producers believe that large-scale gasohol marketing will depress soybean-meal prices because of the competition from distillers' grains. These events would most certainly have a negative impact on food and feed supplies and prices. However, these predictions presume that ethanol fuel will be manufactured from grains. It is virtually certain that grain-based ethanol fuel will be replaced in the 1990s by some of the processes based on biomass cellulosics and, perhaps, coal. Grains are just too valuable and limited to be used as a fuel source for an extended period of time.

Impact on Chemical Manufacture. The availability of large quantities of low cost methanol or ethanol from fuel-manufacturing operations would present chemical manufacturers with new processing schemes that are presently uneconomical. Many of the specialty and heavy organic chemicals could be made from methanol and ethanol because all but ca 10% of the organic chemicals marketed today are made from synthesis gas, ethylene, propylene, butadiene, benzene, and *p*-xylene. Brazil has already taken the lead in this area through the manufacture of ethylene, acetic acid, and octanol from ethanol and is developing other processes.

BIBLIOGRAPHY

1. D. L. Klass, *Energy Topics*, Institute of Gas Technology, Chicago, Ill., April 14, 1980, 8 pp.
2. H. T. Pratt, *Chem. Eng. News* **56**(18), 64 (May 1, 1978); **56**(41), 2, 55 (Oct. 9, 1978).
3. R. K. Duncan, *The Chemistry of Commerce*, Harper & Brothers Publishers, New York, 1907, p. 147.
4. G. Egloff and J. C. Morrell, *Ind. Eng. Chem.* **28,** 1080 (1936).
5. *Use of Alcohol in Motor Gasoline—A Review*, Publication No. 4082, American Petroleum Institute, New York, Aug. 1971.
6. *Alcohols—A Technical Assessment of Their Applications as Fuels*, Publication No. 4261, American Petroleum Institute, New York, July 1976.
7. T. J. Feaheny in *Symposium Papers Nonpetroleum Vehicular Fuels III*, sponsored by Institute of Gas Technology, Washington, D.C., Oct. 12–14, 1982.
8. *Ethyl Alcohol Handbook*, 5th ed., U.S.I. Chemicals, 1981.
9. A. W. Giebelhaus, *Agric. Hist.* **154,** 178 (1980).
10. L. M. Christensen, R. M. Hixon, and E. J. Fulmer, *Power Alcohol and Farm Relief*, Iowa State College, Ames, Iowa, 1934, Chapt. VII.
11. H. F. Wilkie and P. J. Kolachov, *Food for Thought*, Indiana Farm Bureau, Inc., Indianapolis, Ind., 1942.
12. E. R. Riegel, *Industrial Chemistry*, The Chemical Catalog Co., Inc., New York, 1933.
13. J. L. Keller, G. M. Nakaguchi, and J. C. Ware, *Methanol Fuel Modification for Highway Vehicle Use*, HCP/W3683-18, final report for U.S. Department of Energy, Washington, D.C., July 1978.
14. E. Johansson in *Retrofit '79 Proceedings of a Workshop on Air Gasification*, sponsored by Solar Energy Research Institute, Seattle, Wash., Feb. 2, 1979, Chapt. 5.5.
15. T. Powell, *Racing Experiences With Methanol and Ethanol-Based Motor-Fuel Blends*, Paper 750124, Automobile Engineering Congress and Exposition, Society of Automotive Engineers, Detroit, Mich., Feb. 1975.
16. A. W. Nash and D. A. Howes, *The Principles of Motor Fuel Preparation & Application*, Vols. I and II, John Wiley & Sons, Inc., New York, 1935.

17. *Alcohols as Motor Fuels, Progress in Technology Series*, No. 19, Society of Automotive Engineers, Inc., Detroit, Mich., 1980.
18. *Proceedings of Third International Symposium on Alcohol Fuels Technology*, U.S. Department of Energy, Asilomar, Calif., May 29–31, 1979.
19. *Ethanol Fuel Modification for Highway Vehicle Use*, Union Oil Company of California, final report ALO-3683-T1 for U.S. Department of Energy, Washington, D.C., Jan. 1980.
20. J. L. Keller, G. M. Nakaguchi, and J. C. Ware, *Methanol Fuel Modification for Highway Vehicle Use*, final report HCP/W3683-18 for U.S. Department of Energy,, Washington, D.C., July 1978.
21. J. L. Keller, *Hydrocarbon Process.* **58**(5), 127 (May 1979).
22. M. I. Greene, *Chem. Eng. Prog.* **78**(8), 46 (Aug. 1982).
23. P. W. McCallum, T. J. Timbario, R. L. Bechtold, and E. E. Ecklund, *Chem. Eng. Prog.* **78**(8), 52 (Aug. 1982).
24. R. C. Weast, ed., *Handbook of Chemistry and Physics*, 63rd ed., CRC Press Inc., Boca Raton, Fla., 1983.
25. L. M. Gibbs and B. J. Gilbert, *Centra Costo County's One-Year Experience With Gasohol*, Paper 810440, Society of Automotive Engineers, Detroit, Mich., Feb. 1981.
26. W. J. Shadis and P. W. McCallum, *A Comparative Assessment of Current Gasohol Fuel Economy Data*, Paper 800889, Society of Automotive Engineers, Detroit, Mich., Aug. 1980.
27. K. R. Stamper in ref. 18, I-9, pp. 1–14.
28. L. Barker, T. Pucholski, and K. Tholen, *Use of Ethanol in Diesel Engines*, SIM No. 11026, Solar Energy Research Institute, Golden, Colo.
29. J. Strait, J. J. Boedicker, and K. C. Johanson, *Diesel Oil and Ethanol Mixtures for Diesel Powered Farm Tractors*, final report to the State of Minnesota Energy Agency, Minn., 1978.
30. V. Hofman and co-workers, *Sunflower Oil as a Fuel Alternative*, Bulletin No. 13-AENG-5, Cooperative Extension Service, North Dakota State University, Fargo, N.D., 1980.
31. P. S. Berg, E. Holmer, and B. I. Bertilsson in ref. 18, II-29, pp. 1–8.
32. F. F. Pischinger in ref. 18, II-28, pp. 1–11.
33. C. A. Moses in ref. 18, III-51, pp. 1–11.
34. P. A. Boruff, A. W. Schwab, C. E. Goering, and E. H. Pryde, *Trans. ASAE* **25**(1), 47 (1982).
35. T. L. Narasimhan, M. R. K. Rao, H. A. Havemann, *J. Indian Inst. Sci. Sect. B* **28**, 224 (1956).
36. M. R. K. Rao and Y. M. Balakrishna, *J. Indian Inst. Sci. Sect. B* **39,** 17 (1957).
37. J. M. Cruz, W. J. Chancellor, and J. R. Gross in ref. 18, II-27, pp. 1–12.
38. H. Adelman in ref. 17, pp. 267–275.
39. S. Miyawaki and co-workers, *Evaluation of MTBE Gasoline by Japanese Passenger Cars*, Paper 801352, Society of Automotive Engineers, Detroit, Mich., Oct. 1980.
40. J. A. Harrington, D. D. Brehole, and E. H. Schanerberger in ref. 18, III-53, pp. 1–13.
41. L. G. Goran Svahn in ref. 18, I-10, pp. 1–12.
42. F. W. Cox in ref. 18, II-22, pp. 1–14.
43. *Chem. Eng. (N.Y.)* **86**(5), 78 (Feb. 26, 1979).
44. L. Barker and T. Puchalski, SERI Contract No. EG-77-C-01-40-42, Solar Energy Research Institute, Golden, Colo.
45. *Alcohol Production From Biomass in the Developing Countries*, World Bank, Washington, D.C., Sept. 1980.
46. *Alcohol, Tobacco and Firearms Summary Statistics*, ATF P 1323.1, Fiscal Year 1976, pp. 4–77; Transitional Quarters Ending Sept. 30, 1976, pp. 8–77; Fiscal Year 1978, pp. 4–81; Fiscal Year 1979, pp. 7–82, Department of Treasury, Washington, D.C.
47. K. Klier in *Proceedings Biomass-to-Methanol Specialists' Workshop*, sponsored by U.S. Department of Energy, Durango, Colo., March 3–5, 1982, pp. 5–25.
48. E. L. Foo, *Process Biochem.* **13**(3), 23 (March 1978).
49. M. G. Gonikberg, *Chemical Equilibria and Reaction Rates at High Pressures*, 2nd ed., Izdate/'stro Akademii Nauk SSSR, Moscow, 1960.
50. R. S. Liu, M. Iwamoto, and J. H. Lunsford, *J. Chem. Soc. Chem. Commun.*, 78 (1982).
51. W. A. Rains, *paper presented at 1982 Annual Meeting, National Petroleum Refiners Association*, San Antonio, Texas, March 21–23, 1982.
52. *Chemcyclopedia of Chemicals*, Vol. 1, American Chemical Society, Washington, D.C., 1982–1983; *Chem. Eng. News* **58**(50), 10 (Dec. 15, 1980).
53. *Chem. Eng. News* **60**(13), 24 (March 29, 1982).

54. A. Adams, *Large-Scale Alcohol Fuel Plants Directory*, SERI/SP-290-1467, Solar Energy Research Institute, Golden, Colo., Feb. 1982.
55. *Alcohol Week* **3**(41), 1 (Oct. 11, 1982).
56. F. Potter, Information Resources, Inc., Nov. 1982.
57. D. L. Klass in *Symposium Papers Energy From Biomass and Wastes IV*, sponsored by Institute of Gas Technology, Lake Buena Vista, Fla., Jan. 21–25, 1980, pp. 1–41.
58. *Energy Requirements Ethyl Alcohol Production From Beer (Fermented Mash) for Various Processes, Including Proprietary Vulcan, Cincinnati Vapor Reuse Process*, Vulcan Cincinnati, Inc.
59. R. Katzen and co-workers in D. L. Klass and G. H. Emert, eds., *Fuels from Biomass and Wastes*, Ann Arbor Science Publishers, Inc., 1981, Chapt. 21.
60. *Chem. Week*, 42 (Aug. 8, 1979).
61. *Biomass Digest* **1**(11), 5 (Nov. 1979).
62. M. R. Ladisch and K. Dyck, *Science* **205**, 898 (1979).
63. G. F. Fanta, R. C. Burr, W. L. Orton, and W. M. Doane, *Science* **210**, 646 (Nov. 7, 1980).
64. *Alcohol Fuels Program Technical Review*, Solar Energy Research Institute, Golden, Colo., Winter 1981.
65. *Chem. Week*, 46 (July 25, 1979).
66. *Chem. Eng. News* **57**(38), 27 (Sept. 17, 1979).
67. *Chem. Week*, 34 (Sept. 26, 1979).
68. G. H. Emert and R. Katzen in D. L. Klass, ed., *Biomass as a Nonfossil Fuel Source*, American Chemical Society, Washington, D.C., 1981, Chapt. 11.
69. D. L. Marzola and D. P. Bartholomew, *Science* **205**, 555 (Aug. 10, 1979).
70. R. A. Clyde in *Symposium Papers Energy From Biomass and Wastes VII*, sponsored by Institute of Gas Technology, Lake Buena Vista, Fla., Jan. 25–29, 1982, pp. 887–896.
71. *Opportunities for Ethanol-Gasohol Research in Illinois*, Illinois Institute of Natural Resources, Ill., 1981.
72. M. Ichikawa, *CHEMTECH* **12**, 674 (Nov. 1982).
73. *Chem. Eng. News* **60**(38), 41 (Sept. 20, 1982).
74. *Chem. Week*, 40 (Oct. 31, 1979).
75. *Chem. Week*, 38 (Nov. 28, 1979).
76. M. E. Frank, *paper presented at 15th Intersociety Energy Conversion Engineering Conference*, Seattle, Wash., Aug. 18–22, 1980.
77. R. L. Dickenson, A. J. Moll, and D. R. Simbeck, *Oil Gas J.* **80**, 242 (March 8, 1982); *Chem. Eng. News* **60**(24), 14 (June 14, 1982).
78. T. O. Wentworth and D. F. Othmer, *Chem. Eng. Prog.* **78**(8), 29 (Aug. 1982).
79. C. S. Brandon, R. W. Duhl, D. R. Miller, and B. R. Thakker, *paper presented at the Fifth International Alcohol Fuel Technology Symposium*, Auckland, New Zealand, May 13–18, 1982; R. W. Duhl and B. R. Thakker in *Symposium Papers Nonpetroleum Vehicular Fuels III*, sponsored by Institute of Gas Technology, Arlington, Va., Oct. 12–14, 1982, pp. 227–238.
80. P. G. Laux, *paper presented at the International Symposium on Alcohol Fuel Technology*, Wolfsburg, FRG, Nov. 21, 1977.
81. A. Paggini and V. Fattone, *paper presented at the Fifth International Alcohol Fuel Technology Symposium*, Auckland, New Zealand, May 13–18, 1982.
82. *Oil Gas J.* **77**(28), 45 (July 9, 1979).
83. *Chem. Week*, 45 (Oct. 31, 1979).
84. *Alcohol Week* **3**(43), 1 (Oct. 25, 1982).
85. *List of Anhydrous Fermentation Ethanol Fuel Plants Operating with a Capacity of 500,000 Gallons per Year or Greater*, Contract No. 53-3J28-2-00471, report to U.S. Department of Agriculture, Office of Energy, National Alcohol Fuel Producers Association, Sept. 30, 1982.
86. *Alcohol Week* **3**(20), 1 (May 17, 1982).
87. M. Garnero, *Alcohol Fuels in Brazil*, testimony at hearing, Subcommittee on Energy, Nuclear Proliferation and Government Processes, U.S. Senate, March 24, 1981.
88. R. S. Goodrich, *Chem. Eng. Prog.* **78**, 29 (Jan. 1982).
89. *Assessment of Brazil's National Alcohol Program*, Ministry of Industry and Commerce Secretariat of Industrial Technology, Brasilia, Brazil, 1981.
90. B. I. Baer and M. L. Liebranz in *Symposium Papers Nonpetroleum Vehicular Fuels III*, sponsored by Institute of Gas Technology, Arlington, Virginia, Oct. 12–14, 1982.
91. *Chem. Week*, 33 (Aug. 11, 1982).

92. *The Report of the Alcohol Fuels Policy Review*, DOE/PE-0012, U.S. Department of Energy, Washington, D.C., June 1979.
93. R. Katzen and co-workers, *paper presented at the Synfuels Symposium*, American Institute of Chemical Engineers, Houston, Texas, April 8, 1981.
94. E. I. du Pont de Nemours & Co., Inc., Methanol Waiver Application, EPA Docket A-82-33, Number I-A-1, Aug. 25, 1982.
95. Ford Motor Company, submission by H. O. Petrauskas to the Environmental Protection Agency, EPA Docket EN-81-08, Number III-B-1, March 15, 1982.
96. W. H. Kampen, *Hydrocarbon Process.* **59**(2), 72 (Feb. 1980).
97. F. W. Cox, *Physical Properties of Gasoline/Alcohol Blends*, U.S. Department of Energy Publication No. BETC/RI-79/4, U.S. Department of Energy, Washington, D.C., 1979.
98. K. R. Stamper, *Evaporative Emissions from Vehicles Operating on Methanol/Gasoline Blends*, Paper 801360, Society of Automotive Engineers, Detroit, Mich., Oct. 1980.
99. R. Lawrence, *Gasohol Test Program*, Dec. 1978.
100. J. J. Mooney, J. G. Hansel, and K. R. Burns in ref. 17, pp. 237–247.
101. M. D. Gurney, J. R. Allsup, and C. L. Merlotti, *Gasohol: Laboratory and Fleet Test Evaluation*, Paper 800892, Society of Automotive Engineers, Detroit, Mich., Aug. 1980.
102. E. E. Wigg and R. S. Lunt in ref. 17, pp. 71–82.
103. S. E. Moreira Lima, *The Brazilian National Alcohol Program*, Washington Energy News Bureau's Seminar on Alcohol Fuels, Washington, D.C., March 16, 1981.
104. R. J. Nichols, *Experimental Fuel Fleet News Conference*, Dearborn, Mich., April 30, 1981.
105. L. Barker, T. Puchalski, and K. Tholen, *Automobile Gas Engine Conversion to Burn Ethanol*, SIM No. 11025, Solar Energy Research Institute, Golden, Colo., Feb. 1981.
106. G. Pischinger and N. L. M. Pinto in ref. 18, I-8, pp. 1–9.
107. A. McRae, L. Whelchel, and H. Rowland, eds., *Toxic Substances Control Sourcebook*, The Center for Compliance Information, The Aspen Systems Corp., Germantown, Md., 1978.
108. R. J. Lewis, Sr., and R. L. Tatken, eds., *Registry of Toxic Effects of Chemical Substances*, U.S. Department of Health and Human Services, Washington, D.C., Feb. 1982.
109. F. A. Patty, ed., *Industrial Hygiene and Toxicology*, 2nd ed., Vol. II, Wiley-Interscience, New York, 1962.
110. N. Y. Chen and W. J. Reagen, *J. Catal.* **59,** 123 (1979).
111. P. B. Weisz, W. O. Haag, and P. G. Rodewald, *Science* **206,** 57 (Oct. 5, 1979).
112. *Development of an Initial Strategy for Transport Fuels Supply and Gas Utilization on New Zealand, A Second Report to the Minister of Energy From the Liquid Fuels Trust Board*, Report No. LF 502, New Zealand, Oct. 31, 1979.
113. *Chem. Week* **130**(8), 31 (Feb. 24, 1982).
114. *Chem. Eng. News* **60**(32), 6 (Aug. 9, 1982).
115. J. C. Finegold and J. T. McKinnon in *Dissociated Methanol Test Results, Proceedings Biomass-to-Methanol Specialists' Workshop*, sponsored by U.S. Department of Energy, Durango, Colo., March 3–5, 1982, pp. 277–288.
116. J. Haggin, *Chem. Eng. News* **60**(29), 41 (July 19, 1982).
117. R. E. Brown and co-workers, *Economic Evaluation of the Coproduction of Methanol and Electricity with Texaco Gasification-Combined-Cycle Systems*, final report for EPRI AP-2212, Jan. 1982.
118. M. J. Gluckman and B. M. Louks in *Proceedings of the Eighth Energy Technology Conference*, Washington, D.C., March 9–11, 1981, pp. 920–943.
119. R. Noon in *Symposium Papers Nonpetroleum Vehicular Fuels II*, sponsored by Institute of Gas Technology, Detroit, Mich., June 15–17, 1981, pp. 221–232.
120. R. Noon, *Power Grade Butanol Recovery and Utilization*, Energy Research and Resource Development Division, Kansas Energy Office, Topeka, Kan., Feb. 12, 1982.

DONALD L. KLASS
Institute of Gas Technology

ANALYTICAL METHODS

Analytical chemistry is the scientific discipline concerned with the determination of the chemical composition of matter. The term chemical composition may, in practice, refer to elemental, ionic, or molecular composition. Often, a particular basis such as functional-group concentration or molecular concentration is defined for a specific application. In general, it is necessary to define carefully the basis for reporting or for understanding the results of a chemical analysis.

Analytical methods include approaches for ascertaining which chemical species are present in a sample (qualitative analysis) and their relative or absolute concentrations (quantitative analysis). Analytical methods are based upon a wide variety of chemical or physical principles used singly or in combination. A comprehensive summary of analytical methods is given in Vol. 2, pp. 586–683. This article supplements the Volume 2 article and focuses upon recent applications, trends, and analytical methods.

Trends in Analytical Requirements

Analytical methods are used extensively in science, industry, and government. A summary of applications of analytical methods is given in Table 1. For the applications given in Table 1, four trends in analytical requirements are noted.

Determination of trace constituents. The term trace constituents refers to constituents present at very low concentrations, ie, ppm, ppb, and lower. Methods for determining species present at such low levels must be sensitive and selective. Often, sample preparation and analysis must be carried out in clean rooms and may require specially purified reagents and other precautions to minimize contamination (1).

Determination of species information. Most applications of organic analysis now require the determination of individual molecular species information rather than total "class" analyses or carbon, hydrogen, nitrogen, etc, ratios. Increasingly, appli-

Table 1. Summary of Applications of Analytical Methods

process control	resource surveys
feedstock evaluation	criminal investigation
product evaluation	food inspection
effluent monitoring	environmental analysis
toxicological testing	trade specifications
maintenance and materials failure analysis	clinical testing
occupational health and safety	research and development

cations of inorganic analysis require the determination of the chemical form or valence states of atoms or species.

More complete chemical characterization. Increasingly, analysts seek an overall chemical profile of a sample, often requiring the determination of many species contained in the same sample.

Spatial resolution of chemical constituents. Solid materials such as metals, plastics, glass, minerals, and particulate matter are usually chemically inhomogeneous. In such cases, average or bulk concentrations may be poor indicators of properties or performance. Local compositional information (compositional maps) may be required for complete chemical characterization.

Analytical Strategy. Inasmuch as several million (10^6) chemical species are known and a single sample may contain thousands of species, chemical analysis must be undertaken with a clear perspective of the information sought and how the information is to be used. The approach to analysis must take into account analytical considerations such as the physical state of the sample, destructive or nondestructive analysis, qualitative vs quantitative information, trace or principal constituents, selectivity and detection limit, accuracy and precision required, and measurement-quality assurance. In most cases, more than one method can provide the needed chemical information, and the appropriate method may need to be selected on the basis of other criteria such as cost, number of samples, and availability of equipment.

In general, analytical determinations of concentrations are relative and not absolute. Thus, well-characterized materials should be used to prepare a series of standards (2). The signal produced by the unknown solution is compared to those produced by the standards, and the concentration of the specific unknown species is then calculated by interpolation. Care must be taken, however, to determine interferences by other species in complex matrices (3).

Trends in Analytical Methods

Advances in analytical methods, though significant in terms of quality and quantity of data and improved sensitivity and selectivity, are perhaps best described as evolutionary rather than taking place through the introduction of methods based upon newly discovered phenomena. The evolution is stimulated by the requirements associated with the applications listed in Table 1 and is made possible by rapid technological advances such as electronics (computers) and lasers. Developments in signal processing, optimization, and pattern-recognition techniques are affecting not only the approaches to acquisition and interpretation of data but also the underlying experimental design in chemical analysis. Perhaps the most significant and broadly based developments taking place in analytical chemistry are the creation of combined or hyphenated techniques, such as gas chromatography–mass spectrometry, and gas chromatography–Fourier transform infrared spectrometry. These hyphenated techniques afford a multidimensionality to analysis that is particularly well-adapted to the analysis of complex mixtures.

Advances in electronics and computer technologies have led to new generations of automated analytical instruments (4). Electronic devices, computers, and microprocessors have reduced the number of operations that require time-consuming mechanical or operator control. Automation has not only increased sample throughput and decreased operator intervention, but has also made possible elaborate, multistep analyses. The availability of fast electronics and sensitive, discriminating detectors permits the simultaneous determination of many parameters from which compositional information may be derived. The accessibility of on-line data bases (optical or mass spectra) coupled with the capability to perform complex mathematical manipulations and decision-making during data acquisition make automated method development possible. For most analytical instruments, instrument responses are slower than those of the associated computer. Thus, the computer is able to record and display analytical information while simultaneously performing other functions, for example, comparing information with that contained in tables (reference spectra), applying corrections to signals, transforming signals from one domain to another, or determining optimal settings for instruments (5–6). Such capabilities have given rise to a subdiscipline of analytical chemistry called chemometrics (7). Chemometrics is concerned with the use of mathematical and statistical methods not only to design or select optimal measurement procedures and experiments but also to extract maximal chemical information from the analysis of chemical data.

The laser has become a versatile tool in analytical chemistry (8). One or more features of lasers, eg, coherence, high power, tunability, narrow wavelength, and spatial focus, are used to improve existing analytical methods or to provide a basis for new ones (see Lasers). Methods which have been improved through incorporation of lasers include atomic fluorescence, fluorometric analysis, polarimetry, and Raman spectroscopy. New methods include laser-enhanced ionization and resonance ionization spectroscopy. In addition to serving as radiation sources in spectroscopic analysis, lasers are being used to enhance sample introduction (vaporization, ionization), to select small regions of samples for mass or spectroscopic analysis, and to improve selectivity at the sample-introduction stage.

In principle, direct analysis of mixtures can be accomplished by recording molecular information (eg, ir absorption bands or ms peaks) of the sample as a whole and then linking subsets of the observations to quantities of species whose identity is also determined in the analysis. For any but the simplest mixtures, however, it is generally not possible to carry out an unambiguous analysis because of interferences and overlaps in signals. Analysis of a mixture using a second technique based on a different physical or chemical principle may change the pattern of interferences or overlaps and thus help to resolve ambiguities arising with the use of a single technique. Traditionally, complex mixtures were analyzed beginning with separations using acid, base, or solvent extractions, followed by analysis of individual extracts. In recent years, however, multistage techniques have evolved in which the information gathering and separating powers of individual techniques are joined to provide a versatile system capable of analyzing quite complex mixtures. Analytical methods involving the integration of two or more techniques are called combined, or sometimes hyphenated, methods (9) and are designated by a slash representing the interface between the two. To date, the most successful combined methods involve a separation technique as the first stage and a detection technique as the second stage. The success of such combined methods is based upon the fact that the separations stage releases species one at a time into the inlet (sample-introduction) stage of an instrument which then provides characteristic fingerprints of these species.

An excellent example of a combined method based upon the separation–detection principle is gas chromatography (gc) interfaced with mass spectrometry (ms) to produce the hyphenated method gc–ms (or gc/ms). In this method, a sample is inserted onto a gc column along with a carrier gas. Through adjustment of column temperature (temperature programming), individual species present in the sample are swept from the column by the carrier gas into the inlet of the mass spectrometer. Data that are recorded are time of arrival of species and mass spectral intensities. If the gas chromatographic column has succeeded in a time separation of species contained in a sample, the mass spectral intensities will be those associated with single species observed one after the other, free from peak overlaps.

A number of other successful combinations have been demonstrated, eg, tandem mass spectrometry (ms–ms), gas chromatography–Fourier transform infrared spectroscopy (gc–ftir), liquid chromatography–mass spectrometry (lc–ms), and liquid chromatography–fluorescence, ultraviolet, or infrared spectrometry. The trend toward analysis of complex mixtures requires the development of analytical systems capable of yielding much information. Combined methods offer high information potential, and the development of new combinations depends upon the success of efforts to interface instruments which may, in practice, have very different physical requirements.

The analytical chemist has developed or applied other techniques or procedures to cope with the increased demands of sample complexity and vanishingly small quantities of chemicals in samples or for analysis. Ten selected topics not covered in the Analytical Methods article in Volume 2 are discussed in the following pages. Several of the techniques have already been used extensively in chemical analysis, whereas the use of others in analytical chemistry is only now being investigated. The selected topics are not meant to be a comprehensive review of all new applications in analytical chemistry, but represent the wide range of methods used by the analyst to address the problems of today's required chemical analyses.

Flow-Injection Analyses

Flow-injection analysis (fia) is an analytical method that utilizes converging streams of sample and reagent which mix and undergo chemical or physical change that is monitored by means of a flow-through detector (10–11). Common applications of fia include pH measurement, acid–base titration, and electrolyte determination; fia detectors commonly employed are spectrometric (visible or uv) or electrochemical. In concept, fia may be described as a family of sample processing techniques with advantages derived from versatility and high sample throughput. Measurements are generally not made at equilibrium, and the time required for mixing is usually saved. A schematic representation of a fia system is shown in Figure 1.

The basic fia technique involves the injection of sample volumes into a continuously flowing carrier stream. The sample-containing carrier then mixes with the

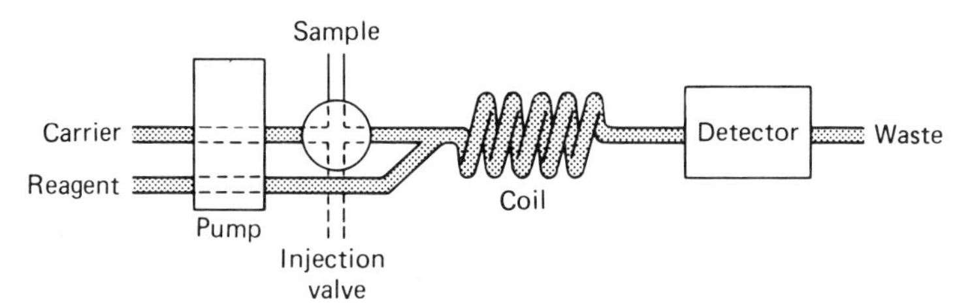

Figure 1. Schematic diagram of a continuous flow analysis manifold (11).

reagent in the fia reactor, in which it immediately begins to undergo chemical or physical change during transit toward the detector. The continuous flow of carrier ejects the reacted sample as waste; after a preset time, the next sample is injected into the carrier.

Flow-injection analysis is a versatile method, adaptable to a wide range of analytical problems. For example, Mn(II) can be determined spectrophotometrically by the reaction with formaldoxime in alkaline medium. The Mn(II) complex formed is oxidized immediately by atmospheric oxygen to form the orange-red Mn(IV) complex $[Mn(CH{=}NOH)_6]^{2-}$ which absorbs at 455 nm. Thus a stream of the unknown sample (or standard solution) is injected into an alkaline stream of formaldoxime and oxidized, and the absorbance of the Mn(IV) complex is measured. The manganese concentrations of the unknowns are calculated relative to the standards. The determinations are fast (ca 1 minute per sample) and use small volumes of samples and reagents (μL and mL, respectively) (12). Most adaptations involve variations of flow rate, reactor design, and choice of detector. Differences in reactor design are intended to control sample-zone dispersion for one or more of the following purposes: optimization of measurement signal, avoidance of dilution, facilitation of mixing, increased residence time when reaction rates are slow, and determining effects of matrix interferences. Flow-injection analysis systems incorporate virtually any flow-through detector, and flows may be adjusted to optimize detection. An important variant of fia is the so-called stopped-flow technique. This technique involves intermittent pumping which lengthens the reaction time without significantly increasing the sample zone. In this technique, the reacting mixture is in residence in the detector, and changes in signal are monitored as the reaction proceeds. An important application of the stopped-flow fia technique is to enzyme-based-rate-reactions analysis.

An advantage of fia is the reproducibility of experimental conditions, eg, flow rates and injection volumes, which lend themselves to ease of automation and high sample throughput. Optimal experimental parameters in the fia system can first be determined and, once established, can be used as the basis for many analyses through microprocessor control.

Inelastic-Electron-Tunneling Spectroscopy

Inelastic-electron-tunneling spectroscopy (iets) was discovered during investigation of the current-voltage behavior of specially prepared metal–metal oxide–metal devices called Josephson junctions (13). Inelastic-electron-tunneling spectroscopy is a branch of vibrational spectroscopy based upon the measurement of the energy losses sustained by electrons tunneling from one metal to another through an insulating barrier containing the sample to be analyzed (14). Energy from the tunneling electrons is transferred to vibrational energy levels in molecules of the sample. After undergoing energy loss, the electron continues across the junction. Figure 2 shows a schematic diagram of a junction.

Experimentally, current is monitored as a function of bias voltage. In a conventional Josephson junction, current is proportional to applied bias voltage. In junctions containing thin layers of sample on the insulating barriers, abrupt changes in current-voltage characteristics of the junction are observed, corresponding to molecular vibrational levels.

An example of a spectrum obtained for benzoic acid on alumina is given in Figure

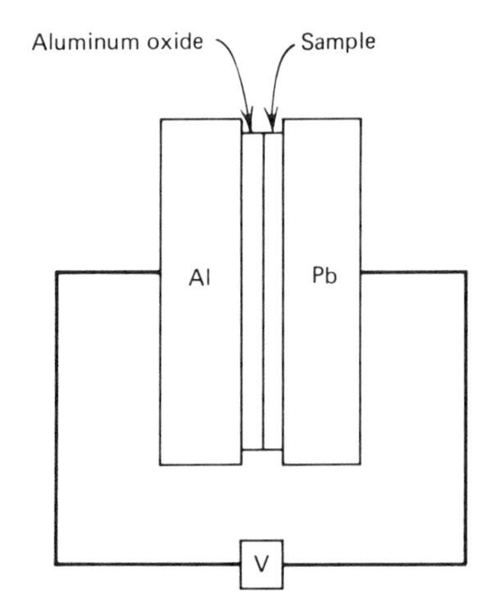

Figure 2. Schematic diagram of an iets junction.

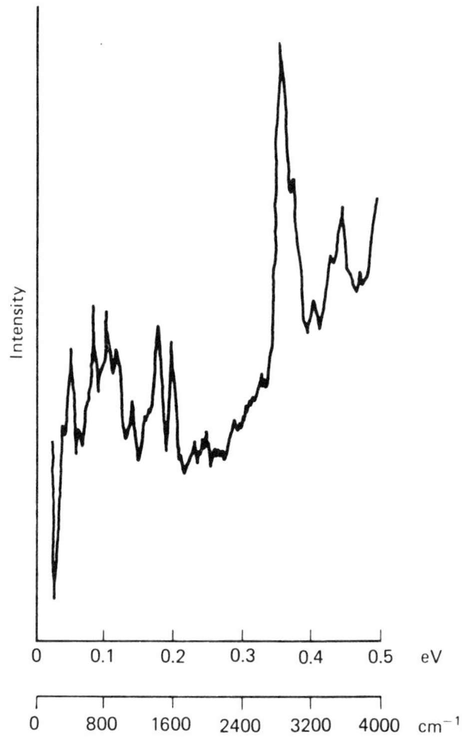

Figure 3. Inelastic-electron-tunneling spectroscopy spectrum of benzoic acid on alumina (1 ev = 8065.7 cm^{-1}) (14).

3. In this case, the carbonyl vibration at 1600 cm^{-1} is also observed in the ir spectrum.

In preparing a junction for iets, the first metal usually is deposited to a thickness of 100–300 nm on a substrate such as a microscope slide. An oxide layer of thickness 2–3 nm is prepared through exposure of the metal to oxygen, with care to avoid contamination by water or organic substances. Doping of the junction is accomplished by absorbing a sample to a depth of 5–10 nm. The junction is operated at liquid helium temperatures to achieve good spectral resolution. Experiments involving iets involve careful investigation of the current-voltage characteristics of the specially prepared junction. Second-derivative techniques are used to enhance sensitivity.

Inelastic-electron-tunneling spectroscopy is a third technique from which vibrational information for molecules can be obtained. However, compared with Raman and ir spectroscopic techniques, iets is nonoptical and is not subject to quantum-mechanical selection rules. That is, Raman-active, ir-active, and optically forbidden vibrational modes may be observed in iets. However, because tunneling electrons preferentially couple to vibrational modes which have their oscillating dipole moment perpendicular to the oxide surface, iets may be used to investigate molecular-orientation effects. The limit of detection for iets is much lower than for Raman and ir spectroscopies. Approximately 2×10^{10} molecules may be detected on the barrier surface.

Inelastic-electron-tunneling spectroscopy has a wide range of applications as a surface analysis–molecular structure technique. It is used to investigate metallic corrosion, surface orientation, radiation damage, sequencing of biomolecules, and mechanisms of catalysis. Quantitative analysis using iets is difficult owing to dependence of intensities upon surface saturation, molecular orientation, junction preparation, and other factors. Assignments of spectral bands in iets are aided by Raman and ir data, chemical substitution, and use of isotopic species.

Extended X-Ray Absorption Fine Structure

Extended x-ray absorption fine structure (exafs) is a technique in which the x-ray absorption spectrum of a material is measured over an energy range that extends above the absorption edge of a selected element (15–16). Fine structure refers to the modulation of the x-ray absorption coefficient at energies ranging from ca 30 to 1000 eV above the absorption edge. Analysis of this fine structure yields information on local structure and short-range order in the vicinity of the absorbing species. A typical x-ray absorption spectrum is shown in Figure 4.

In exafs, an electron is ejected from the atom under investigation by an x-ray photon. The resulting photoelectron wave is scattered by neighboring atoms, producing an interference wave pattern. By varying the energy of the incident x-ray photon, the wavelength of the ejected photoelectron changes and, in turn, changes the interference pattern. The elastic scattering of the photoelectron diminishes with the square of the distance from the absorbing atom. The modulation in the absorption coefficient receives the largest contribution from the nearest neighbor atoms and is thus a probe of local structure. Coordination distances of nearest neighbor atoms can be determined with an accuracy of 1–5 pm. The availability of synchrotron radiation produced by high energy storage rings enables exafs experiments to be carried out more quickly and with much improved signal-to-noise ratios compared with conventional x-ray tubes.

Extended x-ray absorption fine structure is especially well-suited to the analysis of materials which have short-range order but lack long-range order such as disordered solids and amorphous materials. For such cases, the technique is unique in its ability to determine nearest-neighbor distances, coordination numbers, and mean-square

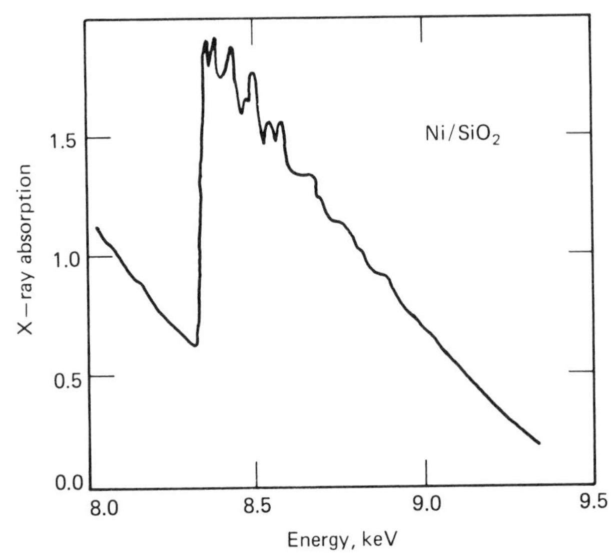

Figure 4. X-ray absorption spectrum of a nickel metal catalyst (16).

relative displacements. Applications include investigation of metal-on-support catalysts, ions in solution, impurities in solids and at surfaces, and structures of metalloenzymes.

Photoacoustic Spectroscopy

Photoacoustic spectroscopy (pas) is based on the fact that optically excited species can lose excess energy through nonradiative processes. These processes produce heat which in turn yield measurable acoustic waves (sound) (17). Since the photoacoustic signal is proportional to the optical radiation absorbed and not to the difference between two intensities as in spectrophotometry, the technique has the potential for much more sensitivity than conventional absorption techniques. A general equation for the photoacoustic signal produced is

$$A = kP_{\text{abs}}\delta c$$

where A is the photoacoustic signal; k is an instrumental and cell constant; P_{abs} is the power absorbed by the sample; δ is the energy conversion efficiency for radiationless processes; and c is a term to describe the time domain over which the sound is produced.

The advent of high power sources (lasers, arcs) and advanced sound detectors (microphones, hydrophones, thin-film piezoelectric detectors) has turned this technique from one of average sensitivity to one that is highly sensitive for the analyses of species in gas, liquid, and solid phases. A typical experimental setup for the analyses of gases is given schematically in Figure 5. The gas is excited with a pulsed or chopped continuous-wave (cw) laser. Excitation energy is lost through nonradiative processes and measured as sound with the acoustic detector. The signal is amplified, processed with a phase-sensitive or gated detector, and fed to a recording device or computer for additional signal processing. Proper microphone and cell designs are critical for high sensitivity. For measurement of solids and solutions, gases are often used to

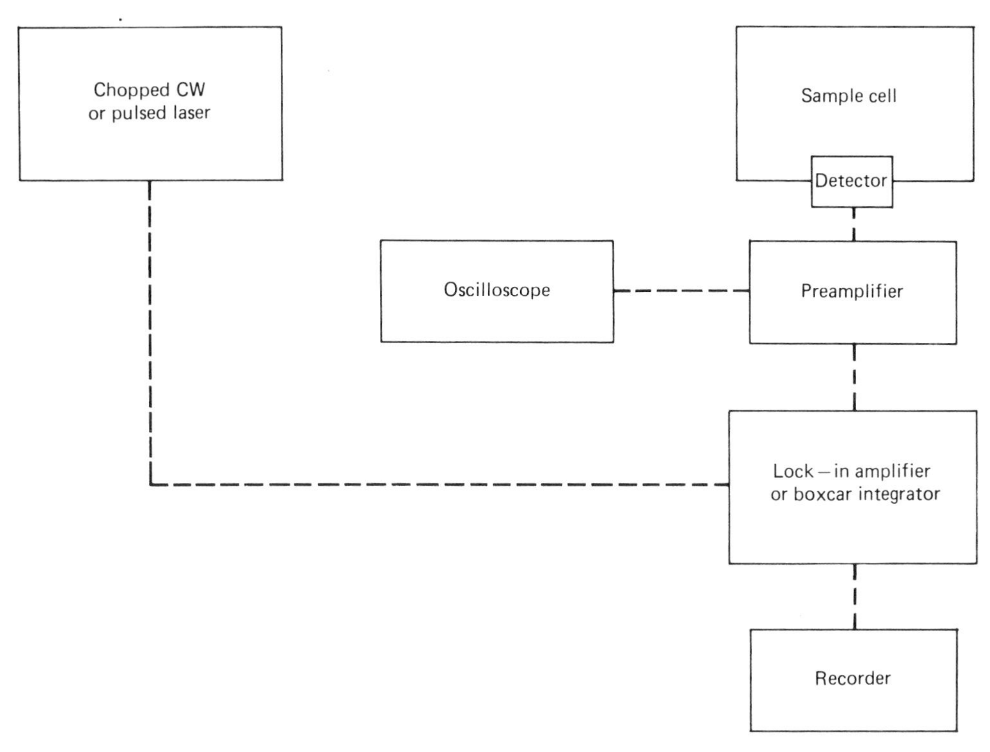

Figure 5. Schematic diagram for photoacoustic-spectrometry experimental system.

transmit the acoustic wave to the detector, although a piezoelectric sensor may be placed in direct contact with the sample. Fourier transform photoacoustic spectroscopy has been developed to exploit Fellgett's (all wavelengths measured simultaneously) and Jaquinot's (no energy-limiting slits) advantages (18).

The applications for pas have been growing tremendously and include the determination of 10–100 ppb levels of gases in pollution studies (19) and spectroscopically difficult-to-measure materials such as opaque or powdered samples. It has also been used in studies of physical properties of materials, depth profiling, and surface or subsurface microscopic imaging looking for defects as a nondestructive evaluation technique and has been recommended as a detector for gas chromatography with higher sensitivity than thermal conductivity or flame ionization detectors (20). In spectroscopy, the pas technique has been used to study low probability transitions and overtone absorptions, energy-level splittings caused by the Zeeman effect, energy-transfer processes, and transitions from excited states.

Inductively Coupled and Direct-Current Plasma-Emission Spectrometry

Flame-atomic-emission spectrometry (faes) and flame-atomic-absorption spectrometry (faas) have been recognized as extremely useful techniques for elemental analysis in a wide variety of matrices since the development of commercial instrumentation in the 1960s. However, the relatively low temperatures of combustion flames (usually $<3000°C$) limit the sensitivity of faes. The development of plasma sources with temperatures approaching 10,000°C has led to a significant revival in the use of emission instrumentation and has increased the potential for sensitive multielement analysis. Although many plasma sources have been developed, including inductively coupled plasma (ICP), direct-current plasma (DCP), capacitively coupled microwave plasma (CMP), microwave-induced plasma (MIP), and laser-induced plasma (LIP), only ICP and DCP are considered, owing to their wide use and the availability of commercial instrumentation. Plasma and flame emission techniques determine atomic species in a sensitive, selective, and easy-to-use manner. The use of plasmas as an excitation source has the added analytical advantages of wide linear dynamic range, minimal chemical and physical interferences, and increased sensitivities for sequential or simultaneous multielement determinations (see Plasma technology).

The general experimental setup is given in Figure 6. The plasma source causes excitation of the sample atoms. Radiation, emitted as characteristic narrow lines by the excited atomic species, is separated by a high resolution dispersion system. The intensity of the resolved analytical line is determined by a detector and is proportional to the concentration of the species in solution. An excellent review of emission spectroscopic instrumentation is given in reference 21, and discussions of ICP and DCP fundamentals and applications are numerous (22–24). Sample introduction into the plasma is similar to that used in faes, ie, pneumatic or ultrasonic nebulization of solutions or slurries.

The ICP is an electrodeless argon plasma operated at atmospheric pressure and sustained as a plasma by inductive coupling to an r-f electromagnetic field (Fig. 7(**a**)). In the DCP, the argon plasma jet is formed by a dc between carbon anodes and a tungsten cathode (Fig. 7(**b**)). The higher excitation temperatures of the plasmas (8,000–10,000 K) compared to flames (2,000–3,000 K) result in more efficient atom excitation (increased sensitivity), especially for the refractory elements such as B, P, W, Nb, Zr, and U. The flame geometries and dynamics (eg, few sample atoms in plasma sheaths) and temperature profiles (especially for the ICP) result in minimal line reversal and matrix interferences with concomitant increases in the linear dynamic ranges and excellent capability to perform simultaneous determinations of principal, minor, and trace constituents in a single sample without dilutions or multiple determinations.

High spectral dispersive systems (spectral bandpasses <20 pm) are required to separate the line emissions produced by the various elements. These systems can be monochromators, polychromators, or echelle-grating spectrometers (23). Multielement analysis can be performed in a sequential or simultaneous mode. The former generally relies upon a scanning monochromator with scanning speeds and integration times at each analysis line controlled by a preset computer program. Simultaneous multielement determination generally relies on wavelength separation by a dispersive element (grating, echelle grating) with premachined/photoetched exit slits to isolate specific element lines for detector measurement. The detectors, placed directly behind

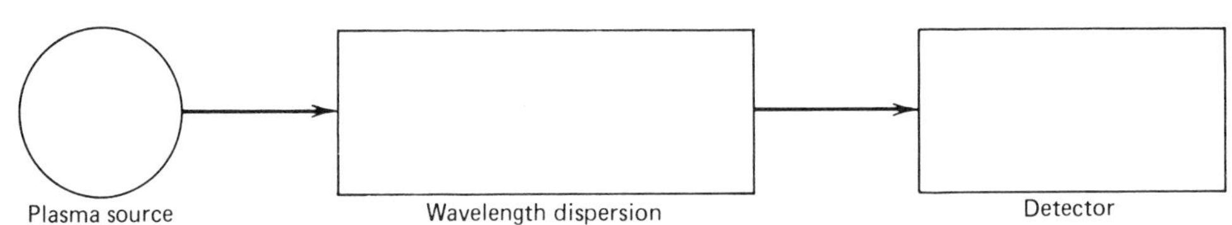

Figure 6. General, simplified schematic diagram of instrumentation used in atomic-emission spectroscopy.

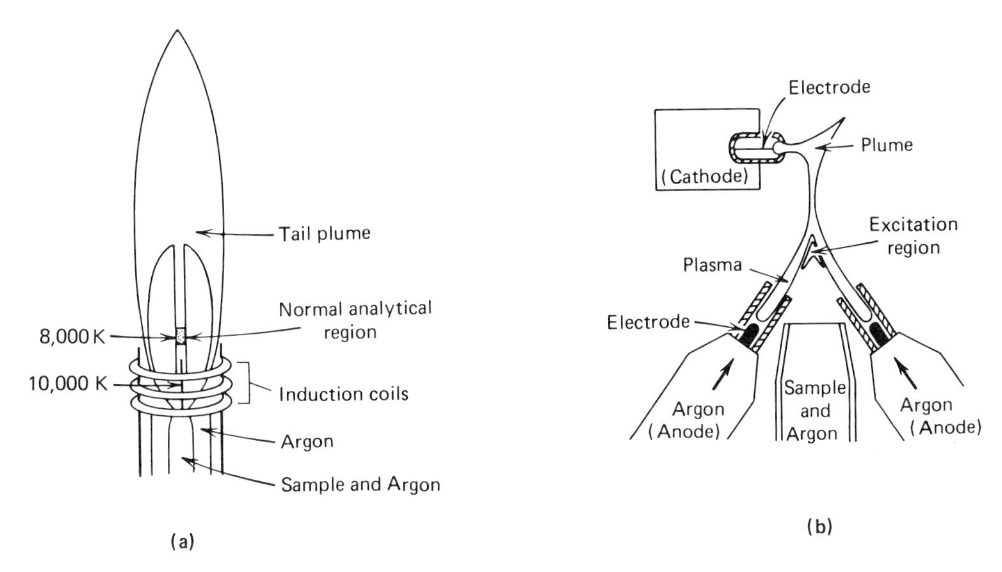

Figure 7. Plasmas, the heart of the aes technique: (**a**) ICP and (**b**) DCP (25) with specific regions identified.

the slits in the mask, can be photomultipliers or recently developed array detectors such as vidicons, reticons, or channel plates which need no mask or exit slits. The electrical signals are amplified, processed, and passed to a computer for data manipulation and comparison. These developments in aes using plasma sources and instrument control and data processing by computers for multielement determinations in a wide variety of matrices are truly indicative of the trends in analytical elemental analyses.

Laser-Enhanced Ionization

In the study of laser-induced atomic fluorescence in a hollow-cathode-discharge source, electrical impedance changes were also noted when the wavelength of a dye laser was tuned to the electronic transition wavelengths of the specific atoms in the discharge. This effect was found to occur as well in atomic-absorption flames used for chemical analysis (8,25) and was interpreted as the production of ions by energy supplied to atoms through a combination of laser photoexcitation and collisional excitation. The total excitation process, called laser-enhanced ionization (lei), is depicted schematically for two elements in Figures 8(**a**) and 8(**b**), and the experimental system is shown in Figure 9.

A pulsed laser is used to pump a dye laser with the wavelength tuned to a resonance electronic transition of the element of interest. The analyte, aspirated as a solution into the flame, absorbs the energy provided by the laser and is raised to an excited state. Additional energy (thermal, kT) is supplied by the flame, and the sum of energies provided results in atom ionization producing an ion and an electron. The change in the current of the external circuit applying voltage across the flame is proportional to concentration. Since the photon wavelength (Fig. 8(**a**)) can be varied, the system is very selective. Stepwise excitation using two different wavelengths improves the lei selectivity dramatically since the probability of both transitions coinciding

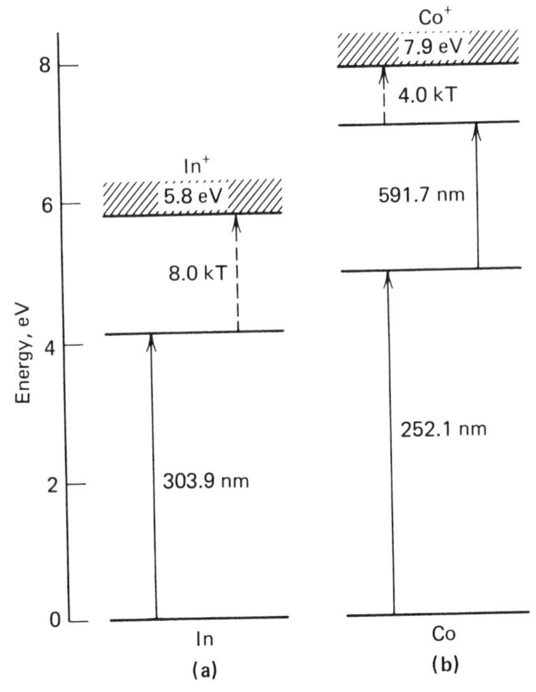

Figure 8. Ionization schemes for In and Co using (**a**) single-photon or (**b**) two-photon laser excitation where the solid arrows represent energy supplied by the laser(s) and the broken arrows represent energy supplied by the flame for which kT = 0.2 eV. Courtesy of J. C. Travis and G. C. Turk, National Bureau of Standards.

simultaneously with two transitions from another element is quite small (26). Additionally, raising atoms to a higher-lying excited state by the second photoexcitation results in more efficient thermal ionization, especially for elements of high ionization potential. Laser-enhanced ionization sensitivities are equivalent to or greater than conventional flame atomic spectrometric techniques (Table 2). Both ionization schemes depicted in Figures 8(**a**) and 8(**b**) show photoexcitation from the ground state. Excitation from low-lying excited states can also be used to enhance atom ionization.

It is anticipated that flame lei will be applicable to all species currently determined by atomic spectrometry with greater sensitivity and selectivity in most cases (27–28).

Resonance-Ionization Mass Spectrometry

Resonance-ionization mass spectrometry (rims) is a combination technique derived from resonance-ionization spectroscopy (ris) and mass spectrometry (ms). The first step, selective production of ions and electrons by tuning narrow-band lasers across atomic resonance absorption lines, is identical to the photon excitation steps in lei. Rather than using the analytical electrical signal produced by the ions and electrons, the ions produced are inserted into a mass spectrometer, providing additional selectivity and sensitivity (29). A schematic diagram of a typical experimental setup is presented in Figure 10. In this case, a Nd–YAG laser is frequency-doubled (to 532

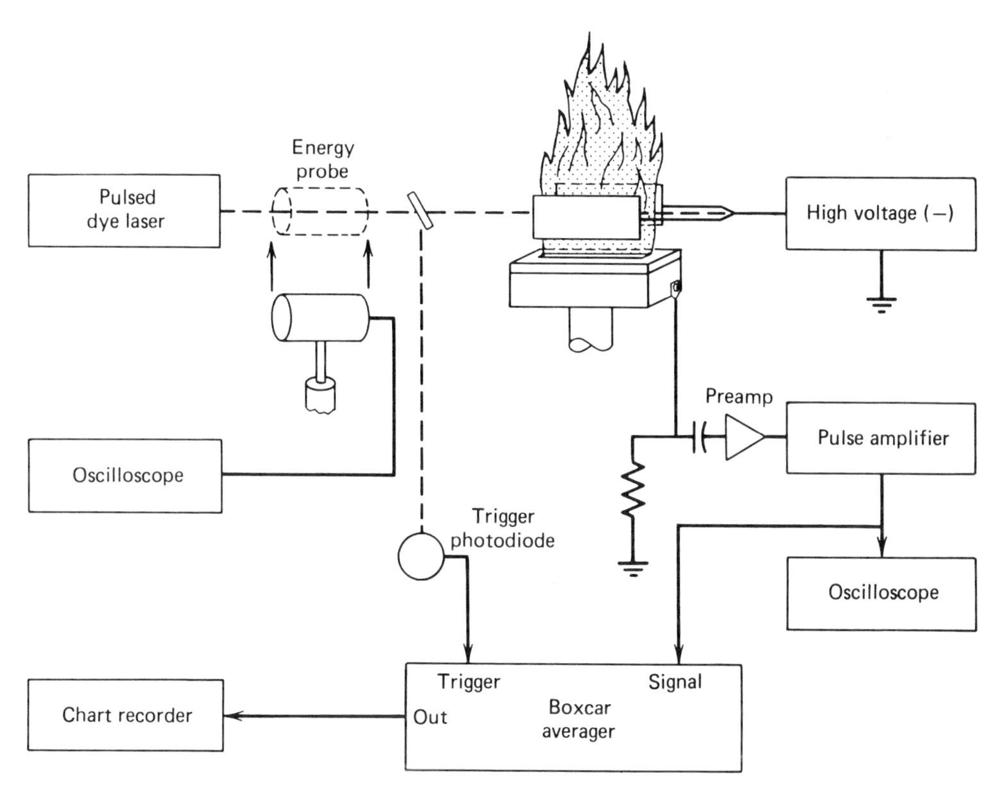

Figure 9. Schematic diagram of experimental system for the lei technique using a pulsed-dye laser source. Courtesy of J. C. Travis and G. C. Turk, National Bureau of Standards.

Table 2. Comparison of Detection Limits for Selected Elements, ng/mL [a]

Element	lei	faa[b]	fae[b]	faf[b]	lif[b]
Ag	1	1	2	0.1	4
Ba	0.2	20	1		8
Bi	2	50	20,000	5	3
Cu	0.07	1	0.1	0.5	1
In	0.008	30	0.4	100	0.2
Li	0.001	1	0.02		0.5
Na	0.05	0.8	0.1		0.1
Pb	0.09	10	100	10	13
Tl	0.09	20	20	8	4

[a] Refs. 8 and 25.

[b] Faa = flame-atomic absorption; fae = flame-atomic emission; faf = flame-atomic fluorescence; and lif = laser-induced fluorescence in flames.

nm) and used to pump a tunable dye laser, the output of which is also frequency-doubled to provide uv radiation. The uv radiation ionizes atoms that have been thermally excited by the heated filament in the mass spectrometer. The resulting ions are introduced into the mass spectrometer for separation, detection, and measurement.

To date, the main use of this technique has been the quantitative determination

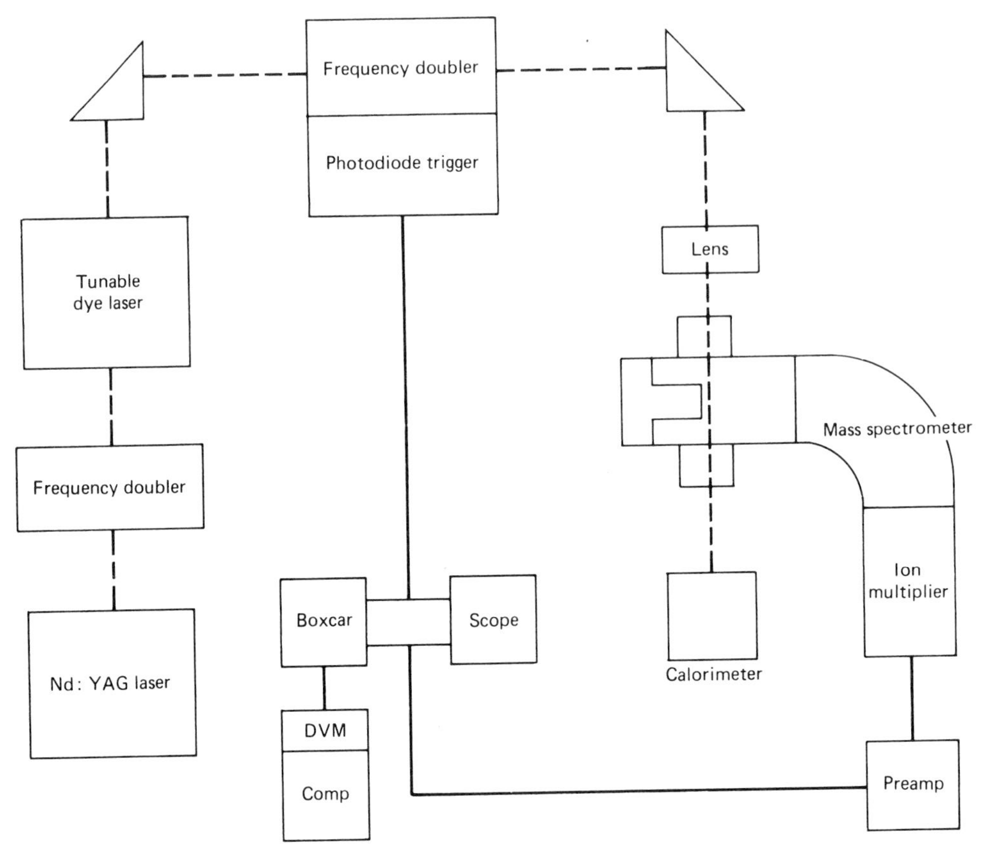

Figure 10. Schematic diagram of the rims experimental system using a Nd–YAG pumped-dye laser for resonance ionization and a mass spectrometer for ion separation–detection. Courtesy of L. J. Moore and J. D. Fassett, National Bureau of Standards.

of rare-earth and actinide elements in the presence of isobaric interferents (29). Recently, the technique has been applied to the formation and qualitative determination of Mo, Re, and V ions (30). Spectral information from this technique is comparable to that from conventional thermal mass spectrometry; however, additional transitions are noted and have been tentatively assigned to transitions originating from excited states.

Excitation schemes have been developed for more than 50 elements, and it is expected that this technique will be used extensively for accurately measuring trace element concentrations in complex matrices without chemical preseparation, an ultimate goal of observable trends in analytical chemistry (31).

Laser-Microprobe Mass Spectrometry

Research in laser-microprobe mass spectrometry culminating in the commercial production of instruments (laser-microprobe mass analyzer (LAMMA) and laser-induced ion-mass analyzer (LIMA)) has added an exciting new dimension to analytical chemistry (32). As with other microprobe instruments, the LAMMA or LIMA can provide qualitative and semiquantitative data on elemental composition. In addition, these instruments have the capability for providing analytical data on organic and inorganic molecular species as well as for determining isotopic abundances. This total analytical capability provides unparalleled potential for yielding compositional data from H to U on samples with 1–2 μm spatial resolutions at sensitivities from 10^{-18} to 10^{-20} g.

The instruments used a Q-switched (pulse duration, 15 ns), doubled (to 530 nm) or quadrupled (to 265 nm) Nd–YAG laser to ablate and ionize portions of a sample (schematic diagram in Fig. 11). A He–Ne laser, with an optical path collinear with that of the Nd–YAG laser, is used to focus on the sample and identify the analytical region. The ionic species formed are introduced into a time-of-flight (TOF) mass spectrometer, focused and accelerated by an ion lens, and detected by an electron multiplier. The analogue signal can be displayed on an oscilloscope, digitized, stored, or transferred to a computer for later data manipulation.

Samples can range from thin films (<1 μm thick) to macrosamples (>20 μm). Ion-introduction geometries to the TOF mass spectrometer vary from 180° for total ablation of thin films, small particles (<2 μm), and large particles (sample grazing) to 0° for front-surface ablation of very large samples.

The formation of ions and atoms is complex and depends on several processes that occur during the sample ablation and lifetime of the microplasma. Thus, interpretation of spectra of organic molecules generally cannot be done by direct comparison with those obtained in conventional mass spectrometry. Typical spectra obtained from

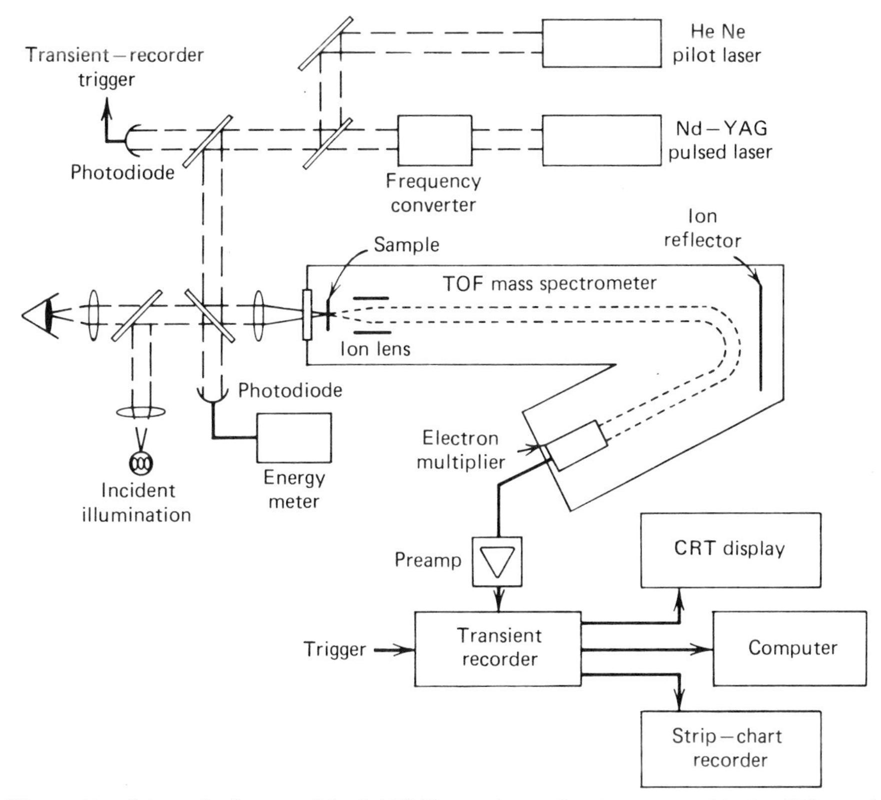

Figure 11. Schematic diagram of the LAMMA experimental system operated in the 180° sample-ion introduction mode. Courtesy of D. S. Simons, National Bureau of Standards.

National Bureau of Standards (NBS) Standard Reference Material (SRM) 610, *Inorganic Ions in Glasses*, and NBS SRM 947, *Uric Acid*, are depicted in Figures 12(**a**) and (**b**), and 12(**c**) and (**d**), respectively. The wealth of information, sensitivity, and selectivity are immediately apparent from these spectra. For example, all identified elements (except Na, Al, Si, and Ca) in Figures 12(**a**) and (**b**) are present in the glass at nominal weight concentrations of 500 ppm.

For inorganic compounds, spectral intensities generally increase with laser power. However, the absence of or production of an elemental or molecular ion does not guarantee that species' presence or absence in the sample since the energy supplied may be below an ionization energy threshhold for specific elements, or as in secondary ion mass spectrometry, the species can be formed by reactions in the microplasma

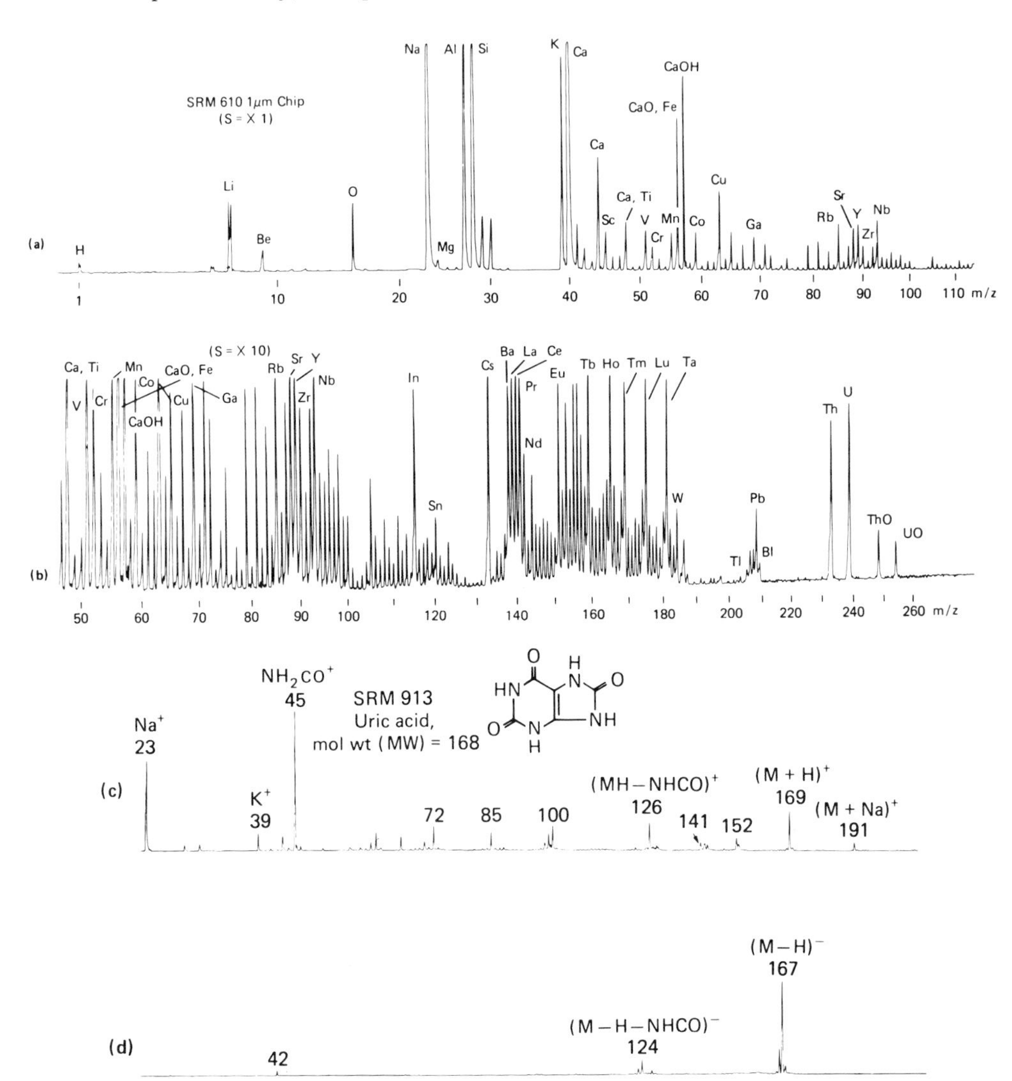

Figure 12. Spectra of NBS SRM 610, *Inorganic Ions in Glasses*, at (**a**) 1 × sensitivity (**b**) 10 × sensitivity, and SRM 947, *Uric Acid*, (**c**) positive-ion spectrum and (**d**) negative-ion spectrum. Figures 12(**a**) and 12(**b**), courtesy of D. S. Simons, National Bureau of Standards; Figures 12(**c**) and 12(**d**), courtesy of G. Byrd, National Bureau of Standards.

during or immediately following the ablation process (33–34). For organic, high molecular weight, nonvolatile compounds, low laser powers may produce parent peak formation, whereas use of high laser powers generally results in extensive molecular fragmentation.

This technique has been used in the determination and identification of various species in a broad variety of biomedical areas including fluorine in teeth, elements and compounds in specific cell sites and tissues, element–compound uptake by algae and bacteria, as well as in the analyses of particles, aerosols, polymers, soils, woods, etc (32–35).

Solid-State Nuclear Magnetic Resonance Spectroscopy

The nmr of solid materials has experienced advances in the last decade (36–39). During this period, experimental techniques such as high power proton dipolar decoupling, magic-angle spinning (mas), and cross polarization (cp) have been developed or applied which increase the sensitivity and resolution of ^{13}C solid-state spectra until they almost approach the resolution of solution spectra. Solid-state nmr spectra of ^{1}H, on the other hand, have one to two orders of magnitude less resolution than solution spectra.

In the past, unlike the nmr spectra of solutions, the spectra of solids were broad, featureless, and of low intensity (Fig. 13(**a**)) owing to low ^{13}C abundance (1.1%) and small magnetic moment, as well as large direct-coupling (dipole–dipole) effects. As a result of low sensitivity and strong dipolar couplings, the information on molecular geometries, internuclear distances, coordination, symmetries, and molecular wave functions available from chemical shielding and spin–spin and J-coupling effects was lost.

The barrier of low sensitivity for ^{13}C solid-state nmr spectra is removed by cross-polarization techniques. In cp, transfer of magnetization from the protons to the ^{13}C nuclei is achieved by r-f pulse sequencing for both nuclei, giving them equivalent overlapping oscillatory components in their rotating frames. Efficient cross relaxation occurs by energy-conserving spin flips, and the sensitivity to ^{13}C is increased. Exacting magnetic-field stabilization and pulse-sequence timing is required, necessitating computer programming and control.

Since most organic species have hydrogen nuclei adjacent to carbon nuclei, decoupling of the strong dipole interactions of ^{1}H and ^{13}C nuclei is necessary and can be accomplished by r-f irradiation at the proton resonance frequency (up to r-f field strengths of at least 45 kHz).

Even after application of cross polarization and decoupling of the dipolar interactions of ^{1}H with ^{13}C, the resonance in the ^{13}C solid spectra, although narrower, is still broadened (Figs. 13(**b**) and (**c**)) because of chemical-shielding anisotropy (csa). In a polycrystalline matrix, the chemical bonds are randomly oriented to the applied magnetic field, resulting in a broad distribution of resonance frequencies which yields, in turn, a broadened spectrum. In the solid state, csa is removed by mas in which the sample rotates rapidly, ie, at a spinning rate of up to 6–7 kHz about an axis that is at 54.7° with the applied magnetic field. Thus, all chemical-shift anisotropies in the solid collapse to an average isotropic shift, and well-resolved ^{13}C spectra are obtained (Fig. 13(**d**)). In solution, an average isotropic shift is attained by rapid molecular tumbling and spectra are quite sharp (Fig. 13(**e**)).

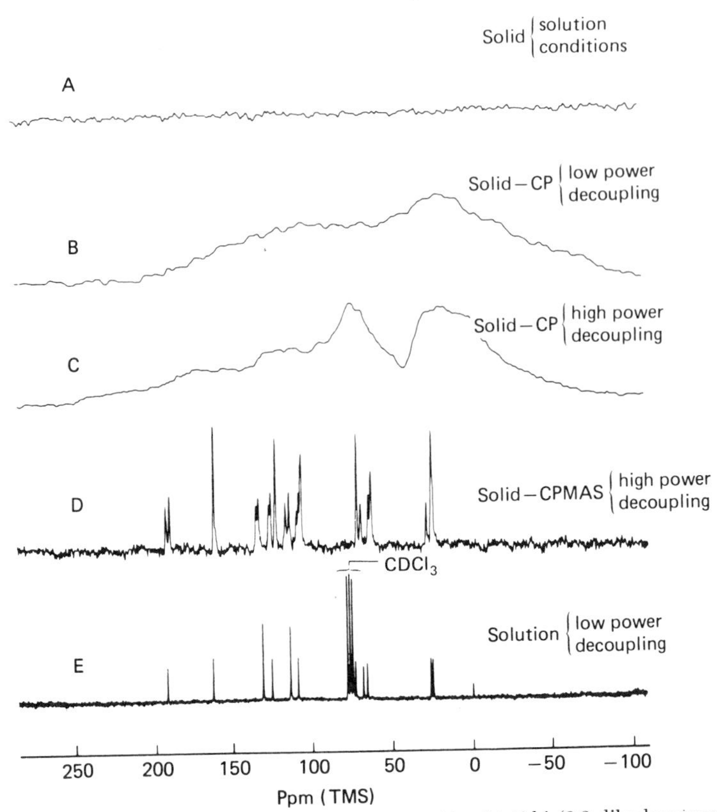

Figure 13. Solid-state ^{13}C nmr spectra of the bisacetonide of 4,4′-bis(2,3-dihydroxypropyl)oxy benzil: A, normal experimental conditions; B, cross polarization and low power decoupling; C, cross polarization and high power decoupling; D, cross polarization, magic-angle spinning, and high power decoupling; E, solution spectrum in $CDCl_3$ with same experimental conditions as in A plus low power decoupling (38).

With these new experimental techniques, it is now possible to obtain information on molecular and local nuclear motions and interactions leading to determinations of various relaxation parameters, chemical-exchange processes, crystal structures, and the presence of reactive intermediates in the solid state. Typical applications include research in resin and polymer chemistry (synthetic and biopolymers), catalysis (surface-attached species), homogeneous micellar and zeolite catalysts, energy sources (shale, coal, and other fossil fuels), and organic solids in general (36–39). The structural and chemical information obtained is complementary to that obtained by x-ray crystallography, and additionally can be used on microcrystalline and amorphous solids to produce well-resolved ^{13}C spectra. As this technique becomes more quantitative, further extensive areas of application are expected.

Biomolecules as Analytical Reagents

During the period 1950–1970, classical wet-chemical analysis decreased in popularity as the analytical instrumentation revolution gained extraordinary momentum. During the last decade, however, wet-chemical analysis has made great strides in reclaiming its previously important position, albeit in a somewhat different context. The development responsible for this reversal is the use of highly selective and sensitive biomolecules as analytical reagents. Two important procedures have developed in this area: enzymatic and immunochemical analyses.

Enzymes are high molecular weight biocatalysts that increase the rate of specific chemical reactions in biological organisms. To carry out enzymatic analyses, the enzymes are separated from the organism, stabilized (if necessary), purified, and used to catalyze a given chemical reaction. In general, the rate of the chemical reaction is monitored and is proportional to the concentration of the analyte of interest (40). A simple model is used to represent the analytical system. It involves an equilibrium between the enzyme (E) and analyte (A, species to be measured) with subsequent reaction to produce the product (P) and release the enzyme:

$$E + A \rightleftharpoons EA \rightarrow E + P$$

In a simplistic sense, the reaction velocity (V or $d[P]/dt$) is linearly proportional to the analyte concentration [A] at low A concentrations, ie, K_M (the Michaelis constant) $\gg$ [A].

$$V = \frac{d[P]}{dt} = \frac{V_{max}[A]}{K_M + [A]}$$

The reactions are affected by temperature, pH, anions, cations, and solvents; but once reaction conditions are established and potential interferences determined and eliminated, the procedures are sensitive, selective, and straightforward. An example of a compound-specific determination, uric acid by the enzyme uricase, is given in the following reaction sequence:

$$\text{uric acid} + O_2 \xrightarrow{\text{uricase}} \text{allantoin} + CO_2 + H_2O_2$$

$$H_2O_2 + \text{leuco dye} \xrightarrow{\text{peroxidase}} P_{dye} + H_2$$

The CO_2 formed could be measured, but usually an additional enzymatic reaction is performed on the peroxide product. The absorbance of the product dye [P_{dye}] formed is measured spectrophotometrically with the rate and/or amount of product formation proportional to uric acid concentration.

A main problem in the use of enzymes is their instability when removed from the organism. In some instances, this problem has been partially circumvented by attaching the enzyme to some support such as polymeric materials, thus producing immobilized enzymes that can be used under experimental conditions normally destructive for the isolated enzyme (40–41).

In immunoassay procedures, antibody molecules (produced by biological organisms after injection or assimilation of specific chemicals–often the analyte of in-

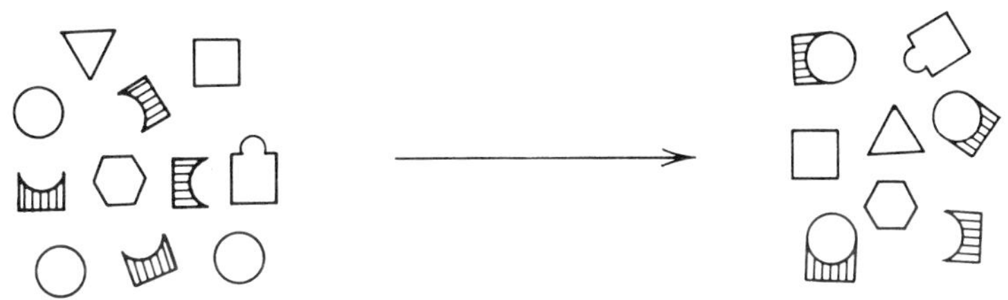

Figure 14. Schematic diagram of the reaction of antibodies (shaded figures) with a specific analyte (unshaded circles) in the presence of other potential analytes (antigens).

terest) have binding sites that react with molecular species called antigens having specific molecular structures. A schematic for this reaction is given in Figure 14.

A known quantity of labeled analyte (A*) and the sample analyte (A) are mixed with the antibodies (B) resulting in a competition for the antibodies by A* and A.

$$nA^* + mA + iB \rightarrow (n - x)A^*B + (m - y)AB + xA^* + yA$$

where $i = (n - x) + (m - y)$.

The unbound analyte ($xA^* + yA$) is separated from the bound species. The concentration of A is determined by measuring signals from bound or free A*, and use of calculations similar to those used in isotope-dilution mass spectrometry (42). The tagged analytes can be radioactive (radioimmunoassay to measure radioactivity signal), fluorescent (fluoroimmunoassay to measure fluorescence signal), on particles or cells (measure scattered radiation), or metals (measure metals by conventional techniques).

All of these techniques are selective and sensitive, measuring as little as 10^{-16} g (100 ag) of analyte in a complex mixture. Biomolecules have been used as reagents in the areas of medicine, clinical and environmental chemistry, forensic science, pharmacology, and food and nutrition. Types of compounds analyzed include macromolecules (proteins or polymeric species), low atomic weight molecules such as gases, urea, amino acids, carbohydrates, and sugars. It is expected that with development of specific enzymes, antibodies, and the appropriate reaction conditions, biomolecules will be used as analytical reagents for many more chemical compounds.

Nomenclature

aes = atomic-emission spectrometry
CMP = capacitively coupled microwave plasma
cp = cross polarization
CRT = cathode-ray tube
csa = chemical-shielding anisotropy
cw = continuous wave
DCP = direct-current plasma
DVM = digital voltmeter
exafs = extended x-ray absorption fine structure
faa = flame-atomic absorption
faas = flame-atomic-absorption spectrometry
fae = flame-atomic emission
faes = flame-atomic-emission spectrometry
faf = flame-atomic fluorescence
fia = flow-injection analysis
ftir = Fourier transform infrared spectroscopy
gc = gas chromatography
ICP = inductively coupled plasma
iets = inelastic-electron-tunneling spectroscopy
LAMMA = laser-microprobe mass analyzer
lc = liquid chromatography
lei = laser-enhanced ionization
lif = laser-induced fluorescence
LIMA = laser-induced ion-mass analyzer
LIP = laser-induced plasma
mas = magic-angle spinning
MIP = microwave-induced plasma
ms = mass spectrometry

NBS = National Bureau of Standards
pas = photoacoustic spectroscopy
rims = resonance-ionization mass spectrometry
ris = resonance-ionization spectroscopy
S = sensitivity
SRM = standard reference material
TMS = tetramethylsilane
TOF = time of flight
YAG = yttrium aluminum garnet ($Al_5Y_3O_{12}$)

BIBLIOGRAPHY

"Analytical Methods" in *ECT* 3rd ed., Vol. 2, pp. 586–683, by Eric Lifshin and Elizabeth A. Williams, General Electric Co.

1. J. R. Moody, *Anal. Chem.* **54,** 1358A (1982).
2. *Standard Reference Materials*, Office of Standard Reference Materials, National Bureau of Standards, Washington, D.C.
3. J. A. Dean and T. C. Rains, eds., *Elements and Matrices*, Vol. 3 of *Flame Emission and Atomic Absorption Spectrometry*, Marcel Dekker, Inc., New York, 1975.
4. C. G. Enke, S. R. Crouch, F. J. Holler, H. V. Malmstadt, and J. P. Avery, *Anal. Chem.* **54,** 367A (1982).
5. P. R. Griffiths, ed., *Transform Techniques in Chemistry*, Plenum Press, New York, 1978.
6. K. Eckschlager and V. Stepanek, *Anal. Chem.* **54,** 1115A (1982).
7. B. R. Kowalski, *Chemometrics: Theory and Applications*, ACS Symposium Series 52, Washington, D.C., 1977.
8. G. M. Hieftje, J. C. Travis, and F. E. Lytle, *Lasers in Chemical Analysis*, The Humana Press, Clifton, N.J., 1981.
9. T. Hirschfeld, *Anal. Chem.* **52,** 297A (1980).
10. J. Ruzicka and E. H. Hansen, *Flow Injection Analysis*, John Wiley & Sons, Inc., New York, 1981.
11. W. E. van der Linden, *Trends Anal. Chem.* **1,** 188 (1982).
12. E. H. Hansen, J. Ruzicka, F. J. Krug, and E. A. G. Zagatto, *Anal. Chim. Acta* **148,** 111 (1983).
13. J. Lambe and R. C. Jaklevic, *Phys. Rev.* **165,** 821 (1968).
14. H. S. Gold and L. J. Hilliard, *Trends Anal. Chem.* **1,** 166 (1982).
15. B. K. Teo and D. C. Joy, eds., *EXAFS Spectroscopy: Techniques and Applications*, Plenum Press, New York, 1981.
16. D. C. Koningsberger and R. Prins, *Trends Anal. Chem.* **1,** 16 (1981).
17. A. C. Tam in D. Kliger, ed., *Ultrasensitive Spectroscopic Techniques*, Academic Press, Inc., New York, 1982; J. F. McClelland, *Anal. Chem.* **55**(1), 89A (1983).
18. M. M. Farrow, R. K. Burnham, and E. M. Eyring, *Appl. Phys. Lett.* **33,** 735 (1978).
19. C. K. N. Patel, *Science* **202,** 157 (1978).
20. L. B. Kreuzer, *Anal. Chem.* **50,** 597A (1978).
21. J. A. C. Broekaert, *Spectrochim. Acta* **36B**(6), 553 (1981).
22. V. A. Fassel, *Science* **202,** 183 (1978).
23. S. Greenfield, *Analyst* **105,** 1032 (1980).
24. J. A. C. Broekaert, *Trends Anal. Chem.* **1**(11), 249 (1982).
25. J. Reednick, *Am. Lab.*, 54 (Mar. 1979).
26. J. C. Travis, G. C. Turk, and R. B. Green, *Anal. Chem.* **54**(9), 1006A (1982).
27. G. C. Turk, J. R. DeVoe, and J. C. Travis, *Anal. Chem.* **54**(9), 643 (1982).
28. S. A. Borman, *Anal. Chem.* **54**(114), 1476A (1982).
29. D. L. Donahue, J. P. Young, and D. H. Smith, *Int. J. Mass Spectrom. Ion Phys.* **43,** 293 (1982); C. M. Miller, N. S. Nogar, A. J. Gancarz, and W. R. Shields, *Anal. Chem.* **54,** 2377 (1982).
30. J. D. Fassett, J. C. Travis, L. J. Moore, and F. E. Lytle, *Anal. Chem.*, (1983), in press.
31. J. C. Travis and L. J. Moore, personal communication, Washington, D.C.
32. R. Stefani, *Trends Anal. Chem.* **1,** 84 (1981).
33. E. Denayer, R. Van Grieken, F. Adams, and D. F. S. Natusch, *Anal. Chem.* **54,** 26A (1982).
34. D. M. Hercules, R. J. Day, K. Balasanmugam, T. A. Dang, and C. P. Li, *Anal. Chem.* **54,** 280A (1982).

35. *Fresenius' Z. Anal. Chem.* **308,** 193 (1981).
36. C. A. Fyfe, L. Bemi, R. Childs, H. C. Clar, D. Curtin, J. Davies, D. Drexler, R. L. Dudley, G. C. Gobbi, J. S. Hartman, P. Hayes, J. Klinowski, R. E. Lenkinski, C. J. L. Lock, I. C. Paul, A. Rudin, W. Tchir, J. M. Thomas, F. R. S., and R. E. Wasylishin, *Philos. Trans. R. Soc. London* **A305,** 591 (1982).
37. M. Reinhold, *Am. Lab.* **14**(2), 164 (1982).
38. C. S. Yannoni, *Acc. Chem. Res.* **15,** 201 (1982).
39. B. C. Gerstein, *Anal. Chem.* **55**(7), 781A (1983).
40. L. D. Bowers, *Trends Anal. Chem.* **1**(8), 191 (1982).
41. *Biochem. Soc. Trans.* **8,** 241 (1980).
42. E. M. Chait and R. C. Ebersole, *Anal. Chem.* **53**(6), 682A (1981).

CURT W. REIMANN
RANCE A. VELAPOLDI
National Bureau of Standards

ANTIBIOTICS, β-LACTAMS

CEPHALOSPORINS AND PENEMS

Cephalosporins

In the last six years, research into cephalosporin analogues has taken a leap forward and produced 11 injectable and three oral compounds, with several others under clinical investigation. The number of pharmaceutical companies interested in cephalosporins has greatly increased since 1976, especially in Japan. The first of the new cephalosporins to be introduced in this time were cefamandole (**1**), cefoxitin (**2**), and cefuroxime (**3**). These compounds were the first to have resistance to Type-1 β-lactamase; cefoxitin (**2**) and cefuroxime (**3**) are also resistant to other lactamases. Sub-

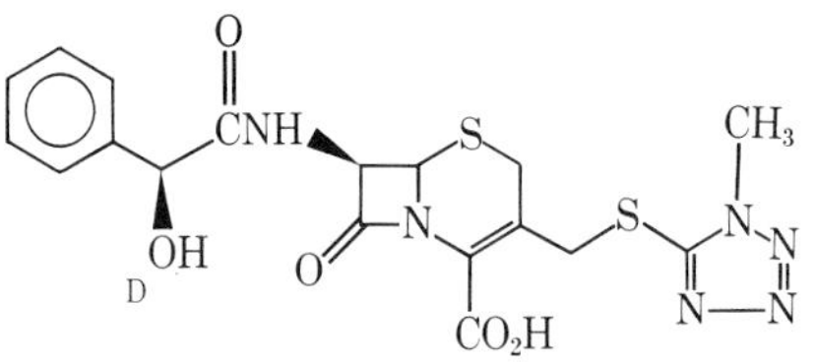

(**1**) cefamandole [*34444-01-4*]
(sometimes available as the formyl ester)

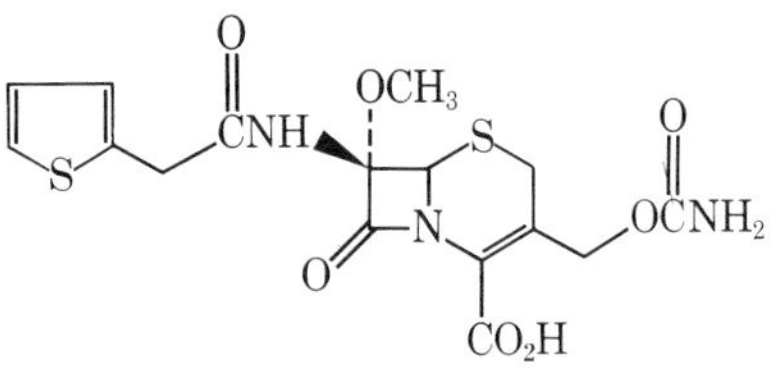

(**2**) cefoxitin
[*35607-66-0*]

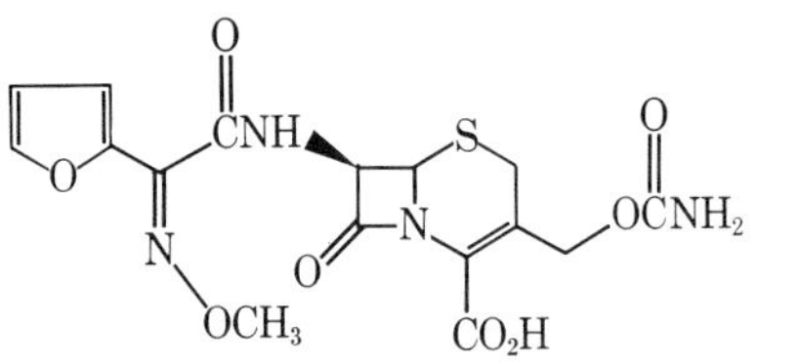

(**3**) cefuroxime
[*55268-75-2*]

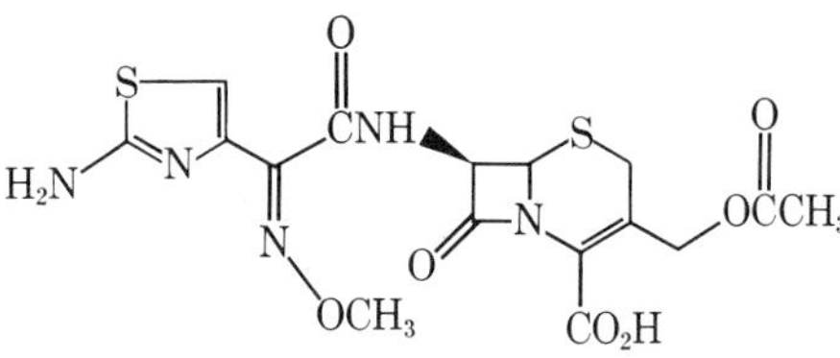

(**4**) cefotaxime
[*63527-52-6*]

sequently, it was observed that cephalosporins with a 2-aminothiazol-4-yl group, and the *syn*-oxime group of cefuroxime (**3**), had β-lactamase resistance and greatly increased potency; cefotaxime (**4**) was the first of these to be marketed.

Ceftazidime (**5**), currently under development, shows good activity against *Pseudomonas aeruginosa* and other gram-negative bacteria. The semisynthetic moxalactam (**6**), where oxygen replaces the sulfur atom of the cephalosporins, also exhibits a high level of potency against gram-negative bacteria including activity against *Ps. aeruginosa.* Other compounds active against *Pseudomonas* are cefoperazone (**7**), the close cephalosporin analogue of piperacillin, and cefsulodin (**8**) (related to sulbenicillin), which is surprising since it has no other noteworthy antibacterial activity.

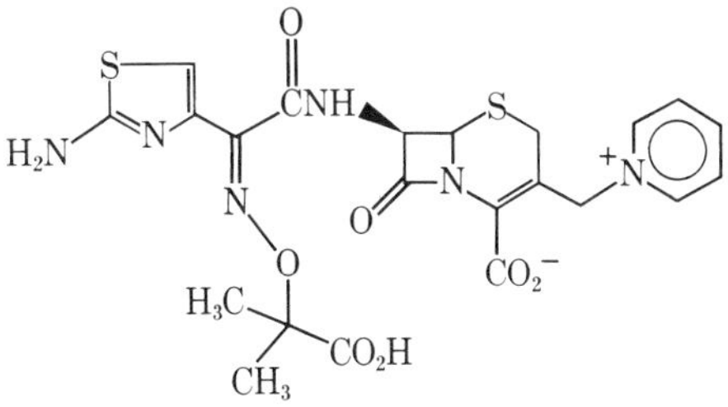

(**5**) ceftazidime
[*72558-82-8*]

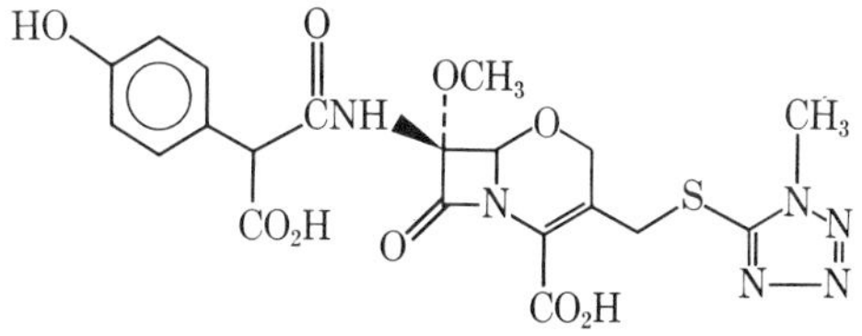

(**6**) moxalactam
[*64952-97-2*]

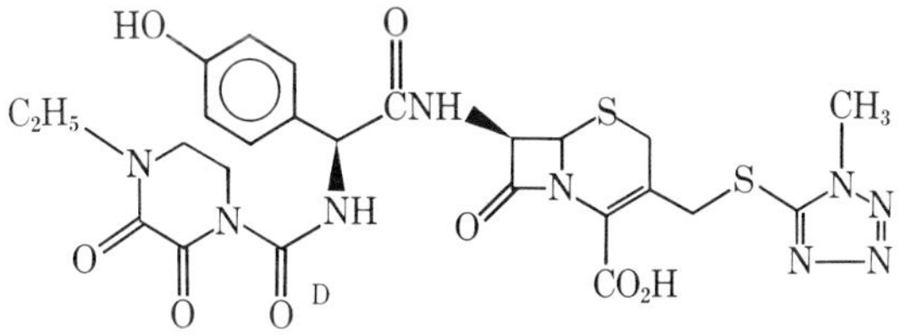

(**7**) cefoperazone
[*62893-19-0*]

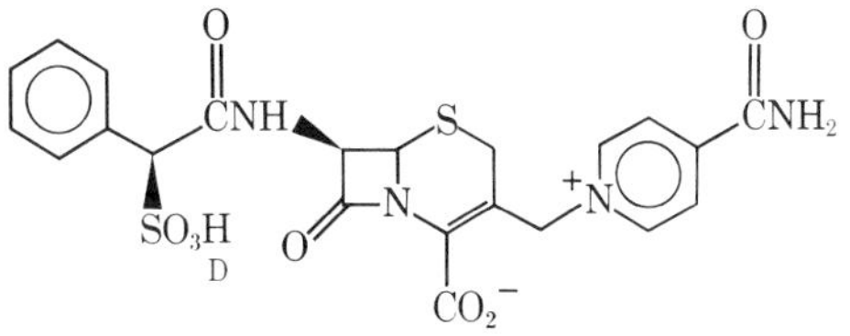

(**8**) cefsulodin
[*62587-73-9*]
Na salt: [*52152-93-9*]

The principal deficiency of the older cephalosporins was their lack of resistance to β-lactamases. The compounds with improved resistance have one of the following characteristics: monovalent substitution on the α-carbon atom in the 7-acyl amino group, eg, cefamandole (**1**) and cefoperazone (**7**); *syn*-oxime substituents, eg, cefuroxime (**3**), and a methoxy substituent on the β-lactam ring at the 7α-position, giving the class of cephalosporins known as cephamycins. These were first discovered in the fermentation liquors of several *Streptomyces* spp and, as in the other naturally occurring cephalosporins, the 7β-amino group is substituted by aminoadipoyl, but at the 3-position they have an α-methoxycinnamoyl group (**9**) or a carbamate group (**10**) (1).

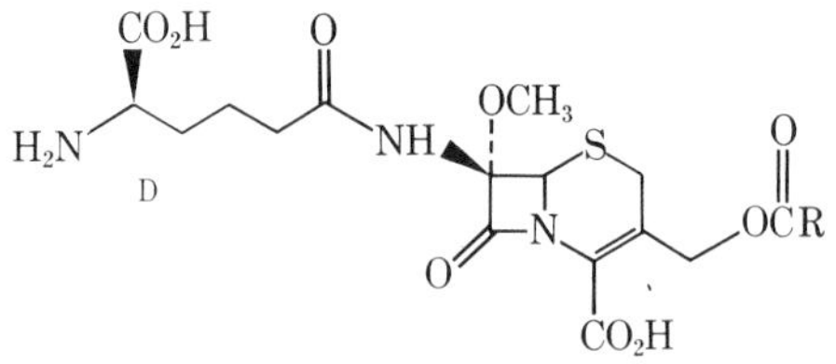

(**9**) R = $C(OCH_3){=}CH$–C_6H_4–OR′

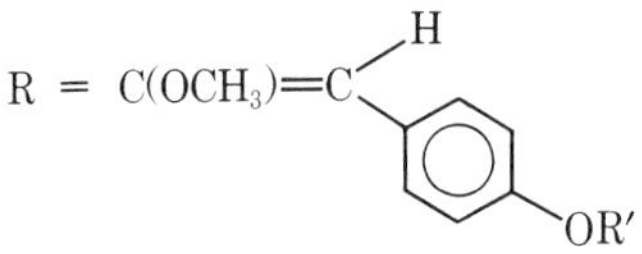

(**a**) R′ = SO_3H cephamycin A [*34279-78-2*]

(**b**) R′ = H cephamycin B [*34279-77-1*]

(**10**) R = NH_2 cephamycin C [*34279-51-1*]

These various structural changes have different effects, and increased resistance to one enzyme does not indicate resistance to all β-lactamases. The 7α-methoxy group is the most effective in imparting a broad-spectrum resistance to β-lactamases, closely followed by acyl side-chain oxime ether substituents. Cefuroxime (**3**), the first oximinocephalosporin, has a resistance spectrum similar to that of cefotaxime (**4**) (2); the less-resistant cefamandole (**1**) is similar to cefoperazone (**7**).

7α-Methoxycephalosporins. Although the antibacterial activity of the naturally occurring 7α-methoxycephalosporins was feeble, it did extend to species that were resistant to the cephalosporins then available, notably *Bacteroides* spp, and most β-lactamase-producing gram-negative organisms. Cephamycin C was converted into the thienylacetyl compound (**2**), cefoxitin (2–4). The conversion of (**10**) into (**2**), via the phosphorus pentachloride route, was not straightforward, because of the reaction of phosphorus pentachloride with the carbamoyl group.

A number of methods have been devised for the stereoselective introduction of a 7α-methoxy group into a cephalosporin derivative (5). Usually, a double bond is introduced between *C*-7 and the attached *N* group, and methanol addition occurs from the α-side of the molecule (see Fig. 1) (5–8). In this way, cefmetazole (**11**) was prepared, which is more potent than cefoxitin (**2**), although retaining its breadth of spectrum (6–9). Cefotetan (**13**), although it is not yet on the market, also has a broad spectrum of good activity against gram-negative organisms (10). It is formed, reversibly, from the isothiazolyl cephalosporin (**12**) (11).

All 7α-methoxycephalosporins have excellent resistance to β-lactamases but they have different levels of antibacterial activity (Table 1). The change in the 7β-substituent from cefoxitin (**2**) to cefotetan (**13**) improves activity, especially with regard to *Enterobacter* spp and *Serratia* spp. However, no such improvement occurred with *B. fragilis* or *Haemophilus influenzae*, against which the activities of the two compounds are virtually identical. Although cefotetan (**13**) is more active against *Ps. aeruginosa*, neither compound has therapeutically useful activity against this organism. The overall antibacterial activity of cefmetazole (**11**) is between those of cefoxitin (**2**) and cefotetan (**13**) (15).

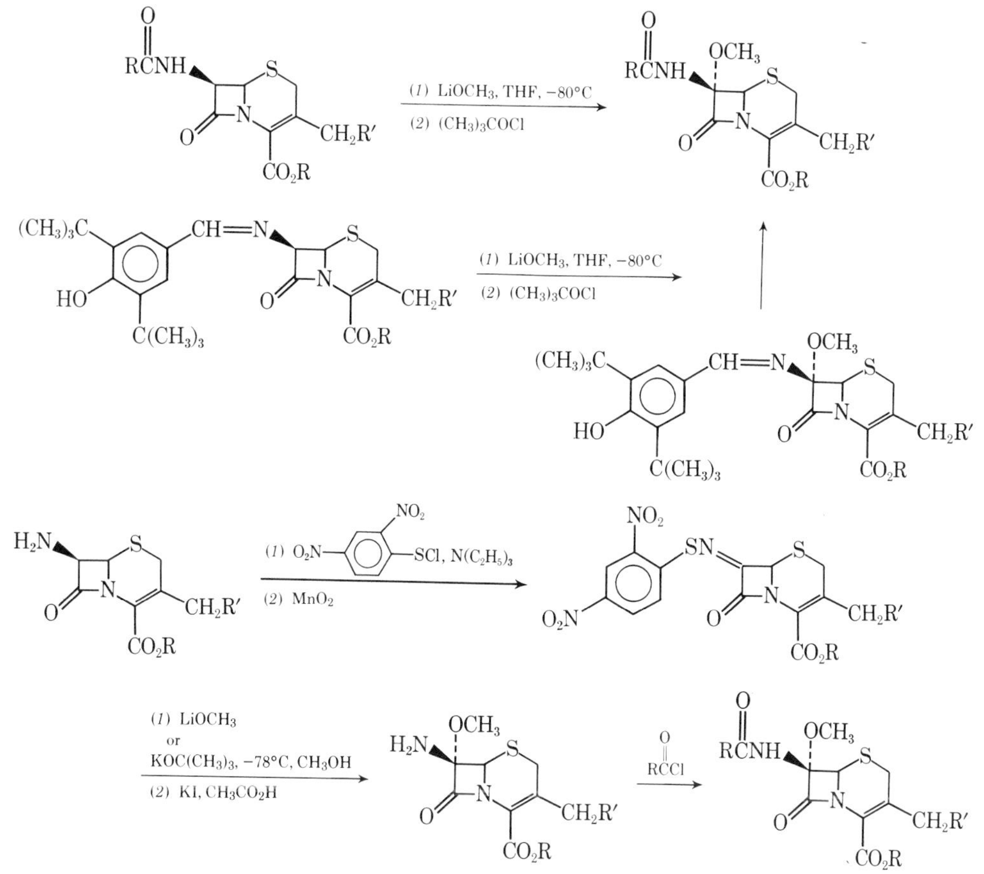

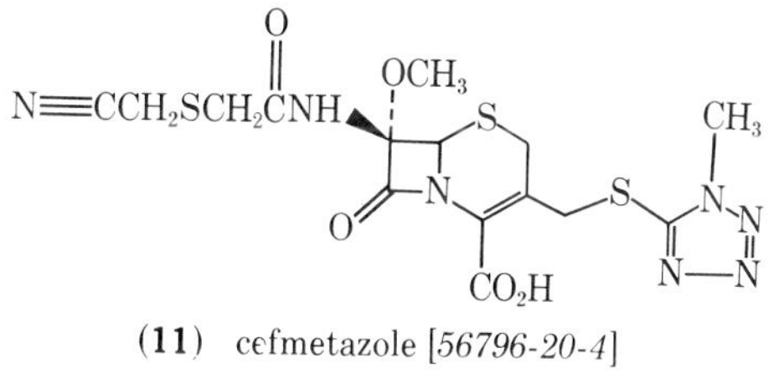

O
RCNH
S
N
O
CH_2R'
CO_2R

(1) $LiOCH_3$, THF, −80°C
(2) $(CH_3)_3COCl$

O
OCH_3
RCNH
S
N
O
CH_2R'
CO_2R

$(CH_3)_3C$
CH=N
S
HO
N
O
CH_2R'
$C(CH_3)_3$
CO_2R

(1) $LiOCH_3$, THF, −80°C
(2) $(CH_3)_3COCl$

OCH_3
$(CH_3)_3C$
CH=N
S
HO
N
O
CH_2R'
$C(CH_3)_3$
CO_2R

H_2N
S
N
O
CH_2R'
CO_2R

NO_2
(1) O_2N–C$_6$H$_3$–SCl, $N(C_2H_5)_3$
(2) MnO_2

NO_2
SN
S
O_2N
N
O
CH_2R'
CO_2R

(1) $LiOCH_3$
or
$KOC(CH_3)_3$, −78°C, CH_3OH
(2) KI, CH_3CO_2H

OCH_3
H_2N
S
N
O
CH_2R'
CO_2R

O
RCCl

O
OCH_3
RCNH
S
N
O
CH_2R'
CO_2R

O
OCH_3
$N{\equiv}CCH_2SCH_2CNH$
S
CH_3
N
O
S
N
N
N
N
CO_2H

(**11**) cefmetazole [*56796-20-4*]

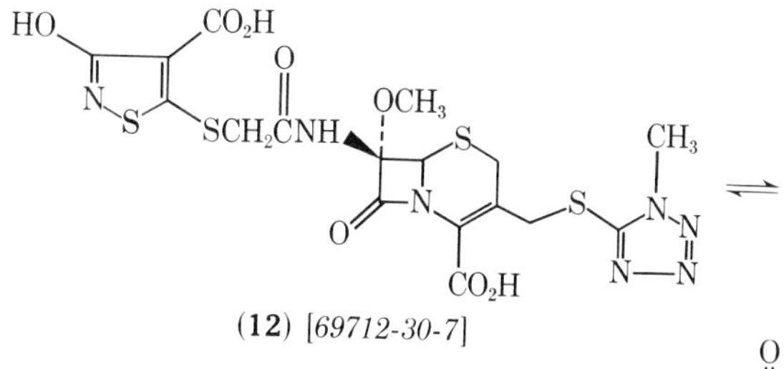

HO
CO_2H
O
N
S
SCH_2CNH
OCH_3
S
CH_3
N
O
S
N
N
N
N
CO_2H
⇌

(**12**) [*69712-30-7*]

O
O
H_2NC
S
CNH
OCH_3
S
HO_2C
S
CH_3
N
O
S
N
N
N
N
CO_2H

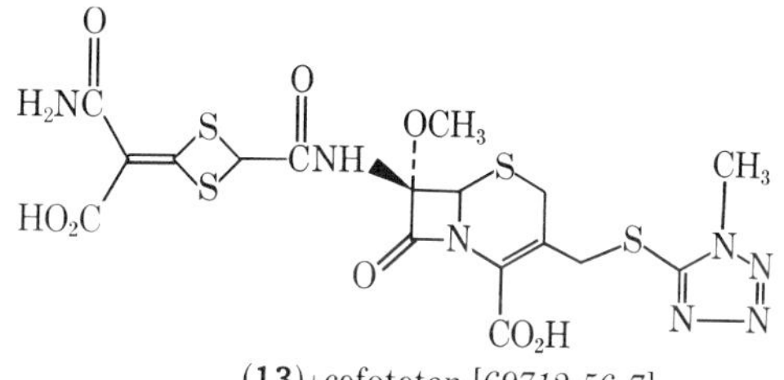

(**13**) cefotetan [*69712-56-7*]

Figure 1. 7α-Methoxylation of cephalosporins.

Table 1. *In Vitro* Antibacterial Activity of Some of the Newer Cephalosporins[a]

Bacteria	CFX[b]	CMZ[c]	CTN[b]	MOX[b]	CXM[c]	CEM[c]	CPZ[d]	CFT[c]	CTX[c]	CAZ[e]
Staph. aureus	2	1.6	4	2	1.6	0.4	3.1	0.8	1.6	6.3
Strep. pyogenes		0.8	0.25		0.05	⩽0.03		0.2	<0.025	0.1
E. coli	4	1.6	0.12	0.12	6.2	1.6	0.8	0.2	0.1	<0.05
Kleb. pneumoniae	2	0.8	0.12	0.12	3.1	1.6	1.6	0.2	0.05	0.05
Pr. mirabilis	4	3.1	0.12	0.015	3.1	1.6	0.8	0.4	0.05	0.05
Proteus spp, indole positive	4	3–6	0.12	0.06	0.8–>400	6.2–>200	0.8–12	0.4–200	⩽0.02–0.2	0.05
Enterobacter spp	128	>100	0.25	0.12	12–100	6.2–50	0.3	1.6–25	0.2–0.8	0.2
Citrobacter sp	2	50	0.06	0.06	6.2	6.2	0.4	1.6	0.4	0.02–0.1
Serratia marcescens	16	25	0.25	0.25	400	200	100	100	1.6	0.4
Ps. aeruginosa	>128	>100	32	8	>100	>100	6.2	>100	50	1.6
H. influenzae	2	3.1	1	0.03	0.8	0.8	0.4	0.8	⩽0.02	0.1
B. fragilis	8	25	4	0.5	12.5	50	50	50	3.1	25

[a] Concentrations in mg/L required to inhibit 50% of strains tested.
[b] Ref. 10. CFX = cefoxitin (**2**); CTN = cefotetan (**13**); MOX = moxalactam (**6**).
[c] Ref. 12. CXM = cefuroxime (**3**); CEM = cefamandole (**1**); CFT = cefotiam (**18**); CTX = cefotaxime (**4**); CMZ = cefmetazole (**11**).
[d] Ref. 13. CPZ = cefoperazone (**7**).
[e] Ref. 14. CAZ = ceftazidime (**5**).

Oxadethiacephalosporins. The first compound in which the sulfur of a cephalosporin was replaced by oxygen was 1-oxa-1-dethiacephalothin (**14**), prepared by total synthesis in the Merck, Sharp, and Dohme Research Laboratories (16). It has, somewhat surprisingly, the same activity as the cephalosporin or, strictly, twice the activity, since the synthetic material was racemic. Subsequently, the Merck team (17) found that the activity of the oxygen analogue of cefamandole (**15**) was as good as or better than cefamandole itself, except for the activity against *Proteus mirabilis.* Investigators at Beecham Laboratories (18), using penicillin V as starting material, again found that oxacephalosporins with the mandelamido side chain, together with a methyl at *C*-3 (**16**) or only hydrogen (**17**), had properties akin to cephalexin [*15686-71-2*] although the phenylglycyl analogue was inactive.

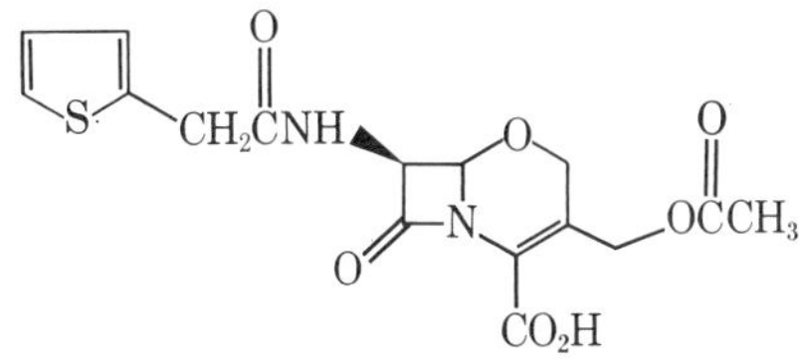

(**14**) (±) 1-oxa-1-dethiacephalothin [*54214-83-4*]

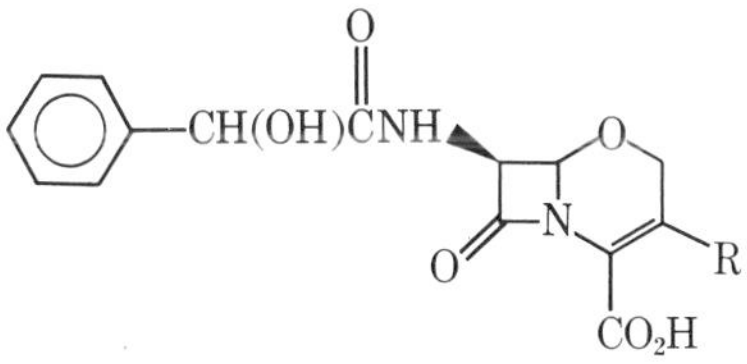

(**15**) R = 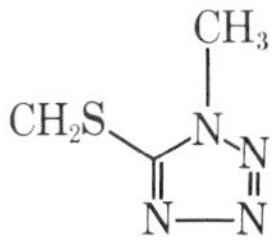 CH_2S–(1-methyltetrazol-5-yl) (±) cephamandole, oxygen analogue [*62504-53-4*]

(**16**) R = CH_3 [*63881-91-4*]

(**17**) R = H [*72777-65-2*]

At Shionogi, the properties of the oxygen analogues of the cephalosporins were studied including the effect of various substituents at *C*-3, *C*-7β, and *C*-7α (OCH_3 or H), and moxalactam (**6**) was prepared. In the original synthesis, the oxygen and the three carbon atoms destined to be *C*-2, *C*-3, and *C*-3a were supplied from external sources, and the overall yield was very low. In the most recently published method, the five carbon chain of the *N*-residue of penicillin is incorporated intact (19–22). The synthesis from penicillin G is shown in Figure 2.

The antibacterial activity of moxalactam (**6**) is very similar in breadth and degree to that of cefotaxime (**4**) and to the 7α-methoxy cephalosporin, cefotetan (see Table 1). However, it shows more activity against *B. fragilis*, *H. influenzae*, and *Ps. aeruginosa* than cefotetan (**13**) (10,23). Moxalactam (**6**) is highly resistant to a broad spectrum of β-lactamases owing to the stabilizing effect of the 7α-methoxy group (24); cefotetan (**13**) shows similar resistance.

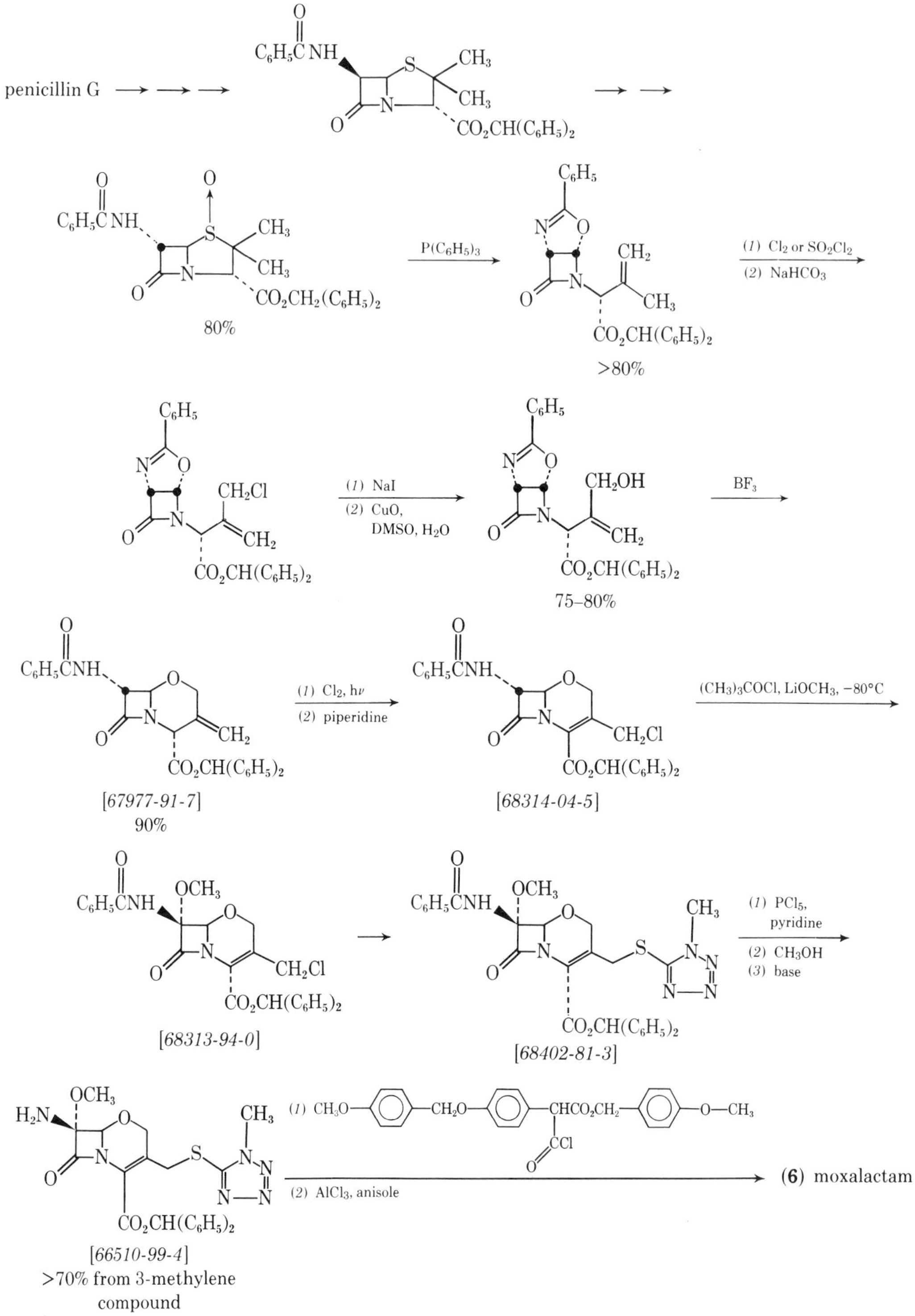

penicillin G → → →

C_6H_5CNH, S, CH_3, CH_3, N, O, $CO_2CH(C_6H_5)_2$ → →

C_6H_5CNH, O, S, CH_3, CH_3, N, O, $CO_2CH_2(C_6H_5)_2$

80%

$P(C_6H_5)_3$ →

C_6H_5, N, O, CH_2, N, O, CH_3, $CO_2CH(C_6H_5)_2$

>80%

(1) Cl_2 or SO_2Cl_2
(2) $NaHCO_3$ →

C_6H_5, N, O, CH_2Cl, N, O, CH_2, $CO_2CH(C_6H_5)_2$

(1) NaI
(2) CuO, DMSO, H_2O →

C_6H_5, N, O, CH_2OH, N, O, CH_2, $CO_2CH(C_6H_5)_2$

75–80%

BF_3 →

C_6H_5CNH, O, N, O, CH_2, $CO_2CH(C_6H_5)_2$

[*67977-91-7*]
90%

(1) Cl_2, $h\nu$
(2) piperidine →

C_6H_5CNH, O, N, O, CH_2Cl, $CO_2CH(C_6H_5)_2$

[*68314-04-5*]

$(CH_3)_3COCl$, $LiOCH_3$, −80°C →

C_6H_5CNH, OCH_3, O, N, O, CH_2Cl, $CO_2CH(C_6H_5)_2$

[*68313-94-0*]

→

C_6H_5CNH, OCH_3, O, N, O, S, CH_3, N, N, N, N, $CO_2CH(C_6H_5)_2$

[*68402-81-3*]

(1) PCl_5, pyridine
(2) CH_3OH
(3) base →

H_2N, OCH_3, O, N, O, S, CH_3, N, N, N, N, $CO_2CH(C_6H_5)_2$

[*66510-99-4*]
>70% from 3-methylene compound

(1) CH_3O–C_6H_4–CH_2O–C_6H_4–$CHCO_2CH_2$–C_6H_4–O–CH_3 (with CCl=O)
(2) $AlCl_3$, anisole → (**6**) moxalactam

Figure 2. Synthesis of moxalactam (**6**) from benzylpenicillin.

Aminothiazolyl Cephalosporins. The high potency of cephalosporins with an aminothiazolyl group against gram-negative bacteria was first demonstrated in 1977 with the announcement of SCE-963, subsequently called cefotiam (**18**), by Takeda Chemical Industries, Ltd. (25). It was synthesized from a 7-(γ-chloroacetoacetamido)cephalosporanic acid [Fig. 3(**a**)] (26) or from desacetylcephalosporin C via the 3-acetoacetoxymethyl derivative [Fig. 3(**b**)] (27).

At the same time, Hoechst-Roussel (28) announced that a group of compounds [(**4**): R = H [*63504-16-5*], CH_3, C_2H_5, $CH(CH_3)_2$, or $CH_2CH{=}CH_2$ [*63504-19-8*]] showed powerful bactericidal activity, 10–100 times that of other cephalosporins against gram-negative bacteria (see Fig. 4). The activity of the anti-isomers was considerably lower than that of the corresponding syn isomers. The methoxime [(**4**): R = CH_3] was found to be the most active (29), and this compound, now known as cefotaxime, is on the market. A method of preparation, not necessarily the best, is shown in Figure 4. The oxime is mostly in the syn configuration if the thiourea reaction is carried out at room temperature.

A number of other oximino-substituted aminothiazolyl cephalosporins have recently come on the market or are in the final stages of development. Examples are ceftizoxime (**19**) (12,30), unsubstituted in the 3-position (Cefizox (SKF)); cefmenoxime (**20**) with the same 3-substituent as cefamandole (**1**) (15); and ceftriaxone (**21**) with a mercaptosubstituted hydroxytriazinone at position 3 (31) (see Fig. 5).

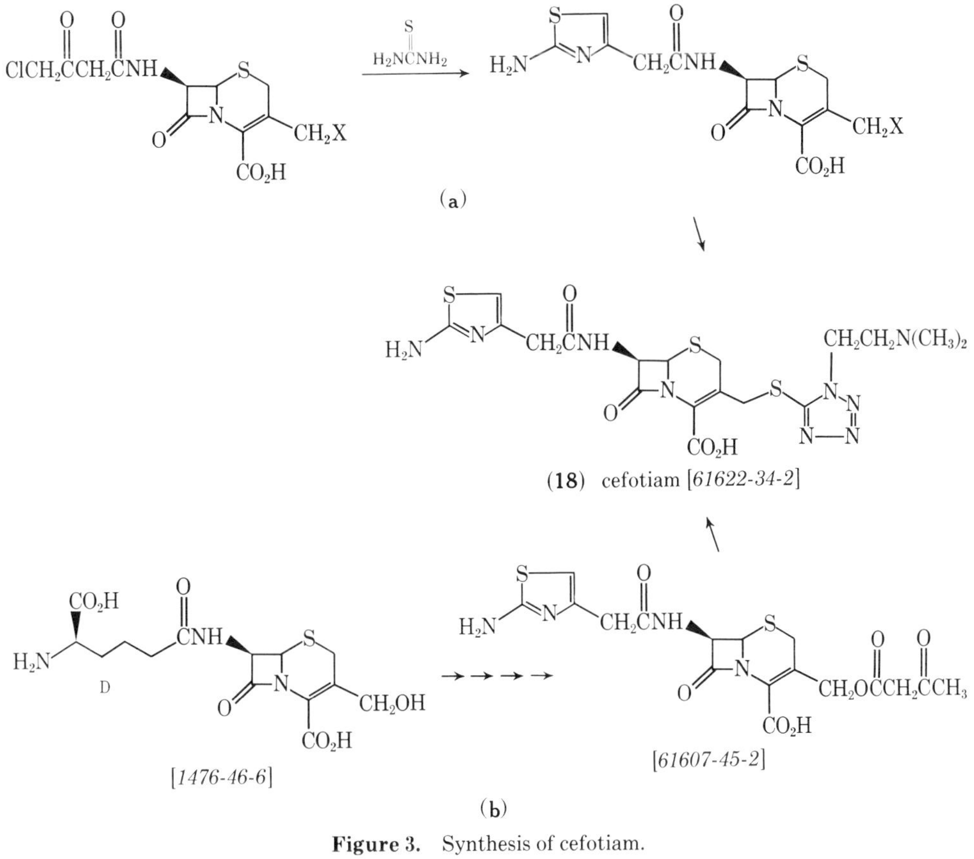

$ClCH_2\overset{O}{\overset{\|}{C}}CH_2\overset{O}{\overset{\|}{C}}NH$ — CH_2X, CO_2H $\xrightarrow{H_2N\overset{S}{\overset{\|}{C}}NH_2}$ H_2N (aminothiazolyl) $CH_2\overset{O}{\overset{\|}{C}}NH$ — CH_2X, CO_2H

(a)

H_2N (aminothiazolyl) $CH_2\overset{O}{\overset{\|}{C}}NH$ — CO_2H, S-tetrazolyl $CH_2CH_2N(CH_3)_2$

(**18**) cefotiam [*61622-34-2*]

H_2N $\overset{CO_2H}{}$ D $\overset{O}{\overset{\|}{C}}NH$ — CH_2OH, CO_2H

[*1476-46-6*]

→ → → →

H_2N (aminothiazolyl) $CH_2\overset{O}{\overset{\|}{C}}NH$ — $CH_2O\overset{O}{\overset{\|}{C}}CH_2\overset{O}{\overset{\|}{C}}CH_3$, CO_2H

[*61607-45-2*]

(b)

Figure 3. Synthesis of cefotiam.

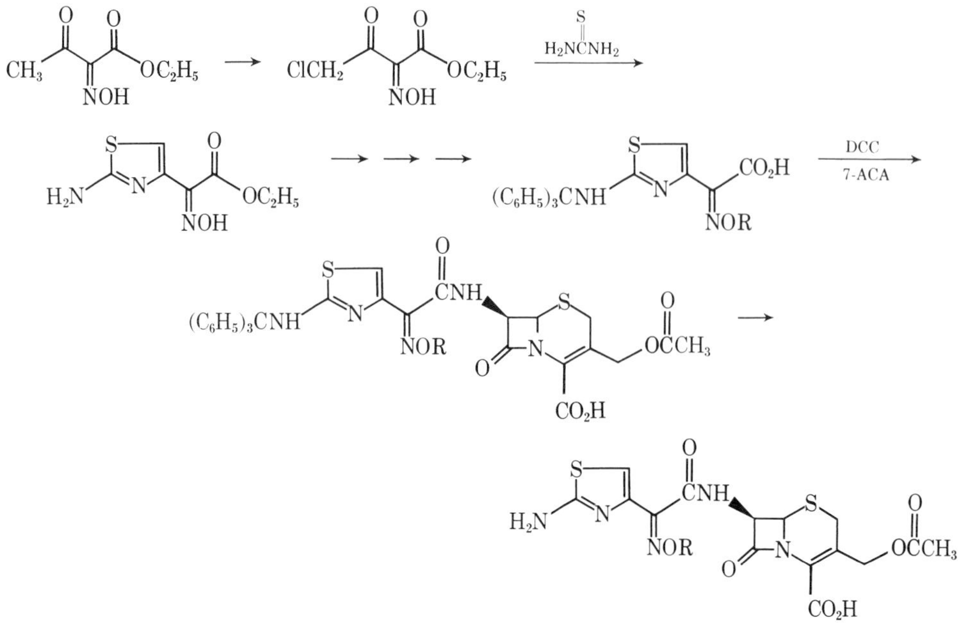

(**4**) R = CH_3 cefotaxime
R = C_2H_5 *[63504-17-6]*
R = $CH(CH_3)_2$ *[63504-18-7]*

Figure 4. Synthesis of cefotaxime and related compounds. DCC = dicyclohexylcarbodiimide; 7-ACA = 7-aminocephalosporanic acid.

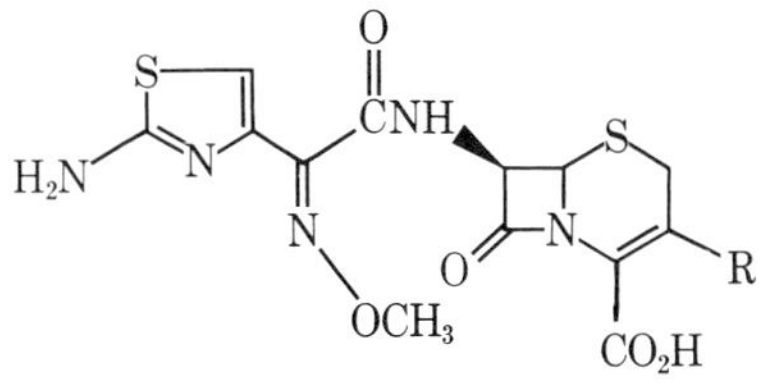

(**19**) R = H ceftizoxime *[68401-81-0]*

(**20**) R = cefmenoxime *[65085-01-0]*

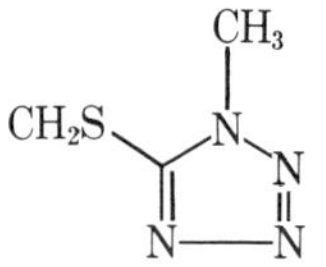

(**21**) R = ceftriaxone *[73384-59-5]*

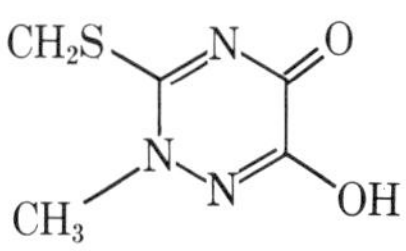

Figure 5. Oximino-substituted aminothiazolyl cephalosporins.

In a related compound, ceftazidime (32) (**5**), the oxime is etherified by a carboxyisopropyl group, and a pyridinium is at position 3. Its activity against *Ps. aeruginosa* is so high that it can be considered a replacement for the aminoglycoside antibiotics. The synthesis is shown in Figure 6 (33).

The antibacterial activities of the four methoxyiminoaminothiazolyl cephalosporins, cefotaxime, ceftizoxime, cefmenoxime, and ceftriaxone, are virtually identical in breadth and degree (12,23,31); the activity of cefotaxime is representative of the

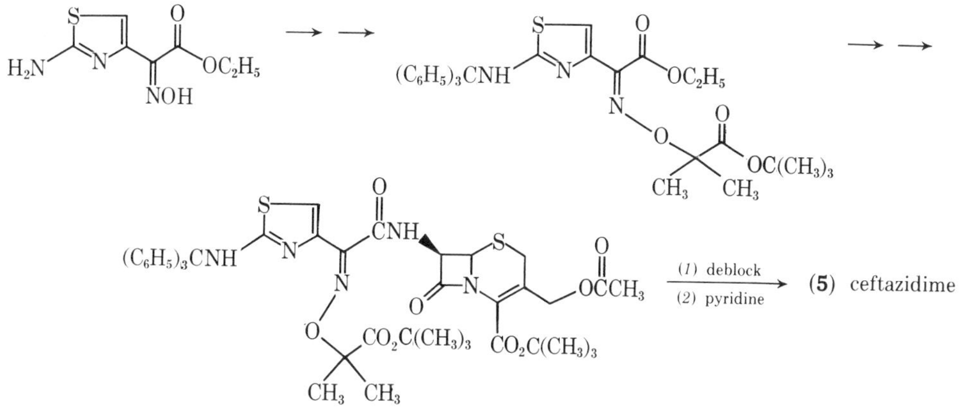

$\rightarrow\rightarrow$ $(C_6H_5)_3CNH$ OC_2H_5 $OC(CH_3)_3$ CH_3 CH_3 $\rightarrow\rightarrow$

$(C_6H_5)_3CNH$ CNH $OCCH_3$ $CO_2C(CH_3)_3$ $CO_2C(CH_3)_3$ CH_3 CH_3 (1) deblock (2) pyridine → (5) ceftazidime

Figure 6. Synthesis of ceftazidime.

group, and is compared in Table 1 with that of cefotiam (**18**), and with ceftazidime (**5**). The oximino compounds are more active and have a broader spectrum than cefotiam (**18**). They also show greater resistance to hydrolysis by β-lactamases. This resistance, in turn, is reflected by the breadth of the antibacterial spectrum of the compounds. Although ceftazidime (**5**) and cefotaxime (**4**) have essentially the same activity against many organisms, and are similar to moxalactam (**6**) against a broad range of bacteria, there are some important differences: cefotaxime is more active against *Staph. aureus* and *B. fragilis*, whereas ceftazidime (**5**) is usually more active against *Acinetobacter* spp and *S. marcescens* and highly effective against *Ps. aeruginosa* (13–14).

Cephalosporins with Special Properties. Until the emergence of compounds with activity against *Pseudomonas* such as the aminothiazoles and moxalactam (**6**), the best results were obtained with cefsulodin (**8**) (21–22), which bears some resemblance to carbenicillin. It has, however, virtually no activity against gram-negative bacteria, other than *Pseudomonas* (34).

Ceftriaxone (**21**) is notable for its remarkably long persistence in the blood (35), partially because of very high binding to serum proteins. In its antibacterial activity *in vitro*, it closely resembles other cephalosporins with the same 7-acyl group, namely, cefotaxime (**4**), cefmenoxime (**20**), and ceftizoxime (**19**), although its activity against *Ps. aeruginosa* is slightly higher. Its rather unusual heterocyclic group attached to the 3-methyl group is synthesized as indicated in Figure 7 (36).

Chromogenic Cephalosporins. A recently described purple cephalosporin, known as PADAC (**22**), loses this color when attacked by β-lactamases (37).

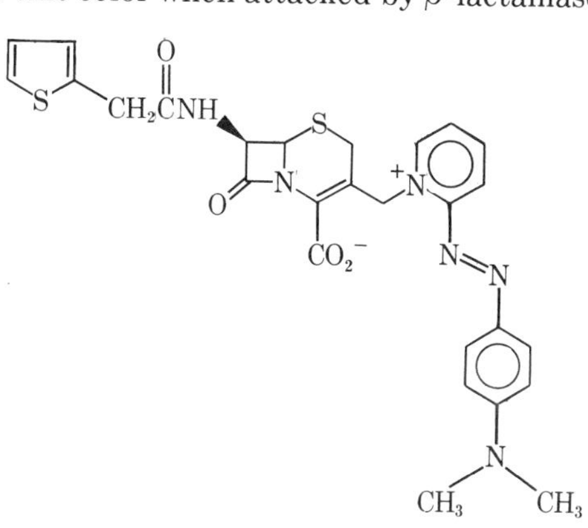

CH_2CNH CO_2^- CH_3 CH_3

(**22**) PADAC purple
[*77449-91-3*]

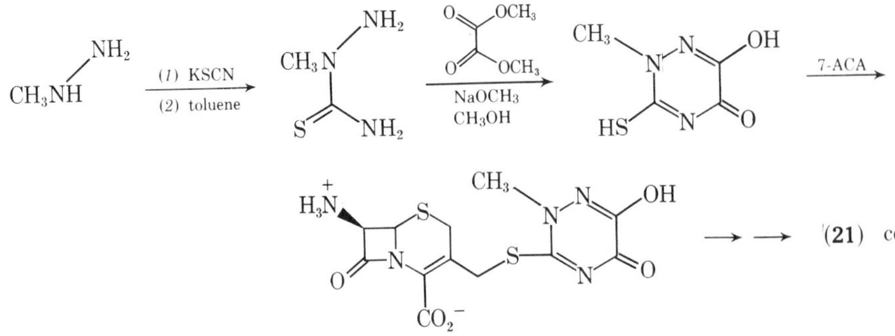

Figure 7. Synthesis of ceftriaxone.

Pharmacokinetic Properties. The pharmacokinetic properties of the cephalosporins depend to a large extent on the substituent at position 3. 1-*N*-Substituted tetrazol-5-ylmercaptomethyl is a widely used group at position 3, eg, in cefoperazone (**7**), cefotiam (**18**), moxalactam (**6**), cefamandole (**1**), cefmenoxime (**20**), cefotetan (**13**), and cefmetazole (**11**) but, as there is always some interaction with the 7-acyl group, their pharmacokinetic properties are not identical. There is a tendency for compounds with this 3-substituent to be excreted via the bile, especially cefoperazone (**7**) (38). Moxalactam (**6**), cefotiam (**18**), and, to a lesser extent, cefamandole (**1**) have also been reported to give high biliary concentrations; the presence of comparatively large amounts of potent antibiotics in the gastrointestinal tract may give rise to digestive upsets. Results *in vitro* and *in vivo* may not correspond.

When this type of 3-substituent bears an acidic group such as that in cefonicid [*61270-58-4*] (39) or the earlier compound ceforanide [*60925-61-3*] (40), the rate of urinary excretion is reduced and higher, longer-lasting serum levels are obtained. Ceftriaxone (**21**), with an acidic group in its 3-substituent, has extremely long-lasting serum levels. Acidic 3-substituents appear to be associated with a very high level of serum-protein binding which may contribute to their long-lasting total serum levels (36).

Orally Active Cephalosporins. In the search for orally active cephalosporins, four compounds have been synthesized (see Fig. 8): cefatrizine (**23**) (41), cefadroxil (**24**) (42), cefroxadine (**25**) (43–44), and cefaclor (**26**) (45–46).

The synthesis of (**25**) and (**26**), which have heteroatoms attached directly to *C*-3, is somewhat unusual (see Fig. 9). Cephalosporins with a 3-acetoxymethyl group can be reduced to the 3-methylene compound (**27**) by a number of reducing agents; (**27**), in turn, can be esterified and ozonized to the 3-hydroxy-ceph-3-em ester (**28**). Treatment with diazomethane gives the 3-methoxyceph-3-em ester (**29**), whereas thionyl chloride or its equivalent gives the 3-chloro compound (**30**). In an alternative route from penicillin to the 3-methylene compound [Fig. 9(**b**)], the penicillin ester sulfoxide is treated with a halogenating agent, such as *N*-chlorosuccinimide, to give the open-chain sulfinyl halide (**31**), which is then cyclized with a Lewis acid to the 3-methylene compound *S*-oxide (**32**). The subsequent steps to cefroxadine (**25**) or cefaclor (**26**) are conventional (47–48).

Cefroxadine (**25**) and cefadroxil (**24**) have virtually the same antibacterial activity as cephalexin (Table 2) and are absorbed orally to a similar extent. Cefaclor (**26**) and cefatrizine (**23**) are more active *in vitro* than cephalexin, but oral absorption is lower and they are not stable in serum (49–50).

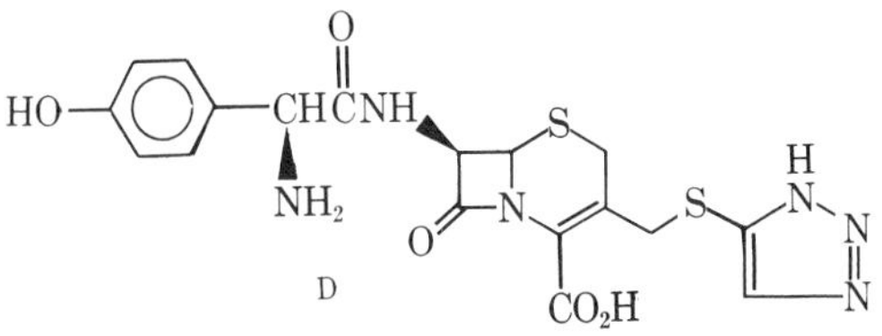

(**23**) cefatrizine
[*51627-14-6*]

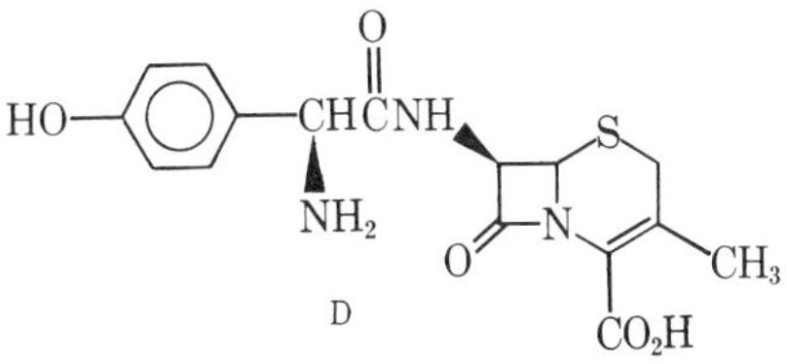

(**24**) cefadroxil
[*50370-12-2*]

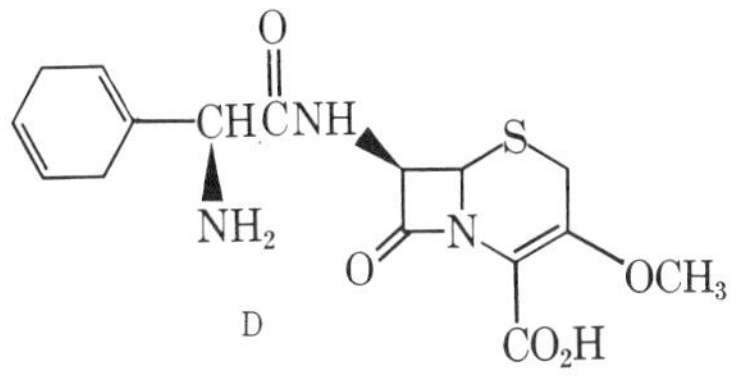

(**25**) cefroxadine
[*51762-05-1*]

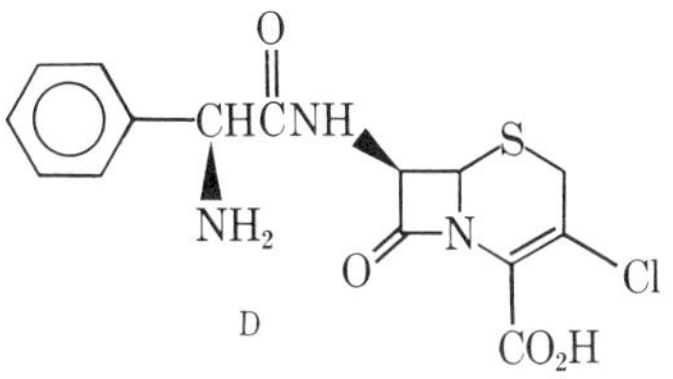

(**26**) cefaclor
[*53994-73-3*]

Figure 8. Orally active cephalosporins.

Table 2. *In Vitro* **Antibacterial Activity of Orally Absorbed Cephalosporins, and Penem Sch 29482**

	Antibacterial activity in μg/mL					
Bacteria	Cephalexin[a]	Cefadroxil[a] (**24**)	Cefroxadine[a] (**25**)	Cefatrizine[a] (**23**)	Cefaclor[a] (**26**)	Sch 29482[b]
Staph. aureus	1.7	2.5	0.6	1.1	2.2	
Strep. pneumoniae	0.9	2.0	0.6	0.2	0.24	0.12
E. coli	2.9	8	2.5	2.3	1.9	0.5
K. pneumoniae	6.4	8	3	2.5	0.8	0.25
Pr. mirabilis	17	16	15	4.2	2.6	1.0
indole positive *Proteus*	91	>100		15	64	2.0
Enterobacter aerogenes	>100	>100	>100	27	27	1.0
H. influenzae	>16	32		4	4	0.5

[a] Ref. 51.
[b] Ref. 52.

In Vitro **Antibacterial Activity.** The antibacterial activity of any particular cephalosporin is a combination of degree and type of activity at the target site, the ease with which it can penetrate, and the ability to resist the attack of destructive enzymes (53). The nature and complexity of the biochemical target for β-lactam antibiotics has now been more clearly elucidated, and the mechanisms of penetration are better understood.

Since many factors are involved, including pharmacokinetic and pharmacodynamic properties, the antibiotic may not perform as predicted even after *in vitro* activity has been determined under the most carefully controlled laboratory conditions.

Intrinsic Activity and Biochemical Targets. All β-lactam antibiotics act by binding to and inhibiting the action of members of the enzyme complex involved in the formation of cell-wall peptidoglycan (54–55). These penicillin-binding proteins (PBPs) are species specific; seven have been identified in gram-negative organisms, but only

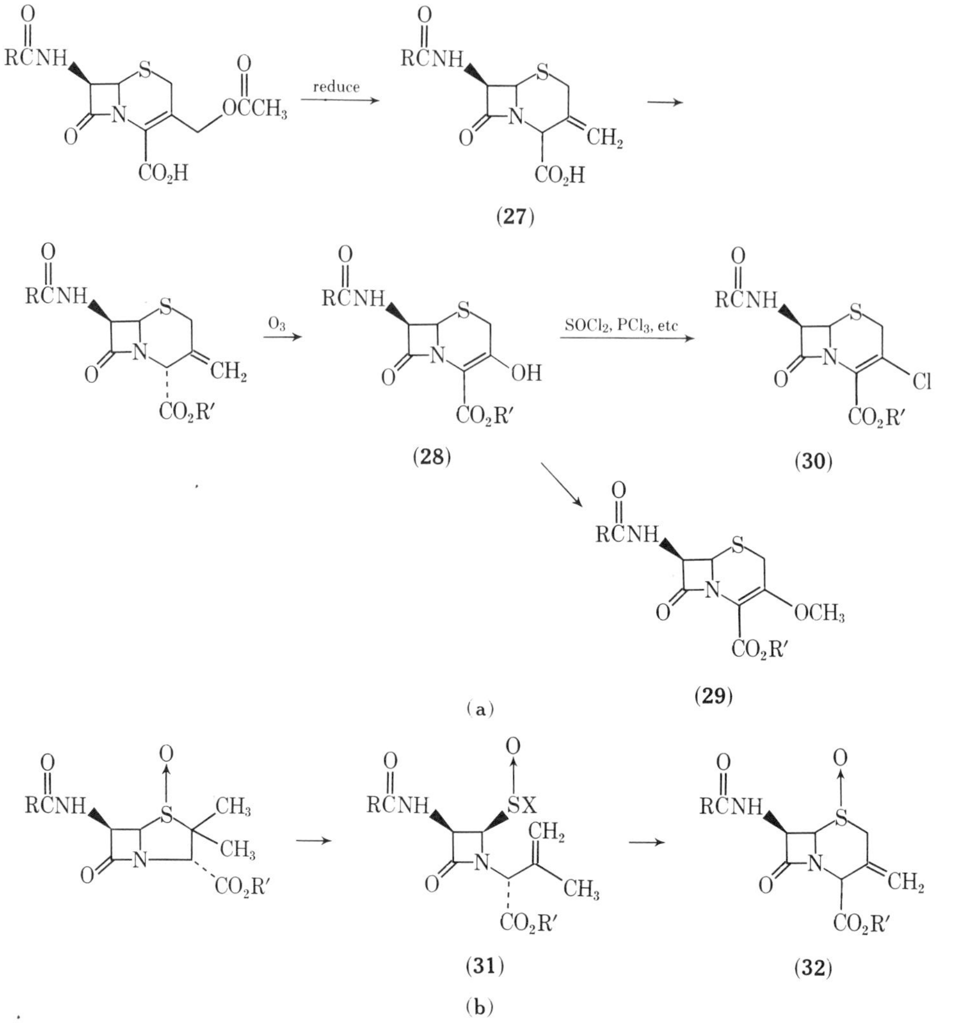

Figure 9. Synthesis of cefroxadine and cefaclor.

four in *Staph. aureus*. The lower molecular weight proteins designated 4, 5, and 6 are not thought to be essential for cell life. In contrast, the higher molecular weight proteins 1A, 1B, 2, and 3 are essential to the life and growth of cells and also govern their shapes.

Although all the cephalosporins are thought to act in the same general way, as shown by competitive binding experiments (56), they differ in their avidities for the various PBPs. Preferential binding to PBP 1B, as with cephaloridine [*50-59-9*], produces rapid lysis, with little malformation of the cells. If PBP 2 is the most susceptible, large round misshapen forms are produced which only lyse slowly (as exemplified by the penicillin mecillinam [*32887-01-7*]). Cross-wall formation is governed by PBP 3 and cephalosporins that primarily inhibit this protein produce highly elongated forms which ultimately lyse. In practice, each cephalosporin inhibits all these proteins to varying extents, and the intrinsic activity of the compound depends largely on the concentration of the compound required to saturate the most susceptible PBP. The

7α-methoxycephalosporins have considerable avidity for the nonessential PBPs 4, 5, and 6, to which cephalosporins without a 7α-substituent bind very little if at all (57).

Cell Penetration. No matter how susceptible the target to the antibiotic, bacterial growth proceeds unimpaired if the cell wall is able to prevent its entry. There can be a considerable difference between the concentration required to inhibit intact cells and the concentrations required to saturate the target enzymes in broken-cell preparations. For example, the concentration of cefoxitin (**2**) (a 7α-methoxy cephalosporin) required to inhibit the growth of whole organisms is ca 100-fold that which saturates the essential PBPs, whereas with, eg, cefuroxime, the minimal inhibitory concentration (MIC) is only fourfold the concentration that saturates the PBPs (58). Recent work has identified proteins in the outer cell wall of several species of bacteria which appear to be connected with the ability of antibiotics to penetrate the cell. Cells deficient in some of these proteins, called porins, appear to be much more impermeable than cells rich in porins. There are indications that porins are, to some extent, specific to certain antibiotics (59–60). Penetration may also be hindered if a cephalosporin binds with high avidity to a superficial cell component. It has been suggested that one mechanism for blocking entry to the cell may be very tight binding to β-lactamases in the periplasmic space.

Resistance to β-Lactamases. The close correlation between enzyme susceptibility and antibacterial activity in the earliest cephalosporins is much less clearcut with the more recent compounds (61). Measurements of conventional β-lactamase resistance with isolated enzyme preparations cannot determine such small losses as may occur during long exposure of low concentrations of the compound to uncontrolled amounts of enzyme, as occurs in the MIC test. In addition, the comprehensive term β-lactamase-resistant is sometimes used to describe a compound which is actually resistant to only a few of the many enzymes now identified.

Biogenesis. The biosynthesis of cephalosporins and penicillins starts from the same amino acids, which form the common intermediate δ-(L-α-aminoadipyl)-L-cysteinyl-D-valine, the LLD-tripeptide (62), but its role was difficult to prove as this compound does not penetrate into intact cells. A cell-free system was developed and LLD was converted into isopenicillin N, ie, the isomer of penicillin N with an L-aminoadipyl side chain; this is believed to be the immediate precursor of the penicillin N produced by *Cephalosporium acremonium*. Since LLD can also be converted to benzylpenicillin by an acyl transferase in *Penicillium chrysogenum*, it appears that isopenicillin N occupies a central place in both penicillin and cephalosporin biosynthesis. Conversion of LLD-tripeptide to isopenicillin N and the conversion of penicillin N to desacetoxycephalosporin C require powerful oxidation and are stimulated by ferrous ion (63–64). However, the steps in the cyclization remain obscure, despite much speculation and some incorporation attempts with postulated intermediates. The further oxidation of the 3-position in cephalosporins and the introduction of the 7α-methoxy group also involve oxygenases, since 3-position oxidation also requires iron (65). The biosynthesis of penicillins and cephalosporins is illustrated in Figure 10 (66).

Economic Aspects. Cephalosporins first entered the antibiotic market in 1964 when cephalothin [*153-61-7*] and cephaloridine, both injectable, were launched. During the next decade, they became established, particularly in hospitals. At this time (1983), ca 114 metric tons of injectable cephalosporins is used annually with the newer com-

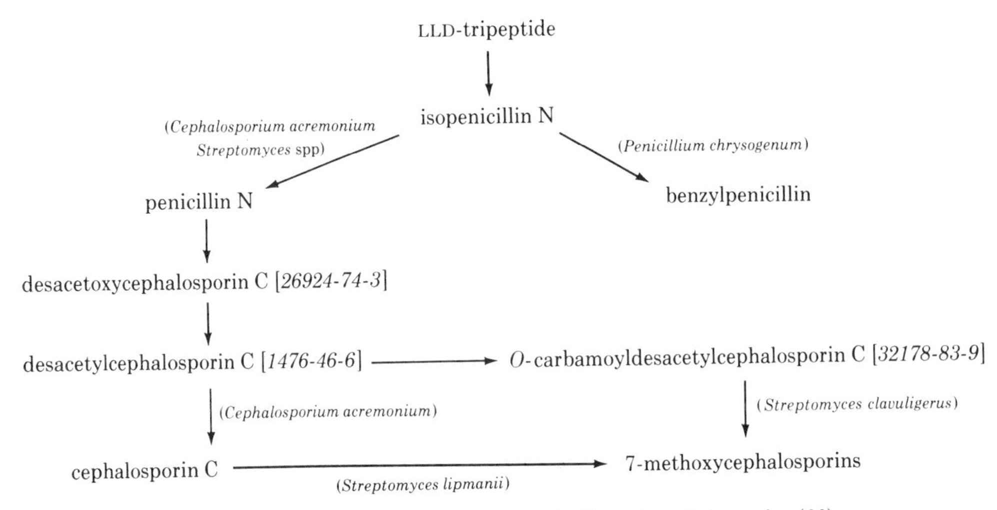

Figure 10. Biosynthetic pathways to penicillin and cephalosporins (66).

pound cefazolin [*25953-19-9*] emerging as a significant product. Although subsequently another six similar compounds appeared, cephalothin and cefazolin were the preferred cephalosporins as recently as 1980. By the late 1970s, the injectable cephalosporins had become important therapeutic agents in hospitals. With the new impetus started at that time by the first β-lactamase-stable products, eg, cefoxitin (**2**) and cefuroxime (**3**), quickly followed by the more active products such as cefotaxime (**4**), cefoperazone (**7**), and moxalactam (**6**), cephalosporins now dominate in many markets with sales in excess of $\$10^9$ per year.

The first oral cephalosporin appeared in early 1964 when cephaloglycine [*3577-01-3*] was launched. By the end of that year, it had already been superceded by cephalexin which remains the preferred oral cephalosporin to the present day. In the intervening period, a series of similar products have entered the market including cephradine [*3882-53-3*], cefadroxil (**24**), cefatrizine (**23**), and cefaclor (**26**) but their combined usage does not seriously challenge cephalexin. Total oral cephalosporin sales now approach $\$9 \times 10^8$/yr.

Penems

Penems were first described by Woodward and co-workers (67–68), who prepared them by an intramolecular Wittig reaction of compounds of type (**33**) (see Fig. 11). These esters were moderately stable but the free acids, apparently, were not. However, compound (**34**) ([*64370-39-4*] R′ = CH_3) was obtained and found to have antibacterial activity resembling that of a cephalosporin, rather than a penicillin. That the instability was associated with the amide side chain was shown when a 6-unsubstituted penem (**35**) was prepared from deshydroxyclavulanic acid. The compound was stable in aqueous solution at pH 4–10. Furthermore, compound (**35**), isolated as the potassium salt [*71355-41-4*], showed good broad-spectrum antibacterial activity *in vitro*, and

S, C_2H_5, N, O, CO_2H

(**35**) (±) [*73761-27-0*]

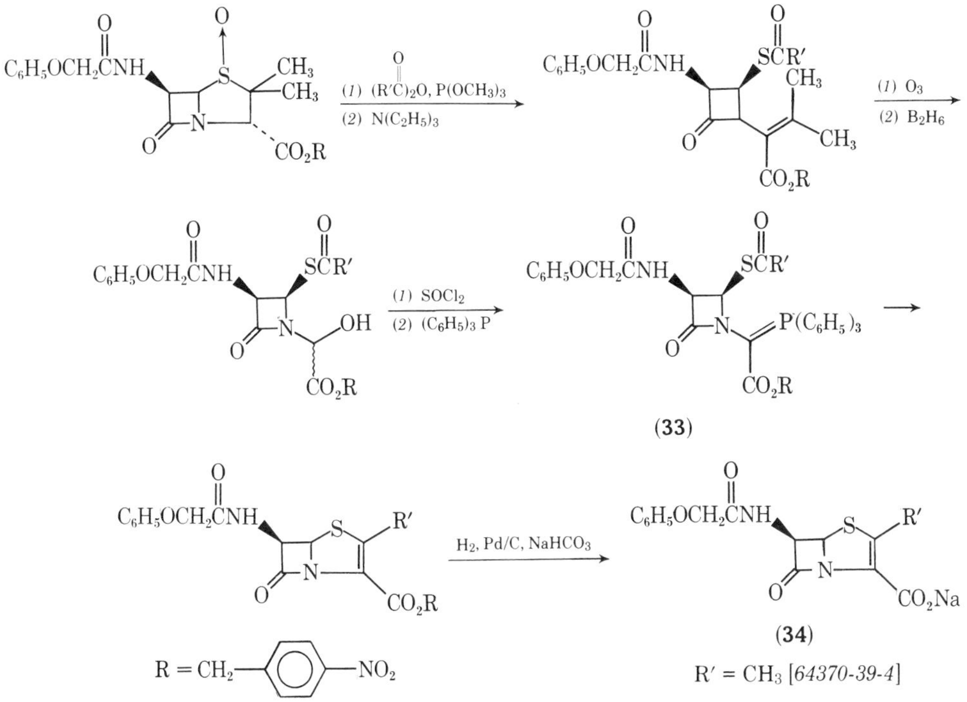

(1) $(R'CO)_2O$, $P(OCH_3)_3$ (2) $N(C_2H_5)_3$

(1) O_3 (2) B_2H_6

(1) $SOCl_2$ (2) $(C_6H_5)_3P$

(33)

H_2, Pd/C, $NaHCO_3$

(34)

$R = CH_2-C_6H_4-NO_2$

$R' = CH_3$ [*64370-39-4*]

Figure 11. Synthesis of 6-acylamino penems from the corresponding penicillin.

was stable to penicillinase from staphylococci and to P99 cephalosporinase; it was, however, destroyed by TEM-1 and K-1 β-lactamases.

The work was followed up to provide a variety of 2-substituted ethyl penem-3-carboxylic acids, with hydroxy, alkoxy, acyloxy (including aminocarbonyloxy), amino, acylamino, alkyureido, and aryl- and acylthio groups as the substituent. In broad terms, their antibacterial activity was similar. Furthermore, their *in vivo* activity was disappointing, apparently not because of instability, high serum binding, or poor pharmacokinetics, but rather because of a lack of activity in presence of serum proteins (69–70).

Substitution at the 6-position of the penem nucleus has been achieved (71) by total synthesis. The 6α-methoxy compound [(**36**): R = OCH_3: R′ = C_2H_5], though less active than the unsubstituted compound, was more active than the 6β-methoxy derivative [(**37**): R = OCH_3: R′ = C_2H_5] and, furthermore, was unaffected by β-lactamases. Again, of the 6-epoxyethyl compounds, one of the two 6α-substituted isomers [(**36**):

R = —CH—CH₂ (epoxide, bridged by O),

R′ = C_2H_5] showed excellent activity against organisms producing the TEM, P99, and K-1 β-lactamases. These results have, broadly, been substantiated by synthesis, using Woodward's cyclization method of a variety of 2-substituted penems unsubstituted at *C*-6 and of 2,6-disubstituted penems (52,72). The activity was little affected by the 2-substituent, but much more by the 6-substituent. The threo-6α-1-hydroxyethyl compound (**38**) was the most active compound synthesized and certainly more active than the erythro-6α-form.

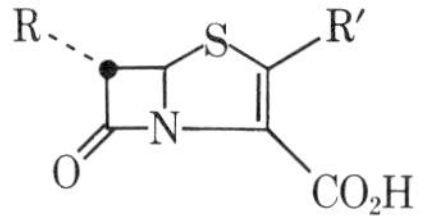

(**36**)
R = OCH_3, R′ = C_2H_5 [*86630-89-8*]

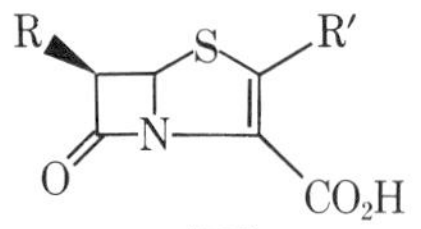

R = —CH—CH_2 (epoxide), R′ = C_2H_5 [*86569-24-6*]

(**37**)
R = OCH_3, R′ = C_2H_5 [*86630-88-8*]

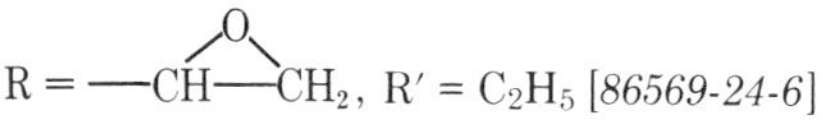

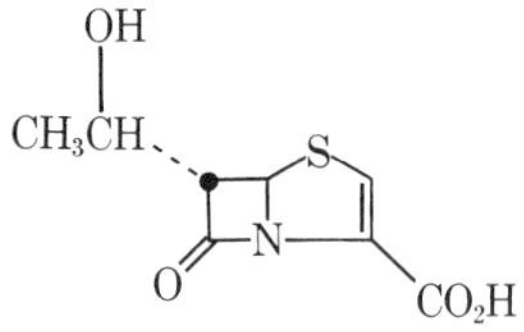

(**38**) (±) [*73845-26-8*]

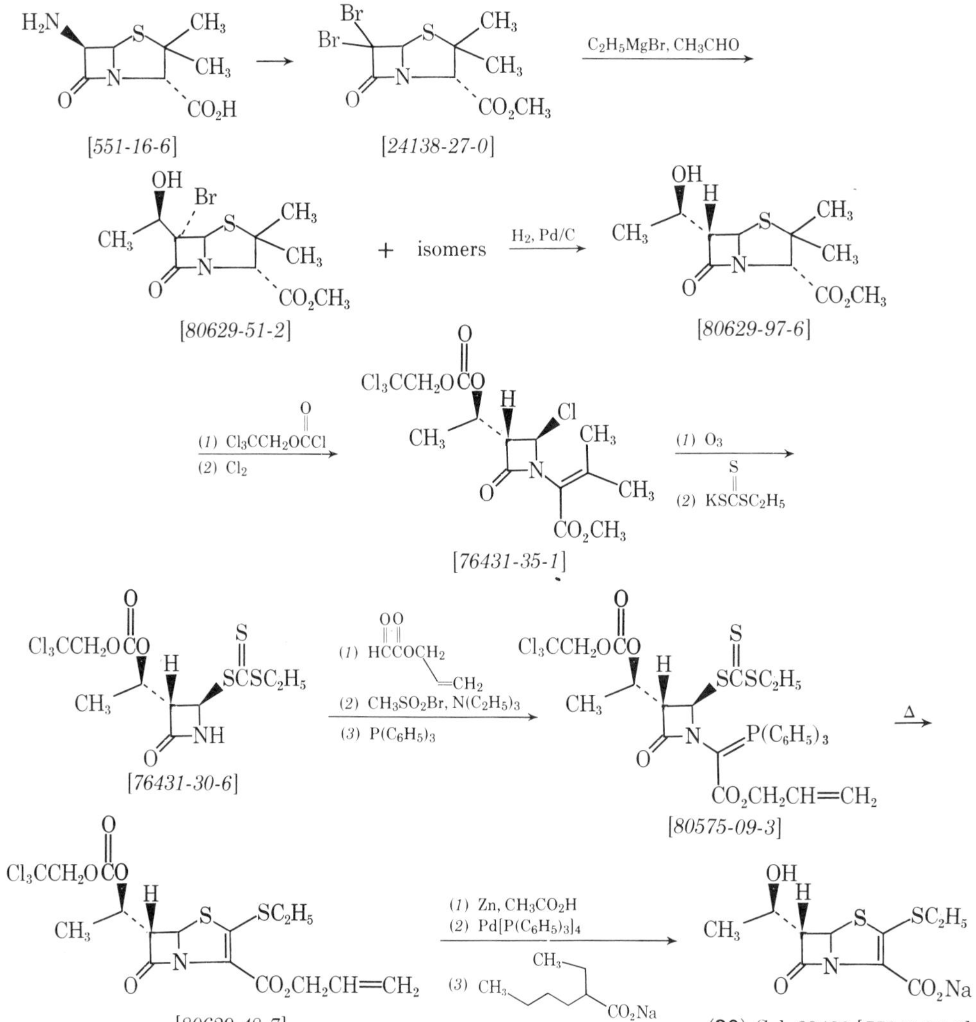

[*551-16-6*] → [*24138-27-0*] —(C_2H_5MgBr, CH_3CHO)→ [*80629-51-2*] + isomers —(H_2, Pd/C)→ [*80629-97-6*] —((*1*) Cl_3CCH_2OCOCl; (*2*) Cl_2)→ [*76431-35-1*] —((*1*) O_3; (*2*) $KSC(S)SC_2H_5$)→ [*76431-30-6*] —((*1*) $HC(O)C(O)OCH_2CH{=}CH_2$; (*2*) CH_3SO_2Br, $N(C_2H_5)_3$; (*3*) $P(C_6H_5)_3$)→ [*80575-09-3*] —Δ→ [*80629-48-7*] —((*1*) Zn, CH_3CO_2H; (*2*) $Pd[P(C_6H_5)_3]_4$; (*3*) sodium 2-ethylhexanoate)→ (**39**) Sch 29482 [*77646-84-5*]

Figure 12. Synthesis of Sch 29482.

It was also demonstrated that all the activity is carried by the form that has the penicillin stereochemistry at *C*-5 (73–75). Workers at the Schering Corporation have synthesized Sch 29482 (**39**) by the method shown in Figure 12 (75).

The spectrum of activity of Sch 29482 differs from that of the newer cephalosporins, and also from the orally absorbable ones. It has a higher activity against *Staph. aureus*, but a similar level of activity to cefotaxime (**4**) against *Haemophilus influenzae* and *Neisseria* spp. Unlike the cephalosporins, it is active against *Streptococcus faecalis*. Against most gram-negative bacteria, it generally has one fifth to one tenth the activity of cefotaxime (**4**) and moxalactam (**6**), but has similar, or higher activity against *Serratia* spp, *Acinetobacter*, and *Bacteroides* spp. Its high resistance to a broad spectrum of β-lactamases is comparable to that of moxalactam and cefoxitin (52). It is generally much more enzyme-resistant and more active than any of the orally absorbed cephalosporins (Table 2). The compound is strongly bound to proteins, and its activity is markedly reduced when it is tested in a medium containing 75% human serum (76–77). Although the activity of Sch 29482 is somewhat lower than is now expected from a new β-lactam, this is not necessarily a disadvantage. The ever-greater *in vitro* activity seen in recent years has not greatly improved clinical results nor significantly lowered dosage regimes (10).

Sch 29482 (**39**) is absorbed orally in humans, and the pharmacokinetics are independent of the administered dose. An oral 1 g dose gives a peak serum concentration of ca 25 mg/L and an elimination half-life of 1.3 h. Most of excreted Sch 29482 appears in the urine within four hours after administration. However, regardless of dose, only ca 2% of the amount administered is recovered (78).

BIBLIOGRAPHY

1. E. O. Stapley and co-workers, *Antimicrob. Agents Chemother.* **2**, 122 (1972).
2. I. N. Simpson, S. J. Plested, and P. B. Harper, *J. Antimicrob. Chemother.* **9**, 357 (1982).
3. T. W. Miller and co-workers, *J. Antimicrob. Chemother.* **9**, 132 (1982).
4. A. K. Miller and co-workers, *J. Antimicrob. Chemother.* **9**, 281, 287 (1982).
5. S. Korady and co-workers, *J. Am. Chem. Soc.* **94**, 1410 (1972); G. A. Koppel and R. E. Koehler, *J. Am. Chem. Soc.* **95**, 2403 (1973).
6. H. Yanagisawa and co-workers, *J. Antibiot.* **29**, 969 (1976).
7. H. Yanagisawa and co-workers, *Tetrahedron Lett.*, 2705 (1975).
8. T. Kobayashi, K. Iino, and T. Hiraoka, *J. Am. Chem. Soc.* **99**, 5505 (1977).
9. T. Kobayashi and co-workers, *Chem. Pharm. Bull. Jpn.* **27**, 2718, 2727 (1979).
10. R. Wise, J. M. Andrews, and J. Hancox, *Antimicrob. Agents Chemother.* **21**, 486 (1982).
11. M. Iwanami and co-workers, *Chem. Pharm. Bull. Jpn.* **28**, 2629 (1980).
12. T. Kamimura and co-workers, *Antimicrob. Agents. Chemother.* **16**, 540 (1979).
13. H. C. Neu and co-workers, *Antimicrob. Agents Chemother.* **16**, 150 (1979).
14. H. C. Neu, *J. Antimicrob. Chemother. Suppl. B* **8**, 131 (1981).
15. K. Tsuchiya and co-workers, *Antimicrob. Agents Chemother.* **19**, 56 (1981).
16. L. D. Cama and B. G. Christensen, *J. Am. Chem. Soc.* **96**, 7582 (1974).
17. R. A. Firestone and co-workers, *J. Med. Chem.* **20**, 551 (1977).
18. C. L. Branch and M. J. Pearson, *J. Chem. Soc. Perkin Trans. 1*, 2268 (1979).
19. W. Nagata, *Phil. Trans. R. Soc. London Ser. B* **289**, 255 (1980).
20. M. Yoshioka and co-workers, *Tetrahedron Lett.* **21**, 351 (1980).
21. H. Nomura and co-workers, *Heterocycles* **2**, 67 (1974).
22. H. Nomura and co-workers, *J. Med. Chem.* **17**, 1312 (1974).
23. P. C. Fuchs and co-workers, *Antimicrob. Agents Chemother.* **20**, 747 (1981).

24. K. Murakami and T. Yoshida, *Antimicrob. Agents Chemother.* **19,** 1 (1981).
25. M. Numata and co-workers, *J. Antibiot.* **31,** 1262 (1978).
26. K. Morita and co-workers, *Phil. Trans. R. Soc. London Ser. B* **289,** 181 (1980).
27. S. Tsushima and co-workers, *Chem. Pharm. Bull. Jpn.* **27,** 696 (1979).
28. R. Bucourt and co-workers, *C. R. Acad. Sci. Ser. D* **284,** 1847 (1977).
29. R. Bucourt and co-workers, *Tetrahedron* **34,** 2233 (1978).
30. H. Kojo and co-workers, *Antimicrob. Agents Chemother.* **16,** 549 (1979).
31. P. Angehrn and co-workers, *Antimicrob. Agents Chemother.* **18,** 913 (1980); *Chem Eng. News*, 18 (Sept. 26, 1983).
32. C. H. O'Callaghan and co-workers, *Antimicrob. Agents Chemother.* **17,** 876 (1980).
33. UK Pat. Specification 2,025,398 (May 1979), C. H. O'Callaghan, D. G. H. Livermore, and C. E. Newall (to Glaxo Group Ltd.).
34. M. P. E. Slack and co-workers, *J. Antimicrob. Chemother.* **5,** 687 (1979).
35. C. D. Findlay and co-workers, *J. Antimicrob. Chemother.* **9,** 57 (1982).
36. R. Reiner and co-workers, *J. Antibiot.* **33,** 783 (1980).
37. P. Schindler and P. Huber in U. Brodbeck, ed., *Enzyme Inhibitors*, Verlag Chemie, Weinheim, FRG, 1980, p. 169.
38. J. Shimizu, *Clin. Ther.* (Suppl. 3), 60 (1980).
39. D. H. Pitkin and co-workers, *Current Chemotherapy and Infectious Diseases* **1,** 252 (1980).
40. R. D. Smyth and co-workers, *Antimicrob. Agents Chemother.* **16,** 615 (1979).
41. F. Leitner and co-workers, *Antimicrob. Agents Chemother.* **7,** 298 (1975); P. Actor and co-workers, *Antimicrob. Agents Chemother.* **9,** 800 (1976).
42. R. E. Buck and K. E. Price, *Antimicrob. Agents Chemother.* **11,** 324 (1977).
43. O. Zak and co-workers, *J. Antibiot.* **29,** 653 (1976).
44. K. Yasuda, S. Kurashige, and S. Mitsuhashi, *Antimicrob. Agents Chemother.* **18,** 105 (1980).
45. J. Santoro and M. E. Levison, *Antimicrob. Agents Chemother.* **12,** 442 (1977).
46. J. Santoro and co-workers, *Antimicrob. Agents Chemother.* **13,** 951 (1978).
47. S. Kukolja in J. Elks, ed., *Recent Advances in the Chemistry of β-Lactam Antibiotics*, Chemical Society Special Publication No. 28, Chemical Society, London, 1977, p. 181.
48. R. R. Chauvette and P. A. Pennington, *J. Med. Chem.* **18,** 403 (1975).
49. H. Lode and co-workers, *Antimicrob. Agents Chemother.* **16,** 1 (1979).
50. J. M. Broughall and co-workers, *J. Antimicrob. Chemother.* **5,** 471 (1979).
51. C. H. O'Callaghan, *J. Antimicrob. Chemother.* **5,** 635 (1979).
52. A. L. Barry and co-workers, *J. Antimicrob. Chemother. Suppl. C* **9,** 97 (1982).
53. M. H. Richmond, *J. Antimicrob. Chemother. Suppl. B* **4,** 1 (1978).
54. B. G. Spratt, *Phil. Trans. R. Soc. London Ser. B* **289,** 273 (1980).
55. J. A. Kelly and co-workers, *Science* **218,** 479 (1982).
56. N. A. C. Curtis and co-workers, *Antimicrob. Agents Chemother.* **16,** 325 (1979).
57. N. A. C. Curtis, *J. Antimicrob. Chemother.* **8,** 85 (1981).
58. M. H. Richmond and W. Wotton, *Antimicrob. Agents Chemother.* **10,** 219 (1976).
59. K. J. Harder and co-workers, *Antimicrob. Agents Chemother.* **20,** 549 (1981).
60. Y. Komatsu and co-workers, *Antimicrob. Agents Chemother.* **20,** 613 (1981).
61. J. L. Ott and co-workers, *Antimicrob. Agents Chemother.* **15,** 14 (1979).
62. P. A. Fawcett, *Biochem. J.* **157,** 651 (1976).
63. M. Kohsaka and A. L. Demain, *Biochem. Biophys. Res. Commun.* **70,** 465 (1976).
64. J. E. Baldwin and co-workers, *Phil. Trans. R. Soc. London Ser. B* **289,** 169 (1980).
65. M. K. Turner and co-workers, *Biochem. J.* **173,** 839 (1978).
66. J. O'Sullivan and co-workers, *Phil. Trans. R. Soc. London Ser. B* **289,** 363 (1980).
67. R. B. Woodward in ref. 47, p. 167.
68. I. Ernest and co-workers, *J. Am. Chem. Soc.* **100,** 8214 (1978).
69. C. E. Newall in G. I. Gregory, ed., *Recent Advances in the Chemistry of β-Lactam Antibiotics*, Chemical Society Special Publication No. 38, Chemical Society, London, 1977, p. 151.
70. P. C. Cherry, C. E. Newall, and N. S. Watson, *J. Chem. Soc. Chem. Commun.*, 663 (1979).
71. M. W. Foxton and co-workers in ref. 69, p. 281.
72. M. Lang and co-workers, *J. Am. Chem. Soc.* **101,** 6296 (1979).
73. I. Ernest, J. Gosteli, and R. B. Woodward, *J. Am. Chem. Soc.* **101,** 6301 (1979).
74. J. R. Pfaendler, J. Gosteli, and R. B. Woodward, *J. Am. Chem. Soc.* **101,** 6306 (1979).
75. V. M. Girijavallabhan and co-workers, *Tetrahedron Lett.* **22,** 3485 (1981).

76. H. C. Neu and P. Labthavikul, *J. Antimicrob. Chemother. Suppl. C* **9**, 49 (1982).
77. R. M. Brown, R. Wise, and J. M. Andrews, *J. Antimicrob. Chemother. Suppl. C* **9**, 17 (1982).
78. R. Gural and co-workers, *J. Antimicrob. Chemother. Suppl. C* **9**, 239 (1982).

CYNTHIA O'CALLAGHAN
J. ELKS
Glaxo Pharmaceuticals Ltd.
Glaxo Group Research Ltd.

CLAVULANIC ACID, THIENAMYCIN, AND OTHERS

The discovery of the cephamycins in the late 1960s and early 1970s led to a resurgance of interest in new β-lactam antibiotics (1–2). During the last decade, the detection, isolation, and characterization of a variety of new β-lactam systems from microorganisms are illustrated by the evaluation of the novel β-lactams, eg, the nocardicins, monobactams, clavulanic acid (**1**), thienamycin (**2**), the closely related olivanic acids, and other carbapenems (see Fig. 1). Although many new penicillins, cephalosporins, and cephamycins with significant improvements in chemotherapeutic properties over the early penicillins and cephalosporins have been developed in the same period, it remains to be seen if members of these new families of β-lactam antibiotics will establish a place as clinically important therapeutic agents. So far, clavulanic acid has reached the market place in a formulation with amoxycillin [*26787-78-0*]; representatives of the thienamycin families are currently being clinically evaluated.

The detection of novel β-lactam systems has resulted from the application of new screening procedures, such as screening for inhibitors of β-lactamase or peptidoglycan synthetase and the use of supersensitive strains of various bacteria as screening organisms. Screening programs have generally involved the examination of soil samples from a wide range of localities throughout the world. This has led to the isolation of new species and strains of fungi, *Actinomycetes*, and bacteria, which produce novel β-lactams, as well as to the isolation of more common strains with the ability to yield new β-lactam metabolites. Culture collections have also served as sources of organisms for screening.

So far, only clavulanic acid has been developed to the clinical stage. An oral formulation of potassium clavulanate (125 mg) plus amoxycillin trihydrate [*26787-78-0*] (250 mg) is now marketed by Beecham Pharmaceuticals in the UK as Augmentin [*74469-00-4*] and is available to hospitals and general practitioners. Clavulanic acid and amoxycillin, in 1:2, 1:4, and other formulations, are being investigated in other European countries and in the United States. Timentin, a combination with ticarcillin [*34787-01-4*], is being developed and is under extensive clinical investigation in a number of countries.

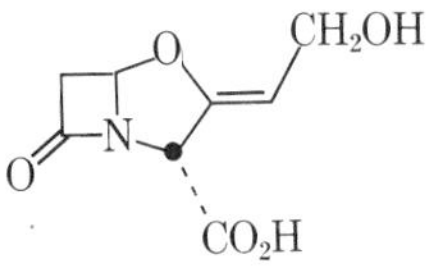

(**1**) Clavulanic acid; MM 14151

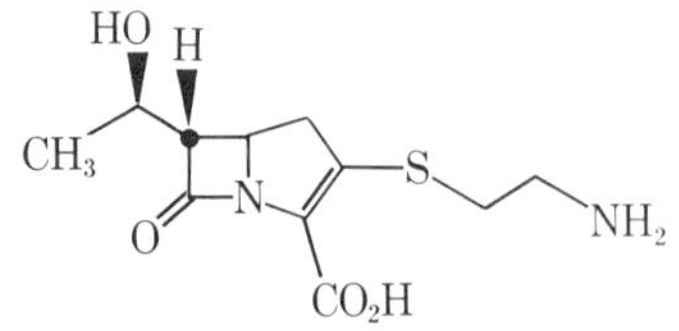

(**2**) Thienamycin [*59995-64-1*]

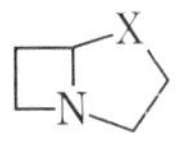

(**3**) X = O
4-oxa-1-azabicyclo[3.2.0]heptane
(**5**) X = CH_2
1-azabicyclo[3.2.0]heptane

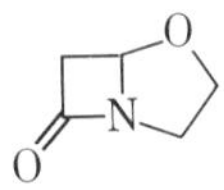

(**4**) Clavam
Oxapenam; 1-oxa-1-dethiopenam,
4-oxa-1-azabicyclo[3.2.0]heptan-7-one
[*64341-46-4*]

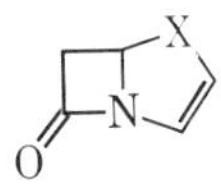

(**6a**) X = CH_2 Carbapenem
1-Carba-1-dethiapenem
1-azabicyclo[3.2.0]hept-2-en-7-one
[*74628-00-5*]
(**6b**) X = S; Penem
4-Thia-1-azabicyclo[3.2.0]hept-2-en-7-one
(**6c**) X = O; Clavem
4-Oxa-1-azabicyclo[3.2.0]hept-2-en-7-one

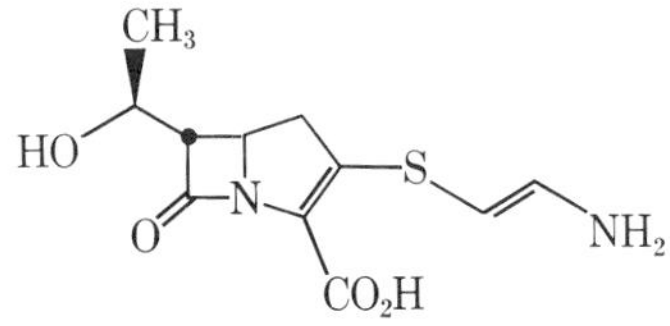

(**7**) Olivanic acid
[*64761-66-6*]

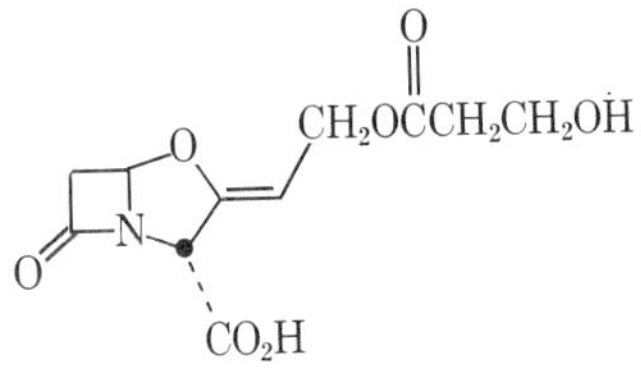

(**8**) β-hydroxypropionylclavulanic acid

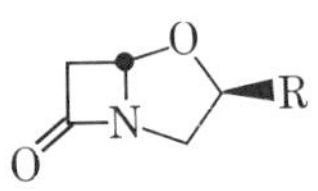

(**9**) R = CO_2H
(**12**) R = $CHCO_2H$ (with NH_2); Ro 22-5417
(**13**) R = CH_2CH_2OH

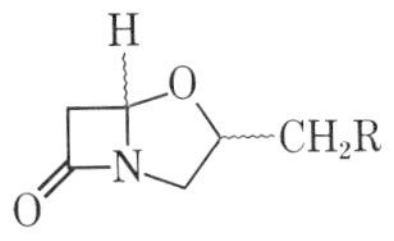

(**10**) R = OH
(**11**) R = OCH(=O)

Figure 1. Clavulanic acid, thienamycin, and olivanic acid and naturally occurring clavam derivatives.

The naturally occurring clavam derivatives are given in Table 1; the structures of clavulanic acid (**1**), its derivatives, and the related naturally occurring compounds are shown in Figure 1. They are derivatives of the 4-oxa-1-azabicyclo[3.2.0]heptane ring system (**3**). Clavulanic acid, *Z*-(3*R*,5*R*)-2-(β-hydroxyethylidene)clavam-3-carboxylic acid, is isolated from *Streptomyces clavuligerus* ATCC 27064 as its sodium or potassium salt (3–4). The parent β-lactam ring, ie, 4-oxa-1-azabicyclo[3.2.0]heptan-7-one (**4**), has been called clavam (12) by analogy with penam (15) and cephem (16). An alternative name for (**4**), though rarely used, is 1-oxa-1-dethiapenam; the designation oxapenam is more commonly encountered.

The same nomenclature is used to describe thienamycin (**2**) produced by *S. cattleya* (17), the epithienamycins, the olivanic acids, and the analogues, which occur as metabolites from a wide range of *Streptomyces* species. They are derivatives of the

Table 1. Naturally Occurring Clavam Derivatives

Compound	CAS Reg. No.	Structure no.	Systematic name	Producing organism	References
clavulanic acid	*[58001-44-8]* [a]	(**1**)	[2*R*-(2α,3*Z*,5α)]-3-(2-hydroxyethylidene)-7-oxo-4-oxa-1-azabicyclo[3.2.0]heptan-2-carboxylic acid	*S. clavuligerus*, ATCC 27064 NRRL 3585	3–6
				S. jumonjinensis, NRRL 5741 FERM-P 1545	7–8
				S. katsurahamanus FERM-P 3944	9
				Streptomyces sp P6621, FERM-P 2804	10
β-hydroxypropionyl-clavulanic acid	*[64675-12-3]* [b]	(**8**)	[2*R*-[2α,3*Z*,5α]-3-[2-(3-hydroxy-1-oxopropoxy)-ethylidene]-7-oxo-4-oxa-1-azabicyclo[3.2.0]-heptan-2-carboxylic acid	*S. clavuligerus*	11
clavam-2-carboxylic acid	*[71657-61-9]* [c]	(**9**)	[3*R*-(3α,5β)]-7-oxo-4-oxa-1-azabicyclo[3.2.0]-heptan-3-carboxylic acid	*S. clavuligerus*	12
2-hydroxymethylclavam	*[66036-39-3]*	(**10**)	3-hydroxymethyl-4-oxa-1-azabicyclo[3.2.0]-heptan-7-one	*S. clavuligerus*	12
2-formyloxymethylclavam	*[66036-40-6]*	(**11**)	3-formyloxymethyl-4-oxa-1-azabicyclo[3.2.0]-heptan-7-one	*S. clavuligerus*	12
2-(3-alanyl)clavam	*[74758-63-7]*	(**12**)	α-amino-7-oxo-4-oxa-1-azabicyclo[3.2.0]-heptan-3-propionic acid	*S. clavuligerus*	13
(−)-2-(2-hydroxyethyl)-clavam	*[79416-52-7]*	(**13**)	[3*S*-(3α,5β)]-3-(2-hydroxyethyl)-4-oxa-1-azabicyclo[3.2.0]heptan-7-one	*S. antibioticus* Tü 1718	14

[a] Na salt *[57943-81-4]*; Li salt *[61177-44-4]*; K salt *[61177-45-5]*.
[b] Isolated as its benzyl ester *[64675-13-4]*; Na salt *[64682-12-8]*.
[c] Na salt *[66002-34-4]*; Li salt *[66036-41-7]*; Mg salt *[66036-42-8]*.

1-azabicyclo[3.2.0]heptane (**5**) ring system. The parent β-lactam is 7-oxo-1-azabicyclo[3.2.0]hept-2-ene (**6a**) which has been defined as 1-carbapenem (18). Alternative names for (**6a**), although rarely used, are des-thiacarbapenem and 1-carba-1-dethiapen-2-em. The systematic name of thienamycin is (5*R*,6*S*,8*R*)-6-(α-hydroxyethyl)-3-(β-aminoethylthio)-7-oxo-1-azabicyclo[3.2.0]hept-2-ene-2-carboxylic acid.

In using the trivial clavam, carbapenem, and oxapenam nomenclature, the numbering of the atoms in the fused bicyclic-β-lactam ring system follows the sequence used for penicillins and cephalosporins (19). On the other hand, with the systematic IUPAC system, the numbering of the azabicyclo[3.2.0]heptane ring commences at nitrogen. Olivanic acid (**7**) and its derivatives are carbapenems that differ from thienamycin in the absolute stereochemistry of *C*-8, ie, the 6-(β-hydroxyethyl) substituent, which is *S*, whereas that of the thienamycin derivatives is *R*; the absolute stereochemistry of the olivanic acids at *C*-6 can be either *R* or *S*. By analogy with carbapenem (**6a**) and penem (**6b**), (**6c**) is designated as a clavem (20).

CLAVULANIC ACID

About 15 years ago, a systematic screening program was undertaken by Beecham Pharmaceuticals whereby fermentation broths were examined for the production of β-lactamase inhibitors. A plate test was devised in which *Klebsiella pneumoniae* ATCC 29665 was grown in agar containing penicillin G [*61-33-6*]. Test samples were added to wells cut in the agar. Because of the production of β-lactamase by *K. aerogenes*, the penicillin G in the agar was inactivated and bacterial growth occurred throughout the plate. If, however, an assay sample containing a β-lactamase inhibitor was added to a well, diffusion of the inhibitor into the agar around the well occurred, inactivation of the penicillin was prevented, and a typical zone of inhibition around the well was obtained (5). Application of this screen to over 1000 actinomycetes isolated from soil samples from a variety of localities throughout the world led to the identification of a number of strains of *S. olivaceus* that gave marked activity in this screen, known within Beecham as the KAG assay. Preliminary studies required that the active metabolites of these *S. olivaceus* strains were compared with the reported cephamycins (1–2) and various cephamycin-producing cultures were purchased from the American Type Culture Collection (ATCC) and examined by KAG assay and biochromatography. It became clear that one of these cephamycin producing cultures, namely, *S. clavuligerus* ATCC 27004 (NRRL 3585), produced a β-lactamase inhibitory metabolite with chromatographic properties distinctly different from the known cephamycins and the β-lactamase inhibitory compounds of *S. olivaceus*. *S. clavuligerus* was known to produce cephamycin C [*34279-51-1*], its analogue lacking the *C*-7 methoxy group (ie, the 3-carbamoyloxymethyl analogue of cephalosporin C), desacetoxycephalosporin C [*26924-74-3*], and penicillin N [*525-94-0*] (1,21). This culture had been isolated from a South American soil sample and described by workers at Eli Lilly (22). Initially, the metabolite responsible for the inhibitory activity was designated as MM 14151, but on characterization it was called clavulanic acid (**1**) (3–4,6,23).

More recently, clavulanic acid has been shown to be produced by *S. jumonjinensis* NRRL 5741 (FERM-P No. 1545) (7), a culture originally isolated by the Sankyo group and noted for its capacity to yield cephamycin C and antibiotic 3008B (8). Clavulanic acid was also obtained from *S. katsurahamanus* FERM-P No. 3944 as disclosed by Takeda in a Japanese patent application (9) and from another organism, *Streptomyces* sp P6621 (FERM-P No. 2804) by the Sanraku-Ocean Co., Ltd. (10). The latter strain also produces cephamycin C (24). Detailed examination of *S. clavuligerus* and mutants has yielded other clavam derivatives. For example, the β-hydroxypropionyl derivative (**8**) of clavulanic acid was obtained (11). Investigators at Glaxo reported (12) the isolation and characterization of clavam-2-carboxylic acid (**9**), 2-hydroxymethylclavam (**10**), and 2-formyloxymethylclavam (**11**) in addition to clavulanic acid (6); 2-(3-alanyl)clavam (**12**) was identified by a Hoffmann-La Roche group (13). At this point, only a tentative comment can be made on the relative stereochemistry at the asymmetric centers in these compounds. Recent studies, however, have shown that the absolute stereochemistry of *C*-5 of clavam-2-carboxylate is opposite to that found in clavulanic acid, ie, is 5*S* instead of 5*R* (25). Similarly, in 2-(2-hydroxyethyl)clavam (**13**) isolated from *S. antibioticus* Tü 1718, the stereochemistry at *C*-5 is the opposite of that in clavulanic acid. The relative stereochemistry between *C*-2 and *C*-5 is the same as for clavam-2-carboxylate (**9**). Clavams (**9**), (**12**), and (**13**) are the only β-lactam examples so far reported where the *C*-5 proton is β-orientated. In contrast to clavulanic acid and its derivatives, clavam-2-carboxylate (**9**), clavams (**10**) and (**11**), and 2-hydroxyethylclavam (**13**) show antifungal activity, particularly against plant pathogens, and were not inhibitors of β-lactamase (12). Clavam (**13**) was detected in an antifungal screen based on morphological effects on *Mucor*, but it was not an inhibitor of chitin synthase (14).

β-Lactams of the clavulanic acid, penicillin, cephalosporin, and cephamycin families are not the only secondary metabolites produced by *S. clavuligerus*. It is also noted for its production of holomycin [*488-04-0*] and MM 19290, a mixture of compounds closely related to the tunicamycins (26).

Physical Properties

Salts of clavulanic acid are colorless crystalline solids with indefinite melting points, decomposing at elevated temperatures. Clavulanic acid itself is an unstable viscous gum with a pK_a of 2.3–2.7. In addition to alkali metal, alkaline earth metal, and ammonium salts, clavulanic acid can be converted into organic base salts via primary, secondary, tertiary or *N*-quaternary amines, and heterocyclic base salts. The salts are soluble in water and other polar solvents; the free acid is readily extracted from acid solution (pH 2–3) into solvents, such as *n*-butanol, ethyl acetate, and methyl isobutyl ketone (6,23).

The ir spectrum of clavulanic acid salts and derivatives contain the characteristic β-lactam carbonyl stretching frequency in the range 1790–1815 cm^{-1} in addition to a band at 1680–1695 cm^{-1} assigned to the strained exocyclic double bond. Absorptions, typical of a primary hydroxyl function and carboxylate or ester functions, are readily identifiable in infrared spectra of clavulanic acid and its compounds. The uv spectrum of clavulanic acid shows only end absorption.

The nmr spectra of clavulanic acid and its derivatives show certain features, again characteristic of the fused β-lactam ring system. Thus, the β-lactam protons generally appear as an ABX system with the *C*-5 proton as a downfield double doublet because of coupling ($J_{5\alpha,6\alpha}$ = ca 2.8 Hz and $J_{5\alpha,6\beta}$ = ca 1 Hz) to the *C*-6α and *C*-6β protons ($J_{6\alpha,6\beta}$ = ca 17.5 Hz). The downfield shift of the *C*-5 proton relative to penicillanic acid derivatives is considered indicative of the effect of replacing sulfur by oxygen in the fused bicyclic ring system, as are the smaller coupling constants (*trans* = ca 0.8 Hz; *cis* = ca 2.8 Hz) for the *C*-5 position.

The stability of clavulanic acid in buffered solutions at 37°C and over several pH values and in human serum and urine has been evaluated (27). At pH 7–8, clavulanic acid had a half-life of ca 6.5 h, whereas at pH 2.6, the half-life was ca 0.7 h. The stability of potassium clavulanate in aqueous solution between pH 2.15 and 10.10 at 35° and anionic strength of 0.5 has been studied (28). The observed rates of degradation at various pH values followed pseudo-first-order kinetics, and the degradation was catalyzed by buffer salts. Rate-constant data indicated that clavulanic acid degradation was 10 times faster than that of penicillin G [*61-33-6*] at pH 6.4, ca 5 times faster under alkaline conditions, and comparable at acidic pH. The degradation rate of compound (**1**) was calculated to be ca 10 times faster than that of amoxycillin.

An evaluation of the reactivity of the clavam nucleus, as exemplified by clavulanic acid, can be obtained by a comparison of its β-lactam carbonyl stretching frequency with those typical of other β-lactam systems. The high frequency (ca 1800 cm^{-1}) of clavams is considered to be related to the greater reactivity of the amide bond between *C*-7 and *N*-4 owing to increased carbonyl character at *C*-7 and greater amine character at *N*-4. The deviation of the β-lactam carbonyl and nitrogen atoms from the superficial amide character is also indicated from x-ray data. Thus, the displacement (ca 0.05 nm) of the nitrogen atom of the clavam β-lactam ring from the plane of the neighboring carbon atoms (*C*-3, *C*-5, and *C*-7) is greater than that found in ampicillin (0.038 nm) or cephaloridine (0.024 nm), and the sum of the angles around the nitrogen atom deviates further from 360°, ie, ca 320° for clavam compounds, ca 339° for ampicillin [*69-54-4*] and ca 351° for cephaloridine. All facts are indicative of the greater pyrimidal nature of the β-lactam nitrogen in clavams. Increased ring strain of the clavam β-lactam system is also shown in the decreased β-lactam carbonyl bond length and increased *C*-7 to *N*-4 bond length. These features explain the greater susceptibility of the clavam β-lactam moiety to nucleophilic attack compared with more stable penicillins and cephalosporins and simpler azetidinones and amides (see Table 2) (29,33).

Table 2. β-Lactam Carbonyl Frequencies and Geometry of the Bridgehead Nitrogen Atom

β-Lactam	CAS Registry No.	ΣN^a	d^b, nm	$\nu_{max}{}^c$	Ref.
sodium clavulanate, anhydrous	[*57943-81-4*]	328	0.047	1785	29
p-bromobenzyl clavulanate (**16**)	[*60297-60-1*]	319	0.05	1800	29
MM 13902 monoethyl monomethyl ester	[*73521-68-3*]	323	0.05	1790	29
N-acetylthienamycin methyl ester	[*68364-11-4*]	326	0.049	1779	30
Carpetimycin A *p*-bromobenzyl ester	[*79244-02-3*]	326	0.051		31
acetonyl penem-3-carboxylate	[*69126-83-6*]		0.050	1785	32
penicillanic acid sulfone (**125**)	[*68373-14-8*]	339	0.038	1769	242
Penicillin V	[*87-08-1*]	337	0.040	1775	33
cephaloridine	[*50-59-9*]	351	0.024	1770	33

[a] Sum of bond angles (°) about bridgehead *N* atom.
[b] Distance of *N* atom from the plane of its three neighboring substituents.
[c] β-Lactam carbonyl stretching frequency, cm^{-1}.

Chemical Properties

Structure Elucidation. The structure of clavulanic acid was deduced from the spectroscopic data obtained on methyl clavulanate (**14**), which was prepared by the esterification of sodium clavulanate with methyl iodide in dimethylformamide solution (3). The rationale behind the structure elucidation has been reported in detail (34–35). Confirmation of structure (**1**) for clavulanic acid was obtained by x-ray analysis of the *p*-nitrobenzyl (**15**) and the *p*-bromobenzyl (**16**) esters which gave the relative and absolute stereochemistry at *C*-5 and *C*-3 and defined the double-bond configuration as *Z*. An analysis of the sodium salt supported the conclusion of these experiments. The x-ray work confirmed that the stereochemistry at *C*-3 and *C*-5 was identical to that of the naturally occurring penicillins.

Clavulanic acid was therefore found to be a novel fused β-lactam which differed dramatically from the known penicillins and cephalosporins. The molecule contained oxygen instead of sulfur, ie, an oxazolidine ring instead of a thiazolidine ring; it possessed no acylamino side chain at *C*-6, characteristic of active penicillins and cephalosporins, but at *C*-2 there was an exocyclic double bond in the form of a β-hydroxyethylidene substituent.

In addition to direct isolation from a culture filtrate by solvent extraction or chromatography, salts of clavulanic acid are available from the hydrogenolysis of benzyl clavulanate (**17**) or by the controlled hydrolysis of methyl clavulanate (**14**). Hydrogenation of methyl clavulanate in ethyl acetate over 10% Pd/C gave an epimeric mixture of dihydro derivative (**18**) that could be separated by high pressure glc and synthesized specifically or as the racemic acetate (**22**) from 4-acetoxyazetidinone (**19**). The route involved the condensation of (**19**) with the bromohydrin (**20**) to give the bromo ester (**21**) that was cyclized with benzyltrimethyl ammonium hydroxide to methyl acetyldihydroclavulanate (**22**), which proved to be identical to the acetate derived by reduction and acetylation of methyl clavulanate (**14**) (36) (see Fig. 2). This sequence of reactions confirmed the structure of clavulanic acid by classical means and this structure was later further confirmed by two total syntheses of (±)-methyl clavulanate (37–38).

Degradation reactions of interest in the derivation of the structure of clavulanic acid concerned ozonolysis and treatment with base. Ozonolysis of benzyl or *p*-bromobenzyl clavulanate yielded the oxazolidinone (**23**), whereas reaction of sodium or benzyl clavulanate with diethylamine in methanol gave products such as methyl β-diethylaminoacrylate (**24**) (39–40).

Reactions. The patent literature on clavulanic acid derivatives is extensive, but relatively few details have been published (35). Almost all the work on the chemistry of structure (**1**) has been described by the Beecham and Glaxo groups, which have ready access to large quantities of the natural product (34,41). Several reaction sequences are shown in Figure 3.

Clavulanic acid and its salts are readily converted (42) into a variety of esters, many of which have been used as the starting materials for further transformations (34); of particular use have been the methyl (**14**), benzyl (**17**), *p*-nitrobenzyl (**15**), phthalidyl (**25**), and phenacyl (**26**) esters. The double bond in structure (**1**) can be reduced (43) to yield dihydroclavulanic acid (**34**) [*59897-42-6*] as a mixture of epimers, oxidized by ozonolysis (39,44) to the substituted oxazolidinone (**23**). Photochemical isomerization of esters produced isoclavulanic acid derivatives (**27**) (45), although the

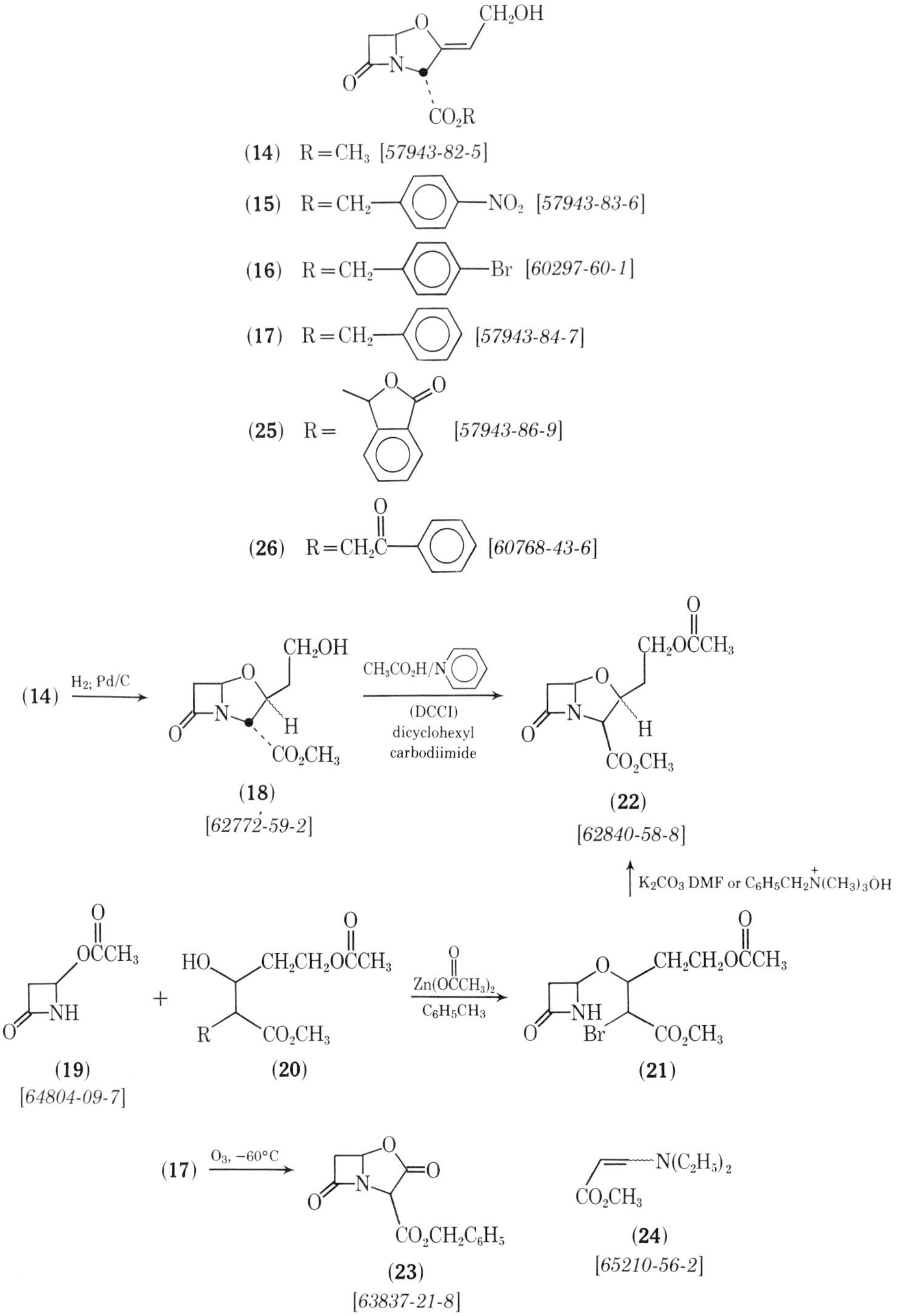

CH_2OH

CO_2R

(**14**) $R=CH_3$ [*57943-82-5*]

(**15**) $R=CH_2-C_6H_4-NO_2$ [*57943-83-6*]

(**16**) $R=CH_2-C_6H_4-Br$ [*60297-60-1*]

(**17**) $R=CH_2-C_6H_5$ [*57943-84-7*]

(**25**) R= [*57943-86-9*]

(**26**) $R=CH_2COC_6H_5$ [*60768-43-6*]

(**14**) $\xrightarrow{H_2;\ Pd/C}$ (**18**) [*62772-59-2*] $\xrightarrow[\text{(DCCI) dicyclohexyl carbodiimide}]{CH_3CO_2H/\text{pyridine}}$ (**22**) [*62840-58-8*]

K_2CO_3 DMF or $C_6H_5CH_2\overset{+}{N}(CH_3)_3\overset{-}{O}H$

(**19**) [*64804-09-7*] + (**20**) $\xrightarrow[C_6H_5CH_3]{Zn(OCOCH_3)_2}$ (**21**)

(**17**) $\xrightarrow{O_3,\ -60°C}$ (**23**) [*63837-21-8*]

(**24**) [*65210-56-2*]

Figure 2. Structure elucidation of clavulanic acid.

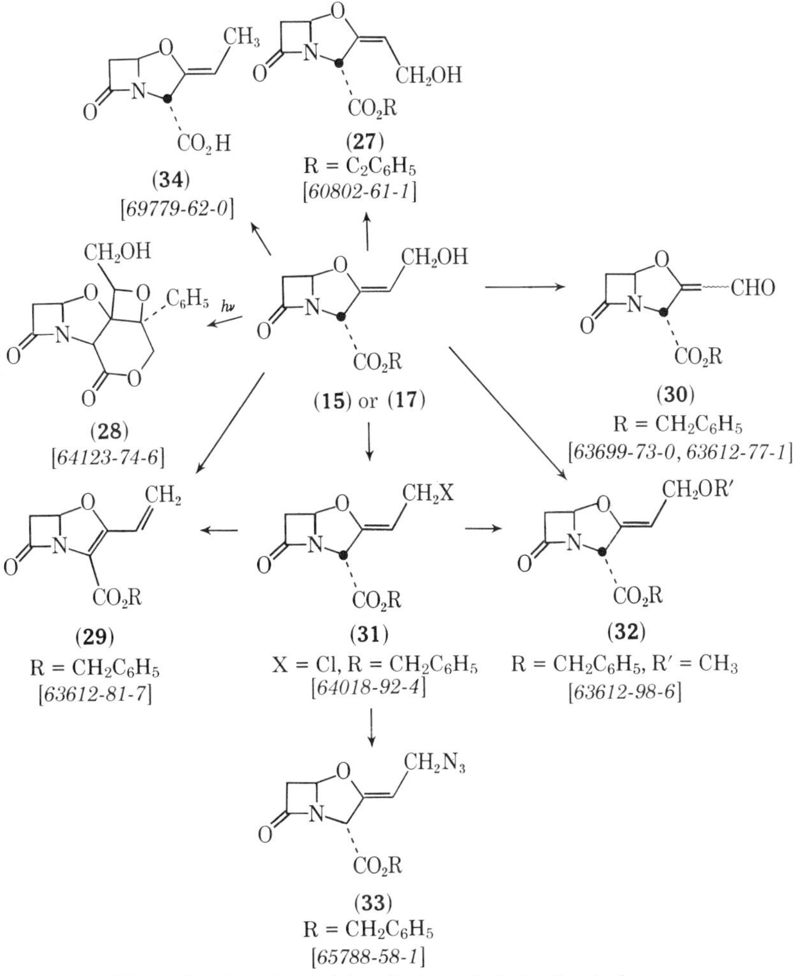

Figure 3. Reactions of clavulanic acid (derivatization).

phenacyl ester also formed the tetracyclic clavam (**28**) (46). The allylic hydroxyl group can be readily acylated, and the resultant allylic esters are important intermediates on route to new clavems and clavams, such as the diene (**29**) and substituted amines (47). Oxidation of the allylic alcohol function with pyridinium chlorochromate gave a mixture of isomeric aldehydes (**30**); however, with dimethyl sulfoxide–dicyclohexylcarbodiimide, the diene (**29**) was formed (48). This compound was also prepared from the halogen derivatives (**31**), obtained by thionyl chloride or bromide treatment of a clavulanic ester in the presence of triethylamine (49). Routes to ethers (**32**) involved use of a Meerwein reagent (50) or alkyl halides in the presence of silver and calcium oxides (51), as did the reaction of compound (**31**) with alkoxide (41). Reaction of clavulanic acid with a number of reagents in the presence of a Lewis acid, such as boron trifluoride etherate, led to a number of clavulanic acid and 9-deoxyclavulanic acid derivatives (52–55). With azide, the intermediate halogen compound (**31**) gave compound (**33**) (56).

Displacement of chloride or bromide from the corresponding halides (**31**) with sulfur nucleophiles gives thiols, sulfoxides, or sulfones (41).

Clavulanic acid forms amides (57); the corresponding primary alcohol is obtained by the borohydride reduction of an ester (58). The azide (**33**) gives triazole derivatives with acetylene (41).

Reaction of clavulanic esters with certain bases and treatment with carbonyl compounds resulted in *C*-6 substituted clavams (**35**) (59). Such clavams have also been claimed via mesylates. The resulting product reacts with azide to form a 6-substituted-9-azido-9-deoxyclavulanate and hence the amine (**36**) on reduction, thus forming an analogue of thienamycin. This sequence, via the mesylate and the azide, has also been proposed for the formation of the preparation of 9-aminodeoxyclavulanic acid (9-ADCA) (60).

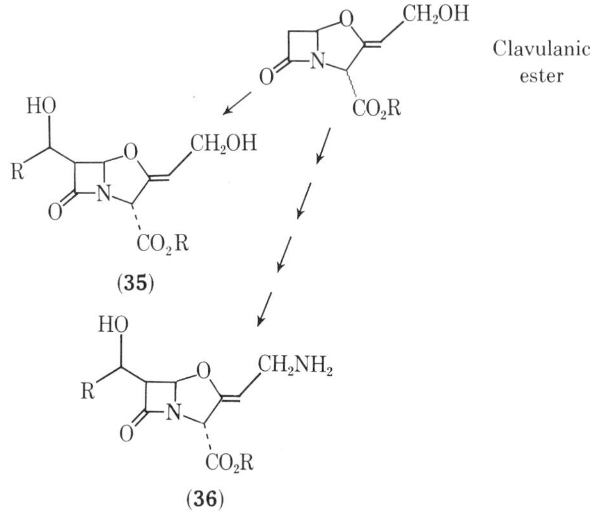

O N CH_2OH CO_2R O

Clavulanic ester

HO R O CH_2OH O N CO_2R

(**35**)

HO R O CH_2NH_2 O N CO_2R

(**36**)

Isoclavulanic acid derivatives are sometimes isolated as by-products of many of the above reactions. Such compounds can also be prepared by direct modification of esters (**27**) or by deliberate isomerization of the clavulanate analogues (61). Under certain hydrogenolysis conditions, benzyl clavulanate (**15**) yielded isoclavulanic acid (**27**) and deoxyclavulanic acid derivatives (**37**) (34,45).

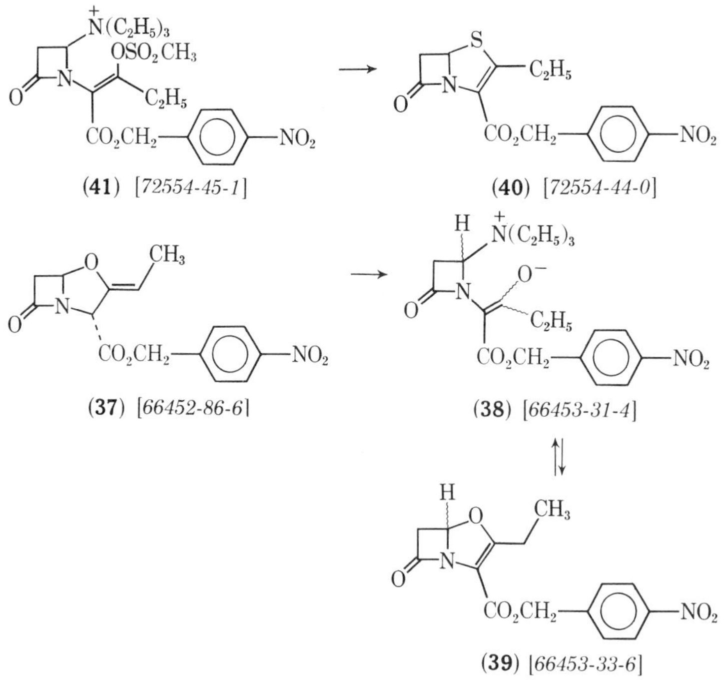

$\overset{+}{N}(C_2H_5)_3$ OSO_2CH_3 O N C_2H_5 CO_2CH_2—NO_2 → S C_2H_5 O N CO_2CH_2—NO_2

(**41**) [*72554-45-1*]

(**40**) [*72554-44-0*]

O CH_3 O N CO_2CH_2—NO_2 → H $\overset{+}{N}(C_2H_5)_3$ O^- O N C_2H_5 CO_2CH_2—NO_2

(**37**) [*66452-86-6*]

(**38**) [*66453-31-4*]

⇅

H O CH_3 O N CO_2CH_2—NO_2

(**39**) [*66453-33-6*]

The base-catalyzed isomerization of certain derivatives of compound (**1**) has given some unexpected results (41,62). Triethylamine treatment of the deoxyclavulanate (**37**) gives the betaine (**38**) which on decarboxylation gave a ketone. The betaine (**38**) rearranges to the clavem (**39**) on heating.

The chemistry of various intermediates and related derivatives has been explored in depth with a view to the formation of substituted penems, eg, the betaine (**38**) has been transformed into the 2-ethylpenem (**40**) via the mesylate (**41**) (41,63).

Catalytic decarboxylation of clavulanic acid (**1**) in the presence of mercury(II) acetate produced the clavams (**42**) and (**43**); these isomers are not interconverted by mercury(II) ion (20). The diene (**44**) was also formed from (**1**) by reaction with triphenylphosphine and diethyl azodicarboxylate and on reduction gave the tetrahydro derivative (**45**) and the two dihydro compounds (**46**) and (**47**).

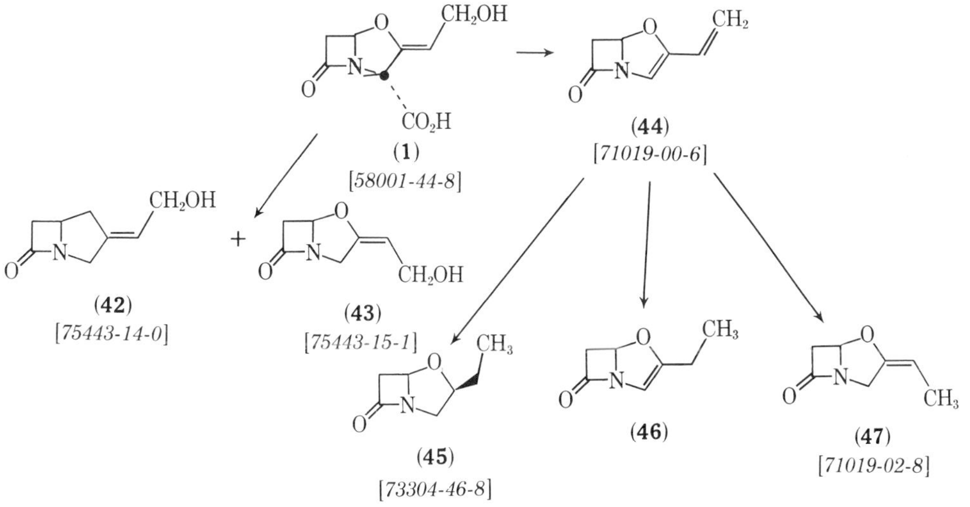

CH_2OH CO_2H (**1**) *[58001-44-8]* → CH_2 (**44**) *[71019-00-6]*

CH_2OH (**42**) *[75443-14-0]* + CH_2OH (**43**) *[75443-15-1]*

CH_3 (**45**) *[73304-46-8]* CH_3 (**46**) CH_3 (**47**) *[71019-02-8]*

Benzyl clavulanate (**15**) is isomerized by triethylamine to the clavem (**48**) which on prolonged treatment with acetic acid gave the β-hydroxypyrroles (**49**) and (**50**). Methanolic cleavage of clavulanic acid gives a pyrrole via a β-aminoacylate (40).

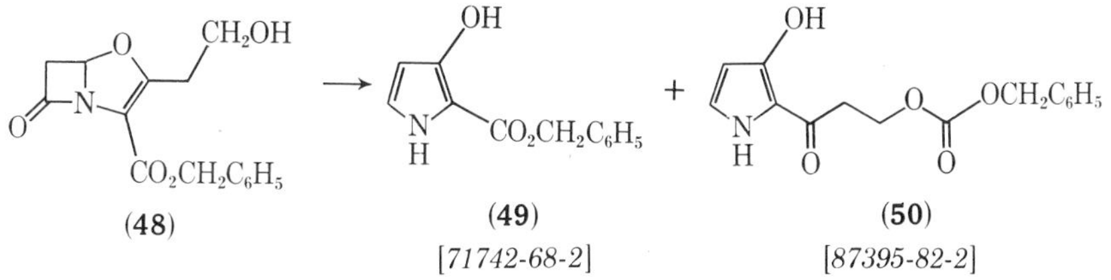

CH_2OH $CO_2CH_2C_6H_5$ (**48**) → OH $CO_2CH_2C_6H_5$ (**49**) *[71742-68-2]* + OH $OCH_2C_6H_5$ (**50**) *[87395-82-2]*

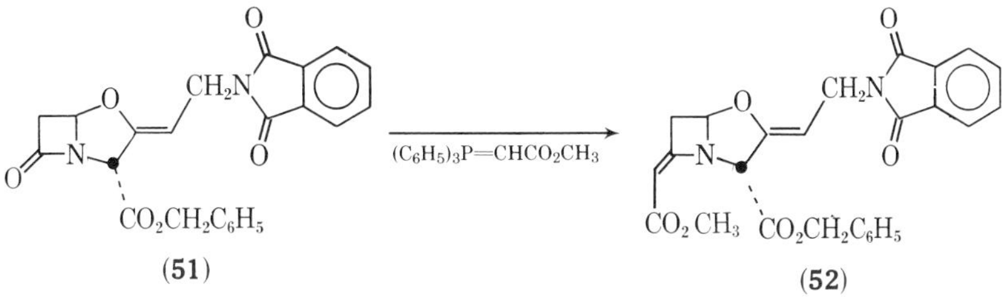

CH_2N $CO_2CH_2C_6H_5$ (**51**) $\xrightarrow{(C_6H_5)_3P{=}CHCO_2CH_3}$ CH_2N CO_2CH_3 $CO_2CH_2C_6H_5$ (**52**)

With phosphoranes, imide (**51**) formed the acrylate (**52**), the configuration of which was confirmed by x-ray analysis and conversion into a bicyclic lactone (64). Similar derivatives can be derived from a penicillin V ester. Such reactions reflect the ketonic properties of the β-lactam carbonyl group.

By the above reactions, it has thus been able to transform structure (**1**) into many derivatives by exploiting the chemistry of the double bond, the allylic alcohol function, the carboxylic acid residue, the *C*-6 position, and even the β-lactam carbonyl. In addition, degradation has resulted in a number of interesting rearrangements and reactions.

Biological Properties

As a potent progressive inhibitor of a wide range of β-lactamases derived from gram-positive and gram-negative bacteria, clavulanic acid differs from the classical β-lactams of the penicillin and cephalosporin series (4,65) (see Table 3). The β-lactamases readily inhibited are mainly of the penicillinase type, ie, of the Class II, III, IV, and V types (66–68); activity against most Class I cephalosporin hydrolyzing β-lactamases was poor. Clavulanic acid is highly effective against plasmid-mediated β-lactamases with limited activity against those that are chromosomally mediated.

In conjunction with certain penicillins and cephalosporins, clavulanic acid produces a pronounced synergistic effect against clinically important β-lactamase-producing organisms. A large number of papers describing the synergistic effect of clavulanic acid plus amoxycillin against β-lactamase-producing strains of *Staphylococcus aureus*, *Klebsiella aerogenes* (including strains producing plasmid and chromosomally mediated β-lactamase), *Proteus mirabilis*, *P. vulgaris*, *Bacteroides fragilis*, *Branhamella catarrhalis*, *Neisseria gonorrhoeae*, and *Haemophilus influenzae*, and certain strains of *Escherichia coli*, *Shigella*, *Salmonella*, *Serratia*, and *Pseudomonas* (ie, those producing plasmid-mediated enzymes, eg, R_{TEM} type) have been published over the last five years (4,69–72). Synergistic effects are not seen against various *Enterobacteriaceae* such as *P. morganii*, *P. rettgeri*, *Providencia*, *P. aeruginosa*, and *S. marcescens* as they produce β-lactamases (chromosomally mediated) against which clavulanic acid has little effect. Good synergistic effects are also found with clavulanic acid plus ticarcillin (71,74), and in combination with other penicillins such as penicillin G (4,75), mezlocillin [*51481-65-3*] (76), and piperacillin [*61477-96-1*] (77). Although

Table 3. β-Lactamase-Inhibitory Activity of Clavulanic Acid

β-Lactamase source	Class[a]	Substrate[b]	I_{50} (μg/mL)[c]
Pseudomonas aeruginosa A (Sabath type)	Id	C	160
Citrobacter freundii Mantio	Ia	C	15
Escherichia coli JT410	Ib	C	56
Enterobacter cloacae P99	Ia	C	10
Escherichia coli JT4[d]	III	P	0.08
Klebsiella aerogenes E70	IV	P	0.02
Klebsiella aerogenes Ba95[d]	III	P	0.07
Proteus mirabilis C889	II	P	0.03
Pseudomonas aeruginosa Dalgleish	III	P	0.1
Staphylococcus aureus Russell		P	0.06
Bacillus cereus (β-lactamase I and II)		P	17

[a] Refs. 66–68.
[b] C = Cephaloridine; P = Benzylpenicillin.
[c] Ref. 4.
[d] Producing plasmid mediated β-lactamase.

cephalosporins such as cephaloridine, cephalothin [*153-61-7*], cefazolin [*25953-19-9*], and cefaperazone [*62893-19-0*] have a considerable degree of stability to β-lactamases, their activity is enhanced in combination with clavulanic acid (71,78–79). However, cephalosporins and cephamycins, such as cefuroxime [*55268-75-2*] and cefoxitin [*35607-66-0*], with pronounced stability to β-lactamases show little or no synergy in conjunction with compound (**1**). Clavulanic acid itself exhibits weak, though broad spectrum, antibacterial activity. It is only slightly effective in combination with some sixteen antimicrobial agents against *Legionella pneumophila*, against which it is, however, the most effective single agent (80). The synergistic properties of clavulanic acid with a β-lactamase labile β-lactam are not adversely affected by inoculum size, media composition at pH 5.5–8, serum, or binding (23%) to human serum proteins. Clavulanic acid alone or with amoxycillin is bactericidal and has greatest affinity for penicillin-binding protein 2 derived from *E. coli* K-12. This was supported by complementary morphological studies showing that against *E. coli* KN 126, lemon-shaped cells formed at low concentrations (30 μg/mL), whereas at higher concentrations (50 μg/mL), spheroplast formation and lysis resulted (81).

Clavulanic acid is well absorbed in animals by the subcutaneous or intramuscular route (in mice, rats, rabbits, dogs, and squirrel monkeys) and orally (in mice, dogs, and squirrel monkeys) (69,82). In combination with amoxycillin (70) or penicillin G (78), it is very effective in mice against experimental infections caused by β-lactamase-producing organisms, whereas both clavulanic acid and penicillin alone were ineffective. Clavulanic acid is well absorbed by the oral route in man; a dose of 250 mg gives a maximum concentration of ca 5.5 μg/mL after 1 h with a urinary recovery of 30–40% (83).

Augmentin (BRL 25000), a formulation consisting of amoxycillin trihydrate (250 mg) and potassium clavulanate (125 mg), has been developed by Beecham and is now marketed in the United Kingdom. It is currently undergoing extensive clinical trial studies and investigation in other countries such as the United States, Japan, the FRG, France, the Netherlands, and South Africa. The proceedings of two symposia on Augmentin (clavulanate potentiated amoxycillin) have been published and contain detailed descriptions of and discussions on the microbiology, pharmacology, and toxicology of Augmentin and accounts of its efficacy against urinary tract, skin and soft tissue, and respiratory tract infections from hospitals and general practice (83–84). Many other papers describe the properties of this formulation (85). Augmentin is active *in vitro* and *in vivo* against a broad range of amoxycillin-sensitive and amoxycillin-resistant bacteria (see Table 4) (86). There is little evidence of clavulanic acid potentiating amoxycillin against non-β-lactamase-producing organisms, and the formulation is ineffective against those *Enterobacter* species producing chromasomally mediated β-lactamases (cephalosporinases). A 30-μg Augmentin-sensitivity disk (20 μg amoxycillin + 10 μg clavulanic acid) gives zone diameters in the range 15–30 mm against β-lactamase-producing strains of staphylococci and gram-negative bacilli; >35 mm zones were obtained against streptococci and penicillin-sensitive staphylococci. Details of the toxicology of clavulanic acid and Augmentin in animals and humans have been reported. The pharmacokinetics of amoxycillin (or clavulanic acid) were not affected by the coadministration of clavulanic acid (or amoxycillin), and both were found to be absorbed orally to the same extent and at a similar rate. The half-life of clavulanic acid is ca 1 h, while that of amoxycillin is ca 1.3 h with ca 30% and 54% urinary recovery, respectively. Food has no effect on the absorption of either amoxycillin or clavulanic

Table 4. Antibacterial Spectrum of Augmentin[a], Amoxycillin, and Clavulanic Acid

Organism	MIC (μg/mL)		
	Augmentin	Amoxycillin	Clavulanic acid
E. coli NCTC 10418	5.0	5.0	50.0
E. coli JT 39[b]	25.0	>500.0	50.0
Salmonella typhi A	1.25	1.25	50.0
Salmonella typhimurium 33	2.5	2.5	50.0
Shigella sonnei A	5.0	2.5	50.0
K. aerogenes A	2.5	125.0	50.0
P. mirabilis 977	1.25	1.25	125.0
P. mirabilis 889[b]	12.5	>500.0	125.0
P. vulgaris E	5.0	>500.0	125.0
Proteus morganii F	250.0	500.0	125.0
Proteus rettgeri I	250.0	125.0	125.0
Providencia stuartii T946	125.0	250.0	125.0
Enterobacter aerogenes T765	125.0	500.0	125.0
Pseudomonas aeruginosa NCTC 10662	250.0	>500.0	125.0
Serratia marcescens US 32	125.0	125.0	125.0
Bacteroides fragilis BC3	1.25	25.0	50.0
Haemophilus influenzae Wy21	0.5	0.25	25.0
H. influenzae VK 38[b]	1.25	50.0	50.0
Neisseria gonorrhoeae 30791	0.05	0.05	1.25
N. gonorrhoeae 1782[b]	1.0	>10.0	5.0
Bacillus subtilis ATCC 6633	0.5	0.25	25.0
Clostridium perfringens 1730	0.05	0.05	25.0
Peptostreptococcus anaerobius 2905	0.25	0.12	25.0
Staph. aureus NCTC 6571	0.12	0.05	12.5
Staph aureus 1555[b]	2.5	250.0	12.5
Staph. aureus W2827[c]	25.0	500.0	500.0
Streptococcus pyogenes CN10	0.02	0.02	25.0
Streptococcus pneumoniae 7	0.05	0.02	25.0
Streptococcus faecalis I	0.5	0.5	250.0

[a] Augmentin is amoxycillin + clavulanic acid (2:1) [BRL 25000].
[b] β-Lactamase-producing strain.
[c] Methicillin-resistant strain.

acid alone or together when the compounds are given at the start of a meal. Clavulanic acid alone or with amoxycillin is well tolerated. On repeat dosing there were predictable minor side effects such as diarrhea or soft stools and slight gastrointestinal upset (83). Parental formulations of clavulanic acid with amoxycillin and ticarcillin are under investigation (87).

The metabolism of clavulanic acid in animals and humans has been studied; the proposed degradative metabolic pathway is outlined in Figure 4. In the rat from radiolabeled studies, ca 30% of compound (**1**) is metabolized into carbon dioxide and 4-amino-3-ketobutan-1-ol, which has been detected in urine (84). No compound, other than clavulanic acid itself, has been detected as a bioactive, ie, β-lactamase-inhibitory, metabolite in serum or urine.

The mode of action of clavulanic acid as a β-lactamase inhibitor has been examined in great detail by a number of groups. References 88–91, especially relevant, show clavulanic acid to be an active site-directed inhibitor. It reacts with the R_{TEM} β-lac-

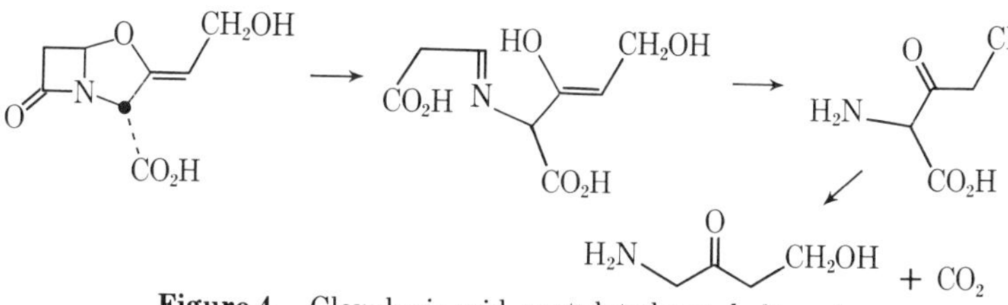

Figure 4. Clavulanic acid–postulated metabolic pathway.

tamase of *E. coli* in three ways: As a substrate for the β-lactamase, it is hydrolyzed at the β-lactam ring; it forms a transient intermediate that can be broken down to yield regenerated enzyme and degraded (**1**); and the β-lactamase can be irreversibly inactivated by compound (**1**). Kinetic studies with structure (**1**) and dihydroclavulanic acid, which does not inhibit β-lactamase, and deoxyclavulanic acid [*69779-62-0*], the use of labeled clavulanate, and the isolation of inactivated enzyme suggest an inactivation pathway illustrated in Figure 5. The three products, formed on irreversible inactivation of the enzyme with excess clavulanate, were treated with hydroxylamine resulting in the regeneration of active enzyme from one of these species. Similar results were obtained by the inactivation of TEM-2 β-lactamase from *Klebsiella pneumoniae* E70 (89). The penicillinase from *Proteus mirabilis* C889 was rapidly inhibited by compound (**1**) giving a moderately stable complex, but the proportion of irreversible inhibition was small compared with that for TEM-2 β-lactamase; the cephalosporinase from *Enterobacter cloacae* P99 reacted only with high concentrations of compound (**1**) because of a low affinity, and the resultant inactivation was not irreversible. With extracellular and intracellular staphylococcal β-lactamase, structure (**1**) formed an inactive complex, and in the absence of excess (**1**), enzyme activity regenerates. With

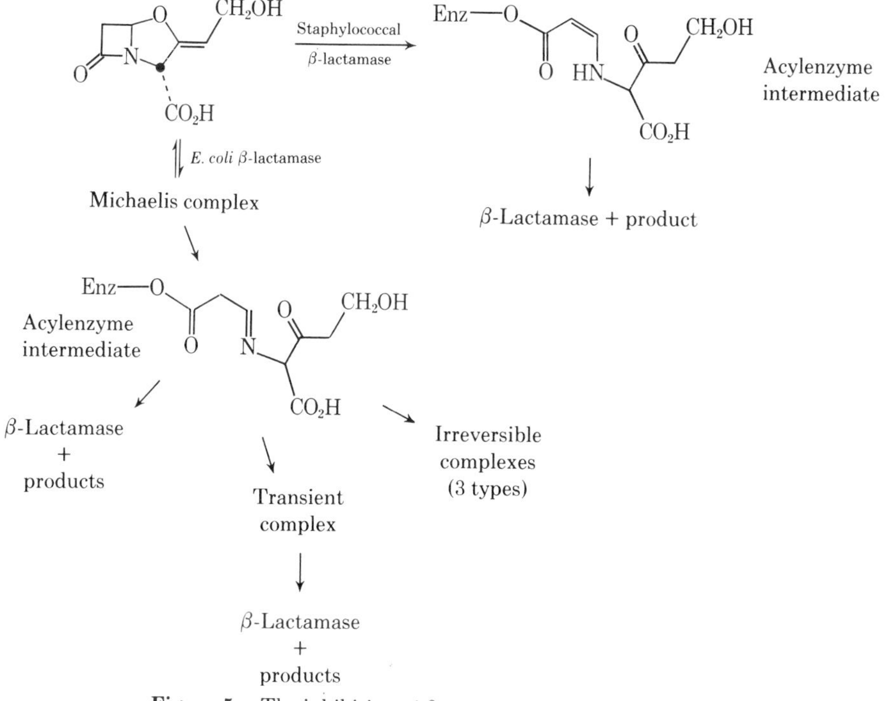

Figure 5. The inhibition of β-lactamases by clavulanic acid.

excess (1) however, complete inhibition of β-lactamase resulted (87). Similar results and conclusions were obtained for the inactivation of β-lactamase I from *Bacillus cereus* 569/H (92).

Derivatives. The β-lactamase inhibitory and synergy data for a number of representative derivatives of clavulanic acid are given in Table 5. For comparison of chemotherapeutic properties, extensive investigation is required (93–94).

Synthesis

The simplest clavam is 4-oxa-1-azabicyclo[3.2.0]heptan-7-one (**4**), which has been synthesized (95) by the cyclization of 4-(2-bromoethoxy)azetidin-2-one (**53**) (see Fig. 6). Clavulanic acid itself has been obtained synthetically by two routes. The first led to the preparation of a mixture of racemic methyl clavulanate (**54**) and (±)-methyl isoclavulanate (**55**) (37), which were separated by chromatography. The ester (**54**) was spectroscopically identical with that obtained from natural (**1**), and ester (**55**) was identical with that obtained from isoclavulanic acid. The second synthesis required the formation of an enolic ester by treatment of the azetidinone (**56**) with vinylacetyl chloride (38). Chlorinolysis and base-catalyzed cyclization and isomerization gave the diene (**57**). Treatment of (**57**) with ozone and catalytic hydrogenolysis of the resultant ozonide gave (±)-methyl clavulanate (**54**). The diene (**57**) could be converted into the analogues (**58**) by reduction and the diol (**59**) by hydroxylation. Under certain conditions, the chloro derivative (**60**) could be cyclized to yield a mixture of compound (**61**) and the isomeric clavem (**62**). Application of the route used to give structure (**4**) led to analogues of type (**63**) which reacted with a variety of nucleophilic agents. This resulted in a synthesis of clavams which by nmr studies were shown to have the same relative stereochemistry as the products (**10**) and (**11**) isolated by Glaxo workers from *S. clavuligerus*.

These synthetic routes were extensions of the earlier procedures devised to yield the clavem and clavam ring systems (96–97) and gave compounds such as the tritylaminoclavem (**64**) and the clavam penicillanic acid analogues (**65**), (**66**), and (**67**). The 2-vinylclavams (**68**) have also been synthesized and their reactions investigated (98). Other syntheses of the substituted clavam and clavem derivatives have been reported (99–100).

Biosynthesis. The biosynthesis of clavulanic acid by *S. clavuligerus* has been studied in detail using ^{13}C and ^{14}C labeled precursors and ^{13}C nmr spectroscopy (39,101). The β-lactam carbon atoms (*C*-5, *C*-6, and *C*-7) are considered to be derived from an intact three-carbon unit based on the incorporation of labeled glycerol and an examination of gluconeogenesis involving triglyceride-based media. Propionate was not considered to be a precursor of the β-lactam carbons. Incorporation of variously labeled ^{13}C-acetate, ^{13}C-propionate, ^{13}C-bicarbonate, ^{13}C-glutamate, and ^{13}C-α-amino-δ-hydroxyvalerate moieties indicated that *C*-2 and *C*-3 of and *C*-8, *C*-9, and *C*-10 on the oxazolidine ring of clavulanic acid were derived from α-amino-δ-hydroxyvalerate via glutamate and 2-oxoglutarate. As all these precursors can be metabolized via the tricarboxylic acid cycle, the label appeared in oxazolidine and azetidinone fragments of the fused β-lactam or clavam ring system when *S. clavuligerus* was grown in a medium based on triglyceride.

Table 5. β-Lactamase Inhibitory Activity (I_{50} $\mu g/mL$) of Clavulanic Acid Derivatives and Analogues, and Antibacterial Activity of Ampicillin in the Presence of These Derivatives (MIC $\mu g/mL$)[a,b]

	R	Organism and β-lactamase from							
		S. aureus Russell		*E. coli* (IIIa, R_{TEM})		*K. aerogenes* E70 (II)		*P. mirabilis* C889 (IV)	*C. freundii* Mantio (I)
		I_{50}	MIC[c]	I_{50}	MIC[c]	I_{50}	MIC[c]	I_{50}	I_{50}
ampicillin			500		>2000		1000		
clavulanic acid	OH	0.06	0.04	0.07	8	0.03	0.8	0.03	10
deoxyclavulanic acid	H	0.12	0.08	0.09	8	0.05	1.6		5
acetate	$OCOCH_3$	0.04	0.2		>500	>0.4	1.6		0.4
carbamate	$OCONHCH_3$	1.5	0.6	2.5	16	2.5	3		0.45
methyl ether	OCH_3	0.05	0.02	0.18	8	0.07	0.6	0.01	8.5
benzyl ether	$OCH_2C_6H_5$	0.005	0.01	0.1	31	0.04	12.5	0.02	4.4
thioether	SCH_3	0.11	0.1	0.04	4	0.13	0.8	0.01	$\gg 10$[d]
amine	$N(CH_2C_6H_5)_2$	0.002	0.4	0.04	62.5	0.08	25	0.01	0.62
isoclavulanic acid		0.6	0.8	1.0	1000	0.45	12.5		5
BRL 20780		<0.08		0.2		3.6		<0.08	>40[d]
BRL 19378		0.11		0.18		4.0		4.0	0.16[d]

[a] Ref. 93.
[b] MIC = minimal inhibitory concentration.
[c] MIC of ampicillin in the presence of 5 $\mu g/mL$ of inhibitor, except in the case of the acetate (R = $OCOCH_3$), where a concentration of 20 $\mu g/mL$ was used.
[d] β-Lactamase from *Enterobacter cloacae* P99.

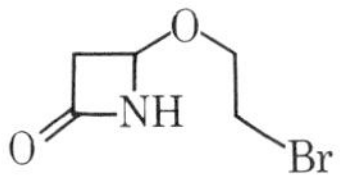

(**53**) [*75995-93-6*]

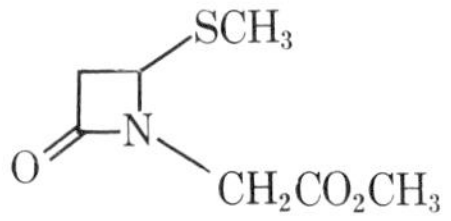

(**56**) [*64455-11-4*]

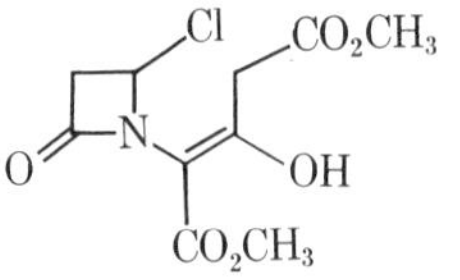

(**60**) [*66314-56-5*]

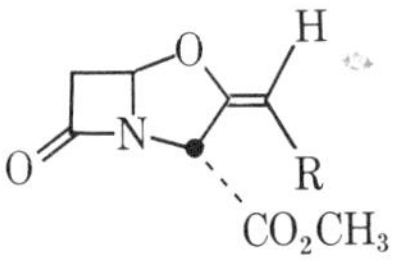

(**55**) R = CH_2OH [*64519-77-3*]
(**61**) R = CO_2CH_3 [*64455-27-2*]

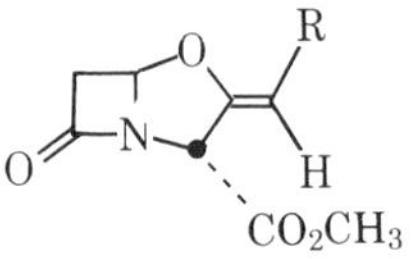

(**54**) R = CH_2OH [*64519-79-5*]
(**57**) R = $CH{=}CH_2$ [*72510-77-1*]
(**58**) R = C_2H_5 [*72510-79-3*]

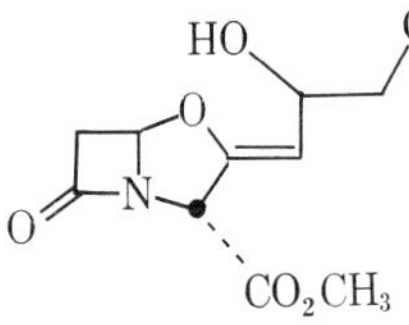

(**59**) [*72510-80-6*]

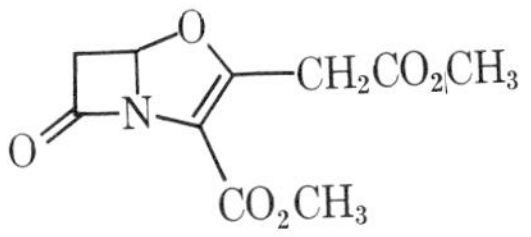

(**62**) [*66314-57-6*]

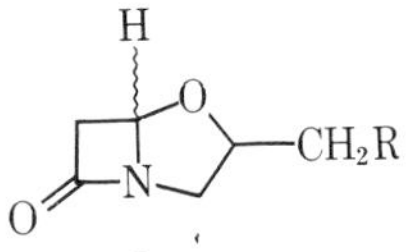

(**63**) R = OCH(=O) [*66036-40-6*] or H

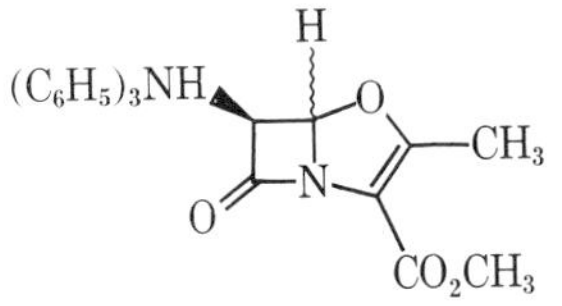

(**64**) *SR* [*66260-27-3*] *SS* [*66289-78-9*]

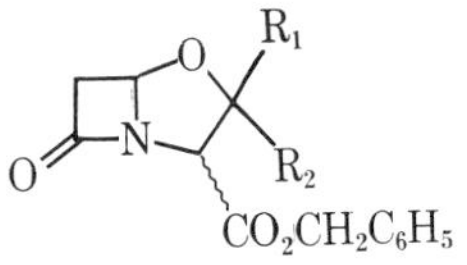

(**65**) $R_1 = R_2 = CH_3$ [*62772-20-7*]
(**66**) $R_1 = H, R_2 = CH_3$ [*62840-53-3*]
(**67**) $R_1 = CH_3, R_2 = H$ [*62772-17-2*]

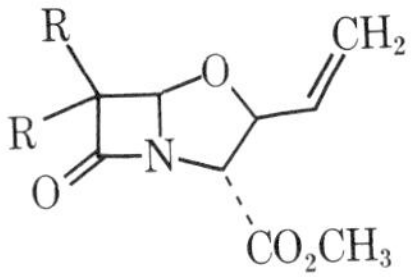

(**68a**) R = H [*68217-55-0*]
(**68b**) R = Br [*62772-51-4*]

Figure 6. Synthesis of clavulanic acid.

Manufacture and Processing

Like penicillin G and cephalosporin C, clavulanic acid is produced commercially by submerged fermentation (4,6–7,23), though it is derived from a streptomycete (a procaryote), whereas the other two are obtained from the fungi (eucaryotes) *Penicillium chrysogenum* and *Cephalosporium acremonium.* Although various processes are proprietary, the manufacture of clavulanic acid has been described in some detail in recent reviews (102–103).

Fermentation. The cultures producing clavulanic acid are given in Table 1. Stock cultures of *S. clavuligerus* preserved in lyophilized form in ampules or in desiccated soil are used to prepare working slant cultures on agar. Isolates for stock cultures are derived from mutation studies on the parent strain ATCC 27064 (NRRL 3585). Roux bottles are inoculated with suspensions of sporulating aerial mycelia from these slants in order to provide a spore suspension for a vegetative seed tank. The spores from a fully grown Roux-bottle culture provide an inoculum for a 100-L stainless steel fermentor containing 75 L of seed medium. The seed fermentation is incubated at 26°C for 72 h under agitation with a flat-bladed turbine impeller (140 rpm) and controlled airflow (75 L/min). Production media in large-scale fermentors are generally based on soybean flour which is a good source of nitrogen. A typical production medium consists of 1.5% soybean flour, 1.0% glycerol, 0.1% KH_2PO_4, and 0.2% (v/v) of 10% Pluronic L81 antifoam in soybean oil. Increased fermentation yields were obtained when lipids from vegetable oils or animal fats, such as soybean, peanut, maize, or lard oils, were used in place of glycerol and antifoam was omitted. Starch, dextrose, and silicone antifoam have also been used. Best results were obtained with Prichem P224, which is a mixture of fatty acid (65% oleic acid) triglycerides. A seed medium consists of 1.5% soybean flour, 1.0% Prichem P224, and inorganic phosphate at pH 7 before sterilization. After ca 90 h fermentation, clavulanic acid reaches a maximum of ca 500 μg/mL. Other media and conditions have been described (6) which claim improved yields at ca pH 6.5. The fermentation steps involved in the large-scale production of clavulanic acid are shown in Figure 7.

Isolation and Purification. Broth, clarified by filtration or centrifugation, can be treated in two ways to provide clavulanic acid as a salt for further purification. These are solvent extraction at acid pH and adsorption on a basic amine-exchange resin. Mycelium is not processed to give structure (1) as almost all clavulanic acid produced is excreted into culture fluid.

Free clavulanic acid is sufficiently stable to allow extraction from clarified broth into *n*-butanol at pH 2.0. It is then back extracted into water at pH 7 with sodium hydroxide. Following initial absorption onto carbon and elution with 90% aqueous acetone, compound (1) can alternatively be extracted into an organic phase consisting of 72% aqueous phenol–*N*,*N*-dimethylaniline–tetrachloromethane (53:5:15 v/v/v) and back extraction with barium hydroxide to pH 6.5 to yield barium clavulanate [*61177-42-2*]. Adsorption onto strongly basic anionic resins, such as Zerolite FFIP SRA61 or Diaion PA306, followed by elution with sodium chloride solution gives a fraction rich in sodium clavulanate. Desalting on Amberlite XAD-4 or Hokuetsu HS columns and concentration by reverse osmosis and lyophilization yields partially purified sodium clavulanate. Chromatography on Zerolite FFIP SRA62 or QAE Sephadex A25 and desalting on Biogel P2 gives the required salt. The crude barium salt is converted into sparingly soluble crystalline lithium clavulanate which, in turn, can be converted into other salts via cation-exchange resins. Sodium clavulanate tetrahydrate [*57943-81-4*] is also prepared by converting partially purified solid sodium clavulanate into benzyl clavulanate using benzyl bromide in dimethylformamide, with subsequent purification of the ester and hydrogenolysis in aqueous ethanol containing sodium hydrogen carbonate and 10% Pd on charcoal as catalyst; the product is recrystallized from aqueous acetone (see Fig. 7).

Assay

Clavulanic acid is assayed in a number of ways. Microbiological assay involves its β-lactamase-inhibitor properties. The principal method used for biological fluids and solutions is the hole-in-plate synergy assay with penicillin G and *Klebsiella aerogenes* NCTC 11228 (*K. pneumoniae* ATCC 29665), defined earlier as the KAG assay. This assay can be used in the presence of other β-lactams such as amoxycillin or ti-

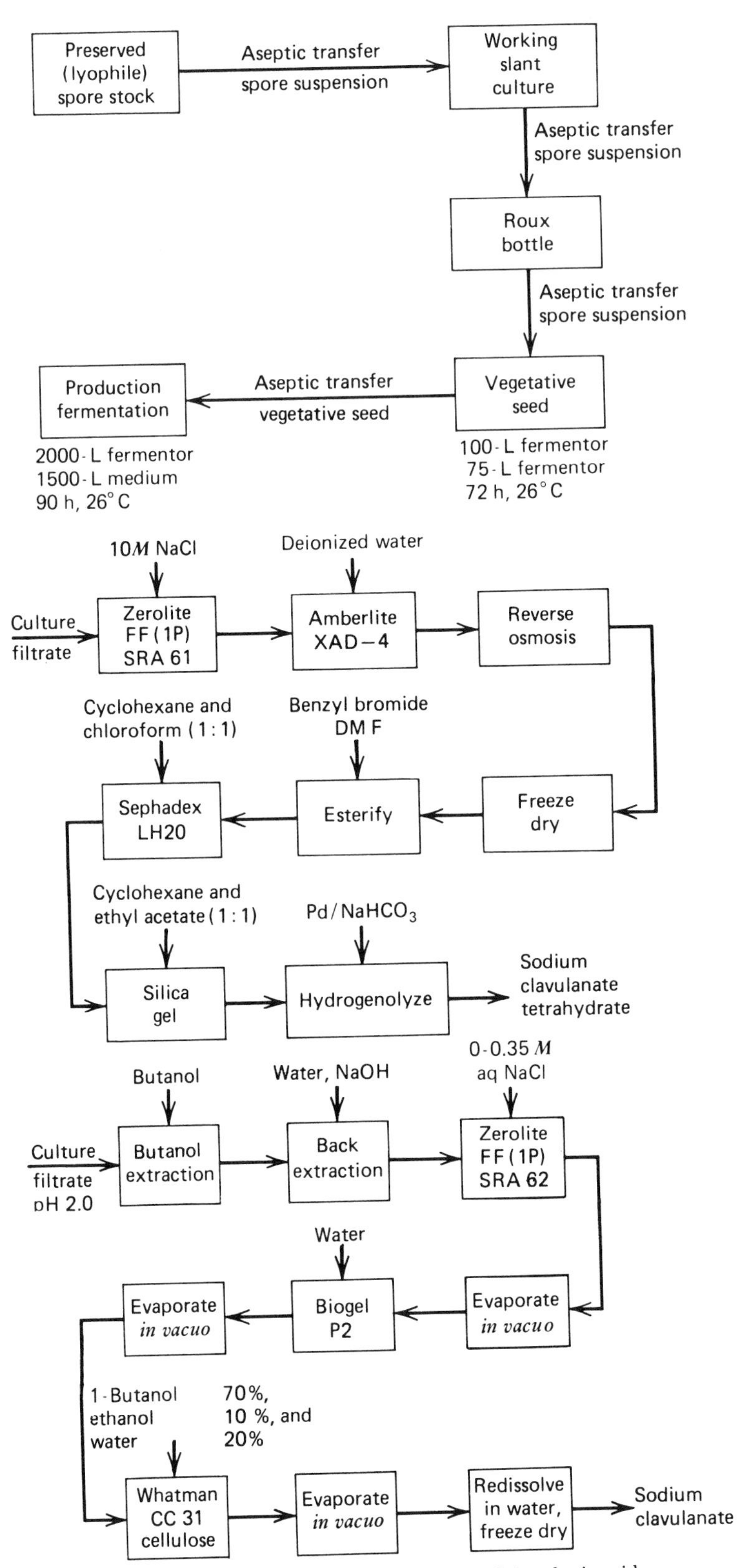

Figure 7. Fermentation and isolation of clavulanic acid.

carcillin. These compounds are readily assayed in the presence of clavulanic acid with organisms such as *Sarcina lutea* NCTC 8340 or *Bacillus subtilis* ATCC 6633 for amoxycillin and *Pseudomonas aeruginosa* NCTC 10701 for ticarcillin. An automated enzyme inhibition assay for structure (**1**) using staphylococcal β-lactamase, has also been described (88). The organism *Comomonas terrigena* ATCC 8461 and an *Acinetobacter* sp have also been used to determine structure (**1**) (6,27). An assay method based on the absorbance at 312 nm of the product formed by the reaction of compound (**1**) with imidazole has been described (104). This procedure can be used for derivatives of compound (**1**) and clavams such as compounds (**9**) and (**12**). Clavulanic acid can be estimated by high performance liquid chromatography (hplc) (C18 μBondapak column; 5% methanol, 95% 0.1 *M* phosphate buffer pH 4.0 as eluant) by detection at 230 nm (105). Combination of these two assay procedures by the precolumn derivatization of clavulanic acid with imidazole at pH 6.8 and room temperature, and hplc using a C18 μBondapak column with 0.1 *M* phosphate buffer and 6% methanol as eluant and monitoring the generated chromophore at 311 nm has been used (104). Hplc utilizing an ion-pair reversed phase has been applied to the assay of compound (**1**) in solution, especially urine (106). The eluant contains tetrabutylammonium bromide which produces a bathochromic shift to a suitable region (λ 220 nm) for detection of clavulanic acid. Solution of compound (**1**) and its salts can also be assayed in dilute alkali (0.1 *M* NaOH) using the absorption maximum which develops at ca 259 nm (ε 16,700) (6). It can be detected on tlc plates by the red color obtained with a triphenyltetrazolium chloride spray (23).

THIENAMYCIN AND OTHER CARBAPENEM ANTIBIOTICS

As indicated under clavulanic acid, application of a β-lactamase inhibition screen—the KAG assay—to soil isolates led to the detection of a number of strains of *S. olivaceus* with β-lactamase inhibitory activity (see under Assays). The family of compounds responsible for this activity was shown to be the olivanic acids (5). Simultaneous to these investigations, Merck workers were investigating soil isolates for the production of inhibitors of peptidoglycan synthesis. Their studies led to the isolation of the thienamycins (17). In addition to these efforts, research groups in Japan were using β-lactamase inhibition screens and supersensitive strains of *Pseudomonas aeruginosa*, *E. coli*, and *Comamonas terrigena* to screen for new β-lactam antibiotics (107–110). These investigations led to the isolation of a large number of novel β-lactams based on the 7-oxo-1-azabicyclo[3.2.0]hept-2-en-2-carboxylic acid or 1-carbapenem (**6a**) ring system (see Table 6).

Thienamycin (**2**) was the first member of the carbapenem group to be reported (17). Until recently, *S. cattleya* was the only organism known to yield compound (**2**); however, *S. penemifaciens* is claimed to produce exclusively (**2**) and without penicillin N and cephamycin C (111). The two latter compounds, together with *N*-acetylthienamycin (**69**) and *N*-acetyldehydrothienamycin (**70**), were also obtained from *S. cattleya* (112,114). Two mutants of *S. cattleya* have given *N*-acetylthienamycin and NS-5 or deshydroxythienamycin (**87**) (113). This last compound has been prepared by deacetylation of PS-5 (**83**) by L- and D-amino acid acylases (128). Northienamycin (**100**) is also considered to be a metabolite of *S. cattleya* (134) as is 8-epithienamycin (140).

Table 6. Naturally Occurring Carbapenems

Carbapenem	CAS Reg. No.	Structure No.	R	R′	Stereochemistry H-5,6	Stereochemistry C-8	Producing organism[a]	Refs.
thienamycin	[59995-64-1]	**(2)**	$CH_3CH(OH)$	$S(CH_2)_2NH_2$	trans	*R*	a, b	17, 111
N-acetylthienamycin	[63701-32-6]	**(69)**	$CH_3CH(OH)$	$S(CH_2)_2NHC(=O)CH_3$	trans	*R*	a, j	17, 112–113
N-acetyldehydrothienamycin	[68738-07-8]	**(70)**	$CH_3CH(OH)$	$SCH{=}CHNHC(=O)CH_3$	trans	*R*	a	17
MM13902 epithienamycin E	[57459-82-2]	**(71)**	$CH_3CH(OSO_3H)$	$SCH{=}CHNHC(=O)CH_3$	cis	*S*	c, d, e, f, g, h	5, 115–121
MM4550 MC696-2Y2A	[12795-21-0] [52012-46-1]	**(72)**	$CH_3CH(OSO_3H)$	$S^{+}(O^{-})CH{=}CHNHC(=O)CH_3$	cis	*S*	c, d, e, f, g, i	5, 115–116, 118–119
MM17880 epithienamycin F	[61036-81-5]	**(73)**	$CH_3CH(OSO_3H)$	$S(CH_2)_2NHC(=O)CH_3$	cis	*S*	c, d, e, f, g, o	115–122
epithienamycin G	[87421-69-0]	**(74)**	$CH_3CH(OSO_3H)$	$S(CH_2)_2C(=O)CH_3$	trans	*S*	e	118
MM22380 epithienamycin A	[63582-78-5]	**(75)**	$CH_3CH(OH)$	$S(CH_2)_2C(=O)CH_3$	cis	*S*	d, e, f, g, o	118–123
MM22381 epithienamycin C	[63599-16-6]	**(76)**	$CH_3CH(OH)$	$S(CH_2)_2NHC(=O)CH_3$	trans	*S*	d, f, e, g	117–121, 123

MM22382 epithienamycin B PS-48	*[68399-26-8]* *[65376-20-7]*	**(77)**	$CH_3CH(OH)$	$SCH{=}CHNHCCH_3$ (C=O)	cis	*S*	d, f, e, g, h	117–121, 123
MM22383 epithienamycin D	*[65322-98-7,* *68399-25-7]*	**(78)**	$CH_3CH(OH)$	$SCH{=}CHNHCCH_3$ (C=O)	trans	*S*	d, f, e, g	117–121, 123
MM27696	*[81542-97-4]*	**(79)**	$CH_3CH(OSO_3H)$	$SCH{=}CHNHCC_2H_5$ (C=O)	cis	*S*	m	124
desacetyl-MM17880	*[68206-95-1]*	**(80)**	$CH_3CH(OSO_3H)$	$S(CH_2)_2NH_2$	cis	*S*	n	125
MM22382 sulfoxide	*[83510-01-4]*	**(81)**	$CH_3CH(OH)$	$S^{+}(O^{-})CH{=}CHN(H)OCH_3$	cis	*S*	o, r	122, 126
MM22383 sulfoxide	*[77448-05-6]*	**(82)**	$CH_3CH(OH)$	$S^{+}(O^{-})CH{=}CHNHCCH_3$ (C=O)	trans	*S*	o	122
MM22744 PS-5	*[67007-79-8]*	**(83)**	C_2H_5	$S(CH_2)_2NHCCH_3$ (C=O)	trans		d, e, g, l	120, 127
PS-6	*[72615-19-1]*	**(84)**	$(CH_3)_2CH$	$S(CH_2)_2NHCCH_3$ (C=O)	trans		d, e, g, l	120
PS-7	*[72615-18-0]*	**(85)**	C_2H_5	$SCH{=}CHNHCCH_3$ (C=O)	trans		d, e, g, l	120
PS-8	*[82837-65-8]*	**(86)**	$(CH_3)_2CH$	$SCH{=}CHNHCCH_3$ (C=O)	trans		g	121
NS-5	*[74806-75-0]*	**(87)**	C_2H_5	$S(CH_2)_2NH_2$	trans		k	113, 128
SF-2103A pluracidomycin A	*[82138-64-5]*	**(88)**	$CH_3CH(OSO_3H)$	SO_3H	cis	*S*	o, p	122, 129

Table 6 *(continued)*

Carbapenem	CAS Reg. No.	Structure No.	R	R′	Stereochemistry H-5,6	C-8	Producing organism[a]	Refs.
pluracidomycin B	[*82138-65-6*]	**(89)**	$CH_3CH(OSO_3H)$	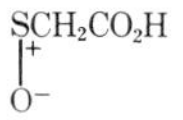 $S^+(O^-)CH_2CO_2H$	cis	*S*	o	122
pluracidomycin C	[*82138-66-7*]	**(90)**	$CH_3CH(OSO_3H)$	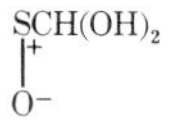 $S^+(O^-)CH(OH)_2$	cis	*S*	o	122
OA-6129A	[*82475-11-4*]	**(91)**	C_2H_5	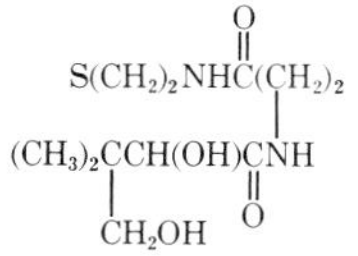 $S(CH_2)_2NHC(O)(CH_2)_2NHC(O)CH(OH)C(CH_3)_2CH_2OH$	trans		q[b]	130
OA-6129B$_1$	[*82510-13-2*]	**(92)**	$CH_3CH(OH)$	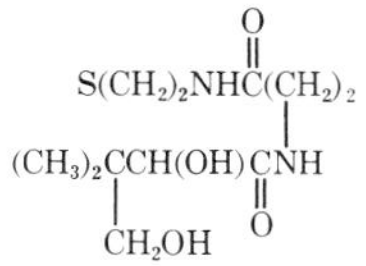 $S(CH_2)_2NHC(O)(CH_2)_2NHC(O)CH(OH)C(CH_3)_2CH_2OH$	cis	*S*	q[b]	130
OA-6129B$_2$	[*82475-10-3*]	**(93)**	$CH_3CH(OH)$	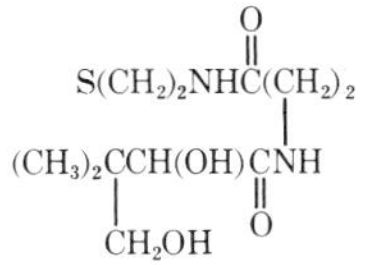 $S(CH_2)_2NHC(O)(CH_2)_2NHC(O)CH(OH)C(CH_3)_2CH_2OH$	trans	*S*	q[b]	130
OA-6129C	[*82475-09-0*]	**(94)**	$CH_3CH(OSO_3H)$	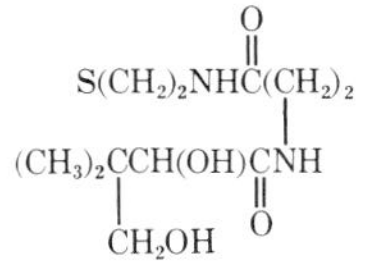 $S(CH_2)_2NHC(O)(CH_2)_2NHC(O)CH(OH)C(CH_3)_2CH_2OH$	cis	*S*	q[b]	130
carpetimycin A C-19393H$_2$	[*77209-15-5*]	**(95)**	$(CH_3)_2C(OH)$	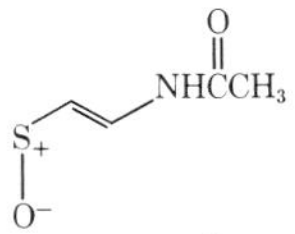 $S^+(O^-)CH{=}CHNHC(O)CH_3$	cis		h, r, s	131–132
carpetimycin B C-19393S$_2$ KA6643-B	[*76094-36-5*] [*77170-94-6*] [*76025-74-6*]	**(96)**	$(CH_3)_2C(OSO_3H)$	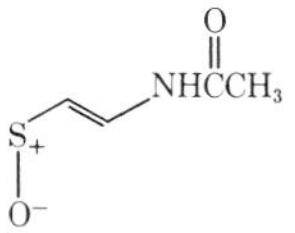 $S^+(O^-)CH{=}CHNHC(O)CH_3$			h, r, s	131–132

asparenomycin A PA-31088-IV	*[76466-24-5]*	**(97)**[c]	$CH_3\overset{\|}{C}(CH_2OH)$	$\overset{+}{S}(O^-)CH{=}CHCCH_3$ (C=O)		h, t	133
asparenomycin B PA-39504-XI	*[79366-72-6]*	**(98)**[c]	$CH_3\overset{\|}{C}(CH_2OH)$	$\overset{+}{S}(O^-)(CH_2)_2NH\overset{O}{\overset{\|}{C}}CH_3$		h, t	133
asparenomycin C PA-39504-X3	*[81018-71-5]*	**(99)**[c]	$CH_3\overset{\|}{C}(CH_2OH)$	$SCH{=}CHNH\overset{O}{\overset{\|}{C}}CH_3$		h, t	133
northienamycin	*[77550-86-8]*	**(100)**	CH_2OH	$S(CH_2)_2NH_2$	trans	a	134
SQ-27860	*[78854-41-8]*	**(101)**	H	H		u, v	135
KA-6643-D	*[82795-82-2]*	**(102)**	$(CH_3)_2C(OSO_3H)$	$S(CH_2)_2NH\overset{O}{\overset{\|}{C}}CH_3$	cis	w	136–137
KA-6643-F	*[82795-81-1]*	**(103)**	$(CH_3)_2C(OSO_2H)$	$SCH{=}CHNH\overset{O}{\overset{\|}{C}}CH_3$	cis	w	136–137
KA-6643-G	*[82889-91-6]*	**(104)**	$(CH_3)_2C(OH)$	$S(CH_2)_2NH\overset{O}{\overset{\|}{C}}CH_3$	trans	w	136–137
KA-6643-X	*[82889-90-5]*	**(105)**	$CH_3C(CH_2OH)$	$S(CH_2)_2NH\overset{O}{\overset{\|}{C}}CH_3$		w	136–137

[a] Producing organisms

a = *S. cattleya* NRRL 8,075
b = *S. penemifaciens* ATCC 31,599 FERM-P5,305
c = *S. olivaceus* ATCC 21,379 to 21,382 and NRRL 11,020
d = *S. olivaceus* ATCC 31365 and 31126
e = *S. flavogriseus* NRRL 8139 and 8140
f = *S. gedanensis* ATCC 4880
g = *S. cremeus auratilis* ATCC 31358
h = *S. tokunonensis* ATCC 31,509 FERM-P4 843
i = *S. fulvoviridis* MC696-SY2
j = *S. cattleya* S-WRI-M459
k = *S. cattleya* S-WRI-M5301
l = *S. fulvoviridis* A933
m = *S. olivaceus* CBS 349.80
n = *S. monjinensis* FERM-P5,467
o = *S. pluracidomyceticus* PA-41,746
p = *S. sulfonofaciens* SF-2103, FERM-5,636
q = *S.* sp OA-6129, FERM-BP-11
r = *S. griseus cyprophilus* ATCC 31486, FERM-P4,774 ISF 13,886
s = *S* sp KC 6643, ATCC 31493, FERM-P5,467
t = *S. argenteolus* ATCC 31589, FERM-P5,265
u = *Serratia* sp SQ 11,482
v = *Erwinia* sp SQ 12,637 and 12,638
w = *S.* sp KC664, FERM BO-58

[b] The following cultures also produce the olivanic acid complex: *S. olivaceus* NCIB 8238 and 8509, *S. argenteolus* ATCC 11009, *S. flavovirens* ATCC 3320, *S. flavus* ATCC 3369, *S. fulvoviridis* ATCC 15863, and *S. sioyalensis* ATCC 13989.

[c] (structure: bicyclic carbapenem ring with R, R′, N, O and CO_2H) is the structure for these compounds.

MM13902 (**71**), MM17880 (**73**), and MM4550 (**72**) were the first olivanic acid derivatives to be reported. Initially they were isolated as a complex and were found to occur in a wide range of *Streptomyces* sp. Detailed investigations (119) of *S. olivaceus* fermentations led to the production and characterization of MM 22380 (**75**), MM 22381 (**76**), MM 22382 (**77**), and MM 22383 (**78**); the first two were isomeric with *N*-acetylthienamycin (**69**), whereas the latter two were isomers of *N*-acetyldehydrothienamycin (**70**). Merck workers have also reported compounds (**75**)–(**78**) from *S. flavogriseus* and called them epithienamycin A, C, B, and D, respectively (117–118). Olivanic acid derivatives have also been produced by *S. gedanensis* (119) and *S. cremeus auratilis* (120). The sulfate MM 4550 (**72**) is considered to be identical to MC-696 SY2A (138), and epithienamycin G (**74**), the sulfate related to epithienamycin C (MM 22381) (**76**), has been detected from *S. flavogriseus* (118). Recently the propionyl analogue MM 27696 (**79**) of MM 13902 (**71**) was isolated from *S. olivaceus* (124). Desacetyl-MM 17880, 8U-207 (**80**) is a metabolite of *S. monjinensis* (125) and sulfoxides (**81**) and (**82**) of MM 22382 and MM 22383, respectively, were derived from *S. pluracidomyceticus* (122); compound (**81**) is also reported from *S. griseus cryophilus* (126).

Compounds PS-5 (**83**), PS-6 (**84**), PS-7 (**85**), and PS-8 (**86**), from *S. cremeus auratilis*, are further examples of carbapenems with trans-β-lactam protons; thienamycin (**2**), epithienamycin C (MM 22381) (**76**), epithienamycin D (MM 22383) (**78**), and NS-5 (**87**) are earlier examples. In PS-5 (**83**) and PS-7 (**85**), *C*-6 has an ethyl substituent, whereas PS-6 (**84**) and PS-8 (**86**) have an isopropyl substituent at *C*-6. Carpetimycin A (C-19393H_2) (**95**) and carpetimycin B (C-19393S_2) (**96**) have been reported from *S. griseus cryophilus*, *S. tokunonesis*, and an unspecified *Streptomyces* sp; both are sulfoxides and possess an hydroxyisopropyl group and the corresponding sulfate ester at *C*-6, respectively (131–133). Metabolites KA-6643D, G, F, and X (**102**)–(**105**) have also been found as a complex (136–137). The most complex *C*-2 substituted olivanic acid analogues yet reported (130) contain a pantetheinyl substituent, ie, OA-6129A (**91**), OA-6129B, (**92**), OA-6129B_2 (**93**), and OA-6129C (**94**). A further carbapenem derivative, perhaps of some biosynthetic significance, is the dihydro-compound (carbapenam) 17927D obtained from a number of *Streptomyces* spp (139). Recently, Squibb reported the production of carbapenem-3-carboxylic acid, SQ 27860 (**101**), from *Serratia* and *Erwinia* species (135). This is the first account of a fused bicyclo-β-lactam from bacteria. Thus, approximately thirty-eight carbapenems have been reported to date. The non-β-lactam metabolites cyclopentenedione G2201-C (**106**) and the azabicyclo[3.3.0]octa-5,7-dien-4-one derivative MM22702 (**107**) were isolated from *S. cattleya* and *S. olivaceus*, respectively, during studies on carbapenems (141).

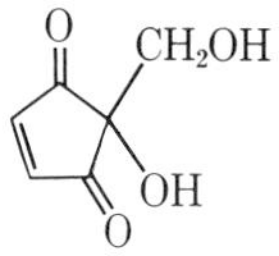

O, CH_2OH, OH, O

(**106**)
G2201-C;
2-hydroxy-2-(hydroxymethyl)cyclopent-4-ene-1,3-dione

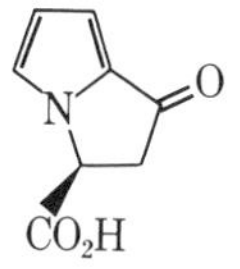

N, O, CO_2H

(**107**)
MM 22702;
(2*S*)-4-oxo-1-azabicyclo[3.3.0]
octa-5,7-diene-2-carboxylic acid

Physical Properties

Thienamycin is a colorless hygroscopic solid, freely soluble in water but sparingly soluble in methanol. As a zwitterion, it has a pK_{a1} of ca 3.1 and a $pK_{a2} > 8$. The uv absorption characteristics of thienamycin and the naturally occurring carbapenems are important criteria in the assay procedures and hence for the evaluation of purity (115,119,123,125,127,131–133). There are distinct differences in maxima, depending on the saturation of the substituent on sulfur at *C*-2 and whether the sulfur atom is at the thiol or sulfoxide level of oxidation. As β-lactams, the carbapenems give an ir absorption in the range of 1760–1792 cm^{-1} and carboxylate anion bands at 1550–1650 cm^{-1}. The sulfates, eg, (**71**)–(**73**), give intense absorption bands at ca 1250^{-1} (116,142–143). The nmr data have certain notable characteristics; Cd spectra have been recorded for the important naturally occurring carbapenems. As might be expected, the data are generally similar for most of these compounds; differences occur between, for example, a thiol derivative at *C*-2 and the corresponding sulfoxide (eg, MM 13902 and MM 4550).

Stability in Aqueous Solution. The stability of thienamycin over a range of pH values and in a variety of buffers has been examined in detail (17). Maximum stability is obtained at pH 6–7 and is comparable to penicillin G at pH 2–5; in phosphate and borate buffers, degradation above pH 7 is very rapid. Stability above pH 7 was enhanced in the presence of buffers based on triethylamine but still poor in those containing tris and ammonium ion. Buffers derived from substituted morpholines, eg, 2-(*N*-morpholino)ethane- and 3-(*N*-morpholino)propanesulfonic acids, gave maximum stability at pH 6–8, at up to molar concentration. Concentrated solutions (>10 mg/mL) of thienamycin decompose rapidly, yielding yellow and brown solutions, with a considerable loss in potency. This has obviously created a problem in the final stages of purification of compound (**2**). Kinetic studies indicate that the β-lactam of thienamycin is degraded by intramolecular aminolysis involving the primary amino function of a second molecule. A dimer may be formed, suggesting a mechanism analogous to the degradation of 6-APA and ampicillin (144).

Although lacking an amino group, the olivanic acid derivatives MM 13902 (**71**), MM 17880 (**73**), and MM 4550 (**72**) are fairly unstable in aqueous solution; MM 4550 has a half-life of 5.5 h at pH 4, compared with 7.5 min and 15 min for MM 13902 and MM 17880, respectively. On the other hand, the latter compounds are more stable than MM 4550 at pH 10.2 (116). Likewise, the sulfoxides carpetimycin A (**95**) and B (**96**) are more stable at acid pH than epithienamycin B (MM 22382; (**77**)), MM 17880, and cephalosporin C (145).

Chemical Properties

Structure Elucidation. Early studies of thienamycin (30) and olivanic acids (142–143) were hampered by the small quantities of partially purified material that were available. Nevertheless, nmr spectra (^{1}H and ^{13}C, plus spin-decoupling experiments), mass spectra (field desorption, low and high resolution data), and elemental analysis on thienamycin, the olivanic acid derivatives, and other carbapenems produced the evidence to establish the structures. The mass spectra of the carbapenem derivatives gave fragmentation patterns analogous to those seen with cephalosporins and penicillins which, in conjunction with ir data, suggested the presence of a β-lactam moiety but with a carbon substituent at *C*-6 rather than an acylamino group. Other fragmentations provided evidence for a partial structure with carboxylic acid/ester and acetylcysteamine or a closely related compound in addition to a β-lactam fragment containing a substituted alkyl substituent.

Methanolysis of methyl *N*-acetylthienamycin in air gave a disulfide and a hydroxypyrrole, both of which were identified. This degradation gave a basis for the proposed structure (**2**), but further information was required to establish relative and absolute configurations. A trans-β-lactam configuration was assigned to thienamycin on the basis of the small vicinal coupling constant, $J_{5,6}$ <3 Hz for the two β-lactam protons; a cis β-lactam, as exemplified by the penicillins and cephalosporins, showed $J_{5,6}$ 5 Hz, and furthermore the trans coupling constant in clavulanic acid derivatives was smaller than the cis. This assignment was further established by x-ray analysis of methyl *N*-acetylthienamycin and by its conversion into a cis β-lactam. The absolute configuration of (**2**) at *C*-5 was determined by converting the enelactam (**108**), obtained from (**2**) via the *N*-benzyloxycarbonyl benzyl ester and a mesylate, into a mixture of aspartic acid and taurine via ozonolysis and oxidation; the aspartic acid had the *R* configuration and as this was derived from *C*-2, 1, 5, and 6, it follows that *C*-6 has the *S* configuration. The *C*-8 of the hydroxyethyl substituent was demonstrated to be *R* by kinetic resolution (146) and nmr configuration correlation (147). The publication (17,30) of the structure of thienamycin established it as the second novel fused β-lactam after clavulanic acid to be reported since the so-called classical β-lactams, ie, the penicillins, cephalosporins, and cephamycins.

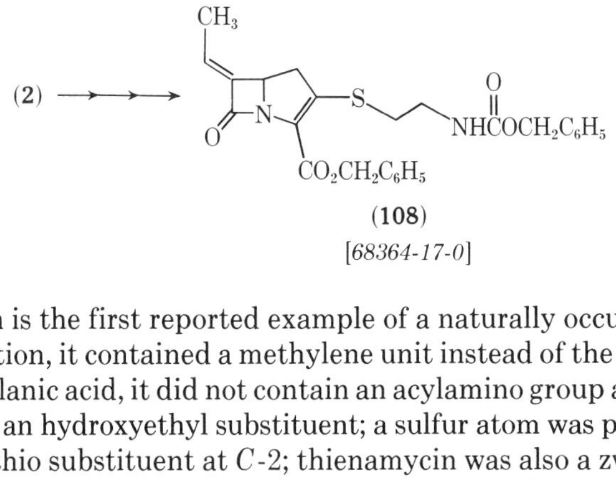

(**2**) ⟶⟶ [structure: CH_3; O; N; S; $NHCOCH_2C_6H_5$; $CO_2CH_2C_6H_5$]

(**108**)
[*68364-17-0*]

Thienamycin is the first reported example of a naturally occurring trans-fused β-lactam. In addition, it contained a methylene unit instead of the traditional sulfur at *C*-1. Like clavulanic acid, it did not contain an acylamino group at *C*-6, but in contrast, it possessed an hydroxyethyl substituent; a sulfur atom was present but as part of an aminoethylthio substituent at *C*-2; thienamycin was also a zwitterion.

The structures of MM 13902 (**71**), MM 17880 (**73**), and MM 4550 (**72**) were established as olivanic acid derivatives (142–143,148); nmr spectral and mass spectral data were used to derive the fused β-lactam ring system. These three compounds had intense bands at 1220–1270 cm^{-1} in the ir, indicative of sulfate ester residues. In addition, field-desorption mass spectra of MM 13902 monoester monosodium salt confirmed the presence of a sulfate residue and gave a fragmentation typical of other β-lactams; cis coupling ($J_{5,6}$ 5.5 Hz) in these compounds was evident.

The results and experience gained in establishing the structures of thienamycin and the sulfated olivanics led to the relatively straightforward characterization of the carbapenems MM 22380, MM 22381, MM 22382, and MM 22383 (epithienamycins A, C, B, and D) as two pairs of isomers; MM 22380 and MM 22382 had the cis configuration, whereas MM 22381 and MM 22383 were trans β-lactams (118,149). The *C*-8 configuration of these compounds and the olivanic sulfates MM 13902 and MM 17880 was studied by a series of stereospecific elimination reactions on esters or monoesters monoquaternary salts to yield *E* and/or *Z* ethylidene derivatives and led to the conclusion that the olivanic acid family, in contrast to thienamycin and its derivatives, possessed 8*S* stereochemistry (148). This result was subsequently confirmed by x-ray analysis of a heavy-atom derivative. The stereochemistry of the *C*-2 acetamidoethenylthio substituent in the metabolites MM 13902, MM 4550, MM 22382, and MM 22383, etc was always found to be *E*, based on the nmr data for a trans-disubstituted double bond (J = 14 Hz) in all these and analogous compounds.

Again nmr and mass spectral data allowed PS-5, PS-6, and PS-7 to be identified as trans-orientated β-lactams (120,150). The absolute stereochemistry of this family was proved by chemical degradation involving reaction with ozone and ultimate conversion of the products into a δ-hydroxy-β-amino acid and hence a lactone, which was assigned the *R* configuration based on Hudson's lactone rule (150). Thus, the *C*-6 of PS-5 was *R* and therefore *C*-5 was *R* because of the trans relationship of the β-lactam protons. The structures of the carpetimycins (**95**) and (**96**) were proposed on the basis of spectral data and on their relationship to MM 4550 (**72**) (31,145). Carpetimycin B (**96**) could be converted by mild hydrolysis into carpetimycin A (**95**). X-ray analysis of the bromobenzyl ester of compound (**95**) defined the absolute stereochemistry of these sulfoxides as the same as MM 13902 at *C*-5 and *C*-6, namely, *R*. The asparenomycins (**97**)–(**99**) were characterized and interrelated on the basis of spectral and x-ray data and comparisons with compounds such as MM 4550 (151). The absolute stereochemistry of the sulfoxide was *R*, whereas chemically the *C*-5 *R* configuration was determined by oxidation of asparenomycin A to D-aspartic acid; its reduction with titanium trichloride gave deoxyasparenomycin A, which was identical to asparenomycin C. The isopropylidene substituent at *C*-6 had *E* geometry because acid hydrolysis of (**97**) gave the dihydropyrrole (152) which does not form a lactone. The lactone (**109**), however, is considered to be true natural product from *S. tokunonensis* and *S. argenteolus*. The structure of the pluracidomycins and the OA-6129 family again followed from their spectral data based on the background of the early carbapenem analogues (122,129–130).

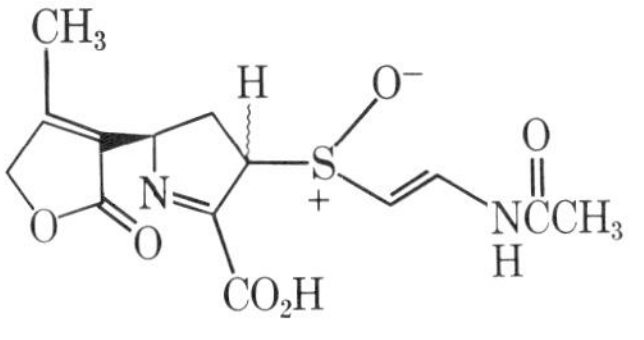

(**109**)

Reactions. The unsaturated *C*-2 side chain present in a number of carbepenems has always been found to occur naturally as a trans double bond. Mercury(II) salt treatment of, for example, MM 13902 (**71**) or MM 22382 (**77**) or the carpetimycins A and B, produced an equilibrium mixture of the *E*- and *Z*-isomer (148,153). The Δ^1-isomer of thienamycin was obtained by conversion of a protected derivative, into a mixture of starting material and isomer using D.B.U. 1,8-diazabicyclo[5.4.0]undec-7-ene and subsequent separation and deblocking (154–155). *N*-Acetyl derivatives of thienamycin are readily obtained under Schotten-Baumann conditions using anhydrides, acid chlorides, and chloroformates or by reaction of acyl halides with silylated thienamycin in anhydrous solvents. Esterified *N*-phenoxyacetylthienamycin can be reductively cleaved to afford the 6-(hydroxyethyl)carbapenem, called descysteaminylthienamycin [*74464-48-5*] (156–157). *N*-Alkylation of thienamycin gives a mixture of secondary, tertiary, and quaternary amines (155). The *O*-methyl ether and *O*-acyl derivatives of compound (**2**) have been prepared after protection of the amine and carboxylic acid functions. In general, thienamycin esters are, as expected, unstable but give stable Schiff bases with salicylaldehyde. Esters of *N*-acyl derivatives are stable (155).

The naturally occurring carbapenems can be transformed into analogues in which the alkyl or alkenyl substituent on the *C*-2 sulfur has been modified. A thiol derivative is alkylated or used in addition reactions (158–159) or a sulfoxide side chain derived by oxidation of the parent metabolite can be displaced by a thiol (160).

The acetamidoethenylthio substituent of olivanic acid derivatives reacts with hypobromous acid to yield a *C*-2 thiol via an intermediate bromohydrin. The thiol is alkylated with alkyl halides or aziridines to afford a *C*-2-alkylthio derivative. Hypobromous acid adds to the acetoxy derivatives of MM 22383 in the expected manner but also furnishes a hydroxyoxazoline.

Carbapenems can be readily converted into their sulfoxides with peracids (161–162). The resultant sulfinyl group may be replaced by different sulfenyl substituents in the presence of mild base.

In order to convert from the olivanic series of metabolites with 8*S* chirality into the thienamycin series possessing 8*R* stereochemistry, MM 22383 and MM 22381 were treated with diethylazodicarboxylate, triphenylphosphine, and formic acid (163). The inverted formate was then hydrolyzed to yield *N*-acetyldehydro- and *N*-acetylthienamycin, respectively. Inversion of stereochemistry at *C*-8 can also be achieved via ethylidene derivatives (148). Ethylidene compounds have been of use to prepare *C*-8 deoxy derivatives (148,164). Reduction of an ethylidene derivative obtained from MM 17880 (**73**) with hydrogen over platinum oxide catalyst gave a mixture of *cis*- and *trans*-deoxyolivanic acids. However, reduction of the ethylidene with sodium borohydride gave almost exclusively the trans isomer and hence PS-5. The cis isomer in this series is prepared by the catalytic reduction of an unsaturated intermediate protected as the *p*-methoxycarbonylbenzyl ester (164).

Both chemical (165–167) and enzymatic methods (167–168) have been used for the deacetylation of *N*-acetyl carbapenems. Enzymatic deacylation of epithienamycin A (**75**), epithienamycin C (**76**), PS-5 (**83**), and *N*-acetylthienamycin (**67**) was achieved by treatment with an amidohydrolase derived from a strain of *Protaminobacter ruber* which was isolated from soil (128,167–169). Epithienamycin A has also been deacylated using an enzyme prepared from hog kidney (118).

N-Formimidoylthienamycin. By far the most interesting derivatives of thienamycin are the amidines, derived by reaction of thienamycin with an imidate ester at pH 8.2 or by reaction of a dialkyliminium chloride with trimethylsilyl thienamycin (155,170). Methyl formimidate and compound (2) give crystalline *N*-formimidoylthienamycin (**110**), designated MK-0787. Amorphous amidines are five to ten times more stable than thienamycin at high concentrations. Furthermore, MK-0787 and the corresponding acetamidine have longer shelf lives than thienamycin. Thus, the conversion of the amino group of structure (**2**) into a more basic, but less nucleophilic, substituent has given a product with dramatically improved stability and, equally important, has resulted in the retention of the good antipseudomonal activity of compound (**2**). Lyophilized samples of MK-0787 are said to contain ca 2% of compound (**2**), suggesting that hydrolysis of the formamidine residue does take place; crystalline MK-0787 contains ≤1% (**2**).

Because of its excellent stability properties, MK-0787 has been selected by Merck as the derivative most suitable for clinical development.

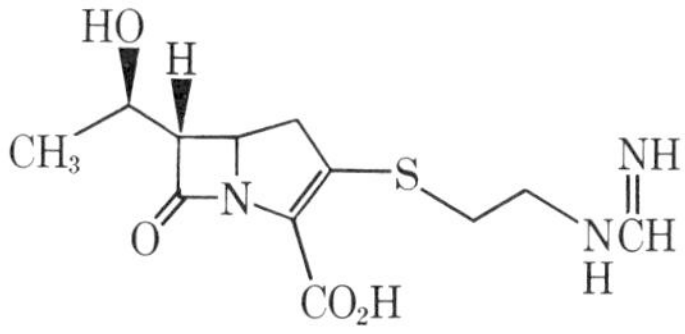

(**110**)
MK-0787 [*64221-86-9*]

Biological Properties

The *in vitro* antibacterial characteristics of thienamycin are summarized on Table 7. It has been shown to be a potent broad-spectrum antibiotic, with pronounced activity against gram-positive cocci and gram-negative bacilli. It is more active than some of the recently developed β-lactams, moxalactam, cefoperazone, piperacillin, mezlocillin, cefamandole, cefoxitin, and ticarcillin, and as active as the aminoglycoside tobramycin; it was particularly effective against multiresistant organisms, and an interesting feature was its promising antipseudomonal activity (171–173). The minimum inhibitory concentration (MIC) of thienamycin against *Bacteroides fragilis* is ≤1 mg/L, and in general, it is more active than clindamycin and metronidazole against anaerobes. Thienamycin is very stable to β-lactamases, and its minimal bactericidal concentration (MBC) is within one or two dilutions of its MIC for most organisms. It has greatest affinity for penicillin-binding protein 2 though it also binds well to proteins 1, 4, 5, and 6, but it has a poor affinity to protein 3 of *E. coli* K-12. Concentrations of 0.1–0.6 mg/L give large osmotically stable round cells; higher concentrations cause lysis with spheroplast formation (79). As an inhibitor of peptidoglycan synthesis, thienamycin is more effective than ampicillin.

The other naturally occurring carbapenems, such as MM 13902, MM 17880, and the epithienamycins (MM 22380, etc), are also broad-spectrum antibiotics but are generally not as potent as thienamycin and all lack significant activity against *Pseudomonas*. For example, whereas MM 13902 has good activity against gram-negative strains, it is less active than compound (**2**) toward gram-positive organisms and *Pseudomonas* (118,174). The hydroxy compounds (MM 22380, etc) are not as active as the sulfates MM 13902 and MM 17880, although the cis β-lactams MM 22380 and MM 22382 are more effective than the trans-β-lactam olivanic acid compounds, MM 22381 and MM 22383. Carpetimycin A, asparenomycin A, OA-6129C, and PS-5 are all similarly active and the most potent of their series (130,132,175–178). Thienamycin affords good *in vivo* protection against a wide range of organisms, although it lacks activity via the oral route (179–180). Of all the naturally occurring carbapenems, thienamycin is the most effective on evaluation in animal models. Detailed studies show that this class of β-lactam antibiotic is extensively degraded by metabolism in the kidney resulting in low and variable urinary recoveries (117,130,181–184). A renal

Table 7. Antibacterial Activity of Some Antibiotics

Strain (10^5 CFU)		MIC (μg/mL) of					
		MK-0787[a]	THM[a]	CFX[a]	CEF[a]	CAB[a]	GEN[a]
Staphylococcus aureus	2985	0.01	0.02	3.2	0.2	0.63	0.32
Staphylococcus aureus	4428[b]	20.0	40.0	>100	>100	80.0	5.0
Enterococcus	2864	0.63	1.3	>100	25.0	40.0	10.0
Enterococcus	2862	1.3	2.5	>100	25.0	40.0	>20
Enterococcus	2863	40.0	40.0	>100	>100	>80	20.0
Escherichia coli	2482	0.32	0.63	6.3	12.5	5.0	0.32
Escherichia coli	2895	0.32	0.63	25.0	>100	>80	1.3
Shigella spp	2880	0.16	0.32	3.2	6.3	5.0	1.3
Salmonella typhimurium	826	0.32	0.32	3.2	6.3	20.0	2.5
Enterobacter cloacae	2647[c]	0.16	0.32	12.5	12.5	20.0	1.3
Enterobacter cloacae	2646	0.63	0.63	>100	>100	>80	1.3
Enterobacter spp	2903	0.63	1.3	>100	>100	5.0	1.3
Enterobacter spp	2902	1.3	2.5	>100	>100	20.0	1.3
Enterobacter aerogenes	2906	2.5	5.0	>100	>100	5.0	1.3
Klebsiella pneumoniae	2921	0.63	0.63	6.3	12.5	>80	1.3
Klebsiella pneumoniae	2922	0.63	0.63	6.3	100	>80	>20
Klebsiella spp	2888	0.63	1.3	>100	>100	10.0	1.3
Serratia spp	2840	0.63	2.5	25.0	>100	10.0	2.5
Serratia spp	2855	0.63	2.5	12.5	>100	>80	1.3
Proteus mirabilis	3125	5.0	10.0	3.2	6.3	1.3	5.0
Proteus mirabilis	2831	2.5	5.0	25.0	50.0	2.5	10.0
Proteus mirabilis	2830	5.0	10.0	6.3	>100	>80	2.5
Proteus morganii	2833	5.0	5.0	12.5	>100	>80	2.5
Proteus morganii	2834	2.5	10.0	12.5	>100	1.3	0.32
Providencia spp	2851	1.3	2.5	1.6	>100	1.3	0.63
Pseudomonas aeruginosa	40	1.6	3.1			50	50
	4294	12.5	50			100	25

[a] Abbreviations: THM, thienamycin; CFX, cefoxitin; CEF, cephalothin; CAB, carbenicillin; GEN, gentamicin.

[b] Methicillin resistant.

[c] Beta-lactamase-negative mutant derived from parent strain 2646.

dipeptidase is responsible for this metabolism; the enzyme is defined as dehydropeptidase-I (DHP-I, EC 3.4.13.11), and there is a close structural resemblance between a carbapenem, eg, (**2**), and a dehydropeptide, eg, glycyldehydrophenylalanine. The degradation products of the carbapenem nucleus by DHP-I are initially derived by rupture of the β-lactam ring (181–182). The *N*-acylated carbapenems are up to fiftyfold more susceptible than thienamycin and MK-0787 to hog renal DHP-I, whereas desacetylepithienamycin A [*63701-77-9*], a 5,6-cis diasteromer of thienamycin, was degraded at a similar rate. Cephalosporins and penicillins are considered to be stable to DHP-I, but penems are readily degraded (185).

Detailed *in vitro* antibacterial studies have shown MK-0787 to possess the pronounced activity of (**2**) (see Table 7) because of the resistance of these two compounds to degradation by a wide variety of β-lactamases (186).

Although DHP-I can be considered to be a mammalian β-lactamase, it is rather ironic that MK-0787, thienamycin, MM 13902, and the other carbapenems that are relatively resistant to most microbial β-lactamases should be susceptible to DHP-I. In order to overcome this rapid metabolism, Merck workers have investigated the inhibition of DHP-I with a number of acylaminodehydroacrylic acid derivatives (187–188); two such compounds, MK-0791 (**111**) and MK-0789 (**112**), have been studied extensively and were reported to have a dramatic effect on the urinary recovery of MK-0787 in animals and man. The plasma half-life of MK-0787 in man is ca 60 min with a mean peak plasma level of 15 μg/mL following a 250 mg dose with a good tolerability on single and repeat dosing. The urinary recovery, however, is 6–31%. When MK-0791 (**111**) was coadministered with MK-0787 up to 1000 mg, up to a 75% recovery of MK-0787 in urine was found (189). It has been inferred that the use of such renal dipeptidase inhibitors reduces the nephrotoxicity of thienamycin or MK-0787 to rabbits (187).

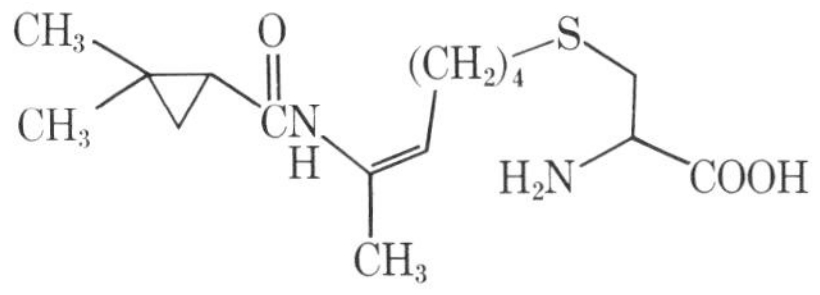

(**111**)
MK-0791 [*78852-98-9*]

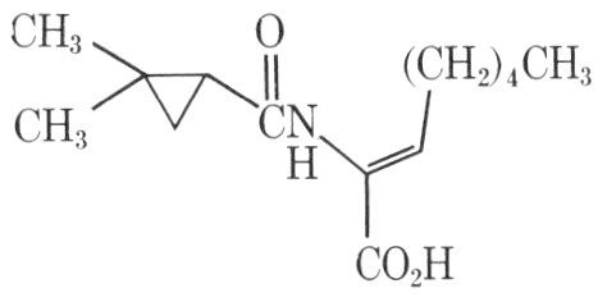

(**112**)
MK-0789 [*74589-65-4*]

In addition to being potent antibiotics, the various members of the carbapenem family are also very effective β-lactamase inhibitors (190). Thus, MM 13902, MM 17880, MM 4550, etc (116,123), the PS-5 group (176–177,191), the carpetimycins (192), the asparenomycins (193–194), the pluracidomycins (122), and SF-2103A (129) have all been shown to inhibit a wide range of β-lactamase types. In general, these carbapenems are better inhibitors of cephalosporinases than clavulanic acid or CP-45,899. Although thienamycin and MK-0787 have good stability to β-lactamase, they are not as effective as the olivanic acids, etc as inhibitors. The inhibition of R_{TEM} β-lactamase by olivanic acid derivatives has been investigated in detail (195).

Synthesis

The synthesis of the carbapenem nucleus and of thienamycin and its naturally occurring and synthetic analogues has been and still is the aim of many academic and industrial groups throughout the world. Various approaches are reviewed in refs. 196 and 197.

1-Carba-2-penem-3-carboxylic acid (**101**), SQ-27860, was synthesized before its isolation from *Serratia* and *Erwinia* spp (135). Racemic (**101**) was obtained as its sodium salt in solution by hydrolysis of various esters (157,197–199). The Beecham route involved the reaction of a glyoxylic ester with 4-allylazetidinone, which was formed from penta-1,4-diene and chlorosulfonyl isocyanate. The resultant hydroxy ester was converted into a phosphorane, and the terminal double bond was oxidized; cyclization gave a carbapenem ester. Other workers showed that the Δ^2-double bond was essential for biological activity (32). Another approach yielded *C*-6 substituted, *C*-2 unsubstituted thienamycin analogues (156–157). A similar route gave *C*-6 unsubstituted, *C*-2 alkyl substituted compounds (200–204). Sulfoxides (analogues of MM 4550) with opposite chirality can be obtained by a thiol addition route, as can (±)-epi-PS-5 when the thiol addition procedure is applied to substituted carbapenem precursors (205). Further application of the thioester–intramolecular Wittig reaction has given a total synthesis of MM 22383 (**78**) and its *C*-8 epimer, *N*-acetyldehydrothienamycin as racemates (206). Here the *C*-6 substituent was introduced via a photolytic or rhodium acetate cyclization of a diazo intermediate followed by reduction and isomer separation (207).

Although the total synthesis of thienamycin has been a prime target for a number of industrial and academic groups, the most important work has been done at Merck, Sharp, and Dohme, who appear to prefer the synthetic approach to large quantities of thienamycin over a fermentation route from *S. cattleya.* This is no doubt because of the problem of producing large titers of thienamycin by strain improvement as encountered with all metabolites in the carbapenem series. The first synthesis of (±)-thienamycin (**2**) also led to (±)-8-epithienamycin (**113**) and (±)-6-epithienamycin (**114**), a compound with stereochemistry not directly available in a naturally occurring carbapenem (208) (see Fig. 8(**a**)).

A stereocontrolled, enantiomerically specific synthesis of thienamycin has also been achieved by Merck workers from an amino acid (152,209–212). L-Aspartic acid (**115**) was converted by a number of steps into the *trans*-3-acetylazetidinone (**116**) and hence by stereocontrolled reactions to the keto esters (**117**) and (**118**). Reaction with a cysteamine derivative gave (**119**), which was readily transformed into (+)-thienamycin (**2**) (Fig. 8(**b**)).

Similarly, 6-APA was converted into thienamycin (213) (see Fig. 9). Diethyl 1,3-acetonedicarboxylate (**120**) has also been utilized as a starting material (214) as shown in Figure 10. A recent patent (215) claims the synthesis of thienamycin by way of intermediates that are derived from D-glucose. The Kametani school has made use of the intermediate (**124**) during the synthesis of thienamycin and analogues (216). The acetal (**123**) is derived from the isoxazoline (**122**) which is prepared by the 1,3-dipolar cycloaddition of the nitrile oxide (**121**) and methyl crotonate and transformed into carbapenem compounds.

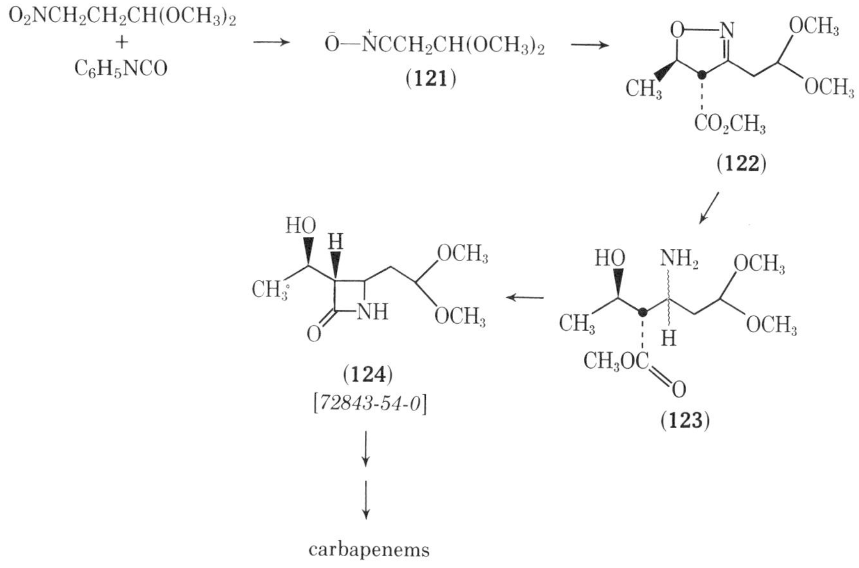

$O_2NCH_2CH_2CH(OCH_3)_2$ + C_6H_5NCO → $\bar{O}—\overset{+}{N}CCH_2CH(OCH_3)_2$ (**121**) → (**122**)

O—N, OCH_3, CH_3, OCH_3, CO_2CH_3

(**122**) → (**123**)

HO, NH_2, OCH_3, CH_3, H, OCH_3, CH_3OC, O

(**123**)

(**123**) → (**124**)

HO, H, OCH_3, CH_3, NH, OCH_3, O

(**124**)
[*72843-54-0*]

↓

↓

carbapenems

The above syntheses result in carbapenems with variations at *C*-2 and *C*-6. Other thienamycin analogues have been claimed (217–223).

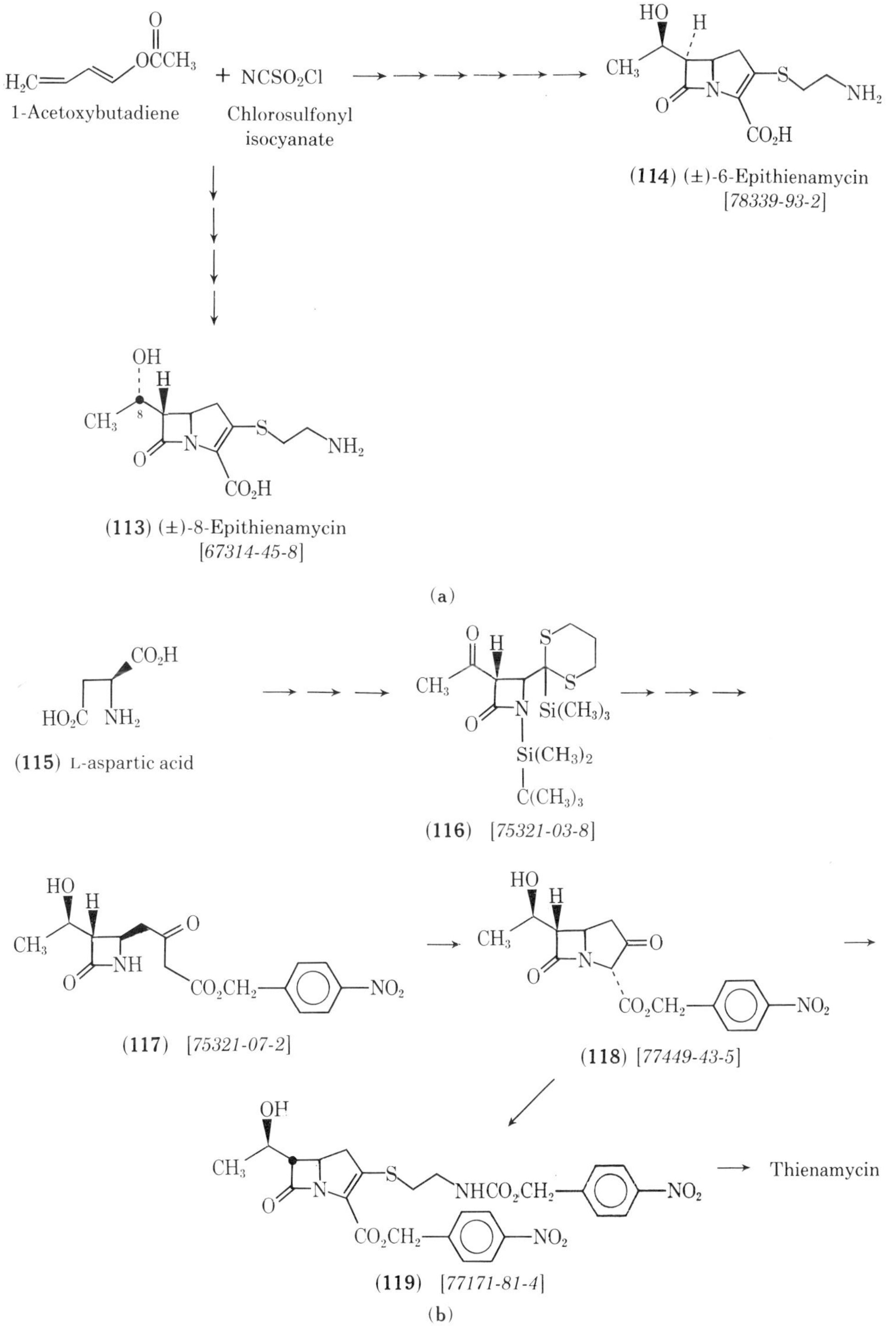

$H_2C{=}CH{-}CH{=}CH{-}OCCH_3$ (C=O)

1-Acetoxybutadiene + $NCSO_2Cl$ Chlorosulfonyl isocyanate → → → → → →

(114) (±)-6-Epithienamycin [*78339-93-2*]

HO, H, CH_3, O, N, S, NH_2, CO_2H

↓ ↓ ↓ ↓

OH, H, CH_3, 8, O, N, S, NH_2, CO_2H

(113) (±)-8-Epithienamycin [*67314-45-8*]

(a)

CO_2H, HO_2C, NH_2

(115) L-aspartic acid

→ → →

O, H, S, S, CH_3, O, N, $Si(CH_3)_3$, $Si(CH_3)_2$, $C(CH_3)_3$

(116) [*75321-03-8*]

→ → →

HO, H, CH_3, O, NH, O, CO_2CH_2—C6H4—NO_2

(117) [*75321-07-2*]

→

HO, H, CH_3, O, N, O, CO_2CH_2—C6H4—NO_2

(118) [*77449-43-5*]

→

OH, CH_3, O, N, S, $NHCO_2CH_2$—C6H4—NO_2, CO_2CH_2—C6H4—NO_2

(119) [*77171-81-4*]

→ Thienamycin

(b)

Figure 8. Synthesis of thienamycin.

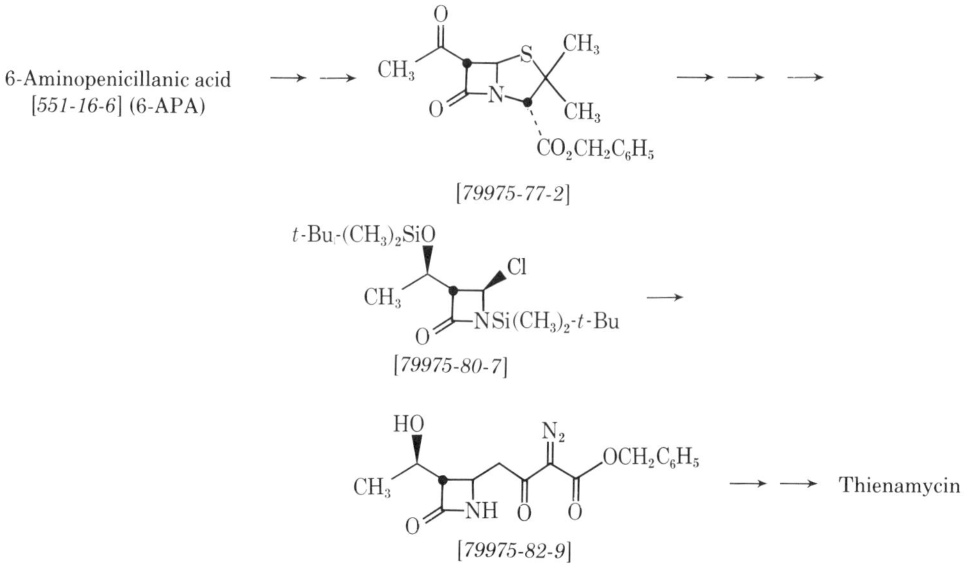

6-Aminopenicillanic acid [*551-16-6*] (6-APA) → → [*79975-77-2*] → → →

[*79975-80-7*] →

[*79975-82-9*] → → Thienamycin

Figure 9. Thienamycin from 6-APA.

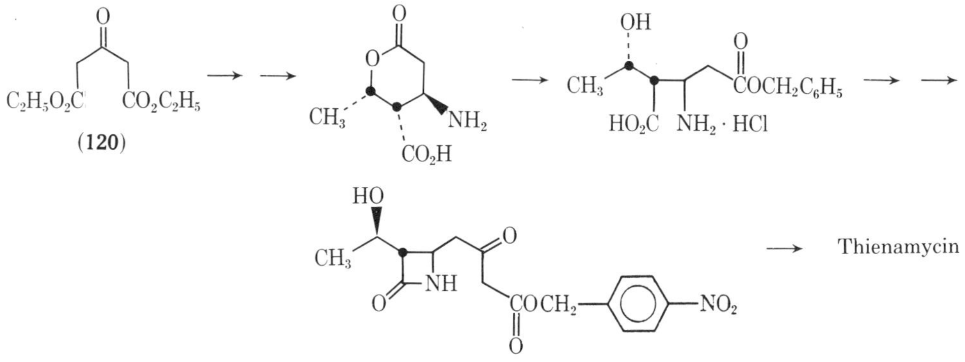

(**120**) → → → → →

→ Thienamycin

(**117**) [*75321-07-2*]

Figure 10. Thienamycin from diethyl 1,3-acetonedicarboxylate.

Biosynthesis. No published information is available on the biosynthesis of thienamycin or other carbapenem. A preliminary report, however, indicates that glutamic acid is a key building block (224). This followed from studies on the incorporation of ^{2}H- and ^{13}C-labeled glutamic acid into thienamycin, but results were not clear. Acetate appears to be the source of *C*-6, *C*-7, *C*-8, and *C*-9. The isolation of compounds such as carbapenem (**101**), the carpetamycins, the PS-5 family, the asparenomycins, in addition to northienamycin (**100**) and the olivanic acid compounds, etc, strongly suggests that variation at *C*-6 is derived by the condensation of aldehyde or ketonic equivalents with a preformed parent carbapenem ring. Addition of cysteamine and further oxidation and acylation accounts for the *C*-2 side-chain varients. The relationship of *cis*-olivanic acid and epithienamycin has been examined by Beecham workers (123). Using certain *S. olivaceus*-blocked mutants, evidence was obtained for the conversion of MM 22380 (**75**) into the sulfate MM 17880 (**73**) and into the unsaturated compound MM 22382 (**77**) and, hence, MM 13902 (**71**). There was no evidence of transformation of MM 17880 into MM 13902 or of MM 13902 into MM

4450 (**72**). *S. cattleya* is said to perform a similar sequence in converting *N*-acetylthienamycin (**69**) into *N*-acetyldehydrothienamycin (**70**) (171).

Manufacture

Fermentation. As shown in Table 6, many species of *Streptomyces* have been reported to produce carbapenem antibiotics; however, only two species, *S. cattleya* (17) and *S. penemifaciens* (111), are capable of producing thienamycin. Again for these compounds, and especially thienamycin and olivanic acids, such as MM 13902, the manufacturing processes leading to the isolation of large quantities of material are proprietary, although limited details are provided in published papers and patents for the compounds given in Table 6. In general, the carbapenems occur as complex mixtures of analogues and in some cases are coproduced with penicillin N and cephamycin C. *Streptomyces* species producing carbapenems give relatively low titers of antibiotics compared with other β-lactam-producing organisms and hence give low yields of desired metabolite. None of these carbapenem-producing strains can be easily mutated to give dramatically increased titers. Although a variety of media have been used, carbapenem fermentation requires ca 0.05% cobalt to obtain optimum antibiotic production. The presence of cobalt seems to be related to glutamine synthetase and hence glutamic acid, which is considered to be a biosynthetic precursor of the carbapenem nucleus. The lower stability of the carbapenems compared to other β-lactams contributes to the low yields. The fermentation and isolation process for thienamycin and the olivanic acids has been reviewed recently (103).

Thienamycin production from a wild-type isolate of *S. cattleya* is described in refs. 17 and 225. Peak titers of up to 4 μg/mL after 140 h fermentation were obtained with the wild-type strain (in shake flask and tank).

Isolation and Purification. The filtered fermentation broth is absorbed on a Dowex 50 Na^+ column, eluted with 2% pyridine, reabsorbed, eluted with a lutidine-based volatile buffer, desalted on XAD- 2, and finally lyophilized to give thienamycin in an overall yield of <2% and a purity of 94% (17). A second procedure employs liquid ion-exchange using *O,O*-dinonylnaphthalenesulfonic acid in butanol (at pH 3) with back extraction into a buffered aqueous solution, followed by a further ion-pair extraction (at pH 11) using Aliquat 336 in butanol and back extraction into potassium acetate solution (226). Final purification via ion-exchange chromatography and desalting gives 90% pure thienamycin in a yield of 80% for the ion-pair extraction steps. This process has to be operated on a continuous-flow basis with minimal holdup volumes because thienamycin is extremely unstable at pHs 3 and 11. The third isolation sequence requires elution from a Dowex 1 × 2 resin with carbon dioxide saturated water and concentration by reverse osmosis (227). Final purification is by further ion-exchange chromatography, another concentration by reverse osmosis, desalting, and lyophilization, to produce 90% pure thienamycin in an overall yield of 25–35%.

Since in the production of the sulfated olivanic acids (**71**)–(**78**) *S. olivaceus* produces several very closely related compounds, it is necessary to influence their relative proportions by using sulfate-free media with certain strains or developing mutants blocked in synthesis of the sulfated compounds. The proportion of sulfates is also affected by pH and addition of citrate. A high oxygen-transfer rate is required for constant titers of olivanic acid derivatives, and in stirred-tank fermentations dissolved oxygen concentrations should not be less than 20% of saturation; oxygen concentration is controlled by agitator speed. The effect of aeration on metabolite production has been illustrated using a sulfation-blocked mutant of *S. olivaceus* ATCC 21379 (228). At high aeration (low volumes in shake flasks), the four hydroxyolivanic acid compounds (epithienamycins) were formed, whereas with lower aeration (high flask volumes), a further metabolite MM 22744 (**83**) identical to PS-5 was obtained.

Ion-pair extraction (liquid–liquid distribution in the liquid ion-exchange mode) has been of great value in the isolation of carbapenems containing sulfate ester groups, eg, MM 13902, MM 17880, and MM 4550 (116). Quaternary ammonium salts such as dimethyldidecylammonium, benzylcetyldimethylammonium, and methyltrioctylammonium chlorides (Aliquat 336) in dichloromethane are particularly effective. Nonsulfated carbapenems are less readily extracted by this procedure, although it has been applied to the isolation of MM 22382 and MM 22383. The metabolite is back-extracted into an aqueous phase containing suitable inorganic salts such as sodium nitrate and iodide, although other cations can be used to give salts other than sodium. In addition, the carbapenems, the pluracidomycins, SF-2103A [*82138-64-5*] and OA-6129C [*82475-09-0*], have been obtained via ion-pair extraction.

Anion-exchange processes to isolate carbapenems have utilized acrylic resins containing quaternary and tertiary ammonium groups; styrene–divinylbenzene resins have also been used, although increased binding, especially in derivatives with unsaturated side chains, can be a complication. QAE Sephadex ion-exchange materials have proved invaluable in the separation of mixtures of carbapenems as has the use of anion exchange resins with hydrophobic characteristics (eg, Diaion HP20 or Amberlite IRA 400). Nonionic styrene–divinylbenzene resins (Amberlite XAD-4, Diaion HP20) are used to desalt, as is gel-filtration (eg, on Biogel P2). In addition, hplc-using reverse-phase systems have been important in providing high quality purified material.

The concentrations in the filtrate are usually very low, at most 20 μg/mL, and large quantities are difficult to obtain. A typical MM 13902 fermentation yields ca 4 g of purified material from 1000 L broth.

Assay

Thienamycin has been assayed by two methods (17). A microbiological disk-diffusion uses *Staphylococcus aureus* ATCC 6538P, whereas a differential spectrophotometric procedure involves reaction of the antibiotic in buffered solution with hydroxylamine at pH 7 for 30 min at 23°C. The difference in absorbance measured at 297 nm between reacted and unreacted antibiotic is related to thienamycin concentration; a ΔA of 1 was equivalent to 36.5 μg/mL antibiotic. A similar differential spectrophotometric assay using cysteine instead of hydroxylamine has also been used. The half-lives of thienamycin to hydroxylamine and cysteine were 5.7 and 2.5 min, respectively; these reactions are considerably faster than for β-lactams such as penicillin G. This assay has also been used quantitatively for the epithienamycins using wavelengths of 308 and 300 nm for unsaturated and saturated side-chain derivatives, respectively. The sulfate esters MM 13902, MM 17880, and MM 4550 can be determined using a β-lactamase-inhibition system based on an agar plate assay with *Klebsiella pneumoniae* ATCC 29655 plus penicillin G (KAG assay), or by an automated assay using penicillin G and R_{TEM} type IIIa β-lactamase, derived from *E. coli* JT4 (116). In addition, antibacterial assays using organisms such as *K. pneumoniae*

ATCC 29655 and *Neisseria catarrhalis* NCTC3622 have been used. The activity of other carbapenems has been measured in broths, in column fractions, and on tlc plates and paper chromatograms using *Bacillus subtilis* ATCC 6633 (123), *Vibrio percolans* ATCC 8461 (117), *Bacillus licheniformis* SC9262 (135), *E. coli* ESS (123), *Comamonas terrigena* B-996 (127), an *E. coli* lacking chromomasomal β-lactamase, and penicillin binding protein 1B (132) and other β-lactam hypersensitive strains (133). Variations of the KAG assay system involving β-lactamase preparations from *E. coli*, *Proteus morganii*, and *Staphylococcus aureus* added to plates finally seeded with *S. aureus* have also been developed (107,115,133). High performance liquid chromatography is employed in the assay of carbapenems and preparation of high purity material; reverse-phase columns have been particularly useful, eg, in the assay of PS-5, PS-6, and PS-7 (120), the asparenomycins (133), and the olivanic family in *S. gedanensis* (119).

PENICILLANIC ACID AND RELATED COMPOUNDS

The discovery of clavulanic acid and the carbapenems, together with their evaluation as β-lactamase inhibitors, has prompted the preparation of a number of β-lactam derivatives as possible alternative inactivators of β-lactamase. Penicillanic acid sulfone ((**125**); CP-45,899; Sulbactam) was the first of such analogues to be studied in detail (27). Subsequently, compounds such as 6α-chloropenicillanic acid sulfone (229), 6β-bromopenicillanic acid (230–232), 6β-iodopenicillanic acid, UK-38,006 (232–234), the 2β-chloromethylpenam, BL-P2103 (**126**) (235), certain penicillin sulfones (236–237), and penamsulfonamines (238) have been described and investigated as mechanism-based or suicide inactivators of β-lactamase.

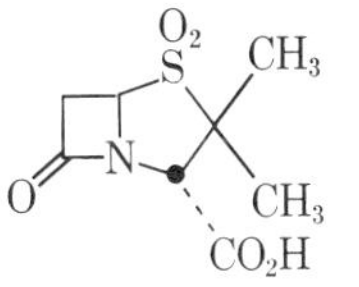

(**125**)
[*68373-14-8*]

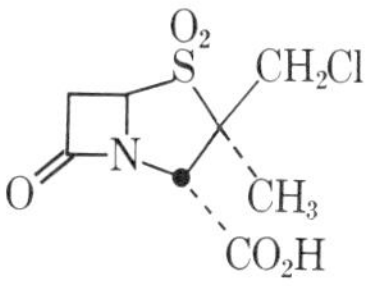

(**126**)
[*79634-05-2*]

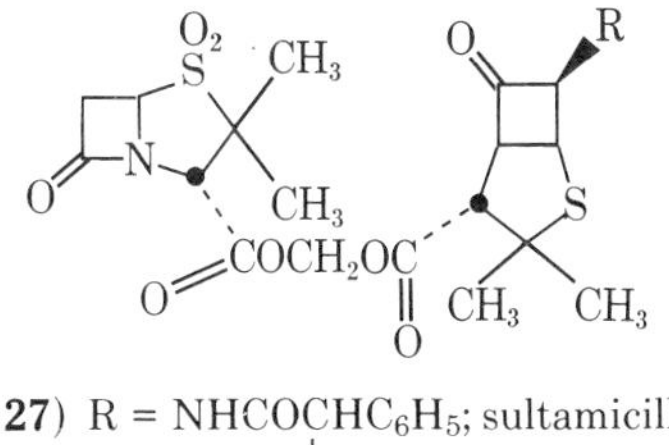

(**127**) R = $NHCOCH(NH_2)C_6H_5$; sultamicillin
[*76497-13-7*]
(**128**) R = $NCHC_6H_{12}$; VD 1825
[*76376-46-0*]

Penicillanic acid sulfone, CP-45,899 (**125**), is obtained by the oxidation of penicillanic acid and the hydrogenolysis of 6,6-dibromo- and 6α-bromopenicillanic acid sulfones (239). It is an irreversible inhibitor of a wide range of β-lactamase types and exhibits marked synergy in combination with a number of penicillins and cephalosporins. Although numerous reports have appeared on its chemotherapeutic potential (77,240–241), it was found to be less effective than clavulanic acid in its capacity to inhibit plasmid mediated β-lactamases of *E. coli* and *K. aerogenes* (73,75,240). Mode of action studies showed that compound (**125**) was both a substrate for and an inhibitor of R_{TEM} β-lactamase (242). It is reported to be poorly absorbed by mouth in humans, although it is well tolerated by the parental route, giving potentially useful blood levels and good urinary recovery (243). A pro-drug form of (**125**) is the pivaloyloxymethyl ester, CP-47,904, which is said to be well absorbed in man (243). Mutual pro-drugs such as (**127**) and (**128**) are prepared from ampicillin and mecillinam, respectively. Esters such as sultamicillin (**127**) are well absorbed from the gastrointestinal tract and are subsequently hydrolyzed to liberate antibiotic and inhibitor (244–245).

6β-Bromopenicillanic acid [*26631-90-3*] has been obtained by equilibration of the 6α-isomer at pH 9.2 (230–233,246) or by the selective reduction of 6,6-dibromopenicillanic acid [*24158-88-1*] with tributyltin hydride (246–248). 6β-Iodopenicillanic acid [*74772-32-0*] was prepared via the 6α-epimer (232) or the corresponding 6α-triflate (233–234). Both compounds have been evaluated as β-lactamase inhibitors in combination with a number of penicillins or cephalosporins and compared with clavulanic acid and CP-45,899 (249). The iodopenicillanate (UK-38,006) was also found to be effective as an inhibitor following oral administration to experimentally infected mice. The mode of action of 6β-bromopenicillanic acid has been examined in some depth and it is found to inactivate *Bacillus cereus* I β-lactamase by reaction with the hydroxyl group of the serine-44 residue (231,246,250). BL-P2013 (**126**) and the halogen penicillanic acids are less active as chemotherapeutic agents but may be of value in detailed mechanistic studies of suicide inhibitors.

BIBLIOGRAPHY

"Antibiotics, β-Lactams" in *ECT* 3rd ed., Vol. 2, pp. 871–919, by John R. E. Hoover and Claude H. Nash, Smith Kline & French Laboratories.

1. R. Nagarajan, D. D. Boeck, M. Gorman, R. L. Hamill, C. E. Higgens, M. M. Hoehn, W. M. Stark, and J. G. Whitney, *J. Am. Chem. Soc.* **93,** 2308 (1971).
2. E. O. Stapley, M. Jackson, S. Hernandez, S. B. Zimmerman, S. A. Currie, S. Mochales, J. M. Mata, H. B. Woodruff, and D. Hendlin, *Antimicrob. Agents Chemother.* **2,** 121 (1972).
3. T. T. Howarth, A. G. Brown, and T. J. King, *J. Chem. Soc. Chem. Commun.*, 266 (1976).
4. C. Reading and M. Cole, *Antimicrob. Agents Chemother.* **11,** 852 (1977).
5. A. G. Brown, D. Butterworth, M. Cole, G. Hanscomb, J. D. Hood, C. Reading, and G. N. Rolinson, *J. Antibiot.* **29,** 668 (1976).
6. Brit. Pat. 1,543,563 (Feb. 7, 1975), I. D. Fleming, D. Noble, H. M. Noble, and W. F. Wall (to Glaxo Laboratories Ltd.).
7. Brit. Pat. 1,563,103 (Oct. 13, 1975), S. Box (to Beecham Group Ltd.).
8. Neth. Pat. 7,603,818 (Apr. 10, 1975), (to Sankyo); Belg. Pat. 804,341 (Aug. 31, 1972), (to Sankyo).
9. Jpn. Pat. Kokai 53-104,706 (Feb. 24, 1977), (to Takeda Chemical Industries Ltd.).
10. Jpn. Pat. Kokai 55-162,993 (Dec. 18, 1980), (to Sanraku-Ocean Co. Ltd.).
11. Brit. Pat. 1,547,222 (Feb. 26, 1976), C. Reading (to Beecham Group Ltd.).
12. D. Brown, J. R. Evans, and R. A. Fletton, *J. Chem. Soc. Chem. Commun.*, 282 (1979).
13. D. L. Preuss and M. Kellett, *Abstr. No.* 163, and R. H. Evans, Jr., and co-workers, *Abstr. No.* 164 in *20th Interscience Conference on Antimicrobial Agents and Chemotherapy*, New Orleans, 1980. U.S. Pat. 4,202,819 (May 13, 1980), M. Kellett and co-workers (to Hoffmann-La Roche).
14. M. Wanning, H. Zähner, B. Krone, and A. Zeeck, *Tetrahedron Lett.* **22,** 2539 (1981); Eur. Pat. 57,644 (Jan. 30, 1981), H. Zahner, M. Wanning, B. Krone, A. Zeeck, and H. Peter (to Ciba-Geigy AG).

15. J. C. Sheehan, K. R. Henery-Logan, and D. A. Johnson, *J. Am. Chem. Soc.* **75,** 3292 (1953).
16. R. B. Morin, B. G. Jackson, E. H. Flynn, and R. W. Roeske, *J. Am. Chem. Soc.* **84,** 3400 (1962).
17. J. S. Kahan, F. M. Kahan, R. Goegelman, S. A. Currie, M. Jackson, E. O. Stapley, T. W. Miller, A. K. Miller, D. Hendlin, S. Mochales, S. Hernandez, H. B. Woodruff, and J. Birnbaum, *J. Antibiot.* **32,** 1 (1979).
18. T. N. Salzmann, R. W. Ratcliffe, and B. G. Christensen, *Tetrahedron Lett.* **21,** 1193 (1980).
19. A. G. Brown, *Antimicrob. Chemother.* **10,** 365 (1982).
20. E. Hunt, *J. Chem. Research* (S), 64 (1981).
21. J. G. Whitney, D. R. Brannon, J. A. Mabe, and K. J. Wicker, *Antimicrob. Agents Chemother.* **1,** 247 (1972); C. E. Higgens, R. L. Hamill, T. H. Sands, M. M. Hoehn, N. E. Davis, R. Nagarajan, and L. D. Boeck, *J. Antibiot.* **27,** 298 (1974).
22. C. E. Higgens and R. E. Kastner, *Int. J. Syst. Bacteriol.* **21,** 326 (1971).
23. Brit. Pat. 1,508,977 (Apr. 20, 1974), M. Cole, T. T. Howarth, and C. Reading (to Beecham Group Ltd.); Brit. Pat. 1,504,425 (Aug. 20, 1975), J. B. Harbridge and T. T. Howarth (to Beecham Group Ltd.); Brit. Pat. 1,561,395 (Feb. 26, 1976), J. B. Harbridge and T. T. Howarth (to Beecham Group Ltd.); U.S. Pat. 4,110,166 (Aug. 29, 1978), M. Cole, T. T. Howarth, and C. Reading (to Beecham Group Ltd.).
24. Jpn. Pat. Kokai 51-110,097 (Sept. 29, 1976), (to Sanraku-Ocean Co. Ltd.).
25. S. Elson and T. J. King, Beecham Pharmaceuticals and Nottingham University, 1982, unpublished results.
26. M. Kenig and C. Reading, *J. Antibiot.* **32,** 549 (1979).
27. A. R. English, J. A. Retsema, A. E. Girard, J. E. Lynch, and W. E. Barth, *Antimicrob. Agents Chemother.* **14,** 414 (1978).
28. J. Haginaka, T. Nakagawa, and T. Uno, *Chem. Pharm. Bull.* **29,** 3334 (1981).
29. T. J. King, Nottingham University, 1982, unpublished results.
30. G. Albers-Schönberg, B. H. Arison, O. D. Hensens, J. Hirshfield, K. Hoogstien, E. A. Kaczka, R. E. Rhodes, J. S. Kahan, F. M. Kahan, R. W. Ratcliffe, E. Walton, L. J. Ruswinkle, R. B. Morin, and B. G. Christensen, *J. Am. Chem. Soc.* **100,** 6491 (1978).
31. M. Nakayama, S. Kimura, S. Tanabe, T. Mizoguchi, I. Watanabe, T. Mori, K. Miyahara, and T. Kawasaki, *J. Antibiot.* **34,** 818 (1981).
32. R. B. Woodward, *Philos. Trans. R. Soc. London Ser. B* **289,** 239 (1980); R. Pfaendler, J. Gosteli, R. B. Woodward, and G. Rihs, *J. Am. Chem. Soc.* **103,** 4526 (1981).
33. R. M. Sweet in E. H. Glynn, ed., *Cephalosporins and Penicillins*, Academic Press, Inc., New York, 1972, p. 297; R. M. Sweet and L. F. Dahl, *J. Am. Chem. Soc.* **92,** 5489 (1970).
34. A. G. Brown, J. Goodacre, J. B. Harbridge, T. T. Howarth, R. J. Ponsford, and I. Stirling in J. Elks, ed., *Recent Advances in the Chemistry of β-Lactam Antibiotics, Special Publication No. 28*, The Chemical Society, London, 1977, 295.
35. R. D. G. Cooper in P. G. Sammes, ed., *Topics in Antibiotic Chemistry*, Vol. 3, Ellis Horwood, Ltd., Chichester, UK, 1980, p. 39.
36. D. F. Corbett, Beecham Pharmaceuticals, unpublished results.
37. P. H. Bentley, P. D. Berry, G. Brooks, M. L. Gilpin, E. Hunt, and I. I. Zomaya, *J. Chem. Soc. Chem. Commun.*, 748 (1977).
38. P. H. Bentley, G. Brooks, M. L. Gilpin, and E. Hunt, *Tetrahedron Lett.*, 1889 (1979).
39. I. Stirling and S. W. Elson, *J. Antibiot.* **32,** 1125 (1979).
40. J. S. Davies and T. T. Howarth, *Tetrahedron Lett.* **23,** 3109 (1982); J. S. Davies, *Tetrahedron Lett.* **23,** 5089 (1982).
41. C. E. Newall in G. I. Gregory, ed., *Recent Advances in the Chemistry of β-Lactam Antibiotics, Special Publication No. 38*, The Chemical Society, London, 1981, p. 151.
42. Brit. Pat. 1,508,978 (Apr. 20, 1974), M. Cole, T. T. Howarth, and C. Reading (to Beecham Group Ltd.).
43. Brit. Pat. 1,501,643 (Oct. 28, 1974), A. G. Brown and T. T. Howarth (to Beecham Group Ltd.).
44. Brit. Pat. 1,509,400 (Dec. 15, 1975), T. T. Howarth and I. Stirling (to Beecham Group Ltd.).
45. A. G. Brown, T. T. Howarth, I. Stirling, and T. J. King, *Tetrahedron Lett.*, 4203 (1976).
46. Brit. Pat. 1,505,859 (Dec. 12, 1975), I. Stirling (to Beecham Group Ltd.).
47. Brit. Pat. 1,587,616 (Feb. 25, 1977), 1,604,822 (Apr. 22, 1977), and 1,604,823 (Apr. 22, 1977), I. Stirling and B. P. Clarke (to Beecham Group Ltd.).
48. D. F. Corbett, T. T. Howarth, and I. Stirling, *J. Chem. Soc. Chem. Commun.*, 808 (1977).
49. P. C. Cherry, G. I. Gregory, C. E. Newall, P. Ward, and N. S. Watson, *J. Chem. Soc. Chem. Commun.*, 467 (1978).

50. Belg. Pat. Appl. 861,716 (Dec. 12, 1976), (to Beecham Group Ltd.).
51. Brit. Pat. 1,582,864 (Mar. 10, 1977), K. Luk, T. T. Howarth, and R. J. Ponsford (to Beecham Group Ltd.).
52. Brit. Pat. 1,565,209 (Oct. 13, 1975), T. T. Howarth and R. J. Ponsford (to Beecham Group Ltd.).
53. Eur. Pat. Appl. 2370 (Dec. 2, 1977); 3254 (Jan. 26, 1978); 13,790 (July 5, 1978); Belg. Pat. Appl. 870,405 (Sept. 16, 1977).
54. Eur. Pat. Appl. 225,287 (Aug. 22, 1979); Eur. Pat. Appl. 18,203 (Apr. 21, 1979); Eur. Pat. Appl. 24,372 (Aug. 3, 1976). Brit. Pat. 1,574,906 (Jan. 31, 1976), R. J. Ponsford (to Beecham Group Ltd.).
55. Brit. Pat. 1,572,259 (Apr. 28, 1977), T. T. Howarth, R. J. Ponsford, and A. J. Eglington (to Beecham Group Ltd.); U.S. Pat. 4,215,128 (Sept. 20, 1977) and U.S. Pat. 4,244,965 (June 15, 1978), (to Beecham Group Ltd.).
56. Brit. Pat. 1,603,208 (Apr. 24, 1977), J. Harbridge (to Beecham Group Ltd.).
57. Brit. Pat. 1,493,197 (July 4, 1975), T. T. Howarth and J. Harbridge (to Beecham Group Ltd.).
58. Brit. Pat. 1,563,427 and U.S. Pat. 4,187,228 (Mar. 4, 1977), K. Luk (to Beecham Group Ltd.).
59. Brit. Pat. 1,582,884 (May 19, 1977), T. T. Howarth and K. Luk (to Beecham Group Ltd.); U.S. Pat. 4,076,826 (Apr. 4, 1976).
60. U.S. Pats. 4,078,067 and 4,078,068 (Apr. 4, 1976), (to Merck & Co. Inc.).
61. U.S. Pat. 4,089,897 (Apr. 4, 1975), (to Beecham Group Ltd.); Brit. Pat. 1,545,467 (Apr. 14, 1975), T. T. Howarth, J. Goodacre, and R. J. Ponsford (to Beecham Group Ltd.); Brit. Pat. 1,587,612 (Oct. 30, 1976), K. Luk (to Beecham Group Ltd.).
62. P. C. Cherry, C. E. Newall, and N. S. Watson, *J. Chem. Soc. Chem. Commun.*, 469 (1978).
63. *Ibid.*, 663 (1979).
64. M. L. Gilpin, J. B. Harbridge, T. T. Howarth, and T. J. King, *J. Chem. Soc. Chem. Commun.*, 929 (1981).
65. M. Cole, *Philos. Trans. R Soc. London Ser. B* **289,** 207 (1980).
66. R. H. Richmond and R. B. Sykes, *Adv. Microb. Physiol.* **9,** 31 (1973); R. B. Sykes and M. Matthews, *J. Antimicrob. Chemother.* **2,** 115 (1976).
67. B. Slocombe in G. N. Rolison and A. Watson, eds., *Augmentin*, Excerpta Medica, Amsterdam, The Netherlands, 1980, p. 8.
68. C. Reading in D. A. Leigh and O. P. W. Robinson, eds., *Augmentin*, Excerpta Medica, Amsterdam, The Netherlands, 1982, p. 5.
69. P. A. Hunter, C. Reading, and D. Witting in W. Siegenthaler and R. Luthy, eds., *Current Chemotherapy*, American Society for Microbiology, Washington, D.C., 1978, p. 478.
70. P. A. Hunter, K. Coleman, J. Fisher, D. Taylor, and E. Taylor, *Drugs Exp. Clin. Res.* **5,** 1 (1979).
71. P. A. Hunter, K. Coleman, J. Fisher, and D. Taylor, *J. Antimicrob. Chemother.* **6,** 455 (1980).
72. R. Wise, *Lancet*, 145 (1977); J. Wust and T. D. Wilkins, *Antimicrob. Agents Chemother.* **13,** 130 (1978); R. Wise, J. M. Andrews, and K. A. Bedford, *Antimicrob. Agents Chemother.* **13,** 389 (1978); H. C. Neu and K. P. Fu, *Antimicrob. Agents Chemother.* **14,** 650 (1978); J. M. Miller, C. N. Baker, and C. Thornsberry, *Antimicrob. Agents Chemother.* **14,** 794 (1978); D. S. Reeves, M. J. Bywater, and H. A. Holt, *Infection* **6**(Suppl. 1), 9 (1978); F. W. Goldstein, M. D. Kitzis, and J. F. Acar, *J. Antimicrob. Chemother.* **5,** 1 (1979); B. Van Klingeren and M. Dessens-Kroon, *J. Antimicrob. Chemother.* **5,** 322 (1979); D. Greenwood, F. O'Grady, and P. Baker, *J. Antimicrob. Chemother.* **5,** 539 (1979); L. Dumon, P. Adriaens, J. Anne, and H. Eyssen, *Antimicrob. Agents Chemother.* **15,** 315 (1979); G. V. Doern, K. G. Siebers, L. M. Hallick, and S. A. Morse, *Antimicrob. Agents Chemother.* **17,** 24 (1980); M. Matsuura, H. Nakazawa, T. Hashimoto, and S. Mitsuhashi, *Antimicrob. Agents Chemother.* **17,** 908 (1980); G. Peters, G. Pulverer, and M. Neugebauer, *Infection* **8,** 104 (1980).
73. R. J. Boon, A. S. Beale, S. J. Layte, C. V. Pierce, A. R. White, and R. Sutherland, *Abstr. No. 442* in *21st Interscience Conference on Antimicrobial Agents and Chemotherapy*, Chicago, Ill., 1981; J. W. Paisley and J. A. Washington, *Antimicrob. Agents Chemother.* **14,** 224 (1978).
74. R. Wise, J. M. Andrews, and K. A. Bedford, *J. Antimicrob. Chemother.* **6,** 197 (1980).
75. R. Wise, A. P. Gillett, and J. M. Andrews, *J. Antimicrob. Chemother.* **5,** 301 (1979).
76. H. C. Neu and K. P. Fu, *Antimicrob. Agents Chemother.* **18,** 582 (1980).
77. M. S. Heerema, D. M. Musher, and T. W. Williams, Jr., *Antimicrob. Agents Chemother.* **16,** 798 (1979).
78. K. P. Fu and H. C. Neu, *J. Antimicrob. Chemother.* **7,** 287 (1980).
79. D. J. Pohlod, L. D. Saravolatz, E. L. Quinn, and M. M. Somerville, *Antimicrob. Agents Chemother.* **18,** 353 (1980).
80. B. G. Spratt, V. Jobamputra, and W. Zimmerman, *Antimicrob. Agents Chemother.* **12,** 406 (1977).

81. L. Mizen, K. Bhandari, J. Sayer, and E. Catherall, *Drugs Exptl. Clin. Res.* **7,** 263 (1981).
82. D. Jackson, D. L. Cooper, T. L. Hardy, P. F. Langley, D. H. Staniforth, and J. A. Sutton in ref. 67, p. 87; J. Haginaka, T. Nakagawa, T. Hoshino, K. Yamaoka, and T. Uno, *Chem. Pharm. Bull.* **29,** 3342 (1981).
83. Ref. 67, entire book.
84. Ref. 68, entire book.
85. G. Ninane, J. Joly, M. Kraylman, and P. Piot, *Lancet* 257 (1978); A. P. Ball, B. G. Davey, A. M. Geddes, I. D. Farrell, and G. R. Brooks, *Lancet* 620 (1980); W. T. Ulmer and I. Zimmerman, *J. Int. Med. Res.* **9,** 372 (1981); H. W. Van Landuyt, M. Pyckaret, and A. M. Lambert, *J. Antimicrob. Chemother.* **7,** 65 (1981); D. A. Leigh, K. Bradnock, and J. M. Marriner, *J. Antimicrob. Chemother.* **7,** 229 (1981); P. Ball, T. Watson, and S. Mehtar, *J. Antimicrob. Agents Chemother.* **7,** 441 (1981); G. A. J. DeKonig, D. Tio, J. F. Coster, R. A. Continno, and M. C. Ansink-Schipper, *J. Antimicrob. Agents Chemother.* **8,** 81 (1981).
86. K. R. Comber, R. Horton, S. J. Layte, A. R. White, and R. Sutherland in ref. 67, p. 19; K. R. Comber, R. Horton, L. Mizen, A. R. White, and R. Sutherland in J. D. Nelson and C. Grassi, eds., *Current Chemotherapy and Infectious Diseases*, Vol. 1, American Society for Microbiology, Washington, D.C., 1980, p. 343; R. J. Boon, A. S. Biele, K. R. Comber, C. V. Pierce, and R. Sutherland, *Antimicrob. Agents Chemother.* **22,** 369 (1982).
87. D. H. Staniforth, D. Jackson, and R. Horton, *Abstr. No.* 443 in *21st Interscience Conference on Antimicrobial Agents and Chemotherapy*, Chicago, Ill., 1981.
88. C. Reading and P. Hepburn, *Biochem. J.* **179,** 67 (1979).
89. C. Reading and T. Farmer, *Biochem. J.* **199,** 779 (1981).
90. J. Fisher, R. L. Charnas, and J. R. Knowles, *Biochemistry* **17,** 2180 (1978); R. L. Charnas, J. Fisher, and J. R. Knowles, *Biochemistry* **17,** 2185 (1978); R. L. Charnas and J. R. Knowles, *Biochemistry* **20,** 3214 (1981); R. L. Charnas, J. Fisher, and J. R. Knowles in N. Seiler, M. J. Jung, and J. Kosh-Weser, eds., *Enzyme-Activated Irreversible Inhibitors*, Elsevier, Amsterdam, The Netherlands, 1978, p. 315.
91. J. Fisher, J. G. Belaseo, R. L. Charnas, S. Khosla, and J. R. Knowles, *Philos. Trans. R. Soc. Lond. Ser. B* **289,** 309 (1980).
92. J. P. Durkin and T. Viswanatha, *J. Antibiot.* **31,** 1162 (1978).
93. P. Hunter, C. Reading, and co-workers, Beecham Pharmaceuticals, unpublished data.
94. A. G. Brown, *J. Antimicrob. Chemother.* **15,** 7 (1981).
95. P. H. Bentley and E. Hunt, *J. Chem. Soc. Perkin I*, 2222 (1980).
96. A. J. Eglington, *J. Chem. Soc. Chem. Commun.*, 720 (1977).
97. A. G. Brown, D. F. Corbett, and T. T. Howarth, *J. Chem. Soc. Chem. Commun.*, 359 (1977); R. G. Alexander and R. Southgate, *J. Chem. Soc. Chem. Commun.*, 405 (1977).
98. P. H. Bentley and E. Hunt, *J. Chem. Soc. Chem. Commun.*, 439 (1978).
99. P. H. Bentley, G. Brooks, M. L. Gilpin, E. Hunt, and I. I. Zomaya, *Tetrahedron Lett.*, 391 (1979); P. H. Bentley, G. Brooks, M. L. Gilpin, and E. Hunt, *J. Chem. Soc. Chem. Commun.*, 905 (1977); E. Hunt, P. H. Bentley, G. Brooks, and M. L. Gilpin, *J. Chem. Soc. Chem. Commun.*, 906 (1977); G. Brooks, T. T. Howarth, and E. Hunt, *J. Chem. Soc. Chem. Commun.*, 642 (1981); L. D. Cama and B. G. Christensen, *Tetrahedron Lett.*, 4233 (1978); S. Oida, A. Yoshida, and E. Ohki, *Chem. Pharm. Bull.* **26,** 448 (1978); T. Kobayashi, Y. Iwano, and K. Hirai, *Chem. Pharm. Bull.* **26,** 1761 (1978); M. D. Bachi and C. Hoornaert, *Tetrahedron Lett.* **22,** 2689, 2693 (1981).
100. P. H. Bentley, P. D. Berry, G. Brooks, M. L. Gilpin, E. Hunt, and I. I. Zomaya in ref. 41, p. 175.
101. S. W. Elson in ref. 41, p. 142; S. W. Elson and R. S. Oliver, *J. Antibiot.* **31,** 586 (1978); S. W. Elson, R. S. Oliver, B. W. Bycroft, and E. A. Faruk, *J. Antibiot.* **35,** 81 (1982); S. W. Elson, *Zol. Bakt. Suppl.* **11,** 436 (1981).
102. D. Butterworth in E. J. Vandamme, ed., *Antibiotic Production by Fermentation Biotechnology*, Marcel Dekker Inc., New York, 1982, in press.
103. D. Butterworth, J. D. Hood, and M. S. Verrall in A. Mizrahi, ed., *Advances in Biotechnological Processes*, Vol. 1, A. R. Liss Inc., New York, 1982, p. 251.
104. A. E. Bird, J. M. Bellis, and B. C. Gasson, *Analyst* **107,** 1241 (1982).
105. M. Foulstone and C. Reading, *Antimicrob. Agents Chemother.* **22,** 753 (1982).
106. J. Haginaka, T. Nakagawa, Y. Nishino, and Y. Uno, *J. Antibiot.* **34,** 1189 (1981).
107. H. Umezawa, S. Mitsuhashi, M. Hamada, S. Iyobe, S. Takahashi, R. Utahara, Y. Osato, S. Yamazadi, H. Omawara, and K. Maeda, *J. Antibiot.* **26,** 15 (1973).
108. K. Kitano, K. Nara, and Y. Nakao, *Japan J. Antibiot.* **30**(Suppl.), 239 (1977).

109. H. Aoki, K. Kunugita, J. Hosoda, and H. Imanada, *Japan J. Antibiot.* **30**(Suppl), 207 (1977).
110. K. Okamura, A. Koki, M. Sakamoto, K. Kubo, Y. Mutch, Y. Fukagawa, K. Kouno, T. Shimauchi, T. Ishikura, and J. Lein, *J. Ferment. Technol.* **57,** 265 (1979).
111. Eur. Pat. Appl. 38,534 (Apr. 15, 1981), K. Tanaka, N. Tsuji, E. Kondo, and Y. Kawamura (to Shionogi and Co., Ltd.).
112. U.S. Pat. 4,165,379 (Aug. 21, 1979) and 4,229,534 (Oct. 21, 1980), J. S. Kahan, F. M. Kahan, R. T. Goegelman, E. O. Stapley, and S. Hernandez (to Merck & Co. Inc.).
113. D. Rosi, M. L. Drozd, M. F. Kuhrt, J. Terminiello, P. E. Cane, and S. J. Daum, *J. Antibiot.* **34,** 341 (1981).
114. U.S. Pat. 4,162,323 (July 24, 1979), J. S. Kahan (to Merck & Co., Inc.).
115. D. Butterworth, M. Cole, G. Hanscomb, and G. N. Rolinson, *J. Antibiot.* **32,** 287 (1979).
116. J. D. Hood, S. J. Box, and M. S. Verrall, *J. Antibiot.* **32,** 295 (1979).
117. E. O. Stapley, P. J. Cassidy, J. Tunac, R. M. Monaghan, M. Jackson, S. Hernandez, J. M. Mata, S. A. Currie, D. Daoust, and D. Hendlin, *J. Antibiot.* **34,** 628 (1981).
118. P. J. Cassidy, G. Albers-Schonberg, R. T. Goegelman, T. Miller, B. Arison, E. O. Stapley, and J. Birnbaum, *J. Antibiot.* **34,** 637 (1981).
119. S. J. Box, G. Hanscomb, and S. R. Spear, *J. Antibiot.* **34,** 600 (1981).
120. N. Shibamoto, A. Koki, M. Nishino, K. Nakamura, K. Kiyoshima, K. Okamura, M. Okabe, R. Okamoto, Y. Fukagawa, Y. Shimauchi, T. Ishikura, and J. Lein, *J. Antibiot.* **33,** 1128 (1980).
121. N. Shibamoto, M. Nishino, K. Okamura, Y. Fukagawa, T. Ishikura, *J. Antibiot.* **35,** 763 (1982).
122. N. Tsuji, K. Nagashima, M. Kobayashi, Y. Terui, K. Matsumoto, and E. Kondo, *J. Antibiot.* **35,** 536 (1982).
123. S. J. Box, J. D. Hood, and S. R. Spear, *J. Antibiot.* **32,** 1239 (1979).
124. S. J. Box, D. F. Corbett, K. G. Robins, S. R. Spear, and M. S. Verrall, *J. Antibiot.* **35,** 1394 (1982); Eur. Pat. Appl. 43,197 (July 2, 1980), S. Box (to Beecham Group Ltd.).
125. Jpn. Pat. Kokai 37-002693 (June 6, 1980), (to Kyowa Hakko Kogyo K.K.).
126. S. Harada, Y. Nozaki, S. Shinagawa, and K. Kitano, *J. Antibiot.* **35,** 957 (1982).
127. K. Okamura, S. Hirata, A. Koki, K. Hori, N. Shibamoto, Y. Okumura, M. Okabe, R. Okamoto, K. Kouno, Y. Fukagawa, Y. Shimauchi, T. Ishikura, and J. Lein, *J. Antibiot.* **32,** 262 (1979); K. Okamura, S. Hirata, Y. Okumura, Y. Rukagawa, Y. Shimauchi, K. Kouno, T. Ishikura, and J. Lein, *J. Antibiot.* **31,** 480 (1978).
128. Y. Fukagawa, K. Kubo, T. Ishikura, and K. Kouno, *J. Antibiot.* **33,** 543 (1980).
129. T. Iot, N. Ezaki, K. Ohba, S. Amano, Y. Kondo, S. Miyadoh, T. Shamura, M. Sezaki, T. Niwa, M. Kojima, S. Inouye, Y. Yamada, and T. Niido, *J. Antibiot.* **35,** 533 (1982).
130. Eur. Pat. Appl. 48,999 (Oct. 1, 1981), M. Okabe, T. Yoshioka, Y. Fukagawa, R. Okamoto, K. Kouno, and T. Ishikura (to Sanraku-Ocean Co., Ltd.); *papers presented at 228th Research Congress of the Japanese Antibiotic Association*, Tokyo, Mar. 1982; M. Okabe, S. Azuma, I. Kojima, K. Kouno, R. Okamoto, Y. Fukagawa, and T. Ishikura, *J. Antibiot.* **35,** 1255 (1982); M. Sakamoto, I. Kojima, M. Okabe, Y. Fukagawa, and T. Ishidura, *J. Antibiot.* **35,** 1264 (1982); T. Yoshioka and co-workers, *Tetrahedron Lett.* **23,** 5177 (1982).
131. M. Nakayama, A. Iwasaki, S. Kimura, T. Mizogushi, S. Taube, A. Murakami, I. Watanabe, M. Okuchi, H. Itoh, Y. Saino, F. Kobayashi, and T. Mori, *J. Antibiot.* **33,** 1388 (1980).
132. A. Imada, Y. Nozaki, K. Kintaka, K. Okonogi, K. Kitano, and S. Harada, *J. Antibiot.* **33,** 1417 (1980).
133. K. Tanaka, J. Shoki, Y. Terui, N. Tsuji, E. Kondo, M. Mayama, Y. Kawamura, T. Hattori, K. Matsumoto, and T. Yoshida, *J. Antibiot.* **34,** 909 (1981); Y. Kawamura, Y. Yasuda, M. Mayama, and K. Tanaka, *J. Antibiot.* **35,** 10 (1982); J. Shoji, H. Hinoo, R. Sakazaki, N. Tsunji, K. Nagashima, K. Matsumoto, Y. Takahashi, S. Kozuki, T. Hattori, E. Kondo, and K. Tanaka, *J. Antibiot.* **35,** 15 (1982).
134. U.S. Pat. 4,247,640 (Jan. 27, 1981), A. J. Kempf and K. E. Wilson (to Merck & Co., Inc.).
135. W. L. Parker, M. L. Rathnum, J. S. Wells, W. H. Trejo, P. A. Principe, and R. B. Sykes, *J. Antibiot.* **35,** 653 (1982).
136. Eur. Pat. Appl. 50,961 (Oct. 22, 1981), M. Okuchi, M. Nakayama, A. Iwasaki, S. Kimura, T. Mizoguchi, S. Tanabe, A. Murakami, H. Ito, and T. Mori (to Kowa Co., Ltd.).
137. S. Tanabe and co-workers, *J. Antibiot.* **35,** 1237 (1982).
138. K. Maeda, S. Takahashi, M. Sezaki, K. Iinuma, H. Naganawa, S. Kondo, M. Ohno, and H. Umezawa, *J. Antibiot.* **30,** 770 (1977).
139. Jpn. Pat. Appl. 5029909 (Aug. 18, 1978), (to Sankyo).

140. U.S. Pat. 4,335,212 (June 15, 1982), K. E. Wilson and A. J. Kempf (to Merck & Co., Inc.).
141. M. Noble, D. Noble, and R. A. Fletton, *J. Antibiot.* **31,** 15 (1978); S. J. Box and D. F. Corbett, *Tetrahedron Lett.* **22,** 3239 (1981).
142. A. G. Brown, D. F. Corbett, A. J. Eglington, and T. T. Howarth, *J. Chem. Soc. Chem. Commun.*, 523 (1977).
143. D. F. Corbett, A. J. Eglington, and T. T. Howarth, *J. Chem. Soc. Chem. Commun.*, 953 (1977).
144. N. H. Grant, D. E. Clark, and H. E. Alburn, *J. Am. Chem. Soc.* **84,** 876 (1962); H. Bundgaard and C. Larsen, *J. Chromatogr.* **132,** 51 (1977).
145. S. Harada, S. Shinagawa, Y. Nozaki, M. Asai, and T. Kishi, *J. Antibiot.* **33,** 1425 (1980).
146. R. Weudman and A. Horeau, *Bull. Soc. Chim. Fr.*, 117 (1967).
147. J. A. Dale and H. S. Mosher, *J. Am. Chem. Soc.* **95,** 512 (1973); G. R. Sullivan and H. S. Mosher, *J. Org. Chem.* **38,** 2143 (1973).
148. A. G. Brown, D. F. Corbett, A. J. Eglington, and T. T. Howarth in ref. 41, p. 255.
149. A. G. Brown, D. F. Corbett, A. J. Eglington, and T. T. Howarth, *J. Antibiot.* **32,** 961 (1979).
150. K. Yamamoto, T. Yoshioka, Y. Kato, N. Shibamoto, K. Okamura, Y. Shimauchi, and T. Ishikura, *J. Antibiot.* **33,** 796 (1980).
151. N. Tsuji, K. Nagashima, M. Kobayashi, J. Shoji, T. Kato, Y. Terui, H. Nakai, and M. Shiro, *J. Antibiot.* **35,** 24 (1982).
152. P. J. Reider and E. J. J. Grabowski, *Tetrahedron Lett.* **23,** 2293 (1982).
153. Jpn. Pat. Appl. 57,011,982 (Jan. 26, 1980), (to Kowa Co., Ltd.).
154. D. H. Shih and R. W. Ratcliffe, *J. Med. Chem.* **24,** 639 (1981).
155. W. J. Leanza, K. J. Wildonger, J. Hannah, D. H. Shih, R. W. Ratcliffe, L. Barash, E. Walton, R. A. Firestone, G. F. Patel, F. M. Kahan, J. S. Kahan, and B. G. Christensen in ref. 41, p. 240.
156. D. H. Shih, J. Hannah, and B. G. Christensen, *J. Am. Chem. Soc.* **100,** 8004 (1978).
157. L. D. Cama and B. G. Christensen, *J. Am. Chem. Soc.* **100,** 8006 (1978).
158. D. F. Corbett, *J. Chem. Soc. Chem. Commun.*, 803 (1981).
159. Eur. Pat. Appl. 44,142 (July 3, 1980), D. F. Corbett (to Beecham Group Ltd.).
160. K. Yamanoto, T. Yoshioka, Y. Kato, K. Isshiki, M. Nishino, F. Nakamura, Y. Shimauchi, and T. Ishikura, *Tetrahedron Lett.* **23,** 897 (1982).
161. Brit. Pat. 1,577,725 (June 30, 1976), A. J. Eglington, T. T. Howarth, and D. F. Corbett (to Beecham Group Ltd.).
162. B. G. Christensen, *North American Medicinal Chemistry Symposium*, Toronto, Can., June 1982.
163. D. F. Corbett, S. Coulton, and R. Southgate, *J. Chem. Soc. Perkin I*, 3011 (1982).
164. D. F. Corbett and A. J. Eglington, *J. Chem. Soc. Chem. Commun.*, 1083 (1980).
165. A. J. Eglington in ref. 162.
166. U.S. Pat. 4,212,807 (Aug. 10, 1978).
167. Eur. Pat. Appl. 2058 (Nov. 16, 1978), K. Okamura, S. Hirata, Y. Okumura, Y. Fukagawa, Y. Shamauchi, T. Ishikura, K. Kouno, and J. Lein (to Sanraku-Ocean Co., Ltd.).
168. U.S. Pats. 4,282,322 and 4,264,734 (Aug. 4, 1981), J. S. Kahan and F. M. Kahan (to Merck & Co., Inc.).
169. K. Kubo, T. Ishikura, and Y. Fukagawa, *J. Antibiot.* **33,** 550, 556 (1980).
170. W. J. Leanza, K. J. Wildonger, T. W. Miller, and B. G. Christensen, *J. Med. Chem.* **22,** 1435 (1979); U.S. Pat. 4,260,543 (Apr. 7, 1981), T. W. Miller (to Merck & Co., Ltd.).
171. P. J. Cassidy, *Dev. Ind. Microbiol.* **22,** 181 (1981).
172. H. Kropp, J. S. Kahan, F. M. Kahan, J. Sundelof, G. Darland, and J. Birnbaum, *Abstr. No. 228* in *16th Interscience Conference on Antimicrobial Agents and Chemotherapy*, Chicago, Ill., 1976.
173. F. P. Tally, N. V. Jacobus, and S. L. Gorbach, *Antimicrob. Agents Chemother.* **14,** 436 (1978); S. S. Weaver, G. P. Bodey, and B. M. LeBlanc, *Antimicrob. Agents Chemother.* **15,** 518 (1979); M. F. Romognoli, R. P. Fu, and H. C. Neu, *J. Antimicrob. Chemother.* **6,** 601 (1980); A. Rodriguez, T. Olay, and M. V. deVincenti in J. D. Nelson and C. Gassi, eds., *Current Chemotherapy and Infectious Disease, American Society for Microbiology*, Washington, D.C., 1980, p. 492; S. D. R. Lang and D. T. Durack, *Clin. Ther.* **3,** 112 (1980); V. Fainstein, B. LeBlanc, S. Weaver, and G. F. Bodey, *Infection* **10,** 50 (1982).
174. M. J. Basker, R. J. Boon, and P. A. Hunter, *J. Antibiot.* **33,** 878 (1980).
175. M. Sakamoto, H. Iguchi, K. Okamura, S. Hori, Y. Fukagawa, T. Ishikura, and J. Lein, *J. Antibiot.* **32,** 272 (1979).
176. M. Sakamoto, N. Shibamoto, H. Iguchi, K. Okamura, S. Hori, Y. Fukagawa, T. Ishikura, and J. Lein, *J. Antibiot.* **33,** 1138 (1980).

177. Y. Fukagawa, K. Okamura, N. Shibamoto, and T. Ishikura in S. Mitsuhashi, ed., *Beta-Lactam Antibiotics*, Japan Scientific Societies Press, Tokyo, 1981, p. 158.
178. F. Kobayashi, Y. Saino, T. Koshi, Y. Hattori, M. Nakayama, A. Iwasaki, T. Mori, and S. Mitsuhashi, *Antimicrob. Agents Chemother.* **21,** 536 (1982); Y. Kimura, K. Motokawa, H. Nagata, Y. Kameda, S. Matsuura, M. Mayama, and T. Yoshida, *J. Antibiot.* **34,** 32 (1982).
179. L. D. Cama and B. G. Christensen in ref. 177, p. 164.
180. H. Kropp, J. G. Sundelof, J. S. Kahan, F. M. Kahan, and J. Birnbaum, *Antimicrob. Agents Chemother.* **17,** 993 (1980); T. Kesado, T. Hashizume, and Y. Asahi, *Antimicrob. Agents Chemother.* **17,** 912 (1980).
181. H. Kropp, J. G. Sundelof, R. Hajdu, and F. M. Kahan, *Antimicrob. Agents Chemother.* **22,** 62 (1982); H. Mikami, M. Ogashiwa, Y. Saino, M. Inone, and S. Mitsuhashi, *Antimicrob. Agents Chemother.* **22,** 693 (1982).
182. N. Shibamoto, M. Sakamoto, H. Iguchi, H. Toni, Y. Fukagawa, and T. Ishikura, *J. Antibiot.* **35,** 721 (1982); N. Shibamoto, M. Sakamoto, Y. Fukagawa, and T. Ishikura, *J. Antibiot.* **35,** 729 (1982); N. Shibamoto, T. Yoshioka, M. Sakamoto, Y. Fukagawa, and T. Ishikura, *J. Antibiot.* **35,** 736 (1982).
183. M. J. Basker, R. J. Boon, S. J. Box, E. A. Prestige, G. M. Smith, and S. R. Spear, *J. Antibiot.* **36,** 416 (1983).
184. Y. Kimura, K. Motokawa, H. Nagata, Y. Kameda, S. Matsuura, M. Mayama, and T. Yoshida, *J. Antibiot.* **35,** 32 (1982).
185. V. M. Giryavallabhan, A. K. Ganguly, S. W. McCombie, P. Pinto, and R. Rizvi, *Abstr. No. 829, 21st Interscience Conference on Antimicrobial Agents and Chemotherapy*, Chicago, Ill.; *J. Antimicrob. Chemother.* **9**(Suppl. C), (1982).
186. V. W. Horadam, J. M. Smilack, C. L. Montgomery, and J. Werringloer, *Antimicrob. Agents Chemother.* **18,** 557 (1980); F. P. Tally, N. V. Jacobus, and S. L. Gorbach, *Antimicrob. Agents Chemother.* **18,** 642 (1980); M. Toda, K. Sato, H. Nakaya, M. Inoue, and S. Mitsuhashi, *Antimicrob. Agents Chemother.* **18,** 837 (1980); M. L. Corrado, S. H. Landesman, and C. E. Cherubin, *Antimicrob. Agents Chemother.* **18,** 893 (1980); W. K. Livingstone, A. M. Elliott, and C. G. Cobbs, *Antimicrob. Agents Chemother.* **19,** 114 (1981); S. Shadomy and R. S. May, *Antimicrob. Agents Chemother.* **19,** 201 (1981); R. Wise, J. M. Andrews, and N. Patel, *J. Antimicrob. Chemother.* **7,** 521 (1981); J. E. Brown, V. E. Del Bene, and C. D. Collins, *Antimicrob. Agents Chemother.* **19,** 248 (1981); L. Verbist and J. Verhaegen, *Antimicrob. Agents Chemother.* **19,** 402 (1981); D. Hanslo, A. King, K. Shannon, C. Warren, and I. Phillips, *J. Antimicrob. Chemother.* **7,** 607 (1981); M. L. Corrado, C. E. Cherubin, M. Shulman, J. Moen, and N. Jhagroo, *J. Antimicrob. Chemother.* **7,** 677 (1981); G. M. Eliopoulos and R. C. Moellering, Jr., *Antimicrob. Agents Chemother.* **19,** 789 (1981); V. A. Tutlane, R. V. McLosky, and J. A. Trent *Antimicrob. Agents Chemother.* **20,** 140 (1981); M. Nasu, J. P. Maskell, R. J. Williams, and J. D. Williams, *Antimicrob. Agents Chemother.* **20,** 433 (1981); C. E. Cherubin, M. L. Corrado, M. F. Sierra, M. E. Gombert, and M. Shulman, *Antimicrob. Agents Chemother.* **20,** 553 (1981); M. V. Borobio, M. C. Nogales, A. Pascual, and E. J. Perea, *J. Antimicrob. Chemother.* **8,** 213 (1981); D. M. Livingstone, R. J. Williams, and J. D. Williams, *J. Antimicrob. Chemother.* **8,** 355 (1981); P. R. Michael, R. H. Alford, and Z. A. McGee, *Antimicrob. Agents Chemother.* **20,** 702 (1981); M. H. Cynamon and G. S. Palmer, *Antimicrob. Agents Chemother.* **20,** 841 (1981); D. A. Martin, C. V. Sanders, and R. L. Marier, *Antimicrob. Agents Chemother.* **21,** 168 (1982); H. C. Neu and P. Labthavikul, *Antimicrob. Agents Chemother.* **21,** 180 (1982); D. L. Cohn, L. G. Reimer, and L. B. Reller, *J. Antimicrob. Chemother.* **9,** 183 (1982); I. Braveny, K. Machka, and R. Elsser, *Infection* **10,** 53 (1982); A. G. Altes, M. D. Enciso, P. P. Garcia, and A. C. Bueno, *Antimicrob. Agents Chemother.* **21,** 501 (1982); J. Butierrez-Nunez, P. T. Harrington, and C. H. Ramirez-Ronda, *Antimicrob. Agents Chemother.* **21,** 509 (1982); P. Patamasucon and G. H. McCracken Jr., *Antimicrob. Agents Chemother.* **21,** 390 (1982); R. L. Thomson, K. A. Fisher, and R. P. Wenzel, *Antimicrob. Agents Chemother.* **21,** 341 (1982); M. A. Miller, J. L. LeFrock, and M. J. Vercler, *Microbiol. Immunol.* **25,** 1119 (1981); E. D. O'Donnell, E. H. Freimer, G. L. Gilardi, and R. Raeder, *Antimicrob. Agents Chemother.* **21,** 673 (1982); A. J. Matzkowitz, A. L. Batch, R. P. Smith, N. T. Sutphen, M. C. Hammer, and J. V. Conroy, *Antimicrob. Agents Chemother.* **21,** 685 (1982); N. Gombert, *Antimicrob. Agents Chemother.* **21,** 1011 (1982); T. Kesado, K. Watanabe, Y. Asahi, M. Isono, and K. Ueno, *Antimicrob. Agents Chemother.* **21,** 1016 (1982); V. I. Ahoukhai, M. F. Sierra, C. E. Cherubin, and M. A. Shulman, *J. Antimicrob. Chemother.* **9,** 411 (1982); M. H. Cynamon and P. A. Cranato, *Chemotherapy* **28,** 204 (1982); S. B. Calderwood, A. Gardella, A. M. Philippon, G. A. Jacoby, and R. C. Moellering, *Antimicrob. Agents Chemother.* **22,** 266 (1982); W. Cullman, W. Opferkuck, M. Stieglitz, and V. Werkmeister, *Antimicrob. Agents Chemother.* **22,**

302 (1982); J. E. Pennington and C. E. Johnson, *Antimicrob. Agents Chemother.* **22,** 406 (1982); R. Auckenthale, W. R. Wilson, A. J. Wright, J. A. Washington, D. T. Durack, and J. E. Geraci, *Antimicrob. Agents Chemother.* **22,** 448 (1982); A. Vrize, *Chemotherapy* **28,** 267 (1982); V. H. Grimm, *Arzneim-Forsch.* **32,** 595 (1982); J. A. Garcia-Rodriguez, M. C. Sainz, J. E. Garcia-Sanchez, and J. Riets, *Drug Exptl. Chim. Res.* **8,** 217 (1981).

187. Eur. Pat. Appl. 28,778 (Oct. 31, 1980), F. M. Kahan and H. Kropp; Eur. Pat. Appl. 10,573 (July 24, 1979), 48,301 (Oct. 24, 1980), D. W. Graham, F. M. Kahan, E. F. Rogers; Eur. Pat. Appl. 6,639 (July 3, 1979), T. W. Miller; Eur. Pat. Appl. 7,614 (July 24, 1979), F. M. Kahan and H. Kropp (to Merck & Co., Inc.).
188. H. Kropp, J. G. Sundelof, D. L. Bohn, and F. M. Kahan, *Abstr. No. 270;* W. T. Ashton, L. Barash, J. E. Brown, R. D. Brown, L. F. Canning, A. Chen, D. W. Graham, F. M. Kahan, H. Kropp, J. G. Sundelof, and E. F. Rogers, *Abstr. No. 271;* and H. Kropp, J. G. Sundelof, R. Hajdu, and F. M. Kahan, *Abstr. No. 272* in *20th Interscience Conference on Antimicrobial Agents and Chemotherapy*, New Orleans, La., 1980.
189. F. Follath, A. M. Geddes, P. Spring, G. D. Ball, K. H. Jones, F. Ferber, J. S. Kahan, and F. M. Kahan, *Abstr. No. 590;* F. Ferber, K. Alestig, F. Kahan, K. Jones, J. Kahan, M. Meisinger, J. Rogers, and R. Norrby, *Abstr. No. 591;* R. Norrby, K. Alestig, B. Bjornegard, L. Burman, F. Ferber, F. Kahan, J. Huber, and K. Jones, *Abstr. No. 592;* and K. H. Jones, K. Alestig, F. Ferber, J. Huber, F. M. Kahan, M. A. P. Meisinger, J. D. Rogers, and R. Norrby, *Abstr. No. 593* in *21st Interscience Conference on Antimicrobial Agents and Chemotherapy*, Chicago, Ill., 1981; R. Norrby, K. Alestig, F. Kahan, J. Kahan, H. Kropp, F. Ferber, and M. Meisinger, *Abstr. No. 338* in *12th International Congress of Chemotherapy*, Florence, Italy, 1981.
190. K. Okonogi, S. Harada, S. Shinagawa, A. Imada, and M. Kuno, *J. Antibiot.* **35,** 963 (1982).
191. K. Okamura, M. Sakamoto, Y. Fukagawa, and T. Ishikura, *J. Antibiot.* **32,** 280 (1979).
192. K. Okonogi, Y. Nozaki, A. Imada, and M. Kuno, *J. Antibiot.* **34,** 212 (1981).
193. K. Murakami, M. Doi, and T. Yoshida, *J. Antibiot.* **35,** 39 (1982).
194. N. Tsuji, E. Kondo, M. Mayama, Y. Kawamura, T. Hattori, K. Matsumoto, and T. Yoshida, *J. Antibiot.* **34,** 909 (1981).
195. C. J. Easton and J. R. Knowles, *Biochemistry* **21,** 2857 (1982); R. L. Charnas and J. R. Knowles, *Biochemistry* **20,** 2732 (1981).
196. T. Kametani, *Heterocycles* **17,** 463 (1982).
197. J. H. Bateson, A. J. G. Baxter, K. H. Dickinson, R. I. Hickling, R. J. Ponsford, T. C. Smale, and R. Southgate in ref. 41, p. 291.
198. A. J. G. Baxter, K. H. Dickinson, P. M. Roberts, T. C. Smale, and R. Southgate, *J. Chem. Soc. Chem. Commun.*, 236 (1979).
199. J. H. Bateson, A. J. G. Baxter, P. M. Roberts, T. C. Smale, and R. Southgate, *J. Chem. Soc. Perkin I*, 3242 (1981).
200. R. J. Ponsford, P. M. Roberts, and R. Southgate, *J. Chem. Soc. Chem. Commun.*, 847 (1979).
201. A. J. G. Baxter, P. Davis, R. J. Ponsford, and R. Southgate, *Tetrahedron Lett.* **21,** 5071 (1980).
202. M. J. Basker, J. H. Bateson, A. J. G. Baxter, R. J. Ponsford, P. M. Roberts, R. Southgate, T. C. Smale, and J. Smith, *J. Antibiot.* **34,** 1224 (1981).
203. A. J. G. Baxter, R. J. Ponsford, and R. Southgate, *J. Chem. Soc. Chem. Commun.*, 429 (1980).
204. J. H. Bateson, P. M. Roberts, T. C. Smale, and R. Southgate, *J. Chem. Soc. Chem. Commun.*, 185 (1980).
205. J. H. Bateson, R. I. Hickling, P. M. Roberts, T. C. Smale, and R. Southgate, *J. Chem. Soc. Chem. Commun.*, 1084 (1980).
206. R. J. Ponsford and R. Southgate, *J. Chem. Soc. Chem. Commun.*, 1085 (1980).
207. *Ibid.*, 846 (1979).
208. D. B. R. Johnston, S. M. Schmitt, F. A. Bouffard, and B. G. Christensen, *J. Am. Chem. Soc.* **100,** 313 (1978); F. A. Bouffard, D. B. R. Johnston, and B. C. Christensen, *J. Org. Chem.* **45,** 1130 (1980); S. M. Schmitt, D. B. R. Johnston, and B. G. Christensen, *J. Org. Chem.* **45,** 1135, 1142 (1980).
209. T. N. Salzmann, R. W. Ratcliffe, B. G. Christensen, and F. A. Bouffard, *J. Am. Chem. Soc.* **102,** 6162 (1980).
210. T. N. Salzmann, R. W. Ratcliffe, F. A. Bouffard, and B. G. Christensen, *Philos. Trans. R. Soc. London Ser. B* **289,** 191 (1980).
211. F. A. Bouffard and B. G. Christensen, *J. Org. Chem.* **46,** 2208 (1981).
212. M. Sletzinger, T. Liu, R. A. Reamer, and I. Shinai, *Tetrahedron Lett.* **21,** 4221 (1980).
213. S. Karady, J. S. Amato, R. A. Reamer, and L. M. Weinstock, *J. Am. Chem. Soc.* **103,** 6765 (1981).

214. D. G. Melillo, I. Shinkai, T. Liu, K. Ryan, and M. Sletzinger, *Tetrahedron Lett.* **21,** 2783 (1980); D. G. Melillo, T. Liu, K. Ryan, M. Sletzinger, and I. Shinkai, *Tetrahedron Lett.* **22,** 913 (1981).
215. U.S. Pat. 4,324,900 (Apr. 13, 1982), P. L. Durette (to Merck & Co., Inc.); N. Ikota, O. Yoshino, and K. Koga, *Chem. Pharm. Bull.* **30,** 1929 (1981); P. L. Durette, *Carbohydr. Res.* **100,** C27 (1982).
216. T. Kametani, S-P Huang, S. Yokohama, Y. Suzuki, and M. Ihara, *J. Am. Chem. Soc.* **102,** 2060 (1980); T. Kametani, T. Nagahara, S. Yokohama, S-P. Huang, and M. Ihara, *Tetrahedron Lett.* **37,** 715 (1981); T. Kametani, S-P. Huang, T. Nagahara, S. Yokohama, and M. Ihara, *J. Chem. Soc. Perkin I,* 964 (1981).
217. U.S. Pats. 4,262,009, 4,262,010, and 4,262,011 (Apr. 14, 1981), B. G. Christensen and D. Shih (to Merck & Co., Inc.).
218. M. Shibuya and S. Kubota, *Tetrahedron Lett.* **21,** 4009 (1980); M. Shibuya, M. Kuretani, and S. Kubota, *Tetrahedron* **38,** 2659 (1982).
219. M. W. Foxton, R. C. Mearman, C. E. Newall, and P. Ward, *Tetrahedron Lett.* **22,** 2497 (1981).
220. Eur. Pat. Appl. 40,494 (May 14, 1980), R. L. Rosati (to Pfizer Inc.); R. L. Rosati, L. V. Kapili, and P. Morrissey, *J. Am. Chem. Soc.* **104,** 4262 (1982).
221. M. Shibuya, M. Kuretani, and S. Kubuto, *Tetrahedron Lett.* **22,** 4453 (1981).
222. E. M. Gordon, J. Plušĉec, and M. A. Ondetti, *Tetrahedron Lett.* **22,** 871 (1981).
223. C. W. Greengrass, D. W. T. Hoople, and M. S. Nobbs, *Tetrahedron Lett.* **22,** 2419 (1982).
224. G. Albers-Schonberg, B. H. Arison, E. Kaczka, F. M. Kahan, B. Lago, W. M. Maiese, R. E. Rhodes, and J. L. Smith, *Abstr. No. 229* in *16th Interscience Conference on Antimicrobial Agents and Chemotherapy*, Chicago, Ill., 1976.
225. F. Foor, B. Tyle, and N. Morin, *Dev. Ind. Microbiol.* **23,** 305 (1982); G. Lilley, A. E. Clark, and G. C. Lawrence, *J. Chem. Technol. Biotechnol.* **31,** 127 (1982).
226. U.S. Pat. 4,168,268 (Sept. 18, 1979), R. Datta and G. Wildman (to Merck & Co., Inc.).
227. L. O. Treiber, V. P. Cullo, and I. Tutter, *Biotechnol. Bioeng.* **23,** 1255 (1981).
228. J. D. Hood in J. D. Bu'lock, L. J. Nisbet, and D. J. Winstanley, eds., *Bioactive Microbial Products: Search and Discovery, Special Publication for the Society of General Microbiology*, No. 6, Academic Press, London, 1982, p. 131.
229. S. J. Cartwright and A. F. W. Coulson, *Nature (London)* **278,** 360 (1979).
230. R. F. Pratt and M. J. Loosemoore, *Proc. Natl. Acad. Sci. USA* **75,** 4145 (1978); M. J. Loosemore and R. F. Pratt, *J. Org. Chem.* **43,** 3611 (1978).
231. V. Knott-Hunziker, B. S. Orlek, P. G. Sammes, and S. G. Waley, *Biochem. J.* **177,** 365 (1979); V. Knott-Hunziker, S. G. Waley, B. S. Orlek, and P. G. Sammes, *FEBS Lett.* **99,** 59 (1979); B. S. Orlek, P. G. Sammes, V. Knott-Hunziker, and S. G. Waley, *J. Chem. Soc. Chem. Commun.*, 962 (1979).
232. W. von Daehne, *J. Antibiot.* **33,** 451 (1980).
233. J. E. G. Kemp in ref. 41, p. 320.
234. J. E. G. Kemp, M. D. Closier, S. Narayanaswami, and M. H. Stefaniak, *Tetrahedron Lett.* **21,** 2991 (1980).
235. W. J. Gottstein, L. B. Crast, Jr., R. G. Graham, U. J. Raynes, and D. N. McGregor, *J. Med. Chem.* **24,** 1531 (1981).
236. J. Fisher, R. L. Charnas, S. M. Bradley, and J. R. Knowles, *Biochemistry* **20,** 2726 (1981).
237. P. S. F. Mezes, A. J. Clarke, G. I. Dmitrienko, and T. Viswanatha, *J. Antibiot.* **35,** 918 (1982); P. S. F. Mezes, R. W. Friesen, T. Viswanatha, and G. I. Dmitrienko, *Heterocycles* **19,** 1207 (1982).
238. E. M. Gordon, H. W. Chang, C. M. Cimarusti, B. Toeplitz, and J. Z. Congoutas, *J. Am. Chem. Soc.* **102,** 1690 (1980).
239. R. A. Volkmann, R. D. Carroll, R. B. Drolet, M. L. Elliott, and B. S. Moore, *J. Org. Chem.* **47,** 3344 (1982).
240. P. A. Hunter and J. P. Webb, *Current Chemother. Infect. Disease* **1,** 340 (1980).
241. N. Aswapokee and H. C. Neu, *J. Antibiot.* **31,** 1238 (1978); K. P. Fu and H. C. Neu, *Antimicrob. Agents Chemother.* **15,** 171 (1979); J. A. Retsema, A. R. English and A. E. Girard, *Antimicrob. Agents Chemother.* **17,** 615 (1980); H. C. Neu, *Abstr. No. 601;* T. Yokota, R. Sekiguchi, and E. Azuma, *Abstr. No. 602;* and R. J. Fass, *Abstr. No. 605* in *20th Interscience Conference on Antimicrobial Agents and Chemotherapy*, New Orleans, La., 1980; A. J. Howard, A. J. Hince, and J. D. Williams, *Drugs. Exptl. Clin. Res.* **5,** 7 (1979); R. G. Washburn and D. T. Durack, *J. Infect. Dis.* **144,** 237 (1981).
242. D. G. Brenner and J. R. Knowles, *Biochemistry* **20,** 3680 (1981); C. Kemal and J. R. Knowles, *Biochemistry* **20,** 3688 (1981).
243. G. Foulds, W. E. Barth, J. R. Bianchine, A. R. English, D. Girard, S. L. Heyes, M. M. O'Brian, and P. Somani in J. D. Nelson and C. Grassi, eds., *Current Chemother.* **1,** 353 (1980).

244. B. Baltzer, E. Binderup, W. von Daehne, W. O. Godtfredsen, K. Hansen, B. Nielsen, H. Sprensen, and S. Vangedal, *J. Antibiot.* **33,** 1183 (1980); UK Pat. Appl. 2044255 (Jan. 25, 1980), W. von Daehne and W. O. Godtfredsen (to Leo Pharmaceutical Products Ltd.).
245. S. Hartley and R. Wise, *J. Antimicrob. Chemother.* **10,** 49 (1982).
246. V. Knott-Hunziker, B. S. Orlek, P. G. Sammes, and S. G. Waley in ref. 70, p. 217; B. S. Orlek, P. G. Sammes, V. Knott-Hunziker, and S. G. Waley, *J. Chem. Soc. Perkin 1,* 2322 (1980).
247. J. A. Aimetti, E. S. Hamanaka, D. A. Johnson, and M. S. Kellog, *Tetrahedron Lett.*, 4631 (1979).
248. D. I. John, E. J. Thomas, and N. D. Tyrrell, *J. Chem. Soc. Chem. Commun.*, 345 (1979).
249. R. Wise, J. M. Andrews, and N. Patel, *J. Antimicrob. Agents Chemother.* **7,** 531 (1981); B. A. Moore and K. W. Brammer, *Antimicrob. Agents Chemother.* **20,** 327 (1981).
250. M. J. Loosemore, S. A. Cohen, and R. F. Pratt, *Biochemistry* **19,** 3990 (1980).

General Reference

R. W. Ratcliffe and G. Albers-Schönberg, "The Chemistry of Thienamycin and Other Carbapenem Antibiotics" and P. C. Cherry and C. E. Newall, "Clavulanic Acid" in R. B. Morin and M. Gorman, eds., *Chemistry and Biology of β-Lactam Antibiotics*, Academic Press, Inc., New York, 1982, Chapts. 4 and 6.

ALLAN BROWN
Beecham Research Laboratories

MONOBACTAMS

The monobactams are monocyclic β-lactam antibiotics characterized by the 2-oxoazetidine-1-sulfonic acid moiety first encountered in gram-negative bacteria. They are the only family of β-lactam antibiotics produced solely by gram-negative bacteria. To date, eight monobactams have been reported as natural products; seven contain a 3*R*-methoxyl group and a single nonmethoxylated compound. Utilizing supersensitive screening strains of different bacteria, groups at Takeda Chemical Industries in Japan and The Squibb Institute for Medical Research in the United States (1) initially encountered different antibiotics characterized by the 2-oxoazetidine-1-sulfonic acid moiety. The term monobactam was coined to distinguish these *mono*cyclic, *bac*terially produced β-lac*tam* antibiotics (2).

During the 1960s and 1970s, complex screening procedures were devised to identify the seemingly limitless number of microbial isolates that produced antibacterial substances. The development of mechanism of action screens, which separated antibiotics according to a particular biochemical mechanism (inhibition of cell-wall synthesis in the case of β-lactams), allowed the sorting of huge numbers of positives from a typical soil-screening program. A second pathway, utilized particularly by several groups in the UK, involved the inhibition of the enzyme β-lactamase as a potential indicator of novel β-lactam structures. A third strategy centered on the development of strains of bacteria that were supersensitive to β-lactam antibiotics but of normal sensitivity to other antimicrobial agents. The latter technique led to the detection of the monobactams in gram-negative bacteria.

Naturally Occurring Monobactams

The penicillins and cephalosporins were first encountered in eukaryotic fungi. Cephamycins, clavulanic acid, and various carbapenem antibiotics are produced by actinomycetes, whereas the monobactams are produced exclusively by various strains of gram-negative bacteria. The naturally occurring monobactams isolated and characterized to date are given in Table 1 along with their producing organisms. The structures of these compounds are given in Figure 1.

Table 1. Naturally Occurring Monobactams

Name	CAS Registry No.	Producing organism	Structure	Refs.
SQ 26,180[a]	[79720-08-4]	*Chromobacterium violaceum*	(1)	2–4
sulfazecin	[77912-79-9]	*Pseudomonas acidophila*	(2)	5–7
SQ 26,445		*Gluconobacter oxydans*	(2)	2
isosulfazecin	[77900-75-5]	*Pseudomonas mesoacidophila*	(3)	5, 8
EM 5400		*Agrobacterium radiobacter*	(4)–(8)	9–10

[a] 3-(Acetylamino)-3-methoxy-2-oxo-1-azetidinesulfonic acid, monopotassium salt.

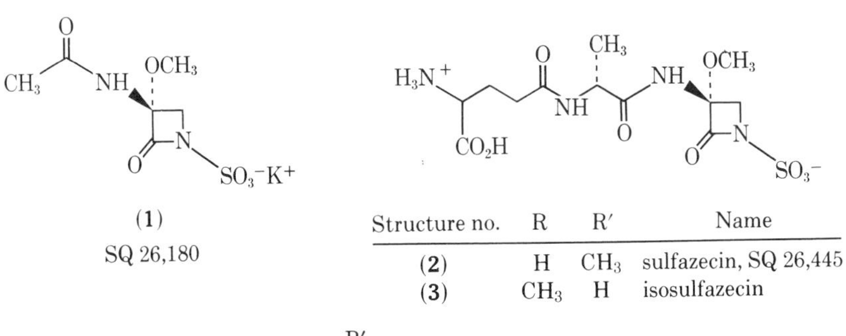

Structure no.	R	R′	Name
(2)	H	CH_3	sulfazecin, SQ 26,445
(3)	CH_3	H	isosulfazecin

EM 5400 components

Structure no.	CAS Registry No.	R	R′	R″
(4)	[79720-13-1]	CH_3O	H	H
(5)	[79720-14-2]	CH_3O	H	OH
(6)	[81919-28-0]	H	H	OH
(7)	[79720-16-4]	CH_3O	OH	$OSO_3^-Na^+$
(8)	[79720-17-5]	CH_3O	$OSO_3^-Na^+$	$OSO_3^-Na^+$

Figure 1. Structures of naturally occurring monobactams.

Of special note is the superficial similarity between monobactams and the nocardicin family of monocyclic β-lactam antibiotics isolated from a species of *Norcardia*. The similarity does not extend beyond the obvious relationship of structure, ie, each family contains a monocyclic 3-acylamino-2-oxoazetidine moiety [see structures (**9**) and (**10**)]. The extremely limited antibacterial spectrum of nocardicin A, the presence of a relatively unactivated β-lactam ring, and the failure of typical penicillin or cephalosporin side chains to elicit useful activity when appended to the nocardicin nucleus stand in stark contrast to the properties of the monobactams discussed below.

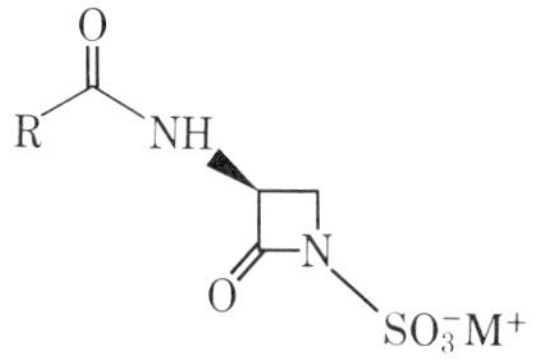

(**9**)

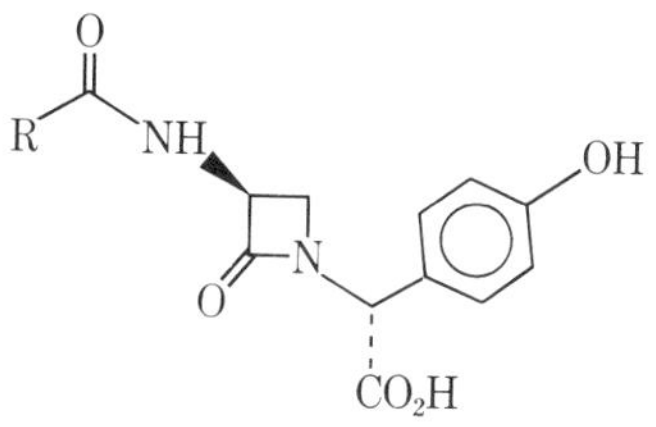

(**10**)

R = CH_3, M = K [*79720-10-8*]
R = $CH_2C_6H_5$, M = Na [*78611-12-8*]
and R = $OCH_2C_6H_5$, M = H [*78611-02-6*]

Comparison with Bicyclic β-Lactams

Structure and Stereochemistry. The simplest monobactams, SQ 26,180 (**1**), sulfazecin (**2**), and isosulfazecin (**3**), were the first to be isolated and characterized. Initial reports in the patent literature (11) were followed by the announcement of the isolation and characterization of the monocyclic β-lactam antibiotics (**1**)–(**8**) from bacteria (2,5). The structure and absolute stereochemistry of (**2**) was unequivocally established by x-ray crystallography (12), whereas (**3**) was characterized by reference to (**2**) (8). The structure and absolute stereochemistry of SQ 26,180 (**1**) was established by synthesis from a precursor of defined structure (3). The structure and absolute configuration of the nonmethoxylated, naturally occurring monobactam (**6**) was also proven by total synthesis (9). The assignment of structures to (**4**), (**5**), (**7**), and (**8**) rests ultimately on spectroscopic comparison to SQ 26,180 (**1**).

The naturally occurring monobactams correspond in configuration to the naturally occurring cephalosporins and cephamycins [see structures (**11**) and (**12**)]. The

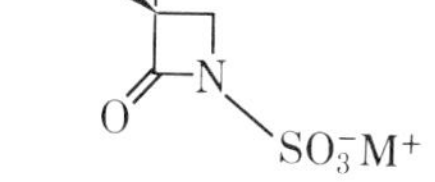

(monobactams)
(**11**) R = H, CH_3O
R′ = acyl

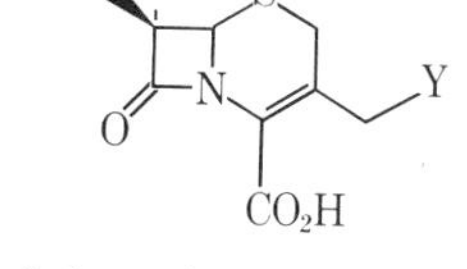

(cephalosporins and cephamycins):
(**12**) R = H, CH_3O
R′ = γ-linked α-aminoadipoyl

parallel does not, however, extend much beyond this point. Monobactams have been encountered in nature with side chains, eg, R (R ⇒ R′), in (**11**) ranging from simple acetyl to complex polypeptides (13).

On the other hand, cephalosporins and cephamycins always contain the γ-linked, D-α-aminoadipoyl [$NH_3^+CH(CO_2^-)$ unit *A* as R′. The presence in (2) and (3) of the D-γ-glutamylalanyl unit *B* suggests a biosynthetic link.

H_3N^+ CO_2^- O *A*

H_3N^+ CO_2^- O NH CH_3 O *B*

Similar to bicyclic β-lactams (14), monobactams of opposite absolute configuration to those encountered in nature are inactive. The enantiomeric pairs of monobactams (13)–(14) and (15)–(16) provide a striking illustration of this specificity (13).

C_6H_5 O NH * N O $SO_3^-K^+$

Structure no.	CAS Registry No.	Asymmetry, *	MIC, μg/mL, for *M. luteus*
(13)	*[80543-46-0]*	*S*	1.6
(14)	*[78612-60-9]*	*R*	>200

S N NH_2 O N O CH_3 NH * N O $SO_3^-K^+$

Structure no.	CAS Registry No.	Asymmetry, *	MIC, μg/mL, for *P. rettgeri*
(15)	*[78626-19-4]*	*S*	<0.05
(16)	*[78612-61-0]*	*R*	>100

β-Lactamase and Chemical Stability. In the penicillins, incorporation of a 6α-methoxyl substituent increases chemical stability three- to fivefold as measured by basic hydrolysis of the β-lactam moiety (15). Similar substitution of a 7α-methoxyl substituent in a cephalosporin results in a more modest 30% increase in chemical stability. Methoxylated monobactams with simple side chains, however, are much less stable than the nonmethoxylated compounds. Stability data for the pair of monobactams (13) and (17) after 24 h at 37°C and pH 7 illustrate this point (13).

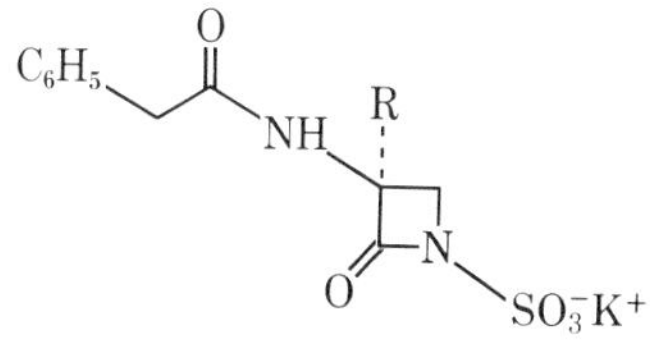

Structure no.	CAS Registry No.	R	% Remaining
(13)		H	>80
(17)	*[78612-58-5]*	CH_3O	<1

As demonstrated for cephalosporins (16), introduction of a 7α-methoxyl group increases the stability to β-lactamases and to chemical agents. The data below clearly show that increased stability to β-lactamase is also demonstrated by methoxylated monobactams when compared to their unsubstituted analogues. The compound SQ 26,180 (1) is much more stable to β-lactamases, including the clinically important RTEM lactamase (17), than its nonmethoxylated counterpart SQ 26,396 *[79720-10-8]* (18).

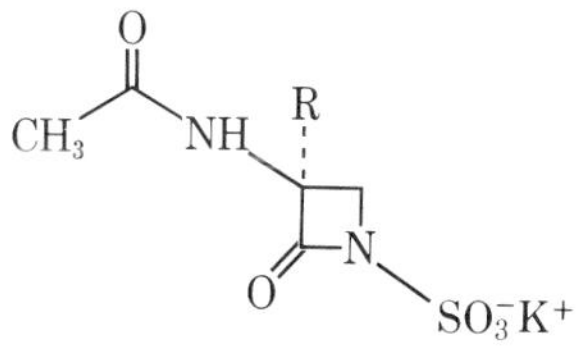

Structure no.	CAS Registry No.	R	Relative V_{max} (RTEM)
(1)		CH_3O	<0.01
(18)	*[79720-10-8]*	H	12
cephaloridine			100

V_{max} = rate at saturating substrate concentration.

This increased β-lactamase stability is thus displayed by the chemically less-stable compound of the pair. This remarkable example of the importance of enzymatic fit, relative to inherent reactivity, is discussed generally in ref. 18.

Mode of Action. The monobactams, like penicillins and cephalosporins, interfere with the synthesis of bacterial cell-wall peptidoglycan. The enzymatic steps involved in the insertion of newly synthesized cell-wall polymers into the expanding peptidoglycan may play a role in the action of penicillin and other β-lactam antibiotics.

The β-lactam antibiotics are known to inhibit DD-carboxypeptidases, peptidoglycan transglycosylases, and peptidoglycan transpeptidases. Peptidoglycan transpeptidase, responsible for cross-linking adjacent peptide strands, has been considered the classic target for penicillin and other β-lactam antibiotics, whereas peptidoglycan transglycosylase is responsible for elongating the glycan strands. The physiological role of DD-carboxypeptidase is less clear. Possibly, it regulates the degree of cross-linking of peptide strands but is seldom a lethal target for penicillin.

In a number of organisms, and especially *Escherichia coli*, these enzymes bind covalently with penicillin and other β-lactam antibiotics and are therefore referred to as penicillin-binding proteins (PBPs).

In accord with the poor antibacterial activity of naturally occurring monobactams, these molecules exhibit a lack of affinity for any essential PBPs of *E. coli* or *S. aureus*. The compounds show good binding to PBPs 1a, 4, and 5/6 of *E. coli* and PBPs of *Staphylococcus aureus*. With the exception of PBP 1a, these PBPs have been identified as DD-carboxypeptidases. In this regard, naturally occurring monobactams (especially methoxylated derivatives) are good inhibitors of the DD-carboxypeptidase from streptomyces R61, which was the model used for the bacterial enzymes in ref. 18.

Synthetic Monobactams. The appropriate modifications of the monobactam nucleus result in compounds with potent antibacterial activity. A large number of monobactams have been prepared synthetically, which carry penicillin- or cephalosporin-like acyl side chains. Compounds with the penicillin-G side chain are predominately active against gram-positive organisms. They exhibit poor binding to the essential PBPs of *E. coli* but good binding to PBPs 1, 2, and 3 of *S. aureus* (19).

Monobactams with piperacillin or azlocillin side chains exhibit broad-spectrum activity and show good binding to PBP 3 of *E. coli* and PBPs 1, 2, and 3 of *S. aureus*. Monobactams with aminothiazole oxime side chains [eg, (**15**)] show potent activity against gram-negative organisms which corresponds with the binding of these compounds to PBP 3. These compounds show poor binding to PBPs 1, 2, and 3 of *S. aureus* (19). Thus, as with penicillins and cephalosporins, antibacterial activity of the monobactams is parallel to essential PBP binding.

Monobactams have been shown to be progressive inhibitors of R61 DD-carboxypeptidase (4,10). As observed with the cephalosporins, 3α-methoxylation increases monobactam binding. Binding to the R61 enzyme is stoichiometric and takes place at the penicillin-binding site (20). However, affinity for the enzyme is generally lower than that of homologous penicillins. Release products of monobactams are less complex than those obtained from penicillins and cephalosporins; release is slow and does not involve secondary fragmentation (20).

Biosynthesis

The monocyclic ring of the naturally occurring monobactams is derived from serine. The biosynthesis has been elucidated by feeding experiments with radiolabeled amino acids. Similar techniques have shown that the methyl moiety of the methoxyl group in 3α-methoxylated monobactams is derived from methionine (21). Synthesis of the 3-acyl substituents and origin of the $—SO_3^-$ group are under investigation.

Synthesis

Naturally Occurring Monobactams. The synthesis of SQ 26,180 (**1**) was reported (3) as part of its structure proof. 7-Aminodesacetoxycephalosporanic acid (7-ADCA) (**19**) was converted to the *N*-1 unsubstituted azetidinone (**20**) and subsequently sulfonated to give SQ 26,180 (**1**). Similarly, 6-APA (**21**) was converted to EM 5400 component (**6**) via the key intermediate (**22**) (9).

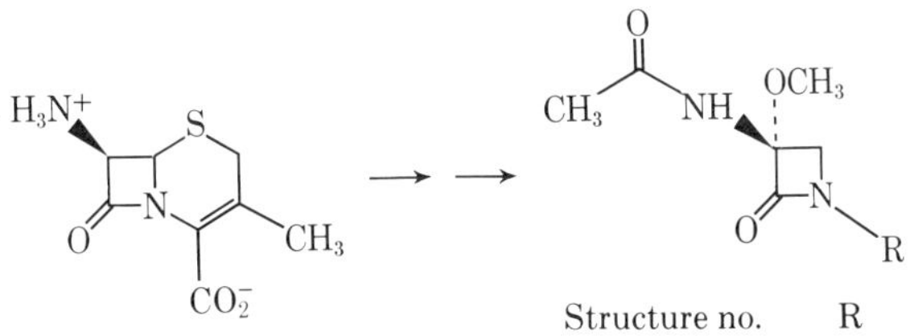

(19)

Structure no.	R
(20)	H
(1)	$SO_3^-K^+$

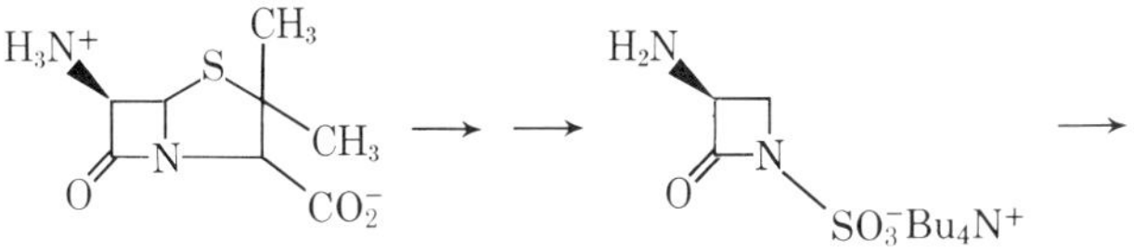

(21)

(22)
[80082-75-7]

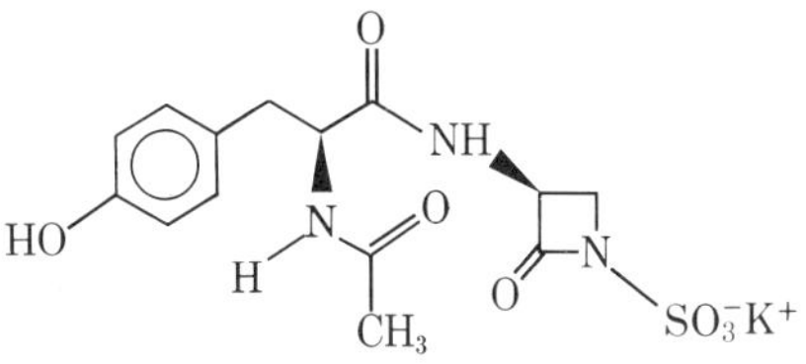

(6)

Each of these syntheses should be regarded as total, since the monobactam moiety was created chemically. Semi-synthesis of monobactams, spectacularly successful for penicillins and cephalosporins, has been only partially successful so far.

A preparation of general intermediates is given in ref. 22.

Semi-Synthetic Routes. The term semi-synthetic, as applied to β-lactams, indicates the chemical modification of a fermentation-derived antibiotic to give another antibiotic. Examples of this strategy include the conversion of penicillin G and cephalosporin C to the nucleii (**21**) and (**19**), respectively, from which literally thousands of new antibiotics have been prepared. A second example is the direct conversion of the cephamycin C derivative (**23**) to cefoxitin (**24**) accomplished by a transacylation process without the isolation of an intermediate (23).

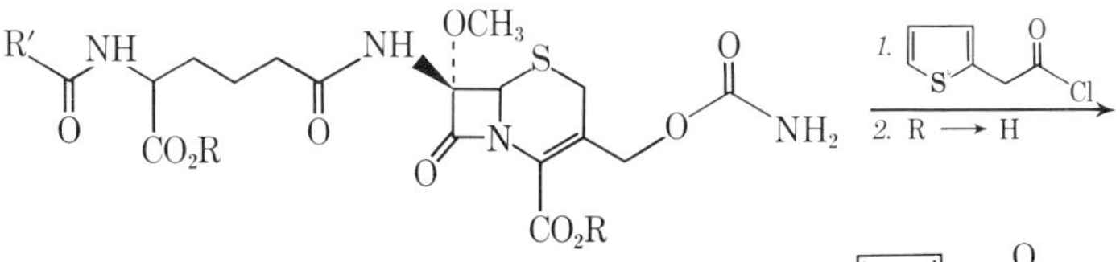

(23)

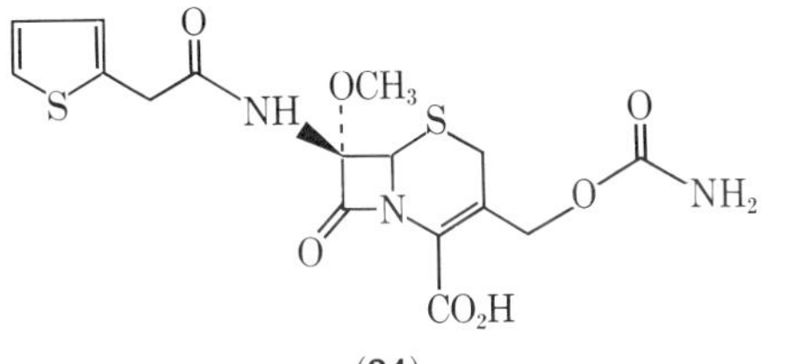

(24)

A striking example of the contrasts between bicyclic β-lactam and monobactam chemistry has been reported (24). Analogy with the chemistry of cephamycin C suggested that conversion of SQ 26,180 (**1**) to (**25**) could be accomplished by iminochloride formation followed either by alcoholysis or reaction with *o*-aminothiophenol. However, the iminochloride (**26**) proved to be unstable and could not be isolated; instead, it readily fragmented to give acetonitrile and the chloroether (**27**) which was identified spectroscopically and by conversion to (**28**) (see Fig. 2).

Nevertheless, this methodology can be applied to nonmethoxylated monobactams. Conversion of the synthetic monobactam ([*78611-47-9*], M = Bu_4N^+) (**29**) via the iminochloride to nucleus (**30**) [*80082-65-1*], an important intermediate to aztreonam (**31**) [*78110-38-0*], has been reported (25).

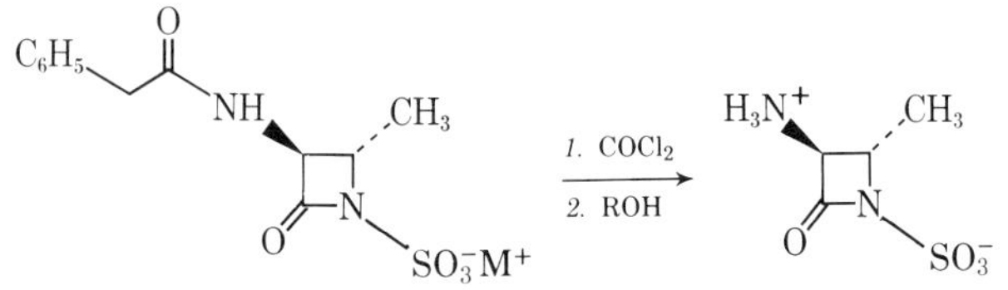

1. $COCl_2$
2. ROH

(**29**) (**30**)

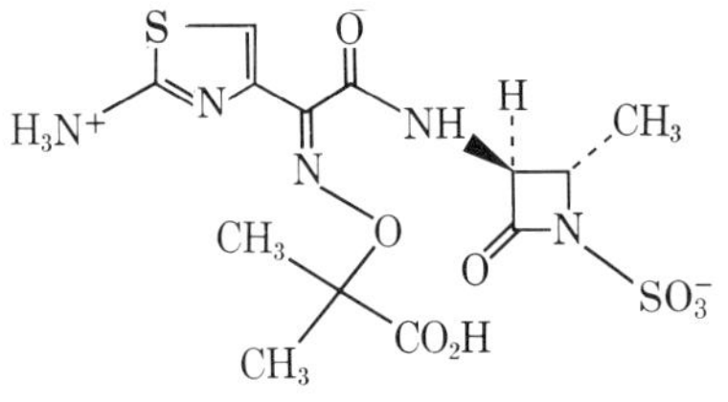

(**31**) aztreonam

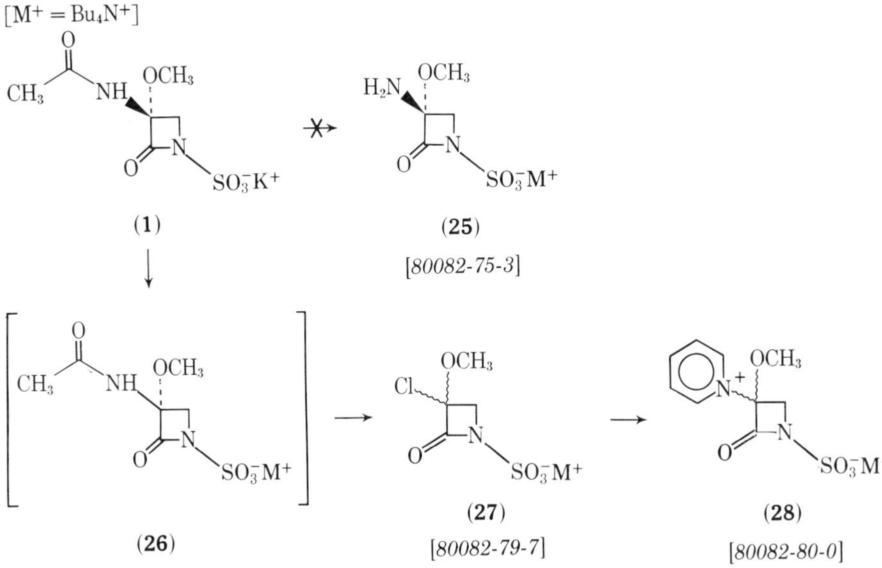

[$M^+ = Bu_4N^+$]

(**1**) (**25**) [*80082-75-3*]

(**26**) [*80544-06-5*] (**27**) [*80082-79-7*] (**28**) [*80082-80-0*]

Figure 2. Products of imidochloride intermediate (**26**) of SQ 26,180 (**1**).

This suggests that nonmethoxylated EM 5400 component (**6**) would serve equally well as a precursor of 3-aminomonobactamic acid [*79720-18-6*] [(**32**), 3-AMA]. However, this transformation has not been reported.

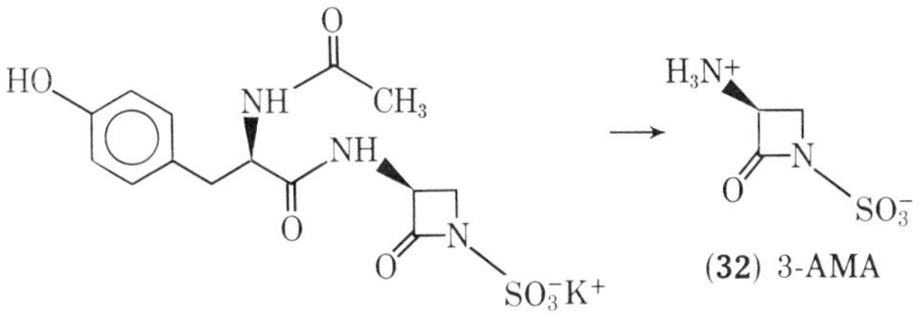

(**6**) → (**32**) 3-AMA

Attempts to apply transacylation to SQ 26,180 (**1**) have been unsuccessful (24).

The acylation of sulfazecin (**2**) to give various acyl derivatives such as (**33**) has been reported (26). Selective removal of the glutamyl moiety from (**2**) (by either iminochloride or nitrous acid methodology with parallels in cephalosporin chemistry) to give alanyl monobactam (**34**) has also been accomplished. Acylation of (**34**), in turn, produces a variety of semi-synthetic monobactams such as (**35**) (27).

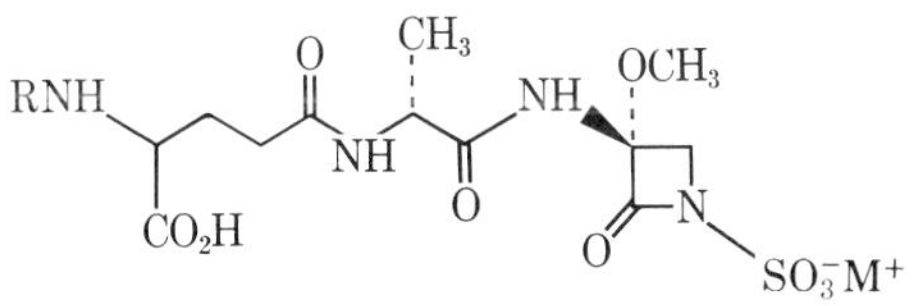

(**2**) R = H
(**33**) R = acyl

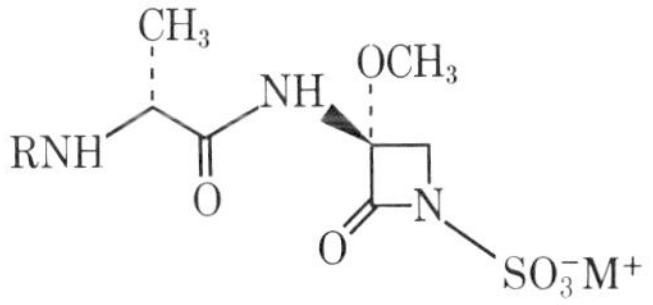

(**34**) R = H
(**35**) R = acyl

Total Synthesis. The inability to convert naturally occurring monobactams to general intermediates led to the early and extensive development of two totally synthetic routes. The first synthesis employs as the key step the sulfonation of an *N*-1 unsubstituted azetidinone. More than 100 monobactam side-chain analogues have been synthesized by the sulfonation of the corresponding *N*-1 unsubstituted derivatives (**36**) → (**11**) (28–29). A specific example of this process is included in Figure 3.

The compound SQ 26,180 (**1**) was synthesized by this procedure (3). A second example of this synthetic procedure involves the sulfonation of *N*-1 unsubstituted azetidinones (**37**) that carry various protecting groups on the nitrogen atom attached to *C*-3. Key intermediates such as (**38**) are readily formed, which could be acylated to a wide variety of side-chain monobactams (see Fig. 4).

A second, conceptually distinct, synthesis of monobactams was also developed (25). Cyclization of the acylsulfamate (**39**) was expected to lead directly to monobactam (**40**) based on analogy to earlier work (30). Consideration of the relative p*Ka* of H_A, H_B, and H_C in (**39**) suggested that neither epimerization nor elimination (H_B removal) or oxazoline formation (H_C removal) would be a competing process under weakly basic conditions. This expectation was realized with aqueous bicarbonate as base; the resulting procedure provides a simple, high yield conversion of L-threonine to 3*S*-(*trans*)-3-amino-4-methylmonobactamic acid (**30**), the key intermediate to aztreonam as indicated in Figure 5.

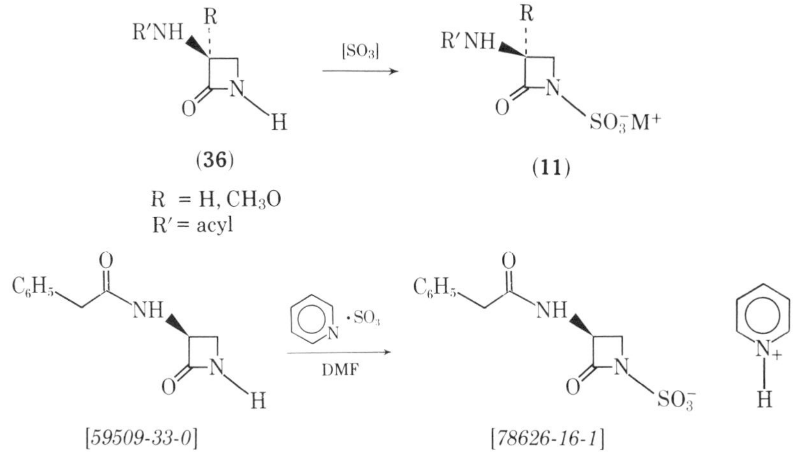
[SO3]
(36)
(11)
R = H, CH3O
R′ = acyl
DMF
[59509-33-0]
[78626-16-1]

Figure 3. Synthesis of monobactam side-chain analogues.

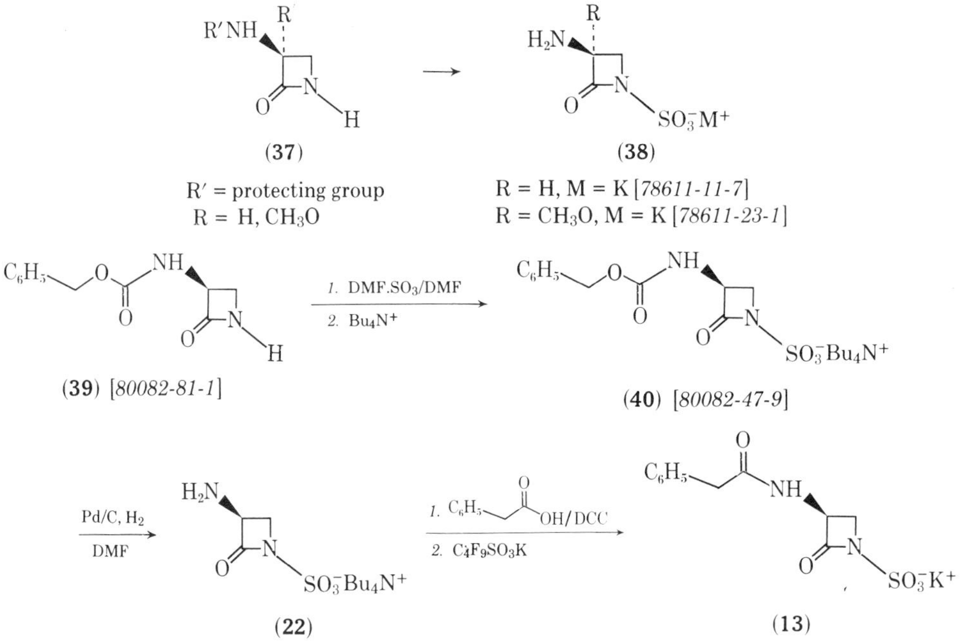
(37)
(38)
R′ = protecting group
R = H, CH3O
R = H, M = K [78611-11-7]
R = CH3O, M = K [78611-23-1]
1. DMF.SO3/DMF
2. Bu4N+
(39) [80082-81-1]
(40) [80082-47-9]
Pd/C, H2
DMF
OH/DCC
2. C4F9SO3K
(22)
(13)

Figure 4. Acylation of intermediates to side-chain monobactams.

Structure–Activity Relationships

Although the activity of methoxylated monobactams could be improved by appropriate side-chain modification (31), difficulty of synthesis and poor chemical stability has focused attention on the nonmethoxylated analogues. High intrinsic activity and excellent β-lactamase stability are exhibited by monobactams that combine aminothiazole oxime side chains and 4-alkyl groups (32). Comparison of the structures of SQ 26,180 (**1**), the simplest naturally occurring monobactam, and aztreonam (**31**), the first monobactam submitted to clinical trials, illustrates these points well.

Exchange of the simple methyl side chain of (**1**) for the complex aminothiazole

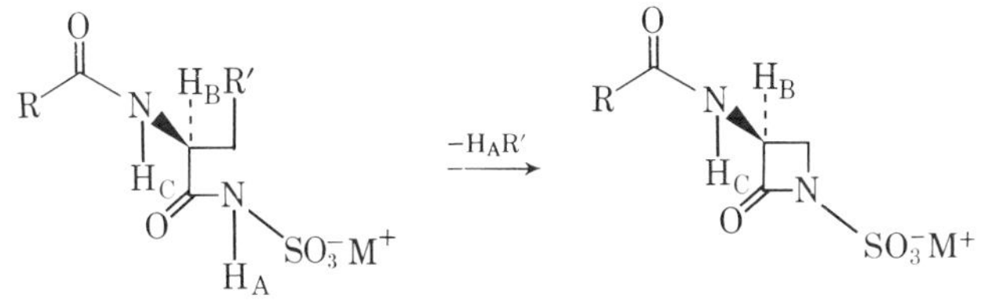

Figure 5. Pathway to aztreonam intermediate.

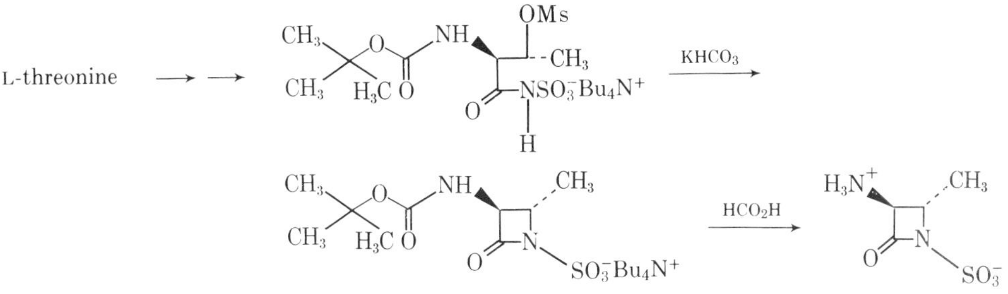

oxime side chain of aztreonam was accompanied by a high increase in intrinsic activity, eg, for a typical *E. coli* strain the increase is 1000-fold. Replacement of the 3α-methoxyl group by a hydrogen atom increased chemical stability markedly. Incorporation of a 4α-methyl group increased β-lactamase stability substantially; relative to its unmethylated parent, the β-lactamase stability of aztreonam is increased ca 100-fold.

Aztreonam

***In Vitro* Properties.** Aztreonam is a totally synthetic β-lactam antibiotic with a unique spectrum of biological activity. The compound exhibits potent and specific activity against a wide range of β-lactamase-producing and nonproducing gram-negative bacteria, including *Pseudomonas aeruginosa*, and displays minimal inhibition against gram-positive organisms and anaerobes.

When tested against 453 strains of *Enterobacteriaceae*, aztreonam inhibited 50% of these isolates at concentrations <0.3 μg/mL (33). Particularly sensitive were members of the *Proteus–Providencia* group, *Escherichia coli*, *Klebsiella pneumoniae*, and *Serratia marcescens*. Similarly, 50% of strains of ampicillin-sensitive and resistant *Haemophilus influenzae* and *Neisseria gonorrhoeae* were inhibited at <0.1 μg/mL. Most dramatic, however, were results with 61 strains of *P. aeruginosa*: 50% were inhibited at 4 μg/mL, whereas 12 μg/mL inhibited 90% of the strains. Cessation of growth is not the sole consequence of exposure of these organisms to aztreonam. The inhibitory concentration is also generally the killing or lethal concentration. In the same study, little or no activity was seen against staphylococci, streptococci, and *Bacteroides fragilis*. Minimal inhibitory concentrations for these organisms were usually >100 μg/mL.

Many of the gram-negative isolates tested were multiresistant β-lactamase-producing strains that are becoming more and more frequent and are gaining increasing clinical importance, particularly in nosocomial infections. Generally resistant to first-

and second-generation cephalosporins and most of the penicillins, these organisms elaborate a diverse array of β-lactamases targeted at destroying β-lactam antibiotics. Aztreonam exhibits a high degree of resistance to enzymatic hydrolysis by most of these common β-lactamases (34). Particularly insidious are the plasmid-mediated TEM β-lactamases carried by a wide range of gram-negative bacteria. The plasmid-mediated genes responsible for enzyme production pass freely among the *Enterobacteriaceae*, pseudomonads, and now *Neisseria gonorrhoeae* and *Haemophilus influenzae*. Other β-lactamases capable of this same promiscuous behavior are the OXA, SHV, and PSE enzymes. Aztreonam, like many of the third-generation aminothiazole–oxime cephalosporins, such as cefotaxime and ceftizoxime, is stable to these β-lactamases. In contrast, compounds like cephaloridine and cefoperazone are readily hydrolyzed. A similar situation exists with the chromosomally mediated β-lactamases which also are widely distributed among specific gram-negative bacteria but are not transferable. These enzymes, which may coexist in the same organisms with a plasmid-mediated β-lactamase, are found in organisms such as *Klebsiella*, *Providencia*, *Enterobacter*, *Serratia*, *Proteus*, and *Bacteroides*.

Aztreonam, cefotaxime, and ceftizoxime are generally stable to these β-lactamases. The only enzyme in this group showing any appreciable destruction of aztreonam is the relatively uncommon K1 β-lactamase produced by some strains of *K. oxytoca*. Cephaloridine and cefoperazone show varying degrees of instability to the chromosomally mediated β-lactamases.

The mode of action of aztreonam has been studied with respect to effects on bacterial morphology, viability, and binding to penicillin-binding proteins (PBPs) (35). As would be expected from its relative inactivity against gram-positive bacteria and anaerobes, aztreonam interacts poorly with PBPs derived from these organisms. With gram-negative bacteria, however, aztreonam has a high and somewhat specific affinity for PBP3. Binding to PBP3 results in filamentation of growing bacteria. A moderate affinity is seen for PBP1a and poor affinity for PBPs 1b, 2, 4, and 5/6. As a consequence of its interaction with susceptible bacteria, cell growth is inhibited and destruction follows. Like other β-lactam antibiotics, the lethal concentration of aztreonam is either the same as or double the MIC.

***In Vivo* Properties.** In experimental animals, aztreonam shows excellent pharmacokinetic properties; ca 40–50% is excreted unchanged in the urine and 15% appears in the bile (33). It penetrates well into the genital tract, heart, kidney, liver, lung, spleen, and meninges at therapeutic concentrations. Aztreonam can protect mice infected systemically with a wide variety of aerobic gram-negative organisms ranging from *E. coli* and *S. marcescens* to *P. aeruginosa* and *H. influenzae*. Aztreonam has been shown to be more efficacious than cefuroxime, cefoperazone, cefotaxime, or piperacillin (35).

The efficacy of aztreonam has also been determined in more complex situations using animal models of human infectious disease. Examples of such models include an infection of the ascending urinary tract in mice caused by *S. marcescens* and *P. rettgeri*. Aztreonam was highly effective. In these *in vivo* situations, it was resistant to hydrolysis by β-lactamase, whereas enzyme-susceptible compounds such as ampicillin, piperacillin, and cefoperazone yielded poor results when the infecting pathogen was a β-lactamase producer.

Although highly effective in eradicating pathogenic bacteria, many of the newer drugs have the capability to destroy the body's microflora. Such disturbances can lead to colonization by potential pathogens, loss of essential nutrients to the host, and the development of intestinal complications as manifested in pseudomembranous colitis. Mass killing can be avoided by directed therapy, ie, antibiotics directed against specific groups of microorganisms. Aztreonam represents such a compound, being directed specifically against the aerobic gram-negative bacteria often implicated in serious hospital-acquired infections.

Clinical studies with aztreonam are ongoing throughout the United States, Europe and Japan. It is anticipated that the drug will become available for general use in 1984.

Outlook

Monocyclic β-lactam antibiotics and the antimicrobial properties of aztreonam were reported in 1981. Their impact should be significant; research in a number of laboratories is already directed at producing monobactams with different properties than those of aztreonam. At the same time, other groups, besides $—SO_3^-$, are under investigation to provide both chemical and biological activation for monocyclic azetidinones.

The ultimate value of the discovery of the monobactams will probably encompass both the clinical efficacy of aztreonam or a subsequent monobactam and the insight into the detailed mechanism of action of β-lactam antibiotics afforded by study of these fascinating monocyclic variants.

BIBLIOGRAPHY

1. C. M. Cimarusti and R. B. Sykes, *Chem. Brit.* **19,** 302 (April 1983).
2. R. B. Sykes, C. M. Cimarusti, D. P. Bonner, K. Bush, D. M. Floyd, N. H. Georgopapadakou, W. H. Koster, W. C. Liu, W. L. Parker, P. A. Principe, M. L. Rathnum, W. A. Slusarchyk, W. H. Trejo, and J. S. Wells, *Nature (London)* **291,** 489 (1981).
3. W. L. Parker, W. H. Koster, C. M. Cimarusti, D. M. Floyd, W. Liu, and M. L. Rathnum, *J. Antibiot.* **35,** 189 (1982).
4. J. S. Wells, W. H. Trejo, P. A. Principe, K. Bush, N. H. Georgopapadakou, D. P. Bonner, and R. B. Sykes, *J. Antibiot.* **35,** 184 (1982).
5. A. Imada, K. Kitano, K. Kintaka, M. Muroi, and M. Asai, *Nature (London)* **289,** 590 (1981).
6. M. Asai, K. Haibara, M. Muroi, K. Kintaka, and T. Kishi, *J. Antibiot.* **34,** 621 (1981).
7. K. Kintaka, Y. Kitano, F. Nozaki, A. Kawashima, Y. Imada, Y. Nakao, and M. Yoneda, *J. Ferment. Technol.* **59,** 263 (1981).
8. K. Kintaka, K. Haibara, M. Asai, and A. Imada, *J. Antibiot.* **34,** 1081 (1981).
9. W. L. Parker and M. L. Rathnum, *J. Antibiot.* **35,** 300 (1982).
10. J. S. Wells, W. H. Trejo, P. A. Principe, K. Bush, N. H. Georgopapadakou, D. P. Bonner, and R. B. Sykes, *J. Antibiot.* **35,** 295 (1982).
11. U.S. Pat. 4,229,436 (Oct. 21, 1980), A. Imada, K. Kitano, and A. Mitsuko (to Takeda Chemical Industries, Ltd.); U.S. Pat. 4,225,586 (Sept. 30, 1980), A. Imada, K. Kintaka, and K. Haibara (to Takeda Chemical Industries, Ltd.).
12. K. Kamiya, M. Takamoto, Y. Wada, and M. Asai, *Acta Crystallogr.* **B37,** 1626 (1981).
13. The Squibb Institute for Medical Research, Princeton, N.J., unpublished results.
14. H. R. Pfaendler, J. Gosteli, and R. B. Woodward, *J. Am. Chem. Soc.* **101,** 6306 (1979).
15. J. M. Indelicato and W. L. Wilham, *J. Med. Chem.* **17,** 528 (1974).
16. J. Birnbaum, E. O. Stapley, A. K. Miller, H. Wallick, D. Hendlin, and H. B. Woodruff, *J. Antimicrob. Chemother.* (*Suppl. B*) **4,** 15 (1978).
17. R. B. Sykes, D. P. Bonner, N. H. Georgopapadakou, and J. S. Wells, *J. Antimicrob. Chemother.* (*Suppl. E*) **8,** 1 (1981).

18. J. M. Frere, J. A. Kelly, D. Klein, J. M. Ghuysen, P. Claes, and H. Vanderhaeghe, *Biochem. J.* **203,** 223 (1982).
19. N. H. Georgopapadakou, S. A. Smith, C. M. Cimarusti, and R. B. Sykes, *Antimicrob. Agents Chemother.* **23,** 98 (1983).
20. N. H. Georgopapadakou, S. A. Smith, and C. M. Cimarusti, *Eur. J. Biochem.* **124,** 507 (1982).
21. J. O'Sullivan, A. M. Gillum, C. A. Aklonis, M. L. Souser, and R. B. Sykes, *Antimicrob. Agents Chemother.* **21,** 558 (1982).
22. D. M. Floyd, A. W. Fritz, J. Plusec, E. R. Weaver, and C. M. Cimarusti, *J. Org. Chem.* **47,** 5160 (1982).
23. S. Karady, S. J. Pines, L. M. Weinstock, F. E. Roberts, G. S. Brenner, A. M. Hoinowski, T. Y. Cheng, and M. Sletzinger, *J. Am. Chem. Soc.* **94,** 1410 (1972).
24. C. M. Cimarusti, H. E. Applegate, H. W. Chang, D. M. Floyd, W. J. Koster, W. A. Slusarchyk, and M. G. Young, *J. Org. Chem.* **47,** 179 (1982).
25. D. M. Floyd, A. W. Fritz, and C. M. Cimarusti, *J. Org. Chem.* **47,** 176 (1982).
26. Jpn. Pat. 56 138,169 (Oct. 28, 1981), M. Muroi, M. Asai, and T. Imada (to Takeda Chemical Industries, Ltd.).
27. Jpn. Pat. 56 139,454 (Oct. 30, 1981), S. Horii and N. Fukase (to Takeda Chemical Industries, Ltd.).
28. Eur. Pat. Appl. 021,678 (Jan. 7, 1981), T. Matsuo, T. Sugawara, H. Masuya, and Y. Kawaro (to Takeda Chemical Industries, Ltd.).
29. Eur. Pat. Appl. 53,387 (June 6, 1982), T. Matsuo, H. Masuya, N. Noguchi, and M. Ochiai (to Takeda Chemical Industries, Ltd.).
30. M. J. Miller, P. G. Mattingley, M. A. Morrison, and J. Kerwin, Jr., *J. Am. Chem. Soc.* **102,** 7026 (1980).
31. H. Breuer, U. D. Treuner, T. Denzel, H. E. Applegate, D. P. Bonner, K. Bush, C. M. Cimarusti, W. H. Koster, W. A. Slusarchyk, R. B. Sykes, and M. G. Young, *paper presented at the 21st Interscience Conference on Antimicrobial Agents and Chemotherapy*, Chicago, Ill., 1981.
32. H. Breuer, C. M. Cimarusti, T. Denzel, W. H. Koster, W. A. Slusarchyk, and U. D. Treuner, *J. Antimicrob. Chemother.* (*Suppl. E*) **8,** 21 (1981).
33. R. B. Sykes, D. P. Bonner, K. Bush, and N. H. Georgopapadakou, *Antimicrob. Agents Chemother.* **21,** 85 (1982).
34. K. Bush, J. Freudenberger, and R. B. Sykes, *Antimicrob. Agents Chemother.* **22,** 414 (1982).
35. D. P. Bonner, R. R. Whitney, C. O. Baughn, B. H. Miller, S. J. Olsen, and R. B. Sykes, *J. Antimicrob. Chemother. Suppl. E.* **8,** 123 (1981).

CHRISTOPHER M. CIMARUSTI
RICHARD B. SYKES
Squibb Institute for Medical Research

AZEOTROPIC AND EXTRACTIVE DISTILLATION

An azeotrope is a liquid mixture that exhibits a maximum or minimum boiling point relative to the boiling points of the components of the mixture and that distills without change in composition. The maximum or minimum boiling point is caused by negative or positive deviations, respectively, from Raoult's law. Application of simple distillation (qv) results in the removal of the azeotrope in the overhead or in the bottoms for a minimum or maximum boiling point azeotrope, respectively. The bottoms or the overhead would be rich in one of the components, depending on the position of the feed composition relative to that of the azeotrope. For the separation of an azeotropic mixture, the properties of the azeotrope must be altered by pressure adjustment; addition of a high boiling point solvent, ie, extractive distillation; formation of a heterogeneous ternary azeotrope, ie, conventional azeotropic distillation; or formation of a heterogeneous binary azeotrope, ie, self-entrained azeotrope distillation. Other processes have been considered for the separation of azeotropes, ie, from the use of membranes to extractive distillation based on the salt effect (1).

Pressure Adjustment

For a given temperature and pressure, an azeotrope defines the condition in which the composition of the vapor phase is the same as that of the liquid phase in equilibrium. If only one liquid phase is present in the azeotrope, the azeotrope is homogeneous. For example, ethanol and water form a homogeneous azeotrope at atmospheric pressure and 78.15°C and at a mole fraction of ethanol of 0.8943. If there are two liquid phases, the azeotrope is heterogeneous. For example, *n*-butanol and water form a heterogeneous azeotrope at atmospheric pressure and 92.25°C and an average mole fraction of *n*-butanol of 0.25. Some of the most common homogeneous and heterogeneous azeotropes are listed in Tables 1 and 2. Some common ternary azeotropes are listed in Table 3. The heterogeneity of certain azeotropes is being exploited in conventional and self-entrained distillations.

The usual thermodynamic equations describe the vapor–liquid and liquid–liquid equilibria for azeotropic mixtures. The fugacity (f) of the liquid phase must be equal to the fugacity of the vapor phase for a homogeneous azeotrope, ie:

$$f_i^L = f_i^V \quad \text{for the component } i \tag{1}$$

resulting in (at low pressure, for the sake of simplicity):

$$\gamma_i x_i P_i^{sat} = y_i P; \quad i = 1, \ldots, c \tag{2}$$

where x = liquid mole fraction, y = vapor mole fraction, and c = number of components. The activity coefficient γ is a function of temperature and compositions, and the vapor pressure P_i^{sat} is a function of temperature only. As indicated in equations 1–2, the azeotropic composition depends on temperature and pressure. Furthermore, equations 1–2 and the equation $\Sigma x_i = 1$ can be solved for the azeotrope at any pressure or temperature.

The dependence of the azeotropes on pressure suggests the use of separation schemes like the one which is used to separate a minimum boiling homogeneous azeotrope, shown in Figure 1. The fresh feed enters the first distillation tower, which

Table 1. Minimum-Boiling-Point Azeotropic Binary Mixtures at 101.3 kPa (760 mm Hg)[a]

System A	System B	A, Mol %	Temperature, °C
water	ethanol	10.57	78.15
	allyl alcohol	54.50	88.20
	propionic acid	94.70	99.98
	n-propyl alcohol	56.83	87.72
	isopropyl alcohol	31.46	80.37
	methyl ethyl ketone	33.00	73.45
	isobutyric acid	94.50	99.30
	ethyl acetate (2 phase)	24.00	70.40
	ethyl ether (2 phase)	5.00	34.15
	n-butyl alcohol (2 phase)	75.0	92.25
	isobutyl alcohol	67.14	89.92
	sec-butyl alcohol	66.00	88.50
	tert-butyl alcohol	35.41	79.91
	isoamyl alcohol (2 phase)	82.79	95.15
	tert-amyl alcohol (2 phase)	65.00	87.00
	benzene (2 phase)	29.60	69.25
	toluene (2 phase)	55.6	84.10
carbon tetrachloride	methanol	44.5	55.70
	ethanol	61.3	64.95
	allyl alcohol	73.0	72.32
	n-propyl alcohol	75.0	72.80
	ethyl acetate	43.0	74.75
carbon disulfide	methanol	72.0	37.65
	ethanol	86.0	42.40
	acetone	61.0	39.25
	methyl acetate	69.5	40.15
chloroform	methanol	65	53.5
	ethanol	84	59.3
	isopropyl alcohol	92	60.8
n-butyl alcohol	cyclohexane	11	79.8
	toluene	37	105.5
isobutyl alcohol	isoamyl bromide	60.0	103.80
	benzene	10.0	79.84
	toluene	50.0	101.15
	α-pinene	96.5	107.90
n-amyl alcohol	isoamyl acetate	96.4	131.3
	isobutyl propionate	85	130.5
isoamyl alcohol	chlorobenzene	42	124.3
	o-xylene	64	128
	m-xylene	58	127
	p-xylene	56	126.8
nitrobenzene	benzyl alcohol	39	204.3
	borneol	60	207.75
	menthol	60	207.9
phenol	*p*-bromotoluene	58	176.2
	carvene	49.5	169.0
	α-pinene	25	152.75
aniline	carvene	48	171.35
benzyl alcohol	guaiacol	38	204.4
	naphthalene	64	204.3
acetic acid	chlorobenzene	72.5	114.65
	benzene	2.5	80.05
	toluene	62.7	105.4
	m-xylene	40	115.38
ethyl alcohol	methyl ethyl ketone	45	74.8
	ethyl acetate	46	71.8
	methyl propionate	67.5	73.2
	n-propyl formate	72	73.5
	benzene	44.8	68.24
	cyclohexane	44.5	64.9
ethyl alcohol	*n*-hexane	33.2	58.68
	toluene	81	76.65
	n-heptane	67	72
allyl alcohol	benzene	22.2	76.75
	cyclohexane	26.6	74
	n-hexane	6.5	65.5
	toluene	61.5	92.4
acetone	methyl acetate	61	56.1
	isobutyl chloride	81	55.8
	diethylamine	43.5	51.5
n-propyl alcohol	ethyl propionate	64	93.4
	benzene	20.9	77.12
	n-hexane	6	65.65
	toluene	60	92.6
isopropyl alcohol	ethyl acetate	30.5	74.8
	benzene	39.3	71.92
	n-hexane	29	61
	toluene	77	80.6
tetrachloroethylene	ethanol	6	77.95
	allyl alcohol	27	94.0
	propionic acid	81	118.95
	n-propyl alcohol	24	94
	isopropyl alcohol	8	81.7
	n-butyl alcohol	47	110
	isobutyl alcohol	40	103.05
trichloroethylene	allyl alcohol	70	80.95
	n-propyl alcohol	69	81.75
	isopropyl alcohol	54	74
	isobutyl alcohol	86	85.4
	tert-butyl alcohol	74	75
	tert-amyl alcohol	83	84
dichloroethylene	allyl alcohol	76	79.6
		77	80
chloral hydrate	cyclohexane	13	76
ethylene bromide	acetic acid	20.7	114.35
	propionic acid	65	127.75
	isobutyl alcohol	22	106.2
	isoamyl alcohol	52	123.2
	ethyl benzene	83.5	131.1
methanol	trichloroethylene	70	60.2
	acetonitrile	84.5	63.45
	ethylene dichloride	62	59.5
	1,1-dichloroethane	28.5	49.05
	ethyl bromide	14	34.95
	chloromethyl methyl ether	57.5	56
	ethyl iodide	52.5	54.7
	acetone	20	55.7
	ethyl formate	30.5	50.95
	methyl acetate	35	54.0
	n-propyl bromide	49	54.1
	n-propyl iodide	88	63.5
	methylal	34.5	41.82
	trimethyl borate	87	59
	ethyl acetate	91.7	62.3
	n-pentane	13	31
	isopentane	9	24.5
	benzene	61.4	53.84
	cyclohexene	63.0	55.9
	cyclohexane	61.0	54.2
	n-hexane	51	50.6
	n-heptane	83	60.5
	D-pinene	98.5	64.5

[a] Ref. 2.

Table 2. Maximum-Boiling-Point Azeotropic Binary Mixtures[a]

A	B	A, Mol %	Temperature, °C	Pressure, kPa[b]
water	hydrofluoric acid	65.4	120	101.3
	hydrochloric acid	88.9	110	
	perchloric acid	32.0	203	
	hydrobromic acid	83.1	126	
	hydriodic acid	84.3	127	
	nitric acid	62.2	120.5	98.0
	formic acid	43.3	107.1	101.3
chloroform	acetone	65.5	64.5	101.3
formic acid	diethyl ketone	48	105.4	
	methyl propyl ketone	47	105.3	
phenol	cyclohexanol	90	182.45	
	benzaldehyde	54	185.6	
	benzyl alcohol	8	206.0	
o-cresol	acetophenone	24	203.7	
	phenyl acetate	42.5	198.6	
	methyl hexyl ketone	97	191.5	
	isoamyl butyrate	80	192.0	
m-cresol	acetophenone	54	209.0	
	isoamyl lactate	60	207.6	
p-cresol	benzyl alcohol	38	207.0	
	acetophenone	52	208.45	
	camphor	38	213.15	

[a] Ref. 2.
[b] To convert kPa to mm Hg, multiply by 7.5.

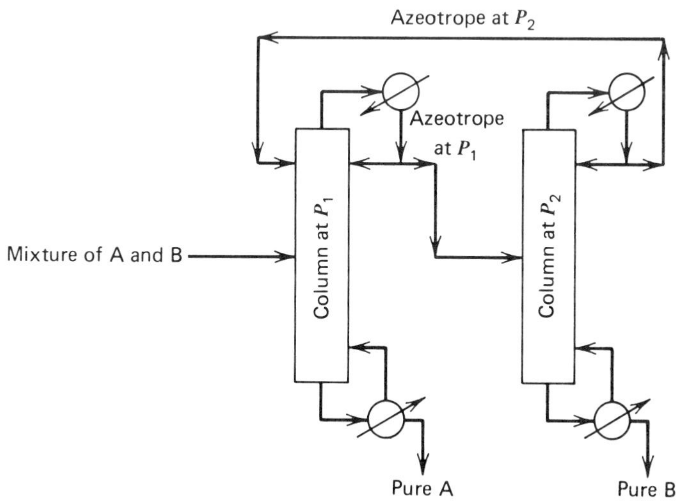

Figure 1. Separation of a pressure-sensitive, minimum-boiling-point, homogeneous azeotrope. P_1 and P_2 = pressures 1 and 2, respectively.

is at pressure P_1. The bottoms product of this tower is pure A and the composition of the overhead approaches that of the azeotrope at pressure P_1. The overhead is fed into the second distillation tower, which is at pressure P_2. The pressure P_2 is chosen

Table 3. Ternary Azeotropic Mixtures[a]

Component A, Mol % A = 100 − (mol % B + mol % C)	Components	Mol %	Temperature, °C
water	B: carbon tetrachloride	57.6	61.8
	C: ethanol	23.0	2 phase
	B: trichloroethylene	38.4	67.25
	C: ethanol	41.2	2 phase
	B: trichloroethylene	49.2	71.4
	C: allyl alcohol	17.3	2 phase
	B: trichloroethylene	51.1	71.55
	C: *n*-propyl alcohol	16.6	2 phase
	B: ethanol	12.4	70.3
	C: ethyl acetate	60.1	
	B: ethanol	22.8	64.86
	C: benzene	53.9	
	B: allyl alcohol	9.5	68.3
	C: benzene	62.2	
	B: *n*-propyl alcohol	8.9	68.48
	C: benzene	62.8	
carbon disulfide	B: methanol	24.1	33.92
	C: ethyl bromide	35.4	
methyl formate	B: ethyl bromide	23.8	16.95
	C: isopentane	31.0	
	B: ethyl ether	7.2	20.4
	C: *n*-pentane	48.2	
n-propyl lactate	B: phenetol	35.2	163.0
	C: menthene	34.1	

[a] Ref. 2.

such that the composition of the azeotrope at this pressure is between that of the fresh feed and that of the overhead from the first tower. The separation in the second tower results in a pure B bottoms product and an overhead similar to the azeotrope at P_2; the latter is recycled. The physicochemical principles underlying this scheme are shown in Figure 2.

Processes similar to the one in Figure 1 are used primarily in the pharmaceutical industry (3). A potential advantage of pressure-adjustment processes is based on the fact that, around the azeotrope, the T–x–y curve is flat. Therefore, vapor recompression can be applied and results in considerable utilities savings. Figure 3 shows a scheme for the separation of ethanol and water by pressure adjustment. The first tower is used to concentrate the ethanol, and operates at atmospheric pressure. The overhead, which is nearly the same as the atmospheric azeotrope, is fed into a low pressure tower which operates at ca 10.7 kPa (0.1 atm). This tower functions only to provide an overhead slightly above the atmospheric azeotrope, which is separated further to pure ethanol in the bottoms and a composition similar to the azeotrope at 101.3 kPa (1 atm) in the third tower. The overhead of the third tower and the bottoms of the second are recycled. The main drawback of such a configuration is the large vapor flow rates, which result in high utilities costs. Since the temperature difference between the top and the bottom of the last two towers is very small, vapor recompression is applied and results in considerable utilities savings.

Configurations similar to those in Figures 1 and 3 or those having a middle tower

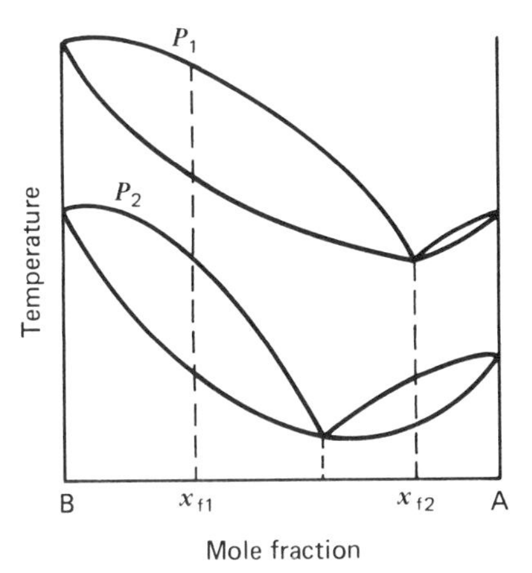

Figure 2. Temperature-versus-mole-fraction diagram of a pressure-sensitive, minimum-boiling-point, homogeneous azeotropic mixture. x_{f1} = mole fraction of feed passed into the column at P_1; x_{f2} = mole fraction of feed passed into the column at P_2.

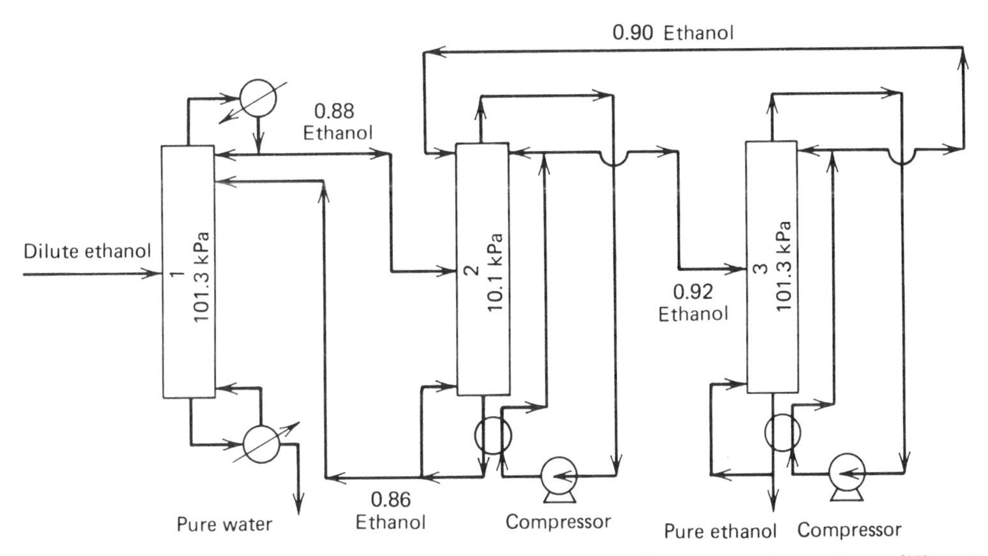

Figure 3. Separation of ethanol and water in a three-towers configuration, each operating at different pressure. Atmospheric azeotrope at 0.894 mole fraction ethanol. Azeotrope vanishes at P = 10.1 kPa (0.1 atm). Vapor recompression is used in towers 2 and 3. 101.3 kPa = 1.0 atm.

operating at above atmospheric pressure may be used to separate maximum-boiling azeotropes. The differences result from the removal of the azeotropes with the bottoms products instead of the overheads (4).

Self-Entrained Systems

The separation of heterogeneous azeotropes is considerably easier than that of homogeneous azeotropes. At the heterogeneous azeotrope, the vapor is in equilibrium with each of the two liquid phases, ie,

$$f_i^V = f_i^1 = f_i^2 \tag{3}$$

or (neglecting the pressure corrections)

$$y_i P = \gamma_i^1 x_i^1 P_i^{sat} = \gamma_i^2 x_i^2 P_i^{sat} \tag{4}$$

Since the vapor pressure is a function of temperature only, the last two equalities of equations 3–4 give

$$\gamma_i^1 x_i^1 = \gamma_i^2 x_i^2 \tag{5}$$

$$i = 1, \ldots, c \tag{6}$$

In that the heterogeneous azeotrope is the average composition of the two liquid phases, equations 5–6 imply that the mole fractions of the two liquid phases in equilibrium lie on the two sides of the azeotrope. Therefore, once the azeotrope has been split into its two liquid phases, the components of the mixture may be obtained by simple distillation. These concepts are applied in the process shown in Figure 4 for the separation of n-butanol and water. The mixture of n-butanol and water at the azeotrope is fed into a decanter. The organic phase of the decanter is passed into a distillation tower for recovery of n-butanol, and the aqueous phase is passed into a stripper for removal of the water. Such systems are said to be self-entrained.

Industrial Processes

In the processes considered in the previous sections, the separation of an azeotropic mixture is accomplished by exploiting the physicochemical properties of the binary mixture. Unfortunately, very few industrially important systems form heterogeneous binary azeotropes. It is not possible in these cases to use processes such as the one shown in Figure 4. Also, not all systems are very sensitive to pressure changes and, therefore, do not lead to cost-efficient designs, eg, those in Figures 1 and 3. When

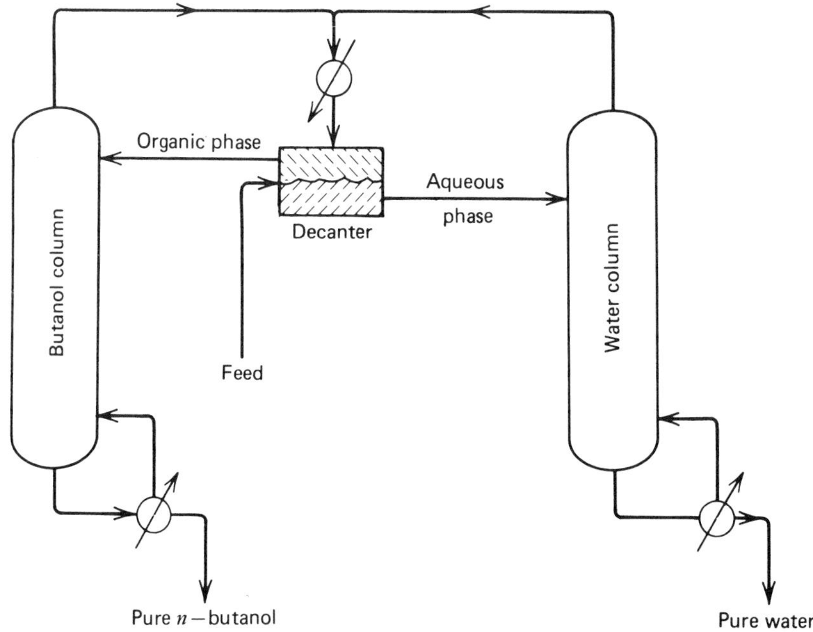

Figure 4. Two-tower separation of n-butanol and water; n-butanol and water form a heterogeneous azeotrope.

the processes discussed in the previous sections are inapplicable, azeotropic or extractive distillation is used to separate close-boiling mixtures.

In both extractive and azeotropic distillation, a third component, called a solvent in extractive distillation or an entrainer in azeotropic distillation, is added to increase the difference in volatility between the key components. The solvent or the entrainer must be present in appreciable concentration in the liquid phase on most of the trays of the tower in order to be effective as a mass-separating agent (5).

The main difference between extractive and azeotropic distillation is in the different methods used to secure appreciable concentrations of the mass-separating agent. In extractive distillation, the solvent is less volatile than the key components; it is fed near the top tray, is almost exclusively removed with the bottoms without formation of an azeotrope, and is recovered in a solvent-recovery column. The solvent is attracted to one or more of the components in the mixture, most probably because of hydrogen bonding. This attraction has also been attributed to the polar characteristics of the resulting mixture, to the formation of unstable chemical complexes, or even to chemical reactions between the solvent and components of the mixture (6).

In azeotropic distillation, the entrainer forms or approaches a heterogeneous ternary azeotrope with the key components in the accumulator, ie, the decanter, at the top of the tower. The entrainer-rich phase is recycled to the azeotropic tower; the other liquid phase is further processed for recovery of the entrainer. Extractive distillation is based on the attraction between the solvent and one or more of the components in the mixture; azeotropic distillation is based on the repulsion between the entrainer and one or more of the components in the mixture. For example, when benzene is used as entrainer to separate the mixture ethanol–water, highly repulsive forces develop between benzene and water, resulting in an immiscible pair. On the other hand, when ethylene glycol is used as the solvent to separate the same mixture in extractive distillation, water is attracted to the solvent and is removed with it from the bottom of the tower. The repulsive forces in heterogeneous azeotropes are attributed primarily to polarity effects.

Separation of Ethanol and Water. The separation of ethanol and water by conventional distillation is not possible because of the presence of a homogeneous azeotrope. However, they can be separated by extractive and azeotropic distillation. Figure 5 is an equipment schematic for the separation of ethanol and water by extractive distillation. Dilute ethanol is fed into a conventional distillation tower at atmospheric pressure. The bottoms product is almost pure water, whereas the overhead composition is near that of the azeotrope. The azeotrope is passed to a second distillation tower. A secondary feed, almost pure solvent, is introduced near the top, and the overhead is almost pure ethanol. The bottoms product consists almost exclusively of solvent and water and is fed into a third tower for solvent recovery. Various solvents have been suggested, including ethylene glycol and gasoline.

Figure 6 shows the production configuration for the separation of ethanol and water by azeotropic distillation. Ethanol is concentrated in a conventional column. The overhead, the composition of which is near the binary azeotrope, is passed to the azeotropic tower. A secondary feed, which is rich in entrainer, is introduced on the top tray. The bottoms product is nearly pure ethanol. The overhead approaches the composition of the ternary azeotrope and is fed into a decanter. The entrainer-rich organic phase of the decanter and a small entrainer make-up stream comprise the secondary feed and are recycled to the azeotropic tower. The aqueous phase of the

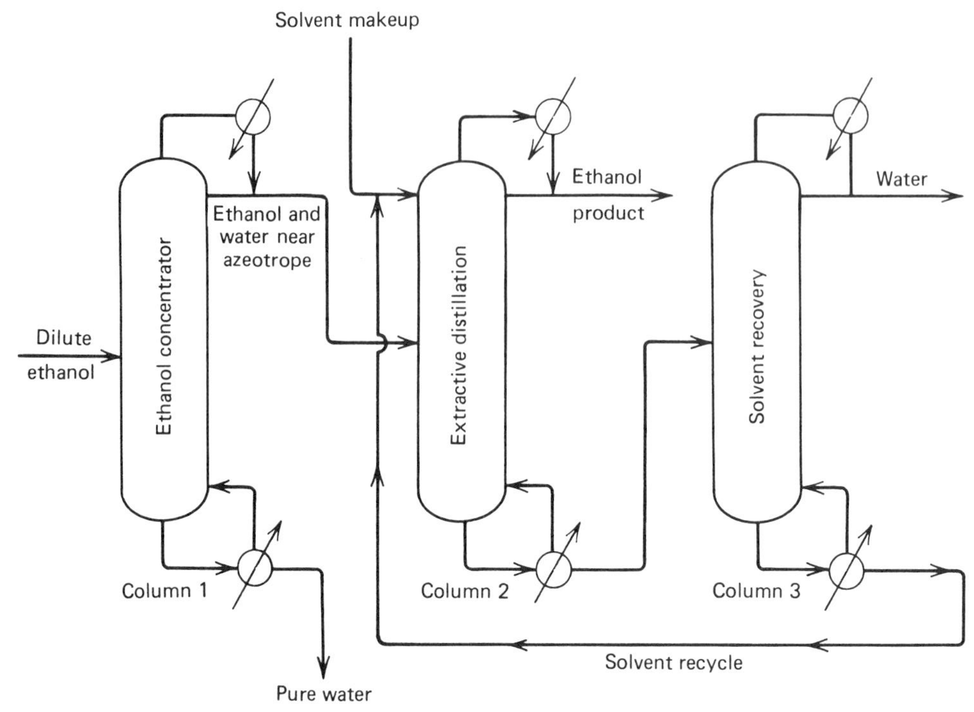

Figure 5. Separation of ethanol and water by extractive distillation.

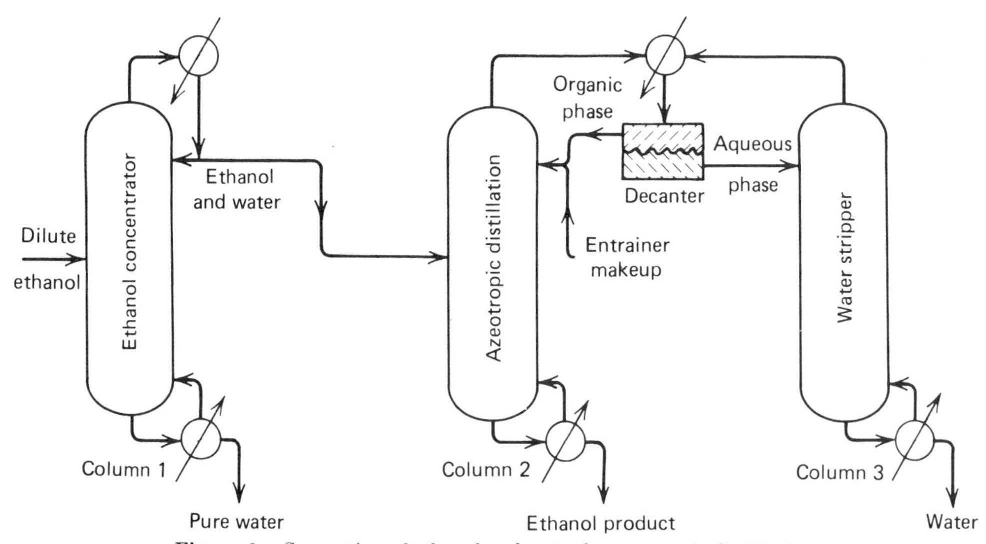

Figure 6. Separation of ethanol and water by azeotropic distillation.

decanter is processed in one or more strippers to recover ethanol and entrainer. Various entrainers have been suggested, including benzene, pentane, cyclohexane, and diethyl ether. The design, operation, and control of an azeotropic separation sequence depends very much on the decanter, where the separation of the two liquid phases of the heterogeneous azeotrope occurs. Also, special care is taken to avoid the formation of the second liquid phase inside the distillation tower. The configurations presented in Figures 5 and 6 are common. Others are described in refs. 7–9.

There is no general or definite answer to the question of which of the two processes, extractive or azeotropic distillation, is more efficient. The two processes in Figures 5 and 6, involving ethylene glycol as solvent and pentane as entrainer, respectively, have been compared (10). It was concluded in this study that azeotropic distillation is a much more economical process and gives better separation. When a suitable entrainer for a given separation is available, azeotropic distillation usually is preferred. However, choosing a good entrainer is not easy for all systems. The principal advantage of extractive distillation is that more solvents can be used, because the precise nature of phase relationships in mixtures of solvent and key components is not critical to the success of the process. For example, it is possible to use gasoline as the solvent for the separation of ethanol and water, if the ethanol is to be used in gasohol (see Gasoline; Ethanol). This significantly reduces both capital investment and utilities cost (9). When high purities are required, azeotropic distillation is the choice.

Three entrainers, ie, benzene, diethyl ether, and *n*-pentane, have been compared in computer-simulation studies (11). For the ethanol and water system, it was concluded that the most suitable entrainer is *n*-pentane, which requires the least number of trays and the smallest heat duties for the same separation. Parametric studies were performed and the key parameter in azeotropic distillation was determined to be the ratio of entrainer to alcohol. An increase in this ratio has a positive effect on the separation, ie, fewer trays and higher purities. However, there also is an increase in the heat load of the reboiler and the condenser and in the internal flow rates without a substantial decrease in the number of trays required for the separation. These findings are summarized in Figure 7.

Separations Based on Combinations of Azeotropic and Extractive Distillation. In certain cases, the desired separation cannot be accomplished by either azeotropic or extractive distillation. In these cases, configurations involving both processes are considered.

Separation of Acetone and Methanol. The configuration for the removal of impurities and separation of a mixture of acetone and methanol is given in Figure 8. The impurities consist primarily of tetramethylene oxide, which forms an azeotrope with acetone. Acetone and methanol also form an azeotrope (see Table 1). It is necessary, then, to separate two azeotropes. To accomplish this, the impure mixture of methanol and acetone is fed into an extraction column; water is the solvent. Being polar, water removes methanol from the bottoms. The overhead, being close to the azeotrope tetramethylene oxide–acetone, is fed into an azeotropic tower; pentane is the entrainer. The oxide is removed with the bottoms stream. The azeotrope of acetone and pentane of the overhead is separated by adding water. In this process, a polar solvent and a nonpolar entrainer are used to break multiple azeotropes (12).

Purification of Methyl Ethyl Ketone. A combination of azeotropic and extractive distillation has been suggested for the purification of methyl ethyl ketone (MEK). The process is shown in Figure 9. A nonpolar entrainer, hexane, removes the oxide impurities in an azeotropic tower. The entrainer is recovered in a water-extraction column; a second water-extraction column removes the acetal impurities. Methyl ethyl ketone and water are finally separated in a pentane-extraction column.

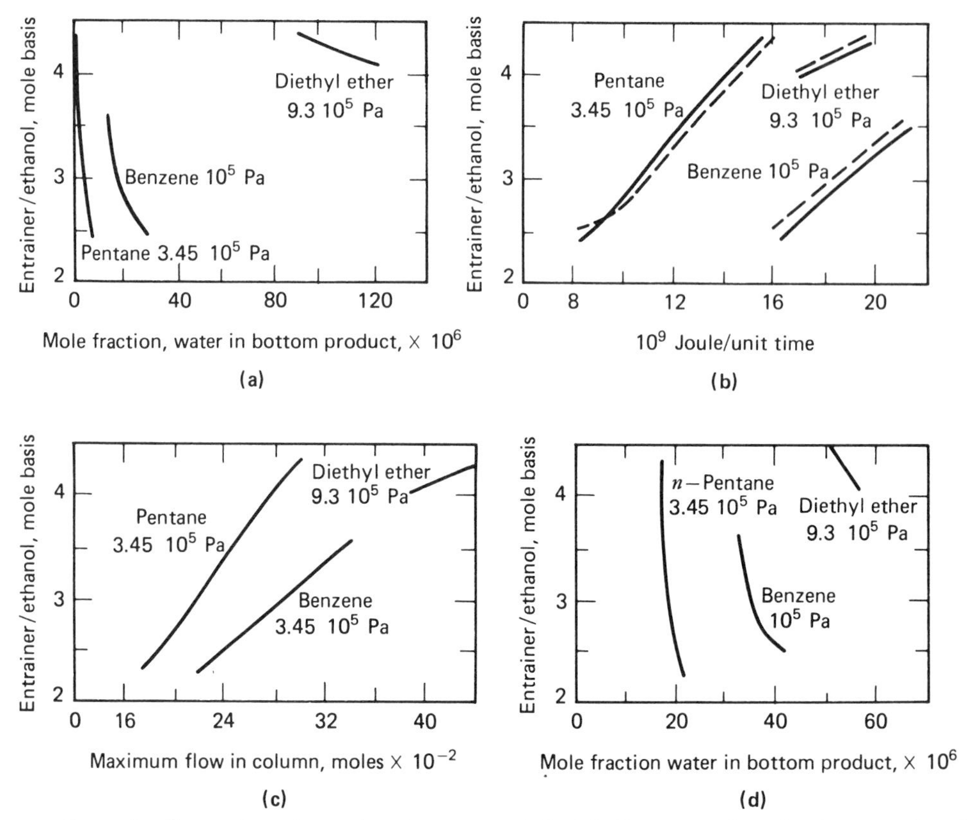

Figure 7. Comparison of n-pentane, benzene, and diethyl ether as entrainers for the separation of ethanol and water by extractive distillation. To convert Pa to mm Hg, divide by 133.3. Effect of entrainer on (**a**) bottoms purity; (**b**) heat duties (solid lines = condenser heat duties, dashed lines = reboiler heat duties); (**c**) internal flow rates; and (**d**) number of trays.

Design of Towers

The basic calculation procedures for extractive distillation towers are the same as for simple distillation towers. An extensive review of these procedures is given in ref. 13.

Prediction of Phase Equilibrium. Perhaps the single most important issue in the design of azeotropic distillation towers is the ability to predict the vapor–liquid and liquid–liquid equilibria. There are many sets of experimental data on azeotropic mixtures available in the literature (14–15). Reference 15 contains sets of fitted parameters with the raw data for many binary and ternary systems. Additional sources, restricted though to UNIQUAC and UNIFAC equations, are references 16 and 17, respectively. The high degree of nonideality of azeotropic mixtures make the simple models for the Gibbs free energy, eg, regular solution, the Margules equation, the Van Laar equation, and the Wilson equation, inadequate. The activity-coefficients model must predict very well the system's behavior as it approaches the composition of the azeotrope and, if possible, the vapor–liquid and liquid–liquid equilibria. Of the available models, only NRTL (nonrandom, two-liquid), UNIQUAC and UNIFAC seem adequate. The various models have been tested in a simulation of a 27-tray tower for the dehydration of ethanol with benzene (18). The results, based on the assumptions

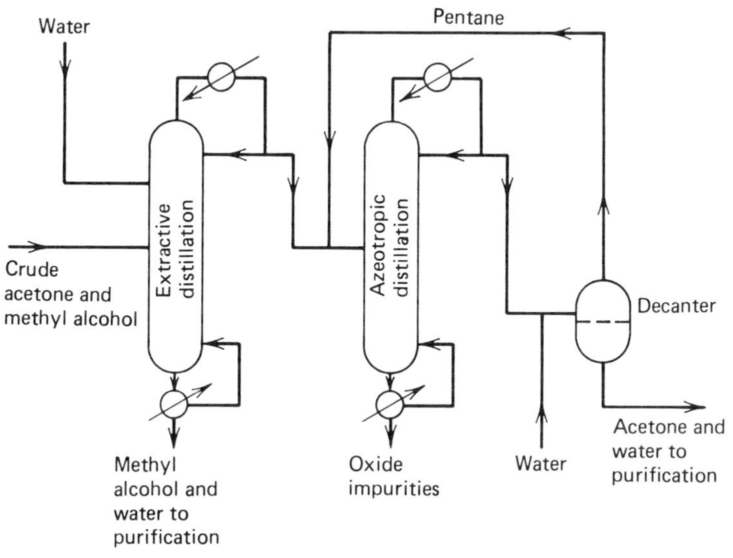

Figure 8. Separation of acetone and methanol by azeotropic and extractive distillation (12).

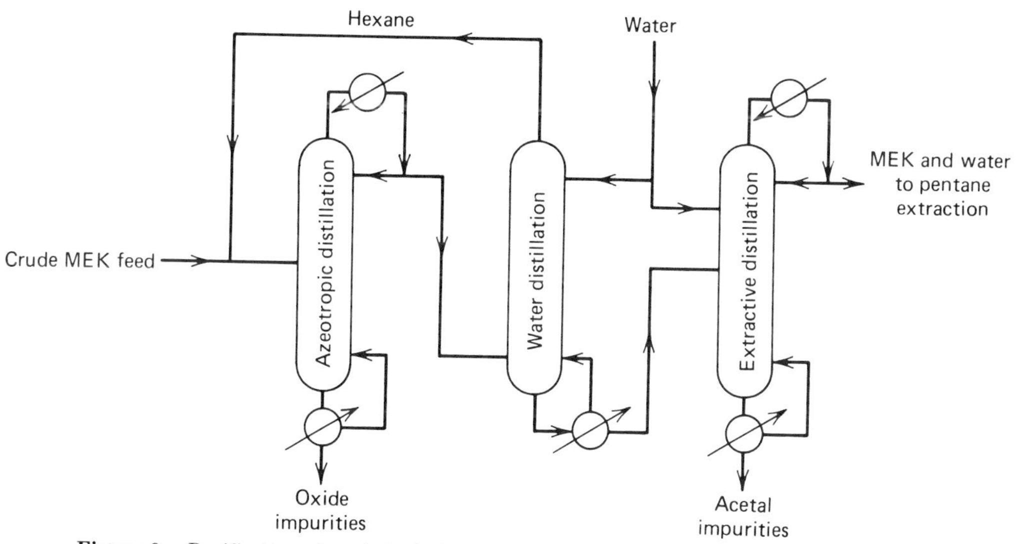

Figure 9. Purification of methyl ethyl ketone by azeotropic and extractive distillation (12).

of constant molal overflow, a single liquid phase on all trays, and no simulation of the decanter, are shown in Figure 10. As shown, even the Wilson equation is inadequate, whereas NRTL, UNIQUAC, and UNIFAC perform equally well. Figure 11 shows the predicted liquid–liquid equilibrium in the decanter as based on the parameters used with the UNIQUAC equation to obtain the liquid-mole-fraction-versus-number-of-tray profile in Figure 10(**c**). The set of parameters taken from ref. 15 predict very well the vapor–liquid equilibrium data reported in ref. 19. However, as shown in Figure 11, these parameters predict an extended, two-liquid-phase envelope and large slopes for the tie lines. These erroneous predictions would lead to the design of an azeotropic tower much smaller than the one needed with a large reflux ratio, and of an unreasonably large stripping tower.

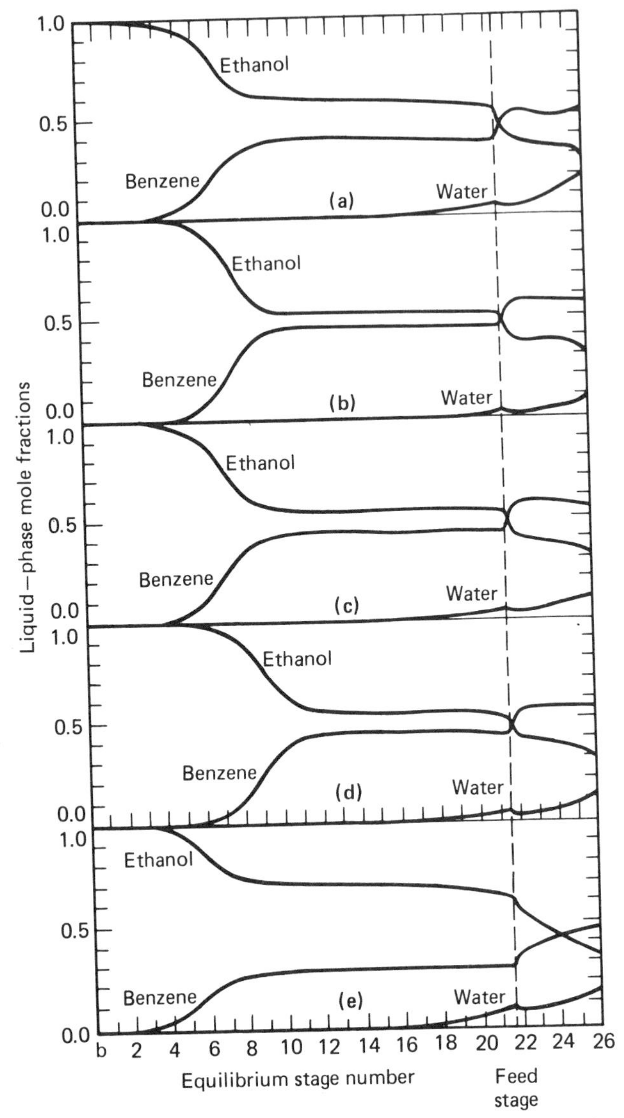

Figure 10. Concentration profiles for the ethanol–benzene–water system, based on various models for the prediction of activity coefficients (18). (**a**) Experimental k values. (**b**) k values from UNIFAC. (**c**) k values from UNIQUAC. (**d**) k values from NRTL. (**e**) k values from the Wilson equation. k value is defined as $k = y/x$.

For a successful design of an azeotropic separation sequence, it is necessary to predict the vapor–liquid–liquid equilibrium (VLLE) of the system. Unfortunately, the available theories do not allow for the determination of a single set of parameters useful for predicting the VLLE. A procedure has been devised for generating parameters to predict satisfactorily VLE and VLLE (16,20).

Design and Simulation. The first algorithm for the design of azeotropic distillation towers is given in ref. 21. The algorithm solves for the material balances tray-to-tray beginning from the reboiler. A similar but more systematic method is published in ref. 11. The latter algorithm includes energy balances as well. In another algorithm,

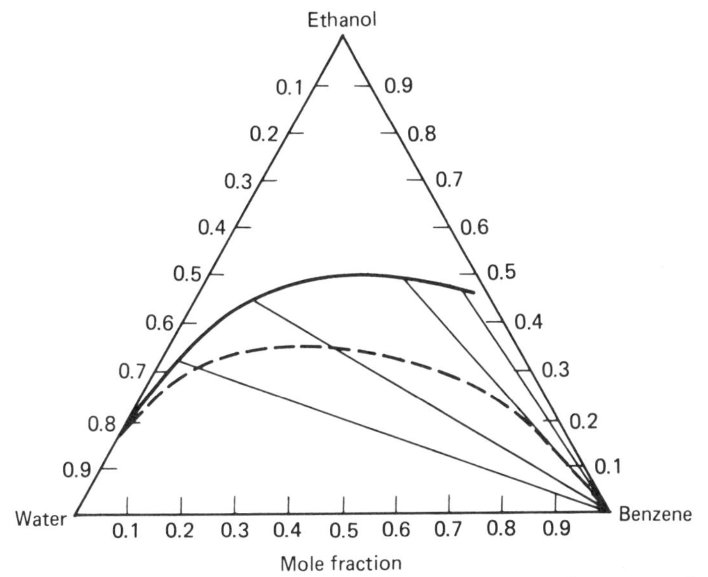

Figure 11. Binodal curve computed from the UNIQUAC equation and parameters from ref. 15 (20). The dashed line shows the experimental curve from ref. 19.

the height of the tower is treated as a continuous variable and a limited number of ordinary differential equations are solved for, rather than many nonlinear algebraic equations (22).

Most of the methods reviewed in ref. 13 are principally suitable for the simulation of azeotropic distillation towers. However, the steep composition profiles (see Fig. 10) cause numerical difficulties in many algorithms. Failures or very slow convergence have been reported (23) when a Boston-and-Sullivan-type algorithm (24) is used. Failures of the algorithms incorporated into the ASPEN software package have also been reported (25) (see Simulation and process design).

The Napthali and Sandholm algorithm (26), the Capital θ Method of Holland (27), and variations of tray-to-tray algorithms seem to be adequate for the simulation of azeotropic distillation towers. A tray-to-tray algorithm has been coupled with an efficient optimization package to obtain conditions that satisfy the material balances and equilibrium relationships of the entire azeotropic distillation tower–decanter–stripper configuration (20).

BIBLIOGRAPHY

"Azeotropes" treated in *ECT* 1st ed., under "Distillation" Vol. 5, pp. 176–179, by E. G. Scheibel, Hoffmann-La Roche Inc.; "Azeotropy and Azeotropic Distillation" in *ECT* 2nd ed., Vol. 2, pp. 839–858, by D. F. Othmer, Polytechnic Institute of Brooklyn; "Azeotropic and Extractive Distillation" in *ECT* 3rd ed., Vol. 3, pp. 352–377, by D. F. Othmer, Polytechnic Institute of New York.

1. W. F. Furter, *Adv. Chem. Ser.* **115,** 35 (1972).
2. *International Critical Tables*, McGraw-Hill Book Co., New York.
3. F. G. Shinskey, *Distillation Control*, McGraw-Hill, New York, 1977, Chapt. 5.
4. C. D. Holland, S. E. Gallun, and M. J. Locket in J. J. McKetta, ed., *Encyclopedia of Chemical Processing and Design*, Vol. 16, Marcel Dekker, Inc., New York, 1982, pp. 96–132.
5. M. Benedict and L. C. Rubin, *Trans. AIChE* **41,** 353 (1945).
6. L. Berg, *Chem. Eng. Prog.* **65,** 52 (1969).

7. R. H. Perry and C. H. Chilton, eds., *Handbook of Chemical Engineers*, 5th ed., McGraw-Hill Book Co., New York, 1973.
8. C. J. King, *Separation Processes*, McGraw-Hill Book Co., New York, 1971.
9. C. Black, *Chem. Eng. Prog.* **76,** 78 (1980).
10. C. Black and D. E. Ditsler, *Adv. Chem. Ser.* **115,** 1 (1972).
11. C. Black, R. A. Golding, and D. E. Ditsler, *Adv. Chem. Ser.* **115,** 64 (1972).
12. W. C. Hopkins and J. J. Fritsch, *Chem. Eng.* **51,** 361 (1955).
13. Y. L. Wang and J. C. Wang in R. H. S. Mah and W. D. Seider, eds., *Computer Aided Chemical Process Design*, American Institute of Chemical Engineers, New York, 1981.
14. L. H. Horsley, *Adv. Chem. Ser.* **116,** (1972).
15. J. Gmehling and U. Onken, *Vapor–Liquid Equilibrium Data Collection*, DECHEMA, Chemistry Data Series, Vols. I and II, Verlag & Druckerei, Friedrick Bischoff, Frankfurt, 1977.
16. J. M. Prausnitz, T. F. Anderson, E. A. Grens, C. A. Eckert, R. Hsieh, and J. P. O'Connell, *Computer Calculations for Multicomponent Vapor–Liquid and Liquid–Liquid Equilibria*, Prentice Hall, Englewood, Cliffs, N.J., 1980.
17. A. Fredenslund, J. Gmehling, and P. Rasmussen, *Vapor–Liquid Equilibria Using UNIFAC*, Elsevier, Amsterdam, 1977.
18. T. Magnussen, M. L. Michelsen, and A. Fredenslund, *Inst. Chem. Eng. Symp. Ser.* **56,** 4.2/1 (1979).
19. W. S. Norman, *Trans. Inst. Chem. Eng.* **23,** 66 (1945).
20. G. J. Prokopakis and W. D. Seider, *AIChE J.* **29,** 49 (1983).
21. C. S. Robinson and E. R. Gilliland, *Elements of Fractional Distillation*, McGraw-Hill Book Co., New York, 1950.
22. D. Van Dongen and M. F. Doherty, *paper presented at the 72nd Annual AIChE Meeting*, San Francisco, Calif., 1979.
23. G. J. Prokopakis, B. A. Ross, and W. Seider in R. H. S. Mah and W. Seider, eds., *Computer Aided Chemical Process Design*, American Institute of Chemical Engineers, New York, 1981.
24. J. F. Boston and S. L. Sullivan, Jr., *Can. J. Chem. Eng.* **52,** 52 (1974).
25. W. Keister, *paper presented at the 1982 Winter Meeting of AIChE*, Orlando, Fla., 1982.
26. L. M. Napthali and D. P. Sandholm, *AIChE J.* **17,** 148 (1971).
27. C. D. Holland, *Fundamentals of Multicomponent Distillation*, McGraw-Hill Book Co., New York, 1981.

GEORGE PROKOPAKIS
Columbia University

B

BLOOD-REPLACEMENT PREPARATIONS

Since 1968, liquid perfluoro compounds and closely related substances have been investigated for biomedical application. This interest is because these nonpolar materials dissolve appreciable quantities of gases, such as oxygen and carbon dioxide, which are of prime importance to living organisms (1). It was demonstrated that the oxygen in solution in F-2-butyltetrahydrofuran (F = perfluoro) is biologically available (2). Mice submerged in this liquid saturated with oxygen were able to extract the oxygen by breathing the liquid in and out. Although it has been shown that animals could survive in oxygenated aqueous solution under pressure, the use of perfluoro compounds allows survival at ambient pressures (3).

It was natural to investigate the possibility that perfluoro compounds could be used for other biological applications where oxygen delivery was crucial. Perfusing organs *in vitro* with perfluoro liquids as such met with limited success because electrolytes and other compounds normally present in perfusion solutions are insoluble. It is also likely that in spite of the low surface tensions and viscosities of these compounds, passage through capillaries was not as facile as might be expected. To overcome these difficulties, an emulsion of F-2-butyltetrahydrofuran was stabilized with bovine serum albumin in an electrolyte solution (4). With this solution, it was possible to maintain the functioning of perfused rat brain for 1–2 h.

In 1968, it was shown that the blood of rats could be almost completely replaced with a protein-free preparation containing emulsified F-tributylamine, electrolytes, and bicarbonate buffer. The emulsifying agent employed was Pluronic F-68, a nonionic material derived from polyoxyethylene and polyoxypropylene (see also Emulsions). In subsequent studies, hydroxyethyl starch was included to furnish additional colloid osmotic pressure (5). It was then possible to replace all of the blood of rats; they survived and thrived. The so-to-say bloodless animals behaved in a normal manner and formed new blood cells and proteins. It was necessary to have the rats breathe 90–100% oxygen since it is carried in solution in the perfluoro compound rather than being bound as with hemoglobin. In other experiments, animals given emulsified perfluoro compounds intravenously were shown to survive in the presence of sufficient carbon monoxide to block oxygen transport by the hemoglobin in their red cells (6–8). The results of these studies proved the efficacy of such blood-replacement preparations.

In spite of the successes with F-2-butyltetrahydrofuran and F-tributylamine, it was necessary to seek other perfluoro compounds that did not remain in the body for a long period. Perfluorodecalin proved to be such a compound with a half-life in the body of ca 7 d (9–10). Unlike F-tributylamine, however, it does not form stable emulsions. Admixture of more easily emulsified perfluoro compounds helps emulsification (11), eg, a commercial product, Fluosol-DA (Green Cross Corp., Osaka), contains a mixture of F-decalin and F-tripropylamine (11–12). This is the only commercially available product that can be used in humans (13–14). After preliminary clinical trials in the FRG, Fluosol-DA is used extensively in Japan and in certain U.S. clinics that have developed protocols for specific uses with FDA approval and in cooperation with the Alpha Therapeutic Corp., the U.S. distributor.

Although Fluosol-DA and a F-tributylamine emulsion (Oxypherol, Green Cross Corp., Osaka) are available to investigators, better preparations are under development. It had been thought that cyclic perfluoro compounds devoid of heteroatoms were necessary for rapid elimination from the body (15). However, this is not the case (16), and a wide variety of fluorinated compounds is under investigation.

Since the perfluoro compounds must be used in emulsified form, the emulsifying agents play an important role in determining the choice of fluoro compounds. Any emulsifying agent destined for intravenous administration must be nontoxic and free of adverse effects on blood components or blood-vessel walls. Currently, only three or four fulfill these conditions.

Because perfluoro compounds carry oxygen in solution, they afford opportunities that cannot be realized with oxygen-binding agents such as hemoglobin. No particular off-loading oxygen pressure is needed. The oxygen freely enters and leaves the perfluoro compounds. Therefore, a high oxygen pressure and small particles can deliver O_2 under conditions not as easily dealt with using red cells. For such reasons, perfluoro-type blood-replacement products are highly versatile. The oxygen-delivery system, however, must be tailored to the individual biological needs. Considerable progress has been made in a relatively short time with a small number of compounds.

The fluorinated compounds used in blood-replacement preparations are variously referred to as perfluorochemicals, perfluorocarbons, fluorocarbons, and fluorochemicals. The compounds used first in this work were F-tributylamine, F-2-butyltetrahydrofuran, and F-decalin. The latter was the only perfluorocarbon.

The perfluoro compounds are colorless and odorless organic compounds in which all hydrogens have been replaced with fluorine. Theoretically, for every hydrogen-containing organic compound there is a perfluoro analogue, and thus the number of possible compounds is enormous. However, because of certain physical properties, the number of those suitable for use in blood-replacement preparations is limited. Consideration of chemical properties is far less important.

Perfluoro compounds vary widely in price, depending upon whether they are common articles of commerce or custom-made specialty products. They should not raise the cost of the blood-replacement preparation above that of a comparable quantity of blood. The commercial product Oxypherol sells for ca $100/500 mL completed mixture (1983 price, Grand Island Biological Co.). The price is expected to fall as sales' volumes increase.

Physical Properties

A number of physical properties are germane to the application of perfluoro compounds in blood-replacement preparations (see Table 1) (17–20). Because there is little intermolecular attraction, molecular packing is minimal, and movement of layers of molecules over one another is largely unimpeded. Thus, surface tensions are very low, and kinematic viscosities are lower than those of the corresponding hydrocarbons (21). The loose packing of the molecules allows considerable space into which gas molecules can enter and leave, and gases are readily soluble in these liquids. This solubility is appreciable, as can be seen from a comparison with the amount of oxygen that is carried by red cells. When the latter are packed by centrifugation, they can hold ca 40 vol % O_2, which is at the lower limit of the vol % O_2 that dissolves in the perfluoro liquids (Table 2) (20,22). Carbon dioxide is even more soluble. Other gases are also soluble, and to date no gases insoluble in perfluoro liquids have been reported; gas mixtures dissolve as well. The amount of gas dissolved depends upon its partial pressure in the external phase. Since Henry's law is followed, a straight line results when the partial pressure is plotted against the concentration of gas in the liquid perfluoro compound. This is different from gas-binding chelates such as hemoglobin that are saturated at a finite external oxygen pressure.

The fluorine atoms densify the perfluoro liquid causing the emulsified particles to settle to the bottom rather than rise to the top like fat emulsion particles. In centrifugation of a mixture of blood and perfluoro emulsion, the blood cells form a layer on top.

Perfluoro compounds are essentially insoluble in water and other polar liquids (Table 3) (22). They dissolve in some other halogenated liquids and usually in each other. Some highly nonpolar solvents dissolve in and are dissolved by the perfluoro compounds. The refractive indices are of some interest because they are extremely low (Table 1). When the refractive index is close to that of water, essentially transparent emulsions are readily formed even though the particle size would indicate the presence of turbidity.

Vapor pressure affects application. Above 3.3 kPa (25 mm Hg), a lethal condition known as bloated lung occurs in which normal respiration becomes impossible. F-2-Butyltetrahydrofuran with a vapor pressure of ca 4 kPa (30 mm Hg) has this effect when its emulsions are injected intravenously. On the other hand, if the vapor pressure is too low, the perfluoro compound remains in the body for too long. F-Tributylamine, for example, has a vapor pressure of 532 Pa (4 mm Hg) and remains in the body almost indefinitely. The most useful perfluoro compounds have vapor pressures between 532 Pa and 3.3 kPa (4–25 mm Hg). Vapor pressure is only one of the physical properties associated with the rate of loss of perfluoro compounds from the body by transpiration through the lungs. It has been postulated that the critical solution temperature of perfluoro compounds correlates inversely with their transpiration rate (18). Thus,

Table 1. Physical Properties of Perfluoro Compounds Suitable for Blood-Replacement Preparations

Perfluoro derivative of	CAS Registry No.	Trade name or abbreviation	Formula	Mol wt	Bp, °C	Vapor pressure at 37°C, kPa	Density at 25°C, g/cm³	n_D^{20}
decalin	[306-94-5]	Flutec PP5 FDC	$C_{10}F_{18}$	462	140–143	1.7	1.946	1.31
1-methyldecalin	[306-92-3]	Flutec PP9 FMD	$C_{11}F_{20}$	512	160	0.64	1.972	1.29
dimethylbicyclo[3.3.1]nonane	[75108-51-9]	FDMBCN	$C_{11}F_{20}$	512	157	0.81		1.31
trimethylbicyclo[3.3.1]nonane	[67711-54-0]	FTMBCN	$C_{12}F_{22}$	562	172–174	0.39	2.025	1.33
7-methylbicyclo[4.3.0]nonane	[75262-86-1]	FMBCN	$C_{10}F_{18}$	462	139	1.79		1.31
bicyclo[5.3.0]decane	[68697-63-2]	FBCD	$C_{10}F_{16}$	462	140	1.73		1.31
tricyclo[5.2.1.0^{2,6}]decane	[307-09-5]	FTCD	$C_{10}F_{16}$	424	135	2.13		
dimethyladamantane	[63267-58-3]	FDMA	$C_{12}F_{20}$	524	174.5	0.25	2.042	
tripropylamine	[338-83-0]	FTPA	$C_9F_{21}N$	521	129–130	2.47	1.82	
tributylamine	[311-89-7]	FC-43	$C_{12}F_{27}N$	671	177	0.15	1.87	1.27
N,*N*-dimethylcyclohexylmethylamine	[82350-91-2]	FMA FDCHMA	$C_9F_{19}N$	483	127–129	2.17	1.84	
4-methyloctahydroquinolidizine	[86563-85-1]	FMOQ	$C_{10}F_{19}N$	495	153	1.43	1.97	
dihexyl ether	[424-20-4]	FHE	$C_{12}F_{26}O$	604	172	0.27	2.025	
Freon E_3	[3330-16-3]	FE3	$C_{11}HF_{23}O_3$	579	143	1.47	1.74	
5*H*,6*H*-5-dodecene	[51249-66-2]	FDD	$C_{12}H_2F_{22}$	545	169	0.17	1.721	
5*H*,6*H*-5-decene	[84551-43-9]	F46E FD F44E	$C_{10}H_2F_{18}$	464	140		1.67	
1-bromooctane	[423-55-2]	FOB	$C_8F_{17}Br$	499	141	1.49	1.94	
2-butyltetrahydrofuran	[335-36-4]	FX80 FC-75	$C_8F_{16}O$	416	103	10.26	1.78	

Table 2. Comparative Solubility at 37°C of Oxygen and Carbon Dioxide in Perfluoro Compounds Emulsified by Pluronic F-68 and Water and Blood

Solvent	Oxygen, vol %	Carbon dioxide, vol %
perfluoro compounds		
tributylamine	41	108
2-butyltetrahydrofuran	58	147
octane [*307-34-6*]	48	180
Freon E_3	56	43
Freon E_4 [*26738-51-2*]	36	77
decalin	45	134
1-methyldecalin	42	126
water	2.3	52.6
blood	23	120

F-tributylamine with a critical solution temperature of 60°C is transpired at a much slower rate than F-tripropylamine that has a critical solution temperature of 44°C.

Chemical Properties

The stability of the C–F bond renders perfluoro compounds chemically inert. Heteroatoms such as nitrogen are usually impossible to detect by classical means. Thus, perfluoroamines show none of the reactions ordinarily associated with a tertiary amine

Table 3. Solubility of Perfluoro Compounds

	Solubility of perfluoro compound in solvent				Solubility of solvent in perfluoro compound			
Solvent	FMD[a]	FDC[a]	FC-47[b,c]	FC-75[b]	FMD[a]	FDC[a]	FC-47[b]	FC-75[b]
acetone			0.9	4.8			0.6	1.2
benzene			0.3	2.6			0.2	3.8
benzotrifluoride			misc	misc			misc	misc
benzyl alcohol			insol	0.2			insol	0.4
carbon tetrachloride	19	31	2.4	20.2	15.0	36.5	24	27
chlorobenzene				1.8			0.4	3.2
chloroform			1.2	7.7			5.4	4.5
cyclohexane			1.8	7.7			2.2	8.4
dioxane			0.1	0.9			insol	2.4
ethyl ether			4.9	misc			5.0	misc
ethyl acetate			2.2	7.5			2.4	6.5
heptane	39	29	6.4	25.5	5.5	8.9	3.4	11.6
isopropyl alcohol			insol	4.1			insol	1.3
methanol			insol	1.0			insol	0.1
petroleum ether			33.2	misc			7.0	misc
Stoddard solvent			1.4	5.9				3.0
toluene	4.6	3.8	0.4	2.9	0.4	3.6		4.1
turpentine			0.9	5.3			insol	1.0
xylene				3.0			1.0	3.0
water			insol	insol			insol	insol

[a] Wt % at 25°C.
[b] g/100 mL at 20°C.
[c] FC-47 = FC-43 = F-tributylamine.

nitrogen. Metallic sodium removes fluorine atoms which forms the basis of an analytical method for determining perfluoro compounds. Refluxing with aqueous KOH or NaOH solutions has no effect and indeed is usually a step in purification procedures. Many of the commercial perfluoro compounds are actually mixtures. In some cases, such as with F-decalin, the chief components are the cis and trans isomers. In general, as supplied, the minor components consist of cleavage and rearrangement products derived during the synthesis steps. Incompletely fluorinated materials should, of course, be removed during processing. However, the desired products of some syntheses may not be true perfluoro compounds but contain residual hydrogens or a double bond which do not affect chemical behavior. The very inertness of the perfluoro compounds is extremely important in blood-replacement preparations. It greatly lessens the chances of adverse reactions and the occurrence of metabolites.

Manufacturing and Processing

Various synthetic pathways for perfluoro compounds are reviewed in refs. 18, 21, 23–25.

Electrochemical Fluorination (Simon Process). The starting material is placed in anhydrous HF in an electrolytic cell equipped with NI and Fe electrodes and cooling provisions (26–27). The cells vary in size from 20 to 50 A at 4–8 V. The starting material must be soluble in HF, ie, a heteroatom must be present such as in amines and ethers. Cyclic and aliphatic hydrocarbons are poor substrates for the Simon process. Since the fluorinated material is heavier than the reactants, it settles to the bottom and can be drained without stopping the operation of the cell. The crude product contains a mixture which is purified by extractions, refluxing with alkali, and fractional distillation.

F-Tributylamine, F-tripropylamine, and F-2-butyltetrahydrofuran are prepared by this process from the corresponding nonfluorinated compound, eg (26):

$$(n\text{-}C_4H_9)_3N \xrightarrow{HF} (n\text{-}C_4F_9)_3N \ (19\%)$$

For F-2-butyltetrahydrofuran, the reaction utilizes the corresponding ether or carboxylic acid (27).

In both cases, a number of other products are also produced that are often difficult, if not impossible, to separate. For example, together with F-*n*-butyltetrahydrofuran, F-2-*n*-propylpentahydropyran is also produced. The mixture is available as the product FC-80 (3M Company). The vigorous reaction occurring at the electrodes causes degradations, and rearrangements may give rise to highly toxic compounds.

Cobalt Trifluoride Catalysis. Fluorine gas with CoF_3 as catalyst is an excellent fluorinating agent at elevated temperature. F-Decalin is prepared in this manner by passing vaporized tetrahydronaphthalene and F_2 over heated CoF_3 (23).

$$\text{(tetrahydronaphthalene)} + 30\ CoF_3 \longrightarrow C_{10}F_{18} + 12\ HF + 30\ CoF_2$$

The resulting product is washed with water, dried, and fractionally distilled. Both cis and trans forms are obtained. Ring cleavage products are formed during the reaction, but these can be fairly effectively removed by fractional distillation procedures.

F-Trimethylbicyclo[3.3.1]nonane is made by this method (18). It is of interest as a component of blood-replacement preparations since it forms clear emulsions.

F-Alkylethenes. F-Alkylethenes have properties that make them of interest as components of blood-replacement preparations. They are produced by the following reactions (24):

F-alkylethenes and ethanes

$$R_FI + CH_2{=}CH_2 \longrightarrow R_FCH_2CH_2I \begin{cases} \nearrow R_FCH{=}CH_2 \\ \searrow R_FCH_2CH_3 \end{cases}$$

$$R_F = C_2F_5, C_4F_9, C_6F_{13}, C_8F_{17}.$$

bis(F-alkyl)ethenes

$$R_FI + R'_FCH{=}CH_2 \xrightarrow[185°C]{} \underset{(60\text{–}80\%)}{R_FCH_2CHIR'_F} \xrightarrow[5°C]{KOH/C_2H_5OH} \underset{(60\text{–}80\%)}{R_FCH{=}CHR'_F}$$

or

$$R_FCH{=}CH_2 \xrightarrow[2.\ KOH/C_2H_5OH]{1.\ Br_2/CCl_4} \underset{(90\text{–}95\%)}{R_FCBr{=}CH_2} \xrightarrow[DMF/150°C]{R'_F I/Cu} \underset{(60\text{–}70\%)}{R_FCH{=}CHR'_F}$$

$$R_F, R'_F = C_2F_5, C_4F_9, C_6F_{13}, C_8F_{17}.$$

tetrakis(F-alkyl)butadienes

$$R_FI + R'_FC{\equiv}CH \xrightarrow[200°C]{} \underset{(80\text{–}90\%)}{R'_FCI{=}CHR_F} \xrightarrow[200°C]{Cu} \underset{(60\text{–}80\%)}{R_FCH{=}C(R'_F)C(R'_F){=}CHR_F}$$

$$R_F, R'_F = C_2F_5, C_4F_9, C_6F_{13}, C_8F_{17}.$$

F-5*H*,6*H*-5-decene has been obtained in 99% purity and has been tested in blood-replacement studies (28).

Emulsion. In high pressure homogenization, equipment such as the Manton-Gaulin homogenizer is used at 13.8 MPa (>2000 psi). Large volumes can be produced and the method is currently used for commercial preparations.

The other method employs ultrasonic equipment such as the Blackstone ultrasonic generator (see Ultrasonics, low power). This procedure is well suited for preparing experimental batches, especially with limited supplies of components. Regardless of the method, it is essential that the heavy perfluoro compound be dispersed as quickly as possible to prevent it from settling. In general practice, the emulsifying agents are dissolved in the water and the mixture is subjected to the action of the dispersing apparatus while the perfluoro compound is fed into the system. Emulsification is continued until the desired particle size is obtained. With improper formulations, a suitable particle size cannot be achieved. In such cases, the perfluoro compound can be recovered, purified, and recycled.

During emulsification, samples are removed for monitoring of particle size, for which several methods are available. Phase microscopy is relatively simple and rapid, and permits constant evaluation of the product. Another method utilizes the Nicoli laser spectrophotometer for quick appraisal of the emulsification progress. Analyses are completed in a few minutes. Readings of optical density are fast and require no complex apparatus. The emulsification is completed when the optical density remains constant. Any method used must be standardized and, when possible, more than one method should be employed. Ultracentrifugation and electron microscopy are not satisfactory for rapid monitoring, but serve to measure final particle size. The equipment, however, is expensive and requires well-trained operators. Because each involves stresses on the emulsified system, it is necessary to guard against artifacts caused by the methods themselves. Both methods also furnish particle-size distributions.

Particle size affects stability and viscosity and the length of time the particles remain in circulation; large particles disappear quicker than small ones. The desired mean particle diameter of $\leq 0.2\ \mu m$ enhances stability and permits the particles to traverse even the smallest capillaries. The emulsification process is greatly influenced by the chemical and physical characteristics of the perfluoro compound. Some are very difficult to emulsify, although they may have other properties that are ideally suited for use in blood-replacement preparations. A combination of perfluoro compounds may facilitate emulsification. For example, this has been used to advantage in stabilizing F-decalin emulsions, eg, addition of F-*n*-dihexyl ether to F-decalin (16). The commercial product Fluosol-DA utilizes a 7:3 mixture of F-decalin and F-tripropylamine.

Processing and Purification. Electrochemical fluorination or CoF_3 catalysis gives partially fluorinated compounds and others formed by ring cleavage, rearrangement, and isomerization. Procedures including several steps are usually less drastic and give rise to fewer impurities.

Impurities may be toxic and may affect emulsification. Chemical, physical, or biological end points may be employed in purification procedures. Biological tests are the most sensitive, especially if a system such as tissue culture is employed (16). Very small traces of highly toxic compounds can escape detection by the most complex instrumental procedure, yet the tissue-culture assay readily discloses their presence. Effects on cells range from complete destruction to modest inhibition of cell multiplication. In some instances, $>1\ \mu L$ of the perfluoro compound in 15 mL of tissue-culture medium causes rapid cell disintegration. These effects occur even though the perfluoro compound may be essentially insoluble in water, and the quantity of toxic substance present extremely small. High resolution gas–liquid chromatography often shows little change in the composition of the original samples before and after they have undergone further purification. When properly purified, perfluoro compounds are not toxic in the tissue-culture-assay systems.

Needless to say, the emulsifying agents must be strictly nontoxic. Pluronic F-68 (BASF Wyandotte Corp.), the most frequently used surfactant for blood-replacement preparations, is purified in several ways. Molecular sieves remove small molecules including polyethylene glycol and polypropylene glycol that form during preparation. Charcoal treatment and column chromatography is another method (16) that gives a product nontoxic to tissue-culture cells. The emulsions made with this material are at least as stable as those made with unpurified Pluronic F-68. In the United States, Pluronic F-68 contains the antioxidant, Ionol (Shell Chemical Corp.), a butylated phenol. It minimizes peroxide formation. The purification procedure given above removes the antioxidant, which in general is not replaced. Purified Pluronic F-68 is stored at $-25°C$ or lower, preferably under nitrogen.

Other most-used emulsifying agents for perfluoro compounds include phospholipids, eg, egg-yolk phospholipids. Phospholipids are a component of the widely used fat emulsion, Intralipid (Vitrum A.B., Stockholm) that is given intravenously as a source of nutritive value and essential fatty acids. The chief constituent of the egg-yolk phospholipids is phosphatidylcholine.

Bovine serum albumin has been employed in animal studies as a stabilizing agent for perfluoro emulsions (4). The albumin keeps the fluoro compound dispersed and, as an oncotic agent, furnishes the necessary colloid osmotic pressure.

As shown in Table 2, the solubility of oxygen in most perfluoro compounds is ca 40 mL O_2/100 mL when equilibrated with 100 wt % O_2. In emulsified blood-replacement preparations, the final concentration of the fluorinated compounds is usually 10–12 vol %, and the oxygen solubility is ca 6 mL/100 mL of product when equilibrated with 95% O_2:5% CO_2. Therefore, a higher oxygen concentration is required.

Administration and Residence Time

When replacing the blood, a small amount is removed first, followed by the administration of the same volume of the artificial product (29). This is repeated until the red cells (hematocrit) and the protein are at the desired concentration. However, the repeated blood-volume losses can have adverse effects.

In another procedure, the blood is removed at the same rate as the administration of the artificial material (30). This so-called isovolemic-exchange method prevents even temporary reductions in blood volume. Changing the internal environment of animals from natural blood to an artificial preparation has little observable effect on their behavior or well-being.

The residence time of perfluoro compounds in the body varies, depending partly on the dose, and comparisons must be made at given doses (see Table 4). The best protocol involves the intravenous administration of the emulsified material. Since the emulsions vary, intraperitoneal injections of the pure compounds should be used for comparisons (18).

Quality Control, Assays

The perfluoro emulsions can be tested for toxicity in both stationary and suspension cultures of human and animal cells. Errors due to turbidity must be avoided.

Table 4. Relative Body Resident Time and Emulsion Quality[a]

Perfluoro compound	Approximate resident time, $t^{1/2}$, days[b]	Emulsion quality[c]
tributylamine	>300	excellent
tripropylamine	63	good
decalin	7	fair
trimethylbicyclo[3.3.1]nonane	30	good
N,N-dimethylcyclohexylmethylamine	20	good–fair
4-methyloctahydroquinolidizine	<10	excellent
F-5*H*,6*H*-5-decene	10	good
Freon E_3	>100	excellent

[a] Refs. 13, 16, 31.

[b] Dose of 4 g/kg body wt in rats weighing ca 100 g.

[c] Pluronic F-68 used as stabilizer at concentration of 2.5% (w/v).

Emulsions that deteriorate during incubation in tissue culture can give erroneous results, especially when electronic particle counters are employed. When testing the fluoro compound in emulsified form, there is no risk of overlaying the cells in stationary cultures such as is possible with the pure perfluoro liquids. When cells are overlaid with the unemulsified material, access to the aqueous medium is cut off, and cell growth or destruction can occur.

Once tissue-culture assays have been successfully passed, animal testing becomes more meaningful. Although it is possible that chemically related toxicity not revealed by *in vitro* assays might be encountered with animal testing, it is more likely that adverse effects *in vivo* are related to physical factors such as particle size, flow characteristics, and viscosity. Specific phenomena like complement activation have to be assayed separately. Adverse effects associated with physical characteristics of the preparation can usually be minimized by adjusting the formulation. In some instances, selective centrifugation and membrane filtration (qv) can prove beneficial.

Animal assays for assessing the toxicity of perfluoro preparations vary widely both in the kind of tests employed and the quantities of materials used in the tests. Doses of 10–40 mL/kg body wt have been given intraperitoneally and intravenously. In the latter case, the increase in blood volume, which is often ignored, may compromise the test results. Tests must be made before or during administration of the product.

A stringent test for such a preparation is to replace most of the animal blood with it (8,32). Suitable preparations allow complete replacement with survival of the animal.

In an even more severe test, the exchange perfusion is repeated daily or every other day for a period of time. These multiple-replacement experiments, however, must be conducted with the animals breathing less than 100% O_2 in order to avoid intoxication from the oxygen. During the perfusions, a number of determinations can be carried out on the animal and on the fluid being removed.

No specifications for perfluoro compounds or blood-replacement products have been established. The compounds should be of reproducible composition as determined by high resolution gas–liquid chromatography and mass spectroscopy as well as by physical constants. Preferably, they consist of a single compound or a set of isomers and are free of traces of reactants, partially fluorinated compounds, residual hydrogen, and free fluoride ions. By definition, perfluorochemicals should contain no residual hydrogen atoms. Infrared spectrometry is routinely used to monitor for the presence of partially fluorinated contaminants.

Materials prepared for injection must be prepared according to pharmaceutical standards, including tests for sterility, safety, and pyrogenicity. They must be chemically and biologically inert.

Health and Safety Factors

There are no known health risks associated with the handling of purified perfluoro compounds or other ingredients of blood-replacement preparations. However, clinicians must be on the alert whenever substances are retained in the body for a long time. When the emulsified compounds or the final complete preparations are administered intravenously, red-cell typing of the recipient is not required. Some reports show that in dogs the blood pressure falls sharply upon administration (33). In humans (34) and animals (35), transient decreases in circulating platelets and leukocytes occur. It is thought that this is related to complement activation caused by the perfluoro compound. A deleterious effect of F-tributylamine on macrophages has recently been suggested (36); more study in this area is needed. Since the different perfluoro compounds remain in the body for different lengths of time, they should not be given in successive doses more frequently than the rate of loss from the body indicates. Excessive accumulation in the liver can lead to temporary changes in liver-function tests. The lungs are the chief organ for the elimination of perfluoro compounds, whereas Pluronic F-68 is removed via the kidneys. Caution should be used in administering these preparations to recipients with impaired functions. Large amounts of the plasma expander, hydroxyethyl starch, may interfere with normal clotting (37); however, no such effect has been reported when this agent is used in perfluoro preparations administered intravenously.

Uses

Because they transport oxygen and carbon dioxide, perfluoro compounds have many potential uses in biology and medicine. In addition, they may be used in industry where gas transport is important, eg, in fermentation (qv) processes (see also Genetic engineering).

As components of blood-replacement preparations, the emulsion particles replace the red cells, and the hydroxyethyl starch replaces the plasma proteins in providing colloid osmotic pressure. The biomedical uses for these oxygen-carrying preparations are both experimental and practical. A number of investigators prepared blood substitutes for animal experimentation, using established or new formulations. Currently, two products manufactured by the Green Cross Corp., Osaka, are being tested for a variety of uses. Oxypherol is designed for animal experimentation. It contains F-tributylamine which is retained too long in the body to be given to humans. However, it serves well for research purposes. Fluosol-DA is employed both experimentally and clinically. It contains both F-decalin and F-tripropylamine with half-life retention values of ca 7 and 63 d, respectively (29). The compositions of these two preparations are given in Table 5.

So far, Fluosol-DA is the only perfluoro product that has been used clinically on a wide variety of hemorrhaging and nonhemorrhaging patients (14,38). For patients who cannot accept blood or its products, Fluosol-DA affords an alternative, as is the case when there are shortages in the supply of compatible blood. Where large amounts of blood are required during operations, especially those utilizing extracorporeal circulation, an artificial preparation saves natural blood for more critical needs. When persons who have received many blood transfusions become sensitized to donor's blood, the artificial product might play an important role. Situations requiring temporary replenishing of the circulatory volume and oxygen-carrying capacity benefit without the risk of hepatitis and other infectious agents. In cases of advanced anemia, perfluoro preparations may be used. Clinical work is underway to determine the usefulness of Fluosol-DA in stroke patients, because the small particles and the increased oxygen pressure may salvage affected brain tissue (39–40). Similar studies are directed toward minimizing permanent damage to heart muscle following failure. Reports to date show that a higher percentage of the tissue at risk can be rescued with perfluoro products than with any other including whole blood (31,41–42). Increased oxygen pressure in

Table 5. Composition of Perfluoro Preparations, w/v %[a]

Constituent	Oxypherol	Fluosol-DA, 20%
perfluorotributylamine	20.0	
perfluorodecalin		14.0
perfluorotripropylamine		6.0
Pluronic F-68	2.56	2.7
yolk phospholipids		0.4
potassium oleate		0.032
glycerol		0.8
NaCl	0.60	0.60
KCl	0.034	0.034
$MgCl_2$	0.020	0.020
$CaCl_2$	0.028	0.028
$NaHCO_3$	0.21	0.210
glucose	0.180	0.180
hydroxyethyl starch	3.0	3.0

[a] Water added to a total of 100 mL.

tissues has been obtained following the administration of these products while the inspired air was enriched with oxygen (43–44). The perfusion of isolated organs such as heart and kidney for transplant purposes or experimental study is an important use for these oxygen-carrying mixtures (45–47).

There are many other actual and potential uses of perfluoro preparations. Since oxygen can be transported even in the presence of carbon monoxide, they may be of therapeutic value in treating CO intoxication. Rats whose blood was completely exchanged with the artificial mixture withstood breathing a 50% CO:50% O_2 mixture for up to 5 h (6–8). Rats given emulsified perfluorochemical and placed in oxygen at a pressure of 193 kPa (28 psi) show convulsions at ca 20 min compared to control values of 4.5 h (48). These results show the effectiveness of products that follow Henry's law.

BIBLIOGRAPHY

1. S. Howlett, D. Dundas, and D. C. Sabiston, Jr., *Arch. Surg. (Chicago)* **91,** 653 (1965).
2. L. C. Clark, Jr., and F. Gollan, *Science* **152,** 1755 (1966).
3. J. A. Klystra, *Fed. Proc. Fed. Am. Soc. Exp. Biol.* **29,** 1724 (1970).
4. H. A. Sloviter and T. Kamimoto, *Nature (London)* **256,** 458 (1967).
5. R. P. Geyer, *Fed. Proc. Fed. Am. Soc. Exp. Biol.* **34,** 1499 (1975).
6. H. A. Sloviter, M. Petkovic, S. Ogashi, and H. Yamada, *J. Appl. Physiol.* **27,** 666 (1969).
7. K. Yokoyama, *Jpn. J. Surg.* **8,** 342 (1978).
8. R. P. Geyer, K. Taylor, R. Eccles, and E. Duffett, *Fed. Proc. Fed. Am. Soc. Exp. Biol.* **35,** 828 (1976).
9. L. C. Clark, Jr., F. Becattini, S. Kaplan, V. Obrock, D. Cohen, and C. Becker, *Science* **181,** 680 (1973).
10. H. Okamoto, K. Yamanouchi, T. Imagawa, R. Murashima, K. Yokoyama, R. Watanabe, and R. Naito, *Proceedings of the IInd Intercompany Conference*, Osaka, Japan, 1973.
11. R. P. Geyer in E. Hasegawa and I. Kishimoto, eds., *Proceedings of the Xth International Congress for Nutrition: Symposium on Perfluorochemical Artificial Blood Substitutes, Kyoto, 1975*, Igakushobo, Osaka, Japan, 1975, pp. 3–19.
12. R. Naito and K. Yokoyama in V. Novakova and L.-O. Plantin, eds., *Research on Perfluorochemicals in Medicine and Biology, 1977 Proceedings*, Gotab, Stockholm, 1978, p. 42.

13. K. Yokoyama, M. Watanabe, and R. Naito in R. Frey, H. Beisbarth, and K. Stosseck, eds., *Oxygen Carrying Colloidal Blood Substitutes: 5th International Symposium on Perfluorochemical Blood Substitutes, Mainz, 1981*, W. Zuckschwerdt, Verlag Munchen, Munich, FRG, 1982, pp. 214–219.
14. K. K. Tremper, A. E. Friedman, E. M. Levine, R. Lapin, and D. Camarillo, *N. Engl. J. Med.* **307,** 277 (1982).
15. L. C. Clark, Jr., E. P. Wesseler, M. L. Miller, and S. Kaplan, *Microvasc. Res.* **8,** 320 (1974).
16. R. P. Geyer in T. Mitsuno, Chair, *Proceedings of the IVth International Symposium on Perfluorochemical Blood Substitutes, Kyoto, 1978*, Excerpta Medica, Amsterdam, The Netherlands, 1979, pp. 3–32.
17. Unpublished data, Green Cross Co., Osaka, Japan, 1982.
18. R. E. Moore and L. C. Clark in ref. 13, pp. 50–60.
19. R. E. Moore, *Final Report: The Synthesis and Biological Screening of New and Improved Fluorochemical Compounds for Use as Artificial Blood Substitutes*, SunTech, Inc., Marcus Hook, Pa., 1979, pp. 7, 61.
20. R. P. Geyer in E. J. Ariens, ed., *Medicinal Chemistry*, Vol. 7, *Drug Design*, Academic Press, New York, 1976, pp. 1–55.
21. J. W. Sargent and R. J. Seffl, *Fed. Proc. Fed. Am. Soc. Exp. Biol.* **29,** 1699 (1970).
22. K. Yokoyama and co-workers, *Fed. Proc. Fed. Am. Soc. Exp. Biol.* **34,** 1480 (1975).
23. D. W. Cottrell in ref. 12, pp. 32–41.
24. J. G. Riess and M. LeBlanc, *Angew. Chem. Int. Ed. Engl.* **17,** 621 (1978).
25. D. D. Dixon and D. G. Holland, *Fed. Proc. Fed. Am. Soc. Exp. Biol.* **34,** 1444 (1975).
26. U.S. Pat. 2,616,927 (Nov. 4, 1952), E. A. Kauck and J. H. Simon (to 3M Company).
27. U.S. Pat. 2,644,823 (July 7, 1953), E. A. Kauck and J. H. Simon (to 3M Company).
28. M. LeBlanc and J. G. Riess in ref. 13, pp. 43–49.
29. *Technical Information Service No. 4*, Green Cross Corp., Osaka, Japan, Dec. 26, 1976.
30. R. P. Geyer, *Bibl. Haematol. (Basel)* **38,** 802 (1971).
31. P. Menasche and co-workers in R. P. Geyer, G. Nemo, and R. Bolin, eds., *Progress in Clinical and Biological Research, Vol. 122: Advances in Blood Substitute Research*, Alan R. Liss, Inc., New York, 1983, pp. 363–373.
32. R. P. Geyer in G. B. Brewer, ed., *Erythrocyte Structure and Function*, Alan R. Liss, Inc., New York, 1975, pp. 565–584.
33. T. Suyama and co-workers in ref. 16, pp. 257–266.
34. G. M. Vercellotti, D. E. Hammerschmidt, P. R. Craddock, and H. S. Jacob, *Blood* **59,** 1299 (1982).
35. P. Lau, V. S. Shanker, L. L. Mayer, H. A. Wurzel, and H. A. Sloviter, *Transfusion* **15,** 432 (1975).
36. R. Bucala, M. Kawakami, and A. Cerami, *Science* **220,** 965 (1983).
37. B. Alexander, K. Odake, D. Lawlor, and M. Swanger, *Fed. Proc. Fed. Am. Soc. Exp. Biol.* **34,** 1429 (1975).
38. T. Mitsuno, H. Ohyanagi, and R. Naito, *Ann. Surg.* **195**(1), 60 (1982).
39. N. J. Zervas and co-workers in *Princeton Conference Proceedings*, Raven Press, New York, 1983.
40. S. J. Peerless, R. Ishikawa, I. G. Hunter, and M. J. Peerless, *Stroke* **12,** 558 (1981).
41. D. H. Glogar, R. A. Kloner, J. Muller, K. W. V. DeBoer, E. Braunwald, and L. C. Clark, Jr., *Science* **211,** 1439 (1981).
42. G. T. Magovern, Jr., J. T. Flaherty, V. L. Gott, B. H. Bulkley, and T. J. Gardner, *Ann. Thorac. Surg.* **34,** 249 (1982).
43. M. Kessler and co-workers in ref. 31, pp. 237–248.
44. R. E. Rude and co-workers, *Clin. Res.* **28,** 617A (1980).
45. M. Iwai and co-workers in ref. 16, pp. 225–236.
46. L. D. Segel and S. V. Rendig, *Am. J. Phys. Heart Circ. Phys.* **242,** H485 (1982).
47. J. Lutz and co-workers in ref. 16, pp. 123–136.
48. R. P. Geyer in A. Cubitt, ed., *Oxygen and Life: Lectures Delivered at the 2nd BOC Priestley Conference*, Royal Society of Chemistry, Burlington House, London, 1981, pp. 132–141.

ROBERT GEYER
Harvard University

C

CARDIOVASCULAR AGENTS

Cardiovascular agents, as described in the article in Volume 4, are drugs that affect the circulatory system of the body. Six types of cardiovascular agents were described in the Volume 4 article, and four additional systems are discussed here: calcium-channel blocking agents, angiotensin-converting enzyme inhibitors, transdermal nitroglycerin systems, and cardiotonic agents.

Calcium-Channel Blocking Agents

Calcium-channel blocking agents or calcium-entry blockers are a group of compounds that are known to alter the availability of calcium by affecting its entry into its receptor sites. The prototype drugs of this group of compounds, such as verapamil, nifedipine, and diltiazem, have been utilized widely for a long period of time for the treatment of angina pectoris and certain peripheral vascular diseases in Europe and Asia before their recent introduction into the United States (1). The significance of the discovery and recognition of the beneficial therapeutic effects of these drugs has been described as a "new orientation to cardiovascular research and therapy" (2). Because of the overwhelming therapeutic improvement in the treatment of angina, considerable interest and research have been generated also for the use of these drugs in arrhythmias and even in the prevention of myocardial infarction (3). Calcium plays a vital role in excitation–contraction coupling, and failure to maintain intracellular calcium homeostasis results in cell death (4). The availability of the calcium-entry blockers is not only useful in the treatment of cardiac diseases but also provides a powerful tool for the basic studies of excitation–contraction coupling, stimulus–excretion coupling, and other specific physiological functions.

The Effect of Calcium on Vascular Smooth-Muscle Contraction. Calcium acts on a number of sites associated with the control of the cytoplasmic-calcium concentration. Vascular smooth-muscle contraction can be initiated by the opening of the calcium-influx channel allowing extracellular calcium to enter the sarcolemma membrane and thus the cytoplasmic compartment. This results in the elevation of the intracellular calcium concentration to 1×10^{-5} M, a threshold concentration to initiate the activation of the contraction (5–6).

In the presence of calcium, the primary contractile protein, myosin, is phosphorylated by the myosin light-chain kinase. This initiates the subsequent actin activation of the myosin adenosine triphosphatase and results in muscle contraction. Removal of calcium inactivates the kinase and allows the myosin light-chain phosphatase to dephosphorylate myosin which results in muscle relaxation (7). Therefore, the general biochemical mechanism for the muscle-contraction process is dependent upon the availability of a sufficient intracellular calcium concentration.

Mode of Action. The calcium-channel blocking agents interfere with the entry of calcium through the membrane channel and therefore prevent the intracellular calcium from reaching the critical concentration necessary to initiate contraction. However, increasing evidence indicates that these drugs may be acting through other mechanisms to prevent the increase of the free calcium ion concentration in the cell (8).

Therapeutic Uses. The calcium-channel blocking agents have been used mainly for the treatment of various types of angina and arrhythmias, but these drugs may also have some potential therapeutic applications in arterial hypertension, left-ventricular failure, acute myocardial infarction, cardiac preservation, cardiomyopathy, cerebral vasospasms, and other vasospastic syndromes (9).

Structure. The chemical structures of a number of calcium-channel blocking agents are given in Table 1. Although some of these compounds are analogues of prototypic compounds, a wide variety of chemical structures is included. Future progress within this class of agents is based on the development of compounds with selectivity for various tissues, eg, myocardial, vascular (cerebral, peripheral, or coronary), blood elements (erythrocytes, lymphocytes), kidney, and so forth.

Verapamil. Verapamil exerts potent negative inotropic effects in isolated cardiac preparations (10). However, little change in contractility occurs in patients (11). The effects of verapamil on the electrocardiogram include prolongation of the P–R and the A–H intervals (12) (the P–R interval is the length of time for an electric impulse to spread across the atrium and through the conduction system to the ventricles; the A–H (atrium to His bundle) interval is the length of time for an electric impulse to spread across the atrium and the atrioventricular node to His bundle). However, the duration of the QRS (ventricular contraction) complex, the Q–T (ventricular contraction and relaxation), and the H–V (His bundle to ventricle) intervals are not affected even at very high doses (12). Thus, it appears that electrophysiological effects of verapamil are selective for atrial-conducting tissues.

The absorption of verapamil after oral administration is nearly complete (92%). The onset of action is 30 min and peak effect occurs 4–5 h later. The main route of elimination is via hepatic metabolism by *N*-dealkylation and *O*-demethylation. About 15% of the pharmacological effect of this drug is due to one of its metabolites, norverapamil. Almost 70% of the drug is eliminated in the urine and 16% is excreted in the feces within 5 d of administration; only ca 3–4% appears unchanged in the urine (9,13). The biological half-life of verapamil is 3.5–7 h.

Table 1. Chemical Structures and Generic and Trade Names of Some Calcium-Channel Blocking Agents in Clinical Use or Clinical or Preclinical Development

Generic name	CAS Registry No.	Structural formula	Trade name
verapamil	*[52-53-9]*	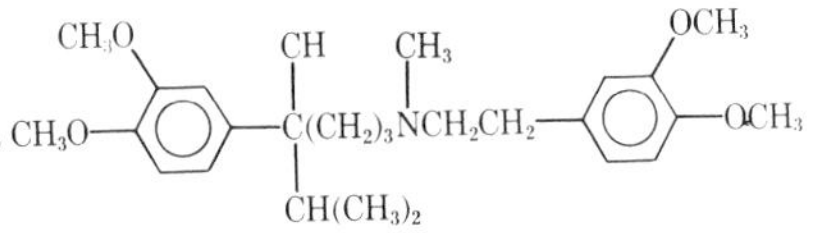 CH_3O; CH; CH_3; OCH_3; CH_3O—; —$C(CH_2)_3NCH_2CH_2$—; —OCH_3; $CH(CH_3)_2$	Isoptin, Iproveratril, Calan
nifedipine	*[21829-25-4]*	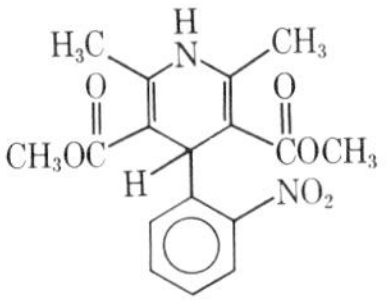 H; H_3C; N; CH_3; O; O; CH_3OC; H; $COCH_3$; NO_2	Procardia, Adalat
diltiazem	*[33286-22-5]*	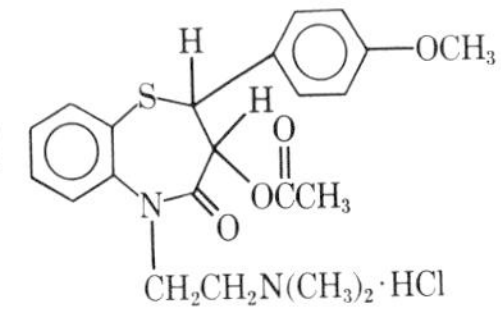 H; —OCH_3; S; H; O; N; O; $OCCH_3$; $CH_2CH_2N(CH_3)_2\cdot HCl$	Herbesser, Anginyl, Cardiem
perhexiline	*[6724-53-4]*	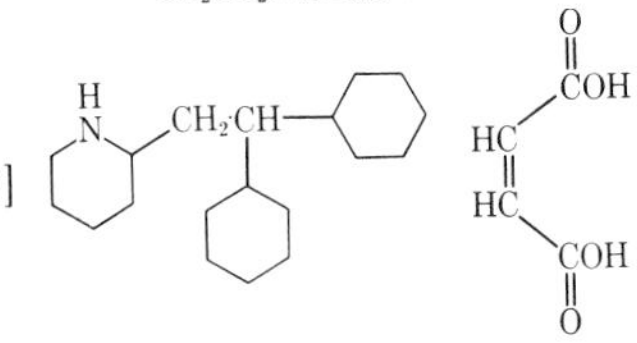 H; N; CH_2CH—; O; COH; HC; HC; COH; O	
lidoflazine	*[3416-26-0]*	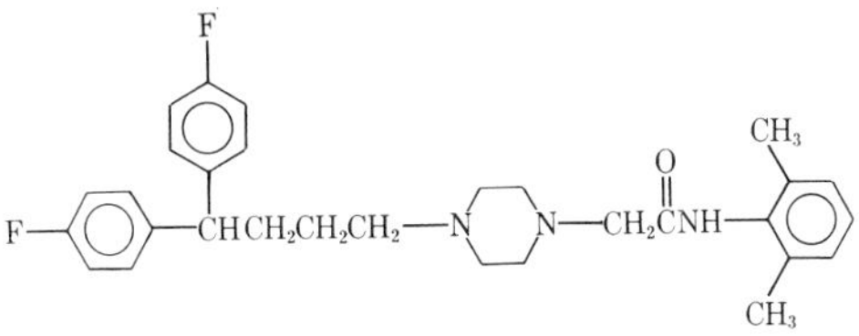 F; F—; —$CHCH_2CH_2CH_2$—N; N—CH_2CNH—; O; CH_3; CH_3	Clinium, Angex
nicardipine	*[54527-84-3]*	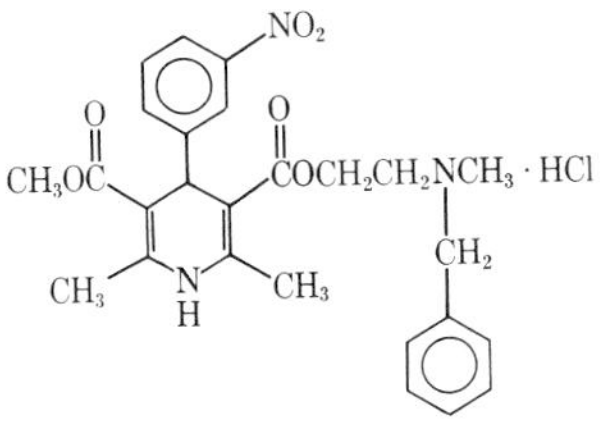 NO_2; O; O; CH_3OC; $COCH_2CH_2NCH_3\cdot HCl$; CH_2; CH_3; N; H; CH_3	
nimodipine	*[66085-59-3]*	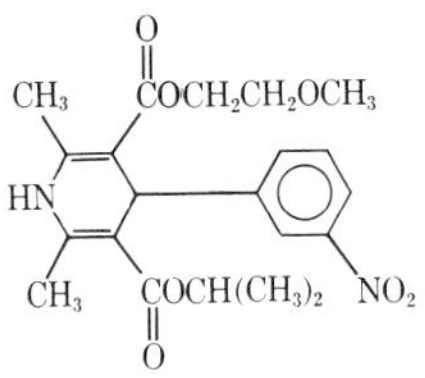 O; CH_3; $COCH_2CH_2OCH_3$; HN; CH_3; $COCH(CH_3)_2$; NO_2; O	
niludipine	*[22609-73-0]*	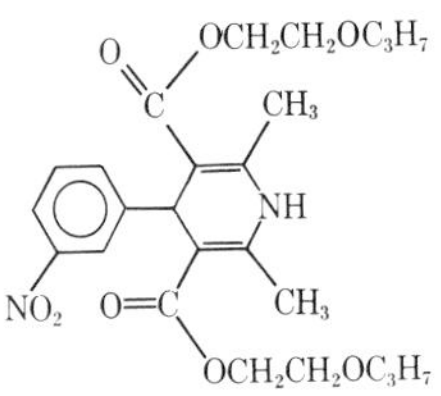 O; $OCH_2CH_2OC_3H_7$; C; CH_3; NH; NO_2; O=C; CH_3; $OCH_2CH_2OC_3H_7$	

Table 1 (*continued*)

Generic name	CAS Registry No.	Structural formula	Trade name
nisoldipine	[*63675-72-9*]	1,4-dihydropyridine ring (HN; H_3C, CH_3), with $C(=O)OCH_3$ and $C(=O)OCH_2CH(CH_3)_2$, and a 2-nitrophenyl group (NO_2)	
nitrendipine	[*80873-62-7*]	1,4-dihydropyridine ring (HN; CH_3, CH_3), with $C(=O)OC_2H_5$ and $C(=O)OCH_3$, and a 3-nitrophenyl group (NO_2)	
flunarizine	[*52468-60-7*]	$(F-C_6H_4)_2CH-N(piperazine)N-CH_2CH{=}CH-C_6H_5$	Sibelium
cinnarizine	[*298-57-7*]	$(C_6H_5)_2CH-N(piperazine)N-CH_2CH{=}CH-C_6H_5$	Sturgeon
bepridil	[*74764-40-2*]	$C_6H_5CH_2-N(C_6H_5)-CHCH_2OCH_2CH(CH_3)_2$, with CH_2-N(pyrrolidine) on the CH	Cerm-1978
prenylamine	[*390-64-7*]	$(C_6H_5)_2CHCH_2CH_2NHCH(CH_3)CH_2-C_6H_5$	
oxatomide	[*60607-34-3*]	$(C_6H_5)_2CH-N(piperazine)N-(CH_2)_3-N$ (benzimidazolone, C=O, NH)	Tinset

Side effects include hypotension, facial flushing, headache, dizziness, constipation, and nausea; frequency is ca 9%. Heart failure and A–V (atrioventricular) block have also been noted; ca 1% of the patients require discontinuation of therapy.

Nifedipine. Nifedipine has more potent negative inotropic effects than verapamil on isolated cardiac tissues (14). Clinically, however, little effect on cardiac contractility in patients has been noted. In contrast to verapamil, nifedipine shortens the P–R interval (15). Nifedipine shows an increase in A–V conduction time only at high doses (16). Nifedipine does not affect the A–H and the H–V interval nor does it have a tendency to precipitate A–V block (17).

Orally administered nifedipine is almost completely absorbed. The onset of action is 20 min and peak effect occurs after 1–2 h. The main route of elimination is via hepatic metabolism by oxidation to a hydroxycarboxylic acid and the corresponding lactone (18). These metabolites are pharmacologically inactive. Almost 70–80% of drug is eliminated in the urine during the first 24 h; ca 15% is excreted in the feces. Its biological half-life is ca 1–2.5 h.

The side effects with nifedipine include hypotension, facial flushing, headache, dizziness, nausea, vomiting, edema, and epigastric pressure. Frequency of occurrence of side effects is ca 17%; in ca 5%, therapy must be discontinued.

Diltiazem. Diltiazem has a slight negative inotropic effect on isolated myocardial tissues (19). The drug slightly prolongs the P–R interval, increases A–V conduction time, and prolongs the A–H interval (20).

The absorption after the administration of one oral dose of diltiazem is nearly complete. The onset of action is ca 15 min and peak effect occurs after 30 min. The main route of elimination is via hepatic metabolism by deacetylation, *N*-demethylation, and *O*-demethylation (21). The primary metabolite, desacetyldiltiazem exhibits some degree of pharmacological activity (20). Almost 60% of the drug is eliminated in the feces and 35% is excreted in the urine. Its biological half-life is 5.5 h.

The side effects noted with diltiazem include facial flushing, headache, dizziness, nausea, abdominal discomfort, and constipation. They occur in ca 4% of the patients and usually do not require discontinuation of therapy. Heart failure, hypotension, A–V block, or edema have not been observed.

Perhexiline. This compound exhibits a slight negative inotropic effect in isolated cardiac preparations (22). Perhexiline prolongs the QRS complex, the P–R interval, and the A–H interval. Orally administered perhexiline is rapidly absorbed and has a relatively long half-life. Upon prolonged administration, it produces serious adverse effects including hepatitis, peripheral neuropathy, and weight loss as well as the side effects exhibited by other calcium-entry blockers (22).

Lidoflazine. This drug has a slight negative inotropic effect in isolated cardiac preparations (23). Lidoflazine is relatively well absorbed after oral administration and has a long half-life. The main route of elimination is through hepatic metabolism via oxidative *N*-dealkylation and hydroxylation; ca 75% is excreted within one week. The adverse effects include headache, flushing, gastric distension, and tinnitus (20).

Nicardipine. This drug exhibits a moderate degree of negative inotropic effect in isolated cardiac preparations (24) and prolongs the A–V conduction time. Nicardipine is completely absorbed after oral administration and has a biological half-life of ca 1 h. The main path of elimination is via hepatic metabolism by hydrolysis and oxidation. The metabolites are relatively inactive and have no pharmacological activities (25).

Nimodipine. This drug blocks the vasoconstrictor effect of a number of agonists in isolated vascular tissues. It dilates smaller arteries and increases cerebral and coronary blood flow and cardiac output. Treatment with nimodipine increases the survival of mice under hypoxic conditions and prevents the occurrence of impaired cerebral reperfusion after total cerebral ischemia in cats (26).

Niludipine. In anesthetized dogs, the intravenous administration of this drug significantly decreased coronary artery resistance, heart rate, and systemic blood pressure. These cardiovascular changes were associated with an increase in coronary blood flow. An increased venous blood return contributed to the increase in coronary blood flow in spite of the presence of the drug-induced hypotension. The increased coronary blood flow is not associated with an increased myocardial oxygen demand (27).

Nisoldipine. Experiments in dogs have shown that this compound effectively increases coronary blood flow and cardiac output and decreases total peripheral resistance and systemic blood pressure. The reflex tachycardia produced is blocked by propranolol which does not affect the fall in blood pressure. At higher doses, the compound exhibits weak tranquilizing and anticonvulsant properties (28).

Other Calcium-Channel Blocking Agents. Other compounds described as having calcium-channel blocking activity are nitrendipine, flunarizine, cinnarizine, bepridil, and prenylamine. Oxatomide appears to possess calcium-channel blocking action in protecting mast cells from degranulation (29) and thus may provide a basis for the therapeutic application of calcium-channel blockers in allergies.

Angiotensin-Converting Enzyme Inhibitors

Over the past decade, the treatment of hypertension has made great strides. Various types of antihypertensive agents with different mechanisms of action are now readily available (see Table 2). The introduction of β-adrenergic receptor blocking agents, such as propranolol, has provided a highly effective modality for the control of hypertension, although its antihypertensive mechanism is still not completely understood. These antihypertensive agents, given alone, in combination, or with a step-care procedure (varying medication) effectively control high blood pressure in most patients.

Captopril, the newly developed converting enzyme inhibitor, has recently been shown to be an effective antihypertensive agent both in experimentally induced hypertension in animals and in hypertensive patients. Because of its unique mechanism of action, the availability of this compound has provided a new weapon to combat hypertensive disease and also serves as a pharmacological tool for studying the role of the renin–angiotensin system in the development of hypertension.

Renin–Angiotensin System and Blood-Pressure Regulation. The normal physiological function of the renin–angiotensin system is the maintenance of both blood pressure and sodium balance in response to decreased renal-perfusion pressure (30). Renin is released from the juxtaglomerular cells in the kidney in response to reductions in renal blood flow and reacts with plasma globulin to produce an inactive decapeptide, angiotensin I. Angiotensin I is then converted within the endothelium of the vascular beds of the lung and other organs by angiotensin-converting enzyme to the active octapeptide hormone, angiotensin II, a potent vasoconstrictor agent. Angiotensin II, in addition to its pressor activity, stimulates the release of aldosterone from the adrenal cortex, which causes sodium retention leading to an increased blood volume. This volume expansion together with the increased blood pressure restores the renal blood flow and, in turn, reduces or inhibits renin release. However, under certain conditions, such as excessive kidney damage, this leads to an uncontrolled activation of the renin–angiotensin system and the initiation of hypertension (31).

Table 2. Chemical Structures and Generic and Trade Names of Various Angiotensin-Converting Enzyme Inhibitors

Generic name	CAS Registry No.	Structural formula	Trade name
captopril, SQ 14225	[*62571-86-2*]	CH_3, CH_2CHCN, SH, O, COH, O	Capoten
enalapril, MK421	[*82009-37-8*]	CH_3, $CH_2CH_2CHNHCHCN$, O, C, O, OC_2H_5, COH, O; HC, HC, COH, O, COH, O	
pivalopril, RHC 3659	[*76963-39-8*]	CH_3, CH_3, CH_3, O, S, O, N, CH_3, COH, O	
SA 446	[*72679-47-1*]	OH, S, N, O=, COH, O, CH_2SH	
CI 906	[*82768-84-1*]	H_3C O, $CH_2CH_2CHNHCHC$—N, COC_2H_5, O, H, COH, O	

Mode of Action. Captopril effectively inhibits angiotensin-converting enzyme *in vivo* (32) and *in vitro* (33). Inhibition of the generation of angiotensin II in plasma was thought to be the main mechanism for the antihypertensive action of captopril. However, recent studies indicate that several factors other than inhibition of angiotensin-II generation may contribute to the antihypertensive activity of captopril.

Systemic or Local Accumulation of Bradykinin. Angiotensin-converting enzyme is identical to kininase II (34) which is responsible for the inactivation of bradykinin. The inhibition of the catabolism of bradykinin markedly potentiates its depressor response. Therefore, captopril has the potential for blocking the generation of angiotensin II, and impairing the degradation of bradykinin. However, the bradykinin levels after captopril treatment reported in the literature are not consistent, ranging from increases (35), to no changes (36), or even decreases (37). It is, therefore, difficult to evaluate the extent of the contribution of bradykinin to the hypotensive effects of captopril.

Release of Prostaglandins. In the anesthetized guinea pig, the angiotensin-converting enzyme inhibitors potentiated bradykinin-induced bronchoconstriction which can be blocked by indomethacin (a cyclooxygenase inhibitor) (38).

Using the blood-bath bioassay technique, it was found that captopril potentiated the release of a bradykinin-induced prostacyclin-like substance whose action was blocked by indomethacin (39). It was also shown that the hypotensive effect of captopril was accompanied by a significant increase in bradykinin metabolites and prostaglandin E_2 (40). Therefore, it was suggested that increased production of prostaglandins may contribute to the antihypertensive effects of captopril (see Prostaglandins in this volume).

Interaction with the Sympathetic Nervous System. Based on preparations of the pithed rat and the rat perfused mesenteric artery, it was proposed that captopril may impair neurogenic vasoconstriction by interfering with the pre- and postjunctional actions of angiotensin on adrenergic neuroeffector systems or by a nonangiotensin-dependent mechanism which results in a decrease in blood pressure (41).

Captopril. Captopril was the first orally active, specific, competitive inhibitor of angiotensin-converting enzyme with an IC_{50} of 2.8×10^{-8} *M* (33). It was shown to be an antihypertensive agent in most animal models of experimentally induced hypertension, such as renovascular and genetic spontaneously hypertensive rats (42). However, captopril had little antihypertensive effect on deoxycorticosterone acetate (DOCA) and salt or in rats made hypertensive by treatment with deoxycorticosterone acetate (DOCA) and salt (43–44). Captopril is an efficacious antihypertensive agent in essential hypertensive patients (45). This effect is greatly increased by concomitant treatment with diuretics (45). The decrease in blood pressure correlates well with the reduction in total peripheral resistance, whereas cardiac output does not change and the stroke index increases (46). Captopril significantly increases plasma-renin activity and decreases aldosterone concentration during the treatment. In long-term treatment, plasma catecholamine concentrations remain the same (36,46).

Absorption after oral administration is rapid and reaches peak blood concentration within one hour. More than 95% of the administered dose is eliminated in the urine within 24 h, with ca 40–50% unchanged. Most of the remainder is eliminated as the disulfide dimer of captopril and captopril–cysteine disulfide. About 25–30% of absorbed captopril is bound to plasma protein and therefore not available for therapy. The plasma half-life is ca 3 h (47).

Few side effects are observed during treatment, most commonly rash and taste disturbances. Less common but more serious are proteinuria and neutropenia (47).

Enalapril. This compound is the first nonsulfur-containing inhibitor of angiotensin-converting enzyme with an IC_{50} of 1.2×10^{-9} *M*, which is ca 17 times more potent than captopril (48). *In vivo* in the rat, the inhibition of the pressor effect of angiotensin I was ca 8.6 times greater than captopril (49). The antihypertensive activity of this compound has been demonstrated in various models of experimental hypertension, such as renovascular and genetic spontaneously hypertensive rats as well as in the perinephretic hypertensive dog (49). Enalapril effectively lowers blood pressure in hypertensive patients with an associated increased plasma-renin activity, and decreased plasma-converting enzyme activity, angiotensin-II and aldosterone plasma concentrations (50).

The absorption after oral administration is rapid and reaches peak blood concentration within 30 min in rats and 2 h in dogs. In rats, about 26% is excreted in urine and 72% in the feces within 72 h. In dogs, 40% is excreted in urine and 36% in the feces (51). In comparison with captopril, enalapril is more potent and has a longer duration of action. The incidence of adverse effects is very rare.

CI 906. This compound is a nonsulfhydryl converting-enzyme inhibitor. It effectively inhibits plasma-angiotensin-converting enzyme in both rats and humans. The onset of action after oral administration is rapid, and the duration of action prolonged. It is effective in reducing blood pressure in several conscious hypertensive animal models such as renovascular and genetic spontaneously hypertensive rats as well as perinephretic hypertensive dogs. This compound has the same antihypertensive activity as enalapril, and has a longer duration of action than captopril (52–53).

Pivalopril. This compound is an orally active converting-enzyme inhibitor, which effectively inhibits the pressor response to exogenous angiotensin I. It has a rapid onset but a relatively short duration of action and is less potent than captopril or enalapril. After oral administration in healthy volunteers, pivalopril increases plasma-renin activity and decreases angiotensin-II and aldosterone plasma concentrations (54). In the sodium-deficient, conscious, spontaneously hypertensive rat, it produces a dose-dependent reduction in arterial blood pressure (55).

SA 446. This compound is an orally active converting-enzyme inhibitor, which bears some structural similarity to captopril. The *in vitro* enzyme inhibitor activity is similar to that of captopril. However, *in vivo* captopril is more potent than SA 446 in its ability to inhibit exogenous angiotensin-I-induced pressor responses (56). In the rat mesenteric bed, SA 446 did not attentuate the vasoconstriction induced by norepinephrine in the way that captopril does (57). SA 446 effectively reduces blood pressure in the renal and genetic spontaneously hypertensive rat; however, in the latter, it is less effective than captopril. On the other hand, the blood pressure of normotensive rats and DOCA–salt sensitive rats was not affected by SA 446 (58).

Nitroglycerin

Glyceryl trinitrate [*55-63-0*], nitroglycerin, was originally synthesized in 1847 (59) and introduced as a drug for the treatment of angina pectoris in 1879 (60).

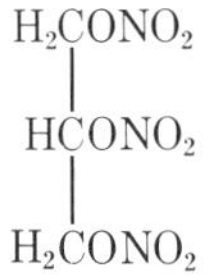

H_2CONO_2
|
$HCONO_2$
|
H_2CONO_2

glyceryl trinitrate,
nitroglycerin

The current use of transdermal nitroglycerin-delivery systems is the culmination of extensive research in dealing with the many problems of nitrate therapy, such as the ease of handling for control of efficacy and side effects, adequate absorption, onset and duration of action, and maintenance of the potency of various pharmaceutical preparations. For example, amyl nitrite [*110-46-3*] was introduced for treatment of angina pectoris over 100 yr ago (61). However, the drug is a highly volatile liquid and must be inhaled, making control of dose and side effects difficult. Nitroglycerin was introduced afterward (60). Adsorption onto lactose for ease of handling and incorporation into a tablet triturate for control of dose and side effects by sublingual administration represented two significant advances. However, action is transient. Various organic nitrates and sustained-release nitroglycerin preparations were introduced to overcome the problems of obtaining a prolonged duration of action but the onset of action is then less rapid. The loss of potency of nitroglycerin tablets (62–63) and the transient actions were largely overcome by the introduction of topically administered nitroglycerin ointments. Ointments are awkward to use and proper dosing erratic. By this route, nitroglycerin must be given several times per day to provide relief from anginal attacks for 24 h. The introduction of transdermal nitroglycerin adhesive or so-called band-aid patches has solved these problems (see below), and nitroglycerin remains the drug of choice for the treatment of angina pectoris.

This disease is manifested as a dramatic, terrorizing substernal pain, often radiating to the left shoulder and over the flexor surface of the left arm to the tips of the medial fingers. The pain of angina is due to cardiac ischemia resulting from acute coronary-artery constriction, or to the inability of the coronary arteries to dilate to meet increased oxygen demands. Other factors, such as decreased oxygen-carrying capacity of the blood (anemia) or reduced aortic pressure may be involved. The attacks may be transient and damage to the ischemic myocardium may be minimal. However, coronary spasm, with or without atherosclerotic changes in the vessel, may lead to sudden death.

Mode of Action. The precise mechanism of action of nitroglycerin is not clearly understood and may be a combination of many factors. The basic pharmacologic action is a relaxation of most smooth muscles, eg, vascular, bronchial, gastrointestinal, ureteral, uterine, etc (63). Vascular smooth-muscle relaxation is a direct effect and does not appear to be mediated via adrenergic receptors (64–66).

It has been suggested that nitroglycerin may release prostacyclin from the vascular endothelial lining and thereby relax vascular smooth muscle (67). Prostacyclin is a potent endogenous vasodilator and inhibitor of platelet aggregation.

Nitroglycerin may be effective in relieving anginal pain by redistributing coronary blood flow to the ischemic myocardium following coronary artery dilatation (68) or by improving the perfusion of the subendocardium (69–70). Reduction of myocardial oxygen demand has been proposed as part of the mechanism for relieving angina. In support of the latter hypothesis are the observations that nitroglycerin also relaxes venous capacitance vessels (71), causes venous pooling, and decreases venous return to the heart (66). These actions on the venous capacitance side of the circulation reduce preload, left-ventricular volume, and ventricular wall tension (72). Combined with its action on arteriolar resistance to reduce afterload and cardiac output (66,72), nitroglycerin reduces external work of the ventricles and, in turn, decreases the oxygen requirements of the myocardium.

Effects of nitroglycerin on subcellular biochemical mechanisms, such as activation of guanylate cyclase, have been demonstrated (63), but their significance in the therapeutic actions of nitroglycerin is not understood.

Sublingual nitroglycerin preparations are rapidly absorbed with peak blood concentrations and pain relief occurring within 1–4 min (63,73–74). Duration of action is 15–20 min (63), although venous plasma nitroglycerin content is not detectable after 10 min (74). Oral nitroglycerin is more slowly absorbed with maximal effects observed in 30–60 min; duration of action orally is 4–6 h (63). Nitroglycerin ointment is absorbed slowly, but its duration of action is 4–8 h (75–76). Site (77) and area of application of the ointment (78) appear to affect the plasma concentrations achieved. With transdermal nitroglycerin systems, detectable blood concentrations are noted in ca 1 h, peak concentration at 4–8 h; these are maintained for 24 h (74).

Once nitroglycerin is absorbed, it is bound to plasma protein (79). Nitroglycerin has a large volume of distribution which suggests that it is widely distributed in the body. (Volume of distribution is a dilution method to determine the volume of fluid in a body compartment.) Nitroglycerin is rapidly metabolized in the liver and other organs (80) and in the blood of some species (81) by glutathione–organic nitrate reductase. The drug is excreted in the urine as the 1,2- and 1,3-glycerol dinitrates and some glycerol mononitrates which are considered pharmacologically inactive (63). It was originally thought that oral nitroglycerin therapy would be inadequate since absorption into portal blood, which passes directly into the liver, would allow hepatic denitration to occur on first pass, clearing the blood of most of the drug before it reached the systemic circulation. However, glutathione–organic nitrate reductase becomes saturated, slowing inactivation and allowing the establishment of therapeutic nitroglycerin concentrations (63). Drugs that increase or decrease liver microsomal enzymes may affect nitroglycerin metabolism, eg, phenobarbital and alcohol, respectively (82).

The principal side effects of nitroglycerin therapy are headache, nausea, flushing, dizziness, and weakness (63). Syncope and increased heart rate can result from postural hypotension produced by nitroglycerin (63). Methemoglobinemia following high doses can also occur because of oxidation of hemoglobin by nitrate ions. Although rashes due to ingestion of nitroglycerin are rare, the reports of incidence of contact dermatitis with nitroglycerin ointment are increasing (83–84).

Decreased sensitivity or tolerance to the vascular effects of nitroglycerin following prolonged exposure to high doses of the drug have been reported (63). Dependence on nitrate has been observed in munitions workers exposed to glyceryl trinitrate (85). Sudden withdrawal of nitrate therapy in patients with coronary insufficiency has resulted in worsening of symptoms (86), and a gradual reduction from prolonged high doses is recommended.

In addition to remaining the drug of choice for the treatment of angina pectoris, nitroglycerin has been successful in the treatment of congestive heart failure (87–89), myocardial infarction (90–91), peripheral vascular disease, such as Raynaud's disease (92), and mitral insufficiency (93), although the beneficial effects on the latter have been questioned (94).

Transdermal Nitroglycerin-Delivery Systems. Marketing approval has been granted by the FDA for three transdermal nitroglycerin systems. Briefly, these are patches which consist of several layers. The outer layer is impermeable. The middle or inner layer consists of a drug reservoir, which delivers the drug in a rate-controlled fashion. The reservoir is bordered by adhesive to attach the unit to the patient's skin. Nitro-Disc (Searle Laboratories) controls the rate of delivery by dispersing nitroglycerin in a solid polymer; Nitro-Dur (Key Pharmaceuticals) by placing the drug in a gel-like polymer matrix, and Transderm-Nitro (CIBA-GEIGY Pharmaceuticals) uses a semipermeable rate-controlling membrane between a reservoir of nitroglycerin in a viscous oil and the skin. The nitroglycerin migrates under the influence of the concentration gradient which is highest in the reservoir and decreases through the layers of the unit to the skin where it is the lowest. Thus, nitroglycerin is constantly released into the systemic circulation (95). The stratum corneum of the skin is the rate determinate for the migration (96).

All products are approved for once-per-day dosing since the units deliver the medication continuously for at least 24 h. It is recommended that the units be applied to a hairless region of the body such as the upper arm or chest and that application

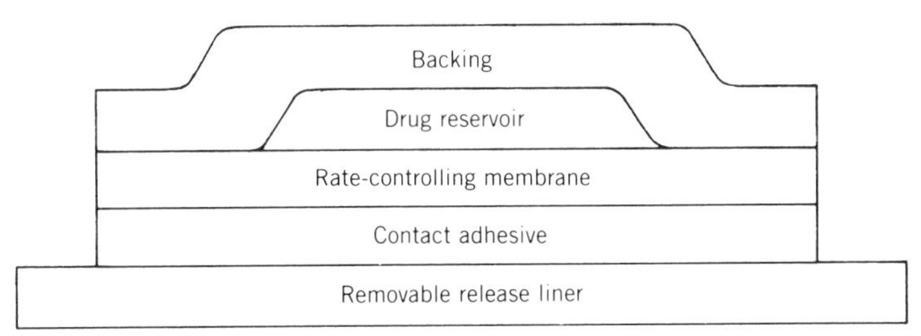

Figure 1. Schematic of the Transderm-Nitro unit.

sites be changed each day. The units may be worn while exposed to or immersed in water (see also Pharmaceuticals, slow release).

Transderm-Nitro. Transderm-Nitro is described as a transdermal therapeutic system (see Fig. 1). The unit consists of an impermeable, aluminized plastic outer layer that holds the drug reservoir consisting of nitroglycerin on lactose in a viscous silicone oil. The reservoir is separated from a hypoallergenic, silicone adhesive layer by a rate-controlling plastic polymer membrane. The adhesive layer is protected by a peel strip which is removed by the patient before the unit is placed on the skin. The rate-controlling membrane, a microporous ethylene–vinyl acetate copolymer film permeable to nitroglycerin, acts as a reservoir; ca 8% of the nitroglycerin from the drug reservoir enters the membrane. The nitroglycerin penetrates the adhesive layer and the stratum corneum of the skin and, in turn, is absorbed into the systemic circulation (95).

Transderm-Nitro is available as Transderm-Nitro 5 and Transderm-Nitro 10. The numbers 5 and 10 correspond to the amount of nitroglycerin delivered in 24 h, although the units contain 25 and 50 mg, respectively. The surface areas of these units are 10 and 20 cm^2 (1.6 and 3.1 $in.^2$), respectively. The drug-delivery rate is 0.5 mg/(cm^2·d) for both units. Plasma concentration of nitroglycerin reaches ca 200–300 pg/mL over 24 h with the 20-cm^2 unit (95). A study of the Transderm-Nitro 5 unit in 87 patients showed that improvement, as judged by significantly fewer anginal attacks and decreased nitrate consumption, was noted in 78%. The patch was preferred by 60% over other delivery systems and no tolerance was noted after six months of administration (97).

Nitro-Disc. Nitro-Disc is described as a microsealed drug-delivery system in which a removable outer impermeable foil layer holds the drug reservoir (see Fig. 2). The drug reservoir is microsealed and contains nitroglycerin evenly dispersed in a solid, silicone rubber polymer matrix. The polymer–drug system is bonded to an aluminum foil disk on an inner layer of an adhesive foam pad. The nitroglycerin dissolves into a liquid compartment and is delivered through the polymer to its surface next to the skin where it is absorbed. Nitro-Disc is available as Nitro-Disc 16 and Nitro-Disc 32 which have surface areas of 8 and 16 cm^2 (1.25 and 2.5 $in.^2$), respectively. The numbers

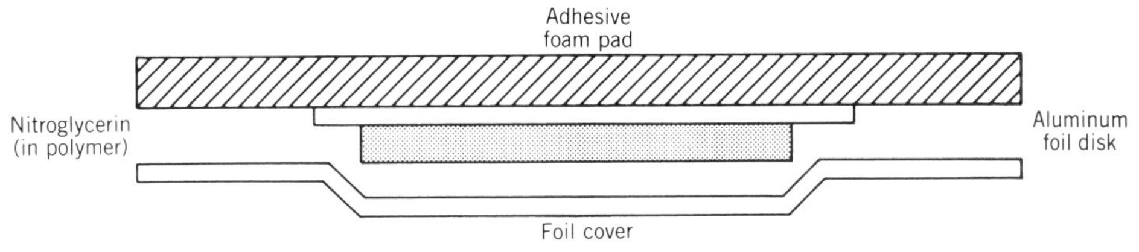

Figure 2. Schematic of the Nitro-Disc unit.

16 and 32 indicate the nitroglycerin content in mg. The units deliver 11.2 and 22.4 mg, respectively, over 24 h, based on a delivery rate of 1.4 mg/(cm^2·d). Plasma concentrations in healthy male volunteers treated with Nitro-Disc 32 were 300 pg/mL over 32 h (95).

Nitro-Dur. Nitro-Dur is described as a transdermal infusion system (see Fig. 3). It is slightly more complex than the other transdermal nitroglycerin units. The outer layer is an impermeable polyethylene film that contains an absorbent bandage to trap moisture. This layer is placed on top of a microporous adhesive tape. The inner layer contains a diffusion matrix bonded to an occlusive aluminum foil base-plate support which is attached to the bottom of the adhesive tape of the outer layer. The diffusion matrix contains nitroglycerin on lactose in equilibrium with nitroglycerin in a glycerol–water phase. Physical stability of the nitroglycerin in the diffusion matrix is assured by enclosing the inner layer with a polyfoil. The polyfoil and a paper-release liner, which protects the stability of the adhesive tape, are removed by the patient before placing the unit on the skin (95).

Nitro-Dur is available in four sizes or strengths: Nitro-Dur 5, 10, 15, and 20. The numbers indicate the surface area in cm^2, but the units contain 26, 51, 77, and 104 mg nitroglycerin, respectively. The release rate is 0.5 mg/(cm^2·d) with the result that 2.5, 5, 7.5, and 10 mg, respectively, are delivered over 24 h.

Venous plasma blood levels were compared in healthy subjects using Nitro-Dur 10, Nitro-Dur 20, and sublingual nitroglycerin tablets (74). The latter had peak blood levels in 1–2 min of 1.2–1.4 ng/mL, which dissipated to nondetectable levels in about 10 min. Subjects taking Nitro-Dur 10 had peak plasma levels in 4 h of 0.84 ng/mL, which remained at that level for 24 h (0.89 ng/mL). Subjects using Nitro-Dur 20 had peak blood levels of 1.83 ng/mL after 8 h and 1.8 ng/mL after 24 h.

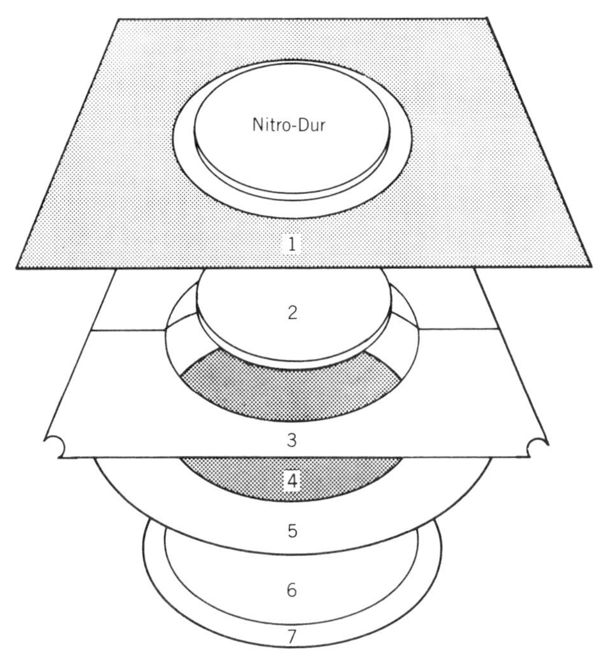

Figure 3. Schematic of the Nitro-Dur unit. 1, Foil coverstrip; 2, drug matrix; 3, release liner; 4, foil base plate; 5, microporous tape; 6, absorbent pad; 7, occlusive overlay.

An improvement in hemodynamic measurements was observed in patients with congestive heart failure 30 min after application of 20–40-cm^2 units of Nitro-Dur. Peak effects were observed at 4–6 h with some effects still present after 24 h (98).

Cardiotonic Agents

The sole function of the heart (ie, the ventricles) is to pump oxygenated blood to perfuse the organs of the body through their arterial supplies and to remove carbon dioxide via their venous outflows. Cardiac performance is mainly regulated by the contractile state of the myocardium, the heart rate, and preload (venous return) and afterload (arterial resistance). The body adjusts for changes in cardiac performance in such a way that the function remains adequate to meet body needs. Thus, the body can compensate for a decrease in pumping capacity caused by an impaired contractile state; it does this by an increase in sympathetic nervous system tone accompanied by an increase in circulating catecholamines. The catecholamines stimulate β-adrenergic receptors in the myocardium and sinoatrial node resulting in an increase in heart rate, contractile force, and cardiac output. In addition, cardiac performance at any given preload (venous return) can be increased by decreasing afterload (arterial resistance or impedance) or by increasing contractile force. Contractile force can be increased by β-adrenergic stimulation (as indicated above) or by an increase in the myocardial fiber length (stretching) up to an optimal fiber length after which further stretching decreases contractile force. When the body's homeostatic mechanisms are no longer adequate, pharmacological therapeutic interventions are required to maintain cardiac performance.

The most important use of drugs that produce positive inotropic effects is in acute heart failure or chronic congestive heart failure. Congestive heart failure is characterized by a decrease in contractile force, dilation of the ventricular chambers (overstretching of the myocardial fibers), and decreasing emptying of the blood from the ventricles resulting in a decrease in cardiac output. In addition, right- and left-ventricular resting or diastolic pressures are increased and an increase in venous pressure (or preload) occurs. As a consequence of these changes in cardiac function, pulmonary and systemic arterial pressures increase and congestion in the lungs and edema of the extremities occurs. The positive inotropic agents have also been used in the treatment of heart failure secondary to myocardial infarction.

The general biochemical mechanism causing muscle contraction is the interaction between the contractile proteins, which depends on the availability of sufficient intracellular calcium ions. The positive inotropic agents, through diverse mechanisms, increase the intracellular calcium ion concentrations to increase the interactions between the cardiac contractile proteins.

Many agents are capable of producing positive inotropic effects, but not all can be used for the treatment of heart failure. For example, the sympathomimetic amines produce positive chronotropic effects or increase blood pressure. Many have relatively short durations of action or lack oral activity. The cardiac glycosides have a relatively small therapeutic index and may produce arrhythmias at doses slightly above the therapeutic range.

At present, two classes of compounds are used as inotropic agents for the treatment of heart failure: the digitalis glycosides and the β-adrenoceptor-agonist drugs.

It has been postulated that the receptor for the inotropic action of the digitalis glycosides is a sodium- and potassium-coupled adenosine triphosphatase ($Na^+ + K^+$ ATPase) situated in the cardiac-cell membrane (99–104). Inhibition of the $Na^+ + K^+$ ATPase by the glycosides leads to an accumulation of intracellular sodium. It is further postulated that as sodium accumulates intracellularly, a sodium–calcium-coupled system reestablishes steady-state sodium exchange at a higher intracellular level (105). Thus, calcium influx is increased and developed tension with each action potential is increased (positive inotropic action) (106). Calcium release from intracellular compartments has also been postulated to contribute to the digitalis glycoside inotropic action (107).

The β-adrenergic agents (dopamine, norepinephrine, isoproterenol, dobutamine, salbutamol, and pirbuterol) stimulate the β-adrenergic receptor on the cardiac cell membrane. Activation of the β-adrenergic receptor stimulates adenyl cyclase in the cardiac cell membrane resulting in increased intracellular levels of cyclic-adenosine monophosphate (AMP). The receptor for cyclic-AMP is a protein kinase which, when activated, phosphorylates proteins that regulate the permeability of the plasma membrane to calcium and calcium transport by the sarcoplasmic reticulum (108).

Amrinone. Amrinone, a bipyridine analogue, is a novel nonglycosidic, non-β-adrenoceptor agonistic positive inotropic agent.

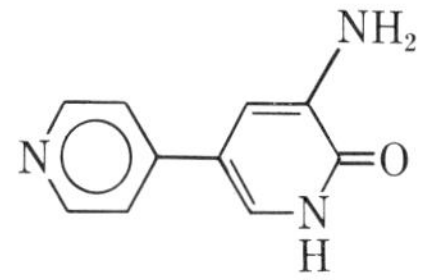

amrinone [*60719-84-8*]

The *in vitro* positive inotropic effects of amrinone have been described in cat atrial and papillary muscle (109), rabbit papillary muscle (110), Langendorff-perfused guinea-pig heart (110–111), and in dog papillary and trabeculae muscle (112). In the Langendorff-perfused rat heart, however, amrinone produces a negative inotropic effect (111). Small insignificant changes in heart rate were noted in cat atrial muscle (109) and in the Langendorff-perfused guinea-pig heart (110–111). The increases in contractile force in cat and dog cardiac muscle were not accompanied by any changes in intracellular action potential configuration (109,112).

In vivo, in the anesthetized (109,113) and unanesthetized dog (109), amrinone produces increases in the force of contraction, cardiac output, and cardiac work with relatively small changes in heart rate or blood pressure except at high doses (109). In several models of heart failure in the dog, eg, pentobarbital-induced (109) or coronary artery occlusion (115), amrinone increases contractile force.

Amrinone, given intravenously or orally, has been shown to increase cardiac output, stroke volume, and stroke work (indices of inotrophy) and to increase exercise tolerance in patients with severe congestive heart failure, with or without ischemic heart disease (115–121). Similar increases in indices of inotrophy were also noted in healthy volunteers (122).

The positive-inotropic effects in animals and man were not accompanied by arrhythmias. However, various arrhythmias were reported in *in vitro* preparations of rabbit papillary muscle and in Langendorff-perfused guinea-pig heart (110). In addition, amrinone, when infused into dogs with acute myocardial ischemia following coronary artery ligation to produce inotropic effects equivalent to those of isoproterenol, would be equally detrimental to the ischemic myocardium (114).

Amrinone is well absorbed following oral administration. The ratio of the intravenous-to-oral dose in the dog is 1:2, indicating good absorption from the gastrointestinal tract (109). Onset of action was rapid and peak effect was 30–45 min (109). Peak plasma concentrations were reached between 0.25 and 1 h (123). In humans, the onset of action was 30–120 min orally (119); intravenously, it was 3–5 min.

The metabolism of amrinone has not been completely elucidated. Urinary and fecal excretion in the dog represent 63 and 26% of the dose, respectively (123). The drug is excreted both unchanged and as a polar metabolite in the urine (123).

Amrinone has a fairly long-acting, positive, inotropic effect. Intravenously, the duration of action in the dog is 30–120 min; orally, it is 4–7 h (109). In humans, the duration of action following intravenous doses has been reported to be as short as 60–90 min (117) and as long as 4–7 h (115); orally, 4–7 h duration has been reported (119). No tolerance to the inotropic actions of amrinone has been observed in humans (118–119).

The exact mechanism of action of the inotropic action of amrinone is not known. An effect on calcium metabolism has been demonstrated in heart muscle and red blood cells (111,124). In the Langendorff-perfused guinea-pig heart, an increase in intracellular calcium turnover has been demonstrated (112), whereas in dog erythrocyte, amrinone stimulated net calcium uptake increasing intracellular calcium content (124). It was suggested that amrinone prolongs the liberation of calcium into the sarcoplasm and increases the rate of calcium sequestration (125). Since amrinone could not reverse the negative inotropic effect of verapamil, it is concluded that amrinone does not act on the slow calcium channel (110).

Other mechanisms associated with the positive inotropic actions of drugs have been ruled out for amrinone. Thus, amrinone does not act by stimulating β-adrenergic receptors since its action is not blocked by the β-adrenergic receptor blocking agent, propranolol (109). Amrinone does not stimulate the release of catecholamines from adrenergic nerve terminals since pretreatment with the catecholamine depletor, reserpine, does not interfere with its inotropic action (109). Amrinone does not inhibit sodium and potassium ATPase (109). Amrinone does not inhibit phosphodiesterase activity or increase cyclic-AMP levels in the heart (109), although in rabbit aorta amrinone has been reported to decrease cyclic-AMP levels (126). The latter effect and decreased availability of calcium in vascular smooth muscle (126) may account for its direct vasodilating activity (109,127–128). Amrinone does not appear to activate histamine receptors in the heart since its action is not affected by histamine H_1 or H_2 receptor blocking agents (109,123).

Although the margin of safety for amrinone is better than that for digitalis (109,123) and no arrhythmias have been noted in humans at doses producing maximal inotropic effects (115–121), the drug has been reported to produce thrombocytopenia and nephrogenic diabetes insipidus in humans (119,121,129).

Amrinone appears to offer some promise as a new potent orally active, positive inotropic agent with a rapid onset of action and a prolonged duration of action. It appears to be less toxic than currently available positive inotropic agents but has some serious toxic effects which may limit its use. Continued surveillance and use will determine its potential as a therapeutic agent. A review of the development of cardiotonic agents to replace the digitalis glycosides in heart failure has been published (130).

BIBLIOGRAPHY

"Cardiovascular Agents" in *ECT* 1st ed., Vol. 3, pp. 211–224, by W. Modell, Cornell University Medical College; "Cardiovascular Agents" in *ECT* 2nd ed., Vol. 4, pp. 510–524, by W. Modell, Cornell University Medical College; "Cardiovascular Agents" in *ECT* 3rd ed., Vol. 4, pp. 872–930, L. Goldman, Lederle Laboratories, American Cyanamid Company.

1. H. J. Sanders, *Chem. Eng. News* **60,** 26 (1982).
2. A. Fleckenstein in G. B. Weiss, ed., *New Perspectives on Calcium Antagonists*, William & Wilkins, Baltimore, Md., 1981, p. 59.
3. A. Schwartz, *Am. J. Cardiol.* **49,** 497 (1982).
4. A. M. Katz and H. Reuter, *Am. J. Cardiol.* **44,** 188 (1979).
5. T. B. Bolton, *Physiol. Rev.* **59,** 606 (1979).
6. R. Casteels, *Chest* **78,** 150 (1980).
7. D. J. Hartshorne, *Chest* **78,** 140 (1980).
8. R. Zelis and S. F. Flaim, *Ann. Intern. Med.* **94,** 124 (1981).
9. P. D. Henry, *Am. J. Cardiol.* **46,** 1047 (1980).
10. A. Fleckenstein, *Arzneim. Forsch. Drug Res.* **20,** 1317 (1970).
11. M. Vincenzi and co-workers, *Arzneim. Forsch. Drug Res.* **26,** 1221 (1976).
12. M. L. Mangiardi and co-workers, *Circulation* **57,** 366 (1978).
13. M. Eichelbaum and co-workers, *Drug Metab. Dispos.* **7,** 145 (1979).
14. G. Gruen and A. Fleckenstein, *Arzneim. Forsch. Drug Res.* **22,** 334 (1972).
15. M. Raschack, *Arzneim. Forsch. Drug Res.* **26,** 1330 (1976).
16. A. Narimatsu and N. Taira, *Naunyn-Schmiedebergs Arch. Pharmakol.* **294,** 169 (1976).
17. N. J. Williams, ed., *Br. J. Clin. Pract. Suppl. 8* **34,** 3 (1980).
18. W. Lochner and co-workers, *Second International Nifedipine Adalat Symposium*, Springer-Verlag, New York, 1976.
19. R. I. Low and co-workers, *Am. J. Cardiol.* **49,** 547 (1982).
20. S. F. Flaim and R. Zelis, *Fed. Proc. Fed. Am. Soc. Exp. Biol.* **40,** 2877 (1981).
21. R. W. Piepho and co-workers, *Am. J. Cardiol.* **49,** 525 (1982).
22. J. V. Talano and C. Tommaso, *Prog. Cardiovasc. Dis.* **25,** 141 (1982).
23. H. M. Einwachter and co-workers, *Eur. J. Pharmacol.* **55,** 225 (1979).
24. K. Satoh and co-workers, *Clin. Exp. Pharmacol. Physiol.* **7,** 249 (1980).
25. S. Higuchi and Y. Shiobara, *Xenobiotica* **10,** 889 (1980).
26. F. Hoffmeister and co-workers, *Acta Neurol. Scand. Suppl.* **60,** 358 (1979).
27. N. Taira and co-workers, *Arzneim. Forsch. Drug Res.* **29,** 246 (1979).
28. S. Kazda and co-workers, *Arzneim. Forsch. Drug Res.* **30,** 2144 (1980).
29. F. Awouters and co-workers in D. L. Temple, ed., *Drugs Affecting The Respiratory System*, American Chemical Society, Washington, D.C., 1980, p. 179.
30. M. J. Peach, *Physiol. Rev.* **57,** 313 (1977).
31. J. H. Laragh, *Prog. Cardiovasc. Dis.* **21,** 159 (1978).
32. F. M. Lai and co-workers, *Life Sci.* **28,** 1309 (1981).
33. M. A. Ondetti and co-workers, *Science* **196,** 441 (1977).
34. E. G. Erdos and H. Y. T. Yang, *Life Sci.* **6,** 569 (1967).
35. R. E. McCaa and co-workers, *Circ. Res.* **43,** I-32 (1978).
36. P. G. Matthews and co-workers, *Clin. Sci. Suppl. 5* **57,** 135S (1979).
37. J. A. Millar and co-workers, *Clin. Sci. Suppl. 6* **59,** 65S (1980).
38. R. Greenberg and co-workers, *Eur. J. Pharmacol.* **57,** 287 (1979).
39. K. M. Mullane and S. Moncada, *Eur. J. Pharmacol.* **66,** 355 (1980).
40. S. L. Swartz and G. H. Williams, *Am. J. Cardiol.* **49,** 1405 (1982).
41. D. P. Clough and co-workers, *Am. J. Cardiol.* **49,** 1410 (1982).
42. R. J. Laffan and co-workers, *J. Pharmacol. Exp. Ther.* **204,** 281 (1978).
43. B. H. Douglas and co-workers, *Clin. Res.* **26,** 510 (1978).
44. B. Rubin and co-workers, *Proc. Cardiovasc. Dis.* **21,** 183 (1978).
45. D. B. Case and co-workers, *Prog. Cardiovasc. Dis.* **21,** 195 (1978).
46. G. Muiesan and co-workers, *Am. J. Cardiol.* **49,** 1420 (1982).
47. *Capoten Product Information*, E. R. Squibb & Sons, Inc., Princeton, N.J., March, 1981.
48. A. A. Patchett and co-workers, *Nature (London)* **288,** 280 (1980).

49. C. S. Sweet and co-workers, *J. Pharmacol. Exp. Ther.* **216,** 558 (1981).
50. H. Gavras and co-workers, *Lancet* **2,** 543 (1981).
51. D. J. Tocco and co-workers, *Drug Metab. Dispos.* **10,** 15 (1982).
52. H. R. Kaplan and co-workers, *Pharmacologist* **24,** 176 (1982).
53. M. J. Ryan and co-workers, *Pharmacologist* **24,** 176 (1982).
54. M. Burnier and co-workers, *Br. J. Clin. Pharmacol.* **12;** 893 (1981).
55. P. S. Wolf and co-workers, *Pharmacologist* **24,** 176 (1982).
56. T. Unger and co-workers, *Eur. J. Pharmacol.* **78,** 411 (1982).
57. T. Saruta and co-workers, *Am. J. Cardiol.* **49,** 153S (1982).
58. Y. Takata and co-workers, *Am. J. Cardiol.* **49,** 1502 (1982).
59. M. Windholz, ed., *Merck Index*, 9th ed., Merck & Co. Publishers, Rahway, N.J., 1976, p. 858.
60. W. Murrell, *Lancet* **i,** 80 (1879).
61. T. L. Brunton, *Lancet* **ii,** 97 (1867).
62. B. Dorsch and R. Shangraw, *Am. J. Hosp. Pharm.* **32,** 795 (1975).
63. P. Needleman and E. M. Johnson, Jr., in A. G. Gilman, L. S. Goodman, and A. Gilman, eds., *Pharmacological Basis of Therapeutics*, 6th ed., MacMillan Publishing Co., Inc., New York, 1980, p. 819.
64. R. F. Furchgott and S. Bhadrakom, *J. Pharmacol. Exp. Ther.* **108,** 129 (1953).
65. D. T. Mason and co-workers, *J. Clin. Invest.* **43,** 1449 (1964).
66. D. T. Mason and E. Braunwald, *Circulation* **32,** 755 (1965).
67. R. I. Levin and co-workers, *Clin. Res.* **28,** 471A (1980).
68. L. D. Horowitz and co-workers, *J. Clin. Invest.* **50,** 1578 (1971).
69. L. C. Becker and co-workers, *Circ. Res.* **28,** 263 (1971).
70. R. E. Goldstein and co-workers, *Circulation* **49,** 298 (1974).
71. R. I. Ogilvie, *J. Pharmacol. Exp. Ther.* **207,** 372 (1978).
72. J. F. Williams and co-workers, *Circulation* **32,** 767 (1965).
73. H. P. Blumenthal and co-workers, *Br. J. Clin. Pharmacol.* **4,** 241 (1977).
74. H. Colfer and co-workers, *J. Cardiovasc. Pharmacol.* **4,** 521 (1982).
75. J. O. Parker and co-workers, *Am. J. Cardiol.* **38,** 162 (1976).
76. P. A. Chandraratna and co-workers, *Cardiology* **66,** 102 (1980).
77. M. S. Hansen and co-workers, *Heart and Lung* **8,** 716 (1979).
78. S. Sved and co-workers, *J. Pharm. Sci.* **70,** 1368 (1981).
79. F. J. DiCarlo and co-workers, *Biochem. Pharmacol.* **18,** 965 (1969).
80. P. Needleman and co-workers, *J. Pharmacol. Exp. Ther.* **179,** 347 (1971).
81. F. J. DiCarlo and M. D. Meglar, *Biochem. Pharmacol.* **19,** 1371 (1970).
82. F. J. DiCarlo, *Drug. Metab. Rev.* **4,** 1 (1975).
83. W. F. Sausker and F. D. Frederick, *J. Am. Med. Assoc.* **239,** 1743 (1978).
84. K. F. Anderson and F. I. Maibach, *Clin. Toxicol.* **16,** 415 (1981).
85. R. L. Lange and co-workers, *Circulation* **46,** 666 (1972).
86. J. A. Franciosa and J. N. Cohn, *Am. J. Cardiol.* **45,** 648 (1980).
87. W. R. Taylor and co-workers, *Am. J. Cardiol.* **38,** 469 (1976).
88. P. A. Chandraratna and co-workers, *Cardiology* **63,** 337 (1978).
89. J. A. Franciosa, *Heart and Lung* **9,** 873 (1980).
90. P. W. Armstrong and co-workers, *Am. J. Cardiol.* **38,** 474 (1976).
91. T. Hardarson and co-workers, *Am. J. Cardiol.* **40,** 90 (1977).
92. A.G. Franks, Jr., *Lancet* **i,** 76 (1982).
93. K. Chatterjee and co-workers, *Circulation* **48,** 684 (1973).
94. L. Gould and co-workers, *Angiology* **31,** 677 (1980).
95. J. F. Dasta and D. R. Geraets, *Am. Pharm.* **22,** 29 (1982).
96. R. J. Scheuplein, *J. Invest. Dermatol.* **67,** 31 (1976).
97. B. Garnier and co-workers, *Praxis* **71,** 511 (1982).
98. M. T. Olivari and co-workers, *Clin. Res.* **30,** 548A (1982).
99. R. Whittam, *Biochem. J.* **84,** 110 (1962).
100. P. C. Caldwell, *J. Physiol. (London)* **152,** 545 (1960).
101. L. E. Hokin and co-workers, *J. Biol. Chem.* **248,** 2593 (1973).
102. J. F. Hoffman, *Am. J. Med.* **41,** 666 (1966).
103. L. E. Lane and co-workers, *J. Biol. Chem.* **248,** 7197 (1973).
104. J. Kyte, *J. Biol. Chem.* **249,** 3652 (1974).

105. G. A. Langer and S. D. Serena, *Ann. N.Y. Acad. Sci.* **242,** 688 (1974).
106. K. Repke in B. Brodie and J. R. Gilette, eds., *Proceedings of the Second International Pharmacology Meeting, Prague, Vol. 4, Drugs and Enzymes*, Pergamon Press, New York, 1965, p. 65.
107. K. S. Lee and co-workers, *J. Pharmacol. Exp. Ther.* **172,** 180 (1970).
108. S. E. Mayer in A. G. Gilman, L. S. Goodman, and A. Gilman, eds., *The Pharmacological Basis of Therapeutics*, 6th ed., MacMillan Publishing Co., New York, 1980, p. 56.
109. A. A. Alousi and co-workers, *Circ. Res.* **45,** 666 (1979).
110. G. Onuaguluchi and R. D. Tanz, *J. Cardiovasc. Pharmacol.* **3,** 1342 (1981).
111. J. Azari and R. J. Huxtable, *Eur. J. Pharmacol.* **67,** 347 (1980).
112. J. E. Rosenthal and G. R. Ferrier, *J. Pharmacol. Exp. Ther.* **221,** 188 (1982).
113. A. Schwartz and co-workers, *Circulation* **59, 60,** II-16 (1979).
114. R. E. Rude and co-workers, *Cardiovasc. Res.* **14,** 419 (1980).
115. J. R. Benotti and co-workers, *New Engl. J. Med.* **299,** 1373 (1978).
116. J. R. Benotti and co-workers, *Circulation* **62,** 28 (1980).
117. T. H. LeJemtel and co-workers, *Circulation* **59,** 1098 (1979).
118. R. F. Malacoff and co-workers, *Am. J. Cardiol.* **45,** 433 (1980).
119. T. H. LeJemtel and co-workers, *Am. J. Cardiol.* **45,** 123 (1980).
120. S. J. Siskind and co-workers, *Circulation* **64,** 966 (1981).
121. J. Wynne and co-workers, *Am. J. Cardiol.* **45,** 1245 (1980).
122. N. T. DeGuzman and co-workers, *Circulation* **58,** Suppl. II, 183 (1978).
123. A. A. Alousi and A. E. Farah, *Trends Pharmacol. Sci.* **1,** 143 (1980).
124. J. C. Parker and J. R. Harper, Jr., *J. Clin. Invest.* **66,** 254 (1980).
125. J. P. Morgan and co-workers, *Fed. Proc. Fed. Am. Soc. Exp. Biol.* **39,** 854 (1980).
126. K. D. Meisheri and co-workers, *Eur. J. Pharmacol.* **61,** 159 (1980).
127. A. Alousi and A. Helstosky, *Fed. Proc. Fed. Am. Soc. Exp. Biol.* **39,** 855 (1980).
128. M. Maruyama and co-workers, *Jpn. J. Pharmacol.* **31,** 1095 (1981).
129. S. A. Rubin and co-workers, *New Engl. J. Med.* **301,** 1185 (1979).
130. A. E. Farah and A. A. Alousi, *Life Sci.* **22,** 1139 (1978).

PETER CERVONI
FONG M. LAI
Lederle Laboratories
American Cyanamid Company

COAL CHEMICALS AND FEEDSTOCKS

INTRODUCTION

Development of the Coal Chemicals Industry

The industrial production of chemicals began more than 100 years ago and was influenced by general technical developments and, particularly, by increasing steel (qv) and coal (qv) production. The steel industry used coke from bituminous coal rather than from charcoal for reduction of ore. The conversion of coal into coke yields by-products, eg, tar, benzene [*71-43-2*], and gas.

Initially, the synthesis of dyestuffs whose quality at first equaled and later exceeded that of natural dyes was based on aromatics, eg, benzene, naphthalene [*91-20-3*], and anthracene [*120-12-7*]; all of these are products from bituminous coal tar (see Dyes). Furthermore, pesticides, drugs, explosives, and plastics, as well as many other materials, were produced from tar (see Insect control technology; Pharmaceuticals; Explosives and propellents; Plastics). The ammonia formed in the coking plants was processed into nitrogenous fertilizers (qv). Only the invention and technical use of the ammonia synthesis from atmospheric nitrogen and hydrogen from coal, however, made it possible to produce nitrogenous fertilizers in sufficient quantities. Thus, coal was the main raw-materials base in the infancy of the chemical industry and remained so until well into the twentieth century.

Current Status

Today, coal and coal-derived products play only a secondary role as chemical raw materials. Petroleum and natural gas were developed and produced in large quantities and proved superior to coal, particularly in two aspects. First, they are richer in hydrogen and thus better basic materials for many products. Second, in most cases, they are easier to transport, handle, meter, and process because of their physical states. Still, the replacement of coal by petroleum differed substantially depending on the amount of raw materials available in the by-products of coking plants, the cost relationships between coal and petroleum, and the development of new production processes.

Since the transition from coal to oil and natural gas as the main chemical raw materials, chemical production has risen enormously worldwide, in particular as a consequence of substantial growth in the fields of plastics and fertilizers.

The amount of monocyclic aromatic compounds, ie, benzene and its homologues, available from tar of bituminous coal became insufficient for the requirements of the chemical industry. The rising demand called for the development of processes for the production of these substances from petroleum. However, the considerably lower demand for naphthalene and anthracene is largely satisfied by bituminous coal tar; although, for these products too, petroleum-based syntheses are known and used. Naphthalene, which used to be the only raw material for the production of phthalic anhydride, has been partly replaced by *o*-xylene. Only small quantities of *o*-xylene are produced in the coking process, but it can be obtained from petroleum relatively easily.

Phenol [*108-95-2*] (qv), which is necessary for the production of artificial resins, occurs in bituminous coal tar only in small amounts; consequently, synthetic processes for industrial phenol production were developed. The industrial demand for pyridine [*110-86-1*] and pyridine bases has decreased, so that tar products still supply a large part of this market (see Pyridine and pyridine derivatives).

Ethylene [*74-85-1*] (qv) forms only in small quantities in the coke oven gas by the coking process of bituminous coal and was occasionally separated from the gas. However, the large amounts of ethylene, as well as propene and butadiene, that are needed can be obtained easily from petroleum. Many products obtained from ethylene, eg, plastics and specifically vinyl chloride, were made from acetylene [*74-86-2*], which is obtainable from calcium carbide [*75-20-7*] from coal or coke (see Vinyl polymers, vinyl chloride and poly(vinyl chloride); Acetylene).

Coke oven gas contains >50 wt % hydrogen, which used to be isolated for ammonia synthesis; alternatively, the hydrogen was obtained by gasification of coal and coke. These processes proved to be too expensive in comparison with cracking of natural gas or oil. Consequently, the synthesis of ammonia [*7664-41-7*] (qv) and methanol [*67-56-1*] (qv) is mainly based on liquid or gaseous hydrocarbons.

By-products of coking, particularly tar and benzene, have maintained their importance as chemical raw materials. Besides its outstanding importance in the production of carbide and coke for the iron industry, coal remains an important chemical raw material in reduction processes for recovery of phosphorus, sodium sulfide, silicon, and other elements and compounds.

Outlook

With decreasing supplies of oil, coal is expected to be used more often as a chemical raw material. All basic chemicals used today can be manufactured from coal, though with greater difficulty in most cases, than from petroleum (see Figure 1). This applies particularly to the products derived from synthesis gas. Synthesis gas, a mixture of hydrogen and carbon monoxide, must be manufactured from petroleum or gas as well as from coal but at higher energy costs. Different gasification processes have been tested on a large scale, and new, high pressure processes are being developed and are running at the prototype stage in various countries. The problem of gas purification and conversion to pure hydrogen or hydrogen–carbon monoxide mixtures has been solved, so that the production of ammonia, methanol, and hydrogen for the various hydrogenation processes or hydrocarbons by Fischer-Tropsch synthesis is possible on a large scale. In South Africa, large amounts of synthesis gas from coal are processed into hydrocarbons by the Fischer-Tropsch synthesis (see Fuels, synthetic).

Coal hydrogenation results in mixtures of hydrocarbons containing large proportions of aromatic compounds, eg, benzene, which can be separated. In the primary step of hydrogenation, phenols are produced in concentrations of $\leq$20 wt %. The

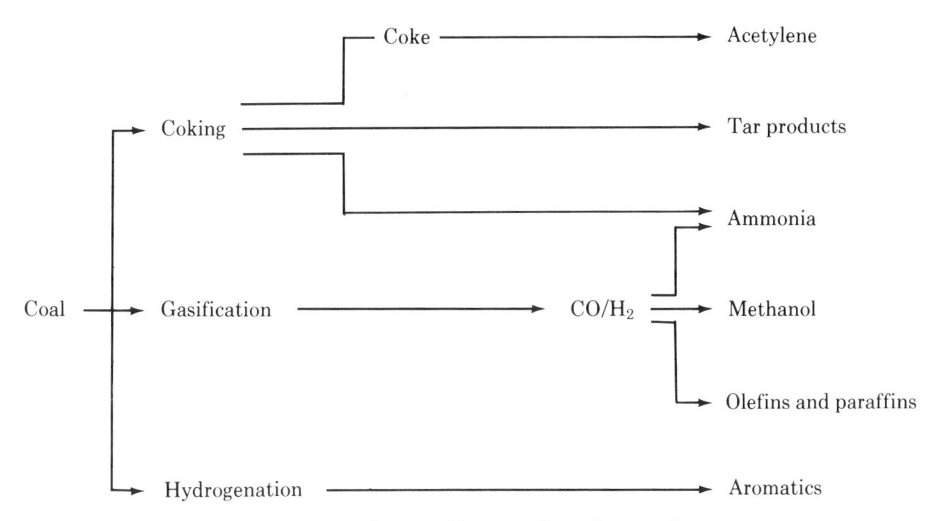

Figure 1. Chemical intermediates from coal.

phenols can be isolated similarly to those from bituminous coal tar and are thus available for direct sale. Special processes that are being developed, eg, hydropyrolysis, lead directly to chemical feedstocks.

Acetylene chemistry may become increasingly important, particularly if it were possible to make the carbide process less expensive or to use other techniques, eg, the thermooxygen or the plasma-based cracking process (see Acetylene-derived chemicals).

It is somewhat more difficult to make large amounts of ethylene and propene from coal as opposed to from petroleum. The following coal-based processes are suitable for this: direct production from synthesis gas based on special versions of the Fischer-Tropsch process; production from methanol involving zeolites (see Molecular sieves); production by cracking of paraffins produced in the Fischer-Tropsch synthesis (see Fuels, synthetic); production from perhydrogenated products of coal hydrogenation; and production by selective hydrogenation of acetylene. The first two processes have given promising results. Methanol, however, is not only a suitable starting material for aromatics and olefins; it can also be converted into ethanol, glycol, acetic acid, and many other compounds. Methanol chemistry will become increasingly important.

BIBLIOGRAPHY

General References

J. Falbe, ed., *Chemical Feedstocks from Coal*, John Wiley & Sons, Inc., New York, 1982.

H. Harnisch, R. Steiner, and K. Winnacker, eds., *Chemische Technologie*, 4th ed., Vol. 5, C. Hanser, Munich, FRG, and Vienna, Austria, 1981.

F. Benthaus and co-workers, *Rohstoff Kohle*, Verlag Chemie, Weinheim, FRG, and New York, 1978.

K. Weissermel and H.-J. Arpe, *Industrielle Organische Chemie*, Verlag Chemie, Weinheim, FRG, and New York, 1976.

Verband der Chemischen Industrie e.V., ed., "Rohstoffsicherung durch Kohlenveredlung," a series of publications in *Chemie und Fortschritt*, Vol. 1, Frankfurt am Main, FRG, 1981.
A. Stratton, *Energy and Feedstocks in the Chemical Industry*, Ellis Horwood Limited, Chichester, UK, 1983.

GEORG KÖLLING
Bergbau-Forschung GmbH

GASIFICATION

As well as some older process variants, the Lurgi, Koppers-Totzek, and Winkler gasification processes are referred to as coal-gasification processes of the so-called first generation (1). All three processes are technically proven and have been used in large-scale plants (2–4) (see Table 1 and Fuels, synthetic).

The main products and by-products of these coal gasification processes are chemical feedstocks. The main product is the chemical-feedstock synthesis gas, which is primarily a mixture of CO and H_2, and is converted in specific synthesis processes to intermediates and products (see Syngas Chemistry). The by-products, ie, methane [*74-82-8*] and sulfur compounds, can, to a certain extent, be processed to provide other chemical feedstocks. Some, eg, tars and phenols, are condensed from the gases while others, eg, phenols and ammonia [*7664-41-7*], are contained in the process water. Special treatment and cleaning steps permit the recovery of all coal-gasification products in principle; however, this is not always economical.

As far as coal gasification as a source of chemical feedstocks is concerned, the process parameters, pressure and temperature, are particularly significant in the gasification processes, whereas the kind of coal is only important as far as the reactivity, throughput, and relative quantities of by-products formed are concerned.

The gasification temperature alone greatly influences the formation of chemical feedstocks. In general, higher reaction temperatures increase carbon conversion and thereby increase the specific yield of synthesis gas as well as reduce the incidence of by-products through *in situ* combustion. The influence of temperature on methane content is diagrammed in reference 5. Methane is also a suitable term of reference to indicate the influence of pressure, as illustrated in Figure 1.

The Lurgi gasification process is the main source of chemical feedstocks from

Table 1. Characteristics of Coal-Gasification Processes

Process	Process principle	Temperature	Pressure	Feed coal
Lurgi	moving fixed bed	low	high	coarse coal
Koppers-Totzek	entrained bed	high	atmospheric	fine coal
Winkler	fluidized bed	medium	atmospheric	medium coal

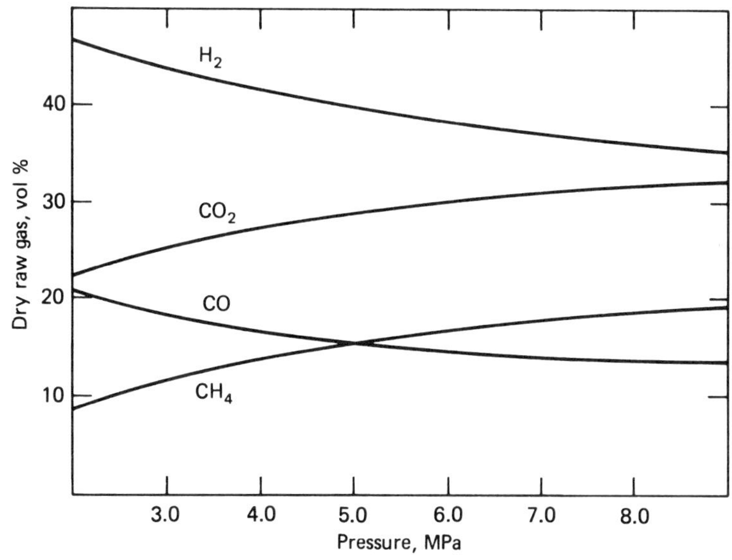

Figure 1. Influence of pressure on the formation of methane in the Lurgi process (6). To convert MPa to psi, multiply by 145.

the by-products of gasification, whereas the Koppers-Totzek and the Winkler gasification processes only provide (other than CO/H_2) elementary sulfur as a final product (7–9). The potential of the Lurgi gasification process as a source of syngas and by-products is shown by the typical gas composition and amount of by-products in Table 2 (3–4).

Coal dust and tars are removed from the raw gas leaving the Lurgi gasifier in a scrubbing cooler, and the gas is then cooled further in a waste-heat boiler. Organic and aqueous condensates also form and make up the gas liquor (10) (see Fig. 2).

Depending on the phase, the by-products of the Lurgi gasification primarily occur in gaseous form with the synthesis gas, ie, as COS and H_2S, or in liquid form, either as liquid condensates of raw gas (tars) or as gas liquors (phenols and ammonia). If the

Table 2. Results of the Lurgi Pressure Gasification Process with Oxygen[a]

Raw gas analysis, vol %	
CO	17–28
H_2	37–41
CH_4	3–12
CO_2 and H_2S	27–33
C_nH_m	0.2–0.6
N_2	0.5–2.0
Product yields, kg/t[b] *coal*	
tars, oils, etc	23–90
raw phenols	4–9
ammonia	ca 8
sulfur	ca 7[c]

[a] Given as the mean values of the gasification of hard coals and lignites (3–4).

[b] Metric ton.

[c] Depends on the sulfur content of the feed coal.

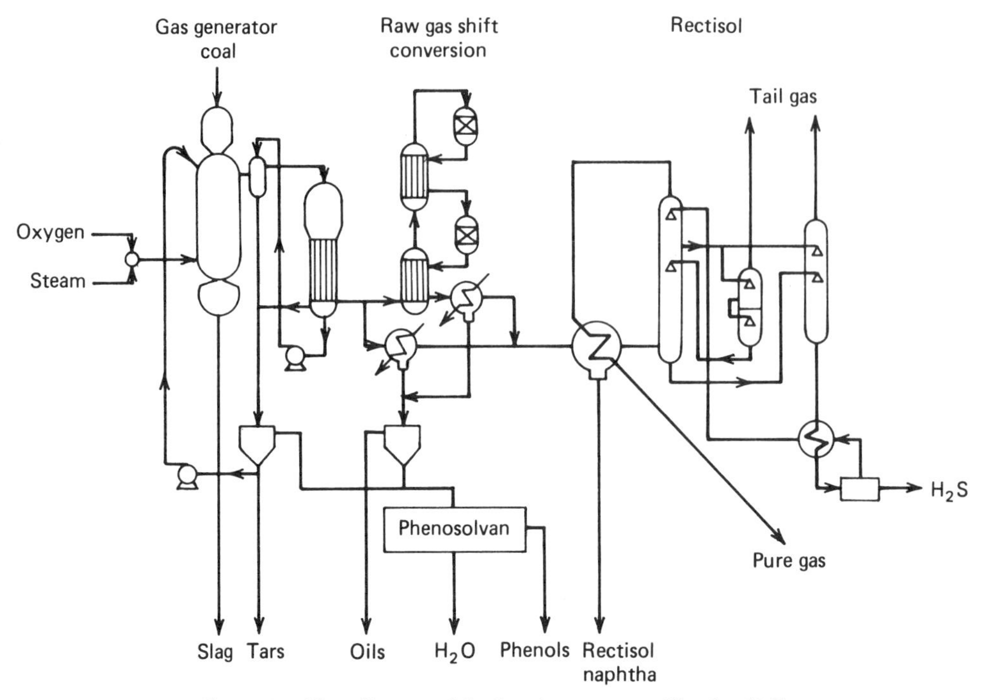

Figure 2. Flow diagram of the Lurgi pressure gasification (1,3).

products of various standard cleaning operations, ie, Rectisol [*67-56-1*], Phenosolvan [*37297-19-1*], etc, are included, five by-product streams can be distinguished (4,11): *tars*, as products of gas cooling or quenching (with fractionated cooling, tars and tar oils of varying densities can be obtained); *Rectisol naphtha* from the precooling stage of the Rectisol wash of the raw synthesis gas, which contains olefins, paraffins, and the aromatics (carbon dioxide also forms); *raw phenol* as a mixture of phenols, heterocycles (eg, pyridines, etc), acids, etc, after treatment of the gas liquor in the Phenosolvan process or after absorptive processes; *ammonia* by the stripping of the wastewater from the Phenosolvan purification or by other processes, eg, CLL or Phosam process, downstream from the phenol removal; and *sulfur* after processing of the acidic gases of the corresponding cleaning processes, eg, Rectisol or Sulfinol [*63665-16-7*], in suitable plants, eg, Claus.

The use of Lurgi pressure gasifiers with downstream synthesis of, for example, Fischer-Tropsch products, syngas, or ammonia requires considerable space and substantial investments for the gasification and by-product processing (12–17). The flow sheet in Figure 3 of the Sasol-I plants shows this clearly.

Subsequent downstream processes, eg, hydrogenative refining, Benzoraffin process, etc, lead to the following derivatives of the five main product streams after conversion or fine separation (4,11):

tar/tar oil and rectisol naphtha ⟶ {
- benzene [*71-43-2*]
- naphthalene [*91-20-3*]
- creosotes
- naphtha
- electrode pitch
- carbon black [*1333-86-4*]

raw phenol ⟶ phenol [*108-95-2*]; cresols (*o*- [*95-48-7*], *m*- [*108-39-4*], *p*- [*106-44-5*])

ammonia ⟶ ammonium sulfate

sulfur ⟶ sulfuric acid

Depending on the size of the downstream plant, by-product formation can be great. For example, in the annual production of ca 1.5×10^6 t motor fuels in the Sasol-II plant in the Republic of South Africa, ca 110,000 t NH_3, 90,000 t sulfur, and 240,000 t phenols, creosotes, pitch, etc also form (12,19–20).

Despite the advanced work-up of by-products, the wastewater of Lurgi gasifiers still contains considerable amounts of pollutants, which also require aftertreatment analogous to that of wastewaters from coking plants (21–22) (see Water, water pollution; Water, industrial water treatment). A comparison of the residues produced in Lurgi gasifiers with those in high temperature gasifiers is given in Table 3.

Development of the Lurgi fixed-bed gasification relies greatly on the reduction of by-product incidence. This is apart from special aims, for example, greater methane yield by increasing pressure or simplification of the technology by liquid-slag removal (6,23–25). To this end, raw recycling and carbonization gas recycling; recycling of tars, phenols, etc; increase of H_2 partial pressure and, thus, the reduction of a portion of condensable components; and special gas-removal devices and raw-gas-splitting processes have been proposed, developed, and tested (23–24,26–27).

Compared to the by-product potential of the Lurgi moving fixed-bed gasification process, the two other coal gasification processes of the first generation are insignificant. An exception is the recovery of sulfur, the yield of which depends on the sulfur content of the feed coal. The sulfur must always be removed for downstream chemical-synthesis processes which, almost without exception, are homogenously or heterogenously catalyzed and require sulfur-free synthesis gases.

Carbon dioxide can also be obtained from the raw gases of all the preceding autothermal coal gasification processes. If required as a starting material, a carbon dioxide concentrate from one of the previous gas-scrubbing steps is particularly suitable.

Syngas Chemistry

Synthesis gas can be produced by the gasification of coal, by the partial oxidation of heavy mineral-oil fractions, ie, by oil gasification, or by the cracking of light hydrocarbons or from methane. Apart from its use as a building block for chemical synthesis, it is also a feedstock for the recovery of pure carbon monoxide and hydrogen.

Depending on the C:H compositions of the starting materials, synthesis gases relatively rich in H_2 form from oil gasification or hydrocarbon cracking ($CO:H_2$ ratios of 0.5 to 1 in partial oxidation). Coal-gasification processes, with the exception of the Lurgi process, generally yield synthesis gas rich in CO. This is especially true for second-generation coal-gasification processes, ie, those involving high reaction temperatures, gasification under pressure, and use of coal fines. This affects the conditioning of the synthesis gas, as subsequent chemical syntheses generally require a feed gas rich in H_2. The theoretical $CO:H_2$ ratio is 1 for oxo synthesis, 0.5 for methanol production, and 0.33 for SNG synthesis (28). Therefore, depending on the method of

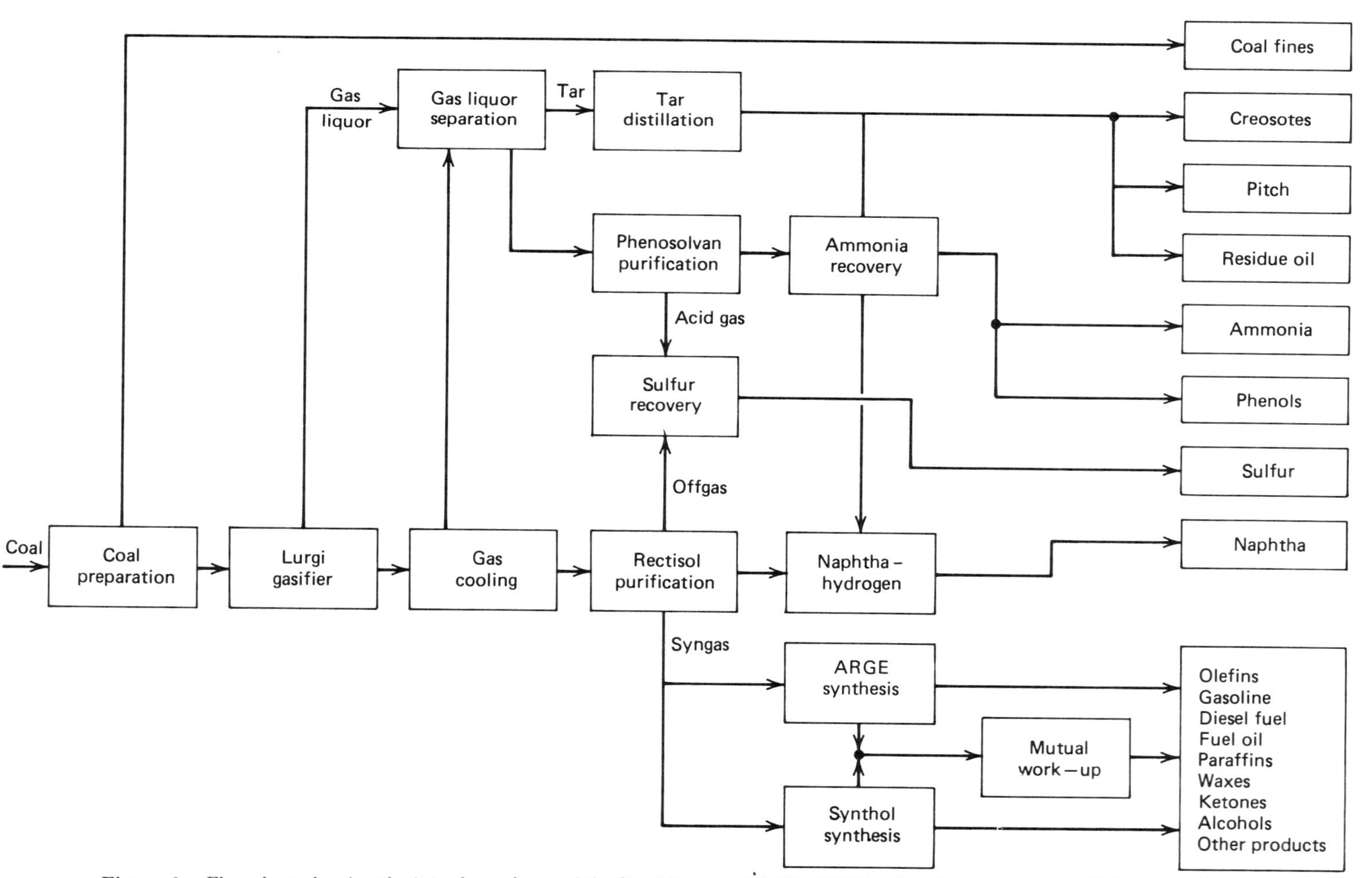

Figure 3. Flow sheet showing the interdependence of the Sasol-I plants (12,18). ARGE = Arbeits gemeinschaft Ruhrchemie/Lurgi.

Table 3. Wastewaters from Coal-Gasification Processes[a]

Pollutants, mg/L	Lurgi gasification	Texaco (high temperature) gasification
total organic carbon (TOC)	≤5,000	≤500
chemical oxygen demands (COD)	≤80,000	≤400
cyanides (as CN)	ca 5	ca 5
phenols	≤5,000	≤0.02
ammonia	≤11,000	≤1,600
carboxylic acids	≤700	≤100

[a] Refs. 4 and 10.

synthesis-gas manufacture, one of the common work-up processes is a shift conversion of CO to produce H.

The potential of synthesis gas as a building block for chemistry is great (see Fig. 4).

The oxo synthesis; methanol (qv) production; Fischer-Tropsch process; production of acetic acid (qv); H_2 recovery for ammonia (qv), urea (qv), and other processes; as well as a wide range of downstream processes are being used industrially (30–32) (see Hydrogen; Oxo process). Synthesis gas from coal is used in technical applications in many countries for the oxo synthesis, methanol production, hydrogen-based ammonia production, and the manufacture of motor fuels and chemical feedstocks (33). After the required synthesis-gas quality has been adjusted, the chemical processes are the same for the production of synthesis gas from coal, oil, and gas.

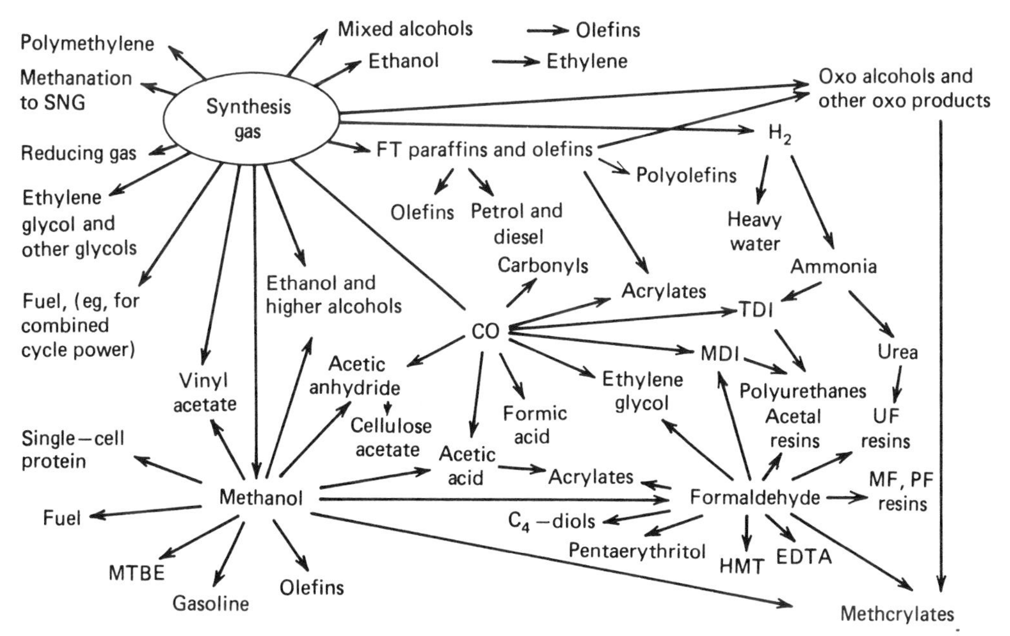

Figure 4. Existing and potential synthesis-gas applications (29). FT = Fischer-Tropsch; TDI = toluene diisocyanate; MDI = diphenylmethane diisocyanate; UF resins = urea formaldehyde resins; MF resins = melamine formaldehyde resins; PF resins = phenol formaldehyde resins; HMT = hexamethylene tetramine; and MTBE = methyl tertiary butyl ether.

The emphasis in the production of synthesis gas from coal is on determining the combination of gasifier and gas-processing stage that is most favorable and energy-efficient, eg, hydrogen recovery from carbon monoxide or from synthesis gas for the production of methanol (31).

Hydrogen Recovery. If pure hydrogen is to be obtained, the synthesis gas is first converted catalytically according to the following equation (31):

$$CO + H_2O \rightleftharpoons CO_2 + H_2$$

In the high temperature shift conversion, an iron–chromoxide catalyst is used at gas inflow temperatures of 280–350°C. As a consequence of the exothermic conversion, a temperature increase to 430°C is permissible; the catalytic material is stable up to 550°C. The conversion efficiency increases with increasing pressure.

With the low temperature shift conversion, a copper catalyst which requires sulfur-free feed gas is used. A final carbon monoxide content of <0.2 wt % in the outflowing gas mixture can be achieved if an upstream high temperature shift conversion occurs. The operating temperature of the low temperature shift conversion is 200–270°C. The low carbon monoxide contents are obtained with the low temperature shift catalyst at technically reasonable reaction rates.

More recently, sulfur-resistant shift catalysts based on cobalt and molybdenum have been developed; these permit the conversion of untreated, sulfur-containing synthesis gas in one heating, provided the gas temperature is correct (34). The catalysts are specially developed for the conversion of sulfur-containing gases at ca 220–475°C. They are only catalytically effective as sulfides; therefore, a certain minimum amount of sulfur must always be present in the feed gas, and this depends on the prevailing operating conditions. The sulfur content in the raw gas is normally high enough if coal is used as the feedstock. In contrast to the high temperature catalyst, sulfur-resistant shift catalysts permit the conversion of gases highly rich in CO, eg, those containing 52 wt % CO, even with a relatively low H_2O–dry gas ratio at the catalyst inlet, without carbon deposition occurring on the catalyst. Since the catalyst operates at low temperatures, a carbon monoxide concentration of ca 0.5 wt % in the converter gas can be achieved.

For many applications, it is essential for COS to be converted to H_2S on the sulfur-resistant conversion catalyst according to the hydrolysis reaction

$$COS + H_2O \rightarrow H_2S + CO_2$$

At high gas temperatures, such as are reached in the second-generation coal-gasification processes (for example, in the Texaco process), a catalyst-free partial conversion can be attained (35).

A carbon monoxide shift conversion in the process route of coal to ammonia is shown in Figure 5 for a Koppers-Totzek coal-gasification process. The catalyst-free conversion with hot wet gas and the use of sulfur-resistant Co/Mo conversion catalysts simplifies the process design appreciably, as shown in Figure 6.

A discussion of the purification of hydrogen, especially of hydrogen to be used for hydrogenation, is given in ref. 7. This involves low temperature separation processes, pressure-swing adsorption (PSA), and the diffusion process (40–41). The pressure-swing adsorption (PSA) process makes use of the loading differences in suitable adsorption agents, eg, molecular sieves (qv), between high pressure for the adsorption and low pressure for the desorption. The pressure adsorption is, therefore,

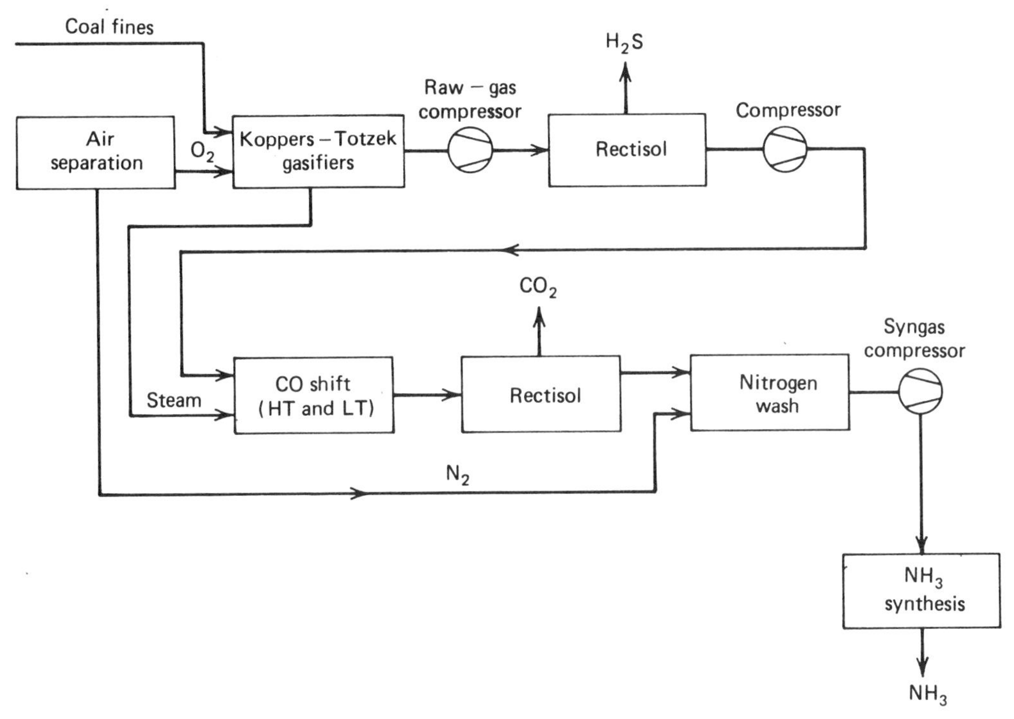

Figure 5. Simplified flow sheet for a Koppers-Totzek ammonia plant (36–38). HT = high temperature; LT = low temperature.

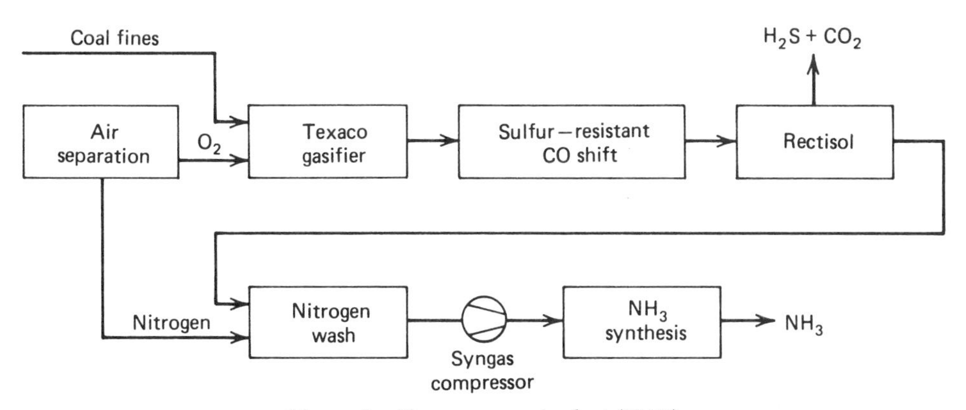

Figure 6. Texaco ammonia plant (36,39).

followed by a regeneration which is accomplished by depressurization of the adsorbent bed. This is followed by a purge of hydrogen obtained from one of the adsorbers undergoing countercurrent depressurization (42–44). The process and flow sheets are shown in Figure 7.

Carbon Monoxide Recovery. The separation of carbon monoxide from a gas mixture containing hydrogen, carbon monoxide, and methane as main components

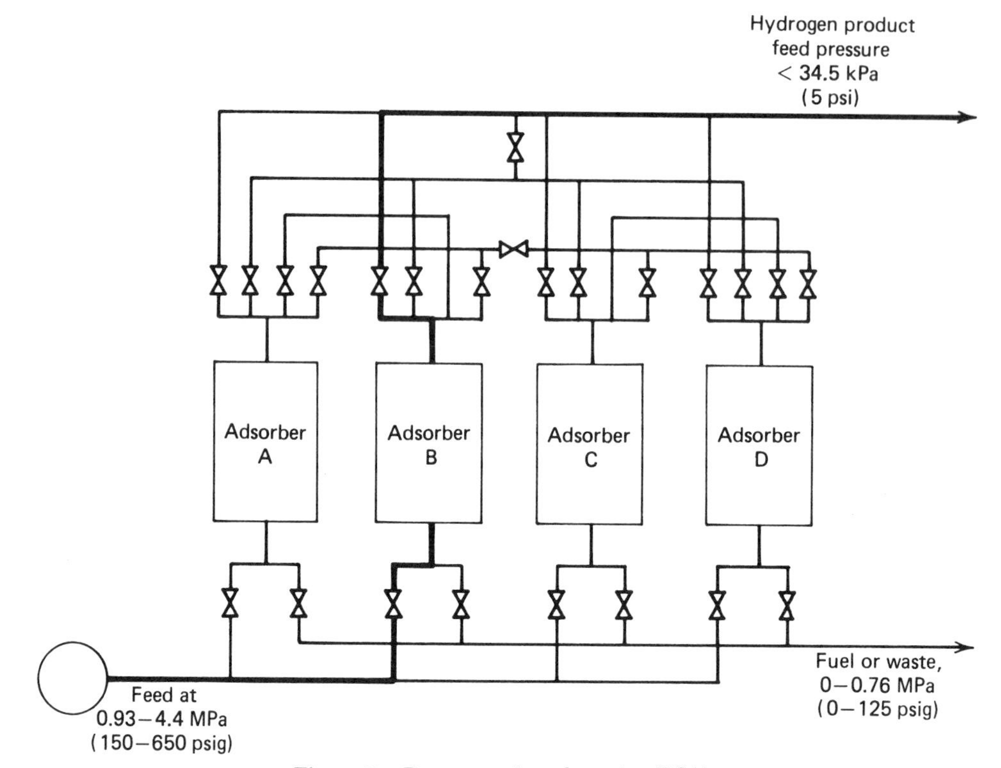

Figure 7. Pressure-swing adsorption (PSA).

can basically be carried out according to two different processes: separation of the components in the liquid phase at low temperatures by condensation and distillation or by selective absorption of the carbon monoxide (31). The purity of the carbon monoxide depends on the application. The impurities are generally hydrogen, carbon dioxide, methane, nitrogen, oxygen, argon, water vapor, oils, higher saturated and unsaturated hydrocarbons, sulfur compounds, nitric oxides, and/or acetylenes. Carbon monoxide (97–99 wt %) that contains hydrogen, methane, or nitrogen generally meets the quality demands for use in chemical processes. Low temperature separation, liquid methane washing, and absorptive methods can be used for separation (45–47) (see Gas cleaning). The process chosen depends on economic aspects (31). Preference is increasingly given to the Cosorb process of Tenneco (48) (see Fig. 8).

In the low temperature separation of synthesis gas involving the isolation of CO, the carbon monoxide and any methane present are first liquefied. This takes place at −100 to −180°C, depending on the pressure, and ending at the sublimation points of methane and CO. In view of the low temperatures, all components that might freeze or become clogged must be removed.

The possible low temperature processes for carbon monoxide recovery from synthesis gases involve partial condensation or liquid-methane scrubbing. With partial condensation, the higher boiling carbon monoxide is condensed on cold walls from the lower boiling hydrogen. The amount of carbon monoxide corresponding to the vapor pressure remains in the gas phase. The cold wall is produced in a heat exchanger

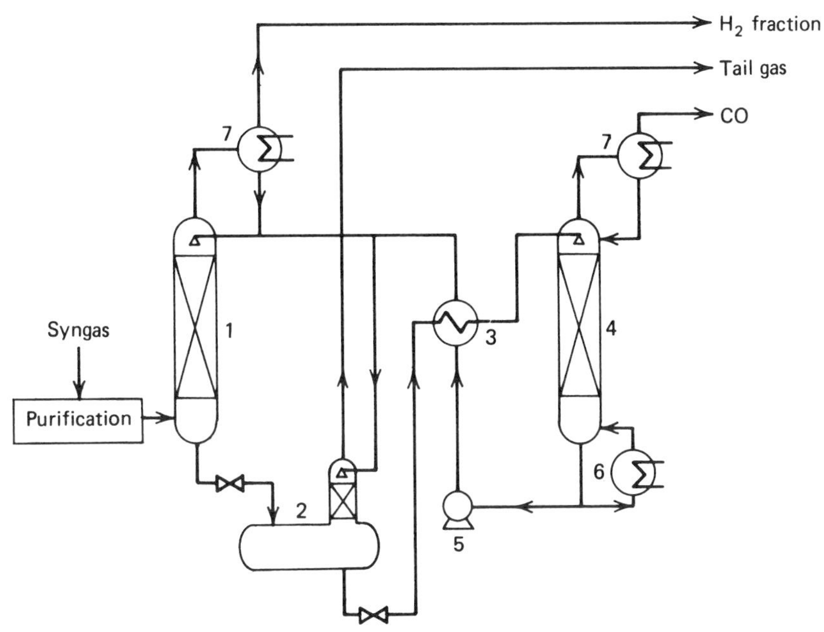

Figure 8. Flow sheet of the Cosorb process. 1, Absorber. 2, Predegassing stage. 3, Heat exchanger. 4, Regenerator. 5, Circulation pump. 6, Reboiler. 7, Reflux condenser (cold trap).

by the liquid carbon monoxide which has been condensed from the gas phase, after flashing to the lowest pressure possible, the CO vaporizes countercurrently.

With the liquid methane washing process, the gas mixture to be separated is scrubbed at just above the freezing point of methane, ie, at ca −180°C. All the carbon monoxide in the hydrogen stream dissolves in the scrubbing methane, so that the hydrogen emitted at the head no longer contains any carbon monoxide. The scrubbing agent is regenerated by recovering the methane by distillation; the carbon monoxide is then flashed overhead. Carbon monoxide is suitable as a circulation medium for this process. It is used according to the heat-pump principle in order to maintain regeneration and to keep the coldness regulated by a refrigerating flash device. The process principle is shown in Figure 9 (33,49).

Synthesis Gas with Different CO:H_2 Ratios. Figures 10 and 11 show examples of various processing combinations in the reaction sequence coal-to-methanol or coal-to-motor fuels. The first technical application of a second-generation coal-gasification process, in which the resulting synthesis gas is used as a chemical feedstock, was by the Tennessee Eastman Corporation in early 1983 (50). In a first reaction stage, acetic anhydride [*108-24-7*] is produced from coal. Figure 12 is a flow sheet of the process combination. Large-scale industrial coal-gasification plants involving chemical syntheses are operated in South Africa (Fischer-Tropsch synthesis and derivatives, ammonia, and methanol [*67-56-1*]); India, Greece, Zambia, and Turkey (ammonia); and the GDR (hydrogen).

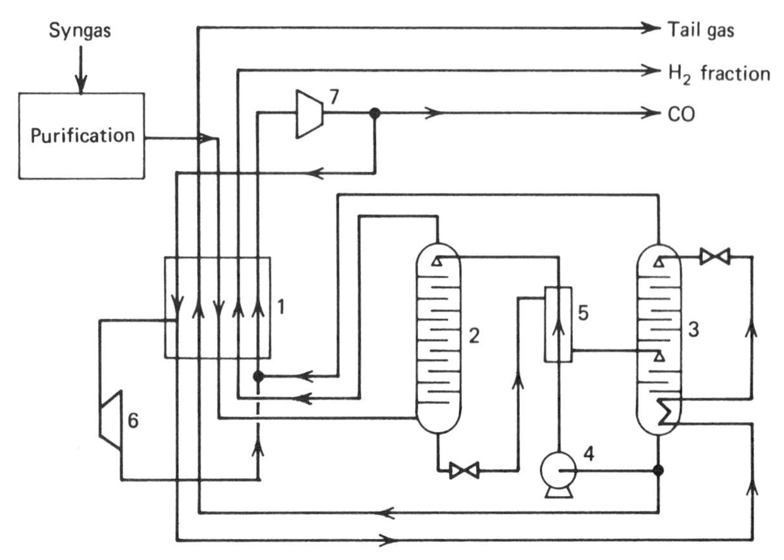

Figure 9. CO separation by liquid methane wash. 1, Heat exchanger. 2, Liquid CH_4 scrubber. 3, CO/CH_4 separator. 4, Liquid CH_4 pump. 5, Liquid CH_4 cooler. 6, Recycle CO decompressor. 7, Recycle CO compressor.

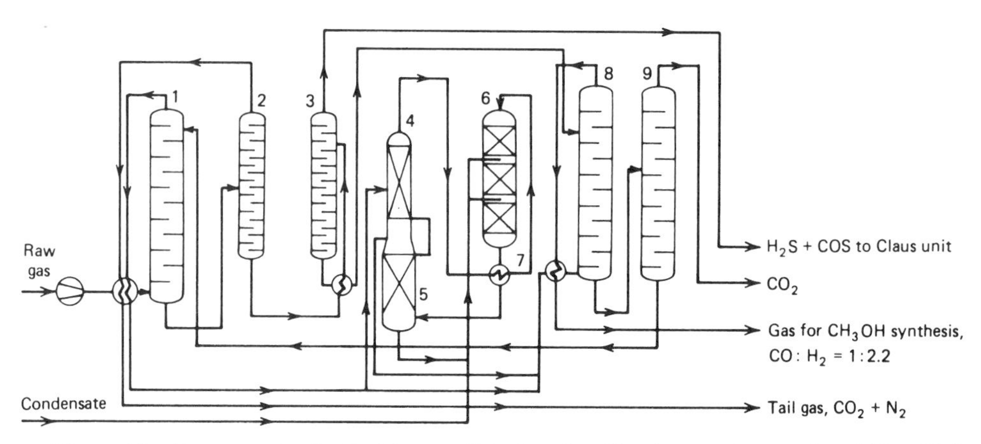

Figure 10. Gas processing for the Koppers-Totzek coal gasification; product: methanol (9). 1, H_2S scrubber. 2, H_2S rectifying section. 3, Regenerating column. 4, Humidifier. 5, Dehumidifier. 6, CO shift conversion. 7, Heat exchanger. 8, CO_2 scrubber. 9, Stripper.

Developments

Coal Gasification. The first of the three principal requirements of second-generation coal-gasification processes is the high gasification temperature, which results in high carbon conversion, pure syngases (ie, low environmental impact), possible slagging, and water decomposition which may serve as an O_2 source. The characteristic high pressures result in lower expenditures for syngas compression and increased space–time productivity. Lastly, the use of finely divided coal implies an increase in specific surface area and carbon conversion rate, the process does not depend on the coal quality, and the trend in the mining industry for increased production of finely divided coal is met.

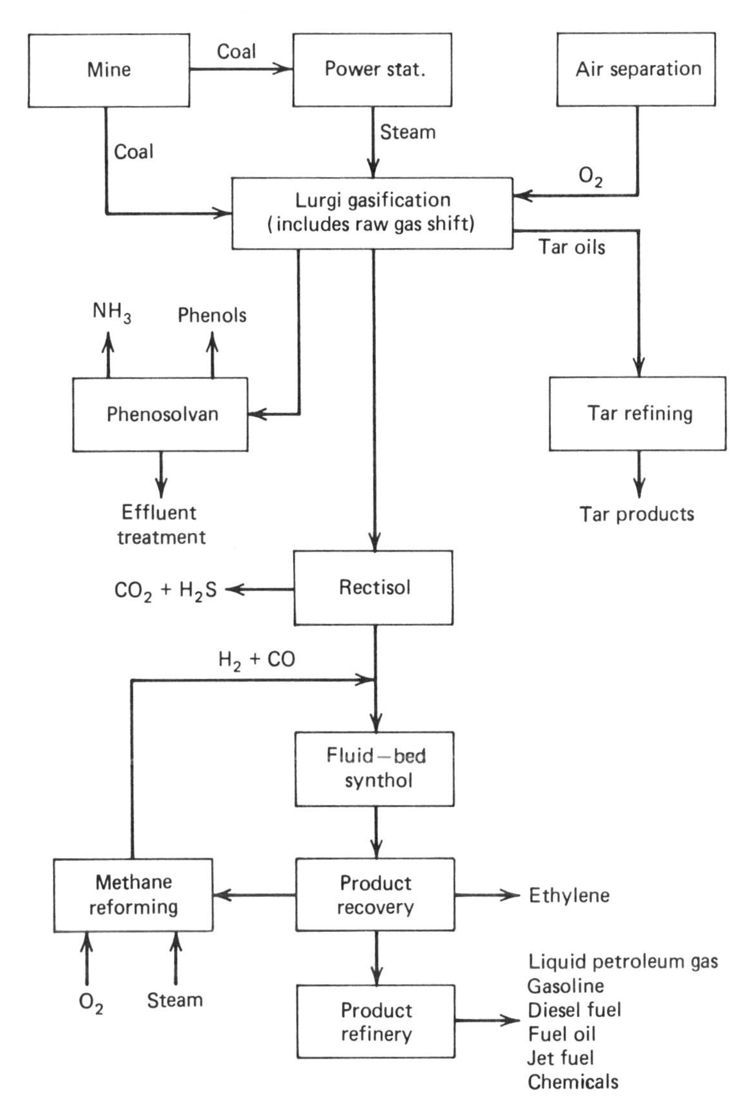

Figure 11. Processing of coal gas from the Lurgi gasification for the Sasol synthesis (4,18).

Second-generation processes are aimed at maximum specific yields of CO and H_2 and minimum by-product incidence by means of the highest reaction temperatures possible (52). The raw-gas compositions of second-generation coal-gasification processes are listed in Table 4.

All entrained-bed processes fulfill the preceding requirements (67). The fluidized-bed gasifications operating at moderately high temperatures result in higher methane concentrations in the raw gas. Consequently, these raw gases are good starting gases for the manufacture of SNG, but they are less suitable than synthesis gas for chemical purposes.

The Texaco coal-gasification process (TCGP) is a particularly suitable and innovative second-generation process; it provides synthesis gas that is practically free of by-products. A flow scheme of the TCGP is shown in Figure 13. As with all gasifi-

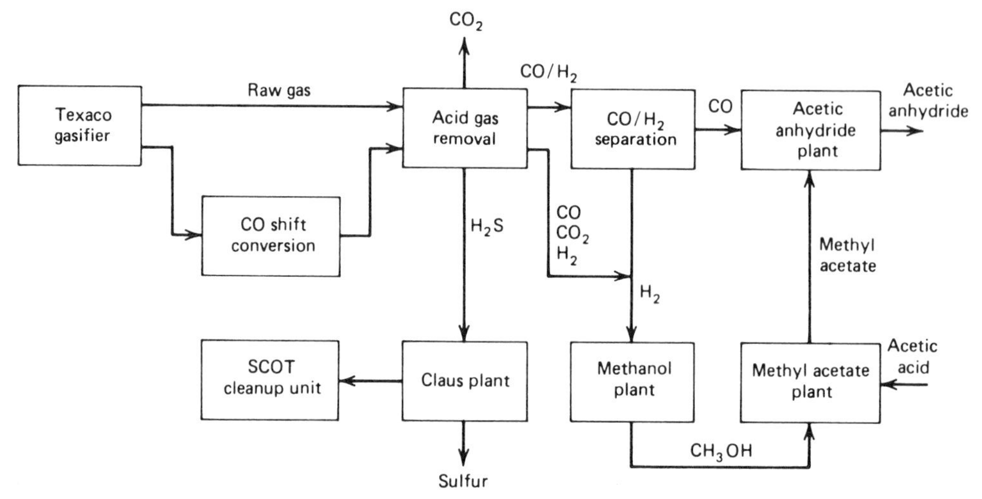

Figure 12. Texaco coal-gasification process: gas processing and associated synthesis in the production of acetic anhydride from coal (51). SCOT = Shell Claus off-gas treating.

Table 4. Raw-Gas Composition of Second-Generation Coal-Gasification Processes

	Fluidized-bed techniques[a]			Entrained-bed techniques[a]				
	U-gas[b]	Westing-house[c]	High tempera-ture-Winkler[d]	Texaco[e]	Shell[f]	Saarberg Otto[g]	Combus-tion Engi-neering[h]	Others KHD[i]
pressure, MPa[j]	0.4	1.6	1.0	8.0	2.0	2.5	0.1[l]	0.1
Typical gas composition, vol %								
CO	34	54	52	55	65	61	8–30	65–70
H_2	41	27	35	33	31	29	5–9.5	25–30
CO_2	19	10	9	11	3	7	3–12.5	0.3
CH_4	3.5	8	3	<0.1	<0.1	0.2	≤3	0.1
N_2 + Ar	0.9	0.5	0.7	0.6	0.7	3	58–74	
sulfur compounds[k]	1.0	<0.02	<0.1	0.3	0.2	<0.1	<0.1	<0.01
C_nH_{2m}				μg range			≤1	
NH_3, ppm				3	500			

[a] Refs. 1–4 and 10.
[b] Utility Gas Process, developed by Institute of Gas Technology (53–54).
[c] Refs. 55–56.
[d] Refs. 57–59.
[e] Refs. 52 and 60.
[f] Refs. 61–62.
[g] Ref. 63.
[h] Refs. 64–65.
[i] Kloeckner-Humboldt-Deutz, FRG (66).
[j] To convert MPa to psi, multiply by 145.
[k] H_2S + COS.
[l] Airblown.

cation processes, whether first- or second-generation, the sulfur content of the feed coals of the TCGP occurs as H_2S + COS in the synthesis gas. Thus, the recovery of elementary sulfur is necessary for every coal-gasification plant.

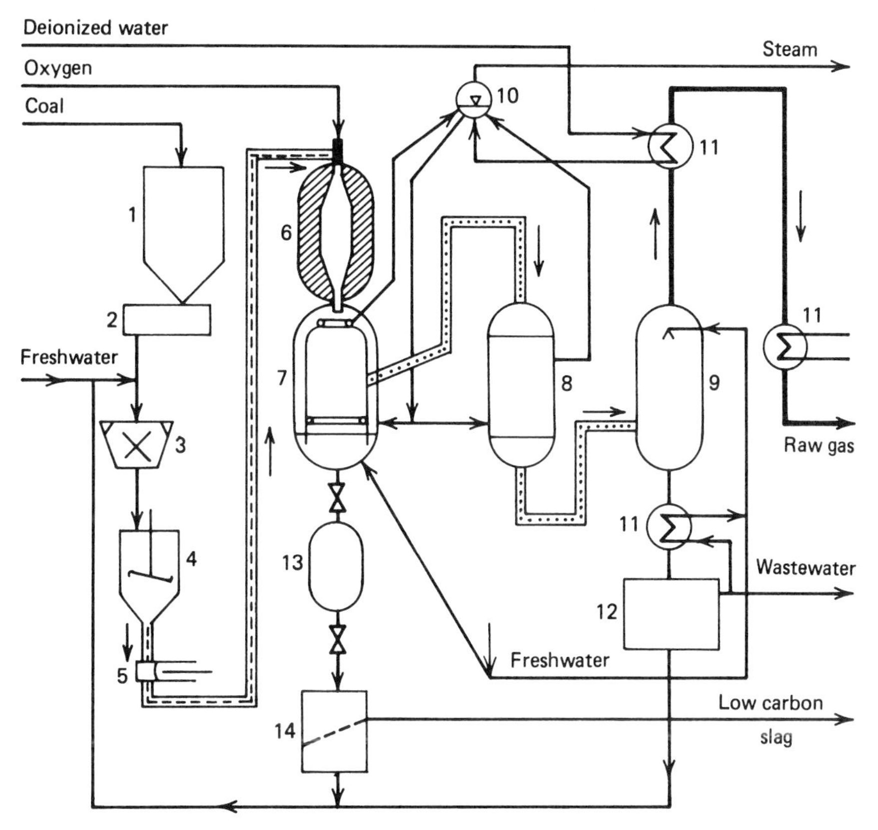

Figure 13. Flow diagram of the Ruhrchemie AG/Ruhrkohle AG plant of the Texaco coal-gasification process (68). 1, Coal bunker. 2, Weigh belt. 3, Mill. 4, Slurry tank. 5, Feed pump. 6, Reactor. 7, Radiant boiler. 8, Convection boiler. 9, Scrubber. 10, Steam drum. 11, Heat exchanger. 12, Settler. 13, Ash lock. 14, Ash separation.

The main goal of processes involving catalysts in the gasifier, eg, Exxon, is a reduction in the required gasification temperature (69–70). The aim is not so much to increase the by-product incidence but rather to secure as high a methane content as possible by hydrogenative coal gasification at low temperatures; the product is SNG.

The second-generation coal-gasification processes are also potential sources of CO_2, which can be further processed to urea, used in the food industry or as refrigerants, etc. Underground coal gasification (UCG) is currently being tested by the Lawrence Livermore National Laboratory (LLNL) near Centralia, Washington. The UCG system uses the controlled-retracting-injection point (CRIP) method (see Fig. 14). The net products are gasoline and a few other associated hydrocarbons. The by-products of the process include elemental carbon dioxide and sulfur (71).

Syngas Chemistry. About 80% of chemicals are derived from petrochemical products. Crude oil and natural gas satisfy >50% of the energy requirements of the chemical industry. Of the petroleum used, ca 70% is used as feedstock and ca 30% as a source of energy.

The use of synthesis gas in the chemistry industry can be roughly divided into three product groups (see also Fig. 5): source of raw material and of energy, eg, in the

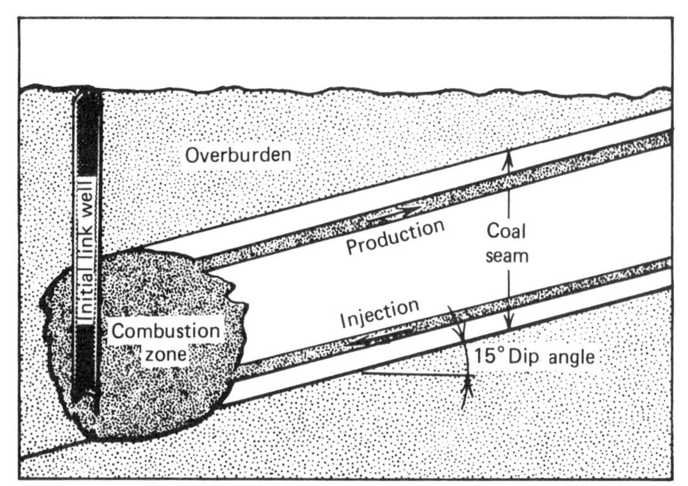

Figure 14. The improved controlled-retraction-injection point (CRIP) method.

synthesis of methane and other hydrocarbons; source of chemical feedstocks, eg, in the synthesis of ethylene, propylene, and aromatic hydrocarbons; and as a component in chemical syntheses, eg, for the preparation of oxygen-containing products (32,58). In the first two of these product groups, synthesis gas from coal or the products made from synthesis gas are economically and effectively competitive with petroleum-based products. Before synthesis gas can be used, two basic requirements must be satisfied: economically favorable prices and technically satisfactory processes for further treatment. The first requirement has been more or less fulfilled; however, the economic potential and technical maturity (see Fig. 15) of the processes for further treatment differ widely.

SNG Preparation. In recent years, methane has been used increasingly in the industrial and private sectors as a source of energy and as a raw material (72–74). In the United States, natural gas satisfies about a quarter of the total primary energy requirements. Although the growth in demand has slowed in the past few years, shortages and price increases are expected in the medium term since there are limited reserves.

The hydrogenation of carbon monoxide to methane is one of the best known heterogeneously catalyzed reactions:

$$CO + 3\,H_2 \rightarrow CH_4 + H_2O \quad \Delta H_R^{500°C} = -214.8 \text{ kJ } (-51.34 \text{ kcal}) \tag{1}$$

$$CO + H_2O \rightleftharpoons CO_2 + H_2 \quad \Delta H_R^{500°C} = -39.8 \text{ kJ } (-9.51 \text{ kcal}) \tag{2}$$

Hydrogenation (eq. 1) is highly selective toward methane, and only small amounts of higher hydrocarbons form. Simultaneous with the methane formation, water–gas conversion (eq. 2) takes place and can be used to meet the entire hydrogen requirement for the reaction. A maximum of 178.6 g CH_4 can be obtained from one m^3 of synthesis gas.

The hydrogenation of carbon monoxide is catalyzed by many metals. In industrial applications, nickel catalysts containing oxidic components as carriers, eg, Al_2O_3 or SiO_2, or as activators, eg, TiO_2, ZrO_2, or Cr_2O_3, are used exclusively. The catalysts must be able to withstand the thermal loads and should be sufficiently resistant to small amounts of catalyst poisons.

Just as in the Fischer-Tropsch synthesis, the heat of reaction, which influences

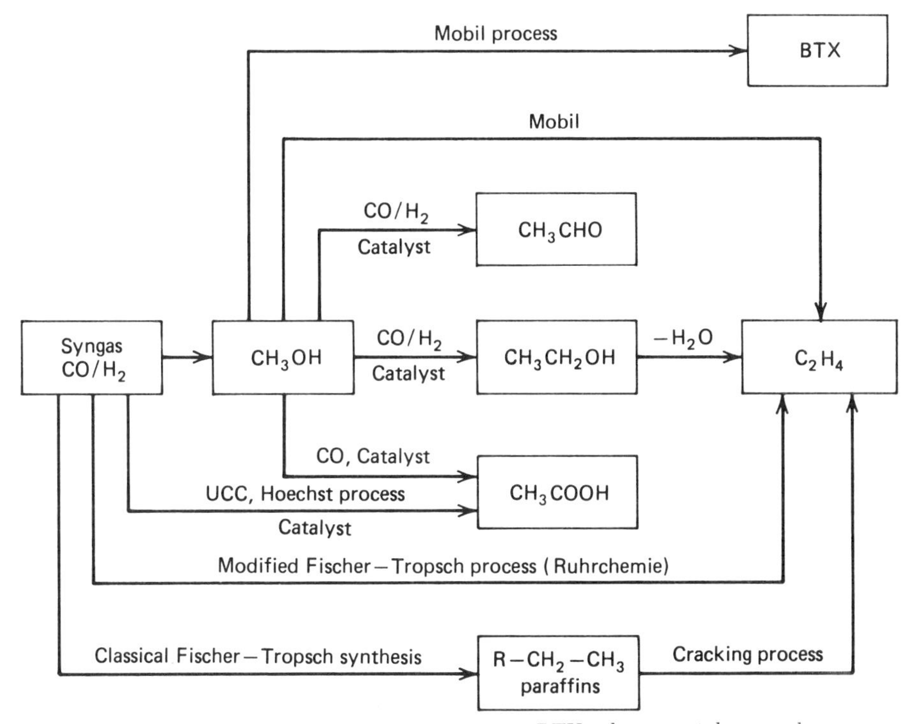

Figure 15. Future applications of synthesis gas. BTX = benzene, toluene, xylenes.

the design of synthesis, must be removed during SNG preparation. Cooled or adiabatic fixed-bed reactors mostly operate on a multiple system. In such a case, the last stage in a cascade arrangement is operated at low temperatures to create a favorable thermodynamic basis for high carbon monoxide conversion. The addition of steam improves heat removal, reduces carbon deposition, and increases the hydrogen concentration through the water–gas equilibrium. The number of reactors can be reduced accordingly by the use of a gas cycle with interim cooling; therefore, two or three stages are sufficient.

A one-stage process is provided in the liquid-phase process, whereby a light hydrocarbon liquid stream is flushed through a granular catalyst. The upward flow of the liquid causes expansion of the catalyst bed by 10–20%, good mass and heat transfer through high linear velocity (10–30 cm/s), and a practically isothermal operation (see Fig. 16). Conversion is high and the stoichiometry corresponds largely to equation 1. After the removal of carbon dioxide, the emitted gas is already of near SNG quality, ie, it contains >95 vol % methane, <2 vol % H_2, and <1 vol % CO. If necessary, it can be postmethanated in a downstream stage.

Another fluidized-bed process also operates on a one-stage principle (74). The characteristics of this process are the construction of the inflow bed in the gas inlet, the arrangement of the cooling aggregate according to the immersion-heater principle, and the jacketed design of the reactor with steam feed (see Fig. 17). The good mass and heat transfer within the fluidized bed and the heat removal through the cooling aggregate and reactor wall result in a low temperature gradient. Thus, a very high conversion of the synthesis gas at relatively high loading can be achieved in one stage, and the residual gas is very rich in methane. A cycle gas flow in the ratio of 1–2:1 to the fresh gas flow enables the load to be increased but is not absolutely necessary.

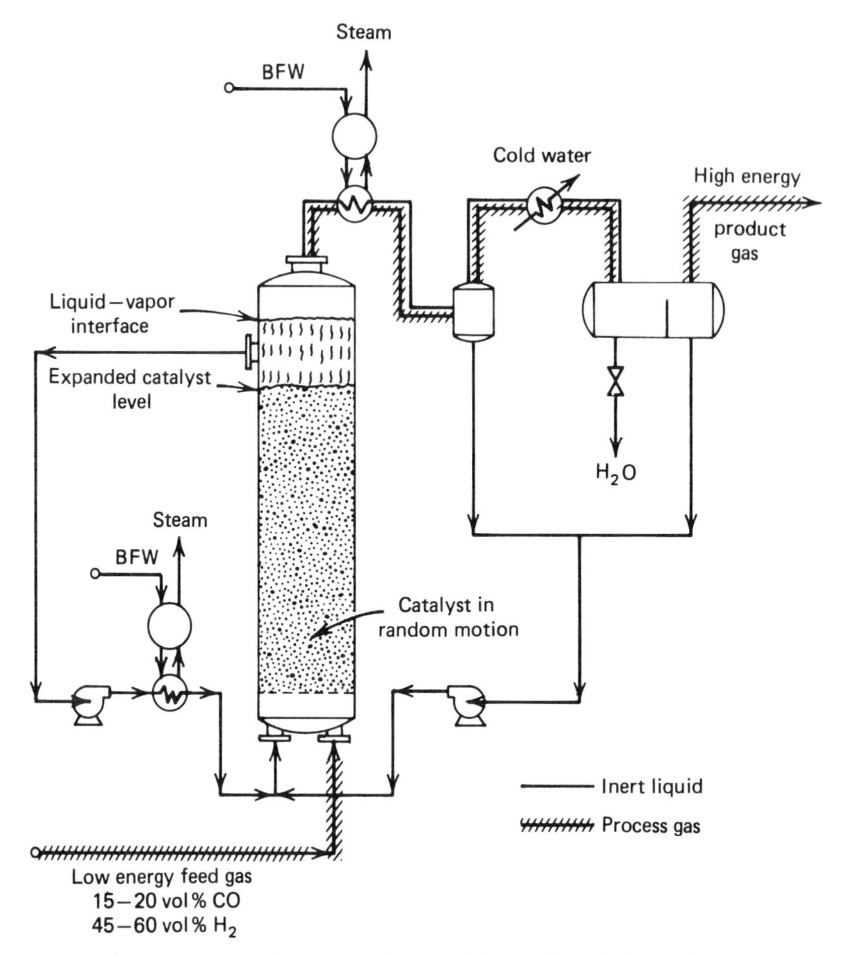

Figure 16. Liquid-phase methanation (72). BFW = Boiler feed water.

The pressures used in the process depend on the composition of the synthesis gas and are 3–6 MPa (435–870 psi). As synthesis gas from coal is usually rich in carbon monoxide, all the processes are designed for a substoichiometric gas composition. The reaction temperatures depend on the procedure: adiabatic, multistage reactors operate at ca 250°C at the inlet and up to 550°C at the outlet; single-stage reactors or post-methanation, at 250–350°C. These and other processes being developed have been tested on a pilot-plant scale (≤4000 m^3/h (5230 yd^3/h) SNG); as of early 1983, no industrial-scale production facilities were being constructed.

Fischer-Tropsch Synthesis. The catalytic hydrogenation of carbon monoxide with catalysts containing iron, cobalt, or ruthenium produces hydrocarbons and, therefore, provides an alternative to the direct hydrogenative liquefaction of coal (75). In the conventional, industrially tested methods, the Fischer-Tropsch synthesis can provide a wide variety of hydrocarbons ranging from methane to waxes with an average molecular weight of ca 1000. Depending on the catalyst and reaction conditions, different proportions of alkenes occur, ie, 40–85 wt %, composed mostly of α-alkenes, as well as alkanes and small quantities of compounds containing oxygen, ie, alcohols, ketones, acids, and esters. The hydrocarbons are predominantly (>90%) linear and contain no sulfur.

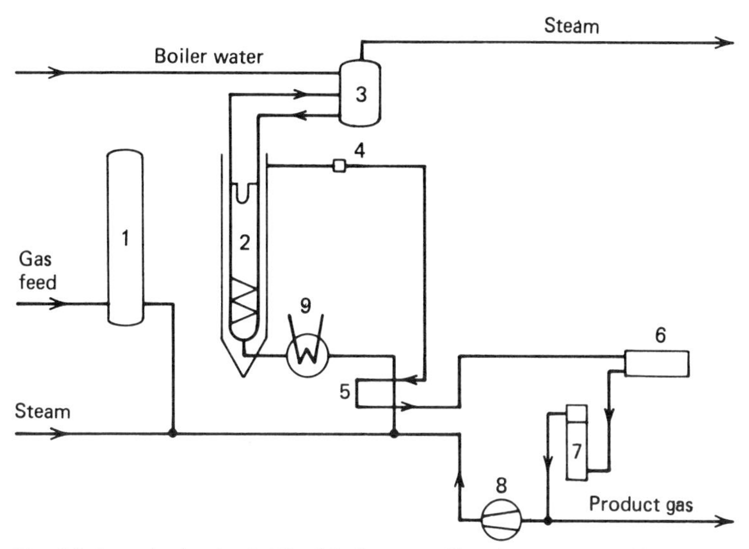

Figure 17. Methanation by the fluidized-bed process (Comflux process of Thyssengas GmbH) (74). 1, ZnO desulfurizer. 2, Reactor. 3, Steam drum. 4, Filter. 5, Heat exchanger. 6, Cooler. 7, Separator. 8, Compressor. 9, Heater.

With increasing prices and mineral oil becoming scarcer, the Fischer-Tropsch synthesis will become economically attractive again if a method is developed for the direct production of the same products as from petroleum, eg, ethylene and propylene. Therefore, the current process variants must be modified to limit the lengths of the hydrocarbon chains and to attain a high alkene–alkane ratio. As all the details of the mechanisms of the Fischer-Tropsch reaction have not yet been clarified, this demand presents some difficulties and only a limited amount of success has been achieved to date.

In Table 5, the reaction conditions and product compositions of two conventional

Table 5. Variants of the Conventional and Modified Fischer-Tropsch Synthesis

Process	ARGE[a] (gas-phase fixed-bed)	Synthol[a] (entrained-bed)	Experimental catalysts: Fe-V-catalyst[b] (gas-phase fixed-bed)	Experimental catalysts: Fe-whisker[b] (gas-phase fixed-bed)
temperature, °C	220–240	320–340	330	320
pressure, MPa[c]	2.6	2.2	1.0	1.5
H_2:CO ratio	1.7	3	1	1
products, wt %				
methane	2.2	10.7	13.7	8.2
C_2–C_4 alkenes	4.4	24.5	36.6	39.3
C_2–C_4 alkanes	5.5	7.8	9.1	5.4
C_5–C_{11} hydrocarbons	17.6	38.2 }	37.6 }	42.1
C_{12}–C_{20} hydrocarbons	14.7	6.6 }		
>C_{20}	50.5	2.9		
others, eg, oxygenates	5.1	9.3	ca 3	ca 5

[a] Ref. 76.
[b] Ref. 77.
[c] To convert MPa to psi, multiply by 145.

Fischer-Tropsch synthesis processes, operated by South African Coal, Oil and Gas Limited (Sasol), are compared with those of more recent processes. The ARGE (Arbeits Gemeinschaft Ruhrchemie/Lurgi) fixed-bed synthesis is used for the production of medium and long chain hydrocarbons, and the synthol entrained-bed synthesis (M. W. Kellogg) is used for the manufacture of gasoline. Relatively small amounts of ethylene and propylene form in either procedure. Selectivity is influenced by the development of modified iron catalysts, and the selection of the reaction conditions must suit the properties of the catalysts. Oxides of vanadium, manganese, chromium, and titanium are suitable promoters. Apart from the composition of the catalyst, the structure of the iron surface is of great significance, as indicated by the results obtained with the Fe-whiskers in Table 5. As of early 1983, the modified versions of the Fischer-Tropsch synthesis have only been tested on a laboratory or pilot-plant scale (77–79).

Oxygen-Containing C_2 Compounds. The conversion of synthesis gas on catalysts containing rhodium or cobalt leads to mixtures of oxygen-containing C_2 compounds. Alkane and alkene occur as by-products. The reaction takes place at 250–300°C at synthesis-gas pressures of 2–10 MPa (290–1450 psi) in the gas phase. The relative proportions of oxygen-containing compounds can be changed by the reaction conditions and catalyst properties, but selectivity is limited and hydrocarbons always form as by-products. The rhodium concentration of the carrier catalysts, which are preferred, is 0.5–5 wt %. Alkali or alkaline earth metal oxides, eg, Li_2O, K_2O, or MgO, and difficultly reducible oxides in groups IV and VI of the periodic table, eg, TiO_2, Cr_2O_3, or MoO_3, have been proposed for use as activating or selectivity-controlling additives. Cobalt-containing catalysts that have been modified with copper yield mainly alcohols, whereas with rhodium-containing catalysts, acetaldehyde and acetic acid form in addition to ethanol. In Table 6, examples of reaction conditions and products from a synthesis on a rhodium catalyst are given (80).

On rhodium-containing catalysts, the hydrogenation of carbon monoxide at raised pressure in the liquid phase leads to a mixture of mono- and polyvalent alcohols, in which ethylene glycol and methanol are the main components. Pressures of 40–60 MPa (5800–8700 psi) are the lower limit, and pressures of 150–200 MPa (21,750–29,000 psi) are recommended. Owing to the high carbon monoxide partial pressure, rhodium is present in the reaction medium, eg, sulfolane, tetraglyme, and reaction products with

Table 6. Conversion of Synthesis Gas to Oxygen-Containing C_2 Compounds[a]

Property	Activator	
	Na_2O/MgO	MgO/ZrO_2
temperature, °C	275	275
pressure, MPa[b]	10	10
H_2:CO ratio	1	1
space–time yield, g oxygen products/($L_{catalyst}$·h)	452	475
selectivity, mol % of converted CO		
ethanol	4	70
acetaldehyde	18	3
acetic acid	32	5
hydrocarbons, C_1–C_4	46	22

[a] Catalyst is Rh/SiO_2 (80).
[b] To convert MPa to psi, multiply by 145.

a high molecular weight, as homogeneously dissolved carbonyl complexes. The temperature ranges from 190 to 240°C, and the rhodium concentration is 0.05–0.3 wt %. The reaction product can consist of up to 65 wt % polyols, of which up to 80 wt % may be ethylene glycol. Glycerol, methanol, and homologues are the most common byproducts (81).

If ruthenium complexes modified with phosphonium or ammonium compounds are used as catalysts and aliphatic carboxylic acids are used as coreactants in the reaction medium, vicinal glycolesters form, as do methyl and ethyl esters of carboxylic acids. The selectivity of the glycol formation is, however, <10%; therefore, even the low pressure of 43 MPa (6235 psi) and the low cost factor resulting from the use of ruthenium do not allow any optimism for industrial application so far.

Despite the favorable results of feasibility studies, the rhodium-catalyzed synthesis of ethylene glycol still must be conducted on a laboratory scale. Several questions still remain to be answered, eg, catalyst recovery and work-up of the products.

Homologation. The reaction of alcohols or aliphatic carboxylic acids with synthesis gas, in which one or several CH_2 groups are inserted, leads to homologues:

$$RCH_2OH + CO + 2\,H_2 \rightarrow RCH_2CH_2OH + H_2O \quad (3)$$

$$RCOOH + CO + 2\,H_2 \rightarrow RCH_2COOH + H_2O \quad (4)$$

Homologation occurs at 25–50 MPa (3625–7250 psi) and at 150–250°C on cobalt- or ruthenium-containing catalysts, which require halogens as cocatalysts. With up to 60% conversion of methanol, as much as 80% ethanol as well as smaller amounts of propanol, butanol, and acetic acid can form with catalyst systems containing Co, Ru, I_2, or phosphines. With up to 70% conversion of acetic acid on Ru/CH_3I, the selectivity to propionic acid is 20–30 mol %; 5–10 mol % to butyric acid; and a few mol % of valeric acids (82–84).

The homologation of methanol or acetic acid provides an access method solely via synthesis gas to functional compounds, which are currently produced from derivatives of mineral oil. The relevant processes are being developed but their future significance is uncertain.

Outlook

The potential of synthesis-gas chemistry shown in Figure 4 has only been exploited to a limited extent. Future research in this field will concentrate on the following types of processes: those that can use CO/H_2 without any other reactants as building blocks, eg, glycol synthesis, modified Fischer-Tropsch syntheses to olefins or waxes, and acetic acid synthesis; those involving CO/H_2 and employed to make additional use of compounds derived from synthesis gas, eg, homologation of methanol to ethanol; those that use a derivative based on CO/H_2, eg, methanol, as a starting material for a further reaction sequence, eg, Mobil's MTG (methanol to gasoline) process or the vinyl acetate process; and energy-efficient processes, eg, those that make syngas for use in combined-cycle plants; and those that use the carbon monoxide for subsequent syntheses, eg, of acetic anhydride, acrylates, Koch acids, etc.

The synthesis gas obtained with the newly developed coal gasification processes are particularly attractive because the infrastructures and pipeline systems that are in many plants that produce syngas from oil and/or natural gas can easily be converted to CO/H_2 from coal. Syngas might well bring about the renaissance of coal chemistry.

BIBLIOGRAPHY

1. J. Falbe, ed., *Chemierohstoffe aus Kohle*, G. Thieme-Verlag, Stuttgart, FRG, 1977; English translation, John Wiley & Sons, Inc., New York, 1982.
2. H. G. Franck and A. Knop, *Kohleveredlung*, Springer-Verlag, New York, 1979.
3. G. Schilling, B. Bonn, and U. Krauss, *Kohlenvergasung*, Verlag Glückauf GmbH, Essen, FRG, 1981.
4. H. Hiller and co-workers in *Ullmann's Encyklopädie der Technischen Chemie*, Vol. 14, Verlag Chemie, 1977, p. 357.
5. B. Cornils, W. Konkol, P. Ruprecht, J. Langhoff, and R. Dürrfeld, *Erdöl Kohle* **35,** 304 (1982).
6. W. Schäfer, G. Heinrich, C. Lohmann, and H. P. Peyrer, *paper presented to the 27th DGMK Discussion*, Aachen, FRG, Oct. 1982.
7. D. Werner, *Chem. Ing. Tech.* **53,** 73 (1981).
8. R. D. Stoll and S. Röper, *Erdöl Kohle* **35,** 380 (1982).
9. H. Staege, *Erdöl Erdgas Z.* **92,** 381 (1976).
10. B. Cornils, J. Hibbel, P. Ruprecht, R. Dürrfeld, and J. Langhoff, *paper presented at the 27th International Gas Turbine Conference, London, April 1982*, Proceedings ASME 82-GT-167.
11. R. Serrurier, *Hydrocarbon Process.*, (9), 253 (1976).
12. H. Schulz and J. H. Cronjé in ref. 4, p. 329.
13. D. Clark, *Hydrocarbon Process.*, (1), 56-C (1979).
14. W. R. Deetus, *Gas Wärme Int.* **31**(6), 265 (1982).
15. A. K. Kuhn, *CEP*, (4), 64 (1982).
16. L. J. Buividas, *CEP*, (5), 44 (1981).
17. *Proceedings of the Symposium "Ammonia from Coal,"* Tennessee Valley Authority, Muscle Shoals, Ala., May 1979.
18. J. C. Hoogendoorn, *Hydrocarbon Process.*, (5), 34-E (1982).
19. J. G. Kronseder, *Hydrocarbon Process.*, (7), 56-F (1976).
20. K. H. Eisenlohr and H. Gaensslen, *Erdöl Kohle* **33,** 201 (1980).
21. I. Neff, *Chemie-Tech.* **11,** 938 (1982).
22. H. Jüntgen, *Haus der Tech. Essen Vortragsveröff.* **453,** 25 (1981).
23. J. A. Lacey, *paper presented to the Executive Coal Gasification Conference/Europe 82*, Amsterdam, The Netherlands, Oct. 1982, p. 2.-7.1.
24. J. A. Lacey and J. E. Scott, *paper presented to the 2nd Annual EPRI Contractors Conference on Coal Gasification*, Palo Alto, Calif., Oct. 1982.
25. K. R. Tart and T. W. A. Rampling, *Hydrocarbon Process.*, (4), 114 (1981).
26. C. Lohmann and G. Röbke, *Gas Wasserfach Erdgas* **121,** 359 (1980).
27. U. Dorstewitz and W. Kaimann, *Haus der Tech. Essen Vortragsveröff* **453,** 67 (1981).
28. Ref. 2, p. 188.
29. C. R. Thorpe, *paper presented to the International Coal Conversion Conference*, Pretoria, Republic of S. Africa, Aug. 1982.
30. G. Lützow and G. Reuss in *Ullmann's Encyklopädie der Technischen Chemie*, Vol. 12, Verlag Chemie, 1976, p. 497.
31. G. Baron, H. Tanz, and K. C. Traenckner in J. Falbe, ed., *Chemical Feedstocks from Coal*, John Wiley & Sons, Inc., 1982, p. 239.
32. J. Falbe, ed., *New Synthesis with Carbon Monoxide*, Springer-Verlag, New York, 1980.
33. *Eur. Chem.*, (2), 72 (1980).
34. I. Dybkjaer in ref. 17, p. 133.
35. J. Hibbel and B. Cornils in ref. 24.
36. M. J. Shires, *paper presented to the COAL CHEM-2000 Conference*, Sheffield, UK, Sept. 1980.
37. L. J. Buividas, *CEP*, (5), 44 (1981).
38. F. Brown, *Hydrocarbon Process.*, (11), 361 (1977).
39. J. A. Burnett in ref. 17, p. 198.
40. U.S. Pat. 4,172,885 (1978), E. Perry (to Monsanto Company).
41. H. A. Stewart and J. L. Heck, *64th National Meeting AIChE*, Mar. 1976.
42. H. A. Stewart and J. L. Heck, *CEP*, (9), 78 (1969).
43. H. Uchida and H. Kyogoku, *Hydrocarbon Process.*, (4), 129 (1977).
44. *Hydrocarbon Process.*, (4), 160 (1982).
45. M. Streich, *DECHEMA Monogr.* **58,** 195 (1960).

46. W. Förg, *Gas Wasserfach* **113**(11), 538 (1972).
47. H. Tanz, *DECHEMA Monogr.* **65,** 293 (1971).
48. D. J. Haase and co-workers, *Chem. Eng.*, (8), 52 (1975); *CEP*, (5), 74 (1974).
49. H. Jockel, E. Supp, *Gas Wärme Int.* **19**(6), (1970).
50. *Chem. Eng.*, 41 (June 13, 1983).
51. H. W. Coover and R. C. Hart, *CEP*, (4), 72 (1982).
52. B. Cornils, W. Konkol, P. Ruprecht, J. Langhoff, and R. Dürrfeld, *Erdöl Kohle* **35,** 304 (1982).
53. J. G. Patel and F. C. Shora in ref. 29.
54. A. Goyal and A. Rehmat, *Proceedings of the 17th IECEC*, Los Angeles, Calif., Aug. 1982, p. 808.
55. K. J. Smith, G. A. Stoops, G. B. Haldipur, and W. J. Havener in ref. 53, p. 814.
56. C. W. Schwartz, K. L. Rath, and M. D. Freier, *CEP*, (4), 55 (1982).
57. H. Teggers, K. A. Theis, and L. Schrader, *Erdöl Kohle* **35,** 178 (1982).
58. K. A. Theis and E. Nitschke in ref. 23, p. 2.-11.1.
59. B. Cornils, J. Falbe, and C. D. Frohning in *Organische Technologie*, Vol. 5 of Winnacker-Küchler, *Chemische Technologie*, Carl Hanser Verlag, München-Wien, FRG, 1981, p. 559.
60. B. Cornils, P. Ruprecht, R. Dürrfeld, and J. Langhoff in ref. 53, p. 801.
61. D. van der Meer in ref. 23, p. 2.-9.2.
62. M. J. van der Burgt, *Hydrocarbon Process.*, (1), 161 (1979).
63. R. Müller, *Haus der Tech. Vortragsveröff.* **453,** 21 (1981).
64. R. B. Knust, *paper presented at the Synthetic Fuels Conference*, San Francisco, Calif., Oct. 1980.
65. R. W. Koucky and H. E. Andrus, *Proceedings of the 16th IECEC*, Atlanta, Ga., Aug. 1981, p. 31.
66. R. Pfeiffer and P. Paschen, *Haus der Tech. Vortragsveröff.* **453,** 43 (1981).
67. H. Jüntgen and K. H. van Heck, *Erdöl Kohle* **34,** 346 (1981).
68. W. G. Schlinger, J. H. Kolaian, B. Cornils, and J. Langhoff in ref. 23, p. 2.-7.1.
69. H. Nie in ref. 23, p. 2.-10.1.
70. K. H. Heck and H. Jüntgen, *Haus der Tech. Vortragsveröff.* **453,** 53 (1981).
71. *Chem. Eng. News*, 15 (July 8, 1983).
72. Chem Systems, *DE-OS 2* **403,** 608 (1977).
73. R. M. Parsons, *DE-OS 2* **345,** 230 (1973).
74. W. Lommerzheim, *Gas Wärme Int.* **29**(4), 171 (1980).
75. C. D. Frohning in ref. 58, p. 518.
76. J. C. Hoogendoorn, *Phil. Trans. R. Soc. London A* **300,** 99 (1981).
77. C. D. Frohning, *Haus der Tech. Vortragsveröff.* **453,** 79 (1981).
78. B. Büssemeier, C. D. Frohning, and B. Cornils, *Hydrocarbon Process.*, (11), 105 (1976).
79. W. D. Deckwer, Y. Serpemen, M. Ralek, and B. Schmidt, *Ind. Eng. Chem. Process Res. Dev.*, 222 (1982).
80. Eur. Appl. 10,295 and 21,241 (1979–1980), F. Wunder, E. I. Leupold, and co-workers (to Hoechst AG).
81. U.S. Pat. 3,957,857 (1974), 4,103,700 (1974), and 4,115,428 (1977), R. L. Pruett and co-workers (to UCC Corporation); B. D. Dombek, *J. Am. Chem. Soc.* **102,** 6855 (1980).
82. H. Bahrmann and B. Cornils, *Chem. Ztg.* **104,** 39 (1980).
83. J. Knifton, *Chemtech*, 609 (1981).
84. H. Bahrmann, W. Lipps, and B. Cornils, *Chem. Ztg.* **106,** 249 (1982).

J. Falbe
C. D. Frohning
B. Cornils
Ruhrchemie AG

HYDROGENATION

The direct coal-hydrogenation processes were developed in Germany during the first quarter of the 20th century and were commercially applied during the 1930s and 1940s (see Coal). Because of the large and low priced petroleum reserves, these technologies had not been further developed. Further development of the traditional processes started at the beginning of the 1960s in the United States and in other countries, eg, the FRG and Japan, during the 1970s. Fears of a future oil shortage and large oil-price increases were the reasons for this development. The most important laboratories and pilot plants in the United States and the FRG are listed in Table 1 (see Fuels, synthetic).

Processes

German Technology. German technology is a development of the Bergius-Pier process (IG process) for the catalytic hydrogenation of coal. Characteristics of the new process are that the reaction products are separated by distillation, the oil used for the slurry preparation consists of an asphaltene-free middle- and heavy-oil distillates in a 40:60 ratio, and the residue is used for the production of hydrogen (1–3). These modifications imply the following technical advantages: the operating pressure is reduced from 70 to 30 MPa (10,200 to 4,400 psi) using an inexpensive iron oxide catalyst, there is up to a 50% increase in specific coal throughput, and there is improved heat recovery.

Figure 1 shows a simplified process flow diagram for coal hydrogenation according to German technology. The coal is ground in two stages to <0.1 mm; then it is dried and mixed with a process-derived solvent after the addition of an iron oxide catalyst

Table 1. Coal Hydrogenation Plants in the United States and the FRG

	United States				FRG		
Company:	Exxon	HRI[a]	Gulf	Lummus	Ruhrkohle/ Veba Oel	Saarberg-werke	Rhein-braun
Process[b]:	EDS	H-Coal	SRC II	LC Fining	German technology	German technology	German technology
plants							
coal throughput, t/d[c]	0.3–1	3	1	0.1	0.5[d]	0.25	0.25
pilot plants							
coal throughput, t/d[c]	200	200–600	30		200	6	
location	Baytown, Texas	Catlettsburgh, Kentucky	Ft. Lewis, Washington		Bottrop	Völklingen-Fürstenhausen	
operation schedule	1980–1982	1980–1982	1977–1982		1981–1985	1982–1984	

[a] Hydrocarbon Research, Inc., of Dynalectron Corporation.
[b] See text for descriptions of processes.
[c] t/d = metric tons per day.
[d] Plant at Bergbauforschung G.m.b.H.

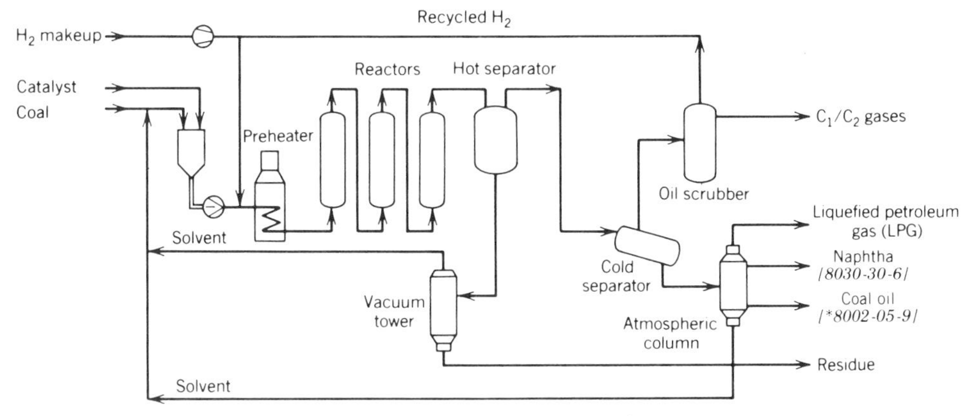

Figure 1. German technology flow diagram.

in a 1:1.5 ratio. The solvent consists of a mixture of ca 40 wt % middle oil (boiling range 200–325°C) and to 60 wt % distillate heavy oil (boiling range >325°C). The slurry pressure is increased to 30 MPa (4350 psi), recycle gas and make-up hydrogen are added and, after heating up to 425°C in the preheater, the slurry is fed into the first reactor. The exothermal reaction takes place in a series of three reactors, through which the reaction mixture ascends. Cold hydrogen is fed into the reactors to prevent temperatures greater than 475°C. Downstream from the third reactor, the hydrogenation products are separated in the hot separator; the gases and the light liquids leave the hot separator at the top and the heavy liquids and solids at the bottom. The separation into gases and coal oil occurs in the cold separator.

In an oil-scrubbing unit, the C_1–C_4 gases are removed from the high hydrogen content gases, which are obtained in the cold separator. Thereafter, the gas is recycled into the process. The liquids from the cold separator are separated by atmospheric distillation into liquified petroleum gas (LPG), naphtha, and coal oil. The solids-containing coal oil from the hot separator is separated by vacuum distillation into vacuum residue and slurrying oil. The vacuum residue is used in a commercial plant to produce the required feed hydrogen, eg, by way of partial oxidation, according to the Texaco process, with subsequent conversion of the synthesis gas produced.

The companies Ruhrkohle A.G., Saarberwerke A.G., Rheinische Braunkohlenwerke A.G., and Veba Oel A.G., as well as the research institute Bergbau-Forschung G.m.b.H., are developing a special method of coal hydrogenation, ie, the so-called German technology.

In 1976, a continuously operating laboratory plant with a throughput of ca 0.5 metric tons per day was constructed on an experimental basis. From 1979 to early 1981, Ruhrkohle A.G. and Veba Oel A.G. jointly constructed a 200-t/d pilot plant in Bottrop (4–6). Operations in this plant began in November 1981. By the middle of 1982, ca 14,000 t of coal was hydrogenated in ca 3000 operating hours.

At Saarbergwerke A.G., the process development started in 1975 in a laboratory plant with a throughput of 0.25 t/d. In 1978, planning started for a 6-t/d pilot plant which began operating in 1982. Both at Ruhrkohle A.G. and at Saarberwerke A.G., hard coal is used for hydrogenation. A laboratory plant for hydrogenation of lignite with a throughput of 0.25 t/d began in 1978 and was owned by Rheinische Braunkohlenwerke A.G.

Exxon Donor-Solvent Process. The Exxon donor-solvent process is a development of the Pott-Broche process; they differ in that a process-derived solvent and additional molecular hydrogen are used in the EDS process (2). The solvent has hydrogen-donating properties. Figure 2 is a block diagram of coal hydrogenation according to the EDS process. The ground coal is mixed with the process-derived donor solvent. The resulting slurry is heated under pressure after the addition of hydrogen and then is fed into the reactors. Hydrogenation takes place at 10.5–17.5 MPa (1520–2540 psi) and 450°C. The separation of the reaction products into gas, oil, solvent, and vacuum residue is carried out by distillation. The solvent is hydrogenated by means of a solid-bed catalyst, thus recovering its donor properties. The hydrocarbon gases are removed from the gaseous products and the purified hydrogen is recycled into the process. In the bottoms recycle, part of the residue is recycled into the hydrogenation process to increase the oil yield.

In 1966, Exxon started coal hydrogenation in laboratories and in technical plants with a throughput of 0.5 t/d. In 1974, a continuously running, 1-t/d plant was set into operation. During 1974–1975, a 200-t/d pilot plant was planned. The Baytown, Texas, plant was operated from the second quarter of 1980 until the middle of 1982, and a total of 90,000 t of coal was processed in 10,692 operating hours. The types of coal hydrogenated were Illinois No. 6, Wyodak, and Martin Lake coals.

H-Coal Process. The H-coal process was developed by Hydrocarbon Research, Inc., and it is based upon the H-oil process for the desulfurization of mineral-oil-distillation residues. It is unique in that coal hydrogenation takes place in an ebullated-bed reactor with an extruded Co–Mo catalyst. In the reactor, the spent catalyst is replaced during the process, thus ensuring a continuous operation.

Figure 3 is a block diagram of coal hydrogenation following the H-coal process. The ground and dried coal is mixed with the process-derived solvent. The slurry is heated after the addition of hydrogen and is fed into the reactor. Hydrogenation takes place at 455°C and 20.7 MPa (3000 psi). Part of the reacted slurry is returned to the reactor inlet and is used to establish the ebullated bed. The hydrogenation reaction products are separated by distillation into naphtha, coal oil, and solid-containing residue.

In 1973–1974, a 3-t/d pilot plant began operation. In 1975, engineering work was started for the construction of a 200-t/d plant (syncrude mode) or a 600-t/d plant (fuel-oil mode). The plant was operated from May 1980 to November 1982 in

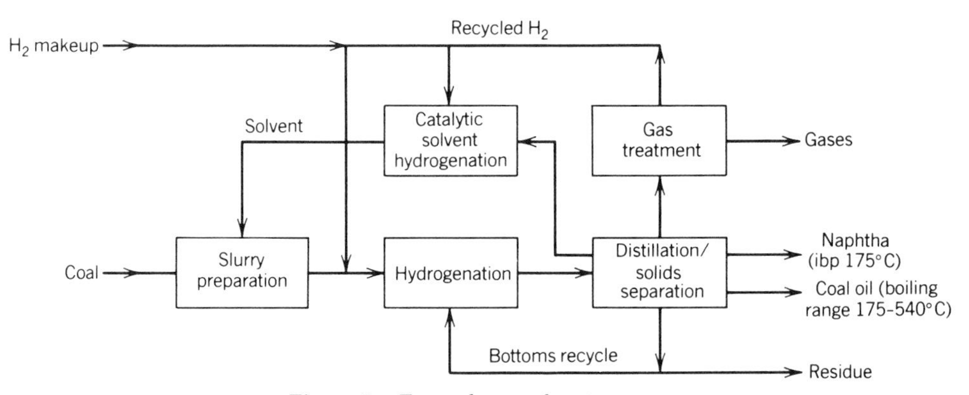

Figure 2. Exxon donor-solvent process.

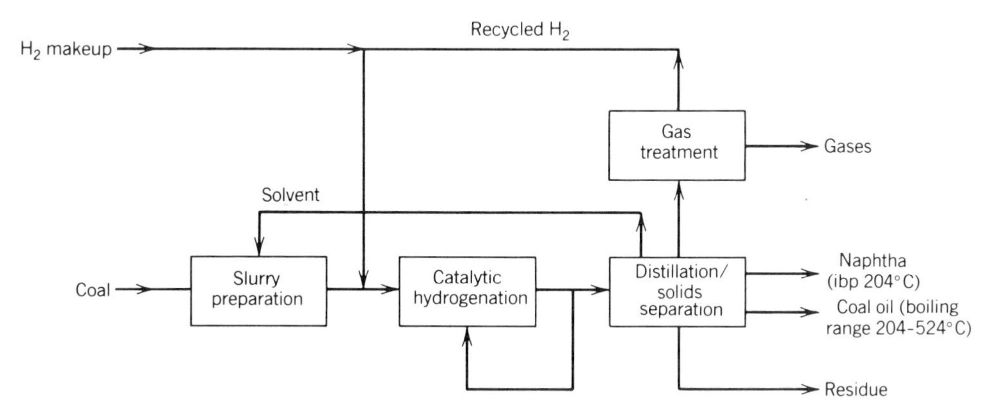

Figure 3. H-coal process.

Catlettsburg, Kentucky. Approximately 52,000 t of coal was processed in ca 7000 operating hours. The types of coal used for hydrogenation were Illinois No. 6, Wyodak, and Kentucky coals.

Gulf Solvent-Refined Coal (SRC) Process. The Gulf SRC II process was developed from the Gulf SRC I process, in which a solid and nearly ash- and sulfur-free extract is produced. During the SRC II process, part of the reacted slurry is recycled for slurrying the coal. This increases the catalytic mineral content in the slurry, which causes increased coal conversion during hydrogenation.

Figure 4 is a block diagram of coal hydrogenation following the Gulf SRC II process. The ground and dried coal is mixed with the solvent from distillation and the recycled reacted slurry. After the addition of recycled and make-up hydrogen, the hydrogenation occurs under the relatively mild conditions of 13.8 MPa (2000 psi) and 460–465°C. The hydrogenation products are separated by distillation into naphtha, coal oil, and solid-containing residue.

During the late 1960s, Gulf started to develop the SRC II process in laboratory plants. A 30-t/d pilot plant, which was near Tacoma, Washington, and originally produced a solid coal extract, was modified at the end of 1976 for SRC II tests. The plant was operated until 1981.

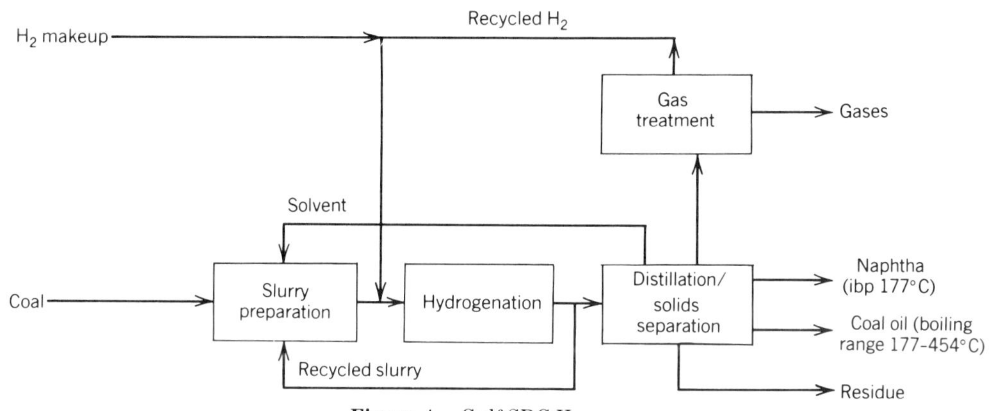

Figure 4. Gulf SRC II process.

LC Fining Process. The LC fining process was developed in the United States by Lummus and is a two-stage hydrogenation process (see Fig. 5). The main process steps involve the short-contact-time hydrogenation of coal by means of a catalytically hydrogenated solvent from the LC finer. The reaction conditions are 460°C and 17.2 MPa (2500 psi). The residence time in the reactor is ca 5–10 min. Downstream from the reactor, the solids are eliminated from the reaction product according to the antisolvent deashing process. The residue obtained in the deashing stage is gasified and forms the make-up hydrogen for the hydrogenation and the LC fining. The coal extract produced without any solids content is then hydrogenated in the LC-fining stage in an ebullated-bed reactor at 400°C and 19.0 MPa (2750 psi) by means of extruded Ni–Mo catalysts. The LC-fining hydrogenation products are LPG, naphtha, and light, middle, and heavy distillates. Part of the products is used as hydrogenated solvent, which has hydrogen-donating properties, and is added to the coal.

The development of this process was started in the United States in 1980. A pilot plant with a coal throughput of 0.3 t/d has operated since 1980. The types of coal tested for hydrogenation purposes are Indiana V and Illinois No. 6.

Japanese Technology. In Japan, research on coal liquefaction was carried on for some time during and after World War II, but then it was stopped for a long time (7). In 1974, the Sunshine Project was started by Miti. It was a research program on coal liquefaction and gasification, solar energy, and geothermal energy, with coal liquefaction an important factor. The first three processes developed for the coal liquefaction project were extractive coal liquefaction, direct hydrogenation, and solvolysis coal liquefaction. Later, the Australian brown-coal liquefaction project was initiated. As of early 1983, four coal liquefaction projects were being developed. These projects are contracted to companies and the total cost of research is borne by the government.

Extractive Coal Liquefaction. Crushed feed coal and a disposable catalyst are slurried in a hydrogen-donor solvent, which is the middle and heavy oil from the fractionator and is catalytically hydrogenated. This slurry and the hydrogen are fed into the reactor from the preheater. The reaction conditions are 450°C and 10–10.5 MPa (1450–1520 psi). Downstream from the reactor, the products are fractionated in atmospheric and vacuum-distillation columns. Most of the middle and the heavy oil is used as recycle solvent. The bottom products from the vacuum column can be used for the gasification to produce hydrogen, for the solids separation to produce more

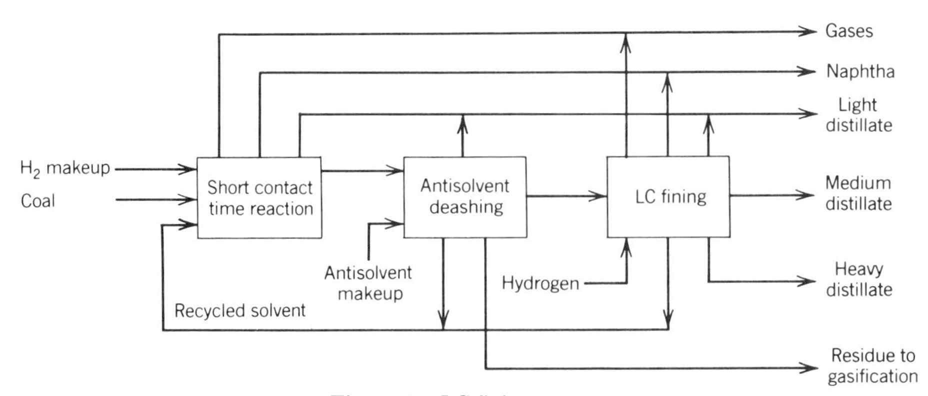

Figure 5. LC fining process.

light oil by a second hydrogenation, and as feedstock for various purposes after solids separation.

Disposable catalysts, eg, iron ore, red mud, and dust, from a molten-iron gasifier were tested, and dust from the molten-iron gasifier was chosen as the most active and inexpensive catalyst. When the molten-iron gasifier is used in the process, the catalyst is recovered after gasification of the bottom products.

Basic research was started in 1978 and, by the end of November 1981, the P.D.U. with a capacity of 1 t/d feed coal was completed in Hasaki, Ibaragi. The P.D.U. test will be continued so as to develop the design basis for the pilot plant.

Direct Hydroliquefaction. Crushed feed coal and a disposable catalyst are slurried in the recycle solvent. This slurry and hydrogen are passed into the reactor from the preheater. The Japanese are trying to reduce the reaction pressure by using a highly active iron catalyst which can be produced at the lowest possible cost. The hydrogenation takes place at 450°C and 30.0 MPa (4350 psi).

Basic research for direct hydroliquefaction started in 1975. A bench-scale unit with a capacity of 0.1 t/d of feed coal was constructed in 1980 and the P.D.U. of 2.4 t/d was completed in Kawasaki by the end of April 1982. In the B.S.U., red mud is used as catalyst. In autoclave tests, the high efficiencies of some new catalysts have been realized.

Solvolysis Coal Liquefaction. Initially, asphalt (qv) from petroleum was used as the solvent. Later, the process was improved. Crushed coal is slurried in the recycled solvent, and the slurry is liquefied in the first step by means of a short-contact-time reaction. The products are deashed and fed into the hydrogenation reactor, where they are catalytically hydrogenated. The products are distilled and the heavy fraction is used as recycle solvent. The reaction conditions in the first stage are 350–450°C and 0.1–3.0 MPa (14.5–435 psi). The catalytic hydrogenation takes place at 350–450°C and 20.0 MPa (2900 psi).

A B.S.U. of 0.1 t/d, which included production from the solvolysis process, was constructed in Hiroshima in 1980.

Brown-Coal Liquefaction. The brown coal of the Latrobe Valley in Victoria, Australia, has a low ash and sulfur content but contains ca 60 wt % water. Raw brown coal and a catalyst are slurried in a recycle solvent. The coal is dried as slurry and is passed to the first hydrogenation stage. The reaction conditions are 410–440°C and 10–15 MPa (1450–2180 psi). The products obtained in the first stage are deashed in a solvent-deashing process and then are catalytically hydrogenated in a second-stage reactor, where various product oils form. Part of the heavy oil is recycled for the slurry preparation.

Basic research has been mainly conducted with a B.S.U. of 0.1 t/d in Kobe, Japan. The P.D.U., with a capacity of 50 t/d of feed coal, has been under construction in Victoria since the end of 1981. The first step in construction was the establishment of the first-stage liquefaction technology in 1983.

As the next step, the New Energy Development Organization (NEDO) will evaluate the experimental results obtained in the extractive coal liquefaction, direct hydroliquefaction, and solvolysis coal liquefaction processes. At present, NEDO is consulting with the agencies concerned about the construction of a pilot plant with a feed-coal capacity of 250–500 t/d. The Australian project will complete the second step of the construction schedule in 1985 (see also Lignite and brown coal).

Large-Scale Plant Projects

The construction of commercial plants for direct coal hydrogenation is possible. The problem, however, is the high amount of capital required and the lack of profitability. As of early 1983, only one definite project for direct hydrogenation was being planned. In the FRG, a reference plant for German technology will be constructed. The planned project for the construction of a 6000-t/d plant based on the Gulf SRC II process was terminated. Several feasibility studies have been made by Ruhrkohle A.G., Veba Oel A.G., and Saarberwerke A.G. for the operation of large-scale plants in the FRG. These would be based on FRG coal and have an annual coal throughput of $2–10 \times 10^6$ t.

Although the commercial coal-hydrogenation plants are not yet profitable, the present stability (in 1983) of the energy markets should be used to develop coal-hydrogenation technologies. This would be possible with a plant processing 10^6 t/yr coal. It could serve as a reference plant with commercial plant components.

Upgrading Oil from Coal

The coal oil produced by hydrogenation is a crude product and must be upgraded into marketable products by further processing. The following processing routes were developed by Ruhrkohle A.G. and Veba Oel A.G. and are similar to the processes developed by other companies for upgrading coal-derived liquids.

Coal oil differs from crude oil, which has a mineral origin. The properties depend on the applied coal-hydrogenation process and on the type of coal used. Table 2 compares the characteristics of coal oil produced by German technology with those of mineral crude oil. The density of 0.950 g/cm^3 and the low H:C atom ratio of 1.25 indicate the high aromaticity of coal oil.

Figure 6 shows the coal-oil upgrading scheme. The crude coal oil is separated by distillation into two fractions: light oil (boiling range up to 180°C) and middle oil (boiling range 185–325°C). These fractions are upgraded separately according to their boiling ranges.

Table 2. Properties of Syncrude from Coal and Arabian Light

Property	Syncrude from coal	Arabian light
density (at 15°C), g/cm^3	0.950	0.856
carbon, wt %	86.60	85.5
hydrogen, wt %	9.05	12.6
sulfur, wt %	0.10	1.7
nitrogen, wt %	0.75	0.2
oxygen, wt %	3.50	
H:C ratio	1.25	1.77
light oil (ibp, <185°C), wt %	20	20
middle distillate (boiling range 185–325°C), wt %	80	27
VGO (boiling range 325–500°C), wt %		28
residue (boiling range >500°C), wt %		25

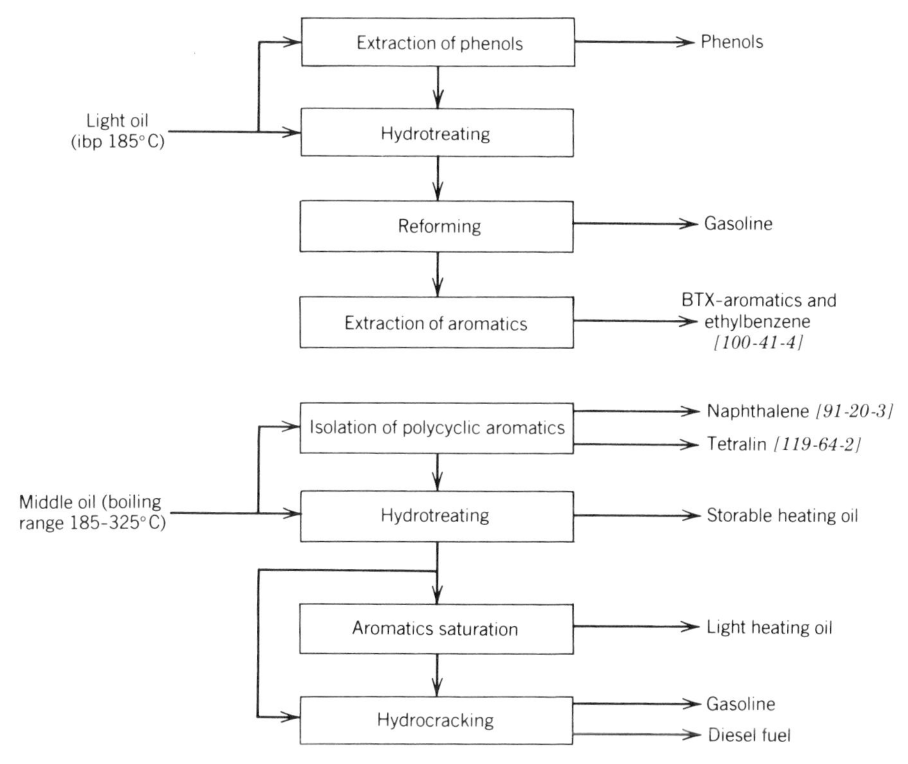

Figure 6. Upgrading of oil from coal. BTX = benzene [*71-43-2*], toluene [*108-88-3*], *o*-xylene [*95-47-6*], *m*-xylene [*108-38-3*], and *p*-xylene [*106-42-3*]; ibp = initial boiling point.

Processes. Coal oil is upgraded in a way similar to that for crude oil. The petroleum-refining processes must be adapted to the specific coal-oil properties. Generally, the reaction conditions applied are more severe than for mineral oil in order to remove sulfur and nitrogen. In cases involving the extraction of phenols from light oil and polycyclic aromatics from middle oil, the processes are the same as those normally used for the tar processing (8).

Equipment. As of early 1983, large-scale plants for coal-oil refining have not been operated. The coal-oil upgrading route described below is based on the results from process-development units operated by Ruhrkohle A.G. and Veba Oel A.G. (9–12). Figure 7 is a flow diagram of a laboratory-scale hydrogenation unit. The basic plant component is a tubular reactor, 2 m in length and 3 L (0.8 gal) in volume. The reactor contains the catalyst bed and is electrically heated from outside. On the left, there is the metering equipment for oil and gas and the preheater; the product cooling system, separator, and let-down valve are on the right. The gases are removed from the hydrogenation products in a stabilizer. The reaction gas is recycled and can be cleaned in an oil-and-water scrubber. The operating conditions can be adapted for the special hydrotreating purposes, with pressures up to 30 MPa (4350 psi) and temperatures up to 500°C.

Products. ***Gasoline and Heating Oil.*** Figure 8 is a schematic of the upgrading route to gasoline and heating oil. The light distillate, ie, naphtha, which has a final boiling point of 185°C, is treated with the naphtha cut obtained from the middle distillate hydrotreatment. The objective of light-oil hydrotreating is the removal of heteroatoms

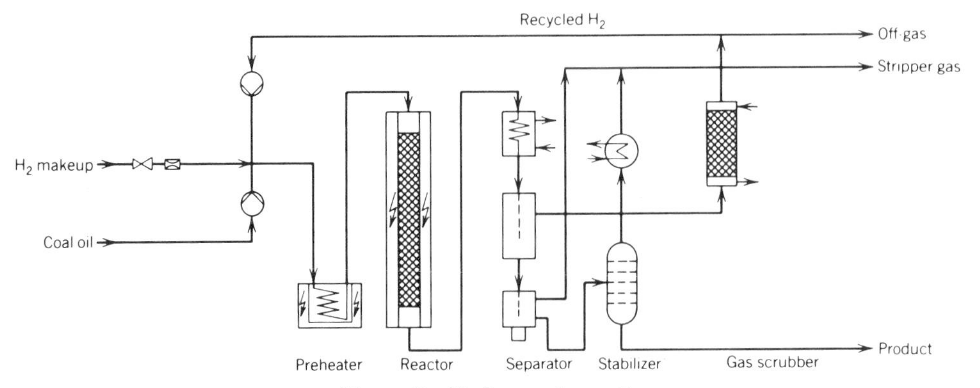

Figure 7. Hydrogenation unit.

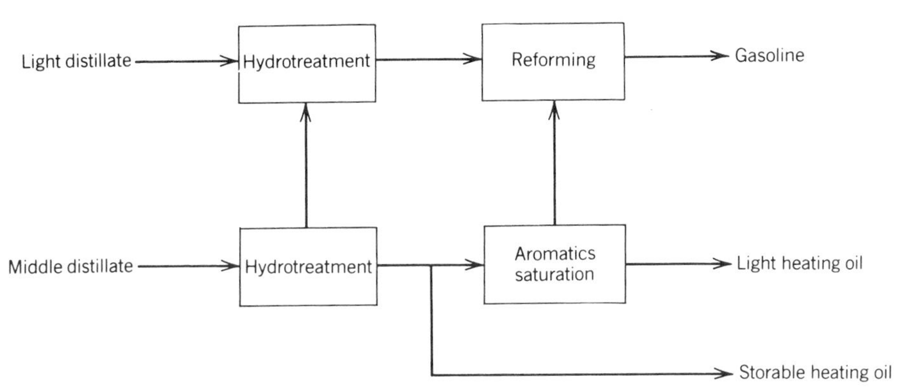

Figure 8. Upgrading route to gasoline and heating oil.

to a concentration of <1 ppm, which meets the reformer feed specification. Hydrotreating in one stage with a cobalt–molybdenum catalyst is sufficient. The reaction conditions are 6–10 MPa (870–1450 psi) and 300–400°C.

The hydrotreated coal-derived naphtha is reformed with a platinum catalyst. The process conditions of the reforming reaction are 2.5 MPa (360 psi) and 490°C. The reformate obtained is gasoline or feedstock for benzene, toluene, xylenes, and ethylbenzene.

For the production of heating oil, two alternatives have been considered and they differ in the degree of hydrogenation. Under relatively mild hydrotreating conditions and in a single-step process, middle distillates can be upgraded to a storable product, which can be used as a blending component for light heating oil burned in special atomizing burners. Satisfying the quality standard of light heating oil requires that most of the aromatic compounds which represent more than 90 wt % of the raw middle distillate be hydrogenated. This can be achieved by an additional process step which is treatment of the storable product from the hydrotreater with a nickel catalyst. The conditions of reaction for the hydrotreatment are 6–10 MPa (870–1450 psi) and 390°C.

Gasoline and Diesel Fuel. Figure 9 shows the upgrading route to gasoline and diesel fuel. The processing of light oil includes hydrotreatment and reforming. The light oil produced in the middle distillate hydrotreatment and hydrocracking is free of sulfur and nitrogen and can be fed into the reformer without further treatment. For hydrotreatment of the raw middle distillate, a sulfur-resistant nickel–molybdenum catalyst is used. The conditions of reaction are 20–30 MPa (2900–4350 psi) and 350–390°C. Varying the reaction conditions in the hydrocracker changes the gasoline yield from ca 60 wt % to total conversion to gasoline (qv).

Table 3 is a listing of the product yields and the hydrogen consumption for different upgrading alternatives based on the feedstock composition of 20 wt % light and 80 wt % middle distillate. The hydrogen consumption varies from 100 m^3/t (130 yd^3/t) to 550 m^3/t (720 yd^3/t), depending on the degree of hydrogenation and the process used. Low hydrotreating of the middle distillate yields 66 wt % of storable heating oil suitable as a blending component for petroleum-based light heating oil. Light heating oil that meets FRG industrial standards is obtainable in a maximum output of 54 wt % of heating oil. Up to 40 wt % diesel fuel can be produced by mild hydrocracking of the middle distillate fraction.

Chemicals. Coal oil is an excellent feedstock for the production of some chemicals, eg, phenols, benzene, toluene, xylenes, ethylbenzene, and naphthalene–tetralin. Their contents in coal oil and in the different fractions are listed in Table 4.

The individual compounds usually can be obtained from the corresponding fractions of crude light oil, reformate, or middle oil according to the process used for tar treatment (see Fig. 6). In the last two columns of Table 4, the potential quantities

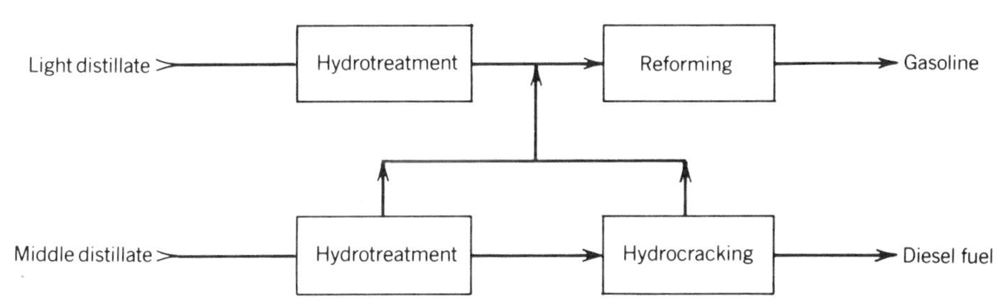

Figure 9. Upgrading route to transportation fuels.

Table 3. Product Yields and Hydrogen Consumption for Different Upgrading Alternatives

	Upgrading routes		
Yields	Gasoline and heating oil (storable)	Gasoline and heating oil[a]	Gasoline and diesel fuel
C_1–C_4, wt %[b]	1.4	2.6	3.8
gasoline, wt %[b]	35	43	57
heating oil, wt %[b]	60	54	
diesel fuel, wt %[b]			40
hydrogen consumption, m^3/t feed oil[c]	100	400–450	450–550

[a] DIN 51603.
[b] Based on feed oil.
[c] To convert m^3/t to yd^3/t, multiply by 1.31.

Table 4. Chemicals from Coal Oil

Chemical	Light oil, wt %	Middle oil, wt %	Reformate, wt %	Potential amount of a 2×10^6-t/yr plant[a]	Consumption in the FRG (1979), 1000 t
benzene	0.8		11.0	34.6	1203
toluene	2.4		19.5	168.6	290
o-xylene	1.5		3.4	23.2	235
m-xylene	1.6		5.5	47.3	
p-xylene	0.6		2.3	19.8	346
ethylbenzene	1.5		6.1	52.5	1030
phenol [*108-95-2*]	11.7			46.8	260
naphthalene		3.5		56.2	95
tetralin		4.4		70.4	
cresols	0.7	4.5		72.4	15.4

[a] The amount of BTX–aromatics is only obtainable if phenols are not extracted.

produced in a commercial 2×10^6-t/yr coal-oil plant are compared with the consumption figures of 1979 in the FRG.

Properties. The FRG industrial standard for gasoline is compared with raw coal-derived naphtha, hydrotreated naphtha, and reformed light oil in Table 5. Because of its content of phenols and aromatics, the raw light oil has a RON/MON of 94.5/84.0. In this respect, it meets the specification for standard gasoline. The hydrotreated light oil shows a remarkable decrease in the octane numbers as a result of the removal of the phenols and aromatics in the hydrogenation process. The reformed product has the highest octane number. The RON clear is higher than the specification for standard gasoline. None of the three light oils meets the maximum standard density of 0.780 g/cm^3.

The reformed light oil is a good blending component for gasoline from petroleum. In addition, the reformed product is an excellent feedstock for the recovery of BTX–aromatics and ethylbenzene (see Table 4). The benzene content is 11 wt %, the toluene content is 19 wt %, and the xylenes and ethylbenzene content is 17 wt %.

Table 6 is a comparison of the properties of middle distillate fractions with those specified for light heating oil. The density and the heating value of the less-severely hydrotreated product do not meet the specified standards, but it can be used as a

Table 5. Properties and Components of Gasoline and of Raw, Hydrotreated, and Reformed Light Oil

Property	Gasoline[a]	Light oil		
		Raw	Hydrotreated	Reformed
density (at 15°C), g/cm^3	0.730–0.780	0.880	0.827	0.865
RON (research octane number)	97.4/91	94.5	76.4	97.5, clear
MON (motor octane number)	87.2/82	84.0	70.0	85.4, clear
gum, mg/100 cm^3	max 5	45	4	1
benzene, wt %		0.8	2	11
toluene, wt %		2.4	7	19
C_8–aromatics, wt %		4.8	9	17

[a] DIN 51600.

Table 6. Properties of Middle Distillate Fractions for Usage as Heating Oil

Property	Light heating oil[a]	Coal oil middle distillate	
		Less-severe hydrotreating	Two-stage hydrogenation
density (at 15°C), g/cm^3	max 0.860	0.930	0.862
chemical composition, wt %			
carbon		89.0	86.65
hydrogen		10.7	13.1
sulfur	max 0.3	16 ppm	4 ppm
nitrogen		186 ppm	6 ppm
heating value, MJ/kg[b]	min 42.0	41.26	42.66
flashpoint, °C	min 55	88	60
viscosity (at 20°C), mm^2/s (= cSt)	max 6.0	3.25	3.16

[a] DIN 51603.
[b] To convert J to cal, divide by 4.184.

Table 7. Properties of the Middle Distillate Fraction from Hydrocracking

Property	Diesel fuel[a]	Hydrocracker product (>185°C)
density (at 15°C), g/cm^3	max 0.855	0.828
distillation, D 86, vol %		
ibp[b], °C		194
ibp[b], <350°C	min 85	100
viscosity (at 20°C), mm/s (= cSt)	2–8	2.4
flashpoint, °C	max 55	73
cetane number	min 45	47
sulfur, wt %	max 0.3	<1 ppm

[a] DIN 51601.
[b] Ibp = initial boiling point.

blending stock. The product of the two-stage hydrogenation of aromatics meets all the requirements of the standard and can be used as light heating oil without any restrictions.

Table 7 is a comparison of the properties of the middle distillate fraction from hydrocracking with the standard for diesel. The hydrocracker bottom product has excellent diesel-fuel properties.

BIBLIOGRAPHY

1. E. Wolowski and O. Funk, *Erdoel Kohle* **33,** 321 (1980).
2. W. Krönig, *Die Katalytische Druckhydrierung von Kohlen, Teeren und Mineralölen*, Springer-Verlag, Berlin, Göttingen/Heidelberg, 1950.
3. J. Falbe in W. Krönig, ed., *Kohlehydrierung*, Georg Thieme Verlag, Stuttgart, FRG, 1977, Chapt. 4, pp. 68–113.
4. J. Langhoff, R. Dürrfeld, and E. Wolowski, *Erdoel Kohle* **34,** 379 (1981).
5. E. Wolowski and O. Funk, *Energie* **34,** 169 (1982).
6. J. Langhoff, E. Wolowski, and O. Funk, *paper presented at International Coal Conference*, Pretoria, Rep. of South Africa, Aug. 16, 1982.
7. Y. Ogisu, *paper presented at COGLAC-Conference 1982*, Pittsburgh, Pa., Aug. 8, 1982.
8. H.-G. Frank and G. Collin, *Steinkohlenteer*, Springer-Verlag, Berlin, Heidelberg, New York, 1968.

9. U. Graeser and A. Jankowski, *Erdoel Kohle* **35,** 74 (1982).
10. W. Döhler, U. Graser, and A. Jankowski, *Proceedings International Conference on Coal Science*, Düsseldorf, FRG, Sept. 7–9, 1981, p. 488.
11. A. Jankowski, W. Döhler, and U. Graeser, *paper presented at Science on Coal Liquefaction*, Lorne, Victoria, Australia, May 24–28, 1982; *Fuel* **61,** 1031 (Oct. 1982).
12. A. Jankowski, U. Graeser, and W. Döhler in W. Classen, ed., *HdT Heft*, Vulkan-Verlag, 1982, p. 453.

JOSEF LANGHOFF
ALFONS JANKOWSKI
Ruhrkohle Oel und Gas GmbH

CARBONIZATION AND COKING

Industrialized countries with significant steel production have a combined output of metallurgical coke of ca 450 × 10^6 metric tons per year. It is used as the reducing agent in the manufacture of pig iron in blast furnaces (1). By the high temperature coking process of bituminous coal at 900–1300°C ca 20 × 10^6 t of the highly aromatic liquid by-products are coal tar and crude benzene (see Fig. 1). Nonaromatic content is as low as 3% in the case of crude benzene and >1% in high temperature coke-oven tar (2–3). Low temperature carbonization processes are characterized by temperatures up to 700°C with bituminous coal or lignite as feedstocks. As a result, the composition of liquid by-products is different for low temperature tars, with lower concentrations of aromatic hydrocarbons but generally higher concentrations of phenolic compounds and paraffins (4–5). Similar liquid by-products are generated in the Lurgi pressure gasification of bituminous coal and lignite. Low temperature carbonization is currently applied in the UK, the USSR, and India, and Lurgi gasification in South Africa, Czechoslovakia, and the GDR. The total of liquid by-products generated by low temperature carbonization and gasification in these countries can be estimated at 2 × 10^6 t from carbonization and coking of lignites and ca 7 × 10^5 t from coal. (see Coal, coal conversion processes).

Coking of Bituminous Coal

A typical coking coal (18–30% volatiles) produces an average of 80% coke, 12% coke-oven gas, 3% tar, and 1% crude benzene in the high temperature batch oven-coking process, based on the weight of water-free coal consumption (6).

Tar and crude benzene are separated by cooling the gaseous products first to ca 100°C to precipitate 60–70% of the tar contained in the crude gas, followed by a secondary precipitation with indirect water condensers and electrofilters. After H_2S and ammonia are removed, the aromatic hydrocarbons are scrubbed with so-called benzene wash oil under 0.8–1 MPa (8–10 atm) pressure from the coke-oven gas to give crude

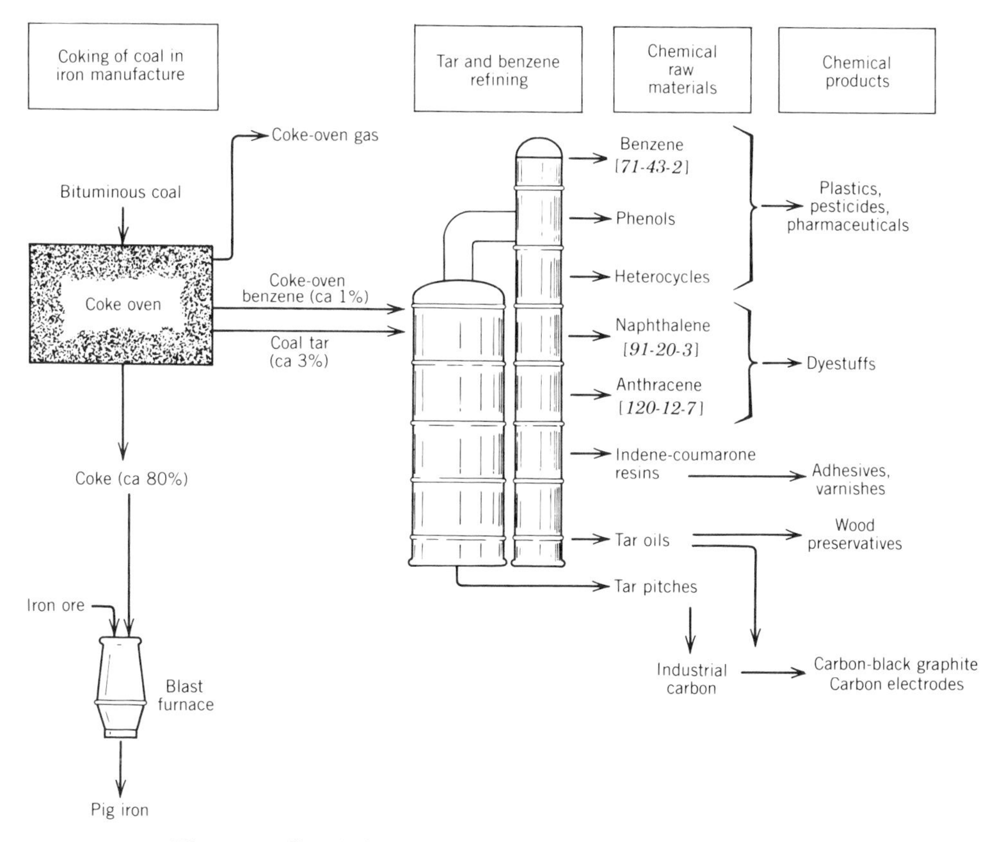

Figure 1. Chemical raw materials from coking of bituminous coal (1).

benzene. Crude tar and benzene are transferred from the coking operations to an aromatics refinery for further separation.

Refining

High Temperature Coke-Oven Coal Tar. According to the physical and chemical properties of high temperature coal tar, various operations are applied (2), namely distillation for primary separation of crude tar and further separation of fractions; crystallization to obtain pure aromatic hydrocarbons (eg, naphthalene, anthracene, acenaphthene [*83-32-9*], and pyrene [*129-00-0*]), nonpolar heterocyclic compounds (eg, carbazole [*86-74-8*]), and filtered oils, such as carbon black and impregnating oils; extraction with base and acid to recover the phenols and nitrogen-containing bases; catalytic polymerization with Friedel-Crafts catalysts to produce indene–coumarone resins from the light-oil fractions (remaining light oils are recovered by distillation after the polymerization and can be refined together with coke-oven crude benzene); thermal polymerization of the primary distillation residual coal-tar pitch to form electrode pitch; and coking of pitch to form low ash electrode coke.

A flow chart of the refining processes of high temperature coal tar is given in Figure 2, a continuous primary tar distillation unit in Figure 3.

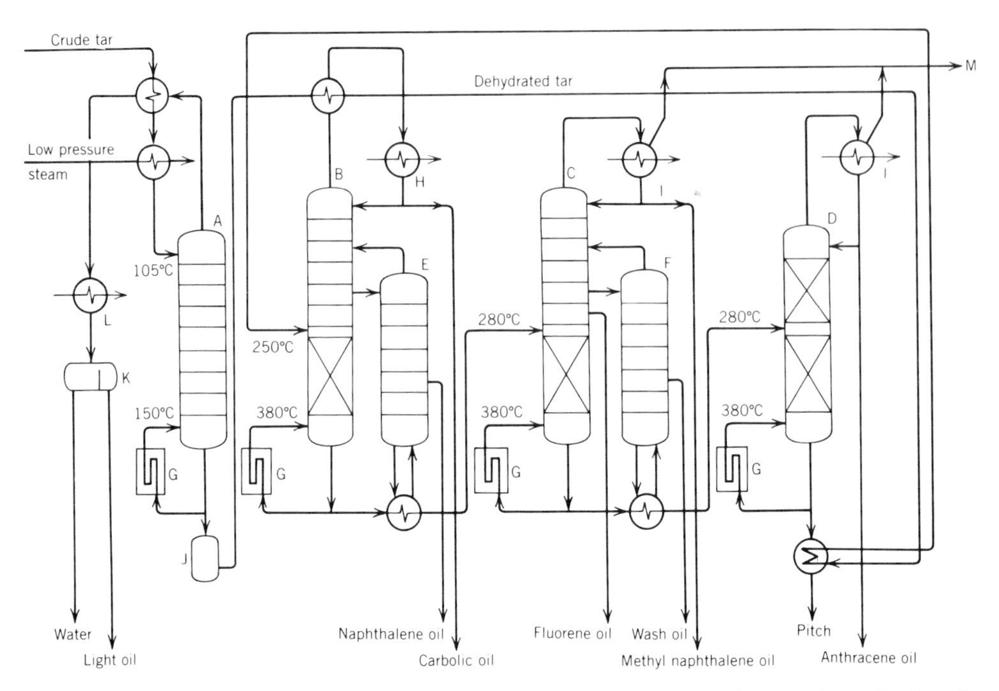

Figure 2. Continuous tar distillation by the Rütgers process (2). A, Dehydration column; B, phenolic oil column; C, methylnaphthalene column; D, anthracene oil column; E, naphthalene oil side column; F, wash oil side column; G, tube furnace; H, condenser; I, condenser (steam generator); J, container for dehydrated tar; K, separator; L, condenser (cooler); and M, to vacuum receiver.

Crude Benzene. After separation of phenols, pyridine bases, and indene–coumarone resins, the remaining tar light oil is refined together with the crude benzene obtained from the coke-oven gas.

Sulfuric Acid Refining. Concentrated sulfuric acid or oleum is used to selectively sulfonate thiophene [*110-02-1*] and to polymerize unsaturated hydrocarbons contained in the crude benzene. Subsequent fractionated distillation gives technically pure benzene with a thiophene content of <1 ppm (7). This method is used in medium-sized plants in both Eastern and Western Europe.

Catalytic Hydrogenation. In catalytic pressure hydrogenation, the sulfur compounds are converted to hydrogen sulfide and the olefins are hydrogenated at 350°C and 2–6 MPa (20–60 atm) pressures over a molybdenum–cobalt–alumina catalyst. The yields are better than with the sulfuric acid process (8–9). This method is used in large plants with capacities of up to 3×10^5 t/yr. The hydrogenated crude benzene is refined by extractive distillation using various solvents, eg, *N*-methylpyrrolidinone (Lurgi-Distapex process), tetrahydrothiophene dioxide (Shell-Sulfolane process), propylene carbonate (Koppers-Propylane process), and *N*-formylmorpholine (Koppers-Morphylane process) (10–12). The purified benzene has a freezing point of 5.5°C. With hydrocracking, extractive distillation is not necessary. In the Houdry-Litol process, at ca 600–630°C, nonaromatic hydrocarbons are broken dow to gases, toluene [*108-88-3*] and xylenes (*o* [*95-47-6*], *m* [*108-38-3*], *p* [*106-47-3*]) are hydrodealkylated, and high benzene yields are obtained (freezing point 5.5°C) (13–15). This process is used in the United States, the UK, and Japan.

The principal aromatic hydrocarbons produced by coke-oven benzene refining

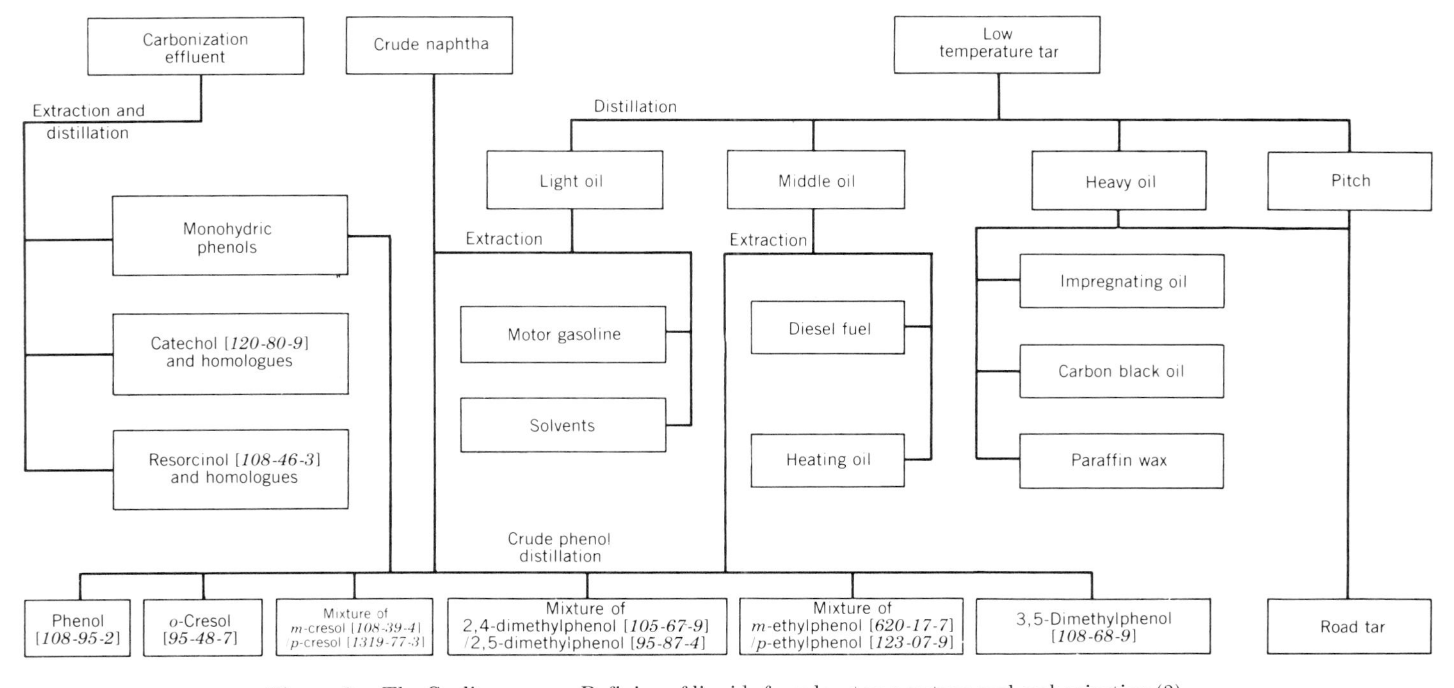

Figure 3. The Coalite process. Refining of liquids from low temperature coal carbonization (3).

are benzene and toluene (see Benzene; Toluene). In addition, ethylbenzene [*100-41-4*] and xylene isomers are recovered in one fraction as blending components for motor gasoline (see Gasoline).

Liquid Products

Low Temperature Carbonization of Coal. In the UK, low temperature carbonization of selected bituminous coals is used to produce solid, smokeless fuels (16). Another source of low temperature coal tar is the liquid by-products of the Lurgi pressure carbonization–gasification process which is used to produce synthesis gas for the Fischer-Tropsch synthesis process in SASOL plants (see also Gasoline and other motor fuels).

The main product of refining low temperature coal tars is phenolic compounds, as opposed to the aromatic hydrocarbons obtained from high temperature coal tars.

In the Coalite process (see Fig. 3) (17–19), the low temperature carbonization tars are recovered from the crude gas in two fractions with different boiling ranges. They are dehydrated similarly to high temperature tars and distilled by a continuous two-stage primary distillation at atmospheric pressure and vacuum. They are usually separated into 4–5 fractions, according to their boiling ranges: ca 2% light oil boiling to 200°C, which contains 30% aromatic hydrocarbons, 35% olefins, 20% paraffins and naphthenes, 10–15% crude phenol, and 2% crude bases; ca 40% middle oil, boiling range 200–300°C, containing 45% crude phenol, 25% aromatic hydrocarbons, 10–15% each of olefins and aliphatic hydrocarbons together with naphthenes; ca 30% heavy oil, boiling range 300–400°C, with a crude phenol content of 35% and a noticeable content of solid paraffinic hydrocarbons, which can be isolated as required; high boiling wax fractions; and ca 25% pitch (distillation residue).

The recovered oil fractions are used as heating oils, in road tar, in fuel oil, and as solvents. The phenols are recovered by extraction of organic fractions with caustic soda, and the Lurgi-Phenosolvan process is used for the recovery of phenols from the aqueous carbonization effluent. The distillation equipment for low temperature tar crude phenols must be of stainless steel because of the corrosive properties of the dihydric phenols. Low temperature carbonization tars can also be refined together with the crude naphtha by direct catalytic hydrogenation to give components for gasoline and diesel fuels.

Low temperature carbonization–gasification tars of bituminous coal as generated in the Lurgi fixed-bed pressure carbonization–gasification process are worked up similarly. The aromatic content is usually higher.

Carbonization of Lignites. A few refineries in Czechoslovakia, the GDR, the USSR, and India recover tars from carbonization and Lurgi pressure gasification of lignites. Process variations reflect the different compositions of lignite crude tars, which generally require a pretreatment, such as partial hydrogenation or treatment with ammonia, before refining because of thermal instability (20) (see Lignite and brown coal).

In comparison with bituminous coal tars, lignite tars contain considerable amounts of paraffinic compounds. They are usually refined by continuous distillation in pipe–still reboil systems (21). Products typically consist of 2% light oil, 18% middle oil (crude diesel fuel), 60% paraffin oil, 17% pitch, 1% coke, and 2% gases and loss.

The paraffin oil fraction contains ca 25–30% of solid recoverable paraffins. They are crystallized in two stages using indirect cooling between 5 and −10°C.

Lignite tar can also be refined by direct catalytic hydrogenation at various temperatures and pressures (22–23).

The crude phenols and aqueous effluent recovered from lignite-tar light oil are worked up in the same way as in low temperature coal-tar processes.

Economic Aspects

At present, ca 130 coal chemical processing refineries are operating worldwide (24). High temperature coke-oven coal tars including crude benzene provide the most significant volume basis for the production of 4×10^6 t/yr of technical carbon products such as carbon electrodes, graphite [*7782-42-5*], and carbon black [*1333-86-4*] products; 4×10^6 t/yr of technically pure benzene and benzene homologue mixtures; 1×10^6 t/yr of condensed aromatic hydrocarbons such as naphthalene and homologues, anthracene, phenanthrene [*85-01-8*], acenaphthene, and pyrene used as chemical intermediates; several thousand tons per year of heterocyclic compounds such as carbazole, quinoline [*91-22-5*], pyridine [*110-86-1*], and indole [*120-72-9*] used in the synthesis of dyes, pesticides, and pharmaceuticals; 4×10^5 t/yr of phenols and their homologues, as well as dihydric phenols which are raw materials for duroplastics, pesticides, disinfectants, and antioxidants; 1×10^5 t/yr of thermoplastic indene–coumarone resins for formulating plastics and adhesives; and $<2 \times 10^6$ t/yr of aromatic oils used as wood preservatives, solvents, and absorbing oils.

Chemicals and feedstocks made from carbonization and coking provide an important contribution to meet the requirements for many key chemicals. On a world scale, they cover 15% of the requirements for pure benzene; 85% of the requirements for naphthalene; 100% of naphthalene homologues, anthracene, phenanthrene, acenaphthene, pyrene, carbazole, quinoline, and indole; 3% of the world demand for phenol; 40% of the world demand for phenol homologues; 25% of the world production of hydrocarbon resins; 90% of the world production of electrode binders for carbon and graphite electrodes; 25% of carbon black feedstocks; and 75% of the wood preservation oils for pressure impregnating processes.

Health and Safety Factors

Tars contain, in various concentration, some compounds which have proven carcinogenic in animal testing, eg, certain polycyclic aromatic hydrocarbons and their homologues, and some *N*-aromatic compounds.

The acute toxicity of tar itself is low.

The prolonged effect of tar on skin may be a brownish patchy pigmentation (melanosis), possibly leading to so-called tar- or pitch-warts. These are usually benign.

The production of skin cancer by certain polycyclic aromatic hydrocarbons and tars has been proven by experiments on rats, mice, hamsters, and guinea pigs (25); on the other hand, the development of lung cancer by the same compounds is not proven or only occurs at very high concentrations (26).

Effects of polycyclic aromatic hydrocarbons have been investigated (27). At workplaces with high exposure, an increased lung-cancer risk cannot be excluded; however, such conditions are rarely found today.

Outlook

The continuing development in the recovery of chemical raw materials and feedstocks from coke-oven tar on an industrial scale during recent decades has provided chemical and allied industries with an important raw-material basis.

The present technology should provide a good economic basis in the foreseeable future since liquid by-products from coking and carbonization of coal are likely to expand in volume and new aromatic liquids from coal gasification and liquefaction processes will become available.

BIBLIOGRAPHY

1. Verband der Chemischen Industrie, ed., "Rohstoffsicherung durch Kohlenveredlung," a series of publications in *Chemie und Fortschritt*, Vol. 1, Frankfurt am Main, FRG, 1981, p. 43.
2. H. G. Franck and G. Collin, *Steinkohlenteer-Chemie, Technologie und Verwendung*, Springer Verlag, New York, 1968.
3. G. Collin and E. Wolfrum in Winnacker-Kuchler, *Chemische Technologie*, 4th ed., Vol. 5, Carl Hanser Verlag, Munich, FRG, 1981, pp. 472–491.
4. G. S. Pound, *Coke Gas* **21,** 395 (1959).
5. R. Kubicka and Z. Kvapil, *Freiberg. Forsch. H.* **A481,** 23 (1970).
6. G. Collin, *Erdöl, Kohle, Erdöl, Kohle, Erdgas, Petrochem.* **35**(6), 294 (1982).
7. H. Ritter, *Erdöl Kohle* **10,** 433 (1957); **16,** 292 (1963).
8. W. Urban, *Erdöl Kohle* **4,** 279 (1951); **7,** 293 (1954).
9. H. Nonnemacher, O. Reitz, and P. Schmidt, *Erdöl Kohle* **8,** 407 (1955).
10. F. Trefny, *Erdöl Kohle* **8,** 874 (1955).
11. M. Stein, *Hydrocarbon Process.* **52**(4), 139 (1973).
12. F. Trefny, *Erdöl, Kohle, Erdgas, Petrochem.* **23,** 337 (1970).
13. A. K. Logwinuk, L. Friedman, and A. H. Weiss, *Erdöl, Kohle, Erdgas, Petrochem.* **17,** 532 (1961).
14. T. Ward, *Erdöl, Kohle, Erdgas, Petrochem.* **26,** 440 (1973).
15. W. Lorz, R. G. Craig, and W. J. Cross, *Erdöl, Kohle, Erdgas, Petrochem.* **21,** 610 (1968).
16. D. McNeil, *Coal Carbonization Products*, Pergamon Press, London, 1966, pp. 3–4.
17. W. A. Bristow, *J. Inst. Fuel* **20**(113), 109 (1947).
18. K. R. Payne, *Yearb. Coke Oven Mgrs. Assoc.*, 154 (1977).
19. S. A. Qader and G. R. Hill, *Chem. Ing. Tech.* **43**(9), 595 (1971).
20. H. Sacher in *Oils and Gases from Coal, General Report Journal of the Symposium on Gasification Liquefaction Coal, Katowice*, 23-27.4.1979, Pergamon Press, New York, 1980, pp. 217–237.
21. W. Köhler, *Chem. Tech.* **4,** 6 (1952).
22. M. Platz, *Energietechnik* **28**(2), 64 (1978).
23. R. Herberling, *Freiberg. Forsch. H.* **A201,** 21 (1961).
24. G. Collin in L. E. St. Pierre, ed., *Proceedings of World Conference, Future Sources Org. Raw Materials Chemrawn I, Toronto*, 10., 13.7.1978, Pergamon Press, New York, 1979, pp. 283–297.
25. W. C. Hueper, *Occupational and Environmental Cancer of the Respiratory Systems*, Springer Verlag, New York, 1966, p. 117.
26. I. Thyssen, J. Althoff, G. Kimmerle, and U. Mohr, *VDI-Ber.* (358), 329 (1980).
27. J. Misfeld in Umweltbundesamt, ed., *Luftqualitaetskriterien fuer ausgewaehlte Polycyclische Aromatische Kohlenwasserstoffe*, Erich Schmidt Verlag, Berlin, 1979, p. 224.

G. COLLIN
Rütgerswerke AG

G. LÖHNERT
Ruetgers-Nease Chemical Company

CARBIDE PRODUCTION

CALCIUM CARBIDE

Calcium carbide [*75-20-7*] is the most important carbide prepared on an industrial scale. It was first described in 1862 (1). Preparation on an industrial scale began in the 1890s (see Carbides, calcium carbide).

Pure calcium carbide, CaC_2, mol wt 64.1, is difficult to prepare and is of scientific interest only (2). Calcium carbide prepared on an industrial scale is generally called carbide. It contains ca 80% CaC_2; the remainder is CaO and other impurities present in the raw materials, eg, SiO_2 and Al_2O_3.

For the manufacture of calcium carbide, lime is treated with a carbon-containing material:

$$CaO + 3\,C \rightarrow CaC_2 + CO \quad + 465\ kJ/mol\ (441\ Btu/mol)$$

The required energy has to be supplied at very high temperatures because the reaction proceeds at an acceptable rate only above 1600°C, and therefore, industrial processes are carried out at 1800–2100°C. This energy is supplied either by combustion of carbon-containing material with pure or enriched oxygen in the presence of CaO in a shaft furnace (3) or by heating and melting the carbon-containing material with CaO in an electrothermal low-shaft furnace.

Despite its promise, development of the former process was discontinued in the mid-1950s because at that time it was less expensive to prepare the large amount of synthesis gas ($CO + H_2$) which formed in this process from natural gas or mineral oil. Although the energy situation has changed since then, no attempt to apply this process has been made (see also Fuels, synthetic (gaseous)).

The development of the latter process began in the 1890s in countries where water power was available at a low price, namely France, Switzerland, and Canada (4). The process has been improved constantly, and today furnaces are constructed with a power consumption of up to 70 MW. At present, carbide is manufactured exclusively electrothermally (see Furnaces, electric).

Calcium carbide owes its importance to two reactions:

$$CaC_2 + N_2 \rightarrow CaCN_2 + C$$

$$CaC_2 + 2\,H_2O \rightarrow C_2H_2 + Ca(OH)_2 \quad + 129.4\ kJ/mol\ (136 \times 10^6\ Btu/mol)$$

The first reaction enabled the nitrogen contained in the air to be chemically bound and to be used as fertilizer (see Cyanamides), whereas the acetylene (qv) obtained according to the second reaction was used as an illuminant around the turn of the century. Later, it became the basis for the development of autogenous welding, and after World War I, it became an important starting material for many chemicals (see Acetylene-derived chemicals).

Small amounts of acetylene are obtained by the gasification of carbide, according to the wet gasification process in which a large excess of water (ca 10 times the weight of carbide) is used. The residue consists of large amounts of lime sludge.

For large amounts, the dry-generation process is generally applied; the heat of reaction is removed by vaporizing a small amount of water. The residual calcium hydroxide can be recycled.

In the Shawinigan generator, the carbide is gasified in three troughs provided with paddles that are arranged one above the other (5). The Knapsack generator comprises several plates that are arranged one above the other in a tower (6). The carbide is deposited in a spiral motion initiated by agitators. Modern dry generators may have a capacity of 3500 m^3 acetylene per hour.

In the last few years, calcium carbide has been used increasingly as a desulfurizing agent in the steel (qv) industry (7), owing to the following reaction:

$$CaC_2 + [S] \rightarrow CaS + 2\ C$$

The phase diagram of CaC_2–CaO is shown in Figure 1 (8).

Manufacture

Calcium carbide is made in the electrothermal low-shaft furnace. Its main components are the furnace vessel with the tapping device and the devices for the introduction of electric power and raw materials (see also Furnaces, electric).

The furnace vessel is made of welded or riveted iron and reinforced against distortions. The bottom of the vessel is covered with carbon bricks or fireproof bricks, and the walls are built in stone or lined with refractory material.

Most furnaces are operated by three-phase-current electric power supplied by three electrodes. These may be arranged in a series, in which case the furnace vessel is rectangular. They may also be arranged at the corners of an equilateral triangle, in

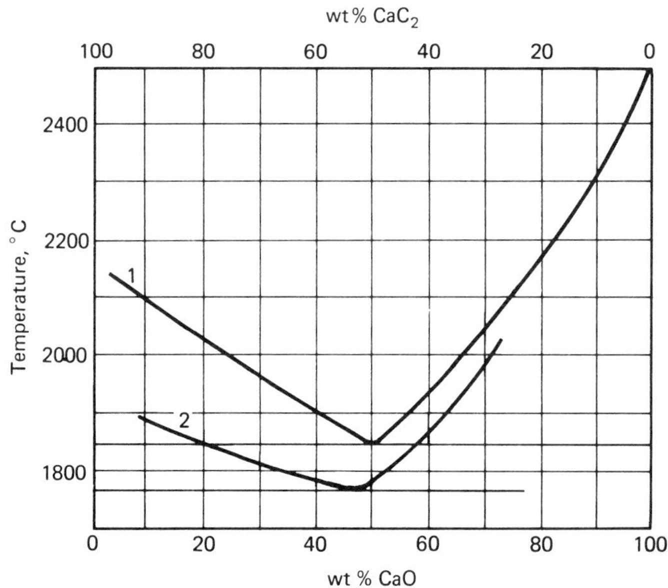

Figure 1. Melting-point diagram for the CaO–CaC_2 system (8). Curve 1 = pure carbide. Curve 2 = technical carbide (contains other impurities than CaO).

which case the vessel is round (Fig. 2(**a**)) or has the form of a triangle rounded off at the corners (Fig. 2(**b**)).

The furnace vessel has three tap holes through which the liquid carbide flows into cast-iron crucibles.

All furnaces are equipped with self-baking Soderberg electrodes. Coal-tar pitch and granular and powdery calcined anthracite and coke are introduced into an iron shell which may be lengthened by means of extendable tubes that are fixed by welding. The mixture burns in the melting zone, and the coal solidifies.

The bases of the electrodes are surrounded by contact plates, usually made of copper, through which the electrodes are supplied with electric power.

The electrode system can be moved up and down in the furnace, which allows regulation of the current consumption. In addition, the electrodes can be pushed separately downward through the contact plates, permitting replacement by molten material.

The raw materials are introduced from various hoppers into the furnace via chutes. Modern furnaces are equipped with so-called hollow electrodes (9) through which the fine-grained material can be blown directly into the melting pot. These electrodes are iron tubes arranged in the center of the electrode shell.

During the reaction, ca 400 m^3 of gas at STP is formed per metric ton of carbide. The gas contains up to 100 g/m^3 of dust, and depending on the nature of the raw materials, 5–15% by volume of H_2 and smaller amounts of methane in addition to CO. To recover this gas, the furnace has to be provided with appropriate devices.

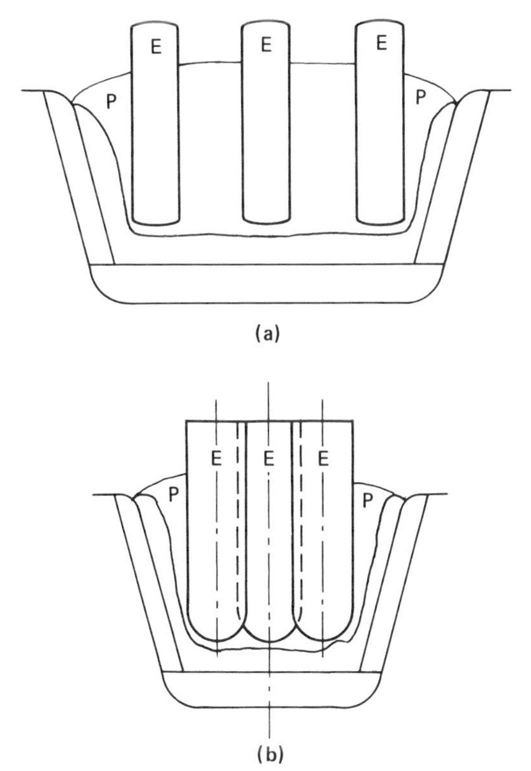

Figure 2. Round (**a**) and triangular (**b**) furnace vessel. E = electrode; P = pool of molten material.

In open furnaces, the gas burns at the surface of the furnace content, and the gas has to be purified. This procedure is expensive because large quantities of gas have to be freed from dust.

In covered furnaces, up to ca 80% of the gas produced is retained by the partial cover of the vessel. In rectangular furnaces, barriers are arranged between the electrodes and the gas is removed at the bottom. Round furnaces are provided with a cupola that is mounted above the furnace vessel and in which annular slots are left free around the three electrodes. The residual gas escapes through these slots and is burned. Thus, less gas has to be freed from dust in the case of covered furnaces than in the case of open furnaces. However, the gas exits in two partial streams, and two purifying apparatuses are required.

In the closed furnace, all the gas that forms is collected and purified (see Fig. 3).

Furnace Operation. The carbide furnace has to be controlled in such a way that the production proceeds continuously and evenly. The open furnace is easiest to control because the materials can at any time be treated mechanically in such a way that the gases are driven out uniformly. Moreover, the materials that have to be added can be conveyed immediately to the location in the furnace where they become effective. The operation of covered furnaces is more difficult. In the case of closed furnaces, the process cannot be controlled while the furnace is in operation. These furnaces can have a power consumption of up to 70 MW, and their operation of the furnace is of particular importance.

Furnace operation is affected by dissociation of the carbide:

$$CaC_2 \rightarrow Ca + 2\,C$$

and by the reaction of carbide with excessive CaO:

$$CaC_2 + 2\,CaO \rightarrow 3\,Ca + 2\,CO$$

Figure 3. A closed-column carbide power furnace.

In normal furnace operation, the calcium forming according to these reactions is converted into carbide in the cooler regions of the furnace by reacting with carbon.

However, the partial pressure of calcium rises rapidly above 2000°C, and above 2200°C, the carbide decomposes by violent eruption. As a result, the operation of the furnace is interrupted and considerable damage may be caused.

For these reasons, the temperature in the melting zone must not increase excessively. This can be avoided by feeding the raw material in such a way that the carbon concentration remains stable and by ensuring that the components, in particular carbon, react rapidly enough at lower temperatures.

Raw Materials. The lime component is usually quicklime with a grain size up to 50 mm (10) (see Lime). Grains smaller than 6–8 mm are separated by screening and conveyed to the hollow electrode. The MgO content should be $<1.5\%$ and total CaO $>90\%$.

The $Ca(OH)_2$ which forms when carbide is gasified is burned and recycled to the furnace. It can also be recycled in the form of briquettes consisting of a mixture of calcium hydroxide and fine coke.

The carbon source is usually coke, generally obtained from hard coal, although coke from brown coal is used where deposits of the latter are nearby. The coke has a grain size of 0–25 mm. The fines having a diameter <2–3 mm (ca 7–10 mesh) are separated by screening and blown through the hollow electrode; ash content should be $<15\%$ and water $<2\%$. The phosphorus contained in the ash forms calcium phosphide which reacts with water giving phosphine. This substance is toxic and responsible for the typical odor of carbide; its content in acetylene must not exceed 0.06 vol % (see also Coal conversion processes, carbonization).

Solidification and Standards. The liquid carbide tapped from the carbide furnace is collected in cast-iron crucibles. In some plants, the blocks are allowed to cool in the crucibles, whereas in others, they are removed after 2–4 h and allowed to cool. In order to avoid losses of acetylene, the blocks are crushed at ca 400°C. Alternatively, the liquid carbide leaving the tap hole is directed into a rotating cooling drum. However, this procedure entails heavy losses and does not permit the granulation required by many standards.

Standards with regard to granulation, C_2H_2 content, and impurities exist in the United States, Australia, Belgium, the FRG, the GDR, France, the UK, India, Israel, Japan, and Mexico.

Economic Aspects

World and U.S. production of calcium carbide was highest from 1955 to 1965 (see Table 1). Because of the competition of petrochemical products and other welding processes using means other than acetylene, carbide consumption has decreased. In

Table 1. Production of Calcium Carbide, 1000 t

Country	1957	1965	1970	1979	1980	1981
United States	922	966	699	244	233	254
Canada	400	120	110	100	100	100
FRG	960	1039	879	453	456	428

communist countries and in South Africa, the production of carbide has remained stable because of the ready availability of raw materials.

Recently, production and capacities have been increased because of the increased demand for carbide for the desulfurization of pig iron and steel.

Health and Safety Factors

Calcium carbide is stable almost indefinitely if stored in the absence of moisture. In contact with water, acetylene, which is inflammable, forms. The limit of explosion in the air is 2.4%, and therefore, handling, storage, and transport of carbide are subject to special safety regulations.

During the production process, gaseous CO forms; its concentration is not permitted to reach 30 cm^3/m^3 air at a place of work.

BIBLIOGRAPHY

1. F. Wöhler, *Liebigs Ann. Chem.* **124,** 220 (1862).
2. S. A. Miller, *Acetylene*, Vol. I, Ernest Benn Ltd., London, 1965.
3. C. Wurster, *Chem. Ing. Tech.* **28,** 4 (Jan. 1956).
4. A. Lang, *Histoire du Carbure de Calcium*, H. Studer S.A., Geneva, Switzerland, 1949.
5. U.S. Pat. 2,343,185 (Feb. 29, 1944), (to Shawinigan Chemicals).
6. Ger. Pat. 530,111 (July 21, 1931), (to AG für Stickstoffdünger).
7. M. A. Palmer and J. S. Beeker, *Desulfurization in the Transfer Ladle in 34th Iron Making Conference, Toronto, Apr. 1975*, Inland Steel Co., Ind.
8. R. Juza and H. U. Schuster, *Z. Anorg. Chem.* **311,** 62 (1961).
9. U.S. Pat. 2,996,360 (Aug. 15, 1961), D. E. Hamby (to A. M. Kuhlman); *Hollow Electrode System for Calcium Carbide Furnaces, Electric Furnace Proceedings*, Union Carbide Corp., 1966, p. 208.
10. E. Shiele and L. W. Berens, *Kalk*, Verlag Stahleisen mbH, Dusseldorf, FRG, 1972.

W. Portz
Hoechst AG

COLLOIDS

Nature and Scope of Colloids

A colloid is a material that exists in a finely dispersed state. It is usually a solid particle, but it may also be a liquid droplet or a gas bubble. Typically, colloids have high surface-area-to-volume ratios, characteristic of matter in the submicrometer-size range. Matter of this size and from ca 100 nm to ca 5 nm, just above atomic dimensions, exhibits physicochemical properties that differ from those of both the constituent atoms or molecules and the macroscopic material. The differences in composition, structure, and interactions between the surface atoms or molecules and those on the interior of the colloidal particle lead to the unique character of finely divided material, specifics of which can be quite diverse (see Flocculating agents). Colloids are encountered in a wide variety of industries, and many important technological problems are related to the behavior of dispersions. Commercial examples are metals, magnetic powders, catalysts, ceramics, minerals, oil recovery, technical glasses, paints and pigments, polymers, pulp and paper, prepared foods, pharmaceuticals, fibers, detergents, and purified water (see also Latex technology; Plasticizers; Printing processes). Natural and biological systems may also depend, to some extent, on the behavior of colloids; phenomenological examples include soil science and plant nutrition, meteorology, the properties of blood, membrane science, and antigen–antibody reactions in medical technology (see also Dental materials). In view of the ubiquity and importance of colloids, the ability to control their formation, destruction, or stabilization is often a critical technological problem. This is true whether the specific colloidal material is deemed desirable, as in the reinforcement of plastics, metals, and ceramics, or undesirable, as is often the case in water treatment, meteorology, and other environmentally important systems (1).

General Properties

Dimensions. Most colloids have all three dimensions within the size range stated earlier. However, if only two dimensions (fibrillar geometry), or even one dimension (laminar geometry), exist in this range, unique properties of the high surface-area portion of the material may still be observed and even dominate the overall character of a system (1). The non-Newtonian rheological behavior of laminar and fibrillar clay suspensions, the reactivity of catalysts, and the critical properties of multifilamentary superconductors are only a few examples of systems controlled by such colloidal materials. For highly dispersed systems that do not exhibit a particularly high surface-area-to-volume ratio, it is common to describe a dispersion factor (2), which is the ratio of the number of surface atoms to the total number of atoms in the particle. Representative values of the dispersion factor for 10-, 100-, and 1000-nm particles would be, respectively, on the order of 0.15–0.30, 0.04, and 0.003–0.02, depending on the specific dimensions of the atoms or molecules that comprise the particles. Since three-dimensional colloidal particles are common, it is useful to compare the effect of particle size on the translational diffusion coefficient D_{tr}, and the distance over which the particle will settle in a fluid. Table 1 contains this type of comparison for spherical particles having a specific gravity ρ_s of 2.0 g/cm^3; the specific gravity of the fluid (water) is taken as unity. For purposes of comparison, the conditions of temperature and time are chosen as 293 K and 1 h, respectively. Dashed lines indicate arbitrary boundaries between which particles are typically considered colloidal since the effect of either Brownian diffusion or sedimentation, ie, relative to the other, is on the order of 1%.

Nomenclature. The properties of colloids greatly depend on their environment, ie, the dispersing medium, as well as their composition and structure. Therefore, colloids must be classified according to their states of subdivision and agglomeration and with respect to the dispersing medium or environment. The possible classifications of colloidal systems are given in Table 2, adapted from refs. 1 and 3–4. From the variety

Table 1. Effect of Spherical Particle Size on Relative Brownian Diffusion and Sedimentation Distances After 1 h in Water at 293 K

Radius (a), nm	D_{tr}/D_{ions}[a,b]	Distance diffused ($\bar{x}$), nm	Distance settled (Δh)[c]	$\bar{x}/\Delta h$[d]	
0.1 (ions)	1	3.9×10^6 ($>\bar{x}$ solvent)	0.08 nm	4.9×10^7	
1 (solvent)	0.1	1.2×10^6	8 nm	1.5×10^5	
10	10^{-2}	3.9×10^5	0.8 μm	4.9×10^2	colloidal dimensions
100	10^{-3}	1.2×10^5	80 μm	1.5	
1,000	10^{-4}	3.9×10^4	8 nm	4.9×10^{-3} (<1%)	
10,000	10^{-5}	1.2×10^4	0.8 m	1.5×10^{-5}	

[a] D_{tr} = translational diffusion coefficient; $D_{tr} = kT/6\pi\eta a$, k = Boltzmann's constant, η = fluid viscosity.

[b] D_{tr}-values are scaled to the value of D_{tr} for ions, D_{ion}.

[c] $\rho_{particle} = 2.0$ g/cm^3; $\rho_{fluid} = 1.0$ g/cm^3.

[d] Dashed lines indicate boundaries between which colloids typically exist.

Table 2. Classifications of Two-Phase Colloidal Systems

Dispersing medium	Dispersed (colloidal) matter	Names[a]
gas	liquid	aerosol, fog, mist
	solid	aerosol, smoke
liquid	gas	foam
	liquid	emulsion
	solid	sol[b], gel, dispersion, suspension
solid	gas	solid foam
	liquid	gel, solid emulsion
	solid	solid sol, alloy

[a] The appropriate descriptive name depends on the specific properties of the system; see refs. 1 and 3–4.

[b] Aqueous sols are sometimes called hydrosols.

of systems represented in this table, it is evident that the problems associated with colloids are usually interdisciplinary in nature and require a broad scientific base to understand completely. This has led to a wide diversity in the theoretical and analytical advances needed to interpret colloidal behavior properly. The following examples represent recent, mid-twentieth century, theoretical and technical developments (1,5) that have occurred because of colloidal behavior: experimental application of the Mie and inelastic theories of light scattering, photon correlation spectroscopy, electrical double-layer models, quantitative formulation of van der Waals interactions, gas adsorption theory, electron microscopy, ultracentrifugation, and various spectroscopic techniques based on electron, ion, and photon probes. Many of these developments can be used, either directly or indirectly, to investigate the changes that a colloidal system undergoes, with the appearance or disappearance of the colloidal state being the most critical event and a change in the tendency to agglomerate (solids) or coalesce (liquids, gases) being the most frequently encountered industrial problem.

Behavior. Among the general physical properties and phenomena that are primarily important in colloidal systems are diffusion, Brownian motion, electrophoresis, osmosis, rheology, mechanics, and optical and electrical properties (1,5–6). Of course, chemical reactivity and adsorption often play important, if not dominant, roles for colloids. Any combination of physical and chemical behavior given here may ultimately determine specific industrial processes and determine product characteristics. An example is the need to pump fluid systems that contain high concentrations of solids, eg, those exceeding 10 vol %. This problem is commonly encountered with extrusion processes in the polymer industry, in processing foods and cosmetic products that are gels, with various suspensions for high performance materials in the ceramics and metallurgical industries, with pigment slurries in the paint industry, and finally, with waste by-products in the minerals and water industrial sectors. A problem shared by all these industries is the ability to predict and to control suspension rheology, in particular, thixotropy and dilatancy. These two rheological properties depend ultimately on the specific interactions (eg, wetting, electrical double layer, and van der Waals forces) among colloidal particles, dispersing medium, and solute additives such as salts, surfactants, and soluble polymers.

Monitoring Techniques. The best choice of physical techniques for investigating or monitoring colloids is usually determined by considering the nature of the colloidal material and the dispersing medium. The purpose for which the information is sought, namely for fundamental knowledge or technological use, must also be considered. In principle, any physical property characteristic of a colloidal system can be used, at least empirically, to monitor changes in dispersed systems. Obviously, the more complex a system is chemically or with respect to its particulate heterogeneity, the less likely it is that a single property uniquely and completely describes changes in the colloidal state.

The preparation and destruction of colloids is often studied using electron and optical microscopy, light scattering, and rheological and surface tensiometric apparatus. Agglomeration or coalescence of the colloidal material can be followed experimentally by various techniques: light scattering, microscopy, rheology, conductivity, filtration, sedimentation, and electrokinetics. Each of these techniques has been successfully employed to monitor changes in colloidal systems.

Physical and Chemical Principles

Formation. Colloid formation involves either nucleation and growth phenomena or subdivision processes (1,7–9). The former case requires a phase change; the latter pertains to the comminution or atomization of coarse particles (solids) or droplets (liquids). Three nucleation and growth processes by which colloids have been synthesized synthetically or in nature are condensation of vapor to yield liquid or solid directly (6), condensation of vapor to form colloidal liquid droplets (aerosols) that subsequently solidify (10), and precipitation from liquid or solid solution (11). Chemical reactions often induce these phase changes. However, this mechanism is not essential for either homogeneously or heterogeneously nucleated colloids. Certainly, growth of a colloid is controlled kinetically and depends fundamentally on the rate at which fresh solute material in the correct orientation approaches a growing nucleus and the bond energy requirements at a nucleus' surface. This last consideration suggests that dislocations in solid colloidal particles significantly influence particle growth. Table 3 summarizes the various mechanisms of colloid formation and resultant powder properties for the industrial production of single component and multicomponent colloidal solids (6). Processes based on subdivision contrast these chemical processes and are widely used (6,12–18). Though it is difficult to reconcile experimental and practical observations of comminution into a single coherent theory, such a theory would clearly have to consider a variety of phenomena including mechanics, particle fracture, and agglomeration. Consequently, subdivision of solids on an industrial scale should be considered more an art than a science at the present time. Nonetheless, a complex technology, if not a specialized one, has developed to conduct and to control comminuting and size-fractionating processes. Mathematical models are available to describe changes in particle-size distribution during comminution, but generally these are restricted to specific processes. Comprehensive reviews of the developments in preparing colloidal solids by subdivision should be consulted for further details (6).

Specific advancements in the chemical synthesis of colloidal materials are noteworthy. Many types of generating devices have been used for producing colloidal liquid aerosols and emulsions (19–23), among them are atomizers and nebulizers of various designs (10,24–30). A unique feature of producing liquid or solid colloids via aerosol processes (Table 3) is that material with a relatively narrow size distribution can be routinely prepared. These monosized colloids are often produced by relying on an electrostatic classifer to select desired particle sizes in the final stage of aerosol pro-

Table 3. Summary of Industrially Produced Colloidal Materials and Related Processes

Mechanism	Examples[a]
vapor → liquid → solid ↓ → → → ↑	oxides, carbides via high intensity arc; metallic powders via vacuum or catalytic reactions
vapor + vapor → solid	chemical vapor deposition, radiofrequency-induced plasma, laser-induced precipitation
liquid → solid	ferrites, titanates, aluminates, zirconates, molybdates via precipitation
solid → solid	oxides, carbides via thermal decomposition

[a] Ref. 6.

duction. Another significant development is the increasing availability of colloidal powders prepared as liquid suspensions. These powders have uniform chemical and phase composition, size, and shape (1). Such colloids of certain elements, eg, sulfur, gold, selenium, and silver, have been obtainable for a long time (8,31–35). Along with the development of aerosol, hydrosol, and dispersion techniques, especially in recent years, the laboratory and industrial preparation of uniform, monosized colloids of complex compositions has advanced rapidly and has the potential for becoming routine. Examples of these materials include many inorganic compounds, such as NaCl, AgCl, $BaSO_4$, $FePO_4$, SiO_2, ZrO_2, V_2O_5, Al_2O_3, α-Fe_2O_3, TiO_2, $Cr(OH)_3$, La_2O_3, Ga_2O_3, and CdS (1,10,36–40), and organic lattices of poly(vinyl acetate), polystyrene, poly(vinyl chloride), styrene–butadiene rubber, and poly(acrylic acid) (1,38,40–44). Practical exploitation of these colloidal materials and related ones is presently limited by problems associated with scale-up. However, monosized powders or monodispersed colloidal sols are used in many industrial and scientific applications, eg, pigments (qv), coatings (qv), and pharmaceuticals (qv) (1,38).

Characterization. The proper characterization of colloids depends on the technological and scientific purposes for which the information is sought, since the complete characterization would be an enormous task (6). The following physical properties are among those to be considered: size, shape, and morphology of the primary particles; surface area; number, and size distribution of pores; degree of crystallinity and polycrystallinity; concentration of defects; nature of internal and surface stresses; and state of agglomeration (6). Chemical and phase composition are required for complete characterization, including information on the purity of the bulk phase and the nature and quantity of adsorbed surface films or impurities.

Particle Size. The most direct methods for determining the particle size, shape, and morphology of colloids are scanning and transmission electron microscopy. Resolution of optical microscopes is usually insufficient to determine primary particle size, although agglomerate size can be effectively evaluated using conventional optical microscopes. Indirect techniques for determining particle size or particle-size distribution include sedimentation and centrifugation, conductometric techniques, light scattering, x-ray diffraction, gas and solute adsorption, ultrafiltration, and diffusiometric methods (3,45). Care must be taken in selecting an indirect method since these require assumptions about either the real size distribution or the process on which the analysis is based. For instance, commercial conductometric equipment, eg, Electrozone and Coulter Counters, relies on the sphericity of particles; light-scattering techniques are reliable only if the particle shape is known or assumed (45); and adsorption analyses rely on model adsorption isotherms, the uniformity of particle size and porosity, and the orientation of adsorbed species. Moreover, as suggested by Table 2, the best analytical choice also depends, to some extent, on the physical properties of the colloid and on the dispersing medium. Typically, more than one technique is employed to develop a realistic description of the particle size.

Surface Area and Permeability or Porosity. These properties are most commonly evaluated using gas or solute adsorption for surface-area determination and mercury porosimetry for permeability in conjunction with at least one other particle-size analysis, such as electron microscopy. Reference 46 gives a rather complete description of the techniques and theoretical models that have been developed, such as the kinetic approach to gas adsorption developed by Brunauer, Emmett, and Teller (BET) (47), which is based on Langmuir's model for adsorption (48), the potential theory of Polanyi

(49) for gas adsorption, the experimental aspects of solute adsorption, and the principles of mercury porosimetry based on the Young–Dupré expression (3,5).

Solid-State Structure. The degree of crystallinity and polycrystallinity of colloids is obtained via x-ray diffraction and the interpretation of Bragg reflections (49). Laue techniques (49) are not routinely employed for studying colloids because they require single crystals of larger size. The structure and concentration of defects in single-crystal colloids can be determined by conventional methods using x radiation and measurement of physical properties down to a particle-size range of ca 1–10 μm. In principle, inherent properties can be studied on finer material. However, in practice, it is often necessary first to conduct analyses of defects on samples of the same material in particles larger than the colloidal size range and then to assume that the determined defect structure also applies to the corresponding colloids. Neutron diffraction is expected to be increasingly used (50), particularly with regard to magnetic structure (51), to complement x-ray studies of colloidal powders.

Internal and Surface Stresses. These stresses in colloidal particulates can be evaluated using the procedures developed for macroscopic samples, namely, electron microscopy, x-ray and neutron diffraction, and physical property measurements (6,49–51). Since many equilibrium crystals are encountered and apparently equilibrated surfaces are not smooth but rather saw-toothed owing to entropic factors (5), the actual planes that predominate at surfaces in practical systems may be determined by experimental or operational conditions. The detailed study of surface stresses in solid colloids is therefore difficult because actual crystal planes are usually incomplete and imperfect at powder surfaces. Moreover, these imperfections may actually represent unequilibrated surface stresses (5) and adsorbed gaseous species or the dispersing medium with its solutes probably affects the energy and atomic configurations responsible for surface stresses. All these factors compound the problems associated with stress analysis. However, some progress in unravelling this problem has been made with the experimental confirmation of surface distortions that result from an imbalance of surface forces, eg, those reported in the recent studies on elemental semiconductors of silicon, germanium, and diamond (52–56). These considerations imply that the practical implications of internal and surface stresses are most important in the formation of solid colloids, particularly during their growth when surface structure changes most radically and in the dispersibility of solids in liquids.

The following example illustrates the impact of surface states on the dispersion of colloidal silicon powder. The process consists of three stages: wetting of powder, breakdown of large agglomerates into colloidal particles or small clusters of primary particles, and stabilization of powder against agglomeration (9). The first two stages involve the replacement of silicon–vapor (air) or silicon–silicon interfaces by a silicon–liquid one and, possibly, the mechanical disintegration of agglomerate structures. These steps constitute complete wetting, for which the change in free energy per unit area is designated as ΔG_{d},

$$\Delta G_{\mathrm{d}} = \gamma_{\mathrm{SL}} - \gamma_{\mathrm{SV}} = -\gamma_{\mathrm{LV}} \cos \theta_{\mathrm{e}}$$

The terms γ_{SL}, γ_{SV}, and γ_{LV} represent the surface tensions or free energies of the solid–liquid, solid–vapor, and liquid–vapor interfacial regions, respectively, when the three phases coexist and establish an equilibrium contact angle θ_{e} measured within the liquid phase. It has recently been demonstrated (57) that the Girifalco-Good theory (58–60) and Fowkes' modification of it (61) adequately describe various dispersibility phenomena observed in a study of this powder (62). ΔG_d, θ_e, and various contributions to the surface free energy of the silicon powder (γ_{Si}) could be estimated using macroscopic contact angles and fundamental physical data of silicon and various liquids. Specifically, γ_{Si} and its dispersion force component were evaluated to be 39.9 mJ/m^2 (= dyn/cm) and 38.0 mJ/m^2 (= dyn/cm), respectively, with the remaining 1.9 mJ/m^2 (= dyn/cm) attributed to the electronic band structure of silicon (53). These results can be, in principle, generally expanded to more complicated solids; however, this step is presently quite difficult to realize.

State of Agglomeration. Nearly all colloidal systems undergo some agglomeration. This leads to a distribution of aggregate size for solid colloids and of droplet size for liquid colloids. Consequently, either the rate or the extent of agglomeration is of practical concern whenever colloidal material is encountered. Ultramicroscopy is the most desired method of evaluating the kinetic or equilibrated aspects of agglomeration; unfortunately, it is often not feasible or too intricate and time consuming. Although it is the only direct method widely available, this technique will be fully exploited once procedures for "wet" or environmental cells are developed. The most common indirect methods for evaluating agglomeration consist of optical techniques (3–4,8), ie, light-scattering or optical-density measurements and photon correlation spectroscopy, for which the magnitude of the optical effect increases as the agglomerates become larger via aggregation or coalescence. Other methods have been devised which are based on the principles of sedimentation, rheology, and electrical and thermal conductivity. These techniques tend to be empirical, qualitative, and specific to a given colloidal system. More than one technique is required to establish the state of agglomeration when a wide range of colloidal dimensions exists. Moreover, if agglomeration is severe, producing clusters exceeding ca 5 μm, aggregate-size distribution can be evaluated using a variety of classification techniques developed for macroscopic solids, eg, sieving (6,63), provided such methods do not themselves induce agglomeration.

The state of agglomeration is intimately related to colloidal stability or the resistance to agglomerate or coalesce. Wetting phenomena play a significant role here, as do various nonwetting, colloidal factors. The stabilizing role of second solvents and surfactants can be exploited if a colloidal solid is not completely wetted by the dispersing fluid and if the interfacial energy in fluid–fluid dispersions such as emulsions and foams is initially unfavorable. This step requires an understanding of the energetics of wetting and adsorption (5,9,64–65) obtained via calorimetry (9) and adsorption isotherms (9,66), or by studying the specific micellization behavior and hemimicelle formation of surface-active agents (66–67). Nonwetting factors in colloidal stability generally pertain to the resistance toward agglomeration. The agglomerating process is induced by particulate collisions arising from diffusion, as in Brownian motion, velocity or shear gradients in a liquid-dispersion medium, and gravitational settling (68). The collisional frequency in the first case depends on the liquid's viscosity, concentration of particles, temperature, and the fraction of collisions that lead to irreversible agglomeration.

This last consideration can be quantified using various models for repulsive or attractive electrostatic (3–5,7–9), London-van der Waals (3,69–71), and steric (72) factors that affect the stabilization of aqueous and nonaqueous colloidal systems. Since these forces originate from different sources, they can be evaluated separately. A comprehensive model of colloid stability, the DLVO (Derjaguin-Landau-Verwey-Overbeek) model has proved to be essentially correct, regarding the roles of electrolytes, dielectric constant, and other physical quantities in colloidal systems, and is considered a cornerstone of modern colloid science (7–8). These references independently describe the interplay of electrostatic and London-van der Waals interaction in the stability of hydrophobic suspensions. This DLVO theory considers the electrostatic interactions between two identically charged, suspended particles to be repulsive and to arise from the overlap of the electrical double layers associated with each particle. London-van der Waals forces are considered in this model to be adequately given by Hamaker's formulas (70), although more appropriate treatments now exist (71). The total interaction is considered, therefore, as the combined effect of surface charges and the London-van der Waals forces. If the particles are sufficiently charged and have an associated surface potential greater than about |25| mV, the binary particle interaction model predicts noticeable stability against agglomeration, owing to a substantial, repulsive potential energy barrier, which inhibits close approach of particles.

Quantitative understanding of the interplay of steric or solvation factors with either the electrostatic or London-van der Waals forces is not part of the DLVO theory, but it has developed since the early 1970s. Long-chain surfactants and high molecular weight polymers have commonly been added to colloidal suspensions to ensure long-term stability. However, the mechanisms of steric stabilization have only recently been elucidated so that these relatively short-range interactions can be used industrially to produce systems with the desired behavior at an economical cost. Molecular architecture, thickness of the adsorbed layer, temperature, and chain or segment solvation are critical parameters in determining the effectiveness of a dispersing agent to provide sufficient steric stabilization (69,72). Depletion stabilization is related to steric stabilization, in that added soluble polymer ensures stability; however, this mechanism relies on the ability of free polymer in bulk solution to provide a thermodynamically metastable barrier to close approach of particles (73). Above a threshold concentration of polymer and on close of approach of particles, the concentration of free polymer segments between the particles is depleted, giving rise to an unfavorable situation and thereby providing colloidal stability.

If velocity or shear gradients are present and are sufficiently large, the frequency of collisions depends on volume fraction of solids and the mean velocity gradient, as well as on the system parameters listed earlier. Several research groups have initiated fundamental studies dealing with colloidal systems in which shear is important (74–79). However, the interpretation of data is sometimes complicated by non-Newtonian flow, stemming from either polydispersity or high solids content. In this regard, various mathematical and rheological techniques have been developed to characterize non-Newtonian behavior (76,80–81).

Assuming that sedimentation is slow compared to the first two collision mechanisms, the overall agglomeration rate is

$$-dN/dt = k_{\mathrm{d}}N^2 + k_{\mathrm{s}}N,$$

where N is the particle-number concentration, k_{d} and k_{s} are the respective rate constants corresponding to diffusion-controlled and shear-induced collision processes, and the minus sign denotes that the number concentration decreases with time t. The constants k_{d} and k_{s} depend on particle properties such as chemical composition of bulk and surface phases, dielectric constant, dipole moment, size, size distribution, shape, surface charge, solid-phase distribution within particles, and particle anisotropy. Properties of the liquid-dispersing medium that contribute significantly to the values of these rate constants are dielectric constant, dipole moment, and the ability to dissolve electrolytes and polymers, in addition to those properties cited earlier. The k_d-term usually dominates in quiescent systems containing submicrometer particles. The full expression for $(-dN/dt)$ and its use are described in more detail elsewhere (3,68). However, chemical reactions can also effect the k_d- and k_s-terms and thereby influence or control colloidal stability (82). Pertinent examples are dissolution, reprecipitation, hydrolysis, precipitation, and chemical complexing. The last reaction may involve either simple species, eg,

$$Al^{3+} + SO_4^{2-} \leftrightarrows AlSO_4^+$$

or complicated solutes such as $Al_8(OH)_{20}^{4+}$, chelated metals (68), synthetic and natural polymers (9,69,72), and a variety (66,83) of surfactants (see also Dispersants). Table 4 lists many of the possible bulk solution chemical reactions that influence colloidal stability, along with specific sample reactions and their general interfacial analogues. The use of natural and synthetic polymers to stabilize aqueous and nonaqueous colloidal suspensions is technologically important. Research in this area has focused on polymer adsorption and steric stabilization (72). Of course, electrolyte flocculation and bridging flocculation by polymers have been addressed experimentally and theoretically. Comprehensive reviews on suspension phenomena related to polymers are available elsewhere (72).

Chemical Properties. Any classical, "wet" chemical analyses or instrumental techniques that are routinely used to analyze the bulk composition of solids and liquids are, in principle, also suitable for colloids. The available instrumental methods range, for example, from spectrographic analysis, which is limited to crude estimates for impurities, to Raman spectroscopy, which provides accuracy to within 1%, depending on the desired accuracy of stoichiometric determinations. Surface-chemical analyses that can not be conducted using conventional methods designed for bulk materials are usually accomplished by optical, diffraction, and spectroscopic techniques, which are often applied under conditions of an ultrahigh vacuum. The spectroscopies measure

Table 4. Sample Solution and Surface Equilibria

Solution	Surface analogue
hydrolysis	
$CH_3C(=O)OCH_3 + H_2O \leftrightarrows CH_3C(=O)OH + CH_3OH$	$M_2O + H_2O \leftrightarrows 2\,MOH$
$PO_4^{3-} + H_2O \leftrightarrows HPO_4^{2-} + OH^-$	$MO^- + H_2O \leftrightarrows MOH + OH^-$
dissociation	
$Al(OH)_2^+ \leftrightarrows Al^{3+} + 2\,OH^-$	$MOH_2^+ \leftrightarrows MO^- + 2\,H^+$
$C_6H_5C(=O)OH \leftrightarrows C_6H_5C(=O)O^- + H^+$	$MOH \leftrightarrows MO^- + H^+$
dissolution	
$ZnC_2O_4(s) \leftrightarrows Zn^{2+} + C_2O_4^{2-}$	
$Al(OH)_3(s) + OH^- \leftrightarrows AlO_2^- + H_2O$	
complexation	
$Cu^{2+} + 4\,OH^- \leftrightarrows Cu(OH)_4^{2-}$	$MO^- + Na^+ \leftrightarrows MO^-\,Na^+$
$n\text{-}C_{12}H_{25}N(CH_3)_2 + HCl \leftrightarrows n\text{-}C_{12}H_{25}\overset{+}{N}H(CH_3)_2Cl^-$	$MOH + HCl \leftrightarrows MOH_2^+Cl^-$

the responses of solid surfaces to beams of electrons, ions, neutral matter, and photons. Each spectroscopy has unique attributes (5,84–96) making it suitable for certain colloids but unsuitable for others. Many of the techniques developed for microchemical analysis of surfaces or thin films are given in Table 5. Classical analytical techniques and instrumental methods can also be used to obtain surface-chemical analyses if the outermost layers of a solid colloid or the adsorbed layers can be quantitatively desorbed and examined. Visible, uv, and, especially, ir spectroscopy (97–102) can be used to examine the nature of adsorbed material. Finally, surface conductance techniques have been developed as qualitative probes into surface-chemical aspects of colloids (8,103).

Hazards of Collodial Systems

Two situations exist for which the chemical reactivity of colloidal materials may be detrimental or even physically harmful. The first one is a result of their high specific surface area. Colloidal chemical reactivity may differ considerably from that of the identical macroscopic material with less surface area. This is particularly important if the colloidal surface is easily and rapidly oxidized. Dust explosions and spontaneous combustion are potential dangers whenever certain materials exist as finely divided dry matter exposed to oxidizing environments. Many of the metals in Table 6 exhibit this property. The second hazard is that many colloidal substances, including fibrillar matter, are often inhaled to the extent that physiological problems may arise if they are retained by bodily tissues such as the lungs. This hazard may be particularly acute if the particulates are ca 1 μm; exceedingly small particulates may be exhaled and particulates that are significantly larger than 1 μm settle and are not often inhaled into the lungs. Short-term allergic conditions are common. Examples include those induced by pollen and household dust, asthma, and hay fever. Long-term effects may be fatal, as is the case with silicosis, asbestosis, and black lung disease. Specific potential hazards have been associated with a diverse spectrum of colloidal materials: chemicals,

Table 5. Diffraction and Scattering Methods for Surface Structural and Compositional Analyses[a,b]

Diffracted (scattered) probe	Incident probe electrons	Ions	Photons
electrons	LEED	INS	ESCA
	HEED		PES
	AES		UPS
	CEELS		XPS
ions	ESD	SIMS	
		NIRMS	
photons	APS	IILE	ellipsometry

[a] Refs. 5, 92.

[b] AES = auger electron spectroscopy, APS = appearance potential spectroscopy, CEELS = characteristic electron energy-loss spectroscopy, ESCA = electron spectroscopy for chemical analysis, ESD = electron-stimulated desorption, HEED = high energy electron diffraction, IILE = ion-induced light emission, INS = ion neutralization spectroscopy, LEED = low energy electron diffraction, NIRMS = noble-gas-ion reflection mass spectrometry, PES = photoelectron spectroscopy; SIMS = secondary ion mass spectroscopy, UPS = ultraviolet photoelectron spectroscopy, and XPS = x-ray photoelectron spectroscopy.

Table 6. Exposure Limits of Selected Colloidal Materials[a]

Substance	Limit, mg/m^3	Substance	Limit, mg/m^3
asbestos	70.6[b]	nickel	1
boron oxide	15	nicotine	0.5
cadmium dust	0.2[c]	oil mist	5
calcium oxide	7	osmium tetroxide	0.002
carbaryl	5	quartz dust	30
carbon black	3.5	phosphoric acid	1
chromium (metal, insoluble salt)	1	phosphorus	0.1
coal dust	24	rhodium dust	0.1
coal tar pitch volatiles	0.2	silver metal	0.01
cobalt dust	1	sulfuric acid	1
copper dust	1	talc	706[b]
cotton dust	1	tantalum	5
diatomaceous earth	80	tellurium	5
DDT	1	thallium	0.2
ferrovanadium dust	1	soluble compounds	0.1
fluoride dust	25[c]	tin compounds[d]	
hafnium	0.5	inorganic	2
iron oxide (fume)	10	organic	0.1
magnesium oxide (fume)	10	tungsten	0.1
mica	706[b]	insoluble compounds	5
mercury	0.1[c]	vanadium oxide	0.1–0.5
molybdenum, insoluble compounds	15	zinc chloride fume	1
		zinc oxide fume	5

[a] Ref. 103.
[b] 10^6 particles per m^3.
[c] 8-h TWA.
[d] Except SnH_4 and SnO_2.

coals, minerals, metals, pharmaceuticals, plastics, and wood pulp, to name a few. Many particulate, hazardous materials are listed in ref. 104. Of course, if the material is a colloidal solid or liquid, the exposure limits expressed in mg/m^3 may be misleading. Large fractions of readily hydrolyzable metals exist as adsorbed species on suspended (colloidal) solids in fresh and marine water systems (105) and can also be anticipated in industrial wastewater. Elements such as lead, zinc, and vanadium that are released into the atmosphere as vapors subsequently condense or are removed as solid particulates by rain (105). Liquid droplets may also constitute a hazard. For instance, smog often contains sulfuric acid-aerosols. Reference 68 should be consulted for further details of colloidal pollutants; Table 7 contains a short list of trace metals commonly found as suspended matter in the chemical form of hydrous oxides and other insoluble materials and their approximate concentrations (105).

Table 8 lists some of the industries that produce waste containing significant amounts of suspended matter. The cost to protect the general population from these solids is, of course, highly variable, owing to transient economic issues and governmental restrictions (106). See refs. 107 and 108 for examples of the latter constraints which set a rigorous, yet largely unenforced, schedule for technologically isolating and containing waterborne and airborne pollutants. Such economic and governmental

Table 7. Metals Commonly Found in Freshwater and Dust Systems[a]

Metal	Aqueous concentration, ppm[b]	Dust concentration, ppm	
		Laboratory dust	Furnace filters
Ag		5	
Al	310		
B	31		
Ba	90		
Ca		70	10,000
Co		19	
Cu	2	38	82
Fe	740	140	5,300
K			3
Mg		720	8,200
Mn	200	140	160
Ni	8	89	160
Pb		14,620	7,400
Sn		110	66
Sr	67		
Ti	7		
V		20	70
Zn			900

[a] Adapted from ref. 105.
[b] Assumed units.

forces have led to quite large water- and air-treatment industries (108–109). The increasing cost of research related to controlling the particulates indicated in Tables 7 and 8 and related hazardous materials may be as great as 60% of the increasing annual cost of research (108,110). The effects of the solid and liquid colloidal particulates that make up smog are widespread and well known. Because of the chemical and physical complexity of smog, health problems associated with exposure to it are often far worse than those associated with exposure to its individual constituents; thus, synergistically detrimental effects can be expected.

Applications of Colloids

General Uses. Although colloids may be undesirable components in industrial systems, particularly as waste or by-products and, in nature, in the forms of fog and mist, they are desirable in many technologically important processes such as mineral beneficiation and the preparation of ceramics, polymers, composite materials, paper, foods, textiles, photographic materials, cosmetics, and detergents. The reasons that colloids are deemed either desirable or undesirable lie in their unique physicochemical properties of diffusion, chemical reactivity with their environment's components, particulate interactions, adhesion/deposition, and electrical, thermal, and magnetic behavior. These properties are responsible for the wide use of colloids. Sample uses are as reinforcement aids in metals, ceramics, and plastics; as adhesion promoters in paints and thermoplastics; as nucleating agents in seeding clouds; as activated powder catalysts; as thickening agents in industrial gels and slurries; and as abrasives in toothpastes (1,3,6,9,63,111) (see also Acrylic-ester polymers; Amino resins and plastics). This list demonstrates the broad applications of (solid) colloids; the remainder of this section focuses on a few specific applications.

Table 8. Summary of Industrial Sources Producing Suspended, Colloidal Waste[a]

Industry	Origin	Principal method of treatment or disposal
food and drug		
canned goods	fruits, vegetables	filtration, lagooning
meat, poultry	bones, grease, fatty residue	filtration, settling, flotation
beet sugar	lime sludge	coagulation, lagooning
yeast	residue	anaerobic digestion, filtration
coffee	pulp	settling, filtration
rice	extractables (starch)	coagulation
soft drinks	equipment washing	municipal sewer
apparel		
textiles	fibers	chemical precipitation, filtration
leather goods	precipitated lime	sedimentation
laundry trades	fabric washing	flotation
chemicals		
detergents	saponified soaps	flotation, precipitation
explosives	metals, oils, soaps	flotation, precipitation
phosphate/phosphorus	clays, slimes	lagooning, coagulation
other materials		
pulp and paper	paper pulp washing	settling, lagooning
steel	coal coking, oils, scale	coagulation
iron-foundry	clay, coal	filtration
oil	drilling muds, sludge	burning
rubber	latex washing	aeration
glass	polishing, cleaning	precipitation
energy		
coal processing	cleaning	flotation

[a] Adapted from ref. 106.

Selected Systems and Applications. Many applications of colloidal materials have been mentioned earlier regarding specific phenomena. Selected examples deserve special mention because they illustrate how colloids affect many technologically important systems in a positive manner and demonstrate the broad range of applications that permeates current synthetic materials.

Colloidal Solids. Used as reinforcement agents in metals, ceramics, and polymers (111–112), particles of colloidal solids may be spherical, angular, fibrillar, or flake-shaped. Examples include aluminum oxide and thoria to reinforce aluminum and nickel, respectively, by providing obstacles to the movement of dislocations or grain boundaries (111). Asbestos, crystalline silicas, and organic solids are added to concrete to improve its strength by providing an interlocking particulate structure within the concrete matrix (113); asbestos (114), various oxides (115), and carbon black (6,115) are added to reinforce polymers by inducing a stiffened or high yield matrix. For instance, maximum strength of natural rubber may be achieved with ca 10 vol % ZnO or ca 22 vol % carbon black and that of TD (thoria-dispersed) nickel by ca 3 vol % ThO_2 (111). The magnitude of the strengthening often depends on particle shape. Fibrillar fillers are commonly used as discontinuous fibers in metals and plastics, eg, for 65 vol % "E" glass in epoxy resin (111). Glass and aluminum oxide are common fillers which are occasionally pretreated with a polymeric or metallic coating; silica and various clays are also used. Unidirectionally oriented continuous fibers are less commonly employed, but they are successfully incorporated into laminated structures so that the fibers in successive layers are orthogonal or otherwise specifically oriented, yielding an alternating fiber structure (111). Finally, the mechanism by which ceramics and metals are reinforced often involves precipitation of colloidal material during thermal treatment of the matrix composition (116–119). This is done in the TiO_2-precipitation in borosilicate glasses designed for enamels (116,119), precipitation in AgCl–NaCl alloy systems (117), $Ni_3(Al,Ti)$ precipitation in Inconel x-750 matrix (118), and Fe_3C-precipitation in tempered martensite (118). Other examples of fillers are listed in Table 9. Table 10 summarizes the mechanisms by which these reinforcement fillers operate (see Fillers).

Fillers can be added not only to improve mechanical properties such as impact strength, fracture toughness, and tensile strength of structural ceramics as indicated for concrete, but also to enhance optical properties, as is done for colored glasses containing colloidal gold or crystalline, chromium-based oxides. Other applications of colloidal solids include the preparation of rigid, elastic and thixotropic gels (111) and surface coatings (6,9). Commercial uses of silica gel and sol-gel processing often

Table 9. Fillers for Reinforcement[a]

Fillers	Substance
nonmetallic	
asbestos	mica
boron on tungsten	nylon 6,6
calcium carbonate	silica
fused silica	tungsten carbide
glass	titanium carbide
glass (SiO_2, Al_2O_3, B_2O_3); "E" glass	titanium dioxide
graphite	zinc oxide
kaolin and other clays	zirconium dioxide
metallic	
copper	titanium
gold	tungsten
nickel	zinc; cermets
whiskers	
alkali halides	iron
Al_2O_3	α-SiC
B_4C	sulfides
graphite	graphite

[a] After refs. 6, 111, 114–120.

Table 10. Strengthening Mechanisms[a]

Reinforcing phase	Mechanism	Toughened materials
metallic	precipitation, dispersion	metals, ceramics
ceramic[b]	dispersion, reinforcement[c,d]	metals, polymers[e]

[a] Ref. 6.
[b] Inorganic, nonmetallic.
[c] For example, by cross-linking.
[d] Higher volume fraction than in dispersion hardening.
[e] For instance, those given in refs. 120–121.

focus on examples of rigid gels having 20–30 vol % SiO_2 (see Silica). Here, the ultimate interparticulate forces are chemical and irreversible, and the solid colloid improves a gel's mechanical strength. Elastic gels are commonly associated with cellophane, rubber, cellulosic fibers, leather, and certain soaps; weak van der Waals dispersion forces operate between particles in these gels, rendering them reversible or elastic. Finally, many thixotropic gels and surface coatings contain colloidal solids such as clays, alumina, ferric oxide, titania, silica, and zinc oxide. Consumer and industrial pastes are in this category, eg, putty, dough, drilling mud, lubricating grease, toothpaste, and paint.

Colloidal Liquids. These fluids are used in the form of emulsions by many industries. Permanent or transient antifoams consisting of an organic material (eg, polyglycol, oils, fatty materials) or silicone oil dispersed in water is one application (5,122–127) that is important to a variety of products and processes: foods, cosmetics, pharmaceuticals, pulp and paper, water treatment, and minerals beneficiation. Other outlets for emulsions are paints, lacquers, varnishes, and electrical- and thermal-insulating materials.

Colloidal Gases. Fluid foams are commonplace in foods, shaving cream, fire-fighting foam, and detergents (5,128). Solid foams such as polyurethane foam and natural pumice also contain dispersed gas bubbles. The production, dispersion, and maintenance of colloidal gas bubbles are essential processes in this technology and related ones.

Status of Colloid Knowledge and Use

Colloids are scientifically and technologically important forms of matter that exhibit unique properties. The fundamentals of preparing and controlling many desirable colloidal systems are known (3–5,7–9) and many products and industries are built upon these principles. Conversely, many of the scientific and technological obstacles have been overcome so that undesirable colloidal materials are routinely prevented, eliminated, or otherwise controlled. However, many problems concerning both favorable and unfavorable colloids still exist because these types of systems are inherently complex and much of the stigma associated with the empirical aspects of colloids remains (1). Viscosity and agglomeration, for example, are complex topics when considered separately; taken together, for instance as in the non-Newtonian behavior of concentrated suspensions, they are subject only to qualitative predictions and explanations. Even when quantitative treatments are possible (75,77), they are usually somewhat empirical, and their application has been restricted to a small number of physically different systems. This situation exists because industrial and natural systems have not often been adequately characterized to apply the developing theories of suspension rheology rigorously. Important problems in colloid science remain to be addressed if the potential of colloids is to be fully exploited—among them, extension of understanding to more concentrated suspensions, testing of predictions using model powders, and examination of relaxation phenomena in ordered colloids (see below). In short, much is known about colloids and their formation and behavior, but considerably more remains unknown. Thus, the full potential to control colloids as materials is not presently realized.

New Developments. Recent work on the structure of regularly arranged colloidal suspensions indicates that ordering of particles in a fluid occurs under restricted solution conditions and solids concentrations exceeding ca 50 vol % (129–133). Experimental advances have spurred theoretical modeling of the liquidlike-to-solidlike phase transition (134–136). Regular patterns corresponding to hexagonal or cubic packing have been identified, along with associated lattice defects, dislocations, grain boundaries, and segregation phenomena (131). The interesting feature of these systems is that liquid separates particles, and there are virtually no particle–particle contacts. Therefore, these stable but ordered suspensions can be considered as precursory structures for ordered, prefired, ceramic components. Potential applications for such systems include various ceramic casting techniques (slip, tape, freeze, pressure, centrifugal, ultrasonic) and isostatic and hot pressing (137). The development of novel techniques based on periodic structures is imminent (138).

BIBLIOGRAPHY

1. E. Matijević, *Chem. Technol.* **3,** 656 (1973).
2. R. van Hardevald and F. Hartog, *Surf. Sci.* **15,** 189 (1969).
3. P. C. Heimenz, *Principles of Colloid and Surface Chemistry*, Marcel Dekker, Inc., New York, 1977; R. D. Vold and M. J. Vold, *Colloid and Interface Chemistry*, Addison-Wesley, Reading, Mass., 1983; H. Sonntag and K. Strenge, *Coagulation and Stability of Disperse Systems*, Halsted Press, New York, 1972.
4. D. J. Shaw, *Introduction to Colloid and Surface Chemistry*, 3rd ed., Butterworth, London, 1980.
5. A. W. Adamson, *Physical Chemistry of Surfaces*, 4th ed., John Wiley & Sons, Inc., New York, 1982.
6. C. R. Veale, *Fine Powders*, John Wiley & Sons, Inc., New York, 1972; J. K. Beddow, *Particulate Science and Technology*, Chemical Publishing, New York, 1980.
7. B. V. Derjaguin and L. D. Landau, *Acta Physicochem. URSS* **14,** 633 (1941).
8. H. R. Kruyt, *Colloid Science*, Vol. I, *Irreversible Systems*, Elsevier Publishing Co., Amsterdam, 1952; E. J. Verwey and J. Th. G. Overbeek, *Theory of the Stability of Lyophobic Colloids*, Elsevier Publishing Co., Amsterdam, 1948.
9. G. D. Parfitt, ed., *Dispersion of Powders in Liquids*, Applied Science Publishers, London, 1981.
10. B. J. Ingebrethsen, Ph.D. thesis, Clarkson College of Technology, Potsdam, N.Y., 1982.
11. H. Füredi-Milhofer and A. G. Walton in ref. 9, Chapt. 5, p. 203.
12. G. C. Lowrison, *Crushing and Grinding*, Chemical Rubber Co., Boca Raton, Fla., 1974.
13. G. E. Agar and P. Somasundaran, *paper presented at 10th International Mineral Processing Congress*, London, 1973.
14. A. Z. Frangiskos, Ph.D. thesis, University of Leeds, Leeds, UK, Aug. 1956.
15. A. G. Evans, *J. Mater. Sci.* **7,** 1137 (1972).
16. J. A. Holmes, *Trans. Inst. Chem. Eng.* **35,** 125 (1957).
17. N. Arbiter and U. N. Bhrany, *Trans. Am. Inst. Min. Metall. Pet. Eng.* **217,** 245 (1960).
18. D. W. Fuerstenau and P. Somasundaran in *Proceedings of the 6th International Mineral Processing Congress, Cannes 1963*, Pergamon Press, Oxford, UK, 1965, p. 25.
19. O. G. Raabe in B. Y. H. Liu, ed., *Fine Particles*, Academic Press, New York, 1975, p. 57.
20. M. Kerker, *Adv. Colloid Interface Sci.* **5,** 105 (1975).
21. N. A. Fuchs and A. G. Sutugin in C. N. Davies, ed., *Aerosol Science*, Academic Press, New York, 1966, Chapt. 1, p. 1.
22. H. Willeke, *Generation of Aerosols and Facilities for Exposure*, Ann Arbor Press, New York, 1980.
23. M. I. Tillery, G. O. Wood, and H. J. Ettinger, *Environ. Health Perspectives* **16,** 25 (1976).
24. T. T. Mercer, M. I. Tillery, and H. Y. Chow, *Am. Ind. Hyg. Assoc. J.* **29,** 66 (1968).
25. M. B. Denson and D. B. Swartz, *Rev. Sci. Instrum.* **45,** 81 (1974).
26. R. N. Bergland and B. Y. H. Liu, *Environ. Sci. Technol.* **7,** 147 (1973).
27. W. H. Walton and W. C. Prewett, *Proc. Phys. Soc. London* **62,** 341 (1949).
28. B. Vonnegut and R. Neubauer, *J. Colloid Sci.* **7,** 616 (1952).

29. A. L. Heubner, *Science* **168,** 118 (1970).
30. E. P. Knutson and K. T. Whitby, *Aerosol Sci.* **6,** 443 (1975); J. K. Agarwal and G. J. Sem, *TSI Quarterly* **6,** 3 (1978).
31. R. Zsigmondy, *Z. Phys. Chem. (Leipzig)* **56,** 65 (1906).
32. E. Wiegel, *Kolloidchem. Beih.* **25,** 176 (1927).
33. H. R. Kruyt and A. E. van Arkel, *Recl. Trav. Chim. Pay-Bas* **39,** 656 (1920).
34. V. K. LaMer and M. D. Barnes, *J. Colloid Sci.* **1,** 71 (1946).
35. V. K. LaMer and R. Dinegar, *J. Am. Chem. Soc.* **72,** 4847 (1950).
36. W. Stöber, A. Fink, and E. Bohn, *J. Colloid Interface Sci.* **26,** 62 (1968).
37. D. Sinclair and V. K. LaMer, *Chem. Rev.* **44,** 245 (1949).
38. E. Matijević, *Acc. Chem. Res.* **14,** 22 (1981).
39. A. Bleier and R. M. Cannon, *Am. Ceram. Soc. Bull.* **61,** 336 (1982).
40. J. H. L. Watson, W. Heller, and W. Wojtowicz, *J. Chem. Phys.* **16,** 997 (1948).
41. S. Hamada and E. Matijević, *J. Colloid Interface Sci.* **84,** 274 (1982).
42. J. W. Vanderhoff, H. J. van den Hul, R. J. M. Tausk, and J. T. G. Overbeek in G. Goldfinger, ed., *Clean Surfaces*, Marcel Dekker, Inc., New York, 1970, Chapt. 2, p. 15; J. W. Vanderhoff, *Pure Appl. Chem.* **52,** 1263 (1980).
43. A. Homola and R. O. James, *J. Colloid Interface Sci.* **59,** 123 (1977); Y. Chung-li, J. W. Goodwin, and R. H. Ottewill, *Prog. Colloid Polymer Sci.* **60,** 163 (1976).
44. V. I. Eliseeva, S. S. Ivanchev, S. I. Kuchanov, and A. V. Lebedev, *Emulsion Polymerization and Its Applications*, Consultants Bureau, New York, 1976.
45. B. H. Kaye, *Direct Characterization of Fineparticles*, Wiley-Interscience, New York, 1981; M. Kerker, *The Scattering of Light and Other Electromagnetic Radiation*, Academic Press, New York, 1969; J. D. Stockham and E. G. Fochtman, eds., *Particle Size Analysis*, Ann Arbor Science, Ann Arbor, Mich., 1979; D. W. Schuerman, ed., *Light Scattering by Irregularly Shaped Particles*, Plenum Press, New York, 1980.
46. S. J. Gregg and K. S. W. Sing, *Adsorption, Surface Area and Porosity*, Academic Press, London, 1967; see also 2nd ed., 1982.
47. S. Brunauer, P. H. Emmett, and E. Teller, *J. Am. Chem. Soc.* **60,** 309 (1938); S. Brunauer, *The Adsorption of Gases and Vapors*, Vol. 1, Princeton University Press, Princeton, N.J., 1945.
48. I. Langmuir, *J. Am. Chem. Soc.* **40,** 1361 (1918).
49. B. D. Cullity, *Elements of X-Ray Diffraction*, 2nd ed., Addison-Wesley Publishing Co., Inc., Reading, Mass., 1978.
50. G. E. Bacon, *Neutron Diffraction*, 3rd ed., Oxford University Press, London, 1975.
51. J. M. Ziman, *Principles of the Theory of Solids*, 2nd ed., Cambridge University Press, Cambridge, UK, 1972, Chapt. 10, p. 329; M. J. Schmank and A. D. Krawitz, *Metall. Trans. A* **13A,** 1069 (1982); A. D. Krawitz and co-workers in *Proceedings of the International Conference on the Science of Hard Materials*, Jackson, Wyo., Aug. 1981.
52. J. A. Appelbaum, G. A. Baraff, and D. R. Hamann, *Phys. Rev. Lett.* **35,** 729 (1975).
53. G. A. Somorjai, *Chemistry in Two Dimensions: Surfaces*, Cornell University Press, Ithaca, N.Y., 1981; G. A. Somorjai and M. A. Van Hove, *Adsorbed Monolayers on Solid Surfaces*, Springer-Verlag, New York, 1979, p. 121.
54. J. A. Appelbaum and D. R. Hamann, *Surf. Sci.* **74,** 21 (1978).
55. J. J. Lander and J. Morrison, *J. Appl. Phys.* **34,** 1411 (1963).
56. J. J. Lander and J. Morrison, *Surf. Sci.* **4,** 241 (1966).
57. A. Bleier, *J. Phys. Chem.* **87,** 3493 (1983); *J. Am. Ceram. Soc.* **66,** C-79 (1983).
58. L. A. Girifalco and R. J. Good, *J. Phys. Chem.* **61,** 904 (1957).
59. R. J. Good and L. A. Girifalco, *J. Phys. Chem.* **64,** 561 (1960).
60. R. J. Good in R. J. Good and R. R. Stromberg, eds., *Surface and Colloid Science*, Vol. 11, Plenum Press, New York, 1979, Chapt. 1, p. 1.
61. F. M. Fowkes, *Ind. Eng. Chem.* **56,** 40 (1964).
62. S. Mizuta, W. R. Cannon, A. Bleier, and J. S. Haggerty, *Am. Ceram. Soc. Bull.* **61,** 872 (1982).
63. C. Orr, *Particulate Technology*, Macmillan, New York, 1966.
64. G. C. Benson, H. P. Scheiber, and F. van Zeggeren, *Can. J. Chem.* **34,** 1553 (1956).
65. A. J. Tyler, J. A. G. Taylor, B. A. Pethica, and J. A. Hockey, *Trans. Faraday Soc.* **67,** 483 (1971).
66. M. J. Rosen, *Surfactants and Interfacial Phenomena*, John Wiley & Sons, Inc., New York, 1978.
67. W. C. Preston, *J. Phys. Colloid Chem.* **52,** 84 (1948).

68. W. Stumm and J. J. Morgan, *Aquatic Chemistry*, 2nd ed., John Wiley & Sons, Inc., New York, 1981.
69. M. J. Vold, *J. Colloid Sci.* **16,** 1 (1961).
70. H. G. Hamaker, *Physica (Utrecht)* **4,** 1058 (1937).
71. I. E. Dzyaloshinskii, E. M. Lifshitz, and L. P. Pitaevskii, *Adv. Phys.* **10,** 165 (1961).
72. T. Sato and R. Ruch, *Stabilization of Colloidal Dispersions by Polymer Adsorption*, Marcel Dekker, Inc., New York, 1980; T. F. Tadros, ed., *The Effect of Polymers on Dispersion Properties*, Academic Press, London, 1982; J. Lyklema, *Adv. Colloid Interface Sci.* **2,** 65 (1968); Y. S. Lipatov and L. M. Sergeeva, *Adsorption of Polymers*, Halsted Press, New York, 1974; B. Vincent, *Adv. Colloid Interface Sci.* **4,** 193 (1974); C. A. Finch, ed., *Chemistry and Technology of Water-Soluble Polymers*, Plenum Press, New York, 1983.
73. R. I. Feigin and D. H. Napper, *J. Colloid Interface Sci.* **74,** 567 (1980); **75,** 525 (1980).
74. A. Okagawa, G. J. Ennis, and S. G. Mason, *Can. J. Chem.* **56,** 2815, 12824 (1978); M. Zuzousky, Z. Priel, and S. G. Mason, *J. Colloid Interface Sci.* **75,** 230 (1980).
75. T. G. M. Van den Ven and R. J. Hunter, *J. Colloid Interface Sci.* **69,** 135 (1979); R. J. Hunter and J. Frayne, *J. Colloid Interface Sci.* **71,** 30 (1979); R. J. Hunter and J. Frayne, *J. Colloid Interface Sci.* **76,** 107 (1980).
76. I. M. Krieger, *Adv. Colloid Interface Sci.* **3,** 111 (1972).
77. B. A. Firth, *J. Colloid Interface Sci.* **57,** 257 (1976).
78. K. Higashitani, S. Miyafusa, T. Matsuda, and Y. Matsuno, *J. Colloid Interface Sci.* **77,** 21 (1980).
79. C. D. Han and R. G. King, *J. Rheol.* **24,** 213 (1980).
80. C. C. Mill, ed., *Rheology of Disperse Systems*, Pergamon Press, New York, 1959.
81. M. R. Rosen, *Polym. Plast. Technol. Eng.* **12,** 1 (1979).
82. E. Matijević, *J. Colloid Interface Sci.* **43,** 217 (1973).
83. *McCutcheon's Detergents and Emulsifiers*, MC Publishing Co., Glen Rock, N.J., 1980.
84. P. F. Kane and G. B. Larrabee, *Anal. Chem.* **49,** 221R (1977).
85. P. F. Kane and G. B. Larrabee, *Anal. Chem.* **51,** 308R (1979).
86. G. B. Larrabee and T. J. Shaffner, *Anal. Chem.* **53,** 163R (1981).
87. S. N. K. Chaudhari and K. L. Cheng, *Appl. Spectrosc. Rev.* **16,** 187 (1980).
88. E. N. Sickafus, *Ind. R&D* **22**(6), 126 (1980).
89. M. H. Koppelman and J. G. Dillard in M. M. Mortland and V. C. Farmer, eds., *International Clay Conference*, Elsevier Publishing Co., Amsterdam, 1979, p. 153.
90. R. J. Blattner and C. A. Evans, Jr., in W. Bardsley, D. T. J. Hurle, and J. B. Mullin, eds., *Crystal Growth: A Tutorial Approach*, North-Holland Publishing Co., Amsterdam, 1979, p. 269.
91. M. Beer, R. W. Carpenter, L. Eyring, C. E. Lyman, and J. M. Thomas, *Chem. Eng. News* **59,** 40 (1981).
92. H. H. Brongersma, F. Meijer, and H. W. Werner, *Philips Tech. Rev.* **54**(11/12), 357 (1974).
93. W. A. Beers, *Res. Dev.*, 18 (Nov. 1975).
94. C. A. Evans, Jr., *Anal. Chem.* **47,** 818A (1975).
95. J. W. Coburn, *Thin Solid Films* **64,** 371 (1982).
96. T. Cosgrove in D. H. Everett, ed., *Colloid Science*, The Chemical Society, Burlington House, London, 1979, Chapt. 7, p. 293.
97. L. H. Little, *Infrared Spectra of Adsorbed Species*, Academic Press, London, 1966; M. L. Hair, *Infrared Spectroscopy in Surface Chemistry*, Marcel Dekker, Inc., New York, 1967.
98. P. P. Yaney and R. J. Becker, *Appl. Surf. Sci.* **4,** 356 (1980).
99. M. R. Basila, *Appl. Spectrosc. Rev.* **1,** 289 (1968).
100. R. P. Eischens, *Acc. Chem. Res.* **5,** 74 (1972).
101. T. A. Egerton and A. H. Hardin, *Catal. Rev.* **11,** 71 (1975).
102. C. H. Rochester, *Powder Technol.* **13,** 157 (1976).
103. R. W. O'Brien, *J. Colloid Interface Sci.* **81,** 234 (1981).
104. R. C. Weast and M. J. Astle, eds., *Handbook of Chemistry and Physics*, 62nd ed., Chemical Rubber Co. Press, Boca Raton, Fla., 1981, p. D-101.
105. A. E. Martell, *Pure Appl. Chem.* **44,** 81 (1975); F. T. Mackenzie and R. Wollast in E. D. Goldberg, ed., *The Sea*, Vol. 6, Interscience Publishers, a division of John Wiley & Sons, Inc., New York, 1977, p. 739.
106. W. F. Echelberger, *Water Pollution Control Technology*, Course Syllabus and Study Materials, Center for Professional Advancement, East Brunswick, N.J., July 1977.

107. Public Law 92-500, Amendment of the Federal Water Pollution Control Act, 92nd Congress, 5.2700, U.S. Government, Oct. 18, 1972.
108. J. Wei, T. W. F. Russell, M. W. Swartzlander, *The Structure of the Chemical Processing Industries*, McGraw-Hill Book Co., Inc., New York, 1979.
109. M. P. Freeman and J. A. Fitzpatrick, eds., *Physical Separations*, Engineering Foundation, New York, 1981; A. J. Rubin, ed., *Chemistry of Wastewater Technology*, Ann Arbor Science, Ann Arbor, Mich., 1978.
110. H. W. Zussman, *Adv. Chem. Ser.* **83,** 116 (1968); *Chem. Eng. News* **60**(30), 41 (1982).
111. Z. D. Jastrzebski, *The Nature and Properties of Engineering Materials*, 2nd ed., John Wiley & Sons, Inc., New York, 1977; A. G. Guy, *Essentials of Materials Science*, McGraw-Hill Book Co., Inc., New York, 1976.
112. S. J. Lefond, ed., *Industrial Minerals and Rocks*, 4th ed., American Institute of Mining, Metallurgical, and Petroleum Engineers, Inc., New York, 1975.
113. J. A. Ames in ref. 112, p. 129.
114. N. Severinghaus in ref. 112, p. 235.
115. L. Mitchell in ref. 112, p. 33.
116. W. D. Kingery, H. K. Bowen, and D. R. Uhlmann, *Introduction to Ceramics*, 2nd ed., Interscience Publishers, a division of John Wiley & Sons, Inc., New York, 1976.
117. R. J. Stokes and C. H. Li, *Acta Metall.* **10,** 535 (1962).
118. K. M. Ralls, T. H. Courtney, and J. Wulff, *Introduction to Materials Science and Engineering*, John Wiley & Sons, Inc., New York, 1976.
119. R. J. Charles in G. Piel and co-workers, eds., *Materials*, W. H. Freeman and Co., San Francisco, Calif., 1967, p. 69; A. Kelly in G. Piels and co-workers, eds., *Materials*, W. H. Freeman and Co., San Francisco, Calif., 1967, p. 97.
120. L. E. Murr, *Interfacial Phenomena in Metals and Alloys*, Addison-Wesley Publishing Co., London, 1975; S. J. Burden, *Ceram. Eng. Sci. Proc.* **3,** 1 (1982); D. W. Richardson, *Modern Ceramic Engineering*, Marcel Dekker, Inc., New York, 1982.
121. Ref. 104, pp. C-740 and C-747.
122. S. Ross, *Chem. Eng. Prog.* **63,** 41 (1967).
123. S. Ross and J. N. Butler, *J. Phys. Chem.* **60,** 1255 (1956).
124. S. Ross and R. M. Haak, *J. Phys. Chem.* **62,** 1260 (1958).
125. J. G. Hawke and A. E. Alexander, *J. Colloid Sci.* **11,** 419 (1956).
126. R. E. Prattle, *J. Soc. Chem. Ind. London* **69,** 363, 368 (1950).
127. P. Becher, *Emulsions*, Reinhold Publishing Co., New York, 1965.
128. J. J. Bikerman, *Foams, Theory and Industrial Applications*, Reinhold Publishing Co., New York, 1952; F. Sebba, *J. Colloid Interface Sci.* **35,** 643 (1971).
129. P. A. Hiltner and I. M. Krieger, *J. Phys. Chem.* **73,** 2386 (1969).
130. A. Kose and S. Hachisu, *J. Colloid Interface Sci.* **46,** 460 (1974).
131. S. Okamuto and S. Hachisu, *J. Colloid Interface Sci.* **61,** 172 (1977).
132. K. Takano and S. Hachisu, *J. Colloid Interface Sci.* **66,** 124 (1978).
133. I. F. Efremov in E. Matijević, ed., *Colloid and Surface Science*, Vol. 8, Plenum Press, New York, 1976, Chapt. 2, p. 85.
134. W. van Meegan and I. Snook, *Faraday Disc. Chem. Soc.* **65,** 92 (1978).
135. E. Dickinson, *J. Chem. Soc. Faraday Trans. 2* **75,** 466 (1979).
136. E. Dickinson in D. H. Everett, ed., *Colloid Science*, Vol. 4, Burlington House, London, 1983.
137. F. Y. Wang, *Ceramic Fabrication Processes*, Academic Press, New York, 1976.
138. A. Bleier in R. F. Davis, H. Palmour III, and R. L. Porter, eds., *Emergent Process Methods for High Technology Ceramics*, Materials Science Research Series, Plenum Press, New York, in press.

ALAN BLEIER
Massachusetts Institute of Technology

COMPOSITES, HIGH PERFORMANCE

High performance composites are formed by combining two or more homogeneous materials in order to achieve a balance of material properties that is superior to the properties of a single material. Increased strength, stiffness, fatigue life, fracture toughness, environmental resistance, and reduced weight and manufacturing cost are some of the common reasons for developing high performance composites. The most common form of high performance composite material is the fiber-reinforced plastic which is the focus of this article (see Laminated and reinforced plastics). High strength, high stiffness, low density fibers are embedded in a plastic matrix to form the fiber-reinforced composite. Most of the strength and stiffness are provided by the reinforcing fibers; the matrix maintains fiber alignment and transfers load around broken fibers. Other forms of high performance composites include fiber-reinforced metals, fiber-reinforced carbon, fiber-reinforced ceramic, whisker-reinforced metals, and particulate-reinforced metals (see Composite materials; Fillers; High temperature composites; Laminated and reinforced metals).

A detailed chronology on the development of fiber-reinforced plastic composites shows that during the last 20 years there has been an explosion of technology involving these materials (1). This was precipitated by the development of high strength glass and high modulus boron fibers in 1960 and the strong desire of the aerospace industry to improve the performance and reduce the weight of aircraft and space vehicles. In 1964 carbon fibers became available in research quantities and ultimately became the most widely used reinforcement in aerospace structural applications. In 1971 aramid (Kevlar) fibers (qv) became available commercially and are now being used extensively in automotive tires and numerous aerospace structures. Specific strength (strength-to-density ratio) and specific stiffness (stiffness-to-density ratio) for reinforcing fibers have continually been increased over the past 20 years. Specific strength and specific stiffness values for glass, Kevlar, boron and carbon fibers are compared with metals in Figure 1 and are as high as 14 times the specific strength and 10 times the specific stiffness of aluminum. Further increases in specific strength and specific stiffness of 20–80% of current values are predicted for reinforcing fibers over the next 20 yr.

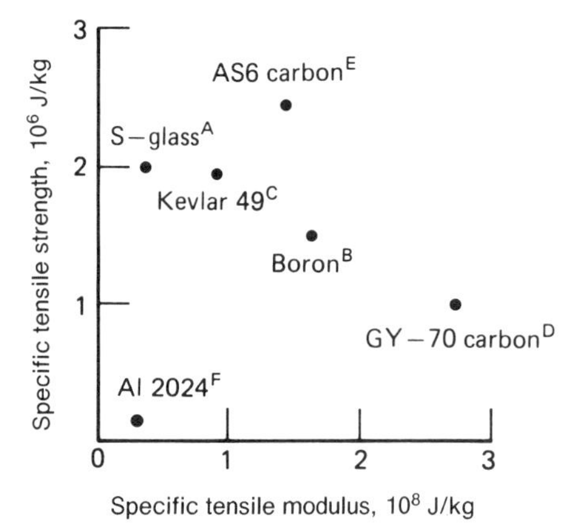

Figure 1. Comparison of specific strength and specific modulus for reinforcing fibers and aluminum (A, ref. 1, B, ref. 2, C, ref. 3, D, ref. 4, E, ref. 5, F, ref. 6).

This article discusses concepts for reinforcing plastics with high strength and high stiffness fibers, properties of the more commonly used fibers and plastic (matrix) materials, and properties of fiber-reinforced plastics. Processing, structural design and applications of fiber-reinforced plastics are also reviewed.

Reinforcing Concepts

Most fiber-reinforced plastics are laminated materials. The fibers in each layer are usually arranged in one of the four configurations shown in Figure 2. Continuous parallel fibers in a unidirectional tape can be used to achieve high values of specific strength and specific stiffness in the direction of the fibers, but exhibit low strength and stiffness in the direction transverse to the fibers. Unidirectional tape is the most widely used reinforcing concept in high performance composites. Many types of reinforcing fibers have been woven into fabrics and used to reinforce plastics. Fabrics are easier to handle and do not exhibit the low transverse strength exhibited by unidirectional tape. A portion of the fibers in a fabric will be crimped and the maximum achievable specific strength and specific stiffness will be lower than the corresponding values for a unidirectional tape. Discrete-length fibers may be aligned or randomly oriented as shown in Figure 2 and used to reinforce plastics. These concepts are usually employed in applications that are controlled by manufacturing considerations and that do not require maximum strength or stiffness.

Reinforcing Fibers

Mechanical and physical properties and costs for glass, boron, carbon, and aramid fibers are listed in Table 1. Specific gravity values for the fibers listed there are ca 50–90% of the value for aluminum; costs range from 30% to 400 times the costs for aluminum sheet.

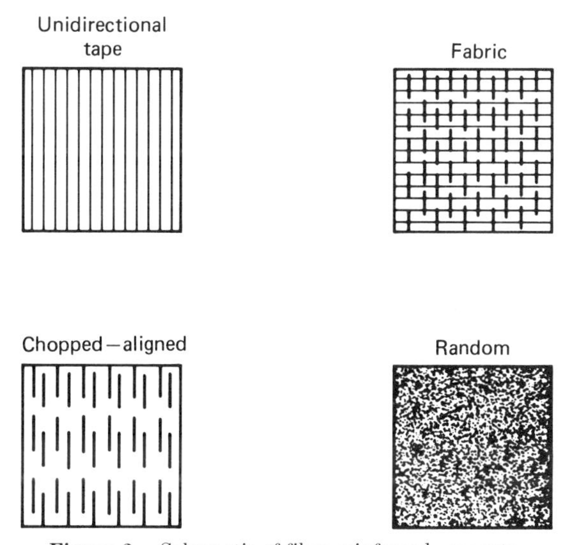

Figure 2. Schematic of fiber-reinforced concepts.

Table 1. Properties and Prices of Reinforcing Fibers

Fiber	Tensile strength, MPa[a]	Tensile modulus, GPa[a]	Specific gravity	Fiber diameter, μm	Coefficient of thermal expansion, m/m °C	Electrical resistivity, ohm·cm	Approximate price, $/kg
glass							
E-glass	3450[b]	72[b]	2.54[b]	7.0[c]	5.0×10^{-6} [c]	10^{15} [c]	2
S-glass	4820[b]	85[b]	2.49[b]	10.0	5.6×10^{-6} [c]	10^{16} [c]	8
boron	3551[d]	386[d]	2.45[d]	140.0[e]	5.4×10^{-6} [d]	10^{14} [d]	200
carbon							
Thornel 300 (Union Carbide)	3448[f]	231[f]	1.75[g]	7.0	$-.5 \times 10^{-6}$ [f]	1.8×10^{-4} [f]	50
AS6	4137[h]	248[h]	1.74	5.5			75
GY-70	1860[i]	517[i]	1.96[i]	8.4[i]		6.5×10^{-4} [i]	1430
P-100S	2241[f]	690[f]	2.15[g]	11.0	-1.6×10^{-6} [f]	2.5×10^{-4} [f]	2750
aramid							
Kevlar 49 (DuPont)	2758[j]	131[k]	1.44[k]	12.0[k]	-2.0×10^{-6} [k]	10^{16} [l]	27

[a] To convert GPa to psi, multiply by 145,000 (MPa × 145).
[b] Ref. 1.
[c] Ref. 7.
[d] Ref. 2.
[e] Ref. 8.
[f] Ref. 9.
[g] Ref. 10.
[h] Ref. 5.
[i] Ref. 4.
[j] Ref. 3.
[k] Ref. 11.
[l] Estd.

S-Glass Fibers. These fibers exhibit higher tensile strength than other fibers that are commercially available. A detailed discussion on the grades of glass fibers available, their properties, the chemical composition of each grade, and the manufacturing process has been prepared (7) (see also Fiber optics). Silica sand, limestone, boric acid, clay, coal, and fluorspar are melted and drawn through a bushing orifice to form glass fibers of 4–13 μm dia. The main component in glass fibers is silicon oxide. Glass fibers offer excellent heat and fire resistance, are not degraded by most chemicals, do not absorb moisture, have a low coefficient of thermal expansion and a high coefficient of thermal conductivity, are electrically nonconductive, and are readily woven into fabrics. The use of glass fibers in structural applications has been limited by their low elastic modulus and high specific gravity compared to other types of fibers.

Boron Fibers. The modulus of boron fibers is nearly twice that of steel; this fiber type has approximately eight times the strength of aluminum and is 10% less dense than aluminum. When boron fibers were introduced in 1960, no other material offered such attractive properties for aircraft structural application (see Boron compounds, halides). Consequently, significant research and development were directed toward exploiting the properties of the fibers in high performance composites and reducing their cost. A gaseous mixture of hydrogen and boron trichloride is vapor deposited on 12.5 μm dia tungsten wire to form 100 μm-, 140 μm-, or 200 μm-dia boron fiber (12).

Boron fibers can also be purchased with a 7 μm B_4C diffusion barrier on the exterior surface. The outlook for producing boron fibers for less than $200/kg does not appear promising and the development of high stiffness, low density carbon fibers has significantly reduced interest in boron fibers (see also Laminated and reinforced metals; Refractory fibers). Reference 8 offers an excellent review of boron fibers and their composites.

Carbon Fibers. Properties for four types of carbon fibers are listed in Table 1 (see also Carbon, carbon and artificial graphite; Novoloid fibers). Fibers currently marketed are made from either polyacrylonitrile (PAN) fibers or pitch-based fibers. PAN fibers are oxidized and stretched simultaneously; pitch fibers do not have to be held under tension during the oxidation process. The chemical carbon yield for mesophase pitch is ca 75%; the yield for PAN is ca 50% (9). The difference in yield has been a major factor in developing pitch-based fibers and continuing efforts to increase the compressive strength of the pitch-based fibers. Table 1 indicates that a wide range of properties has been achieved with carbon fibers. Strength values up to 4137 MPa (600,000 psi) are currently available and a fiber with 5500 MPa (800,000 psi) strength should be in production in 1984 (6). Modulus values up to 690 GPa (100×10^6 psi) are currently available and research on 830 GPa (120×10^6 psi) modulus fiber is underway (9). Carbon fiber's small negative coefficient of thermal expansion can be used to design zero thermal expansion structures. Most carbon fibers can be woven into fabrics. Major producers and/or suppliers of carbon fibers are located in the U.S., UK, and Japan. Costs range from approximately $17/kg for 3–8 cm length fiber with 170 GPa (25×10^6 psi) modulus to $2750/kg for continuous fiber with 690 GPa modulus. Reference 13 is a comprehensive treatment of carbon fibers and their composites that cites 389 sources.

Aramid Fibers. These fibers offer attractive strength, stiffness and density properties, and in addition, they exhibit a high degree of toughness which is attributed to the mode of fracture (see Aramid fibers). An excellent article on aramid fibers and composites has been written (11). Aramid fibers are commercially available from one source, DuPont; they are marketed under the trade name Kevlar. Three grades of fibers are produced: Kevlar, which is made specifically for reinforcing rubber; Kevlar 29, made primarily for use in ropes, ballistics, etc; and Kevlar 49, made for reinforcing plastics in aircraft, aerospace, marine and sporting goods applications.

Kevlar fibers are poly(*p*-phenyleneterephthalamide) [*24938-64-5*]. The procedure for making aramid fibers involves mixing the polymer with a strong acid and extruding from spinnerets (14). Kevlar fibers have excellent resistance to flame and heat, organic solvents, fuels and lubricants; they can be readily woven into fabric. Kevlar fibers can be purchased in large quantities for approximately $27/kg.

Plastic Matrix Materials

Polyester Resin. The most common matrix material in glass-reinforced plastic composites is polyester; the estimated annual usage reached 360,000 metric tons in the early 1980s (15). Polyester contains substantial amounts of several components: resin, catalyst, filler, and accelerator. Most polyester resin is made by combining maleic anhydride with propylene glycol. Peroxide and cobalt naphthenate are frequently used as a catalyst and accelerator, respectively. Typical properties for a rigid polyester matrix material are listed in Table 2. Polyester is perhaps the least expensive matrix

Table 2. Properties and Prices of Matrix Materials

Matrix	Tensile strength, MPa[a]	Tensile modulus, GPa[a]	Tensile elongation at break, %	Specific gravity	Coefficient of thermal expansion, m/m °C	Heat distortion temperature, °C	Electrical resistivity, ohm·cm	Approximate price, $/kg
polyester	73[b]	4.8[b]	1.8[b]	1.22[b]	63×10^{-6} [b]	90[b]	10^{15} [c]	4[d]
epoxy								
RT cure	75[e]	3.2[e]	4.8[e]	1.16[e]	60×10^{-6} [e]	62[e]	3×10^{15}	5[d]
120°C cure	90[e]	2.7[e]	8.5[e]	1.21[e]	68×10^{-6} [e]	121[e]	2×10^{15}	5[d]
177°C cure	59[e]	2.3[e]	3.3[e]	1.15	65×10^{-6}	190[e]	7×10^{16}	5[d]
PEEK[f]	156[g]	3.8[g]	23.0[g]	1.30[g]	63×10^{-6}	150[g]	6×10^{14} [g]	60[g]
PEI[h]	105[i]	3.0[i]	60.0[i]	1.27[i]	56×10^{-6} [i]	200[i]	7×10^{15} [i]	9[d]

[a] To convert GPa to psi, multiply by 145,000 (MPa × 145).
[b] Ref. 16.
[c] Ref. 15.
[d] Ref. 17.
[e] Ref. 18.
[f] PEEK = polyetherketone.
[g] Ref. 19.
[h] PEI = polyetherimide.
[i] Ref. 20.

material and is relatively easy to process. However, the tensile elongation at breaking is too low to exploit the strength potential of high performance fibers. In addition, exposure to solvents, chemicals, and flame and the amount of shrinkage upon cure must be considered for polyester matrices (15–16).

Epoxy Resin. The most popular choice of matrix material for high performance composites containing carbon fibers has been epoxy. A number of attractive characteristics are offered by epoxy; no solvents or condensation products are released during cure, the shrinkage during cure is less than that of polyester and a wide range of properties can be achieved through formulation. Any molecule that contains the oxirane

$$\left(-\overset{|}{\mathrm{C}}\overset{\mathrm{O}}{\diagup\diagdown}\overset{|}{\mathrm{C}}- \right)$$

group is called an epoxy. The backbone of a particular epoxy resin may be built on aromatic and/or aliphatic groups. Aromatic epoxy backbones are stronger than the more flexible aliphatic epoxy backbones. Amine or anhydride agents are used to cure epoxy resins. Curing can be achieved at lower temperatures with the amine agents; anhydride curing agents offer increased thermal stability and moisture resistance. Accelerators such as boron trifluoride or benzyldimethylamine (BDMA) are often used to speed up the cure of epoxies.

Epoxy resin systems (base resin, curing agent, accelerator) are generally formulated to achieve a specific set of properties. Unfortunately, one property is usually decreased when another is increased. For example, room-temperature-curing systems do not possess the moisture resistance and thermal stability required for most aerospace applications. Therefore, a priority ranking must be established for the desired set of properties and then the formulation that best meets the desired properties can be selected. Epoxy–resin systems are often classified by their curing temperature.

Table 2 lists properties for aliphatic amine epoxies cured at room temperature, 120°C, and 177°C. Epoxies are generally more expensive than polyesters. An excellent article that provides a detailed discussion on epoxy resins, and cites 106 references is ref. 18. Researchers are currently seeking methods to increase toughness and resistance to environmental degradation. Construction of two-phase systems that contain small rubber particles is one approach to improving fracture toughness (21).

Other Thermosetting Resins. Polybutadiene, vinyl ester polymers, and polyimide are thermosetting matrix materials that are also used in fiber reinforced plastics. Each material has been discussed in detail elsewhere (22–24). Polybutadiene has a high dielectric constant (4.5–5.0) and is an excellent material for use in aircraft radomes. Vinyl ester has better chemical resistance than polyester and is used in pipe, ducts, tanks, etc. Polyimides can withstand higher temperatures than epoxies but are more difficult to process.

Thermoplastics. Thermoplastic matrix materials are expected to assume a major role in fiber-reinforced plastics in the next few years. Thermoplastics have the potential for reduced fabrication cost, improved repairability, damage tolerance, and chemical resistance. However, development of an inexpensive thermoplastic that adheres well to carbon fibers and has satisfactory resistance to solvents has yet to be achieved. Polyetherketone (PEEK) is a leading candidate but is expensive compared to epoxies (17,25). Polyetherimide (PEI) and polysulfone are being evaluated, but evidence obtained to date indicates that these materials degrade in the presence of many solvents associated with aircraft environments. Properties for PEI and PEEK are listed in Table 2.

Material Properties

The basic building block in high performance fiber-reinforced plastic composites is a single-ply of unidirectional tape. Properties for the single-ply of tape are assumed to equal those for a multi-ply laminate that has all fibers aligned in one direction. Representative moduli and strength values for unidirectional laminates of E-glass–epoxy, S-glass–epoxy, Kevlar 49–epoxy, high strength carbon–epoxy, high modulus carbon–epoxy and boron–epoxy are listed in Table 3. Moduli and strength values in the longitudinal (fiber) direction are outstanding. However, moduli and strength in the transverse direction are $1/40$ to $1/3$ the values in the fiber direction. Both tensile and compressive strength in the longitudinal direction and in the transverse direction are different. In-plane shear strength is generally less than 10% of the longitudinal tensile strength. Analytical methods for predicting the strength and stiffness of unidirectional fiber-reinforced composites are given in References 27–29.

Figure 3 illustrates the influence of fiber orientation on strength and modulus. References 29 and 31 present methods for predicting strength and modulus as a function of fiber orientation. The wide range of properties of unidirectional tapes offers designers the opportunity to tailor strength and stiffness to achieve maximum efficiency. However, designers must exercise care to ensure that fibers are properly aligned to react to the loads.

Fiber-reinforced plastics have excellent resistance to environment degradation. Unidirectional fiber-reinforced epoxy laminates, loaded only in the fiber direction, generally exhibit excellent fatigue behavior. Most matrix materials and some reinforcing fibers absorb moisture. Figure 4 shows the effect of moisture content on flexure

Table 3. Representative Properties of Fiber-Reinforced Epoxy Composites

	Reinforcing fiber					
Property	E-glass	S-glass	Kevlar 49	High strength carbon	High modulus carbon	Boron
longitudinal modulus, GPa[b]	45	55	76	145	220	210
transverse modulus, GPa[b]	12	16	5.5	10	6.9	19
axial shear modulus, GPa[b]	5.5	7.6	2.1	4.8	4.8	4.8
Poisson's ratio, ν_{LT}	0.28	0.28	0.34	0.25	0.25	0.25
longitudinal tensile strength, MPa[b]	1020	1620	1240	1240	760	1240
longitudinal compressive strength, MPa[b]	620	690	280	1240	690	3310
transverse tensile strength, MPa[b]	40	40	30	41	28	70
transverse compressive strength, MPa[b]	140	140	140	170	170	280
in-plane shear strength, MPa[b]	70	80	60	80	70	90
longitudinal tensile strain	2.3	2.9	1.6	0.9	0.3	0.6
longitudinal compressive strain	1.4	1.3	>2.0	0.9	0.3	1.6
transverse tensile strain	0.4	0.4	0.5	0.4	0.4	0.4
transverse compressive strain	1.1	1.1	2.5	1.6	2.8	1.5
fiber-volume fraction	0.60	0.60	0.60	0.60	0.60	0.60
specific gravity	2.1	2.0	1.4	1.6	1.6	2.0

[a] Ref. 26.
[b] To convert GPa to psi, multiply by 145,000 (MPa × 145).

strength for three carbon–epoxy composites. Slightly more than one percent moisture was absorbed by each composite when equilibrium was reached in an outdoor ground-based environment. Figure 4 indicates the room temperature flexure strength was not affected; the 83°C flexure strength was reduced 10–25% at 1% moisture content. The T300–5209 is a composite cured at 120°C and is degraded more by moisture than the other two composites which were cured at 177°C.

Figure 5 illustrates the need to protect epoxy matrix composites from uv radiation. The exposure condition depicted in Figure 5 is severe. Polyurethane paint provided adequate protection; unpainted specimens experienced significant mass loss. The effects of 5 yr exposure of two carbon–epoxy and a Kevlar 49–epoxy composite to fluids common to the operating environment of aircraft appear in Table 4. The carbon–epoxy composites retained at least 89% of the baseline strength whereas the Kevlar 49–5209 retained 77% of the baseline strength in the fuel–water exposure.

Electrical resistivities for glass, boron, and aramid fibers, and polyester and epoxy resins are 10^{14}–10^{16} Ω·cm; the electrical resistivity for carbon fibers is ca 10^{-3}–10^{-4} Ω·cm. Therefore, glass, boron, or Kevlar reinforced epoxy or polyester composites exhibit high values of electrical resistivity in both the longitudinal and transverse directions. Carbon–epoxy composites conduct current in the longitudinal direction and exhibit high electrical resistivity in the transverse direction (33–34).

Processing

The fabrication sequence for most high performance structures, such as aircraft component, is fibers and resin are combined to form tape; sheets of tape are stacked in the specified orientation; resulting laminate is bagged and then cured in an autoclave; cured component is machined and nondestructively inspected.

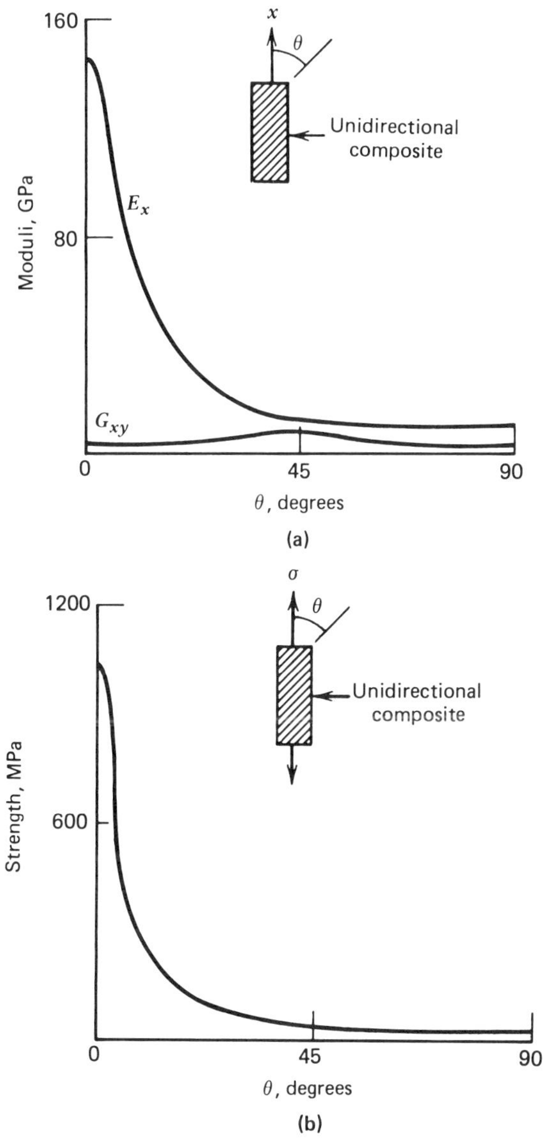

Figure 3. Effect of fiber orientation on moduli and strength; (**a**) extension E_x, and shear, G_{xy}, moduli of high strength carbon–epoxy; (**b**) strength, σ, of S-glass–epoxy. To convert GPa to psi, multiply by 145,000 (MPa × 145) (30).

Prepregging. The process of combining fibers and resin is called prepregging. A schematic diagram of one method for prepregging is shown in Figure 6. Precision equipment is used to control tape thickness, fiber alignment and spacing, resin content and degree of cure in the resin. The nominal thickness for most prepreg tape is 1.5–2.5 mm, however, tape as thin as 0.3 mm may be purchased. Tape is available in widths of 3 mm to 1 m. Resin is usually applied to the fibers in a quantity that produces a tape containing ca 60 vol % fiber. The resin content is maintained within ±2% of the specified value. Degree of cure of the resin is limited to yield a tape that exhibits tack and drape. Most prepreggers have developed specifications, recommended cure cycles and

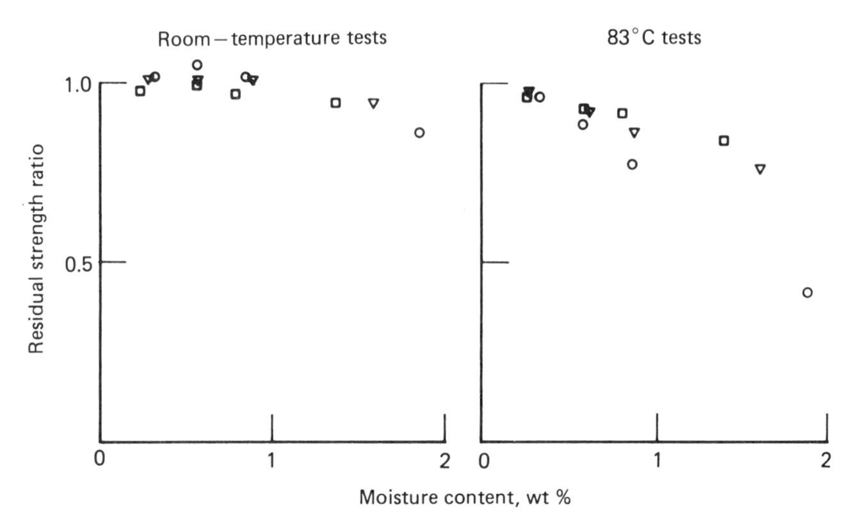

Figure 4. Effect of moisture on flexure strength of carbon–epoxy (32). Materials; □, T300/5208; ▽, T300/1034; ○, T300/5209.

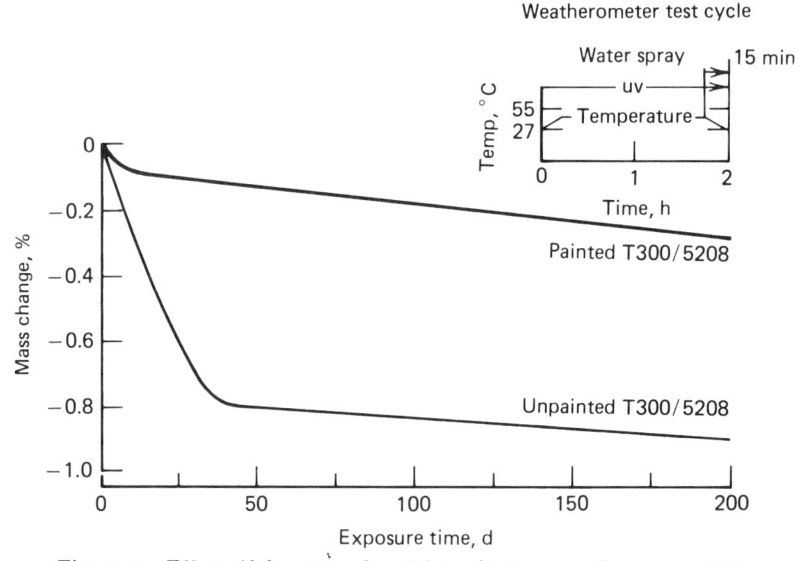

Figure 5. Effect of laboratory ultraviolet radiation on carbon–epoxy (32).

property data for their composite materials. These specifications have matured significantly over the past 15 yr and now often include advanced chemical analyses such as high pressure liquid chromatography.

Stacking, Bagging, and Curing. Sheets of prepreg tape are stacked in a specified orientation and then cured under pressure to fabricate a component. The prepreg tape may be cut and laid-up by hand or by automated tape laying machines. Figure 7(**a**) is a typical bagging diagram for a flat laminate. The tool, which is usually metal, is coated with a release agent or covered with release fabric. The sheets of prepreg, labeled carbon–epoxy layup in Figure 7(**a**), are placed on the tool. The layup is covered with

Table 4. Effect of Aircraft Fluids on Composite Materials after 5 yr Exposure

Composite	T300–5208	T300–5209	Kevlar 49–5209
normalized tensile strength of [±45] laminate	1.00	1.00	1.00
residual strength ratio after 5 yr exposure in			
ambient air	0.98	1.03	0.88
water	1.05		0.83
JP–4 fuel	1.14	0.96	0.85
Skydrol	1.14	1.00	0.84
fuel–water	1.04	0.89	0.77
fuel–air, 24 h cycle	1.02	0.89	0.88

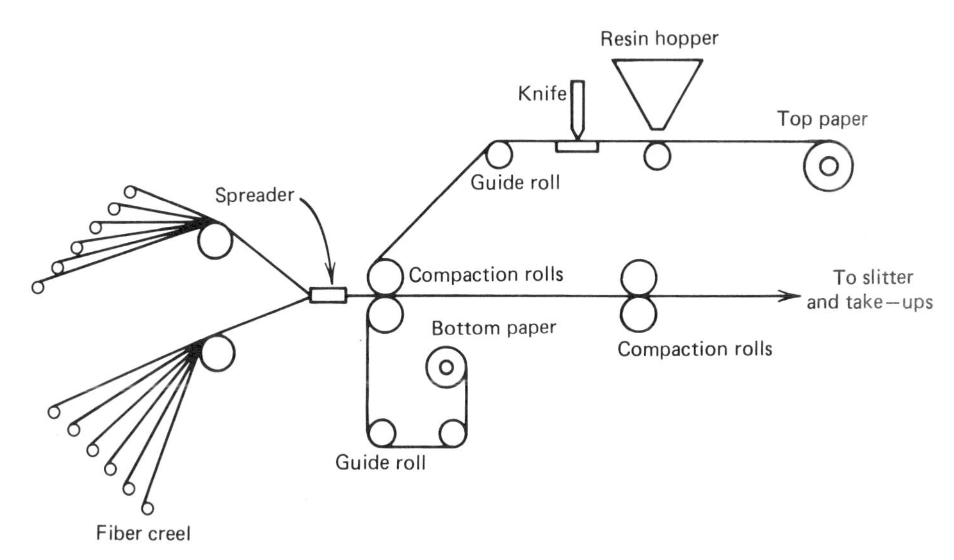

Figure 6. Schematic of prepregging process.

release fabric over which breather plies are placed that provide a path for removal of entrapped air. A vacuum bag is placed over the breather plies and the edges are sealed. The entire assembly is placed in an autoclave. Figure 7(**b**) shows a typical cure cycle at 177°C for a carbon–epoxy composite.

Machining. Practices have been developed for machining all fiber-reinforced epoxy or polyester composites discussed in this article (35–37). Fiber-reinforced composites can be sawed, drilled, countersunk, ground, sanded, chamfered, milled, or turned. Carbon–epoxy is probably the easiest composite to machine whereas Kevlar–epoxy is perhaps the most difficult. Smooth holes and edges can be machined without damaging the composites.

Nondestructive Inspection. Composite components are inspected to ensure the absence of defects. Microcracks, delaminations, and voids are the most common types of flaws; ultrasonic inspection is the most popular method for their detection (38) (see Ultrasonics, low power). Figure 8 shows ultrasonic C-scan records of acceptable and unacceptable quality laminates. The larger dark areas on the C-scan records indicate delaminations (see Nondestructive testing). Other nondestructive inspection methods include visual, sonic, radiography, microwave, infrared, holography, and eddy current (38–39).

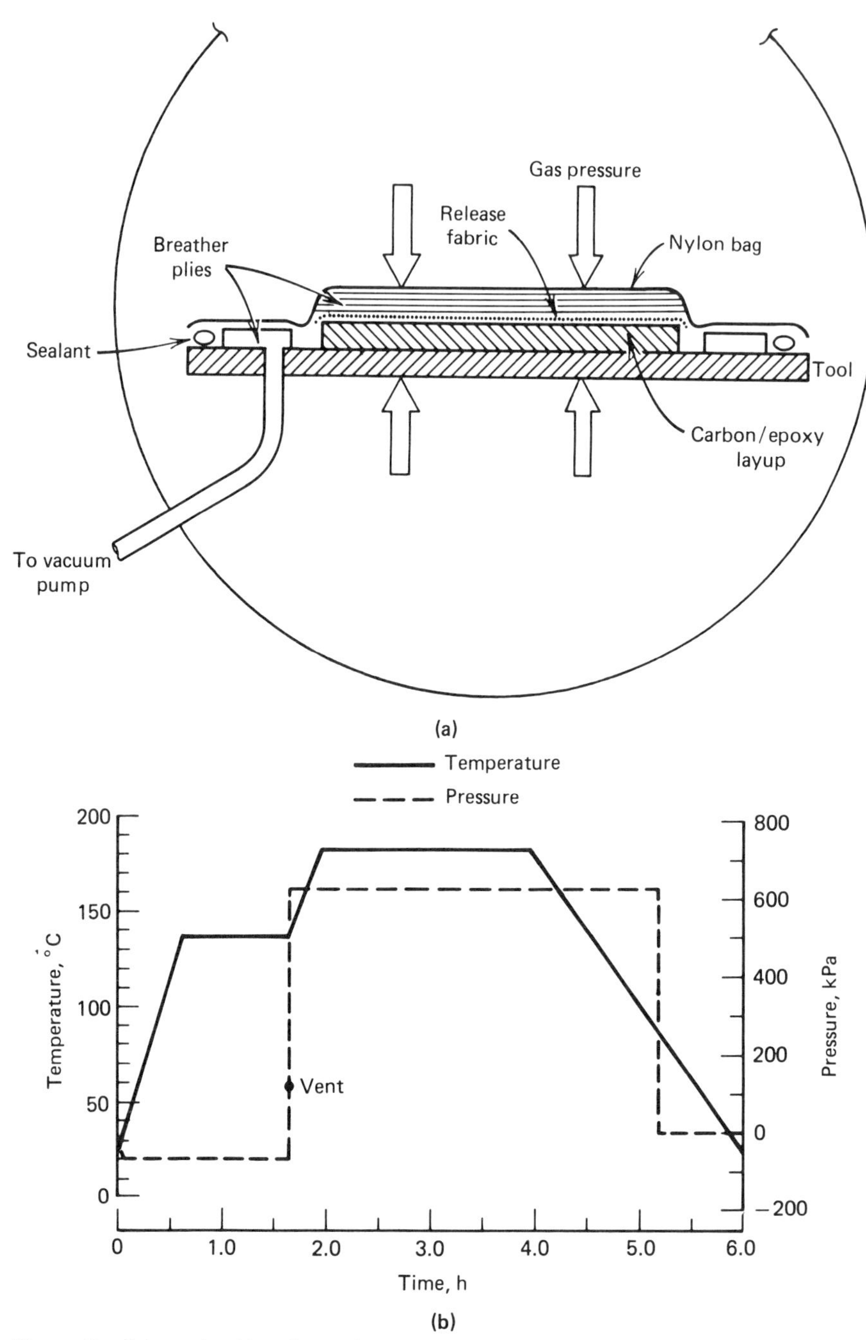

Figure 7. Schematic of bagging and cure cycle for autoclave-curing–resin matrix composites; (**a**) typical bagging diagram; (**b**) typical cure cycle for carbon–epoxy. To convert kPa to psi, multiply by 0.145.

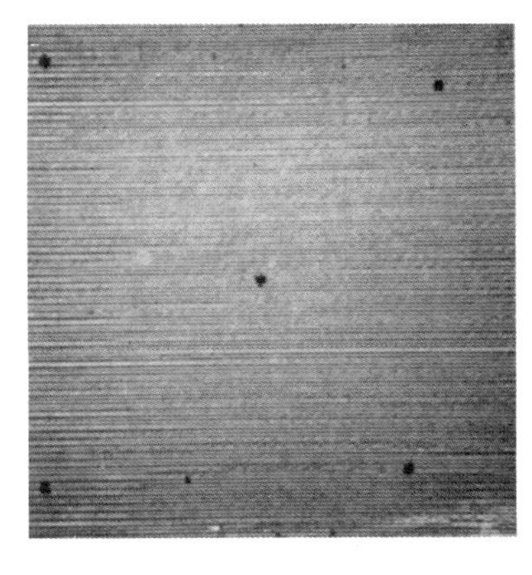

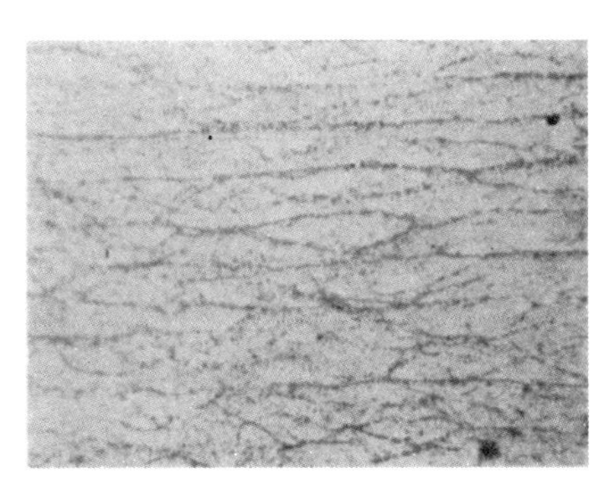

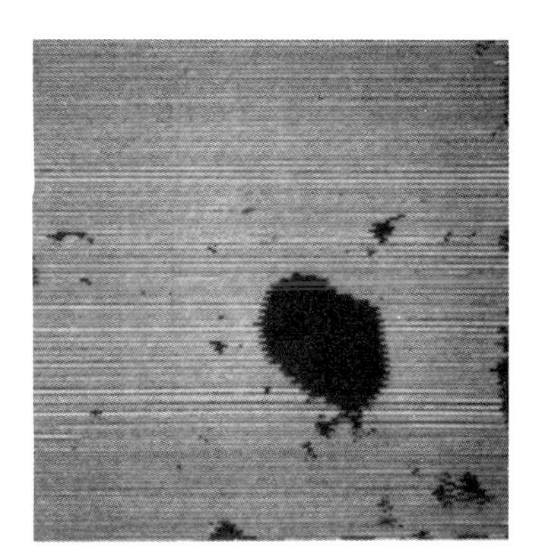

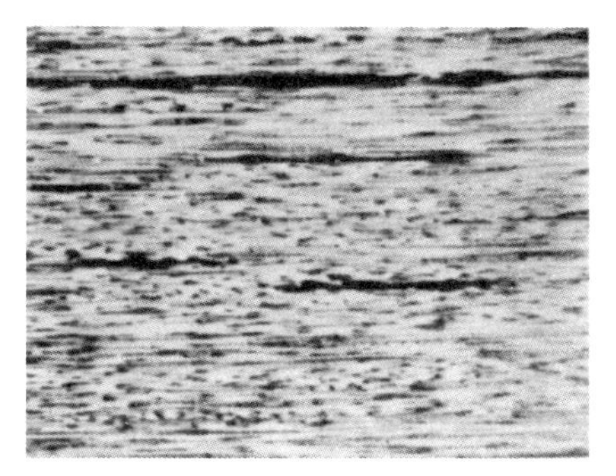

Figure 8. Nondestructive inspection of composite materials.

Automation. Nearly one half the cost of making laminated composite structures may be charged to hand operations (40). Therefore, major efforts are underway at several manufacturers to fully automate the process for using prepreg tape or fabric (sometimes labeled broadgoods) to make composite structures. Figure 9 illustrates one automated approach for cutting and stacking the prepreg material.

Filament Winding and Pultrusion. These methods are more economical fabrication approaches than hand lay-up. Filament winding is commonly used to fabricate bodies of revolution such as rocket–motor cases, high pressure tanks, cylinders, and pipe. The most common practice is to impregnate the filaments with a thermosetting-resin matrix and to wind the filaments over a mandrel in a continuous operation. Cure is usually accomplished by heating to an elevated temperature or at room temperature. Several factors must be considered to successfully use the filament winding process. Resin viscosity in the range 350 to 1500 mPa·s (= cP) is usually required to achieve uniform filament and matrix distribution and to prevent entrapment of air or volatiles in the composite material. Mandrel removal must be taken into account. Aluminum or steel mandrels can readily be used to fabricate open-ended structures such as cylinders, whereas dissolvable or collapsible mandrels are required to fabricate closed

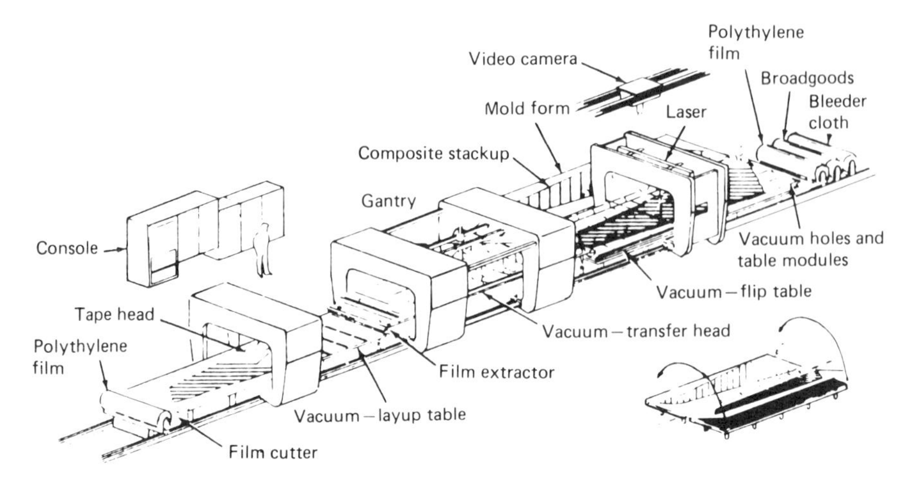

Figure 9. Integrated laminating center.

structures such as pressure vessels. Low melting temperature alloys, eutectic salts, soluble plasters, PVA-bonded sand and inflatable rubber mandrels have been used to fabricate closed structures. Additional information on design and fabrication of filament wound structures is provided in reference 41.

Pultrusion is commonly used to fabricate constant shaped cross-section components such as fence posts, channel-shaped beams and tubing. Bundles of filaments and/or other forms of reinforcement such as fabric are usually consolidated dry on a cylindrical mandrel and then transitioned to the final cross-sectional shape in a continuous operation. A high pressure pump may be used to impregnate the filaments with a rapid-curing resin matrix. Cure may be achieved by conventional heating and/or radio frequency heating in the final shaping die. Production speeds of 400–600 cm/min are achievable. Continuous plate and sheet stock up to 122 cm wide by 38 mm thick have been produced. Reference 42 contains detailed information on pultrusion.

Other Processing Methods. These include spray-up, centrifugal molding, bag molding, resin injection molding and matched-die molding (43–45).

Designing Composite Structures

Fiber-reinforced plastics structures are designed with the same principles of mechanics that are used to design metallic structures. However, the design usually must also be analyzed at one lower level because the composite is composed of a number of individual layers or plies. Because of the anisotropic behavior of each ply, ie, extreme strength and stiffness in the longitudinal direction compared to the transverse direction, care must be taken to ensure that delamination of the plies does not occur. Fiber-reinforced composites do not have as much ductility as some metals and thus the effect of stress concentrations must be included in design.

Figure 10 illustrates the variation in tensile strength and modulus that can be achieved by varying the distribution of plies within a [0°, ±45°, 90°] carbon–epoxy laminate. Similar plots can be developed for all in-plane strength, moduli, and thermal expansion properties for any combination of ply orientations using analyses in References 29 and 31.

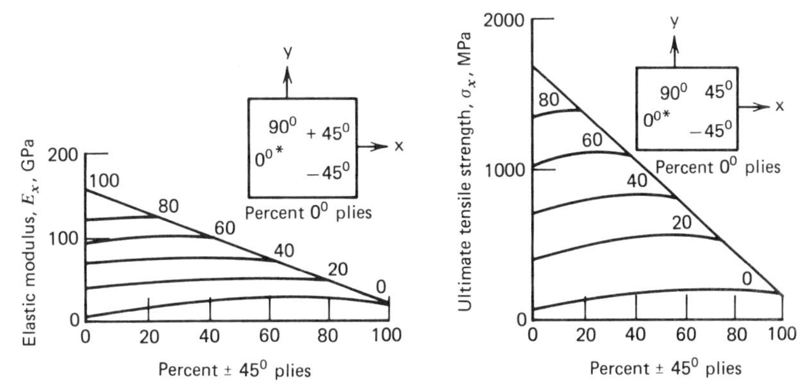

Figure 10. Effect of laminate orientation on modulus, E_x, and strength, σ_x. To convert GPa to psi, multiply by 145,000 (MPa × 145) (46).

Laminated-fiber-reinforced plastics that are designed to fail by breaking fibers in one or more of plies are defined as fiber-dominated composites. This type of composite offers excellent resistance to degradation by fatigue loadings. A comparison of the fatigue behavior of carbon–epoxy and aluminum shown in Figure 11 clearly indicates the superior performance of the fiber-dominated composites. There is good agreement between predicted and measured fatigue life for [0°, ±45°, 90°] carbon–epoxy laminates under spectrum loadings (49). Tensile-fatigue performance of glass–fiber-dominated composites is reported in Reference 50. Laminates that fail in modes other than fiber breaking, such as delamination or matrix cracking, do not exhibit excellent fatigue behavior and designers usually limit interlaminar stresses to low values. Analyses for predicting interlaminar stresses are available (51).

High performance fiber-reinforced composites, loaded to failure in compression, usually fail by delamination or microbuckling of the fibers and do not exhibit local yielding observed in many isotropic metals. Because of these failure modes, the notch sensitivity in compression of composites is quite unlike that of metals (see Fig. 12). Impacting composites with low velocity particles that do not cause visable damage may degrade compressive strength. Figure 13 shows that an isotropic carbon–epoxy laminate which has been impacted may only be strained to 0.003 m/m whereas the virgin laminate can be strained to 0.009 m/m.

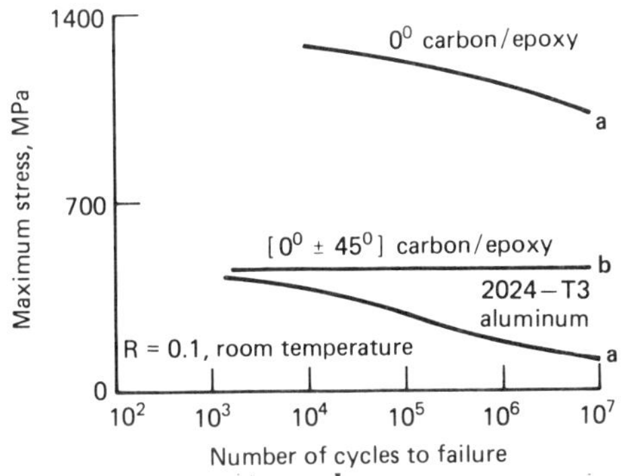

Figure 11. Comparison of fatigue behavior of carbon–epoxy and aluminum. To convert MPa to psi, multiply by 145 (47–48).

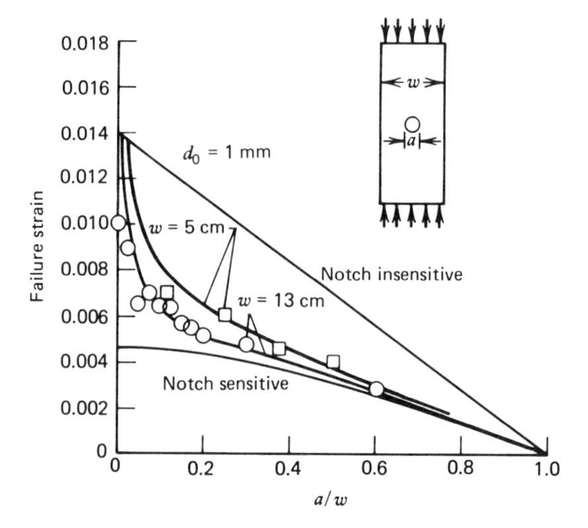

Figure 12. Effect of holes on compressive strength of carbon–epoxy quasi-isotropic laminate (52).

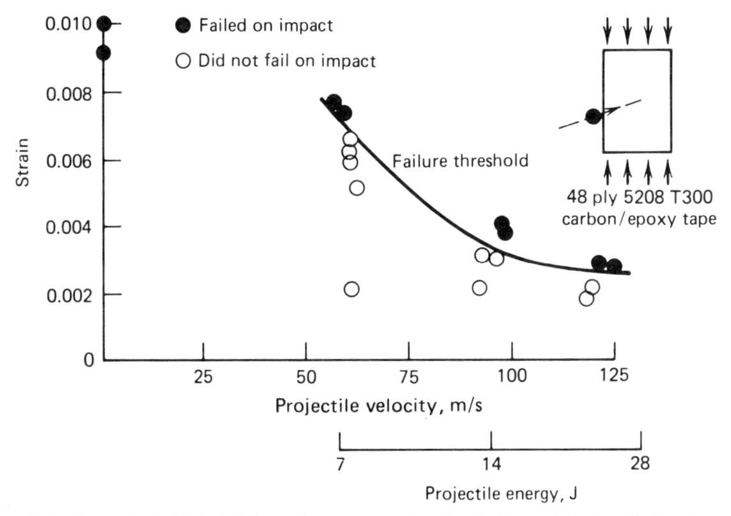

Figure 13. Impact-initiated failure in compression loaded quasi-isotropic laminate (52).

Mechanical fastening and adhesive bonding are the primary methods for joining composite structures. Experimental data and guidelines on joint configurations and proportions for glass–polyester, glass–epoxy, boron–epoxy, aramid–epoxy, and carbon–epoxy are presented in Reference 37. Investigations of bolted joints for carbon–epoxy and hybrid carbon–glass–epoxy laminates have been conducted (53). Figure 14 illustrates the variation in bearing stress that can be achieved by varying the distribution of plies in a [0°, ±45°, 90°] carbon–epoxy composite.

Metals cannot simply be replaced by composites; designers must take into account the unique features of composites. The additional design effort is warranted on the basis of increased performance and/or decreased cost that can be achieved with advanced composites.

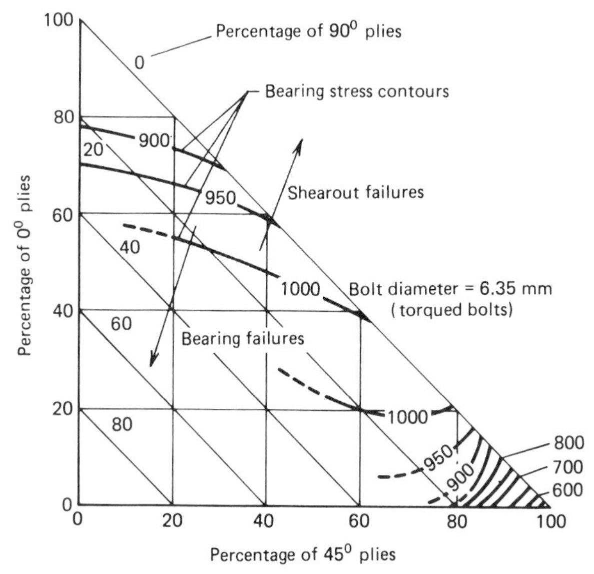

Figure 14. Bearing stress contours for various laminate patterns of MODMOR II–NARMCO 1004 carbon–epoxy composite (53).

Applications

High performance fiber-reinforced composites have been used in space and land vehicles, aircraft, marine vessels, sporting goods, and medical applications. In most applications the use of composite materials offers one or more of the following advantages compared to metal structures: improved performance, lower weight, or lower cost.

Space Vehicles. Several components in the Space Shuttle Orbiter are composite structures. These include carbon–epoxy payload bay doors and skin panels for the orbital-maneuvering system, boron–epoxy reinforced titanium aft thrust structure, boron–aluminum tubular struts in fuselage frames, and Kevlar–epoxy pressure vessels. Figure 15 shows the payload bay doors. Each door is divided into four segments; each segment is approximately 4.6-m long and the arc length between the hinge line and door center-line is approximately 4.0 m. Use of carbon–epoxy payload doors reduced the Orbiter mass by about 400 kg. Other applications in space vehicles include rocket–motor cases which exploit the high specific strength of composites (47), support trusses for optical telescopes which exploit their near zero thermal expansion coefficient (47), and light weight booms which exploit the specific stiffness (54) offered by fiber-reinforced plastics.

Aircraft. Table 5 lists 11 production, 1 prototype, and 3 conceptual aircraft that contain fiber-reinforced composite structures. Table 5 includes the percentage of the total structural weight that is composite structure. In most aircraft applications the composite components are about 25% lighter than metal components and in some applications the weight savings approach 50%. Application of composites to production–fighter aircraft began about 1970 and focused on control surfaces such as the F-14 horizontal stabilizer and F-15 rudder (55). The growth in application of composites to fighter aircraft has increased rapidly. The F-18 contains slightly less than 10%

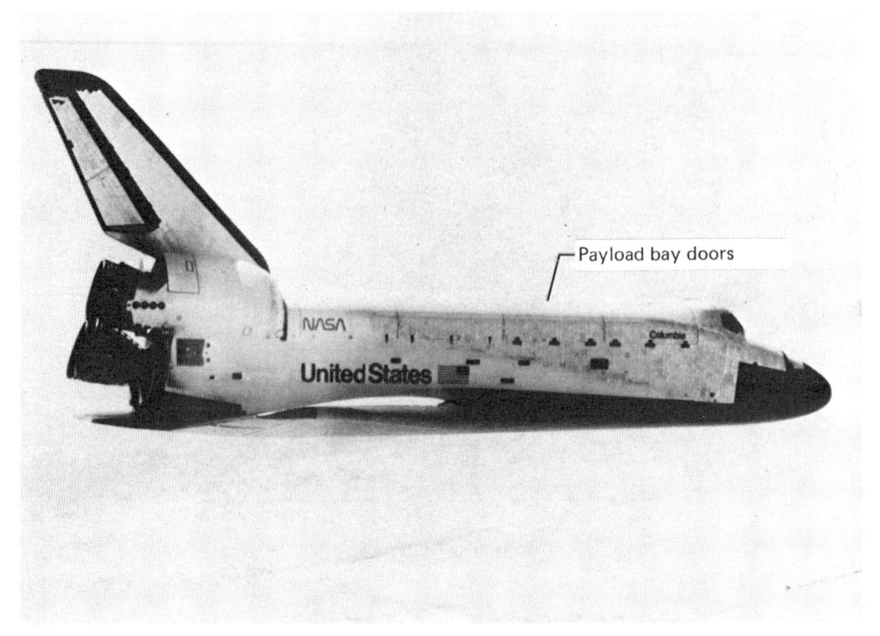

Figure 15. Carbon–epoxy payload bay doors for space shuttle orbiter.

Table 5. Applications of Composites to Aircraft Structures

Aircraft type	Application	Materials used[a]	Composite structure, wt %
F-14	production	B–E, C–E	1.0
F-15	production	B–E, C–E	1.6
F-111	production	B–E, C–E	
F-16	production	C–E	2.5
F-18	production	C–E	9.5
AV-8B	production	C–E	26.0
advanced tactical fighter	conceptual	C–E	45
advanced V/STOL	conceptual	C–E	55
advanced attack helicopter	conceptual	C–E, K–E	ca 100
U.S. army ACAP	prototype	C–E, K–E	ca 100
UH-60	production	C–E, K–E	8.0
S-76	production	C–E, K–E	8.0
Lear Fan 2100	production	C–E, K–E	ca 100
B-757/767	production	C–E, K–E	3.0
L-1011-500	production	K–E, C–E	1.0

[a] B–E = Boron–epoxy; C–E = carbon–epoxy; and K–E = Kevlar–epoxy.

composite structure whereas the AV-8B has a carbon–epoxy wing structure and contains 26% composite structural material (56). Fighter aircraft planned for the next decade are expected to have 45–55% composite structures.

Commercial-transport aircraft have used glass–fiber-reinforced components since the late 1950s. However, the Boeing 757 and 767 aircraft are the first transports to utilize significant amounts of carbon–epoxy and hybrid Kevlar–carbon–epoxy structures (see Fig. 16). Research underway at NASA and transport manufacturers is expected to drastically increase the use of composite structures in future transport

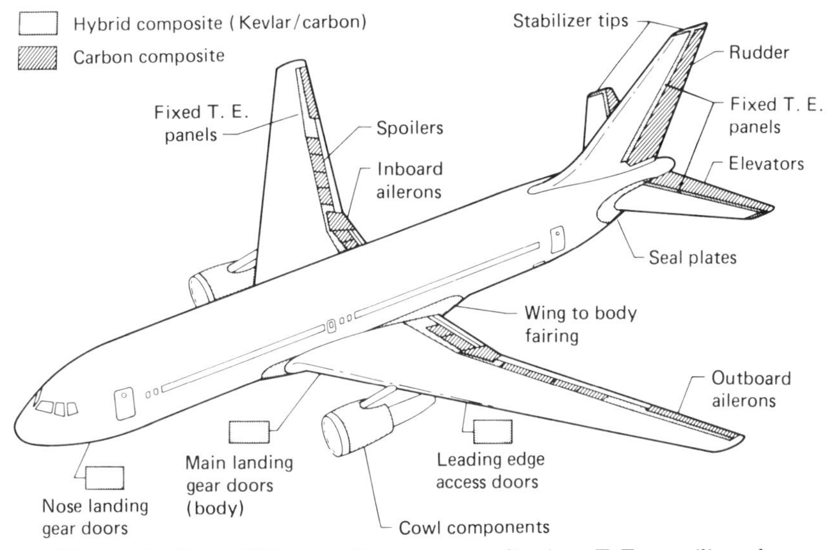

Figure 16. Boeing 767 composite structure applications. T. E. = trailing-edge.

aircraft (57); totally composite wings, and perhaps fuselages, are likely to be developed.

The most aggressive application of composite materials in fixed wing aircraft is the LearAvia Lear Fan 2100 executive aircraft (Fig. 17). Test flights began in 1981 and FAA certification is expected in the near future. The airframe is almost entirely carbon–epoxy or Kevlar–epoxy structure. Lear Fan 2100 is an excellent example of the increased performance that can be achieved through exploitation of advanced composite structures and aeronautics research.

Each Sikorsky S-76 helicopter has over 115 kg of Kevlar–epoxy structure that covers approximately 45% of the exterior wetted surface (57). S-76 Kevlar–epoxy applications include the horizontal stabilizer, doors, fairings, main-rotor blade–tip caps, and internal nonstructural items. The S-76 also uses carbon–epoxy and glass–epoxy in the tail rotor. Approximately 8% of the structural weight is composite materials. Figure 18 shows four composite components that are being evaluated on Bell Model 206 helicopters. The doors and fairing on the 206L are made of Kevlar–epoxy

Figure 17. LearAvia Lear Fan 2100.

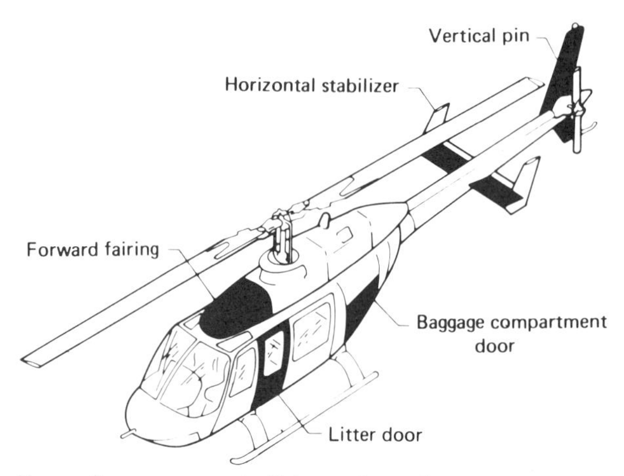

Figure 18. Composite components in flight service evaluation on Bell 206L helicopter.

and the vertical fin is made from carbon–epoxy. Several helicopter manufacturers are producing glass–epoxy main rotor blades. Application of composites in rotorcraft are expected to increase rapidly. The U.S. Army has contracts with two manufacturers to build prototype rotorcraft that use essentially all composite structures, have 22 percent lower mass, and cost 18% less than a metal airframe.

Land Vehicles. Increases in the cost of fuel and international competition have stimulated significant interest among land transportation vehicle manufacturers in the use of composite materials. One example is the 1979 Ford carbon–epoxy prototype vehicle shown in Figure 19. The mass of a production metal vehicle is 1700 kg whereas the mass of the carbon–epoxy prototype is 1135 kg. Approximately 315 kg of the mass reduction is directly attributed to the use of composites. An all-composite automobile does not appear to be imminent. However, the application of composites to selected components such as panels, oil pans, springs, drive shafts and support brackets is under

Figure 19. Composites in automotive applications (58).

serious consideration (59). It is estimated that land transportation industry used 186,000 t of glass–fiber-reinforced plastic in 1980 and the consumption is projected to double by 1990 (60).

Marine Applications. More than 90% of the pleasure boats and approximately 40% of the ≤29-m fishing trawlers in production are made of glass-fiber-reinforced plastics (63). Larger vessels such as a 360-t displacement, 49-m long mine sweeper have also been built from composites. Kevlar-reinforced plastic has experienced limited applications in high performance power and sail boats and canoes. The Kevlar–plastic canoes are 20–35% lighter than comparable glass–plastic or aluminum canoes (61). DuPont analysts project that the added cost for a 10-m commercial fishing boat built from Kevlar–plastic instead of glass–plastic would be recovered in less than two years because the Kevlar–plastic boat would use less fuel and have an increased speed and payload (61). Other marine applications of composite materials are discussed in reference 62.

Sporting Goods. Fiber-reinforced plastics are being used in a number of sporting good applications; for example, most fishing rods are made from glass-fiber-reinforced plastic. Carbon–plastic and hybrid carbon–glass–plastic composites are being used in applications that require stiffness greater than that achievable with glass–plastic. Two examples, a racquetball racquet and golf club, are shown in Figure 20. Other applications of carbon–epoxy and hybrids include fishing rods, skis, ski poles and archer bows.

Medical–Protective Devices. Medical or health-related applications of fiber-reinforced composites in the developmental or prototype stage include: carbon–epoxy patient support tables which permit examination with lower level x-ray exposure, lightweight helmets to protect handicapped people, lightweight wheelchairs, and lightweight prosthesis equipment (see Prosthetic and biomedical devices). The U.S.

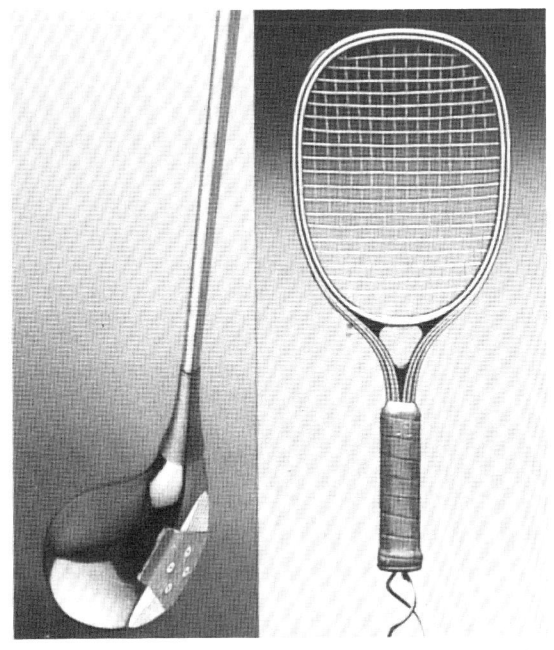

Figure 20. Composites in recreation applications (48).

Army recently ordered 10^6 Kevlar–plastic helmets which are reported to offer improved protection and comfort compared to the M-1 steel helmet (63).

BIBLIOGRAPHY

"High Temperature Composites" in *ECT* 3rd ed., Vol. 12, pp. 459–481, by William B. Hillig, General Electric Company.

1. D. V. Rosato in G. Lubin, ed., *Handbook of Composites*, Van Nostrand Reinhold Company, Inc., New York, 1982, pp. 1–14.
2. J. Economy, *SAMPE J.*, 5 (Nov./Dec. 1976).
3. L. H. Miner, *Proceedings of Technical Symposium III—Design and Use of Kevlar Aramid in Aircraft*, 1981, pp. 3–10.
4. *Material Properties-GY70SE, Bulletin CFM8*, Structural Composites, Celanese Corp., 1981.
5. *Herculines Newsletter* **17**(1), SAE issue (1983).
6. J. R. Kattus, *Aerospace Structural Metals Handbook*, 1981, Sect. 3203.
7. C. E. Knox in ref. 1, p. 136.
8. H. E. DeBolt in ref. 1, p. 171.
9. O. C. Trulson and H. F. Volk, *paper presented at the 14th National SAMPE Technical Conference, Materials and Process Advances '82*, Vol. 14, 1982, pp. 503–512.
10. "Typical Properties of Thornel High Modulus Carbon Yarns and Their Composites," Tech. Bulletin 465-252, Union Carbide Corp., Carbon Products Div.
11. C. C. Chiao and T. T. Chiao in ref. 1, p. 272.
12. C. P. Talley, *J. Appl. Phys.* **30,** 1114 (1959).
13. D. M. Riggs, R. J. Shuford, and R. W. Lewis in ref. 1, p. 196.
14. U.S. Pat. 3,767,756 (Oct. 23, 1973), H. Baldes.
15. I. H. Updegraff in ref. 1, p. 34.
16. M. Rubin in G. Lubin, ed., in *Handbook in Fiberglass and Advanced Composites*, Van Nostrand Reinhold Company, Inc., New York, 1969, p. 31.
17. *Plast. Technol.* **29,** 95 (Sept. 1983).
18. L. S. Penn and T. T. Chiao in ref. 1, p. 75.
19. *Polyether ether ketone: Typical Properties of Crystalline Injection Mouldings*, technical brochure, Imperial Chemical Industries PLC, PK PD5, 1979.
20. *Ultem Polyetherimide Resin*, technical brochure, General Electric Co., Plastics Operations, ULTEM Products Section.
21. R. Drake and A. Siebert, *SAMPE Q.* **11**(6), (1975).
22. M. Stander in ref. 1, p. 50.
23. M. B. Launikitis in ref. 1, p. 38.
24. T. T. Serafinin in ref. 1, p. 89.
25. J. T. Hartness, *paper presented at the 14th National SAMPE Technical Conference, Materials and Process Advance '82*, Vol. 14, 1982, pp. 26–37.
26. C. Zweben, H. T. Hahn, and R. B. Pipes, *Composites Design Guide*, Vol. 1, Mechanical Properties Section 1.7, 1980, pp. 65–71.
27. J. E. Ashton, J. C. Halpin, and P. H. Petit, *Primer on Composite Materials: Analysis*, Technomic Publishing Co., Inc., Westport, Conn., 1969.
28. Z. Hashin, *Theory of Fiber Reinforced Materials*, NASA CR 1974, 1972.
29. R. M. Jones, *Mechanics of Composite Materials*, McGraw-Hill, Inc., New York, 1975.
30. S. W. Tsai, *Strength Characteristics of Composite Materials*, NASA CR 224, 1965.
31. S. W. Tsai and H. T. Hahn, *Introduction to Composite Materials*, Technomic Publishing Co., Inc., Westport, Conn., 1980.
32. H. B. Dexter and A. J. Chapman, "NASA Service Experience with Composite Components," *Proceedings 12th National SAMPE Technical Conference*, 1980.
33. J. T. Kung and M. P. Amason, *SAMPE J.* **23,** 1039 (1978).
34. J. M. Thomson, *The Electrical Properties of Carbon Fiber Composites, AGARD Lecture Series No. 124*, NATO, 1982, pp. 9-1–9-15.
35. *A Guide to Cutting and Machining Kevlar Aramid*, technical brochure, Textile Fibers Department, E. I. du Pont de Nemours & Co., Inc., Wilmington, Del.

36. L. E. Meade in ref. 1, pp. 510–511.
37. S. J. Dastin in ref. 1, pp. 622–632.
38. G. Epstein in ref. 1, pp. 665–675.
39. M. L. Phelps, *In-Service Inspection Methods for Graphite–Epoxy Structures on Commercial Transport Aircraft*, NASA CR 165746, 1981.
40. R. Hadcock, J. Huber, *Manufacturing Processes for Aeronautical Structures, AGARD Lecture Series No. 124*, NATO, 1982, p. 11-9.
41. A. M. Shibley in ref. 1, p. 449.
42. W. B. Goldsworthy in ref. 1, p. 479.
43. C. Wittman and G. D. Shook in ref. 1, p. 321.
44. A. Slobodzinsky in ref. 1, p. 368.
45. P. R. Young in ref. 1, p. 391.
46. *Material Properties of Composites*, Celanese Corp., Advanced Engineering Composites, 1977.
47. *Hercules Composites Structures*, Technical Brochure, Hercules, Inc., Bacchus Works, Magna, Utah.
48. J. R. Kerr and J. F. Haskins, *High-Temperature-Stress Capabilities of Composite Materials for Advanced Supersonic Technology Applications*, NASA CR 159267, 1980.
49. J. N. Yang and D. L. Jones, *Compos. Techno. Rev.* **4,** 63 (1982).
50. J. F. Mandell, D. D. Huang, and F. J. McGarry, *Compos. Technol. Rev.* **3,** 96 (1981).
51. N. J. Pagano, *Stress Fields in Composite Laminates*, AFML-TR-77-114 (1977).
52. M. M. Mikulas, Jr., "Failure Prediction Techniques for Compression Loaded Laminates With Holes," in NASA CP 2142, 1980, p. 1.
53. L. J. Hart-Smith, *Bolted Joints in Graphite–Epoxy Composites*, NASA CR 144899, 1976.
54. J. G. Davis, Jr., *paper presented at the 17th National SAMPE Symposium*, 1972.
55. F. J. Fechek, *AGARD Conf. Proc. No. 288*, 20-1 (1980).
56. R. Hadcock, *AGARD Lect. Ser. No. 124*, 12-3 (1980).
57. H. B. Dexter, *AGARD Conf. Proc. No. 288*, 19-1 (1980).
58. *Profile Newsletter*, Vought Corporation, Vol. 11, No. 4, 1979, p. 1.
59. S. V. Kulkarni, C. H. Zweben, and R. B. Pipes, eds., *Composite Materials in the Automobile Industry*, ASME, New York, 1978.
60. M. Martin and J. F. Dockum, Jr., in ref. 1, p. 695.
61. M. J. Meermans, *Mater. Eng.* **97,** 24 (1983).
62. W. R. Graner in ref. 1, pp. 699–721.
63. *Des. News* **39,** 48 (1983).

JOHN DAVIS
NASA

CONVEYING

This article presents an overview of various types of the more commonly used material-handling devices employed in today's industry and commerce. The Conveyor Equipment Manufacturers Association (CEMA) is the recognized authority in today's material-handling industry, and much of the reference material contained herein incorporates data and information developed and published by this organization. CEMA has prepared several publications relating to the field of material handling, one of which is devoted entirely to terms and definitions of various types of conveying equipment (1). The latest edition defines more than 120 distinct types of conveying units, but only a small fraction is used extensively today.

Conveying equipment can be divided into two broad classifications:

1. Those units that feed material into a system at a predetermined and controlled rate. These devices are generally referred to as feeders.

2. Those units that transport material from point to point. These fall into the general classification of conveyors. Where material must be conveyed straight up, or nearly so, the conveying unit is most often a bucket elevator.

As used throughout the material-handling industry, a feeder is a conveying-type unit, generally of relatively short length, designed to remove bulk material from a storage area (eg, track hopper, truck hopper, surge bin, silo, etc) at a uniform and controlled rate, either fixed or variable, and deliver the material to a process or conveying unit (see Fig. 1 and Weighing and Proportioning). Its rate of removal is usually accomplished on a volumetric basis, as determined by the cross-sectional area of material on or in the feeder, multiplied by the rate of conveying speed. The product, m^3/min (ft^3/min, CFM), is then easily converted into metric tons per hour (short tons per hour, TPH) when material bulk density, in kg/m^3 (lb/ft^3, PCF), is known.

$$t/h = 0.06 \times m^3/min \times \frac{kg}{m^3}$$

$$(TPH = 0.03\ CFM{\cdot}PCF)$$

The cross-sectional area of the material is determined by the width and projected vertical height of the opening in the hopper outlet (3).

The feeder discharges onto or into a succeeding device in the system, eg, a processing unit (mixer, press, furnace, etc) or another conveying unit. If the next unit is in fact a conveyor, it is not considered to be a feeder in material-handling terminology, even though this succeeding unit may be discharging into, or feeding a process unit. Simply stated, a feeder is a volume-controlling device, introducing material into a system, whereas a conveyor is a material transporter.

To begin to make a proper selection of the optimum and most economic types of conveying equipment for use in a given installation, the user or designer must consider the many factors that influence the layout and choice of equipment. Oftentimes, several different types of conveyors are included in a single material-handling system.

Three distinct factors that must be considered in selection of the proper equipment, whether of a single unit or a system, are

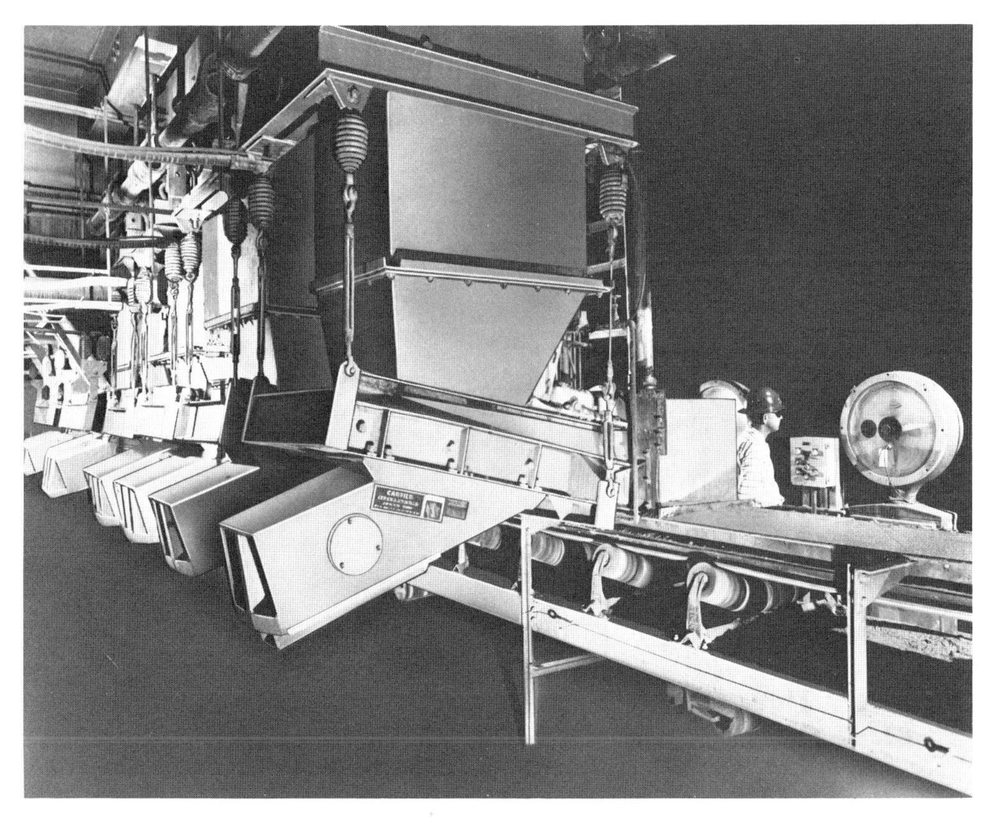

Figure 1. Natural-frequency vibrating feeder Carrier Amplitrol patented design (2). Courtesy of Rexnord Inc.

Material characteristics. Most materials can be handled well by only certain types of conveyors. Before planning a system layout, the designer should determine which types of conveyors are applicable based on this consideration.

Layout. Distances to be traveled, lift required, and space available dictate the practicability of some units and totally rule out others. Any need for intermediate-discharge points, to turn corners, etc, should also be considered at this time.

Economics. When more than one type of conveyor can satisfy these two criteria, the proper choice becomes one of economics.

Other considerations that would generally fall into one of the three categories listed above include such items as material temperature or corrosiveness and any special environmental conditions. High temperature, for example, might dictate the use of a steel pan vibrating conveyor instead of a rubber belt conveyor. Corrosiveness, on the other hand, might direct the user toward a rubber belt conveyor rather than a steel vibrating or apron conveyor. Environmental stipulation such as the need for strict dust control might favor a screw conveyor or covered vibrating conveyor over a rubber belt or steel apron conveyor.

Because of the inherent flexibility of most conveying devices, the layout of the material-handling system is usually left until near the end of the design of a production or processing facility. The design and location of the large and heavy manufacturing or processing equipment must take precedence in the design and layout of the operations. Because of this, choosing the best path or layout of the conveying system may be difficult.

Perhaps the most common problem is attaining a required lift in a restricted distance. Most conveyors are limited in the maximum angle of incline at which they can efficiently convey material. This maximum incline is further influenced by the characteristics of the material being handled. For example, the maximum angle of incline of a vibrating conveyor when handling castings or other similar objects is generally considered to be 5–6° from the horizontal. On the other hand, when handling wood chips or other fibrous interlocking material, the same unit can often attain an incline of up to 25°.

Types of Units

Various designs of feeders, conveyors, and elevators have been developed by equipment manufacturers during the evolution of the industry. Many types and styles were initially marketed as proprietary items but, as patent rights expired, were adopted by the industry as a whole. As a result, most of the units described here are readily available in their basic form from several sources, although most manufacturers have introduced exclusive innovations in the design and construction of particular pieces of their equipment.

The definitions and illustrations that follow are representative of the units that are most generally used for all but very special material-handling requirements and that are considered most reliable when properly applied.

Feeders. *Apron feeder.* A series of apron pans attached to chain or pivotally attached one to another forms the feeding medium beneath a hopper or a bin (Fig. 2(**a**)). These units can be very heavy in construction and are well adapted for handling larger lumps, carrying a considerable headload, or operating with very high temperature materials.

Belt feeder. An endless rubber or metal belt operating over suitable drive, tail, and bend terminals and over belt idlers or a slider bed constitutes a belt feeder (Fig. 2(**b**)). Such a unit can be used to withdraw coarse material from a hopper if the hopper bottom and outlet are designed to minimize the headload on the feeder. Special high temperature rubber belts capable of handling materials up to 200°C are readily available.

Chain-curtain feeder. In this unit, a power-operated curtain of endless lengths of chain rests on and retards the flow of bulk materials in an inclined chute. Chain-curtain feeders are most applicable to handling material with a high percentage of large, lumpy material (Fig. 2(**c**)).

Live-roll grizzly. This device for screening and scalping consists of a series of rotating, parallel rolls with fixed-size spaces between (Fig. 2(**d**)). It is ideally suited for use where a bed of fines is to be laid on a receiving in-line belt conveyor. This is accomplished by closely spacing the openings between the parallel rolls near the charge end of the feeder and gradually increasing this spacing toward the discharge end.

Reciprocating feeder. A reciprocating driven plate or pan operating under a head of bulk material is the principal element of a reciprocating feeder (Fig. 3). The plate operates at a long stroke and a low frequency, and the feeder can handle a sizeable headload of extremely large and heavy lumps. This unit is not self-cleaning; it relies on a headload of material above to push material off the pan.

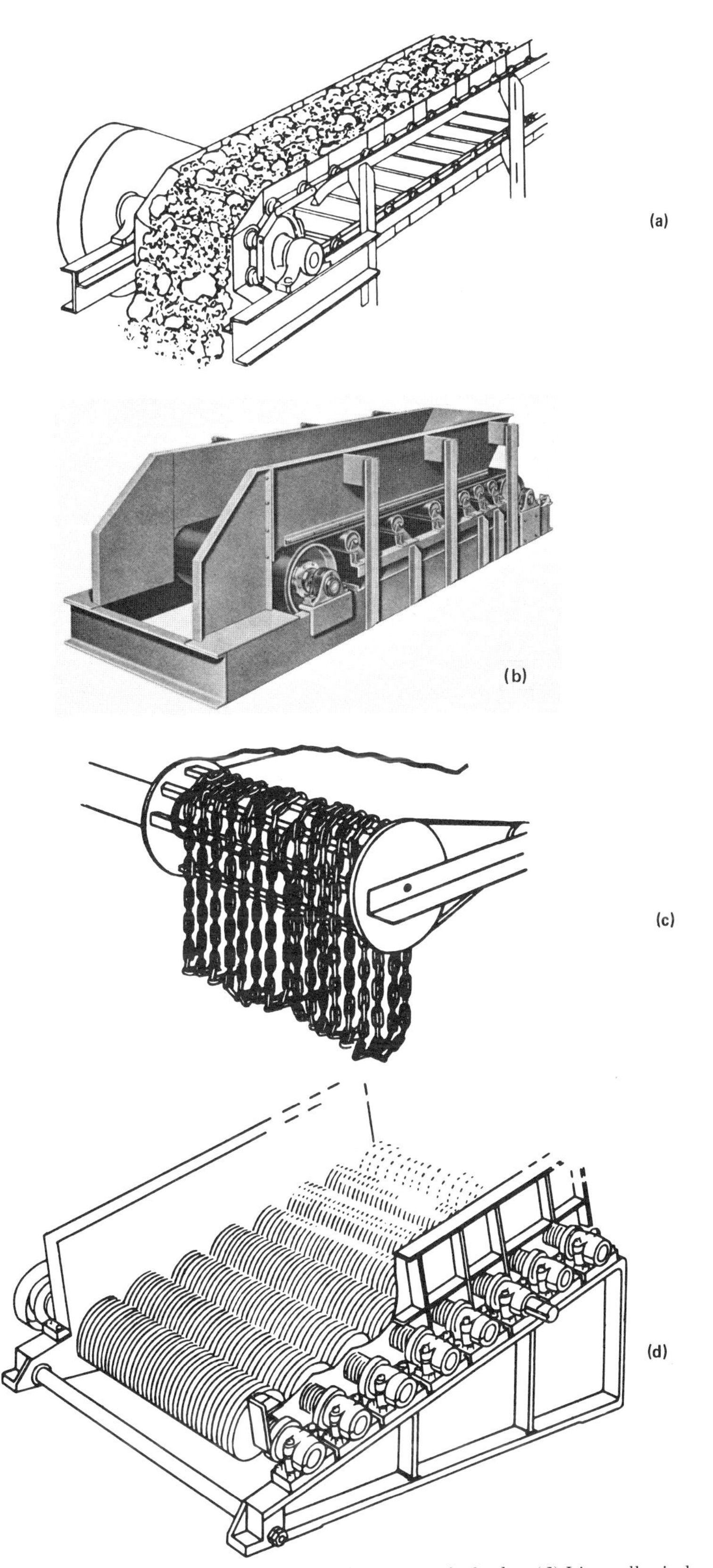

Figure 2. (**a**) Apron feeder. (**b**) Belt feeder. (**c**) Chain-curtain feeder. (**d**) Live-roll grizzly.

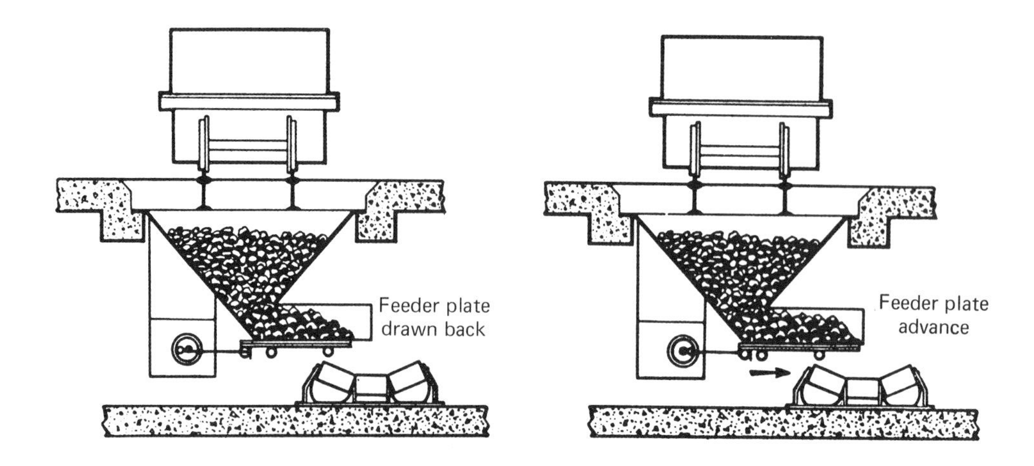

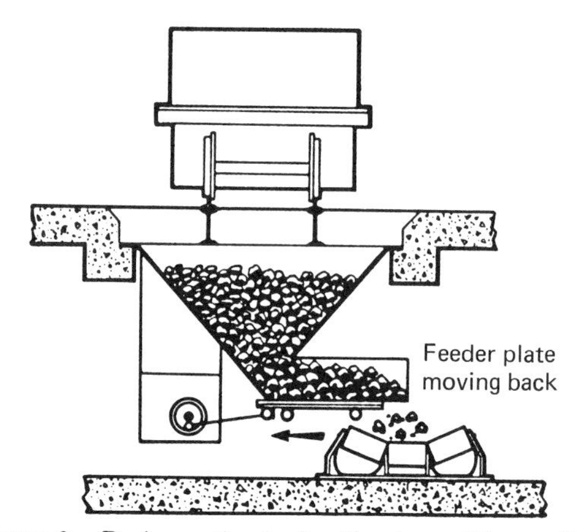

Figure 3. Reciprocating feeder. Courtesy of Rexnord Inc.

Roll feeder. In a feeder of this type, a smooth, fluted, or cleated roll or drum rotates to deliver bulk materials (Fig. 4(**a**)). The unit is limited to handling rather small lumps (<5 cm). However, by adding radial vanes 10–15 cm high across the full width of the drum at ca 60° spacing, the unit is converted to a rotary-vane feeder, which can handle larger volumes of material with more accuracy because of the positive-displacement principle.

Rotary-table feeder. This unit consists of a rotating circular table to which material flows from a round bin or hopper opening and from which it is discharged by a plow (Fig. 4(**b**)). This type feeder is a live-bin bottom and is best applied for withdrawing sluggish materials from large bins or outside storage piles since the plate diameter can be made very large, thereby minimizing the hoppering otherwise required to reduce the size of the discharge opening.

Screw feeder. A screw revolving in a stationary trough or casing fitted with hangers, trough ends, and other auxiliary accessories (Fig. 4(**c**)) makes an excellent volumetric control withdrawal unit with no practical limitation in the angle of incline. A screw feeder is, however, somewhat sensitive to abrasion wear and is not normally recommended for handling materials that lose value from degradation.

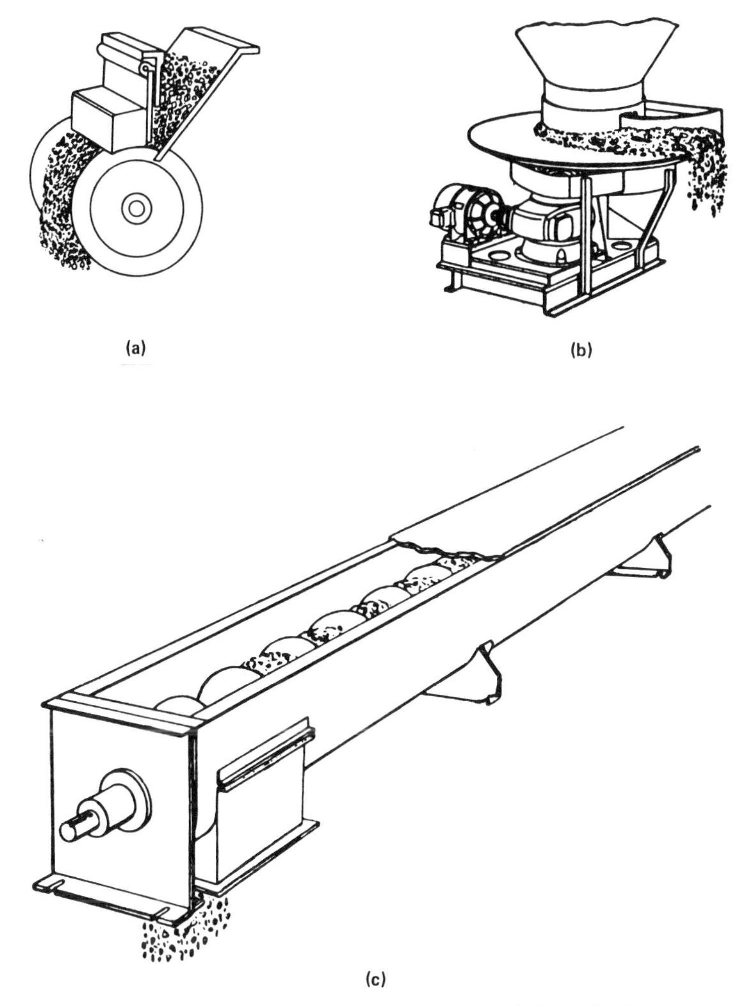

Figure 4. (**a**) Roll feeder. (**b**) Rotary-table feeder. (**c**) Screw feeder–conveyor.

Vibrating feeder. A vibrating feeder consists of a trough or tube, usually supported on or suspended from a steel or rubber spring system that is excited by an electric motor or magnet (see Fig. 1). It generally operates at a relatively low stroke (<1 cm) and high frequency (>15 Hz).

Conveyors. *Apron conveyor.* Sometimes referred to as a pan conveyor, an apron conveyor is a series of apron pans which, when attached to chain or pivotally attached one to another, forms a conveying medium. Apron conveyors are very versatile in contour. They can be straight horizontal, straight inclined (usually to a maximum of 45°), or any combination of the two. They are generally most economical at conveying lengths <18 m and are used for longer runs only when material temperature or size rules out other types of equipment.

Belt conveyor. Like a belt feeder, this conveyor consist of an endless rubber or metal belt operating over suitable drive, tail, and bend terminals and over belt idlers or slider beds. The workhorse of the material-handling industry, the belt conveyor is more flexible, versatile, and economical for more installations than any other conveying medium. Capacities of hundreds of tons per hour have been carried on a single unit for several kilometers. Belt conveyors can operate at inclines approaching 25° to the horizontal, depending upon the characteristics of the material being handled and are available in special fabrics capable of withstanding material temperatures up to ca 200°C.

Drag-chain conveyor. This type of conveyor has one or more endless chains which drag bulk materials in a trough constructed of wood, steel, or concrete (Fig. 5(**a**)). Drag-chain conveyors are often applied in installations handling material having lumps of ≤10 cm at temperatures up to 980°C.

En masse conveyor. An en masse conveyor consists of a series of skeleton or solid flights on an endless chain, cable, or other linkage operating in horizontal, inclined, or vertical paths within a closely fitted casing for the carrying run (Fig. 5(**b**)). The bulk material is conveyed and elevated en masse in an essentially continuous stream having a cross section substantially the same as the conveyor casing. En masse conveyors are most applicable for handling fine, free-flowing nonabrasive materials, and the totally enclosed casing construction provides excellent dust control for fluffy materials.

Flight conveyor. This conveyor consists of one or more endless propelling media, such as chain, to which flights are attached and a trough through which material is pushed by the flights (Fig. 5(**c**)).

Gravity-discharge conveyor–elevator. Gravity-discharge buckets are mounted between two endless chains that operate through troughs and casings in this system (Fig. 6(**a**)). Gravity-discharge conveyors are capable of horizontal, inclined, and vertical paths over suitable drive, corner, and takeup terminals. They are not recommended for highly abrasive materials, but because of their versatility, they can be economically used in place of up to three units where a horizontal–vertical–horizontal layout is required, within reasonable limits of length, lift, and capacity.

Pivoted bucket carrier. A pivoted bucket conveyor is similar to the gravity-discharge conveyor, except that in this case, the buckets remain in the carrying position until they are tipped for discharge (Fig. 6(**b**)). The pivoted bucket carrier, too, can be used in place of as many as three units where horizontal–elevating–horizontal layout is required. Because there is no scraping, dragging, or transfer action involved with this unit, it is applicable for hauling abrasive, dusty, and friable materials.

Pneumatic conveyor. In a pneumatic conveyor, material is charged into an airtight tube and propelled by a continuous blast of forced air. Pneumatic systems are usually self-fed by a surge hopper at the charge points. Materials being conveyed must be free flowing with minimal abrasion and packing tendencies. Conveying tubes are usually relatively small in cross section and readily routed through otherwise restricted plant areas. With proper control, dusting is a negligible problem.

Screw conveyor. A conveyor screw revolves in a stationary trough or casing fitted with hangers, trough ends, and other auxiliary accessories (Fig. 4(**c**)). The screw conveyor is best suited for moderate conveying distances. Its enclosed casing construction affords maximum dust control and weather protection.

Shuttle conveyor. Any conveyor, eg, belt, apron, screw, etc, that is mounted on a self-contained mobile structure, generally operating on rails, parallel to the direction of the flow of material becomes a shuttle conveyor (Fig. 7). In practice, shuttle conveyors are usually belt conveyors, are almost always reversible, and are mounted over a long line of storage bins to provide for multiple loading points into the bins. Shuttles are most often fed at right angles by the preceding conveyor in a system, but in-line

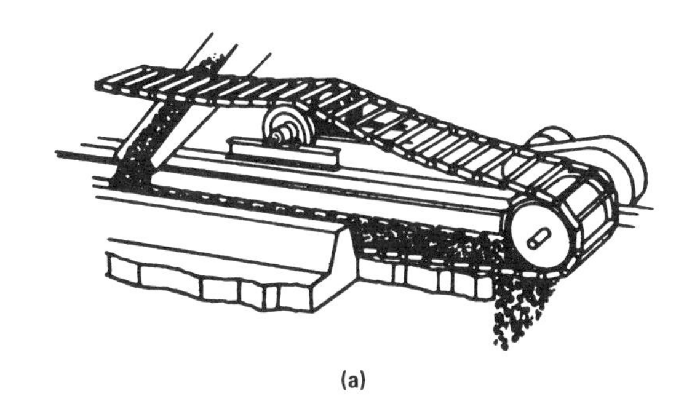

(a)

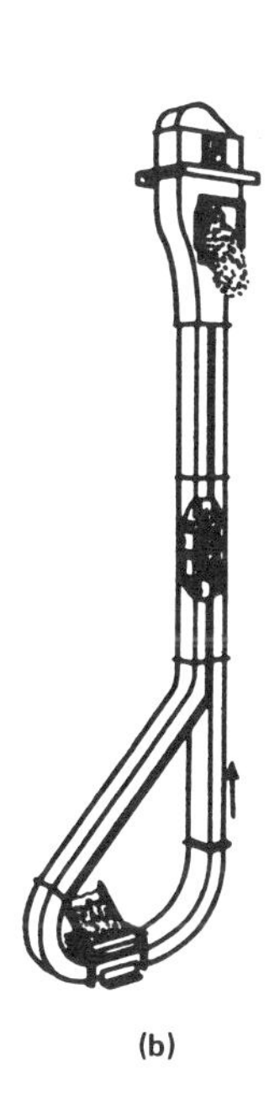

(b)

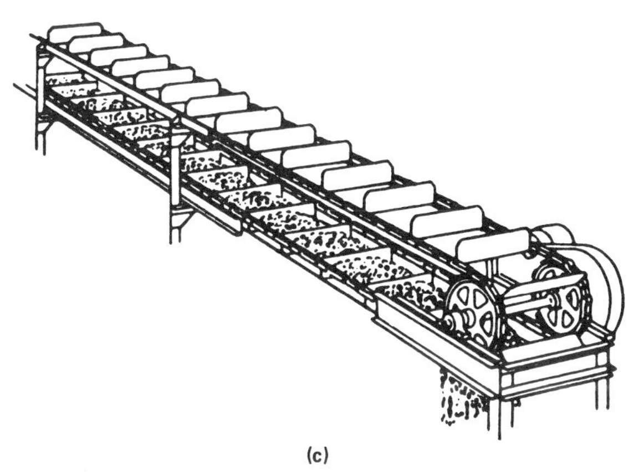

(c)

Figure 5. (**a**) Drag-chain conveyor. (**b**) En masse conveyor. (**c**) Flight conveyor.

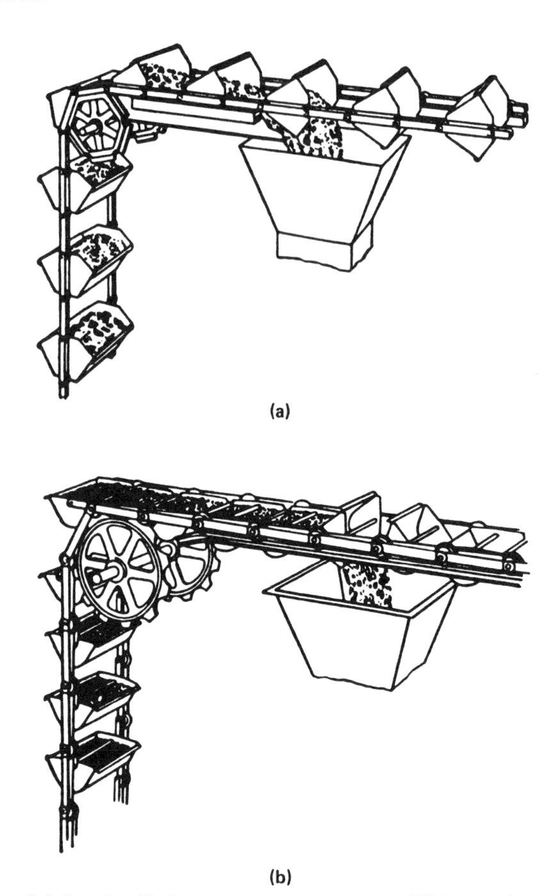

Figure 6. (**a**) Gravity-discharge conveyor–elevator. (**b**) Pivoted bucket carrier.

feeding is feasible where proper clearances are maintained between the shuttle and the preceding unit.

Slat conveyor. A slat conveyor is one or more endless chains to which nonoverlapping, noninterlocking, spaced slats are attached (Fig. 8(**a**)). Because of spacing between adjacent slats, units are not adaptable for handling small or dusty material and are therefore usually limited to unit or package handling.

Sliding-chain conveyor. In a sliding-chain conveyor, one or more endless chains slide on tracks for unit or package handling (Fig. 8(**b**)).

Vibrating conveyor. A trough or tube, usually supported on or suspended from a steel or rubber spring system, is excited by electric motor-driven eccentrics to a relatively high stroke (2–2.5 cm) and low frequency (4–8 Hz). Because there is no conveyor return run, operating and cleanup problems otherwise caused by material adhering to the conveying medium and dropping off on the return run are eliminated. En route processing of conveyed material by addition of grizzly, lumpbreaking, or cooling sections is feasible. Vibrating conveyors are suitable for handling hot materials. They are, however, somewhat imited in angle of incline, depending upon characteristics of the material being handled.

Bucket Elevators. Of the various types of elevators covered in this section, each has features that particularly suit it to certain applications and therefore set it apart from the others. On the other hand, the various types of elevators do have much in

Figure 7. Shuttle conveyor. Courtesy of Rexnord Inc.

common in their general design, features, and components. In working and dealing with bucket elevators, the user should have a broad knowledge of the glossary and definitions in order to better understand their operation.

Principles of Operation. All bucket elevators operate in essentially the same manner. A series of buckets are mounted on a continuous strand of chain or rubber belting and are filled, or loaded, at the bottom (boot or footend) of the elevator casing, and elevated to the top (headend) of the elevator, where material in the buckets is discharged as the buckets are inverted over the headend terminal (headshaft) assembly as they enter the downgoing side (return run) (Fig. 9).

The following are the most common styles of bucket elevators:

A centrifugal-discharge bucket elevator (Fig. 10(**a**)) uses centrifugal-discharge elevator buckets (Fig. 10(**b**)) spaced to permit the free discharge of bulk materials. The bucket is designed to scoop material from the elevator boot and discharge by reason of centrifugal force and gravity. Buckets can be mounted on either chain or belt, depending on the application and characteristics of the material to be handled. Centrifugal-discharge elevators are best suited for handling less coarse (<5-cm lump) free-flowing material.

A continuous bucket elevator uses continuous elevator buckets (Fig. 10(**c**)) mounted consecutively on a single- or double-chain strand. The continuous bucket has sides projecting beyond the front and, when spaced continuously with other buckets, forms a chute for the material discharged by the trailing bucket as it passes over the elevator headend terminal. Since the continuous bucket elevator does not depend upon centrifugal force to cleanly discharge material, the elevator can be operated at a slower speed and is therefore usually recommended for handling of highly abrasive or dusty material.

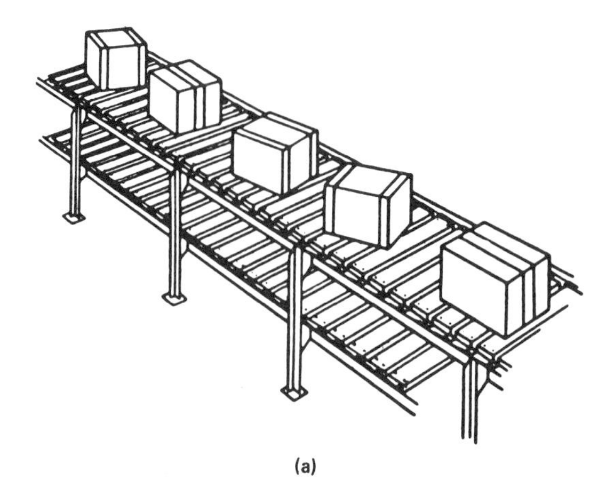

(a)

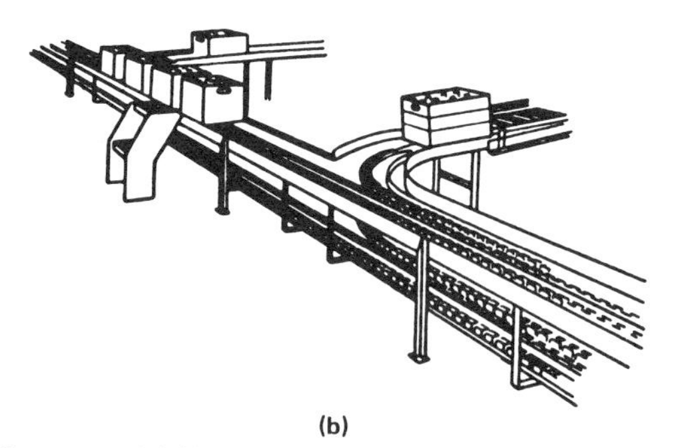

(b)

Figure 8. (**a**) Slat conveyor. (**b**) Sliding-chain conveyor.

A super-capacity bucket elevator operates on the same principle as the continuous bucket elevator, except that the buckets (Fig. 10(**d**)) are much larger and are end-hung between two strands of chain, so that the back of the bucket at the bottom extends backwards into the space between the up and down runs to provide additional capacity without increase in the bucket length or projection. The shape of the super-capacity bucket also enables this unit to handle considerably larger material lumps.

An internal-discharge bucket elevator has continuous buckets abutting, hinged, or overlapping, and is designed for loading and discharging along the inner boundary of the closed path of the buckets (Fig. 10(**e**)).

A positive-discharge bucket elevator is a spaced bucket elevator in which the buckets are held over the discharge chute long enough to permit free gravity discharge of materials. It is used for very sluggish materials.

En masse elevator (see en masse conveyor, Fig. 5(**b**)).

Gravity-discharge elevator (see gravity-discharge conveyor–elevator, Fig. 6(**a**)).

Pivoted bucket elevator (see pivoted bucket carrier, Fig. 6(**b**)).

A mill-duty bucket elevator is a centrifugal discharge elevator, but incorporation of hooded and high-front-lip buckets allows buckets to be spaced almost continuously (Fig. 10(**f**)). This unique bucket design, which also permits higher elevator speeds, results in considerably greater capacity ratings.

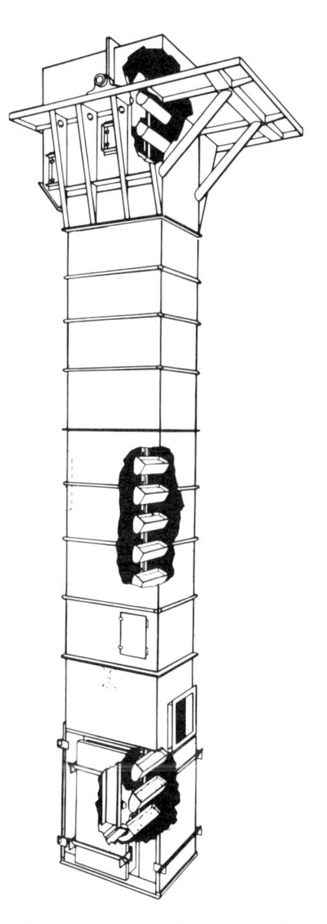

Figure 9. Bucket elevator. Courtesy of Rexnord Inc.

Application Guides

With the many variables that must be considered in the selection of the most efficient and economical type of bulk material-handling unit, selecting the right piece of feeding or conveying equipment to perform a given task may be difficult. A valuable tool in the preliminary selection of a proper material-handling unit is the selection guide (Figs. 11–13).

In the charts, one item in each subheading is selected that most nearly represents the layout requirements and the prevailing conditions for the material to be handled. A vertical line is then drawn in each of the applicable columns. Horizontal lines are drawn for each unit type being considered. If this horizontal line intersects a vertical line in a blank space, that particular type unit is unsuitable for the intended use. When the horizontal line intersects all vertical lines in a space having a solid dot or a properly satisfied provision, the unit being considered is judged to be properly applied. If two or more units are qualified for the same application, the choice is based upon such factors as expected service life, initial cost, and potential maintenance requirements.

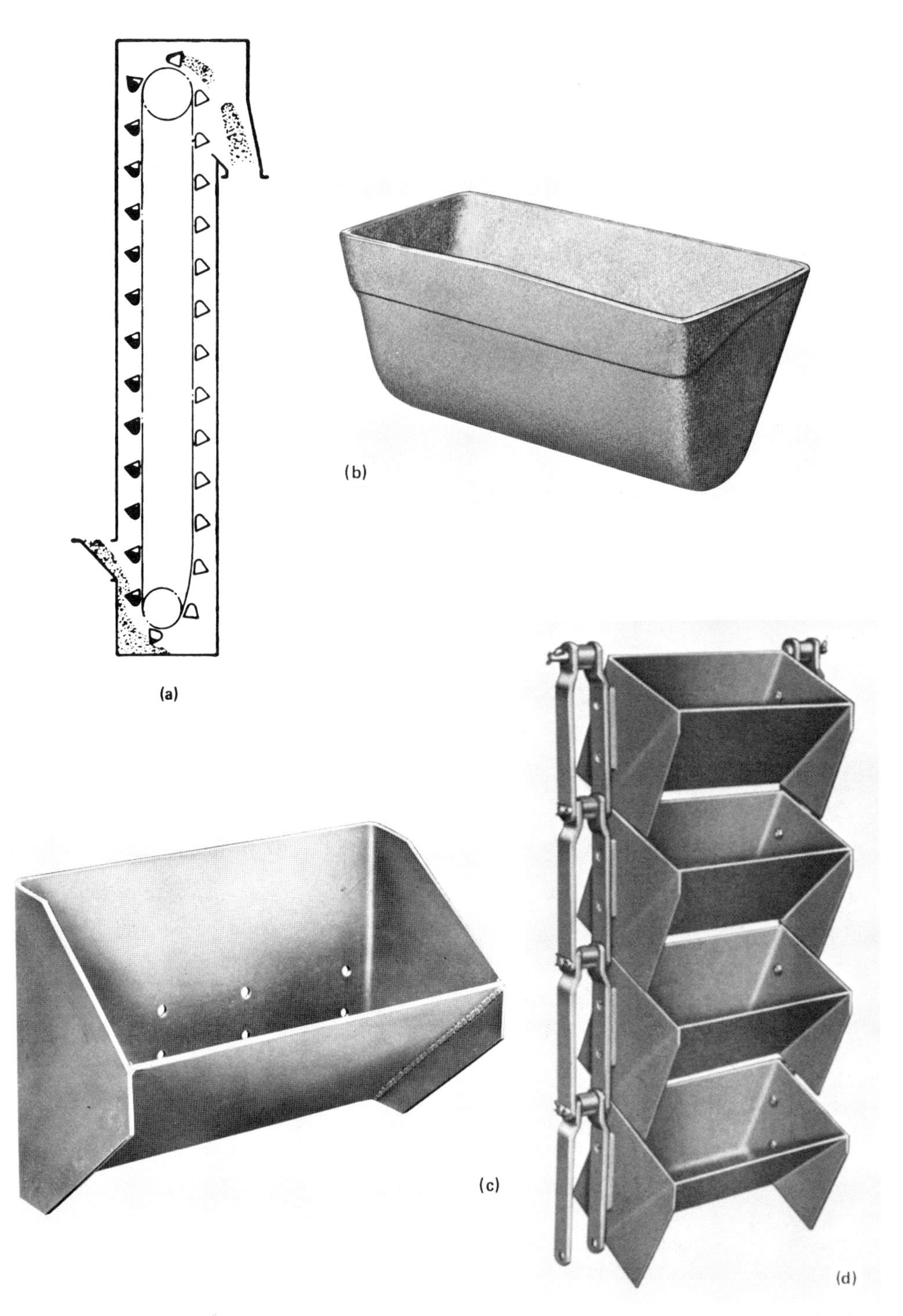

Figure 10. (**a**) Centrifugal-discharge bucket elevator. (**b**) Centrifugal-discharge bucket. (**c**) Continuous-discharge bucket. (**d**) Super-capacity elevator bucket. (**e**) Internal-discharge bucket elevator. (**f**) Mill-duty elevator bucket.

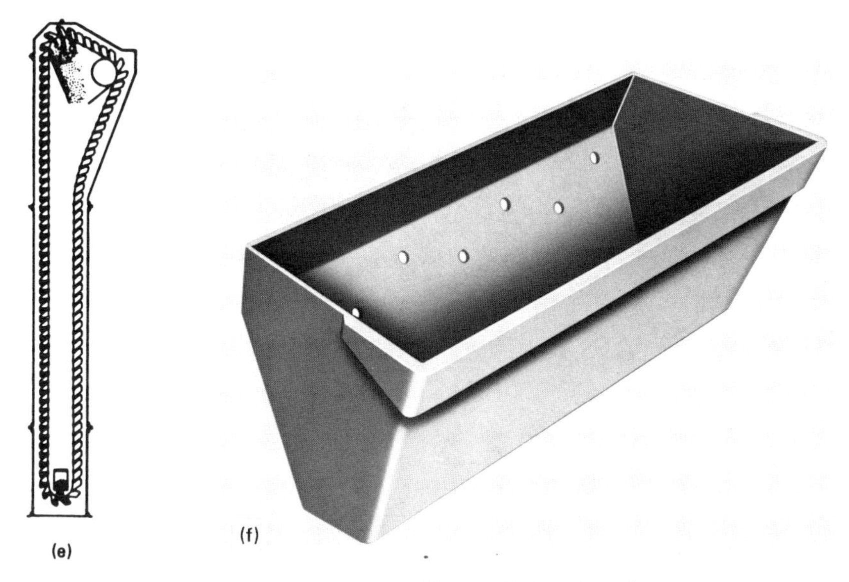

Figure 10. *(continued)*

Proper Unit Size and Speed Selection. In the unit selection guides (Figs. 11–13), capacity to be handled is not a factor in the selection of the most applicable type of material-handling unit. This requirement is usually not taken into consideration until a tentative equipment selection is made, at which time the size and speed of travel of the selected unit is established.

Most equipment manufacturers publish data that provide assistance to the designer in selecting equipment of optimum sizing and speed for the capacity desired. In almost all cases, published allowable-feeding and -conveying speeds have a fairly wide band of minimum to maximum ranges, and the designer needs to be sure only that the speed selected for a particular unit falls within this band.

Bucket Elevators. An exception to the above is found in bucket elevators, where the speed of a particular type of unit has little, if any, allowable variation, since proper discharging or emptying action of the buckets is a function of the centrifugal forces that are created as the bucket enters its dumping position over the elevator head terminal. Operation either above or below the recommended speed creates a condition known as backlegging, in which material discharging from the bucket does not follow its prescribed path into the elevator chute but instead falls back into the elevator boot, where eventual plugging becomes inevitable. When an elevator is operating at less than design speed, discharging material begins to release from the bucket prematurely, ie, before the bucket is in its proper position relative to the elevator discharge chute. On the other hand, should an elevator be operated at a speed in excess of proper design value, the abnormal centrifugal forces generated tend to hold the material in the bucket for a slightly extended period of time. This is often enough to delay the discharge of material until after the bucket has passed the inlet of the elevator discharge chute on its downward run and again can result in material dropping back in the elevator boot section.

Feeder type	Loading (see below)			Feeder profile (see below)					
	Condition A	Condition B	Condition C	Horizontal	10° max	18° max	30° max	45° max	45° to vertical
Apron	●	●	●	●	●	●	●	●	
Belt	●	●		●	●	①	①		
Chain curtain	●			●					
Drag chain	●	●		●					
En Masse	●	●		●	●	●	●	●	●
Flight	●	●		●					
Grizzy, live roll	●		●	●					
Reciprocating plate	●	●		●					
Roll	●			●					
Rotary table		●		●					
Rotary vane	●	●		●					
Screw	●	●		●	●	①	①	①	①
Vibrating	●	①	①	●	●	①			

Feeder type	Material size								Flowability: Angle of repose			
	Silt (>58 μm)	Very fine (58–149 μm)	Granule (149 μm–0.6 cm)	Pebble (0.6–5 cm)	Small cobble (5–10 cm)	Medium cobble (10–15 cm)	Large cobble (15–30 cm)	Boulder (>30 cm)	Flushing (0–15°)	Free-flowing (15–35°)	Sluggish (35–60°)	Sticky (60–90°)
Apron			●	●	●	●	●	●		●	●	●
Belt		①	●	●	●					●	●	①
Chain curtain						●	●	●		●		
Drag chain	●	●	●	●	●					●	●	
En Masse		●	●	●	●					●		
Flight			●	●	●	●				●	●	
Grizzy, live roll						●	●	●		●	●	●
Reciprocating plate			●	●	●	●	●	●		●	●	●
Roll			●	●						●		
Rotary table			●	●	●					●	●	
Rotary vane	●	●	●	●					●	●		
Screw	●	●	●	●						●	●	
Vibrating		①	●	●	●	●	●	●		●	①	

Feeder type	or Specific material						Friability		
	Sludges and filtercake	Setting (trisuper phosphate)	Packing (sodium bicarbonate)	Metal turnings	Wood chips	Sliverstick and pulpwood	Nonfriable	Friable	Dusts excessively
Apron			●	●	●		●	●	①
Belt	●	●	●		●		●	●	①
Chain curtain						●	●		
Drag chain					●	●	●		①
En Masse							●		●
Flight					●	●	●		①
Grizzy, live roll			●				●		
Reciprocating plate			●	●	●		●		①
Roll							●		●
Rotary table				●	●	●	●		●
Rotary vane					●		●		●
Screw	●				●		●		●
Vibrating				●	●		●	●	①

Feeder type	Temperature					Corrosiveness			Abrasiveness		
	Cold (<0°C)	Ambient (0–66°C)	Hot (66–150°C)	Very hot (150–480°C)	High temp (>480°C)	Noncorrosive (>7 pH)	Mildly corrosive (5–7 pH)	Very corrosive (1–5 pH)	Nonabrasive (1–3 Mohs)	Mildly abrasive (3–5 Mohs)	Very abrasive (>5 Mohs)
Apron	●	●	●	●	①	●	●	①	●	●	●
Belt	●	●	①			●	●	●	●	●	●
Chain curtain	●	●	●			●	●	●	●	●	●
Drag chain	●	●	●	●	●	●	●	①	●	●	①
En Masse	●	●	●			●	①	①	●	●	
Flight	●	●	●	●		●	●	①	●	●	
Grizzy, live roll	●	●	●	●		●	●	●	●	●	●
Reciprocating plate	●	●	●	●		●	●	①	●	●	①
Roll	●	●	●	●		●	●	①	●	●	●
Rotary table	●	●	●			●	●	①	●	●	
Rotary vane	●	●	●			●	●		●	●	①
Screw	●	●	●			●	●	①	●	●	①
Vibrating	●	●	●	●	①	●	●	①	●	●	①

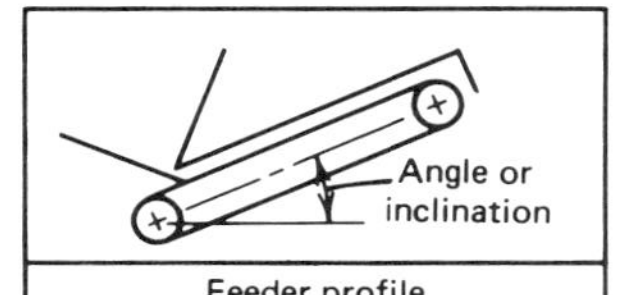

Figure 11. Feeder selection guide. ① Special consideration required; consult manufacturer. Examples include special steels, special belts, special troughs, and enclosures. Courtesy of Rexnord Inc.

Conveyor type	Loading: Controlled feed	Loading: Uncontrolled feed	Discharge: Head end	Discharge: Head or foot end	Discharge: Single intermediate	Discharge: Multiple intermediate	Discharge: Variable intermediate	Angle of inclination: Horizontal	Angle of inclination: 10° max	Angle of inclination: 18° max	Angle of inclination: 30° max	Angle of inclination: Up to 45° max	Complexity: Straight line	Complexity: Compound	Horizontal carry: <30 m	Horizontal carry: <76 m	Horizontal carry: >76 m	Vertical lift: <9 m	Vertical lift: <18 m	Vertical lift: <24 m	Vertical lift: >24 m
Apron conveyor, deep pan	●	●	●	●				●	●	●	●	①	●	②	●	●		●	●	●	
Apron conveyor, hinged pan	●	●	●	●				●	●	●	①		●	②	●	●		●	●		
Apron conveyor, shallow pan	●	●	●	●				●	●	●	①		●	②	●	●		●	●		
Belt conveyor, rubber belt	●		●	●				●	●	③	③		●	②	●	●	●	●	●	●	●
Belt conveyor, shuttle	●		●	●			●	●					●		●	●					
Belt conveyor, steel band	●		●	●				●	●				●	②	●	●	●	●	●	●	●
Belt conveyor with plows	●		●		●	●		●	●				●	②	●	●	●	●	●	●	●
Belt conveyor with mobile tripper	●		●			●	●	●					●	⑤	●	●	●	●	●	●	●
Belt conveyor with stationary tripper	●		●		●	●		●	●	③	③		●	②	●	●	●	●	●	●	●
Drag-chain conveyor	●	●	●	●	●	●		●	③	③		●	●	②	●	●		●			
En masse conveyor	●	●	●		●	●		●	●	●	●		●	●	●	●		●	●	●	●
Flight conveyor	●		●	●	●	●		●	●	●	●	①	●	②	●	●		●	●	●	
Gravity-discharge conveyor	●		●		●	●		●					●	●	●			●	●	●	●
Pivoted-bucket conveyor	●				●	●	●	●	●	●	●	●	●	●	●	●		●	●	●	●
Screw conveyor	●	●	●	●	●	●		●	●	①	①	①	●		●	●	①	●	①		
Slat conveyor	●		●					●	●				●	●	●	●	●	●			
Sliding-chain conveyor	●		●	①	●	●		●	●	①			●	●	●	●	●	●	●	●	
Vibrating conveyor	●	●	●		●	●	●	●	●	①			●		●	●	●	●			

Conveyor type	Material size: Silt (>58 μm)	Material size: Very fine (58–149 μm)	Material size: Granule (149 μm–0.6 cm)	Material size: Pebble (0.6–5 cm)	Material size: Small cobble (5–10 cm)	Material size: Medium cobble (10–15 cm)	Material size: Large cobble (15–30 cm)	Material size: Boulder (>30 cm)	Flowability, angle of repose: Flushing (0–15°)	Flowability, angle of repose: Free-flowing (15–35°)	Flowability, angle of repose: Sluggish (35–60°)	Flowability, angle of repose: Sticky (60–90°)	Flowability, specific material: Sludge and filtercake	Flowability, specific material: Setting (trisuper phosphate)	Flowability, specific material: Packing (sodium bicarbonate)	Flowability, specific material: Metal turnings	Flowability, specific material: Wood chips	Flowability, specific material: Sliverstick and pulpwood
Apron conveyor, deep pan			●	●	●	●	●		●	●	●						●	
Apron conveyor, hinged pan				●	●	●										●	●	●
Apron conveyor, shallow pan			●	●	●	●	●	●		●	●				●		●	
Belt conveyor, rubber belt		①	●	●	●	●	●	●	①	●	●	●	●	●	●	●	●	●
Belt conveyor, shuttle		①	●	●	●	●			①	●	●	●	●	●	●	●	●	
Belt conveyor, steel band		①	●	●	●	●				●	●	●	●	●	●	●	●	
Belt conveyor with plows			●	●						●	●		●		●		●	●
Belt conveyor with mobile tripper		①	●	●	●	●	①	①	①	●	●	①	●		●		●	
Belt conveyor with stationary tripper		①	●	●	●	●	①	①	①	●	●	①	●		●		●	
Drag-chain conveyor	●	●	●	●	●					●							●	●
En masse conveyor		●	●	●	●					●								
Flight conveyor			●	●	●	●				●	●						●	
Gravity-discharge conveyor			●	●	●					●	●		●				●	
Pivoted-bucket conveyor			●	●	●	●			●	●	●					●	●	
Screw conveyor	●	●	●	●	●					●			●					
Slat conveyor					●	●	●	●										●
Sliding-chain conveyor				●	●	●	●	●		●						●	●	●
Vibrating conveyor	①	①	●	●	●	●	●	●		●						●	●	

Conveyor type	Friability: Nonfriable	Friability: Friable	Friability: Dusts excessively	Temperature: Cold (<0°C)	Temperature: Ambient (0–66°C)	Temperature: Hot (66–150°C)	Temperature: Very hot (150–480°C)	Temperature: High temp (>480°C)	Corrosiveness: Noncorrosive (>7 pH)	Corrosiveness: Mildly corrosive (5–7 pH)	Corrosiveness: Very corrosive (1–5 pH)	Abrasiveness: Nonabrasive (1–3 Mohs)	Abrasiveness: Mildly abrasive (3–5 Mohs)	Abrasiveness: Very abrasive (>5 Mohs)
Apron conveyor, deep pan	●	●	①	●	●	●	●	①	●	●	①	●	●	●
Apron conveyor, hinged pan	●	●	①		●	●			●	●	①	●	●	
Apron conveyor, shallow pan	●	●	①	●	●	●	●	①	●	●	①	●	●	●
Belt conveyor, rubber belt	●	④	①	●	●	①			●	●	●	●	●	●
Belt conveyor, shuttle	●	④	①	●	●	①			●	●	●	●	●	●
Belt conveyor, steel band	●		①		●	●			●			●	●	
Belt conveyor with plows	●		①	●	●	①			●	●	●	●	●	●
Belt conveyor with mobile tripper	●		①	●	●	①			●	●	●	●	●	●
Belt conveyor with stationary tripper	●		①	●	●	①			●	●	●	●	●	●
Drag-chain conveyor	●		①	●	●	●	●	●	●	●	①	●	●	①
En masse conveyor	●		●	●	●	●			●	①	①	●	●	
Flight conveyor	●		①	●	●	●	●		●	●	①	●	●	
Gravity-discharge conveyor	●		①	●	●	●			●	●	①	●	●	
Pivoted-bucket conveyor	●	●	①	●	●	●	●		●	●	①	●	●	●
Screw conveyor	●		●	●	●	●	●		●	①	①	●	●	①
Slat conveyor	●	●		●	●	●			●	●	①	●	●	①
Sliding-chain conveyor	●			●	●	●	●		●	●	①	●	●	①
Vibrating conveyor	●	●	①	●	●	●	●	①	●	●	●	●	●	①

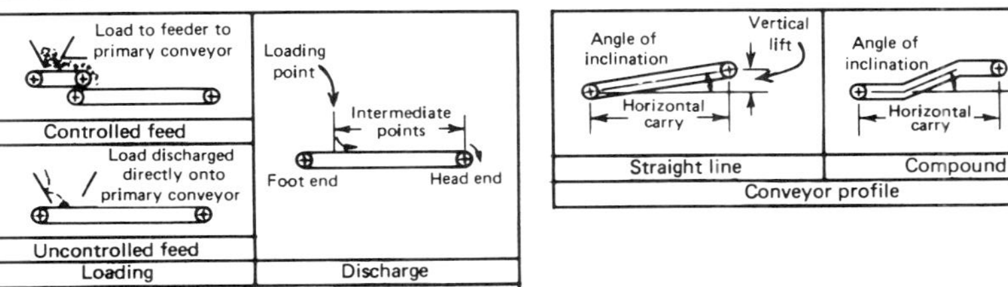

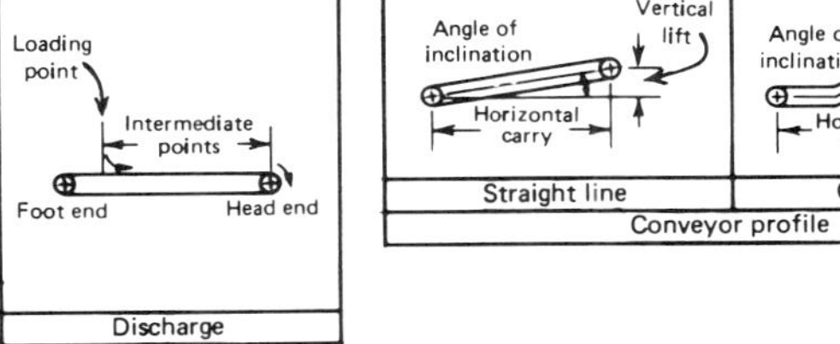

Figure 12. Conveyor selection guide. ① Special considerations required; consult manufacturer. Examples include special steels, special belts, and enclosures. ② In combinations of inclined and horizontal units only. ③ Investigate maximum operating angle for material handled. ④ At controlled belt speeds. ⑤ Tripper must be located on horizontal section. Courtesy of Rexnord Inc.

	Elevator type	Material transfer (see below)							Elevator profile (see below)										
		Loading		Discharge					Angle of inclination				Complexity		Horizontal carry			Vertical lift	
		Controlled feed	Uncontrolled feed	Head end	Head or foot end	Single intermediate	Multiple intermediate	Variable intermediate	45° max	60° max	75° max	Vertical	Straight line	Compound	<30 m	<76 m	>76 m	<24 m	>24 m
Bucket elevators	Centrifugal discharge, chain type	●		●							●	●	●					●	●
	Centrifugal discharge, belt type	●		●							●	●	●					●	●
	Continuous bucket, chain type	●		●							●	●	●					●	●
	Continuous bucket, belt type	●		●							●	●	●					●	●
	Continuous overlapping bucket	●		●								●	●					●	●
	Super capacity buckets, Style F	●		●					●	●			●					●	●
	Super capacity buckets, Style G	●		●							●	●	●					●	●
	Super capacity buckets, Style H or HL	●		●								●	●					●	●
	Positive discharge	●		●								●	●					●	●
	Internal discharge	●		●								●	●					●	●
	En masse conveyor	●	●	●		●	●		●	●	●	●	●	●	●	●		●	●
	Gravity-discharge conveyor	●		●		●	●			●	●	●	●	●	●			●	●
	Pivoted-bucket conveyor	●				●	●	●	●	●	●	●	●	●	●	●		●	●
	Screw conveyor	●	●	●	●	●	●		①	①	①	●	●		●	●	①	●	

Elevator type	Material properties																	
	Material size								Flowability									
									Angle of repose				or Specific material					
	Silt (>58 µm)	Very fine (58–149 µm)	Granule (149 µm–0.6 cm)	Pebble (0.6–5 cm)	Small cobble (5–10 cm)	Medium cobble (10–15 cm)	Large cobble (15–30 cm)	Boulder (>30 cm)	Flushing (0–15°)	Free-flowing (15–35°)	Sluggish (35–60°)	Sticky (60–90°)	Sludge and filtercake	Setting (trisuper phosphate)	Packing (sodium bicarbonate)	Metal turnings	Wood chips	Sliverstick and pulpwood
Centrifugal discharge, chain type		●	●	●	②				①	●	③	③			①		●	
Centrifugal discharge, belt type		●	●	●	②				①	●	③	③			①		●	
Continuous bucket, chain type		●	●	●	●	●				●	●	①				●	●	
Continuous bucket, belt type		●	●							●	●						●	
Continuous overlapping bucket	●	●	●	●	●				●	●	●	①				●	●	
Super capacity buckets, Style F		●	●	●	●	●	●	●	●	●	●	●				●	●	
Super capacity buckets, Style G		●	●	●	●	●	●	●	●	●	●	●				●	●	
Super capacity buckets, Style H or HL		●	●	●	●	●	●	●		●	●	●				●	●	
Positive discharge	●	●	●	●	②				●	●	●	●			①		●	
Internal discharge			●	●						●								
En masse conveyor		●	●	●	●					●								
Gravity-discharge conveyor			●	●	●					●	●		●				●	
Pivoted-bucket conveyor			●	●	●	●			●	●	●					●	●	
Screw conveyor	●	●	●	●	●					●			●					

Elevator type	Material properties													
	Friability			Temperature					Corrosiveness			Abrasiveness		
	Nonfriable	Friable	Dusts excessively	Cold (<0°C)	Ambient (0–66°C)	Hot (66–150°C)	Very hot (150–480°C)	High temp (>480°C)	Noncorrosive (>7 pH)	Mildly corrosive (5–7 pH)	Very corrosive (1–5 pH)	Nonabrasive (1–3 Mohs)	Mildly abrasive (3–5 Mohs)	Very abrasive (>5 Mohs)
Centrifugal discharge, chain type	●		①	●	●	●	①		●	●	①	●	●	①
Centrifugal discharge, belt type	●		●	●	●	①			●	●	①	●	●	①
Continuous bucket, chain type	●	●	●	●	●	●	●	①	●	●	①	●	●	①
Continuous bucket, belt type	●	●	●	●	●	①			●	●	①	●	●	①
Continuous overlapping bucket	●	●	●	●	●	●	●	①	●	●	①	●	●	①
Super capacity buckets, Style F	●	●	●	●	●	●	●		●	●	①	●	●	●
Super capacity buckets, Style G	●	●	●	●	●	●	●		●	●	①	●	●	①
Super capacity buckets, Style H or HL	●	●	●	●	●	●	●		●	●	①	●	●	①
Positive discharge	●	●	●	●	●	●	●		●	●	①	●	●	
Internal discharge	●		●	●	●	●	●		●	●	①	●		
En masse conveyor	●		●	●	●	●			●	①	①	●	●	
Gravity-discharge conveyor	●		①	●	●	●			●	●	①	●	●	
Pivoted-bucket conveyor	●	●	①	●	●	●	●		●	●	①	●	●	●
Screw conveyor	●		●	●	●	●			●	①	①	●	●	①

Controlled feed
Uncontrolled feed
Loading
Head end
Intermediate points
Loading point
Foot end
Head end
Discharge
Material transfer

Angle of inclination
Vertical lift
Horizontal carry
Straight line
Angle of inclination
Vertical lift
Horizontal carry
Compound
Elevator profile

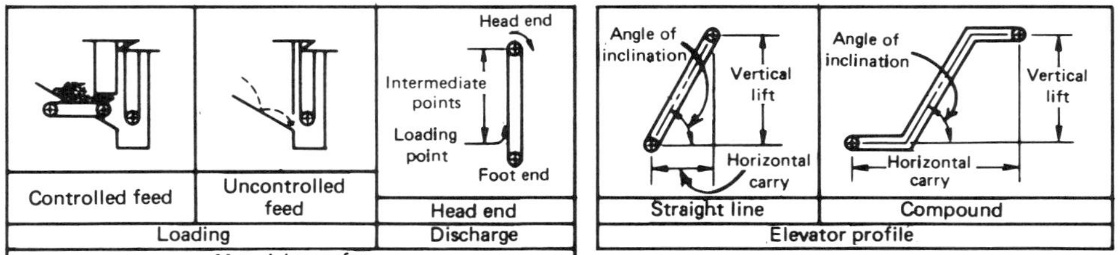

Figure 13. Elevator selection guide. ① Special considerations required; consult manufacturer. Examples include special steels, special belts, and special buckets. ② Satisfactory when material contains large percentage of fines. ③ Satisfactory with special, large-radius buckets. Courtesy of Rexnord Inc.

Proper elevator selection and power requirements can be estimated by first determining the proper elevator type from the selection guide (Fig. 13), and then finding proper elevator size. Elevator capacity can be determined by calculating volume to be handled, from the following formula:

$$\frac{\mathrm{m}^3}{\mathrm{h}} = \frac{\mathrm{t/h}}{D}$$

$$\left(\mathrm{CFH} = \frac{\mathrm{TPH} \times 2000}{\mathrm{PCF}}\right)$$

where D = bulk density in g/cm^3, TPH = required capacity in short tons per hour, and PCF = bulk density of material to be handled in its as-handled condition, expressed in pounds per cubic foot. Refer to data published by most bucket-elevator manufacturers, in catalogue or bulletin form, for tentative elevator sizing and power requirements (4).

Belt Conveyors. Belt conveyors have an extremely wide allowable speed range, anywhere from $\leq$0.3 m/min to >300 m/min. For normal conveying applications, typical belt speeds are 15–180 m/min, depending upon the application. Speeds of <15 m/min are seldom used, except in cases where maximum material retention time becomes a consideration for such reasons as a need for a cooling or setting time during material processing. Belt speeds of >180 m/min are seldom found in in-plant operations, but they are encountered in longer, overland-type conveyors with pulley centers of $\geq$200 m.

The design of units of this length and speed range is usually undertaken by only the more practiced material-handling design engineers because of the need to consider many critical areas of design. For example, loading skirts must be designed to minimize friction between the skirt boards and the material being handled, or the skirt boards and the conveyor belt itself. Also, loading skirts must be long enough to allow the material being loaded onto the belt conveyor to stabilize, particularly where larger lumps are being handled, lest undue material spillage occurs as the material leaves the loading-zone area. At the same time, excessive loading-skirt length should be avoided to minimize unnecessary friction and resultant increase in power requirements. A reasonable rule of thumb dictates that loading-skirt length should be 1.5–2 m for each 100 m/min of belt travel. At the conveyor discharge chute, special care should be taken to accommodate the flatter discharging material trajectory that results from higher speeds.

Type, design, and location of conveyor-belt takeups become important considerations on longer conveyors. Gravity takeups are desirable, if not essential, on any belt conveyor with pulley centers >90 m. On horizontal or nearly horizontal conveyors, gravity takeups are most efficiently located at or near the conveyor head end. On inclined units, the gravity takeup operates more efficiently near the tail end of the conveyor.

Because of the magnitude of the inertia of a longer belt conveyor, some kind of reduced voltage or gradual starting should always be considered to reduce stress on the conveyor belt during the starting cycle, as well as to minimize the possibility of the belt lifting off the conveyor idlers on reverse reflex curves.

Belt-Conveyor Capacity. Upon establishing the feasibility or need of a belt conveyor, the designer next determines optimum belt width and speed for the capacity desired. Most belt-conveyor manufacturers have readily available reference data that show capacities that can be attained for various belt widths, operating at a wide range of speeds (5).

In making the preliminary determination, two requisites must first be established by the designer: the troughing angle of the belt-conveyor idler and the surcharge angle of the material. Most belt-conveyor-machinery manufacturers that make idlers offer three different troughing angles, ie, 20°, 35°, and 45° (Fig. 14).

Selection of troughing angle is made by the designer, but in a large percentage of applications, the 35° idler has become the most widely used. Although the 20° idler preceded the 35° unit in the evolution of belt-conveyor design, later development of more flexible conveyor-belt fabrics allowed the use of the deeper trough idler, with resultant increase in capacity for the same conveyor width.

Introduction of the 45° idler provided an even deeper trough and, hence, still greater capacity. However, some caution must be exercised in the selection of the 45° trough, because the additional flexibility required in the belt fabric can result in problems in load support as the belt spans the space between idlers, especially for heavy material containing large lumps. It is good practice to consult with the belt manu-

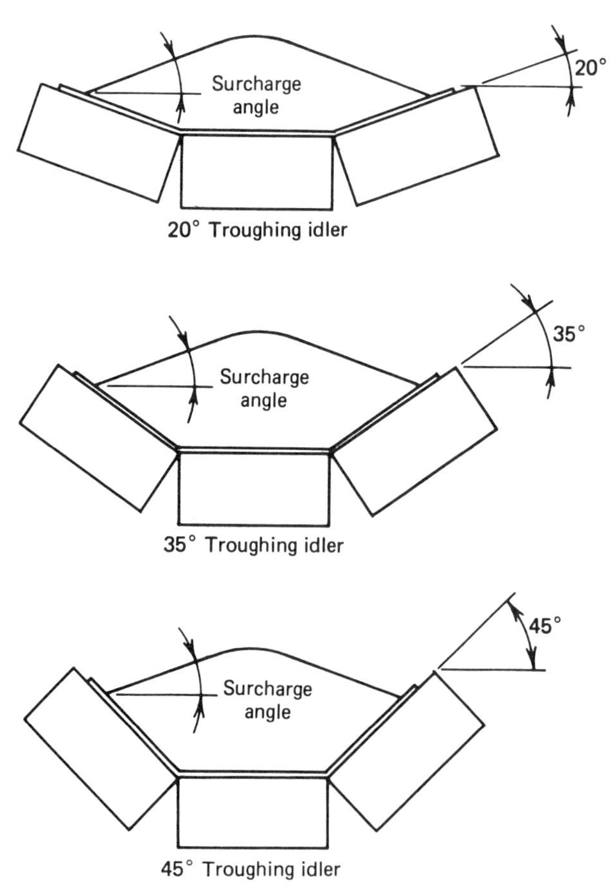

Figure 14. Belt-conveyor idler cross section.

facturer for recommendations regarding belt-fabric construction, optimum trough angle, and appropriate idler spacing.

Surcharge angle of the material as it is carried on the belt conveyor (Fig. 14) is largely a function of the static angle of repose of the material, taking into consideration also the size and shape of the lumps being handled. Although the surcharge angle of the material can vary and is best established by experiment, observation, or experience, the cross-sectional values for the 20° surcharge angle are generally considered to be most representative when the actual value is unknown.

Chain Conveyors. Normal operating speeds of chain-type conveyors are usually considerably less than those of belt conveyors for several reasons.

Conveyor belts are supported by, and roll on, closely spaced rollers, or belt idlers, that are equipped with well-lubricated precision-type roller or ball bearings. The construction of these rolls makes them well suited for operation at speeds of several hundred rpm, if desired. Chain conveyors, on the other hand, generate considerably greater frictional loadings, since in most cases, the chain drags directly on a metallic conveying bed or rails, or, at the very best, is supported by relatively small, nonbushed high frictional chain rollers. The resultant high frictional loadings dictate low maximum speeds.

Longer pitch chains are, size for size, more economical than shorter pitch chains since, for the most part, the bulk of the cost of a conveyor chain is found in the round parts at the chain articulation joints. However, as chain pitch increases, the consequences of chordal action of the chain engaging the head sprocket of the conveyor become more apparent and influential in the operation of the conveyor. As a sprocket tooth engages a chain link and the link begins to seat itself on the sprocket, a slight decrease in the forward speed of the chain occurs. This cyclical motion of the chain is hardly discernible on a short pitch chain, but it increases dramatically as the chain pitch lengthens. It can be compensated for to a great extent by increasing the number of teeth in the sprocket. However, there are certain practical limitations to the use of larger sprockets, since they not only encroach on usually restrictive space limitations but can also considerably increase drive costs because they increase torque requirements.

For reasons such as these, chain conveyors are usually designed with maximum conveying speeds of 30–38 m/min for chain pitches up to 23 cm, and somewhat less (15–23 m/min) for chains with greater pitches.

Feeders. In general, when a unit is serving as a feeder, it is operated at a slower speed than when used as a conveyor. As the feeder withdraws material from a bin or hopper through a restricted opening, the slower operating speed introduces several operational benefits: reduced material agitation; increased accuracy and uniformity of the volume of the material being withdrawn; less dusting of fines; less potential damage to the feeder from turbulent action of the material, especially when hard, lumpy material is being handled; and longer equipment life.

Units such as apron and belt feeders usually operate in a speed range of 3–15 m/min, depending upon the severity of their operating environment and the type of material to be handled. Rotary-table feeders, because of their size and relatively large mass, develop considerable inertia and are usually limited to $\leq$10 rpm. Reciprocating and vibrating feeders are less affected by the inertia of the unit and can operate through a very wide range of strokes and frequencies.

A reciprocating feeder, utilizing a directly connected crank arm, operates at strokes of 7–15 cm and at frequencies of 0.7–1.0 Hz (40–60 cycles per minute). The longer stroke and lower frequency of a reciprocating feeder impart a somewhat cyclical motion to its load; hence, the average conveying speed of the bed of material is determined as the product of the material displacement per stroke and the strokes per minute.

Conveying motion of materials on an electromechanical or electromagnetic vibrating feeder appears uniform to the naked eye, because of the relatively high frequency of the action. The rate of conveying speed of vibrating feeders varies considerably, depending upon the stroke and frequency of the unit as well as the characteristics of the material itself, but speeds of 7.6 m/min are usually readily attainable. Greater speeds can be obtained by slightly declining the slope of the feeder bottom plate.

Operating speed of a screw feeder is measured in revolutions per minute (rpm). Because of the uniqueness of the screw feeder design and construction, it is not practical to list a range of minimum and maximum allowable speeds, since so many variables are involved in the proper selection of this type unit. The designer, after having established the need and practicability of a screw feeder, should rely upon selection guides published by the various manufacturers of this equipment.

Power Requirements. As the cost of energy has increased, the users of material-handling systems have become more cognizant of the need to minimize power consumption from the standpoints of both energy conservation and operating costs.

In an effort to assist both the supplier and the user in units utilizing maximum efficiency, CEMA has assembled test data on uniform and accurate determination of optimum power requirements for material-handling equipment. Almost all conveyor manufacturers have adopted, in whole or in part, the data compiled by CEMA in the marketing of their products.

Generally speaking, power requirements for conveying units can be shown to be the sum of the following components: the power required to move the empty conveyor; the power required to move the load horizontally; the power required to lift the load; and the power consumed by friction or agitation of the load.

The power to move the conveyor itself includes the power to move the conveying medium over its supporting rolls or bed, the power consumed in rotating the terminal-shaft assemblies, and that to overcome the inefficiencies of the drive train. In the case of a screw conveyor, power is used in turning the screw in its bearings; in the case of vibrating conveyors, power is required to overcome the windage and internal-friction losses. In bucket elevators, the power required to move the empty unit consists wholly of that generated in rotating the headshaft and footshaft assemblies. In cases where bucket elevators have internal chain guides, some additional power usage, however slight, may be developed by the chain or buckets bearing against these guides in the upward or return travel.

In the case where material is carried directly on the conveying medium, such as on a belt or pan conveyor, the power required to move material over the horizontal length of the conveyor is obtained by adding the weight of material directly to the weight of the conveying medium. Where material is pushed by the conveying medium, such as a drag or screw conveyor, the material is considered separately, since its friction factor against the conveying bed is likely to be different than that of the conveying medium against its supports.

When evaluating the power required to lift the material, consideration must again be given to whether material is being conveyed on or propelled by the conveying medium because of differences in friction factors.

The power consumed by extraordinary material agitation or friction includes such action as an elevator bucket digging through the material in an elevator-boot section, a belt-conveyor plow-scraping material from a flattened conveyor belt at an intermediate discharge point, or the friction of material within the skirt boards at the conveyor load point. Additional power is also required for certain accessories that are often used with conveyors, such as a belt-conveyor traveling tripper or a belt scraper at a conveyor head end.

A source of significant power consumption common to feeders is the pullout resistance developed as material is literally dragged through a restrictive hopper outlet opening by the feeder medium. Since the function of the feeder is to introduce material into the system at a predetermined rate, the hopper-outlet opening determines the width and depth of the material bed being withdrawn. In establishing hopper-outlet-opening size, maximum lump size of the material being handled must be taken into consideration to prevent material from jamming as it passes through the hopper outlet. A rule of thumb relating to this is that minimum outlet dimensions should be at least 2–3 times the maximum lump size to be handled. This requirement often presents design problems for miniscule capacities of materials having very large lumps, since the high volume–low capacity combination could require slower speeds than are practical on most units.

Approximate power requirement is calculated from the formula

$$\text{power} = \frac{(T_e)(V)}{C}$$

where T_e = force (N or lbf) developed in the conveying medium in overcoming forces of friction and gravity, V = velocity (m/s or ft/min) of conveying medium, and C = constant (1 for watts (745.7 W/hsp) or 33,000 ft·lb/min/hsp). This formula applies for most conveyors and feeders having linear motion such as belt, apron, or chain conveyors and can also be used for inclined and vertical bucket elevators. The formula is not applicable for rotary-motion or reciprocating-type units such as screw conveyors, rotary drums or tables, or reciprocating- and vibrating-type units. Power calculations for units included in the latter category generally entail special design considerations, the derivation of which can usually be obtained from specific manufacturers' published literature.

The effective tension (T_e) used in the power calculation above can be determined when weight of conveying medium, bulk density of material, and friction factors are known. Weight of conveying medium might more accurately be expressed in terms of conveyor moving parts, particularly as in belt conveyors. Here the weight of the rotating idler rolls becomes part of the conveying medium weight, along with the weight of the conveyor belt, even though the belt conveyor idlers are stationary. Bulk density of material must be used in "as conveyed" terms, and can be found in numerous published data.

The value T_e is the sum of the tension developed on the return run (T_{er}), the tension developed on the carrying run (T_{ec}), and other miscellaneous sources (T_{em}):

$$T_e = T_{er} + T_{ec} + T_{em}$$

Referring to Figure 15, since $M_1gL \sin \theta$ acts in the direction of conveying-medium travel, it is considered negative, and

$$T_{er} = -M_1gL \sin \theta + M_1gL \mu \cos \theta.$$

where M_1 = mass of moving parts per unit of length, conveyor return run (kg/m or lb/ft); g = gravitational acceleration (9.80 m/s^2 or 32.16 ft/s^2); L = conveyor centers (m or ft); θ = angle of conveyor incline, relative to horizontal; and μ = friction factor. Where $\tan \theta > \mu$, the return run will tend to move downhill by gravity and T_{er} becomes a negative quantity.

It can be seen that

$$T_{ec} = M_2gL \sin \theta + M_2gL \mu \cos \theta.$$

where M_2 = mass of moving parts per unit of length, conveyor carrying run (kg/m or lb/ft).

Where material is carried on, or in, the conveying medium such as a belt conveyor, apron conveyor, or bucket elevator, the weight of material being conveyed can be added directly to the weight of the conveyor moving parts since the same friction μ will apply to both. However, where material is being moved by (instead of on) a conveyor, such as a drag chain or flight conveyor, the friction factor between the material and trough is usually different than the friction factor of the conveying medium, in which case

$$T_{ec} = M_2gL \sin \theta + M_2gL \mu_c \cos \theta + M_3gL \sin \theta + M_3gL \mu_m \cos \theta$$

or

$$T_{ec} = L[M_2g(\mu_c \cos \theta + \sin \theta) + M_3g(\mu_m \cos \theta + \sin \theta)]$$

where μ_c = friction factor, conveying medium; M_3 = mass of material being conveyed per unit of length (kg/m or lb/ft); μ_m = friction factor, material in trough;

$$M_3g = \frac{2.72 \text{ t/h}}{V}$$

$$\left(\frac{33.3 \text{ TPH}}{V}\right)$$

where V = velocity (m/s or ft/min).

It can be seen that the calculation of T_{er} and T_{ec} involves the use of basic physics and is relatively straightforward. Not all of the components that make up the miscellaneous sources of resistance (T_{em}) are as readily and clearly definable and in some cases require a degree of judgment and entail some assumptions by the designer. These components include such items as: tensions developed by the resistance caused by rotation of conveyor terminals or shaft assemblies; tension developed by friction of material between conveyor skirt boards, as well as the friction of the skirt boards against the conveying medium; tension resulting from conveyor accessories such as belt conveyor trippers, plows, scrapers, etc, or any other device that is propelled by or creates a drag on the conveying medium; drive-train losses including frictional losses of chain or V-belt drives, and efficiency losses through gear reducers; and tensions generated, primarily in feeder applications, by forces developed in the pullout action of material being withdrawn from a hopper (depending on the size and type of material being handled and the configuration of the hopper outlet, these pullout forces plus the shearing forces of material inside the hopper often represent the principal component in the feeder power requirements).

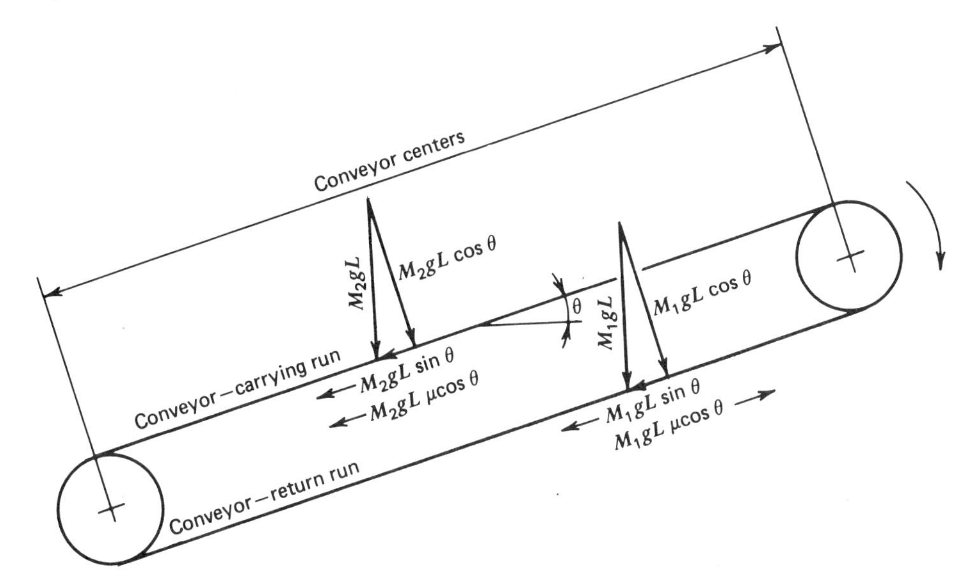

Figure 15. Relationship of T_{er} (tension developed on the return run) and T_{ec} (tension developed on the conveying run).

Values applied to the above factors are often the designer's best estimate. When more exact values are desired, it is recommended that one of the more creditable equipment manufacturers be consulted, since many of these values have already been established, whether empirically or through actual experience.

Pneumatic Conveyors. The principle of pneumatic conveying is simply to fluidize a dry, granular material and direct it through a pipe with a flow of air (6). The various systems on the market accomplish this in different ways, but if they are classified by air flows used, all are either high pressure or low pressure systems.

The high pressure systems require air to be supplied at normal plant air pressures of 590–690 kPa (85–100 psi) and are identified with a pressure vessel, usually called a transporter, into which the material must be loaded by gravity. The transporting, or blow cycle, is accomplished by sealing a gate at the top of the transporter and introducing air of sufficient pressure and volume to aerate the material and force it out the bottom of the transporter through a pipe.

Pressure is maintained until no material remains in the transporter or pipe to offer resistance to the air flow. Pressure drop allows the gate to open, and the transporter is ready to receive another batch. Since the gate must be closed and sealed tightly, it is necessary to interrupt the flow of material to the transporter during the blow cycle.

Low pressure systems operate at 21–69 kPa (3–10 psi) and require larger volumes of air to develop greater transporting velocities. These operate more on the principle of carrying material in suspension in the air stream and are therefore limited to the dry granular materials that can be more easily aerated.

The positive-displacement blower which is an integral part of these systems must be properly sized to suit the installation. Pressure, volume, and power must be calculated to meet requirements of material density, distance, and rated capacity. Transport lines are larger diameter pipe, and all turns must be long-radius sweeps. A suction line is typical with those systems, and operation is usually continuous.

Vibrating Conveyors and Processing Units. Vibrating conveyors are often more than carriers of material or products from one point to another (7–8). In addition to moving, lifting, or feeding materials, vibrating conveyors can be used simultaneously for such physical jobs as sorting, screening, mixing, tumbling, and breaking of material. Because vibrating conveyors keep products in constant motion, air circulation is facilitated, and heat and moisture transfer is efficient. As a result, vibrating units can be used effectively in such production processes as heating, cooling, drying, moisturizing, and washing.

Because vibrating conveyors are not seriously affected by heat, dust, impact, abrasion, heavy loads, and generally bad operating conditions, they offer many operating advantages. They provide longer service life, more reliable operation, and lower maintenance costs than many other types of material-handling equipment.

However, vibrating conveyors do have limitations. They are able to lift most products up an incline of no more than 7–8°. They are inefficient in handling some products such as sticky materials or heavy, fine products like finished cement. Vibrating conveyors generally have a lower carrying capacity because of relatively low conveyor speed, and the initial cost of the equipment is generally higher than other types of powered conveyors.

These limitations do not appear to be restrictive. The number of applications of vibrating conveyors is growing in an expanding group of industries. Food-processing companies, for example, use vibrating conveyors with smooth, sanitary decks for a variety of property-changing functions on such products as bread crumbs, breakfast cereals, snacks, and pet foods (see Food processing). In pulp and paper mills, they are used to feed products directly into chippers, as chip washers, and as refuse reclaimers (see Pulp; Paper). They are unaffected by impact loading and disorientation of long pieces. In the glass industry, vibrating conveyors are used for quench cooling and removal of glass cullet, which is moved in a water-filled trough (see Glass).

The versatility of vibrating conveyors is demonstrated by their use in handling scrap steel. In this extremely rugged duty, vibrating conveyors are used for preheating of scrap to burn out oils and combustible materials. They can be employed in the magnetic separation of ferrous from nonferrous materials. Applications have been developed in the chemical-processing industries for drying and cooling pharmaceuticals, screening fertilizers, and handling explosives.

In natural-frequency vibration, a free-standing spring and weight system can be kept in motion with a relatively small exciting force. This principle is used to advantage in vibrating conveyors. Spring-supported conveyor pans have a drive system tuned to resonance with the natural frequency of the springs and the pan. Most of the force required to vibrate the pan is alternately stored and released by the springs. The drive has only to make up the losses owing to friction and air pressure. Uniform spacing of the springs distributes driving forces over the length of the conveyor.

The conveying action consists of a series of gentle "throws and catches." During each cycle, both pressure and motion are applied between the pan and the conveyed material, but the pressure and motion do not occur at the same time. Thus, even very delicate particles, fragile pellets, cereal flakes, and explosives can be carried and processed without damage or degradation.

Where vibrating conveyors are to be installed in elevated structures or in areas having poor supporting soil conditions, dynamic balancing should be considered. In certain areas, vibration isolation is necessary.

Balancing is achieved through use of separate counterweights consisting of continuous structural members running the full length of the conveyor and a duplicate spring system tuned to resonance with the vibrating-pan frequency. These members are positively vibrated 180° out of phase by separate connecting arms to the common drive shaft so that each spring reaction is counteracted locally by an opposite reaction cancelling out ca 90% of the dynamic forces of the vibrating-conveyor spring system.

Vibration isolation is achieved by mounting the conveyor on a rigid structural base supported by isolation springs designed to reduce the transmission of vibration to the supporting structure by up to 80%. Vibration isolation can be used with either standard or dynamically balanced conveyors. A balanced and isolated installation can remove up to 98% of the transmitted vibrations.

Vibrating conveyors and process units are relatively new in the material-handling industry, but within the past several years, the use of vibrating equipment has become widespread. Their simple construction lends itself to a wide range of applications ranging from handling and orientating very fragile household light bulbs to foundry shakeout operations where mold sand is readily separated from and conveyed away from glowing metallic castings weighing several metric tons.

A unique application of vibrating motion is the relatively newly developed spiral elevator (see Fig. 16) in which material is elevated in an upward helix for distances of ≥12 m. This type of unit is found mainly in chemical and processing plants where

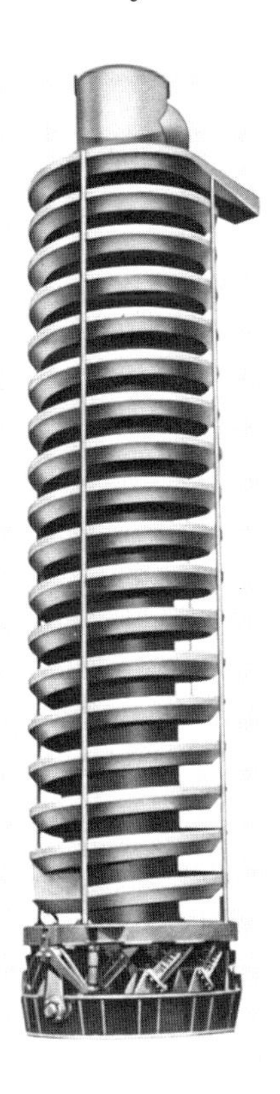

Figure 16. Spiral elevator conveys up while cooling, drying, or combination of processes. Courtesy of Rexnord Inc.

material is processed as it is elevated by having jets of cooling or heating air or water sprayed onto the material as it moves up the trough. It operates with a minimum of dusting, except where high velocity air is blown directly onto a bed of fine material. In this case, dust arresting can be effected by completely jacketing the spiral. It should be noted that, unless severe space restrictions exist or en route processing is desired, the spiral elevator may be cost prohibitive compared to the more common bucket elevators.

Another innovative application for vibrating conveying equipment is in fluid-bed drying and/or cooling (see Fig. 17). The principle of drying and cooling operations is the same in that material is conveyed over a deck having thousands of closely spaced, small-diameter drilled holes. At the same time, a regulated volume of air, entering from an enclosed plenum underneath, is blown through the perforated deck, literally fluidizing the bed of material being conveyed and surrounding each particle of material with a film of air. The spent air rises into a dust-tight hood, where a negative pressure causes the air to enter an exhaust system. There, it is generally blown into a dry-dust collector or wet-scrubber unit for cleansing before being exhausted into the atmosphere.

The only difference between the drying and cooling operations is the temperature of the processing air. For drying operation, the supply air is preheated by passing through heating coils in the ductwork. For most cooling operations, ambient air can be used since the fluid-bed cooler can efficiently reduce temperature of material being handled to within 6–8°C of the air temperature. Where lower temperatures are required, cooling air can be passed through refrigerated coils located in the air duct carrying the cooling air before entering the vibrating unit plenum.

Cooling can be effected either by an evaporative method, where the material being cooled has adequate inherent moisture, or nonevaporatively, where material being

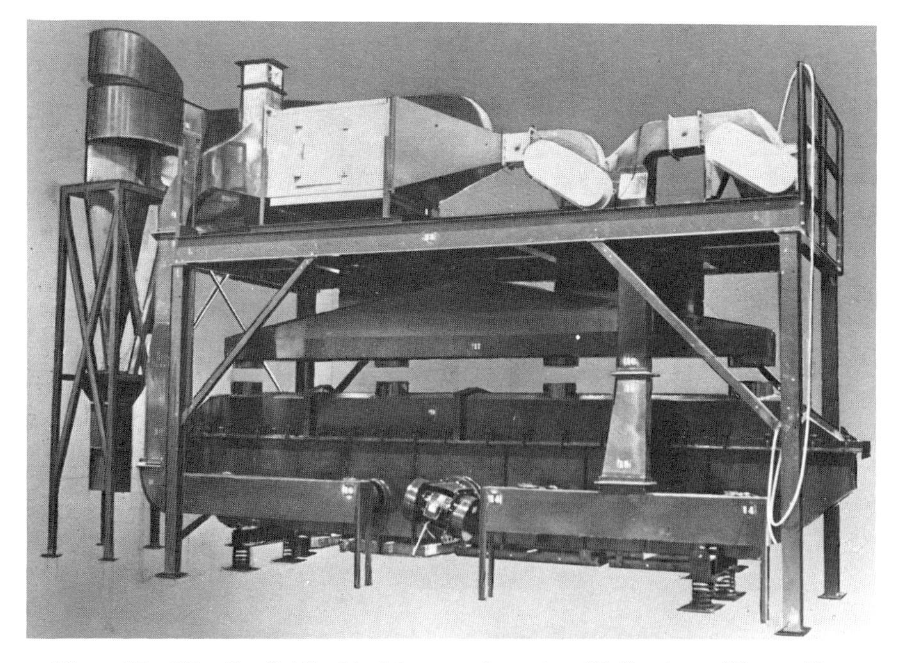

Figure 17. Vibrating fluidized-bed dryer–cooler system (9). Courtesy of Rexnord Inc.

cooled contains little or no moisture. The cooling-deck area for evaporative cooling is usually approximately one-third that of a nonevaporative process. Where addition of moisture to the material is neither harmful nor chemically undesirable, a more economical cooling package can usually be put together by spraying water over and mixing it with the material prior to its entering the fluid-bed cooler.

BIBLIOGRAPHY

"Conveying" in *ECT* 1st ed., Vol. 4, pp. 374–379, by W. G. Hudson, Consulting Engineer; "Conveying" in *ECT* 2nd ed., Vol. 6, pp. 92–121, by Richard R. Towers, Link-Belt Company.

1. *Conveyer Terms and Definitions*, Publication No. 102-1982, Conveyor Equipment Manufacturers Association, Washington, D.C.
2. *Unique Materials-Handling with Vibration*, Bulletin 16912, Rexnord Inc., Louisville, Ky.
3. *Rexnord Inc. Cat. R80*, Rexnord Inc., Milwaukee, Wisc., p. 623.
4. *Ibid.*, pp. 327–332, 599–607.
5. *Ibid.*, pp. 580–582.
6. *Material Handling with Pneumatic Conveyors*, Whirl-Air-Flow, Engineering and Manufacturing Facilities, Minneapolis, Minn., 1982.
7. J. I. Schneider, Jr., *Vibrating Conveyors for Moving and Processing*, Penton Publishing Company, Cleveland, Ohio, 1972.
8. J. R. DeSpain, *Vibrating Conveyors*, Penton Publishing Company, Cleveland, Ohio, 1975.
9. *Bulletin 16504*, Rexnord Inc., Louisville, Ky.

General Reference

M. V. Bhatia, ed., *Solids and Liquids Conveying Systems*, Vol. 4 of Process Equipment Series, Technomic Publishing Company, Inc., Westport, Conn., 1982.

E. J. ANDERSON
Rexnord Inc.

D

DEWATERING

Dewatering is the last mechanical process applied to separate a solid and a liquid. It does not include thermal drying (see Drying). In municipal wastewater treatment, dewatering is regarded as the final process used to achieve a water content of 85% or less in the sludge (1) or to change the behavior of sludge to a "solid" (2). Many other industries would regard 85% water content as feed to a primary liquid–solid-separation device such as a thickener, and would consider 15% moisture (85% solids) an appropriately dewatered product. Dewatering, then, is not defined by moisture content or by the use of a type of liquid–solid-separation equipment (3).

This article treats dewatering very broadly and covers both equipment and processes as well as techniques to improve the performance of existing dewatering equipment. Some of the processes treat the solids before they enter the liquid–solid-separation device, just as flocculants are added before thickening. The resulting products of the pretreatments described in this article probably need no further dewatering beyond the initial liquid–solid separation. A number of dewatering processes are listed in Table 1.

Interaction of Water with Solids

Liquids associate with solids with a range of energies, requiring a range of energy intensities for removal (4). Figure 1 shows a qualitative representation of the energy needed for water removal, and covers a range of interactions of water with a hydrated solid. Most dewatering, however, is concerned with water bound in capillaries, and dewatering processes affect the water and solids by: changing the size distribution of

Table 1. Liquid–Solid Separation in *Encyclopedia of Chemical Technology*[a]

Liquid–solid separation	Kirk-Othmer article[a]
equipment	
filters: pressure, vacuum, cartridge	Filtration, Vol. 10
filter thickening	Filtration, Vol. 10
electrofiltration	Filtration, Vol. 10
centrifugal filtration	Centrifugal separation, Vol. 5
centrifugal dewatering	Centrifugal separation, Vol. 5
expression presses	Fats and fatty oils, Vol. 9
	Filtration, Vol. 10
	Fruit juices, Vol. 11
	Cotton, Vol. 7
thickeners, clarifiers	Sedimentation
dissolved air flotation	Water, sewage, Vol. 24
thickeners	Water, industrial water treatment, Vol. 24
	Flotation, Vol. 10
magnetic separation	Magnetic separation, Vol. 14
	Dewatering
processes and pretreatments	
flocculation, coagulation, sludge conditioning	Water, municipal water treatment, Vol. 24
	Flocculation, Vol. 10
	Sedimentation, Vol. 20
	Filtration, Vol. 10
	Dewatering
sludge drainage on sand beds	Water, sewage, Vol. 24
filter and press admix	Water, sewage, Vol. 24
	Filtration, Vol. 10
	Fruit juices, Vol. 11
	Dewatering
viscosity adjustment	Filtration, Vol. 10
	Dewatering
electroosmotic dewatering	Dewatering
surface tension and wettability adjustment	Dewatering
oil agglomeration, shear flocculation	Dewatering
	Flotation, Vol. 10
oil-phase extraction	Dewatering
solvent extraction of water	Water, supply and desalination, Vol. 24
	Dewatering
thermal treatment	Water, sewage, Vol. 24
	Dewatering
thermal drying	Drying, Vol. 8
	Dewatering
freeze–thaw, freeze crystallization	Water, supply and desalination, Vol. 24
	Dewatering

[a] Third edition.

capillary radii; reducing adhesion of water to the solids; displacing water from the capillaries; and reducing the energy required to cause flow in the capillaries.

Although all of these techniques are effective, in industrial applications there is rarely time to achieve an equilibrium-reduced-saturation state (see Filtration), so variables that affect only the kinetics of dewatering and not the equilibrium-residual moisture are also very important. The most important kinetic variables to more rapidly displace the liquid from the solid are increased-pressure differentials and viscosity reduction.

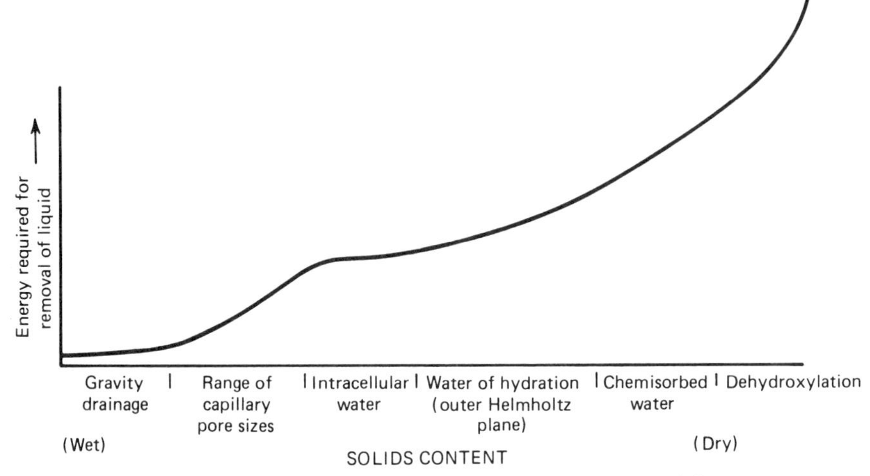

Figure 1. Energy required for removing water from solids.

Filter-Cake Dewatering

The most important function of filtration is the formation of a filter or centrifuge cake from a slurry. The theory of cake formation and the variables affecting it are outlined in Filtration and Centrifugal separation. The theory of dewatering the resulting cakes is also covered in these sections, and in much greater depth in refs. 5–9. Although theory still lags behind practice, it is evolving fairly rapidly. Presently, the theory can identify the principal variables controlling dewatering and can predict both dewatering rate as well as equilibrium-residual moisture in fully defined cakes. Because very few industrial applications of dewatering have consistent or fully characterized cakes, and because no new process in the design stage can fully characterize a filter cake, only the pertinent results of the theory, rather than its development, are included here.

Dewatering of centrifuge cakes is primarily achieved by providing sufficient centrifugal force to expel the liquid from the pores in the cake and through the filter medium. Lowering the final moisture content of the centrifuge cake includes the use of steam, surfactants, or flocculants, as well as pretreatments by pelletizing, oil agglomeration, thermal treatment, and freeze–thaw processes.

To dewater filter cakes, there are more choices of processes and equipment. The final moisture content of a filter cake is dependent on many variables that also control cake formation. Cake dewatering cannot be treated as independent and unrelated to cake formation. The pretreatment processes described in this article, for example, affect both cake formation and dewatering. In contrast, the following discussion covers only those variables and processes for dewatering a filter cake once it is formed. Cake dewatering is achieved by only two mechanisms: compression of the cake or displacement of the residual liquid in the cake with another phase, usually a gas. In order to be effective, each process relies on different properties of the cake. Compression reduces the total void space in the cake by pumping slurry into a fixed chamber that includes a filter medium which lets the liquid pass through. The pores remaining in the cake are completely saturated with liquid. A standard plate-and-frame filter provides compression dewatering. Expression is another method of reducing the void volume in a cake or solid by mechanically pressing on the cake after it is formed (10). Expression dewatering is often used with compressible cakes (11) of particle sizes below 50 μm (7), superflocculated cakes or cakes otherwise containing loosely entrapped liquid, and organic solids that contain liquid-filled cells that must be ruptured for effective dewatering.

Filters that use expression include variable-chamber plate-and-frame filters, belt-filter presses, tube presses, and screw presses. Tube presses can exert pressures up to 14 MPa (2000 psi) (12) which is the highest pressure of any filter-cake expression device. Screw presses are used primarily with fibrous or polymeric materials containing entrapped liquid. Pressures up to 110 MPa (16,000 psi) have been measured in these devices (13).

Expression always reduces the permeability of the cake being compressed, and, as a consequence, the resistance to flow of the liquid increases considerably (14). The effectiveness of expression is governed by cake thickness, specific-cake resistance, consolidation properties, and shear forces. The theory of expression from cakes (15) is somewhat different from the equations developed to describe expression from organic cellular solids (11).

Experimental observation of dilute slurries of paper fibers under mechanical compression shows a regime where the fibers flow with the liquid, and no dewatering is achieved. This can occur with belt-filter presses where the solids may flow out from the edges of the belts as compression is applied. The conditions have been identified for the transition from a flow of fibers in a slurry to a consolidation phase, where the fibers lock into a grid and liquid flows away (16). The limiting rate of expression dewatering of fibers by pressing in rolls has also been studied (17).

Displacement Dewatering. Displacement dewatering is the replacement of the liquid in the voids of a cake by a gas. Air displacement can be accomplished by using a pressure difference to force the liquid from the pores in the cake. The types of filters that provide displacement dewatering include all vacuum filters, rotary pressure disk and drum filters, the Lasta filter press (18), Rosenmund Nutsche-type filters, and the ECLP (12) tube press. In a centrifuge, displacement dewatering is accomplished by applying a body force directly to the liquid by the spinning motion. The factors that control displacement dewatering are shown in Table 2.

The equations developed in the Filtration article show that the rate of displacement dewatering increases by increasing the driving force, increasing bed permeability, increasing filter area, decreasing viscosity, and/or decreasing cake thickness. The dewatered cake becomes drier by increasing the driving force, increasing bed permeability, increasing the contact angle, and/or decreasing the surface tension; a number of techniques to achieve these results are discussed in this article as well as in Filtration.

Table 2. Variables Influencing Displacement Dewatering

Property	Variables
cake properties	particle size and size distribution, shape, packing, dimensions of the cake
fluid properties	density, viscosity
interfacial properties	surface tension (gas–liquid), interfacial tension (solid–liquid, gas–solid)
others	temperature, pressure gradient, rate of displacement

Expression Dewatering of Fibrous Materials. Expression-dewatering techniques used to process materials with liquids trapped in cells are listed in Table 3. These are frequently dewatered in screw, disk, and roll presses, and in batch pot and cage presses. Screw presses are continuous and have replaced batch pot and cage presses in most applications. Traditionally, however, batch presses have been used for squeezing cocoa butter from cocoa beans, requiring pressures up to 41 MPa (6000 psi) (26). A description of many of the types of batch presses is included in ref. 27.

Screw presses (Figs. 2 and 3) do not produce a clear liquid product. Frequently, the "pressate" is further filtered in a filter press to give a clear liquid product. Press aids are added to feed materials that contain fine particles or particles that can deform and plug the slots in the cage of a screw press. Typical press aids include sawdust, rice hulls, perlite, and diatomaceous earth (see Fruit juices).

A disk press, shown in Figure 4, can achieve a compression ratio of about 4:1 and produce a paper pulp with a consistency of 45–50% solids. It is a continuous press, and has also been used on brewers' spent grains and coffee grounds. The two surfaces of the rotating, converging press disks have a screen backing that retains and presses the solids while letting liquid pass through the screen. Slurry is fed at the wide part of the space between the disks, and the slurry is carried through the maximum compression zone before being released as the disks diverge (28).

A different type of press is the Vari-Nip shown in Figure 5 (29). A slurry is forced into the vat at up to 240 kPa (20 psig). The twin perforated rolls form a mat between them and dewater the mat to about 50% moisture. The emerging mat is cut and transported by a serrated screw conveyor shown in the lower figure. This press is somewhat similar to the drilled press rolls used in papermaking. The first steps in papermaking are controlled dewatering of dilute slurries, and of controlling water flow in the pressing rolls, forcing water rapidly into the felt that supports the slurry when it is pressed (see Paper).

Table 3. Applications of Screw, Roll, and Pot Presses[a]

Material	% Liquid in feed	% Liquid in product
paper pulp	97	50
	90	65
wood chips	85	50
sugar cane	68	43
oilseeds		
high oil content[b]	>30	3–7
low oil content[c]	<30	3–6
cocoa (separation of cocoa butter)	53	12
food[d]	60–90	10–30
polymers (elastomeric and thermoplastic)[e]	60	5–8
rendered tissue	20 (fat)	6–10
sewage sludge[f]	98	85

[a] Refs. 19–24.

[b] Includes copra, cottonseed, corn germ, peanuts, flax, safflower, sunflower, sesame, palm kernels, and linseed.

[c] Includes soybean, rice bran, and dry-process corn germ.

[d] Includes apples, carrots, coffee grounds, fish, grapes, pineapples, and tomatoes.

[e] Includes ABS, nitriles, SBR, natural rubber, and EPDM.

[f] The sludge was steam-heated in the press, and treated with CaO and polymer. About 95% of the solids were retained in the cake (25).

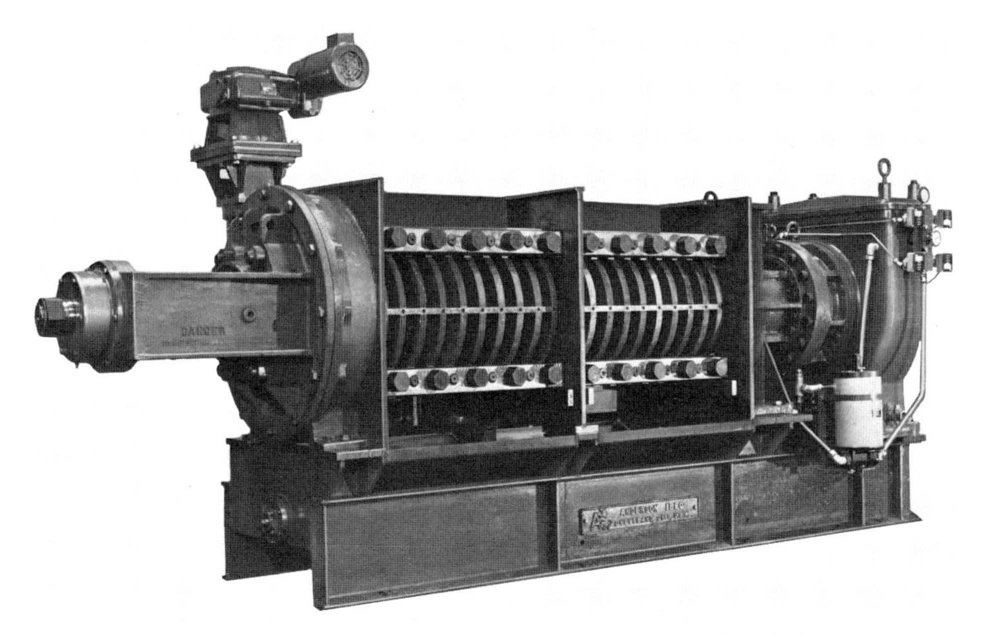

Figure 2. Screw press. Courtesy Anderson International Corp. (23).

Improving Cake Dewatering. ***By Viscosity Reduction.*** Equations relating the rate of liquid flow through a filter cake can be simplified to

$$\frac{V}{A} = \frac{K \cdot \Delta P}{\mu \cdot \ell}$$

where V (m^3/s) is the flow rate through the cake, A (m^2) is the area of the cake, K (m^2) is the permeability, ΔP (P = Pa) is the pressure drop across the cake, ℓ (m) is the thickness of the cake, and μ is the viscosity (Darcy's law). Viscosity is a kinetic variable, and does not appear in equations describing reduced-saturation levels in a cake (30). Since most practical filtration is limited by the time available for the steps of cake forming and dewatering, an increase in the flow rate of filtrate during dewatering translates directly to lower cake-moisture levels. If the liquid is water, the viscosity can drop by a factor of three as the temperature rises from 15°C to 80°C, and consequently the flow rate of water through the cake is tripled. Addition of moderate quantities of salts, polymers, or small amounts of less-viscous miscible liquids (like alcohols) have very little effect on water viscosity (see Water, properties). Temperature is the most important control of this variable.

The usual method of heating to improve dewatering is to apply low pressure steam to the filter cake as it is in the dewatering phase of the filter cycle. Steam has been used on rotary vacuum filters to dewater fine coal, and has been used on horizontal belt filters to dewater pipeline coal. Steaming typically removes an additional 0.7–1.5 kg of water per kilogram of steam applied, and reduces moisture levels in coal filter cakes ca 5% (31–33). The effectiveness of steam depends directly on the permeability of the cake. In highly permeable cakes, up to 90% of the contained moisture can be removed. Generally, steam is ineffective on filter cakes containing particles predominantly smaller than 10 μm (34). However, permeability is the critical factor: one of the largest installations of steam-assisted filtration is on <20 μm nonmagnetic taconites (35).

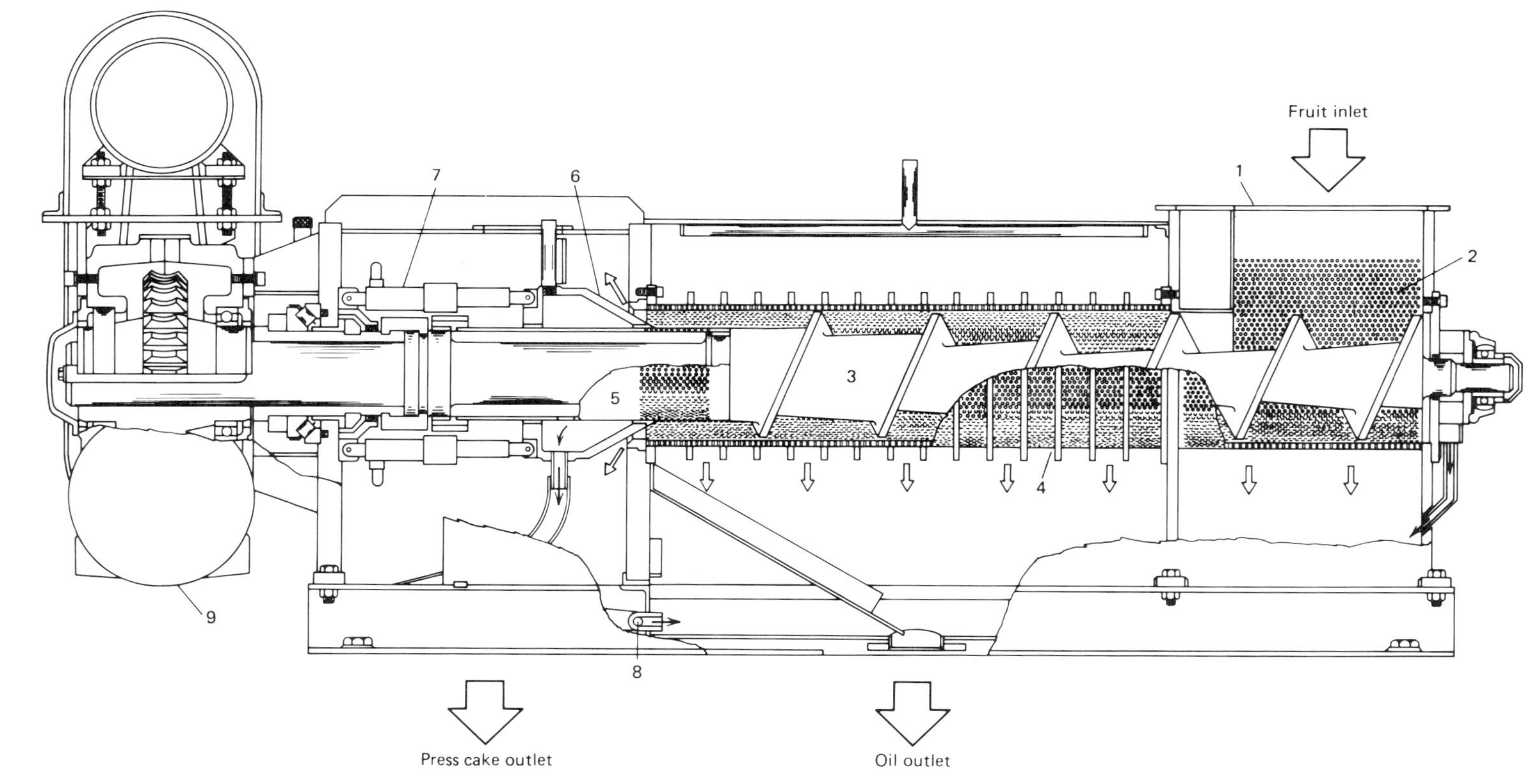

Figure 3. Screw press cross section. Courtesy of The French Oil Mill Machinery Co. (19). 1, Hopper; 2, perforated sheets; 3, main shaft; 4, perforated cage; 5, draining cylinder; 6, cone; 7, hydraulic cylinder; 8, draining cylinder oil; 9, gear box.

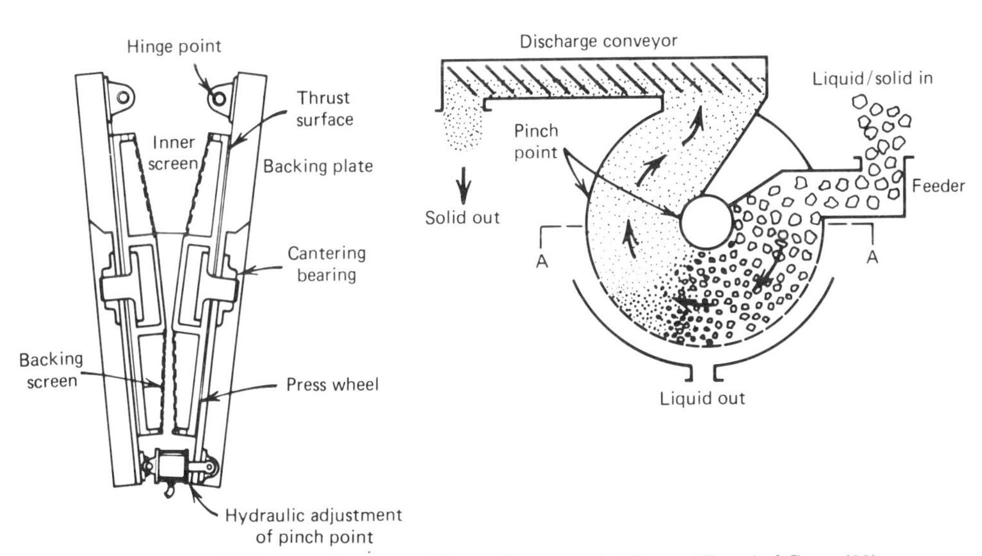

Figure 4. Disk press. Courtesy of Bepex Corp., a subsidiary of Berwind Corp. (28).

By Use of Surfactants. Although the use of steam to improve dewatering is uncontroversial in its effects, the effects of surfactants on residual moisture are highly inconsistent. In this article, those applications where surfactants are effective are covered, and some of the reasons behind the inconsistencies are discussed as a guide to the successful use of surfactants.

Additions of anionic, nonionic, or (sometimes) cationic surfactants of a few hundredths weight percent of the slurry [0.02–0.5 kg/t (metric ton) of solids (36)] are as effective as viscosity reduction in removing water from a number of filter cakes, including froth-floated coal, metal sulfide concentrates, and fine iron ores (Table 4).

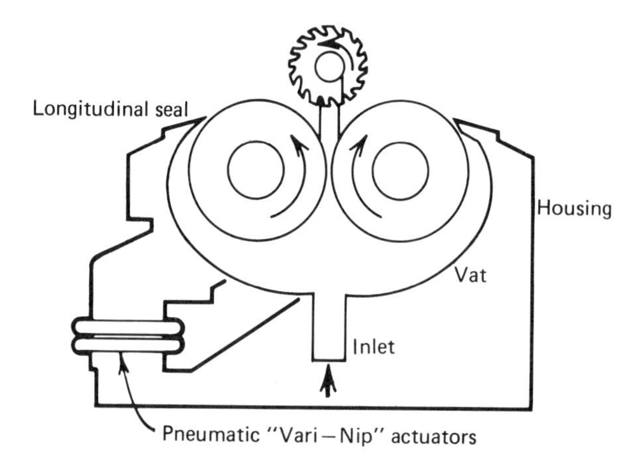

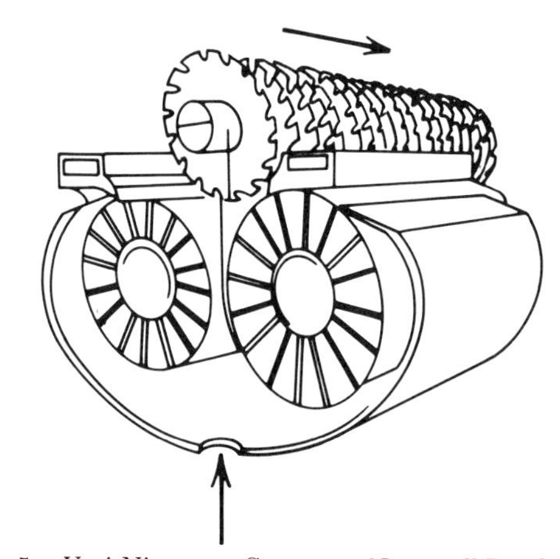

Figure 5. Vari-Nip press. Courtesy of Ingersoll Rand Co. (29).

Table 4. Effect of Surfactants on Residual Moisture in Filter Cakes

Material	Moisture content without surfactant, %	Moisture content with surfactant, %
sulfide flotation concentrates ($CuFeS_2$, MoS_2, ZnS)	15	12
	12	9
iron ore	17	15
	21	17
coal (different coals)	20–22	16
	36–40	30–34
	6	9
silica sand	12	8

A few studies have used both steam and surfactant on coal and iron ore, and found that the effects of each are additive, giving twice the moisture reduction of either treatment alone (30,32,35).

Surfactants aid dewatering of filter cakes after the cakes have formed, and have very little observed effect on cake-formation rate. Equations describing this behavior show that the main effect is to lower capillary pressure of water in the cake, rather than have a kinetic effect. The amount of residual water left in a filter cake is related to the capillary forces holding the liquids in the cake. Laplace's equation relates the capillary pressure (P_c) to surface tension (σ), contact angle of air and liquid on the solid (θ) (a measure of wettability), and to the capillary radius (r_c), or a similar measure applicable to filter cakes.

$$P_c = \frac{2\,\sigma\cos\theta}{r_c}$$

Surfactants lower the surface tension of water [typically from 72 to ca 30–35 mN/m (30–35 dyn/cm)] and most surfactants have a strong effect on the contact angle. Many surfactants act as flotation collectors, and adsorb on solids to make them hydrophobic. If the solids are initially hydrophilic, then, as surfactant is added, the contact angle increases from approximately zero (indicating complete wetting of the solid by water) toward higher values of ≥85° (see Flotation). A moderately hydrophobic coal, for example, can exhibit a contact angle of 54°. If a surfactant adsorbs on the coal, and changes the contact angle to 85°, the value of cos θ decreases more than six times. Recalling Laplace's equation, the lowering of surface tension by a factor of two works in concert with the sixfold effect of a change in contact angle, together having a large potential for reducing residual water. Surfactants adsorbing on a more typical hydrophilic solid such as alumina or silica can change cos θ by a factor of 11.

However, there is a complication. Too much surfactant, near or above the critical micelle concentration (CMC), reverses the effect that the surfactant has at lower concentrations and lowers the contact angle by functioning as a wetting agent (detergent), eg, the CMC for sodium dodecyl sulfate is ca 8 mmol (37). At the higher surfactant concentrations, the contact angle drops back to zero and the surfactant loses this effect on dewatering. Additionally, at or above the CMC, there is no further lowering of surface tension with increased surfactant concentration.

Ineffective applications of surfactants. In addition to the effects of too much surfactant (Fig. 6), a number of filter cakes do not respond to surfactant addition at

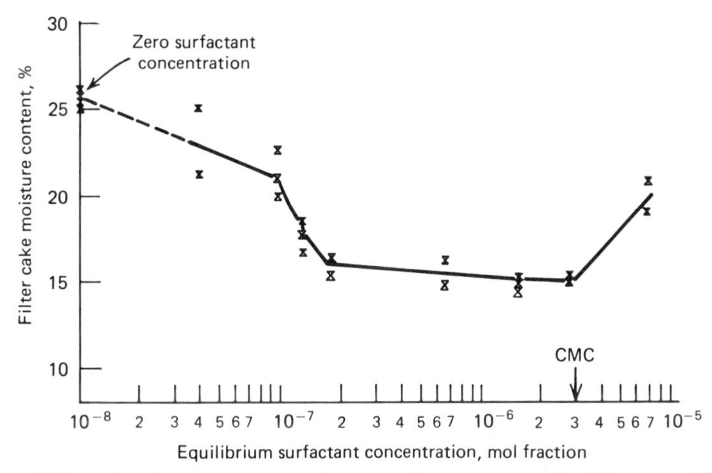

Figure 6. Effect of surfactant concentration on moisture content of <500 μm coal filter cake (38). CMC = critical micelle concentration.

any level. Coal refuse, clay, and hydroxide sludges show no benefits from surfactant additions. Further, some tests on coal have given very mixed results. One study showed that very beneficial effects are caused by large dosages of surfactants but that moisture reduction is insignificant at low surfactant concentrations (30). In Figure 6, the exact opposite behavior is shown: moisture increases as the CMC is reached and exceeded (38–39). Other studies showed beneficial effects of using surfactants to dewater coal, but found no noticeable effect related to approaching or exceeding the CMC (40). Some studies conclude that all surfactants have little commercial value in improving dewatering (41).

It appears that there is an effect that influences dewatering besides those of lowering the surface tension and contact angle. All these conflicting data can be resolved by considering the detergent action that high surfactant concentrations have on fine particles in the cake. Some helpful data are shown in Figure 7. This graph was prepared under unique conditions. The curve shows the time required to pass the same volume of air through a filter cake at a constant pressure drop, rather than passing air through a filter cake for a constant time, as is done commercially and in most other studies. Figure 7 shows that, at concentrations of surfactant above CMC (3×10^{-5} *M*, ref. 37), air passed through the cake with increasing difficulty (see also ref. 38). However, both terms in the Laplace equation suggest that lower pressures and less air resistance should occur with surfactant addition. A conclusion that fits the data is to presume that hydrophobic fine particles are initially adsorbed on coarser particles in the cake (see Oil Agglomeration). These particles are suspended by the detergent in the liquid. The liquid, which moves faster in the large pores of the filter cake, carries the fine particles to the filter medium, partially blocking the larger pores. Migration (consolidation) of filter cakes often occurs (42). This increases the pressure drop across the cake and forces water from the smaller pores, improving dewatering as time passes. For a constant volume of air, more dewatering would occur throughout the cake. Increased residual moisture would be observed if a constant time of drawing air through the cake was used. In this case air flow rate is increasingly impeded by movement of fines (38–39). Of course, after water is displaced from the pores, air flow rate again increases, but only contributes evaporative drying, not water displacement.

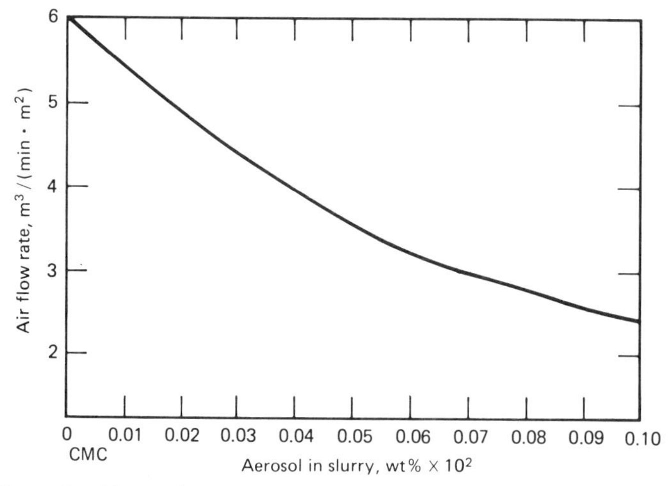

Figure 7. Air rate decreases with increasing surfactant concentration (30).

The beneficial effects of surfactants on dewatering are most pronounced in cakes that have been partially deslimed or in cakes of partially hydrophobic particles (eg, flotation concentrates) which are adsorbed on each other. Surfactants at or above CMC have little practical effect on extremely fine cakes where pores are small and the cake has no opportunity to further consolidate. Under equilibrium conditions, low concentrations of surfactants nearly always have lower residual moisture content.

Summary. For a number of filter cakes, excluding those with a high clay or other submicrometer-particle content, addition of surfactants, steam, or both, can reduce the moisture content 40%, and sometimes higher. Studies on the use of surfactants seem to observe that surfactants have a kinetic effect on dewatering, but unfortunately, have interpreted the effect in thermodynamic terms. The behavior of surfactants in downstream processes and equipment should be considered before the surfactant is added.

By Use of Flocculants and Sludge Conditioning. In the minerals industry and in water treatment (see Water, municipal water treatment; Flocculants), the primary purpose of flocculants is to improve sedimentation rates and overflow clarity in thickening operations. Generally, the flocculants that work best on sedimentation are not the best for filtration. Large, loose flocs are effective in causing rapid settling but trap water in the filter cake. The flocs can deform easily and block the filter media. Any excess polymer sticks to the filter media causing further blinding (43).

However, flocculants have the capability to improve filtration rates up to 100 times, especially of fine clay, sludges, or tailings (44). This is one of their main uses in filtration: to make impossibly slow filtering slurries filterable at reasonable rates. In addition, flocculants are as critical to belt-filter presses and dewatering centrifuges as are electric motors in making the machines effective. There are extensive and helpful reviews on selection of flocculants and their effects on the performance of these dewatering devices (1,45–48). In municipal-sludge processing, where often no flocculant is added to the primary thickening devices, flocculants are subsequently chosen for their effect on dewatering rates. Although filtration rates can be much faster with flocculants, the final cake moisture is often higher in a flocculated cake (40,49–51) (see also Filtration, **10,** 297).

In contrast, if flocculants are used that are optimized for filtration, a vacuum-filter cake of coal and other mineral slurries can be dewatered to moisture contents significantly lower than the untreated cake (52–54). The advantages of rapid filtration rates can also be preserved. Flocculants that provide better filtration tend to form flocs with the characteristics shown in Table 5.

It is generally observed that pumping a polymer-flocculated slurry to a filter degrades or destroys the floccules. To repair the damage, other flocculants can be added that are chosen for their optimum filtration characteristics (54). For example, to filter froth-floated coal (nominally <0.5-mm particle size), a medium molecular weight anionic flocculant (average mol wt of 10×10^6) is used. For sedimentation, much higher mol wt polymers are more effective (52).

In addition to specifying mol wt, the chemical structure of polyacrylamide flocculants has a significant effect on final moisture content. Two references (48,52) show the marked effects of both chemical structure and mol wt on filtration rates and are useful guides to flocculant selection for coal and clay-containing fine slurries (see Fig. 8).

Although there has been significant work tailoring flocculants to optimize clari-

Table 5. Floc Properties that Aid Filtration [a]

Floc characteristics	Beneficial effects
small	reduces intrafloccular water in the late stages of filtration; reduces pick-up problems due to gravity settling in the rotary filter chamber
strong	prevents the floc breakdown due to suspending agitation in the filter tank; resists collapse and premature loss of cake permeability in the early interfloccular stage of filtration
equisized	prevents localized breakthrough of air, prevents cake shrinkage, cracking and early loss of vacuum
good fines capture (into floc structure)	provides good filtrate clarity; prevents cloth blinding and poor discharge

[a] Ref. 52.

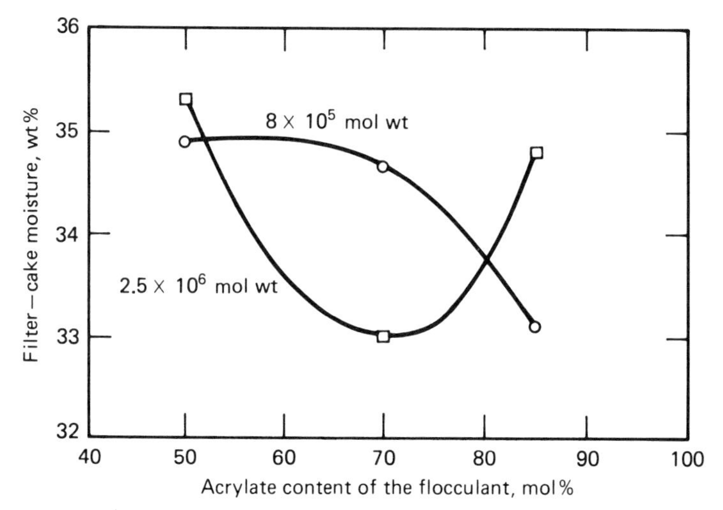

Figure 8. Effect of acrylate content on coal refuse filter-cake moisture (48). Upper curve shows the effect on a longer-chain flocculant. Concentration of flocculant was held constant at 150 g/t.

fication, thickening, and vacuum filtration, there is almost no published information on designing optimized flocculants for the high shear requirements critical to making belt presses work. Poly(ethylene oxide) flocculants have some unusual properties that appear to help increase shear resistance.

The effects of polymer flocculants on dewatering very fine coal, and separately, kaolin in a laboratory pressure filter has shown that small quantities of polyacrylates and polyacrylamide flocculants increase the moisture content of all coal and kaolin cakes, in accordance with generally observed practice (51). However, when air was allowed to break through the cake, the moisture content dropped to levels below the levels found in a cake that was formed and dewatered under the same conditions without flocculants (51).

Further dewatering of a flocculated filter cake can be achieved by using a surfactant dewatering aid, as described previously. The effectiveness of surfactants as dewatering aids seems neither to impair nor be impaired by the flocculant (52).

Sludge conditioning is a specialized branch of chemical treatment for dewatering sludges from water and wastewater treatment (1,55–58). Because sludge handling costs can be 25–50% of the total cost of wastewater treatment, dewatering is critical to cost control. In practice, the art of dewatering has included adding a number of substances to the thickened sludge to increase the permeability of the filter or centrifuge cake. In addition to polymer flocculants, coagulants like ferric and ferrous chloride, alum, and lime are added, which chemically react in the sludge and improve dewatering. Fly ash, ash derived from incinerating dewatered sludge, bark, newspaper, and diatomaceous earth have been added as body feeds to improve sludge dewatering. Typical dosages of lime and ferrous chloride are shown in Table 6. The dewatering results that can be achieved are summarized in Table 7. The relative performance of different filters on a range of waste sludges is shown in Table 8. For recommended test procedures and expected results, consult refs. 1, 55–58.

Comparisons are available on the relative performance and costs for dewatering the same sludge using different equipment. For example, one study (59) compared two belt-filter presses, two different centrifuges, and a rotary vacuum filter on the same sludge. In another study, a variable chamber filter press, fixed-volume filter press, continuous belt-filter press, and a rotary vacuum filter were compared for performance, capacity, and capital and operating costs (55).

Table 6. Chemical Dosages for Various Types of Sludge[a,b]

	Fresh solids, %		Digested solids, %	
Sludge type	$FeCl_2$	CaO	$FeCl_2$	CaO
primary	1–2	6–8	1.5–3.5	6–10
primary and trickling filter	2–3	6–8	1.5–3.5	6–10
primary and activated	1.5–2.5	7–9	1.5–4	6–12
activated only	4–6	ca 0		

[a] Ref. 58.
[b] Chemical dosages are presented as percentage of dry sludge.

Table 7. Solid Content of Different Sludges Dewatered on Different Filters[a,b]

		Dewatered product, %	
Sludge origin	Feed, %	Belt-filter press	Plate-and-frame filter
raw primary	4.6–6.0	22	45
waste-activated sludge (WAS)	2.0–3.0	14	45
mixed primary + WAS	3.5–5.0	20	45
anaerobically digested	5.5–7.8	20	40
aerobically digested	3.0–6.0	14	40
thermally conditioned	4.0–8.6	24	45
paper primary + WAS	3.0–8.0	20	45
petrochemical primary	6.0–8.0	19	
tannery	2.0–5.0	23	48
meat processing	3.0–5.0	17	
metal hydroxides	1.0–2.5		35
water-treatment lime sludge	4.0–5.0	30	48
water-treatment alum sludge	0.5–2.0	14	50

[a] Ref. 56.
[b] Results assume appropriate preconditioning.

Table 8. Comparison of Different Filters on Aerobically Digested Sludges[a]

Property	Belt-filter press		Filter press		Rotary vacuum filter	Centrifuge
cake produced, % solids	12–20[b]	14–25[c]	35–50[d]	35–50[e]	3–6	
operating pressure, MPa[f]	gravity plus pressure rolls		0.7–1.4	0.7–1.4	ca 0.1	2000–3000 g
relative space requirements	2.6	2.6	3.0	3.0	1.0	
approximate relative cost, $	2.2	2.8	3.0	5.2	1.0	
approximate maximum size, m^2	60	60	400	60	100	
usual sludge conditioning	polymer	polymer	chemical and filter aid	chemical and filter aid	chemical and filter aid	polymer, 0–10 kg/t

[a] Ref. 57.
[b] Standard.
[c] Vacuum assisted.
[d] Moveable plates.
[e] Moveable cloth.
[f] To convert MPa to psi, multiply by 145.

In a detailed analysis of municipal sludges, particle size is the most important variable of those shown in Table 9, and is directly related to many of the other variables that were measured and correlated with moisture content (60). Once the dewatering equipment has been selected, the use of sludge conditioning to increase the particle size is the most important variable to change.

Less-Common Commercial Dewatering Processes

When solids dewatering is known to be a problem early in the process-design stage of a plant, or is serious enough to warrant additional capital expense for new equipment, then there is an opportunity to consider a range of dewatering alternatives. The following sections contain brief descriptions of a number of dewatering methods. Two approaches are taken. First, processes that begin the dewatering process while in the original suspension are reviewed. These attempt to exclude liquid and densify the solids into compact and fairly large agglomerates that freely drain when separated. A second approach is to extract water or apply unusual desaturating forces to water present in sludges and filter cakes.

Table 9. Factors Affecting Sludge Dewatering[a]

Composition	Physical properties
cellulose content	pH
organic content	zeta potential, V
bound-water content	filtrate viscosity, mm^2/s (= cSt)
solids concentration	compressibility (ratio of porosities or void fractions)
grease content	particle size, μm

[a] Ref. 60.

Processes for Tightly Agglomerating Suspended Solids. ***Pelleting Precipitation.*** Typical cold soda–limewater softening generates a 5–15% soupy calcium carbonate sludge which is dumped into lagoons for disposal. By controlling reaction conditions of lime with hard water and providing for recirculating seeds of sand or calcium carbonate, precipitation of the carbonate can be controlled to form on the sand and to grow to 1.5-mm tight pellets in a device shown in Figure 9. The pellets dewater by draining to less than 10% moisture (61–62). Control of crystal growth and crystal habit is used to improve dewatering in the production of phosphoric acid and in scrubbing of flue gas with lime (63–64). In each process, a precipitate of gypsum is formed.

Pelleting Flocculation and Superflocculation. Flocculated, thickened municipal sludges can be dewatered by gentle rolling in a mesh-covered (microscreen) drum that allows water to drain as the sludge is reworked and agglomerated by its own weight. A device called a Dual cell gravity sludge dewatering unit (DCG) was used in 1962 in New Jersey (1,65–66). As shown in Figure 10, a DCG consists of two drum frames side by side on parallel axes covered on the outside with a fine screen, similar to a large flat pulley belt between two wheels. Between the drums at the bottom, a third nonporous

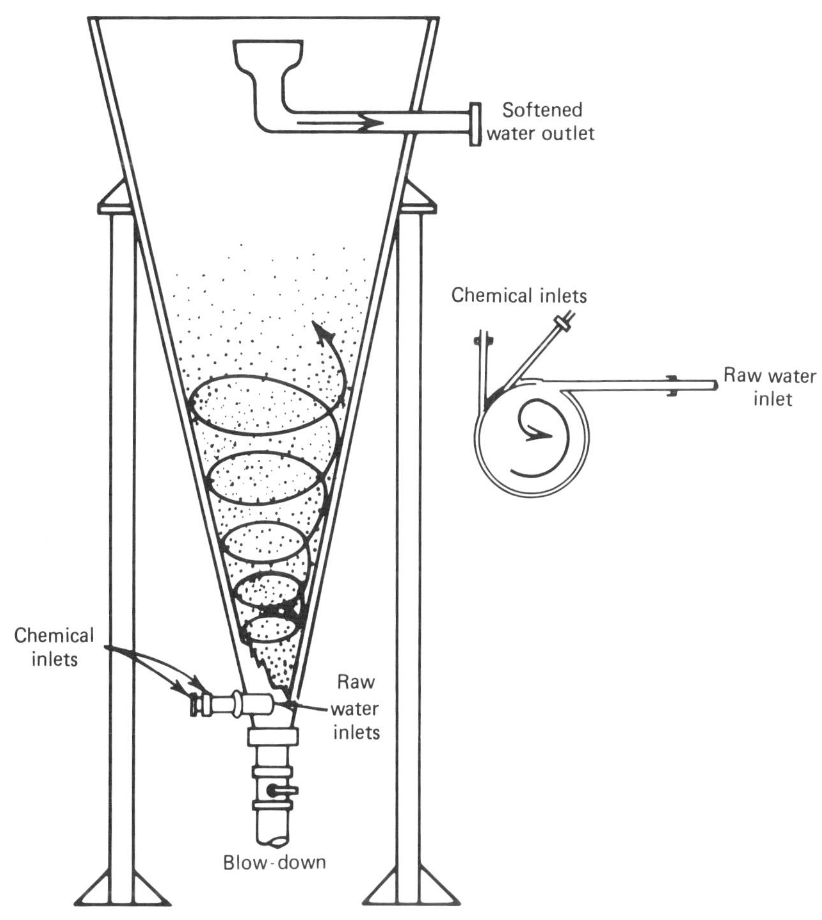

Figure 9. Cross-sectional view of Permutit Spiractor showing flow. Insert gives detail of arrangement of raw water and chemical inlets (61).

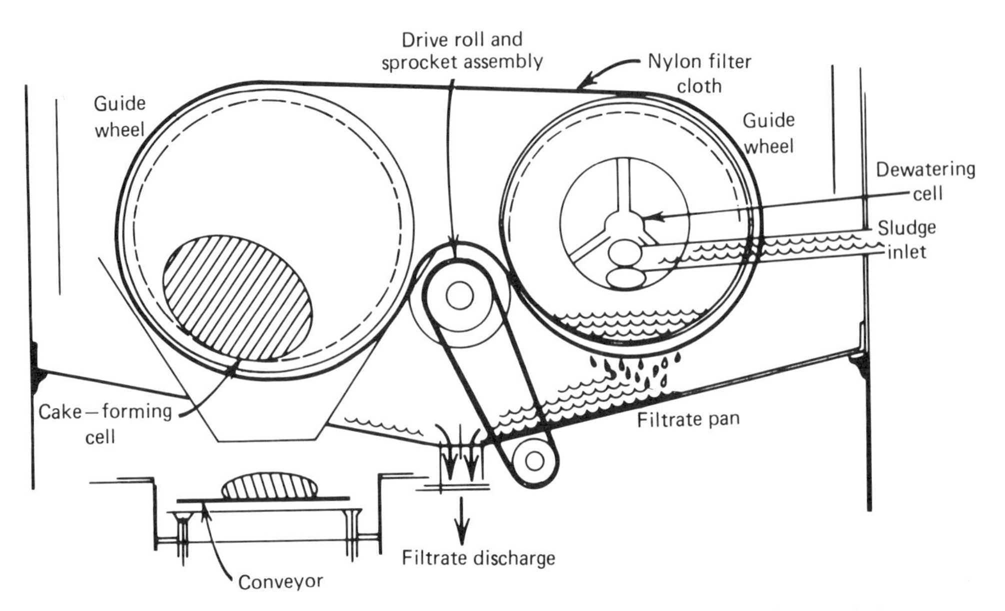

Figure 10. A DCG sludge dewatering machine. Courtesy of Permutit Co., Inc. (66).

pulley cylinder forces the mesh to ride up a hill between the drums. Sludge and flocculant are fed into one of the drums, called the thickening barrel. As the water drains by gravity through the filter cloth, the sludge thickens and moves up over the central hump and into the dewatering barrel where further water removal occurs. As the solids in the dewatering drum build up, they spill over the ends as dewatered product. This device has a low initial capital cost and has been found to operate on many types of sludges. It is reported to be difficult to operate since exact chemical conditioning is required (2). A similar device called a Rotoplug thickener is described in ref. 8. The primary application of these machines is on sewage treatment and on paper-mill effluent (8). The performance of these units has improved and the range of applications has expanded as flocculants have been improved. In municipal sludges, 7–17% solids in the product have been reported.

Additions of new flocculants after conventional thickening produce further dewatering of mineral slimes. As clay was flocculated with polyacrylamides and rotated in a drum the growth of compact kaolin pellets has been reported (67). The pellets could easily be wet-screened and dewatered. This work has resulted in a device called a Dehydrum which flocculates and pelletizes thickened sludges into round, 3-mm pellets. Several units are reported in commercial operation in Japan where they are used to thicken fine refuse from coal-preparation plants. The product contains 50% moisture, compared to 3% solids fed into the Dehydrum from the thickener underflow (68). In Poland, commercial use of the process to treat coal fines has been reported. The pellets are further pressed to complete dewatering of the recovered coal. This process compares favorably both economically and technically to thickening and vacuum filtration (69).

The U.S. Bureau of Mines has run large-scale tests on a similar process for treating <600-μm coal tailings. The material is the tailings of a flotation cell and is treated at a concentration of 3–7% solids. After mixing the slurry with 0.14 kg of poly(ethylene oxide) flocculant per metric ton of dry solids and "working" for less than 30 s, the solids dewater to 50–70% moisture (70). The product is shown in Figure 11. The process is being demonstrated at a rate of 2300 L/min of waste slurry. Waste phosphate slimes have been similarly successfully consolidated from 4 to 40% solids (70).

Other applications include dewatering extremely fine (0.1 μm) laterite-leach tailings (71). These pelletizing processes should be compared in flocculant consumption and operating and capital costs to belt-filter presses.

Oil Agglomeration. Similar pelletizing can be obtained using chemicals which are not flocculants, but instead make hydrophilic particles hydrophobic and agglomerate the particles into tight hydrophobic clusters (72–73). These processes were invented before the flocculant pelletizing processes cited above, and were first used to make selective separations of a relatively minor amount of sulfide minerals (1–2%) in a slurry containing mostly silicates and carbonates. The Cattermole process used addition of oleic and other long-chain fatty acids to the mineral suspension and vigorous stirring of the slurry causes sulfide minerals (of copper, zinc, or lead) to agglomerate selectively to form spheres. These were large enough to be retained on a

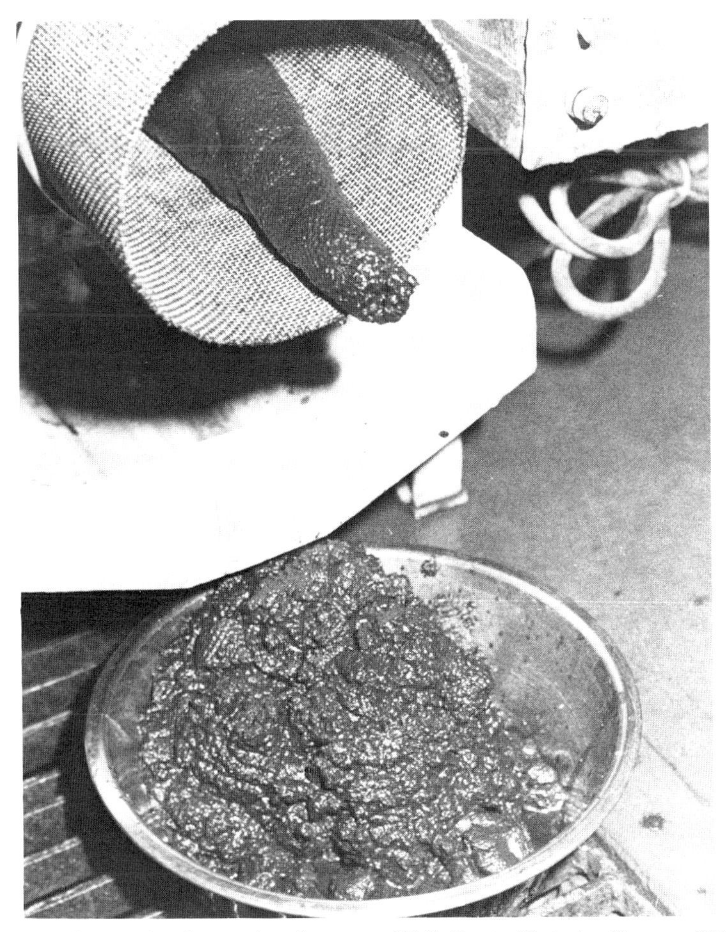

Figure 11. Agglomeration dewatering. Courtesy of U.S. Dept. of Interior, Bureau of Mines (70).

screen, but the finer, dispersed waste rock passed through. This process and a similar one (Murex) that added fatty-acid-coated magnetite to collect the sulfides on the magnetite surface were the precursors to froth flotation. These processes could not compete with the smaller reagent requirements needed for flotation.

However, the recovery and simultaneous dewatering of coal did become commercial about 1920, using fuel oils to agglomerate the coal into pellets. The high cost of oil, relative to coal, made continued operations uneconomical. It is a well-studied, effective, and proven technology for selective agglomeration and dewatering (73).

A new agglomeration process that is also selective for coal adds a chlorofluorocarbon (Freon) to a slurry of <30 μm coal and mineral matter. Similar to oil, the Freon causes the fine coal to agglomerate into large pellets that are recovered from the slurry on coarse static screens. The only remaining water is occluded in the pellets. The Freon, with its low heat capacity and boiling point (24°C), is recovered in a dryer (74).

A successful variation of oil agglomeration was used for removal and dewatering of 1–3% solids suspension consisting of <5-μm soot from refinery process waters (Fig. 12). An amount of heavy oil was added to the dilute slurry and intensely agitated in a multistage mixer. The soot agglomerated with the oil to form 3–5-mm pellets that were easily screened from the water (75). The pellets contained only 5–10% water.

This process has been applied to certain mineral oxides, usually with the objective of selective recovery of one mineral from a mixture of minerals. Examples include recovery of cassiterite (SnO_2) from silicates (76), gold from ores (77), and ilmenite ($FeTiO_2$) from silicates (78). In each case, the normally hydrophilic mineral is treated with a surfactant (often a fatty acid; see Flotation) under conditions that would selectively coat the desired mineral. Oil is then added to the slurry and agitated to agglomerate the mineral into a compact pellet that could be separated by classification (see Gravity concentration) or screening. The technique is most useful when selectivity is desired or when dilute suspensions of minerals would form a very wet and unfilterable sludge if conventionally flocculated in a clarifier (see Sedimentation). Because of the cost of added reagent, the particles should have some intrinsic value. If no excess oil is added, but enough reagent is added and highly agitated, the pelletizing process is known as shear flocculation (79–81). Its main use, too, is for selective separation of a component from a mixture.

Another modification of dewatering methods using oil-agglomeration techniques combines the water exclusion of agglomeration with the ability of froth flotation to thicken. Dissolved air flotation or induced air flotation is used in over 300 municipalities for thickening wastewater solids (65), and is also used at a large number of

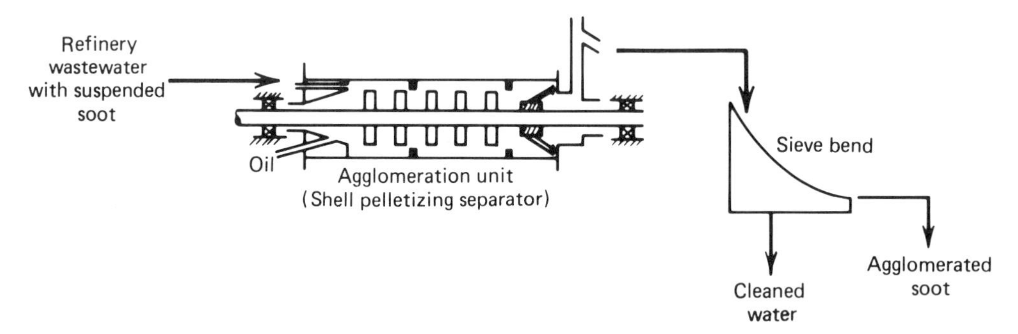

Figure 12. Shell pelletizing separator (72).

oil-well sites, refineries, and rendering plants for removal of oil from wastewaters. Typically, however, the thickened product contains only 2–4% solids. Preceding the operation with oil agglomeration or shear flocculation is particularly useful for inorganic particles finer than 15 μm because it increases the kinetics of flotation and, with that, lessens the amount of water carried into the froth. It allows a much smaller volume of slurry to be treated by further agglomeration or filtration to complete dewatering (82–86).

Extraction Processes. ***Oil-Phase Extraction.*** Among the processes of dewatering by wetting a solid with an immiscible phase, another step in water displacement is possible. The use of very large quantities of immiscible liquid allows extraction or transfer of the particles from one phase into the other and the particles remain in a dispersed state in the new phase. To extract solids (as small as 0.1 μm) from water, a hydrophobic surface on the solid is needed and is usually provided by using flotation reagents (long-chain fatty acids, alkyl sulfonates, amines, xanthates). Conversely, water has been used to agglomerate and extract hydrophilic solids that are dispersed in organic liquids (87). "Crud" is a concentration of solids at the liquid–liquid interface in normal solvent extraction and represents partial extraction (88). Pigment-flushing is a technique used in paint and ink manufacture for transferring paint particles from the aqueous solution where they are formed to a dispersed state in a nonaqueous carrier (89–90). Since the pigments are first filtered to remove soluble salts but are not dried before dispersion in the oil phase, the process function includes dewatering. Similarly, a very important process for manufacturing ferrofluids (stable dispersions of submicrometer magnetite in kerosene) consists of precipitating magnetite from water, adding oleic acid to coat the fine precipitate, and then contacting the wet filter cake with kerosene. The magnetite transfers into the kerosene and forms a stable suspension (91).

The Lummus antisolvent deashing process was developed to overcome very difficult solid–liquid separation of mineral matter and unreacted coal left in the product of solvent-refining coal. Precoat filtration had proven difficult and expensive for removing solids from the very viscous liquid coal product. The Lummus process uses kerosene as a liquid that is immiscible with the coal liquids, yet has an affinity for agglomerating and extracting the solids from the coal liquid. The ash-free coal liquid separates from the kerosene in a settler. Typically, twice as much kerosene as coal liquid is necessary to completely extract the 24% solids from the liquid (92).

Solvent Extraction of the Liquid. Water contained in a cake or slurry can be extracted from the solids by dissolving the water in a solvent that is less expensive to evaporate than the contained water alone. The Institute of Gas Technology has developed a laboratory process called solvent dewatering based on the principle that the solubility of water in selected solvents changes significantly with a change in temperature. In one example, hot solvent is mixed with wet peat and the water–solvent solution then decanted. Upon cooling, the water precipitates as a separate phase (93). So far, this process has not proven economical. The Resource Conservation Company has used chilled triethylamine (TEA) to dissolve water from sludge, and then warmed the extracted liquid to form an immiscible water phase. Note that the effect of temperature on water solubility is opposite to IGT's solvent. With TEA requiring only 309 kJ/kg (133 Btu/lb) to evaporate, the remaining solvent costs less to evaporate than water (73,94). Similar processes have been considered for desalination (see Water, supply and desalination).

A process based on the same principle has been developed for removing solids from thick, viscous, nonaqueous slurries. The Kerr McGee critical solvent deashing process makes use of the controllable dissolving power of solvents at their critical temperatures and pressures to dissolve coal liquids without dissolving or extracting unreacted coal or mineral matter. The coal liquid is extracted away from the solids. The process is, in a sense, the inverse of the Lummus process described above.

Thermal Processes. ***Thermal Drying.*** The solvent-extraction processes discussed so far have progressively included evaporation. In this aspect, the Carver-Greenfield process of multiple-effect evaporation of water from sludges is an important alternative to very-late-stage dewatering. Organic sludges of ca 20% solids are fed to the unit, along with enough oil to keep the sludge moving in the processing equipment. Using three effects in the evaporators, only 700–900 kJ/kg (300–387 Btu/lb) of water evaporated are required [compared to 2.3 MJ/kg (1000 Btu/lb) needed in a single-effect evaporator] (25,95) (see Drying).

Thermal Treatment. A number of the previous dewatering processes alter the interaction of solid with liquid. Of those described, most depended upon making a hydrophilic solid hydrophobic by adding small quantities of surfactants and oils. Many biological sludges cannot be economically treated with reagents to provide hydrophobicity, and such treatments would have no effect on water bound inside the mycelium. Thermal treatment is intended for these organic sludges. It is partial wet-air oxidation (see Water, sewage). It lowers the specific cake resistance of many biological gels and colloidal sludges by 50–100 times. For example, if a municipal sludge that normally thickens to 5% solids is successfully thermally treated, it will thicken to 10–15% solids. Upon filtering the thickened sludge, the cake formed from the untreated sludge will have ca 15% solids. The thermally treated sludge can be filtered to 35–40% solids (96).

Thermal treatment consists of heating the sludge under a pressure of about 2.4 MPa (350 psi) and to temperatures of 150–225°C (96) with or without additional air. Reactions, including partial oxidation, occur which change the nature of the solids and consume 1–5% of the solids. Unfortunately, the process produces significant quantities of acetic acid and other short-chain, soluble organics. In 1979, there were over 100 installations operating on sewage sludge.

At least five related dewatering processes have been applied to peat and lignite. A flow sheet of one of these is shown in Figure 13 (97). Peat and lignite have a high absorbed-moisture content (90% in peat and 40–50% water in lignite) and have a tendency to break down to undesirable fines and to become pyrophoric when dried. Steam drying, as these processes are called, is a family of processes very different from steam dewatering of filter cakes. These processes involve heating the peat or lignite with their initial water content in autoclave to temperatures of 150–200°C under pressures of about 1.3 MPa (189 psi) for ca 15 min. This treatment causes the solids to shrink, eliminating water from pores and removing carboxylic acid and its salts from the surface. The steam treatment itself, considered separately from subsequent evaporative flashing, allows 30–50% of the initial water to drain or be pressure filtered from the product (98). Higher temperature and pressure leads to greater dewatering, and pressures up to 10 MPa (1500 psi) have been tested (99). These high pressures produce a completely dewatered lignite with high stability and with very little tendency to reabsorb moisture.

The only commercially used process in this group is the Fleissner process, de-

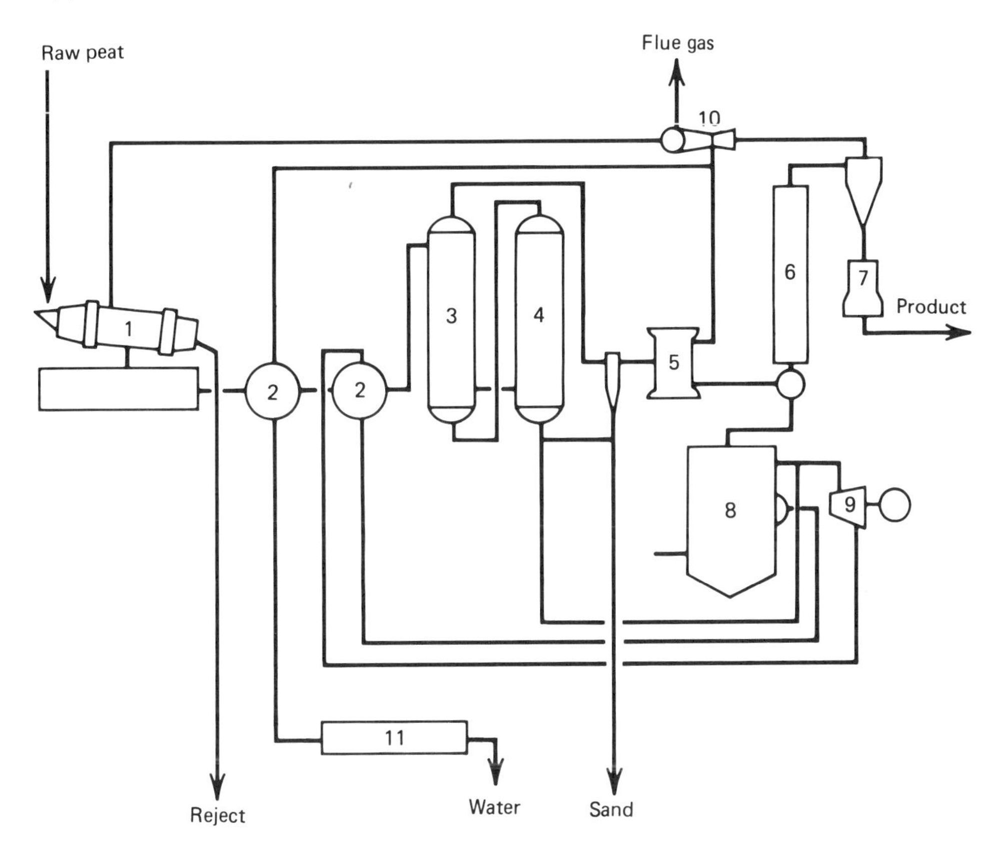

Figure 13. Steam-drying process for peat (Wheelabrator-Frye, Inc.) (97). 1, Raw peat pulper; 2, heat exchanger; 3, preheating tower; 4, reactor; 5, dewatering filter; 6, thermal dryer; 7, briquetting press; 8, steam generator; 9, steam turbine; 10, flue gas cleaner; 11, wastewater purification.

veloped in 1927 for drying lignite. One plant, operating in Austria between 1927 and 1960, achieved a capacity of 1700 t/d. A number of other plants licensed by Fleissner are currently operating (100). There are related processes in the pilot plant stage described in refs. 97 and 99–102.

By raising the pressure, temperature, and available oxygen, virtually all the organic solids can be oxidized to CO_2 and H_2O. "Ignition" occurs at 200–225°C and wet-air oxidation is then autogenous. At 250–300°C, reactions occur rapidly. The vapor pressure of water at 300°C is ca 9 MPa (1300 psi). Rather than needing 25–30% solids to achieve autogenous combustion of a sludge in air, only 0.5–1% organics in water is needed for autogenous wet-air oxidation (103). Methods for recovering power from the process have been devised (103). Raising the temperature and pressure still further to the critical point of water, 374°C and 22 MPa (3190 psi), allows even toxic organics to be completely destroyed (104).

Freeze–Thaw Dewatering. It has been found that slow freezing of hydroxide, clay, and municipal sludges affects the water-retention properties of the solids when the frozen slurry is thawed. Two plants used the process on water-treatment sludges. The sludge is gelatinous aluminum hydroxide with organic and inorganic matter that typically thickens to 1–3% solids. In tests in the UK, the sludge, after a cycle of freezing and thawing, was a sandy, granular material that drained without need for a filter.

The costs were about 20% higher than competitive processing with a filter press (105). In Albany, New York, the sludge was naturally frozen and thawed, allowing complete separation of the solids and the liquids (106). The technical success of freeze–thaw dewatering has been frequently demonstrated on Florida phosphate slimes, allowing a typical thickened 13%-solids sludge to drain to 45%-solids after freezing and melting. The economics have been elusive, but a recent study using improved refrigeration techniques indicates that the technology is competitive with polymer thickening of the slimes (107).

The probable mechanisms of dewatering by freezing have been described, and factors leading to successful application discussed (108). The ability to withstand freeze–thaw cycling is an important consideration in latex-paint manufacture. Some factors leading to such stability have been presented (90). Similar considerations must be important in some frozen-food formulations (see Food processing).

Freeze Crystallization. A final variant of the process is the use of freezing to crystallize pure ice crystals that are then removed from the slurry by screens sized to pass the fine solids but to catch the ice crystals and leave behind a more concentrated slurry. The process has been considered mostly for solutions, not suspensions. However, freeze crystallization has been tested for concentrating orange juice where solids are present (see Fruit juices). An important test of freeze crystallization has concentrated pulp and paper black liquor from 6% dissolved solids to 30% and showed energy savings of over 75% compared to multiple-effect evaporation. Only 35–46 kJ/kg (15–20 Btu/lb) of water removed was consumed in the process (109). There appears to be no large commercial operation yet of freeze crystallization.

Electromagnetic Processes. ***Electrical Enhancement of Dewatering.*** The Filtration article discusses the use of electrophoresis to prevent a filter cake from forming on a filter medium while allowing water to pass through the medium from the slurry. Electrophoresis uses the motion of the particles to swim upstream and prevent blinding of the medium. In this section, the use of electrical currents to withdraw water from a consolidated cake or soil is reviewed. Once a matrix of particles is formed, whether filter cake, thickened underflow, or soil, applying a current to the fluid causes a movement of ions in the water and, with the ions, the water of hydration. The phenomenon is called electroosmosis and the pressure generated on the fluid is given by (110):

$$P = 2\,\zeta ED/\pi r^2$$

where P = pressure, Pa; ζ = Zeta potential, V; E = electric field, V/m; D = dielectric constant; r = radius of capillary, m. The amount of water moved is proportional to the intensity and time that power is applied, proportional to the zeta potential of the solid, and inversely proportional to the conductivity of the fluid (111). Results are often measured in kW·h/t of dewatered product or in kW·h/t of water removed.

High pressures are generated only in small capillary openings where the pressure is needed. But unlike pressure generated on a fluid by an externally applied force, the largest forces are generated at the shear plane of the liquid and the solid in the pores. In Table 10, the effect of particle size on capillary retention force is shown (14). To calculate the pore radius used in the table, it is assumed that pore is a cylinder which can just pass between three monosized spheres. The entries following P_E are the pressures developed by electroosmosis in those same pores, assuming a field of 3000 V/m. To generate that field in an electrolyte, a current of 600 A/m^2 must flow and in

Table 10. Effect of Particle Size on Capillary Retention Force

d, μm	50	20	15	12	1
$r_c = 0.165\ d/2$, μm	4.125	1.65	1.24	0.99	0.0825
P_c[a], kPa	35.5	88.2	117	147	1760
(atm)	(0.35)	(0.87)	(1.15)	(1.45)	(17.4)
P_E[b], kPa				17.2	2430
(atm)				(0.17)	(24)

[a] $P_c = 2\,\sigma \cos\theta / r_c$, capillary pressure.
[b] P_E = electrical "pressure."

this case, it is assumed to be created in a 10^{-4} M, 1:1 electrolyte (0.2 S/m conductivity) (112).

In most applications, far less current and lower voltages are used. For example, in dewatering clay soils to stabilize dams, foundations, or dredged spoil, 20–100 V/m are commonly applied (113–114). In soil stabilization, power is applied for weeks to months.

Over the last decade, the U.S. Bureau of Mines has shown the effectiveness and costs of electroosmotic dewatering on a large number of clay-containing tailings from metallic, nonmetallic, and coal mines (115). The process can be used *in situ* or in a batch dewatering cell. One large test dewatered a very old, stable 50%-solid slime generated by a coal-washing plant. Applying 37 kW·h/t of the final product, moisture was reduced to 19% in 24 h. Using a batch dewatering cell, 1100 t/wk of slimes were dewatered (115).

Magnetic Enhancement of Dewatering. *Liquid–solid separation.* When magnetic forces are considered for liquid–solid separations, it is usually to replace thickening and filtration rather than for dewatering. The commonly used Frantz Ferrofilter (Fig. 14), used for removing suspended ferromagnetic impurities from liquids, is somewhat analogous to a depth filter in its use of multiple collection sites and lack of a definable porous-filter-media surface (116). The Ferrofilter principle has been extended to very large magnetized volumes of 1.7 m^3 (60 ft^3) and at high fields (from 0.15 T (1500 G) for the Ferrofilter to 2.0 T (2×10^4 G) for the new machines). The large, high gradient magnetic filters use a depth filter media of 430 stainless-steel wool to form high gradients in the high field. Commercially, they are widely used for removing paramagnetic impurities from kaolin (117), and less-intense versions are used to separate and dewater magnetic iron ores and nonferromagnetic iron ores (eg, itabirite) and other ores containing specular hematite (Fe_2O_3) (118–119) (see Magnetic separation).

The large-volume magnetic separators have been used to remove 90% of the suspended mill scale solids in steel rolling-mill wastewaters, and for cleanup of steam-boiler water (see Magnetic separation). A much wider range of potential uses of magnetic separation for clarifying, filtering, and dewatering is possible when nonmagnetic impurities are made magnetic. In Japan, plating-wastes with dissolved Cr, Mn, Cu, Zn, Cd, etc, have been precipitated as magnetic ferrites and recovered from the wastewater with a simple magnetic separator. Details of this commercial ferrite precipitation process have been reported (120–121).

A simple separator used to recover the magnetic particles consists of a series of disks mounted on a shaft, which each disk having a number of permanent magnets mounted flush with the surface at its perimeter. The disks rotate into and out of the

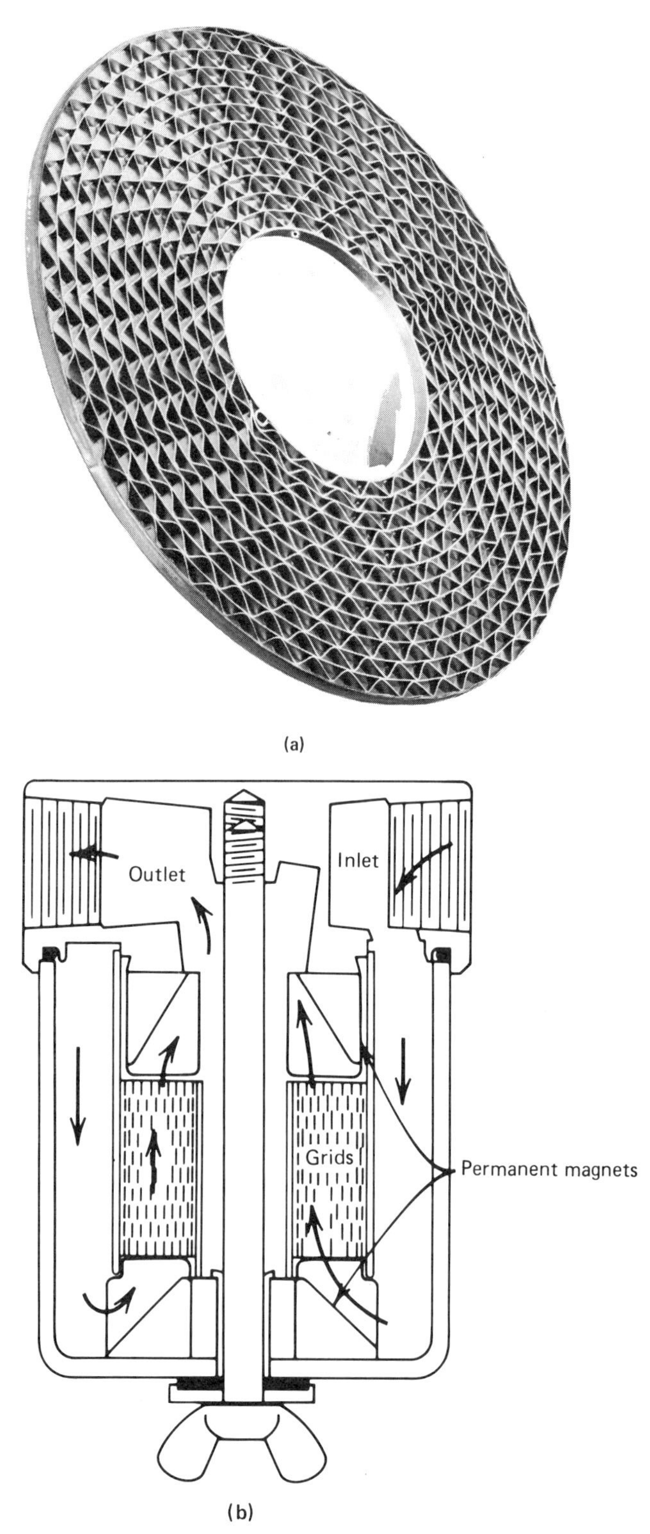

Figure 14. Frantz Ferrofilter (116). (**a**) Grid. (**b**) Assembled unit.

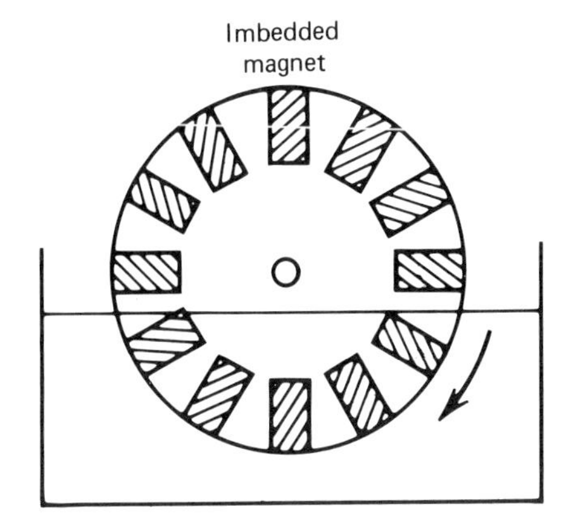

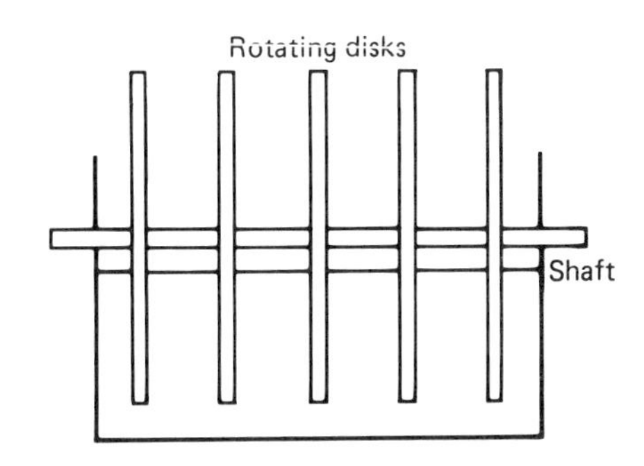

Figure 15. Rotating magnetic-disk separator (122).

liquid containing the suspended magnetic material, and lift the magnetics out of the stream where the magnets are scraped clean (Fig. 15). Very low residence times are needed for removal of the particles compared to settling or flotation (122).

Adsorption of nonmagnetic suspended materials onto magnetic "seeds" has been proposed and tested for removal of suspended solids from drinking water (123) and bacteria from municipal wastewater effluent. Using aluminum sulfate as a coagulant and 100–1000 ppm of magnetite, 80–90% of the suspended solids were removed (124).

There has been at least one study specifically evaluating the use of a high gradient magnetic separator for dewatering of a paramagnetic mineral slurry (malachite, $CuCO_3.Cu(OH)_2$) of an average particle size of 4 μm. Initial slurries of 2–12% solids were passed through the magnetic matrix and allowed to drain. When the field was turned off, the mineral could be successfully washed off the matrix to give a 40% solid slurry (125).

Finally, as briefly described above, selective separation and dewatering of one suspended substance in a slurry containing different minerals or precipitates is possible by selectively adsorbing a magnetic material (usually hydrophobic) onto a solid that is also naturally or chemically conditioned to a hydrophobic state. This process was used on both sulfide ores and on some oxides (126). More recently, hydrocarbon-based ferrofluids were tested and shown to selectively adsorb on coal from slurries of coal and mineral matter, allowing magnetic recovery (126). Copper and zinc sulfides were similarly recoverable from waste-rock slurries (127).

BIBLIOGRAPHY

1. W. E. Stanley, *Sludge Dewatering, Manual of Practice 20*, Water Pollution Control Federation, Washington, D.C., 1969, p. 101 (new edition 1983).
2. P. A. Vesilind, *Treatment and Disposal of Wastewater Sludges*, Ann Arbor Science Publishers, Ann Arbor, Mich., 1974, Chapt. 6.
3. For other definitions procedurally defining dewatering in terms of specific equipment, see: P. W. Thrush, ed., *Dictionary of Mining, Metallurgical and Related Terms*, U.S. Government Printing Office, Washington, D.C., 1968, p. 319; T. C. Collocott, ed., *Chambers Dictionary of Science and Technology*, Barnes and Noble, New York, 1971; D. N. Lapedes, *McGraw-Hill Dictionary of Scientific and Technical Terms*, 2nd ed., McGraw-Hill Book Co., New York, 1978.
4. H. Sato and co-workers, *Filtr. Sep.* **19**, 492 (1982).

5. H. B. Gala and S. H. Chiang, *Filtration and Dewatering: Review of Literature*, Report #DOE/ET/14291-1, U.S. Department of Energy, Washington, D.C., 1980.
6. L. Svarofsky, *Solid–Liquid Separations*, 2nd ed., Butterworths, London, 1982.
7. R. J. Wakeman, *Filtration Post-Treatment Processes*, Elsevier Publishing Co., Amsterdam, 1975.
8. D. B. Purchas, *Solid/Liquid Separation*, Uplands Press, Croyden, UK, 1981.
9. D. A. Dahlstrom in A. L. Mular and R. B. Bhappu, eds., *Mineral Processing Plant Design*, 2nd ed., American Institute of Mining Engineers, New York, 1980, pp. 578–601.
10. M. Shirato and co-workers, *Filtr. Sep.* **7,** 277 (1970).
11. H. G. Schwartzberg, J. R. Rosenau, and G. Richardson, *AIChE Symp. Ser.* **73**(163), 177 (1977).
12. J. Quilter, *Industrial Minerals Magazine, Energy Supplement*, 29 (March 1983).
13. D. K. Bredeson, *J. Am. Oil Chem. Soc.* **60,** 163A (1983).
14. R. J. Wakeman and A. Rushton, *Filtr. Sep.* **13,** 450 (1976).
15. M. Shirato, T. Murase, and K. Atsumi, *Proceedings, Second World Filtration Congress*, The Filtration Society, London, 1979, pp. 39–46.
16. B. K. Steenberg, *Paper Technol. Ind.*, 282 (Oct. 1979).
17. A. K. Mukhopadhyay and H. B. Kingsbury, *Tappi* **63,** 87 (1980).
18. A. F. Westergard, *Eng. Min. J.* **184**(6), 60 (1983).
19. *Elastomer and Polymer Processing Systems*, Bulletin MPR 76 (1976); *Pre-Press*, Bulletin FO 175; *French Dual Cage Screw Press*, Bulletin OP8130 (1981), The French Oil Mill Machinery Co., Piqua, Ohio.
20. L. H. Tindale and S. R. Hill-Haas, *J. Am. Oil Chem. Soc.* **53,** 265 (1976).
21. J. A. Ward, *J. Am. Oil Chem. Soc.* **53,** 261 (1976).
22. D. K. Bredeson, *J. Am. Oil Chem. Soc.* **55,** 762 (1978).
23. *Anderson Duo Crackling Expeller Presses*, Bulletin Duo 375-2, 1981; *Anderson Rubber and Plastic Polymer Dewatering and Drying Equipment*, Bulletin RDD 1173, 1980; *Anderson Expeller Presses*, Bulletin 359 R, 1980, Anderson International Corp., Cleveland, Ohio.
24. *Pressmaster Press*, Bulletin SB82-002B, Beloit Corporation, Jones Division, Dalton, Mass., 1982.
25. K. Ohmiya and S. Takahashi, *J. Water Pollut. Control Fed.* **52,** 943 (1980).
26. *Carver Cocoa Presses*, Bulletins HV-A and FP-1, Fred S. Carver, Inc., Menomonee Falls, Wisc., 1981.
27. R. H. Perry and C. H. Chilton, *Chemical Engineers Handbook*, 5th ed., McGraw-Hill Book Co., New York, 1973, pp. 19-101–19-104.
28. *V-Press Bulletin 64-5*, Bepex Corporation, Rietz Division, Santa Rosa, Calif., 1982; ref. 8, p. 516.
29. *Technical Bulletin 2-2-16/1-B*, Ingersoll-Rand Co., Nashua, N.H.
30. C. E. Silverblatt and D. A. Dahlstrom, *Ind. Eng. Chem.* **46,** 1201 (1954).
31. C. S. Simons and D. A. Dahlstrom, *Chem. Eng. Prog.* **62**(1), 75 (1966).
32. A. F. Baker and A. W. Duerbrouck in A. C. Partridge, ed., *Proceedings of the International Coal Preparation Congress*, 1977; excellent discussion at end of paper shows the confusion present in the role of surfactants.
33. J. H. Brown, *Can. Min. Metall. Bull.* **58,** 315 (1965); *Transactions Can. Inst. Min. Met.* **68,** 105 (1965).
34. F. M. Tiller and J. R. Crump, *Chem. Eng. Prog.* **74,** 65 (Oct. 1977).
35. U.S. Pat. 4,107,028 (Aug. 15, 1978), R. K. Emmett, S. D. Heden, and R. A. Summerhays (to Envirotech Corporation).
36. S. M. Moos and R. E. Dugger, *Min. Eng. (N.Y.)* **31,** 1479 (1979).
37. J. Leja, *Surface Chemistry of Froth Flotation*, Plenum Press, New York, 1982, p. 272.
38. H. B. Gala, S. H. Chiang, and W. W. Wen, *Proceedings World Filtration Congress III*, The Filtration Society, Downington, Pa., 1982.
39. D. V. Keller and co-workers, *Surface Phenomena in the Dewatering of Coal*, DOE Report #FE9001-1, U.S. Department of Energy, Washington, D.C., 1979.
40. S. K. Nicol, *Proc. Australas Inst. Min. Metall.*, 37 (Dec. 1976).
41. J. Hermia in ref. 15, pp. 549–566.
42. F. M. Tiller and co-workers in ref. 15, pp. 1–13.
43. A. Rushton, *Filtr. Sep.* **13,** 573 (1976).
44. P. J. Lafforgue and co-workers, *paper presented at Society of SME-AIME Annual Meeting*, Feb. 1982, preprint 82-22, avail. United Engineering Soc. Library, New York.
45. *Proceedings of the Consolidation and Dewatering of Fine Particles Conference*, University of Alabama, Aug. 1982, available from U.S. Bureau of Mines, Tuscaloosa, Ala.

46. *Proceedings of the Progress in the Dewatering of Fine Particles Conference*, University of Alabama, April 1981, available from the U.S. Bureau of Mines, Tuscaloosa, Ala.
47. F. N. Kemmer and J. McCallion, eds., *Nalco Water Handbook*, McGraw-Hill Book Co., New York, 1979, Chapts. 8–9.
48. M. E. Lewellyn and S. S. Wang in R. B. Seymour and G. A. Stahl, eds., *Macromolecular Solutions Solvent–Property Relationships in Polymers*, Pergamon Press, New York, 1982, pp. 134–150.
49. M. J. Pearse and T. Barnett, *Filtr. Sep.* **17,** 460 (1980).
50. Ref. 8, p. 53.
51. R. Leutz and M. Clement, *Filtr. Sep.* **7,** 193 (1970).
52. M. J. Pearse in ref. 45, p. 41–89.
53. V. P. Mehrotra and co-workers, *Filtr. Sep.* **19,** 197 (1982).
54. R. J. Schwartz in ref. 2, pp. 13–40.
55. A. F. Cassel and B. P. Johnson, *Evaluation of Devices for Producing High Solids Sludge Cake*, NTIS Report No. PB80-111503, National Technical Information Service, Washington, D.C., 1980.
56. G. L. Culp, *Handbook of Sludge Handling Processes*, Garland STPM Press, New York, 1979.
57. D. DiGregorio and J. F. Zievers in W. W. Eckenfelder, Jr., and C. J. Santhanam, eds., *Sludge Treatment*, Marcel Dekker, Inc., New York, 1981, Chapt. 6, pp. 142–207.
58. *Wastewater Engineering*, McGraw-Hill Book Co., New York, 1972, Chapt. 13, 760 pp.
59. B. Sawyer, R. Watkins, and C. Lue-Hing, *Proceedings 31st Industrial Waste Conference*, Purdue University, Lafayette, Ind., 1976, pp. 537–553.
60. P. R. Karr and T. M. Keinath, *J. Water Pollut. Control Fed.* **50,** 1911 (1978).
61. *Spiractor*, Bulletin 5852, Permutit Co., Inc., Paramus, N.J., 1979.
62. *SWA/KW Reactor*, Esmil International, Amsterdam, 1976.
63. D. A. Dahlstrom in M. P. Freeman and J. A. FitzPatrick, eds., *Theory, Practice and Process Principles for Physical Separations*, Engineering Foundation, New York, 1977, pp. 261–273; *EPRI Report F.P. 937*, Electric Power Research Institute, Palo Alto, Calif., 1979.
64. A. D. Randolph and D. Etherton, *Study of Gypsum Crystal Nucleation and Growth Rates in Simulated Flue Gas Desulfurization Liquors*, EPRI Report CS1885, Electric Power Research Institute, Palo Alto, Calif., 1981.
65. O. E. Albertson, *Sludge Thickening, Manual of Practice FD1*, Task Force on Sludge Thickening, Water Pollution Control Federation, Washington, D.C., 1980, p. 33.
66. *DCG for Gravity Sludge Dewatering*, Bulletin 5161, Permutit Co., Inc., Paramus, N.J., 1983.
67. M. Yusa and A. M. Gaudin, *Am. Ceram. Soc. Bull.* **43,** 402 (1964).
68. M. Yusa and co-workers in A. C. Partridge, ed., *Proceedings of the 7th International Coal Preparation Congress*, Australian National Committee, Sydney, Australia, 1976.
69. J. Szczpya in P. Somasundaran, ed., *Fine Particles Processing*, Society of Mining Engineers of AIME, Littleton, Col., 1980, pp. 1676–1686.
70. B. J. Scheiner and A. G. Smelley, *Dewatering of Thickened Phosphate Clay Waste from Disposal Ponds*, Paper A81-6, The Metallurgical Society of AIME, Warrendale, Pa., 1981; J. R. Pederson, ed., *U.S. Bureau of Mines Research 81*, U.S. Government Printing Office, Washington, D.C., 1981, p. 83.
71. R. M. Hoover and P. V. Avotins, *Development of Polymer Pelletization for Enhancing Solid Liquid Separation of Leached Laterite Residue*, Paper A78-13, The Metallurgical Society of AIME, Warrendale, Pa., 1978.
72. V. P. Mehrota and co-workers, *Int. J. Miner. Process.* **10** (1983).
73. V. P. Mehrota and co-workers, *Min. Eng. (N.Y.)* **32,** 1230 (1980).
74. D. V. Keller, Jr., in ref. 45, pp. 152–171.
75. F. J. Zuiderweg and co-workers, *Chem. Engineer (London)*, 223 (July 1968).
76. F. W. Meadus and co-workers, *Can. Min. Metall. Bull.*, 968 (1966).
77. F. W. Meadus and co-workers, *Can. Min. Metall. Bull.*, 1326 (1969).
78. I. E. Puddington and B. D. Sparks, *Miner. Sci. Eng.* **7,** 282 (1975).
79. P. T. L. Koh and L. T. Warren, *13th International Mineral Processing Congress*, Warsaw, Poland, 1979.
80. A. M. Gaudin and P. Malozemoff, *J. Phys. Chem.* **37,** 599 (1933).
81. A. M. Gaudin and P. Malozemoff, *Trans. Am. Inst. Min. Metall. Engrs.* **112,** 303 (1934).
82. I. E. Puddington in P. Somasundaran and M. Arbiter, eds., *Beneficiation of Mineral Fines*, National Science Foundation, Society of Mining Engineers, AIME, Littleton, Col., 1979, Chapt. 28.
83. O. Eidsmo and O. Mellgren, *5th International Mineral Processing Congress*, London, 1960.

84. K. Karjalahti, *Trans. Inst. Min. Metall.* **81,** C219 (1972).
85. K. Li and co-workers, *Trans. Inst. Min. Metall.* **70,** C19 (1960).
86. J. W. H. Chi and E. F. Young, *Trans. Inst. Min. Metall.* **72,** 169 (1962).
87. H. M. Smith and I. E. Puddington, *Can. J. Chem.* **38,** 1911 (1960).
88. G. M. Ritcey and A. W. Ashbrook, *Solvent Extraction Principles and Applications to Process Metallurgy*, Elsevier Publishing Co., Amsterdam, 1979, Part II, p. 669.
89. R. Stratton-Crawley in ref. 82, p. 317.
90. D. Bass, *Paint Manuf.*, 5 (Jan. 1957).
91. G. W. Reimers and S. E. Khalafalla, *Preparing Magnetic Fluids by a Peptizing Method*, U.S. Bureau of Mines Technical Progress Report 59, U.S. Bureau of Mines, Washington, D.C., Sept. 1972; U.S. Pat. 3,843,540 (Oct. 22, 1974), G. W. Reimers and S. E. Khalafalla (to U.S. Department of the Interior).
92. M. Peluso and D. F. Ogren, *paper presented at the 71st Annual AIChE Meeting, Miami 1978*, C-E Lummus, Bloomfield, N.J., 1978.
93. C. L. Tsaros in J. W. White and B. F. Feingold, eds., *Peat Energy Alternatives*, Institute of Gas Technology, Chicago, Ill., 1980.
94. *Chem. Eng.*, 82 (June 4, 1979).
95. S. A. Raksit, *Carver-Greenfield Pilot Demonstration*, LA-OMA Project Los Angeles Department of Public Works, Los Angeles, Calif., 1978.
96. J. Jacknow, *Sludge* **2**(4), 26 (July 1979).
97. *PDF Large Scale Peat Refining*, process licensed by Wheelabrator-Frye, Hampton, N.H., 1981.
98. Can. Pat. 1,010,477 (Nov. 8, 1977), E. J. Wasp (to Bechtel International Corp.).
99. W. H. Oppelt and co-workers, *Drying North Dakota Lignite to 1500 Psi by the Fleissner Process*, Report of Investigations 5527, U.S. Bureau of Mines, Washington, D.C., 1959.
100. B. Stanmore, D. N. Boria, and L. E. Paulson, *Steam Drying of Lignite: A Review of Processes and Performance*, DOE/GFETC/RI-82/1 (DE82007849), U.S. Department of Energy, available from National Technical Information Service, Washington, D.C., 1982.
101. J. B. Murray and D. G. Evans, *Fuel* **51,** 290 (1972).
102. U.S. Pats. 4,052,168; 4,129,420 (1977), E. Koppelman; *Chem. Eng.*, 77 (March 27, 1978).
103. D. F. Othmer, *Mech. Eng.*, 30 (Dec. 1979).
104. J. Josephson, *Environ. Sci. Technol.* **16,** 548A (1982).
105. Ref. 8, pp. 60–63.
106. P. F. Mahoney and W. J. Duensing, *J. Am. Water Works Assoc.* **64,** 665 (1972).
107. A. J. Toerling in ref. 46, 34 pp.
108. G. S. Logsdon and E. Edgerley, Jr., *J. Am. Water Works Assoc.* **63,** 734 (Nov. 1971).
109. H. E. Davis and C. J. Egan, *AIChE Symp. Ser. 207* **77,** 50 (1981).
110. A. W. Adamson, *Physical Chemistry of Surfaces*, 3rd ed., John Wiley & Sons, Inc., New York, 1974, p. 212.
111. N. C. Lockhart in ref. 38, pp. 325–332.
112. M. P. Freeman in G. Hetsrom, ed., *Handbook of Multiphase Systems*, Hemisphere, New York, 1982, Chapt. 9.3, pp. 9-98–9-115.
113. B. A. Segall and co-workers, *ASCE Geotech Engineering Division J.* **GT106,** 1148 (1980), and references therein.
114. C. A. Fetzer, *Proceedings ASCE, Journal of Soil Mechanics and Foundations Division* **93 SM4,** 85 (1967).
115. R. H. Sprute and D. J. Kelsh in ref. 69, pp. 1828–1844.
116. *Frantz Ferrofilter Magnetic and Electromagnetic Separators*, Bulletins EM and PM, S. G. Frantz Co., Inc., Trenton, N.J., 1980.
117. C. Mills, *Ind. Miner.* (*London*), 41 (Aug. 1977).
118. J. E. Lawver and D. M. Hopstock, *Miner. Sci. Eng.* **6,** 154 (July 1974).
119. D. M. Thayer and P. B. Linkson, *Trans. AIME* **270,** 1897 (1981).
120. N. Nojiri and co-workers, *J. Water Pollut. Control Fed.* **52,** 1898 (1980).
121. Y. Tamaura and co-workers, *Water Res.* **13,** 21 (1979).
122. M. Miura and T. M. Williams, *Chem. Eng. Prog.*, 66 (April 1978).
123. B. A. Bolto and co-workers, *J. Polymer Sci. Polymer Symp.*, 211 (1975).
124. R. R. Oder and B. I. Horst, *Filtr. Sep.* **13,** 363 (1976).
125. P. Chakrabarti and co-workers, *Filtr. Sep.* **19,** 105 (1982).

126. T. A. Sladek and C. H. Cox, *Coal Preparation Using Magnetic Separation, Volume 4: Evaluation of Magnetic Fluids for Coal Beneficiation*, EPRI Report CS1517, Energy Electric Power Research Institute, Palo Alto, Calif., July 1980.
127. U.S. Pats. 1,043,851 (Nov. 1912); 1,043,850 (Nov. 1912); 996,491 (Aug. 1911); 993,717 (June 1911), A. A. Lockwood.

BOOKER MOREY
Telic Technical Services

DISPERSION OF POWDERS IN LIQUIDS

The term dispersion refers to the complete process of incorporating a powder into a liquid medium in such a manner that the product consists of fine particles distributed throughout the medium. The dispersion of fine particles is normally termed colloidal if at least one dimension of the particles lies between 1 nm and 1 μm. Such dispersions are classified, in terms of the affinity of the colloidal particle for the medium, as lyophobic (possessing aversion to liquid) or lyophilic (possessing affinity for liquid); and as hydrophobic and hydrophilic, respectively, for aqueous media. Some difficulties may arise with the use of these terms. In the sense of the definition, lyophobic implies no affinity between the colloid and the medium, eg, an insoluble powder, but taken to its logical extreme this would suggest no wetting of the powder by the liquid and hence no dispersion could be formed. But although a metal oxide powder may be insoluble in water, it is nevertheless normally wetted by water and hence the powder surface is hydrophilic, but the dispersion is classified as hydrophobic. Solutions of polymers and surface-active agents are of the lyophilic type and form spontaneously when the components are brought into contact. They are true solutions and are stable in the thermodynamic sense, whereas lyophobic dispersions do not form spontaneously and hence in principle are thermodynamically unstable. The division between solutions and dispersions is without ambiguity.

A powder containing nonporous particles of colloidal dimensions has a high surface area. Consider the change in the surface area of a piece of matter as it is successively subdivided. A 1-cm cube, initially with an area of 6 cm^2, when divided into 10^{12} cubes of 1-μm edge length has an area of 6 m^2, whereas when divided into 10^{18} cubes of 10-nm edge length, it has an area of 600 m^2 or 0.06 ha (ca 0.15 acre). Normally, it is assumed that the surfaces separating individual phases make a negligible contribution to the total energy of a system, since only a relatively small number of the molecules are at the surface. Even if the surface molecules have energies that are substantially different from those in the bulk, they make an insignificant contribution to the macroscopic properties of the system. For colloidal particles, however, the proportion of molecules at the surface may be as high as ca 30%; hence, surface properties become increasingly important the smaller the particle size.

In many practical uses of powders, the primary particle size is sufficiently small for further subdivision to be unnecessary. In the dry state, the powder usually contains some aggregates of primary particles that are attached to other aggregates or primary particles forming agglomerates. Aggregates are groups of primary particles joined at their faces with a surface area significantly less than the sum of the areas of their constituent particles. An agglomerate is a collection of primary particles and aggregates which are joined at edges and corners; its surface area is not markedly different from the sum of the areas of the individual components. Many factors are involved in the powder achieving its particular state, and a knowledge of its manufacture and storage conditions is necessary in order to understand the problem of breaking the bonds between particles in the clusters.

Dispersability is defined as the ease with which the particles in the powder are distributed in a continuous liquid medium in such a manner that each particle is completely surrounded by the medium and no longer makes permanent contact with other particles. The problem is then to maintain the dispersed state since the particles have a natural tendency as a result of Brownian motion to reduce in number with time because of collisions. An attractive force exists between the particles as they approach, the magnitude of which increases significantly as the distance of approach decreases. The reduction in particle number here is to be termed flocculation whatever the mechanism involved. To resist flocculation, a repulsive force is necessary which is usually achieved through the particles being charged or containing adsorbed layers that protect them, or both. The term coagulation, which is also often used to represent this process, has various meanings and is avoided here. Strictly speaking, coagulation involves the formation of compact clusters of particles, leading to the macroscopic separation of a coagulum. Flocculation implies the formation of a loose or open network (floc) which may or may not separate macroscopically. Coagulation is usually irreversible, but flocculation is reversible, ie, the flocs are readily broken down under shear (see also Flocculating agents). Hence, the term stability can have various meanings and should be clearly defined in relation to the type of change under consideration. It could relate to flocculation or coagulation, or to sedimentation both with and without hindered settling, all of which involve a change in the state of dispersion, hence in the properties of the dispersion.

In the early stages of the dispersion process, the solid–air interface is replaced by one between solid and liquid (wetting). The powder consists of aggregates/agglomerates of similar particles, and to disperse these particles into a liquid, mechanical work is performed. The forces that exist at the interface determine the ease with which the process can be brought about. Once dispersed, the particles are free to move in their new environment, and flocculation is prevented by various chemical and physical methods, all of which relate to the character of the solid–liquid interface. This review concerns the forces between primary particles in the powder, the wetting and milling processes, and the stability of the resulting dispersion.

The dispersion process includes incorporation, wetting, breakdown of particle clusters, and flocculation of the dispersed particles. These stages overlap, but one is likely to be dominant with respect to a particular property of the dispersed system. The principles involved in each stage are fairly well established, but because of the overlap it is often not easy to recognize any particular aspect in a practical dispersion system.

Incorporation

The interparticle forces that must be overcome in dispersion may be of several types, such as mechanical forces caused by interlocking of irregular particles; surface tension forces, particularly with powders containing pendular moisture; forces arising from plastic welding caused by contact points between particles coalescing under high loads; electrostatic forces, particularly for substances that easily become charged; van der Waals forces, particularly significant for particles of small diameter; and solid-bridge forces, where crystallization at contact points causes joining of the particles.

All these forces are subject to changes in environment and to the effects of their previous history. The surfaces of freshly ground material, for example, contain a multiplicity of active sites and unsatisfied valency bonds, and such materials do not behave in the same way as aged material. The humidity of the environment can affect the surface moisture on particles and lead to a change in properties. Different methods of particle production give particles of different roughness, leading to changes in the interlock or friction forces between particles.

Finally, the way in which the particles pack together to form the powder must be considered. Even if they were all of the same size and shape, and interparticle attractions were absent, a number of three-dimensional packing structures are possible. Particle size and shape and interparticle attractive forces influence this process, and therefore, the packing structure of fine, irregular particles is extremely complex geometrically.

Liquid Bridges. When appreciable quantities of liquid are adsorbed, the surface film covers the asperities and forms liquid bridges between particles in contact. The curvature of the meniscus determines the vapor pressure above the liquid bridge and the magnitude of the adhesive force due to surface tension. Increasing moisture content of powders influences their strength in diverse ways, depending on the nature of the powders. In some instances strength increases, and in some it decreases as the surface moisture level increases. Considering both the surface tension and the pressure deficiency above the meniscus, a monotonic decrease in strength with increasing liquid content can be predicted for smooth particles (1). Surface roughness reverses this trend, and geometric arguments can explain the opposing effects of alumina (smooth) and silica (rough) powders (2). For two smooth spheres the total force (F_{tot}) is

$$F_{tot} = \frac{2\,\pi r \gamma}{1 + (\tan \phi/2)} \qquad (1)$$

where ϕ is the central angle (Fig. 1), r is particle radius, and γ is the surface tension.

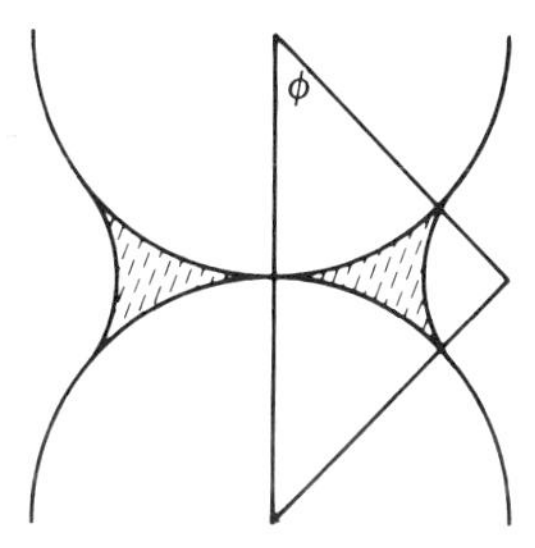

Figure 1. Condensed liquid at the point of contact between two smooth solid spheres, assuming zero contact angle.

As ϕ approaches zero, F_{tot} tends to a limiting value $2\ \pi r\gamma$. As the liquid content increases, the size of the liquid ring increases until a limiting value is reached at $\phi = 53.13°$ and $F_{tot} = 4\ \pi r\gamma/3$. Thus as ϕ increases, F_{tot} falls from $2\ \pi r\gamma$ to $4\ \pi r\gamma/3$, owing to the changing curvature of the liquid–vapor interface.

The asperities on a rough surface are treated as conical contact points, and the force is shown to be

$$F_{tot} = \pi a \gamma \frac{\tan \alpha/2}{\cos \alpha} \qquad (2)$$

(for α and a, see Fig. 2). Thus, F_{tot} increases with a, which is a measure of the size of the meniscus, and also increases with α.

Interaction between Surface Asperities. Particle surfaces are rough when viewed on a macroscopic level, and it seems reasonable to expect surface asperities of adjacent particles to interlock. When two particles come into contact, however, only relatively small parts of their total surface area are actually touching. Only the tips of the asperities come into contact, and if the material is at all compacted, tremendous pressures can be generated over the tiny areas of contact, and plastic flow can occur at contact points as well as elastic deformation within the material. If the asperities are deformable, the real area of contact, and hence the adhesion between the particles, increases with the force pressing them together. Both the surface energy and hardness of the material are important factors. Adsorbed vapor reduces the surface energy; hence, the joint at the asperities is weakened.

Electrostatic Forces. Electrostatic forces may arise from surface charges which are either permanent or acquired from adjacent surfaces. Every particle in an assembly of dry particles may not possess the same sign or magnitude of charge, and the following different types of charged powder have been identified: unipolar powders, in which all particles have charges of similar sign; ambipolar powders, in which particles contain both positive and negative charges; homopolar powders, in which all particles have charges of the same magnitude; and heteropolar powders, in which different particles have charges of different magnitudes.

When two solid bodies are in contact, electrons are transferred from one to the other. In theory, this transfer process should continue until an equilibrium is reached when the electron currents (or work functions) of the solids become equal. When this equilibrium has been reached, there is a potential difference between the two solids; the junction between the surfaces constitutes a capacitance. According to simple theory, there is an attraction between the plates of a simple capacitor, given by

$$E = \frac{\epsilon_o X^2}{2\ H_o^2} \qquad (3)$$

where E is the electrostatic attraction (N^2/m^2); X, the potential difference (V); H_o, the distance between the plates (m); and ϵ_o, the dielectric constant of the medium (C^2/Nm^2). Unfortunately, this simple equation cannot be applied widely because of incomplete knowledge of the electrical properties of the materials involved. Particularly, a serious problem is posed by the effects of adsorbed layers and so-called surface states. The origin of surface states is obscure. They are known to be partly owing to absorbed atoms and imperfections in surface structure. Both of these factors depend on the immediate history of the material, eg, grinding stress and frictional history. Thus, quantitative prediction of electrostatic interparticle forces becomes extremely

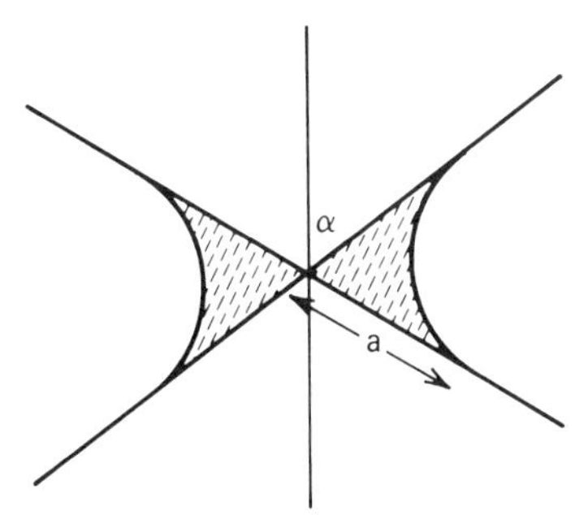

Figure 2. Condensed liquid at the point of contact between two solid cones, representing the asperities on a rough surface; zero contact angle is assumed.

difficult. Calculations indicate that the electrostatic forces arising from excess charge or contact potential are at least one order of magnitude less than the van der Waals force.

Van der Waals Force. The derivation of the attractive force between macroscopic bodies assumes that interactions between individual atoms and molecules in the interacting materials are additive (3). For two equal spheres of radius a, the attractive potential energy V_A is given by

$$V_A = -\frac{A}{6}\left[\frac{2}{s^2 - 4} + \frac{2}{s^2} + \ln\left(\frac{s^2 - 4}{s^2}\right)\right] \tag{4}$$

where A is the Hamaker constant and $s = R/a$; R is the distance between the centers of the spheres. Expressions for other situations are given in ref. 4. For irregular shapes, the force is more difficult to define. In the case of interaction between agglomerates, it can be assumed that, since the van der Waals force is of short range, only the attraction between the particles in the outer surface of the agglomerates is of any consequence.

Adsorbed layers on the particle surfaces affect the magnitude of the attractive force by increasing R and usually reducing A; similarly for liquid bridges.

Solid Bridges. The strength of an agglomerate can be considerably enhanced by the formation of solid bridges between particles. Such bridging can be due to chemical reactions at the surface; crystallization of dissolved substances, eg, water of crystallization may be released on heating which dissolves surface material, and this is redeposited when the temperature is reduced; and formation of bonds of the same material as the particles by surface diffusion at the Tamann temperature (half the surface melting point in Kelvins), and this might occur during frictional heating when adjacent surfaces rub together.

Such a wide variety of mechanisms make it difficult to quantify the effects in any theory of powder strength. However, for practical purposes, the magnitude of the principal forces involved can be estimated (van der Waals and gravity), to illustrate the problem of disagglomeration of the powder during the early stages of incorporation. In general terms, the smaller the particles of a powder, the more important become the interparticle forces in controlling packing and flow behavior. For particles to pack closely together and flow freely, the force of gravity on each particle must be greater than the force holding the particles to each other. For example, it can be shown for titanium dioxide pigment particles that below a radius of ca 100 μm the attractive force exceeds the separation force (Fig. 3). This is in accord with the common experience

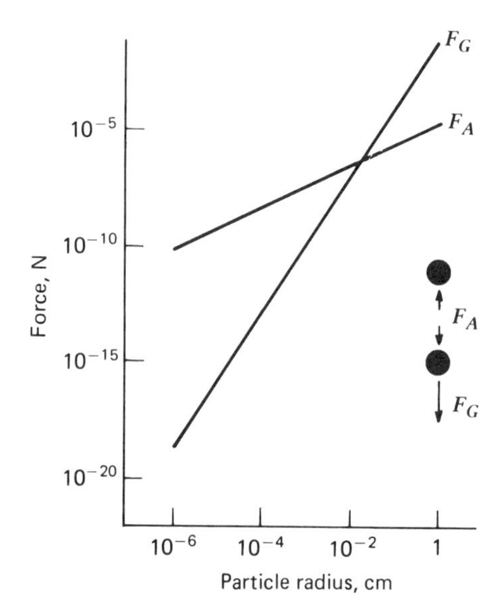

Figure 3. Interparticle attractive force between two titanium dioxide particles compared with the gravity force. F_A = attractive force; F_G = force of gravity. To convert N to dyn, multiply by 10^5.

that powdery materials tend to be free-flowing when their particle size is above this radius, and cohesive when it is less. Particles with a radius of 0.1 μm are therefore intrinsically cohesive. Nevertheless, the pigment flows as loose agglomerates, which under an optical microscope appear to have a diameter of 30–50 μm. If it is assumed that loose agglomerates are loosely packed assemblies of pigment particles with a density of not much more than unity, and that only one or two particles in their surfaces contribute to the attractive force, a similar calculation shows that agglomerates above 30–50 μm would be expected to flow fairly freely (Fig. 4).

Tensile Strength. A simple demonstration of the range of cohesiveness that exists in available TiO_2 pigments is the measurement of tensile strength (5–6). Pigments of high tensile strength are those that tend to be difficult to incorporate, particularly in poorly dispersing media (Fig. 5). Therefore, measurements of tensile strength and cohesiveness clarify the ease of incorporation, ie, on the initial submergence process when pigment and medium are brought into contact and pigment–air interfaces start to be replaced by pigment–liquid interfaces. Pigments of high tensile strength do not break down easily during this process (see also Pigments).

The tensile strength of a material is the maximum stress required to fracture a specimen in simple tension. Most semiempirical approaches linking powder-surface properties with flow properties have attempted to link a defined particle size, the packing fraction, and an interparticle force parameter with tensile strength. In one of the earliest approaches, a clear distinction was made between powders where the bonding is localized at points of interparticle contact and where the space between the particles is filled by some strength-transferring material such as a liquid (7). A basic equation for a model in which the particles are of uniform size (uni-sized) makes four important assumptions: (*1*) there are a large number of bonds in the fracture section, (*2*) the bonds are distributed randomly in space within the fracture section, (*3*) the particles in the compact are randomly distributed in space, and (*4*) a mean microscopic force exists which is the same at any point in the fracture section.

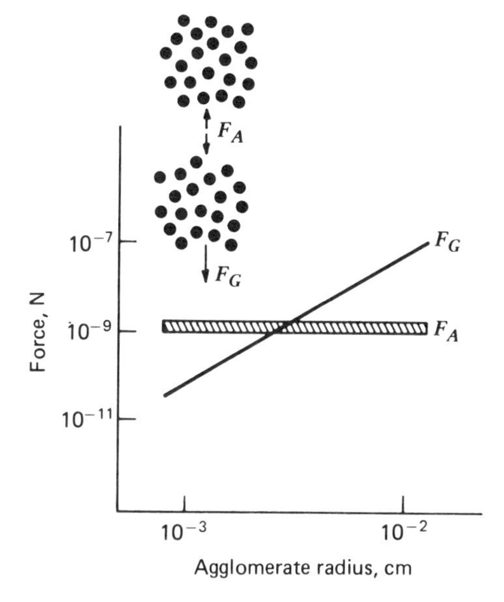

Figure 4. The attractive force between agglomerates of titanium dioxide particles compared with the gravity force. F_A = attractive force; F_G = force of gravity. To convert N to dyn, multiply by 10^5.

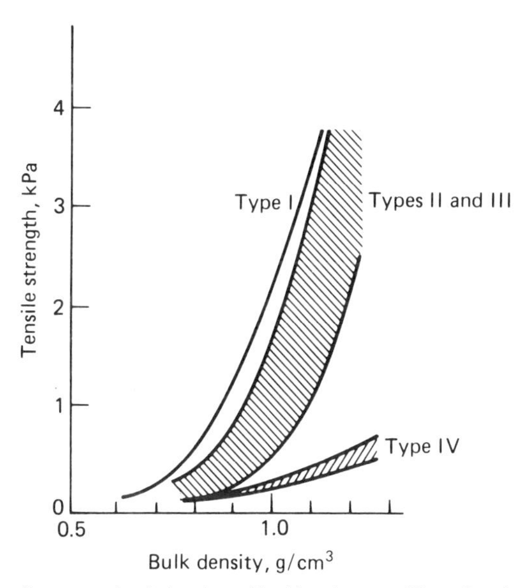

Figure 5. The tensile strength of titanium dioxide pigment: Type I = dry-milled, uncoated; Type II = pigments coated with inorganic oxides; Type III = Type II with an organic treatment; Type IV = Type II with a hydrophobic organic treatment for plastics applications. To convert kPa to psi, multiply by 0.145.

The first three assumptions eliminate packing considerations other than the voidage ϵ and coordination number, and the postulation of uni-sized spheres means that a one-parameter description representing the particle size and shape can be used. The fourth assumption combines all interparticle surface forces known to exist into one composite parameter.

Rumpf's equation is of the form

$$T = \frac{9}{8} \cdot \frac{(1-\epsilon)}{\pi d^2} \cdot cF \tag{5}$$

where T is the tensile strength, d is the diameter of the spherical particles, c is the mean coordination number, and F is the mean microscopic force at the contact face (see assumption (*4*) above). The equation was verified using closely sized limestone particles, which were rotund in shape. The original theory was extended to randomly packed, similar, approximately isometric, convex particles, and ways of treating random particle size distributions were considered (8). It was also shown that molecular forces and plastic welding were particularly important in determining the tensile strength of dry powders and that the strengths of compacts made under vacuum are higher than those made in the atmosphere. Examination of moist powders showed that liquid–solid boundary adhesion is of prime importance in determining compact strength (9). Equations were derived involving the angle of contact of the wetted surface and various relationships plotted between coordination numbers for a network of bridges and powder strength variables.

In another approach, it is assumed that in a powder compact, individual particles are packed together in a more or less random fashion (10). The particles are taken to have an average coordination number, and, on tensile fracture, the tensile strength is taken to be an average over a large number of particle pairs in the fracture plane. Thus, no particular packing structure is attributed to the compact. The fracture plane is not simple, ie, the actual geometry of the surface at the break is somewhat rugged, as illustrated in Figure 6.

Since the tensile strength depends on the number of particle pairs N_{pp} per unit area of the fracture plane, the relation of N_{pp}/area to the coordination number and the bulk density of the compact ρ can be expressed as

$$\frac{N_{pp}}{\text{area}} = \frac{c}{2}\frac{\bar{d}}{\bar{v}}\frac{\rho}{\rho_s} \tag{6}$$

where $\bar{v}$ is the mean volume of the particle of mean effective diameter $\bar{d}$, and ρ_s is the density of the solid.

The second factor determining the tensile strength is the force F_{pp}, which acts between the particles in a pair of particles. Since the particles are not smooth, when they come into contact, the surface protrusions touch and form microcontacts, shown

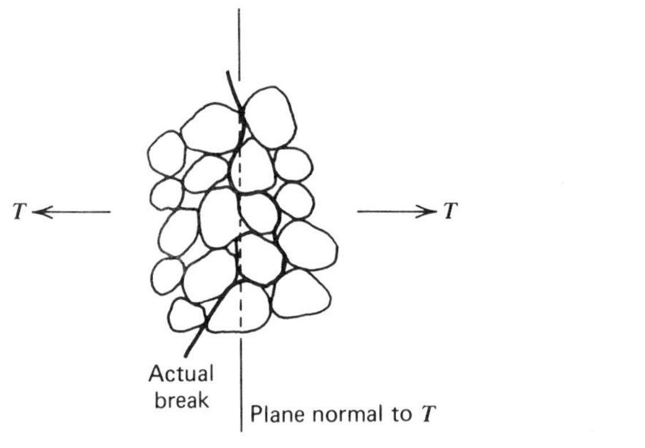

Figure 6. Fracture geometry of a powder compact. T = tensile strength.

as M in Figure 7; N represents protrusions that form microcontacts when the surface separation is reduced.

An area of contact A_{ij} is defined within which the microcontacts lie, and F_{pp} can be defined in terms of this area:

$$F_{pp} = F^{o}_{pp} A_{ij} \tag{7}$$

where F^{o}_{pp} is the interparticle force per unit area of contact.

The tensile strength is given by

$$T = \sum \overline{F}_{pp}$$

where $\overline{F}_{pp}$ is the mean of F_{pp}, and $\overline{F}_{pp} = \frac{1}{2} F_{pp}$ is the interparticle forces are randomly directed at the fracture plane. Thus

$$T = \tfrac{1}{2} F^{o}_{pp} \sum A_{ij} \frac{N_{pp}}{\text{area}} x_{ij} \tag{8}$$

where x_{ij} is the proportion of particle pairs of diameter d_i and d_j.

The equation for the tensile strength of dry powder compacts is

$$T = \tfrac{3}{4} t_o \frac{\overline{ds}}{\overline{v}} \cdot \frac{\rho/\rho_s}{1 - (\rho/\rho_s)} \cdot F^{o}_{pp} \tag{9}$$

where t_o is the surface separation at zero tensile strength, and $\bar{s}$ is the mean surface area per particle (calculated from the particle size distribution assuming spherical particles). The equation is tested by using experimental T vs ρ data,

$$T = \tfrac{1}{2} abc \frac{\bar{s}}{\bar{v}} \frac{\rho}{\rho_s} \left[t_o - \frac{\bar{d}}{3} \left(\frac{\rho}{\rho_o} - 1 \right) \right] F^{o}_{pp} \tag{10}$$

where a is the number of particle pairs per unit area divided by the number of pairs per unit volume, b the true area of contact per particle pair divided by the surface area per particle, and ρ_o the value of bulk density when $T = 0$.

The work on dry powders was extended to powders containing pendular moisture to cause liquid bridges to form between the particles. For these, an expression can be written for T

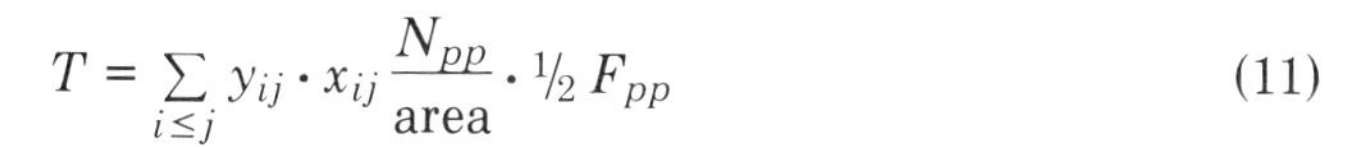

$$T = \sum_{i \leq j} y_{ij} \cdot x_{ij} \frac{N_{pp}}{\text{area}} \cdot \tfrac{1}{2} F_{pp} \tag{11}$$

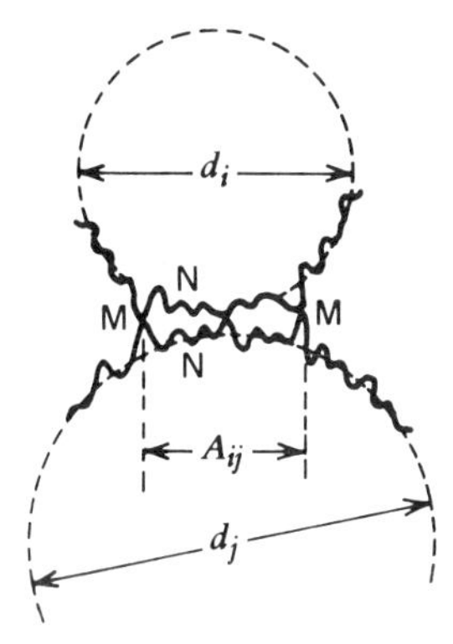

Figure 7. Microcontacts (M) and surface protrusions (N) of two particles with rough surfaces in contact.

y_{ij} is the proportion of particle pairs of diameters d_i and d_j bridged by pendular moisture. Since T is equated with interparticle forces between only those particles joined by moisture, it is implicit in this argument that surface-tension adhesion forces are dominant in powders containing pendular moisture.

Thus, from tensile strength measurements the magnitude of the interparticle forces may be derived and compared, eg, with that calculated from Hamaker's expression for the van der Waals interaction force.

It should be possible to define the theoretical minimum energy ($E_{\min}$), which would be necessary to achieve the dispersion of unit mass of powder by a completely efficient process. The theoretical maximum dispersability would then be defined as $D_{\max} = 1/E_{\min}$. Hence, dispersability has the dimension of mass per unit of energy and is thus a direct measure of the quantity of powder that would be dispersed by the expenditure of one unit of energy. Since no real process can be 100% efficient, it is necessary to incorporate an efficiency factor K to account for the real situation. Hence $D_{\text{real}} = K/E_{\min}$ where $0 < K < 1$. Various factors determine the dispersability of a powder in terms of those that affect $E_{\min}$ and K separately. In principle, $E_{\min}$ can be calculated by integrating the work required to break individual interparticle bonds, although by the very nature of dispersions such a simple integration could be misleading since simultaneous processes take place and, depending on the nature of the system, they may consume or supply additional energy to that expended on the system from external sources. Therefore, apart from the simplest of systems, eg, the dispersion of powders in gases, the calculation of the minimum energy required to separate the particles is difficult.

Wetting

The initial stage of wetting involves both the external surface of particles and the internal surfaces which exist between the particles in the clusters that make up the dry powder. The wetting process is dependent on the nature of the liquid phase, the character of the surface, the dimensions of the interstices in the clusters, and the nature of the mechanical process used to bring together the components of the system. The principles of wetting external surfaces are well established, but extension to internal surfaces involves geometric factors which, for fine particle systems, are not readily defined.

To describe the wetting of a powder it is necessary to be more specific about the kinds of wetting processes that can be involved in the formation of a solid–liquid interface. The three definable distinct types of wetting are designated as adhesional wetting, spreading wetting, and immersional wetting, according to the mechanical process taking place. They are readily defined in terms of the changes in interfacial free energy γ involved.

Adhesional wetting. The formation of 1 m^2 of solid–liquid interface by bringing into contact 1 m^2 of plane solid surface with 1 m^2 of plane liquid surface involves work W_A given by

$$W_A^{\text{SLV}} = \gamma^{\text{SL}} - (\gamma^{\text{S}} + \gamma^{\text{LV}}) \tag{12}$$

where W_A^{SLV} is the work of adhesion, that is, the work required to restore the initial condition, and γ^{SL}, γ^{S}, and γ^{LV} are the interfacial free energies at the solid–liquid, solid–vapor, and liquid–vapor interfaces, respectively, the vapor here being that of the separate phases.

Immersional wetting. If 1 m^2 of solid surface is immersed in a liquid, the work involved is

$$W_W^{SLV} = \gamma^{SL} - \gamma^{SV} \tag{13}$$

where γ^{SV} is the free energy of the surface in equilibrium with the vapor of the liquid.

Spreading wetting. When a drop of liquid spreads over a solid surface, for each unit area of solid surface that disappears, equivalent areas of liquid surface and solid–liquid interface are formed. The work of spreading wetting is given by

$$W_S^{SLV} = (\gamma^{SL} + \gamma^{LV}) - \gamma^{SV} \tag{14}$$

Here γ^{SV} is used, since the solid is assumed to be in equilibrium with the vapor of the liquid.

To apply these concepts to the powder problem, it is useful to consider the particles as cubes (A–D in Fig. 8). If it is assumed that before wetting the solid is in equilibrium with the vapor of the liquid, the energy changes that take place are given by:

A to B adhesional wetting:

$$W_A^{SLV} = \gamma^{SL} - (\gamma^{SV} + \gamma^{LV}) = -\gamma^{LV} (\cos\theta + 1) \tag{15}$$

B to C immersional wetting:

$$W_W^{SLV} = 4\,\gamma^{SL} - 4\,\gamma^{SV} = -4\,\gamma^{LV} \cos\theta \tag{16}$$

C to D spreading wetting:

$$W_S^{SLV} = (\gamma^{SL} + \gamma^{LV}) - \gamma^{SV} = -\gamma^{LV} (\cos\theta - 1) \tag{17}$$

The contact angle θ is introduced into these equations by making use of the Young equation

$$\gamma^{SV} = \gamma^{SL} + \gamma^{LV} \cos\theta \tag{18}$$

For practical purposes, it is useful to consider under what conditions the powder would wet spontaneously. This occurs when W is negative; if W is positive, work must be expended on the system for the process to take place.

For the three separate stages, it may be concluded that adhesional wetting is spontaneous if θ is $<180°$, immersional wetting is only spontaneous if θ is $<90°$, and spreading wetting is only spontaneous when $\theta = 0$.

For the total process

$$W_t = W_A^{SLV} + W_W^{SLV} + W_S^{SLV} = 6\gamma^{SL} - 6\gamma^{SV} = -6\gamma^{LV} \cos\theta \tag{19}$$

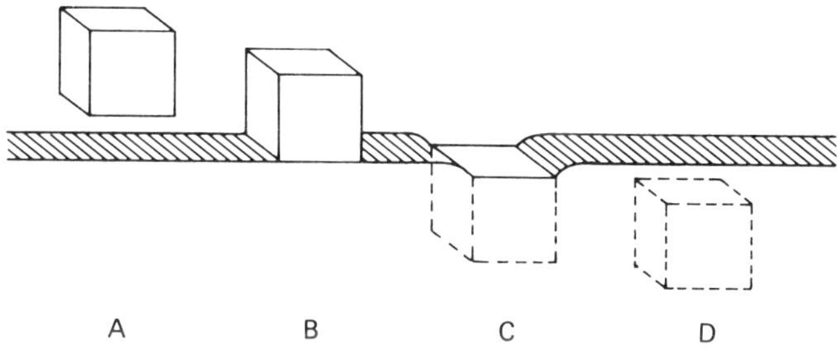

A B C D

Figure 8. The three stages involved in the complete wetting of a solid cube by a liquid: A to B adhesional wetting, B to C immersional wetting, C to D spreading wetting.

and for spontaneity, θ must be <90°. However, since one of the separate stages requires zero contact angle, this must be the condition for spontaneous wetting, for without it the powder would tend to float and some work would be required to bring about the spreading process. Surface-active agents are used to ensure θ is close to zero in the practical wetting of powders. Numerous values of θ are available in the literature, although most have been measured on flat surfaces and problems are caused by surface roughness (11). To measure θ for powders is more difficult, although a number of methods have been reported (12–13).

The penetration of liquid into the interstices of the agglomerates is not easy to define precisely. For the simplest case of a cylindrical tube of uniform cross section, penetration of the liquid is, in the absence of air, only spontaneous when $\theta < 90°$. However, agglomerates are normally filled with air, and once the liquid has penetrated the channel, the air pressure increases and complete wetting seems to be impossible. The structure of the agglomerate is relevant to wetting behavior, and when filled with air, complete wetting can only occur when $\theta = 0$ (14). Therefore, despite the problem of defining the geometry of the system, there are some guidelines available as to how the wetting process may be brought about with the minimum expenditure of energy.

However, an important factor is the rate at which the wetting takes place and, in particular, the rate of penetration of the liquid into the interstices of the agglomerates. This is particularly important since it may decide if the agglomerates are incorporated into the medium with all surfaces wetted or just the external surfaces. The rate of penetration of a liquid into a horizontal capillary of uniform cross section (radius r) is defined by the Washburn equation (15).

$$l^2 = \frac{rt\gamma^{LV}\cos\theta}{2\,\eta} \qquad (20)$$

where l is the depth of penetration in time t, and η is the viscosity of the liquid. For a packed bed of solid particles, the radius may be replaced by a factor K, which contains the effective radius for the bed and a tortuosity factor which takes account of the complex path formed by the channels between the particles. To achieve rapid penetration, high $\gamma^{LV}\cos\theta$, low θ, and low η are desirable, with K as large as possible, eg, a loosely packed powder. High γ^{LV} and low θ are normally incompatible; a low θ is an important requisite, but when $\theta = 0$, further lowering of γ^{LV} reduces the penetration rate.

Hence, the wetting phenomena are directly related to $\gamma_{L/V}$ and θ. In general, the overall process is likely to be more spontaneous the lower θ and the higher $\gamma_{L/V}$, although these two factors tend to operate in opposite senses.

Interactions at the Solid–Liquid Interface

To reduce the surface tension of water, ie, the free energy of the liquid–vapor interface, a large number of compounds are available. They are amphipathic in nature, that is, they contain a hydrocarbon group which is expelled by water and a polar group which is water-soluble and tends to remain in the water. These are normally called surface-active agents and are classified according to the nature of the soluble group into anionic, cationic, and nonionic types (see Surfactants and detersive systems). In all cases, adsorption takes place at the liquid–vapor interface with the hydrophobic part oriented away from the water, and the surface tension is reduced. In the case of a drop of water sitting on a flat hydrophobic surface, subtending an angle of contact θ as described by the Young equation (eq 18), the addition of a surface-active compound leads to adsorption at the liquid–vapor and solid–liquid interfaces; both γ^{LV} and γ^{SL} are reduced, with a consequent reduction in θ since γ^{SV} is hardly changed. Such compounds are referred to as wetting agents, since they improve wetting. It is important that γ^{SL} is significantly reduced by adsorption, even if γ^{LV} is only slightly changed. Therefore, choice of a suitable agent to improve wetting, ie, reduce θ as much as possible, should not be related only to its surface activity at the liquid–vapor interface and its efficiency in reducing surface tension, since that reflects only part of the overall change. In practice, the term dispersing agent is also used with the implication that the surface-active agent assists in dispersion, which could relate to wetting and to stability to flocculation. The latter effect is due to adsorption at the solid–liquid interface, giving rise to a repulsive force associated with charge and steric barriers, as discussed below (see also Dispersants).

The role of surface-active compounds in pigment dispersion in aqueous and nonaqueous media is reviewed in ref. 16, with particular reference to their selection to achieve specific effects. The hydrophile–lipophile balance (HLB) number and the solubility-parameter concept are terms used freely in technology without understanding of their real meaning and significance. The former (16) describes surface activity. A recent paper discusses HLB from a thermodynamic perspective (17) (see Emulsions).

It is thus clear that implicit to an understanding of the dispersion of powders in liquids is a study of the solid–liquid interface and of the variety of interactions that take place at that interface. Of particular importance in many practical systems is the adsorption of the macromolecular components of the liquid phase onto the solid surface; their configuration in the adsorbed state is a vital factor in reducing flocculation.

Practical systems are usually complex; they normally contain three types of components, namely the solid phase (pigment, extender, etc), the continuous medium, and the molecules that are dissolved in that medium. Thus, a number of different kinds of interactions exist in the system, for example, interactions between solid and liquid phases. The work required to separate unit area of interface into the two individual components is called the work of adhesion W_A and is given by $\gamma^{SV} + \gamma^{LV} - \gamma^{SL}$. The magnitude of this work term was defined in terms of the different kinds of interactions that take place across the interface (18). It is expressed as

$$W_A = W_A^d + W_A^h + W_A^\pi + W_A^p + W_A^e \tag{21}$$

where the superscripts refer to the type of interaction; d = dispersion, h = hydrogen bond, π = pi bond, p = other polar interactions, and e = contribution owing to charge separation at the interface giving rise to an electric double layer. Dispersion forces are always present, and the adhesion between dissimilar materials always depends on and is frequently dominated by these dispersion forces. Even in polar molecules, the dipole–dipole interactions are usually insignificant compared with dispersion-force interactions. However, with the exception of aliphatic hydrocarbon solvents, all the other components possess a certain degree of polarity associated with specific chemical groups that are present both in the liquid phase and on the surface of the solid. These groups may be acidic or basic in nature, or even amphoteric, and give rise to those forces other than the dispersive type described above. Specific effects can be identified, such as the chemical interaction or hydrogen-bonding interaction between an adsorbed molecule and a specific group on the solid surface, as well as nonspecific effects, eg, the interaction between the hydrocarbon chain of a polymer with a carbon-black surface.

Many studies have been carried out on the adsorption of molecules from the liquid phase at the solid–liquid interface (see also Adsorptive separation). Reference 19 covers solutions of nonelectrolytes; adsorption from electrolyte solutions has not yet been adequately reviewed although a book covering all aspects of adsorption from solution has been published (20). The technique is simple: solid and solution are equilibrated together and the change in solution concentration is noted. Adsorption isotherms are plots of amount adsorbed as a function of equilibrium concentration. These isotherms are of different shapes and reflect the molecular interactions that take place at the interface (21). Interpretations of data for small molecules is relatively simple except for polymers. The amount adsorbed depends on the configuration of the polymer and how much solvent it carries with it. For example, an alkyd resin of low acidity interacts strongly with the surface of an acidic titanium dioxide pigment and the alkyd molecules interact forming a close-packed layer at the interface. However, for a basic titanium dioxide there are few interactions between alkyd and surface, and the alkyd molecules are extended from the surface in a heavily solvated condition. Hence, the difference between the adsorption behavior of silica-coated (acidic) and alumina-coated (basic) titania surfaces is of particular relevance to the stabilization of coated titanium dioxide pigment surfaces in alkyd paints (22).

The two extreme cases of adsorption involve van der Waals forces on the one hand (physical adsorption, or physisorption) and chemical adsorption (chemisorption) on the other, when a chemical bond is formed between solid surface and adsorbed molecules. Most adsorption processes from the solution phase do not involve a chemical interaction, but the strength of the physical force varies quite considerably from that of the dispersion type to the interaction involving hydrogen bonds.

Analysis of concentration changes caused by adsorption is straightforward. Taking m g of a solid in contact with a solution that initially contained n_1^o mol of component 1 at mole fraction x_1^o, and n_2^o mol of 2 at x_2^o, and after adsorption equilibrium is reached the solution contains n_1 mol at x_1, and n_2 at x_2, then the change on adsorption is given by

$$n^o(x_1^o - x_1)/m = n_1^s x_2 - n_2^s x_1 \tag{22}$$

where n_1^s and n_2^s mol of the components are adsorbed per gram of solid and n^o ($= n_1^o + n_2^o$) is the total number of moles of liquid components present in the system. In equation 22 the measurable quantities are on the left and the unknowns on the right; a plot of $n^o(x_1^o - x_1)/m$ vs x_1 is an adsorption isotherm but with the proviso that it is not an isotherm of actual moles of component 1 adsorbed. Examination of the limiting values of the adsorption function at $x_1 = 0$ and 1 serves to illustrate the point. The measured quantity is an apparent adsorption and represents the change in this quantity as the relative concentrations of the two components are varied across the whole range of mole fraction. The isotherm, being a function of the behavior of two solution components, is commonly called a composite isotherm. In the case of dilute solutions, for which $x_2 \approx 1$ and $x_1 \rightarrow 0$, the measured change in solution concentration on adsorption gives directly the individual isotherm (n_1^s) for the solute.

Equation 22 contains two unknowns n_1^s and n_2^s. To obtain individual isotherms for the two components it is usually assumed that adsorption is confined to a monolayer, and

$$n_1^s/(n_1^s)_m + n_2^s/(n_2^s)_m = 1 \tag{23}$$

is used, where $(n_1^s)_m$ and $(n_2^s)_m$ are the monolayer values for the single components (derived from, for example, molecular models assuming an orientation). This analysis is fraught with difficulties.

Adsorption from dilute solution is analytically simpler, since the measured concentration change gives directly the individual isotherm of the solute. The solvent is adsorbed, but the amount involved does not enter into the simple mathematics. An expression of the Langmuir type, derived from adsorption of vapors, is often found to fit the data

$$\frac{C_2}{n_2^s} = \frac{1}{n^s b} + \frac{C_2}{n^s} \tag{24}$$

where C_2 is the final concentration, n^s the number of adsorption sites in mol/g, and b is a constant. Using a suitable molecular area for component 2 can lead to a value for the surface area, but such analysis involves assumptions on orientation and coverage.

Systems of practical interest usually involve adsorption from dilute solution, and some of the effects that may be observed are illustrated in the following examples taken from the literature.

Normal long chain acids adsorb onto oxide surfaces by hydrogen bonding with the surface hydroxyl groups, and in addition, for unsaturated acids there may be interactions between the surface and the alkene residue of the alkyl chain (23). These two types of interaction may lead to changes in orientation of the adsorbed molecules as the surface coverage is increased (24). Needless to say, the presence of water on the surface of the solid can have a profound effect on the nature of the adsorption, eg, in the case of the adsorption of *n*-octadecanol onto rutile from *p*-xylene solutions, the adsorption on a surface free of water is dominated by a strong alcohol-surface interaction (25). With increasing amounts of preadsorbed water on the surface, the strength of this interaction decreases and the effects of competitive solvent adsorption and molecular interactions in the bulk phase become apparent. At any particular concentration, the octadecanol adsorption is highest on the dry surface.

The adsorption of a wide range of polar organic molecules from organic solvents onto the surface of titanium dioxide, talc, barium sulfate, and a number of other pigments and extenders has been reported (26). The interesting feature of this work is the study of simultaneous adsorption of more than one compound. For a mixture of polar compounds with identical polar groups, the amount adsorbed is proportional to the concentration of the compounds in solution. Relative chain lengths are important; for molecules with similar polar groups the long chain compounds can be desorbed by short chain molecules and vice versa. The effects described are particularly relevant to the situation in obtaining coating systems.

Adsorption from hydrocarbon solutions, as described above, represents a relatively straightforward situation with respect to the mechanism and the nature of the forces involved. However, from aqueous solutions, ionization effects give rise to an electric double layer at the solid–liquid interface. Many substances of practical interest contain ionizable groups. Solid surfaces are also ionized in aqueous media giving rise to a surface charge. The pH of the solution determines the sign and magnitude of the surface charge as well as the acid–base character of dissolved molecules containing acidic and basic groups. The ionic strength of the solution determines the characteristics of the electric double layer as well as the electric potential at the surface of the solid. These factors determine the extent of adsorption of compounds containing ionizable groups; hence, the situation is complex and each case must be considered on its own merit. A useful review of the adsorption behavior of surfactants at solid–water interfaces is given in ref. 27, where the electric double layer and the parameters are described that are important in the physical adsorption of surfactants onto oxide surfaces, namely the role of the hydrocarbon chain, the effect of chain length and chain structure, and the influence of pH and ionic strength. Also presented are some examples of chemisorption, such as the adsorption of sodium oleate and oleic acid on ferric oxide (haematite). The thermodynamic and molecular considerations applied to adsorption of organic molecules are found in ref. 28.

The adsorption of some long chain electrolytes (cationic and anionic) from aqueous solutions on polar and nonpolar adsorbents (alumina, titania, barium sulfate, calcium fluoride, and carbon black) admirably illustrates the effect of ionic strength and pH and shows further that the adsorption reaches a saturation value at the critical micelle concentration (CMC) of the electrolyte, a phenomenon that is commonly observed (29). Orientation effects, at low concentration with the surfactant lying flat and changing to perpendicular orientation on increasing concentration, have been reported (30). Such effects are relevant to the wetting process.

Polymers are important components of many systems, and their adsorption at the solid–liquid interface may be complex and difficult to define precisely (31–34). Numerous adsorption isotherms have appeared in the literature. Adsorption equilibrium is usually reached very slowly, and it may take several days or even weeks. Isotherms nearly always show a rapidly rising region followed by a plateau which suggests the formation of a monolayer, but the quantity of polymer adsorbed in the plateau region often exceeds the monolayer capacity of the adsorbent, from which one might conclude that the polymer is adsorbed in loops and chains. If that is the case, it might be expected that solvent is incorporated in the adsorbed layer. Not much attention has been paid to the fact that the isotherms are of a composite nature, involving both solvent and adsorbate. This phenomenon should give rise to maxima in polymer adsorption isotherms and these have been observed in systems containing alkyd resins (qv) (22) and polyesters (qv) (35).

Several important general findings have been established from polymer adsorption studies. The time taken to reach equilibrium depends on the molecular weight of the polymer, and the amount adsorbed at equilibrium increases with increasing molecular weight. The amount adsorbed also increases with decrease in solvent power of the medium for the polymer. In a poor solvent, polymer–surface contacts are favored, and since the polymer coils are more compact in a poor solvent, the surface can accommodate a greater number. The surface character of the adsorbent also affects the equilibrium adsorption, the extent depending on the nature of the surface groups and the presence of specific substituents in the polymer molecule. A good illustration of acid–base interactions in polymer adsorption is reported in ref. 36, comparing the adsorption of a basic polymer, poly(methyl methacrylate), on an acidic silica and a basic calcium carbonate from acidic, basic, and neutral solvents. Increase in temper-

ature can either increase or decrease the amount of polymer adsorbed, for a variety of reasons which are mostly concerned with the interaction between polymer and solvent.

The distribution of molecular weight fractions during the adsorption process merits attention. It is usually assumed that the polar, low molecular weight fractions of alkyd resins are preferentially adsorbed, based on the fact that the acid number and the hydroxyl number of the adsorbed resin are larger than those of the initial resin (37). As demonstrated by gel permeation chromatography (38), it appears that adsorption of fractions of all molecular sizes can take place on titanium dioxide. The result depends on the affinity of the different fractions for the pigment surface as well as their concentration in the original resin. It is also affected by molecular interactions in solution; solvent–binder interactions in nonaqueous paint systems are described in ref. 39.

The ultimate fate of the alkyd resin fractions is, therefore, a complex function of the various interactions that take place both in solution and at the solid–liquid interface, and a thorough understanding requires application of a variety of techniques. Furthermore, the adsorption takes place in the presence of other nonpolymeric surface-active species, which are commonly used in practical systems. A study of the simultaneous adsorption of an alkyd resin in the presence of fatty acids on titanium dioxide pigments demonstrates the complexity of the adsorption behavior (38). The adsorption of fatty acids is strongly reduced by the presence of the alkyd resin; the relative strengths of the interactions are important and interpretation is difficult.

Breakdown of Agglomerates and Aggregates

In the initial stage of incorporating and wetting, the magnitude of the cohesive forces between the individual particles determines the extent of disagglomeration before the mechanical action is applied. At first, powder is added to the surface of the liquid medium. As the medium penetrates the undersurface of the mass, small agglomerates and crystals are engulfed by the advancing interface and are detached from the bulk. Larger agglomerates tend to be engulfed, and the medium can also penetrate the agglomerate, but at a slower rate than that of the engulfing process because the interparticle spacing is smaller in the agglomerate. At some degree of penetration, the bulk density of the agglomerate exceeds that of the medium and the agglomerate detaches. At this point, the medium continues to penetrate and the air escapes as bubbles, whereas the outer particles detach until finally the whole agglomerate is dispersed. Alternatively, particularly if the binding in the agglomerate is slightly stronger than average, the medium penetrates from all sides until the internal air pressure disrupts the agglomerate and the process starts again with the fragments. Finally, before any milling or stirring, the powder is submerged in the medium, most of the air has been displaced, and the solid surface has largely been wetted by the medium, although some particle–air interfaces may remain in the center of the more strongly bound agglomerates.

Once the particles are wetted, some mechanical energy is required to separate them completely. Aggregates may require considerable energy to be broken down completely to the point where the surface of each primary particle is available to the wetting liquid. Presumably, agglomerates would normally require less energy than aggregates. The ways in which this disruption occurs in various mills are incompletely understood and much remains to be elucidated. The milling action disrupts the agglomerates except those that are strongly bound and which form the grit content. The more intense the milling action, the fewer grits remain. Highly dispersible powders contain few aggregates and therefore have a low grit content. Agitation during wetting increases the extent of breakdown of agglomerates and speeds up the whole process.

Clearly the addition of surface-active agents is important to the wetting and disagglomeration processes both in terms of reducing the work involved and increasing the rate. Surface-active agents play an important role in maintaining the stability of the dispersion, but in the milling stage it is more important to disrupt agglomerates and aggregates than to secure a stable dispersion of primary particles. A millbase should therefore be formulated to obtain the fastest breakdown of oversize particles; the final stability is of only secondary consideration. This means maximizing the speed of wetting and the speed of disagglomeration, provided that the rate of flocculation of primary particles is not so rapid that an unfavorable equilibrium results.

Surface-active agents may also contribute to the mechanical breakdown of the aggregates during the milling stage. The adsorption of surface-active agents at structural defects in the surface considerably facilitates the fine grinding of solids to create new interfaces, and therefore may contribute to the breaking of aggregates during milling (40). The presence of submicroscopic defects in the solid can lead to fracture occurring at only a small fraction of the theoretical strength, and the fracture energy is further reduced by a wetting liquid causing embrittlement of the solid (41).

It is generally accepted that disagglomeration is brought about by shearing or impact. Disruption by shearing relies on viscous drag, whereas the comminution process takes place most easily when unhindered by viscous resistance. Thus, the two mechanisms operate under opposing conditions, and a particular mill works best within fairly close viscosity limits. Comparison of various types of equipment used, for example, for paint-making, shows that mills that work mainly by an impact process require a low millbase viscosity whereas those relying on shearing need a high one (Fig. 9).

Dispersion equipment broadly falls into four classes (42–43):

Low shear equipment employs slow mixing at high viscosity or high solids concentration. The millbase is formulated to give maximum resistance to the movement of the blades within the capabilities of the unit. The motor power is transmitted to the batch through blades which present a large working surface to the batch, and the energy so transmitted is utilized at high shear stress in the normally fairly cohesive mass at extremely low shear rates. In all cases, the blades pass closely to the walls of the container to ensure complete intermixing and impart additional shear at this point. Examples are the Pug mill, planetary mixers, and Z-blade mixers.

High shear equipment uses impellers moving at high peripheral speeds, which generate high shear rates and high rate of interparticle shear by virtue of careful control of millbase rheology, or by the use of a variety of different shear-rate-control baffle arrangements. Examples are disk cavitation mixers, stator–rotor mixers, hynetic or Kady mills, and colloid mills.

Ball mills. This classification covers all types of equipment in which the energies transmitted are applied to the dispersion by use of free-moving members irrespective of size, shape, or means of activation. Ball mills may be of the horizontal, vibratory,

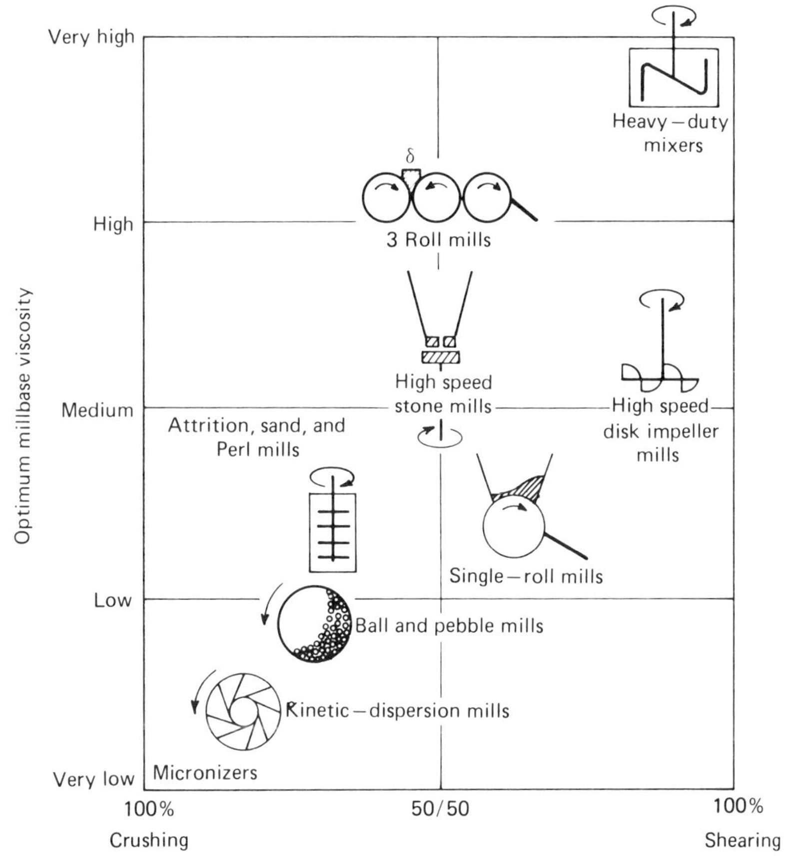

Figure 9. The milling action in dispersion equipment in relation to millbase viscosity.

or planetary type according to the mode of action. Sand and Perl mills, based on sand or small beads, respectively, are similar in principle to ball mills but with smaller grinding media.

Roll mills. Equipment that uses one or more rollers and is designed to carry the millbase between either two rollers or a roller and a blade at a closely controlled gap under pressure. These can be single-roll, two-roll, and triple or multiple rolls.

The definition of dispersability as the ease with which the particles in the powder are distributed in a continuous liquid medium such that each particle is completely surrounded by the medium and no longer makes permanent contact with other particles is too vague. It would be much better if the process could be expressed in terms of the energy necessary to achieve the dispersion of unit mass of powder by a completely efficient process. In practice, the energy has two sources: The first is related to the chemistry of the system, with respect to the chemical nature of the components, in particular the powder surface and the liquid phase, which determine the efficiency of wetting and the breakdown of agglomerates. The second source of energy comes from milling and depends on the type and dimensions of the mill and the characteristics of the millbase. The use of chemical energy can be optimized, eg, by surface treatment of the powder, as is practiced in pigment technology, and selection of the best surface-active agent for effective wetting (low contact angle) and rate of wetting, hence reducing the mechanical energy required which could lead to cost benefit.

Stability to Flocculation

Flocculation is the process that reduces the number of particles by collisions under both static (Brownian motion) and dynamic (under shear) conditions. There are three major types of forces of interaction between particles as they approach each other during the collision process: the van der Waals force of attraction, the Coulombic force (repulsive or attractive) associated with charged particles, and the repulsive force associated with the interaction between adsorbed layers of polymer on the particle surfaces. Most of the theoretical work on flocculation has been directed toward the simple interaction between two particles, and hence the relevant experimental studies have been conducted on very dilute dispersions, many orders of magnitude less concentrated than found in a practical system. Although the basic principles are similar, application to the concentrated system, where two-particle collisions must be influenced by other particles in the vicinity, has received relatively little attention (see also Flocculating agents).

The interplay of the van der Waals and Coulombic forces forms the basis of the classical theory of flocculation of lyophobic dispersions, known as the DLVO (Derjaguin-Landau-Verwey-Overbeek) theory (44–45).

The DLVO Theory. Application of the DLVO theory to flocculation problems has become common practice despite all the limitations. In principle, the theory is correct, but refinements are obviously necessary in many cases.

The theory considers the interaction between two charged particles in terms of the overlap of their electric double layers leading to a repulsive force which is combined with the attractive term to give the total potential energy as a function of distance for the system.

The Attractive Force. The potential energy of attraction V_A between solid spherical particles may be calculated with the help of the Hamaker expression (eq. 4). The Hamaker constant A is given by $\pi^2 q^2 B$, where q is the number of atoms or molecules per cm^3 and B is the London constant. Methods of evaluating A are reviewed in ref. 46, and tables of A values are given in refs. 46–48.

When a medium separates the interacting particles, the attractive force is weaker than that calculated by equation 4 because of the interaction between the molecules of the medium and between the particles and the medium. For the two-phase system, the total A is given by

$$A = A_{pp} + A_{mm} - 2\,A_{pm} \tag{25}$$

The distance function in the brackets of equation 4 remains the same; A_{pp} and A_{mm} are the constants for the particles and medium (*in vacuo*), respectively, and A_{pm} is the constant associated with the particle–medium interaction. In the absence of more precise information, a good approximation is

$$A = (A_{pp}^{1/2} - A_{mm}^{1/2})^2 \tag{26}$$

The value of A is always positive. The values of A_{pp} and A_{mm} are frequently of similar magnitude, and since these values are certain to be in error, it is not possible to estimate A for a system accurately.

For the interaction between dissimilar particles, equation 25 becomes

$$A = A_{12} + A_{33} - A_{13} - A_{23} \tag{27}$$

where the subscripts 1 and 2 refer to the different particle material and 3 to the medium. If $A_{11} < A_{33} < A_{22}$, A becomes negative, ie, a repulsive force develops.

The effect of an adsorbed layer with a value of A different from that of the particle is a decrease in attractive energy (49). The total effect is the sum of two parts: First, the attractive energy is inversely proportional to the minimum surface separation of the two particles; hence, the addition of an adsorbed layer at constant surface separation leads to an increase in separation of the core particles, resulting in a decrease in energy. Second, the change in the A constant for the adsorbed layer leads to a further change in attractive energy, which can be negative or positive, depending on the relative values of the three A constants involved. Under certain conditions a repulsive energy can develop between particles or between particles and plates coated with differing adsorbed layers (48,50).

The Coulombic Repulsive Force. Since in a medium containing ions a charged particle with its electric double layer is electrically neutral, no net Coulombic force exists between charged particles at large distances from each other. As the particles approach, the diffuse parts of the double layers interpenetrate, giving rise to a repulsive force which increases in magnitude as the distance between the particle decreases. The distance at which the repulsive force becomes significant increases with the thickness of the double layer ($1/\kappa$) and the force increases with the surface potential (ψ_o).

For particles that are too far apart to interact, the characteristics of each individual particle double-layer system are determined by the surface charge (or surface potential) and the ionic strength and properties of the medium. When two particles interact, some changes in these parameters must occur, but in the estimation of the interaction energy, either constancy in surface potential (implies rapid establishment of adsorption equilibrium) or in surface charge (implies slow desorption) is assumed.

The treatment of interacting double layers is described in ref. 45. The easiest treatment is that of the interaction of two flat parallel double layers. For a flat double layer,

$$\frac{d^2\psi}{dx^2} = \frac{8\,\pi nze}{\epsilon}\sinh\left(\frac{ze\psi}{kT}\right) \tag{28}$$

where ψ is the potential at distance x from the surface, n the concentration of electrolyte containing ions of valency $z_+ = z_- = z$, e is the electronic charge and ϵ the dielectric constant. Substituting

$$y = ze\psi/kT, \quad w = ze\psi_o/kT, \quad \kappa^2 = 8\,\pi ne^2/\epsilon kT$$

gives

$$d^2y/dx^2 = \kappa^2 \sinh y \tag{29}$$

the solution of which for $\kappa > 1$, ie, at a large distance from the surface, is

$$y = 4\,\gamma \exp(-\kappa x) \tag{30}$$

where $\gamma = \exp[(w/2) - 1]/\exp[(w/2) + 1]$. This solution is a very good approximation for distances down to $\kappa x = 1$ for all values of ψ_o.

To derive an exact expression for the potential distribution between two flat plates separated by a distance $2\,d$, it is necessary to solve equation 27 to satisfy the boundary conditions $\psi = \psi_o$ when $x = 0$ and $x = 2\,d$, and $d\psi/dx = 0$ when $x = d$. The solution involves an elliptic integral which can only be solved numerically. However, when $\kappa d > 1$ (ie, the interaction is not too great), it is sufficiently accurate to assume that the potential may be built up additively from the potentials of the two single double layers. Hence, from equation 30 the potential ψ_d halfway between the plates is given by

$$u = 8\,\gamma \exp(-\kappa d) \tag{31}$$

where $u = ze\psi_d/kT$ and $\kappa d > 1$.

The simplest way to calculate the force of repulsion between two interacting double layers is to consider the force as arising from the osmotic pressure of the excess ions in the space where the double layers overlap. The ionic concentrations (n_+ and n_-) at $x = d$ are given by the Boltzmann equation

$$n_+ = n \exp(-ze\psi_d/kT) \text{ and } n_- = n \exp(ze\psi_d/kT) \tag{32}$$

and the osmotic pressure at this plane is

$$kT(n_+ + n_-) = nkT[\exp(u) + \exp(-u)] = 2\,nkT \cosh u \tag{33}$$

Outside the field of the double layers where ψ and $u = 0$, the osmotic pressure is $2\,nkT$. Hence, the difference between these two pressures is a measure of the force p acting on the unit area of each plate:

$$p = 2\,nkT(\cosh u - 1) \tag{34}$$

The repulsive potential energy is given by the work done against this force when the plates are brought together from a large distance. This work per unit area of plate is given by

$$V_R = -\int_{\infty}^{d} p dx = -2\int_{\infty}^{d} 2\,nkT(\cosh u - 1)dx \tag{35}$$

but in the general case this integral cannot be readily evaluated because the relation between u and x is complex. However, if we again assume small interaction, ie, u is small and $(\cosh u - 1) = u^2/2$, then

$$V_R = -2\int_{\infty}^{d} nkT\, 64\,\gamma^2 \exp(-2\,\kappa x)dx \tag{36}$$

or

$$V_R = 64\,nkT\,\gamma^2[\exp(-2\,\kappa d)]\kappa \tag{37}$$

The exact numerical solutions for the interaction of parallel plates have been computed (51). The corresponding tables for spherical particles are not generally available and in general approximate formulas are used, which are valid for limited conditions. The two equations most commonly used were derived for small ψ_o and prove most useful in many practical situations.

For systems in which $\kappa a \gg 1$ (large particles in aqueous systems with moderate electrolyte concentrations) and interaction is weak (shortest distance H between surfaces of spheres is large compared with $1/\kappa$),

$$V_R = \tfrac{1}{2}\,\epsilon a \psi_o^2 \ln[1 + \exp(-\kappa H)] \tag{38}$$

The equation gives values that are close to those from the exact treatment (using graphical integration) provided $w(= ze\psi_o/kT) \ngtr 2$ and $\kappa a \nless 10$.

For systems in which $\kappa a \ll 1$ (small particles in nonaqueous media or in aqueous systems with very low electrolyte concentration)

$$V_R = \frac{\epsilon a^2 \psi_o^2}{H + 2a} \beta \exp(-\kappa H) \tag{39}$$

This is quite accurate up to $\kappa a = 1$ and ψ_o = ca 50–60 mV; β is a factor to allow for loss of spherical symmetry in the double layer as they overlap (45). The treatment for the intermediate region of κa and for higher potentials is given in ref. 45.

The Total Interaction Energy. Since the double-layer repulsion and the van der Waals attraction operate independently, the total potential energy V_{tot} for the system is given by the sum of the two terms. The form of the resulting potential energy against distance relationship is dependent upon the relative magnitudes of the two forces; V_R decreases exponentially with distance, whereas V_A shows an approximate inverse relationship with the square of the distance. Attraction predominates at short distances other than immediately adjacent to the surface at atomic distances when repulsion dominates. Little is known about this effect, and in colloid problems it is usually ignored. Otherwise, the form of the V_{tot} curve depends to a large extent on the V_R term. Figure 10 illustrates the type of plot expected for particles of radius 0.1–1 μm in an aqueous system containing ca 10 mol/dm^3 of 1:1 electrolyte for which the range of attractive and repulsive forces are similar.

The three important characteristics shown in Figure 10 are directly related to flocculation behavior. The potential energy barrier must be surmounted before the particles make lasting contact in the primary minimum. Provided the barrier is considerably larger than the thermal energy of the particles, relatively few make contact and the system should be stable. If, however, the secondary minimum is of depth $\gg kT$, then the particles flocculate with a liquid film between them in the cluster. Since both the attractive and repulsive forces are approximately proportional to the particle radius, the secondary minimum should become increasingly significant with increasing

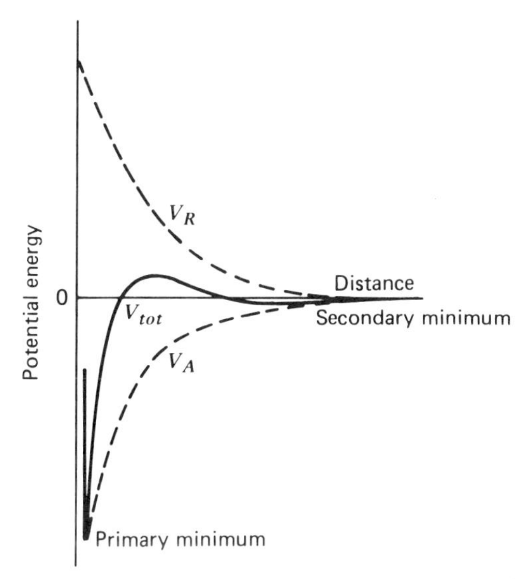

Figure 10. Potential energy curves for the interaction between two spherical particles immersed in an aqueous medium.

particle size, and particularly so with parallel plates. The effect also increases with increasing electrolyte concentration reducing the distance over which the repulsive forces operate, and also the energy barrier, which again would promote flocculation. Systems which have flocculated into the secondary minimum tend to be reversible, ie, they can be readily redispersed (peptized) with shaking and break down under shear, whereas those in the primary minimum (coagulated) need considerably more energy to redisperse. In some cases, particles might on standing pass from the secondary to the primary minimum.

The effect of reducing the electrolyte concentration (increasing $1/\kappa$ at constant ψ_o) on the total potential energy is shown in Figure 11 and illustrates the difference between dispersions in aqueous and nonaqueous solutions of a surface-active agent of the same stoichiometric concentration, eg, sodium di-2-ethylhexyl sulfosuccinate, which dissolves in both water and hydrocarbons but ionizes to markedly different extents in the two solvents. When $1/\kappa$ is large, the secondary minimum disappears since the range of V_R is then considerably greater than that of V_A, and under these conditions, the accuracy of the V_A values becomes much less important.

In making theoretical predictions of stability it is necessary to assign values to the Hamaker constant; the Stern potential (rather than ψ_o), since this potential determines the characteristics of the diffuse part of the double layer, the region particularly relevant to colloid stability; and the particle dimensions.

Since the Stern potential cannot be determined by direct measurement, the zeta potential obtained from an electrokinetic experiment is used and has been successful in a number of cases. Measurement of particle dimensions is itself a considerable problem, hence, the desirability of using monodisperse systems when the validity of the theory is being assessed. Nevertheless, for a dispersion containing only one type of particle, the deviations from monodispersity and spherical geometry may effectively be small enough, and thus, an appreciable error is not involved when the theory for monodisperse spheres is applied. For powders, electron microscopy is the most attractive technique for assessing particle dimensions.

Most of the reported experimental work on flocculation in dilute systems lends strong support to the DLVO theory although, as might be expected, there are deviations in some of the fine detail. Nevertheless, the theory provides an effective approach to flocculation which is universally accepted.

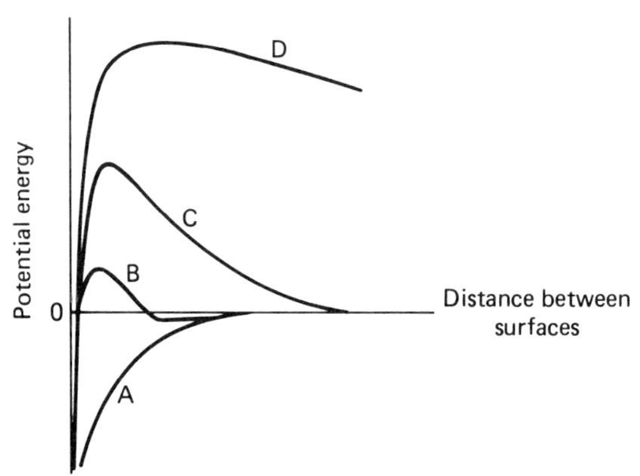

Figure 11. Influence of electrolyte concentration on the total potential energy of interaction of two spherical particles of radius 100 nm in aqueous media. A $1/\kappa = 10^{-7}$ cm, B $1/\kappa = 10^{-6}$ cm, C $1/\kappa = 10^{-5}$ cm, D $1/\kappa = 10^{-4}$ cm.

For practical systems with high particle concentration, the validity of the DLVO theory is open to question. Particle–particle interaction might not begin at infinite distance of separation, ie, one particle is, on average, sufficiently close to another to reduce the effective energy barrier which must be overcome for the particles to come into contact (not considering at this time the secondary minimum effect). A preliminary attempt was made to solve this problem in order to explain flocculation data for the charged droplets of water-in-oil emulsions (52). The two-particle interaction calculations were abandoned and replaced by a model based on twelve nearest neighbors, situated on a spherical shell around a central particle, with all other particles randomly distributed outside this shell. It was shown that the calculation of the interaction energy of the one particle with all other particles leads to an energy barrier, which is significantly lower than that for the simple interaction of two particles; the effect increases with particle concentration. The magnitude of the effect is directly related to the thickness of the double layer, since double-layer overlap is the cause of the repulsion. Hence, it is much more significant in nonaqueous media. The use of the DLVO theory for two-particle interaction is probably useful as a first approximation for concentrated aqueous disperions with moderate to high electrolyte concentrations.

Using an alternative approach, the force between two charged spherical particles of a concentrated dispersion in a hydrocarbon medium was calculated (53–55). It was demonstrated that the electric double-layer force is not sufficient to promote stability to flocculation, in contrast to its effect in dilute dispersions; this is borne out in practice. Furthermore, it was suggested that the electrostatic charge mechanism does not assure stabilization of concentrated dispersions in aqueous or nonaqueous media, and to prevent flocculation, some sort of protective action by polymeric species is required (56). This protective action could arise from the effects associated with adsorption or through the polymer in the bulk forming a gel structure through which diffusion of the particles is restricted. In principle, possible mechanisms range from those directly associated with the properties of the adsorbed layer to those entirely dependent on bulk solution properties, and hence, the whole problem becomes complex and lacks quantitative precision.

Heteroflocculation. When only one type of particle is present, ie, only one value for a, A, and ψ_o is applicable, it is often adequate to assume that the deviations from monodispersity and spherical geometry may be effectively small enough that little error is involved in applying the DLVO theory. When more than one type of particle is present or if there is a large disparity in size, the situation becomes more complex. Interaction between dissimilar particles is termed heteroflocculation.

A theory of heteroflocculation included the interaction of dissimilar double layers on flat plates (51,57). A more recent theoretical treatment (58) is mathematically simpler because it is based on the approximate DLVO theory for spheres with $\kappa a > 10$ and values of $\psi_o < 50$ mV. The potential energy of repulsion for two particles of radius a_1 and a_2 and potentials ψ_{01} and ψ_{02} is given by

$$V_R = \frac{\epsilon a_1 a_2 (\psi_{01}^2 + \psi_{02}^2)}{4\,(a_1 + a_2)} \left[\frac{2\,\psi_{01}\psi_{02}}{(\psi_{01}^2 + \psi_{02}^2)} \ln \frac{1 + \exp(-\kappa H)}{1 - \exp(-\kappa H)} + \ln\left(1 - \exp(-2\,\kappa H)\right) \right] \quad (40)$$

Combining this with the appropriate Hamaker expression for the attraction between two spheres gives the total potential for the interaction from which energy barriers may be evaluated (59).

From the theoretical calculations, two facts emerge: first, the particle with the smaller radius and potential determines the height of the potential energy barrier; and second, although the different particles may have the same sign of surface charge and potential but different magnitude, they may nevertheless attract each other when the double layers overlap. This is a consequence of the relative changes that occur in the charge and potential as the particles approach (60).

The interaction of oppositely charged particles has been recently analyzed (61). Constant charge is assumed, ie, one type has strongly acidic (negative) surface sites and the other strongly basic (positive) sites. When the ratio of the absolute charge densities is significantly different from unity (ca 10:1), stability ratios $>10^4$ are obtained over a significant range of ionic strength. At large ionic strength (ca 10^{-4} mol/L) and very low ionic strength (ca 10^{-7} mol/L), rapid heteroflocculation occurs, whereas in the intermediate range, the system is stable despite the fact that the particles are oppositely charged. Thus, the conditions for initiating flocculation by adding a sol whose particles carry an opposite charge need to be carefully analyzed, and the simple rule of thumb may not apply.

The potential energy of interaction between two dissimilar particles in nonpolar media has been evaluated by applying Maxwell's equations for charged conducting spheres (62). The presence of an electric double layer was ignored, which would seem justified in view of the very small concentrations of ions involved in media of low dielectric constant. Use of the simple Coulombic law leads to significant error in the V_R term. Furthermore, it was demonstrated that it is the particle of lower radius and potential that controls the energetics of the flocculation process, a similar conclusion to that reached for aqueous media, as described above.

All these theoretical treatments relate to dilute dispersions because only the interaction between two particles is considered.

Steric Stabilization. The term steric stabilization embraces all aspects of the stabilization of colloidal particles by adsorbed nonionic macromolecules, and is equally applicable to aqueous and nonaqueous media. Of particular interest are the thickness of the adsorbed layer, the configuration of the polymer chain, the fraction of segments adsorbed, and the segment-density distribution, both normal and parallel to the surface.

Polymers adsorbed at particle–medium interfaces may take up any configuration from lying flat (strong interactions between the surface and many points along the polymer chain) to highly extended into the medium (interaction at few points). For the extreme case of parallel orientation, the only effects on flocculation are the modification of the attractive force between particles, which is likely to be small in most cases, and the effect on the zeta potential, which could be significant, depending on the chemical character of the surface and polymer. When the polymer chains are extended into the liquid phase, a strong effect arises when particles approach sufficiently close for the adsorbed layers to interact. In one extreme case, the polymer chains mix, whereas the other involves no mixing but the layers become compressed on close approach of the surfaces. Both involve a reduction in the configurational freedom of the adsorbed polymer molecules, which results in a repulsive force. Either or both types of interaction could be present in any particular system, depending on the nature of the polymer and solvent and the structure of the adsorbed layer.

The most effective stabilizers are amphipathic block or graft copolymers, which contain anchor groups that are strongly held to the surface and are insoluble in the medium, and stabilizing chains that are soluble in the medium and extended from the surface. High surface coverage is essential. The solvency of the stabilizing chain is the important factor, and instability is readily induced if the solvency is reduced by temperature and pressure changes, or by adding a nonsolvent for the chain to the dispersion medium. When the solvency is reduced, the dispersions often show sharp transitions from long-term stability to rapid flocculation, and the critical flocculation values correlate strongly with the corresponding θ point for the stabilizing moiety in free solution (63). This provides strong evidence of mixing effects as two adsorbed polymer layers interact. Examples of the close correlation between θ temperature and critical flocculation temperature are given in Table 1.

The free energy of the interaction ΔG_R is related to the enthalpy ΔH_R and entropy ΔS_R changes by

$$\Delta G_R = \Delta H_R - T\Delta S_R \tag{41}$$

The sign and magnitude of ΔG_R are dependent on the relative magnitudes of the enthalpy and entropy terms. If ΔG_R is negative, flocculation is promoted; if zero, flocculation proceeds as if no adsorbed layers were involved; and if positive, the particles are protected from flocculation. For ΔG_R to be positive, both the entropy $T\Delta S_R$ and enthalpy terms must be negative with the former being the larger, hence termed entropic stabilization, indicating a dominating effect owing to reduction in the number of configurations of the polymer molecules in the adsorbed layer; or both terms must be positive with the enthalpy change predominating, which is termed enthalpic stabilization; or ΔH_R is positive and ΔS_R negative, and therefore, both terms contribute to stability, a situation termed enthalpic–entropic or combined stabilization.

A characteristic feature of sterically stabilized systems is their different responses to temperature change. Entropically stabilized systems flocculate on cooling, enthalpic stabilized systems on heating, and for combined stabilization there is no accessible temperature for flocculation (64).

Various experiments and calculations have shown that in most nonpolar solvents the primary stabilization effect is the decrease in entropy of the polymer–solvent mixture in the region between the particle surfaces. In many aqueous systems, however, the situation is quite different. The local increase in polymer segments between the particles causes some breakdown in water structure with a consequent increase in system entropy. In this case, stability is a consequence of an increase in potential energy

Table 1. Correlation between θ Temperature and Critical Flocculation Temperature[a]

Polymer stabilizer	Molecular weight	Medium	CFT[b], K	θ[c], K
poly(ethylene oxide)	96,000	3.9×10^{-4} mol/L $MgSO_4$	316 ± 2	315 ± 2
poly(acrylic acid)	51,900	2×10^{-4} mol/L HCl	283 ± 2	287 ± 2
poly(vinyl alcohol)	57,000	2×10^{-3} mol/L NaCl	301 ± 3	300 ± 3
polyisobutylene	23,000	2-methylbutane	325 ± 1	325 ± 2

[a] Ref. 63.
[b] Critical flocculation temperature.
[c] A characteristic value for polymers.

produced by the increase in the number of polymer–polymer interactions and the decrease in the number of polymer–solvent interactions. Such an increase can only occur if polymer–solvent interactions are strong, ie, the solvent is a very good one. Entropically stabilized dispersions tend to flocculate on cooling since entropy effects decrease with increasing temperature. Dispersions stabilized by enthalpic effects, however, tend to flocculate on heating because the stabilizing interaction effects are eventually outweighed by the destabilizing entropic effects (64).

The mixing effects described above are readily interpreted in terms of polymer-solution theory, and apply when the separation between the approaching surfaces is one or two adsorbed layer thicknesses. At closer approach, it is likely that both mixing and elastic compression occurs. The following equation was derived for the free energy of the interaction involving both effects (63):

$$\Delta G_R = \frac{2\,\pi akTn^2\Gamma^2V_s^2}{V_1}(1/2 - \chi)S_{mix} + 2\,\pi akT\Gamma S_{el} \tag{42}$$

This first term refers to the mixing of the absorbed layers, and the second to the elastic compression effect at distances less than one-layer thickness. V_s and V_1 are the volumes of polymer segment and solvent molecule, respectively, n is the number of segments per polymer chain, Γ the number of chains per unit area of surface, χ the Flory interaction parameter, and S_{mix} and S_{el} are geometrical terms. When the stabilizing chain is in a good solvent environment ($\chi < 1/2$) the mixture term is positive, but when θ conditions are exceeded ($\chi > 1/2$) it changes sign and the force is attractive. Attraction between polymer segments in worse than θ solvents is well known (65). However, the elastic term remains repulsive and operates at short distances. Hence, the flocculation behavior depends on the structure of the adsorbed layer (solvency, molecular weight, thickness, segment distribution), and the magnitude of the repulsive or attractive interactions arising from adsorbed layer overlap. For a comprehensive review of steric stabilization, see ref. 34.

An interesting example of the application of the concept of steric stabilization is the interpretation of the optical performance of alkyd paints pigmented with TiO_2 (22). Having established from electrophoresis and opacity measurements that the electric charge on the particles is not the controlling factor in flocculation, a study was made of the adsorption characteristics of the resin using pigments coated with different levels of silica–alumina such that the surfaces created varied from predominantly silica to mostly alumina. Interpretation of the adsorption isotherms indicated that for the acidic silica surface the basic resin molecules interacted strongly with the surface and adopted a parallel orientation, thus contributing little to the prevention of flocculation. On the other hand, for the predominantly alumina-coated surface, the resin only made contact with the surface with its limited number of acid groups; the rest of the molecule is extended into the medium and provides a steric barrier to flocculation.

When the surface coverage of polymer is low, bridging flocculation is possible. This is brought out by the polymer becoming simultaneously adsorbed on two or more particles (66). Since polymer adsorption is usually irreversible, the method of mixing of the components can have a profound effect on the flocculation process. The effect has been demonstrated with silica and aqueous solutions of poly(vinyl alcohol) (67). The effect depends on the pretreatment of silica, hence on the surface hydroxyl population, and on the pH of the solution which determines the particle charge; both factors influence the adsorption sites for PVA adsorption.

Measurement of Dispersion

Perhaps one of the most difficult aspects of the experimental approach to dispersion is the assessment of the state of the dispersion, particularly for particles in the colloidal range where the measurement can often disturb the system, eg, in electron microscopy, rheology, etc.

The state of dispersion of pigments can be studied in various ways, either in the liquid paint or in the dried film. They range from essentially practical methods used in production control to advanced laboratory techniques capable of providing quantitative data on the number, size, and type of particles present in a dispersion, eg, the Coulter counter.

Various electron-microscope techniques have been used for assessing pigment dispersion in coatings. For normal transmission electron microscopy, the paint is diluted which could have an effect on its state of dispersion. A paint film can also be sectioned to produce a suitably thin film for electron transmission, but during the process, pigment particles may be torn out of the film, thus producing an unsatisfactory product for examination. Some attempts have been made to study pigment dispersion by eroding the continuous medium and exposing pigment which can then be examined with a replication technique or by scanning electron microscopy. Since the various stages of incorporation–wetting, disagglomeration, and flocculation overlap in practice, it is difficult to differentiate between the situation when all the agglomerates/aggregates in the dry powder have not been completely dispersed in the initial stages and that in which flocculation has occurred. The clusters of particles commonly seen in electron micrographs of paint films are in most cases impossible to characterize, although carefully designed experiments can give helpful information as to which particular stage is giving trouble in a particular practical dispersion.

The measurement of backscatter of infrared radiation from a dry paint film provides a useful method for studying flocculation. As a result of their large size, scatter preferentially flocculates long wavelength radiation, and the amount of scattered light depends on the size and number of the flocculates, ie, on the degree of flocculation of the paint. The contribution to the long wavelength scattering by well-dispersed particles is small; hence, the measured backscatter is owing to the relatively small number of large particles, whether aggregates, agglomerates, or flocculates. A plot of backscatter against film thickness is linear for thin films, and the slope is termed the flocculation gradient. Measurements on a series of paints covering a range of flocculation levels are readily related to the opacity, gloss, and durability of the films (68–69).

In principle, rheological methods provide a useful way of investigating the state of dispersion although interpretation of the observed phenomena is beset with difficulties (see Rheological measurements). Much of the relevant fundamental work has been carried out with large spheres and at low concentration; such systems usually exhibit Newtonian behavior. With increasing concentration, deviations from ideal behavior are observed, and there are a large number of equations, both empirical and theoretical, that purport to describe experimental viscosity data (70). At volume fractions >0.25, non-Newtonian behavior is frequently observed and this is often owing to flocculation/aggregation of the dispersed particles. A useful illustration of this is the measurement of the rheological behavior of concentrated suspensions of glass spheres (radius ca 15 μm) dispersed in three Newtonian liquids of different polarities

(71). In addition, the surfaces of the spheres were varied by treatment with a polar surfactant and a silane to produce a hydrophobic surface. Suspensions (up to 40 vol %) of the silane-treated beads in the polar liquid (glycerol) were pseudoplastic, and evidence was presented for a significant degree of particle aggregation, arising from incomplete wetting. In other experiments, this low energy surface gave rise to Newtonian behavior provided the liquid adequately wetted the surface.

The influence of adsorbed layers on the rheological behavior of titanium dioxide dispersions (at 10 vol %) was investigated (72). Newtonian behavior was observed with solutions of alkyd resin in toluene, showing that the system was protected from flocculation by the adsorbed resin molecules. With stearic acid, the system was flocculated and exhibited plastic flow with a Bingham yield value. In mixed resin plus acid solutions, the yield values showed a direct relationship with the amount of stearic acid adsorbed.

Dilute systems are frequently used to provide rheological information on degree of dispersion, the thickness of adsorbed layers, etc (73). With monodisperse polymer latices at low volume fraction, the Bingham yield value reaches a maximum at the pH of zero particle charge, and a linear relationship between yield value and (zeta potential)^2 was found, as predicted by the application of the DLVO theory (74). However, most dispersed systems of practical relevance are concentrated and exhibit significant departures from Newtonian behavior; hence, the theories of colloid stability based on two-body collisions can no longer be expected to apply.

Nomenclature

A = Hamaker constant
A_{ij} = area of contact
A_{mm} = Hamaker constant for medium
A_{pm} = Hamaker constant for particle-medium interaction
A_{pp} = Hamaker constant for particles
a = measure of meniscus size; radius; number of particle pairs per unit area divided by the number of pairs per unit volume
B = London constant
b = true area of contact per particle pair divided by the surface area per particle; constant
C_2 = final concentration
c = mean coordination number
D_{max} = maximum dispersability
d = diameter of spherical particles
$\bar{d}$ = mean effective diameter
E = electrostatic attraction
E_{min} = minimum energy
e = electronic charge
F = force; mean microscopic force
F_{pp} = force between particles in a pair of particles
$\bar{F}_{pp}$ = mean of F_{pp}
F^{o}_{pp} = interparticle force per unit area of contact
F_{tot} = total force
H = distance between surfaces of spheres
H_o = distance between plates
K = efficiency factor
l = depth of penetration
M = microcontacts
N = protrusions that form microcontacts

$N_{\text{pp/area}}$ = number of particle pairs per unit area
n = electrolyte concentration
n^o = total number of moles of liquid components present
n^s = number of adsorption sites
n_1^s, n_2^s = moles of components adsorbed per gram solid
p = force acting on the unit area of each plate
q = number of atoms per molecules per cm^3
R = distance between the centers of the spheres
r = particle radius
s = surface; R/a
$\bar{s}$ = mean surface area per particle
S_{mix}, S_{el} = geometrical terms
T = tensile strength
t = time
t_o = surface separation at zero tensile strength
V_1 = volume of solvent molecule
V_A = attractive potential energy
V_R = repulsive potential energy
V_S = volume of polymer segment
V_{tot} = total potential energy
$\bar{v}$ = mean particle volume
W_A = work of adhesion
W_A^{SLV} = work of adhesional wetting
W_S^{SLV} = work of spreading wetting
W_W^{SLV} = work of immersional wetting
W_t = total work
X = potential difference
x = distance
x_1^o, x_2^o = mole fractions
x_{ij} = proportion of particle pairs of diameter d_i and d_j
y_{ij} = proportion of particle pairs of diameter d_i and d_j bridged by pendular moisture
z = valency
α = cone half-angle
β = factor to allow for loss of spherical symmetry in the double layer
Γ = number of chains per unit area of surface
γ = surface tension; interfacial free energy
γ^{LV} = interfacial free energy at liquid-vapor interface
γ^{S} = interfacial free energy at solid-vapor interface
γ^{SL} = interfacial free energy at solid-liquid interface
ϵ = voidage; dielectric constant
ϵ_o = dielectric constant of medium
η = viscosity
θ = contact angle; characteristic parameter for polymers
κ = Debye-Hückel parameter
$1/\kappa$ = thickness of double layer
ρ_o = bulk density when $T = 0$
ρ_s = density of solid
ϕ = central angle
χ = Flory interaction parameter
ψ = potential at distance x from the surface
ψ_o = surface potential

Superscripts

d = dispersion
e = electric charge

h = hydrogen bond
p = polar interaction
π = pi bond

BIBLIOGRAPHY

1. R. A. Fisher, *J. Agric. Sci.* **16,** 492 (1926).
2. B. Elbirli, L. Vergara-Edwards, and R. W. Coughlin, *AIChE J.* **26,** 865 (1980).
3. H. C. Hamaker, *Physica* **4,** 1058 (1937).
4. P. C. Hiemenz, *Principles of Colloid and Surface Chemistry*, Marcel Dekker Inc., New York, 1977, p. 414.
5. H. M. Sutton in G. D. Parfitt and K. S. W. Sing, eds., *Characterization of Powder Surfaces*, Academic Press, London, 1976, Chapt. 3.
6. G. D. Parfitt, *FATIPEC Congr.*, 107 (1978).
7. H. Rumpf in W. A. Knepper, ed., *Agglomeration*, Interscience Publishers, a division of John Wiley & Sons, Inc., New York, 1962, p. 379.
8. H. Rumpf, *Chem. Ing. Tech.* **42,** 538 (1970).
9. W. Pietsch and H. Rumpf, *Chem. Ing. Tech.* **39,** 885 (1967).
10. D. C. H. Cheng, *Chem. Eng. Sci.* **23,** 1405 (1968); *J. Adhes.* **2,** 82 (1970); *Proc. Soc. Anal. Chem.* **10,** 17 (1973).
11. C. Huk and S. G. Mason, *J. Colloid Interface Sci.* **60,** 11 (1977).
12. N. W. F. Kossen and P. M. Heertjes, *Chem. Eng. Sci.* **20,** 593 (1965).
13. J. J. Bikerman, *Ind. Eng. Chem.* **13,** 443 (1941).
14. P. M. Heertjes and W. C. Witvoet, *Powder Technol.* **3,** 339 (1970).
15. E. D. Washburn, *Phys. Rev.* **17,** 374 (1921).
16. W. Black in G. D. Parfitt, ed., *Dispersion of Powders in Liquids*, 3rd ed., Applied Science Publishers, London, 1981, p. 149.
17. R. G. Laughlin, *J. Soc. Cosmet. Chem.* **32,** 371 (1981).
18. F. M. Fowkes, *Wetting*, Monograph 25, Society of Chemical Industry, London, 1967, p. 3.
19. J. J. Kipling, *Adsorption from Solutions of Non-electrolytes*, Academic Press, London, 1965.
20. G. D. Parfitt and C. H. Rochester, eds., *Adsorption from Solution at the Solid/Liquid Interface*, Academic Press, London, 1983.
21. C. H. Giles, T. H. MacEwan, S. N. Nakhway, and D. Smith, *J. Chem. Soc. London*, 3973 (1960).
22. M. J. B. Franklin, K. Goldsbrough, G. D. Parfitt, and J. Peacock, *J. Paint Technol.* **42,** 740 (1970).
23. K. Marshall and C. H. Rochester, *J. Chem. Soc. Faraday Trans. I* **71,** 1754 (1975).
24. R. H. Ottewill and J. M. Tiffany, *J. Oil Colour Chem. Assoc.* **50,** 844 (1967).
25. R. E. Day and G. D. Parfitt, *Powder Technol.* **1,** 3 (1967).
26. L. Dintenfass, *Kolloid Z.* **151,** 154 (1957).
27. D. W. Fuerstenau in M. L. Hair, ed., *The Chemistry of Bio Surfaces*, Vol. 1, Marcel Dekker Inc., New York, 1971.
28. T. W. Healy, *J. Macromol. Sci. Chem.* **A8**(3), 603 (1974).
29. B. Tamamushi and K. Tamaki, *Trans. Faraday Soc.* **55,** 1007 (1959).
30. R. E. Day, F. G. Greenwood, and G. D. Parfitt, *Proc. 45th Int. Congr. Surface Activity* **II,** 1005 (1967).
31. B. Vincent, *Adv. Colloid Interface Sci.* **4,** 193 (1974).
32. A. Silberberg, *Pure Appl. Sci.* **26,** 583 (1971).
33. Y. S. Lipatov and L. M. Sergeeva, *Adsorption of Polymers*, John Wiley & Sons, Inc., New York, 1974.
34. B. Vincent and S. G. Whittington in E. Matijevic, ed., *Surface and Colloid Science*, Vol. 12, Plenum Press, New York, 1982, p. 1.
35. V. T. Crowl and M. A. Malati, *Discuss. Faraday Sci.* **42,** 301 (1966).
36. F. M. Fowkes and M. A. Mostafa, *Ind. Eng. Chem. Prod. Res. Dev.* **17,** 3 (1978).
37. T. Doorgeest, *J. Oil Colour Chem. Assoc.* **50,** 841 (1967).
38. P. M. Heertjes, C. I. Smits, and P. M. M. Vervoorn, *Prog. Org. Coat.* **7,** 141 (1979).
39. A. J. Walton, *Paint Manuf.*, 13 (May 1978).
40. P. A. Rehbinder, *Colloid J. USSR* **20,** 493 (1958).

41. A. T. DiBenedetto, *The Structure and Properties of Materials*, McGraw-Hill, Inc., New York, 1967.
42. I. R. Sheppard in G. D. Parfitt, ed., *Dispersion of Powders in Liquids*, 2nd ed., Applied Science Publishers, London, 1973, p. 221.
43. D. A. Wheeler in ref. 16, p. 327.
44. B. V. Derjaguin and L. D. Landau, *Acta Physiochim. URSS* **14**, 633 (1941).
45. E. J. W. Verwey and J. Th. G. Overbeek, *Theory of the Stability of Lyophobic Colloids*, Elsevier, Amsterdam, The Netherlands, 1948.
46. J. Gregory, *Adv. Colloid Interface Sci.* **2**, 396 (1970).
47. J. Visser, *Adv. Colloid Interface Sci.* **3**, 331 (1972).
48. B. Vincent, *J. Colloid Interface Sci.* **42**, 270 (1973).
49. M. J. Vold, *J. Colloid Sci.* **16**, 1 (1961).
50. D. W. J. Osmond, B. Vincent, and F. A. Waite, *J. Colloid Interface Sci.* **42**, 262 (1973).
51. O. F. Devereux and P. L. de Bruyn, *Interaction of Plane Parallel Double Layers*, M.I.T. Press, Cambridge, Mass., 1963.
52. W. Albers and J. Th. G. Overbeek, *J. Colloid Sci.* **14**, 510 (1959).
53. C. S. Chen and S. Levine, *J. Colloid Interface Sci.* **43**, 599 (1973).
54. G. R. Feat and S. Levine, *J. Chem. Soc. Faraday Trans. II* **71**, 102 (1975).
55. G. R. Feat and S. Levine, *J. Colloid Interface Sci.* **54**, 34 (1976).
56. P. A. Rehbinder and A. B. Taubman, *Colloid J. USSR* **23**, 301 (1961).
57. B. V. Derjaguin, *Disc. Faraday Soc.* **18**, 85 (1954).
58. R. Hogg, T. W. Healy and D. W. Fuerstenau, *Trans. Faraday Soc.* **62**, 1638 (1966).
59. G. D. Parfitt in ref. 16, p. 1.
60. A. Bierman, *J. Colloid Sci.* **10**, 231 (1955).
61. D. C. Prieve and E. Ruckenstein, *J. Colloid Interface Sci.* **73**, 539 (1980).
62. G. D. Parfitt, J. A. Wood, and R. T. Ball, *J. Chem. Soc. Faraday Trans. I* **69**, 1908 (1973).
63. D. H. Napper, *J. Colloid Interface Sci.* **58**, 390 (1977).
64. R. Evans and D. H. Napper, *J. Colloid Interface Sci.* **52**, 260 (1975).
65. P. J. Flory, *Principles of Polymer Chemistry*, Cornell University Press, Ithaca, N.Y., 1971.
66. J. A. Kitchener, *Br. Polym. J.* **4**, 217 (1972).
67. Th. F. Tadros, *J. Colloid Interface Sci.* **64**, 36 (1978).
68. J. Balfour and M. J. Hird, *J. Oil Colour Chem. Assoc.* **58**, 331 (1975).
69. J. Balfour, *J. Oil Colour Chem. Assoc.* **60**, 365 (1977).
70. J. R. Rutgers, *Rheol. Acta* **2**, 305 (1962).
71. S. V. Kao, L. E. Nielsen, and C. T. Hill, *J. Colloid Interface Sci.* **53**, 358 (1975).
72. P. M. Heertjes and C. I. Smits, *Powder Technol.* **17**, 197 (1977).
73. J. W. Goodwin, *Colloid Science*, Vol. 2, Chemical Society, London, 1975, p. 246.
74. J. P. Frend and R. J. Hunter, *J. Colloid Interface Sci.* **37**, 548 (1971).

G. D. PARFITT
Carnegie-Mellon University

F

FIBERS, MULTICOMPONENT

Bicomponent fibers, conjugated, or composite fibers or heterofils have been known for some time. Glass bicomponent fibers were produced in the latter half of the nineteenth century and side–side viscose bicomponent fibers were patented in 1937 (1). Wool (qv), however, is a natural bicomponent fiber. Several acrylic bicomponent fibers were manufactured as self-crimping products in the early 1960s and nylon bicomponent fibers were introduced about the same time for use in hosiery and in nonwoven fabrics (qv) (see Acrylic and modacrylic fibers; Polyamides; Fibers, elastomeric; Fibers, chemical).

These products were all relatively simple with only a single interface between the two components. Today's fibers are more complex; they are manufactured with multiple interfaces and material–air boundaries. Interest in all types of multicomponent, though generally bicomponent, fibers has been revived because of the intense competition of the more conventional synthetic fabrics with the unacceptably high cost of introducing a fiber based on a totally new polymer; the need for some of the unique properties which can only be obtained via a bicomponent route; and improvements in technology which have allowed the manufacture of such sophisticated products.

The most commonly used designation for fibers composed of two polymers is bicomponent fibers, sometimes shortened to bico fibers. Since there are few examples, as yet, of three or more fiber-forming polymer components, bicomponent tends to be used as a general term to include these products. Synthetic fibers always contain materials such as stabilizers, catalyst residues, processing aids, pigments, and other additives. Alternative expresssions are bilaminar filaments, biconstituent fibers, composite fibers, heterofilaments (or heterofils), and sea–island fibers. It is important to distinguish between composite fibers (where every individual fiber or filament has two or more components) and composite yarns, sometimes called heteroyarns. The latter are yarns composed of a mixture of different types of filaments, usually single-component filaments, although a composite yarn may contain a mixture of single-component and bicomponent filaments (or any other combination).

It has been found possible to introduce controlled-composition variations into the cross sections of textile filaments of ca 0.1–0.5 tex (1–5 den), or only a few micrometers in diameter, especially in simpler bicomponent fibers with only a single interface, such as side–side and core–sheath bicomponents. Coupled with an increasing understanding of the viscous flow of fiber-forming polymer melts, and improved spinnerette manufacturing methods, this has led to the development of new biconstituent fibers including multiinterface or matrix–filament fibers (sea–island fibers). Japan, western Europe, and North America have all been involved in this development. In many cases these products are never marketed as bulk fiber but are converted by the producers into materials of higher value.

Bicomponent fibers are classified on the basis of the cross-sectional and longitudinal arrangement of the components (2–3). The simplest class, consisting of fibers with a single interface and with both components possessing an external boundary, is known as side–side fibers irrespective of the cross-sectional shape, which was not necessarily circular. The second class comprised fibers with a single interface but in which only one component had an external boundary, known as sheath–core fibers. These two classes exhibit special properties because of the presence of only a single interface. For example, side–side fibers usually tend to crimp when treated in various ways. There is no generally accepted term for fibers with more than one component–component interface and one or more external boundaries, although they are sometimes termed matrix–filament fibers. This term is used here in preference to sea–island fibers, which implies only a single external boundary, ie, multiple totally enclosed cores.

In the above examples, the cross sections of the components do not vary along the lengths of the fiber. Another class of bicomponent fibers, known as matrix–fibril fibers, exhibits cross-sectional variation with length. The second (minor) component is present as separate fibrils or short lengths of fiber embedded in a matrix of the first (principal) component. The main axis of the fibrils is aligned with the main axis of the fiber. Matrix–fibril fibers are not prepared by the controlled combination of polymer streams as are other bicomponent fibers; they are of little commercial importance.

A great variety of transverse and longitudinal fiber cross sections exists. The scanning electron microscope photograph of a transverse cross section of Diolen ultra matrix–filament fibers is shown in Figure 1. The center contains a complete filament comprising a star-shaped matrix of nylon into which are fitted six triangular filaments of poly(ethylene terephthalate) polyester. This whole cross section is surrounded by split matrix-and-filament components because this particular fiber is designed to split to give very fine fibers.

Materials and Processes

In many respects, the manufacture of bicomponent fibers differs very little from that of single-component fibers (see Acrylic and modacrylic fibers; Olefin fibers;

Figure 1. Transverse cross section of Diolen Ultra fibers (magnification 2500×).

Polyamides; Polyester fibers). Many pairs of fiber-forming polymers may be employed in the manufacture of bicomponent fibers. In fact, it is not necessary for both components to be capable of forming fibers if spun alone. For example, polystyrene, which is not normally used in fiber manufacture, may be used as a temporary component in the manufacture of certain matrix–filament fibers. Clearly, it must be possible to handle both components by the same fiber-forming route, eg, melt-spinning. Thus it would be impossible to manufacture a polyester–acrylic bicomponent fiber because the acrylic component could not be spun by the melt-spinning route; bicomponent fibers with an acrylic component can only include components which can be wet or dry-spun. Similarly, materials which can only be wet-spun cannot be combined with materials that can only be melt-spun or dry-spun. Within these limitations, there is still a large number of possible combinations, especially in the melt-spun group, which today is the most important in technical terms.

The bicomponent fibers or their derived products are given in Table 1. The only products that are not melt-spun are the acrylic fibers (qv), which dominate production in volume, but in value are only a fraction of the unit cost of the most expensive products. Some of these products, such as the synthetic suedes, can cost up to $150/kg. Bicomponent rayon fibers have been reported experimentally, but are not produced commercially by a bicomponent route. The normal wet-spinning process, however, can form nonuniform fibers with self-crimp characteristics (4). Melt-spinning is a better process for bicomponent fibers than wet-spinning because of the reduced hole density and consequent loss of spinning capacity introduced by the assemblies behind the spinnerette required for bicomponent formation. This is particularly important in the case of the very high hole densities normally employed in wet-spinning.

The components employed in bicomponent-fiber manufacture must be compatible in a number of other respects as well as in their spinning route. Most important, their viscosities under the conditions of spinning must not be too dissimilar and high enough to prevent turbulence after the spinnerette. Even if turbulence does not occur,

Table 1. Manufacturers and Trade Names of Multicomponent Fibers or Multicomponent Fiber-Based Products

Manufacturer	Country	Trade name	Type	Composition (if known)
Allied Chemical Company	United States	Source[a]	staple, FY[b]	matrix–fibril nylon 6–polyester
ANIC	Italy	Euroacril	staple	side–side acrylic
Asahi Chemical Industry Co., Ltd.	Japan	Cashmilon GW and Cashmilon H	staple	side–side acrylic
		Lammus	suede	
Akzo N.V.	Netherlands	Diolen Ultra	FY[b]	matrix–filament nylon–polyester
		Diolen Biko	staple	matrix–filament nylon–polyester
		Diolen	staple	sheath–core polyethylene–polyester
		Diolen	staple	sheath–core nylon 6–nylon 6,6
		Enkatron	FY[b]	matrix–fibril nylon 6,6–polyester
American Cyanamid Company	United States	Creslan 68	staple	side–side acrylic
Bayer AG	Federal Republic of Germany	Dralon K/120, 140, 820	staple, FY[b]	side–side acrylic
Courtaulds Ltd.	United Kingdom	Courtelle LC	staple	side–side acrylic
Chisso Corporation	Japan	ES Fiber	staple	side–side polyethylene–polypropylene
		EA Fiber	staple	side–side EVA[c]–polypropylene
Colbond b.v.	Netherlands	Colbond, Colback	spunbonded	sheath–core nylon–polyester
Cyanenka	Spain	Crilenka Bico	staple	side–side acrylic
E. I. du Pont de Nemours & Co., Inc.	United States	Cantrece	FY[b]	side–side nylon 6,6–modified nylon 6,6
		Orlon Types 21, 24; Sayelle	staple	side–side acrylic
Hercules	United States	Herculon 404	staple, FY[b]	side–side polyolefin
ICI Fibers	United Kingdom	Cambrelle PBS3	staple nonwoven	sheath–core nylon 6–nylon 6,6
		Cambrelle PBS5	staple nonwoven	sheath–core modified polyester–polyester
		Epitropic fiber	staple blend	sheath–core modified polyester–polyester
		Bonafill	staple	sheath–core polypropylene–polyester
		Terram	spunbonded nonwoven	sheath–core polyethylene–polypropylene[d]
Japan Exlan	Japan	Exlan	staple	side–side acrylic
Kanebo Ltd.	Japan	Sideria[a]	FY[b]	sheath–core nylon–polyester
		Savina	suede	matrix–filament nylon–polyester
		Belleseime[e]	suede	matrix–filament nylon–polyester

Table 1 (*continued*)

Manufacturer	Country	Trade Name	Type	Composition (if known)
Kanebo Ltd.	Japan	Belina	staple	sheath–core nylon–polyester
		Kanebo Nylon 22	staple	side–side nylon 6–modified nylon 6
Kuraray Ltd.	Japan	Sofrina[f]	apparel leather	matrix–filament
		Amara	suede	matrix–filament
Mitsubishi	Japan	Vonnel V-57	staple	side–side acrylic
Monsanto Textile Company	United States	Monvelle[a]	FY[b]	side–side nylon–polyurethane
		Pa–Quel	staple	side–side acrylic
		Antistatic fibre	FY[b]	matrix–filament polyester–polyethylene
Montefibre	Italy	Leacril BC	staple	side–side acrylic
Rhone-Poulenc-textile	France	Tergal X-403	staple	side–side polyester–modified polyester
Snia	Italy	Velicren	staple	side–side acrylic
Teijin Ltd	Japan	Hilake	suede	matrix–filament nylon–polyester
		Tetoron 88	staple	side–side polyester
		S-28[a]	staple	matrix–fibril nylon 6–polyester
Toho Beslon	Japan	Beslon	staple	side–side acrylic
Toray Ltd.	Japan	Ecsaine[g]	suede	matrix–filament polystyrene–polyester
		Toraylina	suede	matrix–filament polystyrene–polyester
		Crestfil	raised fabric	
		SBII	leather	
		Tapilon	FY[b]	sheath–core nylon 6–nylon 6,6
		Metalian	FY[b] (antistatic)	sheath–core polyester
Unitika	Japan	Lauvest	suede	

[a] Production of these materials is now believed to have been discontinued.
[b] FY = filament yarn.
[c] Ethylene–vinyl acetate copolymer.
[d] Previously sheath–core nylon copolymer–polypropylene.
[e] Or Suede 21, Suede 21L.
[f] Napara for export.
[g] Ultrasuede or Alcantara for export.

if one component is less viscous than the other, the flow patterns designed to give the required filament cross section may not be stable. Even if the filament geometry is correct, if it is without a center of symmetry, and if both components have an external boundary, unstable extrusion can occur leading to flickering filaments and interfilament coalescence in a molten thread line. This can lead to severe problems in a simple side–side geometry even where both polymers are apparently similar; differences in frictional properties, flow behavior, and polymer-to-spinnerette adhesion play a part. As a rule, the viscosities of the two components should not differ by more than a factor of ca four, but ideally should be considerably closer. This can be accomplished by modifying the dope concentrations or the polymer molecular weights, although the effect on downstream processing and fiber properties must be considered. The choice of components depends on the fiber application. If possible, however, a polymer and one of its copolymers are selected because this choice can remove many of the compatibility problems, not only the primary one of viscosity balance, but also the subsequent ones of component adhesion and drawing ratio compatibility.

Frequently, component pairs do not adhere sufficiently to behave as a single fiber through various textile processes or in the final application. Splitting and breakage of components may occur on rolls, guides, or texturing devices resulting in heavy, sticky, or dusty deposits, particularly with matrix–filament fibers. Some matrix–filament fibers, however, are designed to separate during fabric finishing to confer certain properties on the fabrics. In other cases, splitting can be detrimental, for example, in carpets where it may result in matting (interlocking and compaction of the pile). Splitting may be alleviated by mechanically interlocking the components; the ultimate interlocking is a core–sheath fiber where the core component cannot escape. However, if the sheath is very thin or is uneven, it may be stripped or split away from the core. Mutual diffusion of the polymer components at the interface by slight flow disturbance in the spinnerette hole has been suggested for better adhesion in side–side fibers (5). Noninterlocked components may separate at the drawing stage, especially if the natural draw ratio of one component is exceeded.

The orientations, as indicated by the birefringences, of the two components, are always affected by the component structures, which are, by design, different. In some cases, optimal overall fiber strength is not needed because one component is removed at a later stage, or used as an adhesive. This simplifies drawing since the draw ratio can be selected to be suitable for the remaining component. If both components contribute to the fiber properties, a weighted average of the properties of either component results. Thus each property value must be below the value which would have been obtained for that property in a monocomponent fiber prepared under the same conditions from whichever component gives the better level of that property. Bicomponent fibers thus tend to have inferior properties compared to the equivalent monocomponent fiber, but have properties that the monocomponent material does not have. Typical properties of an acrylic bicomponent fiber are compared in Table 2 with those of the equivalent single-component fiber (6). In the case of acrylic fibers, the bicomponent technology is applied only to achieve self-crimping in order to obtain properties similar to those of wool. For self-crimping, the fiber must be asymmetric and the components must possess a differential shrinkage (7). As a rough approximation, the radius r of the helical crimp generated for a fiber of thickness h is given (3) by:

$$\frac{1}{r} \approx 1.6 \frac{\Delta L}{L} \cdot \frac{1}{h}$$

where $\Delta L/L$ is the differential change in length of the two components during the crimp development stage. The crimp may develop spontaneously as drawing tension is removed or it may be latent, ie, it develops as some condition of the fiber is changed. Some bicomponent fibers develop crimp when they are heated in a relaxed state. Some acrylic fibers develop crimp when they are dried, and lose it when they are immersed in water. Such a reversible crimp offers an advantage in that garments made from these fibers regain their original softness and bulk after washing.

No similar simple treatment is available for predicting the extension-recovery behavior or crimping or retraction forces that are associated with bicomponent fiber

Table 2. Properties of Cashmilon Acrylic Fibers [a]

Property	Cashmilon GW, bicomponent	Cashmilon FW, monocomponent
strength, N/tex [b]		
dry	0.25	0.25–0.32
wet	0.21	0.23–0.28
elongation, %		
dry	40–45	28–40
wet	45–50	30–45
loop strength, N/tex [b]	0.31	0.18–0.27
loop elongation, %	20–25	5–20
modulus of tensile elasticity, %	99 ± 2	(90–92) ± 3
specific gravity	1.18	1.188
moisture regain at 20°C, %	1.8–1.6	1.6
shrinkage in hot water, %	4	2–4

[a] Ref. 6.
[b] To convert N/tex to gf/den, multiply by 11.33. N/tex = GPa/density.

crimping (3). Although self-crimping products are the most obvious application for bicomponent fibers, there has been no successful commercial application of the principle in the continuous-filament field, even though producers have tried to enter the textured yarn market. The crimping and retraction forces are too low to be of value in comparison with those obtained by the false-twist texturing or draw-texturing routes. In the case of staple fibers, the relatively weak forces can still be of value where the separated short fibers are concerned. Stronger retractive forces can be obtained if one of the components is elastomeric, but such products are not now commercialized. A number of self-crimping products have been obtained with a crimp potential generated by asymmetric treatment of the filaments after extrusion. Since part of the outside skin of each filament is treated, the maximum possible asymmetry results. Crimping can be achieved by water quenching or hot knife-edge treatments (8–9).

The processes by which multicomponent fibers are produced are largely independent of the polymers used except for the formation of a self-skinning fiber in the wet-spinning process (4). Production processes rely on the stability of the interfaces between the components to maintain the filament cross-sectional geometry as the cross-section area is reduced, perhaps several hundredfold, during extrusion and drawdown. Production methods for multicomponent fibers are given in Figure 2. To obtain the required properties under the best process economics, high filament spinning density is needed with high plant utilization and conversion efficiency, clean startup, and low production of waste fiber. In most production methods the components are combined after they have entered the spinning pack. The design of the spinning pack is more complex than for single-component fibers and this overcrowding reduces the potential filament density. This may not be important in the case of a low-count filament yarn (such as a hosiery yarn), where only a few filaments are required. It is more important in feedstock for staple fiber because the greater the number of filaments produced, the better are the process economics.

Multicomponent fibers can be produced by spinning, under suitable conditions, a blend of immiscible polymers, eg, nylon and polyester. A matrix–fibril fiber is produced in which the minor component forms short individual fibrils dispersed in a fiber

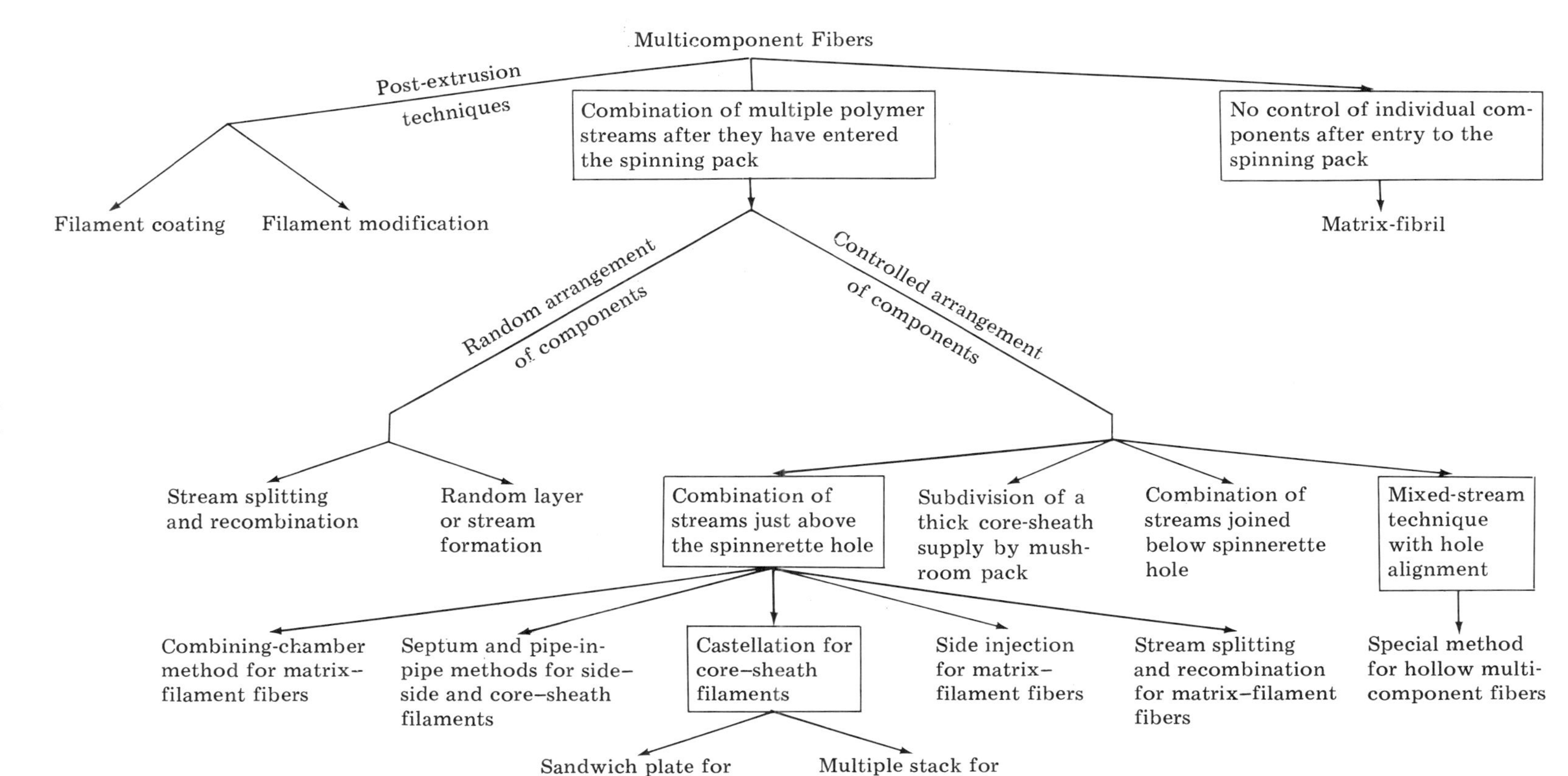

Figure 2. Production of multicomponent fibers.

composed of the principal component. New products were expected from this technique (10), eg, for low flat-spotting tire cords, but commercialization has not materialized. Changes in the ratio of the two components affect the structure of the fibrils. Because of orientation at spinning and drawing, the fibrils always orient parallel to the fiber axes. The components may be blended as chips or powders before extrusion. These fibers are relatively easy to produce and it is unfortunate that they are not of greater interest. However, it is possible that the fibril size and, therefore, the properties may depend upon the degree of mixing during extrusion, which is difficult to control. In order to obtain a fibrous structure, the polymers must be incompatible. For example, polyester is not sufficiently incompatible with polycarbonate to generate fibrils, but polypropylene forms fibrils in a polycarbonate matrix. The properties of the resulting fibers are superior to those of monolaminar polycarbonate filaments. The preparation of multicomponent fibers by post extrusion techniques is not easy. For example, it is difficult to coat fine filaments with a second component. It is possible that Teijin's Metalian antistatic fiber is prepared by vapor deposition of a layer of metal on the fibers (11). Antisoil coatings have been applied by filament-coating techniques to textile fibers, especially carpets (see Textiles, finishing). A perfluoroalkyl ester may be applied in a thin, durable coating to synthetic textile filaments as a spin-finish (12). In this case the sheath is extremely thin. Typical techniques include treatment of the hot filaments with cooling water (8), the passage of filaments over a heated knife edge (9), or the asymmetric heat treatment of the filaments in the spinnerette hole (13). Since the filaments are initially formed from a single component, the modifications change part of it to generate the crimp potential. This may be a change of orientation, crystallinity, molecular weight, or filament cross section.

True multicomponent fibers are prepared by the controlled combination of liquid polymer streams after the individual streams have entered the spinning pack. If only side–side fibers are required and a degree of uncertainty regarding the position of the interface can be tolerated, a conventional spinnerette may be fed with a polymer stream containing alternating layers of the two components (13). Since the mixed-stream flow is not turbulent, the alternating layer pattern is maintained as the stream reaches the spinnerette face and a bicomponent filament is formed whenever a spinnerette hole intersects a layer junction. This intersection takes place randomly and there is no control over the position of the interface in the filament. Depending on the number of spinnerette holes and the number of layers, the interfaces in any filament may vary, resulting in variation in the properties of the individual filaments. Alternatively, the streams can be mixed in one or more static or dichotomic mixers (14). Such mixers have four to nine alternate left- and right-hand helical elements in series. After the polymer components are fed to the opposite sides of the leading edge of the first element of each mixer they are split and recombined to form at the mixer exit a multilaminar stream which is then conducted to the spinnerette. Such mixers may be combined in a suitable configuration in a spinning pack. The principle of mixed-stream formation may also be used to form matrix–filament fibers (15–17). A greater number of streams, ie, fine cores, are normally formed than the number of spinnerette holes and, therefore, the number of interfaces per hole is high and less variable from filament to filament. Furthermore, because a simple spinnerette is employed, the number of filaments which may be produced from a spinnerette by the mixed-stream technique is not limited. In all mixed-stream methods, the number of layers or cores formed in the mixed stream or streams is, in theory, controlled because the streams are formed by the splitting and recombination of the polymers. A truly random number of fine cores is formed in the mixed stream by passage of the laminar stream from a mixer through a layer of gauze screens (18). The gauzes, of carefully selected opening size, thus break up the laminations (of which there may be >1000) to give as many as 150 cores per filament.

Other methods of preparing multicomponent fibers require the controlled flow of the components to the spinnerette holes or beyond. Most of these methods require the components to be combined and formed into the required cross-sectional configuration just above the spinnerette holes. However, three techniques require the controlled arrangement of components away from the spinnerette holes. In the first, the polymer streams are combined before they enter the spinning pack (19). A thick core–sheath supply stream passes through the sand filter and is distributed to a mushroom plate just above the spinnerette (Fig. 3). (Note: letters A and B arbitrarily indicate the positions of the two components. These letters do not necessarily correspond to the sheath and the core, respectively, in all figures.) The distribution channels above the plate are arranged in such a way that their position and areas lead the core and sheath polymers to flow between the underside of the mushroom plate and the upper surface of the spinnerette as an upper layer of core polymer and a lower layer of sheath polymer. Since both polymers are supplied at a fixed rate to the spinnerette pack, fixed quantities of each leave through each spinnerette hole as a concentric core–sheath filament. Despite the simplicity of this technique, it has disadvantages including the difficulty of establishing the correct flow at startup, the almost inevitable presence of some core-only and some sheath-only monocomponent filaments, a limitation to component ratios of between ca 2:1 and 1:1, and the need to design the pack internals for the specific core:sheath ratio needed. The technique is only used for concentric core–sheath spinning; more recent techniques are superior.

Another method is even more specialized; both components must be of the same material but with different orientation. Each filament is formed by the combination of two polymer streams below the spinnerette hole (see Fig. 4). The process is primarily applicable to combined melt spinning and drawing at high speeds (greater than 3500 m/min winding speed). Advantages include simplicity of process and equipment, except for the spinnerette. Disadvantages include limitations in polymer types and cross

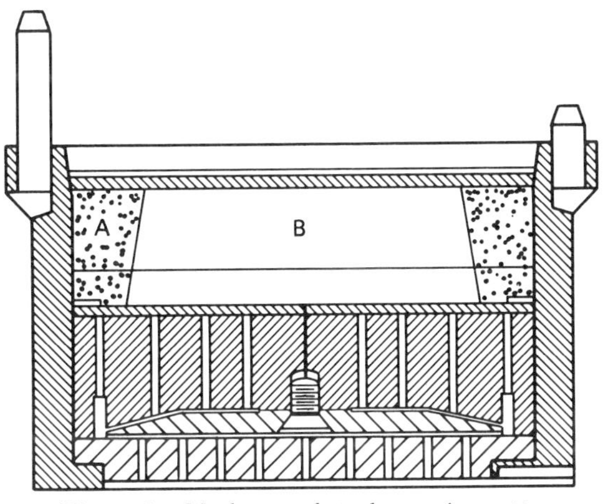

Figure 3. Mushroom plate above spinnerette.

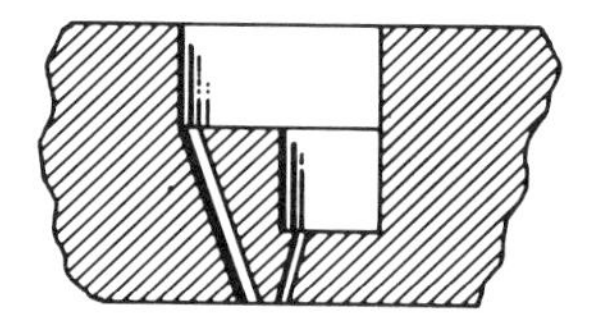

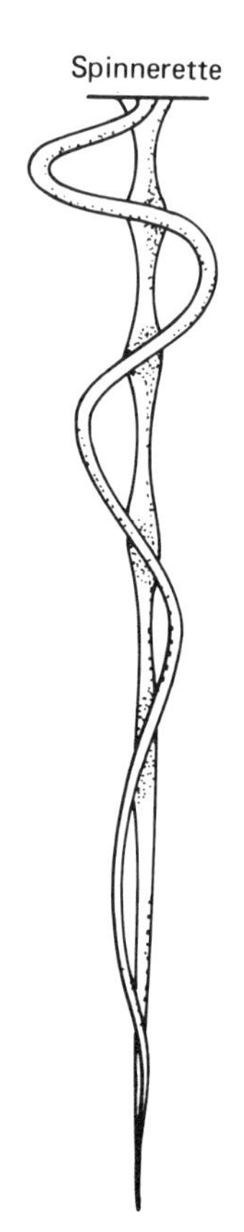

Figure 4. Combination of two polymer streams below spinnerette hole.

sections and low crimping forces. In addition, some spinning conditions can lead to filament oscillation and cross-section variability.

Side–side bicomponent fibers are also produced by feeding a mixed polymer stream to a simple spinnerette with the holes aligned with the stream interfaces. The mixed polymer streams are generated as concentric rings or parallel laminae (7). The pipe–in–pipe method is the simplest for mixed-stream formation (Fig. 5**a**); the concentric plate mixer (Fig. 5**b**) is actually a compact pipe–in–pipe system. At plate 1, polymer A forms a cylindrical stream which is narrowed at plate 2. Plate 3 adds a concentric layer of component B and the core–sheath stream is narrowed at the next plate 2. As many concentric layers may be added as there are pairs of plates. The proportion of the two polymers delivered to each spinnerette hole is determined by the supply rates to the spinning pack. The rings or lines of holes must be carefully aligned with the interfaces and the volume flow of each stream must be in proportion to the number of holes it supplies. If the number of interfaces is large, alignment becomes difficult and random layers form. Special spinnerettes are not needed, but there

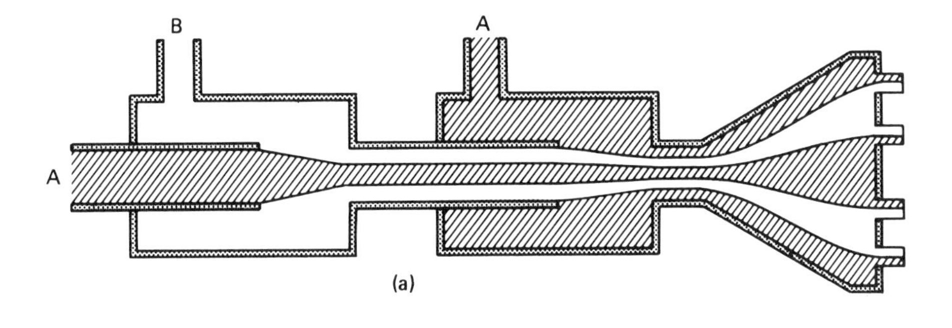

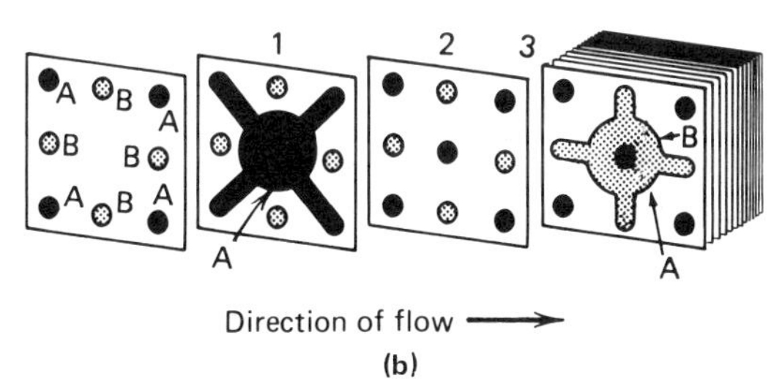

Figure 5. (**a**) Pipe-in-pipe method for mixed-stream formation; (**b**) Concentric plate mixer.

is some loss of control over the components' ratio. The aligned mixed-stream method is not thought to be currently in commercial use. A variant is used in the formation of hollow matrix–filament fibers (20). For each filament, an interrupted, elongated slot meanders back and forth across the interface of a single core–sheath stream. This generates a number of separately spun interfaces, which combine below the spinnerette to form the hollow filament with alternate segments of the two constituents.

In other methods, the polymers are combined just before their passage through the spinnerette hole. The pack designs are more complex than the methods described previously; they are more costly and may be limited in output. On the other hand, the filament cross sections are better controlled and parameters, such as the core:sheath ratio, may be changed without modifying the pack or spinnerette design. In the septum method for side–side fibers the components are kept apart until the very last moment by a partition above the spinnerette holes. The form of the partition, or septum, is not standardized (see Fig. 6); it may be a relatively thin plate (21) or somewhat thicker (22). It may even be a protrusion passing into the spinnerette hole (23). Since the septum generally covers a whole row or ring of holes, the pressure drop to each hole must be the same to obtain fiber uniformity. The ratio of components is determined by the pumping rates. Unless a condensed system is employed (21), the channel system behind the spinnerette severely limits the hole density. The condensed system is less suitable for highly viscous polymer melts because of the high pressures developing. In certain cases (24) a metering plate ensures uniform polymer flows to the spinnerette holes, and the same spinnerette is used to form some monocomponent fibers to yield a composite yarn. To prevent the splitting of filaments, the two components may be combined by a hole through the septum or some other modification (25); this distorts the interface in a controllable fashion. Another method of combining the components is by intermingling at the interface, for example, by the use of a gauze (5).

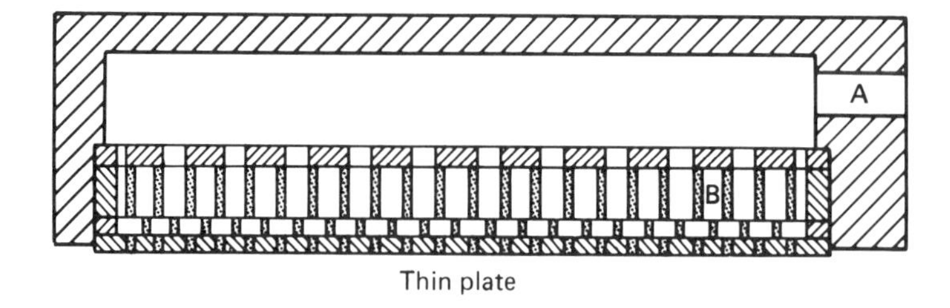

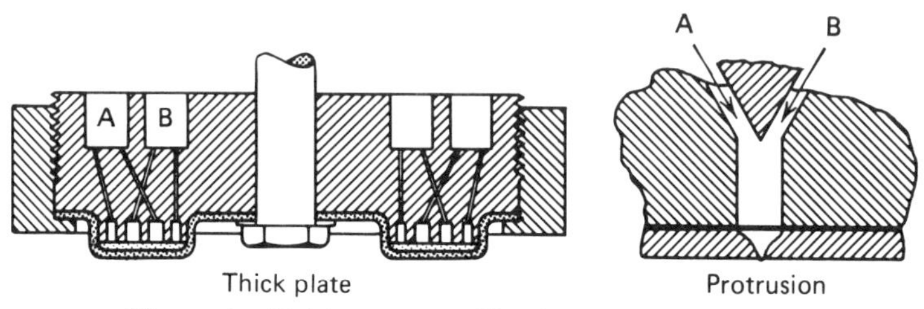

Figure 6. Multicomponent-fiber formation using septa.

The pipe-in-pipe technique is analogous to the septum method. Here, the core stream is inserted directly into the sheath stream via a fine tube. The wall of this tube may be regarded as a circular septum (26). In principle, more than one core could be introduced into a filament by employing a number of injection tubes. In practice, the production of matrix–filament fibers by the pipe-in-pipe technique uses the so-called combining-chamber method in which a number of core–sheath filaments are combined and narrowed in a conical chamber below the spinnerette holes before final extrusion, as shown in Figure 7 (15,27). Various techniques are used to produce tricomponent filaments (28). For example, pipe-in-pipe core–sheath filaments are formed, which pass through a second pipe for addition of a second sheath before a number of such filaments pass into a combining chamber. All pipe-in-pipe designs suffer from the fragility of the injection pipes and the general complexity leading to high cost and low output. Strength may be improved by eliminating the delicate injection tube and merely extruding the core from a hole above the spinnerette hole (29). Larger diameter tubes may be fitted into the spinnerette-hole counterbores in order to receive the core

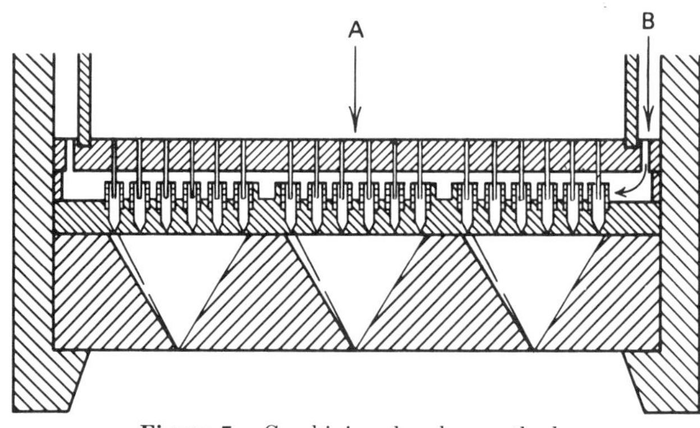

Figure 7. Combining chamber method.

streams from simple holes (30). In still simpler designs, the core is extruded from an internal spinnerette placed above the main spinnerette without fine tubes or passageways. This modification may be used for multisegmented hollow filaments (31) or side–side filaments (32) by suitably offsetting the core holes. This simple hole–above–hole method requires an equal pressure of sheath polymer at each spinnerette hole and an even flow into each hole to ensure uniformity.

The need for control over the sheath polymer has led to the castellation method (Fig. 8), in which the sheath polymer flow is controlled by passage through a fine gap between the upper and lower spinnerettes (33). The upper spinnerette is generally called the distribution plate. The gap is ca 100 μm and extends evenly over the spinnerette; consequently highly uniform fibers are produced. The thickness of the metering gap is influenced by the viscosity of the sheath polymer. If the spinnerette or the distributor plate holes are of noncircular cross section, the filaments, cores, or both can have noncircular cross sections (34–35). The core can also be placed eccentrically by creating an uneven sheath polymer flow (3) by setting it eccentrically relative to the spinnerette hole; making it noncircular; sloping the face; or recessing the distributor plate at one side of the hole. Because of its versatility, reliability, and ruggedness, the castellation method seems likely to increase in importance, especially if the spinnerette area can be reduced. To achieve this, chambers containing the castellations are arranged in more than one plane in an overlapping manner resulting in closer hole spacing (36). Alternatively, an orifice plate containing small holes is aligned over the spinnerette counterbores (37). This permits smaller castellations on the distributor plate, which are closer together than if they meet the counterbore directly. Both techniques approximately double the number of holes per unit area.

Matrix–filament fibers are manufactured by the side-injection technique with a septum device in which a tube connects the chamber containing one component to the spinnerette hole. Fine holes in the sides of this tube communicate with a second chamber containing the second polymer component. By the usual control of supply rates, a segmented matrix–filament is formed. By varying the number of side holes, or introducing fine tubes into the side holes to lead the second component away from the edge of the filament cross section, or introducing a coaxial pin into the tube, complex cross sections may be achieved.

Similar matrix–filament cross sections may be obtained by a version of the plate mixer technique described above. In this case, each spinnerette hole has its own plate stack and thus each filament is a composite stream of the required complexity (38). In practice, many foil plates are connected by webs in such a manner that an assembly for many spinnerette holes can be built up by stacking a few webbed plates. This technique offers the promise of considerable development. As in all matrix–filament spinning, excessive die-swell must be kept in mind.

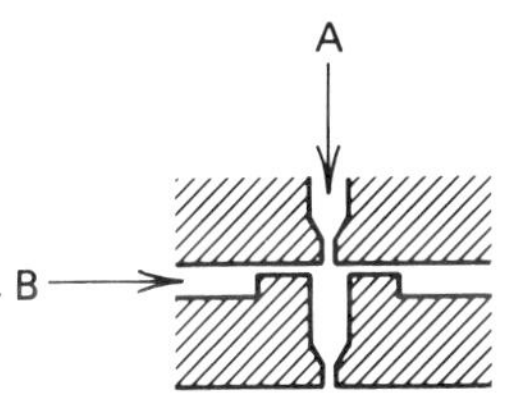

Figure 8. Castellation.

Economic Aspects

The total production of multicomponent fibers throughout the world is difficult to estimate, because the fiber may be used directly by the manufacturer or may not be distinguished in the production statistics from monocomponent fiber of a similar type. It is estimated that world production of bicomponent fibers, excluding acrylics and those for synthetic suede manufacture, amounted to ca 25,000 metric tons in 1981, mainly manufactured in western Europe with a small production in Japan. Japanese synthetic suede capacity was ca 8.4×10^6 m^2 per year (39) by the end of 1981. A small licensed production is based in Italy. Average bicomponent fiber consumption is estimated at 100 g/m^2 of suede. Thus the consumption of bicomponent fiber is 840 metric tons. Suede fabric weights vary between about 150 g/m^2 and 235 g/m^2 of which between 10 and 35% is nonfibrous polyurethane and the remainder may be microdenier (ca 0.1 tex or less per filament) fibers generated from matrix–filament fibers plus woven, knitted, or nonwoven base fabric composed of monocomponent or bicomponent fibers. Up to 80% of the production of the various synthetic suedes may be exported to the United States and western Europe.

Health and Safety Factors

In multicomponent fibers, the individual components contribute to the fibers' properties. Where the components are generally inert, nontoxic, and nonallergenic, the multicomponent fibers behave similarly. Where one or more of the components is not an established textile polymer (eg, a physiologically active agent), special considerations may apply, and each case must be studied individually. Information is usually available from the manufacturers.

Uses

By far the most important application for multicomponent fibers is the creation of desirable textile properties or aesthetic effects. Although initially the aim was to achieve a steric crimp similar to that of wool, much more sophisticated effects can now be achieved. This is often done by using the techniques described above to generate microdenier fibers (<0.1 tex per filament), sometimes of sharply defined cross section. Such fibers are too fine to be spun directly. Other effects may be obtained by drastically modifying the surface of bicomponent fibers with one unmodified component to provide strength. There are hundreds of patents in this area, many from Japan; only a few can be mentioned here. Reversibly crimpable acrylic fibers are described in references 22 and 40–47; reference 3 also cites many earlier patents. Crimpable nylon for hosiery is discussed in references 3, 48, and 49. The use of self-crimping polyester filaments in a feed yarn for draw-texturing is described in reference 24, self-crimping poly(butylene terephthalate)/poly(ethylene terephthalate) filaments in reference 32, and self-crimping products with one elastomeric component in reference 50.

Microdenier (<0.1 tex per filament) fibers formed from matrix–filament fibers may be utilized in various ways. The simplest is in the production of knitted or woven fabrics with appearance and handle different from those made using fibers in the normal 0.1–0.5 tex per filament range. A more silklike appearance is often claimed (15,28,51–57). For textured microdenier yarns it may be necessary to treat the fabrics

with a solvent to dissolve the matrix or to use special procedures to split the polymer junctions. In imitation fur fabric, tufts of microdenier (<0.1 tex per filament) fibers are formed from sea–island pile fabric (58). Flocked fabrics may also be produced (59). The most complicated application of bicomponent fibers is the preparation of washable synthetic suedes which have been very successful commercially (39,60). Such fabrics are based either on conventionally produced textiles or on nonwoven fabrics. The base material is impregnated with a resin, usually a polyurethane, and the matrix material is removed. Typical methods of producing artificial suedes or pile-surfaced fabrics are given in Figure 9 (16,27,31,62–65). An electron microscope photograph of the impregnated suede, Ecsaine, introduced by Toray in 1970, is shown in Figure 10.

In other applications one component serves as an adhesive. Core–sheath fibers are generally used in which the sheath has a melting point ca 20°C lower than the core. The products may be nonwoven fabrics. Applications range from textiles to three-dimensional fillings and foams for aircraft protection (66), to cigarette filters or fiber-tipped pens (67). The fiber–fiber bond may also be employed in the high speed manufacture of staple yarns (68) or in sizing false-twist yarns. If the adhesive action is employed to bond solid particles to fibers, so-called epitropic products are formed. For example, carbon black forms conducting filaments, which can be used for antistatic purposes or current-carrying applications such as heaters (69). Silica particles impart water repellency (70).

In a third group of applications, multicomponent fibers are used to modify the physical properties of textiles, especially for antistatic applications (11,71–80). Many systems attempt to combine both antistatic and humectant or hygroscopic features. Many variations are intended to reduce body cling in women's apparel. Some improve the flat-spotting of nylon tire cords (18), rubber adhesion (81), abrasion resistance (82), and reduce abrasiveness (83). Others change density (with a metal core) (84) and modify tensile properties (85).

A recently developed area is the modification of the chemical fiber properties. For example, one component could impart desirable tensile properties and another, desirable chemical properties. Thus, chemical resistance is improved by coating a poly(ethylene naphthalene-2,6-dicarboxylate) with poly(tetramethylene naphthalene-2,6-dicarboxylate). Similarly, the use of two aromatic polyamides to form a sheath–core fiber with excellent heat-, dyeing-, and flame-resistant properties is claimed (86). Flame resistance may also be imparted by decabromobiphenyl and tin oxide (87). These additives may be incorporated into the sheath after fabric production to avoid spinning problems. Strength and easy dyeing properties may be obtained with cellulose acetate in the partially hydrolyzed sheath (88). A three-layer sheath–sheath–core fiber may possess dyeability and polyester properties by using a polyester core and a saponifiable polyester copolymer sheath (89). Good heat resistance may be achieved in a sheath–core polyamide fiber with a high concentration of metal ions (90).

The inclusion of more than one component in a fiber may confer various optical properties. For example, a bichromic shot effect may be obtained with a fiber of flat cross section with the second component located at the rounded end portions (91). Moire effects are produced with matrix–filament fibers which have been treated to expose the embedded filaments. Unusual color effects may also be obtained by dyeing the microdenier (<0.1 tex per filament) fibers formed by completely removing the matrix from matrix–filament fibers (92). A sheath–core structure in which the sheath

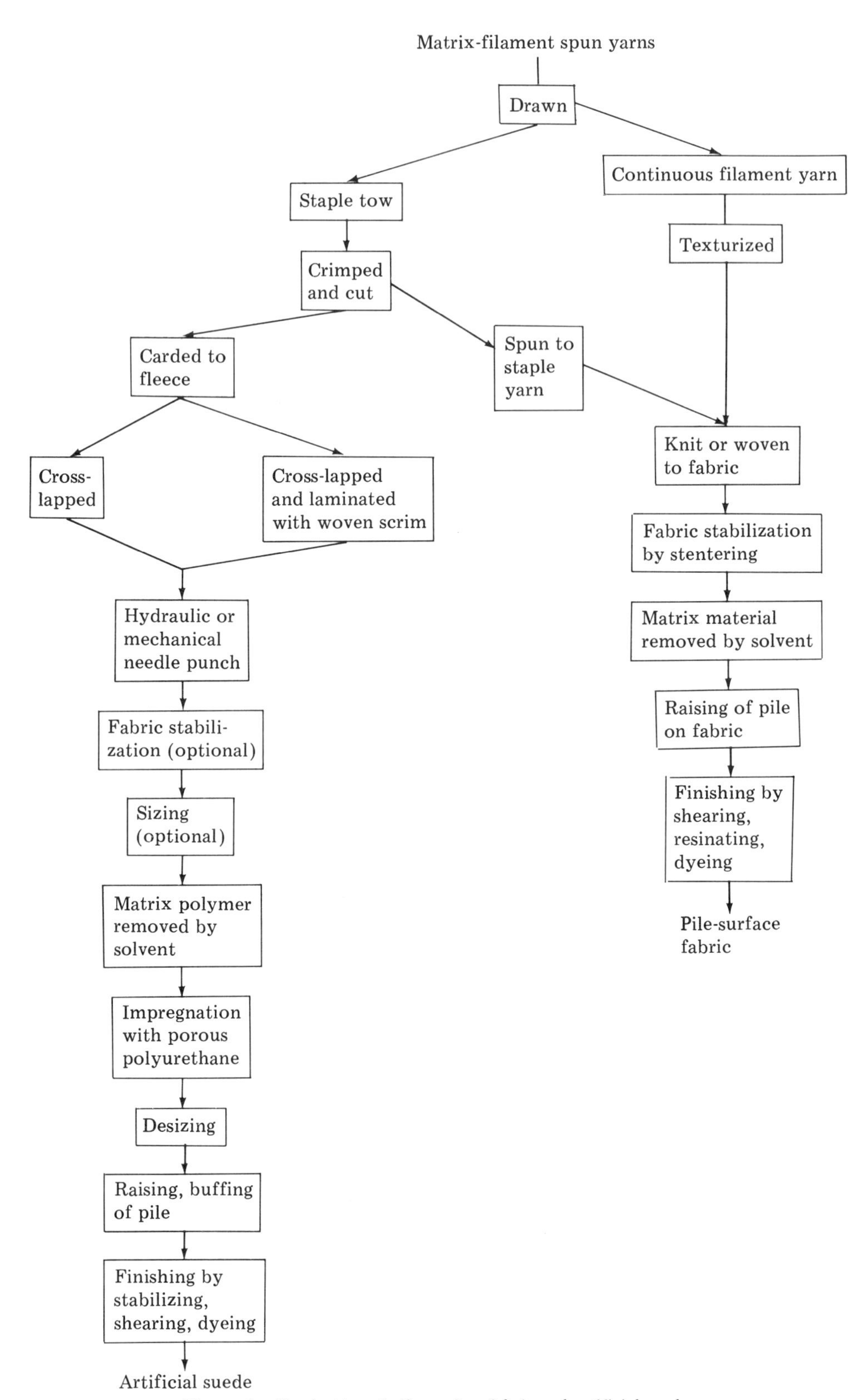

Matrix-filament spun yarns
Drawn
Staple tow
Continuous filament yarn
Crimped and cut
Texturized
Carded to fleece
Spun to staple yarn
Cross-lapped
Cross-lapped and laminated with woven scrim
Knit or woven to fabric
Fabric stabilization by stentering
Hydraulic or mechanical needle punch
Matrix material removed by solvent
Fabric stabilization (optional)
Raising of pile on fabric
Sizing (optional)
Finishing by shearing, resinating, dyeing
Matrix polymer removed by solvent
Pile-surface fabric
Impregnation with porous polyurethane
Desizing
Raising, buffing of pile
Finishing by stabilizing, shearing, dyeing
Artificial suede

Figure 9. Production of pile-surface fabric and artificial suede.

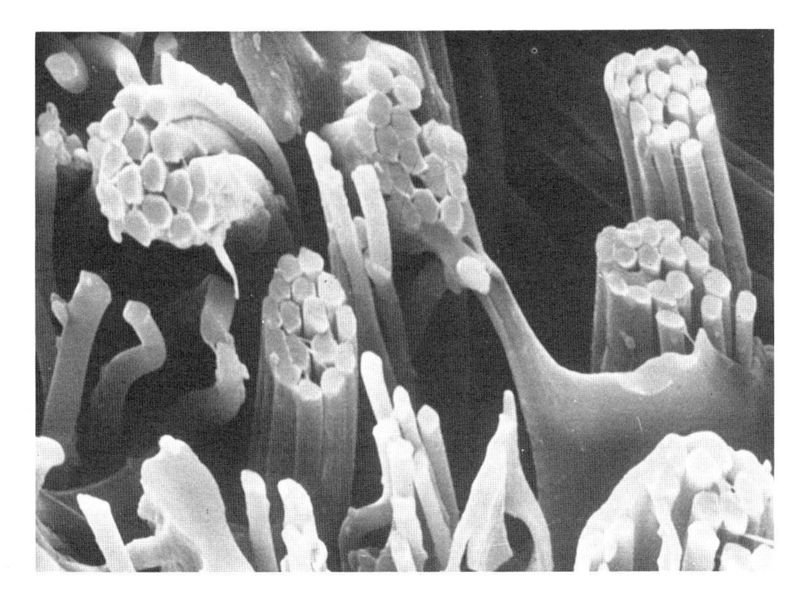

Figure 10. Impregnated suede (magnification 1000×).

possesses many hollow or uneven sections produces a silky luster (93). A nontransparent color effect is obtained from coloring material limited to the core of a sheath–core fiber (94).

More recently, multicomponent fibers have been proposed for use as containers or as diffusion barriers. For example, organic vapors and aerosols may be absorbed by filaments with a microporous sheath and an active sorbent core (95). In a reversal of this process, fibers with a pore-free core and a microporous sheath can be used to store and deliver detergents, wetting agents, bacteriocides, fungicides, etc (96). In a more sophisticated delivery system, sheath–core fibers with a core of progesterone and polycaprolactone and a sheath of polyethylene or polypropylene have been proposed as a long-term (5-year) female contraceptive device (97).

In a possible returning of interest to matrix–fibril fibers, another area of application for multicomponent fibers is in the rheological modification during melt-spinning. It has been recently observed that the presence of a small quantity of microfibers in the molten thread line modifies the drawing process with potential improvements in the processing, economics, or aesthetics of the fiber.

In addition, a glass–resin foam sheath–core fiber has been patented with enhanced aesthetic, chemical, and physical properties. Various three-component fibers are discussed in references 98–100. Bicomponent monofilaments are also proposed for various applications (101). Clearly, now that bicomponent fibers are readily accessible, more applications are likely to be developed using cheaper materials.

BIBLIOGRAPHY

1. P. A. Koch, *Ciba-Geigy Rev.* **1,** 1 (1974).
2. R. Jeffries, *Ciba-Geigy Rev.* **1,** 12 (1974).
3. R. Jeffries, *Bicomponent Fibres*, Merrow Publishing Co. Ltd., UK, 1971, pp. 2–4.
4. W. A. Sisson and F. F. Morehead, *Text. Res. J.* **23,** 152 (1953).
5. Brit. Pat. 1,004,251 (June 10, 1963), (to American Cyanamid Co.).

6. *Technical Information on Cashmilon GW Type (Bi-component)*, Asahi Chemical Industry Co. Ltd., Japan, 1976.
7. W. E. Fitzgerald and J. P. Knudsen, *Text. Res. J.* **37,** 447 (1967).
8. Brit. Pat. 1,425,705 (June 28, 1972), B. Boyes and A. Jones (to Imperial Chemical Industries Ltd.).
9. Brit. Pat. 808,213 (Nov. 30, 1955), R. H. Speakman and R. B. Macleod (to Imperial Chemical Industries Ltd.).
10. P. V. Papero, E. Kubu, and L. Roldan, *Text. Res. J.* **37,** 823 (1967).
11. U.S. Pat. 3,582,448 (Apr. 23, 1968), T. Okuhashi and K. Kumura (to Teijin Ltd.).
12. U.S. Pat. 4,325,857 (May 13, 1980), N. Jivanial and co-workers (to E. I. du Pont de Nemours & Co., Inc.).
13. Brit. Pat. 2,009,032 (Nov. 25, 1977), J. Venot and A. Mottet (to ASA S.A.).
14. Brit. Pat. 2,010,739 (Dec. 22, 1977), P. Chion and co-workers (to Rhone-Poulenc-Textile).
15. U.S. Pat. 3,531,368 (Jan. 7, 1966), M. Okamoto and co-workers (to Toyo Rayon Kabushiki Kaisha).
16. Brit. Pat. 1,514,430 (July 14, 1975), J. Nakagawa, T. Agaki, and K. Hiramatsu (to Kuraray Co., Ltd.).
17. Brit. Pat. 1,045,047 (Feb. 20, 1963), (to Mitsubishi Rayon Co., Ltd.).
18. U.S. Pat. 3,641,232 (Aug. 28, 1967), W. J. Fontijn and K. J. M. van Drunen (to American Enka Corp.).
19. Brit. Pat. 1,100,430 (Dec. 16, 1965), W. G. Parr and A. W. D. Hudgell (to Imperial Chemical Industries Ltd.).
20. Brit. Pat. 2,045,153 (Mar. 27, 1979), J. E. Bromley and J. P. Yu (to Monsanto Co.).
21. Brit. Pat. 1,004,251 (June 10, 1963), (to American Cyanamid Co.).
22. U.S. Pat. 3,038,237 (Nov. 3, 1958), R. B. Taylor, Jr. (to E. I. du Pont de Nemours & Co., Inc.).
23. Brit. Pat. 954,272 (July 14, 1961), (to Japan Exlan Có. Ltd.).
24. Eur. Pat. 0 001 880 (Sept. 16, 1977), M. E. Mirhej (to E. I. du Pont de Nemours & Co., Inc.).
25. Brit. Pat. 1,048,370 (Sept. 26, 1964), (to Kanegafuchi Boseki Kabushiki Kaisha).
26. Brit. Pat. 805,033 (Feb. 26, 1954), (to E. I. du Pont de Nemours & Co., Inc.).
27. Brit. Pat. 1,300,268 (June 4, 1970), (to Toray Industries, Inc.).
28. Brit. Pat. 2,057,344 (Sept. 7, 1979), M. Okamoto and M. Asada (to Toray Industries, Inc.).
29. Brit. Pat. 1,258,760 (Dec. 21, 1967), (to Kanegafuchi Boseki Kabushiki Kaisha).
30. Brit. Pat. 1,302,584 (June 23, 1970), M. Okamoto and co-workers (to Toray Industries, Inc.).
31. Brit. Pat. 1,502,360 (Dec. 12, 1974), (to Teijin Ltd.).
32. Brit. Pat. 1,546,021 (Feb. 27, 1976), G. Barbe and R. Habault (to Rhone-Poulenc-Textile).
33. Brit. Pat. 1,095,166 (Nov. 18, 1964), A. W. D. Hudgell (to Imperial Chemical Industries Ltd.).
34. Brit. Pat. 1,120,241 (July 27, 1965), M. Matsui (to Kanegafuchi Boseki Kabushiki Kaisha).
35. Brit. Pat. 1,066,418 (Nov. 27, 1964), (to E. I. du Pont de Nemours & Co., Inc.).
36. Brit. Pat. 1,565,155 (Nov. 25, 1977), (to Monsanto Co.).
37. Eur. Pat. 0 011 954 (Nov. 30, 1978), P. C. Parkin (to Imperial Chemical Industries Ltd.).
38. Brit. Pat. 2,024,096 (June 19, 1978), J. Bohler (to Akzo, N.V.).
39. *Jpn. Text. News* (327), 56 (Feb. 1982).
40. U.S. Pat. 3,038,236 (Nov. 3, 1958), A. L. Breen (to E. I. du Pont de Nemours & Co., Inc.).
41. U.S. Pat. 3,038,238 (Nov. 20, 1958), T. C. Wu (to E. I. du Pont de Nemours & Co., Inc.).
42. U.S. Pat. 3,038,239 (Mar. 16, 1959), G. M. Moulds (to E. I. du Pont de Nemours & Co., Inc.).
43. U.S. Pat. 3,038,240 (Feb. 2, 1960), F. J. Kovarik (to E. I. du Pont de Nemours & Co., Inc.).
44. U.S. Pat. 3,039,524 (Nov. 3, 1958), L. H. Belck and K. G. Siedschlag, Jr. (to E. I. du Pont de Nemours & Co., Inc.).
45. U.S. Pat. 3,092,892 (Apr. 10, 1961), J. F. Ryan, Jr. and R. L. Tichenor (to E. I. du Pont de Nemours & Co., Inc.).
46. U.S. Pat. 3,470,060 (Feb. 2, 1966), J. Szitz, A. Nogaj, and H. Marzolph (to Farbenfabriken Bayer Aktiengesellschaft).
47. U.S. Pat. 3,473,998 (Aug. 7, 1963), D. R. Spriestersbach and co-workers (to E. I. du Pont de Nemours & Co., Inc.).
48. Brit. Pat. 1,478,101 (July 31, 1974), W. B. Seagraves and K. L. Mullholland (to E. I. du Pont de Nemours & Co., Inc.).
49. Brit. Pat. 2,007,588 (Nov. 10, 1977), C-M. Cerutti and co-workers (to Rhone-Poulenc-Textile).
50. Brit. Pat. 1,518,500 (Nov. 26, 1974), C. M. Bowes (to Courtaulds Ltd.).
51. Brit. Pat. 1,584,313 (Oct. 6, 1976), (to Toray Industries, Inc.).

52. Brit. Pat. 2,015,421 (Dec. 23, 1978), K. Gerlach, N. Mathes, and F. Wechs (to Akzo N.V.).
53. Brit. Pat. 2,043,731 (Mar. 2, 1979), H. W. Bruecher, K-H. Hense, and R. Modtler (to Akzo N.V.).
54. Brit. Pat. 2,062,537 (Nov. 9, 1979), M. Okamoto (to Toray Industries, Inc.).
55. Eur. Pat. 0 44 221 (July 14, 1980), J. T. Summers (to E. I. du Pont de Nemours & Co., Inc.).
56. U.S. Pat. 4,122,658 (May 10, 1977), A. Morioka and T. Nakashima (to Toray Industries, Inc.).
57. Jpn. Pat. 55 116,873 (Feb. 26, 1979), (to Kanebo Kabushiki Kaisha).
58. Eur. Pat. 0 045 611 (Aug. 4, 1980), (to Toray Industries, Inc.).
59. Jpn. Pat. 56 107,080 (Jan. 24, 1980), (to Toray Industries, Inc.).
60. *Jpn. Text. News* (315), 18 (Feb. 1981).
61. Brit. Pat. 1,321,852 (July 1, 1969), (to Toray Industries, Inc.).
62. Brit. Pat. 1,496,369 (Dec. 25, 1974), (to Kabushiki Kaisha Kuraray).
63. Brit. Pat. 1,514,553 (Sept. 13, 1974), T. Kusunose and co-workers (to Asahi Kasei Kogyo Kabushiki Kaisha).
64. Brit. Pat. 1,541,774 (July 12, 1976), (to Mitsubishi Rayon Company Ltd.).
65. Brit. Pat. 2,004,496 (Sept. 6, 1977), K. Ozaki, M. Matsui, and N. Minemura (to Teijin Ltd.).
66. K. Porter, *Melded Fabrics—A Versatile And Unique New Family of Materials*, to be published, *Materials in Engineering*, Scientific and Technical Press Ltd., UK, 1982.
67. Brit. Pat. 2,036,115 (Nov. 15, 1978), T. Sugihara and H. Sonoda (to Chisso Corporation).
68. Brit. Pat. 1,495,546 (Mar. 21, 1974), P. L. Carr and P. M. Ellis (to Imperial Chemical Industries Ltd.).
69. Brit. Pat. 1,417,394 (Feb. 3, 1972), V. S. Ellis and K. W. Mieszkis (to Imperial Chemical Industries Ltd.).
70. Brit. Pat. 1,488,682 (Mar. 11, 1974), B. Walker and D. B. Chambers (to Imperial Chemical Industries Ltd.).
71. Brit. Pat. 1,502,719 (May 27, 1975), (to Monsanto Co.).
72. Brit. Pat. 2,001,901 (Aug. 8, 1977), T. Naruse and co-workers (to Kanebo Kabushiki Kaisha).
73. Brit. Pat. 1,393,234 (July 21, 1972), D. R. Hull (to E. I. du Pont de Nemours & Co., Inc.).
74. Brit. Pat. 1,527,192 (Jan. 12, 1976), (to Fiber Industries, Inc.).
75. Brit. Pat. 1,585,575 (Apr. 29, 1977), G. A. Paton, S. M. Nichols, and J. H. Sanders (to Dow Badische Co.).
76. Brit. Pat. 2,003,084 (July 7, 1977), M. Wandel and co-workers (to Bayer Aktiengesellschaft).
77. Brit. Pat. 2,036,638 (Nov. 23, 1978), S. J. Van Der Meer, W. Peschke, and D. Schilo (to Akzo N.V.).
78. U.S. Pat. 3,582,445 (Nov. 18, 1967), T. Okuhashi (to Teijin Ltd.).
79. U.S. Pat. 3,586,597 (Nov. 20, 1967), T. Okuhashi (to Teijin Ltd.).
80. U.S. Pat. 3,590,570 (Jan. 2, 1969), T. Okuhashi and K. Komura (to Teijin Ltd.).
81. Ger. Pat. 3,022,325 (June 15, 1979), (to Teijin Kabushiki Kaisha).
82. Jpn. Pat. 56 144,218 (Apr. 8, 1980), (to Kureha Chemical Industries Kabushiki Kaisha).
83. Jpn. Pat. 55 158,331 (May 25, 1979), (to Toray Industries, Inc.).
84. Jpn. Pat. 50 118,018 (Mar. 12, 1974), (to Toray Industries, Inc.).
85. Brit. Pat. 1,574,220 (Apr. 7, 1976), S. Yoshikawa, T. Sasaki, and H. Endo (to Kureha Kagaku Kabushiki Kaisha).
86. Eur. Pat. 0 018 523 (Apr. 24, 1979), (to Teijin Ltd.).
87. Brit. Pat. 2,077,310 (May 23, 1980), I. S. Fisher and S. R. Munks (to Imperial Chemical Industries Ltd.).
88. Brit. Pat. 1,601,585 (June 30, 1977), A. L. Heard, P. D. Randall, and D. J. Waters (to Courtaulds Ltd.).
89. Jpn. Pat. 55 132,716 (Apr. 2, 1979), (to Toray Industries, Inc.).
90. Jpn. Pat. 57 047,915 (Aug. 30, 1980), (to Asahi Chemical Industries Kabushiki Kaisha).
91. Jpn. Pat. 56 144,219 (Apr. 7, 1980), (to Kuraray Kabushiki Kaisha).
92. Jpn. Pat. 56 085,474 (Dec. 5, 1979), (to Toray Industries, Inc.).
93. Jpn. Pat. 56 031,015 (Aug. 14, 1979), (to Toray Industries, Inc.).
94. Jpn. Pat. 55 062,210 (Oct. 31, 1978), (to Teijin Kabushiki Kaisha).
95. U.S. Pat. 4,302,509 (June 6, 1980), M. J. Coplan and G. Lopatin (to Albany International Corp.).
96. Eur. Pat. 0 047 797 (Sept. 15, 1980), (to Carl Freudenberg F.A.).
97. R. L. Dunn and co-workers, *Controlled Release Pestic. Pharm., Proceedings of International Symposium*, Plenum, New York, 1980, pp. 125–146.
98. Jpn. Pat. 55 093,813 (Jan. 1, 1979), (to Unitika Kabushiki Kaisha).
99. Jpn. Pat. 55 148,214 (May 1, 1979), (to Asahi Chemical Industries Kabushiki Kaisha).

100. Jpn. Pat. 56 015,416 (July 12, 1979), (to Unitika Kabushiki Kaisha).
101. Ger. Pat. 2,713,435 (Apr. 7, 1976), (to Kureha Kagaku Kogyo).

K. PORTER
ICI Fibres

FINE ART EXAMINATION AND CONSERVATION

SCIENTIFIC EXAMINATION

In few interdisciplinary fields have two so apparently different disciplines been brought together as in the application of physical sciences to the study of art objects (1–3). The scientist, accustomed to think in terms of abstractions, is trained to approach a stated problem by analyzing it for the identification of measurable variables, and devising means to obtain numerical values for the latter; on the other hand, the art historian relies on the trained eye, which enables the visual recognition of stylistic characteristics and the mental comparison of these with observations about numerous other art works, an intellectual process of which the procedural mechanism does not allow for an easy explanation other than "you have to see it."

Practitioners of the two disciplines often have had difficulties in understanding each other's problems and solutions of these, and the lack of communication has in the past sometimes led to wasted effort and mistrust on both sides. Remedying this problem has required mutual efforts, and the fact that this field may now truly be regarded as having come of age must be credited to those art historians who have made themselves familiar with the principles of scientific methodology, as well as to the scientists who have taken an actual, professional interest in the humanities, realizing that this is only possible by learning the language and systematics of their counterparts.

All this does not imply that scientific examination of art objects is a completely new field of study. It is, however, in the twentieth century that the number of technological studies in art have grown virtually exponentially.

One can surmise several reasons for this phenomenon. First, the great museums had to be established, bringing together the large collections for the academic purposes of study and education rather than for the personal aesthetic gratification of a private owner. Once this had happened, it did not take long for the first museum-based laboratories to be established; the first such laboratory was founded in 1888 at the Royal Museum of Berlin, followed by the British Museum during World War I, and the Boston Museum of Fine Arts in 1929.

Second, the explosive development in analytical techniques, especially during the second half of this century, has given birth to many instrumental methods which, through high sensitivities and small sample requirements, are extremely appropriate for the study of art objects. For example, metal-alloy analysis has a long history of application in numismatics, where the evaluation of relative values of contemporary coinages depends on the knowledge of their intrinsic values. With the traditional wet chemical analysis, this often meant the loss of half a coin. Although the analytical results are certainly very accurate and precise, and hence allow meaningful and valuable conclusions, the reluctance of a curator to sacrifice an irreplaceable part of humanity's cultural heritage for this purpose is not only understandable, but justified. In the museum, responsibility for the objects in the collection belongs to the curator, who must constantly be aware of the goals of a museum: to further scholarly study of the material in the collection; to use it for educational purposes through the organization of exhibitions; and, possibly the most important mandate, to preserve it for future generations so that they, too, can see and study the cultural records from the past. Because of the extreme importance of the proper preservation and prevention of damage, a consequence of the irreplaceability of the objects, it is absolutely necessary that any study involving the removal of even the tiniest sample be discussed by the curator and scientist in order to determine the goals of the proposed project; the value of the results, especially in furthering the understanding of the object and its historical context; and the extent of the damage which will be inflicted. The latter has to be balanced carefully against the benefits to be derived from the study. There are several possible reasons why a scientific study of an art work may be desirable. An obvious one is in cases where the authenticity of an object is doubted on stylistic grounds but no unanimous opinion exists. The scientist will identify the materials, analyze their chemical composition, and investigate whether this corresponds to what has been found in comparable objects of unquestioned provenance. If the sources for the materials can be characterized, eg, through trace element composition or structure, it may be possible to determine whether the sources involved in the procurement of the materials for comparable objects with known provenance are the same. Comparative examination of the technological processes involved in the manufacture allows for conclusions whether the object was made with techniques actually used by the people who supposedly created it. Finally, dating techniques may allow the establishment of the date of manufacture.

Several interesting cases of well-known forgeries are discussed in ref. 4. Probably the most notorious case is that of the forger Han van Meegeren, who faked works by Vermeer and de Hoogh. Several of his works were acquired by reputable Dutch museums, and one was sold to Hermann Goering during the German occupation of the Netherlands. When, after the war, van Meegeren was arrested and charged with collaboration, he claimed his innocence, protesting that this painting, and many others, were not originals, but rather forgeries by his hand. These claims were hotly contested, and he had to prove, in captivity, that he could indeed fake Vermeer's style so well as to deceive the stylistic experts. The technical examination which proved these works to be forgeries was the work of Paul Coremans (5). Later, ^{210}Pb dating confirmed the recent date of manufacture (6).

A famous case of fake ceramic objects is that of the large Etruscan Warriors in the Metropolitan Museum of Art. The proof of nonauthenticity was provided by a spectrochemical analysis of the glaze. Although Greek and Etruscan artists applied the decoration using a slip of essentially the same clay as the one from which the object was crafted, and obtained the red and black colors through a special cycle of oxidizing- and reducing-firing conditions, the black color on the Warriors was found to be the result of the use of a manganese black glaze (7).

The Drake Plate is a brass plate, supposedly left behind in 1579 by Sir Francis Drake in California, claiming the land for the British crown. Discovered in 1937, it has since been in the custody of the University of California at Berkeley. Recently, chemical analysis has found that both the principal and trace-element compositions of this plate are incompatible with the attribution (8–9).

The Vinland Map, supposed evidence of the Viking discovery of America long before Columbus, was proven a forgery through the detection of the modern pigment titanium white (10).

The famous bronze Greek horse in the Metropolitan Museum of Art was declared a forgery in 1967, after an examination which led to the conclusion that the object was cast by means of a sand-casting technique, unknown to the ancient Greeks (11). Subsequent technical studies by a team of experts have, however, not only negated all arguments brought against the sculpture, but even authenticated the bronze by means of thermoluminescence dating (12).

Though the cases in which scientific analysis is called upon to assist in the authentication process tend to draw the most attention and have, occasionally, spectacular results, they are by no means the most important applications of the scientific method in the study of art works. Most fruitful and satisfying are those truly interdisciplinary projects in which an object or group of objects, of unquestionable provenance, is studied jointly by the historian and the scientist, in order to better evaluate its historical context. Classifications according to material and technical properties may refine stylistic differentiations, and geographic attributions may lead to the evaluation of past cultural cross-fertilizations (13–15). Without the study of objects of unquestionable attribution, a judgment which generally only the stylistic expert can make, the criteria used in authenticity studies could not be established.

Probably of more direct interest to the scientist than the art historian, but nevertheless of value in the larger framework of the humanities, is the study of the history of technology (16). Generally, this part of the historical record has been overlooked in the past, but affords profound insights into the history of the development of civilization. The earliest existing written records, treatises of craftspeople and artists on the techniques and materials with which they worked, date back to medieval times (17–20). For the millenia of human activities before that time, the record is in the objects which remain, and only through the study of these can knowledge in this regard be furthered.

Last, there is a very practical reason for the need of scientific support in the preservation of art objects (21). In order for the preservation specialist (conservator) to be able to arrest decay processes at work in an object, an understanding of these processes must be reached through study of the material properties and identification of the factors that influence the deterioration process. Subsequently, an evaluation of possible treatments with regard to their potential effectiveness, both immediate and long term, as well as other possible consequences, is made. For this, the active involvement of scientists is necessary and has become, to some degree, standard practice.

Methodologies

Optical Techniques. While stylistic experts perform their observations with the naked eye and under normal illumination, the eye is often aided by optical devices and special illuminations in the laboratory. The most important tool in a museum laboratory is the low power stereomicroscope. This instrument, usually used at magnifications of 3–50 ×, has enough depth of field to be useful for the study of surface phenomena on many types of objects without the need for sample preparation (see Fig. 1).

The information thus obtained can relate to toolmarks and manufacturing

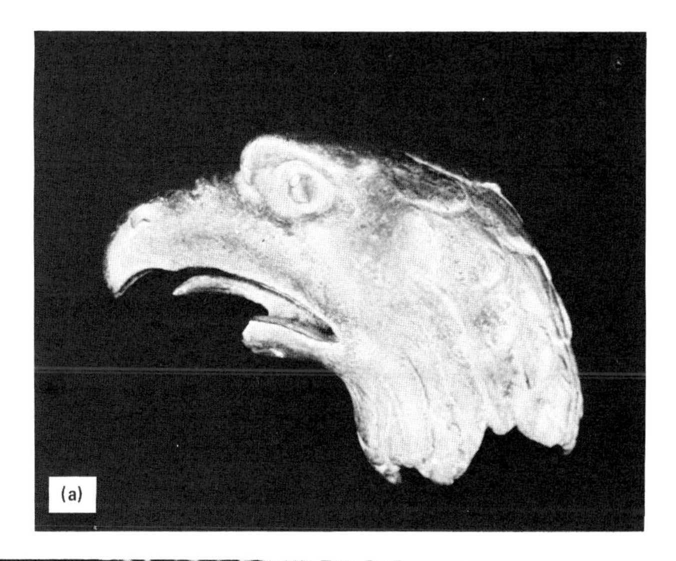

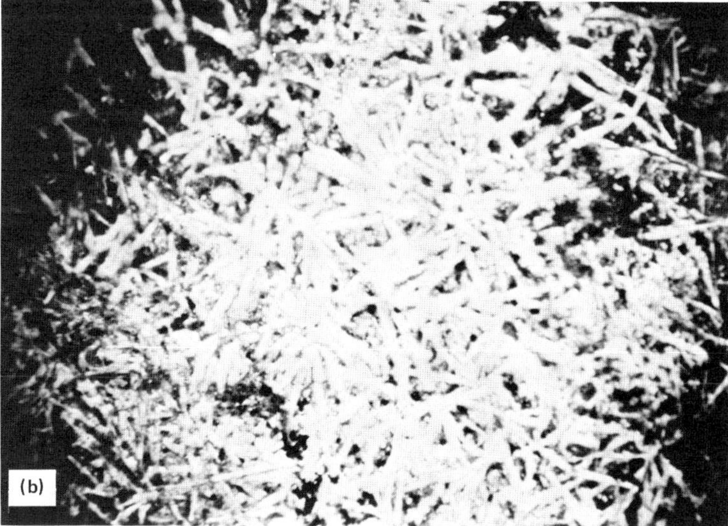

Figure 1. (**a**) A bronze hawk's head, forgery of a Roman finial. (**b**) The patina, studied through a stereomicroscope, shows a structure typical for an artificially induced corrosion, probably effected by pickling. Courtesy of the Research Laboratory, Museum of Fine Arts.

techniques, wear patterns, the structure of corrosion, artificial patination techniques, the structure of paint layers, or previous restorations. Any art object coming into a museum laboratory will, first of all, find its way to this microscope. Of course, it may often be necessary to observe phenomena on a scale too small for the use of these low magnifications. When the higher magnifications obtainable with the polarizing research microscope thus become necessary, samples have to be removed and prepared according to the microscopic technique to be used. Reflected light microscopy, for example, may be useful in the study of a polished metal specimen to determine the in-depth structure of the corrosion or to examine a cross-sectional sample from a painting for the sequence of paint layers (22).

Transmitted light microscopy is the preferred tool in the study of thin sections from stone or ceramic objects, eg, for the examination of the crystal structure of a marble for clues to its geographic provenance, or the technology involved in the forming and decoration of a ceramic vessel.

The research microscope is also of great importance in the morphological identification of materials used in the making of art objects, such as pigments, fibers, or woods.

The very high powers of magnification afforded by the electron microscope, either scanning electron microscopy (SEM) or scanning transmission electron microscopy (STEM), are in great demand for identifications, such as of exact wood species; in technological studies of ancient metallurgical techniques or ceramics; and especially in the study of deterioration processes taking place in various types of art objects. Examination under other than visible light illumination has thus already been introduced. Illuminations with radiations adjacent to those of visible light are frequently used. Ultraviolet light has been in use for a long time in the examination of paintings and other objects, especially for the detection of repairs and restorations (23) (see Fig. 2). Because of the variations in fluorescent behavior of different materials with otherwise similar optical properties, areas of repaint, inpainting, or replacements of losses with color-matched filling materials often can be observed easily. Also, for many materials, the appearance of their fluorescence under uv illumination changes with time as a result of various chemical aging processes, thus providing a means of distinguishing fresh and older surfaces (2,24). Infrared irradiation is often used in the examination of paintings because, owing to the limited absorption by the organic medium, it can penetrate deeply into the paint layers. It is reflected in varying degrees by different pigments. This may enable the detection of changes in composition, called pentimenti; restorations; or underdrawings, ie, working drawings applied by the artist on the prepared ground surface. Infrared-sensitive photographic films or vidicon cameras are used to transform the reflected ir image into a visible one (25–26).

Structural Analysis. Some of the techniques mentioned above qualify for inclusion in this section as well. Microscopic examinations of metallurgical cross sections or of sections through the paint layers of a painting are indeed structural examinations, as is ir reflectography.

The most well-known structural examination technique is probably x-radiography (27), and certainly this represents one of the earliest applications of scientific techniques in the examination of art objects (28–29) (see X-ray techniques).

In the case of paintings, the x rays are primarily absorbed by the heavy metals present in some pigments, ie, mainly by the lead in the pigment lead white, which has historically been used as a principal pigment, both by itself and mixed with other colors (see Pigments). Other pigments containing heavy elements, such as lead–tin yellow

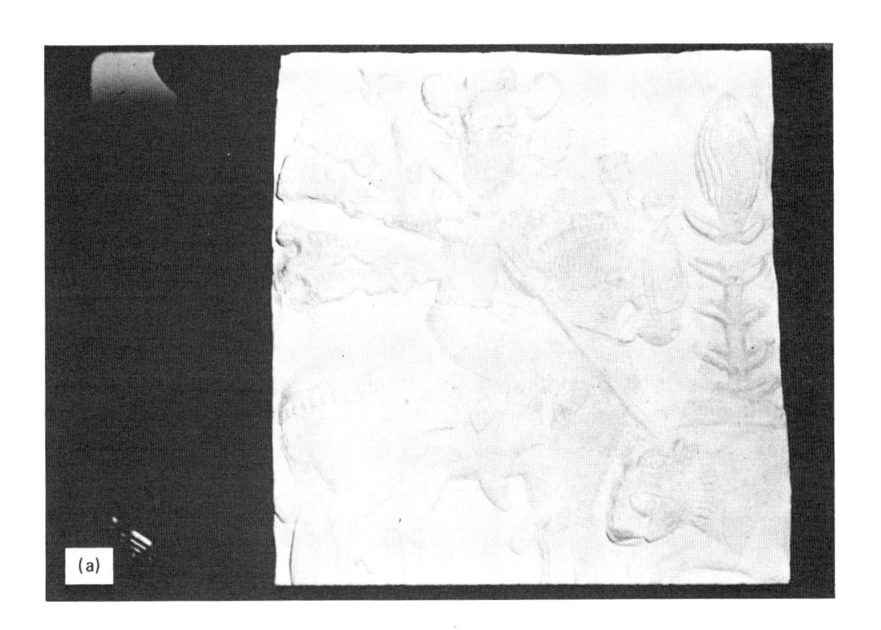

Figure 2. (**a**) A Sassanian stucco relief, photographed under normal light. (**b**) The same object photographed under uv illumination. Extensive repairs and fills show up as dark areas in this figure. Courtesy of the Research Laboratory, Museum of Fine Arts.

and vermilion (mercury sulfide), are also shown in high contrast. Because a contrast between the pigments (the heavy elements) and the support (usually organic materials) is desired, in addition to a large gradation in contrast corresponding to the variation in thickness of the various pigment applications, the conditions under which paintings are radiographed differ distinctly from those normally found in medical applications. This is reflected in the voltages applied over the x-ray tubes: 15–35 kV for canvas paintings and up to ca 50 kV for panel paintings or icons.

X-radiographs of paintings yield information regarding the development of the composition, eg, the blocking out of certain compositional elements, and changes therein, as well as the technique of the artist.

X-radiography is also used extensively in the examination of other types of art objects, in order to obtain information regarding manufacturing techniques or repairs. For example, joining techniques in wooden objects such as furniture or sculpture can be studied; in a ceramic vase, repaired and consecutively overpainted breaks show clearly.

The application of radiography to metal objects is of special interest (30). Using industrial radiographical equipment with tube voltages of 200–300 kV, or even γ-ray sources, information relating to shaping techniques (eg, hammering, raising, repoussé, lost wax, or piece-mold casting) or joining techniques (eg, soldering, welding, or crimping) can be obtained, thereby revealing important details about the technology involved in the fabrication of the object.

In the examination of works of art on paper, the variations in density are often too small to be revealed by conventional x-radiography. Instead, much benefit has been derived from beta-radiography, where the β radiation is provided by an extended radioisotope source, typically a sheet of plastic in which a certain amount of ^{14}C is incorporated homogeneously through labeling on the monomer (see Radioisotopes).

A variation on the conventional technique of x-radiography is xeroradiography, where imaging is obtained by electrostatic rather than photographic means. Because of special merits such as edge enhancement and improved texture rendition, this technique has certain advantages, eg, in the study of ceramics (qv) (31).

A recently developed radiographical technique, applied successfully to paintings, is neutron activation autoradiography (32) (see Fig. 3). A painting is exposed to a flux of thermal neutrons produced in a nuclear reactor. These neutrons, which penetrate the painting completely, interact with a small fraction of the nuclei of the various chemical elements present in the painting. Radioactive isotopes are produced, each with its own characteristic half-life. The radiation, especially β and x rays, produces a blackening in a photographic emulsion exposed to it. Thus, by placing a sheet of photographic film in close contact with the painting, a so-called autoradiograph is obtained which shows the distribution of the predominant activities within the painting at the time of the exposure of the film. Owing to their half-lives, different radioisotopes show a dominant activity level at different times. Therefore, a series of exposures made consecutively after activation yields distribution patterns within the painting of various radioisotopes, and hence of their original elements. A fortunate coincidence is that lead, because of its nuclear properties, does not show on these autoradiographs. Thus, the information obtained relates to a number of pigments other than lead white, and complements rather than duplicates that obtained from x-radiography.

Figure 3. (**a**) "St. Rosalie Interceding for the Plague-Stricken of Palermo" by Anthony van Dyck. The Metropolitan Museum of Art, Purchase, 1871. (**b**) An x-radiograph of this painting reveals the presence of a head under the painting's lower half. (**c**) Autoradiograph, taken two days after activation. Notice the differences compared with the final composition, especially the absence of the putto above the Saint's head. (**d**) Autoradiograph, taken eight days after activation. The underpainting is revealed to be the portrait of a young man. It compares closely with a self portrait of the artist, suggesting that van Dyck may have used an unfinished self portrait as support for the commissioned portrait of the Saint. (**b**)–(**d**) Courtesy of the Metropolitan Museum of Art and Brookhaven National Laboratory. (**e**) Self portrait, by Anthony van Dyck. The Metropolitan Museum of Art, Jules Bache Collection, 1949.

Chemical Analysis. Virtually any technique for chemical analysis, whether qualitative or quantitative, elemental or molecular, organic or inorganic, has been applied in order to solve a wide variety of questions relating to art objects. In authentication studies, the pigments identified in a painting or the alloy composition determined for a metal object may be compared with data obtained from objects of unquestioned provenance, and thus provide evidence with regard to the authenticity of the object under investigation. Similar measurements, when performed on well-provenanced objects, provide the reference information needed for comparison in authentications. They also often lead to inferences with regard to the technology involved in the manufacture of the object, and thus the date and place of introduction of these techniques. Trace-element concentrations of many materials indicate the geographic origins of these materials, and trade and exchange relationships. Identification of deterioration products clarifies questions regarding the decay processes acting on an object, enabling the conservator to design an effective treatment and prevent further deterioration.

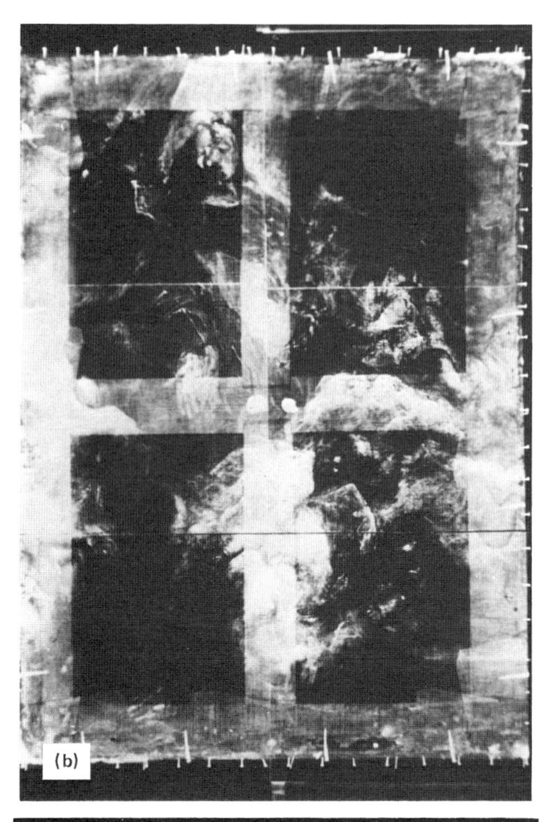

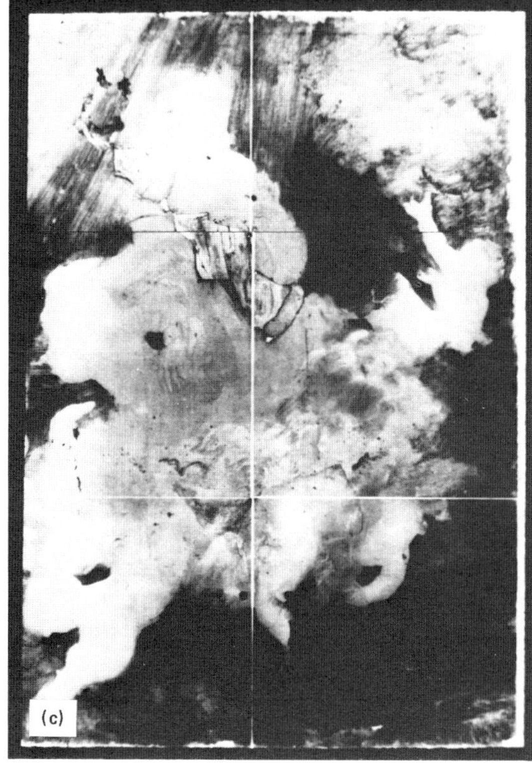

Figure 3. (*continued*)

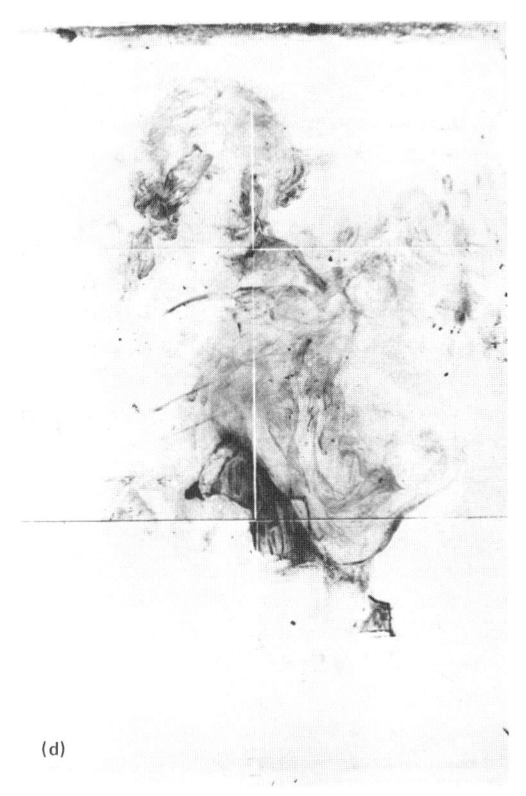
(d)

(e)

Figure 3. *(continued)*

The arguments brought to bear in the selection of an analytical technique for a specific problem are virtually the same as in any other application area of analytical chemistry: information wanted, sensitivity and precision needed, matrix, etc. A very important additional consideration, not often relevant in other fields, is the required sample size. Because works of art and archaeological artifacts must be respected as irreplaceable parts of humanity's cultural heritage, the damage resulting from the removal of a sample for analytical study has to be weighed carefully against the benefits expected from this study. Clearly, techniques that require smaller samples are preferred. The developments in analytical techniques, with the advent of extremely sensitive instrumentation, have been of great significance (see Analytical methods). However, the same problem that accompanies all analyses of very small samples, ie, the accurate representation of the whole object by a small sample, is generally more severe in the study of antiquities. Ancient technologies generally produced materials of a much greater compositional heterogeneity than the results of today's closely controlled industrial processes. The same holds true for many materials used directly as they occur in nature, such as stones, clays, etc. Therefore, often the sample size is dictated not so much by the requirements of the analytical techniques available as by the nature of the material under study.

Another complication that is quite typical for this type of work lies in the lack of proper reference standards. This is, again, because of the age of the materials in question. A typical example is the identification of organic materials such as paint media or natural dyestuffs. In these materials, slow chemical reactions, over a span of time, generate a number of reaction products that do not occur in newly prepared standard materials; thus, the identification of the original material becomes more complicated. In some cases, the chemical behavior of a material may be changed to such a degree that direct comparison with a fresh standard material becomes impossible. One partial solution is in the artificial, accelerated aging of standard materials; however, because such aging rarely yields the identical mixture of reaction products as the natural processes, the value of such standards must be assessed cautiously for each application. Occasionally, chemical alterations have progressed to such an extent that the analysis yields an identification of the degradation product rather than the original material. In such cases, only the experience of the analyst prevents the drawing of misleading conclusions.

In a number of cases, identifications have been extremely difficult, in which the materials were synthetic substances whose existence had actually become forgotten. Thus, several rather commonly encountered synthetic pigments, such as the lead–tin yellow often found in Renaissance paintings, were originally misidentified or left unidentifiable until extensive research, including analyses of elemental composition and chemical and physical properties, and replication experiments, led to proper identification of the material and its manufacturing process.

Dating. The usefulness of dating techniques in the study of art objects, either for authentication purposes or in the evaluation of their historical context, is self evident (33–34).

Radiocarbon dating (35) has probably gained the widest general recognition. Developed by W. F. Libby in the late 1940s, it depends on the equilibrium between the formation of the radioactive isotope ^{14}C and its decay, with a half-life of 5730 yr. After its formation in the upper stratosphere, through nuclear reactions of nitrogen nuclei and neutrons produced by cosmic radiation, the ^{14}C is rapidly mixed into the carbon-exchange reservoir, ie, oceans, atmosphere, and biosphere. Thus, a fairly constant concentration of radiocarbon, ca 1 part ^{14}C per 10^{12} parts ^{12}C, exists in the exchange reservoir. If, however, materials from dead plants or animals are used in the manufacture of an artifact, no exchange takes place, and therefore the radioactive decay process results in a continual decrease of the ^{14}C concentration in the material. The time elapsed since the death of the plant or animal can be determined by comparison of the measured radioactive concentration of the artifact with that of the exchange reservoir.

The concentration of ^{14}C is determined by measurement of the specific β activity. Usually, the carbon from the sample is converted into a gas, eg, carbon dioxide, methane, or acetylene, and introduced into a gas-proportional counter. Alternatively, liquid-scintillation counting is used after a benzene synthesis. The limit of the technique, ca 50,000 yr, is determined largely by signal:background ratio and counting statistics. The application of this technique in dating of art objects has been limited severely by the large samples needed for the traditional variations mentioned above. Even in the case of the most ideal materials, such as wood or charcoal, a sample of several grams is required. In the worst cases, ivory and bone, many hundreds of grams may be necessary. However, recent developments have changed the requirements drastically. The combination of specially designed miniature gas counters and complex electronics that perform elaborate pulse height and shape analysis has resulted in a reduction of the required sample to a few milligrams of carbon (36). Even smaller samples are sufficient when the ^{14}C concentration is determined with a high energy mass spectrometer, which combines a nuclear accelerator and a magnetic mass filter (37).

A fundamental assumption in radiocarbon dating is that the formation rate of ^{14}C is constant. This is only acceptable as a first approximation; in reality, this rate undergoes significant variations over both short and long periods. Hence, for more accurate dating, corrections are in order. These are obtained by dating samples from individual annual growth rings from long-lived trees, such as the bristle-cone pine and the sequoia. Graphs in which the radiocarbon dates for such samples are plotted against their actual age, determined by counting the rings, are used to apply the so-called bristle-cone-pine corrections to the radiocarbon date obtained for an unknown material.

Bristle-cone pines can reach ages of 4000 yr. The use of dendrochronological techniques has enabled the determination of the actual age for rings of trees that died in the past. Thus, correction factors have been established for dates up to ca 8000 yr ago (38).

Dendrochronology depends on the variations in annual growth rates for trees with the climatic conditions during the growing season: wet summers will result in thicker growth rings than dry ones. As all trees in a climatic region are affected to the same degree, the climatic variations in consecutive summers will result in identical patterns of thicker and thinner rings for all trees. Now, the pattern for the outer rings of an older tree will be the same as that of the inner rings of a younger tree with overlapping lifespan. Continuous overlaps to successive younger trees till one presently alive allows then for dating through the counting of the total number of rings.

Dendrochronological sequences have been established for many climatic regions. This is important not only for the corrections thus obtained for radiocarbon dates, but also for the direct dating of the wood used in the manufacture of art objects. For example, the technique is frequently used to date the wood used in paintings on panels (39–40). Measurement of the relative thickness variations in a number of consecutive rings which can be observed on the edge of the panel allows the placing of this pattern in the chronological sequence developed for the area of origin.

Thermoluminescence dating (33–34,41) has, in the short time since its development, had a tremendous impact on the technical study of ceramic materials (see Analytical methods).

Absorption of radiation energy, such as from naturally occurring radioisotopes, in a nonconductor results, via a sequence of primary ionizations and secondary excitations, in the production of electrons and electron-hole pairs with energy levels in the conduction band. With such energies, the electrons and electron-hole pairs are able to move freely through the crystal, which allows them to enter crystal imperfections. At such imperfections, metastable energy levels exist between the valency and conduction bands. One such type of metastable level is the so-called trap: an electron or electron hole which de-excites partially at such an imperfection until it reaches the trap level cannot subsequently de-excite completely because transitions between the trap level and valency band are forbidden. The name of the trap level derives from the consequent restriction in freedom of both movement and energy of the trapped electron or hole. Only transfer of extra energy to the system, as through heating of the crystal, can free them again via re-excitation to conduction band levels. Although most de-excitations to the valency band, followed by recombination, will be nonradiative in nature, some take place in the luminescence centers, which are crystal imperfections of another type.

In a clay, the radiation dose rate is determined, to a large extent, by the concentration of radioisotopes, eg, ^{40}K, and the members of the decay chains of uranium and thorium contained within the constituent minerals. This results in fairly constant rates of ionizations, and hence of trap population. Because the process has been taking place over a geological time period, the trap-population density in a clay is very high. However, firing, the last step in the manufacture of a ceramic object, results in a depopulation, which resets the clock for thermoluminescence dating. Immediately after cooling, repopulation commences, again at a constant rate determined by the dose rate. Thus, at any time after firing, the trap-population density will be a function of the time elapsed since the firing.

In thermoluminescence dating, a sample of the material is heated, and the light that is emitted by the sample as a result of de-excitations of the electrons or holes, freed from the trap, at luminescence centers is measured to provide a measure of the trap-population density. The signal is compared with that obtained from the same sample after a laboratory irradiation of known dose. The annual dose rate for the clay is calculated from determined concentrations of radioisotopes in the material and assumed or measured environmental radiation intensities.

The application range of the technique extends well beyond the first human ceramic production; in fact, it has been used in geological studies such as the dating of lava flows. The accuracy obtained in the dating of ceramics depends largely on knowledge of the environmental history, eg, burial conditions, and physical behavior of the material; the information on burial conditions is needed for the calculation of certain correction factors on the annual dose rate which otherwise have to be estimated. At best, the accuracy of thermoluminescence dating is ca 5–10%. Though for many historical contexts a far more precise dating is possible on stylistical arguments, thermoluminescence dating is very useful in prehistoric archaeology. Accuracv is of less importance in authenticity questions, and it is here that the technique finds extremely widespread use.

A related technique depends on the same principles, only the trap population is measured by means of esr spectrometry (42).

In amino acid racemization dating, the degree to which racemization has progressed since the death of an organism serves as a measure of the time elapsed after that event (43). Because the reaction rates are dependent upon the environmental conditions of the material, eg, average temperature, humidity, and pH of the burial soil, the working range of this technique varies with these locality-bound data, whereas the accuracy depends on their variation in time. The technique is applied, in particular, to the dating of shell, bone, or ivory objects.

Another relative dating technique for fossil materials is known as nitrogen–fluorine dating (44). Bone or ivory will, under burial conditions, lose nitrogen because of groundwater leaching of the amino acids resulting from hydrolysis of proteins, especially collagen. At the same time, fluorine will be absorbed from the groundwater by the hydroxyapatite in the bone, forming fluoroapatite. Thus, the ratio of nitrogen and fluorine concentrations in the bone decreases with time. Another elemental concentration which changes is that of uranium, which can be absorbed by the hydroxyapatite. The reaction rates are dependent on local burial conditions, and the technique only has value as a relative dating for materials excavated or otherwise recovered from the same site.

A few techniques exist that do not provide for direct dating but rather give information as to whether the object is of recent manufacture. One of these is ^{210}Pb dating (45). In the decay series of uranium, the first long-lived member after ^{226}Ra, with its 1622-yr half-life, is ^{210}Pb with a 22-yr half-life. In a lead ore, ^{210}Pb is in radioactive equilibrium with ^{226}Ra, but refining of lead results in a chemical separation in which the greatest fraction of the radium remains in the slag. Thus, the equilibrium is disturbed, with the lead having a much higher ^{210}Pb activity than ^{226}Ra activity. This disequilibrium continues until the much faster decay of the excess ^{210}Pb results in a reduction of this activity to a level where it is in equilibrium again with the ^{226}Ra activity. The time needed to reach this state is of the order of 4–5 half-lives, ie, ca 100 yr. If an object contains refined lead, eg, in the pigment lead-white in a painting, measurement of both ^{226}Ra and ^{210}Pb activities will indicate whether radioactive equilibrium exists between these two radioisotopes and hence whether the lead was refined within the last 100 yr or earlier.

Indirect dating, or dating by inference, is not really dating but rather the reaching of certain conclusions regarding the date of manufacture of an object from other information obtained in its study. The presence on a painting of certain pigments of which the date of invention is known provides a *terminus post quem*, ie, the painting could have been executed at any time after the invention (46). On the other hand, the presence of a material of which the manufacture was discontinued at a certain time can provide a *terminus ante quem*.

Types of Objects—Methods Used

Paintings. Before an expostulation of the procedures used in the examination of paintings, a synopsis of materials and techniques used in their creation is in order (17,47–52).

A painting is a composite of many materials of various types. As a first approach, one can distinguish three groups of components: supports, paint media, and pigments.

The support is the substrate upon which the paint layers are laid down. This can be a specially prepared area on a wall for a wall painting, a wooden panel as in a panel painting, or a fabric in canvas paintings. Paper is a prevalent support in Oriental painting. Other supports are encountered less frequently, eg, metal panels (especially copper sheet).

In the primitive cave paintings, which are the earliest known paintings, the paint was applied directly onto the wall, with little or no preparation. In ancient Egypt, however, as early as the Old Kingdom, wall surfaces were specially prepared with a coating of plaster. In time, the refinement and complexity of the preparation layers increased; in the Renaissance, several layers of different composition and fineness were superimposed. Other preparations used, especially in the Far East, consisted of a clay layer.

The use of wooden panels as support for painting goes back again to Egypt. Although the painting served originally as decoration of a wooden object such as a sarcophagus, later the wooden panel became only a support for the painting, as in the mummy portraits from the Fayum, executed during the Roman period. In Europe's Classical period, wooden panels must have been the supports for portable paintings, and remarks by Pliny indicate so; however, none of these works of art has survived. The tradition was carried through into the Christian era, and wooden panels were virtually the only portable supports used in the Middle Ages and early Renaissance.

The types of wood used in European panel paintings varied with the geographic area. In Italy and southern France, the most popular wood was poplar. In the north, oak was the first choice, followed by pine and linden. Another difference between north and south can be found in the preparation of the panel: whereas the ground layer in the southern paintings consists of real gesso, ie, finely ground burnt gypsum in a glue medium, the northern artists prepared their ground layer with chalk.

The earliest remaining example of painting on a fabric support is from the twelfth dynasty in Egypt. In Europe, although the technique was known, these supports were not frequently used until the Renaissance, when the increasing size of paintings resulted in a tremendous rise in the popularity of canvas supports, a popularity which has lasted until the present. The fabric used almost exclusively by western painters was linen; in the Orient, silk became a frequently used support. The nature and composition of the ground or priming layer on European canvases varied significantly.

The medium is the binder which provides for the adhesion of the pigments. The most important types of medium are the temper media (glue, egg, and gum), the oils, and wax. In addition, there is the true fresco technique, where the pigments are laid down in a fresh, wet plaster preparation layer. Several other media have been used, but much less frequently, eg, casein temper. In modern paints, a number of synthetic resins are used for this purpose.

The use of the various tempera and of wax has been identified on objects dating back to ancient Egypt. The Fayum mummy portraits are beautiful examples of encaustic painting, ie, with molten wax as medium. A rather special variation was the technique used by the Romans for their wall paintings. In these, the medium, referred to by Pliny as "Punic wax," probably consisted of partially saponified wax. In Europe, wax ceased to be used by the ninth century.

The polysaccharides in gum arabic formed a medium used for illuminated manuscripts and inks as well as for painting. Gum is also the binder in watercolors.

The proteinaceous gelatins in the various animal glues were also widely used as paint media, as well as in illuminations. Glues are the traditional media in Oriental painting. They remained the prevalent binders for ground layers in European painting long after oils had become virtually the only medium for the color layers.

Of the various tempera, egg temper was the most important in European painting, both in wall and panel painting. It was little used outside Europe. The main period of its use was in the Middle Ages and early Renaissance. After the sixteenth century, however, it was rarely used, as oils had become the preeminent media.

The earliest written references to the use of oils as paint media date from the twelfth century. The van Eycks, who traditionally have been credited as the inventors of oil painting, did indeed improve the technique to such a degree that oil quickly replaced egg temper as the prevalent medium.

Pigments can be categorized in two different ways: inorganic or organic and natural or synthetic. Organic pigments generally consist of a dyestuff, precipitated on an inorganic substrate. The earlier pigments tend to be inorganic and natural, although important exceptions in both regards exist. Thus, a pigment prepared by the precipitation of madder onto a gypsum substrate was an important color in Egypt during the Graeco–Roman period, and in the Old Kingdom a synthetic pigment, Egyptian blue (copper calcium silicate), was made in large quantities.

For many pigments (qv), a period of time in which they had their widest use can be indicated (47,53). Dates of introduction are known either from documentary sources or from identification on paintings of known dates (46). For some pigments, an approximate date for the discontinuation of their use can be assigned; in some cases, knowledge of the preparation process or even their very existence was lost over an appreciable time span.

A varnish is often applied on top of the paint layers. A varnish serves two purposes: as a protective coating and also for an optical effect that enriches the colors of the painting. A traditional varnish consists of a natural plant resin, dissolved or fused in a liquid for application to the surface. There are two types of varnish resins: the hard ones (the most important of which is copal), and the soft ones (notably dammar and mastic). The hard resins are fossil; to convert them to a fluid state, they are fused in oil at high temperature. The soft resins are dissolved in a suitable organic solvent, eg, turpentine. A typical technical examination of a painting (54), intended to ascertain its period, condition, and the degree of previous restoration would likely entail most of the following steps.

Visual Examination. A close inspection under normal illumination reveals many indications of the condition of the painting and previous repairs. Also, because oil paints become more transparent with age, pentimenti, which originally would have been invisible after the overpainting, can be observed. Raking light illumination is extremely useful at this stage to determine the extent of cracking, distortions of the support, delaminations of the paint layers, etc. This stage of the examination is often done in close cooperation with the stylistic experts; thus, obvious problematic areas can be identified before the other tests are started.

Examination with the Stereomicroscope at Low Magnification. The obtained information relates to the characteristics of the craquelure, the pigments used (grain size, morphology), buildup of paint layers (visible in damages and along cracks), technique of the artist (brush work, use of pure and mixed colors, use of glazes, etc), and condition (eg, damages and losses, amount of inpainting or overpainting).

Examination under Uv Illumination. This, as explained above, principally provides information regarding more recent restorations or changes.

Infrared Reflectography. The ir photograph, or the image from an ir-sensitive vidicon system, gives more evidence of restorations as well as pentimenti. The most important use of this technique is in the revealing of underdrawing: the pigments typically used for underdrawing, such as charcoal or certain earth pigments, do not reflect ir radiation very efficiently. A variation of this technique is transmitted ir photography, where the light source and the camera are placed on the opposite sides of the painting.

X-Radiography. The radiograph reveals evidence of damages and losses in the paint layers and the support. Dimensional changes may also be detected, eg, a cut-down painting on canvas will, on one or more sides, lack the telltale deformations of the canvas caused by the tacking to the stretcher. The radiograph illustrates the artist's compositional approach (eg, blocking out of compositional elements, amount of changes in composition during execution of the painting) as well as the painting technique (eg, preparation of support, brush work, thickness of paint layers, variations in thickness between different areas in the painting) (see Fig. 4). Through the study of x-radiographs, characteristics of a given artist's works can often be established that can serve as benchmarks in the examination of paintings of uncertain attribution.

Pigment Analysis. Identification of pigments present on paintings of unquestioned attribution provides reference information regarding the use of certain pigments in given periods or schools. In the study of paintings with uncertain provenance, the pigments identified in it may either negate the possibility of the proposed attribution or lend more credibility to it. Most pigment analyses are done in one of three ways: microscopy and microchemical tests, x-ray diffraction-powder analysis, or energy-dispersive x-ray fluorescence spectrometry. Although the latter technique does not directly identify the pigment but rather the chemical elements present, this information often allows deduction of the identity of the pigment. The technique can be used nondestructively, without any need for sample removal (see Nondestructive testing). Thus, many areas on a painting can be analyzed, giving better overall information.

Microscopic Examination of Cross Sections through the Paint Layers. This gives definite information regarding the paint-layer sequence in the area from which the sample was taken (22,55). This information illustrates the artist's use of underlayers and glazes, superposition of compositional elements, and changes in composition.

Medium Analysis. The techniques frequently used for identification of the binding media include relatively simple solubility tests, ir absorption spectrophotometry, specific staining techniques (56), tlc (57), gc (58), and ms. Significant work, eg, entailing the identification of the particular oils used in a painting, has recently been made possible through the use of combination gc–ms (59).

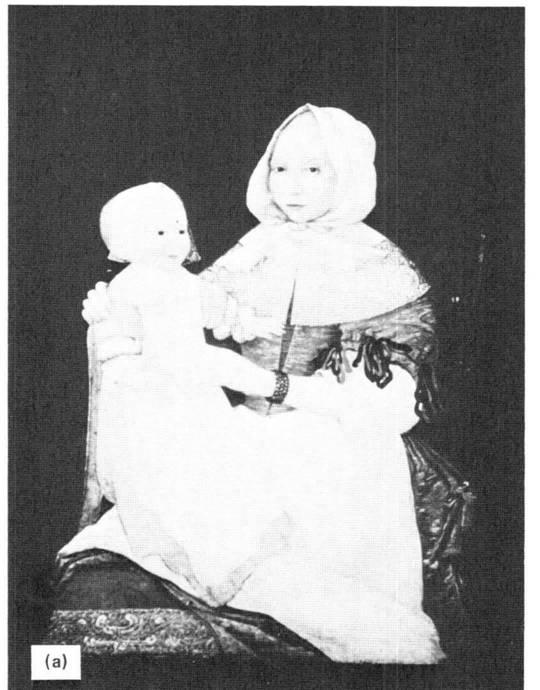

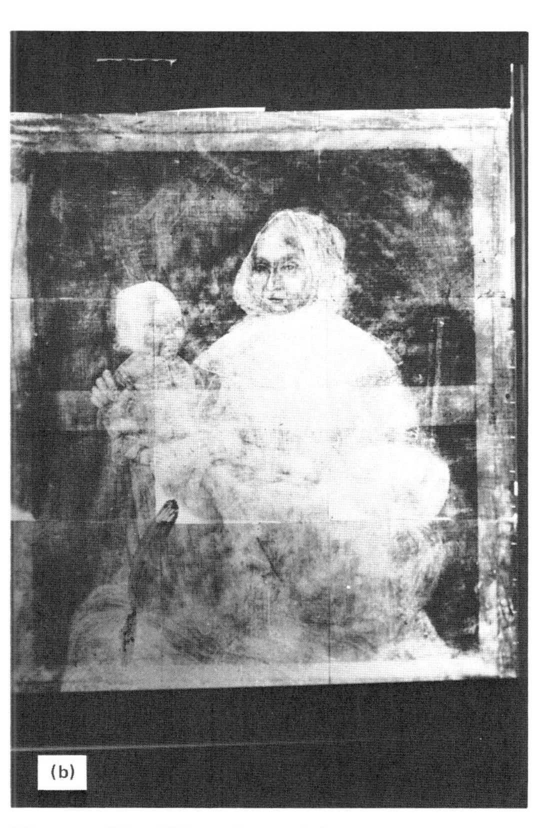

Figure 4. (**a**) "Mrs. Freake and Baby Mary," Boston, 1671–1674. Worcester Art Museum, Gift of Mr. and Mrs. Albert W. Rice. (**b**) The x-radiograph of this painting shows considerable changes in the composition: in an earlier version, the robe of the sitter has a different collar while her right hand does not hold the child but rather lies in her lap, holding a fan. (**c**) Line drawings of the two versions. The baby was not included in the first version of the portrait; when she was added, the mother's dress was also updated (drawing by Carol Greger, copyright 1983). Courtesy of the Worcester Art Museum and the Museum of Fine Arts.

Figure 4. *(continued)*

Wood Identification. In the case of a panel painting, a small sample of the wooden support can be removed, from which a microscopic specimen can be prepared in order to identify the wood used for the panel.

Metal Objects. In the history of technology, the developments of metallurgy probably provide the most complicated and important chapter. Although new archaeological evidence continually necessitates changes in accepted hypotheses regarding the developments in the use of metals in various periods and geographic locations, enough is known to allow the sketching of a general, overall picture (60–61).

Where the earliest developments in metallurgy took place, in particular, the discovery of bronze-making, is a point of intense discussion. The frequent changes in archaeological chronologies for the areas under prime consideration will probably delay a final judgment on this question for a long time. The traditional favorite is Anatolia (Turkey); however, areas in southeast Asia and Europe are frequently considered other prime candidates.

The first use of metals, ca 8000 yr ago, was restricted to those occurring natively, ie, gold and, to a lesser extent, silver and copper. Humans may well have assumed them to be another type of rock, the prevalent material for tool manufacture. At this stage, iron was extremely rare, and obtained from meteorites.

The shaping technique for this new material was beating and hammering, and immediately a difficulty was encountered, especially in the work-hardening of copper. The first real metallurgical breakthrough was the discovery of annealing ca 7000 yr ago.

The invention of the techniques of melting and casting depended on the development of the technology to produce the requisite high temperatures, which may have been first achieved by potters, in ca 4000–3500 BC. Coinciding with this discovery was the development of the technique of smelting copper from oxidized ores, eg, cuprite (cuprous oxide) and malachite (copper hydroxy carbonate). Smelting from these ores is easier than the refinement of copper from sulfidic ores, eg, covellite or chalcopyrite, which involves a roasting step invented ca 2000 yr later.

The first intentional metal alloy, bronze (ca 3000 BC), was probably prepared by the smelting of mixed ores, eg, cuprite or malachite for copper and cassiterite for tin. The metal tin could not yet be refined: the complicated metallurgical technique needed was not developed until ca 1500 BC.

The invention of roasting techniques needed for the refinement of metals from sulfidic ores occurred ca 3000 BC. This discovery enabled the refinement of lead from galena (lead sulfide), and, more important, after the invention of cupellation, the separation of silver present as a significant impurity in lead prepared from argentiferous galena.

During this period, the first iron was refined from ores. This was done by a smelting technique; because the temperatures necessary for the melting of iron were not yet attainable, the product of this operation was a spongy mass of reduced metal with high amounts of slag, collected as residue from the smelting furnace. Forming of this new metal was only possible through mechanical working. In ca 1500 BC, steelmaking was discovered. Thus, ca 1000 BC, humans entered the iron age. It is noteworthy that the development of cast iron did not occur until after 1000 in the West; in China, a natural occurrence of phosphorus in iron ores, causing an appreciable lowering of the melting-point, enabled the production of cast iron ca 1000 yr earlier.

Of the early copper alloys, the tin bronzes were the most important. Another alloy was prepared from copper and lead; the arsenical coppers or bronzes, though in the technical sense alloys, may well have come about through the refinement of copper from arsenic-rich copper ores. Brass, from copper and zinc, was not known until ca 1000 BC and then occurred only rarely; its first large-scale use in the West came as coinage metal by the Romans. Refinement of zinc from its ores posed serious problems through complications owing to the metal's immediate oxidation upon exposure to air and the low sublimation point of the oxide; brass was consequently prepared from copper and zinc ores or, in Roman times, from copper and refined zinc oxide. The refinement of zinc metal was not achieved until medieval times.

The first forming technique was hammering of the metal. This evolved into many different and highly complex cold-working techniques, such as raising, sinking, spinning, turning, and relief application by means of repoussé.

The second class of forming techniques was developed upon the invention of the melting of metals. Originally, casting was done in open stone molds to produce relatively simple forms such as ax heads. The invention of lost-wax casting was the greatest breakthrough in this area. In its simplest variation, a model of the object is made in wax; subsequently, an outer layer (investment), made of refractory clay, is applied to the wax model. Heating of the assembly results in the molten wax running out, leaving a hollow space within the investment, into which metal can be poured. This method allows for the casting of solid objects only, requiring large amounts of expensive metal for the fabrication of sizable objects. A significant improvement was the development of casting around a core in which the model of the object is made by forming the approximate shape of the object out of a clay and sand mixture, upon which a thin wax layer is applied. The final modeling is done in the wax. Again, an investment is applied and the wax removed by melting. A narrow space is left between the core and investment, into which metal is poured.

Lost-wax casting, as described here, destroys the wax model, which makes it impossible to cast more than one piece. Not until the Renaissance were variations on the method developed which allowed the casting of multiple copies of the same model with a lost-wax technique (62). In the meantime, another casting technique had made its entrée in the West: in the Medieval period, the piece mold was introduced. In this technique, the investment is built up around the model in multiple segments which, after drying, can be removed from the model and subsequently reassembled, resulting in an empty mold.

In the West, Near East, and southeast Asia, lost-wax casting was the first development; however, present evidence seems to indicate that, in China, the piece-mold technique existed before the introduction of lost-wax casting (63).

Examinations of metal objects generally include the characterization of the metal, determination of the techniques involved in the manufacture, and study of aging phenomena. Of the latter, the state of corrosion is especially important, both in the examination of an object with the purpose of determining its authenticity as well as when an assessment of its state of conservation is needed (63–64). The layer of corrosion products covering the surface of the metal, the so-called patina, can be studied for its composition and structure. Identification of corrosion products is often performed by means of x-ray-diffraction analysis. The structure of the patina is studied with the low power stereomicroscope, where attention is directed to the growth pattern, crystal size, layering, and adherence to the metal (see Fig. 1). Cross sections prepared from samples can be studied under higher power reflected-light microscopes to determine the extent of penetration of corrosion processes into the interior of the metal, or the inter- and intragranular corrosion (see Fig. 5). The examination with the low power microscope also yields information regarding wear patterns (eg, sharp or rounded edges of incised lines) and manufacturing techniques (tool marks, mold marks, etc). When the crystal structure of the metal is studied through examination of an etched sample under high magnification, conclusions can be drawn regarding metallurgical techniques used in the manufacture such as various types of cold working, annealing, casting with preheated or cold molds, etc (65).

An important tool in the analysis of the structure of the object is x-radiography (30). For metal objects, high energy x rays are needed, obtained from either an industrial radiography instrument or from radioisotope sources (γ-radiography). The radiograph may give direct evidence important in a condition analysis, such as of small cracks, casting defects, failing joints, etc. It also provides information relating to technique, such as the type of joining technique and the forming technique. For example, the methods developed by Italian Renaissance sculptors to enable them to produce multiple copies of one model with lost-wax casting were elucidated with the aid of radiographical examination (62).

Chemical analysis of the metal can serve various purposes. For the determination of the metal-alloy composition, a variety of techniques has been used. In the past, wet-chemical analysis was often used, but the significant size of the sample needed for the purpose was a primary drawback. Today, nondestructive, energy-dispersive x-ray-fluorescence spectrometry is often used when no high precision is needed. However, this technique only allows a surface analysis, and significant surface phe-

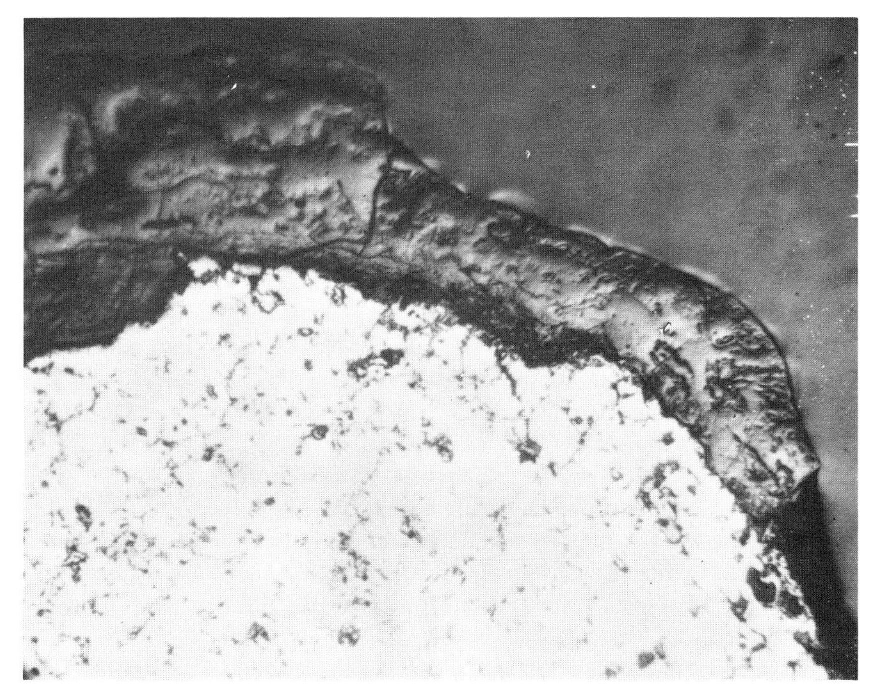

Figure 5. The Roman bronze bust from which this was removed is a forgery. The photomicrograph (100×) reveals how the artificial patina, chemically induced, lies on the metal surface without the presence of a layer structure of this crust or of any penetration into the inner metal. Courtesy of the Research Laboratory, Museum of Fine Arts.

nomena such as preferential enrichments and depletions, which often occur in objects with a burial history, can cause serious errors. For more precise quantitative analyses, samples have to be removed from below the surface, to be analyzed by means of atomic absorption (66), spectrographic techniques (63,67), etc.

Trace-element analysis of metals can give indications of the geographic provenance of the material. Both emission spectroscopy (68) and activation analysis (69) have been used for this purpose.

Another tool in provenance studies is the measurement of relative abundances of the lead isotopes (70). This technique is not restricted to metals, but can be used on any material which contains lead. Finally, for an object cast around a ceramic core, a sample of the core material can be used for thermoluminescence dating.

Ceramics. In the early stages of every ancient civilization, there was a moment where humans achieved the first technological modification of a natural material: an object formed out of wet clay was dried and, subsequently, heated and transformed into a new material with drastically different properties (71–74).

Early pottery is rather simple, with little surface decoration. Because pyrotechnology was still comparatively undeveloped, it was fired at relatively low temperatures. The technology developed, interdependently, in several aspects from this origin. The first aspect is the pyrotechnology. As kilns became available, and these in turn were further perfected, higher firing temperatures became attainable, which resulted in harder, stronger wares. The high point in this development, before the introduction of modern electrical and oil- or gas-fired kilns, was the development in China of the "dragon kiln," a tunnel kiln built against a hillside, with a draft which resulted in the temperatures needed for the firing of porcelain ware.

A second developmental aspect is the knowledge and choice of raw material. Clays differ in both working behavior and firing properties. Choosing the proper clay, cleaning of the clay in order to remove naturally occurring accessory minerals or other impurities which affect its properties, adding other substances to modify its working behavior (tempers or plasticizers), and mixing different clays to obtain a material with the desired properties are developments of the second kind. Deserving special mention are the developments of various artificial-body materials, such as the Egyptian faience (75–76), a paste made out of sand and natron, the latter acting as binder, and flux, or the white-body material developed in the Islamic Near East in imitation of Chinese porcelain, consisting of a mixture of feldspar, quartz sand, and a little white clay.

Several techniques were developed for the forming of the object out of clay. Pots can be made by working a slab of clay into the desired form, by building up with coils, and by throwing on the wheel, though this latter technique was unknown in Precolumbian Central America where the artisans nevertheless succeeded in crafting large, perfectly formed vessels. Sculpture can be hand-formed or, when a production system is devised, pressed in molds, eg, the Greek Tanagra figurines. Clay can also be cast into porous molds, which became the most important manufacturing technique in the porcelain industry (slip casting).

The most spectacular developments took place in the decoration of the surface with paint, slips, and glazes. In painted ware, the pigments are applied to the object after firing. Slips are very fine-grained fractions of clays, which are applied to the object after it has dried but before firing. Slips of different colors can be used to obtain a color effect. A technical and aesthetic high point of slip decoration was reached with the Greek red- and black-painted pottery, where the painting was done with a slip prepared from the same clay as the one used for the buildup of the vessel, and the two colors were obtained by means of a complex firing procedure involving alternating oxidizing and reducing conditions (77).

Glazes stand out through their structure; they are actually glasses. The first glazes were ash glazes, probably initially made by accident when ash from the fire chamber settled on the pots in the kiln. The flux, needed to lower the melting point for the mixture of minerals including quartz, feldspars, and clay minerals, is, in this case, provided by the alkalies from the ash. However, during the Old Kingdom period, the Egyptians used alkaline glazes, in which the alkalies for the flux were added in the form of natron, to glaze objects carved out of steatite. These glazes were already colored through the deliberate addition of metal oxides. Plant and wood ashes have always served as an important source of alkalies (see Potassium compounds). Lead glazes, which were developed around the Roman period, use lead oxide as their fluxing agent.

The ultimate perfection in glaze technology, obtained in Imperial China, depended on highly developed skills, extensive experimentation with glaze formulations, and precisely controlled firing conditions in high temperature kilns (78). Only one color glaze was not developed and perfected by the Chinese potters themselves: the production of a true pink glaze through the introduction of colloidally suspended gold was a Dutch invention, the knowledge of which was brought to China by Jesuit missionaries.

As noted above, glazes are technically glasses, and it is therefore not surprising that the development of glass (qv) as a separate material relates to that of glazes (79). There is a certain amount of debate as to whether the first glass was made in Mesopotamia or Egypt. In Egypt, glass beads were already being made during the fifth dynasty (ca 2500 BC) (76). The greatest development came during the New Kingdom (starting ca 1500 BC), when a multitude of colors was obtained in soda-lime glasses formulated out of quartz sand and natron, the flux which occurs naturally in large quantities in Egypt (80). Calcium, necessary for the chemical stability of the glass, was not added intentionally but happened to be present in large enough quantities in the sand. Indeed, the need for the presence of calcium was not realized until the sixteenth century, as the raw materials had always contained sufficient amounts of this element.

Vessels were formed from this material by application around a sand core. The Egyptian tradition was carried through in the Roman period, when glassmaking technology reached unprecedented heights. During this time, glassblowing was invented. The deservedly famous Venetian glassblowers continued this tradition.

The developments in the Near East are closely related to those in the West; indeed, the two are interdependent, though each has its own characteristics (81).

Natron and ash of seaweeds provided the sodium which served as the flux in all glasses till the Medieval period. Wood ashes then came into use, which changed the glass formulation to such a degree that potassium salts became the principal fluxing alkalies.

Lead glass, though used in Roman mosaics and later in enamels, was otherwise never important in the West until the development of flintglass by Ravenscroft in 1676. In China, however, the first glass was a lead glass, which curiously enough contained barium as the earth alkali instead of calcium. This was probably a consequence of the natural occurrence of barium in the lead ore (82). During the Han period, contact with the West resulted in the introduction of soda-lime glasses with a Roman formulation. Also in this period the barium–lead glasses were replaced by calcium–lead glasses (83).

Examinations of ceramic objects involve a variety of techniques, depending on the type of information sought (84). If an assessment of condition and state of repair is made, the most important tools are the low power stereomicroscope, x-radiography, and examination under uv light. For the identification of glass degradation products, a number of chemical analytical techniques can be used, especially x-ray diffraction.

X-ray diffraction can also be used to great advantage in the analysis of the mineral composition of a ceramic paste; not only can the naturally occurring minerals provide clues to the geographic provenance of the clay, but the presence of certain minerals which are formed at high temperatures can give indications of the firing temperatures.

Elemental chemical analysis provides information regarding the formulation and coloring oxides of glazes and glasses. Energy-dispersive x-ray fluorescence spectrometry is, again, very convenient (see Fig. 6); however, with this technique, the analysis for elements with low atomic numbers is quite difficult, even when vacuum or helium paths are used. The electron-beam microprobe has proven to be an extremely useful tool for this purpose (85). Emission spectroscopy and activation analysis have also been applied successfully in these studies (80).

Trace-element analysis, using emission spectroscopy (86) and, especially, acti-

Figure 6. Energy-dispersive x-ray fluorescence spectrometry is used to identify the elements responsible for the colors of this Chinese porcelain vase, and the impurities which may indicate sources of the glaze materials. Courtesy of the Research Laboratory, Museum of Fine Arts.

vation analysis (87–88) has been applied in provenance studies on archaeological ceramics with revolutionary results. The attribution of a certain geographic origin for the clay of an object excavated elsewhere has a direct implication on past trade and exchange relationships. Microscopic examination of ceramic paste, both at low magnification and at high power with prepared cross sections, can be used for petrographic study of the mineral composition and also for the determination of techniques involved in the manufacturing process. For ware decorated with slips and glazes, microscopic examination of a cross section is irreplaceable as a means of studying the technology. For example, in order to observe the phase separations which are responsible for the effects of some of the unique glazes on Chinese porcelains, the scanning-electron-beam microscope is the only appropriate tool. In archaeology, the ceramic typologies have always formed the basis for the establishment of chronologies. It has been extremely important that an absolute dating technique for ceramic material became available with thermoluminescence dating. As noted in the section on dating, this technique presently does not have enough accuracy to be very useful in areas with historically evaluated chronologies. For authentication purposes, however, the error margin is not quite so critical, and it is here that this technique has had an enormous impact.

Stone Objects. Stone was the first material used by humans to make tools. After it ceased to be used for that purpose, because metals had proven to be superior materials, it remained one of the most important materials for use in architecture and in sculpture.

As with objects of any material, the technical examination of stone objects begins with the use of the low power stereomicroscope. This study yields information regarding toolmarks and, hence, cutting techniques, wear patterns, and wear of toolmark edges. Such information is clearly significant in authenticity studies, but also provides an insight into the skill and the tools of the carver.

When a question exists about whether the carving is ancient or recent, an examination under uv illumination can be very helpful. For example, an aged marble surface exhibits, under a properly filtered mercury lamp, a mellow, brownish fluorescence, whereas a fresh surface, under the same conditions, appears purple. It is not uncommon that old carvings have been sharpened; this is detectable with the microscope and under uv examination.

The first question about a stone object is often which stone was used. Indeed, museums have many labels that carry an erroneous identification. X-ray diffraction provides a relatively easy answer for such questions of identification.

The polarizing microscope is one of the most important tools in the study of stone. Petrographic study of a thin section can be extremely helpful in the identification of a stone, and often yields information that can be used in the assignment of the geographic origin of the materials. Thus, the location of the quarry for many marble sculptures has been deduced. However, this method does not always give conclusive results, eg, the marbles from the Greek islands are notoriously difficult to differentiate by means of microscopic techniques (89). A very interesting application of microscopy is in the structural study of the degree of sulfation in the surface layer of a marble sculpture. Again, this provides an indication whether the surface is the undisturbed original or is of recent origin. The use of an attachment to the microscope allows the study of a specimen by cathodo luminescence, thereby improving petrographic study.

Trace-element analysis provides another approach to these studies, and activation analysis has been applied sucessfully in provenance studies of, eg, limestone sculpture (90). Marble presents a difficulty because of the large degree of inhomogeneity inherent in this material, which is a consequence of its metamorphic genesis. Thermoluminescence studies have also proved applicable to this problem. In several instances, study of the thermoluminescence induced in the stone through artificial irradiation and measurement of the wavelength and intensity of the emitted light as functions of the sample temperature allows the differentiation between stones from various geographic origins.

For marble provenance studies, the most successful technique seems to be the measurement, through mass spectrometry, of the abundance ratios of the stable isotopes of carbon and oxygen (91). Although the inhomogeneity of marble makes trace-element analysis of this material of dubious value for provenance studies, this same property has been used to advantage in establishing whether fragments of broken objects belong together. Multiple sampling along the break planes provides concentration patterns that indicate matching pieces (64).

Other Materials. It is clear that a number of methods are used universally in a museum laboratory. The general approach in the examination of an object is the same regardless of the material: analysis of the materials and techniques used in its manufacture and comparison of the results with data gathered on objects with comparable attribution. In this section, a few techniques, applied specifically on materials not yet discussed, are mentioned.

Some specific questions come up in the study of textiles, a class of objects which includes rugs, gobelins, costumes, weavings, burial shrouds, etc. The first question regards the identification of the textile fiber. These can be collected from animals (eg, wool, silk) or plants (eg, cotton, linen). Identification of the fiber is usually done microscopically. Many fibers can be recognized immediately under the polarizing microscope, but for several fibers the preparation of cross sections is necessary in order to study the cell structure in sufficient detail.

A second common question relates to the dyes used to color the fibers. A small sample is removed, and the dyestuff is stripped from the fiber through treatment with acid followed by further extraction in an organic solvent. This solution can then be used to identify the dye by means of spectrophotometric absorption techniques (92) or tlc; the latter is preferred because of the smaller sample needed (93).

The natural dyestuffs used until the invention of aniline dyes could not be used directly on the fiber, but required the treatment of the fiber with a so-called mordant, a metal salt which is precipitated onto the fibers and facilitates the binding of the dye. Salts of aluminium (alum), iron (ferrous sulfate), tin, and copper have been used as mordants (see Dyes and dye intermediates). In modern dyeing, chromium compounds (qv) are used frequently. The color of the dyed textile depends on the nature of the mordant; for example, cochineal is crimson when mordanted with aluminum, purple with iron, and scarlet with tin. Elemental chemical analysis, such as through emission spectroscopy, is used for identification of the mordant. Undyed fabrics are sometimes sized, a practice which was, for example, used in ancient Egypt. Glues and starches were used for the size. They are identified with chromatographic techniques or through their ir absorption spectra.

Dating of textiles is possible by means of radiocarbon dating. The new developments in this technique, discussed above, will probably greatly increase its application for this purpose. Radiocarbon dating is applicable to any organic material, such as wood, parchment, paper, bone, ivory, etc.

The making of paper (qv) (94) is a Chinese invention, dating back to ca 200 BC. Originally, textile fibers were used as source material, but the traditionally preferred paper in the Orient is mulberry paper, made from the soft inner bark of the mulberry tree. The knowledge of papermaking came to the West via the silk route; it took ca 1000 yr to reach Europe. Until the middle of the nineteenth century, paper was made from rags, and microscopic examination can identify the fibers present in the pulp (95). Later, paper was often made from wood, which causes a very different aging behavior compared to that of rag paper. Because the fiber size is radically reduced, microscopic identification is more cumbersome, but certainly possible.

Inks, watercolor pigments and media, etc, are analyzed similarly to the pigments and media for paintings. Watermarks are studied with the aid of β radiography. The microscopic identification of wood species has already been discussed in the section on paintings, but is, of course, applicable to any wooden object, eg, sculpture. X-radiography is extremely useful not only for condition assessments, but also for the study of joining techniques. Ivory can originate from several animals; elephant, mammoth, or walrus. Examination of the growth-line patterns enables the distinguishing of these sources.

CONSERVATION

Historical Review

The various chemical and physical processes that play a role in the deterioration of art objects are not restricted to the present day, even though the contemporary environment has contributed significantly to the rate of their decay. Hence, people have for a long time been frustrated by seeing their revered masterpieces lose their splendor through losses, discolorations, disfigurations, etc. Indeed, from textual evidence, it is known how artists in the Renaissance restored works of art from Classical times. That restorations were the work of practicing artists was quite logical; they possessed the prerequisite manual skills and were trained in the techniques and use of the materials involved in the creation of the object. These restorers of past centuries attempted to return the object to its original appearance. The fallacy of that idea lies in the fact that they could not know the exact original appearance of the work, ie, immediately after its creation; therefore, they restored the object according to their subjective opinion. This concept of restoration has long been left behind: the modern conservator defines as admissible restoration the compensation of those losses which render the object "unreadable," ie, disfigure the object to an extent that the intent of the artist is obscured. The normal visual effects of age are accepted as such and are not *a priori* subject to modification.

One of the principal problems with restorations in the past was that the effort was directed entirely towards the restoration of the appearance of the object. The awareness that it is necessary to remove the cause of the problem first, and deal with the symptoms secondly, did not often exist. This frequently resulted in an early recurrence of the problems, which were then again treated with the same remedial measures. A spectacular example of this is DaVinci's "Last Supper." This wall painting, which started to exhibit serious problems soon after its completion, has undergone virtually continuous restoration; until recently, the major part of the painting as it appeared to the visitor was the work of restorers rather than the master, whose original was largely buried under the multiple restorations.

In this century, it was realized that a change in approach was necessary, that expertise in chemistry, physics, and materials-sciences should be brought to bear both to define the roots of the problems which plague a work of art and to devise a means by which to arrest the deterioration processes. A new type of specialist originated: the conservator, who possesses the necessary manual skills and talents, and, moreover, has a good grasp of scientific methodology and a sufficient knowledge of chemistry and physics to be able to understand the basic causes of deterioration mechanisms, to devise effective treatments of the objects, and relate with the scientific experts who are called upon to assist in the analysis of the problems or the testing of proposed cures.

A conservator has a thorough training in art history which enables a communication on professional level with the curator, who is responsible for the works in the collection and who must guide the conservator with regard to considerations of an aesthetic nature.

Ethical Considerations

The practitioners of this comparatively new profession have formulated a number of ethical professional standards, many of which are basic to the contemporary approach towards the conservation of art objects and other historic materials (96). Central to all of this is a fundamental respect for the integrity of the object *per se* as a part of humanity's cultural heritage.

No action is allowed that could conceivably carry in it the danger of causing any immediate or future damage to the object. This implies that any proposed treatment must be thoroughly evaluated as to all possible effects it may have besides the intended beneficial ones, not only immediate but also long-term. Many art objects have existed for periods which are very long in terms of human lifespans; they have survived hundreds or even thousands of years and it is for the conservator to see that they are preserved for another such period. Hence, materials that are to be introduced into an object, eg, as structural consolidants, adhesives, etc, have to be tested rigorously for their chemical and physical long-term stability, and for the effects that their failure or decomposition might have in terms of the safety of the object. The need for this knowledge necessitates the testing, often under accelerated aging conditions, by material scientists and chemists, in cooperation with the conservator.

Repairs, restorations, and other treatments should be reversible, ie, it must be realized that it may eventually be necessary to undo today's repairs, just as past restorations often need to be removed by today's conservator. A common reason is that the materials used may have aged with unacceptable visual consequences, but many other reasons occur which could not have been foreseen at the time of application, and it is not unreasonable to assume that this can happen to present efforts as well. If the removal of the proposed repairs should ever be necessary, it should be possible without danger to the object. Consequently, extensive testing of the stability of the materials used for the treatment is needed.

The arrest of the deterioration and the prevention of its recurrence has higher priority than the restoration. Thus, the identification of the causes of the problem and the design of measures to stabilize and consolidate the object are primary considerations; the removal of the symptoms and restoration of the visual appearance comes only after the physical integrity has been safeguarded.

Inducing any changes to the original parts in order to minimize the visual impact of repairs and restorations is prohibited. For example, it used to be a common practice for restorers of porcelains to hide repaired breaks and fills of missing fragments by painting these repairs with a matching color and, in order to minimize the visibility of this restoration, to extend the area of paint application over parts of the adjacent original surface. Although the inpainting of repairs and fills is certainly quite common and often desirable, this overpainting of original surface is not considered to be an ethical practice.

It is imperative that extensive documentation, including photographs during all stages of the work, be kept of any treatment, and that this documentation is stored to be available for consultation in the future, when a renewed conservation assessment of the object may be necessary.

Training and Organization

The extremely high demands put on the knowledge, in a variety of areas, of the modern conservator has necessitated the establishment of special training facilities for this profession. Thus, in the United States, a few universities have established graduate conservation programs. Candidates for admission to such schools must satisfy a number of stringent demands, which generally include an undergraduate degree in art history or archaeology, extensive undergraduate science coursework, and demonstrable manual skills and dexterity, as evident from a portfolio. After a training of three to four years, the students who successfully complete the program are awarded a degree or certificate in conservation, and they are expected to have the theoretical knowledge required of a conservator. An advanced internship with a practicing conservator for another one or two years is generally deemed necessary for the acquisition of the minimum experience needed to be allowed to work independently.

Training programs such as described above have been established in many countries. An alternative method of training is that of the apprenticeship; many experienced conservators are willing to share their knowledge in this way with their future colleagues. Professional organizations for conservators have been established on both national and international scale. Most conservators are members of the International Institute for the Conservation of Historic and Artistic Works (IIC). This organization (in London) organizes biennial international conferences, and also oversees the publication of several professional periodicals (*Studies in Conservation*, *Art and Archaeology Technical Abstracts*). Several countries or geographic areas have their own subdivisions of IIC or, as in the United States, an independent but related organization such as the American Institute for Conservation of Historic and Artistic Works (AIC). This organization holds annual meetings and publishes its own journal. The International Committee of Museums (ICOM) has its own subcommittee on conservation. Headquartered in Paris, this organization sponsors triennial meetings.

Several international organizations have been established which can offer conservation advice or even practical help in areas of the world where such is not readily available, eg, the UNESCO Center for conservation in Rome (ICCROM).

Recently, in the United States, a National Institute for Conservation was established; though presently still in a planning stage, this institute is intended to play a central role in information distribution and research coordination, somewhat analogous to the National Institutes of Health.

Processes Involved in the Deterioration of Art Objects

Metals. Apart from the physical damage that can result from carelessness, abuse, and vandalism, the main problem with metal objects lies in their vulnerability to corrosion (97–100). The degree of corrosion, which depends on the nature and age of the object, ranges from a light tarnish, aesthetically disfiguring as this may be on a polished silver or brass artifact, to total mineralization, a condition not uncommon with slowly corroded archaeological material.

The corrosion processes for various metals are vastly different, and hence the consequences for the object also differ greatly in nature. With bronze, for example, corrosion starts intergranularly, with cuprite (cuprous oxide) formed as primary copper-corrosion product. Because of the relatively large mobility of copper ions, a crust of cuprite forms on the surface and becomes the basis for continued chemical reactions resulting in the formation of products such as malachite and azurite (both basic copper carbonates). This corrosion crust, also known as patina, stabilizes the system to a high degree, resulting in a slowing of the corrosion rate. With the corrosion progressing slowly, and the mobilized copper ions being able to migrate to the surface, the final effect is that the shape of the original surface is preserved, and indeed can be brought to visibility through a careful removal of the corrosion crust. The case with iron is vastly different. Here, the oxides are formed *in situ* and, because of the enormous increase in specific volume penetration of the corrosion into the interior, it leads to serious changes in shape and, ultimately, complete disintegration of the object.

In addition to the metal, the reactive species involved also determine the nature of the corrosion process to an appreciable degree. One primary reactant that all corrosive processes have in common is water, in the absence of which no corrosion of any kind takes place. The presence of other species influences both the reaction rates and the nature of the corrosion products. With bronze, the simplest process involves water, oxygen, and carbon dioxide. When bronze has been buried under conditions that include a sizable chloride content in the groundwater, nantokite (cuprous chloride) forms directly on the metal surface. This compound is highly unstable under conditions of high humidity, which can result in the feared "bronze disease." This notoriously aggressive corrosion process takes place when a bronze with a nantokite layer is exposed to humid air: the cuprous chloride reacts to form copper hydroxy chloride, a light-green, powderous material symptomatic of the condition, and hydrochloric acid. This acid, formed directly on the metal surface, reacts with the metal to produce new cuprous chloride, which in turn hydrolyses: a chain reaction has started which stops only with the destruction of the object, unless arrested by desiccation and treatment (see Fig. 7).

In the present-day environment, the high concentrations of sulfur and nitrogen oxides result in a corrosion with soluble corrosion products and no pacification (see Fig. 8); the results are clearly visible on outdoor bronze sculpture (see Air pollution).

Another air pollutant which can have very serious effects is hydrogen sulfide, which is largely responsible for the tarnishing of silver, but also has played a destructive role in the discoloration of the natural patinas on ancient bronzes through the formation of copper sulfide.

A special vulnerability is created when two metals are in contact: the electromotive force can result in an accelerated corrosion, eg, in bronzes with iron mounting pins.

Stone. An important source of damage to stone objects is mechanical in nature: both breakage and abrasion account for much of the losses on objects made of this relatively fragile material (99,101–102). Although such damages are fairly easy to prevent with careful handling procedures, more difficulties are offered by the processes of a chemical nature which play a role in stone deterioration.

The types of chemical damages are very much dependent on the nature and composition of the stone. Limestones and marbles, for example, are naturally very susceptible to acidic attack. The presence of sulfur oxides in the air and sulfate ions in groundwater has always been high enough to have a marked influence on these

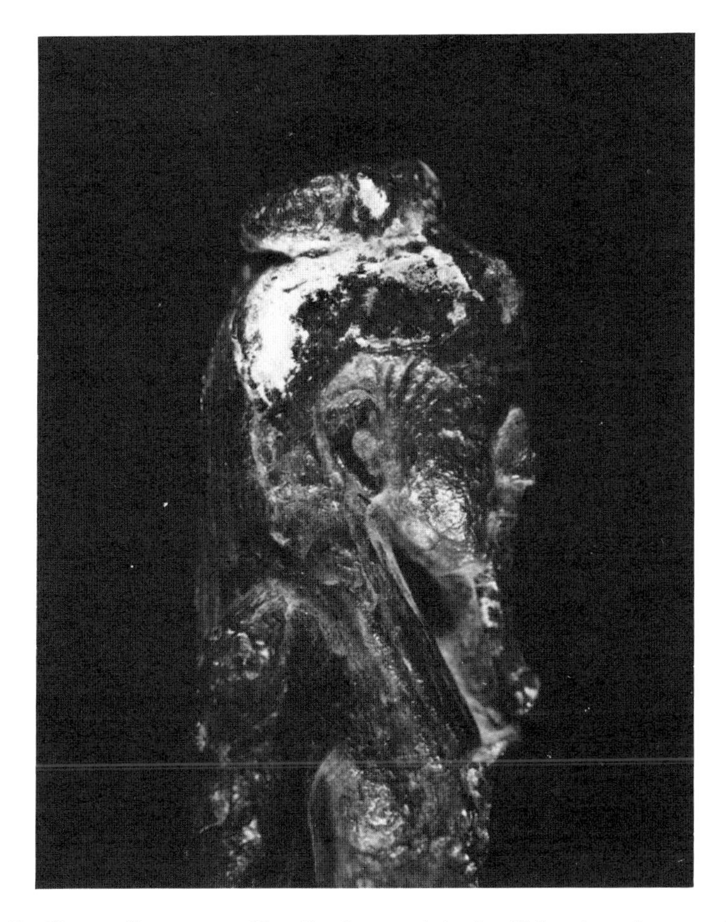

Figure 7. Bronze disease on an Egyptian bronze statuette. This extremely aggressive corrosion, recognizable by its light-green, powdery product, must be stopped immediately through appropriate treatment. Courtesy of the Research Laboratory, Museum of Fine Arts.

stones; with age, objects made out of these materials often acquire an encrustation consisting of gypsum (calcium sulfate) and redeposited calcite (calcium carbonate). The study of the penetration of the sulfation process into the stone through examination of a thin section under the petrographic microscope has been proven to be an aid in solving questions regarding the authenticity of marble sculpture. The effect of the air pollutants in modern urban environments on these stones is self explanatory. In sandstone, the removal of the binder through mobilization of the clay minerals results in a loss of cohesive strength. On the other hand, objects made of several chemically resistant stones, eg, granites and basalts, have survived millenia without much effect.

Exposure to the climatic elements is another important source of decay. Freeze–thaw cycles, in particular, which result in pressures on the pore walls of the stone's interior due to changes in volume during the phase transition of the water, are extremely harmful.

A similar mechanical effect is caused by the presence of water-soluble salts in the stone. Such salts can be present as a result of the genesis of the stone, as in the case of limestone formed from marine deposits, but may also be introduced by groundwater.

Figure 8. Influence of air pollution. The surface of this statue has been almost completely lost through pollution-induced corrosion; the light-colored areas correspond to losses of up to 2-mm thickness. Courtesy of the Research Laboratory, Museum of Fine Arts.

Repeated dissolution and crystallization of these salts, especially those which can contain water of crystallization, results in pressures on the walls of the pores that lead to significant mechanical interior damage. An extra problem arises from the tendency of these salts to percolate toward the exterior. If, as is often the case with aged stone objects, the surface has a decreased permeability, the pressures built up below the surface can lead to losses of entire surface segments through defoliation.

Ceramics. Ceramic objects are, of course, very fragile, and mechanical damages through breakage and abrasions most likely are the principal source of their destruction.

Low fired ceramics can suffer through the rehydration of the body material; this process results in a complete loss of mechanical strength. The presence of soluble salts in porous ceramic bodies has the same disastrous results as in stone.

The deterioration of glasses, including glazes, involves the devitrification of the glass in which the inherently unstable undercooled liquid crystallizes around nucleation centers. The free oxides which are the result of this process are, in turn, prone to removal by water; the alkali oxides in particular cause great problems. The so-called weeping of antique glass is a result of the extremely hygroscopic nature of the potassium oxide which results from the devitrification of glasses made with wood-ash flux; the solution of highly concentrated potassium hydroxide and carbonate which trickles down the objects also acts, through its potent leaching ability, as a stimulator for decay.

On archaeological glass objects, layers of reaction products are formed and the main constituents of these crusts are the less-soluble compounds such as silica and calcium carbonate, which becomes calcium sulfate (see Glass).

Glazed ceramics can be subject to an additional problem: if the glaze does not "fit well," ie, if the thermal-expansion coefficient of the glaze does not match that of the body, consecutive changes in temperature can result in loss of adhesion between glaze and body. On archaeological objects, the glaze, owing to the devitrification process described above, can change so much in thermal-expansion behavior that it becomes extremely unstable under conditions other than a rigidly controlled constant temperature.

Organic Materials. Environmental conditions, especially temperature and relative humidity, above all cause deterioration of objects made from organic (plant or animal) materials (103). Not only extremely high or low values for these parameters, but also large or sudden changes, can often cause irreparable harm. The humidity of the object's environment can be especially critical. For example, wooden objects are subject to warping when exposed to conditions of very high relative humidity. This is a result of absorption of water by the wood fibers and consequent dimensional changes. High humidities also enhance the growth of molds, which destroy the objects by feeding on the organic materials. Conditions of very low relative humidity, on the other hand, are also extremely harmful. Through the desorption of cell water, dimensional changes are caused which can result in serious mechanical damage. Wood and ivory crack and split under such conditions.

At least as harmful as extremes in climatic conditions are changes in them; the resulting thermal expansions and contractions, and dimensional changes due to absorption and desorption of water, especially when occurring in temporal cycles, can lead to total destruction through loss of mechanical strength and deformations. Organic materials can be conditioned to an environment; if this happens slowly and is not followed by consecutive changes, it may result in little harm. Objects made from various organic substances have survived very well under the extremely dry conditions of a desert climate. It is the sudden changes which are often disastrous.

Materials which have been buried underwater cause a special problem. Waterlogged woods and leathers, although quite stable under such burial conditions, are in danger of irreversible damage through drying out upon recovery. Indeed, after excavations from bogs or upon recovery from underwater sites, their only chance for survival rests in being stored underwater until laboratory treatment.

Another source of damage, via chemical-decay processes, is the absorption of electromagnetic radiation, ie, light (104), especially the uv component of the spectrum. Photochemically induced deterioration is a principle cause of damage in textiles, where these processes are responsible for the fading of dyes, as well as the decomposition of several fibers (see Photochemical technology). An illustrative example of the fading of textile dyes can be seen in Figure 25 of Color photography, instant, Vol. 6. The fading effect of light on dyestuffs also causes problems with several organic pigments used in painting; in this regard, light effects the deterioration of several inorganic mineral pigments, notably red lead; the photoinduced dissociation of the green pigments verdigris (basic copper acetate) and copper resinate (a reaction product of verdigris with oil-resin media) cause color changes.

In addition to pigments, the organic binding media used in painting are also light sensitive. Next to dyed textiles, works of art on paper (eg, prints, drawings, photographs, watercolors, and pastels) must be regarded as extremely susceptible to photodegradation processes.

Chemical degradation (105), primarily resulting from depolymerization, oxidations, and hydrolysis, is the third major cause of decay. These reactions are especially harmful in objects made from materials that contain cellulose, such as wood, cotton, and paper. The chemistry of these degradation processes is quite complex; an important role is played by the reaction products, acidic radicals that have a catalytic effect on the degradation.

A special case is the degradation of cellulose nitrate, which for a long time was used as the base for photographic film. This decay process has nitric oxide as one of the reaction products, which serves as an accelerator for the degradation and makes this process autocatalytic. This process can render the material subject to spontaneous combustion, which leads to serious accidents and losses when stored film ignites. Film archives store such films under freezing conditions, and often have extensive programs to copy all old celluloid film onto polyester-based film and destroy the old film. Cellulose nitrate also has widespread use as an adhesive and coating material. Presently, manufacturers add stabilizers to their products, eg, sodium carbonate as neutralizer; nevertheless, many conservators are hesitant to use them because of the inherent instability and the dangers to the object from nitric acid, which may form when the nitric oxide combines with moisture.

Composites. Of this class of objects, paintings are easily recognized as very important representatives. Composite objects have an extremely high vulnerability. First, all the conditions that may negatively effect any of their component materials are therefore harmful to such objects in their entirety. Second, conditions favorable to the survival of one of the component materials may affect the stability of another negatively.

In terms of the natural deterioration, the fact that various materials are, in one way or another, adhered to each other, forms a source of potential trouble. Any dimensional change in one of the components or a differential in dimensional changes between the components as a consequence of changes in environmental conditions results in a strain on the adhesion of the various parts, and finally in failure of this adhesion (see Fig. 9). This is one of the principal causes of losses in panel paintings, where the many dimensional changes in the wooden support cause losses in adhesion between the paint layer and the support. However, paintings on canvas are also subject to this effect, and actually the difference in character between the different paint layers is already enough to cause problems under fluctuating climatic conditions.

A special problem is caused when one of the components promotes chemical-decay reactions in another of the materials. For example, copper pigments promote the deterioration of proteinaceous binding materials such as glue and supports such as silk fabrics (see Fig. 10). Similarly, iron has been found to promote degradation of paper,

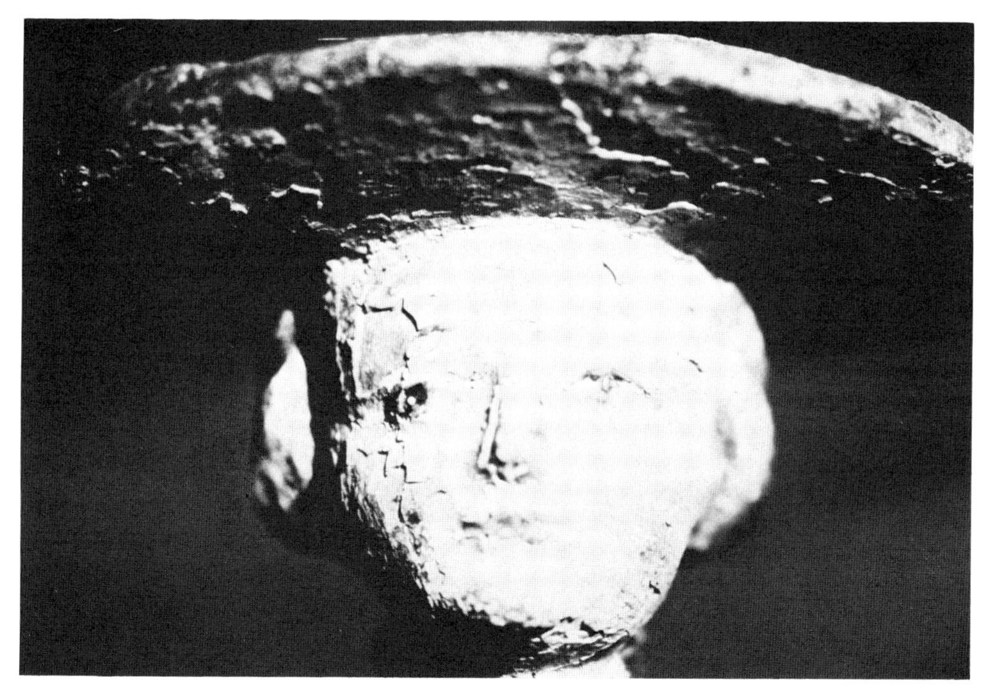

Figure 9. Unfavorable climate conditions have resulted in the paint losses on the face of this sixteenth century Swiss polychrome wooden sculpture. Courtesy of the Research Laboratory, Museum of Fine Arts.

a reason why the historic iron–gall apple ink in many cases has deteriorated the paper onto which it was applied (see Inks).

Preventive Conservation

Temperature and Humidity. No cure can completely undo damages to art objects, which often could have been prevented (106). As discussed in the preceding sections, many deterioration processes are dependent upon environmental conditions, and it is only logical that the conservator first of all directs his or her attention to the optimum environmental conditions.

Of the various environmental factors, temperature is probably the easiest to control. The main concern is that the temperature remains constant, without large or sudden fluctuations, in order to prevent the thermal expansions and contractions that are particularly dangerous to composite objects owing to differential volume changes. Another factor regarding temperature is its inverse relation with relative humidity under conditions of constant absolute humidity, such as exist in closed areas. Extremes in temperature are undesirable, especially very high ones, which can only serve to increase reaction rates. At the areas in which the objects are exhibited and stored must be accessible to visitors and staff, a reasonable temperature setting is room temperature; generally recommended settings are at about 18°C.

Although the temperature can be controlled with a well-designed air-conditioning system, the small fluctuations which most cycling systems cause may be very harmful. The temperature–time record should be a continuous, flat graph. Even more difficulties

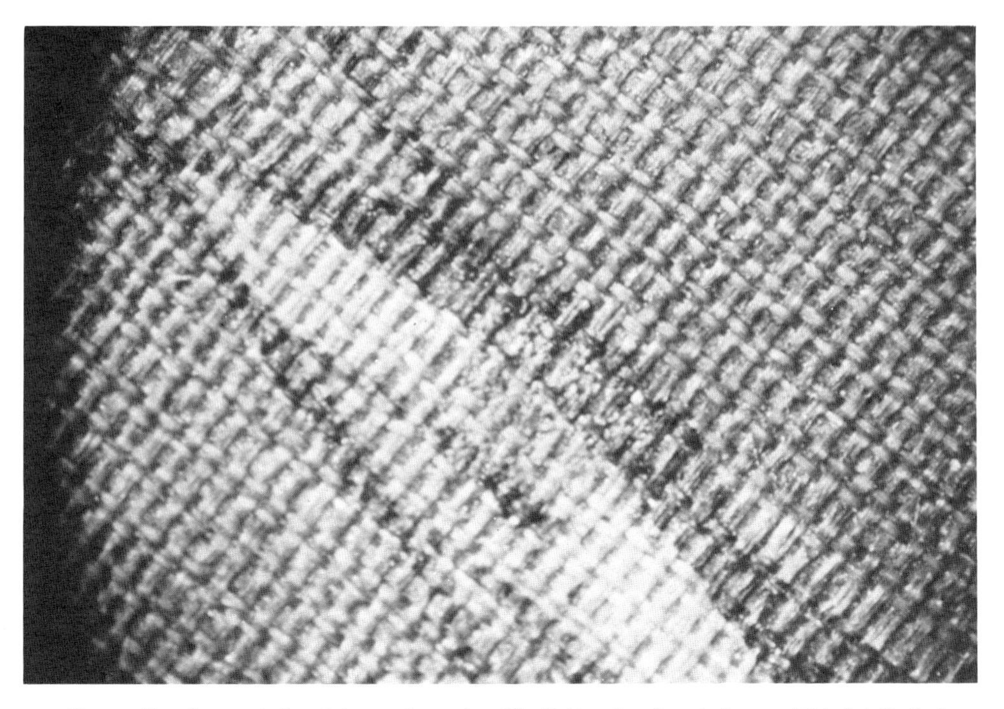

Figure 10. Copper-induced decay of proteins. The light-colored central area of this detail of a Japanese painting on silk was originally painted green with the pigment malachite. Decay of the glue medium, induced by the copper in this pigment, has resulted in the loss of the paint in this area. The silk has also been attacked, resulting in a bleaching. Courtesy of the Research Laboratory, Museum of Fine Arts.

are encountered with the humidity. Not only is it important that the relative humidity remain constant, but also the desirable value depends on the material. Many of the larger institutions have sought to solve the problem through the installation of extensive climate-control systems, which serve to keep both temperature and humidity at a constant value by means of heating, refrigeration, humidification, and dehumidification (see Air conditioning). This is not done without some problems, especially with regard to the humidity. It is a chore to remove excess moisture in humid seasons; for example, because of entrances continuously opening and closing to admit visitors. However, the main difficulties are encountered with humidification in extremely dry winter conditions such as exist along the northeastern seaboard. In existing buildings, the installation of a system capable of efficiently regulating humidity becomes extremely expensive, with the necessary modifications to the building, such as the multiple glazing needed to prevent condensation, adding to the cost. Moreover, the cost of the energy expended to operate these systems is, at present, very high.

The main advantage of wholesale climate control lies in the easy access to the objects, both for visitors and staff, and the absence of differences in conditions between various spaces within the institution, eg, storage areas, conservation laboratories, and exhibition galleries. Whereas the problem of constant humidity can be solved, the actual values for the rh are a matter of compromise; metals, stone, and ceramics are best served with humidities as low as possible, but organic materials generally require higher values. An accepted compromise is the maintenance of the relative humidity at a strictly controlled level of 50–55%.

An alternative to these macroclimate systems is the creation of microclimates; the objects are placed within smaller spaces, such as cases, in which an ideal environment is maintained. One possibility is to install equipment to control the climate in individual cases, or groups of cases with similar materials, by mechanical means.

If the temperature of the space in which the object is placed were absolutely constant, a sealed case with, consequently, a constant absolute humidity would also have a constant relative humidity. Since the temperature is subject to some variations and totally leakproof cases are not easy to build, a second solution is often sought by placing the objects in reasonably well-sealed cases in which the relative humidity is kept at a constant value by means of a buffering agent. Certain grades of silica gel or selected clay minerals are often used as the agent; their appropriateness depends on a suitable relationship between moisture-retention capacity and relative humidity. The buffering material is preconditioned under the selected relative humidity and, after equilibration, installed in the case. This method of microclimate control has proven to be very efficient, not only in exhibition cases and storage spaces, but also in packing crates used for the transportation of sensitive objects. Rigid monitoring of the climatic conditions is an absolute requirement for a successful preventive-conservation program. The desirability of a continuous measurement and a written record favors the use of recording thermohydrographs. Generally, these instruments measure the temperature via the thermal expansion of a metal strip and the relative humidity through dimensional changes induced in humidity-sensitive elements, often a bundle of human hairs. Less-expensive elements are used in dial hygrometers, the smaller of which are appropriate for applications within cases. All of these instruments need regular calibration against dry–wet bulb systems, either sling or motor-driven psychrometers. The use of electronic sensors in sealed cases has been rather limited until recently, both as a result of cost and because of problems with the accuracy of humidity sensors, especially caused by dust contamination.

Light. In the section on deterioration processes, the role of light, and especially its uv component, is discussed. It is imperative that the exposures of especially photosensitive materials is kept to a minimum.

In many museums, the use of natural daylight for illumination is preferred on aesthetic grounds, as its spectral distribution results in an ideal color rendition. As the uv component does not play a role in the visualization of the object and is the most harmful, it must be completely removed. This can be done through the use of appropriate filters, available either in the form of rigid sheets of plexiglass doped with a strongly uv-absorbing organic material or as flexible sheets, adhered to the glass windows.

When artificial light is used, those light sources which produce a significant amount of uv radiation, such as fluorescent bulbs, should also be provided with a suitable filter, such as clear transparent sleeves made from a plastic doped with a uv absorber (see Uv stabilizers). It is generally accepted that the admissible limit of uv light is its proportion in the light emitted by an incandescent tungsten lamp, ca 70 μW uv/lm. There is also a problem in exposure to visible light. The ideal condition, which is no exposure, cannot be maintained continuously. It is, however, clear that storage areas should be kept dark. For the determination of admissible light levels for exhibition galleries, a compromise must be sought between the need for minimum exposure and the desirability of good visibility and color perception. Various international organizations involved with the conservation of art objects have adopted a set of two illuminance levels, with the higher value of 150 lx deemed acceptable for the exhibition of oil and tempera paintings, oriental lacquer, ivory, horn, bone, and undyed leather. For the most photosensitive objects, such as textiles of various kinds, watercolors, prints, drawings, manuscripts, dyed materials, etc, a maximum illuminance of 50 lx is recommended.

One method of reducing the exposure for extremely sensitive objects is to exhibit them only for limited periods and to maintain a regular rotation schedule. In Japan, for example, some extremely important paintings can only be seen a few days per year.

Pollutants. The problems posed by air pollutants have become very serious. Within the museum, measures can be taken to remove harmful substances as efficiently as possible by means of the installation of appropriate filter systems in the ventilation equipment. Without minimizing the difficulties encountered in this endeavor, it is at least not an *a priori* lost proposition. Proposed specification values for museum climate-control systems require filtering systems with an efficiency for particulate removal in the dioctyl phthalate test of 60–80%, and systems able to limit both sulfur dioxide and nitrogen dioxide concentrations to $<10\ \mu g/m^3$, and ozone to $<2\ \mu g/m^3$.

Hydrogen sulfide, a pollutant which has traditionally been a problem in the tarnishing of silver and the discoloration of bronze patinas, can be dealt with in the filters of the climate-control system as well as through the use of proper absorbing agents, as, for example, the paper fabricated especially for this purpose and treated with activated charcoal, within a microclimate. In the preservation of outdoor art objects, the problems caused by air pollution are overwhelming. For metal objects, such as bronze sculpture, a possible solution is the application of a protective barrier layer in the form of a surface coating (see Barrier polymers). Acrylic lacquers, sometimes doped with a corrosion inhibitor such as benzotriazole, and waxes are most frequently used for this purpose (see Waxes). The tremendous cost involved in an effective program of maintenance, necessary in the use of barrier surface layers, is often an insurmountable impediment. With outdoor art objects composed of other materials, eg, stone, the problem is worse because no efficient protection measures are yet available. Surface coatings are often impossible, because they would also seal the surface against moisture passage, which, especially in the presence of water-soluble salts in the stone, would be disastrous. At the same time, the ravages inflicted on especially acid-sensitive stones such as marble and limestone have already resulted in the virtual destruction of innumerable sculptures, and inflicted irreparable damage to architectural monuments.

Physical Safety. Without detail, a discussion on preventive conservation would be greviously incomplete without attention to the physical safety of the objects (107).

The objects should be guarded against acts of vandalism or damage inflicted by well-meaning but ill-informed admirers who want to touch them (in many museums, greasy spots on sculpture can be seen, a result of repeated contact with bare hands). An inordinate amount of damage, however, results from handling by staff, not necessarily through carelessness but rather as a result of unawareness of the mechanical weaknesses of the various materials, especially with ancient objects, where the material properties have been affected by aging processes. Preventive conservation should therefore include a vigorous training and education program for all who handle art objects.

Another area in which the assistance and supervision of a conservator is desirable is in the design and manufacture of mounting devices. Here, again, severe damage has sometimes been inflicted, yet could have been easily prevented with a few simple modifications.

Consolidation and Restoration

The following sections are by no means meant to provide an exhaustive discussion of all possible treatment methods used in the conservation laboratory. Rather, a few of the most common problems and some typical methods of dealing with them will be summarized (108). The conservator, because of the wide variation in conditions caused by the different histories of otherwise similar objects, has to approach each case individually, assessing the problems for each individual object and devising, where necessary, a custom-tailored treatment.

Paintings. Conservation problems in paintings can be considered according to the stratus in which they occur, ie, in the varnish, the paint layers, or the support (109–110).

All natural-resin varnishes exhibit, with age, some or all of the following problems, all related to chemical deterioration caused by autoxidation, association, rearrangement, and loss of volatile oils. First, the surface may become dirty, with the dirt adhering to small blemishes and failures that have developed through the aging of the varnish layer. Second, the darkening or yellowing of the varnish film is the result of oxidation processes. This effect progresses gradually, from a general yellowing which changes the color values of the painting to a darkening which reduces its visibility. The third common problem is the so-called blooming, actually a clouding of the varnish caused by small cracks developing in the coating. Treatment can vary from simple dusting to complete removal of the varnish coat. Two frequently used treatment methods are regeneration and replacement (111). Regeneration, which is only applicable to soft resins, involves the exposure of the varnish to solvent vapors in order to partially redissolve the broken-resin layer and recover the film as a uniform, unbroken mass. More commonly, however, the old resin coat must be removed by means of suitable solvents. Any method requires extreme care and a generous amount of experience on the part of the conservator.

Once removed, the varnish is replaced with a synthetic resin, which is not susceptible to yellowing and which does not change in its solubility properties. The most frequently used plastics are poly(vinyl acetate) and acrylic resins, especially a poly(methyl acrylate-*co*-ethyl methacrylate), which have undergone a good amount of testing (see Acrylic ester polymers; Vinyl polymers). Regrettably, their optical properties are not as good as those of dammar, the most popular natural resin; much work is being done on the development of a polymer that has both satisfactory chemical and optical properties.

The most common problem in the paint layers, which can have a wide variety of causes, is loss of adhesion. Upon drying of the medium, the paint layers develop shrinkage cracks. In itself, this is not a particularly worrisome phenomenon, but, if through any cause the adhesion between paint layers and ground or between ground and support is lost, the paint will begin to flake. First, the flakes curl up, and finally they become completely detached and lost.

Flaking paint is treated by infusion of an adhesive in the areas where needed, followed by resetting the flakes on the substrate; the softening of the paint needed to bend it back is effected through solvent action or heat. Losses can only be filled and inpainted. Inpainting may also be necessary when cracks become so wide as to seriously affect the visual appearance of the painting.

Failure of a canvas support occurs when, due to aging, the fabric becomes hard and brittle and the fibers break; the fabric loses all strength and is unable to function as a support for the paint layers. The common solution, when this has progressed so far as to endanger the safety of the painting, is the so-called lining or relining procedure, which involves the application of a new fabric to the back of the old one. The front of the painting is provided with a temporary facing, then a new canvas is attached to the back with an adhesive, under gentle, uniform pressure applied by means of a low vacuum system. Traditionally, the adhesive was a paste of animal glue and flour. However, the dimensional changes in the various layers frequently induced when this aqueous adhesive was applied to a canvas painting and the difficulties in removal of aged glue paste led to the search for an alternative adhesive. Thus, wax–resin adhesives were formulated. The wax is applied in the molten state (see Adhesives). Wax lining, however, often causes a darkening of the painting, and the adhesive, if it penetrates through the paint layers, is almost impossible to remove. Recent experiments have been performed with synthetic adhesives which can be used at room temperature.

Wooden panels have a tendency to warp and eventually crack (112). If the deformation of these supports becomes dangerous to the paint layers, some measures must be taken. In the past, the restriction of the movement of the support was attempted through the application of rigid cradles, ie, thick wooden strips running both horizontally and vertically across the back. This was not very successful; cracks developed between the strips of the cradle and the cleavage between ground and wood became often more aggravated. There is no easy and safe solution to this problem; the best answer lies in prevention through strict climate control of the painting's environment. In cases where the deformation has progressed too far and immediate danger to the painting exists, a transfer can be done in which the painting is again faced on the front, and the original support is tooled down until the groundlayer is exposed, onto the back of which a new support is applied.

Wall paintings become endangered when the wall which is their support decays. This is often the case with the walls on which the Italian Medieval and Renaissance artists applied their masterpieces. A technique which was specially developed and is now routinely used for the removal of fresco paintings from the wall, the so-called strappo technique, involves the attachment of a canvas facing to the front of the painting with a strong glue. When this glue dries and shrinks, the top layer, which includes the upper preparation layer with the pigments and which is penetrated by the glue, tears off the underlayers and the wall. It is then adhered to a stable support, after which this assembly can be remounted in the original location, yet free of the wall.

Works of Art on Paper. Paper (qv) is a cellulose (qv) product. Until the middle of the nineteenth century, cellulose was obtained from vegetable fibers in linen and cotton rags. This provided a source of relatively pure cellulose. After paper had been cast from the pulp, it was sized separately with gelatin glue. The increasing demand for paper caused the papermakers to seek other sources of cellulose, which they found in wood. When wood is used, however, the cellulose has to be separated from the lignin and natural resins. Strong chemical treatments were devised, such as the sulfite treatment, which had the additional effect of reducing the polymerization degree of the cellulose; this, in turn, has severe consequences for the permanence of the paper. Another source of trouble was also introduced during this period: instead of the separate sizing with gelatin, a rosin–alum mixture was added to the pulp, resulting in intrinsic acidity of the paper (113–116).

The degradation of paper has been recognized as the result of two separate processes. First, there is the hydrolytic degradation of the cellulose. This reaction is acid-catalyzed; hence, the acidity present in aged paper, due to manufacturing techniques, acidic inks, and degradation processes, is extremely dangerous. A second process is oxidation with molecular oxygen; the radicals resulting from this reaction promote depolymerization of the cellulose.

In an attempt to save acidic paper, deacidification treatments have been devised using a variety of reagents, eg, calcium carbonate or magnesium carbonate in water, nonaqueous solutions of magnesium methoxide or methyl magnesium carbonate, and, in order to solve the logistical problems arising in the treatment of complete library holdings, diethylzinc, which has been used for gaseous mass deacidification. Many studies are still ongoing, evaluating the effects of the various deacidification procedures on the rate of the oxidation reactions.

Another common problem with which the paper conservator frequently has to deal is so-called foxing, or the formation of many small, rust-brown-colored spots in the paper. The exact processes involved in foxing are still under investigation, but there seems to be a correlation between this phenomenon and the presence of certain transition metals, such as iron and zinc. The common treatment is bleaching with reagents such as hypochlorite or chloramine-T.

Great danger to paper comes from the growth of molds, which must be prevented through rigorous humidity control. In case of an outbreak, fumigation can be performed with thymol vapor. Access of insects to paper is also a common source of damage. For repairs of tears and damages, in general, three techniques are employed: consolidation by affixing a piece of Japanese paper to the back to bridge the gap; filling in of the damage with a piece of matching paper, a technique which requires extreme skill and a large assortment of papers; and casting a repair into the damage with paper pulp.

Metal Objects. Obviously, the most common conservation problem with metal objects occurs when corrosion processes form a threat to the safety of the object or disfigure its appearance to an unacceptable degree (98,117–118). Treatments are intended to stabilize corrosion processes and to remove aesthetically displeasing corrosion crusts. The latter requires a great deal of thought and discussion as to when a corrosion layer ceases to be a desirable patina and becomes unacceptable.

On highly polished surfaces, as on silverware, the slightest tarnish constitutes a disfiguring effect and must be removed and prevented. Often the cleaning is done by careful polishing with a mild abrasive, such as precipitated chalk. On ancient silver, such polishing results in an undesirable shine; to preserve the soft luster of the metal, cleaning is preferably accomplished through chemical or electrochemical means. Chemical removal of the sulfides and chlorides can be done with formic acid, thiourea, thiosulfate, or thiocyanate. Electrolytic treatment involves the reduction of the corrosion product to metal in an electrochemical cell, in which the object forms the cathode.

In order to prevent recurrence of the corrosion, a lacquer can be applied, or the environment of the object can be controlled strictly with regard to relative humidity and pollutants.

Bronze disease, discussed above, necessitates immediate action to halt the process and remove its causes. For a long time, stabilization was sought by removal of the cuprous chloride by immersing the object in a solution of sodium sesquicarbonate. This process is, however, extremely time-consuming, frequently unsuccessful, and often the cause of unpleasant discolorations of the patina. A localized treatment is the excavation of cuprous chloride from the affected area until bare metal is obtained, and application of silver oxide or silver nitrate which serves to stabilize the chloride by formation of silver chloride. For objects where this is not practical because of widespread affection, the treatment is immersion of the object in a solution of benzotriazole, a corrosion inhibitor for copper.

Removal of disfiguring corrosion crusts can be accomplished in several ways. One approach is mechanical cleaning, the advantage of which is the control it affords the conservator in the degree and extent of cleaning. A second possibility is the use of chemical means. Finally, electrolytic reduction can be used with great advantage in those cases where the original surface needs to be recovered, eg, in order to discover inscriptions or engraved decorations, or where the object has been dimensionally disfigured (see Fig. 11). In the latter case, the metal can, after reduction, be annealed and subsequently reshaped.

Some special problems are encountered in the treatment of iron, especially when recovered from underwater sites. Corrosion stimulators, especially chlorides, must be removed. One treatment used in cases where large amounts of objects are recovered is hydrogen reduction at high temperatures. Other reduction treatments are electrolytic. Attempts to remove the chlorides by washing and leaching procedures have been unsuccessful. Stabilization of corroded iron has been effected by impregnation with polymers, or by treatment with tannic acid. The latter technique is quite historic and has been used for a long time for the protection of iron tools.

Stone Objects. The objectives in the treatment of stone objects are primarily cleaning, stabilization, consolidation, repair, and restoration (99,101–102,119).

Cleaning can vary from a light dusting to the removal of stubborn grime and stains with solvents and detergents. The latter can be used in a poultice method to prevent driving the extraneous material further into the stone.

Stabilization involves the removal of the cause of deterioration, frequently soluble salts present in the stone. If the structural strength of the stone permits it, this can be done through soaking. The object is placed underwater in a tank, and the water is changed regularly. Another method is by application of poultices on the surface.

Consolidation is the introduction of a substance into the interior which will add extra mechanical strength to the stone. Such impregnations have been done with both synthetic resins and inorganic compounds. For objects which will be stored and exhibited in a controlled indoor environment, impregnation with an organic material may be deemed acceptable; however, for stone objects displayed outdoors, the inherent instability of the organic materials is a serious drawback to their use. The main practical problems with resin impregnation are the viscosity of the resin solution and the migration of resin to the surface upon solvent evaporation. Many different resins in various solvents have been tried, most notably poly(vinyl acetate), poly(methyl methacrylate), other similar acrylics, and epoxies. Another approach has been to impregnate the stone with epoxy or acrylic monomers and subsequently instigate polymerization *in situ*.

Apart from the instability of the resins, a negative aspect of such impregnations

Figure 11. Electrolytic reduction. (**a**) A badly corroded Meroitic bronze offering tray. (**b**) A detail of the handle. (**c**) Supported in a brass screen bag, the tray is lowered into a tank filled with a 2 wt % NaOH solution. (**d**) Positioned between two stainless-steel anodes, the tray forms the cathode of an electrolytic cell. The hydrogen evolving at the tray (cathode) induces the reduction of the oxidation products back into their metallic state. (**e**) The same tray, after reduction treatment. (**f**) The handle, after treatment, reveals fine details previously invisible. Courtesy of the Research Laboratory, Museum of Fine Arts.

is that all pores of the stone are completely plugged, which makes moisture movement impossible. If, however, any failure in the resin barrier should develop, moisture can penetrate the interior, which often cannot be impregnated completely owing to the problems alluded to above, and subsequent moisture migration, especially when salts are still present in the stone, could lead to exfoliation at the depth of resin penetration. For this reason, an impregnation with materials which only form a thin coating on the pore walls is preferable. Several processes have been devised wherein inorganic com-

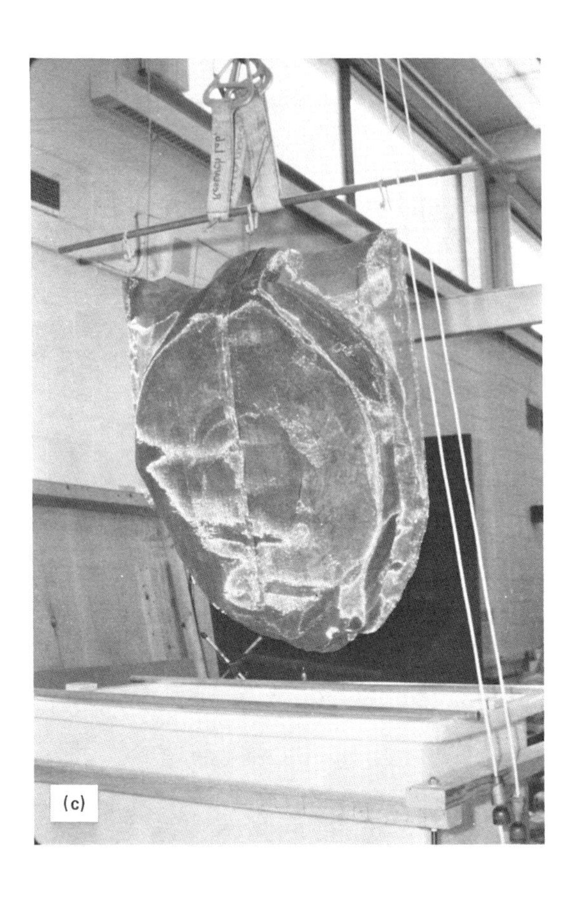

Figure 11. *(continued)*

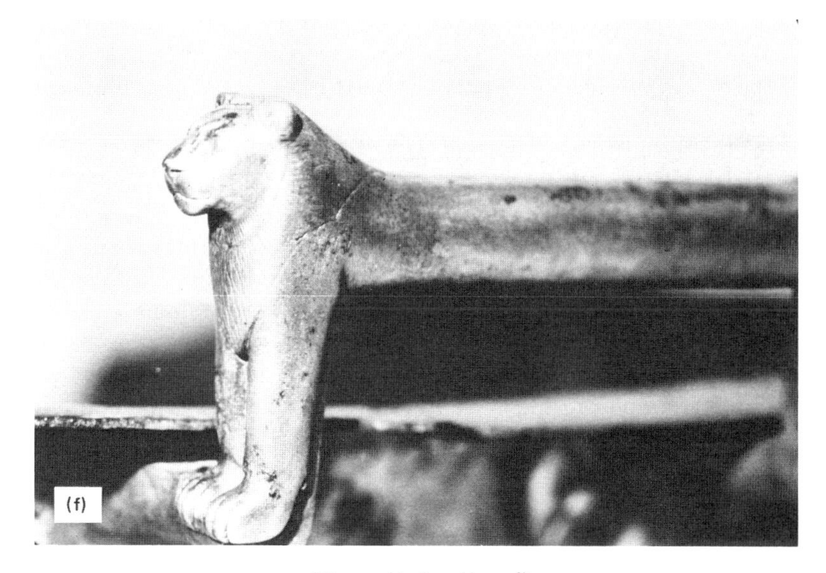

Figure 11. *(continued)*

pounds are thus deposited, such as impregnations with barium hydroxide and urea, which result in homogeneous precipitation of barium carbonate, in turn slowly converted to barium sulfate; with barium ethyl sulfate, resulting in deposition of barium sulfate; with ethyl silicate, causing a silica precipitate; or with alkoxysilanes, resulting in a deposit of silica after curing of the impregnant.

Ceramics and Glass Objects. The adhesives which have been used historically to mend broken ceramics include stick shellac, a paste of lead white (lead hydroxy carbonate) in oil, and cellulose nitrate. None of these is used anymore because of the undesirable properties of these adhesives, notably their discoloration upon aging and the large changes in solubility characteristics for some of them. Most conservators use adhesives based upon poly(vinyl acetate), acrylic resins, some polyesters, and epoxies. Epoxies still have the disadvantage of discoloration, which is especially noticeable in mended glass; for this particular material, however, there is not much choice in adhesives. Low fired, porous ceramics are subject to loss of structural strength, and become sensitive to moisture. Occasionally, an impregnation with poly(vinyl acetate) or acrylic resins becomes necessary. In archaeological ceramics, salts may have been introduced during burial; these must be removed through desalination techniques such as those used for stone.

The layers of decay products and hydrated glass formed on the surface of ancient glass will often start to spall off; in such cases, an impregnation with an acrylic resin may be helpful. The treatment for weeping glass consists of stabilization through the removal of free alkalies with an acid, followed by thorough rinsing and drying, and subsequent storage under dry conditions.

Others. Though the causes of degradation phenomena in textiles (qv) (115–116,120) are many, including pollution, bleaches, acids, and alkalies, the single most important effect is that of photodegradation. Both cellulosic and proteinaceous fibers are extremely photosensitive, with the natural sensitivity of the fibers often enhanced by impurities, remainders of finishing processes, and mordants for dyes. Depolymerization and oxidation leads to decreased fiber strength and to embrittlement.

When a textile has lost its mechanical strength, it must be backed with a lining fabric. In the worst cases, consolidation through application of a polymer may be necessary. When repairs are made, great care must be taken that the mechanical strength of the repair does not surpass that of the textile, otherwise a condition of stress might cause the original to tear.

Light is also the principal cause of damage to the natural dyestuffs that fade through photodegradation (121). Once this has happened, little can be done to remedy the loss. Other major causes of worry to the textile conservator are the dangers of mold growth and of insect damage (122). Strict environmental control is needed as changes in temperature and relative humidity are extremely dangerous for these materials.

Wooden objects (112,123–124) cause several special problems, especially when they are recovered from underwater sites. Waterlogged wood (125), if allowed to dry out, will suffer irreparable damage through warping and cracking. It is, therefore, kept underwater until it arrives in the laboratory, where a few treatments are available. One is the immersion of the object in a tank with water, in which poly(ethylene glycol) (PEG) is subsequently introduced in a slowly increasing concentration, until impregnation of the wood with PEG has been achieved. Another technique involves impregnation with a solution of rosin in acetone.

Furniture is subject to damage by dimensional changes caused by the environment, as well as physical damage through abuse and wear. Insect damage can be very severe; because the infestation is in depth, a penetrating treatment must be used (122). For lesser infestations, an insecticide is often injected into the insect holes; larger infestations, however, require fumigation (see Insect-control technology). Ethylene dioxide is preferred for this purpose, as it has been tested and found to be fairly harmless to the materials of most art objects.

BIBLIOGRAPHY

1. W. J. Young, ed., *Applications of Science in Examination of Works of Art*, Museum of Fine Arts, Boston, Mass., 1959.

2. W. J. Young, ed., *Applications of Science in Examination of Works of Art*, Museum of Fine Arts, Boston, Mass., 1967.
3. W. J. Young, ed., *Applications of Science in Examination of Works of Art*, Museum of Fine Arts, Boston, Mass., 1973.
4. S. J. Fleming, *Authenticity in Art*, Crane, Russak & Co., New York, 1976.
5. P. Coremans, *Van Meegeren's Faked Vermeers and De Hooghs. A Scientific Examination*, Cassell & Co., Ltd., London, 1949; J. M. Meulenhoff, Amsterdam, 1949.
6. B. Keisch, *Science* **160,** 413 (1968).
7. D. von Bothmer and J. V. Noble, *An Inquiry into the Forgery of the Etruscan Terracotta Warriors in the Metropolitan Museum of Art*, Papers, No. 11, The Metropolitan Museum of Art, New York, 1961.
8. H. V. Michel and F. Asaro, *Archaeometry* **21,** 3 (1979).
9. R. E. M. Hedges, *Archaeometry* **21,** 21 (1979).
10. W. C. McCrone, *Anal. Chem.* **48,** 676A (1976).
11. J. V. Noble, *The Forgery of Our Greek Bronze Horse*, The Metropolitan Museum of Art Bulletin, New York, 1968, pp. 253–356.
12. K. C. Lefferts, L. J. Majewski, E. V. Sayre, and P. Meyers, *Journal of the American Institute of Conservation* **21,** 1 (1981).
13. R. H. Brill, ed., *Science and Archaeology*, M.I.T. Press, Cambridge, Mass., 1971.
14. C. W. Beck, ed., *Archaeological Chemistry*, Advances in Chemistry Series No. 138, American Chemical Society, Washington, D.C., 1975.
15. G. F. Carter, ed., *Archaeological Chemistry II*, Advances in Chemistry Series No. 171, American Chemical Society, Washington, D.C., 1978.
16. H. Hodges, *Artifacts*, J. Baker, London, 1968.
17. C. d'Andrea Cennini, *Il Libro dell' Arte;* D. V. Thompson, Jr., translator, *The Craftsman's Handbook*, Yale University Press, New Haven, Conn., 1933; republished Dover Publishers, New York, 1960.
18. G. Agricola, *De Re Metallica*, Basel, 1549; H. C. Hoover and L. H. Hoover, translators, London, 1912.
19. V. Biringuccio, *De La Pirotechnia*, Venice, 1540; C. S. Smith and M. T. Gnudi, translators, M.I.T. Press, Cambridge, Mass., 1942.
20. Theophilus, *De Divers Artibus;* J. G. Hawthorne and C. S. Smith, translators, University of Chicago Press, Chicago, Ill., 1963; republished Dover Publishers, New York, 1979.
21. N. S. Brommelle and G. Thomson, eds., *Science and Technology in the Service of Conservation*, International Institute for Conservation of Historic and Artistic Works, London, 1982.
22. M. H. Butler, *Polarized Light Microscopy in the Conservation of Paintings*, Centennial Volume, State Microscopical Society of Illinois, Chicago, Ill., 1970.
23. J. J. Rorimer, *Ultraviolet Rays and their Use in the Examination of Works of Art*, New York, 1931.
24. E. Rene de la Rie, *Stud. Conserv.* **27,** 1, 65, 102 (1982).
25. J. R. J. van Asperen de Boer, *Appl. Opt.* **7,** 1711 (1968).
26. J. R. J. van Asperen de Boer, Ph.D. Thesis, University of Amsterdam, Central Research Laboratory for Objects of Arts and Science, Amsterdam, 1970.
27. A. Gilardoni, R. A. Orsini, and S. Taccani, *X-Rays in Art*, Gilardoni SpA, Mandello Lario, Como, 1977.
28. A. Burroughs, *Atlantic Monthly* **137,** 520 (1926).
29. A. Burroughs, *Art Criticism from a Laboratory*, Little Brown & Co., Boston, Mass., 1938.
30. P. Meijers in ref. 15, pp. 79–96.
31. R. E. Alexander and R. H. Johnson in J. S. Olin and A. D. Franklin, eds., *Archaeological Ceramics*, Smithsonian Institution Press, Washington, D.C., 1982, pp. 145–154.
32. *Art and Autoradiography*, The Metropolitan Museum of Art, New York, 1982.
33. M. J. Aitken, *Physics and Archaeology*, Clarendon Press, Oxford, 1974.
34. W. J. Young, ed., *Application of Science to Dating of Works of Art*, Museum of Fine Arts, Boston, Mass., 1976.
35. W. F. Libby, *Radiocarbon Dating*, University of Chicago Press, Chicago, Ill., 1952.
36. G. Harbottle, E. V. Sayre, and R. W. Stoenner, *Science* **206,** 683 (1979).
37. R. E. M. Hedges, *Archaeometry* **23,** 3 (1981).
38. E. K. Ralph in ref. 34, pp. 77–79.
39. J. Bauch and D. Eckstein, *Stud. Conserv.* **15,** 45 (1970).

40. E. T. Hall, J. M. Fletcher, and M. F. Barbetti in ref. 34, pp. 68–73.
41. S. J. Fleming, *Dating in Archaeology*, Dent, London, 1977.
42. M. Ikeya, *Archaeometry* **20,** 147 (1978).
43. P. M. Masters and J. L. Bada in ref. 15, pp. 117–138.
44. N. S. Baer, T. Jochsberger, and N. Indictor in ref. 15, pp. 139–149.
45. B. Keisch in ref. 3, pp. 193–198.
46. H. Kuehn in *Code of Ethics and Standards of Practice*, American Institute for Conservation of Historic and Artistic Works, Washington, D.C., 1979, pp. 199–205.
47. R. J. Gettens and G. L. Stout, *Paintings Materials. A Short Encyclopedia*, D. van Nostrand, New York, 1942; republished Dover Publishing, New York, 1966.
48. A. P. Laurie, *The Painter's Methods and Materials*, J. B. Lippincott, Philadelphia, Pa., 1926; republished Dover Publishing, New York.
49. D. V. Thompson, *The Materials and Techniques of Medieval Painting*, Allen & Urwin, London, 1936; republished Dover Publishing, New York, 1956.
50. R. Mayer, *The Painter's Craft*, D. van Nostrand Co., New York, 1948; republished The Viking Press, New York, 1975.
51. R. Mayer, *The Artist's Handbook of Materials and Techniques*, The Viking Press, New York, 1970.
52. C. L. Eastlake, *Methods and Materials of Painting of The Great Schools and Masters* (formerly titled: *Materials for a History of Oil Painting*), Longman, Brown, Green and Longmans, London, 1847; republished Dover Publishing, New York, 1960.
53. R. Harley, *Artist's Pigments c. 1600–1835*, 2nd ed., Buttersworth, London and Boston, Mass., 1982.
54. M. Hours, *Conservation and Scientific Analysis of Paintings*, Van Nostrand Reinhold, New York, 1976.
55. J. Plesters, *Stud. Conserv.* **2,** 110 (1956).
56. E. Martin, *Stud. Conserv.* **22,** 63 (1977).
57. L. Masschelein-Kleiner, J. Heylen, and F. Tricot-Marckx, *Stud. Conserv.* **13,** 105 (1968).
58. J. S. Mills and R. White in N. S. Brommelle and P. Smith, eds., *Conservation and Restoration of Pictorial Art*, Buttersworth, London and Boston, Mass., 1976, pp. 72–77.
59. J. S. Mills and R. White, *Organic Mass Spectrometry of Art Materials: Work in Progress*, National Gallery Technical Bulletin, London, 1981, pp. 3–19.
60. L. Atchison, *A History of Metals*, Interscience Publishers, New York, 1960.
61. R. J. Forbes, *Metallurgy in Antiquity*, E. J. Brill, Leiden, 1950.
62. R. E. Stone, *Metropolitan Museum Journal* **16,** 37 (1982).
63. R. J. Gettens, *The Freer Chinese Bronzes. Vol. II. Technical Studies*, The Freer Gallery of Art, Smithsonian Institution Press, Washington, D.C., 1969.
64. D. Cushing in ref. 2, pp. 53–66.
65. C. S. Smith in ref. 2, pp. 20–52.
66. M. J. Hughes, M. J. Cowell, and P. T. Craddock, *Archaeometry* **18,** 19 (1976).
67. O. Werner, *Spektralanalytische und Metallurgische Untersuchungen an Indischen Bronzen*, E. J. Brill, Leiden, 1972.
68. S. Junghans, E. Sangmeister, and M. Schroeder, *Studien zu den Anfaengen der Metallurgie 2*, Berlin, 1974.
69. P. Meijers, L. van Zelst, E. V. Sayre in ref. 14, pp. 22–33.
70. R. H. Brill, W. R. Shields, and J. M. Wampler in ref. 3, pp. 73–83.
71. F. R. Matson, *Ceramics and Man*, Wenner-Grenn Foundation for Anthropological Research, New York, 1965.
72. A. Shepard, *Ceramics for the Archaeologist*, Washington, D.C., 1956.
73. J. S. Olin and A. D. Franklin, eds., *Archaeological Ceramics*, Smithsonian Institution Press, Washington, D.C., 1982.
74. F. R. Matson in D. Brothwell and E. Higgs, eds., *Science in Archaeology*, Thames and Hudson, London, 1963.
75. P. Vandiver in ref. 73, pp. 167–179.
76. A. Lucas and J. R. Harris, *Ancient Egyptian Materials and Industries*, 4th ed., E. Arnold, London, 1962.
77. J. V. Noble, *The Techniques of Painted Attic Pottery*, Watson-Guptill, New York, 1965.

78. M. Medley, *The Chinese Potter. A Practical History of Chinese Ceramics*, Charles Scribner's Sons, New York, 1976.
79. S. Frank, *Glass and Archaeology*, Academic Press, London, 1982.
80. E. V. Sayre in ref. 2, pp. 145–154.
81. R. W. Smith in F. R. Matson and G. E. Rindone, eds., *Advancess in Glass Technology, Part 2*, Plenum Press, New York, 1963, pp. 283–290.
82. H. C. Beck and C. G. Seligmann, *Nature (London)* **133,** 982 (1934).
83. C. G. Seligmann, P. D. Ritchie, and H. C. Beck, *Nature (London)* **138,** 721 (1936).
84. L. van Zelst in S. L. Hyatt, ed., *The Greek Vase*, Hudson-Mohawk Association of Colleges and Universities, Latham, N.Y., 1981, pp. 119–134.
85. R. H. Brill and S. Moll in ref. 81, pp. 293–302.
86. H. W. Catling and A. Millett, *Archaeometry* **8,** 3 (1965).
87. R. Abascal, G. Harbottle, and E. V. Sayre in ref. 14, pp. 81–99.
88. R. L. Bishop, G. Harbottle, and E. V. Sayre in J. E. Sabloff, ed., *Analyses of Fine Paste Ceramics*, Memoirs of the Peabody Museum of Archaeology and Ethnography 15, No. 2, Harvard University, Cambridge, Mass., 1982.
89. C. Renfrew and J. Springer-Peacy, *Annual of the British School at Athens* **83,** 45 (1968).
90. P. Meijers and L. van Zelst, *Radiochim. Acta* **24,** 197 (1977).
91. K. Germann, G. Holzmann, and F. J. Winkler, *Archaeometry* **22,** 99 (1980).
92. M. Saltzman in ref. 15, pp. 172–185.
93. J. H. Hofenk-de Graaf, *Natural Dyestuffs. Origin, Chemical Constitution, Identification*, ICOM paper 1969, Central Research Laboratory for Objects of Art and Science, Amsterdam, 1969.
94. D. Hunter, *Papermaking. The History and Technique of an Ancient Craft*, A. E. Knopf, New York, 1943; republished Dover Publishing, New York, 1978.
95. W. A. Cote, ed., *Papermaking Fibers*, Syracuse University Press, Syracuse, N.Y., 1980.
96. *Code of Ethics and Standards of Practice*, American Institute for Conservation of Historic and Artistic Works, Washington, D.C., 1979.
97. B. F. Brown and co-eds., *Corrosion and Metal Artifacts*, NBS Special Publication 479, National Bureau of Standards, Washington, D.C., 1977.
98. T. Stambolov, *The Corrosion and Conservation of Metallic Antiquities and Works of Art: A Preliminary Survey*, Central Research Laboratory for Objects of Art and Science, Amsterdam.
99. S. Timmons, ed., *Preservation and Conservation: Principles and Practice*, Smithsonian Institution Press, Washington, D.C., 1976.
100. W. D. Richey in ref. 21, pp. 108–118.
101. R. Rossi-Manaresi, ed., *The Conservation of Stone-I*, Centro per la Conservazione della Sculture all'Aperto, Bologna, 1976.
102. R. Rossi-Manaresi, ed., *The Conservation of Stone-II*, Centro per la Conservazione della Sculture all'Aperto, Bologna, 1981.
103. N. Stolow, *Curator* **9,** 175, 298 (1966).
104. T. B. Brill, *Light: Its Interaction with Art and Antiquities*, Plenum Press, New York, 1980.
105. R. L. Feller in J. C. Williams, ed., *Preservation of Paper and Textiles of Historic and Artistic Value*, Advances in Chemistry Series No. 164, American Chemical Society, Washington, D.C., 1977.
106. G. Thomson, *The Museum Environment*, Buttersworth, London and Boston, Mass., 1978.
107. F. K. Fall, *Art Objects. Their Care and Preservation*, Laurence McGilvery, La Jolla, Calif., 1973.
108. H. J. Plenderleith and A. E. A. Werner, *The Conservation of Antiquities and Works of Art*, 2nd ed., Oxford University Press, London, 1971.
109. G. L. Stout, *The Care of Pictures*, Columbia University Press, New York, 1948; republished Dover Publishing, New York, 1975.
110. H. Ruehemann, *The Cleaning of Paintings*, Faber & Faber, London, 1968.
111. R. L. Feller, E. H. Jones, and N. Stolow, *On Picture Varnishes and Their Solvents*, Intermuseum Conservation Association, Oberlin, Ohio, 1959.
112. N. S. Brommelle, A. Moncrieff, and P. Smith, eds., *Conservation of Wood in Paintings and the Decorative Arts*, International Institute for Conservation of Historic and Artistic Works, London, 1978.
113. A. F. Clapp, *Curatorial Care of Works of Art on Paper*, Intermuseum Conservation Association, Oberlin, Ohio, 1973.
114. F. Dolloff and R. Perkinson, *How to Care for Works of Art on Paper*, 3rd ed., Museum of Fine Arts, Boston, Mass., 1979.

115. J. C. Williams, ed., *Preservation of Paper and Textiles of Historic and Artistic Value*, Advances in Chemistry Series No. 164, American Chemical Society, Washington, D.C., 1977.
116. J. C. Williams, ed., *Preservation of Paper and Textiles of Historic and Artistic Value, II*, Advances in Chemistry Series No. 193, American Chemical Society, Washington, D.C., 1981.
117. R. M. Organ in ref. 97, pp. 107–142.
118. R. M. Organ in S. Doeringer, D. G. Mitten, and A. Steinberg, eds., *Art and Technology*, M.I.T. Press, Cambridge, Mass., 1970, pp. 73–84.
119. R. Rossi-Manaresi and G. Torraca, eds., *The Treatment of Stone*, Centro per la Conservazione della Sculture all'Aperto, Bologna, 1972.
120. J. E. Leene, *Textile Conservation*, Smithsonian Institution Press, Washington, D.C., 1972.
121. T. Padfield and S. Landi, *Stud. Conserv.* **11**, 111 (1966).
122. S. R. Edwards, B. M. Bell, and M. E. King, *Pest Control in Museums: A Status Report (1980)*, The Association of Systematics Collections, Museum of Natural History, University of Kansas, Lawrence, Kan., 1981.
123. N. S. Brommelle and A. E. A. Werner, eds., *Deterioration and Treatment of Wood. Problems of Conservation in Museums*, Editions Eyrolles, Paris, 1969.
124. B. Muehlethaler, *Conservation of Waterlogged Wood and Wet Leather*, Editions Eyrolles, Paris, 1973.
125. W. A. Oddy, ed., *Problems in the Conservation of Waterlogged Wood*, National Maritime Museum, Greenwich, London, 1975.

General References

References 4, 16, 33, 47, 54, 61, 76, 79, 94, 106, 107, 108, 109, 114, and 120 are general references.

Art Archaeol. Tech. Abstr., The Institute of Fine Arts, New York University, New York, for the International Institute for Conservation of Historic and Artistic Works, London, published semi-annually.

D. Brothwell and E. Higgs, eds., *Science in Archaeology*, Thames and Hudson, London, 1969.

R. M. Organ, *Design for Scientific Conservation of Antiques*, Smithsonian Institution Press, Washington, D.C., 1968.

L. VAN ZELST
Museum of Fine Arts, Boston

FIRE-EXTINGUISHING AGENTS

Control and extinguishment of fire has always been a human concern. Water was the first agent to be used; however, the need for agents other than water became apparent as the industries of civilization introduced new and more combustible fire materials. Plastics, chemicals, petroleum products, and combustible metals have become daily fire hazards that have to be dealt with by agents other than water. Along with these products came the need for the classification of the hazards (Table 1) that they imposed and for new agents to effect their extinguishment (see also Plant safety; Flame retardants).

Classification of Fires

Agents and equipment have been classified by the National Fire Protection Association (NFPA) into four categories (Fig. 1) (1–2):

Class A fires are fires in ordinary combustible materials, eg, wood, cloth, paper, rubber, and many plastics.

Class B fires are fires in flammable and combustible liquids, gases, and greases.

Class C fires are fires that involve energized electrical equipment where the electrical nonconductivity of the extinguishing media is of importance. (When electrical equipment is de-energized, extinguishers for Class A or B fires may be used safely.)

Class D fires are fires in combustible metals such as magnesium, titanium, zirconium, sodium, and potassium.

These four classifications are used in Canada and the United States. Those used throughout Europe are shown below (3).

Class A fires involve solid materials normally of an organic nature (compounds of carbon) in which combustion generally occurs with the formation of glowing embers. Class A fires are the most common, and the most effective extinguishing agent is generally water in the form of a jet or spray.

Class B fires involve liquids or liquefiable solids. For the purpose of choosing extinguishing agents, flammable liquids may be divided into two groups: miscible and immiscible with water. Depending on their miscibility, the extinguishing agents include water spray, foam, "Light Water," vaporizing liquids, carbon dioxide, and dry-chemical powders.

Class C fires involve gases or liquefied gases in the form of a liquid spillage or a liquid or gas leak, and these include methane, propane, butane, etc. Foam or dry-chemical powder can be used to control fires involving shallow liquid spills. Water in the form of spray is generally used to cool containers.

Class D fires involve metals. Extinguishing agents containing water are ineffective and even dangerous. Carbon dioxide and the bicarbonate classes of dry-chemical powders may also be hazardous if applied to most metal fires. Powdered graphite, powdered talc, soda ash, limestone, and dry sand are normally suitable for Class D fires. Special fusing powders (eutectic) have been developed for fires involving some, especially radioactive, metals.

Electrical Fires. Combustion of electrical equipment is not necessarily a fire

Table 1. Selection of Agent by Hazard Classification

Fire-extinguishing agent, base description	Hazard Class A	Class B-1[a]	Class B-2[b]	Class C	Class D	Remarks
Aqueous agents						
water	X					
water, spray, fog	X	X	X		X	use only on magnesium and zirconium and with extreme care
wet water	X					
water, thickened	X					both organic and inorganic (slurries) thickener
water, antifreeze	X					calcium chloride and alkali salts
foam	X	X				chemical and air foam
foam, alcohol-resistant	X	X	X			
Liquefied gas						
carbon dioxide		X	X	X		
halogenated hydrocarbon		X	X	X		gas; may have some Class A capabilities in total flooding
halogenated hydrocarbon	X	X	X	X		
nitrogen		X	X	X		
Dry-chemical agents						
sodium bicarbonate		X	X	X		
potassium bicarbonate		X	X	X		
potassium carbonate		X	X	X		limited use in the United States
potassium chloride		X	X	X		
potassium sulfate		X	X	X		used in UK and France
ammonium phosphate, mono basic	X	X	X	X		multipurpose (ABC)
ammonium sulfate	X					normally used in slurry rather than as a dry chemical
Combustible metal agents						
sodium chloride					X	
sodium carbonate					X	sodium, potassium, and Na-K fires only
graphite					X	
trimethoxyboroxine					X	primarily magnesium fires; rarely used

[a] Liquid hydrocarbons.
[b] Water-miscible solvents.

of Class A, B, or D. The normal procedure in such circumstances is to cut off the electricity and use an extinguishing method appropriate to what is burning. When this cannot be done with certainty, special extinguishing agents are required which are nonconductors of electricity and nondamaging to equipment. These include vaporizing liquids, dry-chemical powders, and carbon dioxide, although the cooling and condensation effects of CO_2 may affect sensitive electronic equipment.

Each of the above fires has its own mode of combustion. Class A fires are usually combinations of surface burning (flaming combustion) and combustion within the

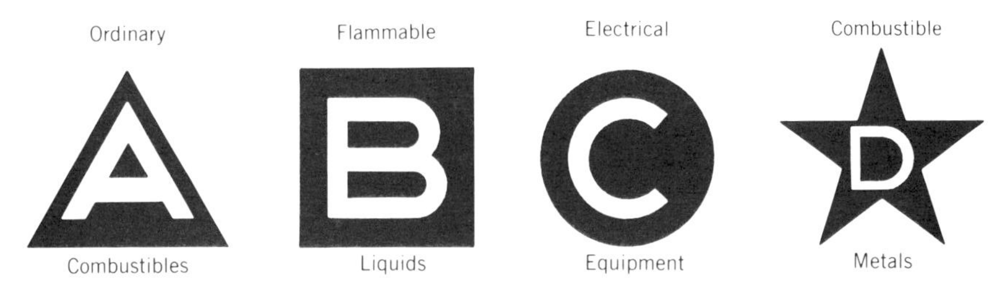

Figure 1. Markings for extinguishers indicating classes of fires on which they should be used. Color coding is part of the identification system. The triangle (Class 1) is colored green, the square (Class B) red, the circle (Class C) blue, and the five-pointed star (Class D) yellow (1).

fuel. Flaming combustion is the rapid vapor oxidation of the gases from the hot fuel and is accompanied by heat that is transferred to the fuel. Deep-seated combustion is characterized by an interior glowing, slower rate of reaction, and a slower rate of heat loss.

Combustion is considered an oxidation reaction; however, unlike the slow oxidation of iron (rust), fire is a rapid oxidation that is self-sustaining and accompanied by the evolution of heat and light.

Typically, the combustion of volatile liquids and gases involves the rapid vapor-phase oxidation of the fuel, subsequent volatilization of more fuel owing to radiative heat transfer, and cracking of the fuel in the vapor phase. The combustion process is characterized by a chain-reaction sequence involving rapidly reacting species (radicals) that are fragments of the fuel or oxidizing-agent molecules. The chain-reaction sequence for a typical hydrocarbon such as methane is (4)

$$O_2 + \cdot H \rightarrow \cdot OH + O$$

$$CH_4 + \cdot OH \rightarrow H_2O + \cdot CH_3$$

$$O_2 + \cdot CH_3 \rightarrow \cdot O\dot{C}H_2 + \cdot OH$$

$$\cdot O\dot{C}H_2 + O \rightarrow H\dot{C}O + \cdot OH$$

$$H\dot{C}O \rightarrow \cdot H + CO$$

$$CO + \cdot OH \rightarrow CO_2 + \cdot H$$

$$2\ \cdot OH \rightarrow H_2O + O$$

$$2\ O + M \rightarrow O_2 + M$$

$$CH_4 + 2\ O_2 \rightarrow 2\ H_2O + CO_2$$

The radicals primarily responsible for sustaining the chain-reaction sequence are ·H, ·OH, and O in the case of a lean CH_4–O_2 flame used as the example. Although plastics are considered Class A hazards, some plastics can decompose into volatile liquids, which are Class B hazards.

From the foregoing, it can be concluded that there are four basic prerequisites (the "Tetrahedron of Fire" (5)) for any combustion process: heat; fuel; oxidizing agent; and suitable chemical-reaction path. From this, it follows that any method for inhibiting a combustion process must involve one or more of the following: remove heat at a faster rate than it is released; separate the fuel and the oxidizing agent; dilute the vapor-phase concentration of fuel and oxidizing agent below that which is necessary for combustion; and terminate the chain-reaction sequence.

Mechanism of Extinguishment

The cooling effect of agents such as water and carbon dioxide is well known, as is the diluting or inerting of oxygen by the latter. Separation of fuel from the air (oxygen) is usually achieved by the foams, although since CO_2 and the Halons are much heavier than air, they, too, can form a barrier between the fuel and air.

The ability for the Halons and dry chemicals to extinguish fires cannot be completely explained by the above concepts. It is felt that their action is owing more to the interruption of the chain reaction. Although the mechanism is not completely understood, there have been a number of possibilities proposed.

Under fire conditions, the halogenated agent, eg, bromotrifluoromethane (Halon 1301), will release a bromine atom (6):

$$CBrF_3 \rightarrow \cdot CF_3 + Br\cdot$$

The Br· atom can react with a hydrocarbon molecule to form hydrogen bromide, HBr:

$$R{-}H + Br\cdot \rightarrow R\cdot + HBr$$

The HBr then reacts with an active hydrogen or hydroxyl radical, releasing the bromine atom for further inhibition reactions:

$$\cdot OH + HBr \rightarrow H_2O + Br\cdot$$

In this way, chain carriers are removed from the system while the inhibiting HBr is continuously regenerated.

Similarly, the following free-radical reactions are probably the means for the Class B extinguishment capabilities of dry chemicals.

$$NaHCO_3 + \text{heat} \rightarrow CO_2 + \cdot OH + Na\cdot$$

$$\cdot OH + \cdot H \text{ (active in flame)} \rightarrow H_2O$$

$$Na\cdot + \cdot H \text{ (active)} \rightarrow NaH$$

$$\cdot OH \text{ (active in flame)} + NaH \rightarrow Na\cdot + H_2O$$

Although the above example is for sodium bicarbonate, it would also be true of potassium bicarbonate, potassium chloride, potassium sulfate, and ammonium phosphate. Potassium carbamate is reported to decrepitate, or break up into finer particles, upon entering the flame front.

The chain breaking is a contributing factor in fire extinguishment. None of the reactions noted take into consideration the anion–cation relationship. There has been no positive work done to explain why ammonium bicarbonate does not extinguish a fire or why ammonium phosphates are the only phosphates that do extinguish a fire.

Other contributions to fire extinguishment requiring further research are radiation shielding, shock waves, critical vibrations, and ion separation from strong magnetic fields.

Approval Agencies and Specifications

There are many agencies controlling agent specifications throughout the world. Most countries have their own, although there is presently an International Standards Organization (ISO) considering international standards for both agents and equipment.

The three main approvals agencies in the United States are the Federal government, Underwriters' Laboratories, Inc. (UL), and Factory Mutual (FM). There are more than subtle differences among the three. The Federal government, for example, issues specifications on the agent and equipment separately. The specifications are basically of the purchase type and for fire extinguishing agents are issued individually on the basis of the base chemical. The tests performed are of the physiochemical type and usually include surface area, particle distribution, chemical composition, etc. Both UL and FM are testing laboratories that evaluate performance characteristics. The former actually rates the combination of agents and hardware (Fig. 2) on fires of varying sizes, whereas the latter only requires a given size fire for a specified extinguisher.

The rating system, as developed by UL, is based as follows (7):

Class A rating is obtained by extinguishing a crib (Figs. 3 and 4), panel, and excelsior fires. The size is determined by the rating the manufacturer is attempting to get (Tables 2 and 3).

Class B rating. In a steel pan, a 5.1-cm layer of gasoline is floated on a 10.2-cm layer of water. The steel pan is square with an area A and a depth of 30.5 cm. A Class B rating (Table 4) is obtained by taking 40% of the largest area consistently extinguished by an extinguisher, ie, a 20-B rating is obtained by extinguishing a 4.65-m^2 (50-ft^2) fire (Fig. 5).

Class C rating is obtained by an electrical conductivity test. A high voltage, 50 Hz, actuating current is impressed between an electrically insulated extinguisher and

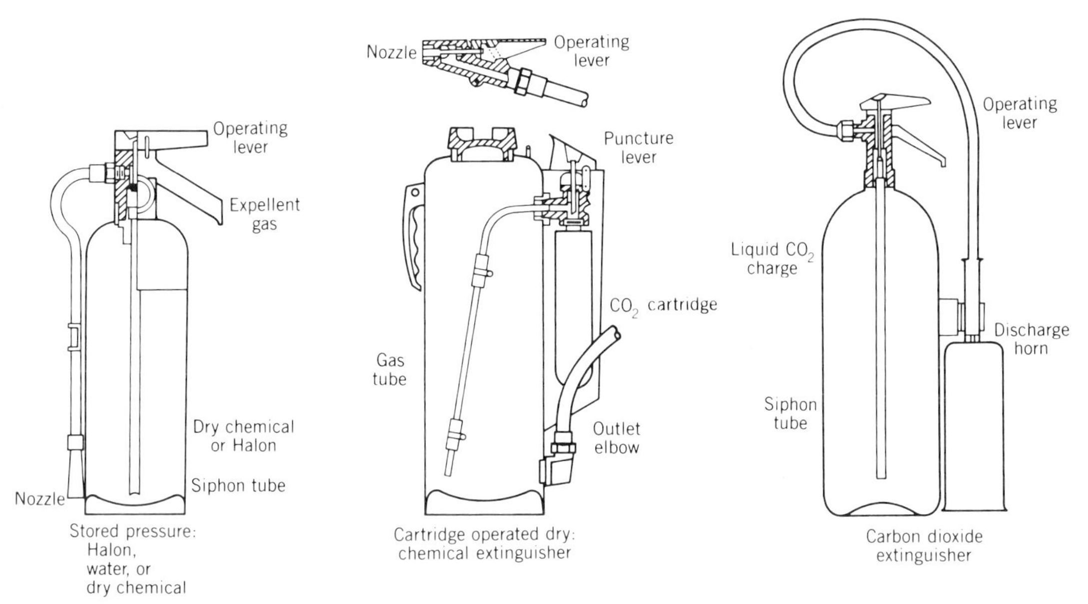

Figure 2. Types of portable extinguishers.

Figure 3. 2A crib fire, one minute prior to attack.

a grounded target. The current flow, if any, is measured through the path of the agent during its discharge. No numbered rating is given since no fire is involved.

Class D rating is usually used for molten metal, castings, or chips, depending on the metal being used (Fig. 6). The fires are a standard size and an agent–fuel ratio is usually determined as in the case of Class C fires. No numerical rating is given; however, the specific fires are noted for a given agent, ie, Met-L-X is listed for sodium, potassium, sodium–potassium alloys, and magnesium. G-1 powder was only listed for magnesium.

Figure 4. 2A crib fire, 8 s into discharge of a 2.27-kg multipurpose extinguisher.

In Canada, the two main agencies, the Canadian Government Specification Board (CGSB) and Underwriters' Laboratories of Canada (ULC), are similar. Testing and specification procedures are similar in nature to those found in the United States. One notable difference is the makeup of the CGSB. It consists of representation from the various branches of the government, as well as representatives of the ULC and the various Canadian manufacturers.

The role of NFPA is not one of specification writing, per se. They do, however, develop consensus standards, and many of these standards have been adopted and made mandatory by some industries and many branches of the government.

Table 2. Comparison of Various Class A Fire Test Cribs

Rating	Wood members	Size and length of wood members, mm	Arrangement of wood members
ULC-S-508			
1-A	72	41 × 38 × 508	12 layers of 6
2-A	112	41 × 38 × 635	16 layers of 7
3-A	144	41 × 38 × 737	18 layers of 8
4-A	180	41 × 38 × 800	20 layers of 9
6-A	230	41 × 38 × 927	23 layers of 10
10-A	324	41 × 38 × 1092	27 layers of 12
20-A	256	38 × 89 × 1400	14 layers of 16 on edge, 4 layers of 8 flat
30-A	328	38 × 89 × 1625	16 layers of 18 on edge, 4 layers on 10 flat
40-A	404	38 × 89 × 1752	18 layers of 20 on edge, 4 layers of 11 flat
UL-711			
1-A	50	38 × 38 × 508	10 layers of 5
2-A	78	38 × 38 × 651	13 layers of 6
3-A	98	38 × 38 × 781	14 layers of 7
4-A	120	38 × 38 × 848	15 layers of 8
6-A	153	38 × 38 × 848	17 layers of 9
10-A	209	38 × 38 × 1207	19 layers of 11
20-A	160	38 × 89 × 1581	10 layers of 15 on edge, 1 top layer of 10 flat
30-A	192	38 × 89 × 1895	10 layers of 18 on edge, 1 top layer of 12 flat
40-A	224	38 × 89 × 2213	10 layers of 21 on edge, 1 top layer of 14 flat

Water

Water is the most widely used agent because of its Class A effectiveness, cost, and availability, and its thermal properties. It requires 2.672 mJ of thermal energy to convert one kilogram of water at 0°C to steam at 100°C. During the process, a volumetric expansion of ×1700 occurs. Two factors account for the extinguishing mechanism of H_2O, ie, cooling of the fuel and dilution of the oxygen concentration. The quantity of water required for extinguishment is dependent on the amount of heat that must be absorbed. Speed of extinguishment then depends on the rate of application, the area of coverage possible, and the mode in which the water is applied. If the water is dispersed in small droplets, such as fog, heat transfer is more rapid and efficient. However, stream pattern must be considered in relationship to the distance the stream of water must be projected. When applied in spray form, water can be an effective agent on flammable liquids with flash point >66°C. Care must be taken around energized electrical nets because of dissolved salts found in water.

The disadvantages of water are its freezing point (0°C), high surface tension (78 N/mm or dyn/cm), and low viscosity (1 mm^2/s or cSt). The relatively high surface tension and low viscosity of water limits its ability to penetrate a burning mass, resulting in a tendency to run off surfaces quickly. In order to increase its effectiveness on fires, water has been made wetter, thicker, fogged, low temperature resistant, and foamed.

Water, because of its versatility, can be applied from portable extinguishers, hose lines, sprinkler systems, trucks with monitor nozzles, and fixed monitor nozzles.

Table 3. Dimensions of Other Class A Test Fires

Designation of test fire	Number of 50-cm wooden sticks in each transverse layer	Length of test fire, cm	Extinguisher capacity, kg	Preburn time[a], min
BFS-5423[b]				
3-A	3	30		
5-A	5	51		
8-A	8	79		
13-A	13	130		
21-A	21	160		
(27-A)	(27)	(270)		
34-A	34	340		
(43-A)	(43)	(430)		
55-A	55	550		
FM[c]				
			4.5	3
			4.5–9.1	5
			9.1–15.4	6

[a] After flame is visible from top of stack.
[b] Cross section of all sticks is 4 ± 0.15 cm; 14 layers for all tests; transverse layers have fixed length of 50 cm; and longitudinal lengths vary according to the table.
[c] Test fires conducted on a single stack of 122 cm × 122 cm × 15 cm pallets.

Table 4. Comparison of Various Class B Fire Tests

Extinguisher capacity, kg	UL-711/ULC S508 Test fire area	Rating-class	Pan size, m^2 (inside)	BS-5423[a] Designation of test fire	Surface area of fire, m^2
1.8	0.5	*Indoor tests*			
1.8–3.2	0.9	1-B	0.23	8-B	0.251
3.2–4.5	1.4	2-B	0.46	13-B	0.409
4.5–6.8	1.9	5-B	1.16	21-B	0.660
6.8–9.1	2.5	10-B	2.32	34-B	1.068
9.1–11.3	3.3	20-B	4.65	55-B	1.728
11.3–13.6	3.9			(70-B)	2.202
13.6–15.9	4.6	*Outdoor tests*		89-B	2.796
		30-B	6.97	(113-B)	3.549
		40-B	9.29	144-B	4.524
		60-B	13.94	(183-B)	5.741
		80-B	18.58	233-B	7.321
		120-B	27.87		
		160-B	37.16		
		240-B	55.74		
		320-B	74.32		
		480-B	111.48		
		640-B	148.64		

[a] BS = British standard.

Wet Water. Water, plus a surfactant, reduces its tension of 73 N/mm (= dyn/cm). This permits the "wetted" water (Fig. 7) to penetrate, or soak in, hazards such as cotton bales, paper rolls, mattresses, etc.

Figure 5. 15.24-m Class B fire test, 2.27-kg cartridge extinguisher for a 20B rating.

Thickened Water. Two types of thickeners generally are used. One is an organic gelling agent and the other is bentonite, which produces a thin gel that many consider a slurry. Both are used in conjunction with either the amonium sulfates or ammonium phosphate. The resultant combination has found great success in forest-fire applications, since either ammonium salts are excellent fire retardants and will rapidly halt flaming or glowing combustion.

Inorganic Freezing-Point Depressants. Originally, calcium chloride was used to lower the freezing point of water to ca −40°C. However, calcium chloride is corrosive to some metals. Inorganic alkali metal salts, such as potassium carbonate and potassium acetate, not only impart freeze protection to water, they also give a modicum of Class B capabilities. These potassium salts have become known as loaded-stream agents. Their main success has been in familiar stainless-steel "water" extinguishers found in stores, warehouses, and the like. A recent and more significant application has been in pre-engineered systems for the protection of restaurant kitchen appliances.

Flow-Property Modifiers. Diluted solutions of poly(ethylene oxide) reduce water to a nonturbulent flow in pipes, fire hoses, etc. The result has been called slippery water and rapid water. It is not widely used.

Foams. The use of foam-making chemicals is another means of modifying the viscosity of water. The foam-making chemical is formed by the reaction of an *A* component (acidic) and a *B* component (basic). For example, the reaction occurs in an aqueous solution using aluminum sulfate as the *A* component and sodium bicarbonate with proteinaceous foam stabilizers as the *B* component. Chemical foams are nearly obsolete because of the ease of handling the more economical liquid-foam-forming concentrates (see Foams).

Figure 6. Extinguishing an aluminum-pigment fire.

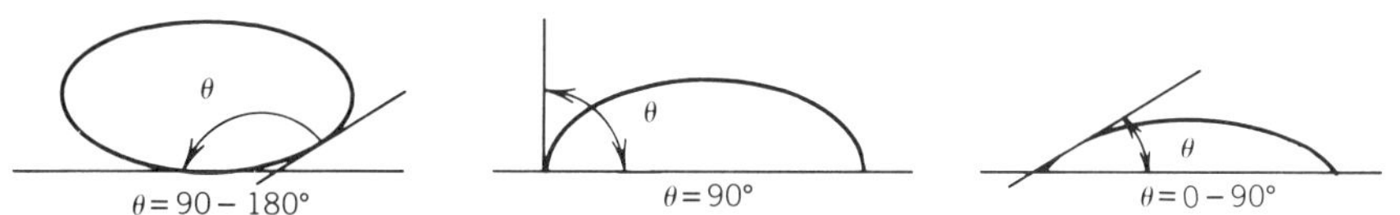

Figure 7. Wetting action. For complete wetting, the contact angle is 0°; for no wetting, the contact angle is 180°; and for partial wetting, the contact angle is between these two values. The smaller the contact angle, the greater the degree of wetting and the higher the free surface area of water.

The liquid-foam-forming concentrates are of two basic types: protein and synthetics. The protein foams include the regular protein and the fluoroprotein. These are protein hydrolysates made from naturally occurring protein (animal or vegetable). The fluoroprotein has a fluorocarbon surfactant added to it to make it more flowable and it also increases its ability to resist fuel attack. The synthetic foam liquids include aqueous-film-forming foam (AFFF), Syndet, alcohol-resistant foams, and high expansion foams.

Foam solution is made by proportioning the liquid concentrates into the water either by aspiration, external pump or pressure heads, or premixing. The foam is generated in one of the following ways: the use of water sprinkler and water spray nozzles will produce a froth (these nonaspirating froths, or foams, rely on air currents and droplet collision for their formation); air-aspirating foam nozzles, foam tubes, and ordinary foam makers generate foam by venturi action and aspirates of air into the foam solution; high back-pressure-foam makers are modified venturi devices that operate at relatively high back pressures for discharging foams down lengths of tubes, or hoses, and are also used for subsurface injection; the high expansion generating devices may be air-aspirating or blow-fan type; and pumped-foam devices in which compressed air is pumped into foam solution under pressure. Since pumped-foam devices are inherently more expensive, they have limited use.

Protein Foam. Protein-foam concentrate is produced by hydrolyzing naturally occurring proteinaceous materials, eg, fish meal, feather meal, and horn and hoof meal. In a typical process, the meal is cooked in an alkaline solution and then neutralized with acid and filtered. The stabilizers and other additives are blended in a large tank and run through a final filter as packaged. Protein-foam concentrate is made in 3% and 6% concentrates. There are some special low temperature blends.

In addition to fire fighting, protein foam has also been applied to airport runways to assist disabled aircraft in safe landing. The foam acts both as lubricant on the runway and as temporary barrier against fires from sparks igniting fuels on the aircraft itself.

Fluoroprotein. The fluoroprotein concentrate is made in the same manner as the regular protein concentrate; however, a fluorocarbon surfactant is added to the blend. The fluorosurfactant makes the foam more flowable and, more importantly, more resistant to the fuel being protected. Fuel resistance makes the fluoroprotein foam suitable for subsurface application in petroleum storage tanks (see Surfactants and detersive systems).

Aqueous-Film-Forming Foam (AFFF). The most important of the synthetic foams is AFFF (see Fig. 8). AFFF is a fluorinated surfactant foam that has been developed for controlling certain flammable liquid fires, in particular, petroleum products. In addition to the fluorinated surfactant, the foam concentrate also contains a foam stabilizer which can be diluted in either fresh or salt (sea) water. AFFF extinguishes the fire by cooling and by releasing a film that both retards vapor formation and excludes air from the fuel surface. It is also used for subsurface application.

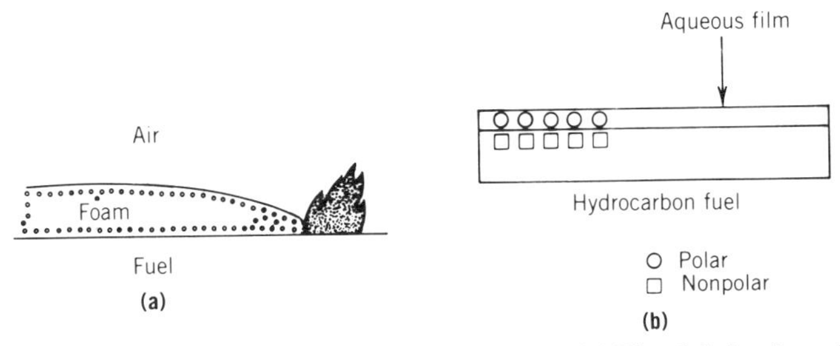

Figure 8. Extinguishment of hydrocarbon fires by foams. (**a**) Fire-fighting foam is a stable mass of small bubbles lighter than oil or water. Foam is a watery suspension of gas (usually air) in forming bubbles separated by films of solution. Foams extinguish fire by separating the air from the burning liquid with a foam blanket. There is some cooling effect from the water contained in the foam bubbles. In addition, the aqueous-film-forming foam (AFFF) is unique, because it allows a film to form on a hydrocarbon fuel surface, which prevents flashbacks by suppressing fuel vapors due to the presence of this film. (**b**) AFFF is obtained from synthetic fluorochemical surfactants. These surfactants are molecules that contain a polar and nonpolar functionality within the same molecule. Basically, the nonpolar end of the molecule is hydrophobic (hydrocarbon-soluble), and the polar end is hydrophilic (water-soluble). The formation of the film is dependent on the relatively rapid breakdown of the foam as it is applied.

AFFF was initially developed and is still used as a "twinned" agent; that is, the fluorinated surfactant foam is discharged separately or simultaneously with potassium bicarbonate dry chemical. This provided the quick knockdown of the dry chemical and prevention of flashovers by the vapor seal formed by the fluorocarbon film floating on the surface of the burning liquid.

Synthetic Foam (Syndet). The synthetic foams, other than AFFF, contain no fluorocarbon surfactants. This foam concentrate is a blend of surfactants and may or may not contain stabilizers. They are used in concentrates of 1–6% and, depending upon formulation, may extinguish the fire either by blanketing the fuel or by emulsification.

Synthetic foams have been called detergent foams, emulsion foams, and more simply Syndet. The latter is becoming the more widely recognized synonym in the United States and abroad.

High Expansion Foam. High expansion (Hi Ex) foam is an aggregate of bubbles generated by the passage of air or other gases through a net, screen, or similarly porous material which is wetted with aqueous solution of surface-action foaming agents. These foams, with expansions of 200:1 to 1000:1, are fluid and stable enough to transport water to inaccessible places for total flooding and for volumetric displacement of vapor heat and smoke.

Hi Ex foam is particularly suited for indoors and confined spaces such as high piled storage areas and in mines. Because of the light density and fragile nature of the foam, it has limited outdoor applications.

Synthetic Foams for Water-Miscible Fuels. Alcohol- (or any polar solvent) resistant foams are formulated synthetic foams that include a polysaccharide that, as a foam drains, forms an insoluble film over water-miscible fuels, such as the alcohols, ketones, and other polar solvents. It is usually used in a 6% concentrate on polar-solvent spills and can be used at a 3% concentration on nonpolar-solvent fires.

Use and Limitations. The principal use of foams is the extinguishment of burning liquids lighter than water. Ignition and fire may be prevented by applying foam blankets to spills or other hazardous areas (8). Foams may also be used to insulate and protect exposures from radiant heat. They also act to prevent ignition of open areas of flammable liquids if spread completely over an exposed surface. It is well to remember, however, that foam breakdown can render such a foam-protective coating of no value to the fire fighter, and frequent renewal may be necessary. Because of the water content, foams may be used to extinguish surface fires in ordinary combustible materials, such as wood, paper, rags, etc.

The limitations of foam are as follows:

Foams are not suitable extinguishing agents for fires involving gases, liquefied gases with boiling points below ambient temperatures, such as butane, butadiene, propane, etc, or cryogenic liquids. Flowing liquid fires, such as overhead tank leakage or pressure leaks, are not readily extinguishable with foams. In addition, foams should not be used to fight fires in materials which react violently with water, such as metallic sodium and metallic potassium, etc. In certain magnesium fires, foams may be judiciously applied to help restrict burning and cool residual metal.

Foam is a conductor and should not be used on energized electrical equipment fires.

The water in foam may cause violent frothing of the contents and even forceful expulsion of a portion of the contents when applied to vessels containing hot oils, asphalts, etc, which are above the boiling point of water, either normally or owing to an exposure fire and to vessels containing high viscosity oils, such as Bunker C fuel oil, which have been burning for extended periods.

Certain wetting agents and some dry-chemical powders may be incompatible with foams. If they are used simultaneously, an instantaneous breakdown of the foam blanket may occur. Most foams are not suitable for water-soluble or polar-solvent liquids. Special foams designed for these materials are available.

Liquefied Gases

Liquefied compressed gases are another class of agents that can be divided into two types. Carbon dioxide is an example of the diluting or inerting type agent because it must be applied in such concentrations in order for it to dilute the oxygen level below that necessary to sustain combustion. The other type of agent in this class is halogenated hydrocarbons which are materials that chemically inhibit the combustion chain reaction sequence. Historically, the chlorinated methanes have been used as agents, but their inherent toxicity resulted in curtailment of their use.

Carbon Dioxide. This agent is the most widely used of the inerting type which includes nitrogen and argon. The latter two are seldom used, but the general comments apply to them as well as CO_2.

The advantages of CO_2 are that it is a clean agent, ie, it leaves no residue; it is inexpensive and readily available; it can be used around energized electrical equipment; and it discharges under its own vapor pressure.

It also has some inherent disadvantages, such as that it requires high concentration for extinguishment; these high concentrations preclude its use in normally occupied areas in which egress is limited. Also, carbon dioxide requires special storage equipment, such as high pressure steel containers or a low pressure (2.1 MPa or 300 psi) unit maintained at low temperatures (−17.8°C) by a refrigeration unit.

Carbon dioxide extinguishes fire by cooling and by the dilution of oxygen. With few exceptions, it requires a concentration of 30–38% to extinguish flammable and combustible liquids and gas fires. It is applied by portable extinguishers or from special total flooding systems.

Halogenated Hydrocarbons (Halons). These extinguishing agents are hydrocarbons (usually methane or ethane) in which the hydrogen has been replaced by a halogen such as fluorine, chlorine, bromine, or iodine. The substitution of one or more of the halogens imparts both nonflammability and flame-extinguishment properties to the compound. The Halons are used in portable fire extinguishers, mobile equipment (specially designed trucks, etc), and total-flooding systems.

The U.S. Army Corp. of Engineers introduced the term Halon and a numbering system that eliminated the need for the unwieldly chemical names and the possible confusion of the alphabet system (Table 5). In the numbered system, the first digit is carbon, the second fluorine, the third chlorine, the fourth bromide, and the fifth iodine (see also Aerosols). Hydrogen is not numbered. Terminal zeroes are not expressed.

The three most used halogens are fluorine, chlorine, and bromine. Their substitution imparts the following properties: fluorine gives the compound stability, reduces toxicity, lowers boiling point, and increases thermal stability; chlorine increases fire-extinguishing effectiveness, increases the boiling point, increases toxicity, and

Table 5. Comparison of Selected Properties of The Halons and CO_2

Property	Methyl-bromide [74-83-9]	Chlorobromo-methane [74-97-5]	Dibromodi-fluoromethane [75-61-6]	Bromotri-fluoromethane [75-63-8]	Bromochloro-difluoromethane [353-59-3]	1,2-Dibromo-1,1,2,2-tetra-fluoroethane [124-73-2]	Carbon-tetrachloride [56-23-5]	Carbon dioxide [124-38-9]
chemical formula	CH_3Br	CH_2BrCl	CFl_2Br_2	CF_3Br	CF_2ClBr	C_2F_4Br	CCl_4	CO_2
Halon number	1001	1011	1202	1301	1211	2402	104	
BFS designation[a]	M.B.	C.B.	D.D.M.	B.T.M.	B.C.F.	D.T.E.	C.T.C.	CO_2
boiling point, °C	3.6	67	24	−58	−4	47.5	77	−118
peak flammability[b] agent concentration, %	9.7	7.6	4.2	6.1	9.3	4.9	11.5	32.0
approx. lethal conc, ppm								
agent	5,900	29,000	54,000	832,000	324,000	126,000	28,000	658,000
decomposition prod.	9,600	4,180	1,850	14,000	7,650	1,600	300	658,000

[a] BFS = British Fire Standard.

[b] Peak flammability data determined by burette method on heptane–air mixture.

reduces thermal stability; and bromine is the same as chlorine, but to a greater degree.

Because the Halons are either compressed gases or liquids that volatilize readily in a fire, they leave no corrosive or abrasive residue. Because of this characteristic, they are called clean agents. In addition, they are nonconductors of electricity and thus suitable for protection in electrical and electronic equipment, air- and ground-vehicle engines, and other areas where rapid extinguishment and equipment damage resulting from cleanup must be minimized.

The two most widely used of the halogenated agents are bromotrifluoromethane (Halon 1301) and bromochlorodifluoromethane (Halon 1211).

Halon 1301. At 21°C, Halon 1301 is a gas with a vapor pressure of 1.5 MPa or 199 psig. Under normal circumstances, this would be adequate to expel the agent; however, the vapor pressure drops rapidly as the temperature is lowered (386 kPa or 56 psi at −17.8°C) and (117 kPa or 17 psi at −40°C). It is usual to increase the pressure by the use of nitrogen. Because of its high volatility, its main use is in total flooding systems. Halon 1301 is seldom used in portable extinguishers. At this time, there is only one (1 kg) such unit listed by UL.

Halon 1211. At 21°C, Halon 1211 is also a gas, but has a vapor pressure of only 253 kPa (22 psig) and a boiling point of −3.9°C. Because of its relatively high boiling point, it can be projected in a liquid stream which makes it suitable for portables and local application systems. Portable extinguishers are pressurized to 584 kPa (70 psig) with nitrogen and as a result, relatively inexpensive hardware can be used. Although Halon 1211 is generally used in portable extinguishers, Halon 1211 total flooding systems (with suitable precautions) are in use outside of the United States.

Toxicity of Halons. The toxicological properties of these two Halons have been studied extensively. Human exposures have shown that Halon 1301 concentrations up to ca 7 vol % and Halon 1211 concentrations of 2–3 vol % have little noticeable effects on the subject. As the concentrate of Halon 1301 reaches 7–10 vol % (Halon 1211 3–4%), the subject began to feel dizziness and tingling of the extremities, indicating mild anesthetics (qv). As concentrations of Halon 1301 exceed 10 vol % (Halon 1211 4–5%), the dizziness becomes more pronounced, and the subject feels loss of manual dexterity, leading to loss of consciousness. In tests where human subjects have been exposed to 7 vol % Halon 1301 concentration for 30 min, the effects appeared within the first 5–10 min of exposure, remained constant throughout the remainder of the test, and quickly disappeared when removed from exposure.

NFPA 12A permits Halon 1301 design concentrations up to 10% in normally occupied areas and up to 15% in areas not normally occupied. Because the required extinguishing concentration of Halon 1211 is near or above its limit for safe exposures, Halon 1211 systems are not recognized in NFPA 12B for normally occupied areas.

Uses and Limitations. Halogenated-agent systems are generally considered useful for the following types of hazards: where a clean agent is required; where live electric or electronic circuits exist; for flammable liquids or flammable gases; for surface-burning flammable solids such as thermoplastics; where the hazard contains a process or objects of high value and where use of conventional extinguishing agents could cause extensive damage or down time; where the area is normally or frequently occupied by personnel; and where availability of water or space for systems using other agents is limited.

There are several types of flammable materials on which halogenated agents are ineffective, such as fuels that contain their own oxidizing agent, eg, gun powder, rocket propellants, cellulose nitrate, organic peroxides, etc; reactive metals, eg, sodium, potassium, NaK eutectic alloy, magnesium, titanium, and zirconium; and metal hydrides, eg, lithium hydride. A more commonly encountered limitation of the capabilities of an agent is its limited effectiveness on deep-seated Class A fires at concentrations <10 vol %. When the concept of control is applied to using halogenated agent systems on Class A fires, rapid response by outside help is a necessity. Otherwise, a reflash potential will exist once the extinguishing concentration is dissipated.

Decomposition Products of the Halons. The breakdown products of the halogenated agents on life safety must be considered since these products have a relatively higher toxicity. Decomposition of these agents takes place in an open flame or where surface temperatures are >482°C. The main decomposition products of Halon 1301 are hydrogen fluoride, hydrogen bromide, and free bromine. The decomposition of Halon 1211 is similar and, in addition, include both hydrogen chloride and chlorine.

The concentration of the decomposition products depends upon size of the room, size of fire, number of hot surfaces present, and the preburn time prior to extinguishment. To keep the decomposition products at a minimum, it is advantageous to attack the fire at an early stage and to release as soon as possible.

Extensive studies on the toxicological effects of halogenated agents indicate that little, if any, risk is attached to the use of Halon 1301 or Halon 1211 when used in accordance with provisions of NFPA standards which recognize the use of these agents.

Dry-Chemical Fire-Extinguishing Agents

Dry-chemical agents are probably the most effective of all agents (Table 6). Their origin can be traced to the time of the Civil War when Solvay began to produce a cheap source of sodium bicarbonate. This appeared for years in long red tubes for extinguishing kerosene lantern fires in railway cars.

The use of sodium bicarbonate agents in pressurized extinguishers went through many stages of development. Initial use of gas as a dispersing agent began in Germany around 1913. The early agents had many drawbacks, especially in moisture pickup and caking problems. In 1948, Dugas introduced a free-flowing agent coated with magnesium stearate that was expelled by the use of carbon dioxide gas. Although

Table 6. Relative Ranking of Agents, Potassium Carbamate = 100

Agent	Relative effectiveness
potassium carbamate	100
potassium bicarbonate base	91
potassium chloride base	90
multipurpose	75
sodium bicarbonate base	57
Halon 1211	53
carbon dioxide	30
Halon 1301	26
AFFF	7

magnesium stearate is still used today, the use of silicones not only proved helpful as a moistureproof barrier but also improved the flow characteristics of the agent and imparted a higher degree of heat resistance to the agent. Along with the silicones came many other agents, such as potassium bicarbonate, monoammonium phosphate, and others.

Most dry chemicals are ground to a particle distribution between 5 and 108 μm, and most have a median particle size of ca 20–30 μm. All existing agents contain additives that make them water repellent, less hygroscopic, and more flowable with a reduced tendency toward packing.

In the United States, the four most common chemical bases used for fire-extinguishing agents are sodium bicarbonate, potassium bicarbonate, monoammonium phosphate, and potassium chloride. European agents also include potassium carbamate (which has limited use in the U.S.) and potassium sulfate.

Sodium Bicarbonate. Many manufacturers still provide what is referred to as a stearated sodium bicarbonate. Magnesium (or possibly zinc or barium) stearate is added to the bicarbonate as a water repellent along with flow promoters, such as mica, clay, starch, colloidal silica, and/or tricalcium phosphate. Sodium bicarbonate can be purchased with a particle distribution suitable for use; therefore, these agents can be simply processed in a ribbon blender. The main drawbacks of the stearated agent are the lack of heat stability and compatibility with protein foam. They are, however, compatible with AFFF. Their use continues because of the ease in which they can be manufactured.

The replacement of magnesium stearate with the silicones, in particular, the cross-linking poly[oxy(methylsilylene)], brought about the need for advanced equipment. The silicones required a special means to introduce the silicone into the blender, and an intimate mixing is required to cure the silicone. State of the art indicates that the ball mill and a horizontal shaft blender equipped with side-entering intensifier blades have been the most suitable for this use. In addition to the silicone, flow promoters are also used, and by adjustments in the formulation, siliconized sodium bicarbonate has been made protein-foam-compatible.

Both the stearate and siliconized sodium bicarbonate are UL and ULC rated for Class B:C fires with very little difference noted in effectiveness.

Potassium Bicarbonate. Products based on potassium bicarbonate are usually referred to as Purple K. The name came about as a result of the purple to violet color that appears in the flame during the extinguishment of a fire. Purple K was developed at the Naval Research Laboratories (NRL), Washington, D.C., in the 1950s because of its effectiveness on Class B fires. In general, it was found that the potassium salt appears to be twice as effective as the sodium salt. Purple K is particularly effective on high pressure gas fires.

Potassium bicarbonate agents are produced by only a few manufacturers. The base raw material has to be imported and ground to the preferred particle distribution either during or prior to blending. The most common means of size reduction are the ball mill, hammer mill, and various types of pin mills. On rare occasions, a turbo mill has been used. After size reduction (qv), processing is the same as for siliconized sodium bicarbonate.

Purple K is classified Class B:C fire-extinguishing agent by both UL and ULC.

Multipurpose Agents. These agents are classified by UL and ULC as Class A:B:C fire-extinguishing agents. They are referred to as multipurpose, all-use, all-purpose, and ABC agents, as well as by some of the brand names. Both of the Underwriters' Laboratories groups refer to them as multipurpose in their listings.

The base chemical used is monoammonium phosphate (MAP) and, in addition to the usual flow promoting and anticaking materials, may contain some amounts of ammonium sulfate, diammonium phosphate, or potassium sulfate. The minimum amount of MAP required to produce an overall effective and reliable agent is ca 80%. Anything less than this can drastically reduce effectiveness on both Class A and B fires.

Like Purple K, the base raw materials for multipurpose agents are ground prior to blending and siliconization. Size reduction is usually done in a hammer mill or a turbo mill. Some manufacturers have had moderate success with pin mills.

Multipurpose agents are acidic and can thus react with the more basic bicarbonate agents. The mixing of these agents is to be avoided because small amounts of contamination can render fire-fighting equipment useless and, in some cases, has caused some portable extinguishers to explode.

To assist in identifying multipurpose agents, all manufacturers in the United States and Canada color their agent yellow. Care must be taken when using imported products because they have no color restrictions, particularly in Europe.

Potassium Chloride. Potassium chloride agent was originally developed as a protein-foam-compatible agent with Class B capabilities and is about as effective as potassium bicarbonate. It can be somewhat corrosive and has a specialized use in aircraft-crash fires where corrosion is not normally a problem.

Potassium chloride, as in the case of the potassium bicarbonate and multipurpose agents, is ground prior to processing. Being neutral, it does not react with either of the bicarbonates or multipurpose agent.

It is rated as a Class B:C agent by UL and ULC and at one time was listed as foam compatible.

Potassium Carbamate. Potassium carbamate is generally assumed to be a reaction mixture of urea and an alkali salt such as potassium bicarbonate. This has been given the empirical formula of KCH_2NO_2 and is generally referred to by its trade name Monnex. It is produced in the UK and is originally reported to be at least three times more effective in Purple K and 15 times more effective than sodium bicarbonate based agents. Most laboratory fire test methods (Table 7) show Monnex to be only 10–15% more effective than potassium bicarbonate, and recent ratings obtained at UL tend to bear this out (Table 8).

Potassium carbamate has limited use in the United States because its superior effectiveness has not been shown and it is expensive. It is listed for Class B:C fires at both UL and ULC. There are no government specifications in either the United States or Canada.

Potassium Sulfate. Potassium sulfate was originally introduced as a protein-foam-compatible agent in Europe. Since potassium sulfate has neither the capabilities nor the efficiency of sodium bicarbonate, its use in the U.S. is presently limited to that of an active filler. Although there are no UL approvals with this agent in either the United States or Canada, it is widely used in Europe and many other parts of the world.

Table 7. Comparative Effectiveness of Dry-Chemical Extinguishing Agents[a]

Chemical name	CAS Registry Number	Weight to extinguish, g Laboratory A[b]	Laboratory B[c]
sodium bicarbonate	[*144-55-8*]	2.5	4.0
potassium bicarbonate	[*298-14-6*]	1.0	1.2
potassium carbamate	[*4366-93-2*]	0.8	0.9
potassium chloride	[*7447-40-7*]	1.2	1.5
potassium sulfate	[*7778-80-5*]	3.5	4.0
monoammonium phosphate	[*7722-76-1*]	3.0	2.2

[a] "Puffer" method (laboratory), 558 cm^2 Class B fire.
[b] Ref. 9.
[c] Ref. 10.

Table 8. Selected Underwriters' Laboratory Ratings[a]

Company	UL file no.	Agent type	Charge, kg	UL rating
Amerex Corp.	EX2764(N)	multipurpose	9.1	20A-120-B:C
		potassium carbamate	4.1	80-B:C
		potassium carbamate	7.7	120-B:C
		sodium bicarbonate	9.1	120-B:C
Ansul Co.	EX1216(N)	potassium bicarbonate	4.5	80-B:C
		potassium bicarbonate	9.1	120-B:C
Chemetron Fire Systems	EX1735(N)	potassium carbamate	5.0	80-B:C
		potassium carbamate	8.6	120-B:C
Fire End & Croker Corp.	EX2703(N)	potassium carbamate	4.1	80-B:C
General Fire Extinguisher Corp.	EX1909(N)	sodium bicarbonate	9.1	120-B:C
Protecto Seal Co.	EX2989(N)	potassium carbamate	4.1	80-B:C

[a] Ref. 11.

Combustible Metal Fire-Extinguishing Agents. Combustible metal or Class D fires occur in lithium, sodium, potassium, sodium–potassium alloy, magnesium, aluminum, titanium, zirconium, and other similar metals. The most commonly used fire-extinguishing agents are based on either sodium chloride or graphite compounds. Dry sand and other oxygen-containing compounds should be used with extreme care, and once applied, the covered fuel should not be disturbed until the mass is completely cool.

Depending on the agent selected, it can be applied by scoop (shovel) from a pail, a portable extinguisher, or systems.

MET-L-X. MET-L-X is a sodium chloride-based agent containing a vinylidene dichloride copolymer to seal off the oxygen from the fuel surface. It is the only agent approved by UL for fires in sodium, potassium, sodium–potassium alloys, and magnesium. It can be applied by any of the three methods noted previously and can control or extinguish fires of aluminum, titanium, zirconium, and other similar metals.

G-1 Powder. G-1 powder is a graphite-base agent containing an organophosphate plasticizer that aids in excluding oxygen. At one time, it had UL approval for magnesium fires. It is suitable for scoop application only and will control or extinguish fires in most all combustible metals.

Lith-X. Lith-X is a free-flowing graphite-based agent specifically developed for lithium fires. It works by rapidly removing heat from the fuel source and smothering the fire. It can be applied by either scoop or portable extinguishers and is also suitable for use in controlling or extinguishing other combustible metal fires.

Dry Sand. Dry sand has been used over the years for metal fires. It has two drawbacks: it must be kept dry, and it should not be mixed with the burning metal since mixing can result in higher intensity fires.

Na-X. Na-X is a special low chloride-containing sodium carbonate agent for use around stainless steel. It was developed for the DOE (formerly AEC) for breeder-reactor use. It has a UL listing for sodium, potassium, and sodium–potassium alloy fires. It can be applied in all three manners and because of its oxygen content should not be used around other burning metals such as magnesium.

Foundry Flux. Foundry flux is a flux used to protect molten magnesium from air. It consists of potassium chloride, barium chloride, magnesium chloride, sodium chloride, and calcium. Although it readily extinguishes fires in chips or shavings, it is not recommended because it is hygroscopic.

Natrix. Natrix is a sodium carbonate-based agent made in Japan. It is similar to Na-X and has the same reactions on metals, such as magnesium.

Pyromet. Pyromet is made in the UK and purported to be a mixture of sodium chloride and either monoammonium or tricalcium phosphate, protein material, clay, and a waterproofing agent. It is effective on sodium or calcium fires and has had some success with magnesium and aluminum chips.

TEC. TEC is an abbreviation of ternary eutectic chloride. It is made in the UK and contains a mixture of potassium chloride, sodium chloride, and barium chloride. Its action on magnesium is similar to that of foundry flux. When laid on, it has been successfully used in glove boxes on small uranium and plutonium fires.

Uses and Limitations. Dry-chemical fire extinguishers (portable or systems) are usually used as follows: when primary use is on flammable liquid fires and high pressure gas fires; where rapid extinguishment is of the utmost importance; and because they are nonconductors, for use on electrical (Class C) fires.

The limitations of dry chemicals are that they do not have a lasting inerting effect over flammable liquid surfaces, and they should not be used around electrical relays and contacts. The insulating properties might render the equipment inoperative. Dry chemicals do not extinguish fire in materials that supply their own oxygen; also, dry chemicals may not be compatible with protein-type air foams. Because of their airborne nature, dry chemicals present a difficult cleanup problem. In areas of high humidity, dry-chemical agents can become electrically conductive or somewhat corrosive. Multipurpose agents should not be used on hot surfaces, such as highly polished rolls. When used in systems, multipurpose agents are not considered effective on deep-seated or burrowing fires.

Toxicity of Dry-Chemical Agents. On the basis of a use factor with people exposed to the various dry chemicals through fire test programs, fire training, and the production of these materials, there is no record of a toxicity problem. When dry chemicals are used on a fire, they do not break down into toxic materials. However, when responding to a fire, many people ignore a most dangerous aspect which is exposure to the products of decomposition produced by the burning material. These products of decomposition can produce species having toxicity levels ranging from a level causing temporary mild illness to lethal. Exposure to these products rather than the dry-chemical extinguishing agent is what, in most cases, causes after-fire illness.

The base chemicals used in dry-chemical fire-extinguishing agents are at least technical-grade materials. Exposure to these chemicals when used as fire-extinguishing agents is not toxic to human beings.

Other additives used in dry-chemical formulations are all inert materials which are considered nontoxic to humans. All of these are present in dry-chemical agents in small quantities only. They are all materials that have been used for a number of years in dry-chemical formulations. There is no record of any toxic or irritating effects from handling any of these in their raw state or from their use in dry-chemical extinguishing agents.

Dry-chemical fire-extinguishing agents are mixtures of chemicals and are, therefore, not subject to the Toxic Substances Control Act (TSCA). However, because dry chemical is a finely divided solid material, it can become suspended in the air causing a mild discomfort comparable to that experienced in any dust-laden atmosphere. It is usually treated as a nuisance; TLV = 15 mg/m^3 (ca. 0.012 oz/ft^3).

Economic Aspects

The overall dollar volume for all fire-extinguishing agents approached $75,000,000 in 1982. Foam agents accounted for 55%, dry chemicals ca 45%, and the Halon agents the remainder. These agents have shown an overall growth rate of 10% per year. The growth rate is expected to continue for at least the next five years. Some of them are shown in Table 9.

Although the foams show the greatest dollar volume, at ca $2.10/L ($2.40/kg or $8/gal) and 22,700 m^3/yr, AFFF accounts for ca 55% of the foam volume and is expected to increase at the expense of the protein and fluoroprotein foams. The principal

Table 9. Selected U.S. Agent Manufacturers' Products and Brand Names

Manufacturer	Agents			
Foam	*AFFF*	*Protein*	*Fluoroprotein*	*High expansion*
The Ansul Company	Ansulite	Ansul Regular	Ansul Fluoro-Protein	Ful-Ex/Mid-Ex
W. Kidde Company				
3M Company	light water			
National Foam	Aer-O-Water	Aer-O-Foam	XL liquid	
Dry chemical	*Sodium bicarbonate stearate/silicone*	*Potassium bicarbonate*	*Monoammonium phosphate*	*Potassium chloride*
Amerex	regular	Purple K	ABC Multi Purpose	
The Ansul Company	+50/+50C	Purple K	Foray	
General Fire Extinguisher	Quick-Aid	Purple K	Triplex	
Pyro Chem	Pyro-BC/Pyro-BCS	Pyro-Pk	Pyro-ABC	Pyro-SK
W. Kidde	regular	Purple K	ABC	Ultra K
Halons	*Type*			
DuPont	Halon 1301			
ICI America	Halon 1211			

markets for the foams are petroleum–petrochemicals, chemical industries, aviation, and Federal government.

The dry-chemical agents account for $20,000,000 or ca 45,000 metric tons. Sodium bicarbonate-based agents account for a solid 45% of this market with multipurpose (MAP-based) agents having a 33% share, Purple K (potassium bicarbonate) 16%, and all other agents 5%. This market could expand considerably more than the anticipated 10% growth if the industry eventually penetrates the domestic consumer market. Several unsuccessful attempts have been made over the past 20 yr with some extinguishers that unfortunately weren't much more than toys. A relatively new market that may affect some of the dry chemicals are the wet-chemical systems for restaurant-kitchen protection, expected to replace the present dry-chemical systems which are based on sodium bicarbonate. The impact of wet-chemical systems will depend on their acceptance, in particular, by the fast-food restaurant chain.

Although the Halons only have a small portion of the present market, they may become the area of greatest growth. High technology, resulting in complete computerization of industry, home, and school, will create a greater demand for a clean agent.

BIBLIOGRAPHY

"Fire Extinguishing Agents" in *ECT* 2nd ed., Supplement Vol., pp. 364–383, by James M. Hammack, National Fire Protection Association.

1. *Fire Protection Handbook*, 14th ed., National Fire Protection Association, Boston, Mass., 1976, p. 16-3.
2. *Hand Portable Extinguishers*, NFPA 10, National Fire Protection Association, Quincy, Mass., 1981, p. 57.
3. R. L. Tuve, *Principles of Fire Protection Chemistry*, National Fire Protection Association, Boston, Mass., 1974, p. 143.
4. *Fire Training Manual*, Wormald U.S., Inc., Marinette, Wisc., 1981, p. 1.
5. W. M. Haessler, *The Extinguishment of Fire*, National Fire Protection Association, Boston, Mass., 1974, p. 2.
6. *Fire Protection Handbook*, 15th ed., National Fire Protection Association, Quincy, Mass., 1981, Sect. 18, Chapt. 2.
7. *Fire Extinguisher, Rating and Testing of*, UL 711, Underwriters' Laboratories, Inc., 1975.
8. *Foam Extinguishing Systems and Combined Agent Systems*, NFPA 11, National Fire Protection Association, Quincy, Mass., 1982.
9. Chemical Concentrates Corporation, Fort Washington, Pa., 1968.
10. The Ansul Co., Marinette, Wisc., 1973.
11. *1983 Fire Protection Equipment Directory*, 1981, pp. 28, 29, 31, and 32.

General References

Portable Fire Extinguishers, NFPA 10 (1981); *Foam Extinguishing Systems and Combined Agent Systems*, NFPA 11 (1982); *Halogenated Extinguishing Agent Systems—Halon 1301*, NFPA 12A; *Halogenated Extinguishing Agent Systems—Halon 1211*, NFPA 12B; *Dry Chemical Extinguisher Systems*, NFPA 17; *Fire Protection Handbook*, 13th, 14th, and 15th eds., National Fire Protection Association, Quincy, Mass.

Dry Chemical Fire Extinguishers, UL 299 (1981); *Fire Extinguisher, Rating and Testing of*, UL 711 (1981); *Air Foam Equipment and Liquid Concentrate*, UL 162 (1975); *2½ Gallon Stored Pressure Water-Type Fire Extinguishers*, UL 626 (1981); *Halogenated Agent Fire Extinguishers*, UL 1093 (1980); *Carbon Dioxide Fire Extinguishers*, UL 154 (1981); *Fire Protection Equipment List* (1981, 1982), Underwriters' Laboratories, Northbrook, Ill.

Portable Fire Extinguishers, Dry Chemical Type, Class Number 5300 Approval Standard, Factory Mutual, 1975.
Fire Extinguishing Agent, Sodium Bicarbonate Dry Chemical, O-F-371a (1966); *Foam Liquid, Fire Extinguishing, Mechanical*, O-F-555-C (1969); *Dry Chemical, Fire Extinguishing, Multi-purpose Phosphate*, O-D-1380a (1972); *Dry Chemical, Fire Extinguishing, Potassium Bicarbonate*, O-D-1407a (1976); *Fire Extinguishing Agent, Aqueous Film Forming Foam (AFFF) Liquid Concentrate, for Fresh and Sea Water*, M/L-F-24385C (1981), Federal specifications.
Standard for Dry Chemical and Dry Powder Hand and Wheeled Extinguishers, ULC-S504-1977; *Standard for Classification, Rating and Fire Testing of Class A, B and C Fire Extinguishers and for Class D Extinguishers or Agents for Use on Combustible Metals*, ULC-S508-1975; *Standard for Halogenated Fire Extinguishers*, ULC-S512-1977; *Standard for Dry Chemical for Use in Hand and Wheeled Extinguishers*, ULC-S514-1978; Underwriters' Laboratories of Canada, Scarborough, Ontario, Can.
Specification for Portable Fire Extinguishing of the Halogenated Hydrocarbon Type, B.S. 1721-1968; *Specification for Portable Carbon Dioxide Fire Extinguishers*, B.S. 3326-1960; *Specification for Dry Powder Fire Extinguishers*, B.S. 3465-1962; *Classes of Fire*, B.S. 4547-1970; *Specification for Portable Fire Extinguishers*, B.S. 5423-1977, British Standards Institute.
Water Type Portable Fire Extinguishers, AS 1840-3-1976; *Foam Type Portable Fire Extinguishers*, AS 1843-5-1976; *Dry Chemical Type Portable Fire Extinguishers*, AS 1846-1976; *Carbon Dioxide Type Portable Fire Extinguishers*, AS 1847-1976; *Halogenated Hydrocarbon Portable Fire Extinguishers*, AS 1848-1976; *Portable Fire Extinguisher—Class, Rating & Fire Testing*, AS 1850-1981, Australian Standards.

E. EUGENE STAUFFER
The Ansul Company

FLOW MEASUREMENT

Flow measurement is a broad field that covers a spectrum ranging from the minuscule viscous flow rates associated with the pharmaceutical industry to the immense volumes involved in river flow measurement. The measurement of flow is an essential part of the production, distribution, consumption, and disposal of all liquids and gases including fuels, chemicals, foods, and wastes. Fluids to be measured may be clean or contaminated, at nearly total vacuums or at high pressure, and range from cryogenic to molten metal temperatures.

This breadth and diversity have led to the development of a multitude of flow-measurement devices. All meet the requirements of certain applications, and some achieve broad utility; none to date, however, comes close to being universal in scope. The field of flow measurement remains application oriented; the specific requirements of the measurement to be made must be analyzed in detail before proper equipment selection can be made.

Flowmeter Selection

A number of considerations should be evaluated before a flow-measurement method can be selected for any application. These considerations can be divided into four general classifications: fluid properties; ambient environment; measurement requirements; and economic considerations.

Fluid Properties. A great variety of equipment exists for measuring clean, low viscosity, single-phase fluids at moderate temperatures and pressures. Any extreme fluid characteristic, such as a high operating temperature, greatly reduces the range of equipment available and so should be given first consideration in any selection procedure. Fluid-related factors that normally should be considered are operating pressure, temperature, viscosity, density, corrosive or erosive characteristics, flashing or cavitation tendencies, and fluid compressibility. Other fluid properties may only be important with certain meter types; ie, heat capacity is important with thermal meters and some fluid electrical conductivity is required for magnetic flowmeter operation. In some cases, particular fluid requirements may limit the metering choices. An example of this is the requirement for the sanitary design of meters used in food processing.

Reynolds Number. One important fluid consideration in meter selection is whether the flow is laminar or turbulent in nature. This can be determined by calculating the Reynolds number, Re, which represents the ratio of inertial to viscous forces within the flow.

$$Re = \frac{\rho VD}{\mu}$$

where ρ = the fluid density; μ = the fluid viscosity; V = the average fluid velocity; and D = the pipe or meter inlet diameter. Since Re is dimensionless, any consistent system of units can be used.

A low Reynolds number indicates laminar flow and a parabolic velocity profile of the type shown in Figure 1(**a**). In this case, the velocity of flow in the center of the conduit is much greater than that near the wall. If the operating Reynolds number is increased, a transition point is reached (somewhere over Re = 2000) at which the flow becomes turbulent and the velocity distribution more even as shown in Figure 1(**b**). This tendency to a uniform velocity profile within the pipe increases at higher Reynolds numbers.

Most flowmeters are designed and calibrated for use on turbulent flow, by far

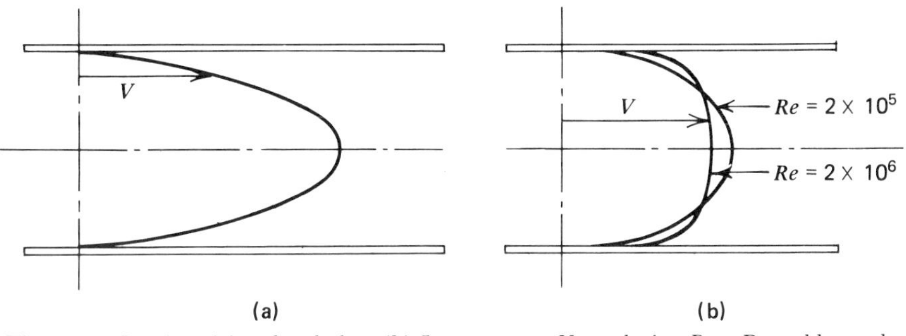

Figure 1. Laminar (**a**) and turbulent (**b**) flow patterns. V = velocity; Re = Reynolds number.

the more common fluid condition. Measurements of laminar flow rates may be seriously in error unless the meter selected is insensitive to velocity profile or is specifically calibrated for the condition of use.

Ambient Environment. The environment around the flow conduit must also be considered in meter selection. Such factors as the ambient temperature and humidity, the pipe shock and vibration levels, the availability of electric power, and the corrosive and explosive characteristics of the environment all may influence flowmeter selection. Special factors such as possible accidental flooding, the need for hosedown or steam cleaning, and the possibility of lightning or power transients may also need to be evaluated.

Enough space must be available to properly service the flowmeter and to install any straight lengths of upstream and downstream pipe recommended for use with the meter. Close-coupled fittings such as elbows or reducers tend to distort the velocity profile and can cause errors in a similar manner to those introduced by laminar flow. The amount of straight pipe required depends on the meter type. For the typical case of an orifice plate, piping requirements are normally listed in terms of the β or orifice:pipe bore ratio as shown in Table 1 (1).

Measurement Requirements. Any analysis of measurement requirements must begin with consideration of the accuracy, repeatability, and range needed for the application. For control applications, repeatability may be the primary criterion; conversely, for critical measurements, the total installed system accuracy should be considered. This includes the accuracy of the flow primary and secondary readout devices as well as the effects of piping, temperature, pressure, and fluid density. The accuracy of the system may also relate to the measurement range that is required.

Table 1. Required Straight Pipe Lengths for Orifice Plates, Nozzles, and Venturis

	On upstream side of the primary device							
β	Single 90° bend or tee (flow from one branch only)	Two or more 90° bends in the same planes	Two or more 90° bends in different planes	Reducer (2D to D over a length of 1.5D to 3D)	Expander (0.5D to D over a length of 1D to 2D)	Globe valve fully open	Gate valve fully open	On downstream side; all fittings included in this table
0.20	10 (6)[a]	14 (7)	34 (17)	5	16 (8)	18 (9)	12 (6)	4 (2)
0.25	10 (6)	14 (7)	34 (17)	5	16 (8)	18 (9)	12 (6)	4 (2)
0.30	10 (6)	16 (8)	34 (17)	5	16 (8)	18 (9)	12 (6)	5 (2.5)
0.35	12 (6)	16 (8)	36 (18)	5	16 (8)	18 (9)	12 (6)	5 (2.5)
0.40	14 (7)	18 (9)	36 (18)	5	16 (8)	20 (10)	12 (6)	6 (3)
0.45	14 (7)	18 (9)	38 (19)	5	17 (9)	20 (10)	12 (6)	6 (3)
0.50	14 (7)	20 (10)	40 (20)	6 (5)	18 (9)	22 (11)	12 (6)	6 (3)
0.55	16 (8)	22 (11)	44 (22)	8 (5)	20 (10)	24 (12)	14 (7)	6 (3)
0.60	18 (9)	26 (13)	48 (24)	9 (5)	22 (11)	26 (13)	14 (7)	7 (3.5)
0.65	22 (11)	32 (16)	54 (27)	11 (6)	25 (13)	28 (14)	16 (8)	7 (3.5)
0.70	28 (14)	36 (18)	62 (31)	14 (6)	30 (15)	32 (16)	20 (10)	7 (3.5)
0.75	36 (18)	42 (21)	70 (35)	22 (11)	38 (19)	36 (18)	24 (12)	8 (4)
0.80	46 (23)	50 (25)	80 (40)	30 (15)	54 (27)	44 (22)	30 (15)	8 (4)

[a] The nonparenthetical values are "zero additional uncertainty" values. The parenthetical values are "±0.5% additional uncertainty" values. All straight lengths are expressed as multiples of the pipe diameter D. They are measured from the upstream face of the primary device.

Depending on the application, other measurement considerations might be the system's speed of response and the pressure drop across the flowmeter.

Economic Considerations. The principal economic consideration is, of course, total installed system cost, including the initial cost of the flow primary, flow secondary, and related ancillary equipment as well as the material and labor required for installation. Other considerations are operating costs and availability of maintenance. Table 2 summarizes these considerations in meter selection.

Flow-Calibration Standards

Flow-measuring equipment must generally be "wet" calibrated to attain maximum accuracy, and principal flowmeter manufacturers maintain extensive facilities for this purpose. In addition, a number of governments, universities, and large flowmeter users also maintain flow laboratories. It has, in fact, been estimated that more money and effort is expended on flowmeter calibration than on the calibration of any other group of measuring instruments (2).

Calibrations are generally performed with water, air, or hydrocarbon fuels using one or more of four basic standards: weigh tanks, volumetric tanks, pipe provers, or master flowmeters. Most standards can be used statically (when the flow rate is quickly started and stopped during the test), dynamically (when readings are taken at the instant the test is initiated and again at the instant it is completed), or in a hybrid

Table 2. Summary of General Considerations in Meter Selection

General classification	Factors to be considered
fluid properties	liquid, gas or multiple phase
	operating temperature and pressure
	viscosity, density, and Reynolds number
	homogeneity, presence of entrained particles
	corrosive, erosive characteristics
	fluid cleanliness, sanitary requirements
	flash point
	cavitation index
ambient environment	temperature and humidity
	availability of electric power
	space and piping limitations, accessibility
	shock and vibration levels
	corrosion characteristics
	explosion hazard
	accidental flooding, hosedown, steam cleaning
	power transients, lightning, radio interference
measurement characteristics	open channel or closed conduit
	size and flow range
	accuracy and repeatability
	measurement or control
	continuous or batch
	pressure drop
	speed of response
	local or remote readout, indication
economic considerations	initial cost including installation operating costs
	cost and availability of maintenance

dynamic start-and-stop, static reading mode. Static systems operate best with flowmeters that have minimal sensitivity to low flow rates. They do not give optimum results with vortex or turbine meters because of errors obtained during the short periods of low flow at the beginning and end of the test. Completely dynamic systems are limited by speed of response considerations and the general difficulties encountered with "on the fly" readings. Because of these limitations, hybrid dynamic start-and-stop static reading weight and volume systems have been developed to provide more accurate liquid calibrations than purely static or dynamic systems. In such systems, the desired test flow rate is first obtained but diverted around the weight or volume flow standard. The test run is initiated by diverting the flow into the standard and completed by diverting it out of the system. The weight or volume is then read after an appropriate settling time. Figure 2 illustrates the basic elements of a standard of this type.

The key to the performance of a dynamic start-and-stop static reading system is the design of the flow-diverter valve that switches the flow in and out of the standard. In a well-designed system, the actual diversion time should be much smaller than the collection time and the flow pattern through the diverter relatively independent of flow rate. Under these conditions, the limiting factor in system accuracy is the basic accuracy and resolution of the weight or volume standard. With care, errors can be reduced to less than 0.1% of reading.

Liquid Displacement Gas Meter Provers. The liquid displacement method is the most prevalent primary standard for the measurement of low-to-moderate gas flow rates. The method consists of displacing a known volume of liquid with gas (see Fig.

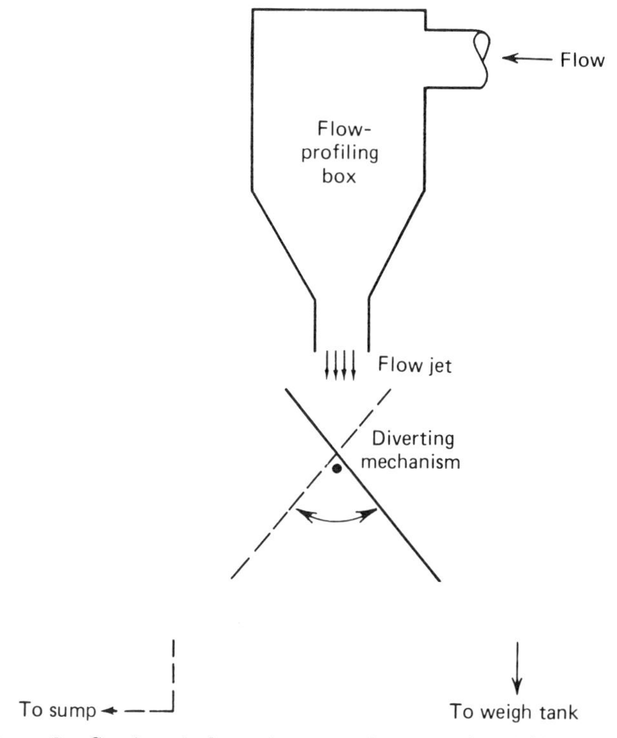

Figure 2. Gravimetric dynamic start-and-stop, static weighing system.

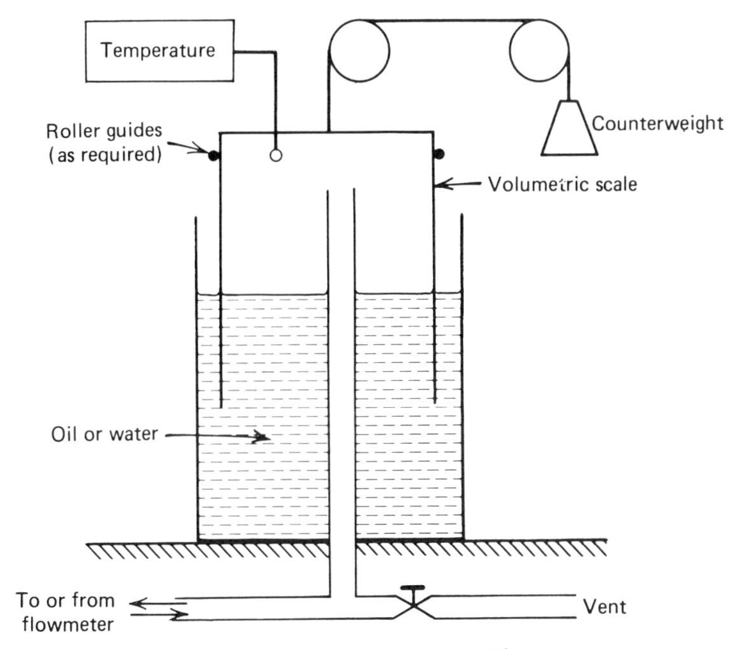

Figure 3. Liquid displacement calibrator.

3). As gas enters the inverted bell, it rises and a volume increment can be timed. Typical proven capacities are 1 m^3 or less, but capacities as large as 20 m^3 are obtainable.

Pipe Provers. In pipe-proving systems, a piston or elastic sphere is driven along a controlled pipe section using the fluid energy. Detectors within the pipe are actuated by the piston so that a known volume of liquid is moved between the detection points while the output of the test meter is simultaneously recorded. Prover systems can be very accurate as well as relatively compact and rugged. For these reasons, portable or truck-mounted systems are widely used for field testing of oil-field meters. Since time runs are relatively short, they work best with high resolution flowmeters such as turbine meters (3).

Master Flowmeters. Perhaps the most common method of flowmeter calibrations is to compare the output of the meter under test with one or more meters of high resolution and proven accuracy, called master flowmeters. This method is both convenient and quick. To obtain optimum accuracy, a system can be set up using multiple masters of overlapping range where the masters are regularly compared and periodically calibrated using weight or volume standards. This combines the convenience of the master meter method with the accuracy of the basic standard (4).

Flowmeter Classification

Flowmeters have traditionally been classified as either electric or mechanical depending on either the nature of the output signal or power requirements, or both. In recent years, improvement in electric transducer technology has blurred the distinction between these categories. Many flowmeters previously classified as mechanical are now used with electric transducers. Some common examples are the electric shaft encoders on positive displacement meters, the electric (strain) sensing of differential pressure, and the ultrasonic sensing of weir or flume levels.

To avoid these electric–mechanical classification problems, the flowmeter discussed in this article are divided based on the method by which their basic flow signal is generated: as category 1, flowmeters in which the signal is generated from the energy of the flowing fluid (commonly called self-generating flowmeters), and category 2, flowmeters that derive their basic signal from the interaction of the flow and an external stimulus (external stimulus flowmeters).

With these definitions, the manner in which the flow signal is transduced, conditioned, or transmitted does not determine the classification. For example, a differential-pressure meter generates a signal from the flow itself; it is part of category 1 regardless of whether the transducer is mechanical or electric in nature. In reviewing the meters in these categories, emphasis will be placed on common flow primaries; devices of a highly specialized nature (such as biomedical flowmeters) are beyond the scope of this article.

Self-Generating Flowmeters

Positive-Displacement Meters. Positive-displacement meters separate the incoming fluid into chambers of known volume. The energy of the fluid is used to advance these chambers through the meter and discharge them into the downstream pipe. The total quantity of fluid passing through the meter is the product of the internal-meter volume and the number of fillings.

Reciprocating Piston Meters. In positive-displacement meters of this type, one or more pistons similar to those in an internal-combustion engine are used to convey the fluid. Capacity per cycle can be adjusted by changing the piston stroke.

Bellows or Diaphragm Meters. Bellows meters use flexible diaphragms as the metering chambers. A series of valves and linkages control the filling and emptying of the chambers. Movement of the flexible walls is regulated for a constant displacement per stroke. Meters of this type were formerly widely used in the gas industry.

Nutating-Disk Meters. In positive-displacement meters of this design, the chambers are formed by a disk mounted on a central ball. The disk is held in an inclined position so that it is in contact with the chamber bottom along a radial section on one side of the ball and in contact with the top at a section 180° away. A radial partition prevents the disk from rotating about its own axis. Inlet and outlet ports are located on each side of the partition. Liquid enters alternately above and below the disk and flows around the conical chamber toward the outlet port. This movement causes the disk to nutate, ie, to undergo a circular nodding motion. This disk motion is coupled to a mechanical meter register. Nutating-disk meters are mechanically simple and rugged. They are widely used as commercial water meters.

Rotary-Impeller-Vane and Gear Meters. One group of positive-displacement meters depends on shaped impellers or gears to form the measuring chambers. Figure 4 illustrates a two-lobed rotary meter of the type used to measure gas flow. The impellers are designed to maintain a continuous seal during rotation. Close tolerances and the use of precision bearings permit these meters to have minimal leakage while keeping overall pressure loss low. Rotating-vane meters are similar in design, but they include a timed gate to isolate the inlet and outlet ports. Figure 5 shows one cycle of a rotating-vane gas meter. The pressure of entering gas rotates the vane assembly counterclockwise and, through timing gears, rotates the gate. In the successive positions, the annular segment of gas is isolated by vanes 1 and 2, rotated through the housing, and discharged by the action of the gate.

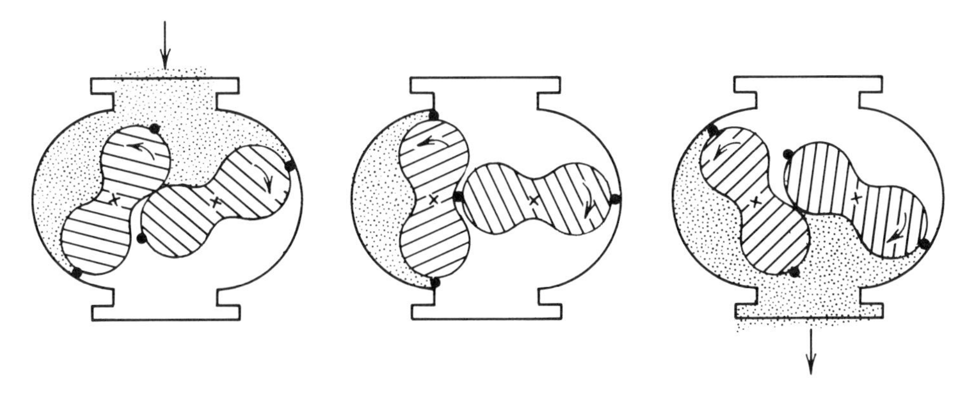

Figure 4. Two-lobed rotary gas flowmeter operating sequence.

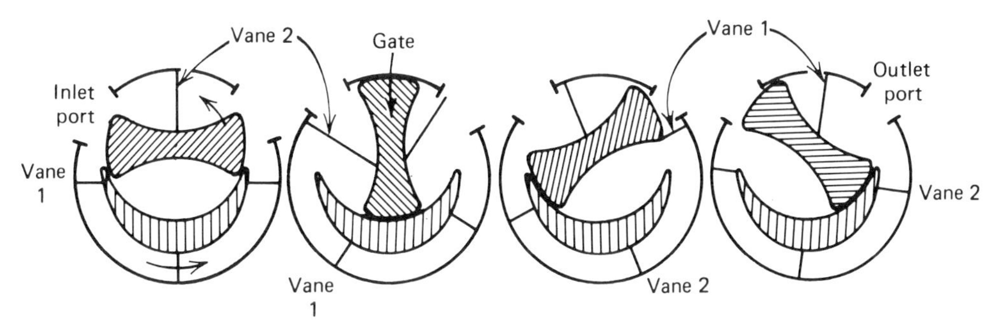

Figure 5. Operating sequence of a rotating-vane positive-displacement meter.

Positive-displacement meters have broad application in the distribution of natural gas because of their completely mechanical nature and their ability to maintain good accuracy over long periods of time. Wear in positive-displacement meters tends to increase leakage so that errors tend to be in the direction of under-registration, the most acceptable mode of error for commercial billing meters within the gas industry. Meters are normally periodically recalibrated and adjusted to read within ±1% of the true flow as required by AGA Report 6 (5) and other standards.

Positive-displacement meters also find broad application in the measurement of viscous liquids since higher viscosities provide lubrication and minimize seal leakage. Positive-displacement designs are inherently insensitive to incoming velocity profile and thus to piping configuration and Reynolds number. Good accuracy can be obtained in transitional Reynolds numbers where many other meters exhibit nonlinearity. Because of their small operating clearances, filters are commonly installed before these meters to minimize seal wear and resulting loss of accuracy (6).

Differential-Pressure Flowmeters. Differential-pressure meters are the oldest and yet still the most common group of flow-measurement devices. This general category includes orifice plates, venturi tubes, flow nozzles, elbow meters, and pitot tubes. All are based on the Bernoulli principle that, in a flowing stream, the total energy (the sum of the pressure head, velocity head, and elevation) remains constant. Differential-pressure devices all create some constriction in the fluid conduit causing a temporary increase in fluid velocity and a corresponding decrease in local head or pressure. For these conditions, the Bernoulli principle can be applied to give a general equation for head meters:

$$q = kA\sqrt{2\,gh}\text{ (in any consistent units)}$$

where q = the volumetric rate of flow; k = the experimentally determined flow coefficient (dimensionless); A = the inside cross-sectional area of pipe; h = the differential produced by the constriction (in height of the flowing fluid); g = the gravitation constant. The basic form of the equation is normally modified so that the differential is expressed in pressure units and the flow coefficient is divided into the product of an experimentally determined discharge coefficient K, and a series of calculated coefficients. In this form, for concentric restrictions:

$$q = K\beta^2 AYF_a\sqrt{2\,g\frac{(P_1 - P_2)}{(\rho)}}$$

where β = the restriction/pipe diameter ratio (d/D); Y = the gas expansion factor (for an adiabatic change from P_1 to P_2); P_1 = the upstream pressure; P_2 = the restriction pressure; F_a = the thermal expansion factor for the restriction; ρ = the fluid density; K = the discharge coefficient.

An outstanding advantage of common differential-pressure meters is the existence of extensive tables of discharge coefficients in terms of β ratio and Reynolds number (7–8). These tables, based on historic data, are generally regarded as accurate to within 1–5% depending on the meter type, the β ratio, the Reynolds number, and the care taken in the manufacture of the device. Better accuracies can be obtained by running an actual flow calibration on the device. Because of the square relationship between differential pressure and flow, the practical range of most differential meters is generally considered to be about 4:1.

Improvements in low differential readout devices and the desire to minimize permanent pressure losses have resulted in a trend toward higher β devices. This is generally at the price of increased sensitivity to upstream piping configuration (see Table 1). To obtain accurate differential-pressure measurements, the installation should conform as closely to reference conditions (1) as possible. The pipe should be inspected with respect to diameter, roundness, roughness, and tap location. The flow primary should be mounted concentric to the pipe internal diameter and the gaskets cut so that they do not protrude into the flow stream. Both pressure taps should be of the same diameter with no roughness or burrs. The differential-pressure transmitter should be as close to the taps as possible and the coupling lead lines sloped to permit condensate or gas-bubble removal.

Orifice Plates. A square-edged orifice plate is a thin, flat plate set perpendicular to the flow with a clean, sharp-edged circular opening. This opening is normally concentric with the pipe centerline, although eccentric plates, with the opening tangent to the pipe axis, are often used with sediment-bearing fluids. Segmented orifice plates, with the opening in the shape of a circular segment, are also available. At Reynolds numbers below 10,000, the coefficient of discharge for a square-edged orifice tends to become nonlinear, and quadrant-edged orifices are commonly used. These are thicker plates on which the inlet edge is rounded, making them somewhat less sensitive to fluid viscosity effects. The various types of orifice plates are shown in Figure 6. The stream issuing from an orifice attains its minimum cross section, defined as the *vena*

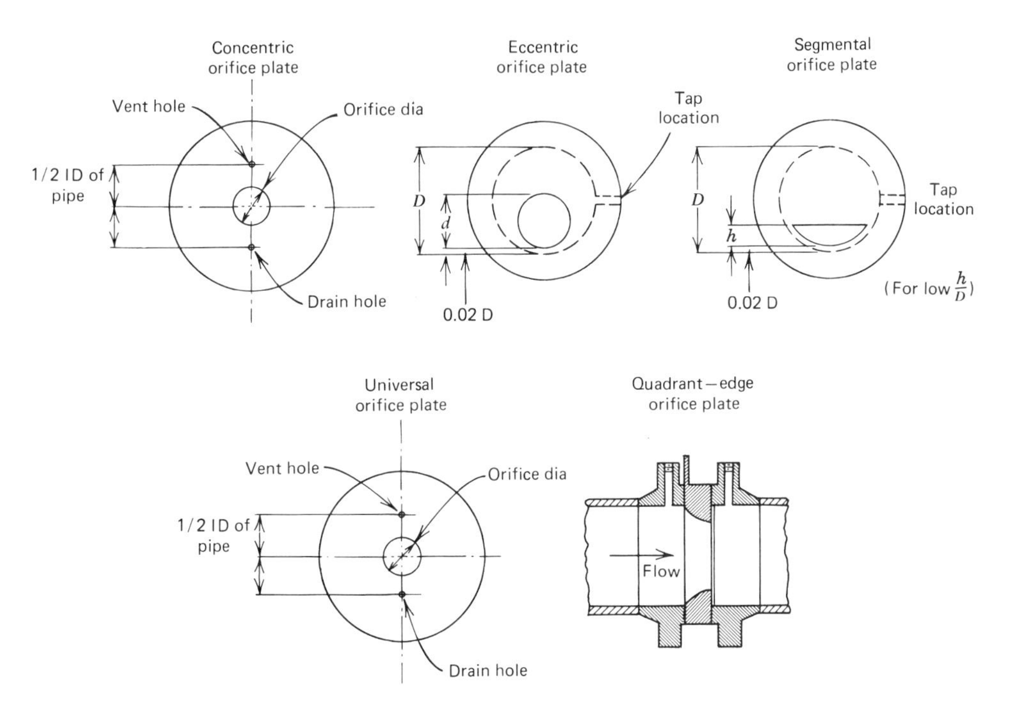

Figure 6. Varieties of orifice plates.

contracta, at a distance downstream of the restriction that is a function of the β ratio. *Vena contracta* pressure taps, commonly used in a steam-flow measurement, are located one pipe diameter upstream of the restriction and at the *vena contracta.* Corner pressure taps are drilled in the orifice mounting flanges, on either side and as close to the plate as possible. They are commonly used for pipe sizes under ca 50 mm. On larger pipe sizes, flange taps, each located 25 mm from the plate, are normally used. Figure 7 illustrates these tap locations. For very low flow rates, the orifice plate is often incorporated into a manifold that is an integral part of the differential-pressure transmitter. This provides a convenient, compact installation.

Two equations for the calculation of square-edged orifice coefficients are currently in general use. The ASME-AGA equation (5) is based on extensive calibration at Ohio State University in the 1930s. The International Standards Association "Universal" equation is based on the work of Stolz (1). In larger pipes, the difference in calculated coefficient is generally small (9).

Orifice plates have the advantages of being simple, hydraulically predictable, readily interchangeable, and reliable. They find wide use in both liquid and gas service where their moderate accuracy and limited range meet the needs of many applications.

Venturi Tubes. A venturi tube (Fig. 8) consists of two hollow truncated cones, the smaller diameters of which are connected by a short circular section known as the throat. Pressure differential is measured between the upstream and throat sections and can be related to flow by equations or tables in a manner similar to orifice plates. The sole purpose of the downstream cone is to recover part of the differential pressure. For an exit cone angle of 7%, the permanent pressure loss is only ca 10% of the differ-

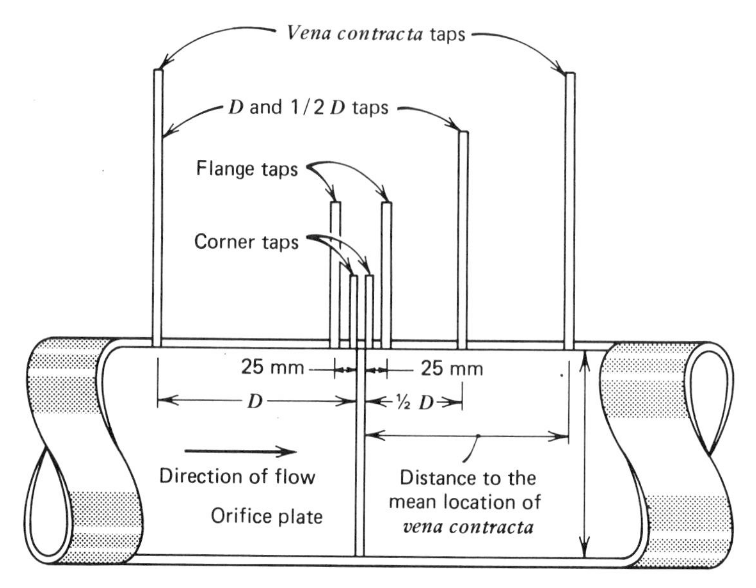

Figure 7. Orifice plate pressure tap location.

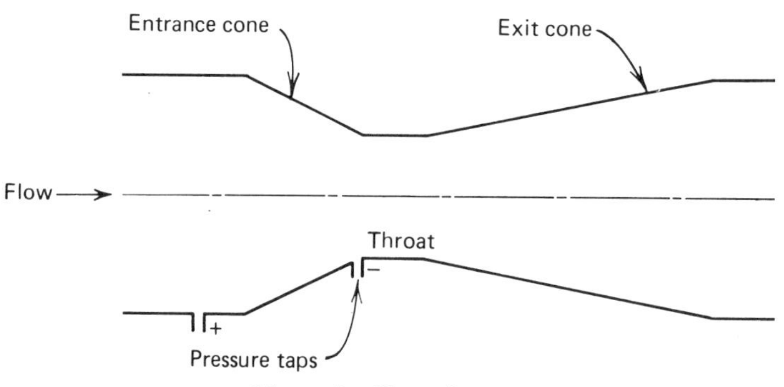

Figure 8. Venturi meter.

ential. A number of short-tube, larger-exit-angle venturi tubes are commercially available; these require less space for installation but have a higher pressure loss.

Venturi tubes have the advantages relative to orifice plates of lower permanent pressure drop and reduced sensitivity to hydraulic conditions. Disadvantages are their greater cost, longer installation length, and lack of easy interchangeability.

Flow Nozzles. A flow nozzle is a constriction with an elliptical or nearly elliptical inlet section. This section blends into a cylindrical throat section as shown in Figure 9. Nozzle pressure differential is normally measured between taps located on pipe diameter upstream and one-half diameter downstream of the nozzle inlet face. A nozzle normally has the approximate discharge coefficient of an equivalent venturi with the pressure drop of an equivalent orifice plate although venturi nozzles, which add a diffuser cone to proprietary nozzle shapes, are available to provide better pressure recovery.

Flow nozzles are commonly used in the measurement of steam and other high velocity fluids where erosion can occur. Nozzle flow coefficients are insensitive to small contour changes and reasonable accuracy can be maintained for long periods under difficult measurement conditions.

Critical Nozzles. As the pressure differential across a nozzle or venturi is increased, the rate of discharge of a gas also increases until the linear velocity in the throat reaches the velocity of sound. Any further increase in pressure differential does not cause an increase in velocity. In this condition, the nozzle is referred to as choked or at critical flow. The ratio of downstream to upstream pressure where critical flow is first obtained is called the critical pressure ratio R_c. For nozzles of the shape shown in Figure 9, R_c is equal to 0.5, but the addition of a venturi-type outlet can provide pressure recovery and values of R_c as high as 0.96 (10). As long as critical conditions are maintained, only the upstream pressure and temperature are needed to determine flow rate.

Because of their simplicity and ruggedness, critical nozzles are coming into increasing use. One principal application area is in the low flow testing of automotive carburetors and emission-control systems. Critical nozzles are also widely used at high gas flow rates where actual nozzles calibration is impractical. For these applications, a theoretical discharge coefficient is used (11).

Elbow Meters. Fluid passing through a common pipe elbow generates a differential centrifugal pressure between the inside and outside of the elbow. This differential can be measured to provide a measurement of flow, on an uncalibrated elbow, to ca ±4% uncertainty. Experimental tests indicate the elbow flow coefficient to be insensitive to changes in relative elbow roughness (7).

Pitot Tubes. The fundamental design of a pitot tube is shown in Figure 10(**a**). The opening into the flow stream measures the total or stagnation pressure of the stream while a wall tap senses static pressure. The velocity at the tip opening V can be obtained (again by the Bernoulli equation) as:

$$V = C\sqrt{2g(P_1 - P_2)}$$

This equation is applicable for gases at velocities under 50 m/s. Above this velocity, gas compressibility must be considered. The pitot flow coefficient C for gas service is very close to 1.0; for liquids, the flow coefficient is dependent on the velocity profile and Reynolds number at the probe tip. The coefficient drops appreciably below 1.0 at Reynolds numbers (based on the tube diameter) below 500.

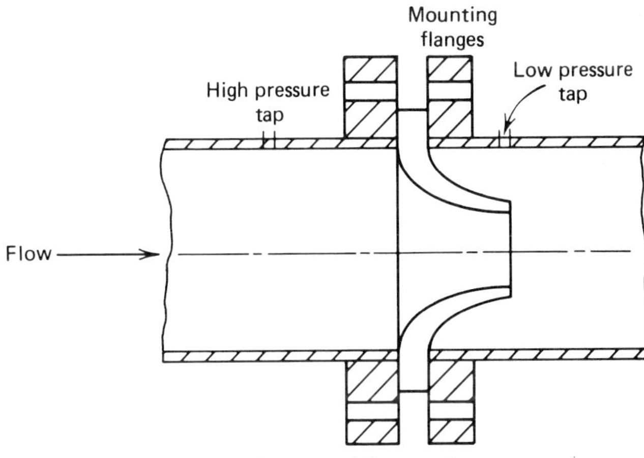

Figure 9. Flow nozzle.

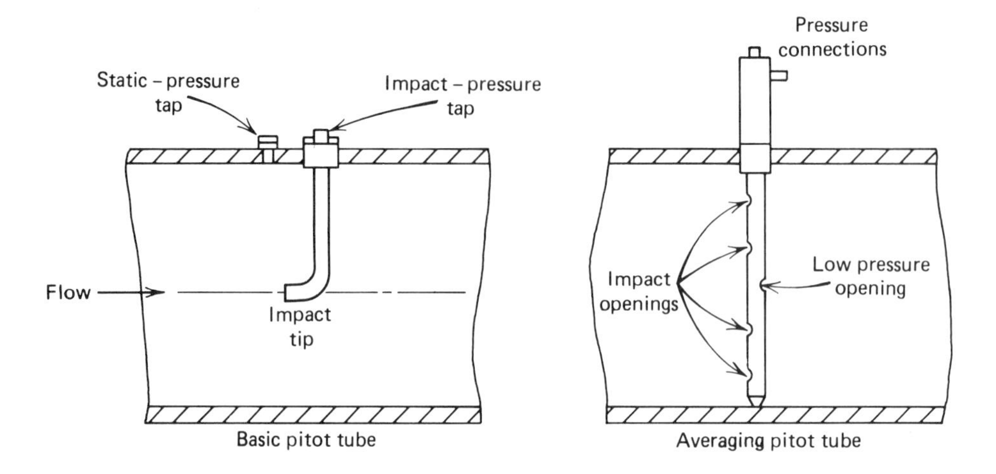

Figure 10. Pitot tube designs.

Standard pitot tubes provide a measurement of point velocity only; any attempt at determining total flow involves the assumption of a velocity profile. To overcome this disadvantage, averaging pitot tubes have been developed. As shown in Figure 10(**b**), these tubes extend across the pipe and use a series of holes at equal annular spacing to obtain an average velocity along the probe length. A single downstream-facing opening senses the downstream pressure.

Advantages of the pitot method of measurement are low pressure loss and easy installation. In some cases, installations in existing lines are made using a flame-cut hole or "hot tap" (12).

Laminar Flowmeters. Each of the previously discussed differential-pressure meters exhibits a square root relationship between differential pressure and flow; there is one type that does not. Laminar flowmeters (Fig. 11) use a series of capillary tubes, rolled metal, or sintered elements to divide the flow conduit into innumerable small passages. These passages are made small enough that the Reynolds number in each is kept below 2000 for all operating conditions. Under these conditions, the pressure drop is a measure of the viscous drag and is linear with flow rate as shown by the Poiseuille equation for capillary flow:

$$q = \frac{\pi}{8} C \left(\frac{D^4}{16}\right) \frac{1}{\mu L} (P_1 - P_2)$$

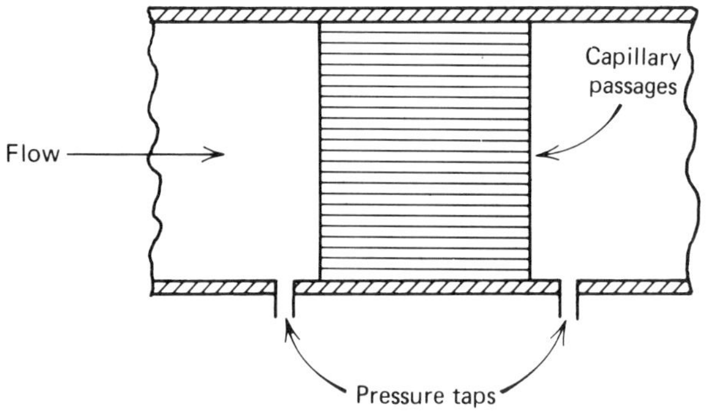

Figure 11. Laminar flowmeter.

where D = the diameter of the capillary tube; L = the tube length; C = the flow coefficient (experimentally determined); μ = the fluid viscosity; $P_1 - P_2$ = the differential pressure over the tube length L. Because of their small passage size and dependence on D^4, laminar flowmeters are suitable for use only with clean fluids.

Variable-Area Flowmeters. In variable-head meters, the pressure differential changes with flow rate across a constant restriction. In variable-area meters, the differential is maintained constant, and the restriction area allowed to change in proportion to the flow rate. A variable-area meter is thus essentially a form of variable orifice.

In its most common form, a variable-area meter consists of a tapered vertical tube containing a float free to move vertically in the tube. When flow is introduced into the small diameter bottom end, the float rises to a point of dynamic equilibrium at which the pressure differential across the float balances the weight of the float less its buoyancy. The shape and weight of the float, the relative diameters of tube and float, and the variation of the tube diameter with elevation all determine the performance characteristics of the meter for a specific set of fluid conditions. A ball float in the conical constant-taper glass tube is the most common design; it is widely used in the measurement of low flow rates at essentially constant viscosity. The flow rate is normally determined visually from a scale etched on the side of the tube. Such a meter is simple and inexpensive but, with care in manufacture and calibration, can provide readings accurate to within several percent of full scale flow (13).

A variety of other float shapes are available, some of which are designed to be insensitive to fluid viscosity changes. Tubes with various tapers are made to give linear of logarithmic scales and long, slow taper tubes are available for higher resolution. Tubes may contain flutes, triangular flats, or guide rods [see Figs. 12(**a**)–(**c**)] to center the float and prevent chatter. Metal tubes are available for high pressure service. In these, a magnetic coupling typically detects the float position and an external indicator or transmitter is used to provide the flow reading. Other, somewhat less common, forms of the variable-area meter use a tapered plug riding vertically within the bore of an orifice. The area of the restriction is controlled to maintain the differential-pressure constant. Flow rate is proportional to the effective restriction area and is derived from the motion of the plug.

Because of their design, variable-area meters are relatively insensitive to the effects of upstream piping and have a pressure loss that is essentially constant over the whole flow range. They have greatest application where direct visual indication of relatively low flow rates of clean liquids or gases are required (14).

Head-Area Meters. The Bernoulli principle, the basis of closed-pipe differential-pressure flow measurement, can also be applied to open-channel liquid flows. When an obstruction is placed in an open channel, the flowing liquid backs up and, with the Bernoulli equation, the flow rate can be shown to be proportional to the head, the exact relationship being a function of the obstruction shape.

Weirs. Weirs are dams or obstructions across open channels that have along their top edge an opening of fixed dimensions and shape through which the stream can flow. This opening is called the weir notch, and its bottom edge is designated the crest. Predictable forms of weirs have been developed which are classified according to the shape of the notch. Figure 13 shows four such forms. The discharge of each can be determined from tables (15) or by actual flow calibration.

Selection of weir type is dependent on the nature of the application. Triangular,

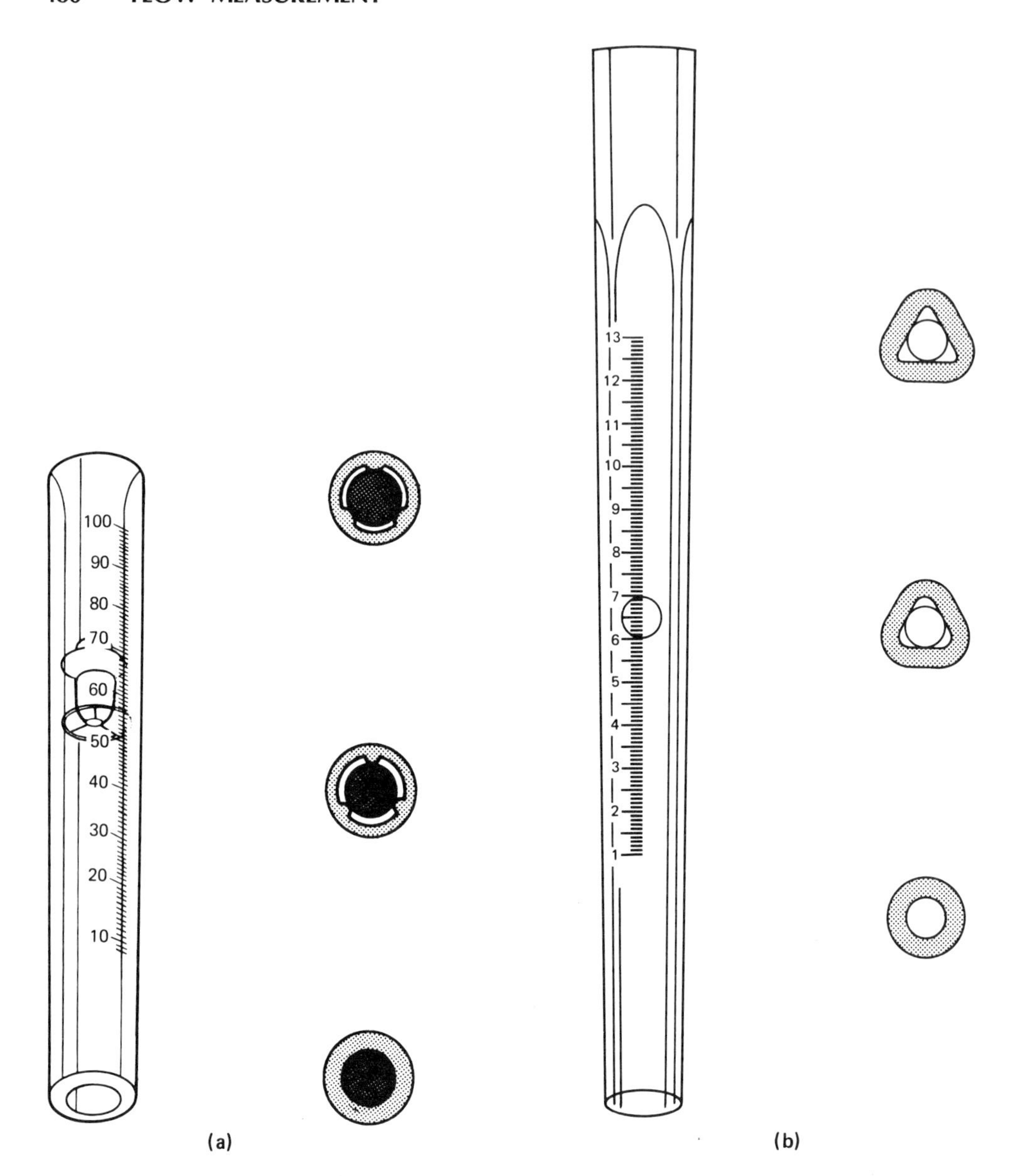

Figure 12. Types of V/A meters: (**a**) beadguide, (**b**) triangular flat, and (**c**) poleguide.

sharp-crested weirs [Fig. 13(**b**)] provide maximum flow range but will not transport floating material. A broad-crested rectangular notch [Fig. 13(**a**)] allows streamline development to pass most floating debris and will work at lower heads than a sharp-crested weir. The trapezoidal notch [Fig. 13(**c**)] is a combination of the rectangular and triangular forms. In the Cippoletti design, the slope of the ends is such that the additional discharge through the triangular portions of the notch exactly compensates for the effects of end contractions. Special forms of notches can be constructed to simplify the flow head relationship. Figure 13(**d**) shows one such form.

For accurate flow measurement, the channel area upstream of the weir should be large enough to allow the flow to develop a smooth flow pattern and a velocity of

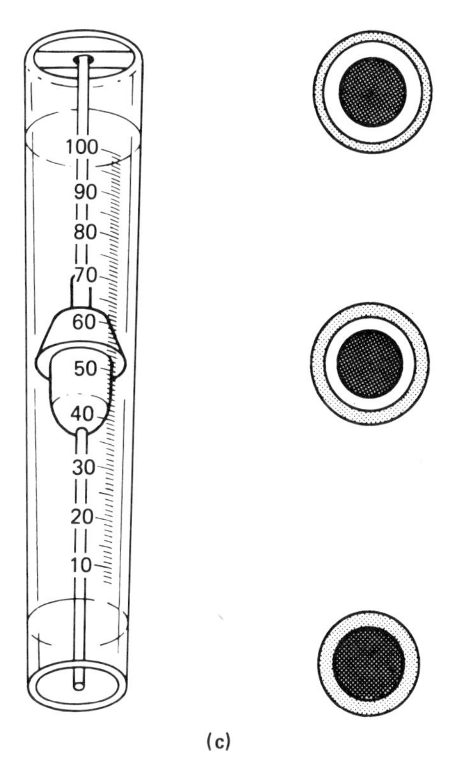

(c)

Figure 12. *(continued)*

0.1 m/s or less. The downstream channel must be large enough to prevent high flow rates from submerging the weir, and the flow over the notch must be sufficient for it to clear the downstream face as shown in Figures 13(**a**)–(**b**). This free discharge mode is the basis of weir capacity tables.

Weirs are commonly used in irrigation, water works, sewage-treatment plants, electricity-generating facilities, and pollution monitoring (16).

Flumes. Flumes are flow channels with gradual rather than sharp restrictions. Flumes for open channels are closely analogous to venturi meters for closed pipes, and weirs are analogous to orifice plates. The flume restriction may be produced by a contraction of the sidewalls, by a raised portion of the channel bed (a low broad-crested weir), or by both. One common design is the Parshall flume shown in Figure 14. This consists of an inlet section with convergent walls and level floor, a throat section with parallel walls and downward sloping floor, and an outlet section with diverging walls and an upward sloping floor.

Dimensions and capacity tables for Parshall flumes can be obtained from ref. 17. They are widely used in measuring irrigation water as alternatives to weirs where their lower head requirements, higher capacity, and reduced sensitivity to silting are advantageous. They are generally considered more expensive and less accurate than sharp-crested weirs.

A number of alternative flume designs have been created specifically for use in partially filled circular conduits such as sewers. They are available in molded fiberglass and can be lowered through a manhole if required (18). As with all open-channel head-area meters, they must be sized to prevent submergence of the restriction.

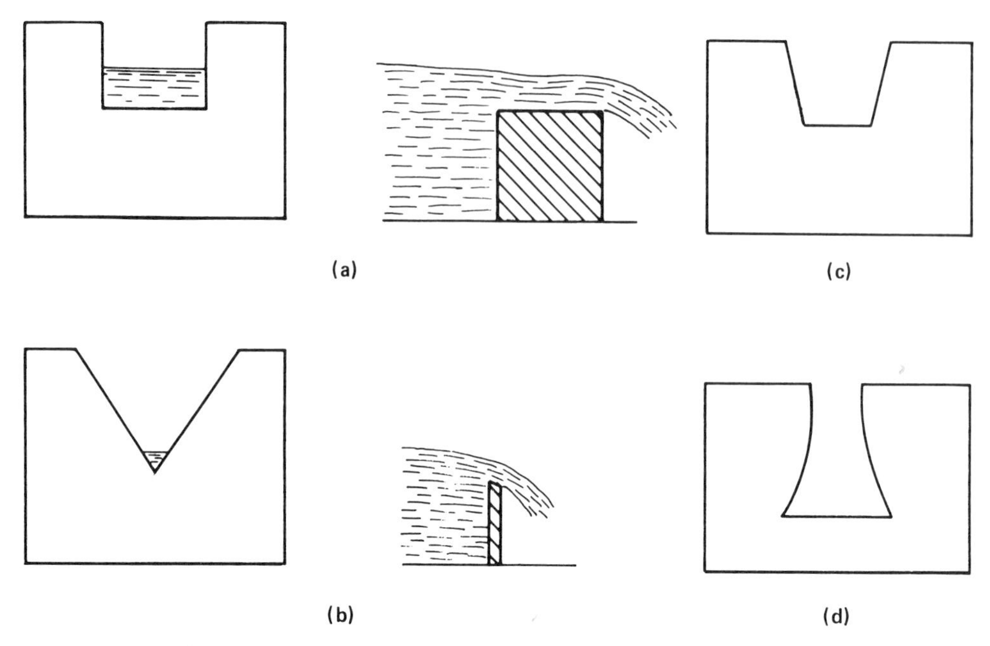

Figure 13. Stream flow over (**a**) a broad-crested, rectangular weir and (**b**) a sharp-crested, triangular weir. (**c**) A trapezoidal-notch weir; (**d**) a hyperbolic-notch weir.

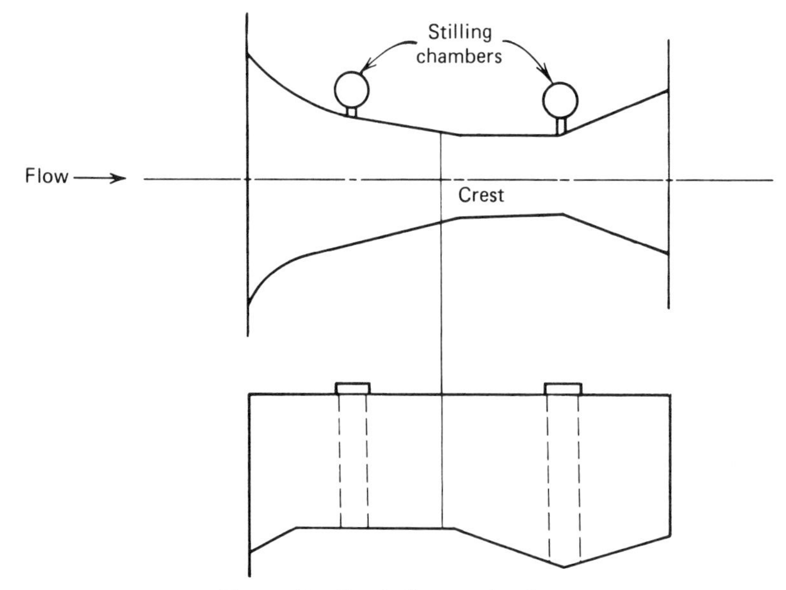

Figure 14. Parshall measuring flume.

Velocity Flowmeters. A number of flowmeter designs use a rotating element that is kept in motion by the kinetic energy of the flowing stream and whose speed is thus a measure of fluid velocity. In general, those used to measure wind velocity are called anemometers; those used for open-channel liquids, current meters; and those used for closed pipes, turbine flowmeters.

Cup and Vane Anemometers. Cup anemometers have shaped cups mounted on the spokes of wheel. The cups, under the action of the fluid forces, spin in a horizontal plane about a vertical shaft mounted in bearings. Vane or propeller types use a multibladed rotor whose axis is parallel to the flow direction as the rotating member. Both designs are commonly used for wind-speed measurement or similar applications such as the velocity in ventilation ducts. Because of their inertia, anemometers are most accurate under steady conditions. Gusty winds cause them to read high.

Current Meters. Various vane designs have been adapted for open-channel flow measurement. The rotating element is partially immersed and rotates rather like a water wheel.

Turbine Flowmeters. The turbine meter represents a refinement of the anemometer or current-meter design for use in a closed conduit. A typical turbine cross section is shown in Figure 15. Flow entering the meter is directed through a flow-straightening section, which shapes the flow and acts as a rotor support. The flow then turns the helically bladed rotor and passes through a rear-support section. A magnetic pick-off coil, or other externally mounted proximity detector, senses the passage of the rotor blades. At a steady velocity, the rotor comes to a speed at which the angle of the fluid striking the blade produces a driving force that is just sufficient to balance the drag forces resisting rotation. This angle of attack is a measure of the total forces on the rotor-resisting rotation. For maximum range, it is essential that these drag forces, and the attack angle, are as small as possible.

The output of a turbine meter is inherently digital, provides a high information rate, and has excellent accuracy and repeatability over a wide range. Typical accuracy is ±0.25% of rate with 0.025% repeatability (19). A typical composite-performance curve is shown in Figure 16. Meter output, in pulses per unit volume, is plotted against operating frequency divided by viscosity (which is analogous to Reynolds number). The curve of Figure 16 was formed by calibration at four different viscosities. The calibrations all overlay down to some low flow rate where either mechanical or magnetic drag forces, or both, become significant. This point defines the lower limit of the predictable-performance range. Because of this excellent curve repeatability, the output can be made linear to low Reynolds numbers using microprocessor-based programmable secondaries.

Initial application of the turbine flowmeter was on clean liquids where high performance was required, as in aerospace and aircraft testing, cryogenic-liquid

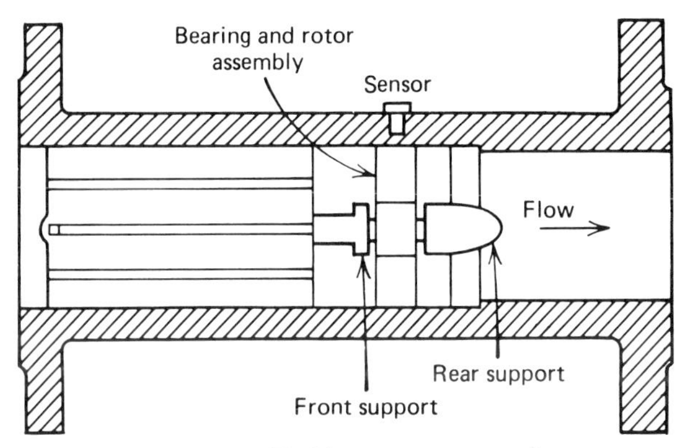

Figure 15. Turbine-meter cross section.

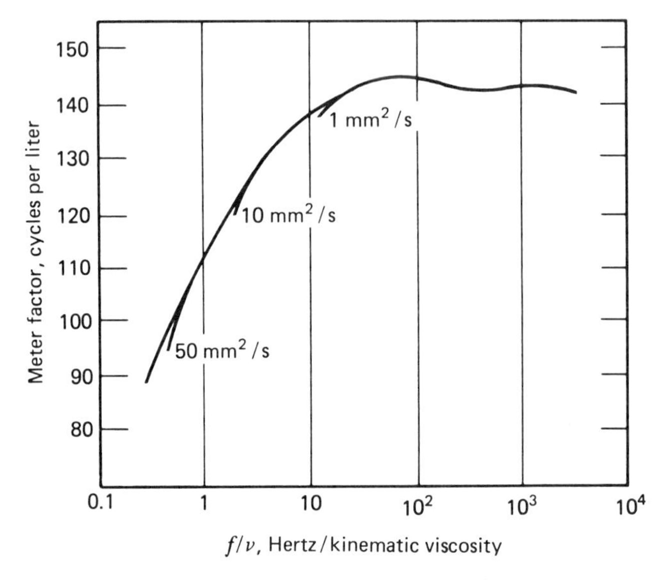

Figure 16. A turbine flowmeter composite-calibration curve. 1 mm^2/s = 1 cSt.

measurement, and the digital-blending of petroleum products. Additional meter developments such as the hydraulic balancing of thrust forces and self-cleaning sleeve bearings have resulted in meters that provide good service life even in the presence of some fluid contamination. Turbine meters are now used in a broad range of industrial environments including liquids, gases, and steam. Their share of the flowmeter market is exceeded only by differential-pressure devices (20).

Target Flowmeters. Target flowmeters use a drag-producing body in the flow stream to generate a force proportional to velocity. This force is sensed using strain gauges or a force balance system. The basic equation governing operation is that for the drag of a body in a flow stream:

$$F = C_D A \frac{\rho V^2}{2g}$$

where F = the drag force; C_D = the drag coefficient; A = the frontal area of the body; ρ = the fluid density; V = the upstream fluid velocity; g = the acceleration of gravity.

In general, the target, commonly a drag disk, is mounted on the pipe centerline to form an annular orifice. For pipe Reynolds numbers above 4000, the drag coefficient of such a design is essentially constant. At lower Reynolds numbers, the drag and meter coefficient depend on the d/D ratio and operating Reynolds numbers (21).

The target meter has the greatest application in the measurement of hot, viscous, or sediment-bearing fluids that would plug or congeal in the pressure taps of a differential-pressure meter.

Momentum Flowmeters. Momentum flowmeters operate by superimposing on a normal fluid motion a perpendicular velocity vector of known magnitude, thus changing the fluid momentum. The force required to balance this change in momentum can be shown to be proportional to the fluid density and velocity, the mass-flow rate.

Axial-Flow Angular-Momentum Flowmeter. In this design, an impeller is rotated at constant speed in the field imparting a constant angular momentum to it. A downstream rotor absorbs this momentum but is restrained from rotating.

By Newton's second law, the torque generated by the charge in momentum is

$$T = WR^2M$$

where T = the fluid torque; W = the fluid angular velocity; R = the radius of gyration of the fluid; M = the mass-flow rate. If a balancing torque is supplied by a closed-loop servo system whose torque is

$$T = KWV$$

where K = a controlled constant, and V = a voltage signal to the servo, then, for T = 0,

$$V = \left(\frac{R^2}{K}\right)^M$$

and the servo voltage is a function of mass-flow rate. Axial-flow angular-momentum meters are commonly used in measuring jet-engine fuel flow as the fuel energy content correlates much more closely with mass than volume.

Coriolis-Type Flowmeters. In Coriolis-type mass flowmeter designs, the fluid angular momentum is changed by causing it to flow around a loop or the curve of C-shaped section of pipe. This pipe section is continuously vibrated at its natural frequency. This subjects each fluid particle to a Coriolis acceleration. The resulting forces angularly deflect or twist the C pipe section. The amount of angular deflection is sensed by optical or magnetic sensors and converted to a digital display directly proportional to mass-flow rate.

Coriolis meters can be used with Newtonian or non-Newtonian fluids, slurries, or multiphase flows. They have found greatest application for low flows in pipe sizes under 25 mm although installations have been made up to 150 mm dia (22).

Oscillatory Flowmeters. Three different oscillatory fluid phenomena are used in flow measurement. They are fluid oscillation, vortex precession, and vortex shedding.

Fluid Oscillation. Fluidic flowmeters are based on the Coanda effect and the technology of bistable fluid oscillators such as those used in fluid logic. The general form of a fluidic meter is shown in Figure 17. It consists of an entrance nozzle section and a diverging section whose walls are designed to permit the flow to attach to one

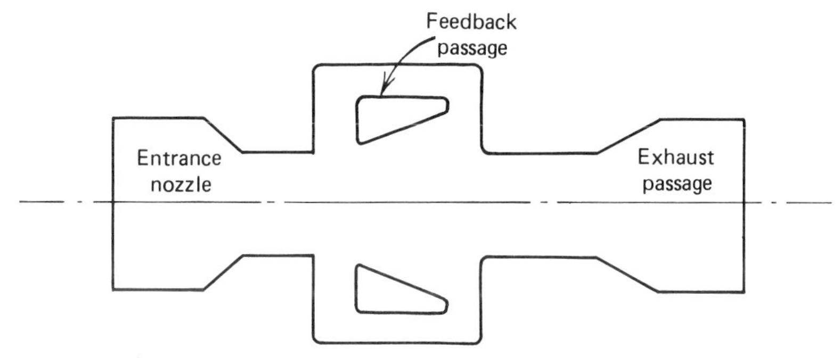

Figure 17. Internal contours of a fluidic-oscillator flowmeter.

side or the other but not to both. Downstream feedback passages connect with control ports upstream of the point of attachment.

In operation, natural turbulence and the Coanda effect cause the flow jet to attach to one side wall. As the flow is biased to this side, a portion of it is directed through the feedback passages to the control port causing the jet to switch to the opposite side where the same feedback action is repeated. The result is a continuous oscillation whose frequency is linearly related to the fluid velocity (23).

Fluidic flowmeters have been applied to the measurement of clean, low viscosity liquids in pipe sizes 80 mm and smaller.

Vortex Precession. When a swirling body of fluid enters a divergent section, the center of rotation precesses; ie, it leaves a straight line and takes up a helical path. Figure 18 shows a meter based on this principle. Entering fluid is given a rotational component of velocity by a set of fixed blades creating a vortex motion whose centerline coincides with that of the meter. This vortex is stabilized and accelerated by a convergent section before it enters an enlarged section that causes precession to take place. The frequency of this precession, detected by a dynamic pressure sensor, is linear with volumetric flow rate over a wide range.

Vortex-precession meters feature no moving parts and a relatively high frequency digital output. They are used in the measurement of gas flows, generally in pipe sizes 80 mm and smaller (24).

Vortex Shedding. When a streamlined body is placed in a flowing stream, the fluid follows the contours of the body without separating from its surface. If, however, the body is bluff or nonstreamlined, the fluid separates at some point from the surface and rolls into a vortex. For two-dimensional symmetrical bodies, the changes in local velocity and pressure associated with the separation on one side interact with the vortex forming on the opposite side. This feedback quickly causes a stable pattern of alternate vortex shedding to establish, so that the downsteam wake becomes a staggered pattern of vortexes commonly referred to as a Karmen-vortex street. This pattern is shown for a cylindrical obstruction in Figure 19. It is this phenomena that causes the flapping of flags and the clear turbulence behind jet aircraft.

The frequency of vortex formation is a linear function of the fluid velocity and the width of the obstruction at the point where shedding occurs. Vortex-shedding flowmeters use various forms of well-defined symmetrical obstructions to optimize vortex formation and detect the vortexes using sensors that respond to local velocity

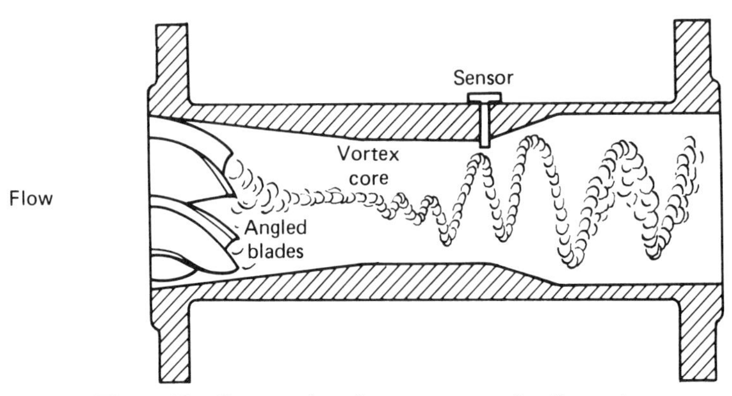

Figure 18. Cross section of a vortex-precession flowmeter.

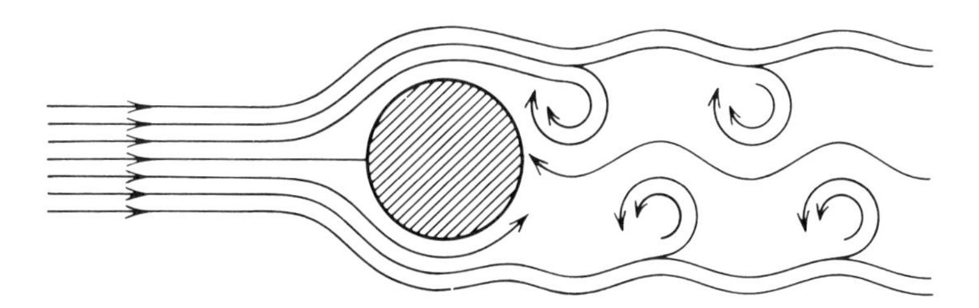

Figure 19. Karmen-vortex pattern behind a circular cylinder.

or pressure changes. Since these meters originally became available in the early 1970s, considerable development in sensing techniques has taken place. Improved sensors have broadened the application range of the vortex-shedding meter in the same way that the development of various sleeve bearings extended the scope of the turbine flowmeter. Current vortex-sensing techniques include differential-pressure-sensing diaphragms (with capacitive or inductive pick-off), strain gauges, piezoelectric elements, or velocity-sensing thermistors. In one design, an ultrasonic beam across the wake is used to detect the vortexes. In other variations, the obstruction consists of two interacting bodies with the torsional forces on the downstream body detected (25) or with the second body serving to simplify the vortexes and improve detection sensitivity.

Vortex-shedding flowmeters provide excellent accuracy and range on liquid, gas, and steam. Their advantage of no moving parts and their digital linear output have resulted in increasing use to the point that vortex shedding is considered the fastest-growing type of flow measurement (26).

External-Stimulus Flowmeters

Flowmeters in this category are generally electric in nature. They derive their signal from the interaction of the fluid motion with some external stimulus such as laser energy, an ultrasonic beam, or a radioactive tracer.

Electromagnetic Flowmeters. Faraday's law of electromagnetic induction states that relative motion, at right angles, between a conductor and a magnetic field induces a voltage in the conductor. The magnitude of the induced voltage is proportional to the relative velocity of the conductor and the magnitude of the magnetic field.

This principle is used to measure the flow of conducting liquids by meter designs similar to that of Figure 20. A pair of coils produces an electromagnetic field through an insulating tube carrying the liquid. Electrodes at a right angle to both the flow and the field sense the induced voltage E, whose magnitude is

$$E = CBdV$$

where C = the meter calibration factor; B = the average magnetic-flux density; d = the distance between electrodes; V = the average fluid velocity.

The history of the electromagnetic flowmeter is one of gradual improvement in techniques for sensing this voltage. Faraday himself realized the flow-measuring potential of the induction principle and attempted to utilize it in measuring the velocity of the Thames. He was unsuccessful as noise and polarization voltages masked any signal. Over a century passed before detection techniques were improved enough to make commercially acceptable electromagnetic meters. These meters used 50/60 Hz coil excitation. Although this sinusoidal excitation eliminated the problem of polar-

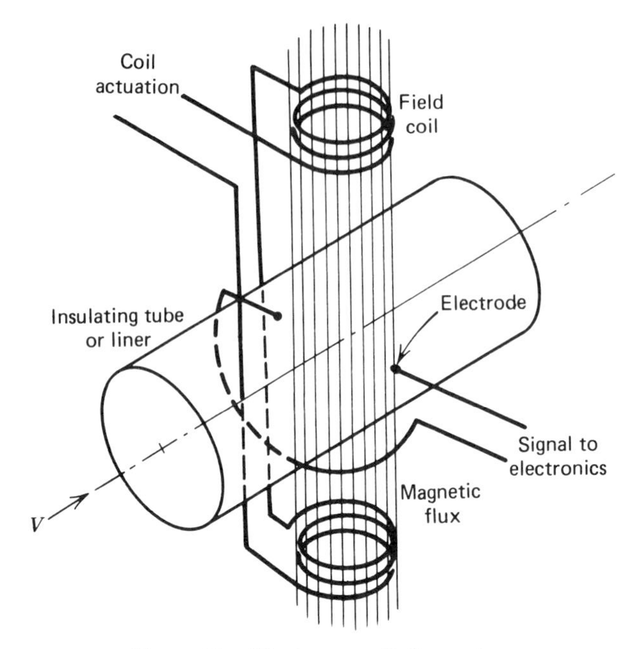

Figure 20. Electromagnetic flowmeter.

ization signals, they created a number of induced voltages within the electrode circuit. These were attenuated by the use of a phase-sensitive amplifier that passed the signal (in phase with the field ϕ) and rejected voltages at 90° to it (the $d\phi/dt$ induced voltages).

In recent years, electromagnetic flowmeters based on a pulsed d-c coil excitation have become available. These use a duty cycle of the form shown in Figure 21. The signal M_1 is measured during steady-state conditions ($d\phi/dt = 0$) with the field coils on, and M_2 similarly measured with the coil off. The on–off period is synchronized at a multiple of the line frequency so any a-c power noise averages out. The difference $M_1 - M_2$ is thus directly proportional to the flow. Meters of this design are claimed to function accurately under conditions where sinusoidal excitation meters do not provide acceptable results (27). Because of their design, pulsed d-c meters have a slower speed of response than do a-c excitation types.

The exact magnitude of the generated voltage of an electromagnetic flowmeter is an integration of the individual velocity and field vectors along the three-dimensional

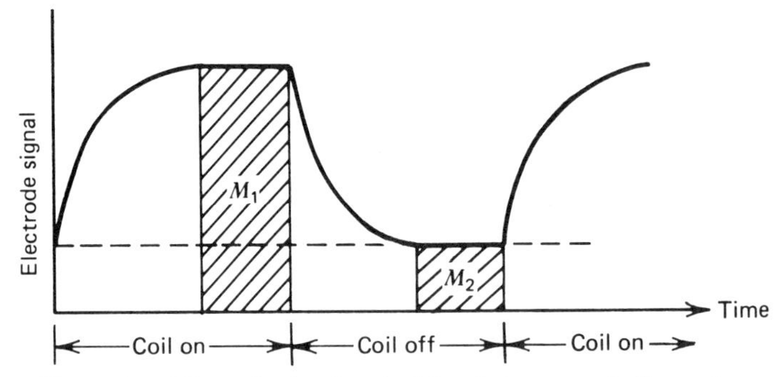

Figure 21. Electrode signal, pulsed d-c electromagnetic flowmeter.

path between the electrodes. Modern designs use characterized fields in order to weigh all velocities equally. In this manner, the meter is made less sensitive to changes in velocity profile than other common meters.

Electromagnetic flowmeters are available with various linear and electrode materials. Selection is based on the liquid characteristics. For corrosive chemicals, a fluorocarbon liner with Hastelloy electrodes is commonly used; polyurethane and stainless-steel electrodes are often used for abrasive slurries. Some fluids tend to form an insulating coating on the electrodes that introduces errors or loss of signal. To overcome this problem, specially shaped electrodes are available which extend into the flow stream and tend to self-clean. In another approach, the electrodes are periodically vibrated at ultrasonic frequencies.

Meters are available in essentially all pipe sizes (2 mm to 3 m) with measurement accuracy of 1% of rate or better over wide ranges. Electromagnetic flowmeters are obstructionless, have no moving parts, and are extremely rugged. Pressure loss is that of an equivalent section of pipe. They are insensitive to viscosity, density, and temperature. Fluid conductivity has no effect on meter performance provided it is above a minimum level that is a function of meter design. Models with thresholds as low as 0.05 μS/cm (tap water is typically 500) are available (28). Electromagnetic flowmeters have the additional advantage of sensing flows with entrained gas or solids on a flowing volume basis provided the flow is well mixed and traveling at a common overall velocity.

Because of these characteristics, electromagnetic flowmeters have been widely applied to the measurement of difficult liquids such as raw sewage and wastewater flows, paper-pulp slurries, viscous polymer solutions, mining slurries, milk, and pharmaceutical products. They are also used in less-demanding applications such as the measurement of large domestic water volumes.

Several special forms of electromagnetic flowmeters have been developed for specific applications. A d-c field version is used with liquid metals such as sodium or mercury. Pitot and probe versions provide low cost measurements within large conduits (29). Another design combines a level sensor with a electromagnetic meter to provide an indication of flow within partially full conduits such as sewer lines.

Ultrasonic Flowmeters. Ultrasonic flowmeters can be divided into three broad groups: passive, or turbulent-noise, flowmeters; Doppler, or frequency-shift, flowmeters; and pulse, or transit-time, flowmeters (see Ultrasonics, low power).

Passive Detectors. Passive, or turbulent-noise, detectors are ultrasonic microphones that are clamped to the flow conduit. These microphones respond to some portion of the frequency spectrum of turbulent noise within the pipe. This noise increases with increasing velocity although the exact relationship is dependent on the particular installation. Passive detectors can be used with liquids, gases, or slurries to activate flow switches or to provide a low cost indication of relative flow. (Since this signal is generated by the flow itself, passive detectors correctly belong in the group of self-generating flowmeters. They are listed here because the other types of ultrasonic flowmeters, the great majority, are not self-generating.

Doppler Flowmeters. Doppler flowmeters sense the shift in apparent frequency of an ultrasonic beam as it is reflected from air bubbles or other acoustically reflective particles that are moving in a liquid flow. It is essential for operation that at least some particles are present, but the concentration can be low and the particles as small as ca 40 μm. Calibration tends to be influenced by particle concentration since higher concentrations result in more reflections taking place near the wall, in the low velocity portion of the flow profile (30). One method used to minimize this effect is to use separate transmitting and receiving transducers focused to receive reflections from an intercept zone near the center of the pipe.

Both wetted-sensor and clamp-on Doppler meters are available for liquid service. A straight run of upstream piping and a Reynolds number of greater than 10,000 are generally recommended to ensure a well-developed flow profile. Doppler meters are primarily used where stringent accuracy and repeatability are not required. Slurry service is an important application area (31).

Pulse Flowmeters. This third type of ultrasonic meter depends on the transit time of an ultrasonic beam through the flow. In most designs, a pair of ultrasonic transducers are mounted diagonally on opposite sides of the pipe section as shown in Figure 22. These transducers may be wetted as well as the clamp-on type shown. In one design, these transducers simultaneously transmit upstream and downstream ultrasonic pulses and measure the time until the leading edge of the pulse is received at the opposite transducer. The transit time for the pulse moving in the direction of the flow is less than for the pulse moving with the flow. Transit times are given by the expressions

$$t_u = \frac{L}{C + V_1} \qquad t_d = \frac{L}{C - V_1}$$

where t_u = the transit time in the upstream direction; t_d = the transit time in the downstream direction; L = the acoustic path length; C = the speed of sound in the fluid; V_1 = the average component of liquid velocity V along the acoustic path.

The difference in transit times is

$$\Delta t = \frac{2\,LV_1}{C^2 - V_1^2}$$

Since $C_2 \gg C_1^2$ and $V_1 = V \cos \theta$:

$$\Delta t = \frac{2\,LV}{C^2 \cos \theta}$$

or

$$V = \frac{C^2 \cos \theta\, \Delta t}{2\,L}$$

The flow velocity is thus proportional to the difference in transit time between the upstream and downstream directions and to the square of the speed of sound in the fluid. Since sonic velocity varies with fluid properties, some designs derive compen-

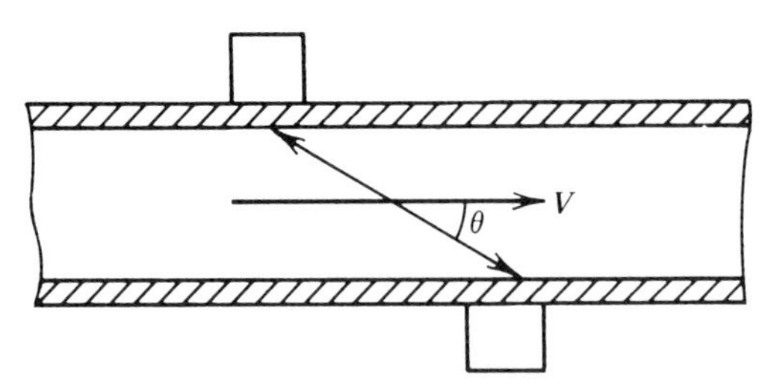

Figure 22. The basic arrangement of a transit-time ultrasonic flowmeter.

sation signals from the sum of the transit times (which can be shown to be proportional to C).

The angle θ can also change because of changes in the refraction angle in accordance with Snell's law. This is primarily a problem in clamp-on designs where the pipe-wall material is not controlled. Thermal gradients in the fluid may also cause problems by distorting the acoustic path. Designs that use a single acoustical path are inherently sensitive to velocity profile and require good upstream piping and fully turbulent flow. They have the greatest application in large pipe sizes where flow profile and fluid properties are relatively constant. Multiple path designs are also available; these provide greater precision where well-controlled conditions cannot be maintained. Both single and multiple path designs are sensitive to swirl.

A variation on the transit-time method is the frequency-difference or sing-around method. In this technique, pulses are transmitted between two pairs of diagonally mounted transducers. The receipt of a pulse is used to trigger the next pulse. Alternatively, this can be done with one pair of transducers where each acts alternately as transmitter and receiver. The frequency of pulses in each loop is given by:

$$f_u = \frac{1}{t_u} \quad \text{and} \quad f_a = \frac{1}{t_d}$$

where f_u = the frequency of pulses in the upstream loop and f_d = the frequency of pulses in the downstream loop.

$$\Delta f = f_u - f_d = \frac{(C + V_1)}{L} - \frac{(C - V_1)}{L}$$

$$= \frac{2\,V_1}{L} = \frac{2\,V \cos \theta}{L}$$

so

$$V = \frac{L}{\cos \theta} \Delta f$$

The flow velocity in this design is therefore proportional to the difference between the frequencies but independent of sonic speed within the fluid.

In practice, Δf is a small number and the sing-around frequencies are scaled up for display. In one example, for a pipe 1 m in diameter and water flowing at 2 m/s, the frequency difference is 1.4 Hz (32).

Transit-time ultrasonic flowmeters require a homogeneous fluid without a high density of reflective particles. (In this sense, fluid requirements are opposite those of a Doppler-type meter although there is an area of overlap.) They have the advantages of being obstructionless with low pressure drop, having a wide range, and being bidirectional. Their accuracy capability depends on the technique selected and the control of flow profile and fluid properties. Typical specifications range from ±5% of full-scale to ±1% of rate (33).

Laser Doppler Velocimeters. Laser Doppler flowmeters have been developed to measure liquid or gas velocities in both open or closed conduits. Velocity is measured by detecting the frequency shift in the light scattered by natural or added contaminent particles in the flow. Operation is generally conceptually analogous to the Doppler ultrasonic meters. Laser Doppler meters can be applied to very low flows and have the advantage of sensing at a distance, without mechanical contact or interaction. The technique has the greatest application in open-flow studies such as the determination of engine-exhaust velocity and ship-wake characteristics (34).

Correlation Flowmeters. ***Tracer Type.*** In this method, a discrete quantity of a foreign substance is injected momentarily into the flow stream, and the time interval for this substance to reach a detection point, or pass between detection points, is measured. From this time, the average velocity can be computed. Among the tracers that have historically been used are salt, anhydrous ammonia, nitrous oxide, dyes, and radioactive isotopes. The most common application area for tracer methods is in gas pipelines where tracers are used to check existing metered sections and to spot check unmetered sections (35).

Cross Correlation. Considerable research has been devoted to correlation techniques where a tracer is not used. In these methods, some characteristic pattern in the flow, either natural or induced, is computer-identified at some point or plane in the flow. It is detected again at a measurable time later at a position slightly downstream. The correlation signal can be electric, optical, or acoustic. This technique is used commercially to measure the flow of paper pulp and pneumatically conveyed solids. Other applications are being tested (36).

Thermal Flowmeters. ***Hot-Wire and Hot-Film Anemometers.*** Hot-wire devices depend on the removal of heat from a heated wire or film sensor exposed to the fluid velocity (see Temperature measurement). The sensor is typically connected in a bridge circuit with a similar sensor that is not exposed to the velocity in the opposite leg of the bridge. This provides compensation for fluid temperature changes.

Hot-wire anemometers are normally operated in a constant-temperature mode. The resistance of the sensor, and therefore its temperature, is held constant at a value slightly over the fluid temperature by a servo amplifier. In this mode, the current to the sensor becomes the flow-dependent variable. Constant-temperature operation minimizes thermal inertia and makes the system capable of sensing rapid changes in velocity.

Hot-wire signals are dependent on the heat transfer from the sensor and thus on both the fluid velocity and density, the mass-flow rate. They are also dependent on the thermal conductivity and specific heat of the fluid and are susceptible to any contamination that changes the heat transfer. For these reasons, hot-wire and hot-film anemometers are primarily used in clean liquids and gases where they can be calibrated for the exact condition of use (37). Principal applications are in the measurement of low air velocities both in the atmosphere and in building ventilation studies.

Differential-Temperature Thermal Flowmeters. Meters of this type inject heat into the fluid and measure the resulting temperature rise or, alternatively, the amount of power required to maintain a constant-temperature differential. The power required to raise the temperature of a flowing stream by an amount ΔT is given by the relation

$$P = Mc_p\Delta T$$

where P = the required power; M = the mass-flow rate; c_p = the specific heat at constant pressure; ΔT = the temperature rise. The thermal meter can therefore measure the mass-flow rate of a particular gas independent of pressure provided the specific heat is constant, a condition that is approximately true for most changes in temperature or pressure.

The original differential-temperature design heated the entire stream via a grid network and measured the temperature upstream and downstream with resistance grids. Because of high power consumption, this design has been supplanted by several forms that retain the essential features but provide lower power consumption and better corrosion protection. In one form, the outside of the meter tube is symmetrically heated by an external coil. Thermocouples are located on the tube wall equidistant from ends of the tube (Fig. 23). Heat sinks placed at the ends of the tube cause a symmetrical temperature pattern at zero flow, and no temperature differential is measured between thermocouples. When fluid flows through the tube, the temperature distribution becomes skewed in the downstream direction and the differential generated between the thermocouples is dependent on the mass flow. Small differential-temperature thermal meters are used to meter corrosive gases such as chlorine (38).

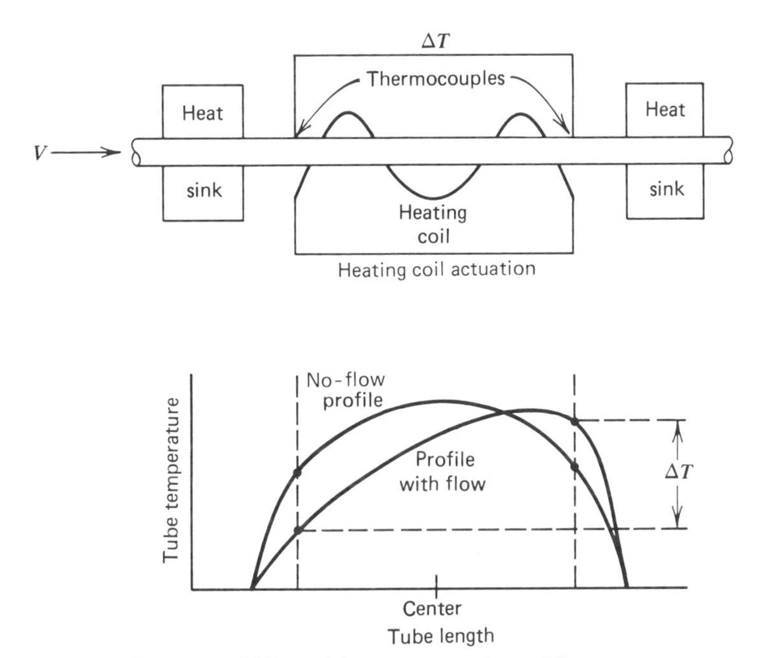

Figure 23. Differential-temperature thermal flowmeter.

Nuclear Magnetic Resonance Flowmeters. The principle of nmr, used in medical-diagnostic scanners, can be applied to liquids containing atoms whose nuclei have nonzero spin and can therefore be magnetically aligned. These atoms are most often hydrogen, but fluorine can also be used. The nmr meter uses a field coil to generate a strong magnetic field. In the presence of such a field, the liquid nuclei align themselves in parallel, spinning like tops. Also like tops, they precess about the axis of spin. The magnetization direction is then modulated and the instantaneous magnetization direction sensed at a slightly downstream point. A feedback circuit changes the modulation frequency to ensure a fixed-constant phase angle between the modulation point and the receiving point. This technique produces a modulation frequency that is linear with liquid velocity (39).

The nmr meter has a straight-through configuration with the measurement system completely external to the flow conduit. It has been applied to a variety of difficult liquid measurements where entrained solids, corrosive or coating tendencies, and lack of fluid conductivity have prevented successful application of other meter types.

BIBLIOGRAPHY

1. *Measurement of Fluid Flow by Means of Orifice Plates, Nozzles and Venturi Tubes Inserted in Circular Cross Section Conduits Running Full*, ISO 5167-1980(E), International Organization for Standardization, Geneva, 1980.
2. A. T. J. Hayward, *Measurement and Control* **10**, 106 (1977).
3. *API Standard 2531*, American Petroleum Institute, New York, 1963.
4. T. H. Burgess and co-workers, *Flow* **2**, 777 (1981).
5. *AGA Gas Measurement Committee Report 3*, 1975.
6. R. C. Pawloski, *Rotary Meter Fundamentals*, Appalachian Gas Measurement Short Course, Morgantown, W.Va., 1970.
7. *Fluid Meters—Their Theory and Application*, 6th ed., American Society of Mechanical Engineers, New York, 1971.
8. R. W. Miller, *Flow Measurement Engineering Handbook*, McGraw-Hill Book Company, New York, 1983.
9. R. W. Miller, *J. Fluid Eng.* **101**, 483 (1979).
10. H. S. Hillbrath, *Flow* **2**, 407 (1981).
11. R. W. Kuzara, *Flow* **2**, 741 (1981).
12. K. O. Plache, *Flow* **1**, 497 (1974).
13. H. H. Phillips, *Measurements and Control*, 136 (Sept. 1981).
14. *Variable Area Flowmeter Handbook*, Fischer Porter Co., Warminster, Pa., 1977.
15. *Liquid Flow Measurement in Open Channels Using Thin Plate Weirs and Venturi Flumes*, ISO 1438-1975(E), International Organization for Standardization, Geneva, 1975.
16. B. L. Thurley, in *Flow-Con 77 Proceedings*, Brighton, UK, 1977, pp. 171–186.
17. P. Ackers and co-workers, *Weirs and Flumes*, John Wiley & Sons, Inc., New York, 1978.
18. K. Komiya and co-workers, *Flow* **2**, 613 (1981).
19. W. O. Strohmeier, *Flow* **1**, 687 (1974).
20. J. Chowdhury, *Chem. Eng.*, 39 (June 28, 1982).
21. D. E. Curran, *Flow* **2**, 263 (1981).
22. K. L. Plache, *Mech. Eng.*, 36 (March 1979).
23. M. P. Wilson, Jr., and co-workers, *J. Basic Eng.* (March 1970).
24. P. J. Herzel, *Flow* **1**, 963 (1974).
25. T. H. Burgess in ref. 16, pp. 321–340.
26. *Chem. Week*, 32 (June 1, 1983).
27. W. Kiene in *FLOMEKO-78 Proceedings*, North Holland Publishing, Amsterdam, 1978, pp. 175–180.
28. A. Krigman, *In Tech*, 29 (Dec. 1982).
29. V. Cushing in ref. 27, pp. 163–169.
30. J. M. Waller, *In Tech*, 55 (Oct. 1980).
31. J. L. McShane, *Flow* **1**, 897 (1974).
32. W. K. Genthe and M. Yomamoto, *Flow* **1**, 947 (1974).
33. T. R. Schmitt, *In Tech*, 59 (May 1981).
34. J. Oldengarm in ref. 27, pp. 443–448.
35. A. E. Uhl, *Flow* **1**, 809 (1974).
36. M. S. Beck and co-workers in *Proceedings International Conference on Modern Developments*, Peter Peregrinus, Ltd., London, 1972.
37. R. B. Suhoke, *Flow Measurement Update, Measurement in Open Channels and Closed Conduits*, NBS Special Publication 484, National Bureau of Standards, Washington, D.C., 1977, pp. 597–620.
38. J. M. Benson, *Flow* **1**, 549 (1974).
39. W. H. Vander Heyden and J. F. Toschik, *Flow* **1**, 857 (1974).

Thomas H. Burgess
Fischer-Porter Company

G

GENETIC ENGINEERING

Recombinant DNA and Associated Technologies

Genetic engineering has generally been equated with recombinant DNA technology. Of the technologies required for a genetic-engineering experiment, the recombinant DNA technology is the most important (see also Genetic engineering in Volume 11). However, genetic engineering depends on or, at least, exploits many other technologies, ie, DNA sequencing and synthesis, protein sequencing and peptide synthesis, monoclonal-antibody generation, microbial genetics, and computer analysis. For the purpose of this article, genetic engineering is considered to be the direct operations performed on genes: the isolation and amplification, the regulable expression, and the directed modification or mutagenesis of DNA segments encoding specific genes.

The utility of being able to isolate a single gene from the complex genome of higher eukaryotic organisms was proven very early in the identification of unexpected genetic elements, ie, the intervening sequences or introns, which interrupt the coding regions of many normal genes. Also, the ability to manipulate genes and regulatory elements for the overproduction of otherwise scarce proteins has been demonstrated with a variety of hormones and blood proteins. Recent discussions of various aspects of genetic engineering are given in refs. 1–7.

Gene Cloning. Microbial systems typically include restriction enzymes, which provide an enzymatic defense against the DNA of invading phage and plasmids. The restriction enzyme makes site-specific cuts in the invading DNA. The same sites within its own DNA are protected by covalent modification of a base within the site, and the protective modification occurs in concert with the DNA replication process. The sizes of the restriction sites typically are 4–6 bases in length. They commonly are palindromic, ie, the sequence of the two strands is the same when read in the 5′-to-3′ direction; and they are frequently cut in a staggered fashion so that there are overhanging, single-stranded, or sticky termini. If fragments taken from different sources but generated by cutting with the same purified restriction enzyme are mixed, they can reassociate by virtue of these complementary termini. These hybrids can be covalently closed by treatment with DNA ligase. If one of the DNAs used in the process is an *Escherichi coli* phage or plasmid, the second DNA segment can be amplified by replicating the hybrid molecule in *E. coli.* The recombination of fragments *in vitro* can be mediated in other ways, including by the addition of synthetic oligonucleotide linkers containing specific restriction sites or of complementary homopolymer tails. DNA fragments may be derived from the chromosomal material of a desired source or by reverse transcription of its messenger RNA (cDNA) (1–4).

In vector technology, there has been increased use of cosmids, which have the elements of both bacteriophage and plasmid vectors and which permit the cloning of large (in excess of 20,000 base pairs or 20 kb) DNA segments (4,8). The *E. coli* phage M13 has become more widely used because of the simplicity of its use in DNA sequencing and mutagenesis protocols (9). Shuttle vectors also have been widely used to move genes between other hosts and *E. coli* where amplification and modification are simpler (4–5,10).

The technical bottleneck in the process tends to be identification and authentication of the desired hybrid, especially if complex mixtures of DNA or RNA are involved. This is typically achieved by nucleic acid hybridization of the hybrid clones with a previously characterized DNA segment that has been radioactively labeled.

A significant development has been the combination of protein sequencing and DNA synthesis to reverse-translate an oligopeptide sequence into an oligonucleotide probe. Techniques permit the sequencing of as little as 10 picomol (ca 10 μg) of protein (7). Approximately 25–35 amino acid residues at the amino terminus of the intact protein or of an internal fragment derived by chemical or enzymatic cleavage can be obtained. Within this sequence, a stretch of ca five amino acids with minimal codon degeneracy is identified. Based on that sequence, a mixture of oligonucleotides is synthesized and collectively represents all the possible sequences that could encode the chosen amino acid sequence. The mixture is labeled radioisotopically and used as a hybridization probe to identify the desired clones. Coupled with techniques developed for the generation of monoclonal antibodies, these procedures provide an access to virtually any gene system. Purified or enriched protein can be used to prepare a monoclonal antibody, which in turn can be used to isolate sufficient amounts of protein for sequencing and initiation of reverse translation.

The identification of hybrids has also been facilitated by refinement of hybrid-selected translation (4). Hybrid DNAs, typically amplified in *E. coli*, are bound individually or in small groups to filter paper, and messenger RNA is hybridized to the filters. After thorough washing, the bound RNA is eluted and translated *in vitro* or in a suitable *in vivo* system, eg, *Xenopus* oocytes. The desired hybrid is identified by its ability to select mRNA, which translates into the target protein. The process is not trivial, but it has been used to select hybrids from populations of cDNA clones as large as 1500 and to confirm the identity of a putative target clone.

Finally, the direct use of antibodies for the selection of specific clones is receiving increasing attention (4,11). DNA fragments are cloned into a vector in order to place the fragment adjacent to an active promoter:ribosomal binding-site system. The host cell subsequently produces whatever polypeptide is encoded in the foreign DNA, and the hybrids producing the protein of interest can be identified immunologically with an antibody specific for the protein whose gene is to be cloned.

Expression of Cloned Genes. The overexpression of an inserted DNA fragment is described in refs. 1, 3, and 12. Besides incorporating the appropriate DNA fragment into the vector, it is necessary that the proper transcription and translation signals be juxtaposed to the gene. This involves linking the vector to genomic or double-stranded cDNA fragments with synthetic pieces designed to optimize the configuration of structural and regulatory elements. This expression is preferably made chemically or thermally inducible, because the constitutive overexpression of a gene is usually detrimental to the host.

E. coli has been the most commonly used host–vector system for overexpression (6,12–13). However, several factors have fueled the search for alternate host–vector systems. First, the commercial-scale production of hormones, for example, in *E. coli*, is complicated by potential pyrogenic responses associated with *E. coli* components. Second, the fermentation (qv) industry and regulatory agencies are fairly unfamiliar with *E. coli*-derived products. Third, *E. coli* does not typically secrete proteins and, thus, the desired protein must be accumulated intracellularly. This can complicate potential schemes for continuous production and recovery of product. Fourth, *E. coli* does not perform many of the post-translational modifications, eg, glycosylation, found in higher eukaryotes.

The yeast *Saccharomyces* has been the second most commonly studied microbial system, especially in the fermentation industry, and it can secrete proteins (14–17). It is not likely to perform accurate glycosylations of mammalian proteins but will undoubtedly be useful for some purposes. *Bacillus subtilis* also is used in fermentation and can secrete proteins (18). However, its phage and plasmids have not been easily incorporated into vector constructions. Many other microbial systems are being studied: *Streptomyces*, *Streptococcus*, and *Pseudomonas*. These are likely to be developed further for use as protein sources in fermentation or as improved microbes in agricultural applications (9,19–20).

Most of the proteins produced to date have not required post-translational modification for biological activity. However, there is much interest in vector systems involving mammalian cells, because they may be required when accurate post-translational modification is needed. Such systems, which are commonly based on mammalian-virus-derived sequences, exist and are likely to be improved (10). Eukaryotic cell:vector transformations typically result in integration of the vector into the host genome. This has been adequate for studies of gene expression and would probably be desirable for achieving permanent changes in mammalian cells and crop plants. However, the desire to use cultured mammalian cells for protein production has led to adenovirus- and SV40-based vectors which, over a short time, exist in a multicopy state and thereby promote high level protein production. Recently, sequences from bovine papilloma virus (BPV), which permit vectors to exist indefinitely as multicopy plasmids, have been used in cultured cells. Other viruses, eg, herpes, are being studied as sources of strong transcription promoters and because of their potential to accommodate large pieces of DNA.

Site-Directed Mutagenesis. Site-directed mutagenesis provides increased genetic diversity and a route to improved protein and cell traits. There are three general routes: a cloned fragment can be subjected to traditional methods of chemical or genetic mutagenesis, one of the strands of the fragment can be nicked at random or at specified sites and subjected to chemical or enzymatic mutagenesis at the site, or a synthetic oligonucleotide encoding the desired change can be used as a primer to direct the mutagenesis.

The second route has been implemented in a variety of ways and is particularly useful for generating a random set of mutants or variants, which can be screened for the improved trait (21). An extension of this approach is to make a double-stranded cut and to use a double-stranded exonuclease to remove variable amounts of DNA. After religation, the resultant population of random deletions can be screened for the desired activity. Since the amino acid code is triplet in nature and the exonucleases are not so cooperative as to only react with triplet units, ca two thirds of the mutants are likely to be undesirably out-of-frame. However, it has been possible to fuse target genes to a second test gene, eg, β-galactosidase, whose activity can be monitored colorimetrically. Then only those transformant clones that exhibit β-galactosidase activity and, therefore, have the β-galactosidase gene as well as the gene of interest in-frame need be picked and characterized.

The third route is a powerful procedure that takes advantage of the chemical synthesis of oligonucleotides and the *E. coli* phage vector M13 (22). M13 produces double-stranded DNA intracellularly but incorporates single-stranded DNA into the phage particle. The gene to be changed is transferred to M13 by means of the vector's intracellular DNA. The single-stranded hybrid phage DNA is subsequently hybridized to a synthetic oligonucleotide. The oligonucleotide contains the desired change but is flanked by DNA sequences that are complementary to the region of the gene in which the change is to be embedded. The complementary regions hold the probe in place on the single-stranded hybrid molecule and act as primers with DNA polymerase for the synthesis of a second strand. A double-stranded, heteroduplex molecule results, a bubble existing where the new DNA sequence is opposite to the region of the gene to be changed. When the heteroduplex is introduced into *E. coli*, the mismatched region is resolved and, approximately half of the time, it is the original DNA strand that is eliminated and the new sequence that is copied. When the desired change can be defined, this is the most direct route to the synthesis.

Cell Transformation. The development of shuttle vectors has been particularly important in permitting the directed and controlled transformation of higher eukaryotic cells (10). In many cases, the host in which a gene is to be expressed is not the one in which the amplification and manipulations are most readily done. Schemes have been devised for moving genes between *E. coli* and the desired host to facilitate such work.

The shuttle vector carries regulatory elements and a selectable marker that permit its replication and selection in the control host, typically *E. coli* or *Saccharomyces*. The vector also carries a set of elements and selectable markers to permit its replication or integration and selection in the second host. Moreover, it carries an expression unit, ie, promoter, etc, peculiar to the second host to direct the expression of the desired gene in that host. As the demands on expression become more complex, elements dictating the time, place, and chemical responsiveness of expression must be incorporated. This is one area in which the most intense work and the greatest progress is likely to occur in the 1980s. An early example involves the use of an intermediate host *Agrobacterium tumefaciens* to move genetic information from an *E. coli* system into the T-DNA region of *Agrobacterium*'s Ti plasmid (23–24). The T-DNA region of Ti is subsequently transferred to the plant genome during transformation. It has been the only system exploited in plants in a scientifically credible fashion and has resulted in the selection of antibiotic-resistant plant cells and their regeneration to whole plants in which the foreign genes are expressed.

Vectors that can be moved directly from *E. coli* to the higher eukaryotic cell have been devised and exploited in animal cells (10,25). These experiments generally involve introduction of genes into cultured cells, but there have been efforts to introduce genes into somatic and germ cells and to return them into the animal. A spectacular example of this was the recent demonstration of enhanced growth in mice that had developed from embryos into which a rat-growth-hormone gene had been introduced (26). A similar result has been obtained with a gene-encoding thymidine kinase (27). Although many ethical and regulatory issues will accompany this class of experimentation, it will be conceivable to introduce genes into human somatic cells, and perhaps germ cells, as a means of gene therapy.

In animal cells, vectors that can move directly between them and *E. coli* have been developed. This is likely to be applied to plant cells as well. Such vectors require an expression unit similar to that needed for overproduction of a protein in *E. coli* or yeast, although in this case peculiar to the second host. These vectors contain regulatory elements that permit their replication in *E. coli* and their replication or genome integration in the second host. They typically require selectable markers for both hosts so that the transformed cells can be readily identified as the information is shuttled back and forth. Finally, to the extent controlled expression is required, elements dictating the time, place, and chemical responsiveness of expression must be incorporated. Progress in this area is expected to be great in the 1980s.

Applied Biotechnology and Genetics

Chemical and Chemical-Process Industries. Chemicals and chemical processes represent one of the largest opportunities for the future application of recombinant DNA technology (28–30). The principal evaluations of this technology have focused on the development of new approaches to the manufacture of existing chemicals at lower costs, reduced energy requirements, and decreased waste-disposal problems. This industry, which is one of the largest in the world and exceeds 40×10^9/yr, has been largely based on the processing and transformation of coal and, more recently, of oil into a great variety of products. Economic, technical, and environmental factors have resulted in a steadily increasing interest in biotechnology and its application to biomass conversion to useful products. Many large-scale production facilities for making antibiotics and organic chemicals, eg, acetone, butanol, and ethanol, were developed in the past few decades. The utility of organic chemical production by fermentation became economically undesirable with subsequent developments in the petrochemicals industry during the 1950s. Although the production of chemicals by fermentation represents a relatively old technology, modern molecular biology has provided a means for precisely tailoring microbes for specific bioconversion tasks, and high petrochemical costs have forced industry to reconsider optional biomass feedstocks for the synthesis of useful chemicals. The combination of fermentation and enzyme technologies will provide a transition from nonrenewable to renewable raw materials and should make fermentation-derived products common (31).

The production of chemicals by biotechnological means and specifically by genetic engineering requires an analysis of the types of compounds that can be produced biologically and the dependence of these compounds on genetic technology. Several variables are the availability of an appropriate organism or enzyme to effect the desired transformation, cost of raw materials, and costs of the process. In general, almost any organic compound can be produced by a biological system at some cost. Proteins are the top of the pyramid of recombinant DNA applications. Table 1 illustrates some of the commercial enzymes that are being or could be produced by recombinant technology. The next level of interest involves the production of small molecules, including vitamins, pesticides, growth regulators, amino acids, solvents, and organic acids (see Table 2). However, the economic considerations for the biotechnological manufacture of specialty chemicals and commodity chemicals are becoming increasingly stringent. Potential representative small molecules to be produced in aerobic and anaerobic biotechnological processes are acetic, acrylic, and adipic acids, and ethanol, ethylene glycol, ethylene oxide, glycerol, and propylene glycol.

Although a large segment of the chemical industry is involved in the manufacture of polymers, it is not likely that present biotechnology will have a significant impact in this area. Most polymers today are derived from simple and relatively inexpensive monomers and, unless the cost of the monomers can be reduced significantly by biotechnology, little impact is expected. However, since biological systems can man-

Table 1. Some Commercial Enzymes Whose Production Can Be Based on Recombinant DNA Technology

Enzyme	Applications
α-amylase	starch hydrolysis, manufacture of alcohol, textile desizing, and human digestive aid
ficin	wound debridement
glucose oxidase	diagnostic for diabetes
glucose isomerase	high fructose corn syrup (HFCS) manufacture
lipase	flavor production in cheese
papain	meat tenderizer
pectinase	clarification of beverages
penicillinase	treatment of penicillin allergic reaction
protease	leather bating
rennin	cheese manufacture

Table 2. Industrial Fermentation Product Categories

Class	Examples
amino acids	glutamic acid, tryptophan, lysine
organic acids	acetic, propionic, acrylic, lactic, malic, citric, gluconic, itaconic
solvents	acetone, methanol, ethanol, isopropanol, *n*-butanol
growth regulator	gibberellic acid
vitamins	riboflavin (B_2), cyanocobalamin (B_{12})
nucleic acids	5′-inosinic acid monophosphate
polyols	ethylene glycol, glycerol

ufacture interesting complex polymers, eg, polypeptides, polysaccharides, and polynucleotides, as well as polyesters, eg, polyhydroxybutyric acid, there is still some interest in this area.

Traditional insecticides, fungicides, and herbicides (qv) have yet to be developed by means of current biotechnological approaches, but the opportunities for the manipulation of microbes to produce specific pesticides has promise (see Fungicides (agricultural)).

The prospects for the application of biological processes to the manufacture of high volume commodity chemicals appear poor with the exception of a few products, eg, ethanol and some short-chain organic acids. Prospects appear much greater for the manufacture of higher priced specialty chemicals or those with no practical alternative manufacture route, eg, alkaloids and other natural products.

Little emphasis has been placed on site-directed mutagenesis as a means for tailoring enzymes for process applications. Enzymes will surely be employed as highly specific catalysts for performing unique chemical steps in a synthetic route, and immobilized microbes that express the desired enzyme activity will probably be used directly. Thus, enzymes with improved physical properties, eg, stability, solubility, etc, and reduced associated costs may provide a means for circumventing fermentative processes.

Reasons for maintaining continuous research involving biotechnology and recombinant DNA applications to the chemical industry include the use of renewable resources, eg, cellulose and lignin, as well as other biomass sources; multistep production, which can be achieved in one-step plants since microorganisms usually carry out a complete series of reactions in a single cell (reactor); the use of biological catalysts and biological processes that facilitate rapid chemical synthesis at less-severe physical conditions, ie, at lower temperatures, pressure, pH, etc; and the specificity of biological reactions which reduce the number of side reactions and by-product production problems.

Because new species of microbes are likely to be utilized, including *Escherichia coli*, *Pseudomonas* sp., and *Acinetobacter*, care must be taken in containment and safety evaluations. The chemical industry has a good record of handling highly toxic and caustic products and should be in a favorable position to provide the necessary checks and balances to protect manufacturing personnel as well as the environment. The containment systems required for large-scale fermentation to protect the organism from being influenced by its environment, eg, by competitive microbes, would provide a useful constraint on environmental impact as well.

Waste and Industrial Pollution. The most widely publicized reports on the first applications of modern biotechnology had to do with the "oil-eating" bacteria developed by Professor A. M. Chakrabarty. This discovery led to the landmark decision by the Supreme Court in 1980 permitting the patenting of new forms of microbial life developed through genetic manipulation. Although no significant commercial results have yet been achieved, the experiment demonstrated the feasibility of engineering microorganisms to degrade potentially toxic chemicals (28–30).

Many natural microbes are capable of digesting almost any product that humans show an interest in and, in fact, many of the waste-disposal ponds utilized for pollution abatement at industrial sites consist of microbes derived from sewage-treatment plants. Areas receiving the greatest amount of attention are the biological degradation of organic substances, eg, pesticides, herbicides, and industrial solvents; the removal of toxic heavy metals from waste streams and the concentration of such heavy metals; and biological denitrification and desulfurization. A number of commercial organizations are working on these problems and some are selling products to companies with biological-waste-treatment plants tailored to degrade their specific chemicals.

Table 3 summarizes some of the pollutants and the organisms which can degrade them. Although efforts are improving in this area, the primary efforts are related to more conventional selection for mutants rather than the specific application of recombinant DNA technology. The limitations have to do with the fact that many molecules, eg, chlorinated hydrocarbons, are recalcitrant to biodegradation and many of the products requiring degradation tend to be inherently toxic to microbes. Also, there is frequently a need for a complex of organisms that can cooperatively metabolize a molecule. The future success in developing useful degradative microbial systems may depend on locating naturally occurring communities of microorganisms that can function cooperatively to destroy mixed-waste streams. However, there are at least two problems that may not be affected at all by developing an improved microbial population. One is the handling of off-site biological or chemical spills and developing an appropriate biotreatment system for spot treatment, and the second has to do with improving the design of on-line waste-stream-treatment facilities. With regard to the latter, many of the biological waste streams from commercial processes are complex mixtures of chemicals and may require simplification by placing bioreactors upstream before the streams become mixed. Also, the streams are frequently too concentrated for organisms to be effective without stream dilution; thus, engineering systems for appropriate dilution of the waste prior to mixing with the microbes may be required.

No unique hazards are foreseen with the application of recombinant DNA technology to the improvement of biodegrading microbes. It is important, however, to minimize aerosol formation.

Pharmaceuticals and Health Care. The application of recombinant DNA technology and genetic engineering has had its greatest impact on the health care and pharmaceutical industry (28–30) (see Pharmaceuticals). Insulin is a hormone made in the pancreas and it is instrumental in the metabolism of glucose (see Insulin and other

Table 3. Microbes Capable of Degrading Pollutants

Potential pollutant	Microbes
petroleum	bacteria (*Acinetobacter*, *Arthobacter*, *Actinomycetes*, *Pseudomonas*); yeasts (*Cladosporium*, *Scolecobasidium*)
insecticides	
cyclodienes, eg, Aldrin	fungus (*Zylerion*)
organophosphates, eg, parathion	bacteria (*Pseudomonas*)
DDT	fungus (*Penicillium*)
herbicides	
2,4-D	bacteria (*Pseudomonas*)
2,4,5-T	bacteria (*Pseudomonas*)
pentachlorophenol	bacteria (*Pseudomonas*)
waste products	
municipal sewage	bacteria (*Pseudomonas*, *Thermomospora*)
pulp-mill lignins	bacteria (*Arthobacter*, *Chromobacter*, *Pseudomonas*); fungi (*Aspergillus*, *Trichosporum*)

antidiabetic agents). Mammalian insulin genes were among the earliest cloned and characterized, and the total chemical synthesis of the gene for proinsulin has been reported (32–33). However, the subsequent production of insulin with cloned genes has presented special problems. The biosynthesis of insulin goes through a single chain form from preproinsulin to proinsulin, but a center fragment C is removed enzymatically to yield the mature form consisting of A and B chains. The genetic engineering of a two-chain molecule is not trivial, and production is further complicated because the chains are too short to be stably accumulated in *E. coli*. One successful approach has been the production of the A and B chains separately as fusion proteins with β-galactosidase in *E. coli* (25). Mature insulin is then obtained by cleavage of the respective fusion products, mixing and reduction of the separate chains, and final reoxidation to combine the two chains. Alternately, proinsulin has been produced in both *E. coli* and mammalian cells and it may be possible to complete its processing enzymatically or chemically *in vitro* (34–35).

Although several genetically engineered proteins are being tested, insulin, produced by the separate-chains approach, is the main product that is being sold. It is sold by Lilly under the trade name Humulin at a price roughly 10% over that for insulin derived from animal sources. Some projections predict that humulin may account for as much as 50% of Lilly's U.S. insulin market or about $\$90 \times 10^6$ by 1985 and an additional $\$50 \times 10^6$ outside the United States. It has been the main advertised target for Lilly's $\$40 \times 10^6$ fermentation facility.

Human growth hormone (somatotropin) represents the second most extensively developed product resulting from recombinant DNA technology (13,36). This pituitary hormone induces growth in many soft tissues, cartilage, and bone and is required for postnatal human growth. Deficiencies result in pituitary dwarfism and other problems related to hormone insufficiency. A spinoff from human growth hormone research has been the cloning and development of host–vector systems for the cloning and production of bovine, porcine, and ovine growth hormones, hormones which are capable of influencing the growth and productivity of dairy cows, pigs, and sheep, respectively (37–38).

The third most extensively exploited area of recombinant DNA technology involves the development of host–vector systems for the production of interferons (39). These glycoproteins are complex and involve at least three classes designated alpha (leukocyte), beta (fibroblast), and gamma (immune). There are more than 15 natural alpha-interferon species and at least three or four beta-interferons. The genetic-engineering aspects of this area are complex but, because of the potential applications to many human viral infections as well as to cancer chemotherapy, there is an enormous interest in the industry to develop sufficient quantities of these materials for clinical testing. Genetic-engineering tools have been used to splice front and back halves of different interferon genes as an approach to novel hybrid molecules.

The generation of viral antigens by genetic engineering for vaccination is another area of active research (see also Vaccine technology). The source of antigens used for vaccinations have generally involved whole killed or attenuated organisms. Adverse effects in vaccinated humans and animals can occur. The availability of pure antigenic proteins or active fractions thereof, eg, those from foot-and-mouth disease virus and hepatitis B virus from recombinant DNA technology could obviate these problems (16–17,40). Other vaccines being studied for production by genetic engineering include influenza virus, hepatitis A, herpes virus, *Gonococcus*, pathogenic *E. coli*, oral bacteria, and malaria parasite.

Another is the development of therapeutic proteins and enzymes for the treatment of blood disorders. The cloning and expression of serum albumin genes has been announced (41). This technology would be useful in the treatment of shock. Serum albumin is the principal product of blood plasma fractionation and, therefore, its availability and cost may encourage its production by genetic engineering. Urokinase, which is used to dissolve unwanted blood clots, has also been cloned and expressed, as has antithrombin III, which is involved in regulating coagulation (42). The gene for Factor IX has been cloned and there is considerable interest in the cloning of Factor VIII, which is used to treat hemophilia.

Human polypeptides of potential interest and the number of amino acid residues (in brackets) are prolactin [198], placental lactogen [192], nerve growth factor [118], parathyroid hormone [84], insulinlike growth factors [67,70], epidermal growth factor, thymopoietin [49], gastric inhibitory polypeptide [43], corticotropin [39], cholecystokinin [33], calcitonin [32], endorphins [31], glucagon [29], thymosin-α_1 [28], secretin [27], and motilin [22]. Calcitonin is manufactured by chemical synthesis. Naturally occurring small peptides being studied by genetic engineers are (amino acid residues in brackets): dynorphin [17], somatostatin [14], bombesin [14], melanocyte-stimulating hormone [13], neurotensin [13], angiotensin I [10], bradykinin [9], vasopressin [9], oxytocin [9], angiotensin II [8], angiotensin III [7], enkephalins [5], and thyrotropin-releasing hormone [3]. The naturally occurring small peptides are particularly interesting since they can be synthesized by direct chemical means, ie, by the Merrifield solid-state synthesis, or by recombinant DNA technology. Somatostatin, a 14-amino-acid hormone which regulates the release of growth hormone, glucagon, and insulin, was among the first hormones made by genetic engineering and was the first for which a totally synthetic gene was used (43). However, it appears now to fall in the group of chemicals with 30–35 amino acids or less for which peptide synthesis is more economical.

Since antibiotics are produced by fermentation, genetic-engineering technology should permit a more focused and rational means for increasing the efficiency of producing and developing antibiotics. The industry has been highly successful in improving the yield of antibiotic-producing microorganisms by mutant-strain selection and has access to an additional powerful tool for gene insertion and gene amplification. These capabilities allow for the enhancement of a particular biological step which regulates the overproduction of the antibiotic or some step which could result in the creation of modified versions of the natural antibiotic. Efforts are being devoted particularly to the study of *Streptomyces* species, since these microbes are important in the production of a series of drugs, eg, streptomycin, erythromycin, and tetracycline.

The isolation and purification of products to be used in human clinical medicine and health care must be carefully controlled and the criteria for quality assurance must be rigorous. Although this is not new to the pharmaceutical industry, it places additional demands on the large-scale production of proteins by microbes which, in some cases, contain potentially biotoxic contaminants. In many of the potential applications, the utility of the protein has yet to be clearly demonstrated and the cost effectiveness of the product and the process to make it have not been clearly established.

Most evidence indicates that *E. coli* and the strain K12 used as a host bacterium for mediating the synthesis of human proteins are quite safe. However, other microorganisms, eg, *Bacillus subtilis* and the brewers' yeast *Saccharomyces cerevisae*, may eliminate the problems of endotoxins and pyrogenic materials associated with *E. coli* and provide potential mechanisms for the secretion of products into the media. The latter would be particularly useful for the isolation and purification of valuable human proteins.

Agriculture. Agricultural genetic engineering involves the isolation and cultivation of cells from plants in tissue culture and the regeneration of whole plants from these cells. Once the tissue-culture system is established, genetic changes can be engineered into the cells with tailored vectors containing the appropriate DNA for the introduction of the desired trait (see, for example, ref. 44). Results obtained with genetic engineering include improved protein quality of seed grain; enhanced protein levels in forage crops; enhanced production of natural products, eg, amino acids, allelopathic chemicals, latex, enzymes, oils, and starch; plant-disease resistance; insect tolerance; drought and flooding tolerance; salt tolerance; metal tolerance, eg, Al^{3+} and Pb^{2+}; herbicide and pesticide tolerance; enhanced photosynthetic carbon fixation; improved response to fertilizers; heat and cold tolerance; development of male sterility; fertility restoration; chemical transformation, eg, $N_2 \rightarrow NH_3$; and improved harvest index.

Early efforts have involved the use of *Agrobacterium tumefaciens*, a bacterium that can introduce genetic information into plant cells creating tumors called crown gall tumors. The Ti plasmid in this bacterium is being engineered to disarm the bacterium's tumorogenicity and for insertion of appropriate genes that confer antibiotic or herbicide resistance to plant cells. However, no significant genetic trait has yet been introduced and expressed in plant cells, but vectors are being developed and successful gene insertion and expression in plants has occurred.

A second application for genetic engineering related to agriculture is the improvement of plant symbionts and associated microorganisms. Nitrogen-fixing bacteria (*Rhizobium* sp) represent one of the most important sources of nitrogen for legumes. The potential for improving the efficiency of nitrogen-fixing bacteria by genetic-engineering technology is being studied.

Finally, the genetic engineering of microbes that can be used to protect plants from insect, fungal, and viral attack is receiving increasing attention. Microbes which secrete allelopathic chemicals capable of killing pathogenic disease organisms and soilborne insects may be developed.

Bacterial, fungal, and viral vectors to be used for the control of insects and plant diseases are receiving renewed attention from molecular biologists. One of the best-known microbial products used for the control of certain leaf-chewing insects is *Bacillus thuringienses*, which produces a protein that is toxic to insects that ingest it. Another strain of this bacterium is useful in the control of mosquito larvae. Such organisms would lend themselves to being developed into more potent microbial forms as well as broader range insecticides.

One of the most serious problems in agricultural genetic engineering is the lack of knowledge regarding gene organization and regulation in higher organisms, eg, plants. Second, the basic biochemical mechanisms that regulate the growth, development, and reproduction of plants are not well understood. Also, the technology of tissue culture must be developed further so that more economically important plants can be regenerated from either protoplast or cell systems. It is expected that the application of genetic-engineering techniques to plant research will expand the knowledge of how genes are organized and regulated temporally as well as spatially, and it is certain that such knowledge will lead to the improvement of crop productivity.

Food Processing. The food industry has begun to apply recombinant DNA techniques to the production of single-cell proteins, enzymes, amino acids, and synthetic sweeteners (qv), eg, aspartame (28). Application of genetic engineering to the production of some natural sweetening agent proteins, eg, monellin and thaumatin, is also being studied.

There appears to be renewed interest in the development of single-cell proteins by genetically engineering microbes to utilize various substrates. For example, the bacterium *Methylophilus methylotrophus* was developed to metabolize methanol as its sole carbon source. This protein-producing bacterium is sold as an animal-feed additive (see Pet and livestock feeds). Genetic-engineering improvements of the organism to alter its nitrogen-assimilating capabilities have resulted in significant yield enhancement and cost effectiveness. Other commercial ventures involving single-cell proteins that develop products for human and animal food use are directed at the reduction of nuclei acid content, toxins, and difficult-to-digest cell walls. Genetic engineering is being studied as a means of improving the functional properties of proteins. Another use of such microbes is in condiments and flavor enhancement.

The genetics of yeast are being studied especially in the baking, brewing, and wine industries. One example has been the development of so-called killer strains of domesticated yeast to be used in the brewing industry. Genetic functions have been transferred to commercially important yeast strains to allow them to kill strains of yeast that may contaminate the cultures. Another example is the development of a yeast strain that produces low carbohydrate-containing beer suitable for diabetics. The development of yeasts tolerant to higher levels of alcohol and capable of completely fermenting grape extracts with unusually high sugar content is being studied. In the baking industry, new strains with improved biological activity and greater storage stability are being developed.

The physical properties of food products are frequently altered or controlled by the use of polysaccharides and proteins, and there is interest in the development of new and alternative sources based on proteins and polysaccharides. Many of these products are incorporated into foods as gelling agents, thickeners, and stabilizers to control ice crystallization and are applied to such food products as salad dressings, sauces, whippings, toppings, processed cheeses, and dairy products. Many of the polymeric sugars are derived from plants and the functional proteins are derived from animals. For example, xanthan gums are produced by *Xanthomonas campestris* in large-scale commercial processes.

Enzymes are important to the food-processing business. Alternative routes to rennin, an enzyme normally derived from the stomach of milk-fed calves, are being studied. This enzyme, which coagulates milk in cheesemaking, is expected to be in short supply as the availability of calves decreases. There are reports of the cloning of the calf rennin gene but there has been no announcement of expression of the enzyme protein. However, the food industry is permitted to use only enzymes obtained from sources approved for food use; thus, only food-grade organisms may be useful as host–vector systems.

The availability of glucose isomerase and amylase has made the production of high fructose corn sweeteners (HFCS) very profitable (see Syrups). Aspartame, which is synthesized from aspartic acid and phenylalanine, has been synthesized by means of a synthetic gene coding for a polypeptide polymer consisting of aspartyl phenylalanyl moieties. Although the economics of this process are not yet available, it illustrates the potential for biotechnological approaches. The genes for monellin and thaumatin, two natural protein products from West African plants, have been isolated and attempts to clone these genes will be made to determine whether a genetically engineered microbe can make these proteins which are thousands of times sweeter than table sugar.

One means to circumvent some of the Federally imposed constraints on the food industry will be to transfer desired genes into organisms that meet FDA standards. Basic knowledge of the characteristics desired to improve food products will also require study before the application of recombinant technology can be successful.

Energy. Biomass fuel sources have several advantages over fossil fuels: they are renewable, their combustion does not result in an increase of the carbon dioxide content of the atmosphere, and the geographical distribution of biomass is more uniform compared to fossil-fuel reserves (see Fuels from biomass). Economic considerations make this approach less attractive in the United States than in places like Brazil where less-expensive or alternative feedstocks to corn are available. Genetic-engineering applications would involve the development of enzymes to break down lignocellulose to small molecules that can be metabolized to products such as ethanol (28–29). In the long run, this will be preferable to the use of foodstuffs, ie, corn and sugarcane, as raw-material feedstocks. Other applications include methane generation by anaerobic (bacterial) digestion of various biomass sources and the generation of hydrogen from water by means of photolytic reactions such as occur in the chloroplasts of green plants.

Most of the approaches represent bench-scale curiosities with some pilot-scale efforts in the production of ethanol or methane from unconventional feedstocks. It is unlikely that food crops will be used ultimately for the generation of energy, but rather more dense sources of cellulosic materials and most likely only those that can be grown as renewable crops, eg, aspen and poplar trees.

Mining. Potential applications of genetic engineering to mining are in tertiary recovery of oil and recovery of minerals associated with rocks (28–29). Polysaccharide lubricants have already been produced from microbial systems. In addition, microbial systems may be useful in the production of xanthan gums to increase the viscosity of water or the production of emulsifiers and surfactants to facilitate the loosening or solubilization of crude oil prior to water flooding (see Emulsions; Surfactants and detersive systems).

A second mining application may be the use of microbes to assist in the leaching, concentration, and recovery of metals from ore deposits and in the removal of heavy metals from pollution effluents. Here, microbes are extremely effective in concentrating metals from dilute solutions.

The direct application of microbes that secrete useful compounds, however, is problematic. Probably the greatest problems are the physical conditions under which the microbes must function, ie, high temperatures, low oxygen, extreme pHs, and the presence of salts and other toxic chemicals. In some cases, the microbes must be supplied a nutrient media at long distances from the soil surface. The concerns over the environmental impact of microbes may inhibit widespread application. The geophysical conditions are varied; thus, these approaches are highly variable with regard to any particular application, be it microbes or the chemicals produced by them. There is not sufficient information to judge the feasibility of these approaches.

Commercial Activities

Companies. There are more than 200 genetic-engineering companies which began business since 1978, and it would appear that new companies will continue to enter the field of genetic engineering at a steady pace (30). Almost all of these companies have used the same basic format for starting their businesses. A business plan backed by a technological plan is prepared and venture-capital financing is sought for the initial capitalization of the company. The companies generally begin by doing both contract research as well as research for their own accounts. In several instances, after a positive track record of accomplishment has been achieved, the companies have gone public and issued stock for future financing of their enterprises. Although a number of the early entry companies initiated genetic-engineering programs for pharmaceutical products, more recently companies have broadened their base by entering into other application areas, eg, animal vaccines, plant genetics, and blood factors.

Of the more than 200 companies, the first and largest in the field are Cetus, Genentech, Biogen, and Genex. Other important firms are Agrigenetics Corp.; Applied Molecular Genetics, Inc.; Bethesda Research Laboratories, Inc.; Bio Logicals, Inc.; Celltech, Ltd.; Collaborative Research Corp.; EnzoBioChem, Inc.; International Plant Research Institute, Inc.; Molecular Genetics, Inc.; and New England Nuclear. Since 1980, there has been increasing commercialization in Canada, the UK, France, the FRG, Switzerland, Sweden, Denmark, the Netherlands, and Israel. The Japanese entry into the field has been mainly through associations with genetic-engineering firms outside Japan, partly because of government restrictions on conducting recombinant DNA research in Japan. However, with changes in regulatory requirements, Japanese efforts in genetic engineering will probably intensify significantly.

In 1982, several companies filed for bankruptcy and others have had to reduce their scientific staffs greatly. A reshaping of the original commercial goals of a number of companies has been announced, and the direction generally tends to be towards specialty chemical products and high value-added products rather than towards commodity chemicals.

Chemical Industry. Numerous chemical and energy-research companies, eg, Exxon, National Distillers and Chemicals, Shell Oil, Chevron, and Arco, and agricultural and food companies, eg, Archer Daniels Midland, A. E. Staley, Campbell Soups, etc, have combined efforts with genetic-engineering firms and have begun their own research and development programs. Some company affiliations are listed in Table 4.

Pharmaceutical Industry. Some of the first pharmaceutical companies involved in recombinant DNA technology and genetic engineering are listed in Table 5. The principal projects undertaken by these companies involve the production of growth hormone, insulin, diagnostics, and interferon; the latter is the largest area of research. A $106 \times 10^6 interferon plant owned by Schering-Plough Corp. and based on technology developed by Biogen is expected to begin production in 1985 (44). Drugs, other hormones, antibiotics, and vaccines are more recent areas of research (see Pharmaceuticals).

Table 4. Chemical Companies Engaged in Genetic Engineering

Company	Affiliations
Allied Corporation	Bio Logicals and Calgene
American Cyanamid Co.	Molecular Genetics
BASF A.G.	
Bayer A.G.	MIT and Miles Laboratories
Celanese Corporation	
CIBA-GEIGY A.G.	
Dow Chemical Co.	Collaborative Research Corporation
E. I. du Pont de Nemours & Co., Inc.	New England Nuclear
Ethyl Corporation	BioTech Research Laboratories
Hoechst A.G.	Massachusetts General/Harvard Medical School
International Mineral & Chemicals Corporation	Genentech and Biogen
Lubrizol Corporation	Genentech
Monsanto Co.	Genentech, Biogen, Genex
National Distillers & Chemical Co.	Cetus Corporation
Rohm & Haas Co.	Advanced Genetic Sciences, Ltd.

Table 5. Pharmaceutical Companies Engaged in Genetic Engineering

Company	Affiliations
Abbott Laboratories	Applied Molecular Genetics
Bristol Myers Co.	Genex
Hoffmann-LaRoche	Genentech
Johnson & Johnson	Bio Logicals
Eli Lilly & Co.	Genentech and International Plant Research Institute
Merck & Co.	
Novell Industries A.S.	Biogen
Schering-Plough Corporation	Biogen
Upjohn Co.	

Regulation and Property Rights

Risk Assessment and NIH Guidelines. Concern among some U.S. scientists that the technology might result in the inadvertant and uncontrollable production of toxic proteins led to the Asilomar Conference in 1975. That conference and subsequent discussions resulted in the elaboration of a set of NIH *Guidelines for Research Involving Recombinant DNA Molecules* and included not only the class of experiments of initial concern but virtually every experiment that involved *in vitro* recombination of DNA fragments from different sources. Levels of containment were identified for individual experiments involving both physical (access, air flow, special equipment) and biological (genetically crippled hosts or vectors) parameters. In the intervening years, review of the *Guidelines* led by a Recombinant DNA Advisory Committee (RAC) has resulted in a document whose focus is on institutional oversight of most recombinant DNA activities.

All experiments resulting in release of recombinant organisms to the environment must be previewed by the RAC. The RAC also acts as reviewer of experiments that are not covered by the *Guidelines* and are referred to them by individual institutional biosafety committees (IBCs). Institutions are explicitly charged with following the intent as well as the letter of the *Guidelines.* Other countries have generally followed the United States, although a wide range in guidelines exists from little or no control in some countries, eg, Belgium, to more stringent controls in others.

The *Guidelines* appear to be adequately controlling the type of experiments that were of initial concern. As the benefits of the technology continue to accumulate and the hazards remain largely theoretical, it is likely that the emphasis will be less on the recombinant DNA technology *per se* and more on the components of the experiment or its product application. For example, the use of productively infecting animal viruses as vectors to produce proteins in mammalian cells must meet containment criteria commensurate with NIH *Guidelines* covering work with animal viruses. Factors for administration to animals also must satisfy appropriate FDA clearances. Hormones produced for human therapy must meet the rigid criteria that cover marketing and administration of ethical drugs and biologicals. It is likely that, in the near term, the RAC and IBCs will serve as monitors of the expanding technology and any potential problems that might develop.

Property Rights. There is a lack of appropriate precedent involving biological systems and their manipulation to clearly define property-right issues. The landmark Chakrabarty case demonstrated that such systems are protectable, although that case did not specifically involve recombinant DNA techniques. As in any other area of invention, discretion will dictate that authentic patent positions be identified and established prior to publication. Even then, it will be interesting to see how the notion of obviousness will be resolved where the level of effort and discussion has been so extensive. Alternatively, protection may be sought in trade secrets. However, in an area where very few ideas are unique and the work force remains highly mobile, such instances will probably remain in the minority over the next few years. This may become more important as general systems for direct cell transformation become established and a feeling for the techniques that incrementally improve a system develops.

Prospects and Problems

Genetic engineering is likely to have a significant impact on any area that involves biological systems or that can be approached with biochemical tools. That is not to say that it will necessarily displace any existing technologies, but rather that it is a powerful tool that is likely to complement many. It should, for example, permit the introduction of genetic information into plant systems which would not readily be available by traditional plant-breeding techniques. Nevertheless, it will be the two together which will permit that information's manipulation and exploitation. Similarly, although economic considerations will sideline genetic-engineering procedures for the production of most commodity chemicals, genetic engineering should permit the more facile execution of one or more steps in many complicated chemical processes by fermentative or enzymatic means.

Undoubtedly, genetic-engineering procedures will continue to make critical contributions to the fundamental understanding of biological systems. This contribution will continue to be essential in systems where proteinaceous components and regulators are available in only limited quantities. With the evolution of techniques for the production of specific protein variants, replacement of normal proteins with variants will permit more complex study of such systems.

Protein hormone overproduction by genetic-engineering procedures is the most direct commercial application insofar as proteins are the primary products of the technology. It is reasonable to assume that the production of pharmaceutically important proteins will continue to be the main commercial application of genetic-engineering procedures in the middle 1980s. Proteins that require post-translational modification for their biological activity will present special problems. A host–vector system that performs the appropriate modification must be identified. Alternatively, the modification might be accomplished chemically or enzymatically. In either case, the procedure has to be cost effective. In some cases, evolving procedures for site-directed mutagenesis may allow identification of simpler variants that completely circumvent the need for the modification. In others, the ability to dissect a biological system with genetic-engineering procedures may make available a totally different protein or chemical alternative that achieves the same purpose.

Some of the problems and their solutions may not involve the technology or may require a component outside the technology, eg, the engineering of equipment and facilities in a cost-effective fashion to satisfy technological and regulatory requirements and the isolation of protein products in adequate amounts and purity. Separations technology may benefit from host–vector systems which are continuous or secrete the product from the host cell. However, a great deal of novel protein biochemistry will be required if separations are to be accomplished successfully. The delivery system or process for the protein must allow the investigator to target or exploit the protein once it is successfully produced by genetic engineering.

Genetic engineering is not likely to have a very broad impact on chemical production, at least of commodity chemicals, and especially not in the near future. It is possible to devise genetic-engineering schemes for most small molecules, but the economics compared to those for traditional chemical approaches will generally be prohibitive. However, fermentation and enzyme technologies have been and will continue to be applied to some chemical processes. It is likely that the optimization of such fermentations or the large-scale production of such enzymes will benefit from genetic-engineering procedures (45).

A third significant area that will be dramatically affected by genetic-engineering procedures is the regulable transformation of whole organisms. In some cases, this will be an extrapolation of processes discussed previously. For example, a soil microorganism that releases a protecting pesticide *in situ* is a logical extension of an *E. coli* that overproduces a protein under fermentation conditions. However, it is reasonable to think that it will be possible to transform crops by genetic engineering to exhibit desirable traits according to a preselected developmental schedule. Similarly, correction of genetic deficiencies will not be a common phenomenon soon, but it may ultimately be possible in some judiciously selected cases. Schemes for the introduction of genes for new proteins into cells from patients suffering from blood disorders and the return of those cells to the patients' marrow have already been proposed. These applications are more speculative than those described in the former areas.

Finally, proposals regarding genetic engineering and debated in Congress will effect the negotiation, funding, and commercialization of bioengineering research (46). Governmental influence on the direction of bioengineering research in the UK includes the offering of grants towards the costs of selected projects and consulting jobs and increased support for centers of biotechnological expertise (47).

Nomenclature

cDNA = complementary DNA to an RNA molecule
2,4-D = 2,4-dichlorophenoxyacetic acid
DNA = deoxyribonucleic acid
IBCs = institutional biosafety committees
kb = 1000 base pairs
mRNA = messenger ribonucleic acid
RAC = Recombinant DNA Advisory Committee
RNA = ribonucleic acid
2,4,5-T = 2,4,5-trichlorophenoxyacetic acid
T-DNA = segment of Ti plasmid transferred to plant genome
Ti = a transmissable plasmid in *Agrobacterium tumefaciens* responsible for inciting tumors in some dicotyledonous plants by mechanisms not fully understood

BIBLIOGRAPHY

"Genetic Engineering" in *ECT* 3rd ed., Vol. 11, pp. 730–745, by A. M. Chakrabarty (University of Illinois at the Medical Center, Chicago).

1. R. W. Old and S. B. Primrose, *Principles of Gene Manipulation*, University of California Press, Berkeley, Calif., 1981.
2. T. Maniatis, E. F. Fritsch, and J. Sambrook, *Molecular Cloning: A Laboratory Manual*, Cold Spring Harbor Laboratory, New York, 1982.
3. R. F. Schleit and P. C. Wensink, *Practical Methods in Molecular Biology*, Springer-Verlag, New York, 1981.
4. R. Wu, ed., *The Recombinant DNA Technology, Methods in Enzymology*, Vol. 68, 1980.
5. J. K. Setlow and A. Hollaender, eds., *Genetic Engineering: Principles and Methods*, Vols. 1–4, Plenum Press, New York, 1979–1982.
6. P. H. Abelson, *Science* **196,** 103 (1977); **209,** 1287 (1980).
7. P. H. Abelson, *Science* **219,** 535 (1983).
8. F. G. Grosveld, T. Lund, E. J. Murray, A. L. Mellor, H. H. M. Dahl, and R. A. Flavell, *Nucleic Acids Research* **10,** 6715 (1982).
9. A. Hollaender, ed., *Genetic Engineering of Microorganisms for Chemicals*, Plenum Press, New York, 1982.
10. Y. Gluzman, ed., *Eukaryotic Viral Vectors*, Cold Spring Harbor Laboratory, New York, 1982.
11. R. A. Young and R. W. Davis, *Proc. Natl. Acad. Sci. USA* **80,** 1194 (1983).
12. A. G. Walton, ed., *Recombinant DNA*, Elsevier Scientific Publishing Co., New York, 1981.
13. J. L. Gueriguian, ed., *Insulins, Growth Hormone, and Recombinant DNA Technology*, Raven Press, New York, 1981.
14. M. V. Olson in ref. 5, Vol. 3, pp. 57–88.
15. J. N. Strathern, E. W. Jones, and J. R. Broach, eds., *The Molecular Biology of the Yeast Saccharomyces*, Cold Spring Harbor Laboratory, New York, 1981.
16. P. Valenzuela, A. Medina, W. J. Rutter, G. Ammerer, and B. D. Hall, *Nature (London)* **298,** 347 (1982).
17. A. Miyanohara, A. Toh-E, C. Nozaki, F. Hamada, N. Ohtomo, and K. Matsubara, *Proc. Natl. Acad. Sci. USA* **80,** 1 (1983).
18. D. A. Dubnau, ed., *The Molecular Biology of the Bacilli*, Vol. 1, Academic Press, New York, 1982.
19. A. Constantinides, W. R. Vieth, and K. Venkatasubramanian, eds., *Biochemical Engineering II, Annals of the New York Academy of Science*, Vol. 369, The New York Academy of Sciences, New York, 1981.
20. D. A. Hopwood and K. F. Chater in ref. 5, Vol. 4, pp. 119–145.
21. D. Shortle, D. DiMaio, and D. Nathans, *Ann. Rev. Genet.* **15,** 265 (1981).
22. M. Smith and S. Gillam in ref. 5, Vol. 3, pp. 1–55.
23. G. Kahl and J. S. Schell, eds., *Molecular Biology of Plant Tumors*, Academic Press, New York, 1982.
24. K. A. Barton and W. J. Brill, *Science* **219,** 671 (1983).

25. D. V. Goeddel, D. G. Kleid, F. Bolivar, H. L. Heyneker, D. G. Yansura, R. Crea, T. Hirose, A. Kraszewski, K. Itakura, and A. D. Riggs, *Proc. Natl. Acad. Sci. USA* **76,** 106 (1979).
26. R. D. Palmiter, H. Y. Chen, and R. L. Brinster, *Cell* **29,** 701 (1982).
27. R. D. Palmiter, R. L. Brinster, R. E. Hammer, M. E. Trumbauer, M. G. Rosenfeld, N. C. Birnberg, and R. M. Evans, *Nature (London)* **300,** 611 (1982).
28. *Impact of Applied Genetics*, Office of Technology Assessment, U.S. Government Printing Office, Washington, D.C., 1981.
29. R. H. Zaugg and J. R. Swarz, *Assessment of Future Environmental Trends and Problems: Industrial Use of Applied Genetics and Biotechnologies*, Contract Grant No. 68-02-3192, EPA No. 600/8-81-020, National Technical Information Service, U.S. Department of Commerce, Springfield, Va., 1981.
30. V. K. Paul, ed., *Genetic Engineering Applications for Industry*, Noyes Data Corporation, Park Ridge, N.J., 1981.
31. D. I. C. Wang, C. L. Cooney, A. L. Demain, P. Dunnill, A. E. Humphrey, and M. D. Lilly, *Fermentation and Enzyme Technology*, John Wiley & Sons, Inc., New York, 1979.
32. A. Ullrich, J. Shine, J. Chirgwin, R. Pictet, E. Tischer, W. J. Rutter, and H. M. Goodman, *Science* **196,** 1313 (1977).
33. R. Brousseau, R. Scarpulla, W. Sung, H. M. Hsiung, S. A. Narang, and R. Wu, *Gene* **17,** 279 (1982).
34. L. Villa-Komaroff, A. Efstratiadis, S. Broome, P. Lomedico, R. Tizard, S. P. Naber, W. L. Chick, and W. Gilbert, *Proc. Natl. Acad. Sci. USA* **75,** 3727 (1978).
35. P. Gruss and G. Khoury, *Proc. Natl. Acad. Sci. USA* **78,** 133 (1981).
36. D. V. Goeddel, H. L. Heyneker, T. Hozumi, R. Arentzen, K. Itakura, D. G. Yansura, M. J. Ross, G. Miozzari, R. Crea, and P. H. Seeburg, *Nature (London)* **281,** 544 (1979).
37. E. Keshet, A. Rosner, Y. Bernstein, M. Gorecki, and H. Aviv, *Nucleic Acids Research* **9,** 19 (1981).
38. P. H. Seeburg, S. Sias, J. Adelman, H. A. de Boer, J. Hayflick, P. Jhurani, D. V. Goeddel, and H. Heyneker, *DNA* **2,** 37 (1983).
39. P. Lengyel, *Annu. Rev. Biochem.* **51,** 251 (1982).
40. H. Kupper, W. Keller, C. Kurz, S. Forss, H. Schaller, R. Franze, K. Strohmaier, O. Marquardt, V. Zaslavsky, and P. H. Hofschneider, *Nature (London)* **289,** 555 (1981).
41. R. M. Lawn, J. Adelman, S. C. Bock, A. E. Franke, C. M. Houck, R. C. Najarian, P. H. Seeburg, and K. L. Wion, *Nucleic Acids Research* **9,** 6103 (1981).
42. S. C. Bock, K. L. Wion, G. A. Vehar, and R. M. Lawn, *Nucleic Acids Research* **10,** 8113 (1982).
43. K. Itakura, T. Hirose, R. Crea, A. D. Riggs, H. Heyneker, F. Bolivar, H. W. Boyer, *Science* **198,** 1056 (1977).
44. *Chem. Eng.*, 36 (March 7, 1983).
45. *Chem. Eng.*, 22 (June 13, 1983).
46. L. Garmon, *Biotechnol.*, 26 (March 1983).
47. R. Stevenson, *Chem. Br.* **19**(1), 17 (1983).

ERNEST JAWORSKI
DAVID TIEMEIER
Monsanto Company

GUANIDINE AND GUANIDINE SALTS

Guanidine [*113-00-8*] (**1**) is a crystalline, strong, organic base discovered by Strecker in 1861 (1) in the course of his work on the constituents of guano, from which the name guanidine is derived.

$$H_2N\overset{\overset{\displaystyle NH}{\|}}{C}NH_2$$

(**1**)

Since that time, it has been found that guanidine is present in combined form in a variety of other natural products, among which are egg albumen, nucleic acids, streptomycin, and folic acid.

Guanidine may be regarded as the imide of urea (qv),

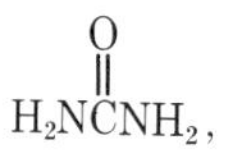

$$H_2N\overset{\overset{\displaystyle O}{\|}}{C}NH_2,$$

or the amidine of carbamic acid (qv),

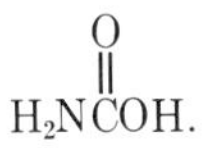

$$H_2N\overset{\overset{\displaystyle O}{\|}}{C}OH.$$

Guanidine is a monoacid base with a strength equivalent to sodium hydroxide. It has been suggested that this marked basicity is owing to the ability of the guanidinium ion (**2**) to resonate among three equivalent structures (**2**).

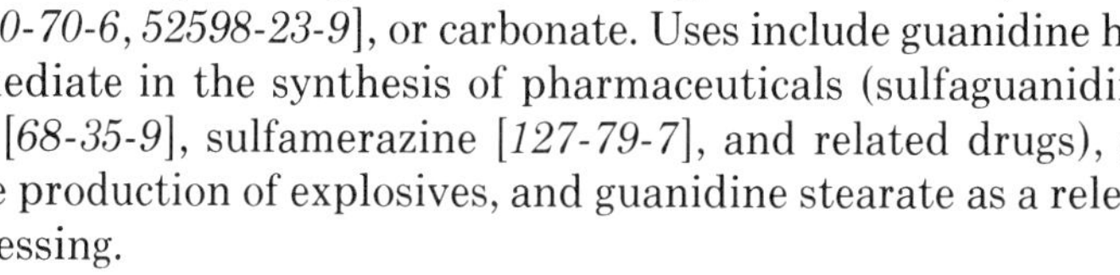

$$H_2\overset{+}{N}{=}C(NH_2)NH_2 \longleftrightarrow H_2N{-}C({=}\overset{+}{N}H_2)NH_2 \longleftrightarrow H_2N{-}C(NH_2){=}\overset{+}{N}H_2.$$

(**2**)

Guanidine is marketed in the form of its salts, usually named as guanidine salts (though more correctly called guanidinium salts), such as nitrate, chloride, stearate, silicate [*39380-70-6, 52598-23-9*], or carbonate. Uses include guanidine hydrochloride as an intermediate in the synthesis of pharmaceuticals (sulfaguanidine [*57-67-0*], sulfadiazine [*68-35-9*], sulfamerazine [*127-79-7*], and related drugs), guanidinium nitrate in the production of explosives, and guanidine stearate as a release agent for plastics processing.

Physical and Chemical Properties

Free guanidine forms colorless, very hygroscopic crystals with a melting point of ca 50°C. It absorbs carbon dioxide rapidly from the air. In aqueous solutions, guanidine hydrolyzes slowly, first to urea and then to carbon dioxide and ammonia. The rate of hydrolysis increases markedly with increased temperature or in the presence of alkali. The ir absorption spectra of guanidine and deuteroguanidine have been described and compared with those of the guanidinium ion (3).

The properties of four available guanidine salts are listed in Table 1 (4–5).

Reactions

Since it is a very strong organic base, guanidine is liberated from its salts only by the use of special procedures such as the treatment of guanidinium carbonate with sodium alkoxide in absolute alcoholic solution. A variety of guanidine salts may be formed by double decomposition or by the treatment of the free base with the appropriate acid.

Like other primary amines, guanidine can be alkylated or acylated and condensed with aldehydes and esters. Since the molecule contains two amino groups, it can be used as a starting material for imidazoles, triazines, and pyrimidines. Typical reactions of guanidine are summarized in Figure 1 (6).

Basic solutions of guanidine or its salts react readily with aqueous formaldehyde to give guanidine-formaldehyde resins. A reaction of special importance is the dehydration of guanidinium nitrate by sulfuric acid or by other dehydrating agents to nitroguanidine [*556-88-7*] (7).

Table 1. Physical Properties of Some Guanidine Salts

Property	Guanidine hydrochloride [*50-01-1*]	Guanidinium nitrate [*506-93-4*]	Guanidinium carbonate [*3425-08-9*]	Guanidinium stearate [*26739-53-7*]
formula	$HN{=}C(NH_2)_2 \cdot HCl$	$HN{=}C(NH_2)_2 \cdot HNO_3$	$[HN{=}C(NH_2)_2]_2 \cdot H_2CO_3$	$HN{=}C(NH_2)_2 \cdot C_{17}H_{35}COOH$
formula wt	95.5	122.1	180.2	343.6
crystal system			tetragonal	
crystal habit	equant to tabular	equant	equant	
mp, °C	184	214.2	241	>76
apparent density of solid at 30°C (±0.004), g/cm³	1.344	1.436	1.251	0.98
refractive index, n_D^{20}	1.520 (N_x) 1.590 (N_y) 1.638 (N_z)	1.350 (N_x) 1.575 (N_y) 1.582 (N_z)	1.4963 (N_o) 1.4864 (N_e)	
soly, g/100 g solvent in				
water at 20°C	215	16	45	
water at 55°C	320	47	62	
ethanol at 78°C	67	13	<0.1	135 (30°C)
acetone at 50°C	<0.05	0.1	0.3	0.2 (30°C)
benzene at 60°C	<0.05	<0.05	<0.05	0.4 (30°C)
hexane at 50°C	<0.05	<0.05	<0.05	
DMF at 30°C	19.5	61	0.02	0.2
pH of aq soln at 25°C				
1 wt %		5.7		
4 wt %	6.4		11.2	
vapor pressure of water (over 4% aq soln at 45°C), kPa (mm Hg)	4.7 (35)		9.3 (70)	

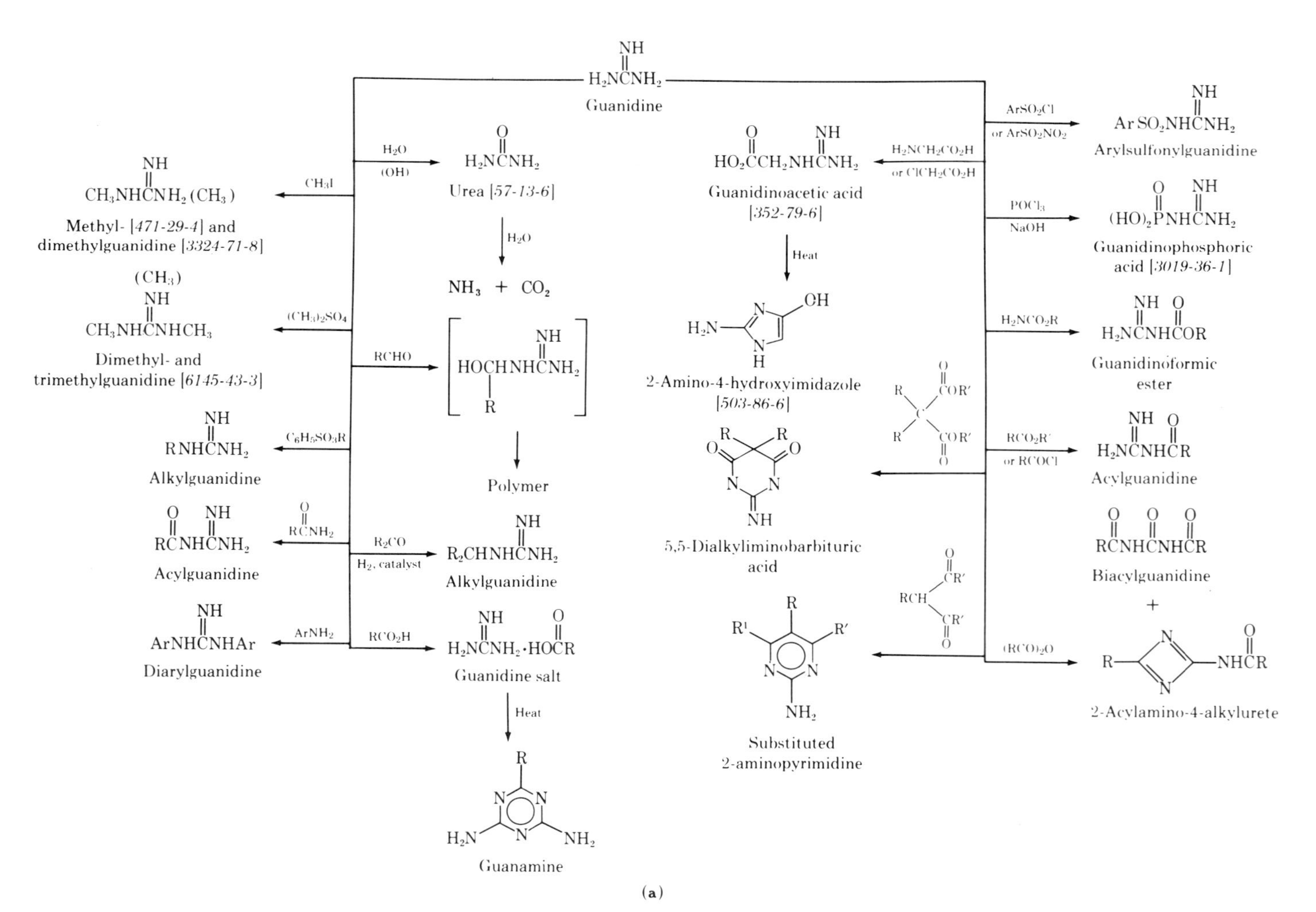

Figure 1. Typical reactions of guanidine.

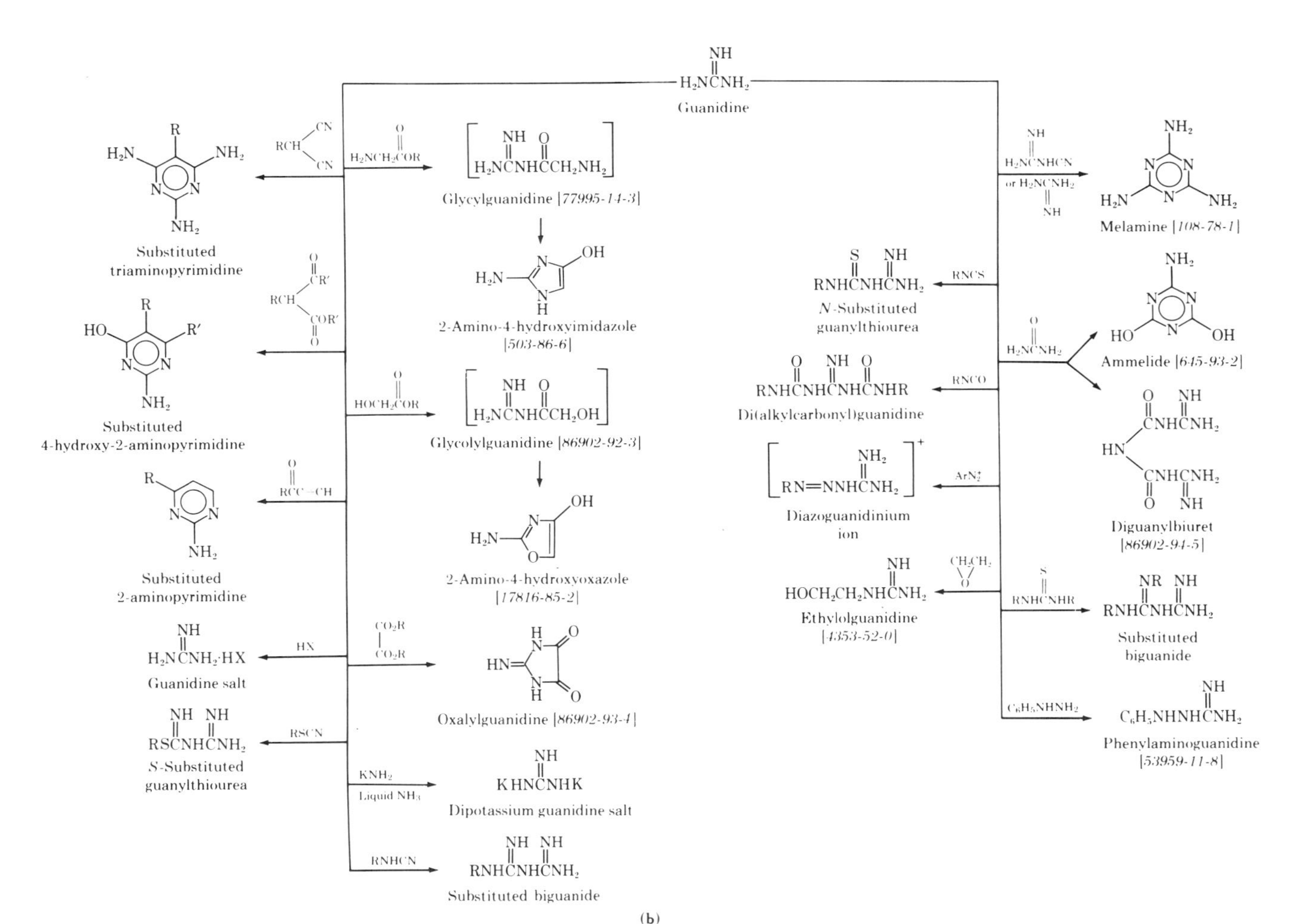

(b)

Figure 1. *(continued)*

Guanidinium sulfate and an equimolar amount of aluminum sulfate react in aqueous solution to form the double salt guanidinium aluminum sulfate, (CH_6N_3)-$Al(SO_4)_2.6H_2O$ [*10199-21-0*], which shows ferroelectric properties (8).

Analytical Methods

The preferred method for the analytical determination of guanidine is potentiometric titration with perchloric acid in ethylene glycol monomethyl ether (half neutral point = 350 mV). Alternatively, guanidine can be precipitated as the picrate by calcium picrate solutions (6). An amperometric method was developed for the estimation of guanidine in the presence of urea and ammonium compounds.

Toxicity

Apparently, guanidine salts do not cause detectable toxic effects in the human body. Cases of allergies are also unknown (4). Toxicity and precautions for handling guanidine hydrochloride are typical of guanidine salts (6).

Guanidine hydrochloride has an acute oral LD_{50} of 1.12 g/kg (rat) and is, therefore, considered to be only slightly toxic in single doses by mouth. A 10% solution produced only mild, transient irritation in the rabbit eye, and repetition of the application for five successive days did not lead to cumulative injury. However, the moist solid is severely irritating to rabbit skin, and systemic symptoms and death were observed following dermal application of dosages from 1 to 8 g/kg. The LD_{50} by skin absorption is ca 2.0 (0.9–4.5) g/kg.

Albino rats were maintained for four weeks on diets containing 10, 100, and 1000 ppm of the compound with no mortality and no recognized evidence of toxic effect. The feeding level of 1000 ppm was equivalent to a daily dosage of 94 mg/kg.

Precautions should be observed against skin or mucous membrane contact with solid guanidine hydrochloride or concentrated solutions. Otherwise, the product should offer no health hazards under ordinary circumstances of industrial handling or application.

Acute oral (rat) and acute dermal (rabbit) LD_{50} values for guanidinium nitrate are 1.26 g/kg and >10.0 g/kg, respectively. Mild primary skin irritation was produced during studies with rabbits (9).

The acute oral LD_{50} (rat) value for guanidine stearate is 7.5 g/kg and for guanidinium carbonate, 1.0 g/kg (5).

Manufacture

The method that has been of greatest industrial importance for the synthesis of guanidine is the reaction of dicyandiamide [*461-58-5*] (cyanoguanidine) with ammonium nitrate in the presence of anhydrous ammonia (7) (see Cyanamides).

$$\underset{\text{dicyandiamide}}{NH_2\overset{\overset{NH}{\|}}{C}NHCN} \xrightarrow{NH_4NO_3} \underset{\text{biguanide nitrate}}{NH_2\overset{\overset{NH}{\|}}{C}NH\overset{\overset{NH}{\|}}{C}NH_2.HNO_3} \xrightarrow{NH_4NO_3} \underset{\text{guanidinium nitrate}}{2\ NH_2\overset{\overset{NH}{\|}}{C}NH_2.HNO_3}$$

Two equivalents of ammonium nitrate are charged into an autoclave with liquid ammonia and one equivalent of dicyandiamide. The mixture is heated at 160°C for one hour, producing pressures up to 13.8 MPa (2000 psi); then, ammonia is flashed off. This process produces guanidinium nitrate in yields of 91–93% and of sufficient purity (91–92%) for the production of nitroguanidine. A single recrystallization from water yields a product of high purity (98.5–99.5%).

Guanidine salts can also be produced by adding ammonia to the nitrile group of cyanamide [*420-04-2*] (10). It is preferable to operate under weakly basic conditions to avoid dimerization of the cyanamide and hydrolysis of the product to urea. Use of a small amount of ammonia with an ammonium salt provides an acid to stabilize the guanidine produced, and ammonia is continuously generated for the reaction. Guanidine salts of all the common acids have been prepared by variations of this process (10).

Guanidinium nitrate can be directly produced by the reaction of ammonium nitrate and urea.

Guanidine salts are produced also by the dry fusion of dicyandiamide and ammonium salts. Although this reaction is exothermic, it has been satisfactory in some instances, particularly if ammonium chloride is the salt used. Other processes involve the use of calcium cyanamide or dicyanodiamide, either in wet or in dry fusion with ammonium salts. The wet fusions have reduced yields because of hydrolysis. The dry fusion with calcium cyanamide requires the use of 3 mol of ammonium salt and is hazardous to operate with ammonium nitrate.

Guanidine may also be prepared by the ammonolysis of ammonium thiocyanate and urea.

Economic Aspects and Specifications

Guanidinium nitrate is manufactured on a commercial scale by Cyanamid Canada Inc. and marketed at ca $1.32/kg. Guanidinium nitrate, guanidine hydrochloride, guanidinium stearate, and guanidinium silicate are produced commercially by SKW-Trostberg A.G. in the FRG (5).

Uses of Guanidine Derivatives

Guanidine salts (in particular guanidine hydrochloride) are used as intermediates in the production of sulfaguanidine,

$$H_2NC_6H_4SO_2NH\overset{\overset{\displaystyle NH}{\|}}{C}NH_2$$

sulfadiazine, and sulfamerazine (see Sulfonamides).

Water-soluble, cationic guanidine-formaldehyde resins, particularly when used in combination with anionic hydrocolloids, impart wet or dry strength to paper, depending on pH and curing conditions (11). Such resin systems are highly efficient for the flocculation of rubber latex, polymer latices, and sewage or industrial wastes (12).

Nitroguanidine is an exceptionally cool and flashless explosive and therefore of value as a propellant for antiaircraft guns and the like (see Explosives and propellants). By suitable reduction, nitroguanidine can be converted to aminoguanidine [*79-17-4*] (which can also be made from hydrazine and cyanamide or one of its salts). Aminoguanidine and substituted aminoguanidines are intermediates for a number of dyes for silk (qv) and wool (qv) and are also of interest because of exceptional properties as reducing agents. Aminoguanidine is also useful for the preparation of lead tetrazide and other explosive materials. Guanidine soaps, made from guanidine salts and fatty acids, are very effective drycleaning agents (13) (see Drycleaning and laundering).

Substituted guanidines are prepared from cyanamide and an amine.

$$RNH_2 + NCNH_2 \longrightarrow RNH\overset{\overset{\displaystyle NH}{\|}}{C}NH_2$$

Dodecylguanidine [*112-65-2*] salts (sold as Cyprex fungicides by the American Cyanamid Co.) are highly effective for agricultural applications, particularly in the control of apple scab and cherry leaf spot (14–15) (see Fungicides). They are also effective algal growth inhibitors (16), useful in the manufacture of microorganism-stable oil and water emulsions (17) and in treatment of saline process water (18).

Aryl derivatives, such as *sym*-diphenylguanidine [*102-06-7*] and *sym*-di-*o*-tolylguanidine [*97-39-2*], are used as accelerators for the vulcanization of rubber (see Rubber chemicals).

N-Substituted guanidine salts have application as antimicrobials (19). Other uses for guanidine salts include the production of an antistatic agent for textiles (20), a stabilizing agent for formalin (21), rubber foam (22), low temperature gas-producing solid fuel (23), a stabilizing agent for synthetic polymers (24), filter additives for cement and alumina suspensions (25), and the propellant triaminoguanidinium nitrate [*4000-16-2*] (26).

BIBLIOGRAPHY

"Guanidine and Guanidine Salts" in *ECT* 1st ed., Vol. 7, pp. 324–328, by J. M. Salsbury and J. M. Affleck, American Cyanamid Company; "Guanidine and Guanidine Salts" in *ECT* 2nd ed., Vol. 10, pp. 734–740, by E. H. Sheers, American Cyanamid Company.

1. C. A. Strecker, *Compt. Rend.* **52,** 1210 (1861).
2. L. Pauling, *The Nature of the Chemical Bond*, 3rd ed., Cornell University Press, Ithaca, N.Y., 1960.
3. W. J. Jones, *Trans. Faraday Soc.* **55,** 524 (1959).
4. *Guanidine Salts*, SKW-Trostberg A.G., Trostberg, FRG, 1976.
5. *Guanidine Products Study*, SKW-Trostberg A.G., Trostberg, FRG, 1976.
6. *Aero Guanidine Hydrochloride*, American Cyanamid Company, New York, 1960.
7. J. H. Paden, K. C. Martin, and R. C. Swain, *Ind. Eng. Chem.* **39,** 952 (1947).
8. J. W. Gilliland, Jr., and H. P. Yockey, *J. Phys. Chem. Solids* **23,** 367 (1962).
9. *Material Safety Data Sheet for Guanidine Nitrate*, American Cyanamid Company, New York, 1982.
10. *Cyanamide*, Cyanamid Canada Inc., Toronto, Can.
11. U.S. Pat. 3,002,881 (Oct. 3, 1961), B. McDonnell and H. E. Jackson (to Consolidated Mining and Smelting Co.).
12. U.S. Pat. 2,745,744 (May 15, 1956), C. L. Weidner and R. Dunlap (to Permacel Tape Corp.).
13. U.S. Pat. 2,978,415 (Apr. 17, 1957), R. J. Chamberlain (to American Cyanamid Co.).
14. R. J. W. Byrde, D. R. Clifford, and D. Woodcock, *Ann Appl. Biol.* **50,** 291 (1962).
15. U.S. Pat. 2,867,562 (Jan. 6, 1959), G. Lamb (to American Cyanamid Co.).
16. U.S. Pat. 3,142,615 (July 28, 1964), D. C. Wehner (to American Cyanamid Co.).
17. Can. Pat. 641,213 (May 15, 1962), C. S. Scanley (to American Cyanamid Co.).
18. U.S. Pat. 2,906,595 (Sept. 29, 1959), E. J. Pelcak and A. C. Dornbush (to American Cyanamid Co.).

19. Ger. Offen. 3,040,993 (June 16, 1982), A. Syldatk, E. Boerner, and K. Disch (to Henkel K.G.a.A.).
20. Jpn. Kokai 81,141,376 (Nov. 5, 1981), (to Nikka Chemical Industry Co., Ltd.).
21. Jpn. Kokai 78 90,209 (Aug. 8, 1978), T. Kamiyama (to Matsushita Electric Works, Ltd.).
22. Fr. Dem. 2,354,360 (Jan. 6, 1978), M. Alicot and A. Tignol (to Produits Chimiques Ugine Kuhlmann).
23. Jpn. Pat. 30,395 (Aug. 8, 1977), T. Shimizu (to Koa Kako K.K.).
24. Ger. Offen. 2,545,647 (Apr. 21, 1977), P. Uhrhan, R. Lantzsch, H. Oertel, E. Roos, and D. Arlt (to Bayer A.G.).
25. Ger. Offen. 2,033,886 (Mar. 4, 1971), H. Stettler (to Sandoz Ltd.).
26. Brit. Pat. 1,428,348 (Mar. 17, 1976), V. E. Haury (to Rockwell International).

PAUL PATTERSON
Cyanamid Canada Inc.

HEAT-EXCHANGE TECHNOLOGY, NETWORK SYNTHESIS

The relatively new discipline of process synthesis has spawned a technology that can be used to solve the problem of efficient energy utilization in process plant design (see Separation systems synthesis). In many plants the viability of the heat-exchange network is the key to economic success because of today's high cost of energy (see Heat-exchange technology). The process of developing a good heat-exchange network is most easily viewed as a multiple-tier optimization problem. Minimum total cost is the objective. This includes capital and operating costs plus some consideration of the flexibility and operability of the selected design. The plant energy network must be developed in concert with the other components of the process design.

In every process plant, heat exchangers are used for process–process-stream heat transfer and heaters or coolers for process–utility heat transfer. Heat-exchange-network synthesis is the art of systematically constructing this network of exchangers. The process engineer must understand the energy, capital, and operating trade-offs involved to confidently arrive at a workable design. Recent literature references use the word resilient to describe a network that has sufficient flexibility, operability, and controllability for use in an actual process design. Once the network has been synthesized, traditional design techniques are used to set stream flows through parallel units and analyze individual heat exchangers, heaters, and coolers. Optimization of these individual units is also considered at this point. The optimization of heat exchangers involves such things as baffle and tube sheet layout as well as pressure drop. This multiple-tier approach to the problem is necessitated by the complexity of the design process which must span the gulf from conceptual network development to detailed mechanical rating of heat exchangers. Figure 1 is a schematic diagram of the network design process. Minimum cost is the overall objective, but other criteria must be used as subobjectives within the various tiers. Iteration between the tiers is used where appropriate; however, any significant change in results from a tier usually requires a completely different solution from the following tiers.

Efficient network-synthesis techniques relieve the process engineer of the burden of accepting designs based on art which cannot be shown to be superior. The opportunities for process improvement are best before the structure of the network is determined. Recent developments in network analysis have progressed to the point that energy use in turbines can also be considered along with heat recovery. Methods for

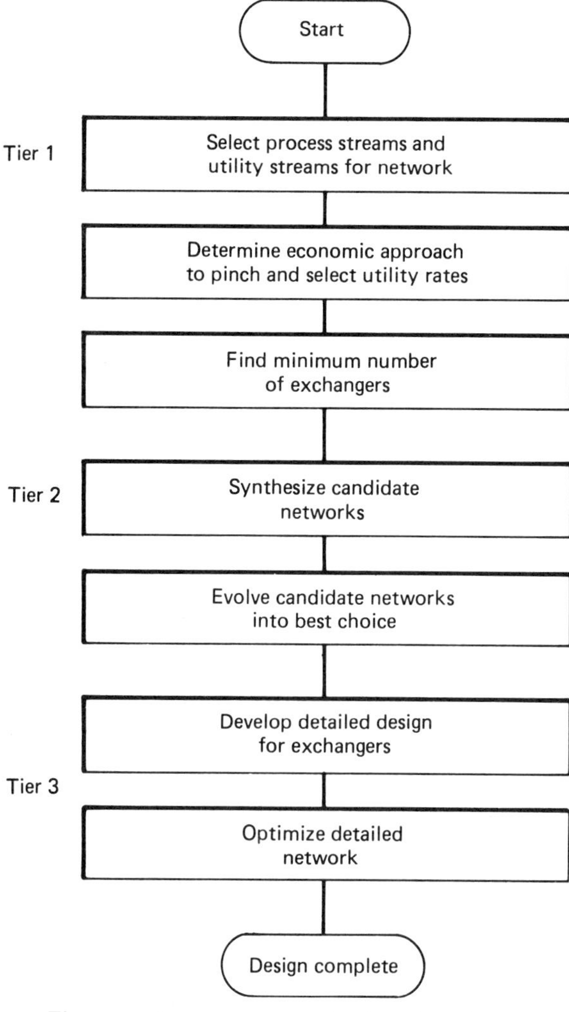

Figure 1. Heat-exchange-network design scheme.

rating individual heat exchangers are well developed and can be used to design new exchangers or simulate the performance of existing units. The approach described here is not limited to new plants but can be used for modification of existing plants as well. Increased heat utilization does not always mean a trade-off. Several studies have shown a reduction in energy consumption as well as capital cost. This is a remarkable recommendation for a systematic approach to network design.

Problem Specification

The problem of heat-exchange-network synthesis can be described as follows. A set of cold streams (i = 1 to M) initially at supply temperature T_{si} is to be heated to target temperature T_{ti}. Simultaneously, a set of hot streams (j = 1 to N) initially at supply temperature T_{sj} is to be cooled to target temperature T_{tj}. Variation in these supply or target temperatures may be permitted in a particular problem if the underlying process conditions are flexible and changes result in significant savings. In addition, some streams may require special alloy heat exchangers or be prohibited from matching with certain other streams. Several hot (heat source) and cold (heat sink) utilities are available for use. The enthalpy vs temperature relationship is known for all these streams. The appropriate physical properties for determining heat-transfer characteristics are also given. The best network of heat exchangers, heaters, and coolers to accomplish the desired temperature changes is desired. Best usually means most economic for the capital cost and utility costs available.

Representation. The impetus for heat-exchange-network design was the development of an adequate means to represent the problem. The temperature and heat (enthalpy) relationship for any process stream can be represented on a temperature–enthalpy diagram. Figure 2 shows four different streams on such a diagram. Stream A is a pure component that is condensing (eg, steam). Streams B and C are streams with constant heat capacity (C_p) that are to be heated (B) or cooled (C). Stream D represents a multicomponent mixture that changes phase as it is cooled. Horizontal distance on this figure represents relative enthalpy. It is possible to slide any of the streams horizontally without changing the amount of energy (enthalpy) required for heating or cooling. Vertical changes are not permitted because they alter process temperatures. This temperature–enthalpy domain representation of the

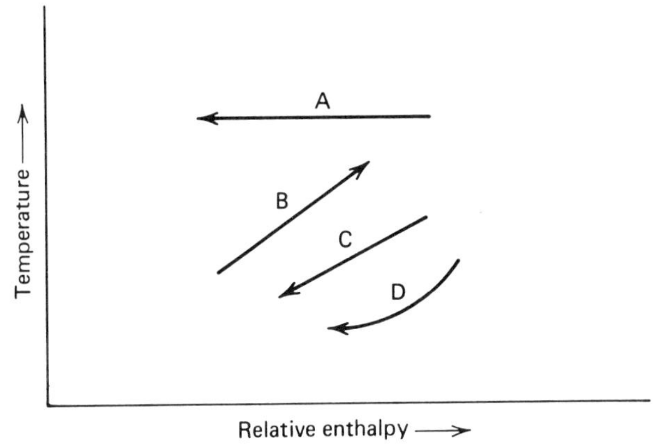

Figure 2. Temperature–enthalpy representation of streams.

network problem dates back to what was called a heat picture (1). The representation was expanded and discussed at length in ref. 2.

An important feature of the temperature–enthalpy representation is that individual streams can be lumped together in an ensemble of single hot and single cold composite streams. The composite representation helps structure the solution to a given heat-exchange-network problem. Figure 3(**a**) shows two hot streams and two cold streams that are to be considered for a heat-exchange network. A typical composite representation of these streams is shown in Figure 3(**b**). If one composite (super) curve is slid towards the other, the curves usually approach each other at a single pinch point. This is the point where the curves would first touch (dashed line) when slid horizontally. The pinch is analogous to the minimum temperature of approach in the design of an individual heat exchanger. The importance of the temperature pinch point to network synthesis cannot be overstated as shown below.

Alternative representations of stream temperature and energy have been proposed. Perhaps the best known is the heat-content diagram, which represents each stream as an area on a graph (3). The vertical scale is temperature, and the horizontal

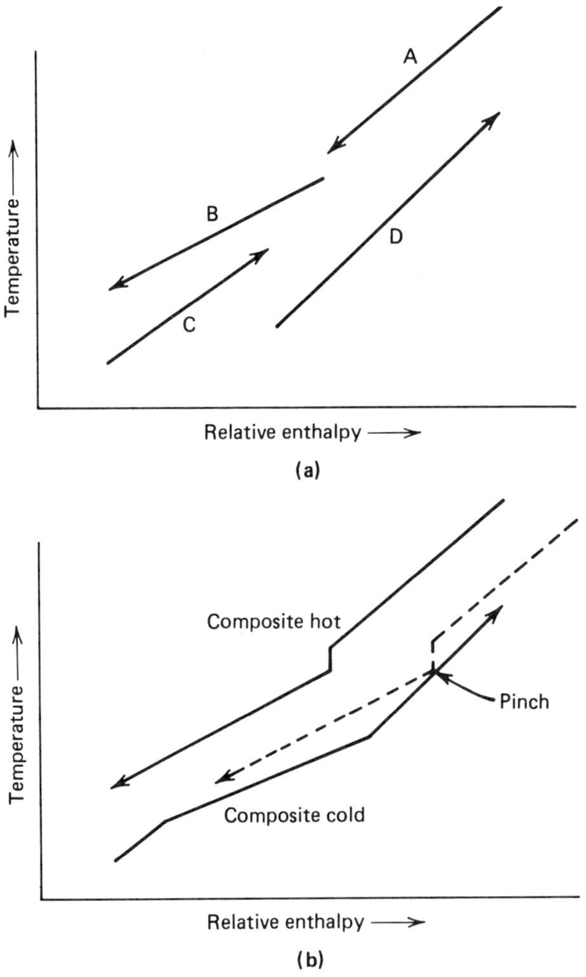

Figure 3. (**a**) Typical process streams in heat-exchange network. (**b**) Superstreams constructed from typical process streams.

is heat capacity times flow rate. Sometimes this quantity is called capacity rate. The stream area (capacity rate times temperature change) represents the enthalpy change of the stream.

In addition to a thermodynamic representation of streams, a means of representing the matching of streams for heat transfer is needed. Process engineers traditionally use the flow-sheet representation. This diagram shows stream flows as both horizontal and vertical lines intersecting at exchangers or nodes represented as circles. This topological representation can also be shown by either of the temperature–energy diagrams described above; however, both are somewhat awkward to use. What is needed is a way to represent an evolving network. A most useful representation was developed (4); its grid method draws streams as horizontal lines with arrows to indicate heat sources (right-pointing arrow) or cold sinks (left-pointing). Heat exchange between two streams, a match or node, is shown as two circles connected by a vertical line. This diagram gives the essential process topology while permitting the viewer to concentrate on the pairing and sequencing of streams. Figure 4(**a**) shows a traditional flow-sheet representation, and 4(**b**) shows the grid representation for a system of five streams.

Idealization. The final rating or design of a heat exchanger requires detailed knowledge of the fluids to be heated and cooled as well as information on the specific type of exchanger. The overall heat-transfer coefficient U usually includes a film coefficient for fluid on the outside of the tubes, a film coefficient for fluid inside the tubes, resistance of the tube wall to heat transfer, and a fouling resistence both inside and outside the tube. These coefficients are correlated with fluid properties, stream-flow velocities, and exchanger type. Hence, they cannot be obtained until a specific network is totally defined. By then of course, the network synthesis is complete and the designer must work with a specific network configuration.

For the purpose of network synthesis, the overall heat-transfer coefficient is usually idealized as a constant value. This independence of the heat-transfer coefficient makes possible the iterations necessary to solve the network problem. Usually, the overall heat-transfer coefficient for each exchanger (match) is defined as

$$U = 1/[(1/h_i) + (1/h_j)]$$

where h_i is the coefficient for the cold stream, and h_j is the coefficient for the hot stream. This simple formulation permits quick calculation of the heat-transfer coefficient for any pair of streams. It does not include the practical consideration of exchanger geometry. In fact, exchangers are often considered to be of the double-pipe variety where the overall exchanger area is given simply by

$$A = Q/(U\Delta T_{\mathrm{LM}})$$

where A is total heat-transfer area required, Q is the heat transferred in the exchanger, U is the overall heat-transfer coefficient, and ΔT_{LM} is the log–mean-temperature driving force. In industrial practice, more complex exchanger geometries are used. The above formula is usually modified to permit these complex geometries by including an F factor to correct ΔT_{LM} as follows:

$$A = Q/(UF\Delta T_{\mathrm{LM}})$$

Frequently, the difference in exchanger type does not influence the desired topology to any significant extent. For industrial problems, however, it is necessary to consider

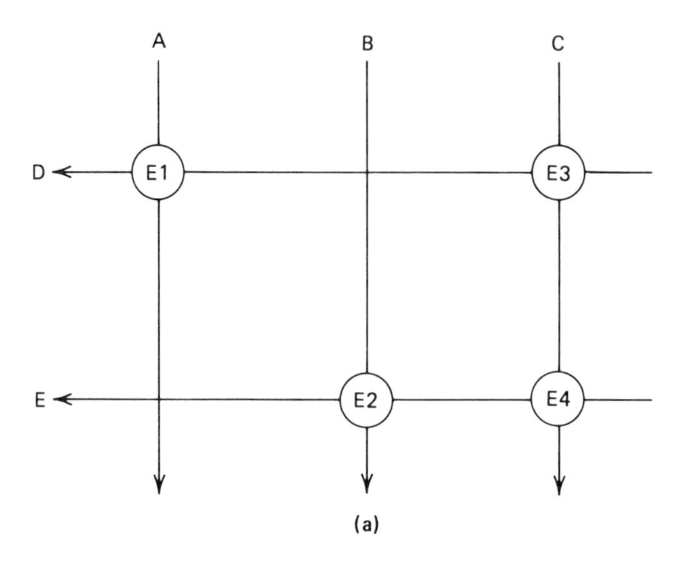

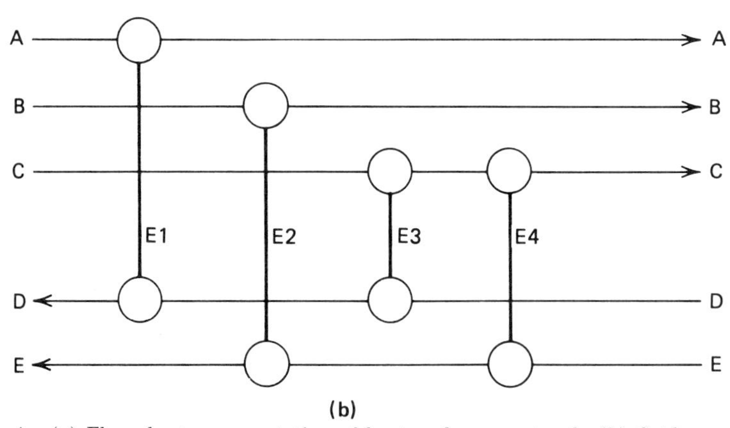

Figure 4. (**a**) Flow-sheet representation of heat-exchange network. (**b**) Grid representation of heat-exchange network.

individual heat-exchanger shells rather than just the match that is called the heat exchanger. If a high level of heat recovery is desired, the effect of the F factor can be important. This problem has been solved but is beyond the scope of this article. With these idealizations, network design may proceed.

Limiting Network Conditions

Heat-exchange-network design has been made easier by the development of various limiting conditions. The technique of limiting conditions is common in many areas of chemical engineering and can provide great insight into a problem. A notable example is the concept of minimum reflux and minimum stages in distillation. Actual distillations must operate between these limits, and economic designs are frequently based on a reflux that is a fixed percentage above the minimum. A detailed understanding of any model requires a clear view of the model at limiting conditions. A variety of limits has been developed and is exploited in heat-exchange-network design.

Feasibility (Heat-Transfer Pinch). A network must be in heat balance, but much more can be said about required heating and cooling utilities. The concept of composite streams permits quick determination of utility requirements. If one composite stream is slid horizontally, it will approach the other (see Fig. 5). The limiting condition of maximum energy recovery (MER_+) occurs when the composite streams just touch

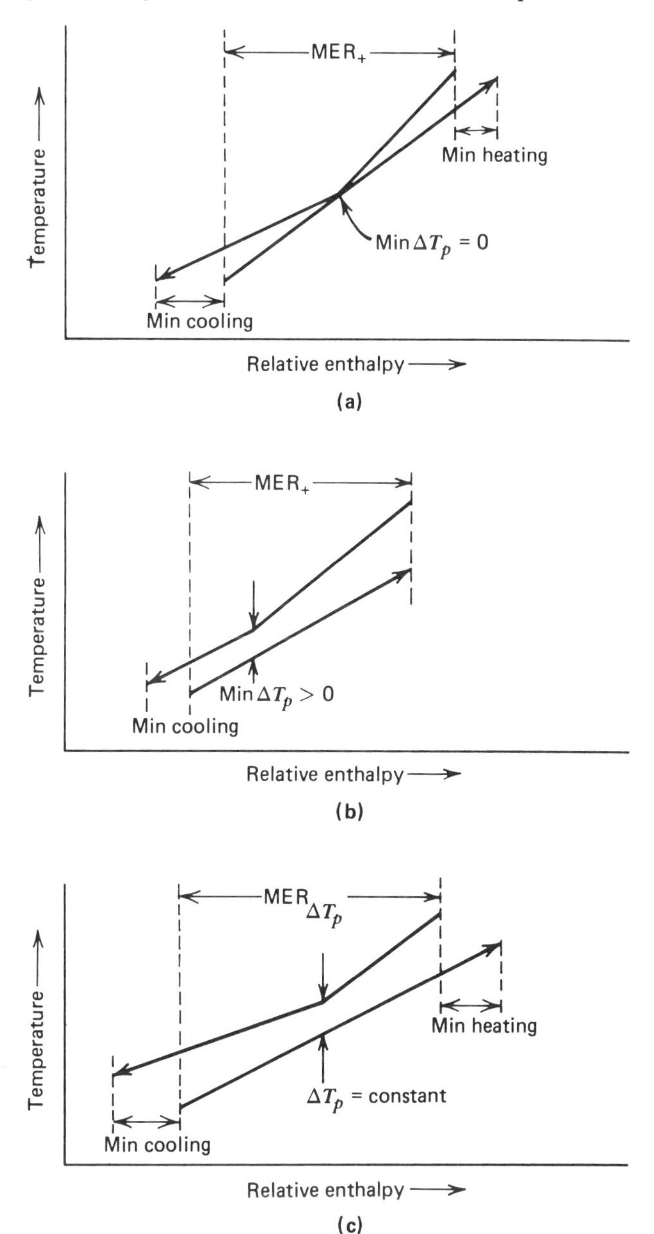

Figure 5. (**a**) Minimum utility requires infinite area. (**b**) Heat balance gives minimum utility. (**c**) Minimum utility requirement at ΔT_p.

(zero-temperature driving force) or when the need for either a hot or cold utility is eliminated. If the streams touch (Fig. 5(**a**)), any possible network with limiting utility rates requires infinite heat-exchange area. On the other hand, if only a heating or cooling utility is required, the limiting case (Fig. 5**b**) might represent a practical solution.

Furthermore, the limiting utility rates for any given temperature of approach at the pinch (ΔT_p) can be found from the temperature–enthalpy diagram. Figure 5(**c**) shows the composite streams for a problem in which a temperature of approach was specified. The composite streams are slid horizontally until the desired ΔT_p is obtained. The maximum energy recovery for this approach temperature would be $\mathrm{MER}_{\Delta T_p}$. Minimum heating and cooling requirements are shown. This procedure is illustrated in a tabular format called a feasibility table (2). A problem table that is used for network synthesis as well as feasibility is shown in ref. 4. Both of these tables are easy to program on a digital computer. Recently, an elegant generalization of these tabular procedures has been published (5). The generalized procedure can take into account restrictions on stream matching and is based on formulating the problem as a classic transportation problem.

A low temperature of approach for the network reduces utilities but raises heat-transfer area requirements. Research has shown that for most of the published problems, utility costs are normally more important than annualized capital costs. For this reason, ΔT_p is chosen early in the network design. Hence, utility rates are determined in the first tier of the solution. An economic network pinch temperature can be chosen in a variety of ways (2,6–7). The temperature of approach (ΔT_p) for the network is not necessarily the same as the minimum temperature of approach (ΔT_e) that should be used for individual exchangers. This difference is significant for industrial problems in which multiple shells may be necessary to exchange the heat required for a given match (7). The economic choice for ΔT_p depends on whether the process environment is heater- or refrigeration-dependent and on the shape of the composite curves (approximately parallel or severely pinched). In crude-oil units, the range of ΔT_p is usually 10–20°C. By definition, $\Delta T_e \leq \Delta T_p$. The best relative value of these temperature differences depends on the particular problem under study.

The concept of a temperature pinch has additional ramifications. If heat passes across the pinch, additional utilities are required (4). This observation can be proven by examining Figure 5(**a**). Suppose one attempted to use heat from the hot composite stream above the pinch to heat the cold composite stream below the pinch. This would displace an equal amount of heating from the hot stream below the pinch. The displaced heat would be available at a temperature suitable for heating only below the pinch, and thus, additional cooling would be required to dispose of the displaced heat. Additionally, the heat from above the pinch would have to be replaced by a utility. An important consequence of the total energy balance around a network is that any change in the hot-utility rate is exactly reflected in the cold-utility rate. Careless design can easily cause a double penalty in utilities. Insightful design can give a double savings in utilities. This savings obviously reduces operating costs but may reduce capital costs as well, since for every two units of utility heat saved, only one additional unit of process–process heat transfer is needed. The unique feature of this analysis is that the thermodynamic feasibility of constructing a network with any given set of process streams and utilities can be determined without resorting to a detailed examination of possible heat-exchange networks.

Minimum Area. The limit of minimum network area is presented in refs. 2–3. If idealized double-pipe exchangers are used, a heat-exchange network with minimum area can quickly be developed for any ΔT_p. In the limiting case, where all heat-transfer coefficients are assumed to be equal, the area for this network can easily be obtained from the composite streams by the following integration:

$$A_{\Delta T_p} = [1/U] \int_{T_{\text{lowest}}}^{T_{\text{highest}}} [dQ/(T_j - T_i)]$$

where A is the minimum possible network area at the given ΔT_p, U is the heat-transfer coefficient, and Q is the heat transferred. The vertical distance between the two composite stream curves $(T_j - T_i)$ is the driving force at each point along the composite streams. This integration is similar to that used for the design of individual heat exchangers. The minimum-area value is not of great interest in itself; however, it does provide a limiting value much like the Carnot cycle for heat engines. Figure 6 shows the feasible solution space for a system of streams (2). The minimum area required for any network is plotted as a function of utility rates. The temperature of approach ΔT_p is shown at selected points.

Methods have been developed that are able to generate a minimum-area network for any given energy recovery (ΔT_p) (2–3). The minimum-area networks developed generally employ many heat exchangers and are not an economic solution. They may be used to start an evolutionary process to develop better networks. Rather than minimum total area, the concern should be with the way in which the area is distributed among the exchangers. The installed cost of exchangers can best be characterized as

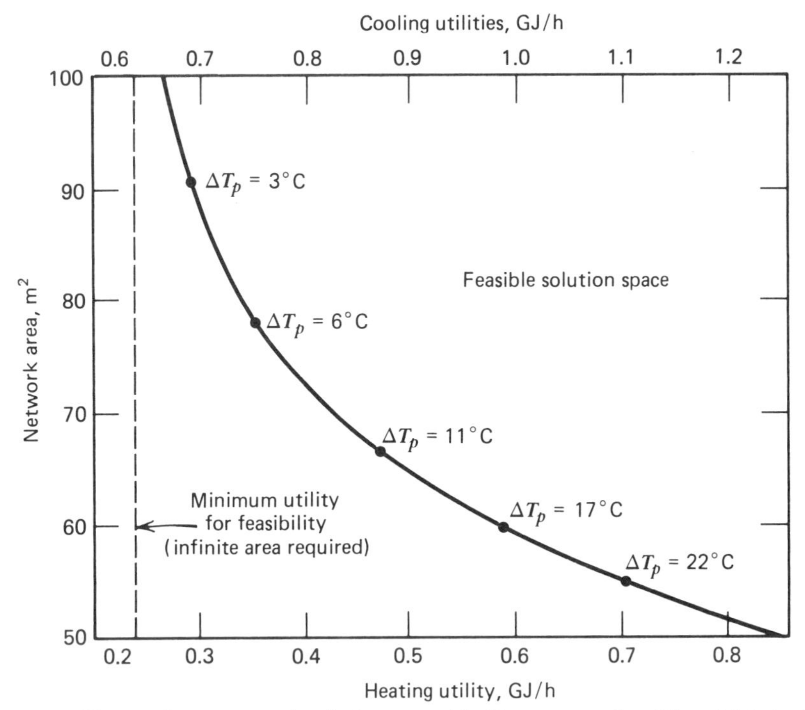

Figure 6. Heat-exchange-network solution space (2). To convert m^2 to ft^2, multiply by 10.76; to convert GJ to 10^6 Btu, divide by 1.054.

$$\text{cost} = a \cdot A^b + c$$

where a, b, and c are constants that depend on economic conditions and materials of construction. The constant b ranges from 0.5 to 1.0, and c is frequently neglected. Recall that

$$[A_1^b + A_2^b + \ldots + A_E^b] \geq [A_1 + A_2 + \ldots + A_E]^b$$

for $0 \leq b \leq 1$. Therefore, the distribution of area among heat exchangers is most important, and minimum area alone is not of paramount importance.

Minimum Number of Exchangers. The fewest number of matches or exchangers that are required in a network can be developed as a limit (2). The number needed (E_{min}) is generally one less than the total number of streams (process and utility) involved in the network:

$$E_{\text{min}} = S_{\text{process}} + S_{\text{utilities}} - 1$$

When this equation holds, each stream match (exchanger) must provide that one of the two streams involved reaches its target temperature. Such a network is called acyclic. In an acyclic network, it is not possible to trace a closed path along stream lines from exchanger to exchanger and return to the starting point without retracing some of the path.

There are exceptions to this simple equation that occur infrequently but nevertheless must be considered. A more complete relationship for the number of exchangers in a network is obtained by applying Euler's network relation from graph theory (8).

$$E = S_{\text{process}} + S_{\text{utilities}} - P + L$$

where P is the number of independent heat loads that can be identified. This means P is the number of possible subproblems, each of which are in heat balance. In practice this is almost always one, but if it exceeds one, the minimum number of exchangers is reduced. Cyclic paths in a network are called heat-load loops. Figure 7 is an illus-

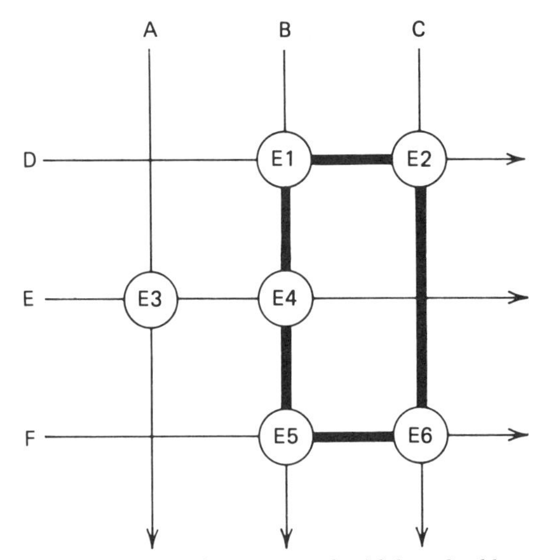

Figure 7. Heat-exchange network with heat-load loop.

tration of a cyclic network (E1, E5, E6, and E2). The number of loops or cycles (L) that exist in a network add to the required number of exchangers. If $P = 1$ (single problem) and $L = 0$ (acyclic), the generalized equation reduces to the equation given for E_{min} above. It is not always possible to construct an acyclic network, and thus the minimum number of exchangers may be increased above the simple minimum (8).

Unfortunately, the minimum number of exchangers is not the same as the number of shells required if conventional shell-and-tube heat exchangers are used. If ΔT_p is large, the difference is not significant. However, as more energy recovery is obtained, ΔT_p must be reduced and exchangers near the pinch may require several shells in series to maintain an acceptable F factor. This problem is discussed, as well as a case-study solution based on the idea of reducing the ΔT_e for individual exchangers while holding ΔT_p constant (7). As with the limit of minimum area, the number of exchangers is not of paramount concern. Rather, the number of exchanger shells and the area distribution among the shells are of primary importance. The cost equation given before should be modified as

$$\text{cost} = a \cdot A_{\text{shell}}^{b} + c$$

to emphasize that the area of each shell must be considered when calculating costs. The two temperatures of approach (ΔT_p and ΔT_e) provide a means of including this effect in the multiple-tiered optimization process. Nevertheless, the minimum number of exchangers represents a worthwhile goal in network development.

Synthesis Algorithms

The synthesis of heat-exchange networks takes place after the feasibility of developing a network has been established for the selected ΔT_p and other constraints. The primary objective is to obtain the lowest cost structure for the utility rates selected. Like many engineering optimization problems, there is not a single distinct solution but rather a set of nearly equal cost solutions. This set of solutions has been called juxta-optimum (2). The selection of the best solution for a given application can be highly dependent on the difficult-to-quantify factors: operability, controllability, and flexibility. Ideally, synthesis methods would lead to all members of the juxta-optimum set without wasting time developing poor networks. The designer could then pick a design based on these secondary factors. Differences in costs of a few percent are below the accuracy of the models used and certainly in the range where nonquantifiable network characteristics would be used to make a selection.

Synthesis methods can be classified in many different ways. The categories chosen here are combinatorial, heuristic, inventive, and evolutionary.

Combinatorial methods express the synthesis problem as a traditional optimization problem which can be solved using powerful techniques that have been known for some time. They may use total network cost directly as an objective function but do not exploit the special characteristics of heat-exchange networks in obtaining a solution.

Heuristic methods rely on the application of sets of rules to lead to a specific objective like network cost. Unfortunately, heuristics cannot guarantee that the objective will be reached.

Inventive synthesis methods develop networks by exploiting selected characteristics of heat exchange among groups of hot and cold streams. These characteristics are used to develop networks that meet certain criteria. One set of characteristics can lead to a minimum network area for a given ΔT_p. Other inventive methods have been able to guarantee networks that recover maximum energy and use the minimum number of heat exchangers.

Evolutionary methods develop new heat-exchange networks from those generated by other means. Obviously, the goal is to improve the starting network according to some specified criteria. This may or may not be network cost.

At this time, the ideal synthesis method does not exist, but some of the recently developed methods provide very good networks. New methods are constantly being developed, and continued improvement is expected.

Combinatorial. Much of the early work in heat-exchange-network synthesis was based on exhaustive search or combinatorial development of networks. Basically, the synthesis problem was transformed into an optimization problem that could be solved by standard programming techniques. A large variety of techniques was used (9–11), but considerable difficulties exist with this approach. The high dimensionality of the problem virtually assures long computation times even with problems carefully formulated. There are many obviously poor networks that must be evaluated. Many of the articles on exhaustive network searching limit the number of possibilities by prohibiting stream splitting. The astonishing number of technically possible networks with more than three or four streams is discussed in ref. 12. For a ten-process-stream example problem (10SP1), the alternative sets of feasible matches are ca 1.55×10^{25} without stream splitting (12).

Heuristic. Another early approach to developing alternative heat-exchange networks was to base network development on heuristics or rules that have been shown to generate good sets of matches. Heuristic development of networks is the procedure that was traditionally used by process engineers, although the synthesis rules were not usually formally stated. With readily understood rules for network construction, repeated matching of the hot stream with the highest supply temperature against the cold stream with the highest target temperature is recommended (13). This procedure leads one from the matches at the highest temperature levels to those at the lowest temperature levels. This simple, quick approach often leads to good networks. Complications arise if two streams are available to supply or receive heat over similar temperature intervals. This situation has been called temperature contention and leads to a dilemma in selecting the stream to match (2). One alternative is to split the heat sink or source stream into parallel branches where each can match with the streams over the temperature interval.

Inventive. Strategies to develop networks that can take advantage of some special characteristic of the problem have been the most successful. These strategies are far more efficient than any previous methods but do not directly utilize network cost as an objective. This is a disadvantage, although networks that meet the subobjectives of maximum energy recovery for a given ΔT_p and a minimum number of exchangers usually seem to be in the juxta-optimum set of networks with minimum cost.

Networks with minimum total area can be quickly and easily developed using the technique of stream splitting to avoid temperature contention (2–3). The difficulty with reaching this subobjective is that the networks usually contain a large number of small heat exchangers. In fact, the number of exchangers will be an effective maximum. Reference 3 shows how to develop such networks and then proposes a method to reduce the number of exchangers. This is called an algorithmic-evolutionary approach. The evolutionary portion of this procedure is discussed below.

A thermodynamic-combinatorial (TC) method follows exactly the opposite tactic with regard to stream splitting (12). The TC method permits construction of all possible networks without stream splitting and using the minimum number of heat exchangers. A prescribed degree of energy recovery ($MER_{\Delta T}$) is specified. The objective to construct all of the networks meeting these criteria without tedious enumeration of many unwanted networks is indeed tantalizing. The method provides for the possibility that no such networks exist for any given problem. The small five-stream problem known as 5SP1 serves as an example of the TC method. The basic stream information for this problem is shown in Figure 8 using the grid representation of ref. 12. The heat exchanger shown connecting streams 1 and 4 is compulsory in all networks that meet the stated synthesis criteria. Since a single heating utility is permitted, the minimum number of heat exchangers is five. Figure 9 is a table used to enumerate the possible stream matches (exchangers) for this problem. The potential exchangers and heaters are shown on the grid in the upper part of the figure. The twenty-four possible networks are listed as rows in the table with five Xs indicating the particular matches in each network. The right hand columns show an N for those matches that would violate any one of three different constraints. The eight selections remaining are indicated with a check mark. When these were examined, three were found to violate the ΔT_p constraint that was applied to each exchanger. The five other networks are shown in Figure 10. Equivalent annual cost was used to select 10(**b**) as the optimum network, although all five had costs within a 1% band.

The temperature interval (TI) method is an alternative to the TC approach of enumerating all networks with maximum energy recovery and minimum number of heat exchangers without stream splitting (4). The TI method was the first to pursue this network goal and is a significant advance in the development of networks. The key to the method is the analysis of a number of subnetworks as synthesis tasks. The original problem is partitioned into subnetworks based on ordered temperature levels. Each level is examined individually for potential matches. Heat is not passed from a higher level to a lower one until all of the cold streams in the higher level have been provided for. In this way, the maximum variety of subnetwork designs is available. The process engineer selects among these designs to arrive at the final network. Subsequently, an evolutionary technique (ED) can be used to improve the networks generated with the TI method. In addition to the technique proposed in reference 4, modifications to the combined TI–ED method have been suggested (7,14).

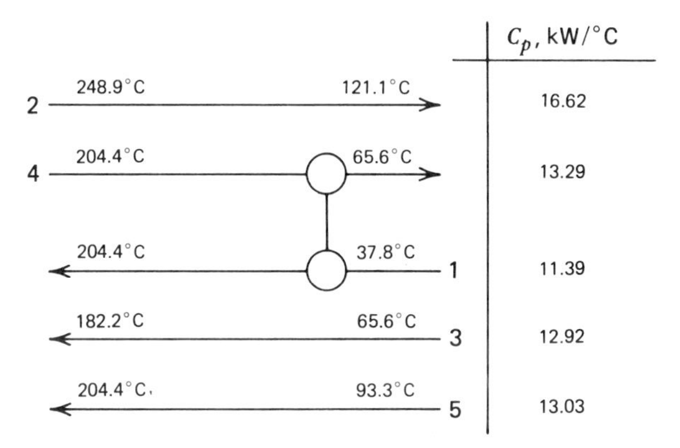

Figure 8. Streams for problem 5SP1 (12).

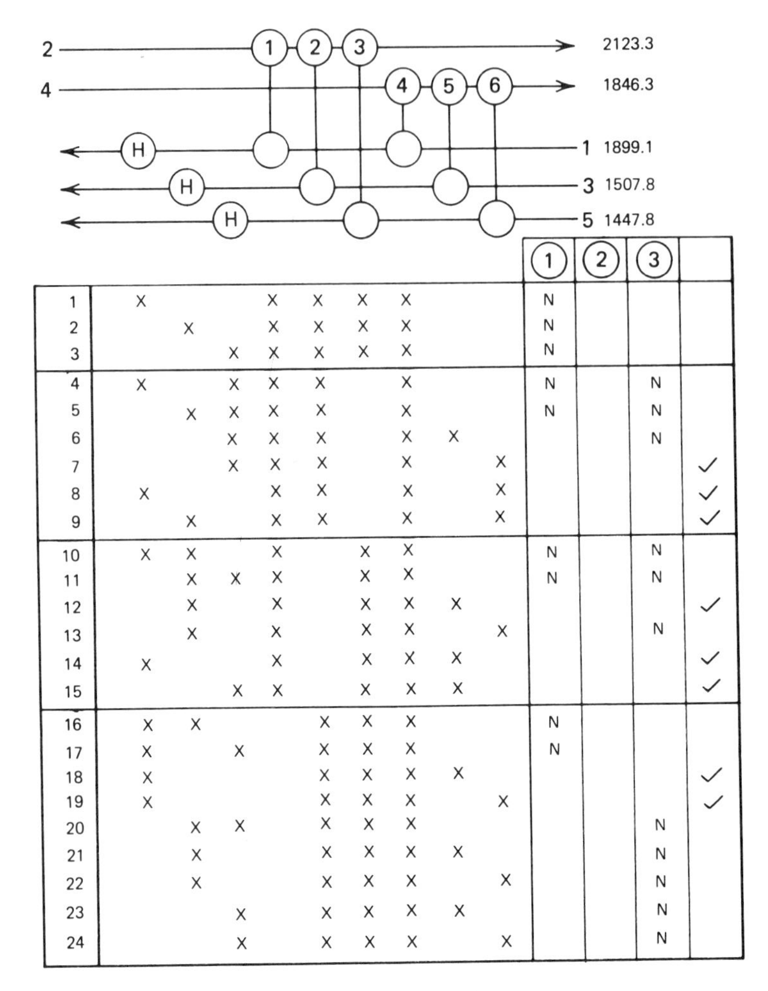

Figure 9. Development of feasible networks (12).

The pinch method for network synthesis gives an important new insight (15). Unlike the other methods discussed, the pinch method starts the synthesis at the network pinch. The pinch is used to split the synthesis problem into two subproblems. These separate problems are solved starting at the pinch and working towards the opposite temperature end of each subproblem. Starting the design at the pinch has the distinct advantage that it is easy to identify whether or not a stream split is required to meet the chosen energy constraint (ΔT_p) for synthesis of a network. Although a design-method summary is presented, the detailed implementation is left up to the engineer. The method cannot be classified as algorithmic. A manual design of the network is advocated. In this way, the engineer can probably guide the design to the best network. A very readable account of the whole network development process using the pinch method for network synthesis is given in ref. 16; however, no specific evolutionary rules for use with the pinch method are shown.

Evolutionary. The evolutionary approach to network development starts with a good network developed by some alternative means and proceeds to change it into an optimal network from the standpoint of cost. The success of any evolutionary

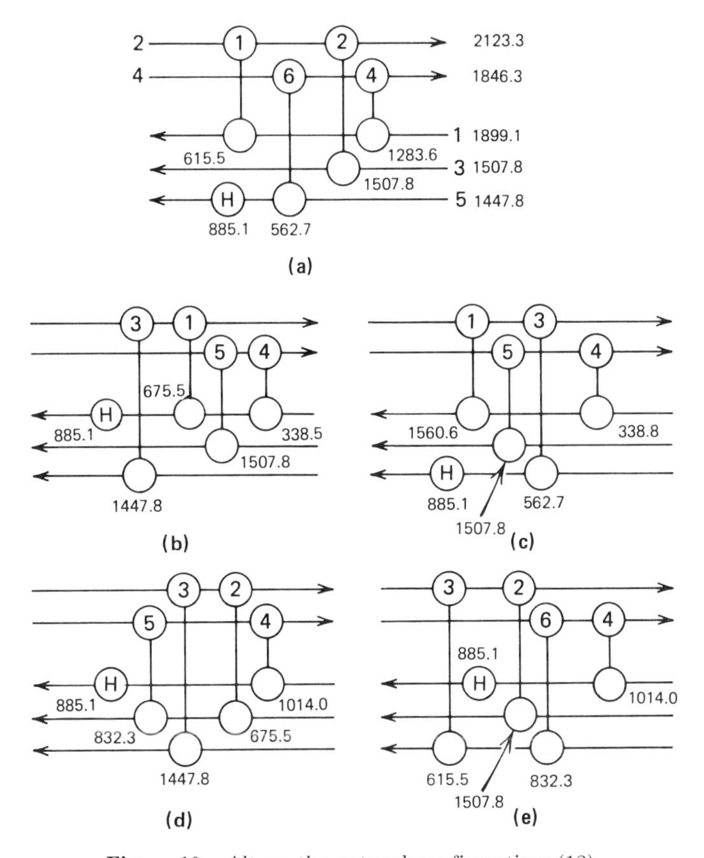

Figure 10. Alternative network configurations (12).

method depends greatly on the starting network. Evolutionary methods are specific to starting network characteristics and are usually matched with the method used to develop the initial network, eg, TI followed by ED. There are two extremes for possible starting networks.

The first is a network that has minimum area but a maximum number of exchangers as proposed by the algorithmic-evolutionary approach (17). The algorithmic part of this method is the development of a minimum area network. The evolutionary part employs a set of rules to modify systematically the initial network. The three rules presented are heuristic in nature and seek to combine exchangers and stream splits to reduce network cost. The problem of reducing stream splits appears difficult to researchers.

An alternative starting network is one without stream splits. The networks from the TI method maximize energy recovery and may introduce heat-load loops (4). Stream splits are not made in the initial steps of network invention. The ED method is proposed to be one in which heuristic rules and strategies would be used to improve the networks developed by the TI method (4). The importance of a thermodynamic base for evolutionary rules is stressed in this proposal, but there is no explicit guidance for the evolutionary process (4).

Some modifications to the TI method and the introduction of a sound loop-breaking strategy that is algorithmic in nature are shown in reference 14. Also shown is the concept of level of loop to indicate the number of source and sink streams involved in any heat-load loop. Loops can range from first level (one source and one sink stream) to Nth level. The smaller of the number of source or number of sink streams is N. The evolutionary procedure searches for loops from the lowest level first and tries to break (eliminate) them to reduce the number of exchangers; the procedure also introduces stream splits (14). Reference 7 includes the concept of using ΔT_p to find utility rates and thereafter using ΔT_e for network synthesis. This indirectly allows minimization of the number of heat-exchanger shells rather than just the number of exchangers. A case-study analysis is necessary to implement this method, which is called the double temperature of approach (DTA).

All these evolutionary methods follow an initial synthesis method. The objective is to develop the best network design prior to the start of detailed and costly calculation of individual network components. In practice, an approximate cost is usually determined for the several candidates before one or more are selected for optimization.

Network Optimization

Process calculations for traditional unit-operations equipment can be divided into two types: design and performance. Sometimes the performance calculation is called a simulation (see Simulation and process design). The design calculation is used to roughly size or specify the equipment. Following the design guideline, a particular piece of equipment is chosen. It is then necessary to calculate the performance of the selected item in the service of interest. The problem of choosing the pipe size required to deliver a fixed amount of fluid illustrates this sequence. The design equation will give a pipe diameter for fixed fluid properties, flow rate, and pressure drop. It is not very likely that the diameter determined will be one of the standard sizes in which pipe is made. Hence, a standard size with a somewhat larger diameter will be chosen. To determine how much fluid will flow through this new pipe, a performance calculation is required. If a pump provides the pressure to drive the fluid through the pipe, an optimization problem exists. Pump-operating costs plus capital cost for the pipe and pump are minimized by proper selection of pipe diameter.

A corresponding problem occurs in the third tier of heat-exchange-network design (Fig. 1). Limits on utilities and number of heat exchangers were set as a first-tier objective. Candidate networks were proposed with the help of synthesis algorithms in the second tier. With the network topology fixed, the detailed design calculations for individual heat exchangers can proceed. Heat exchangers can be designed as single- or multiple-shell units. A low ΔT for a given exchanger may make it necessary to use several shells in series to improve the F factor. In some cases, large flow rates or area requirements may make it necessary to have several shells in parallel. In any case, the design of the various units in the network can be completed. Subsequently, individual equipment is specified, and a performance calculation is needed for the network. Variation in exchanger sizes and stream flows provides an opportunity for further optimization of the network.

The detailed network calculations are time-consuming, and computer-aided methods are virtually required to evaluate the alternatives in the optimization process. The number of variables to consider is large. Parameters that can be varied include exchanger areas, pressure drops, and detailed mechanical aspects of the individual exchangers. These include such items as baffle cut and spacing, tube diameter, gauge, and pitch. If the network under consideration is in an existing plant, currently installed exchangers may be selected for use. Network operation (startup, turndown, and shutdown) should also be of concern at this point.

A flow-sheeting program is needed to effectively handle all of these detailed design calculations. Moreover, a program capable of optimizing the selected variables is desirable. Such programs have been developed and are reported (6,18). The objective function can be any convenient accounting criterion, eg, payout. Most traditional flow-sheeting programs are based on a sequential modular or simultaneous modular approach. This type of program is easy to construct and performs well in situations where various process units are not highly integrated. Unfortunately, networks by nature are very integrated and require a flexible solution approach. Both of the referenced programs use an equation-based approach to the flow-sheeting program. Even small networks involve a relatively large number of equations, and advanced sparse matrix methods must be used. The advantage of the equation-based approach is that virtually any process parameter can be made a variable in the optimization process.

Design Practice

The economic incentive to recover process energy has been discussed at great length in the current literature. Escalation of energy costs during the 1970s was a significant factor in the development of more detailed design techniques. A study recently showed that the relative cost of fuel to heat exchangers has more than doubled in the nine years since 1967 (19). Figure 11 shows that the fuel costs themselves had more than quadrupled during the same period. This large increase in the cost of energy leads to examination of design alternatives. There are certainly many instances where

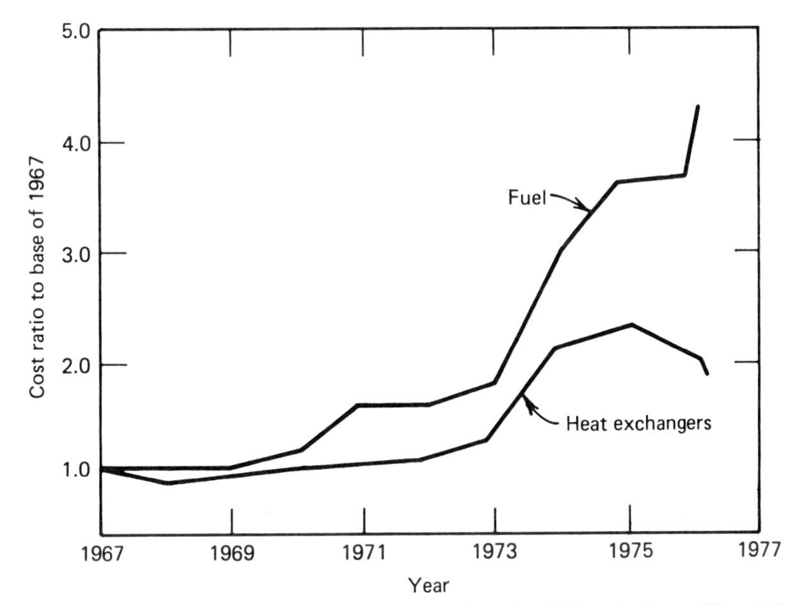

Figure 11. Cost history for fuel and heat exchangers, based on Nelson indexes, *Oil and Gas Journal* (1967 = 1.0) (19).

additional capital for energy-recovery equipment can reduce operating expense. The trade-offs involved in adding insulation to pipes are a classic problem. On the other hand, the double savings (heat and cooling) that result from careful use of energy can lead to both a capital and operating cost savings. Belief in the inevitability of the capital vs operating cost trade-off exists because of the assumption that there are no basic faults in the engineering designs (20). This assumption is not warranted if there has been a reckless energy transfer across the pinch in the heat-exchange-network design. Twelve energy-saving projects at Imperial Chemical Industries have been analyzed using network techniques (20). Energy savings were 6–60% of the original design, and capital savings were as high as 30% of the original design. At least half of the projects resulted in capital savings.

Many of the calculations necessary for heat-exchange-network design can be done by hand. In fact, a clear understanding of the methodology involved can be most easily accomplished with a sheet of graph paper with development of some of the curves and networks discussed above. Emphasis needs to be placed on developing an overall understanding of process-energy use rather than simply adding heat exchangers to recover energy. The practical thermodynamic techniques presented here give an excellent framework with which to do this. Today's computational resources make it attractive for the average process designer to use a computer program to determine and evaluate alternative network configurations. Several companies have developed in-house heat-exchange-network programs. In addition, commercial heat-exchange-network programs are available. These programs typically take the raw stream data available

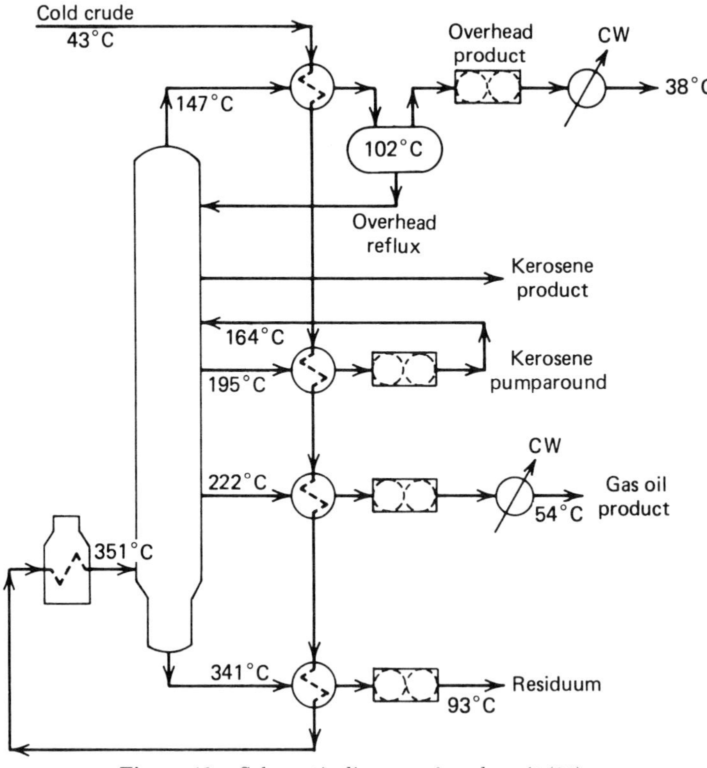

Figure 12. Schematic diagram of crude unit (21).

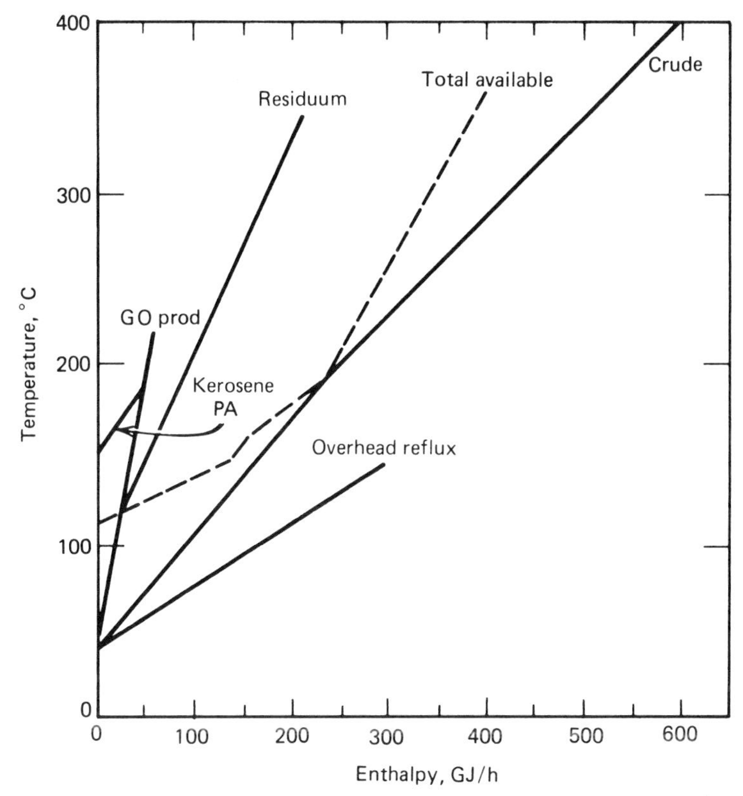

Figure 13. Temperature–enthalpy diagram streams (21). To convert GJ to 10^6 Btu, divide by 1.054.

to the process engineer for the plant and calculate the limiting network conditions. They then determine various alternative networks based on synthesis algorithms. Finally, some of the programs do a more detailed design of individual exchangers and subsequent optimization of the network. Following the accurate determination of flows and other stream conditions, several of the available simulators prepare the input necessary for traditional process simulation programs. However, none of these programs relieves the process engineer of the need to thoroughly understand the energy-integration tools before attempting to apply them.

The earliest use of heat-exchange-network synthesis was in the analysis of crude distillation units (1). The crude stream entering a distillation unit is a convenient single stream to heat while the various side draws from the column are candidate streams to be cooled in a network. So-called pumparounds present additional opportunities for heating the crude. The successful synthesis of crude distillation units was accomplished long before the development of modern network-synthesis techniques. However the techniques now available ensure rapid and accurate development of good crude unit heat-exchange networks.

The heat-recovery opportunities in a 20,700 m^3 (130,000 bbl) per stream-day crude-oil column and some of the practical aspects of optimizing heat recovery in a crude unit have been discussed (21). Most crude units process a variety of crudes and produce a broad range of product slates which requires flexibility in the energy network. A crude-oil feed stream and five hot streams leaving the crude column are shown

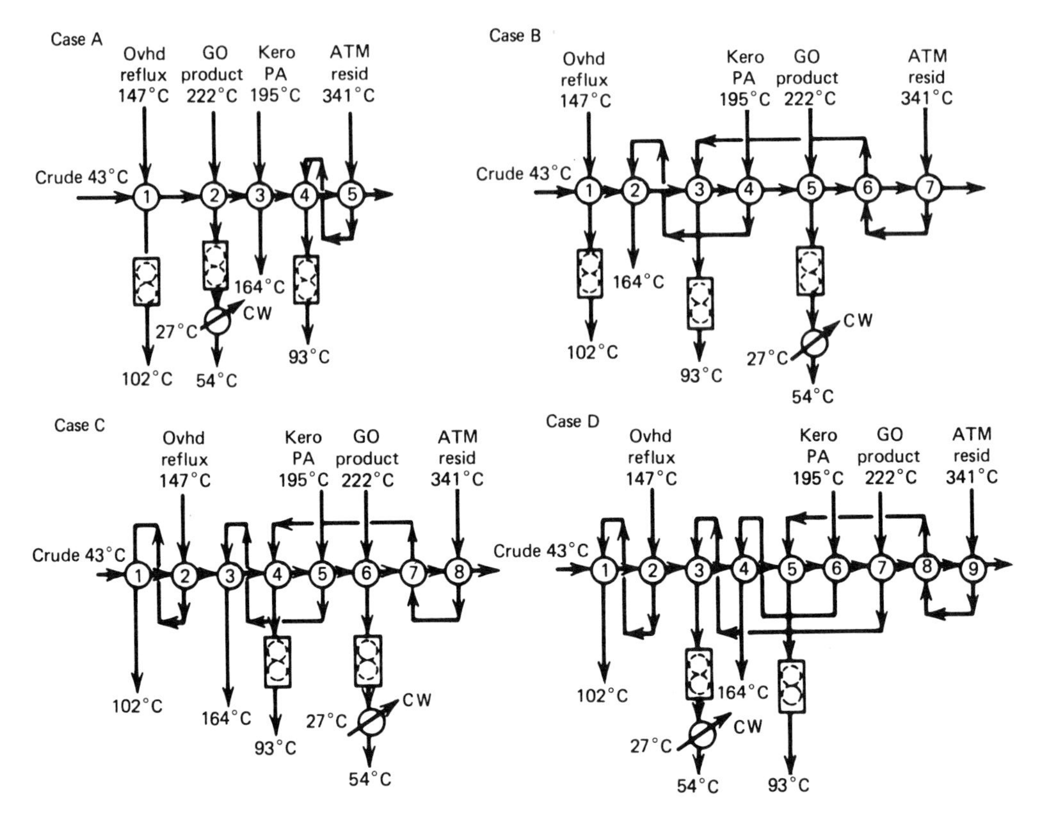

Figure 14. Alternative network configurations (21).

in Figure 12. Side stream strippers and other items are not shown on the column to simplify the drawing. Four of the hot streams are available for heat exchange with the crude: residuum (resid), gas oil (GO) product, kerosene pumparound (Kero PA), and overhead (Ovhd) reflux. The overhead product is not exchanged with the crude because of possible contamination of the naphtha product if an exchanger leaks. The remaining hot stream, kerosene product is not cooled so that it can be used directly in another unit. The crude-oil feed is heated with these various streams, and finally, a fired heater is used to increase the temperature to that required in the column flash zone. The composite enthalpy curve for the hot streams and the crude-stream enthalpy curve are shown in Figure 13. Alternative design cases are shown in Figure 14. Multiple shells are required for several of the heat exchangers. The sensitivity of the various cases to utility and capital cost changes for several different crudes has been investigated (21).

The complexity of problems that can be handled by commercially available programs is illustrated in ref. 22. A 15,900 m^3 (100,000 barrels) per day crude unit consisting of an atmospheric and vacuum crude column, two heaters, and a total of twenty-one heat exchangers were analyzed. Figure 15 shows the process diagram with temperatures and duties indicated. Figure 16 shows the heat-exchange network obtained following the three design steps: feasibility, synthesis, and optimization.

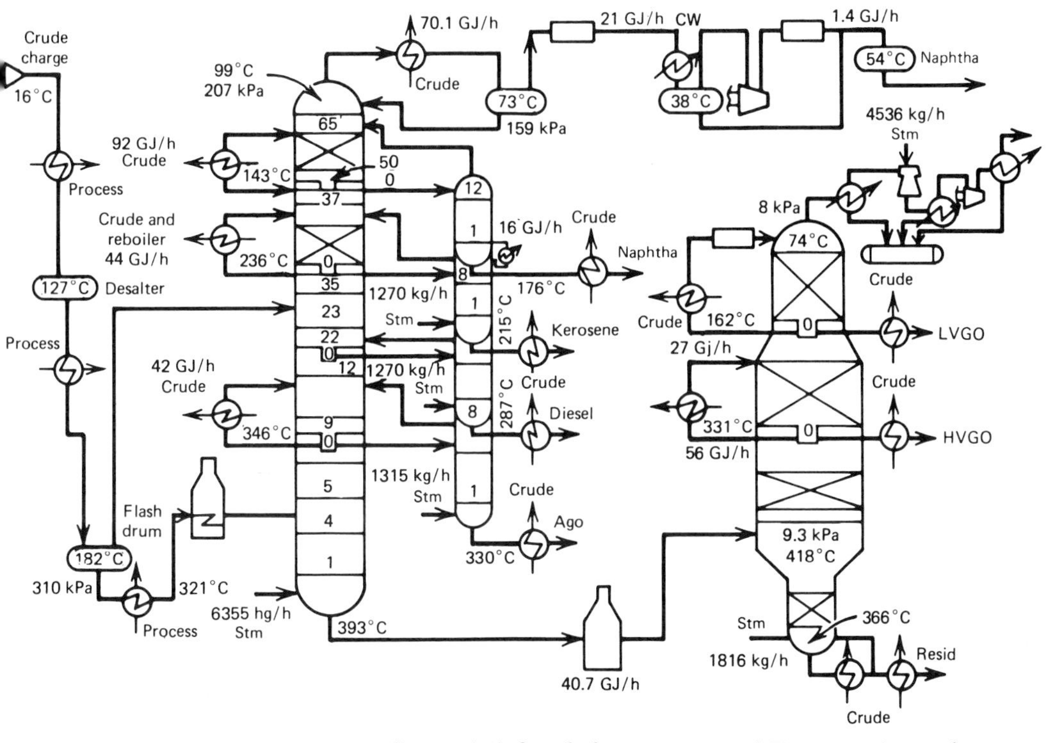

Figure 15. Crude unit-process diagram (22). Overflash set at 3% on feed. To convert GJ to 10^6 Btu, divide by 1.054; to convert kPa to psi, multiply by 0.145.

Future Developments

The topic of heat-exchanger-network design was first explored (1,10) in the industrial environment and subsequently proposed (9,11) as a topic of academic research. Substantial progress has been made in developing techniques that can be used successfully for industrial problems. These techniques are published along with many attempts that are only of academic interest and do not contribute to the solution of the industrial problem. An excellent review of several classes of process-synthesis problems including heat-exchange-network design is given in ref. 23. In a realistic process-design environment, the various synthesis techniques must be used simultaneously. At this time, the literature does not present such an integrated approach.

Acceptance of Techniques. The study of heat-exchange-network design and optimization is a well-established area of interest in the academic community. Viable techniques for industrial work have been developed, and some commercial computer programs utilizing these techniques are available. The acceptance of the techniques by the engineering-design community and plant-operations personnel will eventually follow. The problem is mainly one of education for the academic community and industrial practitioners (20). It is necessary for the academic community to focus on the progress that has been made and translate it into a form that is palatable to the industrial community. Design methods must address the many problems a plant operator has in practice. The economic incentive is such that engineering-design companies

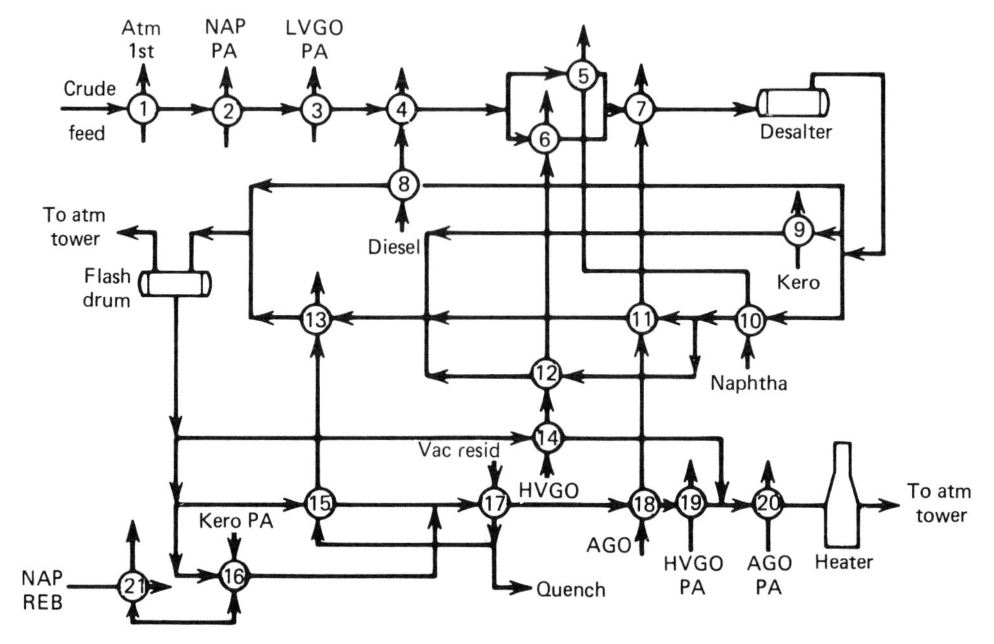

Figure 16. Crude unit heat-exchange network (22).

and others will eventually be forced to utilize these new process-design tools. The general acceptance of the computer will further this end.

Energy Networks. Heat-exchange networks are concerned with just a portion of the total energy in a process plant. As indicated earlier, significant steps have been made toward integrating other parts of the energy system. The essential considerations when turbine–compressor systems and heat-exchange networks are combined are outlined in ref. 24. The temperature–enthalpy representation can still be used to conceptualize the addition or removal of energy from a process. Energy (work) simply enters or leaves via an idealized heat engine that operates over a specified temperature range. The usual operation of a Carnot heat engine is assumed, but the concept of the pinch for a heat-exchange network can be expanded to an energy pinch for the total system. Introduction of a heat engine (eg, letdown turbine) in such a way as to cause energy flow across the pinch wastes process energy. On the other hand, if the engine is introduced properly, "free" energy can be obtained for the capital cost of a machine. Heat-pump placement can be analyzed in a similar manner. Proper placement reduces total energy demand, whereas improper placement may increase it.

Ideally, a process plant should be examined for its total energy consumption. It is likely that other plant energy systems will eventually be included in this type of analysis. This would include not only process thermal energy and shaft energy, but electrical power as well.

Separation Networks. The area of separation networks is important in chemical-process plants. Most separations are conducted by distillation, and there has been considerable interest in the process of designing sequences of distillation columns. Columns are separation-process-run by energy, and the tie-in with heat exchange is obvious. Some progress has been made in the analysis of energy input to and withdrawal from a distillation column (25). The techniques for separation-process synthesis must be integrated with those for energy network synthesis. Several limits have been developed for separation processes similar to those for heat-exchange networks (26) (see Separation systems synthesis). This is the first step toward synthesis of entire flow sheets. Unfortunately, the essential chemical-reactor synthesis lags behind, and it will be several years before significant inroads can be made into entire-plant synthesis owing to the reactor problem.

Nomenclature

A = heat-transfer area
a = constant in the heat-exchanger cost equation
AGO = atmospheric gas oil
b = constant in the heat-exchanger cost equation
C_p = heat capacity
c = constant in the heat-exchanger cost equation
CW = cold water
DTA = double temperature of approach
E = number of heat exchangers in network
$E_{\min}$ = minimum number of heat exchangers in network
ED = evolutionary technique
F = correction factor for ΔT_{LM} in heat exchangers
GO = gas oil
H = relative enthalpy
h_i = heat-transfer coefficient for stream i
h_j = heat-transfer coefficient for stream j
HVGO = high vacuum gas oil
Kero = kerosene
L = number of heat-load loops in a heat-exchange network
LVGO = low vacuum gas oil
M = number of cold process streams in network
MER_+ = maximum energy recovery for given set of process streams
$MER_{\Delta T}$ = maximum energy recovery with network approach of ΔT_n
N = number of hot process streams in network
NAP = naphtha
NAP REB = naphtha reboiled
Ovhd = overhead
P = number of independent stream subsets in a network
PA = pumparound
Q = heat transferred in exchanger
S_{proc} = number of process streams in network
S_{util} = number of utility streams in network
SP = process-stream problem
Stm = steam
T_{si} = supply temperature of cold process streams i
T_{sj} = supply temperature of hot process streams j
T_{ti} = target temperature of cold process stream i
T_{tj} = target temperature of hot process stream j
ΔT_e = minimum temperature of approach for any heat exchanger
ΔT_p = minimum temperature of approach at the network pinch
ΔT_{LM} = log–mean temperature difference for a given heat exchange
TC = thermodynamic-combinatorial
TI = temperature internal
U = overall heat-transfer coefficient

BIBLIOGRAPHY

1. A. M. Whistler, *Pet. Ref.* **27**(1), 57 (1948).
2. E. C. Hohmann, *Optimum Networks for Heat Exchange*, PhD thesis, University of Southern California, Los Angeles, Calif., 1971.
3. N. Nishida, S. Kobayashi, and A. Ichikawa, *Chem. Eng. Sci.* **26,** 1841 (1971).
4. B. Linnhoff and J. R. Flower, *AIChE J.* **24,** 633 (1978).
5. J. Cerda, A. W. Westerberg, D. Mason, and B. Linnhoff, *Minimum Utility Usage in Heat Exchanger Network Synthesis—A Transportation Problem*, DRC-06-25-81, Carnegie-Mellon University, Pittsburgh, Pa., Sept. 1981.
6. T. B. Challand, R. W. Colber, and C. K. Venkatesh, *CEP* **77**(7), 65 (1981).
7. R. W. Colbert, *CEP* **78**(7), 47 (1982).
8. B. Linnhoff, D. R. Mason, and I. Wardle, *Comp. Chem. Eng.* **3,** 295 (1979).
9. C. S. Hwa, *AIChE Int. Chem. Eng. Symp. Ser.* **4,** 101 (1965).
10. M. G. Kesler and R. O. Parker, *Chem. Eng. Prog. Symp. Ser.* **92**(61), 111 (1969).
11. K. F. Lee, A. H. Masso, and D. F. Rudd, *Ind. Eng. Chem. Fundam.* **9,** 48 (1970).
12. J. R. Flower and B. Linnhoff, *AIChE J.* **26,** 1 (1980).
13. J. M. Ponton and R. A. B. Donaldson, *Chem. Eng. Sci.* **29,** 2375 (1974).
14. J. L. Su, *A Loop-Breaking Evolutionary Method for Synthesis of Heat Exchanger Networks*, Masters thesis, Washington University, St. Louis, Mo., 1979.
15. B. Linnhoff and E. Hindmarsh, *Chem. Eng. Sci.* **38,** 745 (1983).
16. B. Linnhoff and J. A. Turner, *Chem. Eng.* **88**(22), 56 (1981).
17. N. Nishida, Y. A. Lir, and L. Lapidus, *AIChE J.* **22,** 539 (1976).
18. J. V. Shah and A. W. Westerberg, *Comp. Chem. Eng. J.* **4,** 21 (1980).
19. T. B. Chilland and S. F. Yang, *Oil Gas J.*, 73 (Feb. 6, 1978).
20. B. Linnhoff and co-workers, *A User Guide on Process Integration for Efficient Use of Energy*, Institute of Chemical Engineers, Rugby, UK, 1982.
21. F. Huang and R. Elshout, *Chem. Eng. Prog.* **72**(7), 68 (1976).
22. F. Kleinschrodt and G. Hammer, *Chem. Eng. Prog.* **79**(7), 33 (1983).
23. N. Nishida, G. Stephanopoulos, and A. W. Westerberg, *AIChE J.* **27,** 3 (1981).
24. B. Linnhoff and B. W. Townsend, *CEP* **78**(7), 72 (1982).
25. B. Linnhoff, H. Dunford, and R. Smith, *Chem. Eng. Sci.*, in press.
26. E. Hohmann, M. Sander, and H. Dunford, *Chem. Eng. Comm.* **17,** 273 (1982).

General References

D. Boland and B. Linnhoff, *Chem. Eng.*, 22 (Apr. 1979).

V. Cena, C. Mustacchi, and F. Natali, *Chem. Eng. Sci.* **32,** 1227 (1977).

J. Cerda and A. W. Westerberg, *Minimum Utility Usage in Heat Exchanger Network Synthesis—A Transportation Problem*, DRC Report No. 06-16-80, Carnegie-Mellon University, Pittsburgh, Pa., 1980.

H. Cohen, G. F. C. Rogers, and H. I. H. Saravanamutto, *Gas Turbine Theory*, Longman, Inc., New York, 1972.

P. R. Cooper, *Royal Society Esso Energy Award*, press release 8(81), Royal Society, London, 1981.

H. A. Dunford and B. Linnhoff, *Cost Savings in Distillation Symposium*, paper no. 10, Institution of Chemical Engineers, Leeds, UK, 1981.

R. V. Elshout and E. C. Hohmann, *Chem. Eng. Prog.*, 72 (Mar. 1979).

R. A. Greenkorn, L. B. Koppel, and S. Raghavan, *Heat Exchanger Network Synthesis—A Thermodynamic Approach*, 71st AIChE Meeting, Miami, Fla., 1978.

L. E. Grimes, *The Synthesis and Evolution of Networks of Heat Exchange That Feature the Minimum Number of Units*, MS thesis, Carnegie-Mellon University, Pittsburgh, Pa., 1980.

I. E. Grossman and R. W. H. Sargent, *Comp. Chem. Eng.* **2,** 1 (1978).

R. W. Haywood, *Analysis of Engineering Cycles*, 3rd ed., Pergamon.

P. Hinchley, *Chem. Eng. Prog.* **73,** 90 (1977); *Chem. Ing. Tech.* **49,** 553 (1977).

E. C. Hohmann and F. J. Lockhart, *Optimum Heat Exchanger Network Synthesis*, AIChE 82nd National Meeting, Atlantic City, N.J., 1976.

E. C. Hohmann and D. B. Nash, *A Simplified Approach to Heat Exchanger Network Analysis*, 85th National AIChE Meeting, Philadelphia, Pa., June 1978.
S. Kobayashi, T. Umeda, and A. Ichikawa, *Chem. Eng. Sci.* **26,** 3167 (1971).
B. Linnhoff, *Thermodynamic Analysis in the Design of Process Networks*, PhD thesis, University of Leeds, Leeds, UK, 1979.
B. Linnhoff, *Design of Heat Exchanger Networks—A Short Course*, University of Manchester Institute of Science and Technology, Manchester, UK, 1982.
B. Linnhoff and K. J. Carpenter, *Energy Conservation by Energy Analysis—The Quick and Simple Way*, Second World Congress of Chemical Engineering, Montreal, Can., 1981.
B. Linnhoff and J. A. Turner, *Chem. Eng. (London)*, 742 (Dec. 1980).
B. Linnhoff and J. A. Turner, *Chem. Eng.*, 56 (Nov. 2, 1981).
R. L. McGalliard and A. W. Westerberg, *Chem. Eng. J.* **4,** 127 (1972).
A. H. Masso and D. F. Rudd, *AIChE J.* **15,** 10 (1969).
R. L. Motard and A. W. Westerberg, *Process Synthesis*, AIChE advanced seminar lecture notes, New York, 1978.
D. B. Nash, E. C. Hohmann, J. Beckman, and L. Dobranski, *A Simplified Approach to Heat Exchanger Network Analysis*, AIChE 85th National Meeting, Philadelphia, Pa., June 1978.
T. K. Pho and L. Lapidus, *AIChE J.* **19,** 1182 (1973).
R. N. S. Rathore and G. J. Powers, *Ind. Eng. Chem. Process Des. Dev.* **14,** 175 (1975).
J. V. Shah and A. W. Westerberg, *Evolutionary Synthesis of Heat Exchanger Networks*, AIChE Annual Meeting, Los Angeles, Calif., 1975.
J. J. Siirola, *Status of Heat Exchanger Network Synthesis*, AIChE National Meeting, Tulsa, Okla., 1974.
R. A. Smith, *Economic Velocity in Heat Exchangers*, ASME/AIChE 20th National Heat Transfer Conference, Milwaukee, Wisc., 1981.
G. Stephanopoulos, *Synthesis of Networks of Heat Exchangers—A Self-Study Block Module*, Project PROCEED, Massachusetts Institute of Technology, Cambridge, Mass., 1977.
J. Taborek in E. U. Schlunder, K. J. Bell, G. F. Hewitt, F. W. Schmidt, D. B. Spalding, J. Taborek, and A. Zukauskas, eds., *Heat Exchanger Design Handbook*, Hemisphere Publishing Corporation, New York, 1982, Chapt. 1.5.
D. W. Townsend and B. Linnhoff, *AIChE J.* (1982).
D. W. Townsend and B. Linnhoff, *Chem. Eng. (London)*, 91 (Mar. 1982).
T. Umeda, T. Harada, and K. Shiroko, *A Thermodynamic Approach to the Synthesis of Heat Integration Systems in Chemical Processes*, Proceedings of the 12th Symposium on Computer Applications in Chemical Engineering, Montreux, Switzerland, 1979, p. 487.
G. L. Wells and M. G. Hodgkinson, *Process Eng.*, 59 (1977).

EDWARD HOHMANN
California State Polytechnic University

I

IONOMERS

Ionomers are a class of important commercial polymers with intriguing scientific characteristics. Typically, ionomers contain a certain number of inorganic salt groups attached to a polymer chain. This article describes ionomers with a maximum of ca 10 mol % ionic groups pendant to a hydrocarbon or perfluorinated polymer chain. Although this definition is arbitrary, it is generally accepted in the polymer field (1). There are also families of water-soluble polymers containing moderate amounts of ionic groups, such as the partially hydrolyzed polyacrylamides. Those systems do not display the ionic interactions typical of the ionomers described in this article. Recent reviews of the subject are included in refs. 1–4.

Unlike homogeneous polymer systems, the pendant ionic groups interact to form ion-rich aggregates contained in the nonpolar polymer matrix. The resulting ionic interactions strongly influence polymer properties and applications, which have made this a fertile area for research and development. A typical ionomer structure can be depicted as follows:

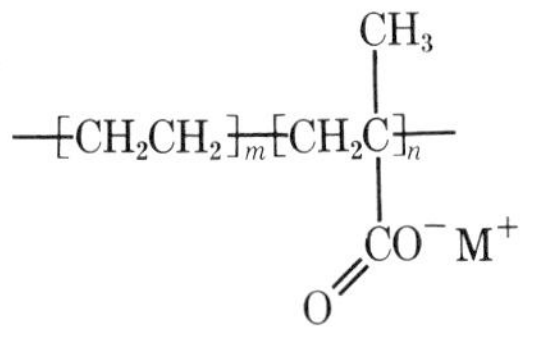

$$-\!\!\left[CH_2CH_2\right]_m\!\!\left[CH_2\overset{\overset{\displaystyle CH_3}{|}}{\underset{\underset{\displaystyle \overset{O\!\!/\!\!/}{}CO^-\,M^+}{|}}{C}}\right]_n\!\!-$$

Typically, the ratio of m/n is on the order of 10 to 100, reflecting a low overall content of ionic groups. The salt groups chemically combined with a nonpolar polymer backbone have a dramatic influence on polymer properties not observed with conventional homopolymers or with copolymers based on nonionic species. It is generally accepted that the ionic groups interact or associate to form ion-rich regions in the polymer matrix, as illustrated schematically in Figure 1.

The ionic interactions and resultant polymer properties are dependent on the type of polymer backbone (plastic or elastomer); ionic functionality (ionic content), generally 0–10%; type of ionic moiety (carboxylate, sulfonate, or phosphonate); degree of neutralization (0–100%); and type of cation (amine, metal, monovalent, or multivalent).

Ionic content, degree of neutralization, and type of cation dominate the properties of the resultant polymer system. With this range of variables, the spectrum of polymer properties within the ionomer family is extremely broad.

In the early 1950s, BF Goodrich introduced one of the first elastomers based on ionic interactions, a poly(butadiene-*co*-acrylonitrile-*co*-acrylic acid) [*25265-19-4*]. These materials can be neutralized with zinc oxide or other zinc salts and plasticized to break ionic association at elevated temperature (5–6). Such ionic elastomers display enhanced tensile properties and improved adhesion compared to conventional copolymers.

A second family of elastomers with a substantial degree of ionic interactions was also introduced in the early 1950s by E. I. du Pont de Nemours & Co., Inc. It was based on the sulfonation of chlorinated polyethylene (7). These materials, suitably cured with various metal oxides, gave rise to a combination of ionic and covalent cross-links and were commercially available under the trade name Hypalon.

A breakthrough occurred in the mid-1960s when DuPont introduced poly(ethylene-*co*-methacrylic acid) [*25053-53-6*] under the trade name Surlyn; these were partially neutralized with sodium and zinc cations. These modified polyethylenes possess remarkable clarity and tensile properties superior to those of conventional polyethylene. This development was an important factor in stimulating research in this area (1,8–16). The Surlyn systems emphasized the versatility of the ionomer structure and the unique properties available by modification of a polyethylene backbone (17–18). A study of the viscoelastic properties of this class of ionomers provided an interpretation of the physical properties based on the existence of hard regions interspersed among soft regions (19).

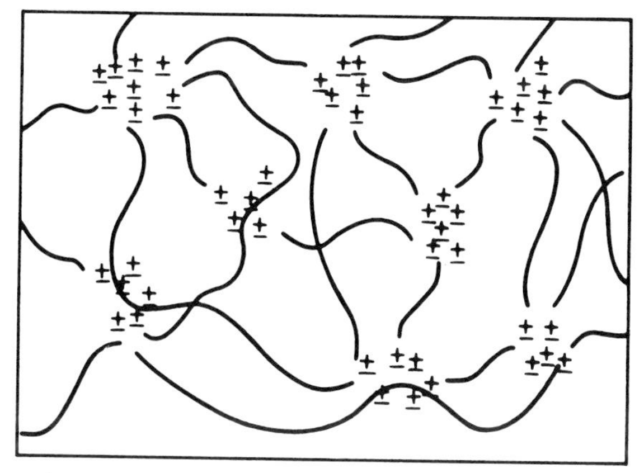

Figure 1. Schematic representation of an ionomer. +, metal ion; −, anion on the polymer; and lines = hydrocarbon polymer chains.

More recently, new families of ionomers have emerged that possess a wide variety of properties leading to different applications. Specifically, DuPont introduced the Nafion class of perfluorosulfonated ionomers (20–21) (see Perfluorinated ionomer membranes), and Exxon the sulfonated ethylene–propylene terpolymer (22–23), a new family of thermoplastic elastomers. Asahi Glass Company in Japan developed perfluorocarboxylate ionomers known as Flemion (24). A number of other polymer systems are still in the laboratory stage.

Structure

A large amount of structural data on ionomers has been accumulated over the past twenty years (1). An important question deals with the state of aggregation of the ionic species in the nonpolar polymer matrix.

Analytical methods applied to the structure elucidation include small-angle x-ray scattering, small-angle neutron scattering, transmission electron microscopy, ir spectroscopy, Raman spectroscopy, Mössbauer spectroscopy, direct measurements of mechanical properties, and dielectric properties.

Above a critical ion concentration, two types of ion aggregates can be defined: multiplets and clusters (2,13). The multiplets are considered to be small isolated polar groups dispersed throughout the polymer matrix. The larger aggregates have a weaker degree of ionic interaction but are significantly larger and are defined as clusters. Raman spectroscopy can be employed in different ionomer systems to elucidate the content of clusters and multiplets at various temperatures for different polymer systems (25–27). In copolymers of low dielectric constant, eg, a styrene-based matrix, the tendency for the ionic species to aggregate at low ionic contents is much more pronounced than in the case of a more polar polymer backbone, such as poly(ethyl acrylate) (28).

In an early theoretical paper (13), multiplets were proposed to result from ion-pair interactions in low polarity media involving contact-ion pairs which associate via electrostatic interactions. Multiplets represent associated ion pairs, triplets, quartets, etc, in which the charges approach each other as closely as possible. The maximum number of ion pairs that can interact to form a multiplet is eight, assuming spherical geometry. Steric factors and the constraint that each ion pair be attached to a hydrocarbon polymer backbone require this limitation.

At higher ion-pair contents, the aggregation of multiplets forms clusters. This aggregation is favored by electrostatic interaction of the ion pairs within the multiplets but opposed by the forces caused by the elastic nature of the hydrocarbon polymer chains. Clusters, therefore, contain ion-pair aggregates interspersed with polymer backbone chains. As the temperature of an ionomer system is increased, the clusters contained in the polymer become unstable, and the multiplets redistribute (see Fig. 2).

Subsequently, the concept of multiplets and clusters was strongly supported (25–27). Specifically, Raman spectra of a family of poly(styrene-*co*-*p*-carboxystyrene) ionomers have been measured and vibrational assignments proposed. By assigning specific Raman bands to multiplets and clusters, the relative concentration of ion pairs in these two different kinds of sites have been determined.

The relative concentration of ions in multiplets or clusters is described in Figure 3 as a function of the total ionic content in poly(styrene–*co*-sodium *p*-styrenecar-

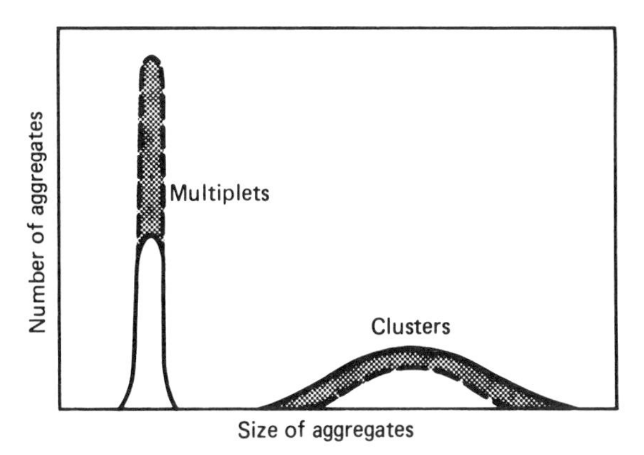

Figure 2. Schematic distribution of ion aggregates in ionomers. —, low temperature; - - -, high temperature (2).

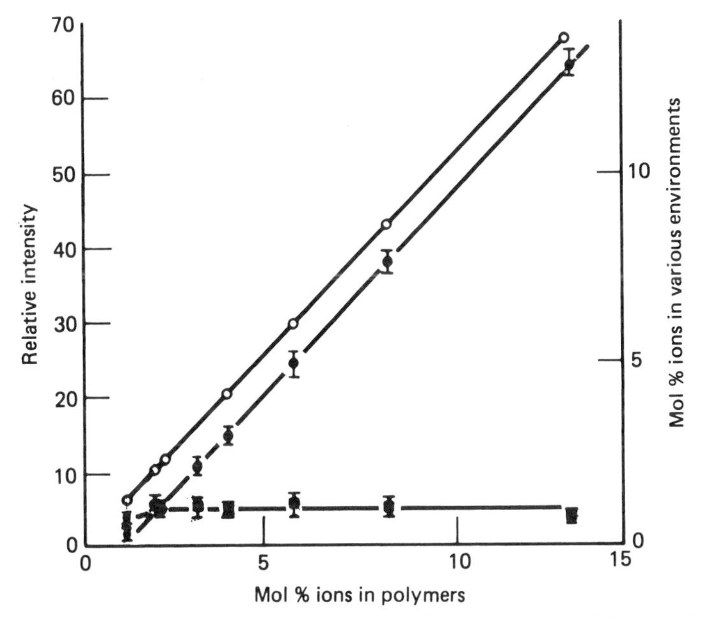

Figure 3. Relative intensities of the bands at 248 cm^{-1} (I_m) and 157 cm^{-1} (I_c) vs c_s (mol %) sodium *p*-styrenecarboxylate for the entire series of poly(styrene-*co*-sodium *p*-styrenecarboxylate)s investigated: ■, I_m; ●, I_c; ○, $I_m + I_c$. The correlation coefficient for the linear $I_m + I_c$ plot is 0.998 (27).

boxylate)s [*77124-41-5*]. The ionic groups contained in multiplets are relatively constant at 1%, whereas, with increasing ionic content, the remaining ionic groups participate in clusters. In effect, these ionomers can be described as three-phase systems composed of hydrocarbon matrix, multiplets, and clusters.

X Ray, Neutron Scattering, and Electron Microscopy. Small-angle x-ray scattering (8) and, more recently, neutron scattering (29–30) have substantially clarified the state of aggregation in ionomers. A typical x-ray diffraction pattern (31) for an ionomer based on a poly(ethylene-*co*-methacrylic acid) ionomer is shown in Figure 4, where three spectra are shown. The first describes the pattern for typical branched polyethylene, the second for a poly(ethylene-*co*-methacrylic acid) in the acid form, and the third describes the pattern achieved with the neutralized ionomer. It is evident

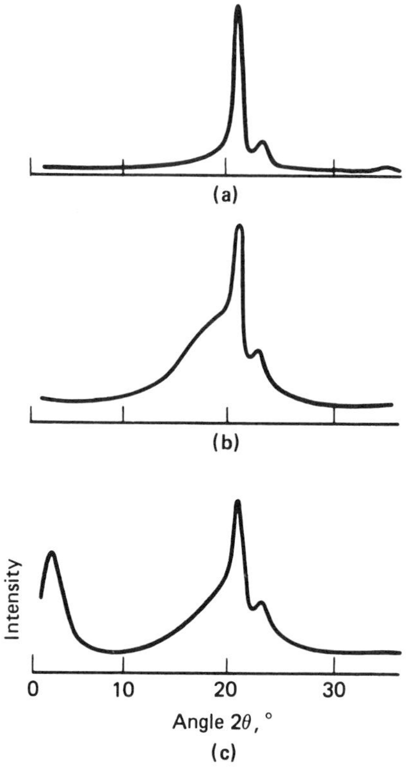

Figure 4. X-ray diffraction patterns of (**a**) branched polyethylene; (**b**) acid copolymer, E-0.06MAA; and (**c**) ionomer, E-0.06MAA—0.90Na (31).

that above a certain ion content, a low-angle peak near 4° is observed for the salts containing cations of a high electron density. This peak is independent of polymer crystallinity and is relatively insensitive to temperature. It is present even above 300°C for poly(ethylene-*co*-methacrylic acid) salts. This low-angle x-ray scattering technique has demonstrated that the ionic species aggregate to form an ion-rich phase in the hydrocarbon matrix. There have been a number of explanations (1) for the appearance of this peak in polyethylene- and polystyrene-based ionomers, but this low-angle peak reflects the existence of scattering centers in the ionomer salts which are not present in the acid form of this copolymer. Explanation for this behavior centers around the belief that microphase-separated regions contain a high concentration of ionic material.

Electron microscopy is potentially the best technique to demonstrate the presence of ionic regions in ionomer systems. A number of studies have shown that profound structural changes occur with a variety of systems upon neutralization of polymer acids (33–36). However, interpretations of the data vary. Therefore, the quantitative aspects of ionic aggregate size are somewhat ambiguous.

Infrared, Mössbauer, and Raman Spectroscopy. Infrared and far-ir spectroscopic data of ionomers are described in refs. 3, 37, and 38. Investigation of the temperature dependence of the relative intensities of the bands at 3540 and 1700 cm^{-1} evident in the acid copolymer gave the dissociation constant K for the carboxyl $\leftrightarrow$ dimer association. Based on this analysis, free acid was detectable only above 30°C.

Fourier transform ir spectroscopy was used to analyze perfluorosulfonate ionomers (39). This study is especially instructive in relating ir spectroscopy to water-solvated ionomers, a state in which these systems are specifically employed. It was observed that the —SO_3^- symmetric stretching mode depended on the degree of hydration and the nature of the cation (Li^+, Na^+, K^+, Rb^+). The peak for this band shifted to higher frequency and broadened as the water content in these film samples decreased. These data were interpreted in terms of increased interaction between the polyanion and the counterion as the water was removed, thereby reducing its shielding action. The magnitude of the frequency shifts was greater for the smaller cations, whereas for Rb^+, the frequency was substantially unchanged at low hydration. The authors provided an interpretation of the width of the hydrated peaks in terms of various states of ionic hydrate association, somewhat analogous to that of simple electrolyte solutions.

Far-ir spectroscopy has been employed as a tool to investigate domain formation in various ionomers. In reference 40, the spectra cover a range of 33–800 cm^{-1}; the metal salts are compared with the acid form of these copolymers at ambient and low temperatures.

In all cases a well-defined band, which was not present in the acid form of the copolymers, was observed for the salts. This band shifted progressively with the change of mass of the cation (450 cm^{-1} for Li^+ to 135 cm^{-1} for Cs^+) and was assigned to cation motion within the anionic field in the copolymer. Models were proposed and analyzed for the vibrational modes of the systems, and vibrational-force constants were determined.

Subsequent studies dealt with copolymers of styrene and methacrylic acid (41). Again, a broad, well-defined band appeared in the spectra for the Cs^+, Ba^{2+}, and Na^+ salts in the region below 300 cm^{-1}; this band was not present in the acid form of the copolymer. As the level of ionic functionality was increased with the Na^+ salt, a band appeared at ca 170 cm^{-1}, which was attributed to the vibrations of aggregates involving many cations and anionic sites close together (higher-order aggregate or clusters). Another band at ca 250 cm^{-1} was assigned to the vibration of an aggregate of a few ions (low-order multiplet). These studies are consistent with the postulate that at low ionic concentrations, small ion aggregates are formed, but at higher concentrations, clustering of such aggregates occurs.

Mössbauer spectroscopy has been applied to several families of ionomers (42–43), eg, to several Fe^{3+} salts of poly(butadiene-*co*-styrene-*co*-4-vinylpyridine) [*26658-72-0*] cross-linked by coordination complexes of Fe(III) and to a series of sulfonated polymers employing Fe^{3+} as counterion. For the most part, these studies revealed the presence of three types of iron species: small units or monomeric species with diameter <3.0 nm; ferric dimers; and aggregates of variable sizes with diameters >3.0 nm. A size distribution of clusters has been presented for the ferric salts of perfluorinated sulfonic acids. The sizes of clusters estimated were found to be in good agreement with those evaluated by alternative experimental means.

Recent studies on various ionomers have employed Raman spectroscopy to characterize the degree of ion aggregation (25–27). Those studies have confirmed that ionomers based on low-polarity hydrocarbon backbones could be viewed as consisting of three phases: matrix, multiplets, and clusters.

Models. The various models proposed for ionic aggregation have been reviewed

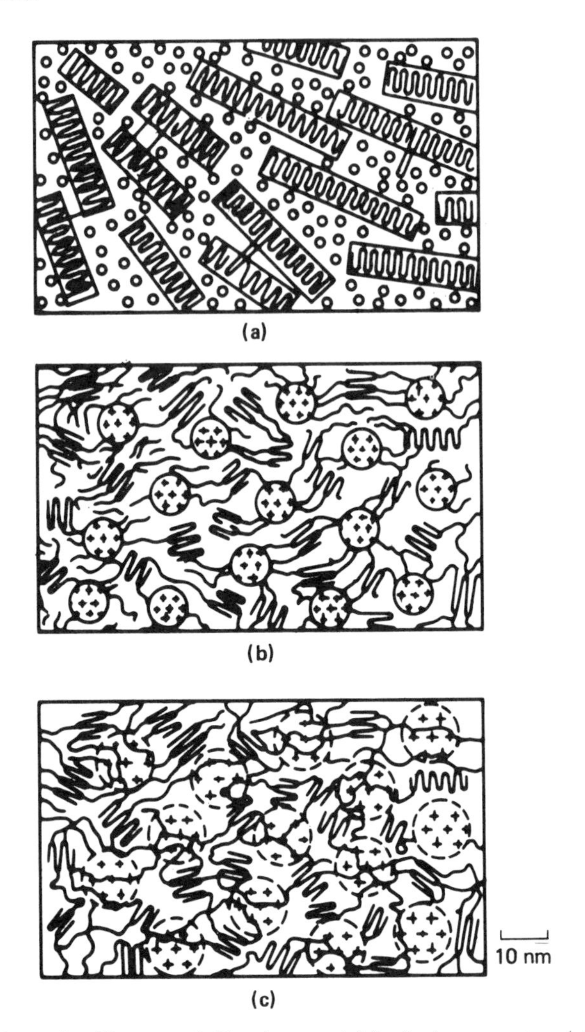

Figure 5. Schematic of Longworth-Vaughan model for ionic aggregates: (**a**) acid copolymer; (**b**) neutralized dry ionomer; (**c**) neutralized wet ionomer (46).

in some detail (1,44). The model shown in Figure 5 schematically represents the structure of poly(ethylene-*co*-methacrylic acid)s and illustrates how this structure is altered by neutralization of the acid (46). These systems are somewhat water-sensitive and, therefore, can be plasticized by the presence of water (Fig. 5(**c**)).

Another model assumes no phase separation, but rather that aggregates of salt and acid groups are homogeneously dispersed in an amorphous phase (47). In principle, this model limits ionic aggregation to a maximum of 6 or 7 ionic groups and, therefore, is inconsistent with clusterlike aggregates.

Currently, the state of the ionic aggregates in both bulk ionomers and hydrocarbon solutions is the subject of intensive research. No single model can adequately describe the wide range of ionomer structures that are now emerging. Most systems are intermediate between the homogenous aggregate and the phase-separated cluster, depending on backbone polarity and ionic functionality.

Properties

The poly(ethylene-*co*-methacrylic acid)s introduced by DuPont in 1964 were the first ionomers whose physical properties were recognized as a consequence of phase-separated ionic-rich domains. Although the DuPont products were based on methacrylic acid, poly(ethylene-*co*-acrylic acid) also behaved similarly over a range of neutralization levels (see Table 1).

Clearly, the tensile strength is enhanced with increasing neutralization, and the melt flow is markedly reduced. Thus, the presence of even monovalent cations neutralizing the acrylic acid moiety gives rise to an ionic association which increases the viscosity markedly and leads to an enhancement of physical properties as progressively more of the acid species is ionized. DuPont's patent describes a range of neutralization from ca 10% to a maximum of 90% neutralized ethylene–methacrylic acid systems (48). At higher levels of neutralization, the resulting product becomes less thermoplastic and, therefore, is not amenable to conventional molding processes.

A growing body of evidence suggests that the characteristic ionomer properties can be attributed to ionic aggregation. Furthermore, the presence of relatively small amounts of metal carboxylate or sulfonate groups pendant to a polymer chain clearly has profound effects on polymer properties. Specific properties, such as the glass transition, the rubbery modulus above the glass transition, dynamic mechanical behavior (49–50), the melt rheology (51), the relaxation behavior, dielectric properties, and solution behavior, are modified by even small amounts of ionic groups (1–2).

Cluster Formation and Physical Properties. In view of the substantial amount of work relating to the morphology of ionomers, the effect of morphology on the physical properties of these systems is of considerable importance. There is little direct ex-

Table 1. Properties of Poly(Ethylene-*co*-Acrylic Acid) Salts[a]

	Percent neutralized	Melt index at 298 kPa, 10 g/min	Melt index at ca 3 MPa, 10 g/min	Secant modulus 1% extension, MPa[b]	Ultimate tensile strength, MPa[b]	Elongation at break, %
control material, 14.8 wt % acrylic acid	0	67	2.57	48.26	14.8	470
sodium salt [*25750-82-7*]	12.0	12.2	256	230	21.7	420
	30.0	3.9	92	540	27.6	330
	47.5	1.0	30	293	31.7	310
	66.0	0.3	7.6	273	33.0	280
potassium salt [*27515-34-0*]	8.0	16.3	360	202	21.0	470
	25.0	4.5	110	362	25.5	410
	51.0	2.7	49	341	30.6	370
	63.0	0.57	15	308	34.4	390
lithium salt [*86472-12-0*]	12.0	18.7	442	181	21.7	410
	28.5	5.2	116	337	26.5	350
	52.5	1.4	38	334	28.2	260
	67.5	0.2	5.4	254	31.7	250

[a] Ref. 33.
[b] To convert MPa to psi, multiply by 145.

perimental evidence correlating multiplet and cluster formation to physical properties. There is, however, increasing indirect evidence relating selected physical properties to the nature of the ionic aggregate.

As mentioned above, the number of ionic groups existing in clusters is strongly dependent on polarity of the polymer matrix, ionic functionality, and temperature. At low ionic content, within a low polarity matrix, only multiplets are favored. However, with increasing ionic content, cluster formation is favored. With an increase in matrix polarity, a progressively higher content of ionic groups is required in order to favor cluster formation (3). As the polarity of the matrix increases, the degree of ionic functionality required for cluster formation increases substantially. The data presented in Table 2 and other data are consistent with that hypothesis but are affected by the nature of the anion and cation and other variables, such as the method by which cluster formation is detected, which may affect the ion concentration required for clustering. Thus, spectroscopic techniques may be more sensitive for detecting clusters than rheological measurements (41).

The rheological behavior of ionomers offers convincing evidence that cluster formation or microphase separation has a significant impact on properties. With poly(styrene-*co*-methacrylic acid) [*9010-92-8*] ionomers, for example, stress–relaxation data at low ionic content (<6 mol %) give satisfactory time–temperature curves. In effect, these results illustrate a low degree of cross-linking which may be attributed to multiplet formation.

Similar stress–relaxation data for samples of higher ion concentration (7.7 mol %) provided a different picture of ion aggregation (3). Attempts to apply conventional time–temperature superposition by shifting the modulus–time curves were unsuccessful. In other words, a single master curve could not be generated which would describe the viscoelastic response over the entire region of time and temperature. The critical ion concentration for a breakdown in time–temperature superposition for the ionomer based on styrene–methacrylic acid is ca 6 mol %. The current interpretation is that superposition does not apply, because at the higher ionic contents a second relaxation mechanism occurs due to a change in the structure.

A significant difference in water absorption is observed for polymers of different ion contents (3,52). For example, in the case of poly(styrene-*co*-sodium methacrylate)s of ion content below 6 mol %, each ion pair accepted ca one water molecule, whereas above 6 mol %, 3–6 molecules of water were absorbed per ion pair. Under those conditions, equilibrium was not achieved even at very long residence times. Similar results were observed for sulfonate ionomers based on a rubbery backbone (23).

Table 2. Critical Ion Concentration for Cluster Formation[a]

System	Critical ion concentration, mol %[b]
polyethylene	1
polystyrene	4–6
poly(ethyl acrylate)	10–15
polyacrylic acid with 20 wt % formamide	10–20
Na_2O–$x SiO_2$	ca 20
Na_2O–$x P_2O_5$	unclustered

[a] Ref. 3.
[b] Approximate values.

Although the concept of clustering at higher ion contents is increasingly supported, others favor a homogeneous distribution of lower-level ion-pair aggregates (3). However, x-ray evidence, rheological data, and water-absorption studies appear to support clustering or microphase separation of some sort.

The impact of clustering or some form of microphase separation upon ionomer properties is seen to be significant. An enhanced degree of water absorption appears conclusive. From a rheological viewpoint, the fact that time–temperature superposition principles do not apply is an indication that multiple-relaxation mechanisms can apply concurrently. There is some evidence that these clusters exhibit an independent broad softening range. Under certain conditions with metal sulfonate ionomers in a low polarity matrix, the elastic modulus is virtually constant up to 250°C (53). The stability of these systems can be attributed to clusters which are resistant to softening. Under such conditions, it appears that the modulus increases with ionic content, owing to a reinforcing-filler effect of the phase-separated ionic-rich region. Another consequence of cluster formation relates to the effect of added polar species. For example, the addition of crystalline zinc stearate (up to 50 wt %) to sulfo-EPDM enhances selected physical properties at ambient temperature. This has been interpreted as the result of the interaction of a phase-separated metal stearate crystal with metal sulfonate clusters to create a combined cluster. These aggregates exhibit sharp melting points and improved melt flow and physical properties (53).

Dynamic Mechanical Properties. The dynamic mechanical properties of ionomer systems provide definitive evidence that the salt forms of these systems are dramatically different from either the acid form or the parent polymers. Specifically, the neutralized forms of these materials generally display a rubbery plateau in the modulus–temperature curves that is not present in the parent materials. This feature can be seen, for example, in the sulfonated ethylene–propylene ionomers in contrast with the behavior of the poly(ethylene-*co*-propylene) from which the ionomer is derived (53). For the sulfo-EPDM systems, the elastic modulus is virtually constant up to 200°C even at very low ionic concentrations, whereas an EPDM polymer displays viscous flow above 50°C. In the latter case, the ionic groups provide a strong temperature-resistent network comparable to that obtained with covalent cross-links.

In other systems, such as polystyrene-based ionomers, dynamic mechanical behavior shows evidence of microphase separation. A single loss-tangent peak is observed for the acid or parent polymer; two such peaks of high intensity appear upon neutralization of the copolymer acid. This phenomenon has been interpreted as the result of two glass-transition temperatures: one is normally observed for the polymer matrix, ie, the T_g of polystyrene of the ionomers based on polystyrene; the other T_g is related to the phase-separated regions containing a high percentage of ionic material. It has been postulated that below a certain ion content, depending upon the polymer matrix, there are small ionic aggregates that behave as cross-links and dominate the physical-property behavior. Above such an ion concentration, large ionic aggregates act to provide phase separation in the ionomer (2).

Preparation and Manufacture

Ionomers are prepared by copolymerization of a functionalized monomer with an olefinic unsaturated monomer or direct functionalization of a preformed polymer. Typically, carboxyl-containing ionomers are obtained by direct copolymerization of acrylic or methacrylic acid with ethylene, styrene, and similar comonomers by free-radical copolymerization (48). The resulting copolymer is generally available as the free acid, which can be neutralized to the degree desired with metal hydroxides, acetates, and similar salts.

The second route to ionomers involves modification of a preformed polymer. Sulfonation of polystyrene, for example, permits the preparation of sulfonated polystyrene (S-PS) with a content of sulfonic acid groups in proportion to the amount of sulfonating agent (54). These reactions are conducted in homogeneous solutions permitting the direct neutralization of the acid functionality to the desired level. The neutralized ionomer is isolated by conventional techniques, ie, coagulation in a non-solvent, solvent flashing, etc (55–56).

The ionomers given in Table 3 are obtained by direct copolymerization of a carboxylate or sulfonate ionomer or by direct functionalization of a preformed polymer. These systems are generally based on carboxylate acids or sulfonic acid systems. The carboxylic acid copolymers are prepared by copolymerization of acrylic or methacrylic acid, whereas the sulfonated polymers are obtained by direct sulfonation of an unsaturated or aromatic preformed polymer. The preparation of ionomers based on phosphonates is described in ref. 1.

Poly(Ethylene-*co*-Methacrylic Acid) Ionomers. The copolymerization of ethylene and acrylic or methacrylic acid is effected at high pressures using a free-radical initiator (33). Typically, 3–6 mol % acid is incorporated in the commercial polymers. Despite the fact that methacrylic acid is much more reactive than ethylene, the resulting polymer chains appear to be reasonably homogeneous with respect to carboxylate distribution. In part, this is probably because the commercial processes are conducted to low conversions (ca 10%).

Although incorporation of low concentrations of methacrylic acid into the ethylene backbone has a modest effect on typical polymer properties, the neutralization of these groups can have a more dramatic influence. In the case of poly(ethylene-*co*-methacrylic acid)s, neutralization can be effectively achieved with bulk polymer on two-roll rubber mills at 150–200°C by addition of sodium hydroxide or other bases in water to the

Table 3. Commercial and Experimental Ionomers

Polymer system	Trade name, if commercial	Manufacturer	Uses
Commercial			
poly(ethylene-*co*-methacrylic acid)	Surlyn	DuPont	modified thermoplastic
poly(butadiene-*co*-acrylic acid)	Hycar	BFGoodrich	high green strength[a] elastomer
perfluorosulfonate ionomers	Nafion	DuPont	multiple membrane users
perfluorocarboxylate ionomers	Flemion	Asahi Glass	chloralkali membrane
telechelic polybutadiene	Hycar	BFGoodrich	specialty uses
sulfonated ethylene–propylene terpolymer	Ionic Elastomer[b]	Uniroyal	thermoplastic elastomer
Experimental			
poly(styrene-*co*-acrylic acid)			model ionomer system
sulfonated polystyrene			model ionomer system
sulfonated butyl elastomer			high green strength[a] elastomer
sulfonated polypentenamer			model ionomer system

[a] Green strength = gum tensile strength (prior to vulcanization).
[b] Development stage.

fluxed copolymer. As the water evaporates, the melt viscosity increases markedly, and at sufficiently high neutralization, the ionomer can be stripped off the mill to yield a tough, flexible sheet (33).

Carboxylated Elastomers. The incorporation of acrylic and methacrylic acid groups into synthetic elastomers via free-radical copolymerization has been commercially practiced for >30 yr. Typical examples of such systems include poly(styrene-*co*-butadiene-*co*-acrylic acid)s [*25085-39-6*], poly(butadiene-*co*-acrylonitrile-*co*-acrylic acid)s [*25265-19-4*], and poly(butadiene-*co*-acrylic acid)s [*25067-26-9*]. Typically <6% of carboxylic monomer is employed in order to preserve the elastomeric properties inherent in these systems. Currently, worldwide production of such systems is estimated to be above 5×10^5 metric tons per year (45).

A second class of carboxyl-containing elastomers is termed telechelic polymers. In these systems, the carboxyl functionality terminates both ends of the polymer chain. Such polymers have molecular weights of 1500–6000 and represent a specialty class of elastomers.

Most commercially available carboxylated elastomers are prepared by emulsion polymerization. Typically, the polymerizations are conducted in acidic emulsion formulations because the free acid copolymerizes much more readily than the neutral salt (45). The low solubility of the monomer salt in the hydrocarbon phase precludes significant monomer incorporation.

A typical procedure for the preparation of a carboxylate elastomer includes the following starting materials: 100 parts butadiene (or combination with other monomers), 100 parts water (deionized), 5 parts methacrylic acid, 1 part sodium alkyl aryl polyether sulfate, and 0.4 parts potassium persulfate. Polymerization is conducted at 30–50°C and can be continued to high conversion. The resulting latex preferably should approach 45–50% solids.

Alternative techniques have been described for preparing such systems including solution copolymerization and grafting procedures on preformed polymer. Owing to the economics and simplicity of the direct polymerization process and the ease of product control, it is the preferred route to elastomers containing random incorporation of carboxyl groups.

Telechelic Carboxylated Elastomers. The synthesis of low molecular weight difunctional carboxyl-terminated butadiene-based polymers has been described in several review articles (57–58). Anionic polymerization or free-radical initiated polymerization processes are usually involved. The first route offers polymers of relatively narrow molecular weight distribution. However, in the molecular weight ranges of interest (1500–6000), substantial amounts of organometallic catalyst are required. The free-radical-initiated process offers a route to polymers of broader molecular weight distribution. This process is utilized for copolymers of 1,3-butadiene and acrylonitrile [*9003-18-3*]. Free-radical initiators, which typically contain carboxyl groups, are used at 70–130°C.

$$(\mathrm{HO\overset{O}{\overset{\|}{C}}CH_2CH_2\underset{CN}{\underset{|}{\overset{CH_3}{\overset{|}{C}}}}{-}N{=})_2} \xrightarrow{\Delta} 2\ \mathrm{HOCCH_2CH_2\underset{CH_3}{\underset{|}{\overset{CH_3}{\overset{|}{C}}}}\cdot} + \mathrm{N_2} \xrightarrow{n\ \mathrm{CH_2{=}CHCH{=}CH_2}}$$

4,4′-azobis(4-cyanopentanoic acid)

$$\mathrm{HO\overset{O}{\overset{\|}{C}}CH_2CH_2\underset{CN}{\underset{|}{\overset{CH_3}{\overset{|}{C}}}}{-}\!\!\left[CH_2CH{=}CHCH_2\right]_n\!\!{-}\underset{CN}{\underset{|}{\overset{CH_3}{\overset{|}{C}}}}CH_2CH_2\overset{O}{\overset{\|}{C}}OH}$$

The selection of appropriate solvents minimizes chain transfer with solvent which has an important effect on the final polymer functionality. Typically *t*-butanol is preferred, although tetrahydrofuran and acetone have also been employed. The liquid polymers are recovered by solvent stripping to yield products with viscosities of 10–40 Pa·s (100–400 P).

These carboxyl-terminated polymers react with suitable metal alkoxides to yield neutralized telechelic polymers (59). Neutralization in toluene gives high molecular weight products; the methanol by-product is removed under vacuum.

$$n\ \mathrm{HOOC{-}P_x{-}COOH} + n\ \mathrm{M^{2+}X_2^{2-}} \longrightarrow (\mathrm{{-}OOC{-}P_x{-}COO^{-}M^{2+}})_n + 2n\ \mathrm{HX}$$

where P_x = polymer.

Sulfonated Ethylene–Propylene–Diene Terpolymers. Sulfonate groups are introduced into ethylene–propylene–diene monomer (EPDM) systems via electrophilic attack of the sulfonation reagent on the polymer unsaturation. A number of different diene termonomers, amenable to sulfonation, can be copolymerized with ethylene and propylene (23,56). A preferred starting material is 5-ethylidene-2-norbornene (ENB). Typically, EPDM based on ENB is dissolved in an aliphatic hydrocarbon, generally hexane, at a concentration of 50–100 g/L. For sulfonation at ambient temperature, acetyl sulfate is generated from acetic anhydride and concentrated sulfuric acid. After 30 min, the sulfonation is terminated by the addition of an alcohol. The resultant polymer sulfonic acid is agitated with an alcoholic solution of metal acetate. The sulfonated metal EPDM is isolated by solvent flashing in boiling water. A 95% conversion of sulfuric acid to polymer-sulfonic acid is typical.

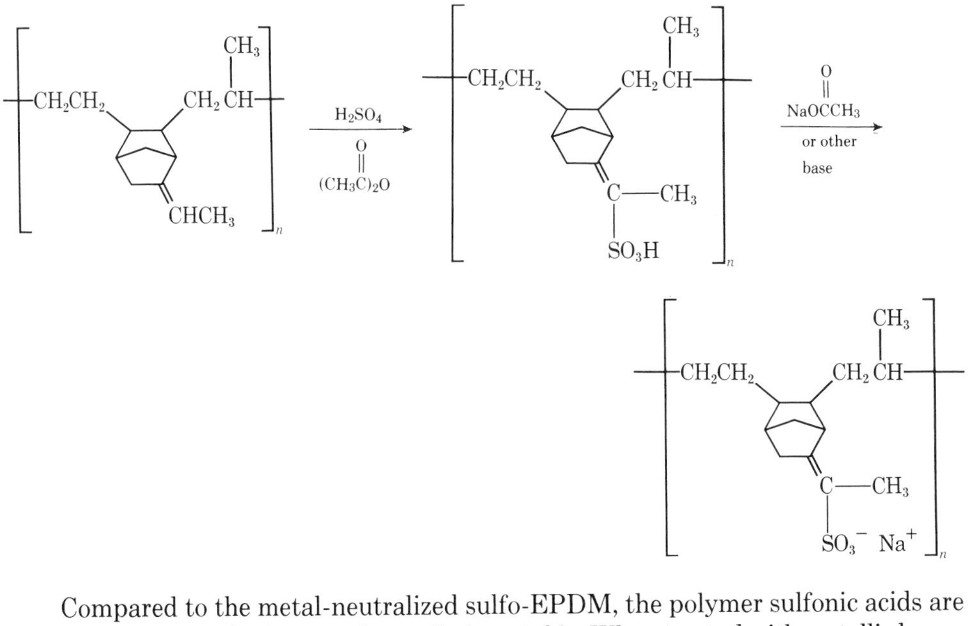

Compared to the metal-neutralized sulfo-EPDM, the polymer sulfonic acids are not highly associated. They are thermally less stable. When treated with metallic bases, bulk and solution properties change markedly. In the absence of a polar cosolvent, such as methanol, the hydrocarbon solutions of the metal salts of sulfo-EPDM are solid gels at polymer concentrations higher than several percent (60–61). With 1–5% alcohol, the polymer solutions have viscosities of 1–5 Pa·s (10–50 P).

Perfluorosulfonate Ionomers, Nafion. Perfluorosulfonate polymers were developed at E. I. du Pont de Nemours & Co., Inc. in the mid-1960s, and these materials were commercialized in the early 1970s (see Fig. 6).

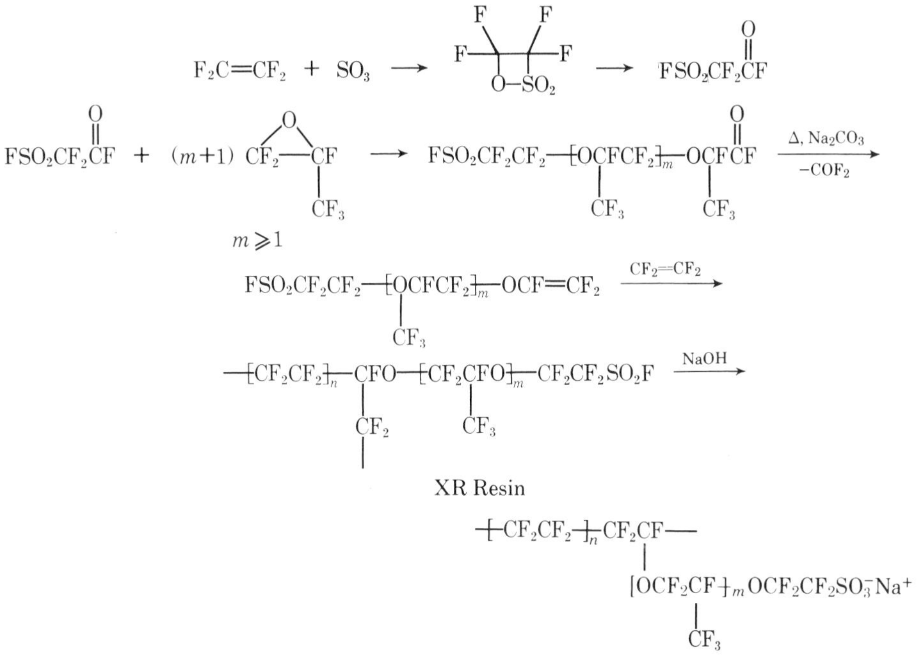

$$F_2C{=}CF_2 + SO_3 \longrightarrow \text{(cyclic F}_2\text{C–CF}_2\text{–O–SO}_2\text{)} \longrightarrow FSO_2CF_2\overset{O}{\overset{\|}{C}}F$$

$$FSO_2CF_2\overset{O}{\overset{\|}{C}}F + (m+1)\ \underset{CF_3}{CF_2\overset{O}{—}CF} \longrightarrow FSO_2CF_2CF_2{-}[OCF(CF_3)CF_2]_m{-}OCF(CF_3)\overset{O}{\overset{\|}{C}}F \xrightarrow[-COF_2]{\Delta,\ Na_2CO_3}$$

$m \geqslant 1$

$$FSO_2CF_2CF_2{-}[OCF(CF_3)CF_2]_m{-}OCF{=}CF_2 \xrightarrow{CF_2{=}CF_2}$$

$$-[CF_2CF_2]_n{-}\underset{\underset{|}{CF_2}}{CFO}{-}[CF_2CF(CF_3)O]_m{-}CF_2CF_2SO_2F \xrightarrow{NaOH}$$

XR Resin

$$-[CF_2CF_2]_n{-}CF_2CF{-}$$
$$\quad [OCF_2CF(CF_3)]_m OCF_2CF_2SO_3^- Na^+$$

Nafion polymer [*39464-59-0*]

Figure 6. Synthesis of Nafion Precursor and Nafion (21)

XR Resin is a high molecular weight polymer which can be melt fabricated and processed into sheet or tubes with standard techniques. The resin is converted into Nafion perfluorosulfonate polymer by hydrolysis.

The sodium counterions are readily exchanged for other metal ions by immersing the polymer in the appropriate aqueous electrolyte. Most commercial Nafion materials have an equivalent weight of ca 1000–1500 per sulfonate group; typically $m = 1$ and n is ca 5–11 (see Perfluorinated ionomer membranes).

Perfluorocarboxylate Ionomers, Flemion. During the mid and late 1970s, Asahi Glass in Japan announced the development of membrane technology based on a family of carboxylated perfluoropolymers. These systems are named Flemion and have the structure shown below.

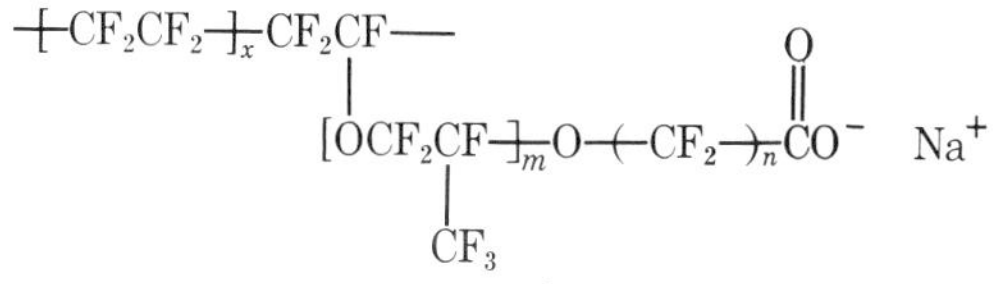

$$-[CF_2CF_2]_x{-}CF_2CF{-}$$
$$\quad [OCF_2CF(CF_3)]_m{-}O{-}(CF_2)_n{-}\overset{O}{\overset{\|}{C}}O^-\ \ Na^+$$

(m = 0 or 1, n = 1–5)

Flemion polymer [*75634-46-7*]

These high molecular weight polymers are based on the copolymerization of the appropriate functional monomers with tetrafluoroethylene and subsequent membrane fabrication. Several synthetic routes to appropriate monomers have been described (24), one of which is shown below.

$$2\ CF_2{=}CF_2 + I_2 \xrightarrow{\Delta} I(CF_2CF_2)_2I \xrightarrow{\text{oleum}} \text{(tetrafluoro lactone: } F_2C{-}CF_2{-}C(=O){-}O{-}CF_2 \text{ ring)} \xrightarrow{CH_3OH}$$

$$FC(=O)CF_2CF_2C(=O)OCH_3 \xrightarrow{+(m+1)CF_2\text{—}CFCF_3 \text{ (epoxide, O)}} FC(=O){-}[CF(CF_3)OCF_2]_{m+1}{-}CF_2CF_2C(=O)OCH_3 \xrightarrow[-COF_2]{\Delta,\ \text{base}}$$

$$CF_2{=}CFO{-}[CF_2CF(CF_3)O]_m{-}CF_2CF_2CF_2C(=O)OCH_3$$

The copolymerization of the fluorinated ester with tetrafluoroethylene can be effected by bulk, solution, or emulsion polymerization. The polymers are fabricated by appropriate extrusion techniques, and the appropriate conversion of the ester group is effected in a suitable electrolyte solution (see Perfluorinated ionomer membranes).

Plasticization

The plasticization of conventional polymers by fugitive and nonfugitive additives is well known. In the case of ionomers, plasticization can have an additional feature since either the polymer backbone or the inorganic salt groups respond to these additives in a selective manner. This is reflected in polymer properties in several ways. For example, the addition of 5 wt % glycerol as a plasticizer in S-PS reduces the melt viscosity by a factor of 1000 at elevated temperatures (54). To obtain the same degree of reduction, 40% dioctyl phthalate is required. This difference is attributed to the preferential solvation of polar glycerol for metal sulfonate groups, whereas dioctyl phthalate plasticizes the polystyrene backbone and affects the glass transition (T_g). Although glycerol has a dramatic effect on melt viscosity, it hardly affects T_g. Conversely, dioctyl phthalate lowers the T_g by 60–70°C, with little effect on melt viscosity. Thus, it is possible to convert S-PS from a rigid, high viscosity, intractable ionomer into a flexible elastomeric composition of controlled viscosity by selective plasticization of polymer backbone and of ionic groups with the appropriate agent (54).

Selective plasticization is particularly applicable in the case of sulfo-EPDM. This ionomer can be plasticized with metal carboxylates to enhance tensile properties and lower melt viscosity. Similarly, the use of paraffinic oils to solvate the poly(ethylene-*co*-propylene) backbone is similar to that described above for S-PS. A variety of thermoplastic elastomers can be created by the addition of plasticizers.

Plasticization With Water. The selective interaction of water and other polar molecules with ionic groups is of special interest for the perfluorinated ionomers such as Nafion (3,21,62). Water interacts with the sulfonic acid groups and thereby facilitates ion transport, which is a key factor in membrane applications. In the acid form, Nafion compounds are extremely hydrophilic and can absorb as much as 30 mol H_2O per sulfonic acid group, which results in doubling the membrane volume (20).

The interaction of water with other ionomers, eg, sulfo-EPDM, is similar (23). The amount of water absorption is highly dependent on the metal cation (see Fig. 7). As much as 10–70 wt % water based on polymer is absorbed even with polymers containing ca 1 mol % sulfonate groups.

Solution Behavior. The behavior of ionomers in dilute solution can be interpreted as a special case of plasticization phenomena where solvents interact to different degrees with the polymer backbone or the ionic groups.

The solution behavior of ionomers has received little attention until recently. Early investigation involved the dilute solution behavior of carboxyl-terminated polybutadienes, revealing a significant amount of end-to-end association in the acidic or partially neutralized ionomers (10). However, the insolubility of many ionomers, such as those based on ethylene–methacrylic acid or those with significant ionic content, precluded a systematic study of the solution behavior of these systems. Some recent studies with sulfonate and carboxylate ionomers, which are readily soluble, have offered additional insight.

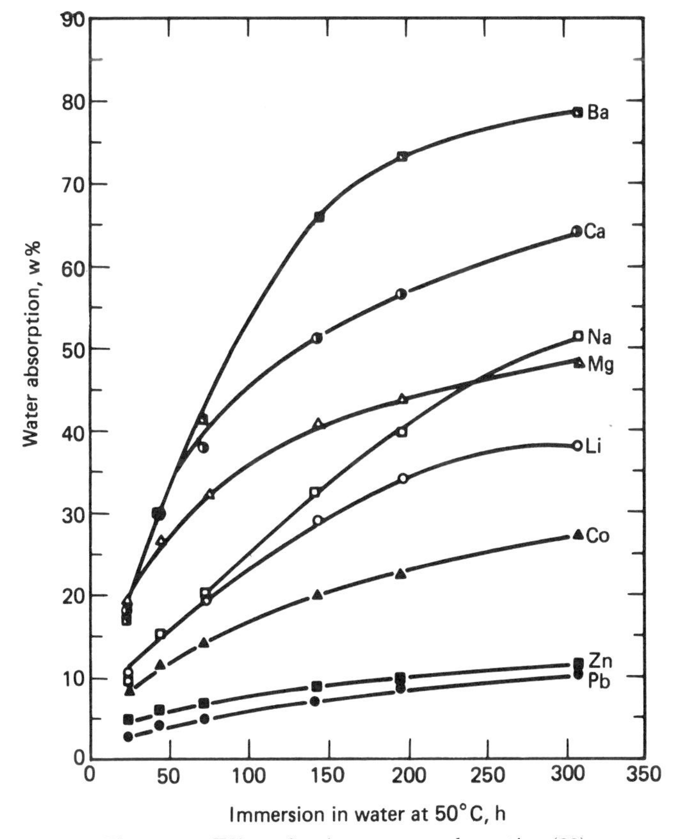

Figure 7. Effect of cation on water absorption (23).

Sulfo-EPDM and sulfonated polystyrene of low functional amounts are effective thickeners for hydrocarbon solvents. Based on dilute solution viscosity measurements, substantial intermolecular association prevails at polymer concentrations ≥0.5 wt %. In more dilute solutions, intramolecular association predominates with the consequence that the ionomers are even less effective thickeners than the unsulfonated polymers (60).

At sufficiently high metal sulfonate levels, sulfonated ionomers are insoluble in hydrocarbon diluents. Under such conditions, strong ionic association precludes solubility in low polarity solvents. However, these polymers can be solubilized by the addition of small amounts of polar cosolvents such as alcohols, amines, acids, etc. Alcohols, for example, solvate the metal cation and weaken the ionic association. This solvation has been interpreted as an equilibrium between alcohol and metal sulfonate groups and is thereby employed to control solution viscosity of oils and hydrocarbons (60–61,63–64).

Typically anionic polymers in aqueous or polar solvents display an increase in reduced viscosity owing to coil expansion as concentration is decreased. The polymer anions expand owing to a polyelectrolyte effect. This behavior is not displayed by typical ionomers in hydrocarbons with or without polar cosolvents (65). These observations, therefore, are compatible with ion-pair association dominating the solution behavior of these polymers in low polarity diluents. As a consequence of these interactions, such solutions exhibit very unusual viscosity–temperature behavior. Viscosity can remain constant or actually increase with an increase in temperature (64–65).

Elastomers

Butadiene–Methacrylic Acid Ionomers. The copolymerization of methacrylic acid with 1,3-butadiene leads to a tougher, less elastic elastomer than is obtained with polybutadiene (5–6,45). This toughness is manifested as an increase in tensile strength

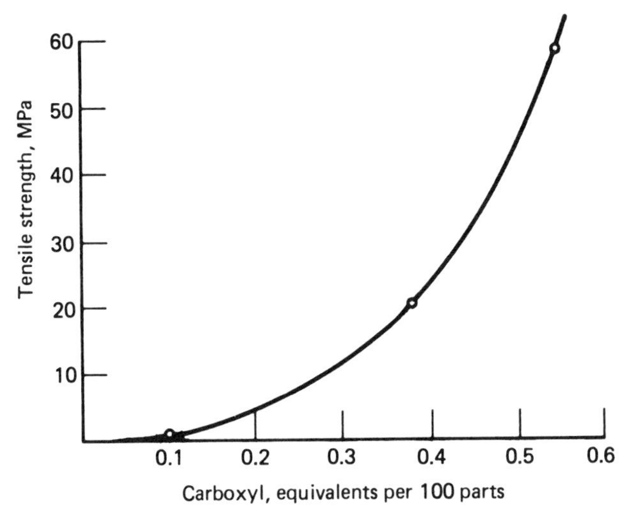

Figure 8. Ultimate tensile strength of raw methacrylic acid–1,3-butadiene copolymers (5). To convert MPa to psi, multiply by 145.

with increasing methacrylic acid content, as shown in Figure 8. If the amount of acid incorporated in the polymer exceeds 40 wt %, the rubbery properties of the material are lost. Although the acid form of such elastomers exhibits superior selected properties and adhesion, significant property changes are observed upon neutralization (or vulcanization) with suitable metal oxides. The incorporation of suitable divalent metal oxides provides rubbers of high tensile strength as compared to sulfur-vulcanized butadiene-based elastomers. The effect of zinc oxide concentration on a methacrylic acid terpolymer is shown in Figure 9. Although such metal oxides neutralize the carboxylate groups and form stable neutral salts, CO_2—Zn—O_2C, the product is a mixture of free acid and mixed salts. The resulting products exhibit poor compression set, high stress relaxation, and fluidity at elevated temperatures. Such behavior is consistent with a combination of chemically bonded cross-links and with secondary bonding owing to ionic aggregation.

Neutralization of these carboxylated rubbers with monovalent salts results in a weaker network than that obtained with divalent salts. However, this weak network is substantially destroyed above 100°C.

Evidence for this thermoplastic network is seen in Figure 10, which illustrates the neutralization of butadiene–methacrylic acid copolymers of varying acid content with lithium counterion (11). The neutralization of the acid results in an increase in the glass transition and an enhancement of the rubbery modulus at elevated temperatures. However, above 75°C viscous flow is evident. Therefore, the monovalent salts of these carboxylated rubbers exhibit a weak ionic association consistent with some degree of multiplet and cluster formation in these systems.

Sulfonated Ethylene–Propylene Terpolymers. Recent patents (56) and publications (22–23) have described a new type of sulfonated ionomer based on the sulfonation of EPDM. EPDM is a commercially available elastomer with excellent oxidation and weathering resistance owing to a saturated polymer backbone. The sulfonation of EPDM can be conducted in hydrocarbon diluents at low concentration without substantial change in the EPDM molecular weight. The polymer sulfonic acid is neutralized with a suitable base, preferably a metal acetate, to yield a fully neutralized

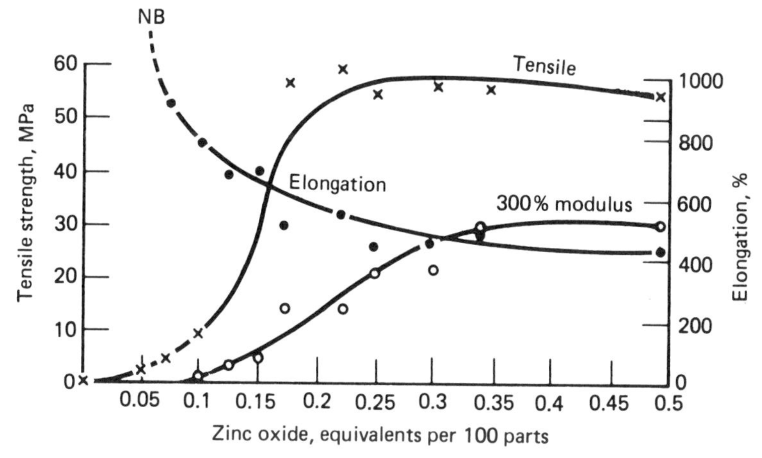

Figure 9. Effect of ZnO concentration on poly(butadiene-*co*-acrylonitrile-*co*-methacrylic acid) polymer. 55:33:10 copolymer of 73% conversion, 0.099 ephr COOH (6). To convert MPa to psi, multiply by 145.

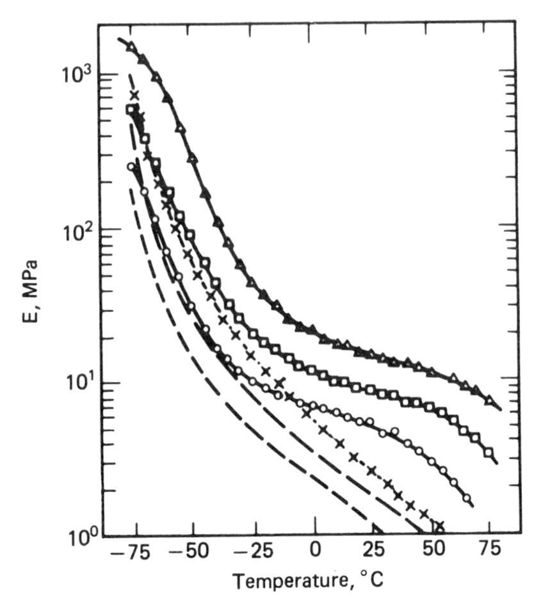

Figure 10. Modulus temperature behavior for butadiene–methacrylic acid copolymers and their lithium salts: ○, 4.7% salt; □, 7.7% salt; △, 11.6% salt; - - -, 4.7% acid; – –, 7.7% acid; and -x-, 11.6% acid (66). To convert MPa to psi, multiply by 145.

metal sulfonated EPDM. The sulfonic acid intermediate is described as thermally unstable and, therefore, is not isolated.

Neutralized sulfo-EPDMs are thermally stable, strongly associating ionomers with properties approaching those of covalently cross-linked EPDM with few metal sulfonate groups (53). Because the ionomer is obtained by direct modification of the base polymer, these systems offer an excellent opportunity to compare the ionomer properties with those of the base EPDM.

The effect of varying the sulfonate content on melt viscosity is shown in Figure 11; zinc is the metal cation. Even at zinc sulfonate contents as low as 20 meq/100 g polymer (or ca 0.6 mol %), the melt viscosity is substantially enhanced over that of the base EPDM. Similarly, the effect of sulfonate content on tensile strength is shown in Figure 12. Above ca 0.5 mol %, the tensile strength increases markedly in noncrystalline EPDM. In the presence of some crystallinity (samples E-55 and E-70), tensile strengths approach 27.6–41.4 MPa (4000–6000 psi) at ca 1 mol % ionic content.

Based on these and other data, it appears that the metal sulfonated EPDM ionomers offer an ionic network that is more persistent (ie, more temperature resistant) at a lower degree of ionic functionality than is obtained with the carboxylated elastomers described above. The effect of metal cation on the melt viscosities and physical properties of sulfo-EPDM is shown in Table 4 (23). Only zinc and lead significantly increase melt flow at 200°C; other cations, both mono- and divalent, effect very high melt viscosities. These data suggest that sulfonate ionomers based on EPDM exhibit an exceptional degree of ionic association, which is resistant to flow at low ionic contents. Nevertheless, these polymers can be dissolved in suitable mixed solvents, demonstrating that they are not covalently cross-linked.

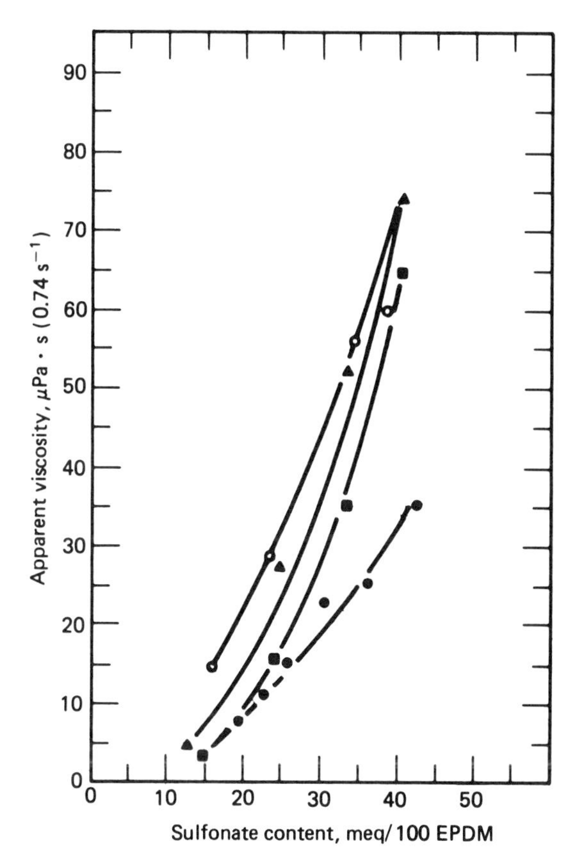

Figure 11. Effect of sulfonate content on melt viscosity at 200°C and 0.74 s^{-1} (metal cation is zinc). The different curves represent different EPDM compositions: ○, E-55; ▲, E-70; ■, CR-709-A; and ●, CR-2504 (23). To convert μPa·s to centipoise, divide by 1000.

Thermoplastics

Crystalline Ethylene–Methacrylic Acid Ionomers. Ionomers based on the polyethylene backbone exhibit a number of characteristic properties including excellent tensile properties, good clarity, and high melt viscosities (48). These properties are typically manifested at 20–80% neutralization with copolymers containing 3–10 wt % methacrylic acid. The effect of various metal cations on ionomer properties are shown in Table 5. Monovalent, divalent, and trivalent cations enhance the physical properties. The improvement in tensile properties is accompanied by a marked decrease in melt flow as shown by the melt index.

The reduction in ionomer melt flow, compared with that of the starting methacrylic acid copolymer, demonstrates that the ionic interactions persist to a substantial degree even at very high temperatures (see Fig. 13). The high melt viscosities, characteristic of ionomer melts, make such materials especially attractive in extrusion or blow molding applications. Materials of commercial interest exhibit high melt viscosities primarily at low shear stress; at high shear stress the viscosities are substantially reduced, approaching those of the acid precursor.

Another ionomer property of practical interest is the marked reduction in haze

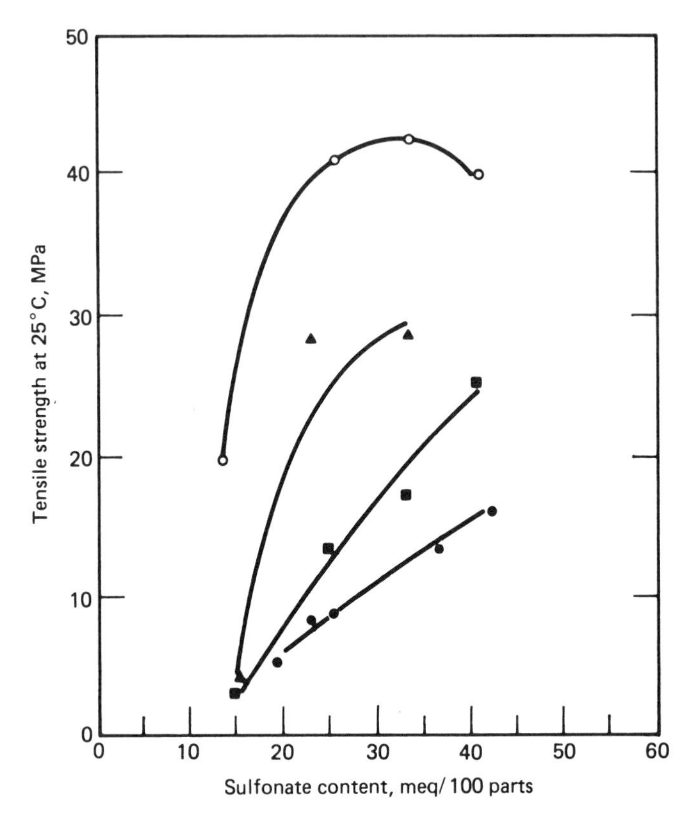

Figure 12. Effect of sulfonate content and EPDM on tensile strength at 25°C (metal cation is zinc). The different curves represent different EPDM compositions: ○, E-70; ▲, E-55; ■, CR-709-A; and ●, CR-2504 (23). To convert MPa to psi, multiply by 145.

after neutralization of ethylene–methacrylic acid copolymers (see Table 5). In the case of conventional low density polyethylene, the polymers are translucent due to spherulitic crystallinity. In the case of ionomers, the degree of crystallinity is comparable to that of conventional polyethylene. However, the presence of ionic interactions are postulated to effect two changes: nucleation of crystallites; and increase in viscosity which slows down the growth of crystallites into spherulites (33).

The physical properties of ionomers are compared with those of conventional low density polyethylene in Table 6 (67). The increase in tensile properties and improvement in clarity and other properties, eg, abrasion resistance, make this class of materials a versatile family of thermoplastics which complement the conventional polyethylenes.

Amorphous Polystyrene Ionomers. ***Carboxylated Derivatives.*** The properties of poly(styrene-*co*-methacrylic acid) containing various amounts of metal carboxylate groups are described in refs. 3, 15, and 16. Those studies clearly show that above the polystyrene glass transition, a second phase emerges above a critical ion concentration. Below ca 6 mol % ionic functionality, it is postulated that simple multiplets of ions occur, forming transient cross-links; however, for ion contents greater than 6 mol %, cluster formation results, leading to microphase separation.

In a comparison of the behavior of poly(styrene-*co*-sodium methacrylate), ie,

Table 4. Effect of Cation on Flow and Physical Properties of Sulfo-EPDM[a,b]

Metal	Apparent viscosity[c], μPa·s[d]	Melt fracture at shear rate, Hz	Melt index (190°C, 3.3 MPa[e], 10 g/min	Room temperature Tensile strength, MPa[e]	Room temperature Elongation, %
Hg			disintegrated		
Mg	55.0	<0.88	0	2.2	70
Ca	53.2	<0.88	0	2.8	90
Co	52.3	<0.88	0	8.1	290
Li	51.5	<0.88	0	5.2	320
Ba	50.8	<0.88	0	2.3	70
Na	50.6	<0.88	0	6.6	350
Pb	32.8	88	0.1	11.6	480
Zn	12.0	147	0.75	10.2	400

[a] Ref. 23.
[b] Sulfonate content: 31 meq/100 EPDM.
[c] At 200°C and 0.88 s^{-1}.
[d] To convert μPa·s to centipoise, divide by 1000.
[e] To convert MPa to psi, multiply by 145.

Table 5. Influence of Cation on Ethylene–Methacrylic Acid Ionomer Properties, 10 mol % Methacrylic Acid[a]

Property	Methacrylic acid	Na^+	Li^+	Ba^{2+}	Mg^{3+}	Zn^{2+}	Al^{3+}
anion		CH_3O^-	^-OH	^-OH	$CH_3CO_2^-$	$CH_3CO_2^-$	$CH_3CO_2^-$
cross-linking agent, wt %		4.8	2.8	9.6	8.4	12.8	14
melt index, 10 g/min	5.8	0.03	0.12	0.19	0.12	0.09	0.25
yield point, MPa[b]	6.1	13.2	13.1	13.4	15.0	13.2	7.1
elongation, %	553	330	317	370	326	313	347
ultimate tensile strength, MPa[b]	23.4	35.8	33.9	33.9	40.4	29.7	22.0
stiffness, MPa[b]	68.9	190.3	206.8	223.4	164.1	208.0	103.4
visual transparency	hazy	clear	clear	clear	clear	clear	clear

[a] Ref. 48.
[b] To convert MPa to psi, multiply by 145.

where the shear storage modulus temperature profile of this family of copolymers of varying ionic content is compared with that of polystyrene, at low ionic contents, the glass transition increases, consistent with multiplet formation. At ionic contents >6 mol %, a pronounced rubbery plateau appears. Such curves are consistent with multiphase polymer systems; in other words, the glassy polymer phase and the ionic-rich phase act independently of each other.

Sulfonated Derivative. In recent years, a number of patents and publications (54,68) have described lightly sulfonated polystyrene (S-PS), generally as the sodium salt, derived from sulfonation of polystyrene. In this process, the polymer backbone molecular weight does not change significantly. Many of the features that apply to polystyrene–carboxylate ionomers also apply to the sulfonated species, except that ionic association appears to be considerably stronger in the case of the sulfonate derivatives. The melt viscosity of fully neutralized S-PS is extremely high compared to polystyrene, the unneutralized polymer sulfonic acid, or even in comparison to an

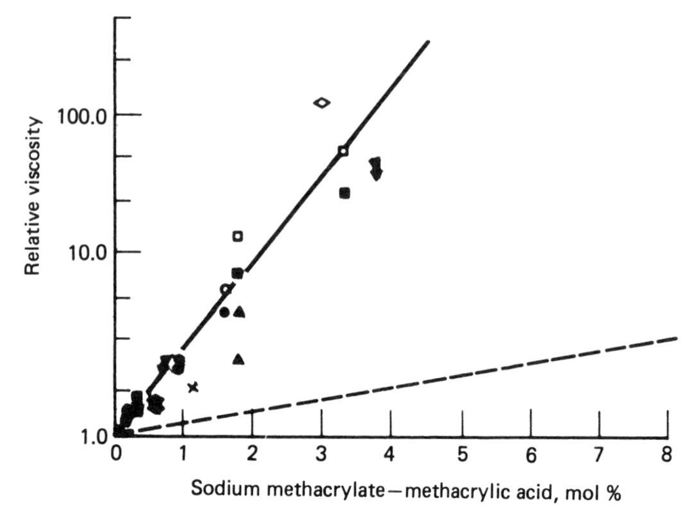

Figure 13. Increase in viscosity relative to polyethylene of ethylene–methacrylic acid ionomers (Na). Mol % MAA: ○, 1.7; ◇, 3.1; □, 3.5; ×, 5.85; ▽, 7.5; and △, 17.8. Solid symbols, 120°C; open symbols, 140°C; - - -, line for unionized copolymers (33).

Table 6. Comparison of Properties of Low Density Polyethylene and Typical Ionomers[a]

Property	Low density polyethylene	Ionomers
molding ease	excellent	excellent
injection molding temperature, °C	135–176	148–293
specific gravity	0.910–0.925	0.93–0.96
tensile strength, MPa[b]	4.1–15.8	24.1–34.4
elongation, %	90–800	350–450
impact strength, Izod test, J/m[c] per notch	no break	320–800
thermal conductivity, kW/(m·K)	33.5	24.3
heat resistance continuous, °C	82–100	71–104
clarity	transparent to opaque	transparent
transmittance, %	0–75	75–85
water absorption, 24 h, $\frac{1}{8}$ in therm.	0.01	0.1–1.4

[a] Ref. 67.
[b] To convert MPa to psi, multiply by 145.
[c] To convert J/m to ft·lbf/in., divide by 53.38.

analogous sodium carboxylated polystyrene (54). In this respect, the sulfonated polystyrenes appear analogous to sulfonated EPDMs since extremely strong, temperature resistant ionic networks are created at low (0.5–3 mol %) ionic concentrations. It is suggested that the observed sulfonate–ionomer interactions are stronger than corresponding carboxylate interactions in a low polarity polymer matrix owing to the higher degree of polarity inherent in the sulfonate system (55) and a consequent increase in the strength of the ion-pair associations.

Health and Safety Factors

Specific information relating to toxicity of specific ionomers can be obtained from the manufacturers.

Ethylene–methacrylic acid based ionomers (Surlyn) produce no hazardous by-products upon heating to normal processing temperatures (160–300°C). The combustion products are carbon dioxide and water. In oxygen-poor combustion environments, thermal cracking can produce carbon monoxide, unsaturated hydrocarbons, and trace amounts of acrolein (69). Typical ionomers based on ethylene and methacrylic acid comply with FDA regulations permitting their use in contact with foods.

The safety aspects of the perfluorosulfonated product, Nafion, are discussed in some detail in DuPont bulletins (70). No skin irritation occurs in normal handling, non-garment, commercial uses. Ingestion of sulfonic acid polymers by young adult rats shows LD_{50} >20,000 mg/kg of body weight, suggestive of very low acute toxicity. Exposure to the thermal decomposition products of Nafion may cause a temporary fluelike condition (polymer fume fever). This condition can occur following exposure to vapors evolved on heating of polymer to temperatures above 250°C.

Uses

Most ionomer applications exploit several key characteristics which can be attributed to ionic aggregation or cluster formation or the interaction of polar groups with ionic aggregates. Changes in physical properties caused by ionic aggregation in elastomeric systems or in polymer melts are most readily detected. Therefore, the marked enhancement in elastomeric green strength is a general characteristic of ionomer-based systems and provides a new class of thermoplastic elastomers. The ionic aggregation is also apparent in the enhanced melt viscosity. In the case of polyethylene-based metal carboxylate ionomers, the high melt viscosity is utilized in heat sealing. It also provides a particular processing advantage during extrusion operations. Under some conditions, however, high melt viscosity is a limitation, eg, in injection molding. Other properties attributable to ionic aggregation include toughness and outstanding abrasion resistance, as well as oil resistance in packaging applications.

The interaction of various polar agents with the ionic groups and the ensuing property changes are unique to ionomer systems. This plasticization process is especially important in membrane applications. The use of perfluorinated ionomers in a variety of separation processes depends on the specific solvation characteristics of these materials as well as their capability to absorb substantial volumes of an aqueous phase.

A different application of ionic cluster plasticization involves the interaction of metal stearates with sulfo-EPDM to induce softening transitions. This plasticization process is required to achieve the processability of thermoplastic elastomers based on this technology.

Surlyn. The outstanding characteristics of the ethylene–methacrylic acid-based ionomers (Surlyn, zinc or sodium salts) are toughness, clarity, adhesion, and melt strength during processability. As a result of these properties, these materials are finding broad application in packaging films, including composite structures employed as heat-seal layers, vacuum packaging for processed meats, and skin packaging for electronic and hardware items (see also Film and sheeting materials; Packaging materials). The high melt strength of these ionomers provides good extrusion performance for paper and foil coatings of multiwall bags. The toughness and abrasion resistance of Surlyn have led to widespread acceptance in sporting goods as diverse as golf-ball covers, roller-skate wheels, and bowling-pin coatings. Similar properties are useful in shoe soles, ski boots, and foam-molded applications. Automotive applications depend on the impact toughness and paintability of ionomers and include bumper pads and similar exterior-trim applications (69).

Sulfo-EPDM. Ionic Elastomer is a metal sulfonated EPDM available in powdered form (71). It can be compounded with fillers, rubber-processing oils and selected polymers into a variety of elastomeric materials. Initially, two grades of Ionic Elastomers are available which can be formulated into a wide variety of compounds of interest in rubber applications. These include adhesives, impact modifiers, footwear applications such as unit soles, calendered sheet, garden hose, and similar rubber goods (72).

Ionic Elastomer differs from conventional vulcanized rubbers by the presence of metal sulfonate groups which provide a strong cross-link at ambient temperatures. However, when a suitable polar additive, termed ionolyzer, such as zinc stearate, is incorporated, the elastomer becomes thermoplastic at elevated temperature permitting melt processing. Thus, by combining materials such as zinc stearate along with other formulation ingredients, a wide variety of products can be made based on several ionic elastomer gums. Typical physical property ranges of ionic elastomer compounds are shown in Table 7.

Perfluorinated Ionomers. The perfluorosulfonate ionomers marketed by DuPont as Nafion products exhibit outstanding chemical and thermal stability. These characteristics, combined with the capacity to absorb substantial amounts of water, have led to a number of membrane applications for these systems such as films and tubing (see also Perfluorinated ionomer membranes). Specific applications for these membranes are based on the electrolytic production of chlorine and caustic (see also Alkali and chlorine products). They are also actively employed as membranes in inorganic and organic electrochemical processes such as fuel cells, electrodialysis, spent acid regeneration, and selectively permeable separations in chemical processing. In addition, Nafion is interesting as an ion-selective membrane to reduce metal ions in aqueous waste. Thus, reductions of undesirable metal ions to limits below those established by the EPA is readily achieved. In addition, changes in conditions permit the concentration of valuable metals from dilute solutions to permit their recovery (70).

A number of metals-removal processes have been proposed, including removal of zinc from textile wastes, softening of brackish water, recovery of copper from industrial wastes and low grade ore, and similar selective recovery processes.

Table 7. Typical Property Range of Ionic Elastomer Compounds[a]

Property	Typical range
Shore A hardness	45–90
100% modulus, MPa[b]	1.17–6.9
tensile strength, MPa[b]	3.4–17.2
elongation, %	350–900
tear strength, MPa[b]	0.89–2.3
specific gravity at ambient temperature	0.95–1.95
compression set, %	30–35
brittle point, °C	−57 to −46
processing temperature, °C	93–260

[a] Ref. 71.
[b] To convert MPa to psi, multiply by 145.

Recently, Asahi Glass has announced the availability of perfluorocarboxylate ionomers for membrane processes such as chlor-alkali processes (24). The water absorption of these membranes at a given ion-exchange capacity is lower than that of Nafion membranes. As a consequence, a higher fixed ion concentration is achieved within the membrane. This condition provides a high current efficiency for electrolysis and thereby a strong caustic solution.

Styrene Ionomers. Several patents have described the unique high temperature properties inherent in rigid foams based on metal sulfonated polystyrene (61). Low content (2–5 mol %) of ionic groups provides improved dimensional stability at high temperatures and improved solvent resistance.

Sulfonated polystyrenes are rigid plastics with softening points above 100°C. However, a flexible, elastomeric product is prepared by the addition of 50–100 parts of the appropriate low polarity, nonvolatile plasticizer. The tensile properties and behavior of these systems are similar to those of plasticized poly(vinyl chloride) (54). When polystyrene itself is plasticized in a similar manner, a viscous liquid of no significant strength is obtained.

Recent Developments. A new family of ionomers is based on polypentenamer (1). Phosphonate, sulfonate, and carboxylate derivatives have been prepared over a broad range of ionic contents.

$$-\!\!\left[CH_2CH_2CH_2CH{=}CH\right]_x\!\!\left[CH_2CH_2CH_2CH{=}\underset{\displaystyle SO_3Na}{\underset{|}{C}}\right]_y\!\!-$$

sodium salt of sulfonated poly pentenamer

Recent studies describe the physical properties and solution behavior of model ionomers with terminal functionality (59). Typically, they are low molecular weight polybutadienes terminated by carboxyl groups and neutralized by various cations. A broad range of systematic modifications can be applied to these materials. A key to the synthesis of these systems was the observation that apparently very high molecular weights could be achieved by complete removal of the water or alcohol evolved during preparation. Viscoelastic, small-angle x-ray and dynamic-mechanical studies have been conducted on these polymers, and the results are consistent with a phase-separated ionic morphology.

A novel approach to ionomers involves the chemical modification of asphaltic bitumen. Treatment of asphalt with maleic anhydride or sulfur trioxide trimethylamine complexes gives a chemically modified product. Reaction with suitable oxides or bases yields asphalt ionomers. The resulting products give composites when mixed with aggregate fillers that retain a very high fraction of their strength when wet. In contrast, unmodified asphalt loses most of its strength upon exposure to water. These asphalt ionomers have been suggested as road-paving materials (73).

Nomenclature

ENB = 5-ethylidene-2-norborane
EPDM = ethylene–propylene–diene monomer
ephr = equivalents per hundred parts
MAA = methacrylic acid
NaMA = sodium methacrylate

S-MAA = sulfonated methacrylic acid
S-PS = sulfonated polystyrene

BIBLIOGRAPHY

1. W. J. MacKnight and T. R. Earnest, *J. Macromol. Rev.* **16,** 41 (1981).
2. C. G. Bazuin and A. Eisenberg, *Ind. Eng. Chem. Prod. Res. Dev.* **20,** 271 (1981).
3. A. Eisenberg and M. King, *Ion-Containing Polymers*, Academic Press, Inc., New York, 1977.
4. L. Holliday, ed., *Ionic Polymers*, Halsted Press, a division of John Wiley & Sons, Inc., New York, 1975.
5. H. P. Brown, *Rubber Chem. Technol.* **35,** 1347 (1957).
6. *Ibid.*, **36,** 931 (1963).
7. R. R. Warner, *Rubber Age* **71**(205), 2 (1952).
8. B. W. Delf and W. J. MacKnight, *Macromolecules* **2,** 309 (1969).
9. W. J. MacKnight, *Polym. Prepr. Am. Chem. Soc., Div. Polym. Chem.* **11,** 504 (1970).
10. E. P. Otocka, *J. Macromol. Sci.* **C5,** 275 (1971).
11. E. P. Otocka and F. R. Eirich, *J. Polym. Sci. Pt. A-2* **6,** 921 (1968).
12. E. P. Otocka, M. Y. Hellman, and L. L. Blyler, *J. Appl. Phys.* **40,** 4221 (1969).
13. A. Eisenberg, *Macromolecules* **3,** 147 (1970).
14. A. Eisenberg, ed., *J. Polym. Sci. Polym. Symp.*, 45 (1974).
15. A. Eisenberg and M. Navratil, *Macromolecules* **6,** 604 (1973).
16. *Ibid.*, **7,** 90 (1974).
17. R. W. Rees and D. J. Vaughan, *Polym. Prepr. Am. Chem. Soc. Div. Polym. Chem.* **6,** 287 (1965).
18. *Ibid.*, 296 (1965).
19. T. C. Ward and A. V. J. Tobolsky, *Appl. Polym. Sci.* **11,** 2403 (1967).
20. T. D. Gierke in *Perfluorinated Ionomer Membranes*, ACS Symposium Series, No. 180, 1982.
21. H. L. Yeager and A. Eisenberg in ref. 20, pp. 386–388.
22. H. S. Makowski and R. D. Lundberg, *Adv. Chem. Ser.* (187), 37 (1980).
23. H. S. Makowski, R. D. Lundberg, L. Westerman, and J. Bock, *Adv. Chem. Ser.* (187), 3 (1980).
24. H. Ukihashi and M. Hamabe in ref. 20, Chapt. 17, p. 427.
25. A. Neppel, I. S. Butler, and A. Eisenberg, *Macromolecules* **12,** 948 (1979).
26. A. Neppel, I. S. Butler, and A. Eisenberg, *J. Polym. Sci. Polym. Phys.* **17,** 2145 (1979).
27. A. Neppel, I. S. Butler, N. Brockman, and A. Eisenberg, *J. Macromol. Sci.* **B19,** 61 (1980).
28. A. Eisenberg, H. Matsuura, and T. Tsutsui, *J. Polym. Sci. Polym. Phys.* **18,** 479 (1980).
29. T. R. Earnest, Jr., J. S. Higgins, and W. J. MacKnight, *Polym. Prepr. Am. Chem. Soc. Div. Polym. Chem.* **21**(1), 179 (1980).
30. M. Pineri, R. Duplessix, and F. Jolino in ref. 20, Chapt. 12, p. 249.
31. F. C. Wilson, R. Longworth, and D. J. Vaughan, *Polym. Prepr. Am. Chem. Soc. Div. Polym. Chem.* **9,** 505 (1968).
32. E. J. Roche, R. S. Stein, and W. J. MacKnight, *J. Polym. Sci. Polym. Phys.* **18,** 1035 (1980).
33. R. Longworth in ref. 4, Chapt. 2.
34. M. Pineri, R. Dupplessix, S. Gauthier, and A. Eisenberg, *Adv. Chem. Ser.* (187), 283 (1980).
35. P. J. Phillips, *J. Polym. Sci. Polym. Lett. Ed.* **10,** 443 (1972).
36. E. L. Thomas, *Bull. Am. Phys. Soc.* **25,** 350 (1980).
37. W. J. MacKnight, L. W. McKenna, B. E. Read, and R. S. Stein, *J. Phys. Chem.* **72,** 1172 (1968).
38. E. P. Otocka and T. K. Kwei, *Macromolecules* **1,** 244 (1968).
39. S. R. Lowry and K. A. Mauritz, *J. Am. Chem. Soc.* **102,** 4665 (1980).
40. A. T. Tsatsas, J. W. Reed, and W. M. Risen, *J. Chem. Phys.* **55,** 3260 (1971).
41. G. B. Rouse, W. M. Risen, A. T. Tsatsas, and A. Eisenberg, *J. Polym. Sci.* **17,** 81 (1979).
42. M. Pineri, C. T. Meyer, A. M. Levelut, and M. J. Lambert, *J. Polym. Sci. Polym. Phys.* **12,** 115 (1974).
43. C. Heitner-Wirguin, E. R. Bauminger, A. Levy, F. Labinsky de Kanter, and S. Ofer, *Polymer* **21,** 1327 (1980).
44. W. J. MacKnight, W. P. Taggart, and R. S. Stein, *J. Polym. Sci. Polym. Symp.* **45,** 113 (1974).
45. D. K. Jenkins and E. W. Duck in ref. 4, Chapt. 3.
46. R. Longworth and D. J. Vaughan, *Nature (London)* **218,** 85 (1968); *Polym. Prepr. Am. Chem. Soc. Div. Polym. Chem.* **9,** 525 (1968).

47. C. L. Marx, D. F. Caulfield, and S. L. Cooper, *Macromolecules* **6,** 344 (1973).
48. U.S. Pat. 3,322,734 (Aug. 2, 1966), R. W. Rees (to E. I. du Pont de Nemours & Co., Inc.).
49. T. R. Earnest, Jr., J. S. Higgins, and W. J. MacKnight, *Polym. Phys.* **16,** 143 (1978).
50. K. Sakamoto, W. J. MacKnight, and R. S. Porter, *J. Polym. Sci. Pt. A-2* **8,** 277 (1970).
51. E. Shohamy and A. Eisenberg, *J. Polym. Sci. Polym. Phys.* **14,** 1211 (1976).
52. H. Matsuura and A. Eisenberg, *J. Polym. Sci. Polym. Phys.* **14,** 773 (1976).
53. P. K. Agarwal, H. S. Makowski, and R. D. Lundberg, *Macromolecules* **13,** 1679 (1980).
54. R. D. Lundberg, H. S. Makowski, and L. Westerman, *Adv. Chem. Ser.* (187), 67 (1980).
55. R. D. Lundberg and H. S. Makowski, *Adv. Chem. Ser.* (187), 21 (1980).
56. U.S. Pat. 4,221,712 (1980), H. S. Makowski, J. Bock, and R. D. Lundberg (to Exxon Res. and Eng. Co.).
57. S. E. Reed, *J. Polym. Sci. Pt. A-1* **9,** 2147 (1971).
58. D. N. Schulz, J. C. Sanda, and B. G. Willoughby, *ACS Symposium Series 16.6*, 1981, Chapt. 27, p. 427.
59. G. Broze, R. Jerome, and Ph. Teyssie, *J. Polym. Sci. Polym. Lett.* **19,** 415 (1981).
60. R. D. Lundberg, *J. Appl. Polym. Sci.* **12,** 4623 (1982).
61. U.S. Pat. 3,867,319 (1980), R. D. Lundberg (to Exxon Res. and Eng. Co.); U.S. Pat. 3,947,387 (1976), R. D. Lundberg (to Exxon Res. and Eng. Co.).
62. R. Duplessix, M. Escoubez, B. Rodmacq, F. Volino, E. Roche, A. Eisenberg, and M. Pineri, *Polym. Prepr. Am. Chem. Soc. Div. Polym. Chem.* **20**(2), 670 (1979).
63. R. D. Lundberg, *Polym. Prepr. Am. Chem. Soc. Div. Polym. Chem.* **19**(1), 455 (1978).
64. R. D. Lundberg and H. S. Makowski, *J. Polym. Sci. Polym. Phys. Ed.* **18,** 1821 (1980).
65. R. D. Lundberg and R. R. Phillips, *J. Polym. Sci. Polym. Phys.* **20,** 1143 (1982).
66. M. Navratil and A. Eisenberg, *Macromolecules* **7,** 84 (1974).
67. *Handbook of Plastics and Elastomers*, McGraw-Hill, Inc., New York, 1975, pp. 3–25, 29.
68. U.S. Pat. 3,870,841 (1975), H. S. Makowski, R. D. Lundberg, and G. S. Singhal.
69. *Modern Plastic Encyclopedia*, 1981–1982, p. 27.
70. *Nafion: Safety in Handling and Use*, DuPont Company Bulletin E24084, Wilmington, Del.
71. *Ionic Elastomer*, Uniroyal Technical Information Bulletin, 1982.
72. B. M. Walker, ed., *Handbook of Thermoplastic Elastomers*, Van Nostrand-Reinhold, New York, 1979.
73. L. Ciplijauskas, M. R. Piggott, and R. T. Woodhams, *Adv. Chem. Ser.* (187), 171 (1980).

General Reference

A. Eisenberg, ed., *Ions in Polymers, Advances in Chemistry Series*, ACS Monograph 187, 1980.

ROBERT D. LUNDBERG
Exxon Research and Engineering Company

O

OPIOIDS, ENDOGENOUS

"Opiates in our heads" was a common newspaper headline recently which made the names of the endogenous opioids household words. The actual discovery of the endogenous opioid peptides culminated decades of research into opioid analgesia (see Analgesics). The first half of this century saw a great deal of synthesis, based on the structure of morphine, aimed at discovering the ideal analgetic. The growing body of opioid analgetic SAR (structure–activity relationships), evidence of stereospecificity of action, and reversal by the narcotic antagonists led to speculation on the nature of receptor-based structural requirements for analgesia (1–2). A conceptual model that accounts for most (but not all) of the structural alterations is shown in Figure 1.

This model is presented primarily as a way to rationalize most opioid SAR, whereas other, more refined models take into account multiple modes of interaction with the opioid receptors (3), the nature and structure of antagonist interactions (4–7), and different aromatic recognition sites on opioid receptors (8).

The concept of the opioid receptor remained hypothetical until 1973, the year the modern era of opioid research was considered to have begun. In 1973, the presence of stereospecific opioid binding in rat brain was independently identified in three separate laboratories (9–11). These demonstrations relied on opioid stereospecificity (eg, levorphanol [*77-07-6*] and its inactive enantiomer, dextrorphan [*125-73-5*], differ by four orders of magnitude in their ability to displace ^{3}H-ligand) and on the use of ligands with high specific radioactivity ($37–148 \times 10^{10}$ Bq/mmol (10–40 Ci/mmol)). Pioneering attempts to demonstrate specificity met with only marginal success, largely because the researchers were limited to high ligand concentrations resulting from the low specific activity of opioid ligands available at that time (12–13).

The presence of opioid receptors in all vertebrates and the obvious absence of a phylogenetic relationship between vertebrates and the poppy suggested an important physiological role for the opiate receptor. It also suggested the presence of an as-yet-unidentified endogenous ligand. The existence of such a ligand received considerable support from workers who were able to demonstrate that electrical stimulation of specific brain sites in the rat produces profound analgesia (14–15), which is naloxone-[*465-65-6*] reversible (16), and is subject to tolerance development and to cross-tolerance to morphine (17). These results are most readily explained by the electrically induced release of an endogenous substance with morphinelike properties. Further support for the existence of an endogenous opioid was the close correlation found between those brain areas most sensitive to stimulation-produced analgesia and regions containing a high density of opioid receptors.

Evidence emerged from three laboratories that these endogenous opioids were peptides rather than simple morphinelike molecules (18–21). The first direct evidence for the existence of such a morphinelike substance in extracts from pig brain was provided late in 1976 in a report that the endogenous ligand was not just one peptide, but two peptides (20), each with five amino acids and differing only in the carboxyl terminal amino acid (Fig. 2). The peptides were called methionine- (Met-) and leucine- (Leu-) enkephalins, the term enkephalin being coined from the Greek word meaning "in the head." Shortly thereafter, these findings were confirmed, but with one important difference (21). Whereas Hughes and Kosterlitz (20) had found four times more Met- than Leu-enkephalin in porcine brain, Simantov and Snyder found that the ratio was reversed in calf brain (22). The common presence of a tyrosine residue in both the enkephalins and in morphine quickly underscored the relevance of these findings to opioid SAR.

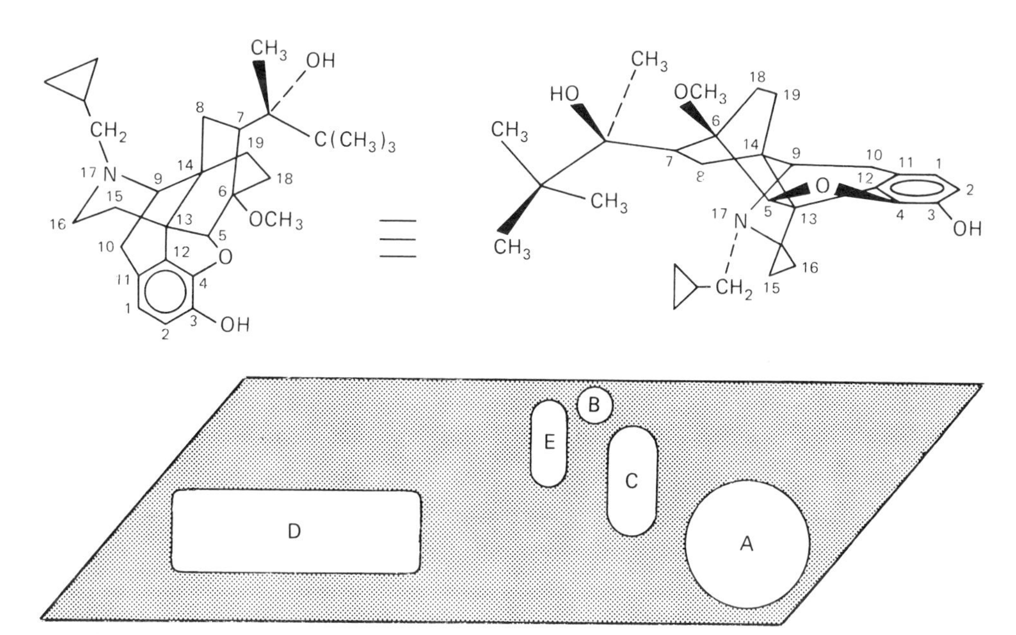

Figure 1. Conceptual model of the opioid receptor. A, flat surface for aromatic ring; B, anionic site; C, cavity for *C*-15 and *C*-16; D, lipophilic site; and E, antagonist site.

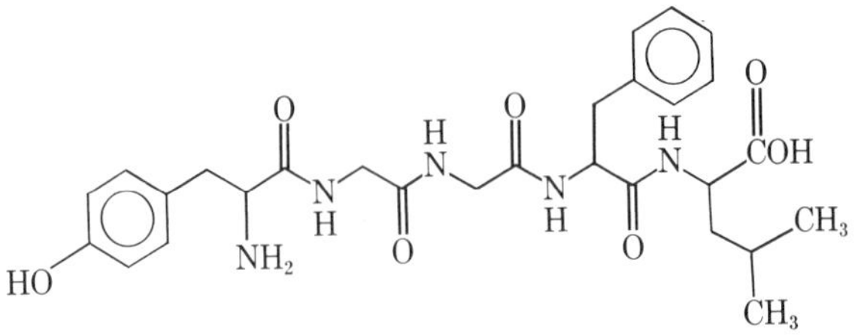

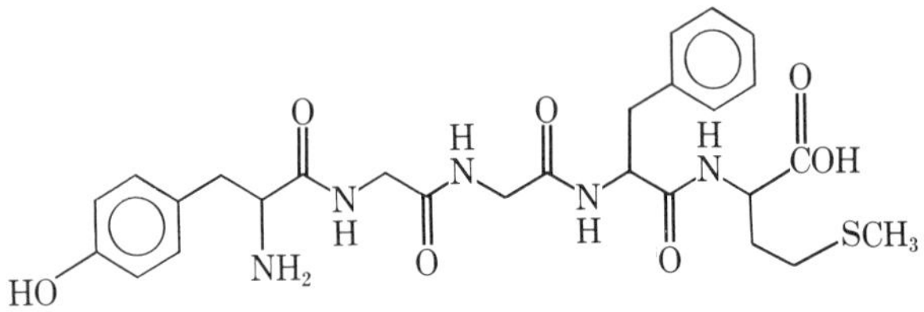

Leu-enkephalin

Met-enkephalin

Figure 2. Structures of leucine and methionine enkephalin.

Nomenclature and Origin of the Opioid Peptides

The term endorphin (a contraction of endo and morphine) as a general term for endogenous opioids was adopted. The term enkephalin is retained as a specific pentapeptide subset.

At the time of the discovery of Met-enkephalin [*58589-55-4*], the sequence of Met-enkephalin was observed to be identical to that of residues of 61–65 contained in the pituitary *C*-fragment hormone β-lipotropin (β-LPH_{61-65}), first isolated by C. H. Li in 1964 (23) (see Hormones, anterior-pituitary). Subsequent studies have shown that indeed the *C*-fragment possesses potent opioid activity (24–25).

Recently, another very potent opioid peptide in the pituitary was discovered (26). This 17-amino-acid peptide was named dynorphin and the first five amino acids ($dynorphin_{1-5}$) are identical to the Leu-enkephalin [*58822-25-6*] sequence.

The remarkable structural homology between β-LPH, β-endorphin [*60617-12-1*], and Met-enkephalin suggested that Met-enkephalin might be formed by the proteolytic cleavage of β-endorphin or β-LPH. Likewise, it was generally thought that dynorphin [*72957-38-1*] and neoendorphins [α-*77739-20-9*, β-*77752-00-2*] [two other peptides with properties and structures similar to dynorphin (27–28)] were precursors for Leu-enkephalin. Since the anatomical distribution of these precursors is different from that of Met- and Leu-enkephalin (29), other explanations for their similarities were sought. The precursor relationship of the various opioid peptides has been somewhat clarified by recent work using cloned (c)-DNA techniques. Three distinct precursors, all with molecular weights around 28,000, have been characterized. The structures of the opioid peptides arising from these precursors is shown in Table 1. Group I contains β-endorphin and its fragments and arises from pro-opiomelanocortin [*66796-54-1*] which is found predominately in the pituitary. The Group II enkephalins come from adrenal pro-enkephalin [*72877-67-9*]. The last and most recently discovered entry comes from pro-dynorphin and contains dynorphin and neoendorphin (30).

Table 1. Structure of Opioid Peptides

Group I. Pro-opiomelanocortin-derived	
α-endorphin	H–TyrGlyGlyPheMetThrSerGluLysSerGlnThrProLeuValThr–OH
γ-endorphin	H–TyrGlyGlyPheMetThrSerGluLysSerGlnThrProLeuValThrLeu–OH
β-endorphin	H–TyrGlyGlyPheMet ThrSerGluLysSerGlnThrProLeuValThr LeuPheLysAsnAlaIleValLysAsnAlaHisLysLysGlyGln–OH
Group II. Pro-enkephalin-derived	
Leu-enkephalin	H–TryGlyGlyPheLeu–OH
Met-enkephalin	H–TyrGlyGlyPheMet–OH
octapeptide	H–TyrGlyGlyPheMetArgGlyLeu–OH
heptapeptide [*73024-95-0*]	H–TyrGlyGlyPheMetArgPhe–OH
Group III. Pro-dynorphin-derived	
dynorphin (1–17)	H–TyrGlyGlyPheLeuArgArgIleArgProLysLeuLysTrpAspAsnGln–OH
dynorphin (1–8)	H–TyrGlyGlyPheLeuArgArgIle–OH
α-neo-endorphin	H–TyrGlyGlyPheLeuArgLysTyrProLys–OH
β-neo-endorphin	H–TyrGlyGlyPheLeuArgLysTyrPro–OH

Opioid Receptor

Concept of Multiple Receptors and A Simple Analgesic Model. The concept of multiple receptors for hormones and neurotransmitters is well documented. For example, the classical neurotransmitter acetylcholine is believed to interact at two distinct cholinergic receptors, the nicotinic and muscarinic sites, producing very different pharmacological effects (see Neuroregulators). Other chemical transmitters such as epinephrine, norepinephrine, histamine, dopamine, serotonin, and adenosine also act through multiple-receptor sites. The concept of multiple opioid receptors was first postulated in 1976 (31). Three distinct opioid receptors were proposed, termed μ, κ, and σ. A fourth type of opioid receptor, the δ receptor, was later postulated in 1977 (32) after discovery of the endogenous opioids, enkephalins, and endorphins. The prototype agonists (Fig. 3) for these receptors are morphine (μ) [*16206-77-2*], ketazocine (κ) [*36292-69-0*], *N*-allylnormetazocine (SKF-10,047) (σ) [*14198-28-8*], and D-Ala²-D-Leu⁵-enkephalin (δ) [*63631-40-3*]. The pharmacological effects associated with agonists interacting at the μ, κ, and σ receptors can be generalized as shown in Table 2. Leu- and Met-enkephalin are believed to interact selectively at the δ receptor, whereas β-endorphin is thought to interact at both the μ and δ receptors. A fifth receptor type, the ϵ receptor, found in rat vas deferens, may also be a binding site for β-endorphin (33).

Evidence has recently begun to accumulate indicating that the σ receptor is the

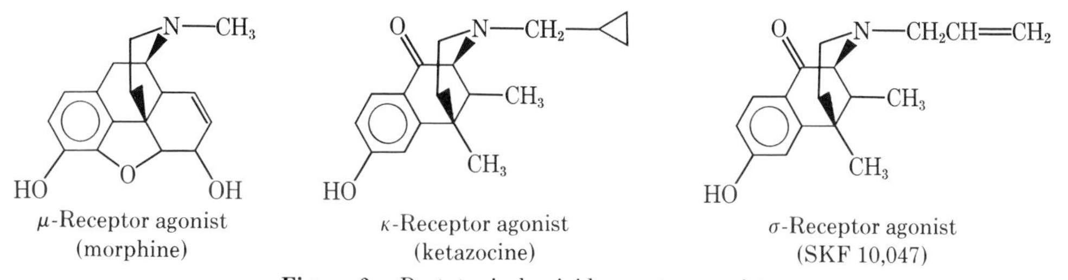

Figure 3. Prototypical opioid-receptor agonists.

Table 2. Characteristic Pharmacological Actions of μ, κ, and σ Agonists

μ	κ	σ
brain analgesia	spinal analgesia	disphoria
respiratory depression	myosis	hallucinations
euphoria	sedation	respiratory and vasomotor stimulation
physical dependence		
gastro-intestinal (GI) effects		

same as the 1-(1-phenylcyclohexyl)piperidine [*77-10-1*] (PCP), "angel dust," receptor (34) since σ agonists exhibit pharmacological effects similar to PCP (35–36) and only the cyclazocinelike opioids displace [^{3}H]-PCP from this binding site (37–38). The σ receptor is no longer considered to be an endogenous opioid receptor.

The true physiological function of each distinct opioid receptor is not known. A hypothesized model (39) for the interactions of the μ and δ receptors in analgesic processes is shown schematically in Figure 4. Although overly simplified, this model is a useful tool to discuss the interactions of these two receptors. The basic features of this model include the following: distinct μ and δ receptors coexist in an opioid-receptor complex; Leu-enkephalin is a δ-agonist; β-endorphin is both a μ and δ agonist; the interaction between the two receptors is symbolized by the hatched arrows and this interaction is not considered to be a thermodynamic equilibrium or two inter-converting forms of the same receptor; the solid arrow connecting the μ receptor to a hypothetical membrane-bound effector symbolizes a coupling mechanism in analogy to the adenyl cyclase model; and the solid arrow connecting the δ receptor to the

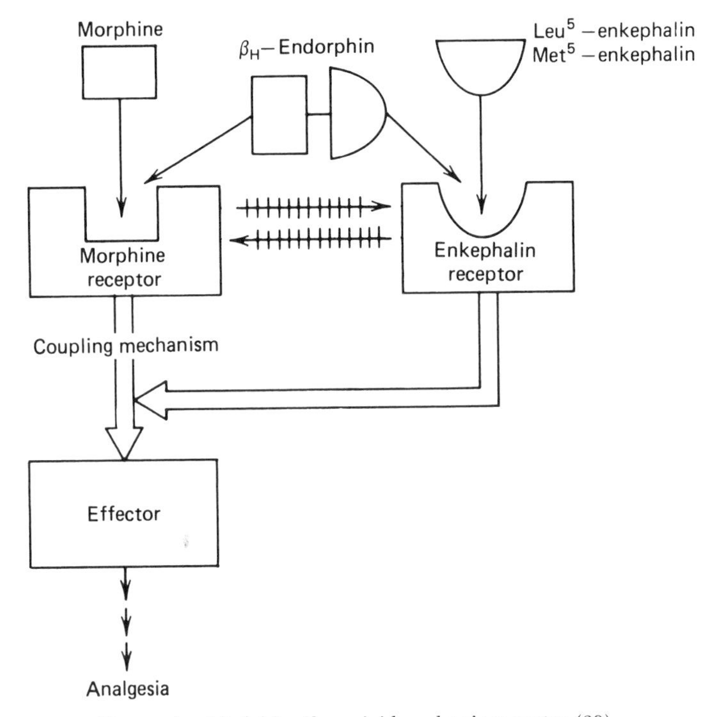

Figure 4. Model for the opioid analgesic receptor (39).

coupling mechanism represents the ability of the δ receptor to modulate coupling of the μ agonists to the effector. This model provides a working hypothesis that is consistent with many experimental observations. Analgesia is produced by occupation of the μ receptor by an agonist (morphine) leading to a receptor conformational change which allows coupling of the receptor to its effector. μ-Receptor antagonists such as naloxone do not induce such a conformational change. Occupation of the δ receptor by an agonist (Leu-enkephalin) is hypothesized to induce a conformational change in this receptor which results not in analgesia but a facilitation of the coupling of the μ receptor to its effector. Leu-enkephalin thus potentiates μ-agonist analgesia whereas Met-enkephalin does not. β-Endorphin binds to both receptors and thus potentiates its own analgesic effects. The endogenous ligand for the μ receptor is still undefined.

Recent advances have led to further characterization of the κ receptor. Dynorphin has been suggested as the endogenous κ-receptor ligand (40–41). In addition, dynorphin$_{1-8}$ and dynorphin$_{1-9}$ are proposed as the neurotransmitters at the κ receptor, whereas dynorphin acts hormonally at this site. The physiological functions of endogenous opioids acting at the κ receptor are under investigation by several groups. The relationship between localization of κ receptors in deep layers of the cerebral cortex and possible sedative effects may be of importance. The possible role of endogenous κ-receptor ligands in the regulation of feeding and drinking should also be noted.

Biological Function of Endorphins

Since the structures of Met- and Leu-enkephalin were proposed (20), there has been an explosion in the studies concerning the nature and physiology of opioid peptides and their receptors. The endogenous role of endorphins quite naturally focused on pain modulation because of the early relationship between opioid peptides and pain. Experimental evidence for such a role relies on a lowering of the pain threshold by the opioid antagonist naloxone. The shortened response latencies after naloxone were shown with a mouse hot plate (42). However, such demonstrations have proved to be very sensitive to experimental variables including diurnal variations reflective of fluctuations in endorphin release (43). The concept that the endorphins make up an endogenous pain-control mechanism has further support in reports that naloxone reverses the analgesia produced by acupuncture (44) and electrical stimulation (45) but, interestingly, not hypnosis (46). Surgical stress apparently also produces its own analgesia based on the notation (47) of a curious hyperalgesic effect of naloxone in patients following surgery, an observation that has also been noted in patients following dental surgery (48). In a more controversial report (49), it has been postulated that among normal patients, only placebo responders are naloxone sensitive. This suggests an interesting and still-to-be-confirmed physiological rather than psychological explanation for the analgesic placebo response.

The enkephalins are extremely weak analgetics in laboratory animal tests owing to their short (2–3 min) biological half-life. Thus, only transient analgesia has been found in rats, mice, and cats (50–51) for Met-enkephalin even when administered by intracerebroventricular (ICV) injection. Not surprisingly, massive doses (320 mg/kg) of Leu-enkephalin administered systemically (iv, intravenous) produced only weak analgetic activity in the mouse-tail-flick procedure (53). On the other hand, β- endorphin, with a half-life of 2–3 h, produces potent, long-lasting (1–2 h) analgesia in a variety of tests (54–56), including humans (57–58). In comparison to morphine, β-endorphin is about 20 times as potent following ICV administration in rodents and about four times as potent following iv injection.

The finding of analgetic activity for the endorphins created a renewed but short-lived hope that these or related peptides might lead at last to an analgesic devoid of dependence liability. Both Met-enkephalin and β-endorphin produce symptoms of physical dependence (59–60) and evidence of tolerance and morphine cross-tolerance in animals and *in vitro* (61). Furthermore, self-administration experiments in rats indicate the development of dependence on both enkephalins (62). Drug generalization experiments indicate that Met-enkephalin is perceived by rats as subjectively comparable to the narcotic fentanyl (63). In support of these findings, β-endorphin is also reported to support morphine abstinence in humans (64). Rapid metabolic inactivation may partially explain why humans are not addicted to their endogenous endorphins, but part of the explanation must also be that their role is to provide intermittent rather than tonic neuronal modulation.

Although an analgetic role was the first and most logically examined function for the endogenous opioids, the endorphins also influence a wide range of behavior. Since intraventricular administration of Leu- and Met-enkephalin produces a suppression of operant behavior maintained by food reinforcement (65), it may be that endorphins are important in regulating biological-drive mechanisms. The feeling of well being attained by reward is probably mediated by opioid receptors and thus the endogenous opioid peptides act as a natural euphoriant or reward transmitter (66). Many studies have linked opioids and the memory process. Thus, it has been found that Leu- and Met-enkephalin can facilitate memory consolidation (67) and may influence, either directly or indirectly, memory processes (68) (see Memory enhancing agents and antiaging drugs). Although it is an open question whether endorphins modulate sexual behavior in humans, studies in rodents with β-endorphin (69–70) and an enkephalin analogue (71) have shown that copulatory behavior was completely suppressed by these substances. Numerous other actions have been found for endorphins. They exhibit a wide variety of analgetic activity in animals and humans, but the modulation of pain mechanisms may not be the sole or even primary function of the endorphins. Only time and intensive research will explain why the human body produces substances that physiologically resemble a plant alkaloid.

Metabolic Inactivation

Role of Enzymes in Controlling Biological Functions. One explanation for the difference in analgesic effects between the enkephalins and β-endorphin is their relative stabilities. The half-life of the enkephalins in the presence of synaptic membranes is only 2–3 min compared to 2–3 h for β-endorphin (72–73). Several enzymes are known which cleave Leu- and Met-enkephalin at various peptide bonds (Fig. 5). The first enkephalin-degrading enzyme identified was an aminopeptidase which cleaves the amino terminal Tyr–Gly bond (74). This aminopeptidase is the most abundant (ca 70% of overall activity) enkephalin-degrading enzyme and is a cytoplasmic enzyme which is uniformly distributed throughout the brain and is therefore not thought to be specific for enkephalins. The increased analgetic activity of synthetic enkephalins substituted by D-amino acids at position 2, eg, [D-Ala^2]-Met-enk, has been attributed to increased stability towards this aminopeptidase (75). A second enkephalin-de-

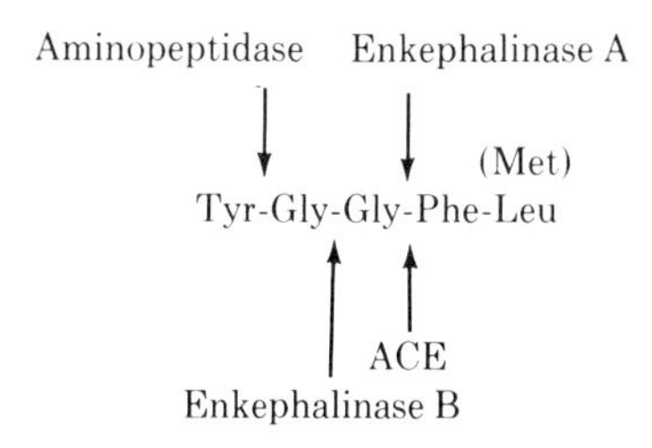

Figure 5. Enzymic cleavage of enkephalins.

grading enzyme, enkephalinase B, is a dipeptidyl aminopeptidase which cleaves the Gly–Gly bond (76). This membrane-bound enzyme has the least abundant overall activity in crude brain homogenates and is uniformly distributed throughout the brain. Enkephalinase A is a dipeptidyl carboxypeptidase which cleaves the Gly–Phe bond (76–79). This membrane-bound enzyme accounts for ca 20% of the overall enkephalin-degrading activity in the brain and its distribution parallels that of the opioid receptors. This latter observation has led to several suggestions that enkephalinase A is responsible for *in vivo* inactivation of the enkephalins. Analogues with increased stability towards enkephalinase A have been synthesized (*N*-methylation of Gly–Phe peptide bond or amidation of the carboxy terminus) which show even more enhanced analgesic activity than the D-Ala^2 analogues.

Intensive research efforts have focused on the discovery of potent and specific inhibitors of the enkephalin-degrading enzymes as novel analgetic agents. Since enkephalinases A and B as well as aminopeptidase are metallo (Zn) enzymes (80), a series of protected amino acid hydroxamates which are transition metal chelators was tested as enkephalinase inhibitors. The most potent analogue (*Z*-Leu-NHOH) possesses a K_I of 40 n*M* against enkephalinase A (81) (Fig. 6). Another potent and specific enkephalinase A inhibitor (K_I of 4.7 n*M*), thiorphan (82), has been reported to potentiate [D-Ala^2]-Met-enk-analgesia in the tail-withdrawal test and to elicit naloxone-reversible analgesia in the mouse hotplate-jump tests (83).

Finally, angiotensin-converting enzyme (ACE) also shows enkephalinaselike activity and, similarly to enkephalinase A, cleaves the Gly–Phe bond. This enzyme is not uniformly distributed throughout the brain, but its distribution does not correlate with that of the opioid receptors. In addition, several specific ACE inhibitors did not significantly alter the overall enkephalin-degrading activity in brain tissue (84).

Compared to enkephalin metabolism, little is known about metabolic inactivation of β-endorphin. Enkephalinase A is only weakly active against β-endorphin (78). Enzymes are known which degrade β-endorphin *in vitro* under nonphysiological

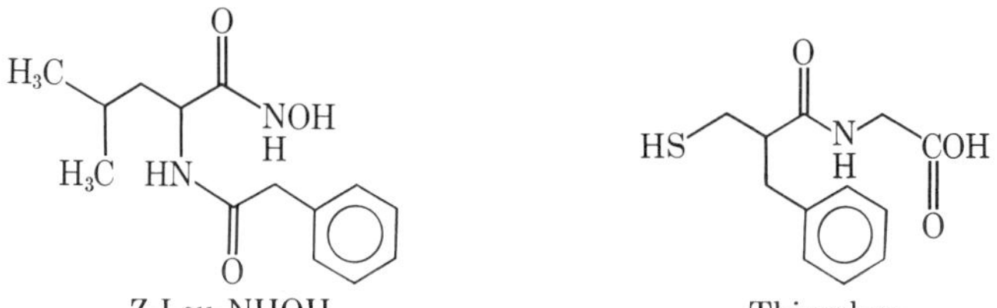

Z-Leu-NHOH

Thiorphan

Figure 6. Enkephalinase inhibitors.

conditions (85) or which inactivate β-endorphin by *N*-acetylation (86). Because of its long half-life, circulating levels of β-endorphin are measurable in plasma.

Structure-Activity Relationships

As was the case with morphine, the identification and structural elucidation of Met- and Leu-enkephalin gave rise to the syntheses of a large number of analogues in the attempt to identify a metabolically stable derivative with superior analgetic properties and fewer side effects. The common presence of a tyrosine [β-(4-hydroxy-phenyl)ethylamine] moiety in both the alkaloid and peptide opioids was quickly noted (87–90), as was the presence of a lipophilic binding region for Phe[4] comparable to that postulated for the oripavine analogue PET [*13965-63-4*] [Fig. 7 (87,91–93)]. It is important to note that several of these early structural comparisons employed incorrect absolute stereochemical assignments (93–94). In the correct assignment, the tyrosine–glycine amine linkage of enkephalin assumes the stereochemical position of the hydrogen at *C*-9 of morphine noted in Figure 7 (95).

Thus, some of the early work was directed towards the similarity between the tyrosine of position 1 and Met-enkephalin and the phenethylamine moiety of morphine. Since this is an essential piece in the synthetic opioids, it is not surprising that most analogues were considerably less active than Met-enkephalin. Removal of the amino group of tyrosine (96–97) and masking of the phenolic group (98) lead to relatively inactive analogues. The presence of an *N*-methyl group (compare morphine and normorphine) usually confers a potency increase in synthetic opioids. Thus, the addition of a methyl group on the tyrosyl nitrogen increased activity in a number of test systems, as expected (99–100). The link to classic opioid SAR was further strengthened by lack of activity for D-Tyr stereospecificity or amino acid replacements (Ala, Gly, Ser, His, Trp). Since *N*-alkyl alkaloids are well known for their narcotic antagonist properties, it was somewhat surprising that the substitution of *N*-allyl tyrosine for tyrosine yielded a compound devoid of antagonist activity.

Structural simplification of the morphine has lead to retention of useful analgetic activity (eg, meperidine, methadone, and pentazocine). However, synthesis of enkephalin analogues containing three or fewer peptide units were found to be inactive

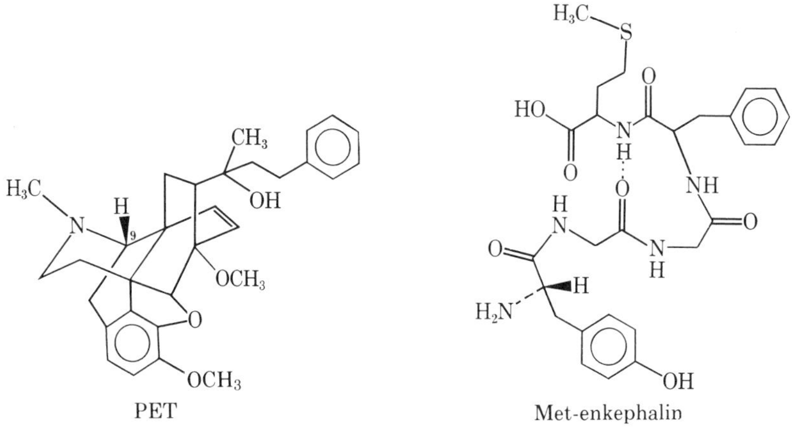

PET

Met-enkephalin

Figure 7. Structural comparison of an alkaloid and peptide opioid.

or much less potent than Met-enkephalin (96,98,101–104). One possible exception is a dipeptide described in ref. 105. Called kyotorphin [*70904-56-2*], this dipeptide (Tyr–Arg) was about four times as potent as Met-enkephalin in the tail-pinch test. The contribution of a D-amino acid at the 2 position not only increases metabolic stability but might also be important in an improved conformational fit to the opioid receptor. However, the analgetic activity of this seemingly unique dipeptide may be due to its ability to stimulate the release of endogenous opioids rather than a direct effect on the opioid receptor.

Many different substitution patterns in the full pentapeptide are allowed. A few of these are outlined in Table 3. It is interesting that methionine and leucine can be modified and replaced by a number of functions. Unlike the *N*-terminal amine, the terminal carboxyl is not essential for activity and can be extensively modified. Conversion of the acid to the corresponding primary amide results in increased potency, presumably owing to increased metabolic stability.

From the numerous synthetic peptides made, several exhibit systemic activity and have been extensively examined for their potential use in humans (Fig. 8). FK-33-824 is a potent analgetic in animals (120); however, an early clinical study (121) failed to produce analgesia at doses below a series of symptoms that were not reversed by naloxone.

Met-kephamid (LY-127,623) is another systemically active analogue of Met-enkephalin which possesses analgesic potency comparable to meperidine in rodents (122). Preliminary clinical data indicate that it has analgesic activity in humans (123) without the limiting side effects observed for FK-33-824. These differences in the clinical pharmacology may be due, in part, to differences in receptor selectivity (123–124). A third analogue, FW-34-569 does not appear to have therapeutic potential as an analgesic agent in humans (125).

Table 3. Allowed Substitutions of the Enkephalins

Substitutions	Ref.
H–Tyr	
$NHCH_3$	106
NH CH_2	107
NHArg	108
NHLys	109
HO NH_2	
NHGly	110
NHTyr	110
Gly	
D-Ala	111
D-Thr	112
D-Nle	113
D-Met	114
D-Ser	114
$(CH_3)_2CH$	115
Gly	
N for CH	116
Phe	
N-CH_3	106
Trp	117
Met–OH	
D-Leu	118
Thz	112
$COHN_2$	106
CH_2OH	106
CO_2R	114
$S \rightarrow O$	106
Pro–NHC_2H_5	114
Pro–NH_2	114
O O	114
N-allyl	119

H-Tyr-D-Ala-Gly-Phe-MeMet-NH_2
Met-kephamid [*66960-34-7*]

MeTyr-D-Ala-Gly-MePhe-Met(O)-ol
FW-34-569 [*70021-31-7*]

H-Tyr-D-Ala-Gly-MePhe-Met(O)-ol
FK-33-824 [*64854-64-4*]

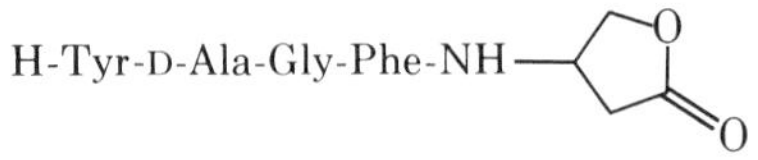

H-Tyr-D-Ala-Gly-Phe-NH— (2-oxotetrahydrofuran-4-yl, ring O, =O)

ICI-121,444 (344) [*66575-22-2*]
H-Tyr-D-Ala-Gly-Phe-D-Leu-OH
BW-180c (347) [*63631-40-3*]

Figure 8. Structures of some enkephalin analogue studies in humans.

Biologically Related Endogenous Substances

Substance P. The endogenous opioids are members of a rapidly growing family of peptides that exhibit neurotransmitter or hormonal properties (Table 4). Several of these peptides have been shown to interact with the opioids in pain/analgetic pathways. Perhaps the best candidate as a neurotransmitter of sensory pain information in the spinal cord is substance P (SP) [*33507-63-0*], an undecapeptide first isolated in the intestines over 50 years ago. Substance P may act as a neurotransmitter in the spinal cord since SP is released from the spinal cord after depolarization by K^+ or electrical stimulations in the Ca^{2+} dependent manner; peripheral nerve stimulation of pain fibers elicits release of SP into the cerebrospinal fluid; application of SP results in prolonged excitation of dorsal horn neurons which are also activated by painful stimuli; capsaicin causes release and depletion of SP from nerve terminals in the spinal cord which results in a short-term pain response and a long-term increase in peripheral thresholds to pain; long-term administration of capsaicin leads to degeneration of small-diameter sensory neurons that contain SP; and SP is found in synaptic vesicles at primary sensory neuron synapses in the substantia gelatinosa area of the spinal cord. Several pieces of data link SP and the endogenous opiates. A striking overlap of SP and opioid receptors exists in the spinal cord, and opioids inhibit the release of SP in the dorsal horn presumably by interaction at presynaptic opioid receptors. Intracerebroventricular administration of SP also produces a short-lived analgesia hypothesized to result from the release of Met-enkephalin.

Substance P

Arg–Pro–Lys–Pro–Gln–Gln–Phe–Phe–Gly–Leu–Met–NH_2

Neurotensin, CCK-8, and Somatostatin. Neurotensin is a 13-amino-acid peptide that may also function as a neurotransmitter. Neurons containing neurotensin localized within the brain show a similar location-mapping to enkephalins, and neurotensin is also highly localized in the dorsal gray and substantia gelatinosa areas of the spinal cord, implying a role in pain perception. Neurotensin produces potent analgesia by an unknown mechanism which is not naloxone-reversible. CCK-8 is an octapeptide (CO_2H terminal portion of cholecystokinin) found, among other brain locations, in the periqueductal gray area, and it may be involved in pain-integration functions. Injection of CCK-8 into the brain or spinal cord produces a long-lasting analgesia which may be opioid-mediated since it is naloxone-reversible. Somatostatin is a cyclic te-

Table 4. Peptides Tentatively Identified and Measured in Human Cerebrospinal Fluid[a]

Peptides	Refs.
16 κ fragment	127
β-lipotropin	127
adrenocorticotrophic hormone (ACTH)	127–131
β-endorphin	127, 132–137
Leu-enkephalin	138
Met-enkephalin	136, 139
α-melanocyte-stimulating hormone	140–141
β-melanocyte-stimulating hormone	142–144
calcitonin	145–148
oxytocin	149
vasopressin (AVP)	150–153
neurophysin	154
anginine vasotocin (AVT)	149, 155–156
angiotensin I	157–159
converting enzyme[b]	160–161
angiotensin II	159, 161
substance P	162–163
neurotensin	164
somatostatin (SRIF)	165–169
gastrin	170–171
cholecystokinin (CCK)	171–173
vasoactive intestinal peptide	153, 174–175
bradykinin	176
bombesin	153, 173
thyrotropin-releasing hormone (TRH)	177–180
thyroid-stimulating hormone (TSH)	181
luteinizing-hormone releasing hormone (LHRH)	182–183
luteinizing hormone (LH)	177
follicle-stimulating hormone (FSH)	177
growth hormone (GH)	181, 184
prolactin	177, 185–188
insulin	189–191

[a] Ref. 126.
[b] First peptide processing enzyme reported in CSF.

tradecapeptide that inhibits secretion of pituitary growth hormone. This peptide is also found in small-diameter sensory ganglion cells and may be a principal transmitter of pain perception. Somatostatin and substance P have both been shown to inhibit aminopeptidase degradation of Leu-enkephalin.

neurotensin

*p*Glu–Leu–Tyr–Glu–Asn–Lys–Pro–Arg–Arg–Pro–Tyr–Ile–Leu–OH

CCK-8

Asp–Tyr–Met–Gly–Trp–Met–Asp–Phe–NH_2

somatostatin

Ala–Gly–Cys–Lys–Asn–Phe–Phe–Trp–Lys–Thr–Phe–Thr–Ser–Cys–OH

BIBLIOGRAPHY

"Analgesics and Antipyretics" in *ECT* 1st ed., Vol. 1, pp. 851–861, by A. W. Ruddy, Sterling-Winthrop Research Institute, Division, Sterling Drug, Inc.; "Analgesics and Antipyretics" in *ECT* 2nd ed., Vol. 2, pp. 379–393, by G. deStevens, Ciba Pharmaceutical Company; "Analgesics, Antipyretics, and Anti-Inflammatory Agents" in *ECT* 3rd ed., Vol. 2, pp. 574–586, by W. F. Michne, Sterling-Winthrop Research Institute.

1. A. H. Beckett and A. F. Casy, *J. Pharm. Pharmacol.* **6,** 986 (1954); A. H. Beckett, *J. Pharm. Pharmacol.* **8,** 848 (1956).
2. P. S. Portoghese, *J. Pharm. Sci.* **52,** 865 (1966) and references cited therein.
3. P. S. Portoghese, *Acc. Chem. Res.* **11,** 21 (1978).
4. B. Balleau, P. Morgan, T. Conway, F. R. Amhed, and A. O. Hardy, *J. Med. Chem.* **17,** 907 (1974).
5. V. M. Klob, *J. Pharm. Sci.* **67,** 999 (1978).
6. R. H. B. Galt, *J. Pharm. Pharmacol.* **29,** 711 (1977).
7. D. M. Zimmerman, R. Nicklander, J. S. Horiig, and D. T. Wong, *Nature (London)* **275,** 332 (1978).
8. P. S. Portoghese, B. D. Alreja, and D. L. Larson, *J. Med. Chem.* **24,** 782 (1981).
9. L. Terenius, *Acta Pharmacol. Toxicol.* **32,** 317 (1973).
10. C. B. Pert and S. H. Snyder, *Science* **179,** 1011 (1973).
11. E. J. Simon, J. M. Hiller, and I. Edelman, *Proc. Natl. Acad. Sci. U.S.A.* **70,** 1947 (1973).
12. N. A. Ingoglia and V. P. Dole, *J. Pharmacol. Exp. Ther.* **175,** 84 (1970).
13. A. Goldstein, L. I. Lowney, and B. K. Pal, *Proc. Natl. Acad. Sci. U.S.A.* **68,** 1742 (1971).
14. D. J. Mayer and J. C. Liebeskind, *Brain Res.* **68,** 73 (1974).
15. J. C. Liebeskind, D. J. Mayer, and H. Akil, *Advan. Neurol.* **4,** 261 (1974).
16. H. Akil, D. J. Mayer, and J. C. Liebeskind, *Science* **191,** 961 (1976).
17. D. J. Mayer and R. Hayes, *Science* **188,** 941 (1975).
18. J. Hughes, *Brain Res.* **88,** 295 (1975).
19. L. Terenius and W. Wahlström, *Acta. Physiol. Scand.* **94,** 74 (1975).
20. J. Hughes, T. W. Smith, H. W. Kosterlitz, L. A. Fothergill, B. A. Morgan, and H. R. Morris, *Nature (London)* **258,** 577 (1975).
21. G. W. Pasternak, R. Goodman, and S. H. Snyder, *Life Sci.* **16,** 1765 (1975).
22. R. Simantov and S. H. Snyder, *Proc. Natl. Acad. Sci. U.S.A.* **73,** 2515 (1976).
23. C. H. Li, *Nature (London)* **201,** 924 (1964).
24. B. M. Cox, A. Golstein, and C. H. Li, *Proc. Natl. Acad. Sci. U.S.A.* **73,** 1821 (1976).
25. A. F. Bradbury, D. G. Smyth, C. R. Snell, N. J. M. Birdsall, and E. C. Hulme, *Nature (London)* **260,** 793 (1976).
26. A. Goldstein, W. Fischli, L. I. Lowney, M. Hunkapiller, and L. Hood, *Proc. Natl. Acad. Sci. U.S.A.* **78,** 7219 (1981).
27. N. Minamino, K. Kangawa, N. Chino, S. Sakakibara, and H. Matsuo, *Biochem. Biophys. Res. Commum.* **99,** 864 (1981).
28. K. Kangawa, N. Minamino, N. Chino, S. Sakakibara, and H. Matsuo, *Biochem. Biophys. Res. Commun.* **99,** 871 (1981).
29. J. Rossier, *Proc. Natl. Acad. Sci. U.S.A.* **74,** 5162 (1977).
30. *Science* **220,** 395 (1983).
31. W. R. Martin, C. G. Eades, J. A. Thompson, R. E. Huppler, and P. E. Gilbert, *J. Pharmacol. Exp. Ther.* **197,** 517 (1976).
32. J. A. H. Lord, A. A. Waterfield, J. Hughes, and H. W. Kosterlitz, *Nature (London)* **267,** 495 (1977).
33. A. Herz, J. Blasig, H. M. Emrich, C. Cording, S. Pirée, A. Kölling, D. V. Zerssen, *Adv. Biochem. Psychopharmacol.* **18,** 333 (1978).
34. R. S. Zukin and S. R. Zukin, *Life Sci.* **29,** 2681 (1981).
35. J. J. Teal and S. G. Holtzman, *J. Pharmacol. Exp. Ther.* **212,** 368 (1980).
36. S. G. Holtzman, *J. Pharmacol. Exp. Ther.* **214,** 614 (1981).
37. R. S. Zukin and S. R. Zukin, *Mol. Pharmacol.* **20,** 246 (1981).
38. S. R. Zukin and R. S. Zukin, *Proc. Natl. Acad. Sci. U.S.A.* **76,** 5372 (1979).
39. J. L. Vaught, R. B. Rothman, and T. C. Westfall, *Life Sci.* **30,** 1443 (1982).
40. J. Pablo, H. Tolo, K. Yoshimura, N. M. Lee, H. H. Loh, and E. L. Way, *Eur. J. Pharmacol.* **72,** 265 (1981).
41. C. Chaukin, I. F. James, and A. Goldstein, *Science* **215,** 413 (1982).
42. J. J. Jacob, E. C. Tremblay, and M. C. Colombel, *Psychopharmacologia* **37,** 217 (1974).
43. R. C. A. Fredrickson, V. Burgis, and J. D. Edwards, *Fed. Proc. Fed. Am. Soc. Exp. Biol.* **36,** 965 (1977).

44. B. Pomeranz and D. Chiu, *Life Sci.* **19,** 1757 (1976); D. J. Mayer, *Neurosci. Res. Progr. Bull.* **13,** 98 (1975).
45. J. Traber, K. Fischer, S. Latziri, and B. Hamprecht, *FEBS Lett.* **49,** 260 (1974).
46. A. Goldstein and E. R. Hilgard, *Proc. Natl. Acad. Sci. U.S.A.* **72,** 2041 (1975).
47. L. Lasagna, *Proc. R. Soc. Med.* **58,** 978 (1965).
48. J. D. Levine, N. C. Gordon, R. T. Jones, and H. L. Fields, *Nature (London)* **272,** 826 (1978).
49. J. D. Levine, N. C. Gordon, and H. L. Fields, *Lancet* **ii,** 654 (1978).
50. A. F. Bradbury, D. G. Smyth, C. R. Snell, J. F. W. Deakin, and S. Wendlandt, *Biochem. Biophys. Res. Commun.* **74,** 748 (1977).
51. H. H. Loh, L. F. Tseng, E. Wei, and C. H. Li, *Proc. Natl. Acad. Sci. U.S.A.* **73,** 2895 (1976).
52. W. Feldberg and D. G. Smyth, *J. Physiol.* **260,** 30P (1976).
53. H. H. Büscher, R. C. Hill, D. Romber, F. Cardinaux, A. Closse, D. Hauser, and J. Pless, *Nature (London)* **261,** 423 (1976).
54. J. M. vanRee, D. DeWied, A. F. Bradbury, E. Hulme, D. G. Smyth, and C. R. Snell, *Nature (London)* **264,** 792 (1976).
55. L. F. Tseng, H. H. Loh, and C. H. Li, *Nature (London)* **263,** 239 (1976).
56. C. H. Li, D. Yamashiro, L. F. Tseng, and H. H. Loh, *J. Med. Chem.* **20,** 325 (1977).
57. D. H. Catlin, K. K. Hui, H. H. Loh, and C. H. Li, *Commun. Psychopharmacol.* **1,** 493 (1977).
58. Y. Hosobuchi and C. H. Li, *Commun. Psychopharmacol.* **2,** 33 (1978).
59. E. Wei and H. Loh, *Science* **193,** 1262 (1976).
60. J. Bläsig and A. Hertz, *Naunyn Schmiedebergs Arch. Pharmakol.* **294,** 297 (1976).
61. A. A. Waterfield, J. Hughes, and H. W. Kosterlitz, *Nature (London)* **260,** 624 (1976).
62. J. D. Belluzzi and L. Stein, *Nature (London)* **266,** 556 (1977).
63. F. C. Colpaert, C. J. E. Niemegeers, P. A. J. Janssen, and J. M. vanRee, *Eur. J. Pharmacol.* **47,** 115 (1978).
64. D. H. Catlere, K. K. Hi, H. H. Loh, and C. H. Li in E. Costa and M. Trabucchi, eds., *Advances in Biochemical Pharmacology*, Vol. 18, Raven Press, New York, 1978.
65. J. D. Belluzzii and L. Stein, *Nature (London)* **266,** 556 (1977).
66. L. Stein and J. B. Belluzzi in E. Costa and M. Trabucchi, eds., *The Endorphins*, Raven Press, New York, 1978, p. 299.
67. H. Rigter, H. Greven, and H. vanRiezen, *Neuropharmacology* **16,** 545 (1977).
68. H. Rigter, *Science* **200,** 83 (1978).
69. B. J. Meyerson and L. Terenius, *Eur. J. Pharmacol.* **42,** 191 (1977).
70. B. J. Meyerson and M. Berg, *Acta Pharmacol. Toxicol.* **41,** 64 (1977).
71. B. P. Quarantotti, M. G. Corda, E. Paglietti, G. Biggio, and G. L. Gessa, *Life Sci.* **23,** 673 (1978).
72. M. Knight and W. A. Klee, *J. Biol. Chem.* **253,** 3842 (1978).
73. R. J. Miller, K-J. Chang, P. Cuatrecasas, and S. Wilkinson, *Biochem. Biophys. Res. Commun.* **74,** 1311 (1977).
74. J. M. Hambrook, B. A. Morgan, M. J. Rance, and C. F. C. Smith, *Nature (London)* **262,** 782 (1976).
75. C. R. Beddell, R. B. Clark, G. W. Hardy, L. A. Lowe, F. Ubatuba, J. R. Vane, S. Wilkinson, R. J. Miller, K-J. Chang, and P. Cuatrecasas, *Proc. R. Soc. London Ser. B* **198,** 249 (1977).
76. C. Gorenstein and S. H. Synder, *Life Sci.* **25,** 2065 (1979).
77. F. B. Craves, P. Y. Law, C. A. Hunt, and H. H. Loh, *J. Pharmacol. Exp. Ther.* **206,** 492 (1978).
78. B. Malfroy, J. P. Swerts, A. Guyon, B. P. Roques, and J. C. Schwartz, *Nature (London)* **276,** 323, 523 (1978).
79. S. Sullivan, H. Akil, and J. D. Barchas, *Commun. Psychopharmacol.* **2,** 525 (1978).
80. C. Gorenstein and S. H. Snyder, *Pro. R. Soc. London Ser. B* **210,** 123 (1980).
81. S. Blumberg, Z. Vogel, and M. Altstein, *Life Sci.* **28,** 301 (1980).
82. C. Llorens, G. Gacel, J.-P. Swerts, R. Perdrisot, M.-C. Fournie-Zaluski, J.-C. Schwartz, and B. P. Roques, *Biochem. Biophys. Res. Commun.* **96,** 1710 (1980).
83. R. P. Roques, M.-C. Fournie-Zaluski, E. Soroca, J. M. Lecomte, B. Malfroy, C. Llorens, and J.-C. Schwartz, *Nature (London)* **288,** 286 (1980).
84. A. Arregui, C.-M. Lee, P. Emson, and L. L. Iversen, *Eur. J. Pharmacology* **59,** 141 (1979).
85. B. M. Austen, D. G. Smyth, and C. R. Snell, *Nature (London)* **269,** 619 (1977).
86. S. Zakarian and D. Smyth, *Proc. Natl. Acad. Sci. U.S.A.* **76,** 5972 (1979).
87. A. F. Bradbury, D. G. Smyth, and C. R. Snell, *Nature (London)* **260,** 165 (1976).
88. A. S. Horn and J. R. Rodgers, *J. Pharm. Pharmacol.* **29,** 257 (1977).

89. B. P. Roques, C. Garbay-Jaureguibery, R. Oberlin, M. Anteunis, and A. K. Lala, *Nature (London)* **262,** 778 (1976).
90. C. R. Jones, W. A. Gibbons, and V. Garsky, *Nature (London)* **262,** 779 (1976).
91. A. P. Feinberg, I. Creese, and S. H. Snyder, *Proc. Natl. Acad. Sci. U.S.A.* **73,** 4215 (1976).
92. G. D. Smith and J. F. Griffen, *Science* **199,** 1214 (1978).
93. P. W. Schiller, C. F. Yam, and M. Lis, *Biochemistry* **16,** 1831 (1977).
94. B. E. Maryanoff and M. J. Zelesko, *J. Pharm. Sci.* **67,** 591 (1978).
95. S. Bajusz, A. Z. Ronai, J. I. Szekely, Z. Dunaikovacs, I. Berzetei, and L. Graf, *Acta. Biochim. Biophys. Acad. Sci. Hung.* **11,** 305 (1976).
96. A. R. Day, M. Lujan, W. L. Dewey, L. S. Harris, J. A. Radding, and R. J. Freer, *Res. Commun. Chem. Pathol. Pharmacol.* **14,** 597 (1976).
97. B. A. Morgan, C. F. C. Smith, A. A. Waterfield, J. Hughes, and H. W. Kosterlitz, *J. Pharm. Pharmacol.* **28,** 660 (1976).
98. N. Ling and R. Guillemin, *Proc. Natl. Acad. Sci. U.S.A.* **73,** 3308 (1976).
99. B. A. Morgan, J. D. Bower, K. P. Guest, B. K. Handa, G. Metcalf, and C. F. C. Smith in M. Goodman and J. Meienhofer, eds., *Peptides: Proceedings of the Fifth American Peptide Symposium*, John Wiley & Sons, Inc., New York, 1977, pp. 111–114.
100. A. F. Bradbury, W. F. Feldberg, D. C. Smyth, and C. R. Snell in H. Kosterlitz, ed., *Opiates and Endogenous Opioid Peptides*, Elsevier/North Holland, Amsterdam, 1976, pp. 9–17.
101. W. A. Klee and M. Nirenberg, *Nature (London)* **263,** 609 (1976).
102. H. Y. Meltzer, R. J. Miller, R. G. Fessler, M. Simonovic, and V. S. Fang, *Life Sci.* **22,** 1931 (1978).
103. H. H. Buscher, R. C. Hill, D. Romer, F. Cardinaux, A. Closse, D. Hauser, and J. Pless, *Nature (London)* **261,** 423 (1976).
104. F. R. Beddell, R. B. Clark, G. W. Hardy, L. A. Lowe, F. B. Ubatuba, J. R. Vane, and S. Wilkinson, *Proc. R. Soc. London Ser. B* **198,** 249 (1977).
105. H. Takagi, M. Satoh, H. Shiomi, A. Akaike, H. Ueda, S. Kawajiri, M. Yamamoto, and H. Amano in E. L. Way, ed., *Endogenous and Exogenous Opiate Agonists and Antagonists*, Pergamon Press, Elmsford, N.Y., 1980, pp. 201–204.
106. J. Pless, W. Bauer, F. Cardinaux, A. Vlosse, D. Hauser, R. Huguenin, D. Romer, H. H. Buscher, and R. C. Hill in I. MacIntyre and A. G. E. Pearse, eds., *Proceedings in Endocrinology*, Elsevier Publishing Co., Amsterdam, 1978.
107. E. F. Hahn, J. Fishman, Y. Shiwaku, F. F. Foldes, H. Nagashima, and D. Duncalf, *Res. Commun. Chem. Path. Pharmacol.* **18,** 1 (1977).
108. P. Y. Law, E. T. Wei, L. F. Tseng, H. H. Loh, and E. L. Way, *Life Sci.* **20,** 251 (1977).
109. A. S. Dutta, J. J. Gormley, C. F. Hayward, J. S. Morley, J. S. Shaw, G. J. Stacey, and M. T. Turnbull, *Life Sci.* **21,** 559 (1977).
110. G. Gacel, M.-C. Fournie-Zaluski, E. Fellion, B. P. Roques, B. Senault, J.-M. Lecomte, B. Malfroy, J.-P. Swerts, and J.-C. Schwartz, *Life Sci.* **24,** 725 (1979).
111. C. B. Pert, A. Pert, J.-K. Chang, and B. T. Fong, *Science* **194,** 330 (1976).
112. L-Fu Tseng, H. H. Loh, and C. H. Li, *Life Sci.* **23,** 2053 (1978).
113. S. Bajusz, A. Z. Ronai, J. I. Szekely, Z. Dunaikovacs, I. Berzetei, and L. Graf, *Acta. Biochem. Biophys. Acad. Sci. Hung.* **11,** 305 (1976).
114. A. S. Dutta, J. J. Gormley, C. F. Hayward, J. S. Morley, J. S. Shaw, G. J. Stacey, and M. J. Turnbull, *Br. J. Pharmacol.* **61,** 481P (1977); J. S. Shaw and M. J. Turnbull, *Eur. J. Pharmacol.* **49,** 313 (1978).
115. R. Nagaraj and P. Balaram, *FEBS Lett.* **96,** 273 (1978).
116. J. S. Morley, *lecture given at the EMBO Workshop on Hormone Fragments and Diseases*, Leiden, The Netherlands, 1977.
117. P. W. Schiller, C. F. Yam, and M. Lis, *Biochemistry* **16,** 1831 (1977).
118. C. R. Beddell, R. B. Clark, L. A. Lowe, S. Wilkerson, K.-J. Chang, P. Cuatrecasas, and R. Miller, *Br. J. Pharm.* **61,** 351 (1977).
119. K. B. Mathur, B. J. Dhotre, R. Raghubir, G. K. Patnaik, and B. N. Dhawan, *Life Sci.* **25,** 2023 (1979).
120. D. Roemer, H. H. Buescher, R. C. Hill, J. Pless, W. Bauer, F. Cardinaux, A. Closse, D. Hauser, and R. Hugenin, *Nature (London)* **268,** 547 (1977).
121. B. von Graffenried, E. del Pozo, J. Roubicek, E. Krebs, W. Poldinger, P. Burmeister, and L. Kerp, *Nature (London)* **272,** 729 (1978).
122. R. C. A. Frederickson, E. L. Smithwick, R. Shuman, and K. G. Bemis, *Science* **211,** 603 (1981).

123. R. C. A. Frederickson, E. L. Smithwick, and D. P. Henry in C. A. Marsan and W. Z. Traczyk, eds., *Neuropeptides and Neural Transmissions*, Raven Press, New York, p. 227.
124. K.-J. Chang, *Psychopharmacol. Bull.* **17,** 108 (1981).
125. T. Lindeburg, V. Larsen, H. Kehlet, and E. Jacobsen, *Acta Anaesthesiol. Scand.* **25,** 254 (1981).
126. R. M. Post, P. Gold, D. R. Rubinow, J. C. Ballenger, W. E. Bunney, Jr., F. K. Goodwin, *Life Sci.* **31,** 1 (1982).
127. Y. Hosobuchi in J. H. Wood, ed., *Neurobiology of Cerebrospinal Fluid, Vol. 11*, Plenum Press, New York.
128. S. Finkielman, C. Fischer-Ferraro, A. Diaz, D. I. Goldstein, and V. E. Nahmod, *Proc. Natl. Acad. Sci. U.S.A.* **69,** 3341 (1972).
129. M. Kleerekoper, R. A. Donald, and S. Posen, *Lancet* **i,** 74 (1972).
130. R. M. Jordan, J. W. Kendall, J. L. Seaich, J. P. Allen, C. A. Paulsen, C. W. Kerber, and W. P. Vanderlaan, *Ann. Intern. Med.* **85,** 49 (1976).
131. J. P. Allen, J. W. Kendall, R. McGilvra, and C. Vancura, *J. Clin. Endocrinol. Metab.* **38,** 586 (1974).
132. W. J. Jeffcoate, L. H. Rees, L. McLoughlin, S. F. Ratter, J. Hope, P. J. Lowry, and G. M. Besser, *Lancet* **ii,** 119 (1978).
133. K. Nakao, Y. Nakai, S. Oki, S. Matsubara, T. Konishi, H. Nishitani, and H. Imura, *J. Clin. Endocrinol. Metab.* **50,** 230 (1980).
134. M. M. Wilkes, R. D. Stewart, J. F. Bruni, M. E. Quigley, S. S. C. Yen, N. Ling, and M. Chretien, *J. Clin. Endocrinol. Metab.* **50,** 309 (1980).
135. H. Akil, D. E. Richardson, J. Hughes, and J. Barchas, *Science* **201,** 463 (1978).
136. H. Akil, D. E. Richardson, J. D. Barchas, and C. H. Li, *Proc. Natl. Acad. Sci. U.S.A.* **75,** 5170 (1978).
137. F. E. Bloom and D. S. Segal in J. H. Wood, ed., *Neurobiology of Cerebrospinal Fluid, Vol. 1*, Plenum Press, New York, 1980, p. 651.
138. Y. Sarne, R. Azov and B. A. Weissman, *Brain Res.* **151,** 399 (1978).
139. A. Dupont, A. Villeneuve, J. P. Bouchard, R. Bouchard, Y. Merand, D. Rouleau, and F. LaBrie, *Lancet* **ii,** 1107 (1978).
140. T. L. O'Donohue, G. E. Handelmann, T. Chaconas, R. L. Miller, and D. M. Jacobowitz, *Peptides* (1982).
141. T. L. O'Donohue, C. G. Charlton, N. B. Thoa, C. J. Helke, T. W. Moody, A. Pert, A. Williams, R. L. Miller, and D. M. Jacobowitz, *Peptides* **2,** 93 (1981).
142. D. Rudman, A. E. Del Rio, B. M. Hollins, D. H. Houser, M. E. Keeling, J. Sutin, J. W. Scott, R. A. Sears, and M. A. Rosenberg, *Endocrinology* **92,** 372 (1973).
143. S. Shuster, A. Smith, N. Plummer, A. Thody, and F. Clark, *Br. Med. J.* **1,** 1318 (1977).
144. A. G. Smith and S. Shuster, *Lancet* **i,** 1321 (1976).
145. K. L. Becker, O. L. Silva, R. M. Post, J. C. Ballenger, J. S. Carman, R. H. Synder, and C. F. Moore, *Brain Res.* **194,** 598 (1980).
146. D. M. Pavlinac, L. W. Lenhard, J. G. Parthemore, and L. J. Deftos, *J. Clin. Endocrinol. Metab.* **50,** 717 (1980).
147. J. S. Caraman, R. M. Post, J. C. Ballenger, K. L. Becker, and O. Silva, *Soc. Biol. Psychiatry Abstr.* **73,** 105 (1980).
148. J. Carman, R. Post, J. Ballenger, and F. Goodwin in B. Jansson, C. Perris, and G. Struew, eds., *Proceedings of the 3rd World Congress of Biological Psychiatry*, Elsevier Publishing Co., Amsterdam.
149. P. W. Gold, D. Fisher, D. Rubinow, J. C. Ballenger, R. M. Post, and F. K. Goodwin, personal communication, 1981.
150. T. G. Luerssen and G. L. Robertson in J. H. Wood, ed., *Neurobiology of Cerebrospinal Fluid*, Plenum Press, New York, 1980, p. 613.
151. P. W. Gold, F. K. Goodwin, J. C. Ballenger, H. Weingartner, G. L. Robertson, and R. M. Post in D. de Wied and P. A. van Keep, eds., *Hormones and the Brain*, MTP Press, Ltd., Lancaster, UK, 1980, p. 241.
152. J. S. Jenkins, H. M. Mather, and V. Ang, *J. Clin. Endocrinol. Metab.* **50,** 364 (1980).
153. A. Gjerris, J. Fahrenkrug, M. Hammer, O. P. Schaffalitzky De Muckadell, S. Bojholm, P. Vendsborg, and O. J. Rafaelsen, *Abstracts of the 12th Collegium Internationale Neuro-Psychopharmacologicum*, Gothenborg, Sweden, June 1980, p. 115.
154. A. G. Robinson, E. A. Zimmerman, *J. Clin. Invest.* **52,** 1360 (1976).

155. S. Pavel, D. Psatta, and R. Goldstein, *Brain Res. Bull.* **2,** 251 (1977).
156. S. Pavel, *J. Clin. Endocrinol. Metab.* **31,** 369 (1970).
157. D. Ganten and G. Speck, *Biochem. Pharmacol.*, in press.
158. D. Ganten, G. Speck, P. Schelling, and T. Unger in J. B. Martin, S. Reichlin, and K. L. Bick, eds., *Neurosecretion and Brain Peptides: Implications for Brain Function and Neurological Disease*, Raven Press, New York, 1981, pp. 359–372.
159. P. Gold, D. Ganten, R. M. Post, F. K. Goodwin, and J. C. Ballenger, 1981.
160. J. Saavedra, personal communication, 1981.
161. P. Schelling, D. Ganten, R. Heckl, K. Hayduk, J. Hutchison, G. Sponer, and U. Ganten in J. P. Buckley and E. M. Ferrario, eds., *Central Actions of Angiotensin and Related Hormones*, Pergamon Press, New York, 1977, p. 519.
162. J. G. Nutt, E. A. Mroz, S. E. Leeman, A. C. Williams, W. K. Engel, and T. N. Chase, *Neurology* **30,** 1280 (1980).
163. J. G. Nutt, E. A. Mroz, T. N. Chase, W. K. Engel, and S. E. Leeman, *Neurology* **26,** 359 (1978).
164. C. B. Nemeroff, *Biol. Psychiatry* **15,** 283 (1980).
165. Y. C. Patel, K. Rao, and S. Reichlin, *N. Engl. J. Med.* **296,** 529 (1977).
166. S. Kronheim, M. Berelowitz, and B. L. Pimstone, *Clin. Endocrinol.* **6,** 411 (1977).
167. D. R. Rubinow, P. W. Gold, R. M. Post, J. C. Ballenger, F. K. Goodwin, and S. Reichlin in R. M. Post and J. C. Ballenger, eds., *Neurobiology of the Mood Disorders*, Williams & Wilkins, Baltimore, Md.
168. K. V. Sorenson, S. E. Christensen, E. Dupont, A. P. Hansen, E. Pedersen, and H. Orskov, *Acta. Neurol. Scand.* **61,** 186 (1980).
169. H. Orskov, A. P. Hansen, B. De Fine Olivarius, E. Pedersen, K. Sorenson, E. Dupont, and J. Ingerslev in ref. 148.
170. C. Kruse-Larsen and J. F. Rehfeld, *Brain Res.* **176,** 189 (1979).
171. A. Gjerris and J. Fahrenkrug in ref. 148.
172. J. F. Rehfeld and C. Kruse-Larsen, *Brain Res.* **155,** 19 (1978).
173. R. H. Gerner in ref. 167.
174. A. M. Ebeid, R. R. Attia, P. Sundaram, and J. E. Fischer, *Am. J. Surg.* **137,** 123 (1979).
175. J. Fahrenkrug, O. B. Schaffalitzky De Muckadell, and A. Fahrenkrug, *Brain Res.* **124,** 581 (1977).
176. M. Barrie and A. Jowett, *Brain* **90,** 785 (1967).
177. M. T. Hyyppa and J. Liira, *Med. Biol.* **57,** 367 (1979).
178. C. Oliver, J. P. Charvet, J. L. Codaccioni, and J. Vague, *Lancet* **i,** 873 (1974).
179. G. E. Schambaugh, J. F. Wilber, E. Montoya, H. Ruder, and E. R. Blonsky, *J. Clin. Endocrinol. Metab.* **41,** 131 (1975).
180. C. Kirkegaard, J. Faber, L. Hummer, and P. Rogowski, *Psychoneuroendocrinology* **4,** 227 (1979).
181. C. Schaub, M. T. Bluet-Pajot, G. Szikla, C. Lornet, and J. Talairach, *J. Neurol. Sci.* **31,** 123 (1977).
182. E. Rolandi, T. Barraca, P. Masturzo, R. Gianrossi, A. Polleri, and C. Perria, *Lancet* **i,** 1080 (1976).
183. A. Gunn, H. M. Fraser, S. L. Jeffcoate, D. T. Holland, and W. J. Jeffcoate, *Lancet* **i,** 1057 (1974).
184. J. A. Linfoot, J. F. Carcia, W. Wei, R. Fink, R. Sarin, J. L. Born, and J. H. Lawrence, *J. Clin. Endocrinol. Metab.* **31,** 230 (1970).
185. G. Sedvall, G. Alfreddson, L. B. Jerkenstedt, P. Eneroth, G. Fyro, C. Harnryd, C. G. Swahn, F. A. Wiesel, and B. Wode-Helgodt in M. Airaksinen, ed., *Proceedings of 6th International Congress of Pharmacology, Vol. 3, Central Nervous System and Behavioral Pharmacology*, Pergamon Press, New York, 1975, p. 255.
186. J. Assies, A. P. Schellekens, and J. L. Touber, *J. Clin. Endocrinol. Metab.* **46,** 576 (1978).
187. B. Wode-Helgodt, B. Fyro, B. Gullberg, and G. Sedvall, *Acta Psychiatr. Scand.* **56,** 129 (1977).
188. D. C. Jimerson, R. M. Post, J. Skyler, and W. E. Bunney, Jr., *J. Pharm. Pharmacol.* **28,** 845 (1976).
189. B. R. Brooks, J. H. Wood, M. Dias, C. Czerwinski, L. P. Georges, J. Sode, M. H. Ebert, and W. K. Engel in ref. 137, p. 113.
190. A. V. Grew, G. Ghirlanda, G. Feddi, and G. Gambassi, *Eur. Neurol.* **3,** 303 (1970).
191. L. P. Ratzman and R. Hampel, *Endokrinologie Band.* **76,** 185 (1980).

General References

Reference 126 is also a general reference.
B. E. B. Sandberg and L. L. Iversen, *J. Med. Chem.* **25,** 1009 (1982).
D. R. Brown and R. J. Miller, *Annu. Rep. Med. Chem.* **17,** 271 (1982).

R. Couture and D. Regoli, *Pharmacology* **24,** 1 (1982).
T. M. Jessell, *Nature (London)* **295,** 551 (1982).
R. Schwyzer, *Naturwissenschaften* **69,** 15 (1982).
S. H. Buck, J. H. Walsh, H. I. Yamamura, and T. F. Burks, *Life Sci.* **30,** 1857 (1982).
T. M. Jessell, *Neurosecretion and Brain Peptides* **189,** (1981).
D. J. Krieger and J. B. Martin, *N. Engl. J. Med.* **876,** 944 (1981).
R. J. Miller, *Pharmacol. Ther.* **12,** 73 (1981).
S. E. Leeman and R. Gamse, *TIPS*, 119 (1981).
W. H. Sweet, *J. R. Soc. Med.* **73,** 482 (1980).
S. H. Synder, *Science* **209,** 976 (1980).
T. Hökfelt, O. Johansson, A. Ljungdahl, J. M. Lundberg, and M. Schultzberg, *Nature (London)* **284,** 515 (1980).
T. M. Jessell and J. S. Kelly, *Nature (London)* **285,** 131 (1980).
J. L. Marx, *Science* **205,** 886 (1979).
E. A. Mroz and S. E. Leeman, *Vitam. Horm. (N.Y.)* **35,** 209 (1977).
J. B. Malick and R. M. S. Bell, eds., *Endorphins: Chemistry, Physiology, Pharmacology, and Clinical Relevance*, Marcel Dekker, Inc., New York, 1982.
Current Contents, 5 (April 18, 1983).

M. ROSS JOHNSON
DAVID A. CLARK
Pfizer Inc.

P

PERFLUORINATED-IONOMER MEMBRANES

The term perfluorinated-ionomer membranes generally refers to ion-exchange membranes composed of perfluorinated polymeric backbones (see Membrane technology).

An overview of both the fundamental properties and the technological aspects of perfluorinated-ionomer membranes is given in ref. 1.

The first perfluorinated-ionomer membranes, the Nafion membranes, were developed and commercialized by DuPont in the early 1970s. They are made of the perfluorinated sulfonic acid ionomer called XR resin (2).

Nafion was first applied to a separator of a fuel cell which was necessarily used in space exploration, and then opened the way to the innovative electrolytic process for chlor-alkali production by using perfluorinated ion-exchange membranes (3) (see Alkali and chlorine products; Electrochemical processing, inorganic). Companies with chlor-alkali processes with Nafion membranes in the United States use technology devised by Eltech Systems Corp. and Occidental (Hooker) Chemical Corp. Three such companies are Fort Howard Paper Co., Georgia-Pacific Corp., and Vulcan Materials Co.

Since the electrolytic performance of the Nafion series available at the early stage was unsatisfactory from the industrial point of view, extensive efforts have concentrated on the improvement of current efficiencies of plain sulfonic acid membranes by their surface modification with chemical reactions (4–6) and with film lamination (7).

A new type of perfluorinated-ionomer membrane, a carboxylic acid type, Flemion, was developed in 1975 and has been commercially produced since 1978 by Asahi Glass (8–10). The different ion-exchange groups greatly affect the physical and electrochemical properties of the membranes.

The development of these high performance membranes has promoted the progress of electrolytic-cell technologies, and now the membrane chlor-alkali process is recognized worldwide as energy-saving and pollution-free, compared with the mercury or the diaphragm processes.

Structure and Properties

Typical polymeric structures for both sulfonic acid and carboxylic acid membranes are shown below:

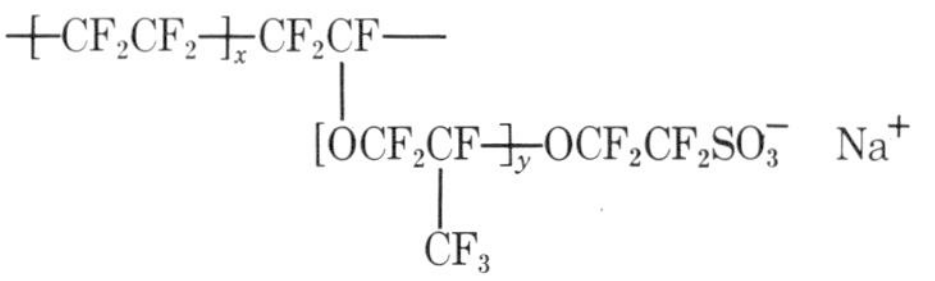

$$\text{-}\!\!\left[CF_2CF_2\right]_x\!CF_2CF\!-\!\!-\qquad \text{side chain: } [OCF_2CF(CF_3)]_y OCF_2CF_2SO_3^- \; Na^+$$

Nafion polymer [*39464-59-0*]

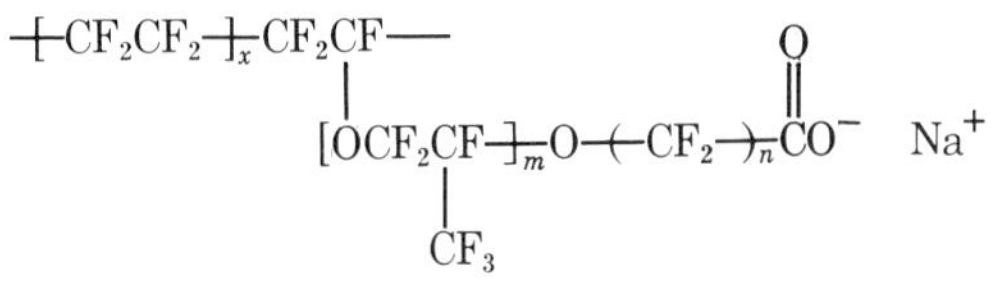

$$\text{-}\!\!\left[CF_2CF_2\right]_x\!CF_2CF\!-\!\!-\qquad \text{side chain: } [OCF_2CF(CF_3)]_m O\!-\!(CF_2)_n\!-\!\overset{O}{\overset{\|}{C}}O^- \; Na^+$$

(m = 0 or 1, n = 1–5)

Flemion polymer [*75634-46-7*]

Both polymers are melt-processable and fabricated into film by means of extrusion-molding. The resultant film can be easily converted to the corresponding ion-exchange membrane by alkaline hydrolysis.

Tensile properties of these membranes drastically change before and after hydrolysis (9–10) and the ionic interaction or the formation of ionic clusters has been extensively investigated by means of various spectroscopies such as small angle x-ray spectroscopy, nmr, ir, Mossbauer, and neutron-activation studies, and the results have been summarized in ref. 1 (see also Analytical methods). These studies provide different aspects of structural features of the membranes, but their ion-clustered morphologies remain unsolved because of their structural complexities.

Electrochemical behaviors, such as electric conductivity and current efficiency of the membrane, mainly depend upon the following inherent properties of polymer materials: ion-exchange capacities, water contents, and fixed-ion concentrations.

The ion-exchange capacity represents the concentration of ion-exchange groups in a polymer and the high conductivity of a membrane can be achieved by incorporating as many ion-exchange groups as possible without impairing polymer strength. Relationships between ion-exchange capacities and electric conductivities are shown in refs. 5, 10–11.

These membranes have lower ion-exchange capacities than other ionomers (see Ionomers) in that ions present in the perfluorinated ionomers tend to aggregate and form domains. The superior performance of these membranes is thought to be due to the presence of these domains. The diffusion of redox ions in Nafion films has been discussed in connection with the two-phase structure of Nafion (12).

The water content or water uptake is determined by the balance of power based on the contraction of the polymer backbone and the swelling by water with which the counterion and the ion-exchange group in the polymer are accompanied. Consequently, the concentration of ion-exchange groups in the water phase inside the membrane is determined and expressed as the fixed ion concentration, which is an important factor governing the current efficiency in the electrolysis. When the fixed-ion concentration is sufficiently high, the introduction of an ion with the same charge as that of the ion-exchange group is efficiently suppressed by the theory of Donnan's equilibrium (see Gelatin; Membrane technology; Polyelectrolytes).

The carboxylic acid group attached to perfluoroalkylene group ($—CF_2CO_2^-$) shows weaker acidity than the sulfonic acid group ($—CF_2SO_3^-$), associated with the weak hydration force of the carboxylic acid group. Therefore, the carboxylic acid membrane retains lower levels of the water content and higher fixed-ion concentrations than the sulfonic acid membrane through a wide range of ion-exchange capacities, taking advantage of high current efficiencies in the production of concentrated caustic soda (13).

Preparation

The general procedure of manufacturing perfluorinated-ionomer membranes includes synthesis of a perfluorovinyl ether moiety with a functional group, its copolymerization with tetrafluoroethylene, and the formation of a membrane from the resulting copolymer. Usually, membranes are reinforced by fabrics or some other ingredients to ensure the reliability of their mechanical properties over long-term use.

Monomer Synthesis. Industrial preparation of various perfluorovinyl ethers has become possible through DuPont's excellent work on fluorine chemistry with hexafluoropropylene oxide (HFPO) (see Fluorine compounds, organic, perfluoroepoxides).

The synthetic procedures for the key monomers of both Nafion and Flemion polymers are described below (2,13):

$$CF_2{=}CF_2 \;[116\text{-}14\text{-}3] \xrightarrow{SO_3} \underset{[697\text{-}18\text{-}7]}{F_2C{-}CF_2 \text{ (}O{-}SO_2\text{ ring)}} \xrightarrow{F^-} \underset{[677\text{-}67\text{-}8]}{FC({=}O)CF_2SO_2F} \xrightarrow{2\,HFPO} \underset{[4089\text{-}58\text{-}1]}{FC({=}O)CF(CF_3)OCF_2CF(CF_3)OCF_2CF_2SO_2F}$$

$$\xrightarrow[-COF_2]{\Delta,\, Na_2CO_3} \underset{[16090\text{-}14\text{-}5]}{CF_2{=}CFOCF_2CF(CF_3)OCF_2CF_2SO_2F}$$

$$CF_2{=}CF_2 \xrightarrow{I_2} \underset{[375\text{-}50\text{-}8]}{I(CF_2CF_2)_2I} \xrightarrow{oleum} \underset{[702\text{-}35\text{-}2]}{\text{tetrafluoro-}\gamma\text{-butyrolactone (F}_6\text{ ring with O and C=O)}} \xrightarrow{CH_3OH} \underset{[63425\text{-}24\text{-}1]}{FC({=}O)CF_2CF_2CO_2CH_3} \xrightarrow{(x+1)\,HFPO}$$

$$FC(CF(CF_3)OCF_2)_{x+1}CF_2CF_2CO_2CH_3 \xrightarrow[-COF_2]{\Delta,\, base} CF_2{=}CFO(CF_2CF(CF_3)O)_xCF_2CF_2CF_2CO_2CH_3$$

$x = 1$ [64033-91-6] $x = 1$ [62361-02-8]
$x = 0$ [64725-41-3] $x = 0$ [19190-61-5]

Recently, new sulfonic-acid-type perfluorovinyl ethers with shorter chain length have been proposed (5,14):

$$CF_2{=}CFO(CF_2)_nSO_2F$$

$n = 2$ [29514-94-1]

$n = 3$ [77416-84-3]

These monomers are considered to be useful in preparing a membrane with higher ion-exchange capacities. Furthermore, new synthetic methods of some perfluorinated vinyl ethers have been disclosed (8,15).

Polymer Synthesis. The functionally substituted vinyl ethers described above can be polymerized with tetrafluoroethylene in the presence of radical initiators in an aqueous medium or in an inert organic solvent (16–18).

A copolymer composition curve shows random copolymerization of tetrafluoroethylene with carboxylated perfluorovinyl ethers (10).

Fabrication of Ionomer Membranes. The crystallinity of the copolymer depends upon the content of the functional comonomer and the polymer structure changes from crystalline to amorphous with the increase of the comonomer content (10). Amorphous or partly crystalline copolymers are fabricated into films with designed thickness (usually 100–250 μm) by means of conventional extrusion techniques. Two kinds of films with different ion-exchange capacities (19–20) or even with different ion-exchange groups (7) can be laminated, if necessary.

The polymer films are generally reinforced by Teflon cloth, but, in some cases, a reinforcement of membranes can be attained by using a premixed copolymer with small amounts of fibrous materials (21).

The resulting films are converted to sulfonic- or carboxylic-acid-type ion-exchange membrane by alkaline hydrolysis.

Chemical Conversion of Functional Groups. The conversion of a sulfonic acid group to a carboxylic acid group can be achieved by the following chemical reactions:

$$\cdots\cdots{-}O(CF_2)_nSO_2X \rightarrow \rightarrow \cdots\cdots{-}O(CF_2)_{n-1}\,CO_2H$$

$$(X = Cl \text{ or } F)$$

The sulfonyl halide group in the membrane is converted to sulfinic acid by reduction and then the carboxylic acid group, having one CF_2 less than the original chain of sulfonic acid, is formed through a desulfonylation reaction (5). Otherwise, the direct conversion of sulfonyl chloride groups is performed by treating the membrane with hot air in the presence of alcohol vapor (6).

Applications

Reference 1 has summarized various expected applications of perfluorinated-ionomer membranes, among which fuel cells, solid polymer electrolyte (SPE), battery separators, metal-winning, and permeation-distillation are expected to have marketing potential (22–23).

In this article, the description of applications is focused on the innovative chlor-alkali process with combined technologies of high performance membranes and cells.

Chlor-Alkali Process. The principle of electrolysis of brine using a cation-exchange membrane is schematically shown in Figure 1.

The membrane permits sodium cations to pass through into a cathode compartment selectively and to form caustic soda. Ordinary hydrocarbon-type membranes cannot be used for this purpose, because they deteriorate under such a strongly oxidative atmosphere.

The membrane material is required to satisfy at least the following properties: high selective permeability of sodium cation and low electrical resistance; exceptional chemical stabilities against chlorine and concentrated caustic soda even at a high temperature of electrolysis; and strong mechanical properties to endure handling and electrolysis under severe conditions.

DuPont's Membrane. DuPont's Nafion has a long history of improving its electrolytic performance. The initial Nafion 400 series was composed of plain sulfonic acid membranes that were not suitable for caustic soda production because of low current efficiencies.

The Nafion 300 series was then introduced for the purpose of producing 10–20%

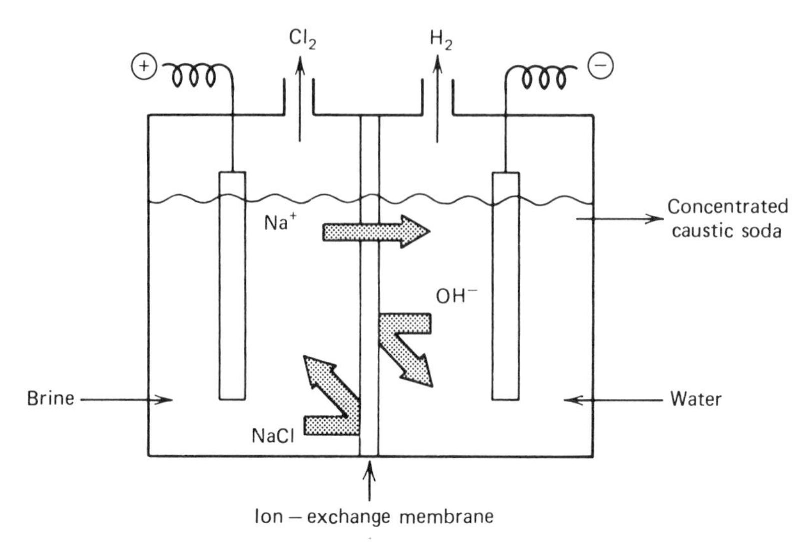

Figure 1. Role of ion-exchange membrane in electrolysis.

caustic soda. These are laminated membranes with two sulfonic-acid-type polymers having different ion-exchange capacities, one with ca 0.7 meq/g and the other with 0.91 meq/g (4).

For the production of 20–28% caustic soda, the Nafion 200 series was developed, which is characterized by having chemically modified surfaces treated with diamines (4).

The latest Nafion 900 series membranes have been designed to be carboxylate–sulfonate two-layer membranes with unique reinforcement, eg, ca 95% current efficiency at 33% caustic soda with Nafion 901 (7).

Asahi Glass' Membrane. Based on Asahi Glass' own technology of producing carboxyl-substituted perfluorovinyl ethers, various types of Flemion membranes have been commercialized in response to users' requirements.

For the production of 35% caustic soda, a standard Flemion 230 is used advantageously with a current efficiency of 94%. On the other hand, Flemion 430 is suitable for the case where a low concentration caustic solution of 20–25% may be utilized directly from a electrolysis plant on site. A specially designed Flemion 330 is recommended for the production of 35% caustic potash with a current efficiency of 97–98% (13).

Recently, the Flemion 700 series, which has organic and inorganic composite texture, has been developed. With these types of membranes, gas bubbles can be removed easily from the membrane surfaces. This indicates the feasibility of the use of the Flemion 700 series with zero-gap between the membrane and electrode.

A new electrolytic process with a zero-gap cell, called the AZEC system, has attained drastic reduction in energy consumption combined with Flemion 723 or 753 and a new electrode system, which requires ca 2000 kW·h (dc) to produce one metric ton of NaOH. This means ca 30% less energy consumption than that required for the mercury process.

Flemion 753 is a composite membrane based on lamination of two films having different ion-exchange capacities (20).

Asahi Chemical's Membrane. Asahi Chemical has developed a two-layer membrane which is characterized by introducing a carboxylic acid layer on the cathode side of a sulfonate membrane. Also, a new sulfonate membrane with a shorter side-chain length was prepared as a starting material of such two-layer membranes. These membranes exhibit ca 96% of current efficiency in the production of 21–30% caustic soda (5).

By roughening the membrane surface on the side facing the catholyte, the accumulation of hydrogen bubbles on the surface is prevented and this allows an anode–cathode gap close to zero with virtually no increase in voltage, resulting in the reduction of the energy consumption. For example, at a current density of 4.0 kA/m^2 and a caustic soda concentration of 25%, 2100–2200 kW·h (dc)/t NaOH has been achieved in a commercial plant with a cell size of 1.5 m height and 3.6 m width (24).

Tokuyama Soda's Membrane. Tokuyama Soda has also succeeded in improving the electrolytic performance of Nafion-type membranes by chemical modification of their cathode-side surface to the carboxylic acid type.

Neosepta-F series (C-1000 for caustic concentration of 20–25% and C-2000 for that of ca 30%) has been developed (6).

Outlook

Besides the membranes described above, a large number of attempts to develop a perfluorinated membrane with high electrolyte performance have been made.

At present, ca 2.8×10^6 t caustic soda in combination with 2.5×10^6 t chlorine are produced in the world by brine electrolysis.

In the United States, the diaphragm process covers ca 60% of the total production. In Europe and Japan, the mercury process has been predominant, but the Japanese government has decided to change from the mercury process to nonmercury ones by the end of 1984 as a solution to the pollution problem.

Although the principal soda producers in Japan have already converted ca 60% of the total production to the diaphragm process, the rest will be converted to the membrane process, which is far superior to the diaphragm process both in quality and in energy-saving.

The total world production of caustic soda by the membrane process reached ca 430,000 t/yr in 1982, ca 230,000 t of which was produced in Japan and ca 200,000 t in the rest of the world.

The use of the membrane process will increase and it is estimated that the chlor-alkali producers in noncommunist countries will add ca 10×10^6 t/yr of new chlorine capacity by 1990 (25).

A wide variety of research work both on membrane material and on cell technologies continues in order to meet the requirements of users.

BIBLIOGRAPHY

1. A. Eisenberg and H. L. Yeager, eds., *Perfluorinated Ionomer Membranes*, ACS Symposium Series 180, American Chemical Society, Washington, D.C., 1982.
2. D. J. Vaughan, *DuPont Innovation* 4(3), 10 (1973).
3. *Chem. Week*, 35 (Nov. 17, 1982).
4. C. J. Hora and D. E. Maloney, *paper presented at the 152nd National Meeting, The Electrochemical Society*, Atlanta, Ga., 1977.

5. M. Seko, *paper presented at the 159th National Meeting, The Electrochemical Society*, Minneapolis, Minn., 1981.
6. T. Sata and Y. Onoue in ref. 1, pp. 411–425.
7. S. M. Ibrahim, E. H. Price, and R. A. Smith, *paper presented at the 25th Chlorine Plant Operations Seminar of the Chlorine Institute*, Atlanta, Ga., 1982.
8. H. Ukihashi and T. Asawa, *paper presented at the 151st National Meeting, The Electrochemical Society*, Philadelphia, Pa., 1977.
9. H. Ukihashi, *paper presented at the American Chemical Society/the Chemical Society of Japan Chemical Congress, Polymer Division*, Honolulu, Hawaii, 1979.
10. H. Ukihashi, *CHEMTECH*, 118 (Feb. 1980).
11. W. G. F. Grot, G. E. Munn, and P. N. Walmsley, *paper presented at the 141st National Meeting, The Electrochemical Society*, Houston, Texas, 1972.
12. D. A. Buttry and F. C. Anson, *J. Am. Chem. Soc.* **105,** 685 (1983).
13. H. Ukihashi and M. Yamabe in ref. 1, pp. 427–457.
14. U.S. Pat. 4,358,412 (Nov. 9, 1982), B. R. Ezzel, W. P. Karl, and W. A. Mod (to The Dow Chemical Co.).
15. U.S. Pat. 4,138,426 (Feb. 6, 1979), D. C. England (to E. I. du Pont de Nemours & Co., Inc.).
16. U.S. Pat. 3,528,954 (Sept. 15, 1970), D. P. Carlson (to E. I. du Pont de Nemours & Co., Inc.).
17. U.S. Pat. 4,138,373 (Feb. 6, 1979), H. Ukihashi, T. Asawa, M. Yamabe, and H. Miyake (to Asahi Glass Co., Ltd.).
18. U.S. Pat. 4,116,888 (Sept. 26, 1978), H. Ukihashi, T. Asawa, M. Yamabe, and H. Miyake (to Asahi Glass Co., Ltd.).
19. E. H. Price, *paper presented at the 152nd National Meeting, The Electrochemical Society*, Atlanta, Ga., 1977.
20. M. Nagamura, H. Ukihashi, and O. Shiragami, *paper presented at the Symposium on Electrochemical Membrane Technology in 1982 AIChE Winter Meeting*, Orlando, Fla., 1982.
21. U.S. Pat. 4,225,523 (March 10, 1981), H. Ukihashi, T. Asawa, and T. Gunjima (to Asahi Glass Co., Ltd.).
22. R. S. Yeo in ref. 1, pp. 454–473.
23. B. Kippling in ref. 1, pp. 475–487.
24. M. Seko, A. Yomiyama, and S. Ogawa, *paper presented at the Ion Exchange Membrane Symposium 1983, Society of Chemical Industry*, London, 1983.
25. S. C. Stinson, *Chem. Eng. News*, 22 (March 15, 1982).

MASAAKI YAMABE
Asahi Glass Company, Ltd.

PLASMA TECHNOLOGY

The correspondence between members of the four states of matter and the ancient "four elements" is often noted: solid = earth, liquid = water, gas = air, and plasma = fire. Most of the matter in the universe is in the plasma state. Plasmas consist of mobile, positively and negatively charged particles that interact because of Coulomb forces. Individual atoms and molecules have some internal properties that are plasmalike. Furthermore, liquids and solids exhibit the collective behavior of charged particles that is characteristic of the plasma state. Even matter that is not in what usually is described as the plasma state can be studied and described using the tools and concepts of plasma physics.

A widely accepted idea, called the Big-Bang theory, holds that, ca $(10\text{–}15) \times 10^9$ yr ago, all matter and energy in the universe existed in one superhot, dense plasma (1). Explosion, expansion, and cooling followed. Matter accumulated gravitationally to the point where nuclear-reaction reheating occurred. Natural re-formation of gaseous plasmas also results from electric fields, eg, lightning, and dissipation of kinetic energy, eg, meteorite impact (see Space chemistry).

According to some theories, life originated on earth by processes in which plasmas are ascribed a central role. The atmosphere of the early earth is thought to have contained significant amounts of H_2, H_2O, NH_3, CH_4, and H_2S. Laboratory experiments have shown that electric sparks, simulating early lightning bolts, produce heavier molecules from these simple gases; uv radiation from the solar plasma produces similar effects (2). These processes involving plasmas may have been the initial steps toward production of the amino acids and proteins on which terrestrial life is based.

Synthetic plasmas are produced in diverse ways, eg, in fluorescent tubes and nuclear explosions. Laboratory plasma physics has its roots in the study of gas discharges by Faraday in the 1830s and by Crookes and others in the 1870s. Electrical discharge devices were used almost exclusively in plasma studies until around 1940. Thereafter, high power r-f generators also were used to generate plasmas. The study of plasmas was given tremendous impetus in the 1950s by efforts to achieve controlled thermonuclear reactions, following the development of hydrogen-plasma thermonuclear weapons (see Fusion energy). The beginnings of space exploration in the 1950s gave further urgency to the study of gaseous plasmas. In the early 1950s, the collective behavior of electrons in conductors was studied, and the similarities and differences between gaseous and solid-state plasmas were enumerated and examined.

Types

The multitude of natural and synthetic plasmas is classified according to plasma temperature and density. Temperature is expressed in degrees kelvin and, equivalently, in electron volts. One eV is equal to 11,600 K, according to $E = kT$, where k is Boltzmann's constant. Densities of electrons and ions are measured by the number of charged particles per cubic centimeter of charged particles and they are determined by the number of particles of each type and the volume accessible to them.

Plasmas are most broadly classified as either condensed-state or gaseous plasmas. The former refers both to systems of electrons that are condensed onto, and bound to, nuclei in atoms and molecules, and to electron-containing systems of atoms or molecules that are condensed to liquids or solids. The latter refers to collections of unbound electrons and ions. This distinction is useful, although it is somewhat artificial since the densities of compressed gaseous plasmas can exceed those of liquids and solids. Figure 1 shows the temperature and density regions applicable to condensed-state and gaseous-state plasmas (3). The wide ranges, ie, factors of 10^7 in temperature and 10^{26} in density, are indicative of the pervasive nature of plasmas.

The electrons around nuclei in atoms and molecules, including ions, are mobile, although they are bound by the central potential. They have velocities, and equivalent kinetic temperatures and densities, that are estimated easily from the binding energies and radii of orbitals (4–5). The range of values is indicated in Figure 1. The number of electrons in free or condensed atoms (<100) and in most molecules ($\leq 10^6$) is small, as is the volume available to them. Hence, the densities of electrons in atoms and molecules are high. Such electrons have well-defined velocities or momentum and spatial distributions that are unlike those in other condensed and gaseous plasmas. Atomic and molecular plasmas are characterized by near equality of the interaction length for collective behavior and the dimensions of the region of confinement. Nonetheless, plasma behavior in atoms and molecules manifests itself in photoabsorption cross-sections in the vacuum uv and soft x-ray regions and in the efficiency with which energetic particles are slowed down when penetrating and exciting matter (6). A recent review of collective aspects of atomic dynamics is given in ref. 7.

Plasmas composed of carriers in metallic and semimetallic, and semiconducting liquids and solids also vary widely in their characteristics. Electron plasmas in liquid and solid metals under ordinary conditions have equivalent temperatures of ca 10^4 K and densities of ca 10^{22} particles/cm^3, with semimetals having lower equivalent densities and temperatures (8). The temperatures of electron and hole plasmas in semiconductors are from ca 0 K to vaporization points. They also vary widely in density, 10^{12}–10^{20} particles/cm^3, depending on the number of their free carriers (9). Plasmas in metals and semimetals usually have equal numbers of electrons and positive charges. Some plasmas in semiconductors are neutral and others are characterized by an excess of either charge, depending on the dopant concentration.

The temperatures, densities, and sizes of synthetic gaseous plasmas vary widely (3). Their temperatures range from flames, in which ionization may occur at ca 10^3 K, to those of plasmas produced in fusion-energy experiments with associated temperatures of ca 10^8 K. Their densities vary from ca 1 nPa (10^{-11} mm Hg or 10^6 particles/cm^3) to densities in excess of normal solid values that are produced in experiments using photon, electron, or ion beams to compress matter.

Temperatures of natural gaseous plasmas vary widely because of their very great differences in ionizations and sizes (see Fig. 1) (3). Cold, sparse plasmas occur between stars, whereas plasmas between planets are hotter and denser. Ionospheric plasmas are relatively cool but are denser. The density of the solar corona is similar to that of the ionosphere but the former is much hotter. The bow shock, where the solar wind encounters the earth, is substantially hotter than the interplanetary plasma, but the former is farther from the earth and less dense than the ionosphere. At the opposite extreme, the plasmas of stellar interiors are heated to many 10^6 K and condensed to densities comparable to those of liquids and solids.

Synthetic and natural plasmas most commonly consist of electrons and ions. However, finely divided charged particles, ie, colloids or dust, also exhibit plasma behavior (10). Usually, the average density of opposite charges in a gaseous plasma

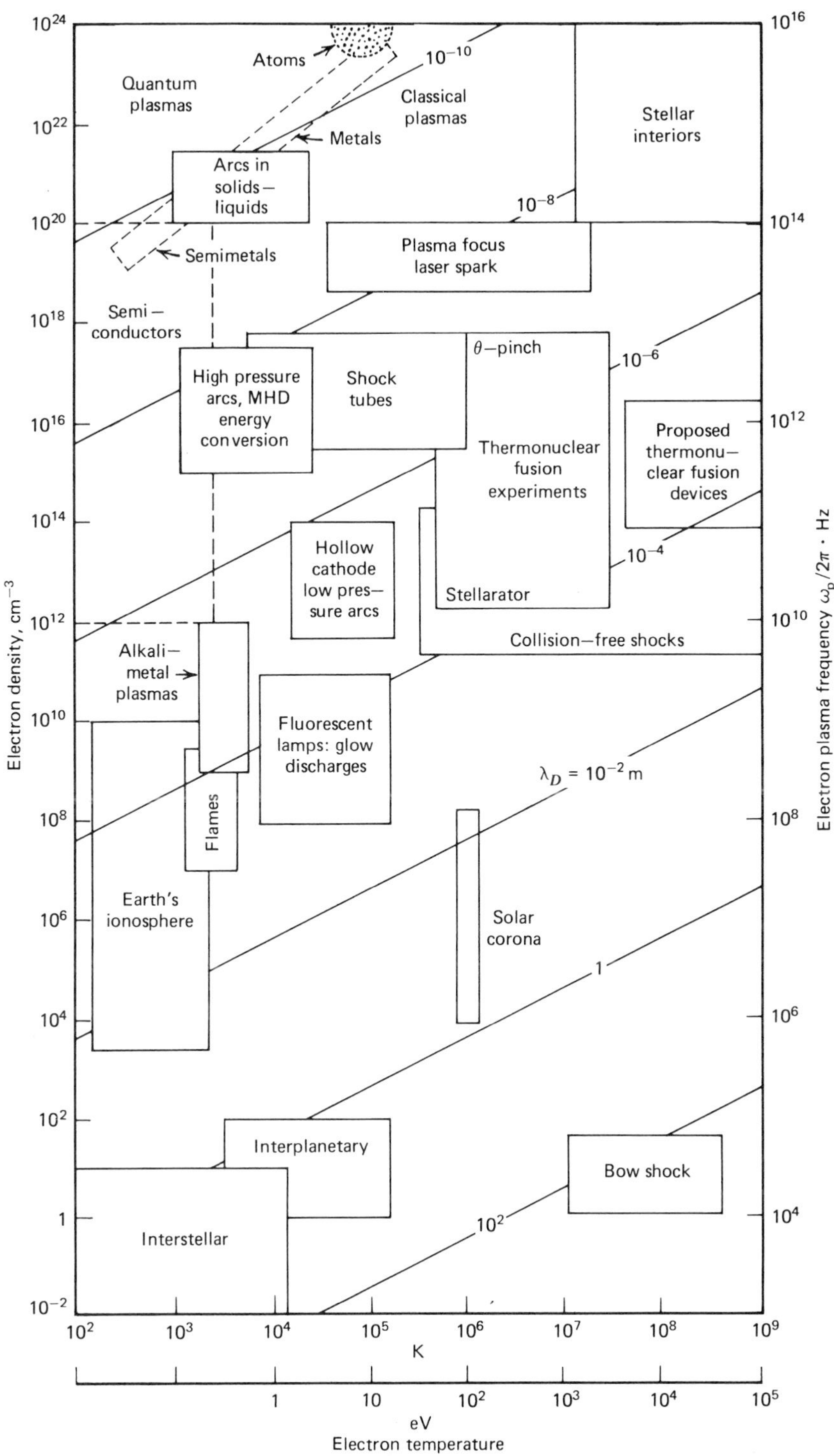

Figure 1. Electron temperature and density regions for plasmas (3). The Debye screening length (λ_D in meters) is given by the diagonal lines and the plasma frequency is given on the right-hand axis. The atomic regime is shaded. Condensed-state plasmas, indicated by dashed lines, tend to have relatively low temperatures and high densities. Plasmas in metals and semimetals fall along the line separating degenerate quantum plasmas from nondegenerate classical plasmas. Gaseous plasmas, shown by solid lines, have characteristics that vary widely. MHD = magnetohydrodynamic.

is equal and the plasmas are neutral. However, nonneutral plasmas are not uncommon in the laboratory; they result from preferential escape from the plasma of particles of one sign, usually fast electrons or, conversely, the injection of either electrons or ions into a gaseous plasma.

Principles

Gaseous Plasmas. The study of plasmas has much in common with the microscopic kinetic theory of gases and the macroscopic theories of fluid dynamics. For example, attention must be paid to both the particle collisions within a plasma and to its overall fluid behavior (11–13). However, the existence of mobile charges in plasmas and their responses to each other and to applied fields complicate the study of plasma behavior on microscopic and macroscopic levels.

The distribution of equilibrated particle velocities for atoms and molecules in gases and for ions and electrons in plasmas is Maxwellian:

$$n(v) = 4\,\pi n \left(\frac{m}{2\,\pi kT}\right)^{3/2} v^2 \exp\left(-\frac{mv^2}{2\,kT}\right)$$

where particles of mass m and density n (particles/cm^3) have temperature T (K). The most probable velocity is $v_o = (2kT/m)^{1/2}$, with the mean velocity $\bar{v} = (8kT/\pi m)^{1/2}$. The particle velocities and the frequencies and violence of collisions in a gas and plasma increase as the square root of the temperature. As the temperature of a gas increases, collisions become sufficiently energetic to knock electrons off the constituent particles. Conventionally, a fractional ionization of a few percent is required before a hot gas is termed a plasma. The equilibrium degree of ionization depends primarily on the plasma temperature, as indicated in Figure 2 (14). Higher temperatures and the associated higher particle velocities make it possible to ionize more tightly bound electron orbitals. The electron and ion temperatures determine ionization rates and the time required for the electron and ion distributions to equilibrate internally and with each other.

Two lengths are important for ordinary gases: the average interparticle spacing $d = n^{-1/3}$ and the mean free path between collisions $\lambda = (\pi n D^2)^{-1}$, where D is the atomic or molecular particle size. The important length parameter for plasmas is the Debye length λ_D, which is given by $\lambda_D = (kT/4\pi ne^2)^{1/2}$, where e is the charge on the electron. The number of electrons within a Debye sphere around an ion is $4\pi\lambda_D^3 n/3$. The Debye length has various interrelated physical interpretations. It can be thought of as the distance over which the electrons shield or screen the positive ionic charges from the remainder of the plasma; the approximate distance at which the attractive electric potential is balanced by the thermal kinetic energy; and the distance below which an ionized gas acts as individual particles and above which it exhibits collective behavior. Since collective response is characteristic of the plasma state, λ_D must, by definition, be smaller than the dimensions of the plasma; then, collective behavior prevails. If λ_D is larger than the size of the plasma, then ordinary individual-particle gaseous behavior occurs.

The primary characteristic frequency of an ordinary gas is the rate of collision $f = \bar{v}/\lambda = \pi \bar{v} n D^2$. A wide variety of special frequencies exists in plasmas, most notably

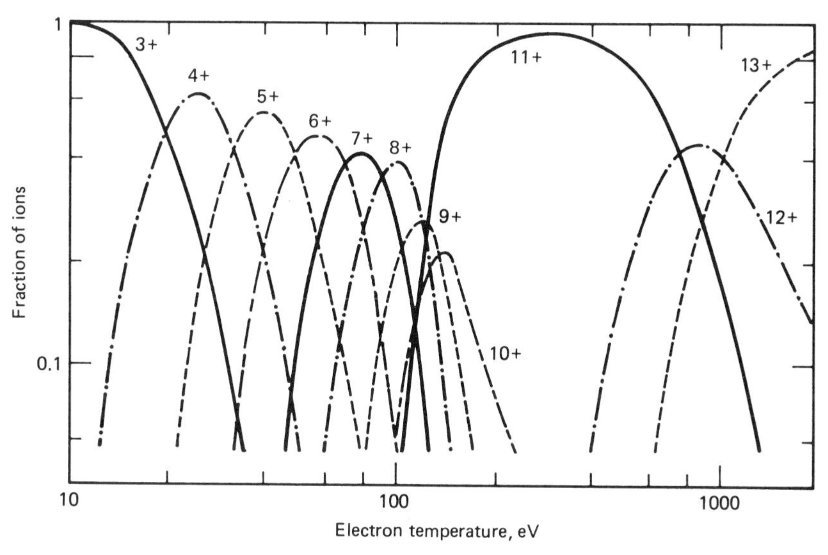

Figure 2. Calculated distribution of ionization states versus temperature for an aluminum plasma (14). The higher electron velocities at higher temperatures result in greater ionization.

the plasma frequency $\omega_P = (4\pi ne^2/m)^{1/2}$. This frequency is a measure of the vibration rate of the ions and electrons relative to each other. For true plasma behavior, ω_P must exceed the particle-collision rate f. The plasma frequency plays a central role in the interactions of electromagnetic waves with plasmas. The frequencies of electron plasma waves depend on the plasma frequency and the thermal electron velocity. They propagate in plasmas, because the presence of the plasma oscillation at any one point is communicated to nearby regions by the thermal motion. The frequencies of ion-plasma waves, also called ion acoustic or plasma sound waves, depend on the electron and ion temperatures as well as on the ion mass. Both electron and ion waves, ie, electrostatic waves, are longitudinal in nature; that is, they consist of compressions and rarefactions along the direction of motion.

The velocities, lengths, and frequencies are intrinsic to gases and plasmas, independent of incident radiation or existing fields. However, the richness of plasma phenomena only appears in the presence of static or dynamic fields. External static electric fields tend to separate and accelerate the plasma charges. Such fields and the resultant electron motions produce and heat plasmas. Magnetic field effects in plasmas are so important that, at one time, plasma physics and magnetohydrodynamics (MHD) were practically synonymous. However, some plasmas are not magnetized, and some MHD processes do not involve plasmas, eg, the fluid motions in the earth's core which produce the terrestrial magnetic field.

Transverse electromagnetic waves propagate in plasmas if their frequency is greater than the plasma frequency. For a given frequency ω, there is a critical density $n_c = m\omega^2/4\pi e^2$ above which waves do not penetrate a plasma. The propagation of electromagnetic waves in plasmas has many uses, especially as a probe of plasma conditions.

The presence of a static magnetic field within a plasma affects microscopic particle motions and microscopic wave motions. The charged particles execute cyclotron motion and their trajectories are altered into helices along the field lines. The radius of the helix, or the Larmor radius, is given by $mcv_{\perp}/qB$ for a particle of mass m and charge q, the velocity $v_{\perp}$ normal to the magnetic field of strength B. The cyclotron frequency, which is introduced by presence of the field, is $\omega_c = qB/mc$. The particle–field interaction is the means by which magnetic fields can exert pressures on plasmas and vice versa.

The presence of static magnetic fields does not alter the propagation of longitudinal electrostatic electron or ion waves if the propagation direction is parallel to the field. However, propagation that is orthogonal to the field involves new frequencies that depend on the field strength. In contrast, the propagation of transverse electromagnetic waves in plasmas is altered by a magnetic field regardless of the relative geometry of the direction of motion and the field vector.

Magnetic fields introduce hydromagnetic waves, which are transverse modes of ion motion and wave propagation that do not exist in the absence of an applied field. The first of these are Alfven waves and their frequency depends on B and ρ, the mass density. Such waves move parallel to the applied field with velocity $v_A = B/\sqrt{4\pi\rho}$ and are similar to the waves that travel along a string. Magnetosonic waves are a second type of hydromagnetic wave, and they propagate perpendicular to the magnetic field; their frequency depends on the Alfven and the acoustic velocities. Hydromagnetic waves are electromagnetic waves. Even though the applied B field is static or nonoscillatory, the waves are transverse and are characterized by oscillatory electric (E) and magnetic (B) components.

A lucid introductory discourse on plasma waves and a tabulation of their characteristics are given in ref. 12. Useful plots of the dispersion relations for various frequencies, field conditions, geometries, and detailed mathematical relationships are given in ref. 15.

Gaseous plasmas are far from equilibrium and, therefore, exhibit microscopic or particle instabilities, and macroscopic or hydromagnetic instabilities (15–16). Microscopic instabilities are caused by departures from equilibrium Maxwellian distributions for the electrons or ions. Such departures can result from lack of equilibrium, eg, when a plasma expands while cooling, to anisotropies in the velocity distribution caused by applied magnetic fields, or the motion or streaming of a particle beam through a plasma. Macroscopic instabilities produce the motion of the plasma as a whole. Causes include pressure or density gradients or magnetic field curvature. All instabilities represent the tendency of plasmas to reach equilibrium more quickly than is possible by ordinary collisions alone. Instabilities can reduce plasma confinement times by many orders of magnitude; this is a significant problem in fusion research. The conduction and diffusion of energy through plasmas, and the way in which these processes are influenced by magnetic fields are described in refs. 12 and 13.

The high energy densities of many gaseous plasmas raise safety concerns. The sources of energy used to produce and heat plasmas, eg, steady-state, high voltage, and high current generators, and capacitors for pulsed electron-discharge heating and laser beams, can be hazardous (17–18). Work with plasmas usually requires careful attention to proper electrical safety cautions and to eye hazards. Even in the absence of lasers (qv), plasmas can pose a threat to vision because they often are very bright and can emit dangerous levels of uv radiation. X radiation from plasmas usually is not a safety concern, since most energy from multimillion (10^6)-degree Kelvin plasmas is soft and does not escape from the experimental chamber or traverse significant distances in air. The hard x rays emitted by most plasma sources usually are of very low intensity. However, high voltage ($>10^4$ V), plasma-generating machines and some fusion-energy research devices emit unsafe x-ray emissions. Shielding, eg, lead or concrete, and distance from the source reduce exposures to acceptable levels.

Production. Sources of matter and energy are necessary for the production of gaseous plasmas, and such plasmas serve as sources of matter and energy in their applications; ie, gaseous laboratory plasmas can be viewed as transducers of matter and energy. The initial and final forms of the material that enters a plasma and the requisite energy vary widely, depending on the particular plasma source and its utilization.

The molecules that are disassociated and the atoms that are ionized during plasma production can be in any state at the start. Steady-state plasmas are formed most often from gases, although solid sources of matter also are used. Gases and solids routinely serve as sources of material in pulsed plasma work.

The energy for plasma formation may be supplied in a variety of ways. The source may be internal, eg, the release of chemical energy in flames. A second energy source involves electric discharges through a plasma (19). The externally applied d-c or a-c field internally heats the plasma, which is part of the electric circuit. Electromagnetic fields, eg, in the microwave range (20), can be used to form plasmas that interact with or feed back into the source. The fourth kind of energy source includes externally produced beams of photons, eg, laser beams or energetic particles, that create a plasma by their impact and absorption independent of the source. Plasmas may be produced in a fifth manner by strong shock waves. Chemical, discharge, and high frequency sources often produce steady-state plasmas, whereas beam and shock heating usually produce pulsed plasmas.

In the production of plasmas by steady-state electric discharge heating, the current across two electrodes in the gas can vary widely, as indicated in Figure 3 (21). At low values, externally induced ionization is needed to maintain current flow. Milliampere currents in gases with pressures at ca 100 Pa (10^{-3} atm) produce a glow discharge that is sustained by electrons produced by positive-ion bombardment of the cathode. If the current is increased to ca 10 A, a self-sustaining arc forms and exists even at pressures above 101 kPa (1 atm).

Modification and Dissipation. Changes in the composition or energy of a plasma after it is formed often are desired. For example, materials can be introduced into a plasma and excited, thereby producing information for their spectrochemical analysis. Plasma heating is a more common modification. Many energy sources and coupling mechanisms can be employed to heat plasmas (12). Plasma formation and heating often are driven by the same energy source. However, an entirely separate second source can be used for heating. For example, laser-beam and particle-beam sources often are used to heat plasmas produced by electric discharges. Energetic neutral, ie, atomic, beams are employed to heat plasmas that are formed initially by discharges.

Restraining a gaseous plasma from its tendency to expand, and compressing it

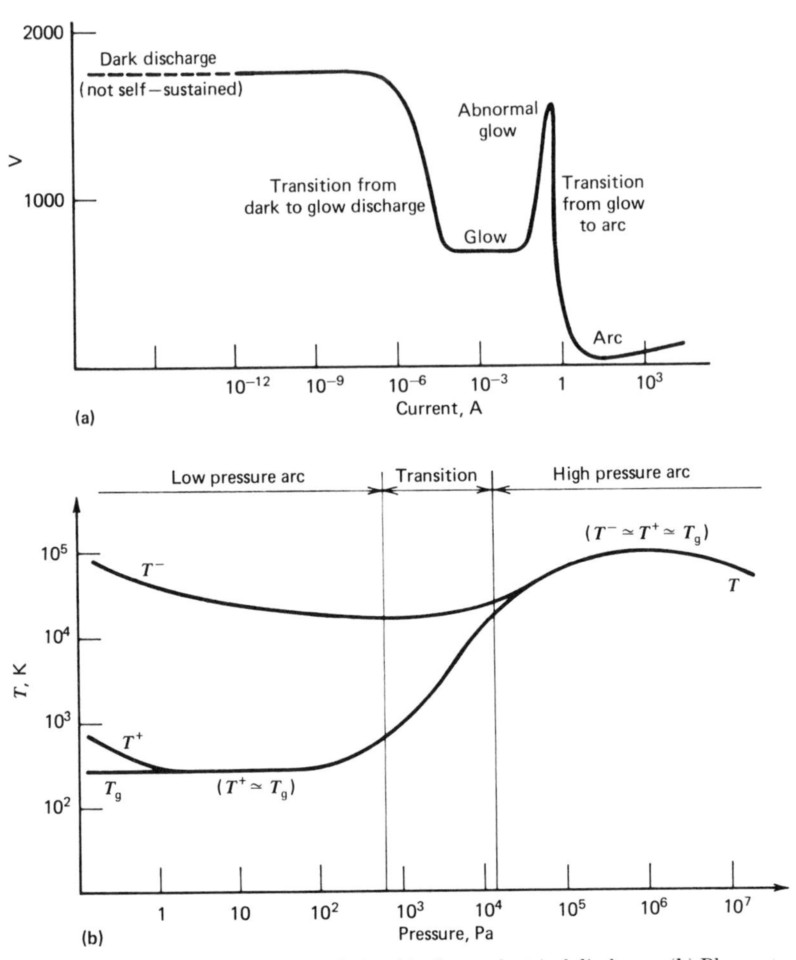

Figure 3. (**a**) Typical voltage-current relationship for an electrical discharge; (**b**) Plasma temperature versus pressure, showing the low pressure nonequilibrium cold-arc and the high pressure equilibrium hot-arc regimes (21). Temperatures are indicated for the plasma electrons (T^-) and ions (T^+) and for the neutral gas (T_g).

also are forms of plasma modification. The two reasons for plasma confinement are maintenance of the plasma and exclusion of contaminants. Plasmas may be confined by surrounding material, eg, the technique of wall confinement (22). The second approach to confinement involves the use of magnetic fields. The third class of confinement schemes depends on the inertial tendency of ions and associated electrons to restrain a plasma explosion for a brief but useful length of time, ie, forces active over finite times are required to produce outward particle velocities. This inertial confinement is usually, but not necessarily, preceded by inward plasma motion and compression.

Low density plasmas are confined magnetically by a variety of field configurations that are designed to prevent particle losses and overall fluid instabilities. Low temperature plasmas used for sputtering and high temperature, fusion-research plasmas are subjected to magnetic confinement, as shown in Figure 4 (23). Magnetron sputtering sources include a variety of magnetic-field configurations designed to restrain

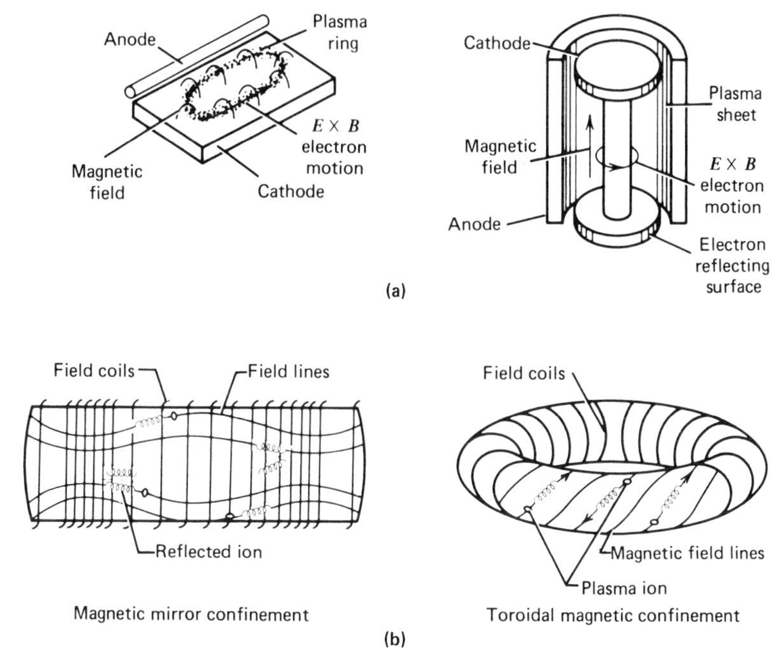

Figure 4. Schematics of magnetic-confinement geometries. (**a**) Planar and cylindrical geometries for magnetron sputtering sources (23). (**b**) Open-ended mirror and closed toroidal geometries for containing fusion-research plasmas (24).

plasma particles near the cathode from which atoms are sputtered by impact of ions (23). The two basic magnet configurations for fusion-research plasmas are open and closed (24). Increasing strength along magnetic field lines produces magnetic mirrors, which exert a retarding force on particles that tend to escape the plasma. However, the plasma in a simple mirror is not stable against overall fluid motion. For stability, the B field must increase in all directions from the plasma outward, ie, the plasma must be in a minimum B location to be magnetohydrodynamically stable against gross translational motion. Curvature of the magnet geometry for mirror systems produces such stability. Nevertheless, sufficiently energetic particles can penetrate the mirror, producing end losses. Closure of the field lines to form loops counteracts such losses, but the field strength decreases radially from the center, and particles tend to drift out of the plasma normal to planar, closed field lines. Addition of out-of-plane fields, which yields helical fields in a toroidal configuration, produces both particle confinement and MHD stability. The history and physics of magnetic confinement of fusion plasmas are discussed in ref. 25 (see also Coal, magnetohydrodynamics).

High density plasmas can be confined and compressed magnetically by fields produced by strong electric currents flowing in and heating the plasmas, as well as by externally applied fields (26). The radial force produced by the magnetic field, which affects the flowing plasma electrons (the $v \times B$ force), produces inward plasma motion and compression. Such confinement is not stable and is short-lived, but it produces plasmas with high energy densities. High density plasmas are studied primarily as x-ray and neutron sources.

Inertial confinement depends on plasma heating outpacing plasma expansion. When a target is struck by a short, $< \mu s$, pulse of laser photons, electrons, or ions, it is heated on a time scale comparable to, or shorter than, that on which the resulting plasma expands and cools; this is true for targets of any geometry. If a cylindrical or spherical target is simultaneously irradiated from two or more sides, the outer surface separates violently, producing a rocket effect that accelerates the remainder of the target inward, thereby compressing and heating the central materials (27).

A plasma loses matter and energy through particle escape and radiation of photons. When matter or energy are no longer supplied to a plasma, it expands and cools before interacting with the surroundings and totally dissipating. Such processes must be offset during the lifetime of a plasma since the material and energy from a plasma often produce undesirable effects. Energetic particles impinging on solids near plasmas can profoundly alter the material properties. Plasma effects on materials are especially important in proposed thermonuclear fusion reactors. Materials in the reactors must withstand high temperatures and stresses for an economically useful period (28).

Diagnostics. Plasma diagnostics is the determination of conditions within gaseous plasmas. The term also refers to the broad collection of experimental techniques and the associated calibration and analytical methods used to assess the characteristics of hot plasmas. Noteworthy properties include the identities and distributions of the various particle species, ie, neutrals, electrons, and ions, and their velocity distributions as functions of space and time. Quantities, eg, plasma flow velocities, turbulence, instabilities, and flow of energy into and out of plasmas by various means, also are desired. All diagnostic methods have limited resolution spatially, temporally, and spectrally. Therefore, plasma characteristics that are derived from measurements always are averaged. Measured quantities in plasma diagnostics usually are integrated along a line of sight through the plasma to the instrumentation, yielding spatially integrated results. No one method for assaying plasmas is universally applicable. A variety of diagnostic tools is needed to characterize a plasma empirically and to compare its empirical and theoretical characteristics.

Diagnostic methods can be categorized most broadly according to those that involve external probes of plasmas or those that rely only upon plasma self-emission. Probes of plasmas include solid instruments inserted into gaseous plasmas and beams of quanta, ie, photons or charged particles, which are shot through plasmas. Electric voltage and current measurements for discharge-heated plasmas provide useful information; thus, the energy source also is a plasma probe. Self-emission includes radiated fields, photons, electrons, ions, and neutrons, ie, any plasma effect or constituent. Gaseous-plasma diagnostic methods are described in refs. 29 and 30. Useful compilations of gaseous plasma characteristics and the diagnostic techniques employed to measure them are given in refs. 31 and 32.

Electrodes or Langmuir probes may be inserted into plasmas that are large enough (>1 cm) and relatively cool (<10^4 K). The net current to the probe is measured as a function of the applied voltage. Electron temperatures, electron and ion densities, and space and wall potentials may be derived from the probe signals. Interaction of plasmas with solid probes tends to perturb plasma conditions.

Monochromatic light from short-pulse lasers may be focused into plasmas that are large enough (≥1 cm), regardless of their temperatures, with little alteration of plasma conditions. Elastic, ie, Thomson, scattering of the light at right angles to the incident direction is measured. Spectral or Doppler broadening of the light, which results from motion of the scattering electrons, yields the electron temperature. The overall scattered intensity is a measure of the electron density. Thomson scattering is a primary diagnostic method for magnetic-fusion research plasmas.

Diagnostic techniques that involve natural emissions are applicable to plasmas of all sizes and temperatures and, clearly, do not perturb the plasma conditions. They are especially useful for the small, high temperature plasmas employed in inertial fusion-energy research. Their small sizes (≤1 cm) and brief lifetimes (<1 μs) preclude the use of most probe techniques, although laser-pulse imaging of such plasmas yields valuable spatial information. Diagnostic methods involving plasma emissions give the most useful results, if the methods can be used to resolve spatial and temporal nonuniformities in plasma emission. However, useful information can be obtained without space and time resolution, if adequate spectral resolution is available. Figure 5 is an example of a space- and time-integrated photon spectrum that yielded temperature and density data (33). Diagnostic methods based on x-ray emission are especially useful, since the x rays are preferentially emitted from the hottest and densest parts of plasmas.

Natural Gaseous Plasmas. Lightning is the most common atmospheric, plasma-related phenomenon (34–35). The separation of charges in clouds, and between clouds and the ground, produces potentials as high as 100 MV. The currents of the discharges are as high as 100 kA. Spectroscopic data have yielded considerable information on plasma conditions within a lightning discharge (36).

Meteors produce atmospheric plasmas as their kinetic energy is converted to thermal energy (37). Most particles from space are consumed before they reach an

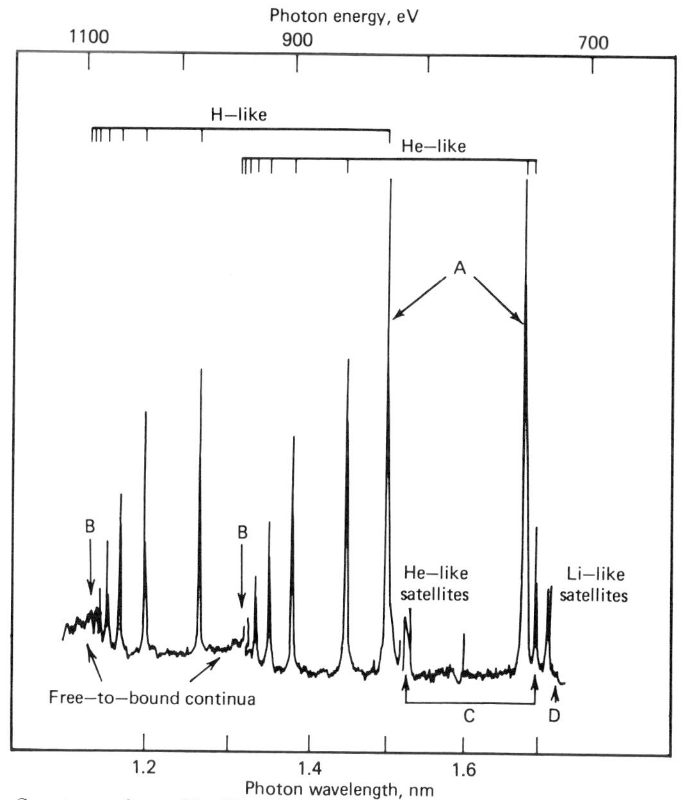

Figure 5. Spectrum of a multimillion-degree laser heated plasma from one-electron (H-like) and two-electron (He-like) fluorine ions as a function of photon energy (top) and wavelength (bottom) (33). Lines labeled A reflect the plasma temperature, B and C yield density values, and D indicates the degree of equilibrium.

altitude of 50 km. Meteors are of little practical use, although radio waves can be bounced off the plasmas left in their wakes.

Auroras are observed primarily at polar latitudes near the geomagnetic poles and resut from impact on the atmosphere of energetic particles that are guided by the earth's field (38). The emissions are produced at altitudes of one to several hundred kilometers by electrons, with energies up to 10 keV, or by protons, with energies as high as 200 keV. Ionization densities that are produced by auroras occasionally reach plasma levels, following intense solar flare activity.

Absorption of solar uv radiation high in the atmosphere produces a tenuous but important plasma, the ionosphere (39). Many physical processes occur in the ionosphere during the creation and loss of free electrons (40). As shown in Figure 6, the electron density can exceed 10^6 e/cm^3 and the electron temperature generally is less than 1000 K (0.1 eV) (41). Atmospheric motions below, and solar activity above the ionosphere influence its characteristics, which vary markedly with the time of day and with the solar cycle and emission. Understanding of the ionosphere has led to the increasingly effective use of it in long-range communication (see Communications and Space Travel).

The ionosphere is part of the larger magnetosphere, which is a cavity in the stream of particles from the sun; the cavity is produced by the earth's magnetic field (42). The ionosphere and the Van Allen radiation belt lie within the plasmasphere, which extends to a maximum distance of ca 15,000 km above the earth's surface.

Magnetospheric plasmas are produced and heavily influenced by solar emissions and activity and by magnetic fields of the planets. Plasmas between planets result from solar-emission processes alone. Protons in the solar wind have low densities (10–100/cm^3) and temperatures below 10^4 to more than 10^5 K (1–10 eV). Their average outward kinetic energy from the sun is ca 400 eV (43). The various zones and phenomena from the sun's visible surface to the upper atmosphere of the earth have been reviewed (44–46).

Classical astronomy is largely concerned with the classification of stars without regard to the details of their constituent plasmas (47). Only in the recent past have satellite-borne observations begun to yield detailed data from the high temperature regions of other stellar plasmas (see also Space chemistry). Cosmic plasmas of diverse size scales have recently been discussed (48).

Plasmas in Condensed Matter. In contrast to gaseous plasmas, which are closely related to kinetic and fluid theories, plasmas in condensed matter are intimately related to the theory of solids (9,49–50). The formation of a diatomic molecule is a good example of the freedom and high velocities that electrons experience in the condensed state. Electrons that initially are confined to either atom bond the atoms; therefore, they are free to range over the larger molecular volume. In the buildup of larger aggregates of condensed matter, addition of more atoms similarly expands the accessible volume. Bonding electrons increase their velocities when atoms form into molecules and condensed matter.

The energy levels of bonding electrons in a conductor are shown in Figure 7 (51). Since electrons are Fermions, they cannot occupy the same spatial and energy states simultaneously, according to the Pauli exclusion principle. Energy bands form as more atoms agglomerate, forming a solid. They may be continuous, eg, for metals, or have an energy gap, eg, narrow for semiconductors or wide for insulators. The Fermi distribution or the distribution of electrons in metals is uniform with energy. The bands

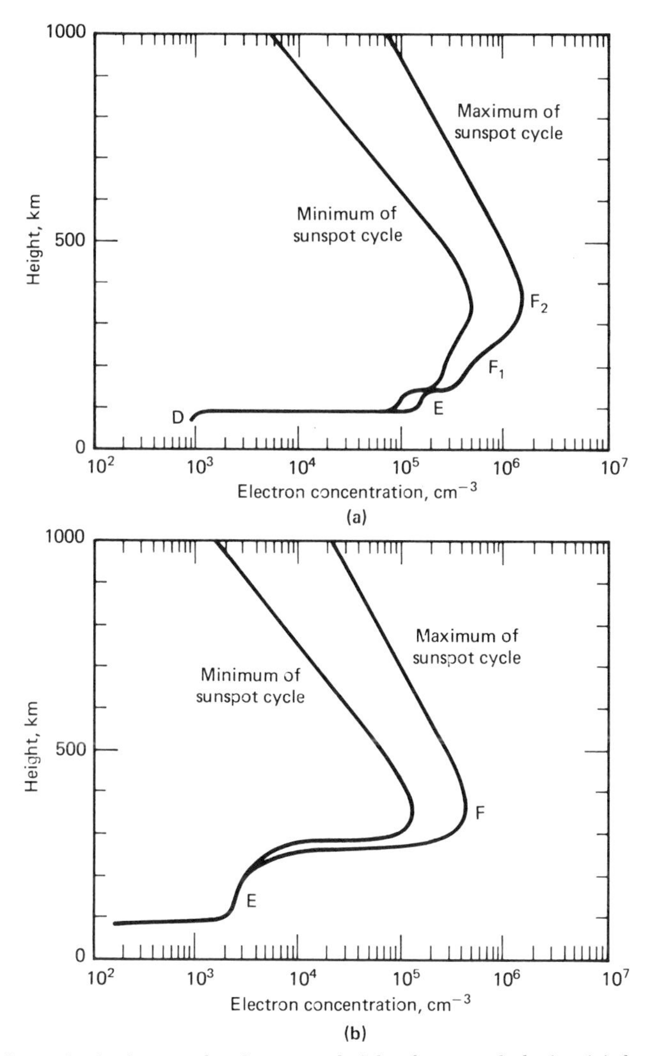

Figure 6. Ionospheric electron density versus height above earth during (**a**) day and (**b**) night, at the extremes of the 11-yr sunspot cycle (40). E, F, F_1, and F_2 are conventional labels for the indicated regions of the ionosphere.

and the distribution are filled to an energy called the Fermi level E_F. Electrons at E_F move with velocities near 10^8 cm/s, which is equivalent to temperatures near 1 eV, even though the lattice is cold (ca 300 K). Densities of mobile electrons in metals are ca 10^{22} particles/cm^3. The presence of plasmas in metals is readily shown by the excitation of electron plasma waves or plasmons. The characteristic energies of plasmas introduce additional peaks into electron and x-ray spectra from metals (50).

In perfect semiconductors, there are no mobile charges at low temperatures. Temperatures or photon energies high enough to excite electrons across the band gap, leaving mobile holes in the Fermi distribution, produce plasmas in semiconductors. Thermal or photoexcitation produces equal numbers of electrons and holes, similar to the situation in gaseous plasmas where the charges of electrons and ions usually

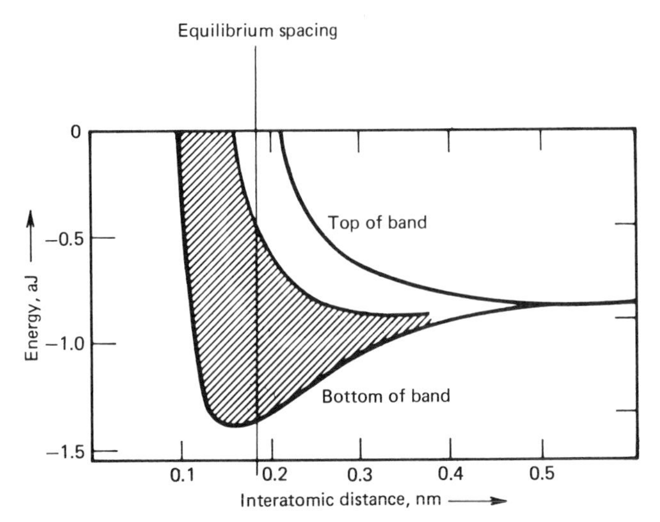

Figure 7. Schematic energy levels of a solid as a function of interatomic distance (51). A band of states obeying Fermi distribution is required by the Pauli principle. High electron velocities and equivalent temperatures exist in conductors even when the lattice is at ordinary temperatures. To convert aJ to Rydberg energy, multiply by 0.4587.

balance. Noteworthy are the photoexcited plasmas, which are characterized by particle densities of ca 10^{17}–10^{18}/cm^3 and which exist stably in Si at temperatures below 20 K and in Ge below 8 K (52–53).

Impurity-produced plasmas in semiconductors do not have to be compensated by charges of the opposite sign. They can be produced by introduction of either electron donors or electron scavengers, ie, hole producers, into semiconductor lattices (see Semiconductors, fabrication and characterization). Their densities range from a lower limit set by the ability to produce pure crystals (ca 10^{12} particles/cm^3) to values in excess of 10^{20} particles/cm^3. Plasmas in semiconductors generally are dilute, so that the Pauli principle cannot be used to determine their energy distribution. They are Maxwellian and have the same temperature as the lattice in equilibrium. The uncompensated nature of plasmas in semiconductors and, especially, their unique and very wide ranges of density and temperature, make them particularly interesting and potentially useful. Numerous plasma effects, including waves and instabilities, are studied in semiconductor plasmas (54–55). Effects peculiar to solids, eg, phonon scattering of plasma particles and anisotopic effective masses, also are unique to semiconductor plasmas.

Plasmas in gaseous and condensed states are related by more than the principles that govern them. For gaseous matter, there is a continuum of behavior from a low density, Maxwellian plasma to a high density, Fermion plasma. The boundary region in the density–temperature plane is given by the equality of the thermal energy kT_F, and the Fermi energy E_F, namely $kT_F = E_F = \bar{h}^2/2m_e(3\pi^2 n_F)^{2/3}$, where h is Planck's constant (56). Densities less than n_F indicate classical, nondegenerate behavior, whereas those above n_F imply quantum, degenerate conditions, such as can be found in condensed-state plasmas (see Fig. 1).

Contrasts between gaseous and condensed plasmas are summarized in refs. 8, 9, and 57. Fundamental differences between gaseous and condensed plasmas include their states of excitation and their characteristic lengths. Gaseous plasmas are produced by classical, ie, collisional, effects. They always are out of equilibrium thermodynamically and are accompanied by photons. In contrast, solid-state metallic plasmas are produced by quantum effects. They can be in or near the ground state and can exist in the absence of photons. Instabilities are common in gaseous plasmas but can be avoided or excited as desired in condensed-state plasmas. Long, mean free paths, compared to the plasma size, are common in gaseous plasmas. However, in solids, scattering usually produces short mean-free paths in comparison to the millimeter sizes of ordinary samples. Scattering often interferes with plasma behavior in semiconductor plasma experiments, a fact that dampens prospects for technological use of solid-state plasma devices.

Uses

Radiation Sources. Ordinarily, electron beams are produced from solids *in vacuo* by thermal or field-assisted processes. Plasmas also serve as electron sources, but they are more uniquely used as ion sources. Ions can be produced by sputtering and field-assisted processes in the absence of plasmas. However, most ion sources involve plasmas (58).

Pulsed plasmas containing hydrogen isotopes can produce bursts of alpha particles and neutrons as a consequence of nuclear reactions. The neutrons are useful for radiation-effects testing and for other materials research. A dense plasma focus filled with deuterium at low pressure has produced 10^{11} neutrons in a single pulse (59). Intense neutron fluxes also are expected from thermonuclear fusion-research devices employing either magnetic or inertial confinement.

Plasmas frequently are used as sources of incoherent and coherent, ie, laser, electromagnetic radiation. Infrared radiation is emitted by plasmas, but they usually are not employed expressly as ir sources. Visible light sources involving steady-state plasmas, in which the electrons are hot compared to the ions, are common. Fluorescent lights and carbon-arc searchlights are examples. Recently developed plasma displays that are flat and may replace bulky cathode-ray tubes also depend on emission of visible light. Direct-current plasmas emit visible and uv radiation for chemical analysis. Pulsed sources of visible and uv light for high speed photography usually involve plasmas. Plasma uv sources are used commercially for production of microelectronics circuits by lithography. Multimillion-degree ($\times 10^6$ K) plasmas provide uniquely short pulses of incoherent uv and soft x radiation for spectroscopy, materials analysis, and other applications.

Lasers act as sources and, sometimes, as amplifiers of coherent ir–uv radiation. Excitation in lasers is provided by external particle or photon-pump sources. The high energy densities required to create inverted populations often involve plasma formation. Certain plasmas, eg, cadmium, are produced by small electric discharges, which act as laser sources and amplifiers (60). Efforts are being directed to the improvement of the energy-conversion efficiencies at longer wavelengths and the demonstration of an x-ray laser in plasma media (61).

Chemistry. The material and energy available in plasmas is used to excite materials and drive chemical reactions. The unique characteristics of plasmas, especially their abundance of energetic quanta, are exploited in plasma-chemical applications (62–63).

The analysis of existing materials, and the production of new chemicals and materials, involve a gamut of gaseous plasma sources (see Fig. 3). Nonequilibrium or cold plasmas, in which the ion or gas temperature is much less than the electron temperature, are widely used. In these, the electrons provide energy that induces excitations and reactions without excessive heating of the desired products or the surroundings by the heavier particles in the plasmas. Equilibrium or hot plasmas, in which the electrons and ions are characterized by approximately the same temperature, also are used in plasma chemistry, especially for spectrochemical analysis and the processing of refractory materials (64). Plasma-electron temperatures below 50,000 K ($\leq$5 eV) are most useful for plasma chemistry.

Chemical Analysis. Plasma oxidation and other reactions often are used to prepare samples for analysis by either wet or dry methods. Plasma excitation is commonly used with atomic emission or absorption spectroscopy for qualitative and quantitative spectrochemical analysis (65).

Samples to be analyzed may be collected on filter papers, either directly from the air, water, or other carriers or after wet chemical separation or concentration. The filters, which are composed of organic materials, interfere with subsequent analysis steps; thus, if the sample is to be heated or excited in steps subsequent to filtration, the filter must be removed or destroyed. Ashing to rid the samples of organic materials can be accomplished in oxidizing or nonoxidizing plasmas (66). Excitation is provided by r-f energy coupled either inductively or capacitively to the plasma. Ashing also is used in the preparation of samples for microscopic examination (67).

Many sources of energy are used to excite samples to emit characteristic wavelengths for chemical identification and assay (68). Very high temperature sources can be employed but are not necessary. All materials can be vaporized and excited with temperatures of only a few electron volts. The introduction of samples to be analyzed into plasmas and their uniform excitation often are problematic.

Use of glow-discharge and the related, but geometrically distinct, hollow-cathode sources involves plasma-induced sputtering and excitation (69). Such sources are commonly employed as sources of resonance-line emission in atomic absorption spectroscopy. The analyte is vaporized in a flame at 2000–3500 K. Absorption of the plasma source light in the flame indicates the presence and amount of specific elements (65).

Pulsed spark sources, in which the material to be analyzed is part of one electrode, are used for semiquantitative analyses. The numerous and complex processes involved in spark discharges have been studied in detail by time- and space-resolved spectroscopy (70). The temperature of d-c arcs, into which the analyte is introduced as an aerosol in a flowing carrier gas, eg, argon, is ca 10,000 K. Numerous experimental and theoretical studies of stabilized plasma arcs are available (62,71).

Plasmas can be produced by radio frequencies that are inductively coupled to aerosol samples in argon streams. A schematic and photograph of such a plasma analysis arrangement is shown in Figure 8 (72). Inductively coupled plasma devices are operated at ca 30 MHz. Temperatures of 10,000–15,000 K are produced with electron densities of ca 10^{15} e/cm^3 (68). These conditions tend to ensure vaporization, disassociation, and excitation. Microwaves near 2 GHz also can be used to excite plasmas for analysis. The analyte is carried in an argon–gas flow which passes through a cavity excited by a magnetron or other source of microwaves. Temperatures of ca 6000 K and electron densities of 10^{12}–10^{15} e/cm^3 prevail in microwave-excited analytical plasmas (73) (see Microwave technology).

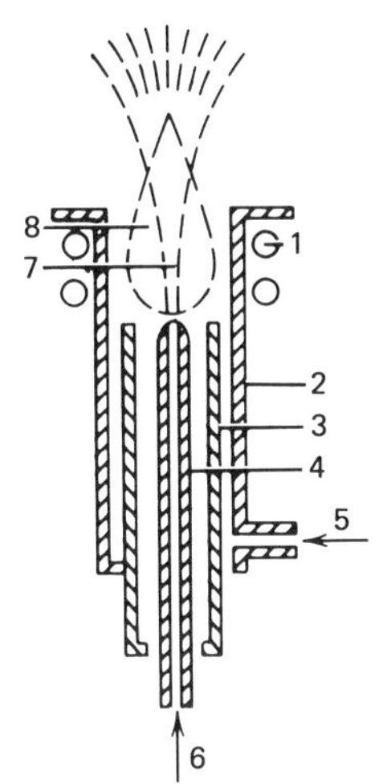

Figure 8. Schematic of an inductively coupled plasma used as an excitation source for chemical analysis (72). The components are 1 = r-f induction coil, 2–4 = silica tubes, 5 = coolant gas, 6 = aerosol carrier gas, 7 = high temperature region, and 8 = low temperature region.

High power pulsed lasers are used to produce plasmas and, thus, to sample and excite the surfaces of solids. Improvements in minimum detectable limits and decreases in background radiation and in interelement interference effects result from the use of two lasers (74).

In plasma chromatography, molecular ions of the heavy organic material to be analyzed are produced in an ionizer and pass by means of a shutter electrode into a drift region. The velocity of drift through an inert gas at ca 101 kPa (1 atm) under the influence of an applied electric field depends on the molecular weight of the sample. The various ionic species are separated and collected every few milliseconds on an electrode. The technique has been employed for studying upper-atmosphere ion–molecule reactions and for chemical analysis (75).

Factors involved in each method and the source for plasma chemical analysis are reviewed biennially (76).

Processing. Plasma processing is described in refs. 62, 77, and 78. Plasmas can accelerate reactions that are otherwise slow to the point of impracticality. Substances not producible by conventional means can be made with plasmas. Plasmas also offer several operational and cost advantages, eg, replacement of batch processes with flow processes. Plasma equipment often is smaller than conventional process hardware, with associated savings in capital expenditure and floor space. The lack of large, hot masses with thermal inertia leads to rapid startup and quenching. Plasma processing based on hydroelectrical power can be less expensive than processes that derive energy from fossil fuels. Plasma sources that are used to drive chemical reactions range from relatively low pressure and temperature glow-discharge and inductively coupled sources to dense, hotter arcs.

Organic and other complex molecules that are exposed to nonequilibrium plasmas can be affected in terms of polymerization, rearrangements (isomerizations), elimination of constituent parts, and total destruction of the original molecules with the generation of atoms and ions (79). The production of polymers and other heavier molecules from gaseous monomers is an attractive application of plasmas in inorganic chemistry (80–81) (see also Radiation curing). A wide variety of chemically inert, adherent films, eg, plasma-polymerized fluorocarbon films, can be produced with simple apparatus (see Fig. 9) (80). Radio-frequency discharge polymerization of organic chemical monomers is employed to produce organosilicon films, which are useful as light guides for integrated optics (82) (see Fiber optics; Film deposition techniques). Semipermeable membranes for hyperfiltration also are prepared with the use of plasmas, eg, poly(vinylene carbonate) deposited on a Millipore filter (see Ultrafiltration; Membrane technology). Such composite filters are easy to prepare, dense, pinhole-free, and provide high salt rejection (83).

Inorganic substances with small molecules also are produced in glow discharges and r-f-induced plasmas, ie, by chemical vapor deposition (cvd). The molecules are introduced in the gaseous state and the products usually deposit on a chosen substrate. Collisional processes relevant to cvd have been tabulated (84). A cylindrical radial-flow reactor for deposition of silicon nitride is illustrated schematically in Figure 10 (84). Amorphous silicon containing high concentrations of hydrogen can be produced by decomposition of SiH_4 in a glow discharge (85).

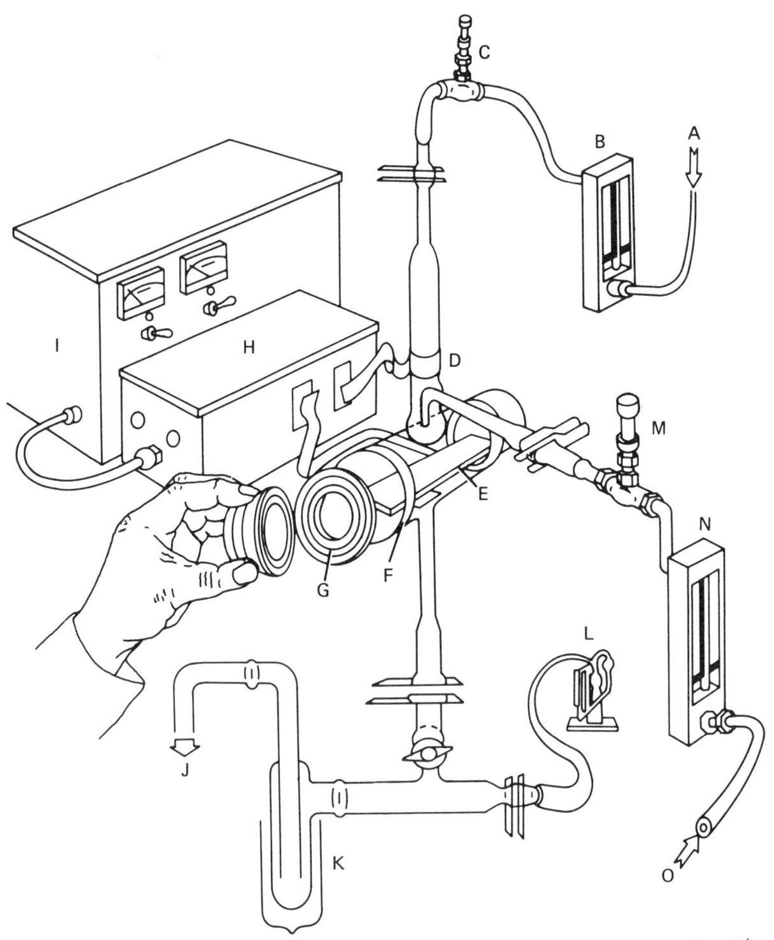

Figure 9. Flow system for coating substrates with plasma-polymerized fluorocarbon films (80). The components are A = gas inlet, B = rotometer, C = flow control valve, D = high voltage, E = substrate, F = ground, G = O-ring, H = r-f coupling unit, I = r-f generator, J = exhaust to pump, K = cold trap, L = pressure gauge, M = flow control valve, N = rotometer, and O = gas inlet.

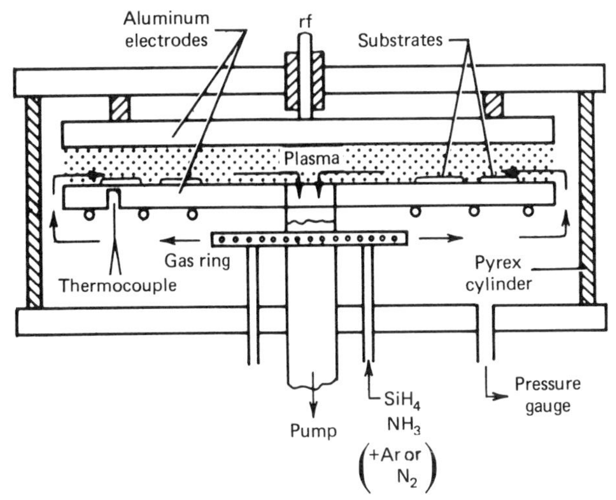

Figure 10. Schematic of a cylindrical radial-flow reactor for deposition of silicon nitride (84). The reactive gases SiH_4 and NH_3, mixed with Ar or N_2 carrier gases, flow from the gas ring, are excited by radio-frequency (r-f) energy, and exhaust to a pump. The temperature of the plate holding the substrates onto which the silicon nitride is deposited is monitored with a thermocouple.

Simple organic and inorganic molecules are produced in reactions driven by plasma arcs, eg, the fixation of nitrogen to hydrogen, carbon, or oxygen (86). The numerous fluorine compounds that have been produced by plasma chemistry are tabulated in ref. 87. Reactions involving many other elements also are accomplished by plasma stimulation (62).

Materials Production and Modification. Plasma materials production and modification embraces processes such as production of thin coatings, heat treatment, and the joining of materials. Plasma sources utilized for the production of materials and their modification are similar to those used to effect chemical reactions. Low temperature, glow-discharge, and r-f devices are employed to coat and heat the surfaces of solids and to alter them by ion-bombardment sputtering. Higher temperature plasmas are used in materials processing. Plasma torches are produced by confining the heating by r-f fields or arcs to a chamber through which gas flows at high velocity. Temperatures in excess of 10,000 K are attained in the plasma, which cools as it is swept along to form a jet.

The reduction of high melting and other inorganic compounds to produce elemental solids also is achieved with plasmas. Extractive metallurgy (qv) requires breaking up ores and similar compounds. The decomposition of MoS_2 to the metal and sulfur has been demonstrated with an induction plasma torch (88). Plasma arc-powered furnaces have been used for gaseous, eg, hydrogen, reduction of metal-containing materials (89).

Thin surface films also can be produced by plasma-sputtering deposition, without chemical reaction. Both single- and multiple-layer materials are produced. Especially important is the use of sputtering to produce multilayered microstructures consisting of dozens of layers of two elements or compounds (90). The microstructures are used as x-ray reflectors (91). Bulk as well as thin materials can be produced, albeit at low rates, by vapor deposition (see Magnetic materials). For example, plasma-driven cvd can be used as a source for growth of refractory crystals, eg, TiN (78). Plasma deposition and etching techniques are being used to modify materials to produce fine patterns, eg, diffraction gratings and Fresnel lenses for use with uv and soft x radiation (92–93).

Plasmas are used extensively to melt materials for a variety of purposes. In many cases, the materials are introduced as a powder into the gas stream in a plasma torch. The molten droplets can be used to grow crystals of refractory materials, eg, Nb (94).

More common and important is spray coating of materials with plasma-melted substances (95). Plasma torches heated by d-c arcs can be hand-held for spray coating (96). Plasma spraying is employed to apply oxidation-resistant coatings to metals, for example, ceramic coatings of aircraft engine components and a proprietary cobalt–chromium–aluminum–yttrium coating for gas-turbine blades (97–98). High temperature, self-lubricating coatings have also been applied to materials with plasma techniques (99). In addition to coating existing structures by plasma spraying, it also is possible to build composite materials by spraying fibers or whiskers with a binding substance (95). The molten droplets produced in a plasma torch can be cooled without impacting a surface in order to obtain fine particles. Refractory metals and oxides can be made into powders of spherical particles using plasma torches. The resulting materials are used, for example, to produce filters and electrodes (89).

Direct-current arcs into which no material is introduced have many applications as heat sources. Industrial processing of metals in the USSR with the use of plasma torches is described in ref. 89. Plasmas also are used in surface and heat treatment of materials. Metals are hardened by exposure to heat from plasmas, and natural fibrous materials, eg, wool (qv), can be made shrink-resistant by such treatment (100). Alteration of surface properties, eg, wettability, by use of plasmas has been reviewed (101). The forming of materials using plasmas is widespread, eg, plasma torches are employed to cut thick metal plates (96). Also common is the use of d-c arcs for joining metals by welding (qv).

Energy Production. ***Fusion.*** The fusion or joining of two light nuclei, generally isotopes of hydrogen, leads to the formation of a heavier nucleus, eg, helium, with conversion of mass to energy. High energy is needed to overcome the mutual electrical repulsion experienced by two positive particles when they are near each other. The high velocities required for nuclear contact and reaction can be produced in plasmas if the ion temperature is greater than a few thousand electron volts. Because of the far greater number of nuclei in plasmas compared to beams of light nuclei, net energy production by nuclear fusion may occur in a plasma. Many exothermic fusion reactions are known, the easiest to initiate being the deuterium–tritium reaction: $D + T \rightarrow He + n, + 17.6$ MeV. Deuterium is abundant in and easily separated from water. There is enough deuterium on earth to provide power for geological time scales. In contrast, tritium is not available in nature, but it can be produced from n + Li reactions. Natural lithium is exhaustible, but sufficient tritium can be provided from it until fusion-energy production is efficient enough to involve only D–D reactions: $D + D \rightarrow T + p, + 4.0$ MeV or $D + D \rightarrow He + n, + 3.3$ MeV. Deuterium–deuterium reactions are harder to ignite and yield less energy than D–T reactions, but they eventually should be the basis of fusion energy production (102). Research into the production of fusion power has a history of more than 30 years (103–105) (see Fusion energy).

High temperature is an important requirement for the attainment of fusion reactions in a plasma. The conditions necessary for extracting as much energy from the plasma as went into it is the Lawson criterion, which states that the product of the ion density and the confinement or reaction time must exceed 10^{14} s/cm^3 in the most favorable cases (103). If the collisions are sufficiently violent, the Lawson criterion specifies how many of them must occur for breakeven. Conventional magnetic confinement involves fields of as much as 10 T (10^5 G) with large (1 m^3) plasmas of low densities ($<10^{14}$ particles/cm^3) and volumes and reaction times of ca 1 s. If the magnetic flux can be compressed to values above 100 T (10^6 G), then a few cm^3 of plasma can be confined at densities of ca 10^{19} particles/cm^3 for correspondingly shorter times (ca 10 μsec). Inertial confinement requires compression of minute plasmas ($<10^{-3}$ cm^3) to densities in excess of those of solids (ca 10^{25} particles/cm^3) for very short reaction times (ca 10 ps). The goal of all fusion devices is to produce high ion temperatures, which usually are in excess of the electron temperature, in contrast to low energy plasmas in which the electrons are often much hotter than the ions (see Fig. 3). Fusion-energy research costs about $\$10^9$ annually worldwide (106–109).

Guns and Missiles. The rapid burning of powder in a gun barrel produces relatively cold plasmas which eject the projectile on a ballistic trajectory. Missiles carry a propellant which burns during flight; this generates motion by high velocity ejection of mass (see Explosives and propellants). Modern missiles contain liquid or solid propellants with high energy densities. The chemical reactions that occur during their burning produces plasmas in the reaction chamber and exhaust nozzles.

High velocities can be reached by a variety of means (110). The highest muzzle velocities of conventional guns are ca 10^6 cm/s. Rockets can attain higher speeds and are limited by the weight of the nonfuel parts of the rocket. Laser-plasma ablation can propel small masses at over 10^7 cm/s but the technique is not likely to be useful for weaponry; a more promising plasma-based launcher is the rail gun (111). In a rail gun, a plasma created by an electric discharge provides pressure to accelerate the projectile. Rail guns have the potential of speeds in excess of 10^7 cm/s, although motion of bullets at such speeds in the atmosphere may be problematic.

Impacts and Explosives. The collision of high velocity bullets or other projectiles with solids causes rapid conversion of kinetic to thermal energy. Plasmas result incidentally, whereas the primary effects of impact are shock and mechanical effects in the target. Impact-produced plasmas are hot enough to cause thermonuclear burn (112).

Most modern projectiles and virtually all missiles contain explosives. The plasmas that result from explosives are intrinsic to operation of warheads, bombs, mines, and related devices (see also Chemicals in war).

Nuclear weapons and plasmas are intimately related. Plasmas are an inevitable result of the detonation of fission and fusion devices and they are fundamental to the operation of fusion devices. Compressed pellets, in which a thermonuclear reaction occurs, would be useful militarily for simulation of the effects of nuclear weapons on materials and devices.

Conventional chemical explosives have many peaceful uses, eg, mining and construction. The potential for peaceful uses of nuclear devices is also evident. A U.S. program, called Plowshare, explored the use of nuclear explosives as earth movers, eg, for canal building (113). Peaceful employment of nuclear explosives for gas production, mining, and scientific purposes, eg, neutron-impact studies, also was studied in the Plowshare program. Concern about radioactive contamination from such tests and related possible projects have led to international treaties banning such work.

Directed-Energy Weapons. Modern weapons mostly involve the propulsion of masses to inflict damage. Energetic quanta from lasers and accelerators have potential as directed-energy weapons (114). They would be based on much of the same pulsed-power technology employed to produce high temperature plasmas, and their impact on targets could produce surface plasmas. Potential laser weapons are likely to be produced within a few decades and could be useful in the atmosphere, weather permitting, and in space.

Communications and Space Travel. Electromagnetic waves in and near the r-f region permit communications between points on or near earth or in space (115). The F layer of the ionosphere has the greatest electron density (see Fig. 6) and reflects radio waves of the greatest frequency. The plasma frequency, which is determined by the electron density and the angle of incidence, determines the frequency cutoff, below which reflection occurs and above which r-f waves penetrate and are partially absorbed by the ionosphere. Low frequencies (long wavelengths), which bounce off the F layer, are partially absorbed in the lower D and E regions of the ionosphere.

The ionosphere is subject to sudden changes resulting from solar activity, particularly from solar eruptions or flares that are accompanied by intense x-ray emission. The absorption of the x rays increases the electron density in the D and E layers, so that absorption of radio waves intended for F-layer reflection increases. In this manner, solar flares disrupt long-range, ionospheric-bounce communications. Time histories of solar x-ray emission as measured by a satellite and the frequencies passed by an ionosphere-bounce radio link are provided in Figure 11 (116).

Earth to space (satellite) to earth communication links are relatively insensitive to ionospheric disturbances. Communications between earth and manned space vehicles are barely affected by plasmas when the spaceships are well away from the atmosphere, eg, in orbit or in a translunar trajectory. However, during reentry of a spaceship, a low temperature plasma forms around the vehicle and interrupts communication with it (117). Plasmas are incidental to the performance of present-day rockets used to explore the solar system.

Spaceships capable of reaching the regions of stars will be more directly involved with plasmas than are contemporary spaceships, in terms of their motion through the interstellar plasmas and their propulsion. Very high velocities will be required for travel to other stars, eg, travel to Proxima Centauri, which is 4.3 light years distant, would require 43 yr at one-tenth the speed of light.

Most schemes that have been proposed to propel starships involve plasmas. They differ both in the selection of matter for propulsion and the way it is energized for ejection. Some proposals involve onboard storage of mass to be ejected, as in present-day rockets, and others consider acquisition of matter from space or the picking up of pellets, and their momentum, which are accelerated from within the solar system (118–119). Energy acquisition from earth-based lasers also has been considered, but most interstellar propulsion ideas involve nuclear fusion energy; both magnetic, ie, mirror and toroidal, and inertial, ie, laser and ion-beam, fusion schemes have been considered (120–124).

Outlook

The applications of plasmas to promote chemical reactions and to produce and modify materials are rapidly growing areas. If fusion-energy production becomes

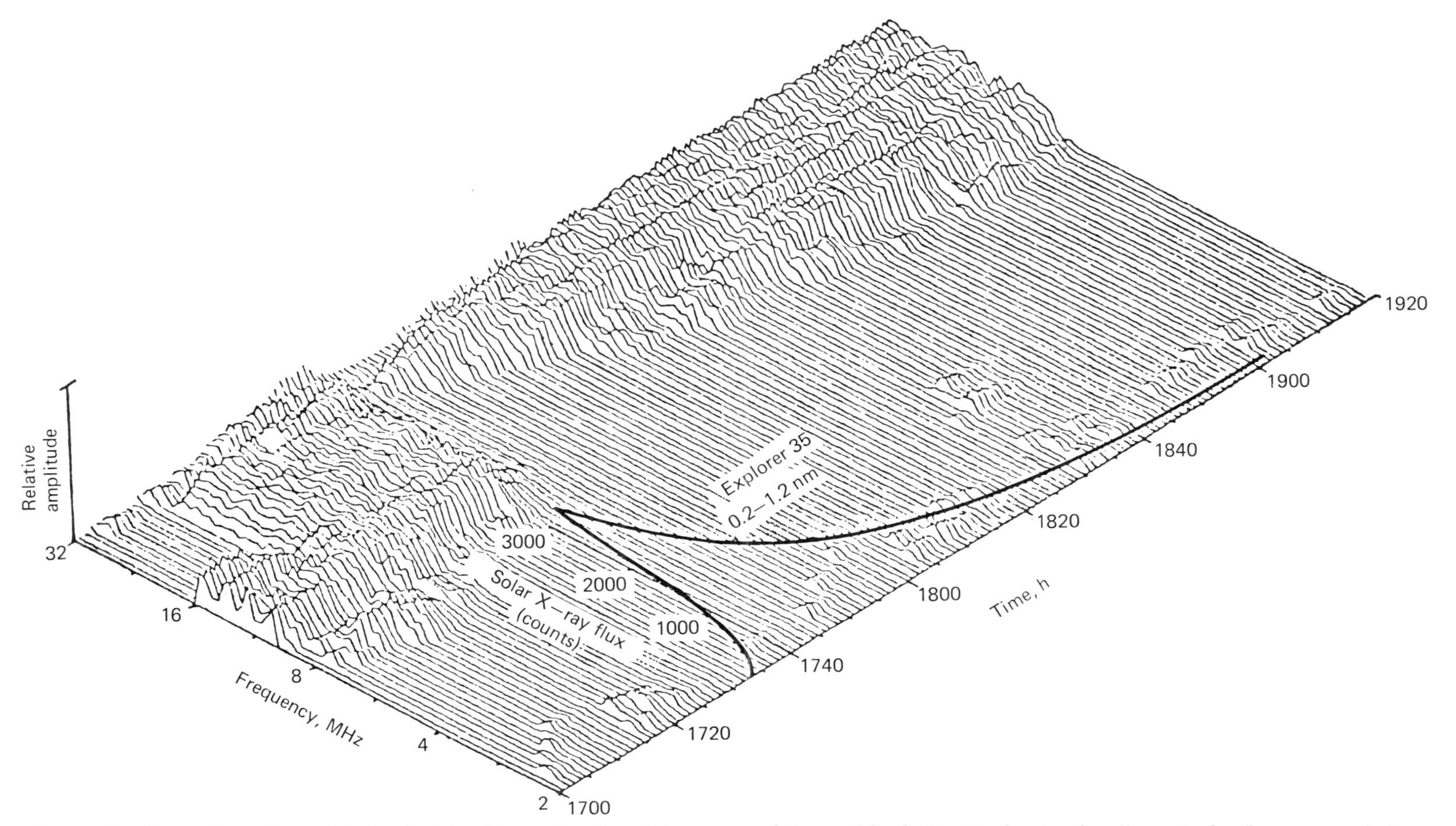

Figure 11. Three-dimensional plot of radio intensity as a function of frequency and time on March 12, 1969 showing the effects of solar-flare x-ray emission on communications (116).

technically and economically feasible, it could dominate all other areas of plasma technology in one or two centuries. There seems to be no end to the growth of the types and expense of weapons in the foreseeable future; thus, application of plasmas in this area will expand.

Nomenclature

B	= magnetic field strength
c	= velocity of light, 3×10^8 m/s
cvd	= chemical vapor deposition
D	= particle size; region of ionosphere
d	= average interparticle spacing
E	= electric field strength; region of ionosphere
E	= oscillatory electric
e	= charge on electron
E_F	= Fermi energy
F	= region of ionosphere
f	= particle-collision rate
h	= Planck's constant, 6.626×10^{-34} J·s
$\bar{h}$	= $h/2\pi$
k	= Boltzmann's constant, 1.38×10^{-23} J/K
m	= mass
m_e	= mass of electron
MHD	= magnetohydrodynamic
n	= density, particles/cm^3
n_c	= critical density
q	= charge
T	= temperature, K or eV (1 eV $\sim$ 11,600 K)
T_F	= Fermi temperature
v	= velocity, m/s
$\bar{v}$	= average velocity
v_A	= Alfven velocity
$v_\perp$	= velocity normal to magnetic field
λ	= mean free path between collisions
λ_D	= Debye length
ρ	= mass density, g/cm^3
ω_P	= plasma frequency
ω_c	= cyclotron frequency

BIBLIOGRAPHY

1. S. Weinberg, *The First Three Minutes*, Basic Books, Inc., New York, 1977.
2. C. Ponnamperuma in W. O. Milligan, ed., *Proceedings of Robert A. Welch Foundation Conferences on Chemical Research XXI: Cosmochemistry*, Robert A. Welch Foundation, Houston, Texas, 1978, p. 137.
3. S. C. Brown and co-workers, *Am. J. Phys.* **31**(8), 637 (1963).
4. J. A. Bearden and A. F. Burr, *Rev. Mod. Phys.* **39,** 125 (1967).
5. J. T. Waber and D. T. Cromer, *J. Chem. Phys.* **42,** 4116 (1961).
6. J. J. Monaghan, *Aust. J. Phys.* **26,** 597 (1973); **27,** 169 (1974); **27,** 667 (1974).
7. S. Lundquist and G. Mukhopadhyay, *Phys. Scr.* **21,** 503 (1980).
8. P. M. Platzman and P. A. Wolff, *Waves and Interactions in Solid State Plasmas*, Academic Press, Inc., New York, 1973.
9. M. F. Hoyaux, *Solid State Plasmas*, Pion Ltd., London, UK, 1970.
10. M. S. Sodha and S. Guha, *Adv. Plasma Phys.* **4,** 219 (1971).
11. M. B. Gottlieb, *Int. Sci. Technol.*, 44 (Aug. 1965).
12. F. F. Chen, *Introduction to Plasma Physics*, Plenum Publishing Corp., New York, 1974.

13. G. Schmidt, *Physics of High Temperature Plasmas*, Academic Press, Inc., New York, 1979.
14. C. Breton and co-workers, *Association Euratom Report EUR-CEA-FC-948*, Fontenay-aux-Roses, France, 1978.
15. J. C. Ingraham in E. U. Condon and H. Odishaw, eds., *Handbook of Physics*, 2nd ed., McGraw-Hill, Inc., New York, 1967, pp. 4, 188, 216.
16. A. Hasegawa, *Plasma Instabilities and Nonlinear Effects*, Springer-Verlag, Berlin, FRG, 1975.
17. W. H. Bostick, V. Nardi, and O. S. F. Zucker, eds., *Energy Storage, Compression and Switching*, Plenum Press, Inc., New York, 1976.
18. D. Sliney and M. L. Wolbarsht, *Safety with Lasers and Other Optical Sources*, Plenum Press, Inc., New York, 1980.
19. J. M. Meek and J. D. Craggs, *Electrical Breakdown of Gases*, John Wiley & Sons, Inc., New York, 1978.
20. A. D. McDonald, *Microwave Breakdown in Gases*, John Wiley & Sons, Inc., New York, 1966.
21. M. F. Hoyaux, *Arc Physics*, Springer-Verlag, Inc., New York, 1968.
22. R. A. Gross, *Nucl. Fusion* **15,** 729 (1978).
23. J. L. Vossen and W. Kern, eds., *Thin Film Processes*, Academic Press, Inc., New York, 1978.
24. H. T. Simmons, *Sciquest* **53**(7), 16 (1980).
25. K. Miyamoto, *Plasma Physics for Nuclear Fusion*, The MIT Press, Cambridge, Mass., 1979.
26. D. E. Evans, ed., *Pulsed High Beta Plasmas*, Pergamon Press, Inc., Oxford, UK, 1976.
27. H. Motz, *The Physics of Laser Fusion*, Academic Press, Inc., New York, 1969.
28. G. M. McCracken, P. E. Stott, and M. W. Thompson, eds., *Plasma Surface Interactions in Controlled Fusion Devices*, North-Holland, Amsterdam, The Netherlands, 1978.
29. R. H. Huddlestone and S. L. Leonard, eds., *Plasma Diagnostic Techniques*, Academic Press, Inc., New York, 1965.
30. W. Lochte-Holtgreven, ed., *Plasma Diagnostics*, North-Holland, Amsterdam, The Netherlands, 1968.
31. K. Bockasten and co-workers, *Controlled Thermonuclear Fusion Research*, International Atomic Energy Agency, Vienna, Austria, 1961.
32. C. B. Wharton in T. P. Anderson, R. W. Springer, and R. C. Warder, Jr., eds., *Physico-Chemical Diagnostics of Plasmas*, Northwestern University Press, Evanston, Ill., 1964.
33. D. J. Nagel, *Adv. X-Ray Anal.* **18,** 1 (1975).
34. M. A. Uman, *Lightning*, McGraw-Hill, Inc., New York, 1969.
35. R. H. Golde, ed., *Lightning*, Academic Press, Inc., New York, 1977.
36. R. E. Orville in R. H. Golde, ed., *Lightning*, Vol. 1, Academic Press, Inc., New York, 1977, p. 281.
37. L. Kresak and P. M. Millman, eds., *Physics and Dynamics of Meteors*, D. Reidel, Dordrecht, The Netherlands, 1968.
38. A. Omholt, *The Optical Aurora*, Springer-Verlag, Inc., New York, 1971.
39. H. Risbeth and O. K. Garriot, *Introduction to Ionospheric Physics*, Academic Press, Inc., New York, 1969.
40. M. J. McEwan and L. F. Phillips, *Chemistry of the Atmosphere*, Halsted Press, a division of John Wiley & Sons, Inc., New York, 1975.
41. C. T. Russell in K. Knott and B. Battrick, eds., *The Scientific Programme During the International Magnetospheric Study*, Academic Press, Inc., New York, 1969, p. 9.
42. E. R. Dyer, ed., *Critical Problems of Magnetospheric Physics*, National Academy of Sciences, Washington, D.C., 1972.
43. R. S. White, *Space Physics*, Gordon & Breach, Science Publishers, Inc., New York, 1970.
44. E. N. Parker, C. F. Kennel, and L. J. Lanzerotti, eds., *Solar System Plasma Physics*, Vols. I, II, and III, North-Holland, Amsterdam, The Netherlands, 1979.
45. *Space Plasma Physics*, (3 Vols.), National Academy of Sciences, Washington, D.C., 1978.
46. *Solar System Space Physics in the 1980s*, National Academy of Sciences, Washington, D.C., 1980.
47. S. Mitton, ed., *The Cambridge Encyclopedia of Astronomy*, Crown Publishers, Inc., New York, 1977.
48. H. Alfven, *Cosmic Plasma*, Kleuver, Boston, Mass., 1981.
49. R. Bowers, *Sci. Am.* **209**(5), 46 (1963).
50. M. Glicksman, *Solid State Phys.* **26,** 275 (1971).
51. J. C. Slater, *Solid-State and Molecular Theory*, John Wiley & Sons, Inc., New York, 1975.
52. T. M. Rice, *Solid State Phys.* **32,** 1 (1977).
53. J. C. Hensel, T. G. Phillips, and G. A. Thomas, *Solid State Phys.* **32,** 87 (1977).

54. A. C. Baynham and A. D. Boardman, *Plasma Effects in Semiconductors: Helicon and Alfven Waves*, Taylor and Francis, London, UK, 1971.
55. J. Pozhela, *Plasma and Current Instabilities in Semiconductors*, Pergamon Press, Inc., Oxford, UK, 1981.
56. J. L. Delecroix, *Plasma Physics*, John Wiley & Sons, Inc., New York, 1965.
57. S. J. Buchsbaum and A. G. Chynoweth, *Int. Sci. Technol.* 40 (Dec. 1965).
58. D. J. Clark, *IEEE Trans. Nucl. Sci.* **NS-24,** 1064 (1977).
59. A. Bernard in D. E. Evans, ed., *Pulsed High Beta Plasmas*, Pergamon Press, Oxford, UK, 1976, p. 69.
60. W. T. Silfvast, L. H. Szeto, and O. R. Wood II, *Appl. Phys. Lett.* **36**(8), 617 (1980).
61. D. J. Nagel, *Naval Research Laboratory Memorandum Report* **4465** (1982).
62. R. F. Baddour and R. S. Timmins, eds., *The Application of Plasmas to Chemical Processing*, The MIT Press, Cambridge, Mass., 1967.
63. *Chem Week*, 24 (Nov. 2, 1983); *Chem. Eng.*, 14 (Dec. 26, 1983).
64. V. A. Fassel, *Science* **202,** 183 (1978).
65. W. G. Schrenk, ed., *Analytical Atomic Spectroscpy*, Plenum Publishing Corp., New York, 1975; J. W. Carnahan, *Am. Lab.*, 31 (Aug. 1983).
66. J. R. Hollahan in ref. 63, Chapt. 7.
67. R. S. Thomas in ref. 63, Chapt. 8.
68. S. Greenfield, H. McD. McGeachin, and P. B. Smith, *Talanta* **22,** 1 (1975); **22,** 553 (1975); **23,** 1 (1976).
69. P. J. Slevin and W. W. Harrison, *Appl. Spectrosc. Rev.* **10,** 201 (1976).
70. J. P. Walters, *Science* **198,** 787 (1977).
71. C. D. Keirs and T. J. Vickers, *Appl. Spectrosc.* **31,** 273 (1977).
72. P. W. J. M. Boumans in E. L. Grove, ed., *Analytical Emission Spectroscopy*, Part II, Marcel Dekker, New York, 1972, pp. 1–254.
73. R. K. Skogerboe and G. N. Coleman, *Anal. Chem.* **48,** 611A (1976).
74. R. M. Measures and H. S. Kwong, *Appl. Opt.* **18,** 281 (1979).
75. F. W. Karaske, *Anal. Chem.* **43,** 1982 (1971).
76. R. M. Barnes, *Anal. Chem.* **50,** 100R (1978); W. J. Boyko, P. N. Keliher, and J. M. Malloy, *Anal. Chem.* **52,** 53R (1980); W. J. Boyco, P. N. Keliher, and J. M. Patterson, III, *Anal. Chem.* **54,** 188R (1982).
77. P. H. Wieks, *Pure Appl. Chem.* **48,** 195 (1976).
78. S. Veprek, *Pure Appl. Chem.* **48,** 163 (1976).
79. H. Suhr in ref. 63, Chapt. 2.
80. M. Millard in ref. 63, Chapt. 5.
81. M. Shen and A. T. Bell, *Plasma Polymerization*, American Chemical Society, Washington, D.C., 1979.
82. P. K. Tien, G. Smolinsky, and R. J. Martin, *Appl. Opt.* **11,** 637 (1972).
83. T. Wydeven and J. R. Hollahan in ref. 63, Chapt. 6.
84. M. J. Rand, *J. Vac. Sci. Tech.* **16,** 420 (1979).
85. M. H. Brodsky, M. Cardone, and J. J. Cuomo, *Phys. Rev.* **16B,** 3556 (1977).
86. R. S. Timmins and P. R. Ammann in ref. 62.
87. B. R. Bronfur in ref. 62, Chapt. 7.
88. R. J. Munz and W. H. Gauvin, *AIChE J.* **21,** 1132 (1975).
89. N. N. Rykalin, *Pure Appl. Chem.* **48,** 179 (1976).
90. T. W. Barbee, Jr. and D. C. Keith, *Stanford Synchrotron Radiation Laboratory Report No. 78/04*, pp. III–26 (May 1978).
91. J. V. Gilfrich, D. J. Nagel, and T. W. Barbee, Jr., *Appl. Spectrosc.* **36,** 58 (1982).
92. J. M. Ballantyne, ed., *Proceedings of NSF Workshop on Opportunities for Microstructure Science*, National Science Foundation, Washington, D.C., 1978.
93. N. M. Ceglio in D. T. Attwood and B. L. Henke, eds., *Low Energy X-Ray Diagnostics—1981*, American Institute of Physics, New York, 1981, pp. 210–222.
94. T. B. Reed, *Int. Sci. Technol.*, 42 (June 1962).
95. N. N. Rykalin and V. V. Kudinov, *Pure Appl. Chem.* **48,** 229 (1976).
96. B. Gross, B. Gryca, and K. Miklossy, *Plasma Technology*, Iliff Books Ltd., London, UK, 1969.
97. D. L. Ruckle, *Thin Solid Films* **64,** 327 (1979).
98. *Chem. Week* (Sept. 3, 1980).

99. H. E. Sliney, *Thin Solid Films* **64,** 211 (1979).
100. A. E. Pavlath in ref. 63, Chapt. 4.
101. M. Hudis in ref. 63, Chapt. 3.
102. D. J. Rose and M. Clark, Jr., *Plasmas and Controlled Fusion*, The MIT Press, Cambridge, Mass., 1961.
103. K. Miyamoto, *Plasma Physics for Nuclear Fusion*, The MIT Press, Cambridge, Mass., 1979.
104. F. F. Chen, *The Sciences*, 6 (July/Aug. 1979).
105. T. J. Dolan, *Fusion Research*, Pergamon Press, New York, 1982.
106. R. S. Pease, *Phys. Technol.*, 144 (July 1977).
107. *Fusion Energy Update*, U.S. Dept. of Energy, P.O. Box 62, Oak Ridge, Tenn. 37830.
108. D. A. Dingee, *Chem. Eng. News*, 32 (Apr. 2, 1979).
109. D. E. Thomsen, *Science News* **117,** 92 (1980).
110. D. J. Nagel, *IEEE Trans. Nucl. Sci.* **NS-26,** 122B (1979).
111. D. E. Thomsen, *Science News* **119,** 218 (1981).
112. A. T. Peaslee, Jr., ed., *Los Alamos Scientific Lab Report* **LA- 8000C** (Aug. 1979).
113. D. B. Lombard, *Phys. Today* **14,** 24 (1961).
114. W. J. Beane, *Naval Inst. Proc.*, 47 (Nov. 1981).
115. J. A. Ratcliff, *Sun, Earth and Radio*, Wiedenfeld and Nicolson, London, UK, 1970.
116. B. E. Erickson, Sr., and V. E. Hildebrand, *Observations of the Effects of Solar X-Ray Radiation on HF Ionospheric Propogation*, Naval Weapon Center, China Lake TP885C, 1969.
117. J. J. Martin, *Atmospheric Reentry*, Prentice-Hall, Inc., Englewood Cliffs, N.J., 1966.
118. R. W. Bussard, *Astronaut. Acta* **6,** 179 (1960).
119. E. C. Singer, *J. British Interplanetary Society* **33,** 107 (1980).
120. G. H. Miley, *Fusion Energy Conversion*, American Nuclear Society, Washington, D.C., 1976.
121. A. A. Jackson IV and D. P. Whitmire, *J. British Interplanetary Society* **31,** 335 (1978).
122. A. R. Martin and A. Bond, *J. British Interplanetary Society* **32,** 283 (1979).
123. F. Winterberg, *J. British Interplanetary Society* **32,** 403 (1979).
124. A. Martin, *J. British Interplanetary Society Supplement* **S1-S192** (1978).

D. J. NAGEL
Naval Research Laboratory

PLASTICS, ENVIRONMENTALLY DEGRADABLE

The environmental degradability of polymeric materials has been studied extensively for many years, mainly to increase the useful life of products such as textiles, coatings, electrical insulation, boats, house sidings, and many other durable goods made from polymers. The protection and stabilization of polymeric materials is very important economically and often is vital to the public health and safety.

Since the 1970s, scientists in several countries have investigated polymer degradability with another goal in mind, ie, to produce plastic products that will degrade in a predictable manner when exposed to environmental stresses. Three approaches have been used: to develop polymers that will degrade with sunlight exposure; to develop polymeric structures that will be degraded and assimilated by microorganisms; and to develop polymers that will be absorbed by the body for use in medical and veterinary applications (see Sutures; Prosthetic and biomedical devices).

This article discusses biodegradable and photodegradable polymers, including the effect of polymer structure and environmental factors on the ease of degradation. Proposed mechanisms for polymer degradation are presented as well as methods for measuring the extent of degradation. Applications for biodegradable and photodegradable polymers in the areas of agriculture, surgery, and consumer packaging are described. Technical information pertaining to the stabilization and protection of polymers from environmental damage is presented elsewhere in the *Encyclopedia* (see Antioxidants and antiozonants; Heat stabilizers; Uv stabilizers).

Definitions

Although most scientists who work with degradable plastics make accurate distinctions among different types of degradation, there is considerable confusion among laymen and legislators as to the meaning of various terms in common usage.

Environmental Degradation. Environmental degradation of polymeric substances results from exposure to the combined environmental effects of sunlight, heat, water, oxygen, pollutants, microorganisms, insects and animals, and from mechanical forces such as wind, blown sand, rain, wave action, vehicular traffic, etc. This term is often mistakenly called biodegradation.

Photodegradation. Photodegradation refers to the degradation of polymeric substances and other organic compounds when exposed to sunlight and other intense sources of light. The ultraviolet wavelengths are primarily responsible for the observed damage.

Oxidative Degradation. Oxidative degradation is the degradation of polymeric chains through attack by oxygen and ozone. In many polymers, oxidative degradation occurs by a chain mechanism catalyzed by ultraviolet light, catalyst residues, or both. Oxidative degradation is also accelerated by higher ambient temperature.

Biodegradation. Biodegradation is the degradation and assimilation of organic polymers and other compounds by the action of living organisms. For samples exposed to the environment, the principal organisms are fungi and bacteria. Microbiological degradation often is facilitated by hydrolytic and oxidative breakdown of the polymer owing to environmental exposure. For polymers implanted in an animal body, biodegradation involves some or all of the body's catabolic processes, including hydrolysis and oxidation processes which usually involve enzyme mediation. Synthetic polymers are usually more difficult to degrade by either animals or microbes than natural polymers.

Biodegradation of Synthetic Polymers

Measuring Biodegradation. Several general methods have been developed for measuring the biological decomposition of organic polymers. These measurements provide such information as rate of polymer weight loss, loss of polymer physical properties such as tensile strength, or rate of increase in colony size of the microbial culture in contact with the polymer.

Growth Ratings. In one procedure formerly called ASTM D 1924-63 (1) and since 1980 designated ASTM G 21-70, the specimens are placed in or on a solid agar growth medium that is deficient in available carbon. After inoculation with the test microorganisms, the medium and samples are incubated for three weeks. Any growth of the colony is dependent on utilization of the polymer specimen as a source of carbon. Standard test organisms commonly used are the fungi *Aspergillus niger*, *Aspergillus flavus*, *Chaetomium globosum*, and *Penicillium funiculosum* in a mixture, although other organisms have been used, such as *Pullaria pullulans*.

After various exposure times up to three weeks, the samples are examined for evidence of colony growth on the polymer sample surface and ratings assigned as follows:

0 = no growth visible
1 = <10% of the surface shows growth
2 = 10–30% covered (light growth)
3 = 30–60% covered (medium growth)
4 = 60–100% covered (heavy growth)

The method can be used with plastic specimens of any thickness. Semisolid waxes and greases are tested by depositing them on fiberglass cloth.

Another test, formerly ASTM D 2676T (2) and since 1980, ASTM G 22-76, follows the same procedure just described, except that a bacterium such as *Pseudomonas aeruginosa* is employed. Generally, fungi are preferred over bacteria for this type of testing because fungi show greater activity and pose less threat of infection to personnel.

Both of these ASTM tests are capable of detecting the presence of minute amounts of available carbon and are extremely sensitive to the presence of biodegradable additives or impurities in or on the sample, such as plasticizers (qv) and cling agents, sebum from fingerprints on the surface of plastic film and sheeting (see Plastics, processing). Some investigators have failed to differentiate between the biological activity of such additives and that of the polymer itself. Positive tests for biodegradability under these circumstances are likely to be inconclusive. Negative results in this test signify that a source of available carbon is not accessible to the microorganism.

If these agar tests are continued past 3–4 weeks, results are compromised because of gradual depletion of the mineral nutrients, possible dehydration of the agar, and gradual breakdown of the agar.

Petri Dish Method, Quantitative. A polymer sample is deposited on the inside of a Petri dish by solvent casting or melting, and the dish is dried and weighed. Nutrient agar is poured over the polymer, which is then inoculated and incubated. After the test period (3–4 wk), the agar and culture are washed off and the plates redried and reweighed to determine weight loss. Physical properties can also be measured (3).

Clear Zone Method. The polymer in finely powdered form is suspended in a nutrient–agar medium in Petri dishes. After inoculation and incubation of the hazy suspension, colonies of cells are observed growing on the gel if the polymer is being assimilated. In some instances, a clear zone may occur in the medium surrounding the colony, which indicates that the polymer in the vicinity of the colony has been assimilated by the colony after degradation by the extracellular enzymes (3).

Soil Burial and Retrieval. This method can be performed either under laboratory conditions or in a natural setting. Uniformly sized samples are buried in a mixture of equal parts of sand, garden soil, and peat moss in laboratory containers which are kept at high humidity away from sunlight (4). Outdoor testing affords a means of testing plastic samples for rate of breakdown in a given soil type containing a native microbe population and in variable geographic regions. Inoculation of soil with test organisms is usually not successful because the native microbe population prevents establishment of new organisms (5). Samples retrieved from soil burial may be tested for weight loss or deterioration of mechanical properties, or they may be examined by scanning electron microscopy for evidence of attack.

Plate-Count Method. The polymer sample is finely ground, suspended in a shaker flask in nutrient broth which has been inoculated with the bacteria of choice, and incubated in the presence of air. Plate counts are performed on aliquots removed at intervals in a procedure similar to that performed in blood analysis (6). It is not useful for studying fungal cultures, which undergo dendritic growth, and is used exclusively for bacterial cultures.

This test is often performed under conditions where other carbon sources are also present, such as potato starch or soybean flour. Plate counts from such tests are misleading unless adequate controls are also carried out. Most synthetic plastics of high molecular weight that are not biodegradable when used as the sole source of carbon are also not degraded in the presence of nutrients such as potato starch. However, in some instances, added nutrients enhance hydrolysis of functional groups along the polymer chain. Polymer samples containing significant quantities of oligomers or fragments resulting from photodegradation may show enhanced assimilation rates in the presence of added carbon source nutrients.

Radioactive ^{14}C Tracer Studies. One of the earliest uses of ^{14}C as a tracer for measuring the rate of biodegradation of plastics was in a study of the biodegradation of ^{14}C-labeled polyethylene (7). The labeled plastic was ground to a powder and mixed with fresh garden soil. Water-saturated air, free from CO_2, was passed through the pot containing the sample and then through a solution of 2 *M* KOH to absorb both $^{12}CO_2$ and $^{14}CO_2$ generated in the pot by microbial action. After $\geq$30 d, the KOH solution was titrated with 1 *M* HCl to pH 8.35, and the total amount of CO_2 formed was calculated. A portion of the titrated solution of KOH was added to a scintillation counter and the amount of ^{14}C determined as counts per minute. By comparison to the original radioactivity of the labeled polymer, the percent by weight of carbon in the polymer which was decomposed to CO_2 can be determined. Although this method is not subject to interference by biodegradable impurities or additives in the polymer or in the soil, it is very sensitive to the presence of labeled low molecular weight polymer molecules which may be present in the sample.

Environmental Factors. Whether an organic polymer undergoes biodegradation depends on three factors: molecular size and structure; the microbial population; and various environmental factors such as temperature, pH, humidity, and availability of nutrients. The effect of polymer morphology, chemical structure, and molecular weight will be discussed later in this article. Here, the effect of environmental factors on the activity of microorganisms and the enzymes they elaborate are discussed.

Microbiological degradation is favored by the absence of light and the presence of high humidity, adequate minerals, and sources of available carbon. Organisms differ with regard to their pH, temperature, and O_2 requirements. Fungi require O_2, a pH above 4.0, and temperatures of 20–45°C to grow. Yeasts have similar requirements but need high levels of carbohydrates to grow. Actinomycetes are also aerobic and prefer a more neutral pH range. Bacteria are either aerobic or anaerobic and function best in the pH range of 5–7. Some bacteria and actinomycetes are thermophilic (heat loving) and function at temperatures as high as 70°C, although 50–55°C is optimum. Compost heaps, which often get quite hot from microbiological action, contain thermophilic organisms (8).

A detailed discussion of the effects of various species of organisms on polymers is beyond the scope of this article. For synthetic polymers, the available evidence indicates that polymeric structure, morphology, molecular weight, and in some cases, surface-to-volume ratio are more important factors than types of organisms used for testing. Thus far, commensality among microorganisms has not played a significant role in causing synthetic polymers to become biodegradable (5). As more biodegradable polymers are synthesized, this form of synergism may become more important.

Enzymes in Polymer Biodegradation. The microbiological degradation of polymeric material occurs because of the presence and activity of various enzymes which are produced by the microorganisms surrounding the polymer. Because they are proteins, most enzymes are soluble in water, dilute salt solution, and dilute solutions of alcohol in water. They are precipitated by concentrated salt solutions, heavy metal ions, organic solvents, and strongly acidic or basic solutions (see Enzymes).

Because of the pronounced water insolubility of most synthetic plastics, it is not surprising that enzymes, which have an affinity for water, have difficulty in interacting with plastics. It was thought likely that increasing the hydrophilicity of such polymers would increase their susceptibility to biodegradation. However, this has not proved to be an effective way to increase the ease of biodegradation of such plastics because of the specificity of enzyme action (9).

The reaction rate of enzyme-catalyzed reactions increases with temperature until the enzyme activity begins to decrease owing to heat-activated denaturation. Enzyme activity is also greatly decreased during winter months in cold climates and is adversely affected by exposure to ionizing radiation, uv light, ultrasound, and high hydrostatic pressures.

Biodegradability of Additives. Various organic compounds are added to synthetic polymers as oxidation inhibitors, plasticizers, lubricants, colorants, slip agents, uv light stabilizers, and antistatic agents, and for other purposes. Some of these compounds are quite biodegradable, some are resistant, and a few are biocides. Some, such as plasticizers, are used in large proportions, and others, such as slip agents, are present in very low concentrations. It is absolutely essential that the effect of such additives on the results of biodegradability testing be recognized and understood. In many situations, the possible ambiguity can be resolved by removal of the additive from the plastic by solvent extraction, followed by testing of the purified polymer.

Examples of the effect of additives on the growing rating (ASTM D 1924-63) of commercial plastics are given in Table 1 (10). A thin layer of cling agent on the film surface produces the fungal susceptibility of sample 1. This additive is removed by toluene. Sample 3 contains a highly biodegradable plasticizer which is responsible for its +3 rating.

The fungal susceptibility of many chemicals used as additives has been discussed (11). Esters of phthalic acid are generally inert, as are most organic phosphates. Derivatives of tricarboxylic acids, such as tricarballylic, citric, and aconitic, are somewhat resistant, whereas mono esters, such as acetates, butyrates, caprylates, and oleates, are very active. Oils containing fatty acids such as tung, linseed, castor, and cottonseed are quite susceptible. Generally, glycol and glycolic acid derivatives containing aliphatic chains below 10 carbon atoms are resistant.

An investigation of 127 compounds, mostly plasticizers, using 24 species of fungi, showed that oxalic acid was not utilized, but that higher dibasic aliphatic acids from malonic to sebacic were all readily assimilated by the microorganisms (12). The diesters of the saturated, aliphatic dibasic acids containing 12 or more carbon atoms support fungal growth. Also, normal or straight-chain hexyl adipates are more active than branched chain isomers. Table 2 summarizes the results of a study on the susceptibility of additives that are widely used in commercial plastics formulations (13). Many of these additives are readily assimilated by fungi. A distinction must be made between the susceptibility of a given synthetic polymer and that of additives it may contain.

Long-term soil-burial studies have demonstrated that blends of biologically inert polymers, such as polyethylene with up to 50 wt % of biodegradable fillers such as starch or sugar, undergo removal of the biodegradable filler by microbial action, which leaves the plastic sample riddled with holes. The hole diameters depend on the particle size of the fillers (14). These studies also showed that blends of sugar or starch with biodegradable polymers, such as aliphatic polyesters, undergo biodegradation of the entire sample. Another study demonstrated that when blended with starch, copolymers of ethylene and vinyl acetate with high vinyl acetate contents undergo greater weight losses than can be accounted for by the amount of starch incorporated in the sample (15). Starch–poly(vinyl acetate) mixtures give similar results. Because increasing exposure leads to greater water sensitivity and there is no evidence of chain scission, it seems likely that esterase cleavage of side-chain ester groups occurs and produces vinyl alcohol groups along the polymer chains (see also Soil chemistry of pesticides).

Table 1. Fungal Susceptibility of Plastic Films before and after Extraction[a]

Sample number	Sample description	ASTM D 1924-63 growth rating
1	polyethylene film before toluene extraction	2
2	polyethylene film after toluene extraction	1
3	poly(vinyl chloride) film plasticized with epoxidized soybean oil	3
4	poly(vinyl chloride) film extracted with toluene	1

[a] Ref. 10.

Table 2. Growth Rating of Additives Commonly Used in Plastics[a]

Identification	Chemical name or type	ASTM D 1924-63 growth rating
Antioxidants		
butylated hydroxytoluene	hindered phenol	0
Santonox R	hindered phenol thioether	0
Topanol CA	hindered phenol	0
Irganox 1010	hindered phenol	0
dilauryl thiodipropionate	thioether ester	4
distearyl thiodipropionate	thioether ester	4
Polygard	nonyl phenyl phosphate	ZI[b]
Slip or antiblock agents		
erucamide	C_{22} unsaturated primary amide	4
oleamide	C_{18} unsaturated primary amide	4
stearamide	C_{18} saturated primary amide	4
behenamide	C_{22} unsaturated amide	4
HTSA-1	olealyl palmitamide	2
zinc stearate	metal salt	4
Plasticizers		
Flexol DOP	di(2-ethylhexyl) phthalate	0
Flexol TCP	tricresyl phosphate	1
Flexol EPO	epoxidized soy bean oil	4
Rucoflex 2, STM	tris(2-ethylhexyl) trimellitate	0
Plastolein 9765	aliphatic polyester	4
Flexol A 26	di(2-ethylhexyl) adipate	0
Lubricants and processing aids		
Mecon white wax	microcrystalline wax	2
Hoechst wax E	hydrocarbon wax	2
Acryloid K izon	acrylic polymer	0
Heat stabilizers		
Vanstay HTA		4
Vanstay SD	phosphite	0
dibutyl tin dilaurate	tin compound	4
Ultraviolet absorbers		
Eastman DOBP	2-hydroxy-4-dodecyloxy benzophenone	0
Eastman OPS	*p*-octyl phenyl salicylate	0

[a] Ref. 13.
[b] ZI = zone of inhibition; indicates compound is acting as a fungicide.

Biodegradability of Hydrocarbon Polymers. ***Hydrocarbon Oxidation.*** The biological degradation of fatty acids occurs as a result of the catalyst action of oxidase enzymes by a process known as β-oxidation. This mechanism was first described in 1905 (16). The biological oxidation of hydrocarbons to fatty acids or derivatives has been established by several investigators (17–20). Cetyl palmitate is a product of the growth of a gram negative coccus grown in hexadecane as the sole source of carbon (21). Also, $^{18}O_2$ from the atmosphere is incorporated into cetyl palmitate in this process. Cetyl alcohol and palmitic acid are the expected products of the oxidation of hexadecane. The formation of methyl ketones from the oxidation of hydrocarbons by *Pseudomonas methanica* and *Mycobacterium* species has also been reported (22–23). It was suggested that methyl ketones arise from attack on the penultimate carbon atom of the hydrocarbon.

Linear vs Branched Chains. Among low molecular weight paraffins, the linear molecules are more easily biodegraded than the branched molecules. Linear paraffins up to *n*-eicosane were found to be readily available sources of carbon for micro-organisms (24). *Pullularia pullulans* assimilates *n*-alkanes (25), and paraffin wax and gas-oil samples are utilized by *Candida lipolytica* and *C. intermedia* (26). The maximum efficiency of alkane assimilation has been noted for nonadecane and tetracosane.

Table 3 gives the results of a study of the biodegradability of pure linear and branched hydrocarbons in the mol wt range of 170–620 (10). Linear molecules up to ca 450 mol wt are readily assimilated, while none of the branched hydrocarbons are active. Above a molecular weight of 500, even the linear molecules were inactive in the ASTM test. Microbial degradation of *n*-alkanes up to a mol wt of 620 was measured in terms of the biological oxygen demand of a mixture of the hydrocarbon with Hudson-Collamer silk loam soil with mineral nutrients (27). A molecular weight effect was also noted; the O_2 uptake in the presence of tetratetracontane and tetracontane was appreciable only after 20 d of exposure.

Polyethylene. Pure compounds with mol wts above 620 have not been tested for biodegradability because of their unavailability; however, several studies show that the ease of biodegradability of polyethylene decreases with increasing mol wt. The earliest of these examined the bacterial susceptibility of polyethylene over the mol wt range 4,800–41,000 using strains of *Pseudomonas aeruginosa*, *Nocardia*, and *Brevibacterium* (6). The plate-count method was used. The greatest increase in plate count was observed for the paraffin wax control, which after 185 d was 3.5×10^9 counts per mL. The plate counts for the 4800 mol wt sample were also high initially, but after 185 d, the plate counts of these samples as well as those of the higher mol wt polymer samples were hardly different from that of the carbon-free control (ca 5×10^6 counts per mL). The high activity of the paraffin wax sample was attributed to the presence of a large amount of biodegradable material in the sample. The marked decline in activity of the polymer samples with incubation time was evidence that the supply of low molecular weight, biodegradable oligomer molecules in these samples had become exhausted.

More recent studies of high and low density polyethylene, summarized in Table 4, also show greater activity of lower molecular weight samples (28). Even more striking

Table 3. Branching and Mol Wt Effects on Pure Hydrocarbon Biodegradability[a]

Compound	Mol wt	No. of branches	ASTM D 1924-63 growth rating
dodecane	170	none	4
2,6,11-trimethyldodecane	212	3	0
hexadecane	226	none	4
2,6,11,15-tetramethylhexadecane	282	4	0
tetracosane	338	none	4
squalane	422	6	0
dotriacontane	450	none	4
hexatriacontane	506	none	0
tetracontane	562	none	0
tetratetracontane	618	none	0

[a] Ref. 10.

Table 4. Molecular Weight Effects on Polyethylene Degradability[a]

Density, g/cm^3	Mol wt viscosity average	ASTM D 1924-63 growth rating
0.96	10,970	2
0.96	13,800	2
0.96	31,600	0
0.96	52,500	0
0.96	97,300	0
0.88	1,350	1
0.95	2,600	3
0.92	12,000	2
0.92	21,000	1
0.92	28,000	0

[a] Ref. 28.

proof of the importance of molecular weight in controlling degradability is found in experiments summarized in Table 5, in which high molecular weight, initially biologically inert, high and low density polyethylenes are pyrolyzed to lower molecular weight polymer molecules. As the severity of pyrolysis increases, the molecular weight of the polymer decreases and the concentration of small biodegradable molecules increases (29–30).

Linear polyethylene usually contains a few percent by weight of wax having a molecular weight below 5000. Experiments have been reported in which this wax fraction was removed from the polymer by cyclohexane extraction and exposed to fungi for periods up to a year to determine the extent of biodegradation (31). After 16 weeks of exposure, the average weight loss of 9 samples was 8.4 ± 0.04%. This increased to 10.6% after 22 wk and leveled off at ca 13.5% between 30 and 52 wk. The fact that the wax samples lost no more weight after 30 wk of exposure was explained by exhaustion of the biodegradable fraction from the wax. These results are useful in assessing the extent to which biodegradation might affect underground polyethylene pipe used to carry water, etc. If the wax content of the pipe compound was 5%, and 13.5% of the

Table 5. Effect on Biodegradability of Pyrolytic Degradation of Polyethylene[a]

Pyrolysis temperature, °C	Viscosity average, mol wt	ASTM D 1924-63 growth rating
High density polyethylene		
control (no pyrolysis)	123,000	0
400	16,000	1
450	8,000	1
500	3,200	3
535	1,000	3
Low density polyethylene		
control (no pyrolysis)	56,000	0
400	19,000	1
450	12,000	1
500	2,100	2
535	1,000	3

[a] Refs. 29–30.

wax biodegraded, the maximum expected weight loss from biodegradation would be ca 0.67%. This should have a negligible effect on the performance characteristics of such a pipe.

In a study of the rate of CO_2 evolution from a commercial polyethylene sample and a ^{14}C-labeled sample buried in soil enriched by the addition of composted garbage, it was found that when the ^{14}C-labeled polymer was protected from light, only traces of ^{14}C-labeled CO_2 were evolved, probably from low molecular weight oligomers originally present in the polymer (7). When the polymer contained a photodecomposition accelerator and was exposed to light for 26 or 42 days prior to soil burial, the rate of evolution of $^{14}CO_2$ decreased steadily with time of burial, indicating that the low molecular weight, biodegradable fragments generated by photodecomposition were being depleted. After 6 months of soil burial, there was no difference between the rate of evolution of $^{14}CO_2$ in samples containing accelerator and those free of accelerator. A similar study, with refinements, has been done (32). The biological oxidation of photodegraded polypropylene and polyethylene was investigated by measuring O_2 uptake from garden soil containing pulverized, severely oxidized polymer (33). Films 25.4 m thick (1 mil) were photooxidized under uv light until friable and the M_n was 2200. The photodegraded polyethylene consumed more O_2 than did the photodegraded polypropylene; both had a higher O_2 consumption rate than the plain garden soil. Polyethylene oxidized with 60% HNO_3 produces excellent growth of *Aspergillus fumigatus* (34). One of the oxidation products is succinic acid, which is biodegradable.

Copolymers of Ethylene. The following ethylene copolymers of stipulated composition are essentially not assimilated when tested by ASTM D 1924-63: ethylene–vinyl acetate (67:33), ethylene–vinyl alcohol (3:7), ethylene–acrylic acid (17:3), ethylene–ammonium acrylate (18:7), ethylene–ethyl acrylate (41:9), ethylene–carbon monoxide (47:3), ethylene–aconitic acid (41:9), ethylene–itaconic acid (79:21), and ethylene–lauryl acrylate (3:1) (35). It appears that increasing polarity and hydrophilicity of polyethylene by adding short, polar side branches does not increase the ease of biodegradation.

Copolymers of ethylene with various unsaturated vegetable oils such as castor, linseed, safflower, soybean, neats-foot, peanut, rape, olive, corn, and oleic acid also give negative results in ASTM D 1924-63 (36). None of these ethylene copolymers is biodegradable, which supports the previously discussed inhibiting effect of chain branching on polymer degradation. The pendant groups which result from copolymerization of ethylene with these monomers interfere with β-oxidation of the hydrocarbon backbone of the polymers. Thus, there appear to be effects of both branching and molecular weight on the biodegradability of polyethylene and ethylene copolymers.

Polystyrene and Styrene Copolymers. It is generally recognized that polystyrene is resistant to microbial attack. Nonetheless, efforts have been made to modify the polymer in order to increase its susceptibility to biological degradation, although without much success. Efforts to make polystyrene biodegradable by attaching biologically active end groups (37), by copolymerizing with other monomers (34,38–41), and by pyrolysis (31) have been unsuccessful. Blends of polystyrene and styrene copolymers with various rubbers are also not active, nor was a thermoplastic block copolymer elastomer. These results are at variance with earlier reports which claimed that synthetic rubbers were biodegradable (42–46).

Other Addition Polymers. ***Poly(Vinyl Chloride) and Vinyl Chloride Copolymers.*** Poly(vinyl chloride) is generally regarded as being immune to attack by fungi and bacteria and is widely used for outdoor house sheathing, gutters, and window frames. It must be protected against sunlight deterioration with stabilizers and coatings. Plasticized poly(vinyl chloride) is a flexible product used as automobile and furniture upholstery, garden hoses, swimming pool liners, wearing apparel, and shower curtains. If the plasticizer used is biodegradable, the film may become brittle if exposed to fungi (see Table 6) (47).

Poly(vinyl chloride) containing dioctyl phthalate, a common plasticizer, was not materially affected by the bacteria; the sample containing R_2H was slightly embrittled; the sample containing H707 was severely embrittled. There is no evidence that the poly(vinyl chloride) itself was attacked by the microorganisms.

Previously, it was stated that dibutyltin dilaurate was biodegradable (see Table 2). If this poly(vinyl chloride) stabilizer is removed from a plasticized film by bacterial action, the polymer will be subject to dehydrohalogenation, which will result in an unsaturated polymer of increased oxidative and microbial susceptibility.

A number of commercial vinyl chloride copolymers and blends have been tested for biodegradability and found to be inactive (41). These are Cycovin (ABS–PVC blend), Barex 210, Lopac, Kydene (PVC–PMMA blend), Saran (vinyl chloride–vinylidene chloride copolymer), and poly(vinyl acetate-*co*-vinyl chloride).

Elastomers. Apparently, poly(*cis*-isoprene) is readily biodegraded (48–49). In one observation, natural poly(*cis*-isoprene), in the form of coagulated latex, was decomposed by three weeks of soil burial (49). Early workers reported that synthetic rubbers such as styrene–butadiene, polybutadiene, and acrylonitrile–butadiene–styrene (ABS) rubber were also biodegradable (42–46). However, in these tests of synthetic rubbers, no account was taken of the presence of biodegradable additives such as emulsifying agents, vulcanizing agents, and plasticizers. In the case of the poly(*cis*-isoprene) study, the entire sample was consumed, so that there is no question about the polymer's biodegradability.

Other tests have shown that the following synthetic rubbers are not susceptible to microbial attack (41): ABS rubber, styrene–butadiene block copolymer (thermoplastic elastomer), butadiene–acrylonitrile rubber, polyisobutylene, and chlorosulfonated polyethylene (Hypalon rubber).

Vinyl and Acrylic Latexes. Vinyl and acrylic latexes made by emulsion polymerization are used in latex paints, adhesives, paper and foil coating, carpet backing, textile finishing, and other similar applications. Many polymers and copolymers are available in latex form, including poly(vinyl acetate), poly(vinyl chloride-*co*-ethylene), poly(vinyl

Table 6. Effect of Plasticizer Loss on Physical Properties of Plasticized Poly(Vinyl Chloride) after Exposure to Bacteria[a]

	Tensile modulus of elasticity, MPa[b]		
Plasticizer	Unexposed	Exposed	Change, %
R_2H polyester	16.33	19.91	22
H707 Hercules	9.46	33.73	258
dioctyl phthalate	7.60	8.04	6

[a] Ref. 47.
[b] To convert MPa to psi, multiply by 145.

chloride-*co*-acrylic acid), poly(butadiene-*co*-styrene), and many copolymers containing acrylate or methacrylate esters. Although these polymers are resistant to attack by microbes, the aqueous latex emulsions are subject to spoilage by microbial attack on various substances in the emulsion formulation, such as monomers, thickening agents (cellulosic and poly(vinyl alcohol)), stabilizers, surfactants, and defoamers (see Latex Technology).

The causes and prevention of microbiological spoilage of latex emulsions have been discussed (50). Spoilage results from the presence of substances susceptible to microbial attack and microbes such as bacteria, fungi, and yeast in the formulation and environmental conditions favorable to microbial growth. Loss in viscosity on storage results from microbial degradation of polymeric thickening agents, such as water-soluble cellulose ethers and poly(vinyl alcohol). Substitution of nondegradable additives for degradable ones minimizes microbial spoilage, as do sanitary manufacturing and handling procedures. The use of broad-spectrum biocides in the emulsion formulation is strongly recommended.

Miscellaneous Addition Polymers. Other addition polymers which were reported not to be susceptible to microbial attack are as follows (41): poly(methyl methacrylate) (Lucite), rubber modified poly(methyl methacrylate), poly(4-methyl-1-pentene) (TPX), poly(vinyl butyral), poly(vinyl ethyl ether), poly(vinyl acetate), and partially hydrolyzed poly(vinyl acetate) (50% hydrolyzed).

Poly(vinyl alcohol), both in film form and in aqueous solutions, has been found to be degraded by microbes. Poly(vinyl alcohol) [*9002-89-5*] films are degraded by sewage sludge (51). High temperature soluble films show a weight loss of 30–50% at 20–30°C. Films soluble in cold water are biodegraded in about four days. Films with a degree of polymerization of $\leq$2500 are biodegradable. Another study reports the biodegradability of poly(vinyl alcohol) in aqueous solutions in the concentration range of 0.65–1.44% (50).

Condensation Polymers. Because synthetic condensation polymers have the same ester, amide, and ether linkages found in naturally occurring polymers such as proteins, fats, and carbohydrates, it is reasonable to expect that condensation polymers are more easily biodegraded than carbon chain polymers. However, it has been observed that most commercially available high molecular weight condensation polymers such as nylons and aromatic polyesters are also resistant to biological attack. This permits their use in applications such as rope, textiles, carpets, and molded parts where resistance to biodegradation may be important.

Certain types of condensation polymers, such as aliphatic polyesters and polyurethanes derived from ester diols, have been found to be susceptible to microbial degradation. Also, the introduction of amino acids and other similar linkages seems to increase the susceptibility of certain condensation polymers to attack by enzymes and microbes. Because of their anticipated high cost, applications for such products may be restricted to use as biomedical implants and surgical sutures where resorption by the body is desired. The effect of structural variations in condensation polymers on their microbial susceptibility follows.

Polyamides. Although these polymers contain amide linkages similar to those in proteins, high molecular weight crystalline polymers such as nylon-6 (poly(ϵ-caprolactam)), nylon-6,6, and nylon-12 have been found to be quite resistant to microbial attack (10–11,52–54). However, a bacterium (*Corynebacterium aurantiacum* B-2) isolated from an effluent water line of a nylon-6 plant was found to be capable of splitting the cyclic trimer, tetramer, and pentamer of ϵ-caprolactam as well as the linear dimer, trimer, and tetramer (53–56).

The possibility of enhancing the biodegradability of polyamides by inserting amino acids into the polymer chain has been investigated (57). Attempts at random insertion resulted in considerable blocking of the amino acid groups. An alternating copolymer of glycine and ϵ-aminocaproic acid was degradable by soil microorganisms. The water-soluble copolymer of serine and ϵ-aminocaproic acid was more degradable than that based on glycine.

Other workers prepared benzylated nylon-6,3 (mp 140–145°C, $\overline{M}_n$ 2000) from achiral benzylmalonic acid and found it to be hydrolyzed by both chymotrypsin and buffer solution (58). However, for nylons (2,6), (4,6), (6,6), and (8,6) derived from DL-α-benzyladipic acid, very little degradation was observed. It was speculated that the D isomers of the DL polymers acted as enzyme inhibitors. A series of polyamides containing methyl and/or hydroxy substituents from the polymerization of 1,2-diaminopropane and/or 1,3-diamino-2-propanol with diacid chlorides were prepared (59). These polymers ranged in $\overline{M}_n$ from 9,470 to 20,000 and melted in the range 180–245°C. All of these polymers supported the growth of *A. niger* and *A. flavus*, whereas nylons prepared from ethylenediamine and 1,3-diaminopropane were resistant to fungal attack.

Polyureas. These are made by the reaction of polyisocyanates and polyamides. Linear polyureas are crystalline and resistant to degradation, but substituted polyureas prepared from the methyl and ethyl esters of L-lysine support the growth of *Aspergillus niger* (59). The methyl ester becomes water soluble after exposure to buffered chymotrypsin and subtilisin, apparently because of side-chain hydrolysis. Also, poly(esterurea) derived from the amino acid phenylalanine is degraded by the enzyme chymotrypsin (58). The polymer was prepared by esterifying ethylene glycol with phenylalanine followed by reaction of the amino groups with 1,6-diisocyanate hexane to give a product of mp 194–198°C with M_n between 1930 and 2640. Based on the small change in amino acid group concentration, it was concluded that ester linkages are preferentially degraded. A similar polymer prepared using glycine in place of phenylalanine is not degraded. Poly(amide–urethane)s prepared from mandelic acid and 1,6-dicyanatohexane (mp 110–115°C, $\overline{M}_n$ 7500) are biodegradable, whereas a similar polyester prepared from glycolic acid is not biodegradable by either enzymes or *Aspergillus niger*.

Aromatic Polyesters. Polyesters containing significant amounts of aromatic constituents are generally found to be resistant to biodegradation. For example, it was found that Arnite G [poly(ethylene terephthalate)], Vitel PE 100 [poly(ethylene–terephthalate–isophthalate)], Kodel [poly(cyclohexane dimethanol terephthalate)], Mylar [poly(ethylene terephthalate)], and poly(bisphenol A carbonate) are resistant to degradation in ASTM D 1924-63 (41). Poly(ethylene terephthalate) fibers show good resistance to biological attack after seven years exposure to seawater (60). Dacron (a copolyester of terephthalic acid, methyl terephthalate, and ethylene glycol) in the form of cloth shows good resistance to fungal attack and no loss of strength after soil burial (61). Terylene [poly(ethylene glycol terephthalate)] and polycarbonates are also resistant (11).

Low molecular weight polyketoesters were prepared from 4,4′-bis(chloroacetyl) diphenyl ether, 4,4′-bis(bromoacetyl diphenyl ether, and 4,4′-bis(2-bromopropionyl) diphenyl ether and aliphatic diacids of 6, 8, 9, 10, and 12 carbon atoms (62). These polymers were estimated to have molecular weights of ca 2000 by end-group analysis. The polyester made from $(CH_2)_{10}$ diacid gave an ASTM rating of 3, those from $(CH_2)_8$ and $(CH_2)_6$ gave ASTM ratings of 2, whereas the others gave 0 or 1 ratings. High molecular weight versions of these polymers were not tested.

Aliphatic Polyesters. There are many reports describing aliphatic esters and polyesters as biodegradable. The fungal susceptibilities of three polyester prepolymers, poly(ethylene glycol adipate) of $\overline{M}_n$ = 2390, poly(1,3-propanediol adipate) of $\overline{M}_n$ = 5240, and poly(1,4-butanediol adipate) of $\overline{M}_n$ = 1950, were found to be pronounced as shown by growth ratings of 4 using ASTM 1924-63 (63). Seven organisms were tested: *A. niger*, *A. flavus*, *A. versicolor*, *P. funiculosum*, *Pullularia pullulans*, *Trichoderma*, and *Chaetomium globosum*.

Another study tested four polyesters: poly(propylene sebacate), sebacic acid polyester, and two unidentified polyesters (Code 90, 91) (12). All of these commercial plasticizers were utilized readily by all 24 fungi tested, with one exception. *A. niger* gave no colony growth when the sebacic acid polyester was the sole source of carbon. The following ester plasticizers, which are unavailable to mesophilic organisms, are utilized by thermophilic soil microorganisms at 48°C (64). These are tri-*n*-butyl citrate, dioctyl sebacate, poly(ethylene glycol) 200, tricresyl phosphate, dioctyl phthalate, and triethylphosphate. Benzyl benzoate, di-*n*-butyl tartrate, and glycerol triacetate are quite inactive to thermophilic organisms, and polyester plasticizers incorporating structures similar to these three should be resistant to microbiological attack.

When the relative aggressiveness of fungi vs bacteria toward butylene glycol polyadipate plasticizer as the sole source of carbon was compared, this polymer was found to be susceptible to attack by 51 fungal isolates with no other carbon source present (65). All the bacterial cultures tested required the presence of a supplemental source of carbon such as yeast extract.

The biodegradability of a number of aliphatic polyesters of different structure and molecular weight was investigated by other workers using ASTM D 1924-63 as the test method (66). The results are shown in Table 7. Sample 1, a linear polyester obtained by the ring-opening polymerization of ϵ-caprolactone, has a molecular weight of ca 40,000. This polymer is readily utilized by both fungi and bacteria. Sample 2, a branched polyester obtained by the ring-opening polymerization of pivalolactone, is not utilized, although it is much lower in molecular weight (reduced viscosity = 0.1). Polyesters based on fumaric acid, which is unsaturated, are less readily utilized than

Table 7. Biodegradability of Aliphatic Polyesters[a]

Sample number	Description	Reduced viscosity	ASTM growth rating
1	poly(ϵ-caprolactone)	0.7	4
2	polypivalolactone	0.1	0
3	poly(ethylene succinate)	0.24	4
4	poly(tetramethylene succinate)	0.59	1
5	poly(tetramethylene succinate)	0.08	4
6	poly(hexamethylene succinate)	0.91	4
7	poly(hexamethylene fumarate)	0.25	2
8	poly(hexamethylene fumarate)	0.78	2
9	poly(ethylene adipate)	0.13	4

[a] Ref. 66.

those based on saturated dibasic acids such as succinic and adipic acids. A marked dependence of ease of degradation on molecular weight is observed in samples 4 and 5, poly(tetramethylene succinate).

A good correlation between biodegradability and melting point exists for a series of aliphatic polyesters, as shown in Table 8 (67). Biodegradability decreases markedly with increasing melting point for these polyesters. Copolymers prepared by transesterification between aliphatic polyester poly(ϵ-caprolactone) and various aromatic polyesters show the same trend. Only the aliphatic part of the copolyesters appear to be biodegraded.

In an examination of the degradability of polyesters from DL-tartaric acid and diols having 2, 6, 8, 10, and 12 carbon atoms, the C_2 polyester did not support the growth of *A. niger* in ASTM D 1924-61T (68). The longer chain diol polyesters were degraded, with a maximum effect occurring with C_6 and C_8 diols. Blocking of the pendant OH group of the tartaric acid moiety caused greatly reduced activity. Polyesters containing 50% mandelic acid were degraded; however, those containing ≤20% mandelic acid were inactive, as were all of the glycolic acid polyesters tested.

Block and Graft Polymers. Several graft or block copolymers were prepared using polyethylene, polystyrene, or poly(ethylene terephthalate) as backbone polymers and poly(ϵ-caprolactone) as the biodegradable polymer (69). Table 9 shows the ASTM growth rating as a function of the composition of ethylene–caprolactone graft polymers. It is apparent that the sample must contain at least 50 wt % caprolactone in order to show appreciable growth in ASTM D 1924-63; the caprolactone polymer is selectively degraded.

Table 8. Biodegradability of Saturated Aliphatic Polyesters as Functions of Melting Point[a]

Polyester	Biodegradability, total organic carbon formed, 10^3 ppm	Mp, °C
poly(ethylene adipate)	9.4	48
poly(ethylene azelate)	3.7	51
poly(ethylene suberate)	1.6	64
poly(ethylene sebacate)	1.0	74
poly(ethylene decamethylate)	0.3	86
poly(butylene sebacate)	3.3	65
poly(butylene adipate)	3.0	74
poly(butylene succinate)	0.2	112
poly(ϵ-caprolactone)	3.6	59
polypropiolactone	1.7	95
poly(hexamethylene sebacate)	1.3	73

[a] Ref. 67.

Table 9. Biodegradability of Graft Polymers of Polyethylene and Caprolactone[a]

Sample number	Ethylene, %	Caprolactone, %	ASTM growth rating
1	24	76	4
2	48	52	4
3	60	40	2
4	88	12	0

[a] Ref. 69.

The graft polymers of styrene and caprolactone give good growth at caprolactone concentrations as low as 35%. This probably reflects the greater incompatibility of polystyrene and polycaprolactone compared to polyethylene and polycaprolactone, which results in more of the polycaprolactone chains being available on the sample surface. No evidence of attack on the polystyrene part of these samples was observed. For poly(ethylene terephthalate)–poly(ϵ-caprolactone) block copolymers, owing to the screening action of the PET blocks, a high growth rate was observed only at the highest PCL concentrations. The PET blocks appeared not to have been degraded.

These results for block and graft copolymers suggest that attaching biodegradable polymer molecules to nondegradable polymer molecules does not enhance the degradability of the inert component and that the latter component may encapsulate the active component if present in large proportion.

Miscellaneous Condensation Polymers. A study of the effect of polymer structure on the fungal susceptibility of polyurethanes, using the mixed-culture Petri dish method, revealed that polyether-linked polyurethanes are, in general, much less susceptible to attack by fungi than are the polyester-linked polyurethanes (63). This is a very important consideration in selecting polymeric structures for use in spandex-type garments which are exposed to fungus attack in normal service and in military applications. Polyurethanes made from polypropylene glycols in the molecular weight range 1000–1300 were quite active, but not as active as the polyester-based urethanes.

Other polymers reported to be resistant to microbial attack are phenolformaldehyde and urea–formaldehyde resins (in cross-linked form); cellulose triacetate, butyrate, and propionate; chlorinated polyethers; and silicone resins (11). Another report indicates that cellulose acetate and butyrate are not attacked, nor is polyformaldehyde (Celcon) (41). Polysulfone and poly(2,6-dimethylphenylene oxide) are generally regarded as being resistant to attack because of their similarity to bisphenol A polycarbonate and other aromatic polyesters.

A. fumigatus grows on nitrocellulose suspended in a nitrogen-deficient, carbon-containing medium (70). This organism utilizes nitrogen from nitrocellulose without attacking the carbon backbone and relies on a hydrolysis mechanism to obtain the nitrogen. This process is useful for the removal of nitrate groups from the surface of propellant grains in order to control propellant burning rate.

Poly(ϵ-caprolactone) Biodegradation Studies. Because of its ease of polymerization to high molecular weights and its commercial availability, poly(ϵ-caprolactone) [*24980-41-4*] has been the subject of a number of studies pertaining to its biodegradability (3–4,71–74). The synthesis of ϵ-caprolactone and its conversion to high molecular weight polymers are well documented (75–76), and poly(ϵ-caprolactone) is available from Union Carbide Corporation. The structure is

$$-\!\!\left[(CH_2)_5\overset{\overset{\displaystyle O}{\|}}{C}O\right]_n\!\!-$$

The polymer is very crystalline, melts at around 60°C, and begins to decompose to monomer at temperatures above 250°C. It resembles medium density polyethylene in stiffness (ca 350 MPa, stiffness modulus) and has a waxy feel.

Using the quantitative Petri dish method and the organism *Pullularia pullulans*, a sharp dependence of weight loss per cm^2 on initial molecular weight of the polymer has been observed for poly(ϵ-caprolactone) (3). After 40 d exposure, a 1250 $\overline{M}_n$ sample lost 15 mg/cm^2, a 2000 $\overline{M}_n$ sample lost 5 mg/cm^2, while samples of 14,700 and 27,500 $\overline{M}_n$ lost only ca 2 mg/cm^2. The $\overline{M}_n$ of the remaining polymer was essentially unchanged after three weeks of exposure, indicating surface attack. The study also demonstrated that the observed weight loss was not because of simple hydrolysis.

The effect of soil burial on bars molded from PCL-700, a commercial poly(ϵ-caprolactone) (Union Carbide) of ca 40,000 $\overline{M}_n$ has been determined (77). The bars were buried in a mixture of equal parts of New Jersey garden soil, Michigan peat moss, and builder's sand for intervals up to 12 months. Samples were removed periodically to determine weight loss and measure tensile properties. The results appear in Table 10. Although the weight loss was only 16% after 4 mo of burial, the ultimate tensile strength had decreased by 80% and the elongation at break by 99%. Figure 1 illustrates the appearance and extent of degradation of 56-mL molded containers after burial for time intervals up to one year. Because of the greater surface-to-volume ratio in these containers compared to the test bars, the weight loss was more rapid (2 mo, 12%; 4 mo, 29%; 6 mo, 48%; 12 mo, 95%). Figures 2 and 3 are scanning electron micrographs

Table 10. Effect of Soil Burial on High Molecular Weight Poly(ϵ-Caprolactone)–PCL-700[a]

Burial time, mo	Tensile strength, MPa	Elongation at break, %	Weight loss, %
0	18 ± 0.7	369 ± 59	0
1.25	13 ± 1.5	9 ± 1.4	
2.0	11 ± 1.2	7 ± 2	8
4.0	3.06 ± 1.5	2.6 ± 1.1	16
6.0	0.7	negligible	25
12.0	negligible	negligible	42

[a] Ref. 77.

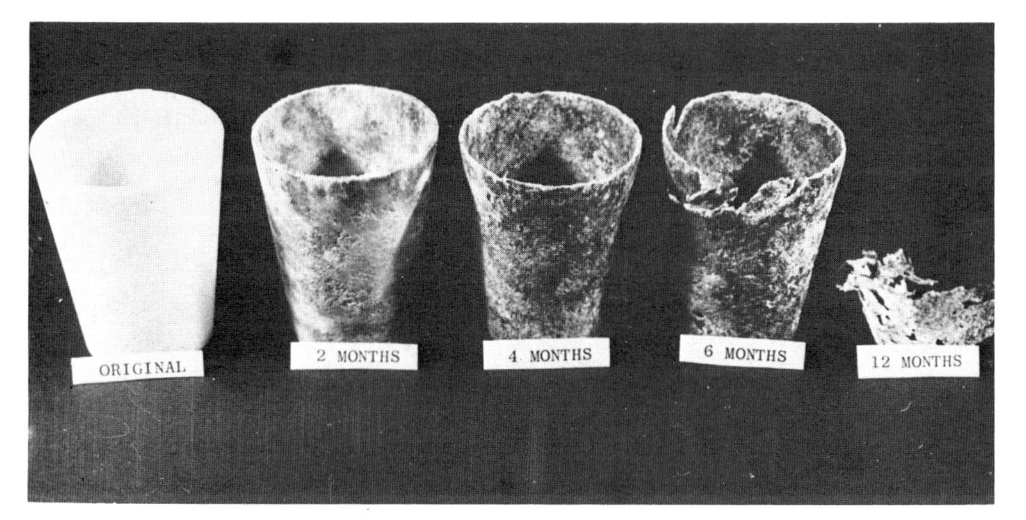

Figure 1. Biodegradation resulting from soil burial for periods up to one year of PCL-700 grade of poly(ϵ-caprolactone) (10). Courtesy of Union Carbide Corporation.

Figure 2. Scanning electron micrograph of surface of molded poly(ϵ-caprolactone) test bar before soil burial. Magnification 1770× (10).

of the surface of molded poly(ϵ-caprolactone) test bars before and after two months of soil burial. Figure 2 (before) reveals microscopic scratches on the surface, while Figure 3 (after) reveals the cavernous surface markings after microbiological attack (71). In other scanning electron microscopic studies of poly(ϵ-caprolactone), surfaces were attacked by *A. flavus* and *Penicillium funiculosum*, and care was taken not to disturb the mycelial growth on the plastic surface (78). Figure 4 shows diffuse biode-

Figure 3. Scanning electron micrograph of surface of molded poly(ϵ-caprolactone) test bar after 2-mo soil burial. Magnification 1770× (10).

gradation at the amorphous areas exposing spherulite arms. Figure 5 reveals early signs of crystalline areas being biodegraded while amorphous areas are severely degraded.

Water-Soluble Polymers. The biodegradability of water-soluble polymers is dependent on their molecular structure, the presence of specific enzymes, and in some instances on the mol wt of the polymer. An old review of the properties of water-soluble polymers includes their biodegradability, as measured by the biological oxygen demand (BOD) test (79).

Water-soluble synthetic polymers which are resistant to biodegradation by the BOD test are poly(ethylene glycol)s, carboxyvinyl polymer (Carbapol), poly(ethylene oxide) (Polyox), poly(vinyl pyrrolidinone), poly(vinyl methyl ether), poly(acrylic acid) salts, polyacrylamide, and poly(vinyl alcohol). However, poly(vinyl alcohol) has been found to be biodegraded by other tests (48–49). Poly(ethylene oxide) has been found to be biodegradable at molecular weights below 500 but is resistant at higher molecular weights.

Most water-soluble cellulose derivatives, such as Klucel (a cellulose ether), and other ether derivatives, such as ethyl 2-hydroxylethyl cellulose, ethyl methyl cellulose, 2-hydroxyethyl cellulose, 2-hydroxyethyl methyl cellulose, 2-hydroxypropyl methyl

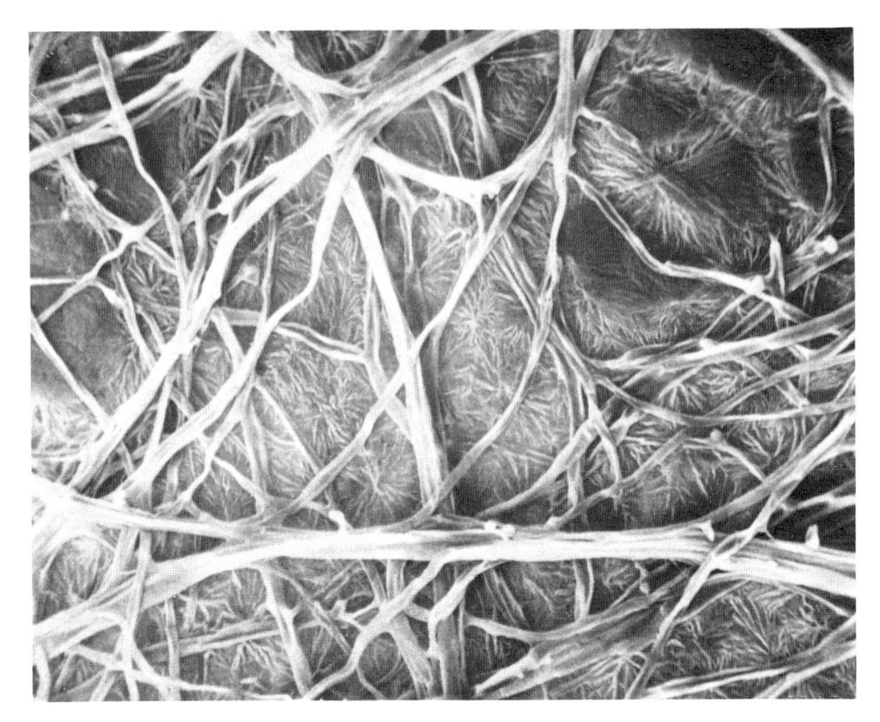

Figure 4. SEM visualization at 140× magnification of surface of poly(ϵ-caprolactone) after attack by *A. flavus* and *P. funiculosum*. Biodegradation of amorphous areas exposing spherulite arms. Mycelial growth on surface not disturbed (78).

cellulose, methyl cellulose, sodium carboxymethyl cellulose, sodium carboxymethyl 2-hydroxyethyl cellulose, and sodium cellulose sulfate, are biodegradable. Naturally occurring water-soluble polymers such as gum arabic, gum karaya, gum tragacanth, guar gum, locust bean gum, starch and various starch derivatives, casein, gelatin, various alginates, and carrageenan (Irish moss) are considered to be biodegradable (79). Agar-agar degrades slowly enough so that it is widely used as an inert microbiological culture medium.

Applications of Biodegradable Polymers. Many intriguing and significant applications for biodegradable polymers are being investigated; some of these have resulted in commercial products. The three application areas which have received the most attention are medical, agricultural, and consumer goods packaging. Because of their specialized nature and greater unit value, medical device applications have progressed faster than the other two.

Medical Applications. Biodegradable plastics have been developed for use as surgical implants (in vascular and orthopedic surgery), as implantable matrices for the controlled, long-term release of drugs inside the body, as absorbable surgical sutures, and for use in the eye. The advantage of biodegradable or absorbable plastics for implantation is that they obviate the need for surgical removal.

In vivo degradation involves hydrolytic, oxidative, and enzymatic processes. Many polymers which are stable when buried in the soil are found to undergo degradation and erosion inside the body of a human or animal. However, in many applications inside the body, degradation rates higher than that observed for conventional polymers are needed. Polymers and copolymers containing biodegradable chain segments have been designed to accelerate the *in vivo* degradation rate of the polymer (74).

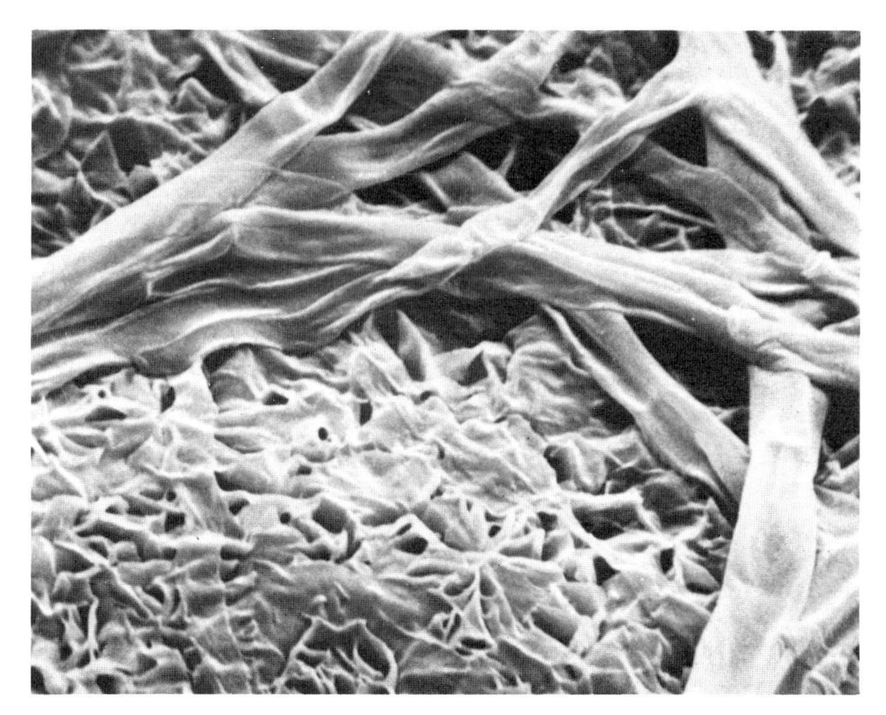

Figure 5. SEM visualization at 533× magnification of surface of poly(ϵ-caprolactone) after attack by *A. flavus* and *P. funiculosum*. Mycelial growth preserved. Amorphous areas severely degraded; crystalline areas slightly degraded (78).

Absorbable surgical sutures have been made by the extrusion of polylactide polymers and copolymers (80). These commercially available polymers are useful for internal sutures to be absorbed by the body after healing has occurred. These products are also claimed to be useful as absorbable pins, screws, plates, and fabric for surgical repair.

Biodegradable polyesters have been studied as possible drug delivery systems for contraceptives and narcotic antagonists implanted under the skin of a person or animal (81). Eye inserts of biodegradable hydroxypropyl cellulose polymers can produce artificial tears (82). Copolymers of L-glutamic acid and γ-ethyl L-glutamate are useful as dispensers of drugs such as steroids, narcotic antagonists, peptide hormones, antimalarials, and anticancer drugs (83). The controlled release of biologically active compounds from biodegradable polymers has been reviewed (84–85).

Agricultural Applications. Perhaps the largest commercial application of degradable polymers is the use of urea–formaldehyde polymer in slow-releasing fertilizers, which permits the application to lawns of up to 30% nitrogen fertilizer without burning. The use of biodegradable polymers as matrices for the slow release of both fertilizers and pesticides has been proposed. In some instances the active chemical is chemically attached to the polymer, while in others the polymer acts as a carrier.

A potentially very large market is believed to exist for biodegradable containers for growing and transplanting tree seedlings, annual flowers, vegetables, and ornamental shrubs. Containerized growing and transplanting enable large growers to reduce costs through automation, to lengthen the transplanting season, and to obtain higher survival rates. If the container is not biodegradable, it must be removed prior to transplanting to permit rapid root growth. However, the bare root plug is very fragile and difficult to plant by machine. The biodegradable container protects the root plug during machine planting and protects the roots against transplanting shock, thus increasing the survival rate. The degradation rate of the container must be adjusted to match the environmental conditions encountered during the greenhouse growing phase and the outplanting phase.

Various aspects of biodegradable-container technology have been described in patents (86–87) and publications (4,31). The biodegradable containers described are fabricated from polymer blends containing poly(ϵ-caprolactone) as the principal biodegradable polymer; there are other additives, both biodegradable and inert. Various fabrication methods have been explored, including injection molding, blow molding, and thermoforming. Blow molding is preferred for lightweight thin-walled containers. Figure 6 depicts a long leaf pine tree seedling grown in a greenhouse in a blow-molded container. Containerized tree seedlings grown in these biodegradable containers have been outplanted extensively in British Columbia, Can., Washington, New Mexico, and several southern states.

Other containers were fabricated from polymeric blends containing a photodegradable ethylene copolymer or homopolymer with photosensitizing additives as well as biodegradable thermoplastic polymers such as poly(ϵ-caprolactone) or poly(ethylene adipate) (87–89). After activation by 8-h exposure to an ultraviolet light source, these containers were buried in soil. After three months, the containers were removed and found to have undergone embrittlement, biodegradation, and disintegration.

Figure 6. Long leaf pine tree seedling grown in greenhouse in biodegradable PCL-700 container by U.S. Forest Service, Southern Forest Experiment Station, Pineville, La. Courtesy of Union Carbide Corporation.

Biodegradable Packaging Plastics. Because of the large increase in the use of plastic film and containers in discardable packaging material, it has been proposed to make throwaway materials from biodegradable plastics to alleviate waste-disposal problems.

There are a number of reasons why this technology has not developed and why it is unlikely to be developed except in very special situations (90–95). The low cost, high volume packaging plastics such as polyethylene, polystyrene, polypropylene, and poly(ethylene terephthalate) are not biodegradable; efforts to make them so by blending with biodegradable fillers or other additives have not been successful. Existing biodegradable plastics are deficient in properties required in most packaging applications and are more expensive than commonly used packaging plastics. Degradable plastics are more difficult to successfully recycle than are nondegradable plastics. Nondegradable plastics are preferred in most landfill sites as are other inert products such as glass bottles, ceramics, and bricks because they do not generate noxious or toxic gases or water-leachable degradation products.

Photodegradability of Synthetic Polymers

Unlike biodegradation, to which almost all types of synthetic polymers are quite resistant, most synthetic polymers are subject to weathering degradation, which is caused by the actions of sunlight, heat, oxygen, and moisture in various combinations. The ease of weathering of polymers is markedly dependent on the presence of photochemically active functional groups and structures that form peroxides.

In most outdoor applications of polymers, such as house paints, plastic signs, plastic glazing, and structural uses, in which photodegradation is undesirable, chemical additives act as uv absorbers, antioxidants, and acid scavengers. Packing plastics are also stabilized to protect the contents of the package against spoilage and to increase shelf life. The stabilization of polymeric systems is discussed extensively in the literature (96–100).

Interest in the development of polymeric systems that degrade rapidly when exposed to sunlight was catalyzed by the enormous increase in the use of convenience packaging for food, beverages, and other consumer goods and by pressure from the ecology movement, which objected to the proliferation of convenience-packaging litter. Other application areas for degradable plastics which have been explored include their use as agricultural mulch film, fumigant retention film, trash and shopping bags, and traffic paint.

Measuring Photodegradability. The most direct way to evaluate the photodegradability of a given polymer is to subject samples to outdoor weathering in a representative location and measure some physical property as a function of exposure time. The desirable rate of degradation varies with the application. For example, a degradable mulch film may be required to maintain most of its strength during the growing season and degrade rapidly thereafter. A degradable loop carrier used to hold together cans of soda may be required to degrade within two months of outdoor exposure.

Degradable polymers become brittle within a specified exposure time and then disintegrate from the action of natural forces such as wind and rain. For flexible film and sheet materials, the brittle point is often reached at or below 20% elongation at break. Rigid degradable polymers such as polystyrene formulations become quite brittle at tensile strengths below 3.5 MPa. A plot of the measured physical property vs exposure time indicates how rapidly the physical-strength property is decaying.

Changes in measured properties should be correlated with other evidence that the plastic is becoming extremely brittle, such as by hand tests.

Laboratory methods of accelerated weathering are widely used to predict long-term outdoor durability of plastics and paints. These methods allow simulation of a variety of climatic and geographic conditions. Polymer degradation is usually accelerated by conditions of high humidity, high temperature, and atmospheric pollution by acid rain, and is retarded by such factors as suspended dust in the atmosphere which reduces the intensity of radiation incident on the plastic object. Devices such as the Xenon Weatherometer expose samples to light that has spectral characteristics similar to sunlight, under selected conditions of temperature and humidity. It is estimated that 100-h exposure of a sample in the Atlas Type XI Weather-Ometer is equivalent to ca 1 yr of outdoor exposure. This type of accelerated weathering testing is discussed elsewhere (101).

The measurement of surface oxidation by attenuated total reflectance infrared spectroscopy (ATR) can be used to follow the extent of photooxidative degradation of polyolefin samples, provided the sample thickness is $\leq 500\ \mu m$ (ca 20 mils) (102). For thicker samples, the extent of surface oxidation may not correlate well with loss of physical properties. Also, for polymeric systems in which the primary mode of degradation is photochemical, the measurement of extent of oxidation may not be a reliable indication of the extent of embrittlement of the plastic sample.

Because the embrittlement which develops in degradable plastics is usually accompanied by surface crazing and cracking, photomicrography, scanning electron microscopy, and conventional photography are useful tools for documenting the changes in the appearance of plastic samples with increasing exposure time.

Uv Catalyzed Polymer Degradation Mechanisms. When exposed to uv radiation, polymers containing carbonyl or hydroperoxide groups have been observed to undergo photochemical reactions which result in chain cleavage. These chemical groups absorb radiation in the 290-nm region, slightly above the shortest uv wavelength to reach the earth's surface through the atmosphere (270 nm). Carbonyl groups and hydroperoxides usually develop in polymers as a result of oxidation of the polymer during processing or from atmospheric aging. One of the most frequently used techniques for enhancing the photodegradability of polymers such as polyolefins is to add catalysts and prooxidants which will stimulate the formation of carbonyls and hydroperoxides and lead to autooxidative chain degradation reactions (103–105). The other approach is to introduce ketone groups into the polymers by copolymerization with monomers such as carbon monoxide and alkyl or aryl vinyl ketones (106–109).

Photochemical Degradation Mechanisms. Polymers containing internal ketone groups undergo chain cleavage by Norrish I and Norrish II mechanisms, as illustrated below (110).

Norrish type I reaction

$$\sim CH_2CH_2\underset{\underset{O}{\|}}{C}CH_2CH_2\sim \xrightarrow{h\nu} \sim CH_2CH_2\overset{\overset{O}{\|}}{C}\cdot + \cdot CH_2CH_2\sim$$

$$\sim CH_2CH_2\overset{\overset{O}{\|}}{C}\cdot \longrightarrow CO + \sim CH_2CH_2\cdot$$

Norrish type II reaction

$$\sim CH_2CH_2\underset{\underset{O}{\|}}{C}CH_2CH_2CH_2\sim \xrightarrow{h\nu} \sim CH_2CH_2\underset{\underset{O}{\|}}{C}CH_3 + CH_2{=}\underset{\underset{H}{|}}{C}\sim$$

Both Norrish I and II reactions produce main-chain cleavage when the ketone group is in the main chain, as in ethylene–carbon monoxide copolymers. For these copolymers, the Norrish type I reaction accounts for ca 10% of the observed chain scission at room temperature, and ca 50% at 120°C. The Norrish type II reaction has been found to be independent of temperature and O_2 concentration above the T_g of the polymer (111).

The photochemical cleavage of copolymers containing side-chain ketones, as for example, poly(ethylene-*co*-methyl vinyl ketone) and poly(styrene-*co*-phenyl vinyl ketone), also undergo Norrish I and II types of degradation (106,112). However, only the Norrish II reaction produces main-chain cleavage directly, as shown below.

Methyl vinyl ketone copolymer

$$\text{\textasciitilde CH}_2\text{CH}_2\underset{\underset{\text{CH}_3}{|}}{\underset{\text{C=O}}{|}}{\overset{\text{R}}{\overset{|}{\text{C}}}}\text{CH}_2\text{CH}_2\text{CH}_2\text{\textasciitilde} \xrightarrow{\text{Norrish I}} \text{\textasciitilde CH}_2\text{CH}_2\overset{\text{R}}{\overset{|}{\dot{\text{C}}}}\text{CH}_2\text{CH}_2\text{\textasciitilde} + \text{CH}_3\dot{\text{C}}\text{=O}$$

$$\xrightarrow{\text{Norrish II}} \text{\textasciitilde CH}_2\text{CH}_2\underset{\underset{\text{CH}_3}{|}}{\underset{\text{C=O}}{|}}{\overset{\text{R}}{\overset{|}{\text{CH}}}} + \text{CH}_2\text{=}\overset{\text{H}}{\overset{|}{\text{C}}}\text{CH}_2\text{\textasciitilde}$$

Phenyl vinyl ketone copolymer

$$\text{\textasciitilde CH}_2\underset{\text{C}_6\text{H}_5}{\underset{|}{\overset{\text{H}}{\overset{|}{\text{C}}}}}\text{CH}_2\underset{\underset{\text{C}_6\text{H}_5}{|}}{\underset{\text{C=O}}{|}}{\overset{\text{H}}{\overset{|}{\text{C}}}}\text{\textasciitilde} \rightarrow \text{\textasciitilde CH}_2\underset{\text{C}_6\text{H}_5}{\underset{|}{\overset{\text{H}}{\overset{|}{\text{C}}}}}\text{CH}_2\overset{\text{H}}{\overset{|}{\dot{\text{C}}}}\text{\textasciitilde} + \underset{\text{C}_6\text{H}_5}{\underset{|}{\dot{\text{C}}\text{=O}}} \quad \text{(Norrish I)}$$

$$\rightarrow \text{\textasciitilde CH}_2\underset{\text{C}_6\text{H}_5}{\underset{|}{\overset{\text{H}}{\overset{|}{\text{C}}}}}\text{CH}_2\underset{\text{CO}\cdot}{\underset{|}{\overset{\text{H}}{\overset{|}{\text{C}}}}}\text{\textasciitilde} + \cdot\text{C}_6\text{H}_5 \quad \text{(Norrish I)}$$

$$\rightarrow \text{\textasciitilde CH}_2\underset{\text{C}_6\text{H}_5}{\underset{|}{\text{C}}}\text{=CH}_2 + \underset{\underset{\text{C}_6\text{H}_5}{|}}{\underset{\text{C=O}}{|}}{\text{CH}_2}\text{\textasciitilde}$$

Quantum yield for the Norrish type II reaction depends on chain length and on whether the ketone carbonyl group is in a side chain or in the main chain of the polymer (113). For structures with the carbonyl group in the main chain, such as di-*n*-butyl ketone or poly(ethylene-*co*-carbon monoxide), the quantum yield decreases from 0.11 to 0.025 as the length of the carbon chains on either side of the carbonyl group increases from 4 to 100 carbon atoms. For methyl or phenyl alkyl ketones, the quantum yield is almost independent of increasing alkyl group chain length; it varies from 0.2 to 0.3. The inference is that less methyl or phenyl vinyl ketone would have to be introduced into a copolymer to obtain a given ease of degradation than if carbon monoxide were used. However, in judging the relative merits of these two ways of introducing carbonyl groups into polymers, one must consider the greater cost of vinyl ketone monomer compared to carbon monoxide, the lesser effect on modulus of introducing the same number of CO groups into the polymer chain, and the greater ease of removal of unreacted CO from the copolymer.

Photooxidative Degradation Mechanisms. Because of the tendency of polymers such as polyethylene, polypropylene, polybutylene, and polybutadiene to undergo photooxidative degradation unless protected with antioxidants and other stabilizers, their degradability can be enhanced by the addition of catalysts and prooxidants rather than by the introduction of carbonyl groups via copolymerization (103–105,114–116). The mechanism of photooxidative degradation begins with the formation of a polymer-chain radical. This event is initiated by exposure to uv, ir, or ionizing radiation, by mechanical shear, or by chemicals such as peroxides, singlet oxygen, or ozone.

$$\mathrm{PH \rightarrow P\cdot + H\cdot}$$

The radical then forms polymer peroxides and hydroperoxides by reaction with oxygen.

$$\mathrm{P\cdot + O_2 \rightarrow POO\cdot}$$

$$\mathrm{POO\cdot + P'H \rightarrow POOH + P'\cdot}$$

$$\mathrm{P'\cdot + O_2 \rightarrow P'OO\cdot}$$

Polymer hydroperoxides are highly susceptible to decomposition, most often through the following mode at wavelengths greater than 290 nm (117):

$$\mathrm{POOH \rightarrow PO\cdot + \cdot OH} \qquad E_{\mathrm{diss}} = 42\ \mathrm{kcal/mol}$$

Alkoxy radicals (PO·) can undergo chain cleavage to give aldehyde or ketone groups as shown below (118–119):

$$\mathrm{{\sim}CH_2C(R)(O\cdot)CH_2C(R)(H){\sim} \rightarrow {\sim}CH_2C(R){=}O + \cdot CH_2{-}C(R)(H){-}CH_2{\sim}}$$

$$\mathrm{{\sim}C(R)(H)C(H)(O\cdot)C(R)(H)CH_2{\sim} \rightarrow {\sim}C(R)(H)C(H){=}O + \cdot C(R)(H)CH_2{\sim}}$$

There are also several reactions in which polymer hydroperoxide groups decompose to give ketone groups without chain cleavage resulting (120–121).

Stabilizers. Antioxidants stabilize polymers by donating hydrogen atoms to polymer radicals. The resulting antioxidant radical is usually not active enough to abstract hydrogen from polymer chains but will react with polymer radicals by a termination reaction. Antioxidants react faster with alkyl peroxy radicals (POO·) than with alkyl radicals (P·).

$$P\cdot + AH \rightarrow PH + A\cdot$$

$$POO\cdot + AH \rightarrow POOH + A\cdot$$

$$POO\cdot + A\cdot \rightarrow POOA \text{ (stable)}$$

$$A\cdot + A\cdot \rightarrow A\text{—}A \text{ (stable)}$$

Metal salts such as ferric dibutyl dithiocarbamate act as stabilizers of polyolefins at high concentrations and act as photodegradation catalysts at low concentrations (122). Transition metal ions, such as Fe^{2+} and Co^{2+}, undergo an oxidation–reduction cycle in the process of catalyzing the decomposition of hydroperoxides in polymers, as shown below (123).

$$M^{n+} + POOH \rightarrow M^{(n+1)+} + PO\cdot + OH^-$$

$$M^{(n+1)+} + POOH \rightarrow M^{n+} + POO\cdot + H^+$$

The overall reaction is, therefore,

$$M^{n+} + 2\,POOH \rightarrow PO\cdot + POO\cdot\, H_2O$$

Reports indicated that when metal dialkyl dithiocarbamates decompose, they generate Lewis acids such as SO_2 or SO_3 which act as catalysts for the destruction of hydroperoxides, and that all the transition metal dithiocarbamates destroy hydroperoxides at the same rate (122,124). Dithiocarbamates are claimed to provide more heat-stabilizing activity to the polymer during hot milling than other metal salts such as stearates or acetonylacetonates but decompose more rapidly when exposed to uv light, thus leading to more rapid photodegradation of the polymer.

Sensitizers. Many organic compounds, such as aromatic ketones, diketones, quinones, nitroso compounds, and dyes, are known to sensitize the photoreactions of polymers that lead to degradation, cross-linking, and grafting (125–127). Benzophenone, in its photochemically excited first triplet state, abstracts hydrogen from polymers such as polyethylene, poly(vinyl chloride) and poly(vinyl alcohol) (118). Low concentrations of the ketone and long-term exposure to sunlight lead to photodegradation of polyethylene, whereas higher ketone concentration (1–2%) and exposure of the polymer to higher intensity ultraviolet fluxes from artificial light sources lead to cross-linking, presumably from the higher polymer radical concentrations and the reduced exposure to oxygen. Whereas *p*-hydroxybenzophenones act as photodegradation sensitizers, *o*-hydroxy compounds act as photostabilizers (125–126).

Photodegradable Addition Polymers. The science and technology of polymer degradation and stabilization have developed to the point that almost any addition polymer could be formulated to degrade rapidly when exposed outdoors if necessary or desirable. However, degradable plastic products have been restricted to those plastics which are widely used in packaging or in agricultural uses such as mulch film. These include high and low density polyethylene, ethylene copolymer, polypropylene, poly(butene-1), polystyrene, and poly(vinyl chloride).

Photodegradation Additives. Table 11 lists twenty companies that have developed technology for making degradable plastics using additives to catalyze photooxidation. Products developed by Gerald Scott, Akerlund and Rausing, Amerplast, and Biodegradable Plastics, Inc., were marketed in North America or Europe during the 1970s but have not been promoted during the 1980s thus far.

Because of the proprietary nature of this field, it is often difficult to determine the composition of various formulations and evaluate the relative efficacy of various

Table 11. Polyolefins Containing Additives to Promote Photodegradation

Company	Polyolefin	Additive	Ref.
Union Carbide Corp.	polyethylene–ethylene copolymers	transition metal salts, polypropylene antioxidants, oxyalkanoyl or dialkanoyl polymer	128
Eastman Kodak	polyethylene, polypropylene, ethylene–CO–CO polymers	opaque pigment or dye, metal salts	129
Texas Oil Co.	polyolefin	cation-exchange coke, polypropylene, fertilizers, fungicides, nematocides	130
Dow Chemical Co., Inc.	75% polyethylene, 25% polypropylene	benzophenone	131
DeBell and Richardson	polyethylene, polypropylene	metal oxide particles, eg, silica, dispersed in fatty acids	132
Princeton Polymer Laboratories, Inc.	polyethylene, polypropylene, polystyrene	metal salts or chelates, aromatic ketones	133
Princeton Chemical Research, Inc.	poly(butene-1) and blends (Ecolan)	filled with fertilizer for mulch film; stabilized carbon black filled mulch film; control of history and antioxidant level	134
Biodegradable Plastics, Inc.	polyethylene, polystyrene	aromatic ketones, metal salts, and hindered phenols	135
Gerald Scott (UK), Dvorkovitz Licensee	polyethylene	ferric *N*,*N*-dibutyldithiocarbamate	136
Imperial Chemical Industries	polyolefins	benzophenone, iron salts	137
Ethylene Plastique (France)	polyolefins	Fe salts with aromatic ketones, dialkyl polysulfides or sulfur	138
AB Akerlund & Rausing (Sweden)	polyethylene	soluble metal salts and aromatic ketones, colloidal metal complexes supported on silica, benzophenones and metal stearates dispersed in fillers and pigments	139
Amerplast (Finland)	Ecoten polyolefin	ferric *N*,*N*-dibutyl dithiocarbamate	[a]
Mitsubishi Petrochemicals (Japan)	polyolefins	cumene derivatives, alkyl or aryl ketones, benzoanthrone	140
Sumitomo Chemical Co., Ltd.	polyolefins, polystyrene, poly(vinyl chloride)	anthraquinone, transition metal, transition-metal compound	141
Chisso Corp. (Japan)	polypropylene	mercaptobenzothiazoles	142
Sekisui Chemical Co. (Japan)	polystyrene	benzophenone and derivatives	143
Mitsui Petrochem Industries Ltd. (Japan)	polyethylene, polypropylene	diphenylacetaldehyde and acid, tetrabromophthalic anhydride, dipyridyl glycol, aromatic bromides, ketones, esters, carboxybenzaldehydes, anthraquinones	144

[a] Licensed from G. Scott.

additives. However, some information is available. Table 12 shows the effect of varying the concentration of ferric *N*,*N*-dibutyldithiocarbamate in low-density polyethylene on the degradation rate of sheet stock 500 μm (20 mils) thick. Exposure was in an Atlas Weatherometer (Model XW, Carbon Arc, with a Corex filter to remove wavelengths below 280 nm). At concentrations >0.05% of the additive, the compound acts as a

Table 12. Effect of Ferric *N,N*-Dibutyldithiocarbamate (FDTC) on Polyethylene Weathering[a]

FDTC in LDPE, wt %	Elongation to break, %, XW-Weatherometer, h: 0	80	180	200	220	240
0.00	430	460	80	70	40	40
0.01	450	380	35	40	3	2
0.05	430	280	50	5	3	3
0.10	430	470	60	15	30	8
0.20	430	470	130		50	
0.50	440	450	400		440	

[a] 500 μm (20 mil) thick films tested. 37.5-μm (1.5 mil) thick films slightly faster, ≤20% elongation required.

stabilizer rather than as an oxidation accelerator. At 0.05% of the iron salt present, 200 hours of exposure in the XW Weatherometer are required to obtain a brittle sample. Although this degradation rate is unacceptably slow for certain products which are subject to littering, it may be acceptable for certain degradable mulch-film applications, where the film must remain intact during an entire growing season.

In a study using plaques of poly(butene-1) 762 mm (30 mils) in thickness, it was found that various concentrations of cobalt octanoate gave rapid degradation of the polymer in 12–30 d (145). Stabilizers such as Irganox 1010 (tetrakis[methylene-3-(3′,5′-di-*t*-butyl-4′-hydroxyphenyl)propionate]methane) and dilauryl thiodipropionate (DLTDP) were necessary ingredients to ensure against premature decomposition. Carbon black effectively neutralized the destabilizing effect of the cobalt salts. Benzophenone and PTP (an unsaturated hydrocarbon prepolymer) were less active than cobalt octoate. Those formulations which contained antioxidants were stable in the shade for 45 days and possibly longer.

Samples of polyethylene and ethylene copolymers containing various transition-metal salts, autooxidizable polymers such as polypropylene and poly(ethylene oxide), and stabilizers were exposed to sunlight; some were subjected to prior ionizing radiation to activate them (87–88). Polypropylene, either crystalline or amorphous, was found to be the preferred autosusceptible polymeric additive. Table 13 compares the efficacy of soluble salts of cobalt, iron, manganese, cerium, zinc, lead, zirconium, and calcium in catalyzing the photoinduced decomposition of 0.922 density polyethylene containing 2% isotactic polypropylene. The concentration of metal salts was 0.05% (as metal) in the formulation. These formulations were pressed into plaques and weathered in an Atlas XW Weatherometer (carbon arc–Corex D filter). The exposure times in hours required for embrittlement to occur, for the surface carbonyl ratio R to equal or exceed 1.7, and for the ultimate elongation to drop below 20% are given in Table 13. R is the ratio of carbonyl absorbance at 5.8 μm to the methylene absorbance at 7.3 μm, as measured by attenuated total reflectance ir spectroscopy. The most active metal salts were cobalt, iron, manganese, and cerium.

The photooxidative degradation of polystyrene has been investigated extensively (146–152). Because of the ca 260-nm phenyl ring uv absorption band in polystyrene, many investigators used 253.7-nm wavelength light in their photolysis studies. However, solar radiation reaching the earth contains very little ultraviolet below 270 nm; consequently, these studies are not directly relevant to the photodegradation that

Table 13. Effect of Metal Salts on the Embrittlement and Surface Oxidation of Polyethylene[a,b]

Salt[c]	Exposure time required, h		
	Embrittlement	FMIR index, $R \geq 1.7$[d]	20% elongation
Co octanoate	150	60–150	60
Fe octanoate	100	60	60
Mn octanoate	150	150	150
Ce naphthenate	150		150–250
Zn octanoate	250		150
Pb octanoate	250–500		≥150
Zn octanoate	500		
Ca octanoate	500		

[a] Ref. 88.
[b] Polymer composition contained 2% polypropylene.
[c] Metal salt conc 0.05% as metal in polymer composition.
[d] R = absorbance at 5.8 μm/absorbance at 7.3 μm. Attenuated total reflectance ir spectroscopy.

occurs when polystyrene is left outdoors. The poor weathering characteristic of polystyrene is believed to be caused by impurities, such as acetophenone groups, formed in the thermal oxidation of the polymer (152).

For degradable applications, polystyrene does not embrittle fast enough; photodegradation and photooxidative degradation is enhanced by methods similar to those described for polyethylene and poly(butene-1). Increased polystyrene photodegradation rates at wavelengths >280 nm require that ketone or other carbonyl groups be introduced into the molecule by copolymerization or grafting. Photooxidative degradation of polystyrene is accelerated by the presence of compounds which become activated by sunlight, such as benzophenone, acetophenone, and anthrone, and by metal salts which catalyze the decomposition of hydroperoxides. In one mechanistic study, thin films of polystyrene containing 3 wt % benzophenone decreased in ultimate tensile strength from 3.0 to ca 0.5 kg/mm^2 after 50 h of exposure to 365-nm wavelength light in air (153). Oxygen was absorbed linearly with increasing film thickness, but the diffusion of oxygen into the film was not the rate limiting step.

The effects of aromatic ketones, with and without cobalt octanoate, on the photooxidative degradation of 0.254-mm (10-mil) thick plaques of clear polystyrene were compared (154). Samples were exposed in an XW Weatherometer for as long as 600 h, after which the polymer solution viscosity was measured at a concentration of 0.2 grams per 100 grams benzene at 30°C. Table 14 shows the percent decrease in original viscosity from weathering. Anthrone was the most effective of the three photosensitizers.

Exposure of polypropylene to uv radiation in air can produce rapid photooxidation with loss of physical strength through the formation of hydroperoxide and carbonyl groups. Traces of transition metal catalysts used in making the polymer contribute to its ease of photooxidation (155). Of course, polypropylene can be stabilized by additives, particularly certain pigments. Polypropylene has been used as an additive to enhance the photooxidative degradation of other polymers for certain degradable polymer applications (87–88), but with the exception of the brief interest of the Chisso Corporation of Japan, there has been no marketing of photodegradable polypropylene.

Table 14. Effect of Aromatic Ketones and Metal Salt on Photooxidation of Polystyrene[a]

	Percent original viscosity[c] after hours exposure in XW Weatherometer					
Photosensitizer[b]	0	150	300	400	500	600
control	100	69	60	48	53	45
with cobalt	100	72	49	45	38	37
benzophenone	100	56	47	43	42	41
with cobalt	100	55	37	36	38	34
xanthone	100	49	34	37	35	28
with cobalt	100	47	41	41	33	32
anthrone	100	38	32	31	32	24
with cobalt	100	33	28	24	24	25

[a] Ref. 153.
[b] 2% added ketone; 0.1% cobalt added as cobalt octanoate.
[c] Solution viscosity relative to control.

The Japan Synthetic Rubber Co. has made degradable products based on syndiotactic poly(1,2-butadiene), but their commercial development appears to be at a standstill (116). The polymer photodegrades by forming hydroperoxides and also undergoes cyclization reactions which lead to brittleness.

The photochemistry of many other addition polymers has been investigated, including polyacrylonitrile (156), poly(vinyl chloride) (157), poly(vinyl alcohol) (158), poly(vinyl acetate) (159), and various polyacrylates and polymethacrylates (160–161). The photooxidation of polyacrylonitrile leads to the formation of a ladder polymer with a minimum of degradation (156). Poly(vinyl chloride) undergoes dehydrochlorination that is accelerated by the HCl evolution. The unsaturated polymer so formed is easily oxidized. Because of the similarity in thermal and photodegradation mechanisms, it is difficult to develop a photodegradable PVC with satisfactory hot processing stability and good shelf life. The ease of photodegradation of poly(vinyl alcohol) is greatly influenced by the aldehyde content of the vinyl acetate monomer used to prepare the precursor polymer (158).

Incorporation of Photosensitive Groups. The alternative to the additives approach for enhancing polymer degradability is to incorporate photosensitive groups into the polymer molecule during polymerization. Copolymerization with carbon monoxide or a vinyl ketone is the usual method.

Although carbon monoxide will not homopolymerize, it combines with ethylene in proportions up to a 1:1 ratio; these copolymers have been made by patented processes for some time (162–164). The crystal structure of poly(ethylene-*co*-carbon monoxide) is the same as that for polyethylene over a wide range of composition (165), and copolymers containing small percentages of CO are quite similar to polyethylene in appearance and physical properties. The presence of ketone groups along the polymer chain causes absorption of uv light in the 270–290-nm wavelength region. When exposed to sunlight, the copolymers undergo photodegradation, principally by the Norrish II mechanism (107–108). Carbon monoxide apparently copolymerizes with propylene, isobutylene, butadiene, allyl esters, vinyl acetate, vinyl chloride, acrylonitrile, and tetrafluoroethylene (166), but only the ethylene copolymer is presently available as a commercial product. Copolymers of ethylene and carbon monoxide have been commercialized by E. I. du Pont de Nemours & Co. and Union Carbide Corporation.

Figure 7 shows the effect of increasing the percent carbon monoxide incorporated into the copolymer on the time required for the polymer to become embrittled when exposed to a fluorescent sunlamp (167). Only low concentrations of CO are required to obtain rapid photodegradation and embrittlement of the copolymer. Blends of low density polyethylene with the copolymer did not decrease in percent elongation to the brittleness point as rapidly as did the pure copolymer. Although the blends of polyethylene and carbon monoxide copolymers are able to cocrystallize (165), the predominant photodegradation mechanism for the copolymer does not generate free radicals which might attack the polyethylene chains. Similar results are obtained with blends of polyethylene with copolymers of ethylene and methyl vinyl ketone.

Methyl or phenyl vinyl ketone will copolymerize with a variety of vinyl monomers, including styrene, ethylene, methyl methacrylate, acrylonitrile, methacrylonitrile, and propylene (168). The resulting polymers are readily degraded, even with small amounts of the ketone-containing monomer incorporated. Apparently, the only polymers of this type available commercially are those from Van Leer-Eco Plastics, Ltd., Ecolyte-E (poly(ethylene-*co*-methyl vinyl ketone)), and Ecolyte-S (poly(styrene-*co*-phenyl vinyl ketone)). Ecolyte-E degrades very rapidly when exposed to sunlight; it becomes embrittled in two weeks. Blends of Ecolyte-E and polyethylene lose their strength more slowly and are not embrittled after 4–6 wk.

Ecolyte-S and blends of it with polystyrene have also been studied (169). The tensile strength of a 0.6-mm thick Ecolyte-S film decreases with increasing exposure time in a uv accelerator as shown in Table 15. Because of the efficiency of the Ecolyte-S photodegradation reaction, blends can be made with styrene homopolymer which have different degradation rates, as shown by Figure 8 (170). The master batch of Ecolyte-S degrades rapidly, reaching essentially zero tensile strength within two weeks of summer

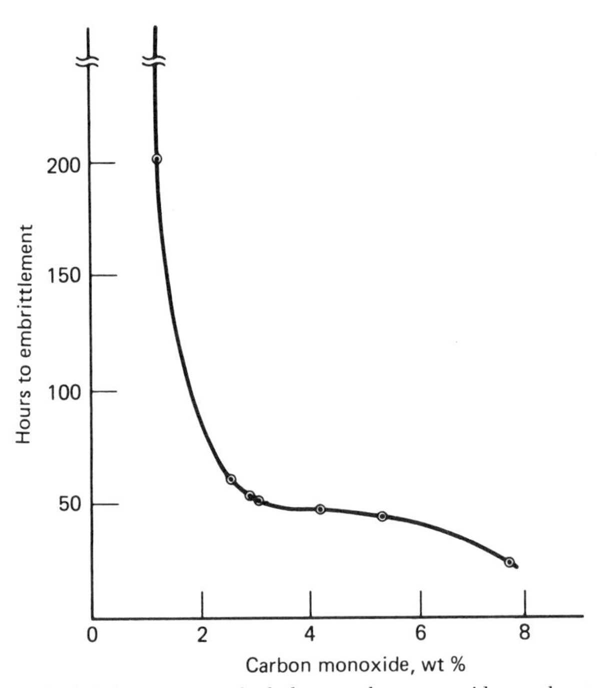

Figure 7. Embrittlement rate of ethylene–carbon monoxide copolymers (167).

Table 15. Tensile Strength of Ecolyte-S vs Hours Exposure in Uv Accelerometer[a]

Exposure in uv accelerometer, h	Ultimate tensile strength, MPa[b]
0	50.3
3	48.3
8	47.6
21	37.2
34	20.0
43	14.5
52	6.2

[a] Ref. 169.
[b] To convert MPa to psi, multiply by 145.

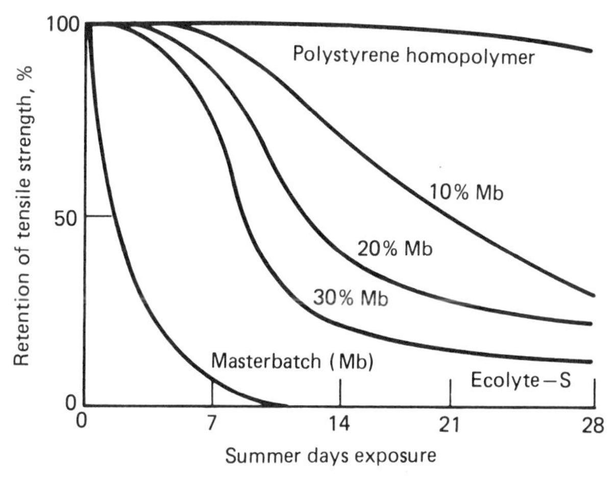

Figure 8. Degradation of Ecolyte-S–polystyrene blends.

exposure. The blends containing 10, 20, and 30% master batch degrade at rates directly related to the amount of Ecolyte-S. Thus, blending permits adjustments of degradation rate to suit the application for which the product is intended.

Random copolymers from styrene and the following monomers, methyl vinyl ketone, phenyl vinyl ketone, isopropenyl methyl ketone, and benzalacetophenone have been prepared using emulsion polymerization (171). Benzalacetone produces alternating copolymers with styrene; these copolymers photolyze rapidly. The degradation rates of the vinyl and isopropenyl ketones are even faster than those of benzalacetophenone and benzalacetone. A Norrish II degradation mechanism is proposed.

Photodegradable Condensation Polymers. Most patents pertaining to degradable polymers have been issued for addition polymers rather than condensation polymers (172). This is partly because of cost, but mainly because applications suitable for degradable plastics usually involve addition polymers. Table 16 lists a variety of patented condensation polymers. None of these products have been commercialized.

The most widely used polyester is poly(ethylene terephthalate); the ester carbonyl groups absorb light in the wavelength region 280–400 nm. In the presence of air, photooxidative degradation involving hydroperoxide formation and cleavage takes place to form hydroxyl radicals and chain alkoxyl radicals which undergo several re-

Table 16. Patented Photodegradable Condensation Polymers

Polymer type	Photodegradable structural feature or additive	Ref.
polyamides	condensation of diamines with acetyl dicarboxylic acid; ketone carbonyl 0.1–3% as side chain	173
polyesters	condensation of 1,4-butanediol with γ-ketopimelic acid; up to 15 wt % of ketone carbonyl in main chain	174
polyesters	ketodicarboxylic acids used to incorporate up to 3 wt % ketonic carbonyl into polyesters	175
polyacetals	anthrone additives	176
polyurethanes	polyalkylene oxide–toluene diisocyanate condensation product is uv degradable	177
cellulose acetate	aromatic ketone and metal salt additives	178
polyurethanes	polyurethane with butadiene units; made in presence of aromatic ketones and metal salts	179
epoxides	photodegradable condensation product of propylene oxide, epichlorhydrin, and styrene oxide plus 30% mineral fillers	180
polycarbonates	photodecomposable poly(oxycarbonyl disulfide)s	181

actions leading to chain scission, the formation of monohydroxy terephthalate groups, and CO and CO_2 as volatile products (182). Polyesters containing keto groups degrade by a Norrish I reaction (183). The chain radicals formed during degradation initiate free radical polymerization in the presence of suitable monomers. Unsaturated polyesters, for example, those containing maleic acid groups, undergo chain scission and some cross-linking when exposed to uv radiation (184). Aromatic polyesters contain chromophoric groups which absorb in the 280–400-nm range. The incorporation of additional photolabile groups in these polyesters may lead to problems with indoor light stability, thermal stability during fabrication, and chemical interaction with foods and beverages if the products are used as bottles or packaging films.

Aliphatic polyamides (nylons) are used to some extent in coextruded films between layers of polyethylene to obtain good grease resistance in food packaging applications. The photolysis of nylon-6, nylon-6,6, and nylon-6,10 have been studied extensively (185). These polymers undergo photodecomposition by breaking of C—N bonds and eventual formation of ethylene and CO. Photooxidation results in formation and decomposition of hydroperoxides. Exposure of nylon-6,6 sheet to a 450 watt uv lamp for 96 h changed the sample from white, opaque, and flexible to straw-colored and brittle (186). Cellulose esters have also been studied in this way (186). The example in Table 17 represents the only effort to prepare degradable nylons.

Cellulose paper products, cellophane, viscose rayon, cellulose acetate, and cellulose nitrate are all subject to photodegradation and photooxidative degradation. Chain-scission reactions generate free radicals, hydroperoxide groups form, and volatile products such as aldehydes, esters, ketones, alcohols, and hydrocarbons are produced in addition to cellulosic monomers and oligomers (187). Photodegradable cellulose acetate containing aromatic ketones and metal salts has been made (178).

The thermal and photooxidation properties of three condensation polymers made from bisphenol A [2,2-bis(4-hydroxyphenyl)propane] have been studied (188). Polycarbonate is made from bisphenol A and phosgene, polysulfone from bisphenol A and 4,4-dichlorodiphenyl sulfone, and phenoxy resins from bisphenol A and epi-

Table 17. Outdoor Weathering of Materials Used to Package Beverage Cans[a]

Material	Thickness, mm[b]	Not exposed	1 mo	2 mo	5 mo
Plastic		*Ultimate elongation after exposure, %*			
low-density polyethylene	0.43	584	625	535	460
ethylene–CO copolymer	0.43	550	10	0	0
polyethylene shrink wrap	0.05	563	380	40	350
polypropylene	0.25	527	380	40	0
Paper		*Ultimate tensile strength*, MPa[c]			
six-pack carton		43.2	18.9	24.5	8.07
corrugated box		5.36	5.49	4.07	1.16

[a] Ref. 191.
[b] To convert from mm to mils, divide by 2.54×10^{-2}.
[c] To convert from MPa to psi, multiply by 145.

chlorohydrin. Polysulfone is outstanding among thermoplastic polymers in thermal oxidation stability and is exceeded in this property only by polyethylene terephthalate and fluorinated polymers. Polycarbonate also has good thermal oxidative stability whereas phenoxy resins, with a low T_g, suffer by comparison. The photooxidation stabilities of the three polymers are compared in Figure 9. Polycarbonate has by far the best photostability of the three. Chain scission at the sulfone link in polysulfone is thought to be the main photodegradation step and is followed by production of a highly oxidized short-chain disulfonic acid (189). Phenoxy photooxidation leads to chain scission and cross-linking (188). Polycarbonate is used as a glazing material because of its toughness and good light stability. Phenoxy is used in zinc-rich primer coatings for steel and magnetic-tape coating. Polysulfone is used as a molding resin in applications where outstanding thermal stability is required. It seems unlikely that these polymers will ever be used in degradable applications.

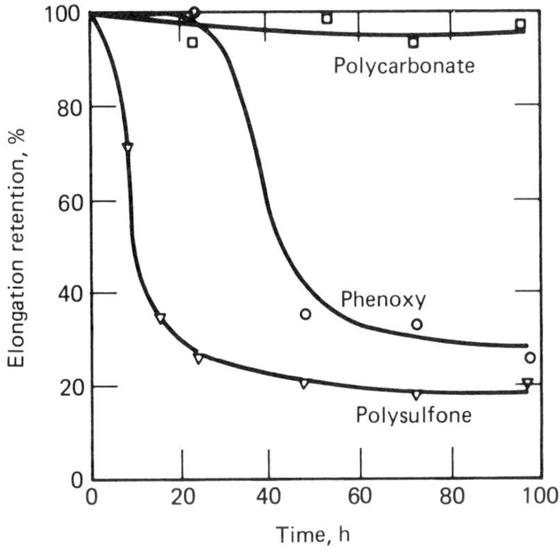

Figure 9. Effect of RS sunlamp exposure on elongation of polycarbonate, phenoxy, and polysulfone (188).

Oxidative Breakdown of Polymer Films During Soil Burial. Polymer films buried in the soil under aerobic conditions are subject to oxidative degradation to varying degrees depending on polymer structure and morphology, the prior history of thermal oxidative abuse, and the presence of additives, such as stabilizers or oxidative accelerators. The embrittlement of polyethylene (LDPE and HDPE), polypropylene (PP), nylon-6,6, and poly(ethylene terephthalate) (PET) in film form under soil burial conditions for periods up to 32 mo has been studied (190). Based on decrease in elongation at break, the films were ranked in order of increasing ease of embrittlement as follows: PET < PP < LDPE < HDPE < nylon-6,6. The embrittlement of the nylon and polyethylene films were followed by ir luminescence spectroscopy, SEM, percent elongation at break and hydroperoxide concentration changes. The rapid deterioration of nylon was similar to that observed in oxygenated water slightly above room temperature.

Applications for Photodegradable Plastics. All currently available photodegradable plastics have been developed to facilitate disposal of the plastic materials after their purposes are served. They can be divided into two categories: those degradable products, mostly packaging materials, that are designed to degrade as rapidly as possible once they are placed outdoors; and plastics that are designed to degrade only after having served a useful function outdoors for a definite time span. Adding degradability requirements to the many other performance specifications that packaging materials must satisfy increases product cost and may interfere with performance specifications vital to the application. The control of photodegradability in products that must function outdoors before degrading is even more difficult to achieve.

Disposable packaging applications for degradable film which have been proposed or developed include shopping bags, trash bags, garbage bags, produce and fruit bags, bread bags, frozen-food bags, snack bags, loop carriers for six-packs of beer or soft drinks, and overwraps for paper products, clothing, stationery, and newspapers and other periodicals. Applications for molded, extruded, or vacuum-formed degradable plastics include plastic plates, eating utensils, drinking cups, foamed plastic trays for use in supermarkets, foamed plastic egg cartons, and containers for dairy products. These are packaging materials that are used once and discarded. The degradable plastics most often used are formulated from polyethylene, polystyrene, or their degradable copolymers.

In all these applications, degradability of the plastic object or container is not important to its functional utility. If the packaging material or container is disposed of properly, ie, by being incinerated or buried in a landfill, the photodegradability feature is not utilized. Only when the packaging material is improperly discarded as litter is the photodegradability feature of societal benefit. Fast food drive-ins became interested in degradable plastics to alleviate litter, but it was soon realized that degradable plastics would have to disappear almost immediately after being discarded to accomplish the desired purpose. Most litter problems were solved instead by providing indoor eating facilities, plenty of trash receptacles, and more effective policing of parking lots. Also, consumers are reluctant to pay a premium price for degradable plastics, which is necessary to cover the incremental cost of manufacture. In some of the applications for which degradable plastics have been promoted, such as lawn trash bags, photodegradability can interfere with the proper functioning of the bag by causing its premature rupture when left in the sun.

Most degradable packaging applications have not become commercial or have not been profitable. On the other hand, the degradable loop carrier for six-packs of beer and soft drinks has increased in volume since 1975. The total resin volume consumed in making loop carriers has not been published but is estimated to be 25,000–35,000 metric tons per year, of which 20% is believed to be of the degradable type. The degradable plastic used in greatest volume for loop carriers is probably a copolymer of ethylene with carbon monoxide. Demand for degradable loop carriers is a result of legislative action in various states banning conventional polyethylene loop carriers but permitting degradable types. Such legislation has been enacted in Alaska, California, Maine, Massachusetts, Oregon, and Vermont. New York State has banned conventional loop carriers and is expected to allow degradable carriers; Delaware may enact similar legislation. If many more states enact such legislation, industry will probably find it advantageous to use degradable loop carriers in all states.

The outdoor disintegration rates of the most commonly used packaging materials for six-packs of canned beer were determined (191). The results are shown in Table 17. The samples were exposed outdoors in New Jersey during the months of Nov. 1975 through Mar. 1976 inclusive, and their ultimate elongation and tensile strength were then determined. The conventional polyethylene samples lost 20–40% of their original elongations but did not disintegrate. The ethylene–carbon monoxide copolymers retained only 10% elongation after one month exposure and disintegrated in four months. The polypropylene sheet had no elongation after five months but was intact and not brittle to the touch. The paper products lost much of their tensile strength, but were still intact. Thus, ethylene–carbon monoxide copolymer apparently disintegrates more rapidly than other widely used packaging materials when exposed outdoors, even during the colder months of the year. However, it is unlikely that properly engineered degradable plastics will continue to replace conventional polymers in other packaging applications where littering has been conspicuous. More probably, packaging materials of greater recycle value than plastic loop carriers, such as plastic beverage bottles, will become subject to the same deposit legislation as glass beverage bottles and cans.

Photodegradable plastic applications in the second defined category are those in which the plastic must perform outdoors for days, weeks, or months without failing and subsequently must disintegrate fairly rapidly. Examples of outdoor applications in the construction field include film to wrap lumber, mill work, sheetrock, Portland cement and other palletized building materials; film to protect partially completed buildings from the weather; film to cover concrete roads and bridges while curing; and netting used to stabilize soil on roadside embankments until grass cover can be established. In the field of agriculture, applications include temporary covers for storing grain outdoors, netting for protecting berry and fruit crops from birds, perforated sleeves for the protection of seedlings from browsing animals, and mulch films used in horticulture, particularly for high value crops such as tomatoes, cucumbers, peppers, pineapples, strawberries, etc. The most compelling reason for the use of degradable plastics in large volume outdoor applications is the savings in the cost of collecting and disposing of conventional film after use. Formerly, the used film could be burned on the site, but this practice has been banned in most states because of air pollution problems and fire hazards.

Estimates are that the U.S. consumption of polyethylene film for agricultural mulch purposes grew from ca 11,400 t in 1968 to 13,600 t in 1971 and to 45,000 t in 1974 (192–193). Figures after 1974 have not been published, but the growth rate is believed to have leveled off because of increases in raw material costs and in the cost of collection and disposal of the used film. The consumption of mulch film in Europe and Japan is greater because of more intensive farming methods practiced there. In 1982 the total U.S. and Japanese consumption of plastic mulch was ca 159,000 t (194). Several brands of degradable mulch film have been field tested in Japan, Israel, Canada, and the U.S., but in most cases, the details have not been published. The annual consumption of photodegradable mulch film is not known but is believed to be at the experimental level. The expected increase in the cost of removal and disposal of nondegradable mulch film makes degradable mulch film an increasingly attractive possibility.

Agricultural mulch films are manufactured in various types for a variety of applications. Black or white opaque films are used to conserve heat or to protect against excessive heat. Embossed films are made to keep the film from slipping out of the soil which holds it in place. Clear films are used to construct row tunnels held up by U-shaped wire supports; these serve as hothouses for starting plants when frost is still prevalent. White films repel aphids, and fumigant retention films retard the evaporation of volatile soil sterilants. Opaque films suppress the growth of weeds, and all mulch films retard the loss of moisture from the soil. In order for photodegradable films to compete successfully with conventional agricultural mulch film, they must be available in the same variety of formulations and must also meet certain other requirements:

Films must be stable in storage for extended periods, and must not be thermally degraded by the temperature encountered in warehouses and in the sun.

Films with different induction periods, during which they maintain adequate strength, must be available for various vegetable crops which grow at different rates.

After the induction period, films must disintegrate into small pieces in order to reduce litter and avoid interference with soil preparation for a new crop.

Disintegrated films must not release harmful substances in the soil or interfere with soil drainage.

Biodegradability is not a present requirement of photodegradable mulch film, but it would alleviate the problems of the expected increase in the concentration of film fragments from repeated use of the product over many years. Ethylene polymers or copolymers that have been thermally or photolytically degraded to sufficiently low molecular weights become linear waxes and are easily biodegraded. Polypropylene that has been degraded in like manner resists biodegradation because branches remain on the chain, and it is likely that photodegraded poly(butene-1) would also resist bioassimilation.

Poly(butene-1) plastics have received the most attention as photodegradable mulch films. Mobil Chemical Company produced this polymer and evaluated it in various degradable formulation (165) but then sold their interest. Princeton Chemical Research developed poly(butene-1) for degradable mulch-film applications and promoted it under the trade name Ecolan (see Table 11) (134). Shell Chemical Company has recently described a photodegradable polybutylene designed to degrade in sunlight in two months (195).

Clear Ecolan has been evaluated as a degradable agricultural mulch film for growing summer squash, muskmelon, eggplant, and tomatoes (196). These plants were selected because their growth habits provide different types of foliage cover on the mulches. Ecolan was compared to nondegradable black and clear polyethylene, and plastic-coated (polyethylene) black paper; nonmulched controls were also included in the study. The effects of the mulches on soil temperatures, soil moisture retention, movement of soil nutrients, and yield and maturity of the indicated vegetable crops were tested. In general, the mulches significantly increased total marketable yield, number of fruit per plant, and advanced maturity by about two weeks. Clear Ecolan 60-day degradable film was equivalent to the other films tested and did not have to be removed and disposed of after harvest with few exceptions.

Although poly(butene-1) has been used more than degradable polyethylene for degradable mulch film, the economic feasibility of producing degradable polyethylene for this purpose should not be overlooked. The rate of photodegradation of a mulch film consisting of polyethylene and radiation-modified atactic polypropylene as a photosensitizer was compared indoors (greenhouse) and outdoors (197). The films were used as plastic mulch for strawberry beds. Both the yield strength and percent of elongation at break of the films used indoors for 6 months were within 70–80% of their initial values, but those exposed outdoors were found to have lost 50% of initial yield strength and 70% of initial elongation at the end of the same period. The photooxidized film was exposed to bacteria found in soil around petroleum wells and found to undergo a weight loss which was attributed to biodegradation. In another study, polyethylene film sensitized with 0.05–0.1% biferrocene was degraded at a high rate outdoors and appeared suitable for use as an agricultural mulch film (198).

Conclusions

With a few notable exceptions, synthetic plastics are resistant to microbiological degradation, but naturally occurring polymers are susceptible to breakdown by fungi and bacteria. Biological degradation of polymers is facilitated by linearity of the polymer chains, by the presence of unhindered aliphatic ester and peptide linkages in the polymer chain and by low molecular weights (below 1000 $\overline{M}_n$). These conclusions are supported by laboratory tests in which polymer samples are exposed to mixtures of known microorganisms and by soil burial tests, where commensality among naturally occurring strains of microorganisms is possible. Applications for biodegradable plastics in agriculture include slow release urea–formaldehyde fertilizers and degradable containers and matrices for the sustained release of pesticides and herbicides. Absorbable sutures are prominent examples of surgical applications for biodegradable plastics. Absorbable plastics have also been developed for use in orthopedic surgery in the form of screws, pins, and plates, and have been tested as implantable matrices for the sustained release of hormones and various drugs in veterinary medicine.

Plastics manufacturers have developed a number of techniques for enhancing the photodegradability of synthetic plastics. The most widely investigated methods involve the use of additives, such as transition metal salts, prooxidants, and photosensitizers, and the incorporation of photolabile groups into the polymer molecules during polymerization. The first method can be carried out by plastic formulators while the second is restricted to primary producers of plastics. The latter method results in fabricated plastics with superior shelf life and more dependable degradation characteristics.

Many applications for photodegradable plastics have been explored and a few have been commercialized. The photodegradable loop carriers for packaging beer and soda cans are used in large volume in those states where mandated by law in an effort to mitigate the litter problem. Agricultural applications for photodegradable plastics such as degradable mulch film appear promising but are in the experimental stage.

If the recycling of throwaway plastic beverage containers becomes well established, it is doubtful if degradable plastics will be used in such applications. Deposit legislation has been enacted in some states as an incentive to recycling and as a litter control measure.

BIBLIOGRAPHY

1. *1970 Annual Book of ASTM Standards*, Pt. 27, ASTM D 1924-63, American Society for Testing and Materials, Philadelphia, Pa., p. 593.
2. *Ibid.*, ASTM D 2676T, p. 758.
3. R. D. Fields, F. Rodriquez, and R. K. Finn, *Am. Chem. Soc. Div. Polym. Chem.* **14,** 1244 (1973).
4. R. A. Clendinning, S. Cohen, and J. E. Potts, *Great Plains Agric. Counc. Pub.* **68,** 244 (1974).
5. M. Alexander, *Microbial Ecology*, John Wiley & Sons, Inc., New York, 1971, Chapt. 12.
6. L. Jen-Hao and A. Schwartz, *Kunststoffe* **51,** 317 (June 1961).
7. N. B. Nykvist, *Proceedings of the Conference on Degradable Polymer Plastics*, The Plastics Institute, London, Nov. 1973, Paper 18, pp. 1–13.
8. K. H. Walhausser in ref. 7, Paper 17, p. 5.
9. Ref. 5, Chapt. 16, p. 415.
10. J. E. Potts, R. A. Clendinning, and W. B. Ackart, *Contract No. CPE-70-124, An Investigation of the Biodegradability of Packaging Plastics EPA-R2-72-046*, Aug. 1972, p. 22.
11. C. J. Wessel, *SPE Trans.*, 199 (July 1964).
12. S. Berk, H. Ebert, and L. Teitell, *Ind. Eng. Chem.* **49,** 1115 (1957).
13. Ref. 10, pp. 61–65.
14. J. E. Potts, R. A. Clendinning, and W. B. Ackart, experimental results, Union Carbide Corporation, Bound Brook, N.J., 1975.
15. G. J. L. Griffin and H. Mivetchi, *Proc. 3rd Int. Biodegr. Symp.*, 807 (1976).
16. F. Knoop, *Beitr. Chem. Physiol. Pathol.* **6,** 150 (1905).
17. J. R. Sokatch, *Bacter. Physiol. Metab.*, Academic Press, London, 1969, p. 157.
18. J. W. Foster and A. van Leeuwenhoek, *J. Microbiol. Serol.* **28,** 241 (1962).
19. E. J. McKenna and R. E. Kallio, *Am. Rev. Microbiol.* **19,** 183 (1965).
20. A. C. vander Linden and G. J. E. Thysse, *Adv. Enzymol.* **27,** 469 (1965).
21. J. E. Steward and co-workers, *J. Bacteriol.* **78,** 441 (1959).
22. E. R. Leadbetter and J. W. Foster, *Arch. Biochem. Biophys.* **82,** 491 (1959).
23. H. B. Lukens and J. W. Foster, *J. Bacteriol.* **85,** 1074 (1963).
24. P. K. Barua and co-workers, *Appl. Microbiol.*, 657 (Nov. 1970).
25. E. Merdinger and R. P. Merdinger, *Appl. Microbiol.* **20**(4), 651 (1970).
26. T. C. Miller and M. J. Johnson, *Biotech* **8,** 567 (1966).
27. J. R. Haines and M. Alexander, *Appl. Microbiol.* **28,** 1084 (1974).
28. Ref. 10, p. 34.
29. J. E. Potts, *Prepr. Pap. Nat. Meet. Div. Water Air Waste Chem. Am. Chem. Soc.* **10**(2), 229 (1970).
30. J. E. Potts, *Ind. Water Eng.*, 32 (1970).
31. J. E. Potts in H. H. G. Jellinek, ed., *Aspects of Degradable and Stable Polymers*, Elsevier, 1978, Chapt. 14, p. 636.
32. A. C. Albertsson and B. Rånby, *Proc. 3rd Int. Biodegr. Symp.*, 743 (1976).
33. P. H. Jones, D. Prasad, M. Heskins, M. H. Morgan, and J. E. Guillet, *Environ. Sci. Technol.* **8,** 919 (1974).
34. J. Mills and H. O. W. Eggins, *Int. Biodeterior. Bull.* **6,** 13 (1970).
35. Ref. 31, p. 640.
36. Ref. 31, p. 641.
37. J. E. Potts, W. B. Ackart, R. A. Clendinning, and W. D. Niegisch in ref. 7, Paper 12, p. 6.
38. *Ibid.*, p. 7.
39. Ref. 31, p. 643.
40. J. E. Guillet, T. W. Regulski, and T. B. McAnenly in ref. 33, p. 923.

41. Ref. 10, p. 24
42. C. E. Zobell and J. D. Beckwith, *J. Am. Water Works Assoc.* **36,** 439 (1944).
43. L. R. Snoke, *Bell Syst. Technol. J.* **86,** 1095 (1957).
44. M. G. Crum, R. J. Reynolds, and H. G. Hedrick, *Appl. Microbiol.* **15,** 1352 (1967).
45. M. G. Crum, R. J. Reynolds, and H. G. Hedrick, *Dev. Ind. Microbiol.* **8,** 260 (1967).
46. W. M. Heap and S. H. Morrell, *J. Appl. Chem.* **18,** 189 (1968).
47. Ref. 10, p. 67.
48. K. W. H. Leeflang, *J. Am. Water Works Assoc.* **55,** 1523 (1963).
49. D. H. Taysum, *Soc. Chem. Ind. London Monogr.* **23,** 105 (1966).
50. J. A. Jakubowski, S. L. Simpson, and J. Gyuris, *J. Coat. Technol.* **54**(685), (1982).
51. K. Nakagawa and Y. Nakagawa, *Fukuyama Sen'i Kogyo Shikenjo Hokoku* **6,** 96 (1979).
52. F. Rodriguez, *Chem. Technol.*, 409 (July 1971).
53. N. N. Pavlov and Z. A. Akopdzhanyan, *Soc. Plast.* (5), 62 (1967).
54. N. N. Pavlov and Z. A. Akopdzhanyan, *Eng. Transl. Plast. Massy* (5), (May 1966).
55. T. Fukumura, *Plant Cell Physiol.* **7,** 93 (1966).
56. T. Fukumura, *J. Biochem.* **59,** 531 (1966).
57. W. J. Bailey, Y. Okamoto, W. C. Kuo, and T. Norita, *Proc. 3rd Int. Symp. Biodegr.*, 200 (Aug. 1975).
58. S. J. Huang and co-workers, *Proc. 3rd Int. Biodegr. Symp.*, 731 (1967).
59. S. J. Huang, M. Bitritto, K. W. Leong, J. Pavlisko, M. Roby, and J. R. Knox, *Adv. Chem. Ser.* **169,** (1978).
60. R. A. Connolly, *Mater. Res. Stand.* **3**(3), 93 (1963).
61. W. H. Stahl and H. Pessen, *Mod. Plast.* **31**(11), 111 (1954).
62. S. J. Huang and C. A. Byrne, *J. Appl. Pol. Sci.* **25**(9), 1951 (1980).
63. R. T. Darby and A. M. Kaplan, *Appl. Microbiol.*, 900 (June 1968).
64. H. O. W. Eggins, J. Mills, A. Holt, and G. Scott, *Soc. Appl. Bacteriol. Symp. Ser.* **1,** 267 (1971).
65. R. E. Klausmeier, *Monogr. Soc. Chem. Ind. London* **23,** 232 (1966).
66. Ref. 10, p. 47.
67. Y. Tokiwa and T. Suzuki, *J. Appl. Polym. Sci.* **26,** 441 (1981).
68. M. M. Bitritto, J. P. Bell, G. M. Brenckle, S. J. Huang, and J. R. Knox, *J. Appl. Polym. Sci. Symp.* **35,** 405 (1979).
69. Ref. 10, pp. 57–59.
70. B. W. Brodman, M. P. Devine, *J. Appl. Polym. Sci.* **26,** 997 (1981).
71. Ref. 10, pp. 46–60.
72. J. E. Potts, R. A. Clendinning, W. B. Ackart, and W. D. Niegisch, *Polym. Prepr. Am. Chem. Soc. Div. Polym. Chem.*, 13 (1972).
73. J. B. Titus, *Environmentally Degradable Plastics, A Review*, Plastec Note N-24, Plastics Technical Evaluation Center, Picatinny Arsenal, Dover, N.J., Feb. 1973.
74. R. D. Gilbert, V. Stannett, C. G. Pitt, and A. Schindler in N. Grassie, ed., *Developments in Polymer Degradation*, Vol. 4, Applied Science Publishers, London, 1982, Chapt. 8, pp. 259–293.
75. The Collected Papers of W. H. Carothers, Vol. 1 of *High Polymers*, Interscience Publishers, New York, 1940, pp. 81–140.
76. U.S. Pats. 3,021,309–3,021,317 (Feb. 13, 1962), E. F. Cox and F. Hostettler (to Union Carbide Corp.).
77. Ref. 10, pp. 50–51.
78. W. J. Cook, J. A. Cameron, J. P. Bell, and S. J. Huang, *J. Polym. Sci. Polym. Lett. Ed.* **19,** 159 (1981).
79. R. L. Davidson and M. Sittig, *Water Soluble Resins*, 2nd ed., Van Nostrand Reinhold Co., 1962, pp. 8–9.
80. U.S. Pat. 3,636,956 (Jan. 25, 1972), A. K. Schneider (to Ethicon).
81. C. G. Pitt, T. A. Marks, and A. Schindler, *Symposium of the 6th International Meet Controlled Release Soc.*, Academic Press, N.Y., 1980.
82. D. W. Lamberts, D. P. Langston, and W. Chu, *Ophthalmology* **85,** 794 (1978).
83. K. R. Sidman, A. D. Schwope, W. D. Steber, S. E. Rudolph, and S. B. Poulin, *J. Membrane Sci.* **1**(3), 277 (1980).
84. J. Heller, *Biomaterials* **1**(1), 51 (1980).
85. J. Heller and R. W. Baker in ref. 83, pp. 1–7.

86. U.S. Pat. 3,844,987 (Oct. 29, 1974), U.S. Pat. 3,850,862 (Nov. 26, 1974), U.S. Pat. 3,850,863 (Nov. 26, 1974), U.S. Pat. 3,852,913 (Dec. 10, 1974), U.S. Pat. 3,867,324 (Feb. 18, 1975), U.S. Pat. 3,919,163 (Nov. 11, 1975), U.S. Pat. 3,923,729 (Dec. 2, 1975), U.S. Pat. 3,921,333 (Nov. 25, 1975), U.S. Pat. 3,929,937 (Dec. 30, 1975), U.S. Pat. 3,931,068 (Jan. 6, 1976), and U.S. Pat. 3,932,319 (Jan. 13, 1976), R. A. Clendinning, J. E. Potts, and W. D. Niegisch (to Union Carbide Corporation).
87. U.S. Pat. 3,935,141 (Jan. 27, 1976), J. E. Potts, S. W. Cornell, and A. M. Sracic (to Union Carbide Corporation).
88. U.S. Pat. 4,067,836 (Jan. 10, 1976), J. E. Potts, S. W. Cornell, and A. M. Sracic (to Union Carbide Corporation).
89. U.S. Pat. 3,867,324 (Feb. 18, 1975), R. A. Clendinning, J. E. Potts, and S. W. Cornell (to Union Carbide Corporation).
90. *Recycling Plastics*, Society of the Plastics Industry, New York, 1972.
91. W. H. Clifford, *Technological and Sociological Significance of Degradable Plastics*, Michigan State University, E. Lansing, Mich., 1974.
92. *Legislative Bulletin, No. 2*, March 8, 1978, Society of the Plastics Industry, New York.
93. *U.S. Municipal Solid Waste Management*, The Plastic Bottle Institute, Society of the Plastics Industry, New York, Nov. 1978, pp. 13–14.
94. Ref. 10, pp. 14–17.
95. *Plastics and Biodegradability*, Public Affairs Council, Society of the Plastics Industry, New York, 1976.
96. B. Rånby and J. F. Rabek, *Photodegradation, Photooxidation and Photostabilization of Polymers*, John Wiley & Sons, Inc., New York, 1978.
97. L. Reich and S. S. Stivala, *Elements of Polymer Degradation*, McGraw-Hill, Inc., New York, 1971.
98. D. L. Allara, *Adv. Chem. Ser.* **169,** (1977).
99. W. O. Lundberg, *Autooxidation and Antioxides*, Interscience Publishers, a division of John Wiley & Sons, Inc., New York, 1961, Vol. 1, 1962, Vol. 2.
100. W. L. Hawkins, *Polymer Stabilization*, Interscience Publishers, a division of John Wiley & Sons, Inc., New York, 1972.
101. Ref. 96, pp. 493–500.
102. N. J. Harrick, *International Reflection Spectroscopy*, Interscience Publishers, a division of John Wiley & Sons, Inc., New York, 1967.
103. Ref. 98, Chapt. 1, pp. 3–4; Chapt. 4, pp. 30–35.
104. G. Geuskens, ed., *Degradation and Stabilization of Polymers*, John Wiley & Sons, Inc., New York, 1975, Chapts. 5–6.
105. J. Guillet, ed., *Polymers and Ecological Problems*, Vol. 3 of *Polymer Science and Technology Series*, Plenum Press, New York, 1973, Chapts. 1–4.
106. F. J. Golemba and J. E. Guillet, *SPE J.* **26,** 88 (1970).
107. G. H. Hartley and J. E. Guillet, *Macromolecules* **1,** 165 (1968).
108. M. Heskins and J. E. Guillet, *Macromolecules* **1,** 97 (1968).
109. Y. Americk and J. E. Guillet, *Macromolecules* **4,** 375 (1971).
110. Ref. 96, p. 144.
111. Ref. 97, p. 32–35.
112. M. Kato and Y. Yoneshige, *Macromol. Chem.* **164,** 159 (1973).
113. Ref. 105, p. 14.
114. A. W. Birley and D. S. Brackman in ref. 7, Paper 1.
115. V. T. Kagiya in ref. 7, Paper 8.
116. Y. Takeuchi, Y. Harita, and A. Sekimoto in ref. 7, Paper 9.
117. S. W. Benson, *J. Chem. Educ.* **42,** 502 (1965).
118. H. C. Beachell and G. W. Tarbet, *J. Polym. Sci.* **45,** 451 (1960).
119. A. Charlesby and R. H. Partridge, *Proc. Roy. Soc. A* **283,** 312 (1965).
120. Ref. 96, pp. 100–101.
121. *Adv. Chem. Ser.* **9,** 4 (1977).
122. Ref. 105, Chapt. 2.
123. C. E. H. Brown and S. A. Chaudhri, *Polymer* **9,** 81 (1968).
124. J. D. Holdsworth, G. Scott, and D. Williams, *J. Chem. Soc.*, 4692 (1964).
125 J. F. Rabek, *Chem. Stosow.* **11,** 53 (1967).
126. E. D. Owen and R. J. Bailey, *J. Polym. Sci.* **A-1,** 10, 113 (1972).
127. Ref. 96, Chapt. 6.

128. U.S. Pat. 3,935,141 (Jan. 27, 1976), UK Pat. 1,412,396 (Nov. 5, 1975), Fr. Pat. 2,154,596 (May 11, 1973), U.S. Pat. 3,867,324 (Feb. 18, 1975), and U.S. Pat. 3,901,838 (Aug. 26, 1975), J. E. Potts, S. W. Cornell, A. M. Sraic, and R. A. Clendinning (to Union Carbide Corporation).
129. U.S. Pat. 3,454,510 (July 8, 1969), U.S. Pat. 3,592,792 (July 13, 1971), and U.S. Pat. 3,676,401 (July 11, 1972), G. C. Newland, G. R. Greear, J. W. Tamblyn, and J. W. Henry (to Eastman Kodak).
130. U.S. Pat 3,707,056 (Dec. 26, 1972), E. L. Cole, H. V. Hess, and R. C. Pomatti (to Texas Oil Co.).
131. U.S. Pat. 3,825,626 (July 23, 1974), M. C. McGaugh (to Dow Chemical Co., Inc.).
132. U.S. Pat. 3,847,852 (Nov. 12, 1974), R. A. White and co-workers (to DeBell and Richardson).
133. U.S. Pat. 3,830,764 (Aug. 20, 1974) and U.S. Pat. 3,981,856 (Sept. 21, 1976), D. E. Hudgins and co-workers (to Princeton Polymer Laboratories, Inc.).
134. U.S. Pat. 3,590,528 (July 6, 1971), U.S. Pat. 3,886,683 (June 3, 1975), U.S. Pat. 3,896,585 (July 29, 1975), and U.S. Pat. 3,984,940 (Oct. 12, 1976), T. Shepherd, D. Hudgins, M. Reich, and R. Miller (to Princeton Chemical Research Inc.).
135. U.S. Pat. 3,888,804 (June 10, 1975) and UK Pat. 1,385,497 (Feb. 26, 1975), C. E. Swanholm (to Biodegradable Plastics, Inc.).
136. UK Pat. 1,356,107 (June 12, 1974), G. Scott (to Gerald Scott).
137. UK Pat. 1,382,062 (Jan. 29, 1975), D. S. Brackman (to Imperial Chemical Industries).
138. U.S. Pat. 3,882,058 (May 6, 1975) and Fr. Pat. 2,236,896 (Feb. 7, 1975), G. Lebrasseur (to Ethylene Plastique).
139. U.S. Pat. 3,865,767 (Feb. 11, 1975), UK Pat. 1,396,451 (June 4, 1975), and Belg. Pat. 805,088 (Jan. 16, 1974), A. Boberg (to AB Akerlund & Rausing).
140. Jpn. Pats. 7,138,686, 7,138,687, 7,138,688, 7,239,568, and 7,239,569 (Oct. 6, 1972), A. Ohnishi (to Mitsubishi Petrochemicals).
141. Jpn. Kokai 50 34045 (Apr. 2, 1975), (to Sumitomo Chem. Co. Ltd.).
142. Jpn. Kokai 47 42730 (Oct. 28, 1972), H. Aob (to Chisson Corp.).
143. Jpn. Kokai 50 21079 (Mar. 6, 1975), (to Sekisui Chemical Co.).
144. Jpn. Kokai 48 76938 (Oct. 16, 1973), Jpn. Kokai 48 103643 (Dec. 26, 1973), Jpn. Kokai 49 10945 (Jan. 30, 1974), Jpn. Kokai 48 76938 (Oct. 16, 1973), Jpn. Pat. 7,211,654, and Jpn. Pat. 7,236,011, T. Takeno and M. Daikichi (to Mitsui Petrochem Ind. Ltd.).
145. P. J. Canterino and D. J. Dibiasi, Plast, *Programmed Life of Plastics*, U. of Massachusetts, Amherst, Mass., Aug. 22, 1972.
146. H.C. Beachell and L. H. Smiley, *J. Polym. Sci.* **A1,** 5, 1635 (1967).
147. Ref. 96, pp. 165–184.
148. R. B. Fox and T. R. Price in ref. 111, Chapt. 8, pp. 98–108.
149. N. Grassie and N. A. Weir, *J. Appl. Polym. Sci.* **9,** 963 (1965).
150. *Ibid.*, 975 (1965).
151. *Ibid.*, 987 (1965).
152. G. Geuskens and C. David in ref. 104, Chapt. 6.
153. G. Geuskens and C. David in ref. 104, pp. 126–134.
154. J. E. Potts and S. W. Cornell, unpublished data.
155. Ref. 96, pp. 128–141.
156. J. Brandrup and L. H. Peebles, *Macromolecules* **1,** 64 (1968).
157. Ref. 96, p. 192.
158. M. K. Lindeman in N. M. Bikales, ed., *Encyclopedia of Polymer Science and Technology*, Interscience Publishers, a division of John Wiley & Sons, Inc., New York, 1971, Vol. 14, Fig. 22.
159. G. Geuskens, M. Borsu, and C. David, *Eur. Polym. J.* **8,** 883 (1972).
160. F. J. Golemba and J. E. Guillet, *J. Paint Technol.* **41,** 315 (1969).
161. Japan Tokkyo Koho 76-35218, 75-33908, 76-35217, 76-27478 (to Japan Agency of Industrial Science & Technology).
162. Ger. Pat. 863,711 (1941), F. Bullauf, O. Bayer, and L. Teichmann (to Bayer A.G.).
163. U.S. Pat. 2,495,286 (Jan. 24, 1950), M. M. Brubaker (to E. I. du Pont de Nemours & Co., Inc.).
164. M. M. Brubaker, D. D. Coffman, and H. H. Haihn, *J. Am. Chem. Soc.* **74,** 1509 (1952).
165. Y. Chatani, T. Takizawa, and S. Murahashi, *J. Polym. Sci.* **62,** 527 (1962); *J. Polym. Sci.* **55,** 811 (1961).
166. G. Peiper in N. M. Bikales, ed., *Encyclopedia of Polymer Science and Technology*, Vol. 9, Interscience Publishers, a division of John Wiley & Sons, Inc., New York, 1968, pp. 397–402.
167. Private communication, J. R. Leech, Union Carbide Corporation, Bound Brook, N.J., to J. E. Potts, 1978.

168. Ref. 105, Chapt. 1, p. 20.
169. Ref. 105, Chapt. 1, pp. 1–26.
170. Ecolyte, sales literature, Van Leer-Eco-Plastics, Ltd. (for further information contact J. E. Guillet, University of Toronto).
171. E. Nenkov, T. Georgieva, A. Stoyanov, and V. Kabaivanov, *Angew. Makromol. Chem.* **91**, 69 (1980).
172. W. Sedriks, *Environmentally Degradable Polymers*, private report to clients, Stanford Research Institute, Menlo Park, Calif.
173. Brit. Pat. 1,362,173 (July 31, 1974), J. E. Guillet (to University of Toronto).
174. Brit. Pat. 1,372,830 (Nov. 6, 1974), J. E. Guillet (to University of Toronto).
175. U.S. Pat. 3,878,169 (Apr. 15, 1974), J. E. Guillet (to University of Toronto).
176. U.S. Pat 3,835,096 (Sept. 10, 1974), Cherdron and co-workers (to Hoechst).
177. U.S. Pat. 3,590,527 (July 6, 1971), J. P. Pijst (to Hollandsche Draad-en Kabelfabriek).
178. Jpn. Tokkyo Koho 82 26531 (1982), (to Daicel).
179. Jpn. Kokai 50 18596 (Feb. 27, 1975), J. Niizeki and co-workers (to Nippon Soda).
180. Jpn. Kokai 5 20510 (Sept. 23, 1975), M. Ohme (to Osaka Soda).
181. Jpn. Kokai 49 117600 (Nov. 9, 1974), T. Fujisawa and co-workers (to Sagami Chemical Research Center).
182. D. M. Wiles in ref. 104, Chapt. 7, pp. 142–146.
183. J. Dhanraj and J. E. Guillet, *J. Polym. Sci. C* (23), 433 (1968).
184. J. Voigt, *Makromol. Chem.* **27,** 80 (1958).
185. Ref. 96, pp. 237–239.
186. S. M. Cohen, R. H. Young, and A. H. Markhart, *J. Polym. Sci.* **A1**(9), 3263 (1971).
187. Ref. 96, pp. 245–251.
188. B. D. Gesner and P. G. Kelleher, *J. Appl. Polym. Sci.* **13,** 2183 (1969).
189. *Ibid.*, **12,** 1199 (1968).
190. G. Colin, J. D. Cooney, D. J. Carlsson, and D. M. Wiles, *J. Appl. Polym. Sci.* **26**(2), 509 (1981).
191. G. M. Harlan, personal communications, Union Carbide Corp., Bound Brook, N.J.
192. *Mod. Plast.*, 10 (Jan. 1969).
193. *Mod. Plast.*, 56 (May 1972); 37 (Nov. 1975).
194. D. Carnell, *American Vegetable Grower*, 54 (May 1983).
195. *Plast. World*, 62 (May 1982).
196. C. A. Burga-Mendoza and B. L. Pollack, *Proceedings of the 11th National Agricultural Plastics Conference, Nov. 7–9, 1973*, Rutgers University, New Brunswick, N.J.
197. H. Omichi and M. Hagiwara, *Polym. Photochem.* **1**(1), 15 (1981).
198. N. J. Kondrashkina, T. N. Zelenkova, M. Z. Borodulina, and B. I. Sazhin, *Plast. Massy* (6), 11 (1982).

JAMES E. POTTS
Union Carbide Corporation

PROCESS ENERGY CONSERVATION

Between 1973 and 1983, the ratio of energy price to capital price increased by a factor of 5–10. As a result, the old rules for optimum reflux, pressure drops, and temperature differentials no longer apply. Yet, the basic concepts and processes of the past are still sound. They must, however, be applied in a new economic framework.

As Figure 1 shows, failure to change in the direction the energy price:capital price ratio suggests, can cause a large increase in costs (1). The figures also suggest some very large changes. The pressure drop in piping should decrease by a factor of 3.4 and heat loss to the environment by a factor of 2. But the rise in energy prices has also increased the value of good engineering which makes it possible to save capital as well as energy. In the following cases, for example, appropriate design may save both energy and capital cost: shift to a continuous process; selection of a higher yield catalyst; choice of a tighter design and/or a less flexible design; development of a lower energy product form; bettering process heat interchange; and improvement of mechanical efficiency in moving equipment.

Many plants also offer opportunities for significant energy conservation by various cogeneration processes (see Energy management; Power generation).

Energy Balance

An energy balance has historically been prepared for components of a process, primarily to assure that heat exchangers and utility supply are adequate. Often, an overall process energy balance was not developed, but today, with the increased emphasis on energy cost, the energy balance for the overall process has become a document almost as important as the material balance.

Its purpose is to serve as an "evergreen" framework during design to highlight the areas with greatest potential for improvement and to serve as a tool for plant-operating personnel after startup, to aid optimization of energy use.

The energy balance should analyze the energy flows by type and amount (ie, electricity, fuel gas, steam level, heat rejected to cooling water, etc). It should include realistic loss values for turbine inefficiencies and heat losses through insulation.

Exergy, Lost Work, and Second-Law Analysis. When energy is critically important to process economics, the simple energy balance is sometimes carried into an analysis of lost work. This compares the actual design against the theoretical ideal at each step and defines where the true energy use (or lost work) is occurring (see Thermodynamics).

In the following discussions of reaction, separation, heat exchange, compression, refrigeration, and steam systems, the importance of this concept is illustrated. A few terms are defined below.

Exergy (E) is the potential to do work. Thermodynamically, this is the maximum work a stream can deliver by coming into equilibrium with its surroundings.

$$E = (H - H_0) - T_0(S - S_0)$$

where E = exergy, the maximum theoretical work potential; H, S = enthalpy and entropy of the stream at its original conditions; H_0, S_0 = enthalpy and entropy of the

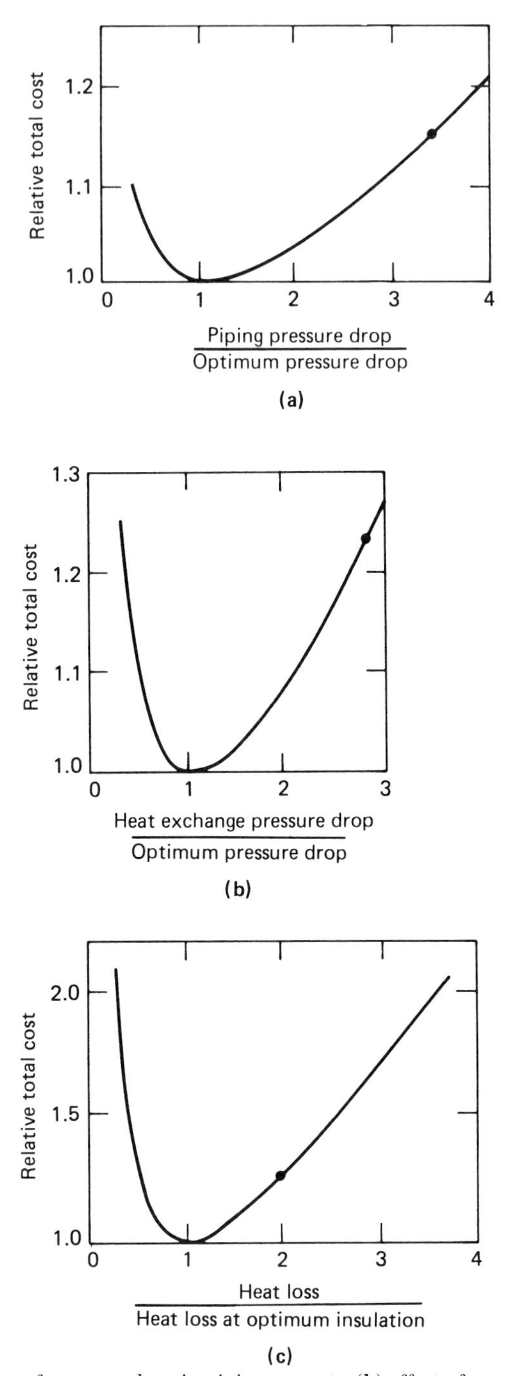

Figure 1. (**a**) Effect of pressure drop in piping on costs; (**b**) effect of pressure drop in exchangers on costs; (**c**) effect of heat loss through insulation on costs; (**d**) effect of reflux use on costs; and (**e**) effect of energy recovery through waste-heat boiler use on cost. Total cost is the sum of capital plus energy costs (for the lifetime of the plant, discounted to present value). Point marked on each graph is the design point if energy price is low by a factor of 4.

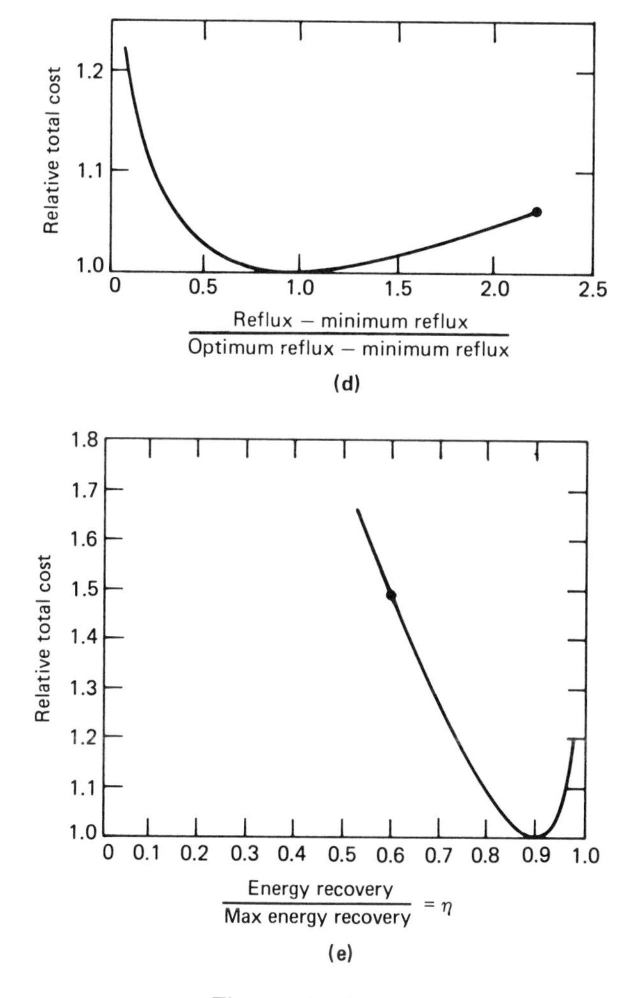

Figure 1. *(continued)*

same stream at equilibrium with the surroundings; and T_0 = temperature of the surroundings (sink).

Exergy is also sometimes called availability or work potential.

Free energy (G) is a related thermodynamic property. It is most commonly used to define the condition for equilibrium in a processing step.

$$\Delta G = \Delta H - T\Delta S$$

It is identical to ΔE if the processing step occurs at T_0.

Lost work (LW) is the irreversible loss in exergy that occurs because a process operates with driving forces or mixes material at different temperature or composition.

$$LW = E_{\text{in}} - E_{\text{out}}$$

Second-Law Analysis looks at the individual components of an overall process to define the causes of lost work. Sometimes it focuses on the efficiency of a step and ratios the theoretical work needed to accomplish a change (eg, a separation) to that actually used. This has been done to entire processes such as NH_3, HNO_3, ethylene, and methanol manufacture (2–4).

The main limit to this type of analysis is that it addresses only to the best possible energy efficiency and ignores the economic optimum in the balance between capital and energy. Sometimes it is more cost effective simply to compare design against a second-law-violation checklist covering items such as (4): mixing streams at different temperatures and compositions; high pressure drops in control valves; reactions run far from equilibrium; high temperature differentials; and pump-discharge recirculation.

Reactor Design for Energy Conservation

How close a designer comes to minimum energy is heavily determined by the raw materials and catalyst system chosen. However, if the designer only has the freedom to choose reaction temperature, residence time, and diluent, he or she still has a tremendous opportunity to influence energy use via their effect on yield. But even given none of these, there is still wide freedom to optimize the heat interchange system.

Maximizing Yield. Often the greatest single contribution to reduced energy cost is increased yield. High yield reduces the amount of material to be pumped, heated, and cooled while also simplifying downstream separation. This says nothing about the indirect energy reduction achieved through reduced raw material use. On average, the chemical industry uses almost as much energy in its raw materials as it does in direct purchases of electricity and fuel.

Minimizing Diluent. The case concerning diluent is less clear. A careful balance needs to be made of the benefits it gives in higher yield against the costs in mass handling and separation.

Optimizing Temperature. Temperature is usually dictated by yield considerations and often this opposes the simple dicta of energy: in an endothermic reaction, put the heat in at the lowest practical temperature; in an exothermic reaction, operate (and remove the heat) at the highest possible temperature.

In an exothermic reaction, there is an inevitable loss of work potential which is proportional to the free energy change (ΔG_{T_r}). With all reactants preheated to reaction temperature (T_r) and all heat recovered at T_r (5),

$$\text{lost work} = \Delta G^0_{T_r} \left(\frac{T_0}{T_r}\right)$$

For highly exothermic reactions, eg, oxidations, this is typically the dominant loss in the entire process.

Heat Recovery and Feed Preheat. The objective, to bring the reactants to and from reaction temperature with the least utility cost and to recover maximum waste heat at maximum temperature is generally achieved by the criteria given below under Heat Exchange. Sometimes, control and safety conditions prevent these criteria from being completely followed. The impact of feed preheating merits a particularly careful look. In an exothermic reaction, it permits the reactor to act as a heat pump, ie, "to buy low and sell high." The most common example is combustion-air preheating for a furnace.

Batch vs Continuous Reactors. Usually, continuous reactors yield much lower energy use because of increased opportunities for heat interchange. Sometimes the savings are even greater in downstream separation units than in the reaction step itself.

Especially on batch reactors, the designer should critically review any use of refrigeration to remove heat. Batch processes often have evolved little from the laboratory-scale glassware where refrigeration was a convenience.

Separation

About one third of the chemical industry's energy is used for separation. It is not surprising, therefore, that a correlation exists between selling price and feed concentration (Fig. 2) (6).

This in turn can be traced to the minimum work of separation.

$$W = RT_0 \, \Sigma N_i \ln (x_i \gamma_i)$$

where T_0 is the sink temperature; N_i is the number of moles of a species present in the feed; $x_i = N_i/\Sigma N_i$; and γ_i is an activity coefficient.

This looks complicated, but actually, it provides a target that is easily calculated and approachable in practice. For example, work calculated from this expression closely approaches the performance of a real-world distillation after inefficiencies for driving forces are taken into account.

For ideal solutions ($\gamma_i = 1$) of a binary mixture, this simplifies to:

$$W = RT_0[x_1 \ln x_1 + (1 - x_1) \ln (1 - x_1)]$$

This applies whether the separation is by distillation or any other technique.

When a separation is not completed, less work is required. For x_1 equal to 0.5:

Product purity, %	*Relative work*
100	1
99.9	0.99
99.0	0.92
90.0	0.53

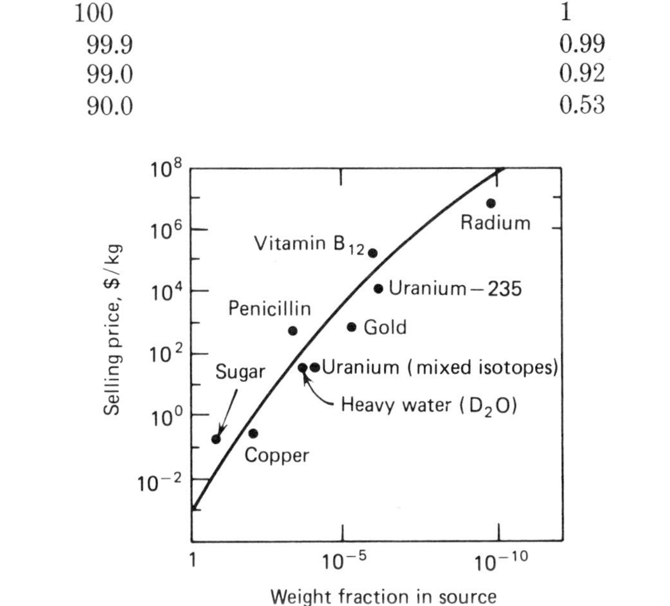

Figure 2. Commercial selling prices of some separated materials (6). Courtesy of McGraw-Hill Book Company.

Note that it takes only a little work to move from 99% separation to 100%, but a great deal to move from 90% to 100%. This is important to recognize when comparing separation techniques. Some leave much of the work undone, as for example, in crystallization involving an unseparated eutectic mixture.

Distillation. Distillation is overwhelmingly the most common separation technique because of its inherent advantages: phase separation is clean; it is relatively easy to build a multistage, countercurrent device; and equilibrium is closely approached in each stage.

Minimum work for an ideal separation at first glance appears unrelated to the slender vertical vessel with a condenser at the top and a reboiler at the bottom. The connection becomes evident when one calculates the work embedded in the heat flow which enters the reboiler and leaves at the condenser. An ideal engine could extract work from this heat.

$$\text{work potential} = QT_0\left[\frac{1}{T_{\text{condenser}}} - \frac{1}{T_{\text{reboiler}}}\right]$$

Comparison of actual use of work potential against the minimum allows calculation of an efficiency relative to the best possible separation:

$$\eta = \frac{RT_0[x_1 \ln x_1 + (1 - x_1)\ln(1 - x_1)]}{QT_0\left[\frac{1}{T_{\text{condenser}}} - \frac{1}{T_{\text{reboiler}}}\right]}$$

There is still no obvious reason to believe that the efficiency of separating a mixture with an α (relative volatility) of 1.1 will be related to that for an α of 2; however, it is known that when α is small, the required reflux and Q are large, but ($T_{\text{condenser}} - T_{\text{reboiler}}$) is small (see Distillation).

The two effects almost cancel to yield an approximation for the work potential used in a distillation (6–7).

$$\text{work potential} = RT_0(1 + [\alpha - 1]x_1)$$

When this is combined with the definition of minimum separation work, an approximation for distillation efficiency can be obtained:

$$\eta = \frac{x_1 \ln x_1 + (1 - x_1)\ln(1 - x_1)}{1 + (\alpha - 1)x_1}$$

This efficiency is high and shows only minor dependence on α over a broad range of α. For $x_1 = 0.5$,

$\alpha = 1.1$	$\alpha = 1.5$	$\alpha = 2$
$\eta = 0.66$	$\eta = 0.55$	$\eta = 0.46$

The dependence on x_1 is greater:

	$\alpha = 1.05$	$\alpha = 2$
η for $x_1 = 0.1$	0.32	0.30
η for $x_1 = 0.01$	0.056	0.053

This suggests that distillation should be the preferred method for feed concentrations of 10–90% and that it is probably a poor choice for feed concentrations of less than 1%.

This matches experience. Techniques such as adsorption, chemical reaction, and ion exchange are chiefly used to remove impurity concentrations <1%.

The high η values above conflict with the common belief that distillation is always inherently inefficient. This belief arises mainly because past distillation practices utilized such high driving forces (pressure drop, reflux ratio, and temperature differentials in reboilers and condensers). A real example is instructive:

	C_2 *splitter* (*relative numbers*)
Theoretical work of separation	1.0
Net work potential used	1.4

$$\eta = \frac{\text{Theoretical work}}{\text{Net work potential used}} = \frac{1.0}{1.4} = 0.7$$

Losses for driving forces:	
Reflux above the minimum	0.1
Exchanger ΔT	2.1
ΔP in tower	0.5
ΔP in condenser and tower	0.8
Total losses	*3.5*

$$\eta_{\text{Including losses}} = \frac{1.0}{1.4 + 3.5} = 0.2$$

These numbers show first that the theoretical work can be closely approached by actual work after known efficiencies are identified, and second, that the dominant driving force losses are in pressure drop and temperature difference. This is a characteristic of towers with low relative volatilities.

What Does Optimum Design Look Like Today? ***Condenser and Reboiler Δ T's.*** As shown by this example, the losses for ΔT typically are far greater than those for reflux beyond the minimum. As shown below under Heat Exchange, the economic optimum for temperature differential is typically under 15°C. This contrasts with the values over 75°C often used in the past. This is probably the biggest opportunity for improvement in the practice of distillation.

Adjusting the Process to Optimize Δ T. First glance may show only three or four utility levels (temperatures) to choose from. These might well be 100°C apart. Some ways to increase the options are to consider multieffect distillation, which spreads the ΔT across two or three towers; to use waste heat for reboil; and to recover energy from the condenser. Often to make some of these possible, the pressure in a column may have to be either raised or lowered.

Reflux Ratio. A number of studies have shown that at today's energy prices, the optimum reflux ratio is generally below 1.15 minimum and often below 1.05 minimum. At this point, excess reflux is a minor contributor to column inefficiency. When designing to this tolerance, correct vapor–liquid equilibrium (VLE) and adequate controls are essential.

State-of-the-Art Control. Modern microprocessor or computer control with feed-forward capability can save 5–20% of a unit's utilities (8). It does this by reducing the margin of safety. This sounds hard to believe, particularly for systems designed to operate within 10% of minimum reflux; however, unless the discipline of a controller forces this to happen, operators typically opt for increased safety. They are probably right to do so unless the proper set of analyzers and controllers is provided and maintained.

Right Feed Enthalpy. Often it is possible to heat the feed with a utility considerably less costly than that used for bottom reboiling. Sometimes the preheating can be directly integrated into the column-heat balance by exchange against the condensing overhead or against the net bottoms from the column. Simulation and a careful look at the overall process are required to assess the value of feed preheating accurately.

A vapor feed is favored when the stream leaves the upstream unit as a vapor or when most of the column feed leaves the tower as overhead product. The use of a vapor feed was a key component in the high efficiency cited above for the C_2 splitter where most of the feed goes overhead.

Low Column-Pressure Drop. The penalty for column-pressure drop is an increase in temperature differential

$$\Delta T = \left(\frac{dT}{dP}\right)\Delta P$$

$$\frac{dT}{dP} = \frac{R}{\Delta H}\frac{T^2}{P}$$

As this suggests, the penalty becomes very large for low vapor pressure materials, ie, for components that are distilled at or below atmospheric pressure. The work penalty associated with this ΔT is approximately defined by the ratio:

$$\frac{\Delta T \text{ for pressure drop}}{T_{\text{reboiler}} - T_{\text{condenser}}} = \text{fraction of work potential for } \Delta P$$

This penalty is severest for close-boiling mixtures. The most powerful technique for cutting ΔP is the use of packing. Conventional packings such as 5-cm (2-in.) pall rings can achieve a factor of four reduction over trays.

In applying this analysis, one must consider the overhead line and condenser pressure drop as well. (Note the high loss in the C_2 splitter example.)

Intermediate Condenser. As shown by Figure 3, an intermediate condenser forces the operating line closer to the equilibrium line, thus reducing the inherent inefficiencies in the tower. With the use of intermediate condensers and reboilers, it is

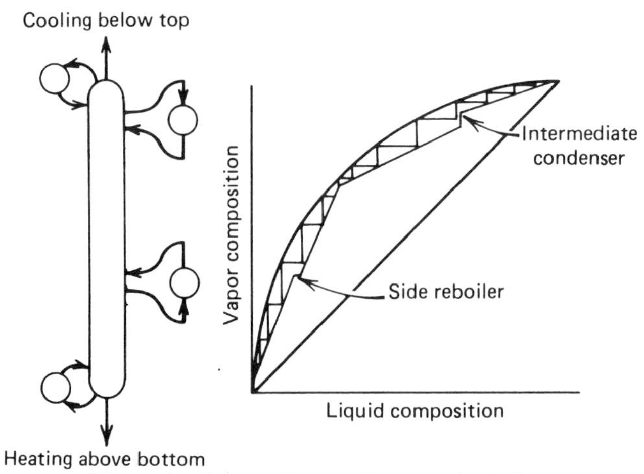

Figure 3. Intermediate condenser and reboiler.

possible to exceed the efficiency given above, particularly when the feed composition is far from 50:50 in a binary mixture.

	Max efficiency 50% of heavy component in feed	*Max efficiency 95% of heavy component in feed*
1 condenser, 1 reboiler	67	20
2 condensers, 1 reboiler	73	47
3 condensers, 1 reboiler	77	62

The intermediate condenser is most effective when the feed leaves the column, mainly as bottoms. This approach is economical only if a less costly coolant can be substituted for refrigeration, or if it permits reuse of the heat of condensation.

Intermediate Reboiler. Inclusion of an intermediate reboiler moves the heat-input location up the column to a slightly colder point. It can permit use of waste heat for reboil when the bottoms temperature is too hot for the waste heat.

Heat Pumps. Because of added capital and complexity, heat pumps have been used rarely, but with the shift to cheaper electricity relative to gas and oil, they may become more common.

Lower Pressure. Usually, relative volatility increases as pressure drops. For some systems, a 1% drop in absolute pressure cuts required reflux by 0.5%. Again, if operating at reduced pressure looks promising, it can be evaluated by simulation (see Simulation and process design).

Steam Stripping. Steam (or other stripping gas) and vacuum are largely interchangeable. Steam stripping allows more tolerance for pressure drop, but this comes at the penalty of much higher energy use.

In a complete study of distillation processes, some other questions also need to be asked:

Is the separation necessary?
Is the purity necessary?
Are there any recycles that could be eliminated?
If the distillation is batch, why?
Can the products be sent directly to downstream units, thereby eliminating intermediate heating and cooling?

Other Separation Techniques. Under some circumstances, distillation is not the best method of separation. Among these instances are the following: when relative volatility is <1.05, as often happens with isomers or compounds of different chemical type but close boiling points; when ca 1% of a stream is removed, as in gas drying (adsorption or absorption) or C_2H_2 removal (reaction or absorption); when thermodynamic efficiency of distillation is <5% (for whatever reason); and when a high boiling point pushes thermal stability limits. A variety of other techniques may be more applicable in these cases.

Reaction. Purification by reaction is relatively common when concentrations are low (ppm) and a high energy but low value molecule is present. Some examples are hydrogenation of acetylene and oxidation of waste hydrocarbons:

$$C_2H_2 + H_2 \rightarrow C_2H_4$$

$$\text{waste hydrocarbon} + O_2 \rightarrow H_2O + CO_2$$

Absorption (Extractive Distillation). As a separation technique, absorption starts with an energy deficit because it mixes in a pure material (solvent) and then separates it again. It is nevertheless quite common, primarily because it shares most of the advantages of distillation. Extractive distillation separates by molecular type and hence can be tailored to obtain a high α. The following ratios are suggested for equal costs (9):

$\alpha_{distillation}$	$\alpha_{extraction\ distillation}$	$\alpha_{extraction}$
1.2	1.4	2.5
1.4	1.9	5
1.6	2.3	8

Feed concentration also should be a key variable. Most of the applications have come where a small part (<5%) of the feed is removed. Examples include H_2S/CO_2 removal and gas drying with a glycol (see Azetropic and extractive distillation).

Extraction. The advantage of extraction (qv) is that it purifies a liquid rather than a vapor, allowing operation at lower temperatures and removal of a series of similar molecules at the same time, even though they differ widely in boiling point. An example is the simultaneous extraction of aromatics from hydrocarbon streams.

The disadvantage of extraction relative to extractive distillation is the greater difficulty of getting high efficiency countercurrent processing.

Adsorption. Adsorbents can achieve even more finely tuned selectivity than extraction. The most common applications is the fixed bed with thermal regeneration, which is simple, attains essentially 100% removal, and carries little penalty for low feed concentration. An example is gas drying. A variant is pressure-swing adsorption. Here, regeneration is attained by a drop in pressure. By use of multiple stages, high impurity rejection can be achieved, but at the expense of also losing part of the desired product.

Another approach is the simulated moving-bed system described in Adsorptive separation, liquids (Vol. 1, p. 570). It has large-volume applications in normal-paraffin separation and *para*-xylene separation. Since its introduction in 1970, it has largely displaced crystallization in xylene separations. The unique feature of the system is that the bed is fixed but the feed point shifts to simulate a moving bed.

Melt Crystallization. Crystallization (qv) from a melt is inherently attractive in competition to distillation because the heat of fusion is so much lower than that of evaporation. It also benefits from lower operating temperature. Typically, organic crystals are virtually insoluble in each other so that a pure product is possible in a one-stage operation.

However, crystallization has a unique set of disadvantages that often outweigh its virtues and have sharply limited its application. Industry practice suggests the use of a workable alternative, if one exists. The disadvantages of melt crystallization are as follows:

Physical separation is difficult. Impure liquid is trapped as occlusions as the crystals grow, and liquid wets all surfaces.

It requires a second separating process to separate the eutectic mixture. In some senses, it is akin to formation of two liquid phases: little energy is required to get the two phases, but a great deal is required to finish the purification.

It is difficult to add or remove heat because of the thermal resistance of the crystal.

It is difficult to move the liquid countercurrent to the crystals.

Thermodynamic efficiency is hurt by the large ΔT between the temperatures of melting and freezing. In an analogy to distillation, the high α comes at the expense of a big spread in reboiler and condenser temperature. From a theoretical standpoint, this penalty is smallest when freezing a high concentration (ca 90%) material.

One process is shown in Figure 4. It is a semibatch operation in which liquid falls down the walls of long tubes. This permits both staged operation and sweating of crystals. Typically, the sweating and staged operation require melting 5 kg of material for each kilogram of product (10).

Membranes. Patented separators are used in purification of hydrogen. The working principle is a membrane that is chemically tuned to pass a molecular type (see Fig. 5).

Liquid separation via membranes (reverse osmosis (qv)) is used in production of pure water from seawater. The chief limit to broader use of reverse osmosis is the high pressure required as the concentration of reject rises.

Mole fraction of reject	*Min* ΔP, MPa (psi)
0.05	7.6 (1100)
0.10	15.2 (2200)
0.20	31.8 (4600)

For most processes, the probability of finding a membrane with requisite strength, as well as selectivity and permeability, is low (see Membrane technology).

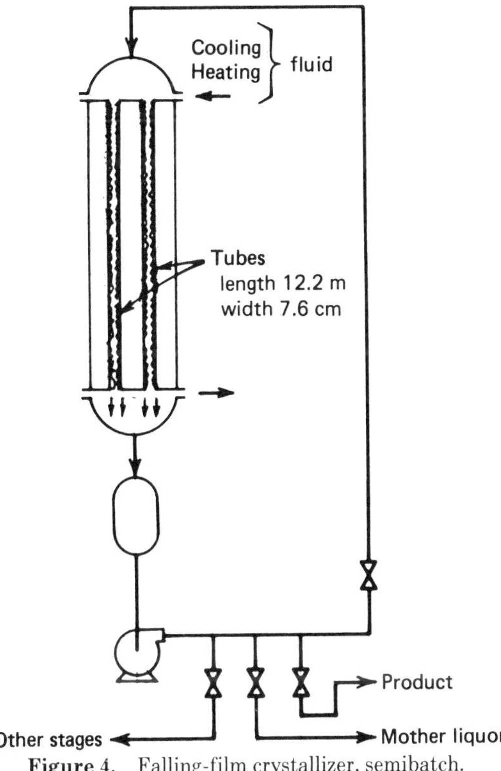

Figure 4. Falling-film crystallizer, semibatch.

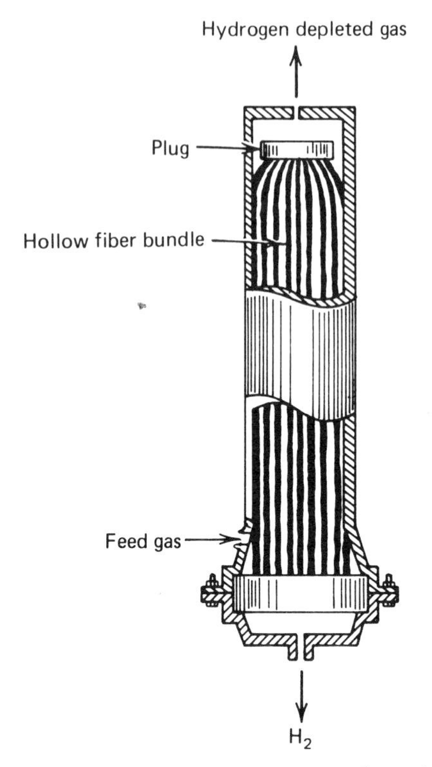

Figure 5. Prism (Monsanto trademark) separator (membrane hydrogen purifier).

Heat Exchange

Most processing is thermal. Reaction systems and separation systems are typically dominated by their associated heat exchange. Optimization of this heat exchange has a tremendous leverage on the ultimate process efficiency (see Heat-exchange technology).

Heat exchangers use energy two ways: as frictional pressure drop and as the loss in ability to do work when heat is degraded.

$$\text{lost work} = QT_{\text{sink}}\left(\frac{1}{T_{\text{cold}}} - \frac{1}{T_{\text{hot}}}\right)$$

The selection of design numbers for ΔP and ΔT is frequently the most important decision the process designer makes, but often this goes unrecognized.

	Typical design time, d	*Typical economic impact, % of unit cost*
ΔT	1/8	25
ΔP	1/16	15
process detail	5	10

The above data show that the designer commonly becomes lost in the detail of tube length and baffle cut in an effort to optimize the hardware to meet a target and spends far too little time on choosing that target.

Before guidelines for the ΔT and ΔP targets are provided, heat-exchange networks and overdesign are discussed briefly.

Heat-Exchange Networks. A basic theme of energy conservation is to look at a process broadly, ie, to look at how best to combine process elements. The heat-exchange network analysis can be a useful part of this optimization (see Heat-exchange technology network synthesis, Supplement Volume). Typical applications have been crude still preheat trains, multistage exothermic reactors, and furnace-convection sections. However, big networks containing more than four shells tend to be relatively rare because of limits imposed by plot plan as well as process restraints. Figure 6 illustrates a simple example of the basic concept of what network analysis does; it builds cumulative heating and cooling curves and merges them until a minimum ΔT is reached.

This example also illustrates one of the key targets of network analysis, ie, obtaining the minimum number of shells:

$$\begin{aligned}\text{minimum shells} &= \text{no. of streams} - 1\\ &= 3 \text{ process streams} + 2 \text{ utilities} - 1 = 4\end{aligned}$$

For more complex problems, there are two available approaches: use of a standard software package or use of the manual method described in ref. 11. For small networks (those with two to four exchangers), the key problem is defining the optimum ΔT.

Overdesign. Overdesign also has a great impact on the cost of heat exchange and sometimes is confused with energy conservation, through lower ΔT and ΔP. The best approach is to define clearly what the objective of overdesign is and then to explicitly specify it. If the main concern is a match to other units in the system, a multiplier is applied to flows. If the concern is with the heat balance or transfer correlation, the multiplier is applied to area. If the concern is fouling, a fouling factor is called for. But

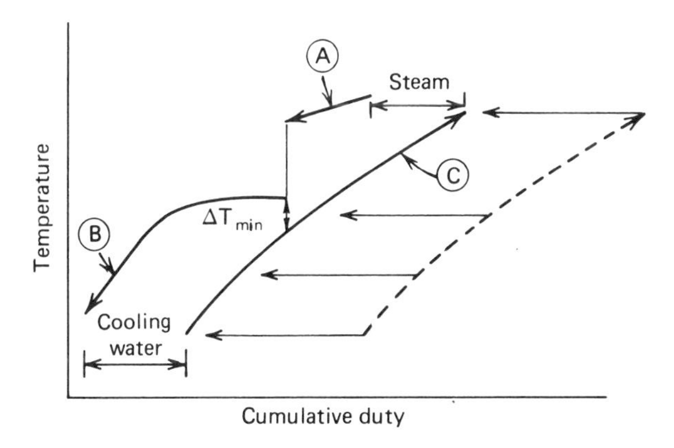

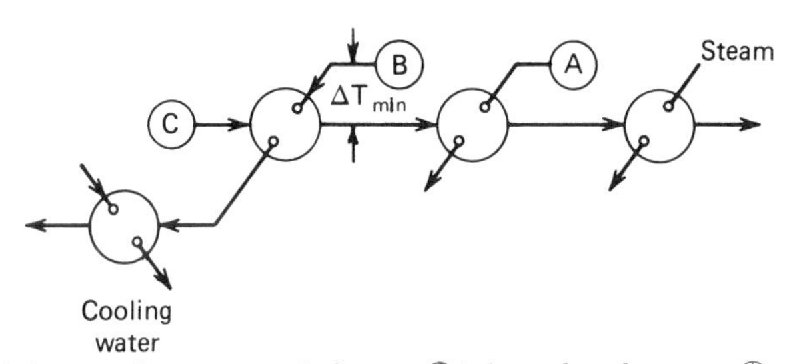

Figure 6. Simple heat-exchange network. Stream Ⓒ is heated, and streams Ⓐ and Ⓑ are cooled.

if low ΔT or ΔP is the principal concern, that should be specified. Adding extra surface saves energy only if the surface is configured to do so. Doubling the area may do nothing more than double the ΔP, unless it is configured properly.

ΔT and ΔP Optimization. Ideally, ΔT and ΔP are optimized by trying several values, making preliminary designs, and finding the point where savings in utility costs just balance incremental surface costs. Where the sums at stake are large, this should be done. However, for many cases the simple guidelines given below are adequate. The primary focus is the impact of surface and utility prices; a secondary focus is the impact of fluid properties on heat-transfer coefficient.

Optimum ΔT. There are three general cases of high importance: the *waste-heat boiler*, in which only one fluid involves sensible heat transfer, ie, a temperature change; the *feed–effluent exchanger*, in which both fluids involve sensible heat transfer and are roughly balanced, ie, undergo essentially the same temperature change; and the *reboiler*, in which neither fluid involves a temperature change, ie, one fluid condenses and the other boils.

Waste-Heat Boiler. In a waste-heat boiler (Fig. 7), the optimum ΔT occurs when

$$\Delta T_{\text{approach}} = \frac{K_1}{K_v}\frac{1.33}{U}$$

where K_1 = annualized cost per unit of surface, \$/(m²·yr); K_v = annualized cost per unit of utility saved, \$/(W·yr); and U = heat transfer coefficient, W/(m²·K). The factor 1.33 includes the value of the pressure drop for the added surface.

For example, the optimum $\Delta T_{\text{approach}}$ is computed

$$K_1 = \frac{\$215/\text{m}^2}{2\ \text{yr}} = \frac{\$107.5}{(\text{m}^2\cdot\text{yr})}$$

$$K_v = \frac{0.017}{\text{kW}\cdot\text{h}}\cdot 8322\ \text{h/yr}$$

$$= \frac{\$142}{(\text{kW}\cdot\text{yr})} = \frac{\$0.142}{(\text{W}\cdot\text{yr})}$$

$$\Delta T_{\text{approach}} = \frac{107.5/(\text{m}^2\cdot\text{yr})}{0.142/(\text{W}\cdot\text{yr})}\frac{1.33}{56.8\ \text{W}/(\text{m}^2\cdot\text{K})} = 17.7\ \text{K}$$

where U = 56.8 W/(m²·K) (10 Btu/(h·ft²·°F)); surface cost = \$215/m² (\$20/ft²); payout time = 2 yr; energy price = \$0.017/kW·h (\$5/10⁶ Btu); and onstream time = 8322 h/yr. This case underlines a dramatic change in process design. Note that $\Delta T_{\text{approach}}$ varies

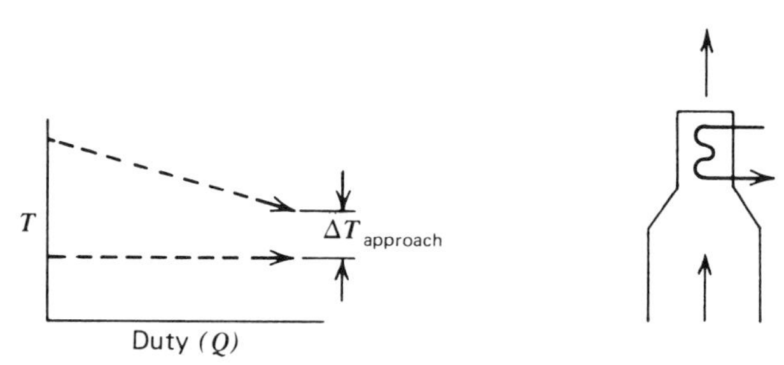

Figure 7. ΔT in a waste-heat boiler.

to the first power of the ratio of surface price to energy price. The most visible result has been a change in typical heater design efficiency from 65–75% to 92–94%. A secondary result has been appearance of waste-heat recovery units in many processes at the point where air coolers were once used.

Feed–Effluent Exchanger. The detailed solution for the optimum ΔT in a feed–effluent exchanger (Fig. 8) involves a quadratic equation for $\Delta T_{\text{approach}}$, but within the restrictions

$$0.8 < \frac{\Delta T_{\text{hot}}}{\Delta T_{\text{cold}}} < 1.25$$

$$\frac{T_{\text{hot}_{\text{in}}} - T_{\text{cold}_{\text{in}}}}{\Delta T_{\text{log mean}}} < 10$$

An excellent approximation is given by

$$\Delta T_{\text{log mean}} = \left[\frac{K_1}{K_v}\frac{1.33}{U}(T_{\text{hot}_{\text{in}}} - T_{\text{cold}_{\text{in}}})\right]^{1/2}$$

For example, the optimum $\Delta T_{\text{log mean}}$ for a feed–effluent exchanger is computed

$$K_1 = \frac{\$107.5/\text{m}^2}{2\ \text{yr}} = \$53.8/(\text{m}^2\cdot\text{yr})$$

$$K_v = \frac{\$0.027}{\text{kW}\cdot\text{h}}\cdot 8322\,\frac{\text{h}}{\text{yr}}\cdot 1000\,\frac{\text{kW}\cdot\text{h}}{\text{W}} = \frac{\$0.227}{\text{W}\cdot\text{yr}}$$

$$\Delta T_{\text{log mean}} = \left[\frac{53.8}{0.227}\frac{1.33}{284}(200 - 100)\right]^{1/2} = 10.5°\text{C}$$

where $T_{\text{hot}_{\text{in}}} = 200°\text{C}$; $T_{\text{cold}_{\text{in}}} = 100°\text{C}$; $U = 284\ \text{W}/(\text{m}^2\cdot\text{K})$ (50 Btu/(h·ft²·°F)); surface cost = \$107.5/m² (\$10/ft²); payout time = 2 yr; energy price = \$0.027/kW·h (\$8/10⁶ Btu); onstream time = 8322 h/yr; and $\Delta T_{\text{hot}}/\Delta T_{\text{cold}} = 1.20$.

The Reboiler. The case shown in Figure 9 is common for reboilers and condensers on distillation towers. Typically, this ΔT has a greater impact on the excess energy

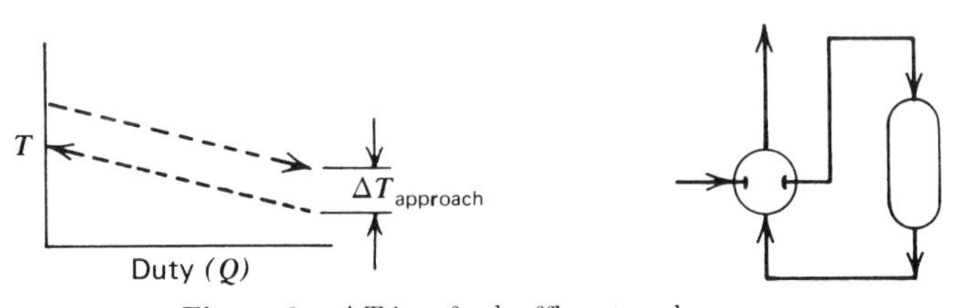

Figure 8. ΔT in a feed–effluent exchanger.

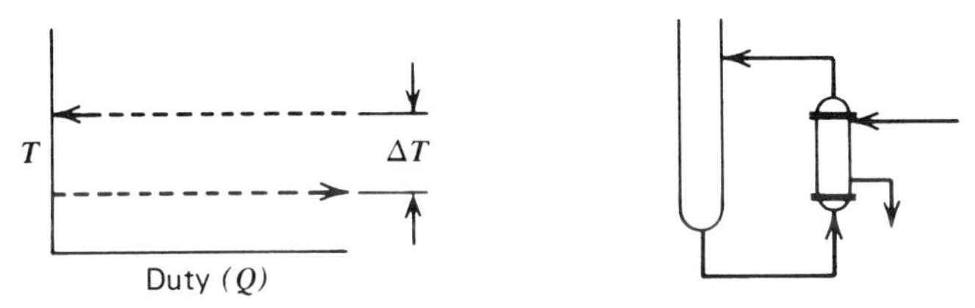

Figure 9. ΔT in the reboiler.

use in distillation than does reflux beyond the minimum. The capital cost of the reboiler and condenser is often equivalent to the cost of the column they serve.

The concept of an optimum reboiler or condenser ΔT relates to the fact that the value of energy changes with temperature. As the gap between supply and rejection widens, the real work in a distillation increases. The optimum ΔT is found by balancing this work penalty against the capital cost of bigger heat exchangers.

If the Carnot cycle is used to calculate the work imbedded in the thermal flows with the assumption that the heat-transfer coefficient (U) is constant and that process temperature is much greater than ΔT,

$$\Delta T_{\text{optimum}} = T_{\text{p}} \left[\frac{K_1}{K_{\text{v}} U T_{\text{sink}}}\right]^{1/2}$$

where T_{sink} = temperature (absolute) at which heat is rejected; T_{p} = process temperature (absolute); K_1 = annualized cost per unit of surface; and K_{v} = annualized value of power.

For utilities above ambient temperature,

$$K_{\text{v}} = K_{\text{p}}\ (\text{turbine efficiency})$$

where K_{p} is the annualized cost of purchased power. The above relations will typically give ΔT values in the 10°C to 20°C range.

One strong caution is that the assumption of a constant U is usually inaccurate for boiling applications. Simulation is generally needed to fix ΔT accurately, particularly at ΔT values below 15°C.

Optimum Pressure Drop. For most heat exchangers, there is an optimum pressure drop. This results from the balance of capital costs against the pumping (or compression) costs. A common prejudice is that the power costs are trivial compared to the capital costs. The total cost curve is fairly flat within ± 50% of the optimum (see Fig. 1**b**), but the incremental costs of power are roughly one third of those for capital on an annualized basis. This simple relationship can be extremely useful in quick design checks.

The best approach is to have a computer program check a series of pressure drops and see how energy requirements decrease as surface increases. If this option is not available, the simple method shown below can be used to obtain specification sheet values.

Start with a pressure drop of 6.9 kPa (1 psi), and apply three correction factors ($F_{\Delta T}$, F_{cost}, F_{prop}):

$$\Delta P_{\text{opt}} = 6.9\,(F_{\Delta T})(F_{\text{cost}})(F_{\text{prop}})$$

The correction for temperature difference is given by

$$F_{\Delta T} = \frac{T_{\text{in}} - T_{\text{out}}}{(T_{\text{hot}} - T_{\text{cold}})_{\text{mean}}}$$

This term is a measure of the unit's length. Sometimes it is referred to as the number of transfer units (see Mass transfer).

The correction for costs is

$$F_{\text{cost}} = 0.017\left(\frac{\$/(\text{m}^2\cdot\text{yr})}{\$/\text{kW}\cdot\text{h}}\right)^{0.75}$$

The correction for physical properties is

$$F_{\text{prop}} = \left(\frac{c}{c_w}\right)^{0.6} \left(\frac{k_w}{k}\right)^{0.6} \left(\frac{\mu}{\mu_w}\right)^{0.3} \left(\frac{\rho}{\rho_w}\right)^{0.5}$$

where c = specific heat; k = thermal conductivity; μ = viscosity; ρ = density; and c_w, k_w, μ_w, and ρ_w are the same properties for water at 25°C.

From these equations, the optimum ΔP for a feed–effluent exchanger where the fluid has the physical properties of water and:

$$T_{\text{in}} - T_{\text{out}} = 20°\text{C}$$

$$\Delta T = (T_{\text{hot}} - T_{\text{cold}})_{\text{mean}} = 10°\text{C}$$

$$\left.\begin{array}{l}\text{surface cost} = \$215/\text{m}^2 \\ \text{payout time} = 2\text{ yr}\end{array}\right\}\$107.5/(\text{m}^2\cdot\text{yr})$$

$$\text{power cost} = \$0.03/\text{kW}\cdot\text{h}$$

is calculated:

$$F_{\Delta T} = \frac{20}{10} = 2$$

$$F_{\text{cost}} = 0.017\left(\frac{107.5}{0.03}\right)^{0.75} = 7.8$$

$$F_{\text{prop}} = 1$$

$$\Delta P_{\text{opt}} = 6.9(2)(7.8)(1) = 107.67\text{ kPa (15.6 psi)}$$

If all else remains the same as above except a gas is obtained with the following properties: $\mu = 0.02\ \mu_w$; $\rho = 0.00081\ \rho_w$; $c = 0.25\ c_w$; and $k = 0.066\ k_w$; then,

$$F_{\text{prop}} = (0.25)^{0.6}\left(\frac{1}{0.066}\right)^{0.6}(0.02)^{0.3}(0.00081)^{0.5} = 0.0194$$

$$\Delta P_{\text{opt}} = 6.9(2)(7.8)(0.0194) = 2.1\text{ kPa (0.3 psi)}$$

The great impact of density in this example and in Table 1 should be noted. Probably the most common specification error is to use the large ΔPs characteristic of liquids in low density gas systems.

Fired Heaters. The fired heater is first a reactor and second a heat exchanger. Often, in reality, it is a network of heat exchangers (see Burner technology; Furnaces, fuel-fired).

Table 1. Impact of Fluid Density on Optimum ΔP

	$\frac{\rho}{\rho_w}$	$\left(\frac{\rho}{\rho_w}\right)^{0.5}$	=	Relative optimum ΔP
water	1	$(1)^{0.5}$		= 1
oil	0.8	$(0.8)^{0.5}$		= 0.9
high pressure gas	0.05	$(0.05)^{0.5}$		= 0.22
atm pressure gas	0.001	$(0.001)^{0.5}$		= 0.03
vacuum	0.0002	$(0.0002)^{0.5}$		= 0.014

The Fired Heater as a Reactor. When viewed as a reactor, the fired heater adds a unique set of energy considerations:

Can the heater be designed to operate with less air by O_2 and CO analyzers?
How does air preheat affect fuel use and efficiency?
How could a lower cost fuel (coal) be used?
Can the high energy potential of the fuel be used upstream in a gas turbine?

CO control. Control of excess air by carbon monoxide is one of the most significant new energy technologies. The key is that CO is highly sensitive to excess air as shown by Figure 10.

Coal and low Btu gas. The much lower cost of coal has caused a rapid resurgence of coal-firing for steam generation. Its direct use in process heaters has been negligible because of historical prejudice and the space and capital demands of coal-handling facilities. Concern about attack on high alloy tubing by ash also precludes its direct use in applications such as ethylene furnaces.

Low Btu gas appears to be an attractive way of utilizing coal in process heaters for a number of reasons: Its technology is old and well established; it keeps the ash out of the process heater; and it requires only mild retrofitting, chiefly of burners. The flame temperature and total flue-gas generation for low Btu gas are both tolerably close to natural gas as shown in Table 2 (12).

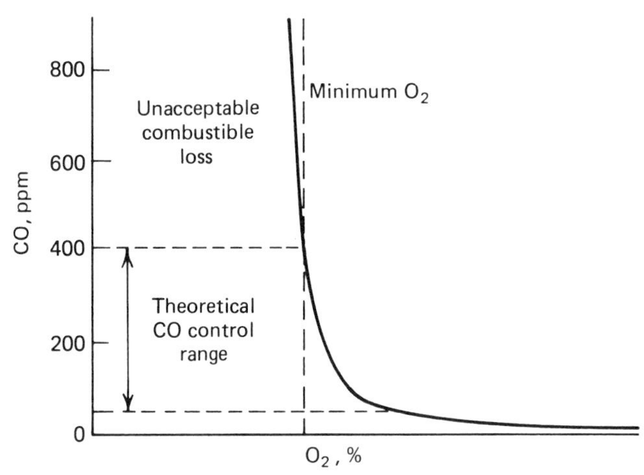

Figure 10. Relationship of CO concentration to O_2 concentration, in a fired heater.

Table 2. Comparison of Combustion Characteristics of Natural and Low Btu Gas

	Natural gas	Low Btu gas
relative heating value per volume		
fuel	1	0.17
fuel and air mixture	1	0.78
relative vol air:vol fuel	1	0.13
relative flue-gas flow rate/unit of energy	1	1.11
adiabatic flame temperature, °C	1927	1760

Air preheating. Use of unpreheated air in the combustion step is probably the biggest waste of thermodynamic potential in industry (see Table 3). It is not practical to preheat to the flame burst temperature, the optimum thermodynamic situation, but some preheating is invariably profitable. Air preheating has the unique benefit of giving a direct cut in fuel consumed. It also can increase the heat-input capability of the firebox because of the hotter flame temperature.

The most common type of air preheater on new units is the rotating wheel. On retrofits, heat pipes or hot-water loops are often more cost-effective because of ductwork costs or space limits.

Upstream firing of fuel, gas turbines. Limitations in the material of construction make it difficult to use the high temperature potential of fuel fully. This restriction has led to insertion of gas turbines into power-generation steam cycles and even to use of gas turbines in preheating air for ethylene-cracking furnaces.

The Fired Heater as a Heat-Exchange System. Improved efficiency in fired heaters has tended to focus on heat lost with the stack gases. When stack temperatures exceed 149°C, that attention is proper, but other losses can be much bigger when viewed from a lost-work perspective. For example, a reformer lost-work analysis gave the breakdown shown in Table 3.

The losses for ΔT in the convection section are almost twice those for the very hot exit flue gas. Furnace optimization is the clearest illustration of the benefits of lost-work analysis. If losses from a stack are nearly transparent, the losses imbedded in an excessive ΔT in a convection section are even harder to identify. They do not show up even on the energy balance that highlights the hot stack. These losses can be cut by adding surface to the convection section and shifting load from the radiant section, as well as by looking at the overall process (including steam generation) for streams to match the cooling curve of the flue gases.

Concern over corrosion from sulfuric acid when burning sulfur-bearing fuels often governs the temperature of the exit stack gas. However, the economics of heat recovery are so strong that flue gases are being designed into the condensing range of weak sulfuric acid. It is not a forbidden zone, but the designer should recognize that tube replacement is necessary.

Simple heat losses through the furnace walls are also significant. This follows from the high temperatures and large size of fired heaters, but these losses are not inevitable. In an optimized system, losses through insulation are roughly proportional to

$$\left(\frac{\text{refractory price}}{\text{energy price}}\right)^{1/2}$$

This means that if the price ratio has decreased by a factor of 9, then losses should be

Table 3. Lost-Work Analysis for a Fired Heater

	Lost-work potential, %
combustion step	54
radiant section ΔT	7
convection section ΔT	24
stack losses (exit temp 225°C)	13
wall losses	2

down by a factor of 3. If the optimum allowed a 3% loss in 1973, today's optimum would be closer to 1%.

Dryers. A drying (qv) operation needs to be viewed as both a separation and a heat-exchange step. When it is seen as a separation, the obvious perspective is to cut down the required work. This is accomplished by mechanically squeezing out the water. The objective is to cut the moisture in the feed to the thermal operation to less than 10%. In terms of hardware, this requires centrifuges and filters and may involve mechanical expression or a compressed air or superheated steam blow. In terms of process, it means big crystals.

When the dryer is seen as a heat exchanger, the obvious perspective is to cut down on the enthalpy of the air purged with the evaporated water. Minimum enthalpy is achieved by using the minimum amount of air and cooling as low as possible. A simple heat balance shows that for a given heat input, minimum air means a high inlet temperature. However, this often presents problems with heat-sensitive material and sometimes with materials of construction, heat source, or other process needs. All can be countered somewhat by exhaust-air recirculation.

Minimum exhaust-air enthalpy also means minimum temperature. If this cannot be attained by heat exchange within the dryer, preheating the inlet air is an option. The temperature differential guidelines of the feed–effluent interchange apply.

Like the fired heater, the dryer is physically large, and proper insulation of the dryer and its allied ductwork is critical. It is not uncommon to find 10% of the energy input lost through the walls in old systems.

Optimum Design of Pumping, Compression, and Vacuum Systems

Pumping. Is piping pressure drop optimal? Many companies have optimum-pipe-sizing programs, but in the absence of one, a good rule of thumb is that in an optimized system the annualized cost for pumping power should be one seventh the annualized cost of piping (1) (see Piping systems; Pumps in Supplement Volume).

Is exchanger pressure drop optimal? Similarly, for an optimized heat exchanger the annualized cost for pumping should be one third the annualized cost of the surface for the thermal resistance connected with that stream.

Is the pump specified for the right flow? As Figure 11 shows, a 50% overdesign factor will increase power by 35% in a combination of higher head and lower efficiency.

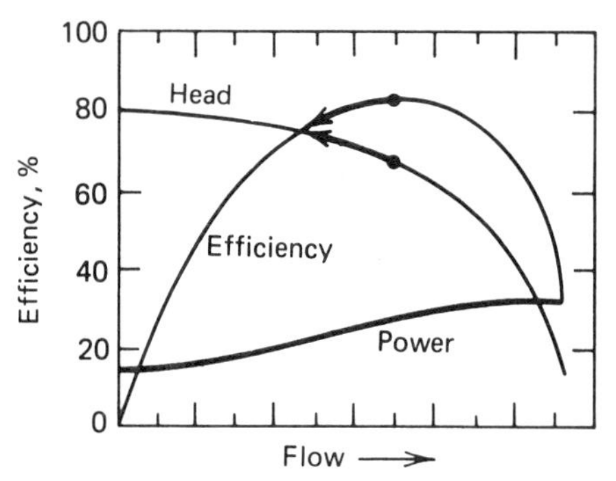

Figure 11. Impact of excess design capacity on pump energy use.

Can the allowance for control be reduced? One option is the use of a variable-speed drive. This eliminates the control valve and its pressure drop and piping. Its best application is where a large share of the head is required for friction and where process demands cause the required flow to vary.

What can be done to get a more efficient pump? Sometimes a higher available net positive suction head (NPSH) permits a more efficient machine.

Compression. The work of compression is typically compared against the isentropic-adiabatic case.

$$\eta_{comp} = \frac{W_{min}}{E_{out} - E_{in}}$$

For an ideal gas, this can be expressed in terms of temperatures

$$\eta_{comp} = \frac{W_{min}}{W_{actual}} = \frac{T_{in}\left[\left(\frac{P_{out}}{P_{in}}\right)^{R/cp} - 1\right]}{T_{out} - T_{in}}$$

where R/cp is the ratio of gas constant to molar specific heat. Minimum work is directly proportional to suction temperature. This means that cooling-water systems should be run as cold as possible. Simply measuring temperature rise permits monitoring efficiency for a fixed pressure ratio and suction temperature.

Sometimes, W_{min} for compression is expressed for the isothermal case, in which it is always lower than for the adiabatic. The difference defines the maximum benefit from interstage cooling.

Efficiencies should always exceed 0.6, and 1.00 is approachable in reciprocating devices. Their better efficiency needs to be balanced against their greater cost, greater maintenance, and lower capacity.

The guidelines on pressure drop in piping and exchangers discussed above also apply here. The opportunity for variable-speed drive is possibly even greater, as is the importance of tight control of minimum flows.

Thermocompressors. A thermocompressor is a single-stage jet using a high pressure gas stream to supply the work of compression. The commonest application is in boosting waste-heat-generated steam to a useful level. An example is shown in Figure 12. Thermocompressors can also be used to boost a waste combustible gas into a fuel system by use of high pressure natural gas. The mixing of the high energy motive stream with the low energy suction stream inherently involves lost work, but as long as the pressures are fairly close, the net efficiency for the device can be respectable (25–30%). Here, efficiency is defined as the ratio of isentropic work done on the suction gas to the isentropic work of expansion that could have been obtained from the motive gas. The thermocompressor has the big advantage of no moving parts and low capital cost.

Vacuum Systems. The most common vacuum system uses the vacuum jet. Because of the higher ratio of motive pressure to suction pressure, the efficiency of vacuum systems is lower than thermocompressors. Generally, it is 10–20%. The optimum system often employs several stages with intercondensers. Steam use in this range varies roughly as $(1/P)^{0.3}$, where P is absolute suction pressure (see Vacuum technology).

Because of the low efficiency of steam-ejector vacuum systems, there is a range of vacuum above 13 kPa (100 mm Hg) where mechanical vacuum pumps are usually

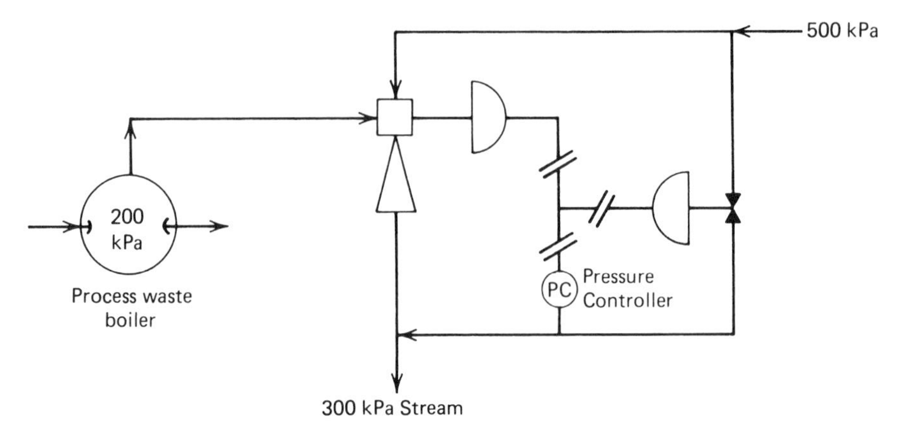

Figure 12. A thermocompressor.

more economic (13). The capital cost of the vacuum pump goes up roughly as (suction volume)$^{0.6}$ or $(1/P)^{0.6}$. This means that as pressure falls, the capital cost of the vacuum pump rises more swiftly than the energy cost of the steam ejector which increases as $(1/P)^{0.3}$. Usually below 1.3 kPa (10 mm Hg), the steam ejector is more cost-effective.

Other factors that favor the choice of the steam ejector are the presence of materials that could form solids and high alloy requirement. Factors that favor the vacuum pump are credits for pollution abatement and high cost steam. The mechanical systems require more maintenance and some form of backup vacuum system, but they can be designed with adequate reliability.

Refrigeration

Refrigeration (qv) is a very high value utility. The value of heat in a hot stream is the work it can surrender:

$$\frac{W}{Q} = \left(\frac{T - T_{\text{sink}}}{T}\right) \eta_{\text{turbine}}$$

And the value of refrigeration is the work required to heat pump it to the sink temperature:

$$\frac{W}{Q} = \left(\frac{T_{\text{sink}} - T}{T}\right) \frac{1}{\eta_{\text{compressor}}} \frac{1}{\eta_{\text{fluid}}}$$

The value of refrigeration is compared to heating in Figure 13 for $\eta_{\text{turbine}} = \eta_{\text{compressor}} = 0.7$ and for $\eta_{\text{fluid}} = 0.8$. Here, η_{fluid} accounts for cycle inefficiencies such as the letdown valve shown by Figure 14.

Because of its value, refrigeration justifies thicker insulation, lower ΔTs in heat exchange, and generally much more care in engineering (14). Some questions that the designer should ask are

Is refrigeration really necessary? Could river water or cooling-tower water be used directly? Could they be used for part of the year? Could they replace part of the refrigeration?

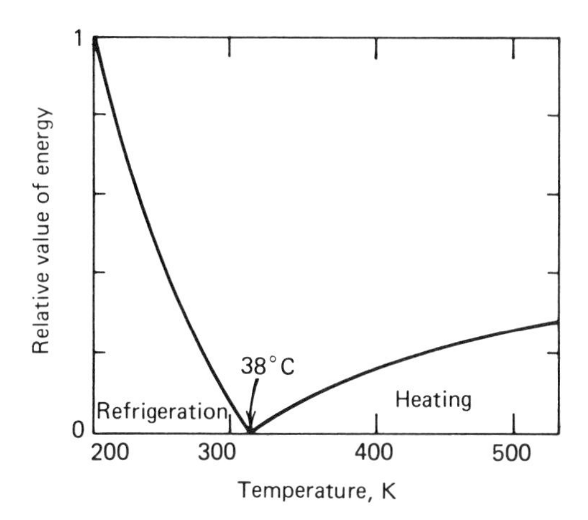

Figure 13. Relative value of energy at various temperatures.

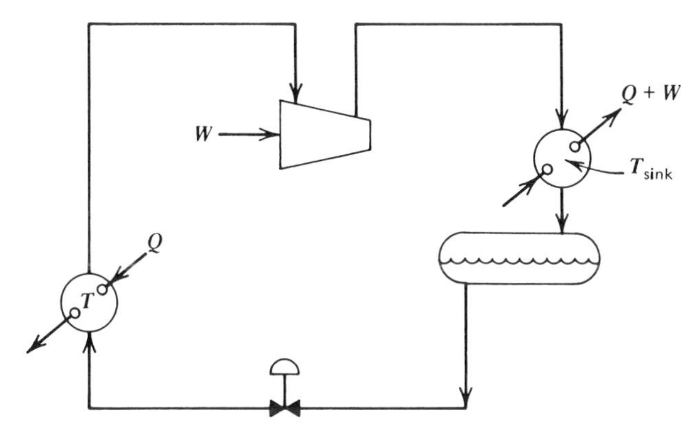

Figure 14. Compression refrigeration.

Can the refrigerant-condensing temperature be reduced? Could it be reduced during part of the year?

Can the system be designed to operate without the compressor during the cold weather?

Is the heat-transfer surface the economic optimum?

Is a central system more efficient than scattered independent systems?

Does the control system cut required power for part-load operation? (Multiple units could yield increased reliability as well as efficiency.)

Are enough gauges and meters provided to monitor operation?

Is there an abundance of waste heat available from the plant (above 90°C)? If so, refrigeration could be supplied by an absorption system.

Absorption chiller units (Fig. 15) need 1.6–1.8 J of waste heat per joule of refrigeration. Commercially available LiBr absorption units are suitable for refrigeration down to 4.5°C.

For low level waste heat (90–120°C), absorption chillers utilize waste heat as efficiently as steam turbines powering mechanical refrigeration units. Absorption refrigeration using 120°C saturated steam delivers 4.5°C refrigeration with an efficiency of 35% where efficiency is referred to the work potential in the steam.

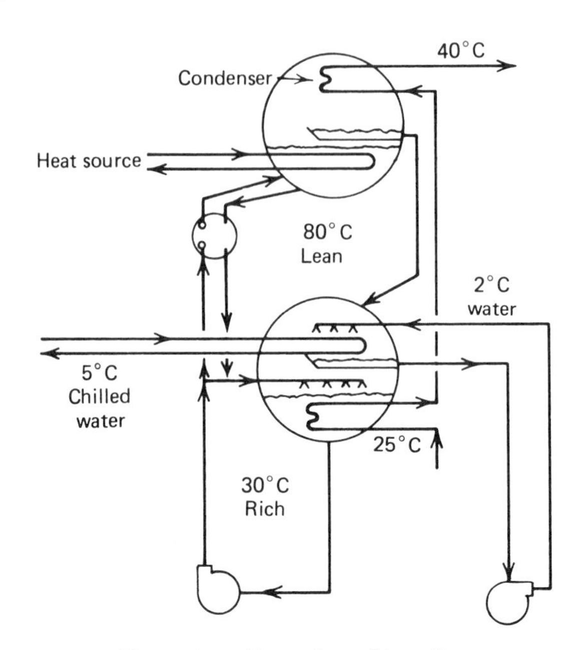

Figure 15. Absorption refrigeration.

Steam and Condensate Systems

In the process industry, steam (qv) serves much the same role as money does in an economy, ie, it is the medium of exchange. If its pricing fails to follow common sense or thermodynamics, strange design practices are reinforced. For example, many process plants employ accounting systems where all steam carries the same price regardless of temperature or pressure. This may be appropriate in a polymer or textile unit where there is no special use for the high temperature; clearly, it is wrong in a petrochemical plant.

Some results of the constant-value pricing system are typically the following: generation in a central unit at relatively low pressure, <4.24 MPa (600 psig); tremendous economic pressure to use turbines rather than motors for drives; lack of incentive for high efficiency turbines; excessively high temperature differentials in steam users; tremendous incentive to recover waste heat as low pressure steam; and a large plume of excess low pressure steam vented to the atmosphere.

A number of alternative pricing systems have been proposed that hinge on turbine efficiency and the relative pricing of fuel and electricity (15–16). Perhaps the best system relates the value of steam to that at the generation pressure by its work potential (exergy content).

$$\frac{\text{value at pressure}}{\text{value at highest generation pressure}} = \frac{\text{exergy at pressure}}{\text{exergy at highest generation pressure}}$$

Design of a central power–steam system is beyond the scope of this discussion, but the interaction between the steam system and the process must be considered at all stages of design. There is a long list of factors to consider in designing a steam-using system:

Can steam-use pressure be lowered? (If ΔT in the heater is above 20°C, the steam pressure is probably above the economic optimum.)

Are there any turbines under 65% efficiency? (Today, turbines are being limited to large sizes above 500 kW, where good efficiencies can be obtained; they are used only for those small drives essential to the safe shutdown of the unit.)

Can a gas turbine be utilized for power generation upstream of the boiler?

Are there any waste streams with unutilized fuel value?

Is there a program to monitor turbine efficiency by checking temperatures in and out?

Is condensate recovered?

Is the flash steam from condensate recovered?

Is feedwater heating optimized?

Is there any pressure letdown without power recovery?

Has enough flexibility been built into the overall condensing–turbine system? (The balance changes over the history of a unit as a process evolves, generally in the direction of less condensing demand.)

Is steam superheat maintained at the maximum level permitted by mechanical design?

Can a thermocompressor be used to increase steam pressures from waste heat?

Are all users metered?

Is low level process heat used to preheat deaerator makeup?

Are ambient sensing valves used to turn off steam tracer systems?

Are steam traps appropriate to the service?

Cooling-Water Systems

Cooling water is a surprisingly costly utility. On the basis of price per unit energy removed, it can cost one fifth as much as the primary fuel. Roughly half of this cost is in delivery (pump, piping, and power). This fact has several important implications for design.

Heat exchangers should be designed to use the available pressure drop. A heat exchanger that is designed for 10 kPa when 250 kPa is available will have five times the design flow.

If an exchanger cannot be economically designed to use available ΔP, orifices should be provided to balance the system. This can be done without compromising the guidelines that no unit should be designed for less than 0.8 m/s on tubeside or less than 0.3 m/s on shellside.

If temperature requirements permit, the system will cost less to operate with exchangers in series.

An installed measuring element is usually justified. Typically this takes the form of a low-pressure-drop, averaging, Pitot tube.

If only part of the system requires a high head, this should be supplied by a booster pump. The whole system should not be designed to use or waste the high head.

Other energy considerations for cooling towers include use of two-speed or variable-speed drives on cooling-tower fans, and proper cooling-water chemistry to prevent fouling and excess ΔT in users (see Water, industrial water treatment).

Air coolers can be a cost-effective alternative to cooling towers at 50–90°C (just below the level where heat recovery is economic).

Special Techniques

Heat Pumps and Temperature Boosting. A heat pump is a refrigeration system that raises heat to a useful level. The most common application is the vapor recompression system for evaporation (Fig. 16). Its application hinges primarily on low cost power relative to the alternative heating media. If electricity price per unit energy is less than 1.5 times the cost of the heating medium, it merits a close look. This tends to occur when electricity is generated from a cheaper fuel (coal) or when hydroelectric power is available.

Use in distillation systems is rare. The reason is a recognition that almost the same benefits can be achieved by integrating the reboiling–condensing via either steam system (above ambient) or refrigeration system (below ambient).

In an optimized system,

$$\frac{Q}{W} = \frac{T_{\text{hot}}\eta_{\text{compressor}}}{T_{\text{hot}} - T_{\text{cold}}}$$

where T_{hot} and T_{cold} are in absolute units, K.

This provides another criterion for testing whether a heat-pump system may be cost-effective. A power plant takes three units of Q to yield one unit of W; therefore, to provide any incentive for less overall energy use, Q/W must be far in excess of 3.

Organic Rankine Cycles. For the same reason that steam is not an ideal medium of exchange at low temperature (low vapor density), it is also not an ideal working fluid for converting waste heat to power. A number of cycles using organic fluids (eg, butane, Freons) have been developed, primarily for geothermal brines. There have also been a limited number of applications in the process industries, but in the process industry there is generally a more attractive use for this low level heat, such as direct interchange or absorption refrigeration.

Energy-Management Systems. The considerable reduction in computing costs has made it possible to do a wide range of routine monitoring and controlling. One can, for example, continuously monitor a distillation system and compare energy use against an optimum and display the cost-per-hour deviation from optimum setpoint. The computer can also test specific actions to achieve the optimum. A computer can

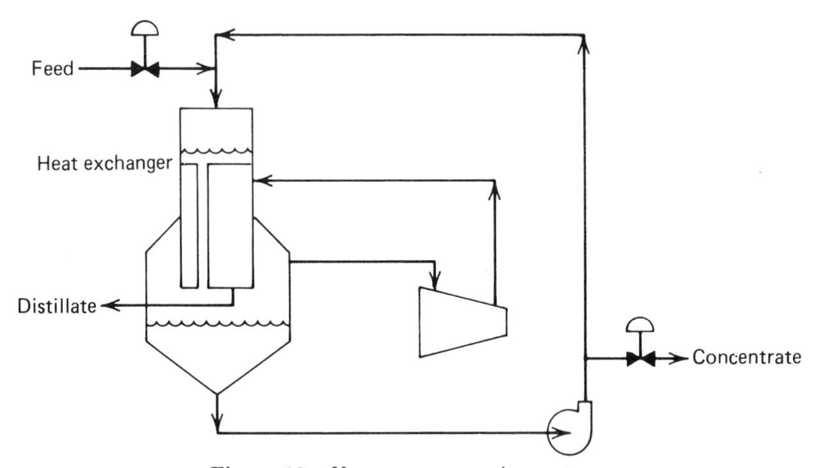

Figure 16. Vapor-recompression system.

monitor a steam system, advise how best to load a set of boilers and choose which turbines to run. Another frequent use of energy-management systems is to turn off lights and cut back heating, ventilating, and air conditioning (HVAC) in building operations.

The Existing Plant

How do existing plants differ from new plants? Good ideas for new plants are also good ideas for existing plants, but there are three basic differences: *1*. Because a plant already exists, the capital vs operating cost curve differs; usually, this makes it more difficult to reduce utility costs to as low a level as in a new plant. *2*. The real economic justification for changes is more likely to be obscured by the plant accounting system and other nontechnical inputs. *3*. The real process needs are measurable and better defined.

An example in support of the first of these is the case of optimum insulation thickness. Suppose a tank was optimally insulated ten years ago. If the value of heat quadrupled in the interim, this change would justify twice the old insulation thickness on a new tank. However, the old tank may have to function with its old insulation. The reason is that there are large costs associated with preparations to insulate. This means that the cost of an added increment of insulation is much greater than assumed in the optimum insulation thickness formulas (Fig. 17). An example of the second difference is that many things appear to be strongly justified by savings in low pressure steam if the steam is valued artificially high. A designer of a new plant has the advantage of focusing attention on savings in the primary budget items, ie, fuel and electricity at the plant gate, rather than on cost-sheet items such as steam at battery limits. The third difference is that many process details are relatively uncertain when a plant is designed. For example, inert loading for vacuum jets is rarely known to within 50%. Although the first two differences are negative, the third provides a unique opportunity to measure the true need and revise the system accordingly.

There has been an evolution in plant programs since the early 1970s.

$$\begin{array}{ccccc} \text{Heat balance} & \rightarrow & \text{Energy audit} & \rightarrow & \text{Energy survey} \\ 1974 & & 1978 & & \text{Today} \end{array}$$

The change of title is more than semantics. The focus today is sharper. For example, a heat balance would certainly pinpoint steam lost through purges, leaks, etc, but it

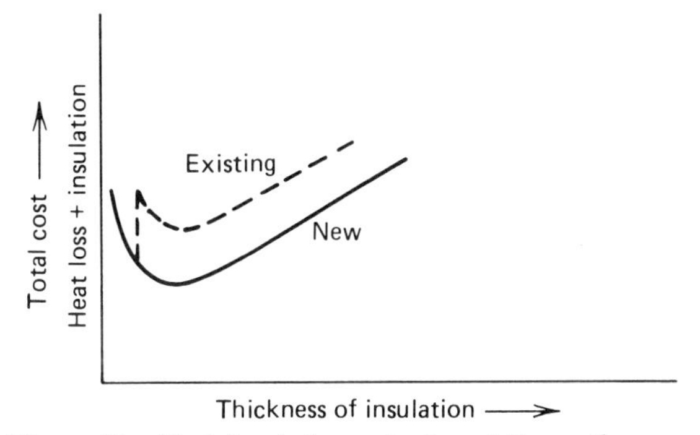

Figure 17. Tank-insulation costs, the existing vs the new.

might indicate nothing about an inefficient turbine, an atmospheric-pressure gas cooler using 35 kPa (5 psi) pressure drop, or a 600°C ΔT in a convection section. Also, an energy survey breaks the 'audit' mentality of steam leaks, etc, and focuses on opportunities to make important process improvements, and integration with other units at the site.

The energy survey has seven components: *"as-it-is" balance; field survey; equipment tests; check against optimum design; idea-generation meeting; evaluation* and *followup.*

As-It-Is Balance. This is a mandatory first step for the energy survey. It permits targeting of principal potentials; check of use against design; check of use against optimum (how a new plant would be designed today); definition of possible hot or cold interchanges; definition of the unexpected uses (eg, the large steam purge to process or high pressure drop exchanger); and contribution from specialists not familiar with the unit.

Field Survey. This is often done by a team of two: one who knows what to look for and one who knows the process. On field surveys, it is as important to talk to the operators ("What runs when the unit is down?"; "What happens when you cut reflux?"; "Where are the guidelines for steam–feed ratio? How close do you usually run?") as it is to record items like air leaks, high pressure drops across values, frost on piping, lights of the wrong type or at the wrong time, steam plumes (a reason to climb to the top of the unit), or minimum-flow bypasses in use. The field survey should develop detailed "fix" lists for leaking traps, uninsulated metal, lighting, and steam leaks.

Equipment Tests. Procedures for rigorous, very detailed, efficiency determination are available (ASME Test Codes) but are rarely used. For the objective of defining conservation potentials, relatively simple measurements are adequate. For fired heaters, stack temperature and excess O_2 in stack should be measured; for turbines, pressures (in and out) and temperatures (in and out) are needed.

Check Against Optimum Design. This attempts to answer the question, "Need a balance be as it is?"

In the large view, the first thing to compare against is the literature claims. For the large-volume chemicals, this is often available. Chemicals where state-of-the-art discussions exist include NH_3, HNO_3, CH_3OH, and phthalic anhydride (2–4,17–18).

The second thing to do is look for the obvious: stack temperatures >149°C; process streams >121°C, cooled by air or water; process streams <65°C, heated by steam; $\eta_{turbine}$ <65%; reflux ratio >1.15 times minimum; and excess air >10% on clean fuels.

Idea-Generation Meeting. The idea-generation meeting is most productive if three important guidelines are followed: *1*. Get the right people. A good guiding principle is to get "two wise old Turks" for each corporate expert. It is extremely important to ask, "Given free choice, whom would you choose to attend?," and then get them. *2*. Discuss the as-it-is balance for each area and record all ideas. *3*. Assign follow-up responsibility.

Evaluation. The evaluation of each idea should include a technical description and its economic impact and technical risk. The ideas should be ranked for implementation. A report should provide a five-year framework for energy projects.

Follow-up. If no savings result, the effort has been wasted. The survey leader needs to be sure that the potential of the good ideas is recognized by management and the project-generation channels of his company.

What Do You Not Do? Often, what looks like negligence can be a tried and proven practice, and one should be cautious about experts bearing lists and offering huge savings. The process has to work, and present utility saving may or may not compensate for future repair bills or lost products. Some examples are the following: an idling turbine may be necessary to permit a safe plant shutdown if a power failure occurs; a cooling-water flow that is throttled to below 0.6 m/s (in winter) will likely assure a heat-exchanger cleaning in late spring; a furnace that runs too low on excess air may run into after-burning; and a column run too close to the minimum reflux ratio without adequate controls runs a risk of off-specification product.

One also does not trust the plant accounting system unquestioningly. All energy is not created equal. The energy that is recovered from flashed steam or that is shaved off a reboiler's duty may not be worth its cost-sheet value. The meters that matter are the primary meters at the plant gate. Only if the recovered energy reduces the meter settings does it save the plant money.

One does not accept the first solution to an energy-waste problem without seeing how this problem fits into the overall plant-energy balance. The sudden rise in energy prices has lifted many options into the justifiable range. There might be better alternatives available.

BIBLIOGRAPHY

1. D. E. Steinmeyer, *Chemtech*, 188 (Mar. 1982).
2. A. Pinto and P. L. Rogerson, *Chem. Eng. Prog.*, 95 (July 1977).
3. J. M. Dealy and M. E. Weber, *Appl. Energy* **6**, 177 (1980).
4. W. F. Kenney, *Proceedings 1981 Industrial Energy Conservation Technology Conference*, Texas Industrial Commission, p. 247.
5. K. G. Denbigh, *Chem. Eng. Sci.* **6**, 11 (1956).
6. C. J. King, *Separation Processes*, 2nd ed., McGraw-Hill, Inc., New York, 1980.
7. C. S. Robinson and E. R. Gilliland, *Elements of Fractional Distillation*, 4th ed., McGraw-Hill, Inc., New York, 1950.
8. A. J. Andrews and D. E. Griffin, *Oil Gas J.*, 91 (Dec. 8, 1980).
9. M. Souders, *Chem. Eng. Prog.* **60**(2), 75 (1964).
10. D. Carter, personal communication, Monsanto Corp., St. Louis, Mo., 1982.
11. B. Linnhoff and J. Turner, *Chem. Eng.*, 56 (Nov. 2, 1981).
12. C. J. Nebeker, *Proceedings 1982 Industrial Energy Conservation Technology Conference*, Texas Industrial Commission.
13. G. A. Huff, *Chem. Eng.*, 83 (Mar. 15, 1976).
14. W. F. Furgerson, *Conserving Energy in Refrigeration*, Manual 12 of *Industrial Energy—Conservation*, MIT Press, Cambridge, Mass., 1982.
15. W. V. L. Campagne, *Hydrocarbon Process.*, 117 (Aug. 1981).
16. H. R. Null, A. C. Pauls, R. J. Larson, and D. E. Steinmeyer, *Process Design for Energy Conservation Course Notes* in *AIChE Today Series*, AIChE, 1976.
17. G. Keunecke and C. Mitchem in ref. 12.
18. J. R. Leblanc, D. O. Moore, and R. V. Schneider in ref. 12.

D. STEINMEYER
Monsanto Co.

PRODUCT LIABILITY

Product liability is the legal term used to describe an action in which an injured party (plaintiff) seeks to recover damages for personal injury or loss of property from a seller (defendant) when it is alleged that the injuries resulted from a defective product. The term has also been used when a consumer or business enterprise has suffered commercial loss owing to the breakdown or inadequate performance of a product. For the most part, however, commercial loss cases have been dealt with under the law of contracts and are governed by the Uniform Commercial Code. The product liability explosion of the past decade has arisen primarily from personal injury cases in which a product is implicated in the injury (1).

Theories of Liability

An injury occurs and a product is implicated as a possible cause of the injury. Where does one begin in assessing whether the law will impose liability on either the manufacturer or seller of the product? The mere occurrence of an injury in which a product played a role is not, *ipso facto*, grounds for imposing liability. The manufacturer or seller is not an insurer. He or she does not normally guarantee that no ill will befall the user of his or her product. Product liability is not absolute liability. It must be established that the manufacturer violated a legal responsibility to the user or consumer.

The following three basic legal principles can be used in most states as the framework within which the plaintiff can bring an action in product liability: negligence, which tests the conduct of the defendant; strict liability and implied warranty, which test the quality of the product; and express warranty and misrepresentation, which test the performance of products against the explicit representations made on their behalf by the manufacturer and sellers.

Negligence. Negligence has been defined as "conduct which involves an unreasonably great risk of causing damage (2). Alternatively, it has been described as conduct "which falls below the standard established by law for the protection of others against unreasonably great risk of harm" (3). To make out a cause of action in negligence, it is not necessary for the plaintiff to establish that the defendant either intended harm or acted recklessly in bringing about the harm. The test requires that the conduct of the defendant be measured against a norm. The norm that has been established by the courts is the "reasonable person" test.

All human activity involves an element of risk. The defendant's conduct is deemed negligent only when it falls below what a "reasonable person" would have done under similar circumstances. It could be argued that such a standard would be too vague for courts to use in everyday litigation, but it provides a flexible tool for judging behavior. For example, a soda-pop bottle that exploded owing to a flaw in the glass may be defective. It does not necessarily follow that the defendant manufacturer of the bottle was negligent. If the manufacturer took "reasonable care" as demonstrated by the level of quality control techniques, the manufacturer's conduct was not unreasonable, even though the product that came off the assembly line was in fact defective and caused injury to the plaintiff.

The flexibility that is the hallmark of the negligence standard is achieved by permitting the interplay of three factors. In every negligence action the court must consider the probability of the harm's occurring; the gravity of the harm, if it should occur; and the burden of precaution to protect against the harm, ie, what the manufacturer might have done to minimize or eliminate the harm. The more remote and improbable that the defendant's conduct will result in harm, the less reason he or she has to protect against it. If, however, the harm can result in serious injury, the actor must consider the serious implications of even improbable or unlikely harm. Finally, the two aforementioned factors must be weighed against the "burden of precaution," ie, the cost of preventing harm. Thus, in the exploding soda-pop bottle case, on the assumption that more exacting inspection techniques existed at the time of manufacture, it still must be weighed against the increased probability of discovering flawed bottles through the more sophisticated inspection techniques, as well as against the seriousness of the injuries that can occur from exploding soda-pop bottles. Thus, the negligence principle admits the existence of real risks associated with product use. The question that must be answered in each case is whether the risk is justified.

In assessing the manufacturer's conduct, the negligence test focuses on the technology and information available to him or her at the time of manufacture of the product. Note, however, that a plaintiff can allege a cause of action based on what a reasonable manufacturer should have known about products manufactured at the time the defective product was produced. Thus, the reasonable manufacturer is expected to maintain an awareness of industry-wide standards and improving technology. At the same time even an entire industry may be found negligent for the continuation of unacceptable standards of manufacture. Thus, negligence may be imposed for the failure of an industry to adopt or undertake technological improvements. An example of the operation of this principle is found in Marsh v. Babcock and Wilcox (4). The plaintiff, who had suffered injury as a result of an exploding boiler, successfully established negligence on the part of the boiler manufacturing industry in their continued use of a hydrostatic test to establish boiler integrity. The court found that it was reasonable for a jury to determine that a more sophisticated test that had recently been developed by a professor of metallurgy at the University of Wisconsin was a feasible alternative to the hydrostatic test and had a higher probability of discovering flaws in the metal.

Since the determination of negligence arises on a case-by-case basis, initiated by a plaintiff who seeks recovery for harm done by a defective product, it is possible that the same conduct may be judged both negligent and nonnegligent in two different legal proceedings. Although this is somewhat unsettling, there is a core concept of "reasonableness" that governs negligence cases. The issue is not left to the vagaries of the jury. If a judge reaches the conclusion that the core concept of "reasonableness" has been met, based on precedents set in similar decisions, the judge will direct a verdict for the defendant manufacturer; ie, the judge will not let the jury decide the case. Thus, a manufacturer may take guidance from the general flow of product cases about whether his future conduct will be judged substandard. Based on the results of litigated cases, a sense can be developed about the considerations that determine an acceptable level of risk.

Evidence that a manufacturer has met self-imposed standards, voluntary consensus standards, or even statutory standards is significant in a negligence action, but such evidence alone does not establish that a manufacturer is nonnegligent. These standards are often used as a floor, not as a ceiling, for establishing the "acceptable" conduct. A 1971 Pennsylvania case illustrates this principle (5). A helicopter manufacturer had met Federal Aviation Administration (FAA) standards for the time available to the pilot for engaging the autorotation control after engine failure. This manufacturer could not raise as an absolute defense to liability the fact that he had conformed to governmental standards. The court held that a jury could well find that a reasonable manufacturer would design the controls to allow the pilot more time after the engine failure to activate the autorotation control than is required by FAA standards. The lesson is clear: regardless of existing standards, either industrial or governmental, the final responsibility for reasonable behavior rests with the manufacturer.

Strict Liability and Implied Warranty. ***Strict Liability.*** In 1964, the concept of strict liability emerged as an alternative avenue for establishing an action in products liability. The clearest expression of the strict liability concept is part of the Restatement (Second) of Torts, Section 402A (1965):

(*1*) One who sells any product in a defective condition unreasonably dangerous to the user or consumer or to his property is subject to liability for physical harm thereby caused to the ultimate user or consumer, or to his property if
  (a) the seller is engaged in the business of selling such a product, and
  (b) it is expected to and does reach the user or consumer without substantial change in the condition in which it is sold.

(*2*) The rule stated in Subsection (*1*) applies although
  (a) the seller has exercised all possible care in the preparation and sale of his product, and
  (b) the user or consumer has not bought the product from or entered into any contractual relation with the seller.

The differences between the negligence and strict liability theories are significant and far-reaching. Fundamentally, in deciding whether a product is or is not unreasonably dangerous, the focus in strict liability is on the product and not on the conduct of the manufacturer. The shift from negligence to strict liability requires, if nothing else, that the inquiry be focused on the product and its use and away from what the manufacturer should or should not have done or perceived.

Recall that, in regard to the exploding soda-pop bottle, it was established that under the negligence theory a manufacturer could claim that the quality-control techniques met the standard of reasonableness and that the manufacturer should not be found liable for a flawed bottle that exploded. Under the theory of strict liability it is no defense that the manufacturer acted reasonably. If the product is in fact unreasonably dangerous and it caused the plaintiff's injury, the manufacturer can be held liable. Liability will attach, even though "the seller has exercised all possible care in the preparation and sale of his product," and it will be of no avail to the defendant in a production-defect case to argue that better quality-control procedures are prohibitively expensive.

If the plaintiff's claim is, however, that the defendant's product is defective because of a design feature, it is essential, as in the theory of negligence, to weigh the burden of precaution to protect against the harm together with the probability and gravity of the harm. In testing a design defect, then, these basic considerations, balancing the probable risks inherent in the use of the product against its utility, remain common to both the negligence and strict liability causes of action.

Use of this balancing process is crucial; for example, in a Texas case (6) the plaintiff, who was attending a party, walked from the living room toward the patio. Unbeknown to her, a clear glass door separated the living room and the patio. The glass was cleanly polished and the plaintiff mistook the glass for open space. The result was a plaintiff with a bruised face. In prior years plaintiff would have gone home and applied a steak to her black eye. But the year was 1971, and so the plaintiff sued the glass company for manufacturing an "unreasonably dangerous" door. The court rejected the plaintiff's claim. In doing so it decided that the value of clear glass doors, which give home dwellers an outdoors feeling, exceeded the danger level of the occasional situation in which someone mistakes the door for open space. To reach this conclusion, the court was deciding whether it ought to establish a standard for glass doors that would lessen the risk of injury. By implying that the existing criteria for the manufacture and installation of glass doors were adequate and that no design alterations or warnings were necessary, the court was stating that the product was reasonably safe and not defective.

The need for an external standard is obvious because the design of a product cannot be measured against the manufacturer's own internal standard. By definition in a design-defect case the manufacturer has met his own internal standard. The difference between negligence and strict liability in a design defect case is thus a subtle one. The key to understanding the difference is again the distinction between defendant conduct and product performance. Negligence questions the reasonableness of the product in the environment of its use (see Design Defects and Failure to Warn).

Another principal distinction between negligence and strict liability theory arises from the potential liability of defendants in the distributive chain. Under strict liability, any seller who has sold the product in a defective condition will be held liable to the plaintiff for his injury. Thus, potential liability exists for the retailer, the wholesaler, and the distributor, as well as the manufacturer. In addition the manufacturer of a component part of the product is also potentially liable. The only caveat is that the product that caused injury must have left the hands of the particular defendant containing the "defective" condition complained of. Thus, if the plaintiff is able to establish that a component part of a product left the hands of the component-part manufacturer in a defective condition and was later assembled into the product, the component-part manufacturer can be held liable, as well as the assembler of the product. The distributor, the wholesaler, and the retailer, will be held liable, even though there is no duty upon them to inspect the product and even when there is no way in which they could have learned of the defective condition. It should be clear that under the negligence theory many of the members of the distributive chain would not be at fault, since there would be no practical way for them to inspect and discover the defect. They would thus have been acting "reasonably" in failing to inspect. Since strict liability focuses, not on the conduct of any party, but rather on the condition of the product in its use environment, liability may attach to any member of the distributive chain regardless of the reasonableness of his conduct.

Implied Warranty of Merchantability. A parallel guarantee of product quality, similar to that of strict liability, is given to the user or consumer by the Uniform Commercial Code, Section 2-314. The code has been adopted in every state except Louisiana. Section 2-314 provides as follows:

(*1*) "Unless excluded or modified, a warranty that the goods shall be merchantable is implied in a contract for their sale if a seller is a merchant with respect to goods of that kind . . . ," and

(2) "Goods to be merchantable must be at least such as . . .

(c) are fit for the ordinary purposes for which such goods are used. . . ."

Although the interpretation of this section of the Uniform Commercial Code by the courts is somewhat unclear, there is general agreement that Section 2-314, which defines an implied warranty of merchantability, provides the same consumer protection as the Restatement requirement that a product be reasonably safe. There may be understandable confusion when it appears that there are two alternative theories of law that accomplish almost the same purpose. Without delving into subtle legal complexities, suffice it to say that, in proceeding under the Uniform Commercial Code (a statute designed primarily for commercial entrepreneurs and not directed specifically to consumers), special care must be taken to comply with all the requirements set forth by the complex provisions of the Uniform Commercial Code. Thus, for example, failure of an injured party to give notice of the defect in the product to a manufacturer within a reasonable time after his injury may preclude his ability to bring suit. In addition, the common practice of manufacturers to disclaim the implied warranty of merchantability must be given special consideration. When the plaintiff proceeds under the Uniform Commercial Code, he may be faced with an argument that the disclaimer prevents his recovery. Courts have differed sharply on whether a disclaimer of the implied warranty of merchantability should be effective in a case where the plaintiff has suffered personal injury. Further, the validity of a disclaimer between parties of unequal bargaining power has been seriously questioned. The Code is ambiguous on this subject. There is, however, no question that strict liability admits of no disclaimers of any kind.

Express Warranty and Misrepresentation—Implied Warranty of Fitness. The previous discussion focuses on legal theories requiring that the plaintiff establish some form of product defect. There is, however, an alternative theory for recovery that does not require the establishment of a defect; ie, it is not necessary to establish that the product is unreasonably dangerous. Where a seller expressly warrants or represents that a product has certain characteristics or will perform in a certain manner and the product fails to meet the seller's own warranty or representations, then, if the buyer can prove that his or her injury resulted from the failure of the product to meet the warranty, liability is established.

There is often a question about whether a certain statement is a warranty or a representation, or whether it is merely "puffing." Thus, at what point a representation of the qualities of a used car becomes a warranty rather than puffing depends on the type and age of the car and the nature and strength of the representations. The statement that a car is in "A-1 shape" may not mean a great deal if the car is seven years old with 160,000 kilometers on the odometer. A somewhat stronger representation (eg, "Factory New") applied to a demonstrator model may qualify as a warranty. No hard and fast rules exist about where to draw the line. It is all a question of good common sense.

The caution to manufacturers not to oversell the product is important. If a manufacturer markets beyond the capability of the product to perform, liability may result. This will be true even though the manufacturer honestly believes in the truth of the representations and is reasonable in this belief. An excellent example of a court's holding a manufacturer to his or her own rhetoric is *Crocker v. Winthrop Laboratories*

(7). The defendants had developed Talwin, a fairly strong painkiller that they advertised to the medical profession as nonaddictive (see Analgesics, antipyretics, and anti-inflammatory agents). The plaintiff, who had been injured in an industrial accident, took Talwin regularly, as prescribed by his physician, to relieve the pain and became addicted to it. Sometime thereafter the plaintiff died as a result of taking a large enough dose so that it was toxic. In finding for the plaintiff, the court did not declare Talwin to be an "unreasonably dangerous" drug. Instead it held the defendant to its promise that Talwin was "nonaddictive." It was irrelevant that the defendant drug company believed the truth of its own assertion at the time of marketing the product. Thus an express warranty is a form of absolute liability.

Applying The Theory To Product-Defect Types

Production or Manufacturing Defects. A production-defect case is one in which the product that came off the assembly line did not meet the manufacturer's internal specifications with regard to the product. It generally results from the failure of quality control to discover an imperfection in the product. An improper weld, a screw that is not sufficiently fastened, the missing vent hole in a cap all exemplify this form of product defect.

In the early years of product liability law, the plaintiff was required to establish negligence in order to make out a case against the seller of a product which had a manufacturing defect. The litigation focused on the adequacy of the defendant's quality-control techniques. Today, in almost all jurisdictions, courts will impose liability if the product was defective. This is the most common application of the much-heralded strict-liability doctrine. The plaintiff must still establish that the product was defective when it left the hands of the manufacturer and that the defect was the cause of the manufacturer's harm; it is simply no longer possible to defend a manufacturing defect on the theory that "we did our very best."

Design Defects and Failure to Warn. In sharp contrast to the production defect case, the design defect case does not provide a built-in internal standard for establishing defect. The product by definition meets the manufacturer's own internal standard. In attempting to define an external standard for establishing defect in design, it is clear that a balancing process must help determine the standard; that standard must essentially be one of reasonableness.

When the test is the conduct of the defendant, the reasonableness standard is fully understandable. When, however, it is the product itself that must be considered to establish defect (as it must be in strict liability), then the standard of "unreasonable danger" requires some explanation.

Products are not capable of reasoning. Discussion of the "unreasonably dangerous" product, begins with the premise that all products present risks to the consuming public. Some risks, when balanced against the important functions the product performs and the cost of providing for greater safety, are deemed "reasonable." This means that a reasonable person who had actual knowledge of the product's potential for harm would conclude that it was proper to market it in that condition.

How does this differ from negligence? Strict liability theory, unlike negligence, is not concerned with the conduct of the defendant that brought about the unreasonably dangerous condition. Thus, for example, assume that a defendant manufacturer has acted reasonably in designing a product and has adequately tested it before it is marketed. After the product is in actual use in the marketplace, however, it is discovered that the testing and design process failed to account for certain dangers, and even though it was reasonable not to anticipate those dangers, the design is in fact substandard. It is not a defense in this case that the defendant acted reasonably. If the product can be proclaimed not reasonably safe, that a reasonable person who had knowledge of the danger would have decided against marketing without the design alteration or additional warning, then the product is unreasonably dangerous.

The articulation of a standard of liability which holds a manufacturer liable for failing to design a product against risks which were *not* reasonably foreseeable or to warn against risks which were not discoverable at the time of manufacture has caused great conflict among courts and scholars. Some courts have flatly stated that even in cases based on defective design, strict liability will apply. In a landmark opinion the New Jersey Supreme Court has gone a step further. In *Beshada v. Johns–Manville Products Co.* (8) defendants argued that they should not be held liable for failing in the 1930s to warn about the dangers of asbestos (qv) causing mesothelioma since medical knowledge was not available until the 1960s that relatively low level exposure to asbestos insulation products could cause the deadly disease. The court took the position that even though it was technologically impossible for the defendants to have been aware of the problem in the 1930s they would be held liable nonetheless if, at the time of trial, it was clear that such a warning should have been given.

Other courts have taken the opposite position. If the plaintiff's claim is based on defective design or failure to warn, the test to be applied is that of negligence. If the defendant acted reasonably at the time it designed and marketed the product there is no liability because technology has advanced or risks have been discovered at a later date.

There is no magic formula which will assure a designer that his or her judgment in favor of a given design is certain to avoid legal liability. However, it is clear that courts are serious about balancing the risks created by a product against the benefits of the products design. The more carefully these considerations are consciously weighed by the designer or engineer, the greater the likelihood that the design will meet the approval of the courts.

Torts Concerning Toxic Substances

A fertile area for lawsuits has developed in the field of harms resulting from toxic substances. The actions that injured claimants can bring are many and varied. When courts have been faced with pollutants from industry they have had to give consideration to the benefits that industry bestows upon society as well as the harm caused by the pollution. The range of judicial opinions on this subject varies greatly. Some have imposed harsh sanctions on industry and have enjoined continuation of the pollution. Others have refused to cripple industry unless the danger caused by the pollution is very serious.

Product liability law has dealt with products that have toxic qualities. Two cases have received great notoriety. The asbestos litigation spawned thousands of claims arising from the toxic qualities of asbestos which cause asbestosis and mesothelioma (see Asbestos). The long latency period masked the symptoms of these diseases for many years. When the claims did materialize, they came in such numbers that they threatened the financial solvency of the main asbestos producer, Johns-Manville

Company, which has sought the approval of the bankruptcy court for Chapter 11 reorganization. The toxic-product case that has become headline news is the Agent Orange litigation (see Herbicides). In this case, a class action brought by veterans who served in Vietnam, the contention is that the chemical companies that sold the Agent Orange (dioxin) defoliant to the U.S. Government permitted contamination of the product with traces of the deadly chemical dioxin. The plaintiffs allege that the U.S. Government was not made fully aware of the degree of contamination, nor of the serious health hazards that could result from exposure to even minute traces of dioxin. The defendant chemical companies refute these contentions and also argue that the level of dioxin contamination was not sufficient to cause health hazards to humans (see Toxicology, Supplement Volume).

The Balancing Process—Some Practical Applications

To illustrate how the standard of "reasonableness" is determined after an evaluation of the trade-offs of product risk versus the utility of the product, consider the case of McCormack v. Hankscraft Company (9).

McCormack v. Hankscraft Co. A major manufacturer of steam vaporizers placed an inexpensive model on the market. One of its features was a lift-off cap to which was attached the heating element that produced the steam. The container with the lift-off cap was placed on a kitchen stool in a child's bedroom. In the middle of the night the child awoke and walked through the darkened room, tripping over the electrical cord attached to the vaporizer, causing the vaporizer to tip over on its side. The scalding water flowing from the overturned vaporizer severely burned the child. An action brought against the manufacturer on behalf of the child claimed that the design of the vaporizer was defective.

Using the balancing process for reasonableness, it is possible to identify the factors necessary for determining whether or not the design of the vaporizer was unreasonably dangerous. The utility of a vaporizer is generally unquestioned. It is entirely foreseeable, however, that steam vaporizers will be used in close proximity to children, and often in darkened bedrooms. Placing the vaporizer on a stool or chair so that the steam is directed toward the child in bed is also a common and expected practice. Thus there is a good chance that a vaporizer will be inadvertently tipped by a child or an adult. Given this chance, one must consider the implications of heated water flowing out of the tipped container.

Since the water in the container was hot enough to cause third-degree burns, the question arises whether it is possible to eliminate the danger without seriously impairing the usefulness of the product or making it unduly expensive. It seems reasonable to examine possible alternatives to the design of this inexpensive vaporizer. The plaintiff, through the use of expert testimony, established that a screw-on cap, in place of the lift-off cap, was both feasible and relatively inexpensive. The defendant countered that the screw-on cap raised other design problems, since there would be a possible buildup of pressure within the vaporizer in addition to the increased cost of adding this feature. The plaintiff's experts countered with testimony that a small vent hole in the cap would serve to relieve the pressure buildup. While small dribbles of water might flow from the vent hole of a tipped vaporizer with a screw-on cap, there would be a significant lessening of the danger of extensive burns arising from such small amounts of water. In other words the risk of injury, although not completely eliminated, would be substantially reduced. In considering the acceptability of the manufacturer's design, the court also took into account the fact that there was no real way for the user, in this case the mother, to know with clarity that the water in the container was scalding hot, and she thus could not realistically assess the danger that could arise from a tipped vaporizer.

The danger of the vaporizer was assessed in the context of its use environment. Unlike negligence, where the emphasis is on the conduct of the manufacturer, strict liability assesses the issue of unreasonable product danger from the vantage point of the environment of its use by a consumer. It is thus clear that the conclusion of whether a given design is, or is not, unreasonably dangerous is an expression, through the litigation process, of society's willingness to accept or reject that design. When the cost of an alternative to make the product safer outweighs the probable danger, then the product will be declared "reasonably safe" as is. When the risk of the product exceeds its inherent utility, the product will be declared "unreasonably dangerous."

Uloth v. City Tank Corp. A recent Massachusetts case demonstrates the complexity of the balancing process. In *Uloth v. City Tank Corp.* (10), the plaintiff was a town-employed general laborer. He had helped with trash collection on several occasions and twice had been assigned to a Loadmaster 316 garbage truck. The design of the Loadmaster included a rear step running the full width of the truck on which the workers rode. Above the step was a "trash hopper" area into which trash was loaded. A packer blade swept through this trash hopper area during the compaction cycle, coming in contact with the loading sill, and pushed the trash into the storage area of the truck.

The compaction cycle on the Loadmaster 316 was activated by placing the truck in neutral, operating several switches in the cab, and pulling a lever on the side of the truck, after which the packer blade would descend. The compaction cycle could be interrupted at any time by disengaging the lever at the side of the truck.

On the day of the accident, the plaintiff operated the lever once without incident. Ten minutes after starting work, Uloth signaled the driver to put the truck in neutral and activate the switches in the cab, after which the plaintiff operated the lever on the side of the truck and lit a cigarette. When he heard the engine noise increase and saw the truck move slightly, the plaintiff assumed that the truck was about to move ahead, and leapt onto the rear step of the truck. He lost his balance; the descending panel caught his left foot, dragged it into the trash hopper, and severed it from his leg.

Uloth is of particular interest because the packer was purchased by a municipality which had the option to purchase competing models equipped with safety features that might well have prevented the accident. The plaintiff contended that the Loadmaster 316 was defectively designed because of its continuous cycle-compaction mechanism, and pointed to other companies using safety devices such as interlocking hopper doors, a stop bar, an interrupted cycle or a "deadman control." In addition, the plaintiff argued that had the steps on the garbage truck been placed on the side rather than in the rear the accident would not have occurred.

The issue of design defect was sharply disputed. The defendant argued that the competitor's compaction chambers equipped with the safety devices had a much faster compaction cycle, whereas the Loadmaster's slower cycle permitted the user to become aware of the slowly descending panel and avoid danger. Furthermore, the defendant argued that the more sophisticated safety devices were unusually dangerous in that sanitation workers seeking to find shortcuts around time-consuming safety features might bypass them so that they could complete their work more quickly. Thus, according to the defendant, in this instance the simpler design that was not subject to tampering was safer than the more complex design which offered a surface appearance of greater safety. The court acknowledged that there was conflicting evidence but held that the controverted issue was for the jury to resolve. A jury verdict for the plaintiff was upheld since there was a strong argument that a safer design could have been instituted.

Wilson v. Piper Aircraft Corp. An excellent example that the courts will demand that a plaintiff introduce strong evidence on the risk–utility trade-offs is found in *Wilson v. Piper Aircraft Corp.* (11). In *Wilson*, the plaintiffs alleged that the airplane crash in which their decedents died was caused by engine failure resulting from carburetor icing and that the carburetor's susceptibility to icing was inherent in the basic design of the engine.

The court's reasoning in reversing the jury verdict for the plaintiffs is instructive. The court ruled that there was ample evidence to support each of the plaintiffs' allegations concerning the design of the aircraft, the likelihood of this design contributing to carburetor icing, and the causal connection between such icing and the crash of the aircraft; the evidence was nonetheless insufficient to support a jury finding that the airplane was "dangerously defective" in design. The court based its decision primarily upon the impracticality of requiring all aircraft to be manufactured with fuel-injection engines in order to avoid the dangers of icing. The court noted that 80–90% of comparable small aircraft were manufactured with carburetor-rather than fuel-injection engines and that the plaintiffs had provided no evidence as to the effect of the proposed design alternative on such matters as "cost, economy of operation, maintenance requirements, over-all performance, and safety" of such planes.

Wilson cannot be dismissed simply as a case in which the court believed the quantum of evidence to be insufficient on the issue of defective design. The plaintiffs had clearly introduced evidence that the alternative design they were proposing was feasible. The fuel-injection engine was not an untested hypothetical design; it had been in actual use and had apparently performed well. There is no question that it did not present the danger of carburetor icing that the carburetor engine exhibited. Why then did the court find the evidence insufficient to go to the jury on the issue of design defect?

The court held that the case should have been withheld from the jury because there was insufficient evidence on the cost effectiveness and overall performance of the fuel-injected engine. However, in offering an alternative design to a court, it is virtually impossible to present the kind of evidence that the court was professing to seek in *Wilson*. In most design-defect cases, the plaintiff's proposed alternative will be a new and relatively untested design which does not have a track record on such matters as cost, overall performance, and maintenance.

The *Wilson* court did, however, tip its hand, indicating that the carbureted engine was extremely popular. It was used widely and considered safe by FAA standards in spite of the icing problem. Such a widely accepted design could not be held defective on evidence showing only that an alternative design could eliminate a fairly remote danger without some demonstration that the costs to society of declaring the design defective were negligible.

In short, *Wilson* demonstrates that courts can take their policy-making role very seriously. A court that perceives it is making an important policy decision by letting a design defect case go to a jury will transcend the simple counting or balancing of the elements of a risk–utility analysis in order to consider the overall social and economic impact of a possible jury finding of defect.

Causation

Even if the plaintiff establishes that the product was defective, a causal relationship between the defect and the malfunction of the product must be established. This issue is often litigated in automobile accident cases. Typically, the car crashes and upon examination of the wreckage a defective part is discovered. Thus, for example, after an accident the motor mounts may be found to have been made from seriously flawed metal. The question is whether the motor mounts failed, thus causing the accident, or if they cracked upon impact and the accident resulted from some other cause. The battle of the experts becomes of great significance on this issue. The physical evidence may legitimately be subject to differing opinions. The judge and jury must ultimately decide this crucial issue.

The role of causation and the ways in which it should impact the products litigation process are illustrated in Figure 1. It is evident that the causation issue can short-circuit a lawsuit at the very outset. It is thus an issue that deserves the attention of counsel and experts at the earliest stages of litigation. The causation question has become of considerable importance in the "crashworthiness" cases (Fig. 1). In these cases brought against the manufacturers of automobiles, the contention is not that the design of the car caused the accident but rather that the car was not adequately designed to withstand a collision. Since the accident was not caused by the automobile but rather by driver error of some kind, the manufacturer should not be held liable for the damages that would have occurred even if the automobile had been well de-

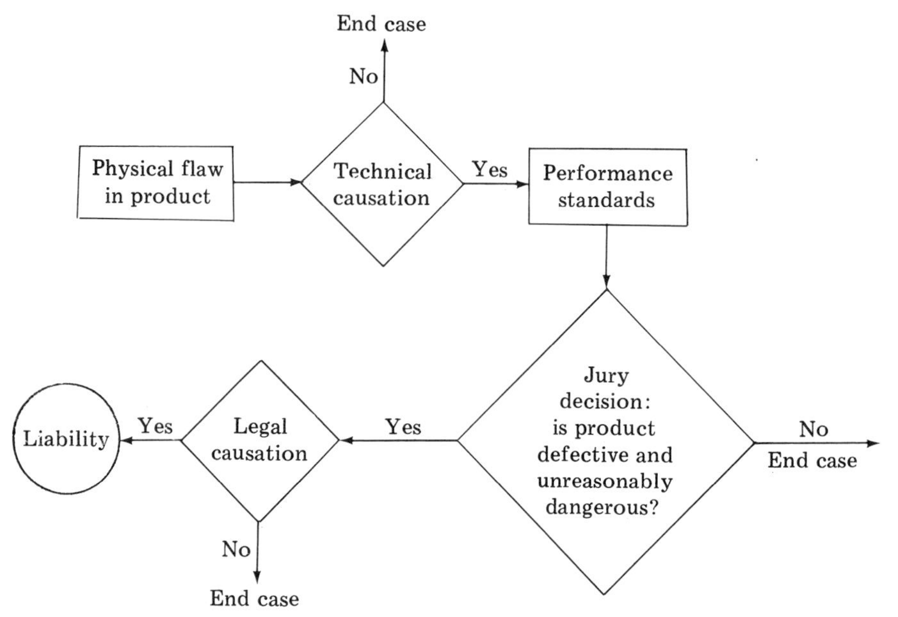

Figure 1. Flow diagram highlighting technical and legal issues in products liability litigation.

signed. Many courts have required that the plaintiff establish by expert testimony the degree of injury aggravation caused by the fact that the automobile was not sufficiently crashworthy. Other courts have recognized that to place such a burden on the plaintiff may make it impossible for recovery and have switched the burden of proof onto the defendant manufacturer to establish that the "second collision" injuries were less than the totality of the plaintiff's entire injuries. Either way, experts are called upon to hypothecate as to the injuries which could have been avoided had the proper design been utilized.

Finally, there are cases of defective products which cause injuries that do not bear the onus of liability. Products can be significantly altered and misused and thereby cause injury. The original defect of manufacture or design may have contributed to the injury and yet the courts may take the position that the defect was not the "proximate cause" of the harm. Thus, a mechanically activated punch press, poorly designed because it did not have a safety guard that would prevent the worker's hands from invading the machine at the point of operation, could be made far more dangerous by changing the activation to an electrically operated push pedal. The machine was dangerous as originally designed, but made many times more dangerous by the alteration. Here too, what is considered "proximate cause" is a matter of differing interpretation by courts. But as one court has noted:

> "It is difficult to know where to draw the line on a manufacturer's liability in the sale of products. There must be a limit to such liability somewhere. We think this case lies outside the area where liability exists. The manufacturer here has done all anyone could reasonably expect it to do in making the chattel safe for the use for which it was intended. The responsibility for maintaining it in the safe condition in which it was sold rested with the user."

The Role of the User

In most personal-injury actions, if the plaintiff failed to act reasonably, the recovery will be reduced under the doctrine of comparative fault. A jury will be asked to assess whether the plaintiff's conduct contributed to the injury and to apportion fault on a percentage basis between the parties. In a minority of jurisdictions, the contributory fault of the plaintiff will prevent the recovery of any damages at all from the defendant.

In product-liability actions the courts have yet to definitively establish how to deal with the plaintiff whose fault has contributed to his or her own harm. The clear trend, however, is in the direction of utilizing the comparative-fault principle and reducing the recovery of the claimant in proportion to the fault which the jury attributes to his conduct. Several courts have indicated that when the fault of the manufacturer has been in failing to design a safety feature onto a piece of industrial machinery they will not reduce the plaintiff's recovery by the percentage of his fault. Their reasoning is that the safety feature should have been available to have prevented the very kind of conduct which occurred. To reduce the plaintiff's recovery for an injury which clearly should have been rendered impossible by an adequate design would be unjust. These cases reflect a concern for the inability of the worker to control the environment of the workplace. These courts believe that the most efficient place to control the risk is at the point of machine design. They thus refuse to apply the comparative-fault principle to these cases.

Product Liability Reform

The explosion of liability in the field of product liability has fostered an atmosphere of crisis among manufacturers. Many contend that insurance costs are so high as to price them out of the market. Others argue that they are so paralyzed with fear of product liability that it has affected their willingness to be creative or innovative in design.

Many states have passed legislation seeking to protect manufacturers. Some states have passed "statutes of repose" that prevent plaintiffs from bringing product liability actions after a given number of years have passed following the original sale. These statutes, which range from six to twelve years, thus place an outside cap after which there can no longer be product liability exposure. Other states have sought to protect manufacturers as long as they have complied with the "state of the art" at the time of manufacture or sale. Still others have ordained that if the injury was occasioned by product alteration and misuse that no liability will attach.

The differing statutes have given rise to a movement calling for a Federal product-liability statute. Several statutes have been proposed and have been given serious consideration by the Congress. It remains questionable whether a sufficient consensus exists as to the problems which should be covered by such legislation to make significant product-liability reform at the Federal level a real possibility.

BIBLIOGRAPHY

1. A. W. Weinstein, A. T. Twerski, W. D. Donaher, and H. P. Piehler, *Products Liability and the Reasonably Safe Product*, John Wiley & Sons, Inc., 1978. A significant portion of the material presented herein is taken from this reference. This book is intended as a guide to management, design, and marketing personnel. It should be referred to for a full discussion of materials which are treated in this presentation in a summary fashion.
2. Terry, "Negligence," *Harvard Law Review* **29,** 40, 1915.
3. Restatement (Second) of Torts, §291, 1965.
4. *Marsh Wood Products Company v. Babcock & Wilcox*, 207 Wisconsin 209, 240 Northwestern 392 (1932).
5. *Berkebile v. Brantly Helicopter Corp.*, 219 Pennsylvania Superior 479, 281 Atlantic 2d 707 (1971).
6. *Metal Window Products Co. v. Magnusen*, 485 Southwestern 2d 355 (1972).
7. *Crocker v. Winthrop Laboratories Division of Sterling Drug, Inc.*, 514 Southwestern 2d 429 (1974).
8. *Beshada v. Johns-Manville Products Co.*, 90 N.J. 191, 447 Atlantic 2d 539 (1932).
9. *McCormack v. Hankscraft Co.*, 278 Minn. 322, 154 Northwestern 2d 488 (1967).
10. 376 Mass. 874, 384 Northeastern 2d 1188 (1978).
11. 282 Oregon 61, 577 Pacific 2d 1322 (1978).

AARON TWERSKI
Hofstra Law School

PROSTAGLANDINS

Prostaglandins are members of a class of organic compounds that are biosynthetically derived from polyunsaturated fatty acids and that possess as a common structural feature the prostan-1-oic acid (**1**) skeleton. Arachidonic acid (**2**) is the predominant fatty acid precursor of the prostaglandins in humans, and it can be obtained directly from the diet or by anabolic conversion of the dietary essential fatty acid, linoleic acid.

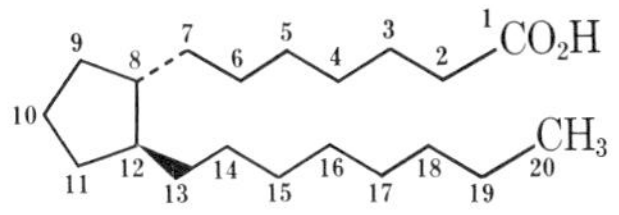

(**1**) 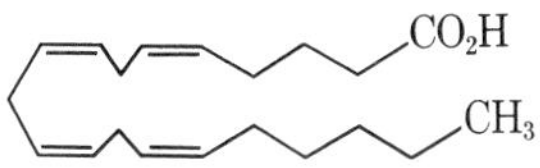 (**2**)

The prostaglandins, including prostacyclin (prostaglandin I_2), exhibit a wide range of biological activities and are biosynthesized by most mammalian tissues. Biosynthetically related substances include the thromboxanes, which have as a common structural feature the thromboxan-1-oic acid (**3**) skeleton, and the more recently discovered leukotrienes (see Table 1).

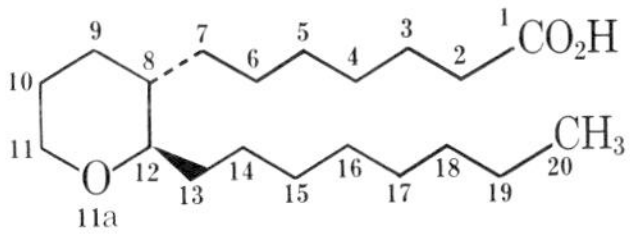

(**3**)

The scientific investigation of the prostaglandins, thromboxanes, and leukotrienes (collectively referred to as eicosanoids) has resulted in important advances in the fields of synthetic organic chemistry, biochemistry, pharmacology, and physiology (see Contraceptive drugs). Notable among contemporary prostaglandin researchers are Nobel prizewinners S. Bergstrom, B. Samuelsson, and J. R. Vane (6). Clinical applications of these substances are expected to increase. The name prostaglandin was given in 1935 to a factor in human seminal fluid which causes contraction of a variety of smooth muscles and lowers the blood pressure of laboratory animals when injected into their systemic circulation (7). In 1960, the isolation from sheep prostate glands of two crystalline prostaglandins, ie, PGE_1 (**4**) and $PGF_{1\alpha}$ (**5**), was reported (8). Subsequent gas chromatographic–mass spectrometric analysis and chemical manipulation led to the structural elucidation of PGE_1, $PGF_{1\alpha}$, and $PGF_{2\alpha}$ in 1962 (9). By 1964, the

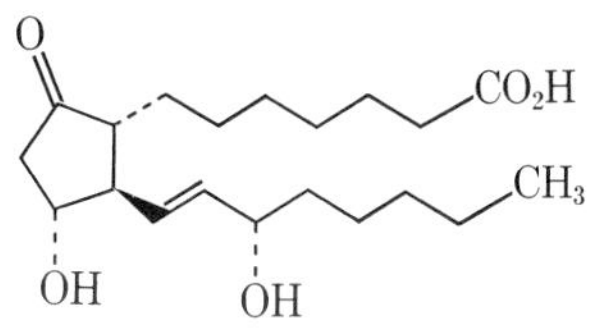

(**4**)
Prostaglandin E_1 (PGE_1) [*745-65-3*]

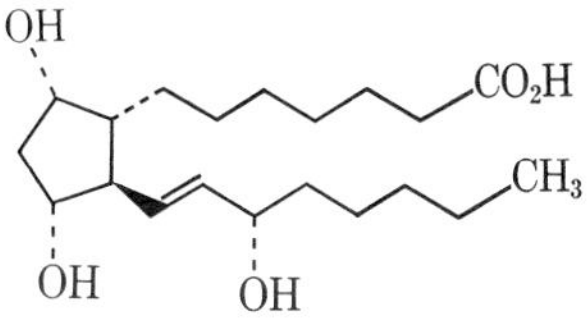

(**5**)
Prostaglandin $F_{1\alpha}$ ($PGF_{1\alpha}$) [*745-62-0*]

Table 1. Nomenclature for Eicosanoids Derived from Arachidonic Acid

Semisystematic name[a]	CAS name[b]	CAS Registry No.	Structure
arachidonic acid	(*Z*,*Z*,*Z*,*Z*)-5,8,11,14-eicosatetraenoic acid	[*506-32-1*]	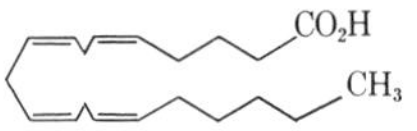
prostaglandin A_2 (PGA_2)	(5*Z*,13*E*,15*S*)-15-hydroxy-9-oxoprosta-5,10,13-trien-1-oic acid	[*13345-50-1*]	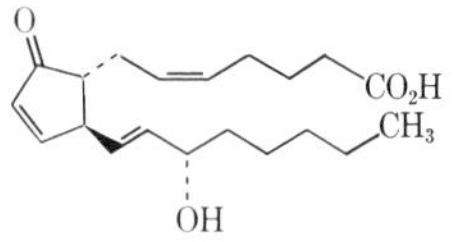
prostaglandin B_2 (PGB_2)	(5*Z*,13*E*,15*S*)-15-hydroxy-9-oxoprosta-5,8(12),13-trien-1-oic acid	[*13367-85-6*]	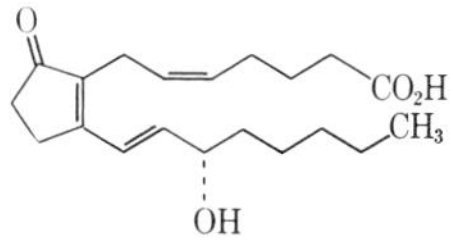
prostaglandin C_2 (PGC_2)	(5*Z*,13*E*,15*S*)-15-hydroxy-9-oxoprosta-5,11,13-trien-1-oic acid	[*49825-91-4*]	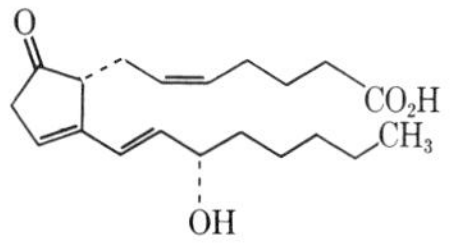
prostaglandin D_2 (PGD_2)	(5*Z*,9α,13*E*,15*S*)-9,15-dihydroxy-11-oxoprosta-5,13-dien-1-oic acid	[*41598-07-6*]	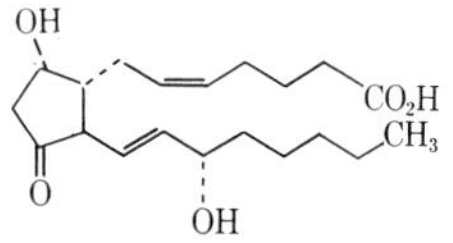
prostaglandin E_2 (PGE_2)	(5*Z*,11α,13*E*,15*S*)-11,15-dihydroxy-9-oxoprosta-5,13-dien-1-oic acid	[*363-24-6*]	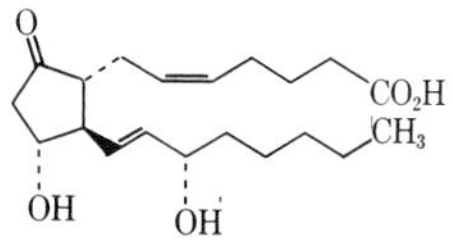
prostaglandin $F_{2\alpha}$ ($PGF_{2\alpha}$)	(5*Z*,9α,11α,13*E*,15*S*)-9,11,15-trihydroxyprosta-5,13-dien-1-oic acid	[*551-11-1*]	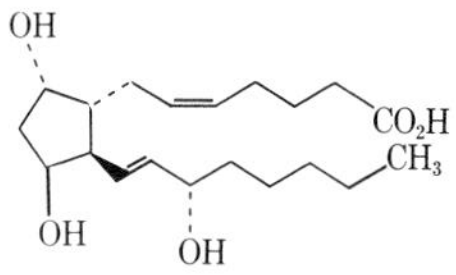
prostaglandin G_2 (PGG_2)	(5*Z*,9α,11α,13*E*,15*S*)-9,11-epidioxy-15-hydroperoxyprosta-5,13-dien-1-oic acid	[*51982-36-6*]	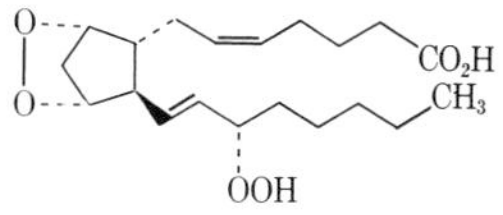
prostaglandin H_2 (PGH_2)	(5*Z*,9α,11α,13*E*,15*S*)-9,11-epidioxy-15-hydroxyprosta-5,13-dien-1-oic acid	[*42935-17-1*]	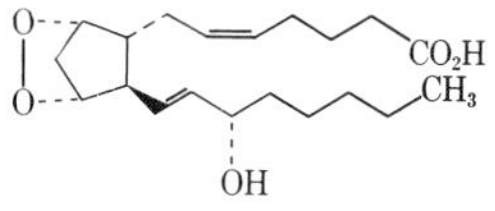
prostaglandin I_2 (PGI_2); prostacyclin	(5*Z*,9α,11α,13*E*,15*S*)-6,9-epoxy-11,15-dihydroxyprosta-5,13-dien-1-oic acid	[*35121-78-9*]	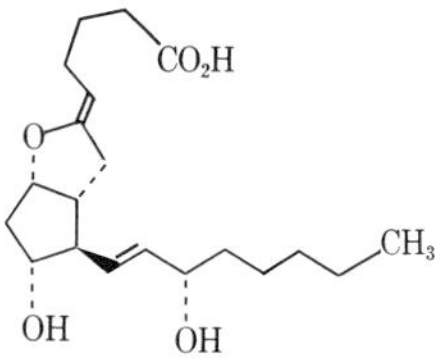
6-oxoprostaglandin $F_{1\alpha}$ (6-oxo-$PGF_{1\alpha}$); 6-ketoprostaglandin $F_{1\alpha}$ (6-keto-$PGF_{1\alpha}$)	(9α,11α,13*E*,15*S*)-6-oxo-9,11,15-trihydroxyprost-13-en-1-oic acid	[*58962-34-8*]	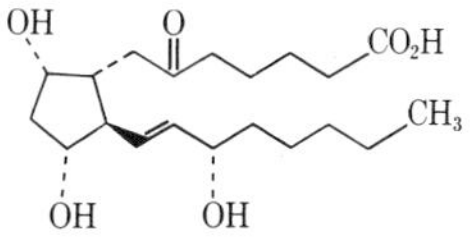

Table 1.

Semisystematic name[a]	CAS name[b]	CAS Registry No.	Structure
thromboxane A_2 (TXA_2)	($5Z,9\alpha,11\alpha,13E,15S$)-9,11-epoxy-15-hydroxythromboxa-5,13-dien-1-oic acid	*[57576-52-0]*	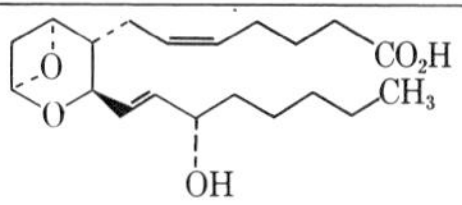
thromboxane B_2 (TXB_2)	($5Z,9\alpha,13E,15S$)-9,11,15-trihydroxy-thromboxa-5,13-dien-1-oic acid	*[54397-85-2]*	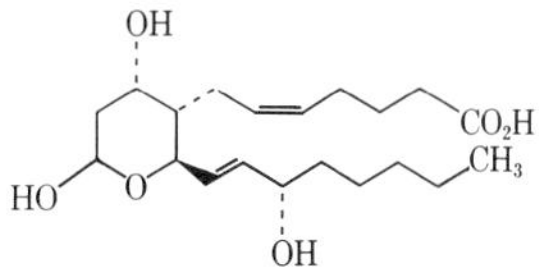
leukotriene A_4 (LTA_4)	[$2S$-[$2\alpha,3\beta(1E,3E,5Z,8Z)$]]-3-(1,3,5,8-tetradecatetraenyl)-oxiranebutanoic acid	*[72059-45-1]*	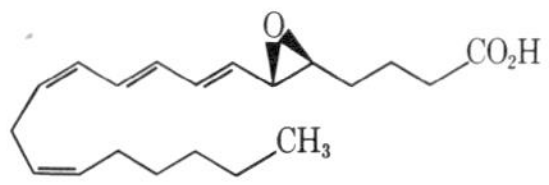
leukotriene B_4 (LTB_4)	[S-[R^*,S^*-(Z,E,E,Z)]]-5,12-dihydroxy-6,8,10,14-eicosatetraenoic acid	*[71160-24-2]*	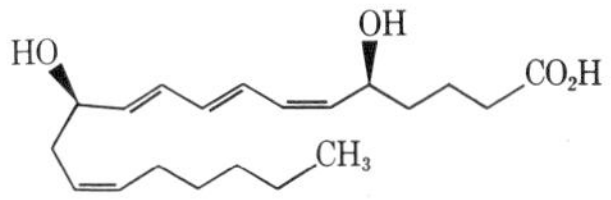
leukotriene C_4 (LTC_4)	[R-[R^*,S^*-(E,E,Z,Z)]]-N-[S-[1-(4-carboxyl-1-hydroxybutyl)-2,4,6,9-pentadecatetraenyl]-N-L-γ-glutamyl-L-cysteinyl]-glycine	*[72025-60-6]*	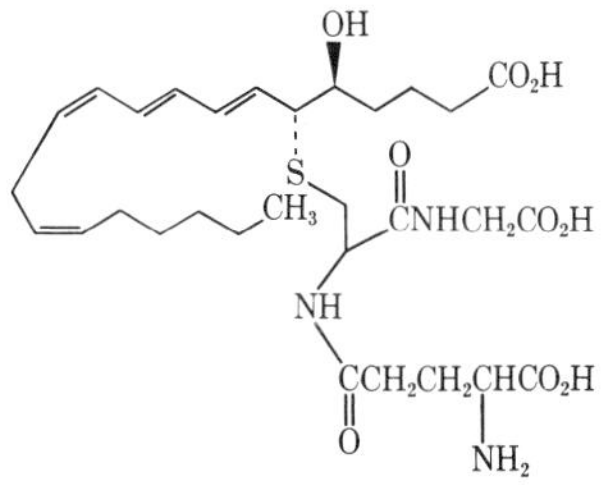
leukotriene D_4 (LTD_4)	[R-[R^*,S^*-(E,E,Z,Z)]]-N-[S-[1-(4-carboxy-1-hydroxybutyl)-2,4,6,9-pentadecatetraenyl]-L-cysteinyl]glycine	*[73836-78-9]*	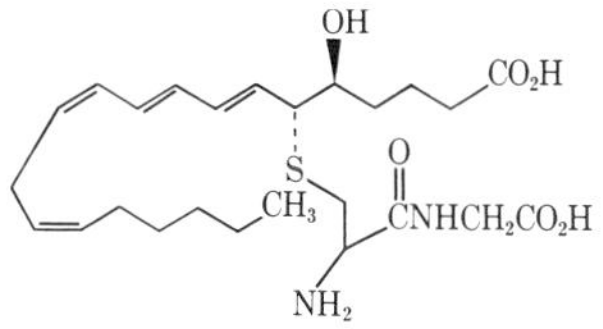
leukotriene E_4 (LTE_4)	[$5S$-[$5R^*,6S^*(S^*),7E,9E,11Z,14Z$]]-6-[(2-amino-2-carboxyethyl)thio]-5-hydroxy-7,9,11,14-eicosatetraenoic acid	*[75715-89-8]*	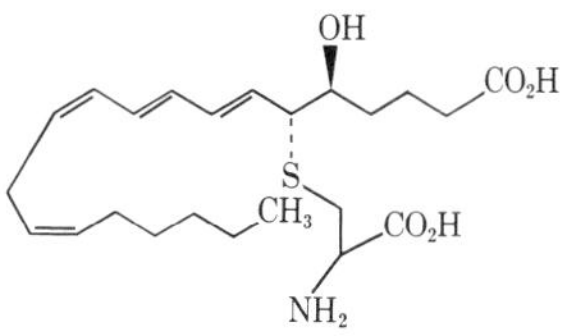

[a] Refs. 1–4.
[b] Ref. 5.

structures of the three remaining primary prostaglandins, ie, PGE_2, PGE_3 (**6**), and $PGF_{3\alpha}$ (**7**), were known (10–13).

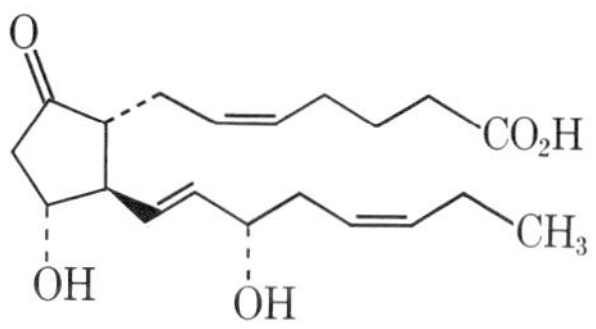

(**6**)

PGE_3 *[802-31-3]*

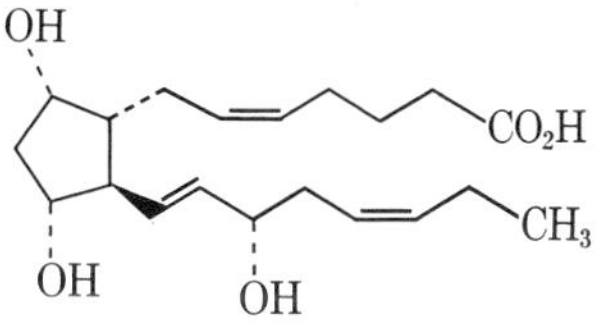

(**7**)

$PGF_{3\alpha}$ *[745-64-2]*

The stereochemical subtleties of the primary prostaglandins were determined by x-ray crystallographic analyses of the bromo- and iodobenzoates of $PGF_{1\beta}$ *[10164-73-5]* and the correct absolute stereochemical configuration was determined by oxidative ozonolysis experiments (14–16).

Nomenclature

Chemical Abstracts currently uses prostane (**8**) and thromboxane (**9**) as stereoparents or index-heading parents.

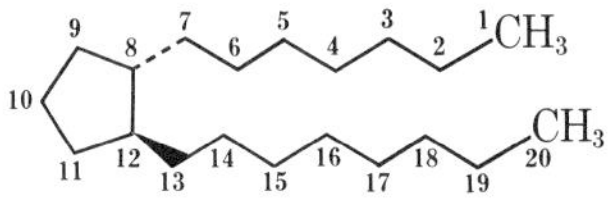

(**8**)

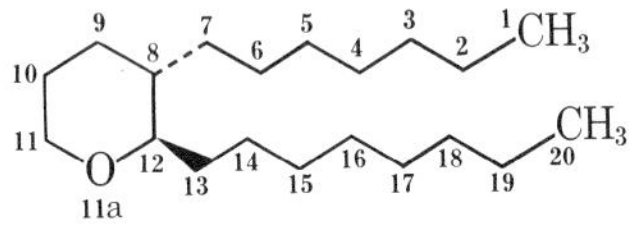

(**9**)

Implicit in these names are the absolute configurations at carbons 8 and 12 and the numbering systems illustrated in structures (**1**), (**3**), (**8**), and (**9**). Derivatives of these parent structures are named based on the rules of terpene and steroid nomenclature (5). *Chemical Abstracts* names the leukotrienes derived from arachidonic acid either as derivatives of eicosatetraenoic acid, eg, leukotrienes B_4 and E_4; as derivatives of glycine, eg, leukotrienes C_4 and D_4; or as derivatives of oxiranebutanoic acid, eg, leukotriene A_4. The cumbersome nature of this systematic nomenclature for the prostaglandins, thromboxanes, and leukotrienes has encouraged the development of semisystematic nomenclatures for each of these subclasses of compounds, and these nomenclatures are used throughout the remainder of this article.

Since 1964, research on the prostaglandins has increased enormously and has proceeded along many different directions. Advances in the total chemical synthesis of the naturally occurring prostaglandins (including prostacyclin or prostaglandin I_2), thromboxanes, and leukotrienes, as well as analogues of these substances; the biosynthesis and metabolism of these eicosanoids; and the development of sensitive analytical instrumentation and procedures for identifying and assaying these eicosanoids have fueled the evolution and growth of the arachidonic acid cascade. Key scientific events during this period are reported in refs. 17–58. Several accounts of the early history of the prostaglandins have also been published (59–60).

Arachidonic Acid Cascade

Biosynthesis. Detailed accounts of the biosyntheses of the prostaglandins, thromboxanes, leukotrienes, and associated hydroxylated fatty acids are reviewed in refs. 41 and 61–67. Although it is possible to biosynthesize prostaglandins of the *one* series, eg, PGE_1, and the *three* series, eg, PGE_3, from (Z,Z,Z)-8,11,14-eicosatrien-1-oic acid and (Z,Z,Z,Z,Z)-5,8,11,14,17-eicosapentaen-1-oic acid, respectively, these unsaturated fatty acids are not present in significant quantities *in vivo* (68–71). Arachidonic acid is the most abundant 20-carbon precursor fatty acid *in vivo*, which accounts for the predominance of the two-series prostaglandins, eg, PGE_2. The many avenues of arachidonic acid metabolism are illustrated in Figure 1, which is known as the arachidonic acid cascade.

The naturally occurring two-series prostaglandins, thromboxanes, and leukotrienes are not stored as such in cells but are biosynthesized on demand from arachidonic acid released from cell-membrane phospholipids (72–73). The precursor arachidonic acid is generally esterified at the two position of phospholipids, eg, phosphatidylcholine (lecithin) (**10**), phosphatidylinositol, phosphatidylethanolamine, and phosphatidylserine.

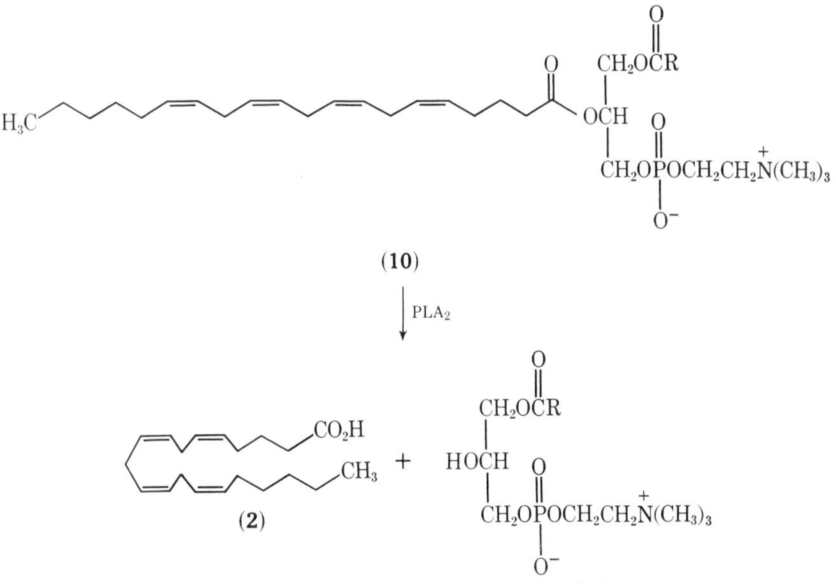

H_3C CH_2OCR OCH $CH_2OPOCH_2CH_2\overset{+}{N}(CH_3)_3$ O^-

(**10**)

PLA_2

CO_2H CH_3 + CH_2OCR HOCH $CH_2OPOCH_2CH_2\overset{+}{N}(CH_3)_3$ O^-

(**2**) (**11**)

It is released by the action of a phospholipase A_2 (PLA_2), which is specific for the two position of the phospholipid molecule (74). In the case of (**10**), lysophosphatidylcholine (lysolecithin) (**11**) is the by-product of this enzymatic hydrolysis. Triglycerides have also been reported to be sources of precursor arachidonic acid (75).

Arachidonic acid is converted into endoperoxides PGG_2 and PGH_2 by the enzyme complex prostaglandin cyclooxygenase [*39391-18-9*] (61). The proposed mechanism of this transformation involves the selective abstraction of the pro-*S* hydrogen at *C*-13 of arachidonic acid (**2**) and oxygen trapping to yield a hypothetical intermediate (**12**).

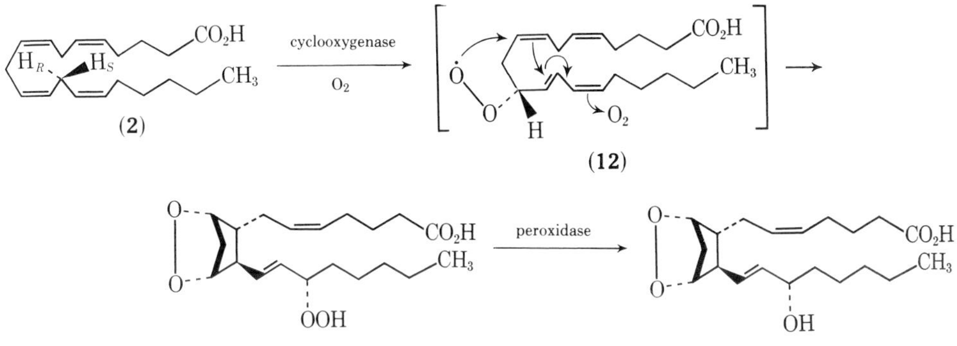

H_R H_S CO_2H CH_3 cyclooxygenase O_2 CO_2H CH_3 O O H O_2

(**2**) (**12**)

CO_2H CH_3 OOH peroxidase CO_2H CH_3 OH

(**13**) PGG_2 (**14**) PGH_2

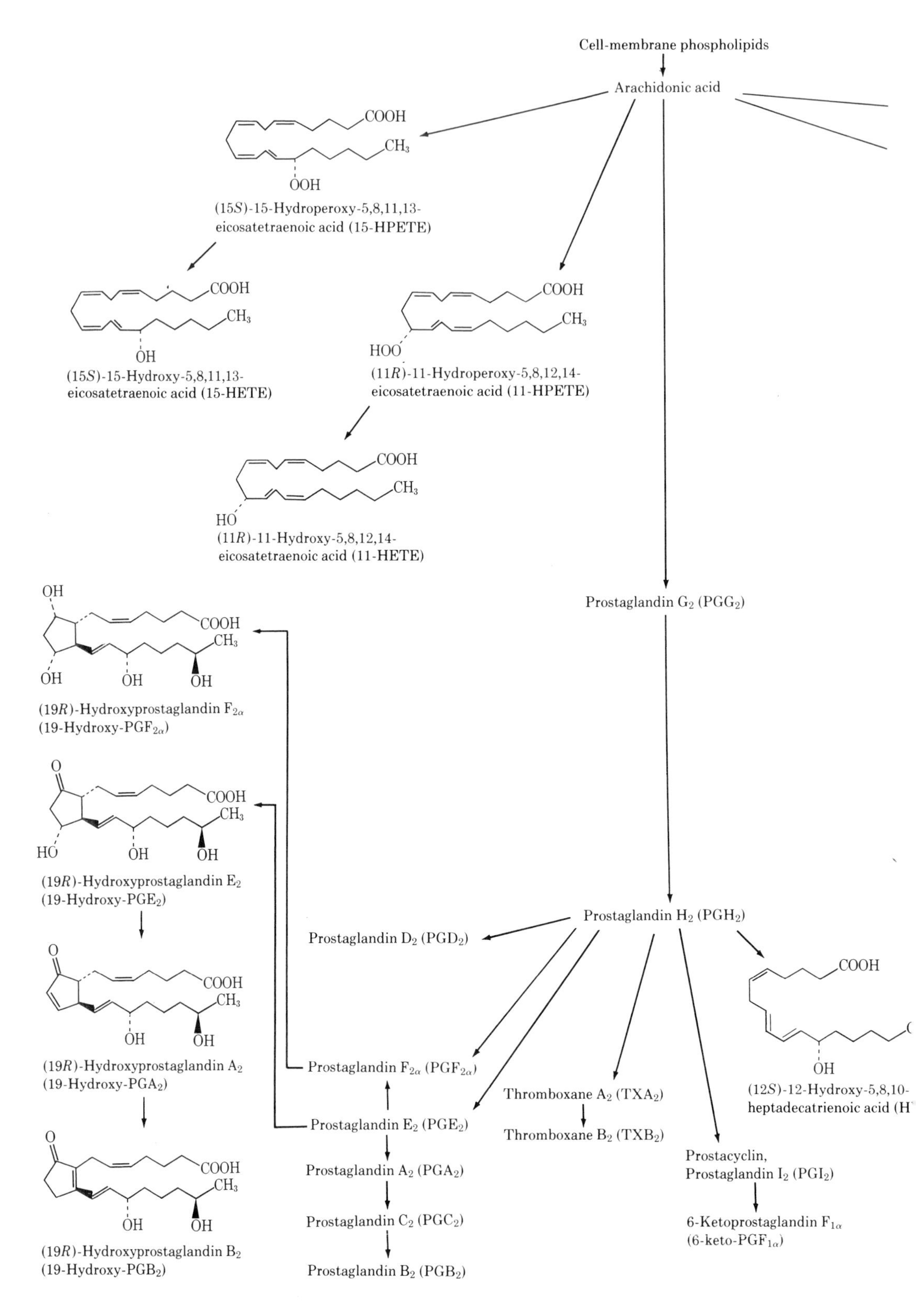

Figure 1. Arachidonic acid cascade.

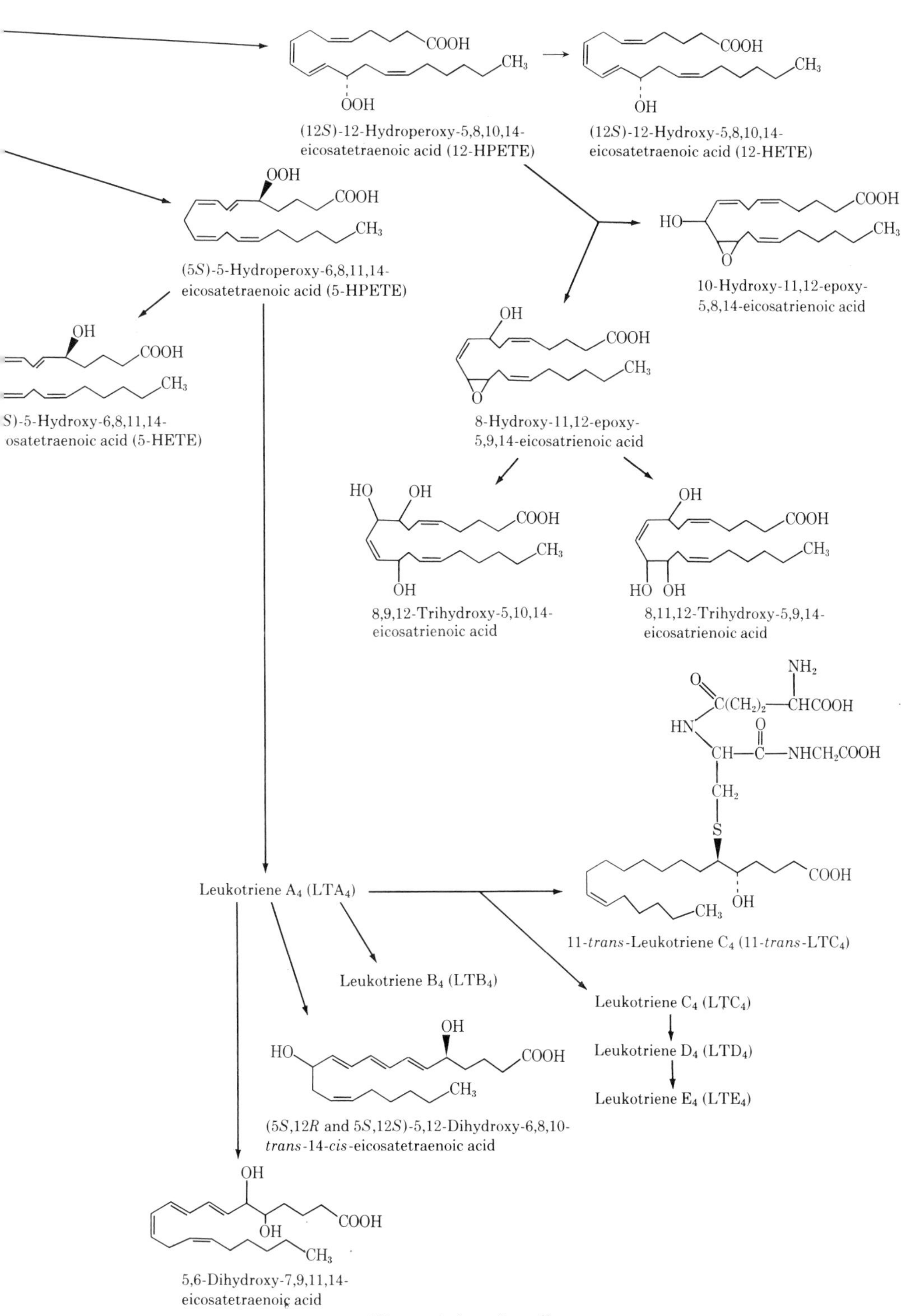

Figure 1. (*continued*)

Cyclization with the incorporation of a second molecule of oxygen produces PGG_2. The cyclooxygenase is associated with a peroxidase which further transforms PGG_2 into PGH_2 (61). The endoperoxides PGG_2 and PGH_2 are extremely important intermediates in the metabolism of arachidonic acid. In addition to having pronounced biological activity, eg, aggregation of blood platelets and constriction of vascular tissue, they serve as substrates for several other enzymes in the cascade, including those which lead to the classical prostaglandins, eg, PGE_2, $PGF_{2\alpha}$, and PGD_2, and those which lead to thromboxane A_2 and to prostaglandin I_2 (31). Collectively, these enzymatic processes, which begin by the conversion of arachidonic acid to PGG_2 or PGH_2, or both, are known as the cyclooxygenase branch of the arachidonic acid cascade (41,61–67).

A second main branch of the arachidonic acid cascade is that comprised of the lipoxygenase pathways. Arachidonic acid is a substrate for a number of lipoxygenases including an 11-lipoxygenase which produces (5*Z*,8*Z*,11*R*,12*E*,14*Z*)-11-hydroperoxy-5,8,12,14-eicosatetraen-1-oic acid (11-HPETE), a 15-lipoxygenase which produces (5*Z*,8*Z*,11*Z*,13*E*,15*S*)-15-hydroperoxy-5,8,11,13-eicosatetraen-1-oic acid (15-HPETE), a 12-lipoxygenase which produces (5*Z*,8*Z*,10*E*,12*S*,14*Z*)-12-hydroperoxy-5,8,10,14-eicosatetraen-1-oic acid (12-HPETE), and a 5-lipoxygenase which produces (5*S*,6*E*,8*Z*,11*Z*,14*Z*)-5-hydroperoxy-6,8,11,14-eicosatetraen-1-oic acid (5-HPETE) (34,76–79). Each of these hydroperoxides is further converted to its corresponding hydroxyl derivative by peroxidases.

The hyperoxide 11-HPETE produced by the 11-lipoxygenase is very similar to the hypothetical intermediate (**12**) in the conversion of arachidonic acid to PGG_2 (**13**). It might serve as an alternative substrate for the cyclooxygenase; however, this has not been proven (79).

The 5-lipoxygenase pathway of the arachidonic acid cascade has recently been the focus of intensive research (61,80). The initial biosynthetic product in this pathway, 5-HPETE, is enzymatically transformed by several cell types to leukotriene A_4 and, from this labile epoxide, into leukotrienes B_4, C_4, D_4, and E_4. Leukotrienes C_4, D_4, and E_4 are main constituents of the slow-reacting substance of anaphylaxis (SRS-A) and potent constrictors of human bronchi, whereas leukotriene B_4 is a potent chemotactic agent for leukocytes and causes leukocyte adhesion to vascular endothelium in post-capillary venules.

There is increasing emphasis on the modulation of the arachidonic acid cascade. Nonsteroidal anti-inflammatory drugs, eg, aspirin and indomethacin, inhibit the cyclooxygenase which transforms arachidonic acid into PGG_2 or PGH_2, or both (29–30). It has been further postulated that the anti-inflammatory effect of aspirin results from its ability to inhibit the biosyntheses of proinflammatory prostaglandins, eg, PGE_2. Anti-inflammatory steroids (qv) are effective inhibitors of arachidonic acid release from phospholipids and, thereby, inhibit prostaglandin biosynthesis (81–87). More recent studies have concluded that this inhibition of substrate release in various cells results from the induction of a phospholipase-A_2-inhibiting protein (74,88–90). Research efforts have led to the identification of selective inhibitors of thromboxane synthetase, which transforms PGG_2 or PGH_2, or both, into the platelet-aggregatory and vascular-constrictive substance thromboxane A_2 (TXA_2). Inhibitors of this enzyme are being studied in a variety of pathological conditions wherein abnormal platelet function is believed to be an underlying cause (61,91). Finally, research efforts are also being directed toward the identification of selective inhibitors of the enzymes of the 5-lipoxygenase branch of the arachidonic acid cascade (92–93).

Metabolism. The naturally occurring prostaglandins and thromboxanes have a short biological half-life because of their rapid metabolism and excretion. In general, these naturally occurring substances are subject to four metabolic transformations, as illustrated in Figure 2 for PGE_2: *C*-15 prostaglandin dehydrogenase (15-PGDH); 13,14-reductase; β-oxidation of the carboxylic acid side chain; and ω- and $(\omega - 1)$-oxidation of the aliphatic side chain (94–96). The 15-PGDH step occurs most rapidly in humans and most other mammals, and it converts the parent prostaglandin molecule into its corresponding *C*-15 ketone structure. The lung is especially enriched in 15-PGDH, and this enzymatic transformation is so rapid that 90–95% of a prostaglandin is enzymatically transformed into its essentially inactive *C*-15 ketone metabolite on a single pass through the lungs. Urinary metabolites isolated from animals treated with natural prostaglandins usually reflect the combined results of several degradative pathways, as exemplified for PGE_2.

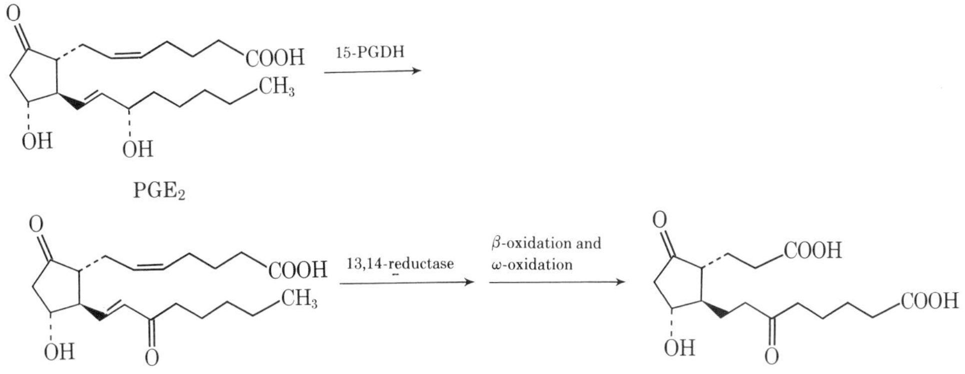

O COOH CH₃ OH OH PGE_2 —15-PGDH→

O COOH CH₃ OH O —13,14-reductase→ —β-oxidation and ω-oxidation→ O COOH COOH OH O

main PGE_2 urinary metabolite

The metabolism of the recently identified leukotrienes is being examined, and initial results have been reported (97).

A kit for the determination of PGE_2 levels in biological samples is marketed by New England Nuclear.

Physical Properties of Selected Eicosanoids. The melting points, optical rotations, and uv spectral data for selected arachidonic-acid-derived eicosanoids are listed in Table 2. Additional physical properties for the prostaglandins of the A, B, E, and F

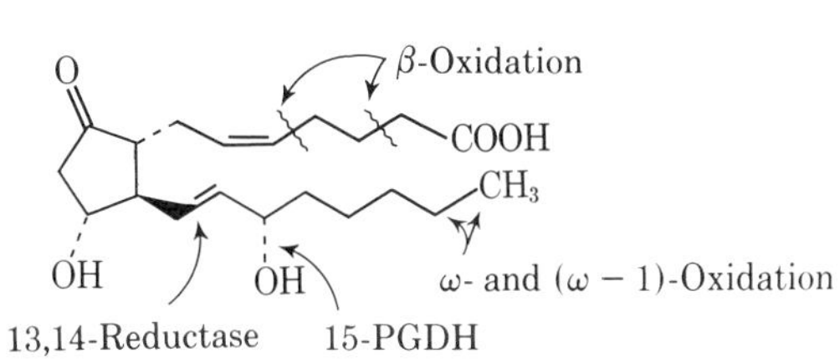

O COOH CH₃ OH OH β-Oxidation ω- and (ω − 1)-Oxidation 13,14-Reductase 15-PGDH

PGE_2

Figure 2. Principal metabolic transformations of PGE_2.

Table 2. Physical Properties of Selected Eicosanoids Derived from Arachidonic Acid

Compound	CAS Registry No.	Mp, °C	$[\alpha]_D$ (solvent)	Uv ($\lambda_{max}^{solvent}$ nm (ϵ))	Refs.
PGA_2		pale yellow oil	+145° (trichloromethane)	$\lambda_{max}^{C_2H_5OH}$ 217 (ϵ 10,300)	98
PGB_2		30–34	+16° to +18° [a]	$\lambda_{max}^{C_2H_5OH}$ 278 (ϵ 26,000)	99–100
PGC_2		oil		$\lambda_{max}^{CH_3OH}$ 234 (ϵ 17,000)	101
PGC_2, methyl ester	[51172-26-0]	oil		$\lambda_{max}^{CH_3OH}$ 229 (sh), 234	102
PGD_2		62.8–63.3	+13° (trichloromethane)		103–104
PGE_2		65–66	−61° (tetrahydrofuran)		105
		62–64	−52° (tetrahydrofuran)		106
		65.0–67.5			98
$PGF_{2\alpha}$		30–35	+26° (ethanol)		98, 107
			+23.8° (tetrahydrofuran)		105
PGI_2, sodium salt	[61849-14-7]	166–168	+88° (trichloromethane)		108
		116–124	+97° (ethanol)		108
PGI_2, methyl ester	[61799-74-4]	30–33	+78° (trichloromethane)		108
6-oxo-$PGF_{1\alpha}$		75–78			108
TXB_2		91–93	+57.4 (ethyl acetate)		109
		92–94			38
		89–90			36
LTA_4, methyl ester	[73466-12-3]	28–32		$\lambda_{max}^{CH_3OH}$ 270 (ϵ 43,900), 278 (ϵ 56,700), and 290 (ϵ 43,100)	110
LTB_4				$\lambda_{max}^{CH_3OH}$ 260 (ϵ 38,000), 270.5 (ϵ 50,000), and 281 (ϵ 39,000)	54
LTC_4				$\lambda_{max}^{CH_3OH}$ 270 (ϵ 32,000), 280 (ϵ 40,000), and 290 (ϵ 31,000)	50
				$\lambda_{max}^{CH_3OH}$ 272 (ϵ 31,000), 278 (ϵ 38,000), and 288 (ϵ 30,000)	111
LTD_4				$\lambda_{max}^{CH_3OH}$ 272 (ϵ 31,000), 278 (ϵ 38,000), and 288 (ϵ 30,000)	111
				$\lambda_{max}^{CH_3OH}$ 271 (ϵ 39,200), 280.5 (ϵ 48,600), and 290 (ϵ 38,400)	110

[a] Solvent unspecified in literature.

series have been summarized in the literature, and the physical methods in prostaglandin research have been reviewed (112–114). The molecular conformations of PGE_2 and PGA_1 have been determined in the solid state by x-ray diffraction, and special ^1H and ^{13}C nuclear magnetic resonance (nmr) spectral studies of several prostaglandins have been reported (115–121). A recent letter in *C&EN* (122), reports an explosive reaction of arachidonic acid at STP, probably owing to peroxide formation. Several excellent reviews of analytic methods for the naturally occurring prostaglandins and thromboxanes and their metabolites have been published (123–127).

Biological Activities

The naturally occurring prostaglandins, thromboxanes, and leukotrienes exhibit a large variety of biochemical and pharmacological actions; excellent reviews have been published (10,61,67,112,128–129). A list of biological activities that are generally associated with selected naturally occurring eicosanoids is presented in Table 3.

Synthesis of Naturally Occurring Eicosanoids

Recent syntheses of prostaglandins and thromboxanes are described in ref. 132.

Total Syntheses. The total chemical synthesis of the naturally occurring prostaglandins, thromboxanes, and leukotrienes has intrigued and challenged organic chemists largely because of the stereochemical complexities and sensitive functional group arrays in these substances. Also, eicosanoids are in great demand for biological, pharmacological, and clinical investigation; consequently, many chemical approaches to these molecules have been developed since the early 1960s. Comprehensive review of all the total syntheses of the naturally occurring eicosanoids are given in refs. 133–136. The following syntheses are commercially important and illustrate unusually novel and creative approaches to the target molecule in question.

Classical E and F Prostaglandins. The synthesis of classical prostaglandins PGE_2 and $PGF_{2\alpha}$ developed by E. J. Corey and co-workers at Harvard University (Fig. 3) is one of the most widely utilized procedures, both commercially and in research lab-

Table 3. Biological Activities Generally Associated with Naturally Occurring Eicosanoids[a]

Eicosanoid	Associated biological activity
PGD_2	inhibitor of blood-platelet aggregation
PGE_1	inhibitor of blood-platelet aggregation; stimulates release of erythropoietin from renal cortex; relaxes bronchial and tracheal muscle (bronchodilator); inhibits basal rate of lipolysis from adipose tissue; inhibitor of allergic responses; and is a vasodilator
PGE_2	stimulates release of erythropoietin from renal cortex; relaxes bronchial and tracheal muscle (bronchodilator); contracts uterine muscle; inhibits gastric acid secretion; and protects gastric mucosal lining[b]
$PGF_{2\alpha}$	contracts bronchial and tracheal muscle (bronchoconstrictor); contracts uterine muscle; luteolytic in some mammalian species; and is a vasoconstrictor
PGH_2	vasoconstrictor; inducer of blood-platelet aggregation; and is a bronchoconstrictor
PGI_2	inhibitor of blood-platelet aggregation; slight bronchodilator; vasodilator; and protects gastric mucosal lining
TXA_2	vasoconstrictor; bronchoconstrictor; and is an inducer of blood-platelet aggregation
LTB_4	stimulator of leukocyte behavior and function, eg, aggregation, chemokinesis, chemotaxis, and release of lysosomal enzymes
LTC_4/LTD_4	bronchoconstrictor; and stimulates airway mucus secretion and plasma leakage from venules

[a] Ref. 130.
[b] Ref. 131.

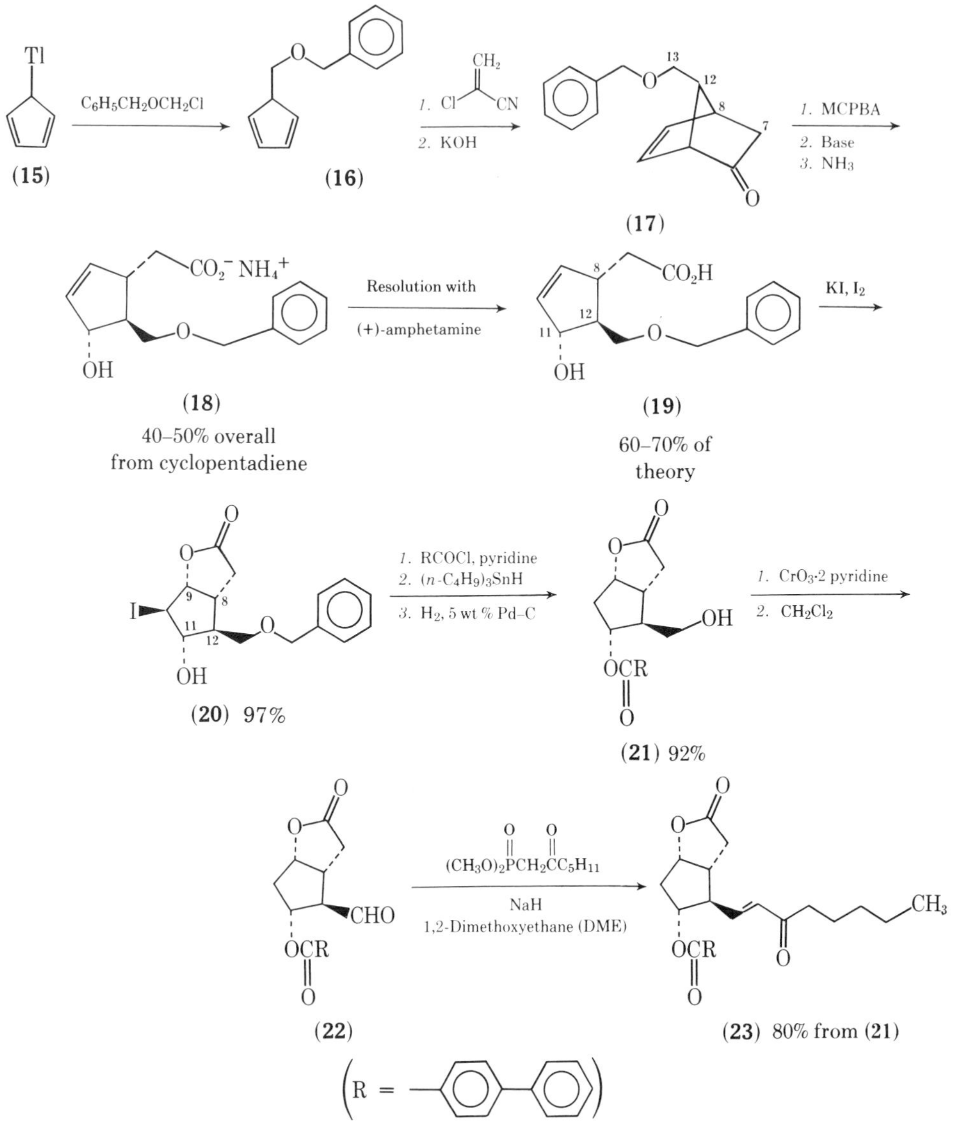

Figure 3. The Corey process.

oratories (105,134–137). In addition to providing the natural products PGE_2 and $PGF_{2\alpha}$, various intermediates in this process, eg, lactone aldehyde (**22**) and lactone diol (**26**), are extensively used in the synthesis of prostaglandin analogues. This stereocontrolled synthesis starts from thallous cyclopentadienide (**15**), which is alkylated with benzyl chloromethyl ether to give the substituted cyclopentadiene (**16**). This reactive and somewhat unstable diene is subjected to a Diels-Alder cycloaddition with α-chloroacrylonitrile, and the resulting cycloadduct is treated with base to release the latent ketone function in (**17**). It should be noted that the rigid bicyclic structure of (**17**) has the desired trans relationship between the eventual side chains at carbons

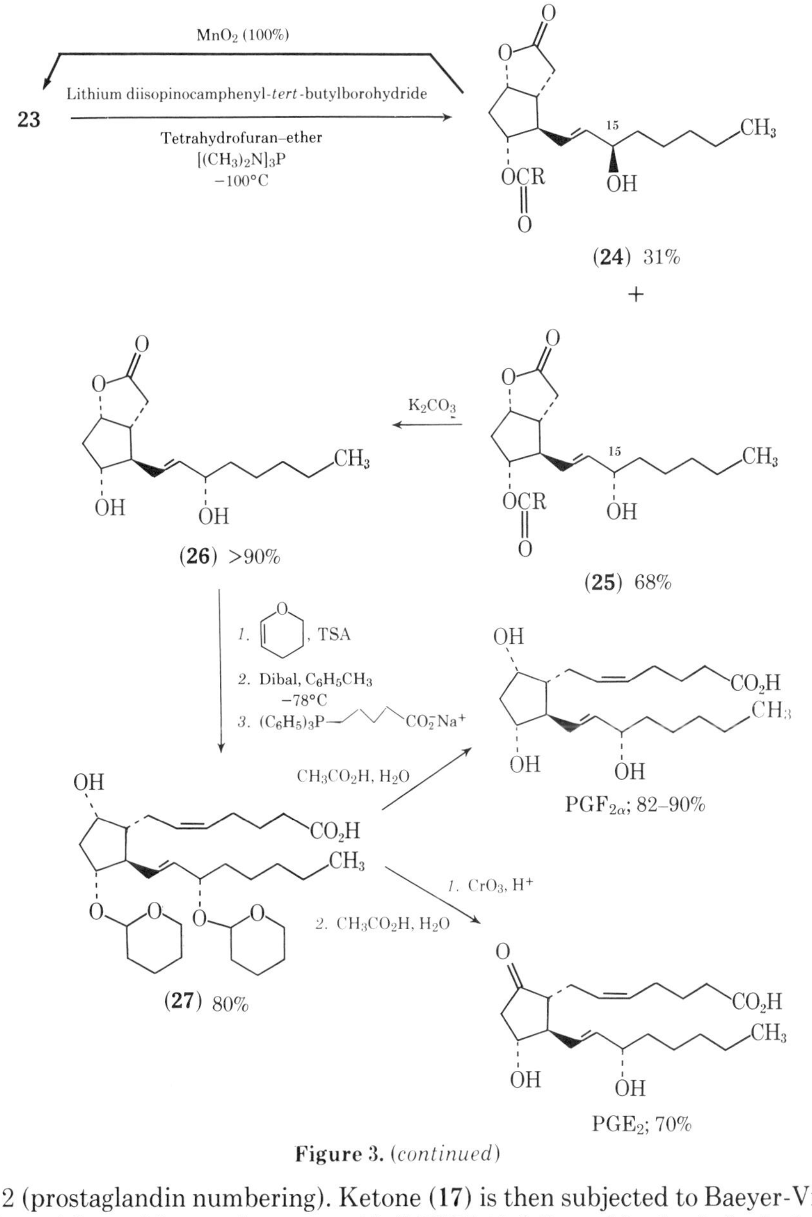

Figure 3. (*continued*)

8 and 12 (prostaglandin numbering). Ketone (**17**) is then subjected to Baeyer-Villiger oxidation with *m*-chloroperbenzoic acid (MCPBA) followed by basic hydrolysis to give, initially, racemic acid (**19**) which is isolated as its ammonium salt (**18**). Racemic acid (**19**) is resolved by way of its (+)-amphetamine salt to give material whose absolute configuration at carbons 8, 11, and 12 is shown in structure (**19**). Subsequent reaction of (**19**) with potassium iodide and iodine results in the formation of iodolactone (**20**), which has the correct absolute configurations of $PGF_{2\alpha}$ at carbons 8, 9, 11, and 12. The hydroxy function at *C*-11 is then acylated with *p*-phenylbenzoyl chloride, the iodine atom at *C*-10 is removed, and the benzyl ether is reductively cleaved to afford lactone alcohol (**21**). This alcohol is then oxidized by the *in situ* modification of the Collins oxidation, ie, with the bispyridine complex of chromium trioxide, to produce the lac-

tone aldehyde (**22**), which is commonly referred to as the Corey aldehyde. The lower side chain is stereospecifically attached by way of the Wadsworth-Emmons modification of the Wittig reaction to give the *trans*-enone (**23**), and the ketone group at *C*-15 of (**23**) is stereoselectively reduced with lithium diisopinocamphenyl-*tert*-butylborohydride to a 1:2 mixture of 15(*R*):15(*S*) alcohols (**24**) and (**25**). The unnatural *R* isomer (**24**), which is separated from (**25**) by silica gel chromatography, is efficiently recycled to (**23**) by manganese dioxide oxidation, whereas the natural *S*-isomer (**25**) is hydrolyzed to lactone diol (**26**). The hydroxyl functions of (**26**) are protected with 1,2-dihydropyran in the presence of *p*-toluenesulfonic acid (TSA), the lactone is reduced to the corresponding lactol with diisobutylaluminum hydride, and Wittig reaction of this lactol with the ylid derived from (4-carboxybutyl)-triphenylphosphonium bromide affords bis-protected $PGF_{2\alpha}$ (**27**). Deprotection of (**27**) with aqueous acetic acid gives $PGF_{2\alpha}$, whereas oxidation of (**27**) with chromic acid followed by aqueous acetic acid deprotection leads to PGE_2.

Since the Corey synthesis has been extensively used in prostaglandin research, improvements on the various steps in the procedure have been sought, identified, and exploited. These improvements include *inter-alia* improved procedures for the preparation of norbornenone (**17**), alternative methods for the resolution of acid (**19**), improved procedures for the deiodination of iodolactone (**20**), alternative methods for the synthesis of Corey aldehyde (**22**) or its equivalent, and improved procedures for the stereoselective reduction of enone (**23**) (138–195). An alternative synthesis of lactone alchol (**21**) (R = CH_3) has been developed and is illustrated below (196). Reaction of (*S*)-(−)-2-acetoxysuccinyl chloride (**28**), which is prepared in 80% yield from (*S*)-(−)-malic acid, with the bis(magnesiobromide) salt of methyl hemimalonate (**29**) affords (*S*)-dimethyl 4-acetoxy-3,6-dioxosuberate (**30**). Cyclization of (**30**) with magnesium carbonate results in cyclopentenone (**31**) which is subsequently transformed in three steps to the lactone ester (**32**) and then to the lactone alcohol (**21**) (R = CH_3).

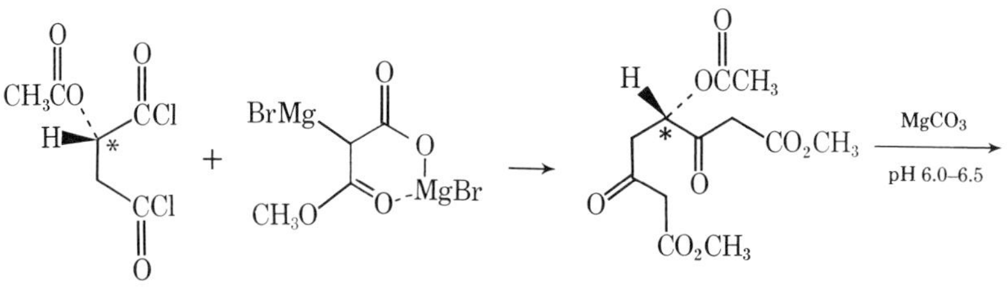

(**28**) + (**29**) → (**30**) 73%

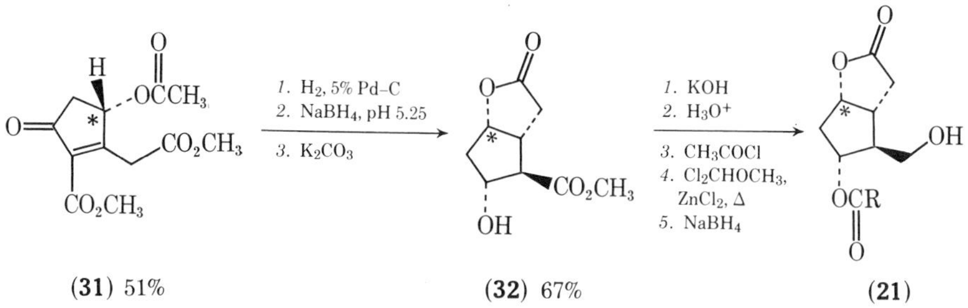

$MgCO_3$, pH 6.0–6.5 → (**31**) 51%

1. H_2, 5% Pd–C
2. $NaBH_4$, pH 5.25
3. K_2CO_3

→ (**32**) 67%

1. KOH
2. H_3O^+
3. CH_3COCl
4. Cl_2CHOCH_3, $ZnCl_2$, Δ
5. $NaBH_4$

→ (**21**) (R = CH_3); 90%

The chiral carbon atom in (**28**) becomes C-9 (prostaglandin numbering) in (**21**) and directly or indirectly controls the stereochemistry in each of the chemical steps leading to (**21**). Also, this process starts from commercially available and optically pure (S)-(−)-malic acid and, therefore, does not involve an optical-resolution step.

The Corey process (Fig. 3) is also useful for the synthesis of the one-series prostaglandins (26). Catalytic reduction of the bis-protected $PGF_{2\alpha}$ (**27**) results in selective saturation of the 5,6-double bond to afford (**33**). Subsequent transformations, as shown below, lead to PGE_1 and $PGF_{1\alpha}$.

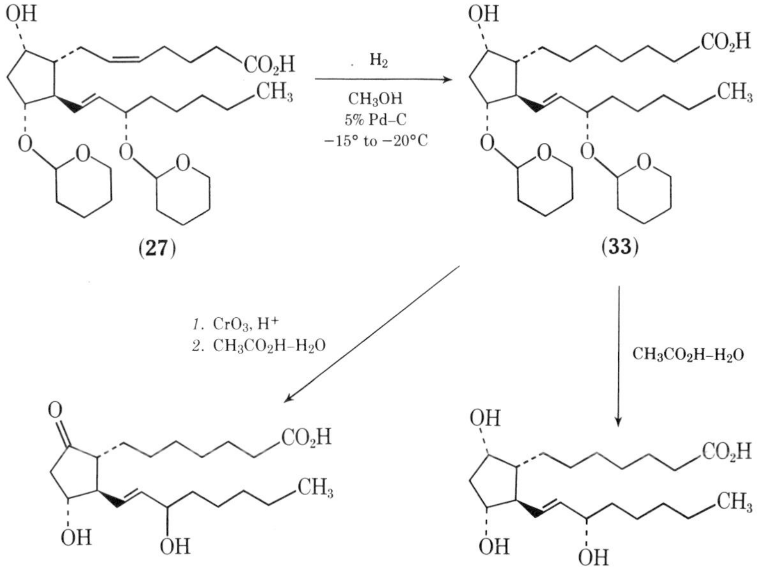

(**27**) (**33**)

PGE_1; 64% $PGF_{1\alpha}$; 80%

Similar catalytic reductions have been reported, and total syntheses of one-series prostaglandins involving reversal of side-chain attachment have been developed (197–201). A more recent example of a total synthesis involving reversed side-chain attachment is reported in ref. 202.

Intermediates in the Corey synthesis (Fig. 3) are also useful for the synthesis of three-series prostaglandins (203). The key step in this synthesis is the Wittig reaction

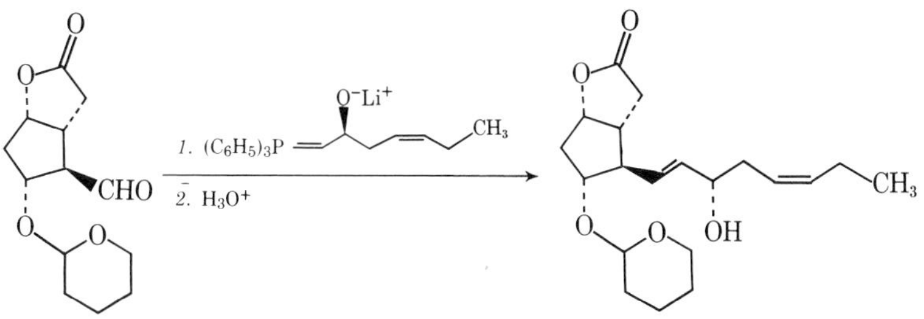

(**34**) (**35**) 35%

of aldehyde (**34**) with a β-oxidoylid to yield stereospecifically the lactone alcohol with a *trans*-13,14-double bond. This lactone is subsequently converted to $PGF_{3\alpha}$ and PGE_3 by methods analogous to those outlined in Figure 3.

In addition to the Corey synthesis, another commercially important total synthesis is that outlined below (204–205).

(36) → 1. CH_3CO_3H, Na_2CO_3; 2. H^+; 3. $HOCH_2C(CH_3)_2CH_2OH$, H^+ → **(37)** 61%

(37) → 1. $Cl_2CHCOCl$, $(C_2H_5)_3N$; 2. Zn, NH_4Cl, C_2H_5OH → **(38)** 81%

(38) → 1. resolution with *l*-ephedrine; 2. MCPBA → **(39)** 63% of theory

(39) → 88% HCO_2H → **(40)** 95%

(40) → $Br_2C(CN)(CH_2)_4CH_3$, $[(CH_3)_2N]_3P$ → **(41)**

(41) → 1. HCO_2H; 2. dilute H_2SO_4; 3. $NaHCO_3$ → **(42)** 61% from **(40)**

This efficient and short synthesis starts with norbornadiene (**36**), which is monoepoxidized and rearranged under acidic conditions and then reacts with 2,2-dimethyl-1,3-propanediol to afford the bicyclo[3.1.0]hexane (**37**). This intermediate is then converted to cyclobutanone (**38**) by means of a ketene–cycloaddition reaction, and (**38**) is resolved by diastereomeric oxazolidine formation with *l*-ephedrine. Hydrolysis of the acetal group in (**39**) followed by condensation of the resulting aldehyde with 1-cyano-1,1-dibromohexane affords glycidonitrile (**41**), which is subsequently solvolyzed to enone (**42**). This enone can then be acylated to produce the lactone enone (**23**) from the Corey process (Fig. 3). As noted in a recent review, this synthesis has produced the equivalent of more than 50 kg/yr of $PGF_{2\alpha}$, and it has effectively reduced the cost of synthesizing prostaglandins to less than one hundredth of that based on bioconversion of arachidonic acid.

A third synthetic procedure for PGE_2 and $PGF_{2\alpha}$, which is potentially commercially useful, is shown (206).

In this synthesis, racemic bicyclo[3.2.0]hept-2-en-6-one (**43**) is the starting material. Upon reduction with bakers' yeast, it affords alcohols (**44**) and (**45**). These al-

R = t-butyldimethylsilyl

(**43**) bakers' yeast 71% conversion → (**44**) 53% + (**45**) 25%

DBDMH acetone–H_2O → (**46**) 77%; DBDMH acetone–H_2O → (**47**) 77%

(**46**) 2 steps → (**48**) 84%; (**47**) 3 steps → (**49**) 50%

(**48**) 50–56% 3 steps → (**50**); (**49**) 2 steps 93% → (**50**)

cohols are separable by chromatography or careful distillation, and each then reacts with 1,3-dibromo-5,5-dimethylhydantoin (DBDMH) in aqueous acetone to yield bromoketones (**46**) [from (**44**)] and (**47**) [from (**45**)]. These ketones are each, in turn, converted to the mixture of lactone diols (**50**) in comparable overall yield. Chromatographic separation of (**50**) then provides lactone diol (**26**) from the Corey synthesis (Fig. 3), and the unnatural R isomer may be recycled. The appealing aspect of this synthesis is that an enantioconvergent approach is used, ie, both enantiomers (**46**) and (**47**) are utilized, which obviates the need for optical resolution and its associated loss of at least one half of the resolved product. Other total synthetic procedures which lead to the Corey lactone diol (**26**) or its equivalent include those developed at Imperial Chemical Industries, Ltd.; Sagami Chemical Research Center; and the University of Chicago (207–213).

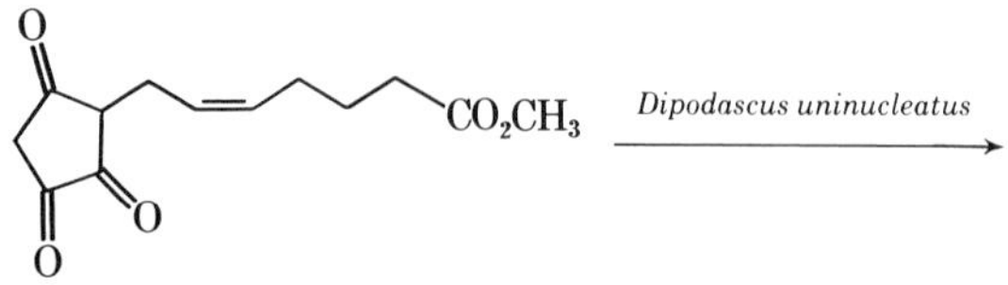

CO_2CH_3 *Dipodascus uninucleatus* →

(**51**)

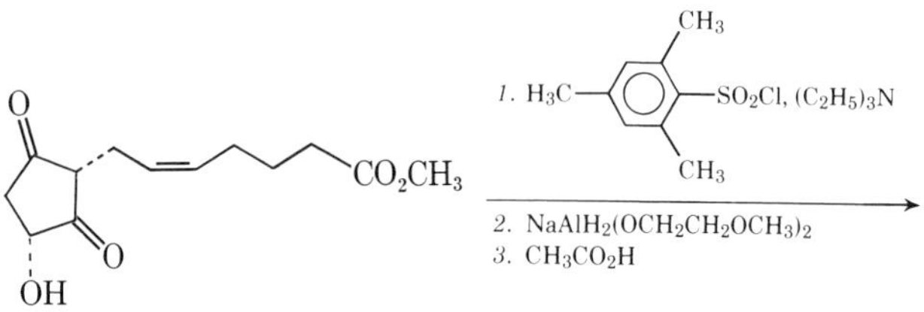

1. H_3C–(mesityl)–SO_2Cl, $(C_2H_5)_3N$
2. $NaAlH_2(OCH_2CH_2OCH_3)_2$
3. CH_3CO_2H

CO_2CH_3, OH

(**52**) 47%

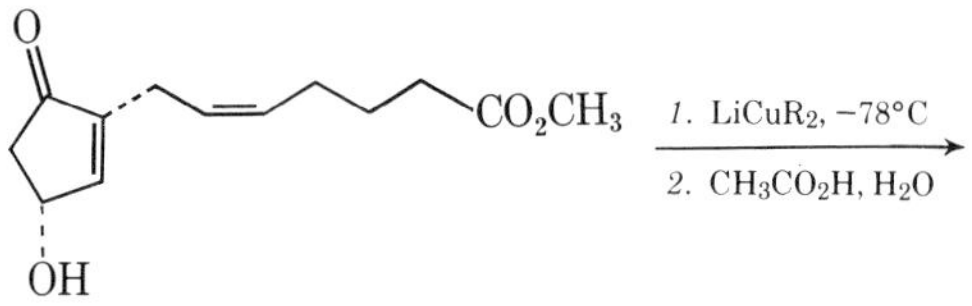

1. $LiCuR_2$, −78°C
2. CH_3CO_2H, H_2O

CO_2CH_3, OH

(**53**) 50%

R = [CH_3, O, CH_3, OC_2H_5]

CO_2CH_3, CH_3, OH, OH *Rhizopus oryzae* → PGE_2; 55%

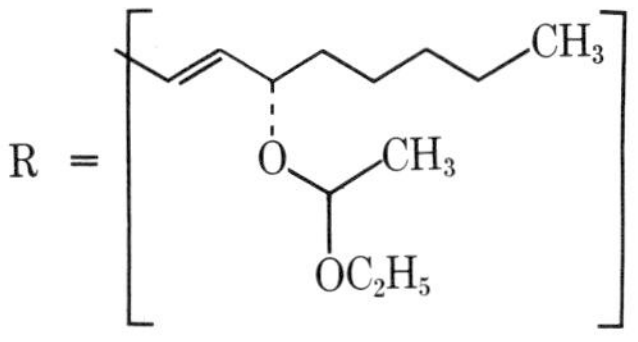

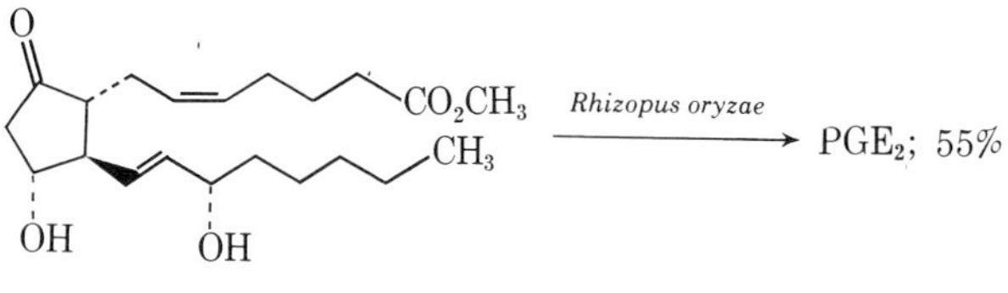

(**54**) 50%

Another important category of total synthetic procedures leading to the prostaglandins includes those that involve, at some stages in their sequences, the 1,4-addition of an organometallic reagent to an α,β-unsaturated ketone function (214–221). This type of approach is exemplified by the synthesis shown below (216).

Trione (**51**) is microbiologically reduced to the 11(R)-hydroxydione (**52**), which is further converted to the cyclopentenone (**53**) by the indicated transformations. Reaction of this enone with the homocuprate derived from (1E,3S)-1-iodo-3-hydroxyoct-1-ene, ethoxyethyl ether gives the methyl ester of PGE_2 (54), which is then converted to PGE_2 microbiologically. This latter transformation may also be effected by a crude enzyme preparation from the soft coral *Plexaura homomalla* or by crude pancreatic lipase (218,222). The conjugate addition has been modified with the use of a cis homocuprate, eg, (**56**).

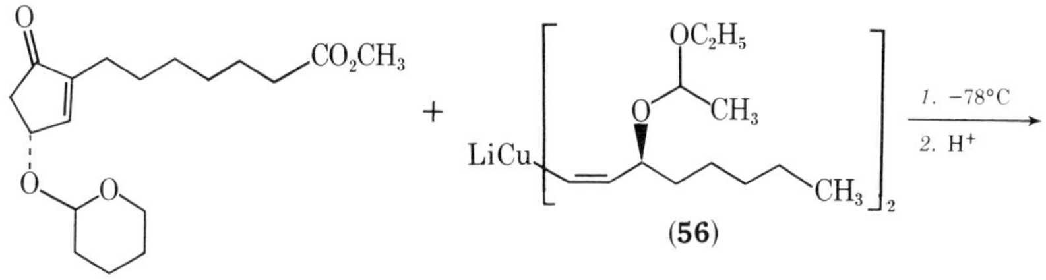

O, CO_2CH_3, O, O (**55**) + LiCu[OC_2H_5, O, CH_3, CH_3]$_2$ (**56**) $\xrightarrow[\text{2. } H^+]{\text{1. } -78°C}$

O, CO_2CH_3, OH, CH_3, O, O (**57**) 70% $\xrightarrow[\text{2. } (CH_3O)_3P \quad \text{3. } CH_3CO_2H, H_2O]{\text{1. } p\text{-}CH_3C_6H_4SCl, (C_2H_5)_3N}$ O, CO_2CH_3, CH_3, OH, OH (**58**) 64%

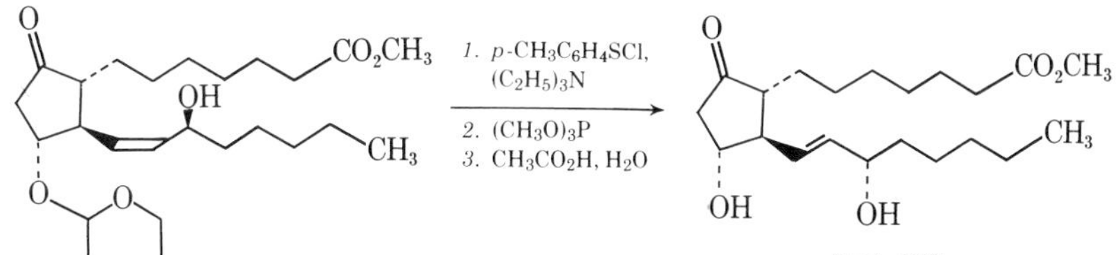

PGE_2, methyl ester [*3434-33-1*]

The advantage of this procedure is that cis homocuprates, eg, (**56**), give higher yields of adduct, ie, (**57**), than the corresponding trans homocuprates. Further, the 1,4-addition of the cis homocuprates occurs with a higher degree of stereoselectivity than with the corresponding trans homocuprates. The intermediate adduct with the *cis*-13,14-double bond (**57**) is stereospecifically converted to the PGE_1 methyl ester, as shown above.

Numerous articles have been published describing improvements in the conjugate addition approach or alternative synthetic procedures for various intermediates (**51**)–(**53**). For example, alternative syntheses of trione (**51**) have been reported as have alternative synthetic procedures for enone (**53**) or its equivalent (214,223–251). An especially creative application of the conjugate addition approach to the synthesis of $PGF_{2\alpha}$ is shown below (252).

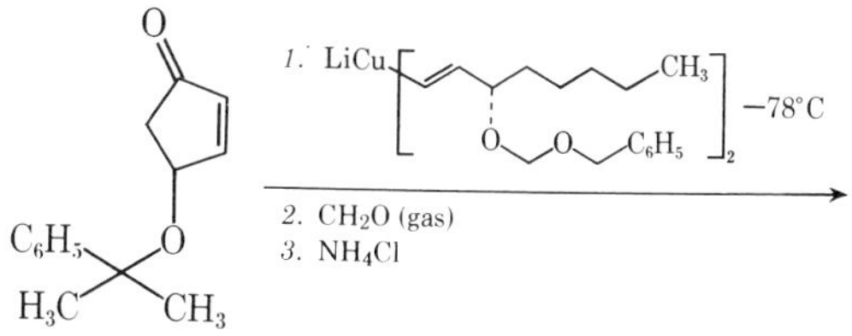

O, C_6H_5, O, H_3C, CH_3 (**59**) $\xrightarrow[\text{2. } CH_2O \text{ (gas)} \quad \text{3. } NH_4Cl]{\text{1. } LiCu[CH_3, O, O, C_6H_5]_2, -78°C}$

O, CH_2OH, CH_3, C_6H_5, O, H_3C, CH_3, O, O, C_6H_5 (**60**) 50–60% $\xrightarrow[\text{2. } (C_2H_5)_3N]{\text{1. } CH_3SO_2Cl, \text{ pyridine}}$

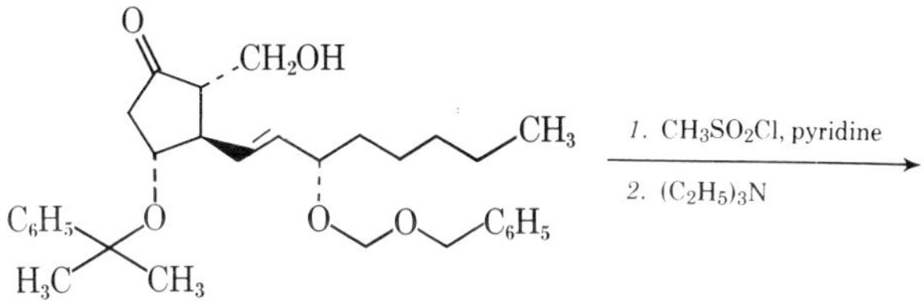

1. LiCu
2. CH_3CO_2H, H_2O
3. CrO_3, H^+

(**61**) 80%

2 steps → $PGF_{2\alpha}$

(**62**) 78%
[*16763-79-4*]

Conjugate addition of a trans divinylcuprate to cyclopentenone (**59**) followed by trapping of the resulting enolate with formaldehyde gas affords β-hydroxyketone (**60**). The hydroxyl group is esterified with methane sulfonyl chloride and eliminated with base to produce the exomethylene cyclopentanone (**61**). A second conjugate addition is then carried out on (**61**) to attach the upper side chain and, after selective *C*-1 protecting group removal and oxidation, the diprotected PGE_2 (**62**) is obtained. Related and very creative conjugate addition–enolate trapping sequences have been reported by others (253–256). The procedure described in ref. 256 is especially interesting, the key step being the addition of vinyl lithium reagent (**64**) to cyclopentene (**63**) followed by alkylation of the resulting anion intermediate with iodide (**65**) to produce (**66**). From (**66**), natural PGE_2 is produced in 54% overall yield after six steps.

1. Li (**64**)
2. I (**65**)

(**63**)

(**66**) 67%

Lastly, an elegant total synthesis of $PGF_{2\alpha}$ from D-glucose is described in ref. 257.

Manufacturers. Prostaglandin $F_{2\alpha}$ is manufactured by the Upjohn Company (Kalamazoo, Michigan); Ono Pharmaceutical Company, Ltd. (Osaka, Japan); Chinoin Chemical and Pharmaceutical Works, Ltd. (Budapest, Hungary); and Toray Industries, Inc. (Tokyo, Japan). The Upjohn process is related to that illustrated in (**36**)–(**41**) and Ono and Chinoin presumably use modified Corey processes. Toray Industries,

Inc., manufactures $PGF_{2\alpha}$ by biosynthesis from arachidonic acid, based on technology licensed from Unilever Research Laboratories. Prostaglandin E_1 and E_2 are manufactured by The Upjohn Company and Ono Pharmaceutical Company, Ltd., and prostaglandin E_2 is manufactured by Toray Industries, Inc. In general, specific information on the manufacturing processes used by these pharmaceutical firms, as well as information on production quantities and costs, are proprietary. Registered trademarks and indications for the naturally occurring prostaglandins have been summarized (59).

Classical D Prostaglandins. Prostaglandin D was first synthesized in 1973 and the procedure is summarized below (258).

1. dihydropyran, H^+
2. diisobutylaluminum hydride
3. $(C_6H_5)_3P{=}CH(CH_2)_3CO_2^-Na^+$

(25) R = CH_3

(67) 65%

CrO_3, H^+, −30°C

(68) 19–25%

(69) 50–56%

CH_3CO_2H, H_2O → PGE_2

CH_3CO_2H, H_2O → PGD_2 45%

The key step in this process is the selective functionalization of the *C*-15 hydroxyl in Corey intermediate **(25)** (Fig. 3) to produce **(67)**. This same intermediate has also been used in other total syntheses of PGD_2 (259–260). Additional separate synthetic procedures leading to PGD_2, PGD_1, and PGD_3 are described in refs. 206, 261–264.

Classical A, B, and C Prostaglandins. In general, the classical prostaglandins of the A, B, and C series are prepared by partial syntheses from prostaglandins of the E series (see Partial Syntheses). Total synthetic procedures leading to these substances have been reviewed (265). Since 1976, descriptions of total syntheses of prostaglandins of the A and C series have been published (206,265–271).

Prostaglandin Endoperoxides. The endoperoxides PGH_1 and PGH_2 have been prepared in multimilligram quantities by biosyntheses, and PGH_2, its methyl ester, and PGG_2, have recently been prepared by total syntheses (56–58,272). The key ring-forming step in each of these procedures involves the double displacement of a 1,3-dihalocyclopentane, eg, **(70)** as shown for the synthesis of PGG_2 (57).

Br Br Cl (70) CO_2H CH_3 $AgOCCF_3$ (O ||) 90% H_2O_2 ether → O- O- OOH CO_2H CH_3 PGG_2; 15–20%

I-Series Prostaglandins. The total syntheses of prostaglandins of the I series have been extensively reviewed (133,273). The first total synthesis of the PGI_2 methyl ester and the PGI_2 sodium salt (also known as epoprostenol sodium) was completed in 1976 (44,274–275). The process is still used to synthesize multigram quantities of these derivatives of PGI_2.

OH OH OH CO_2CH_3 CH_3 (71) $PGF_{2\alpha}$, methyl ester → I_2, KI, Na_2CO_3, H_2O

H I O CO_2CH_3 CH_3 OH OH (72) >85% → 1,5-diazabicyclo[4.3.0]non-5-ene (DBN), $C_6H_5CH_3$, 40°C

O CO_2CH_3 CH_3 OH OH (73) 80–90% → NaOH, CH_3OH, H_2O → O $CO_2^-Na^+$ CH_3 OH OH (74) 77%

Iodoetherification of (**71**) produces two isomeric compounds (**72**), ie, the 5*R*,6*R* and 5*S*,6*S* diastereomers, which are dehydroiodinated without separation to afford the PGI_2, methyl ester (**73**). Since the parent carboxylic acid PGI_2 is very unstable with respect to hydrolysis of the reactive enol ether double bond and can only be isolated in semipure form with great care, the methyl ester (**73**) is converted to its sodium salt, which is a white, free-flowing powder (274). Although a natural PGI_1 has not been identified, the syntheses of (6*R*)-PGI_1 [*62777-90-6*] and (6*S*)-PGI_1 [*62770-50-7*] have been reported by several laboratories, as has the synthesis of PGI_3 (133,274,276).

Thromboxanes. As of March 1983, the successful synthesis of thromboxane A_2 (TXA_2) had not been reported; however, several synthetic procedures for its stable hydrolysis product, ie, thromboxane B_2 (TXB_2), have been described (36–38,109,277–282). Two of these procedures have been successfully used to synthesize multigram quantities of TXB_2. In the first of these, the Corey aldehyde (**22**) (Fig. 3, R = C_6H_5) is converted to enal (**75**) by Florisil (36).

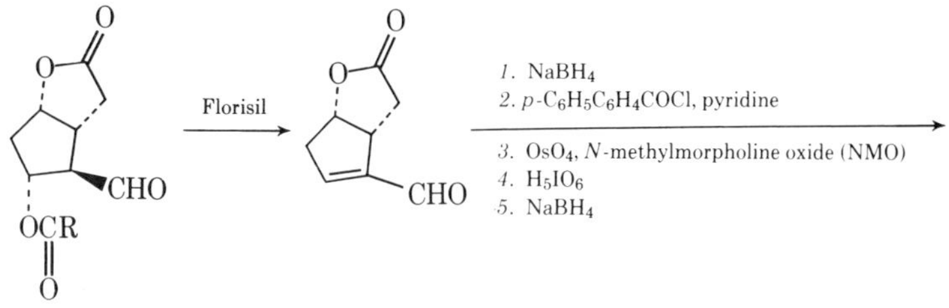

Florisil

1. $NaBH_4$
2. p-$C_6H_5C_6H_4COCl$, pyridine
3. OsO_4, N-methylmorpholine oxide (NMO)
4. H_5IO_6
5. $NaBH_4$

(**22**) R = C_6H_5 (**75**) 74%

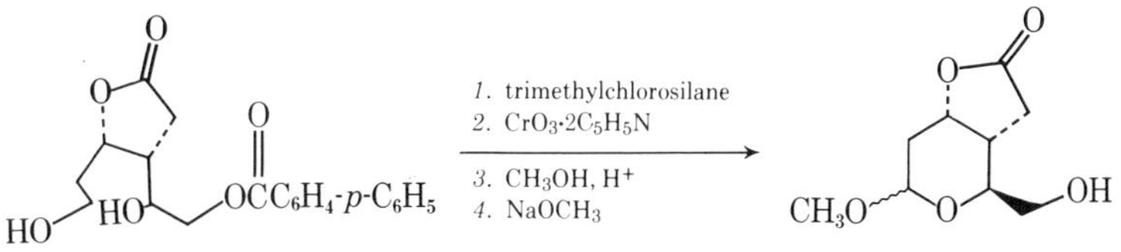

1. trimethylchlorosilane
2. $CrO_3{\cdot}2C_5H_5N$
3. CH_3OH, H^+
4. $NaOCH_3$

(**76**) 56% (**77**) 43–48%

This intermediate is then converted to diol (**76**) by a series of transformations, and (**76**) is cyclized in four steps to the isomeric acetals (**77**). By means of conventional prostaglandin chemistry developed by Corey (Fig. 3), (**77**) is then further transformed to TXB_2. A second synthesis of TXB_2 is quite short and starts from the 9,15-diacetate of the $PGF_{2\alpha}$ methyl ester (**78**) (38). Treatment of (**78**) with lead tetraacetate results in ring cleavage to the rather unstable aldehyde (**79**).

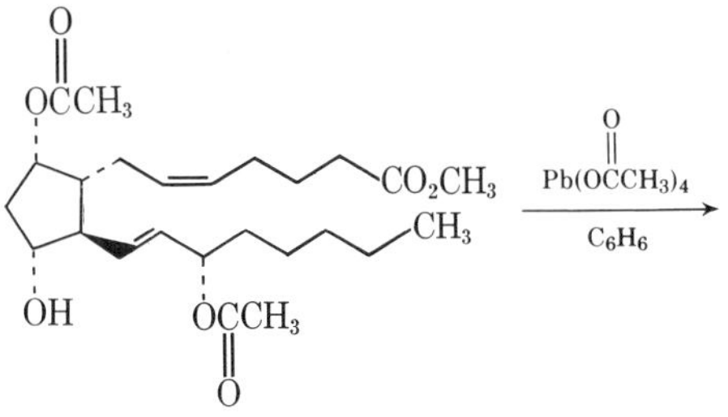

$Pb(OCCH_3)_4$, C_6H_6

(**78**)

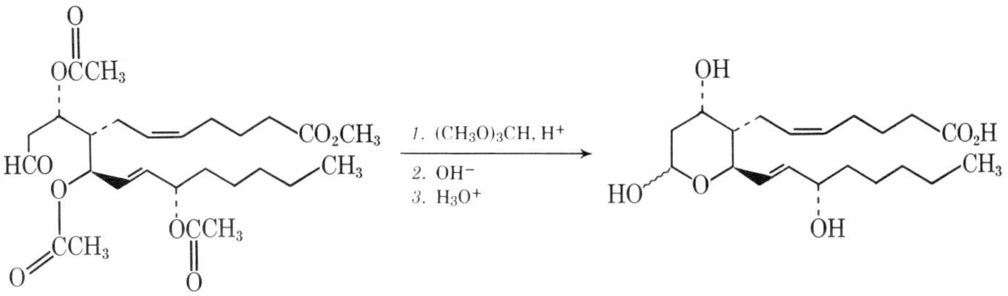

1. $(CH_3O)_3CH$, H^+
2. OH^-
3. H_3O^+

(**79**) TXB_2; 25% from (**78**)

This aldehyde is then protected as its dimethyl acetal; the acetate groups at *C*-9, *C*-12, and *C*-15 and the methyl ester are hydrolyzed with aqueous base; and the dimethyl acetal is cleaved with aqueous acid to yield TXB_2. It should be noted that five syntheses of TXB_2 starting from commercially available sugars have been developed, which is not surprising in view of the nature of the six-membered ring in TXB_2 (109,277,279–281).

Leukotrienes. The first stereospecific total synthesis of leukotriene C_4 (LTC_4) is summarized, in part, below (50,134). The optically active epoxy alcohol (**80**), which is prepared in several steps from D-(−)-ribose, is oxidized to the corresponding aldehyde with the Collins reagent ($CrO_3{\cdot}2C_5H_5N$).

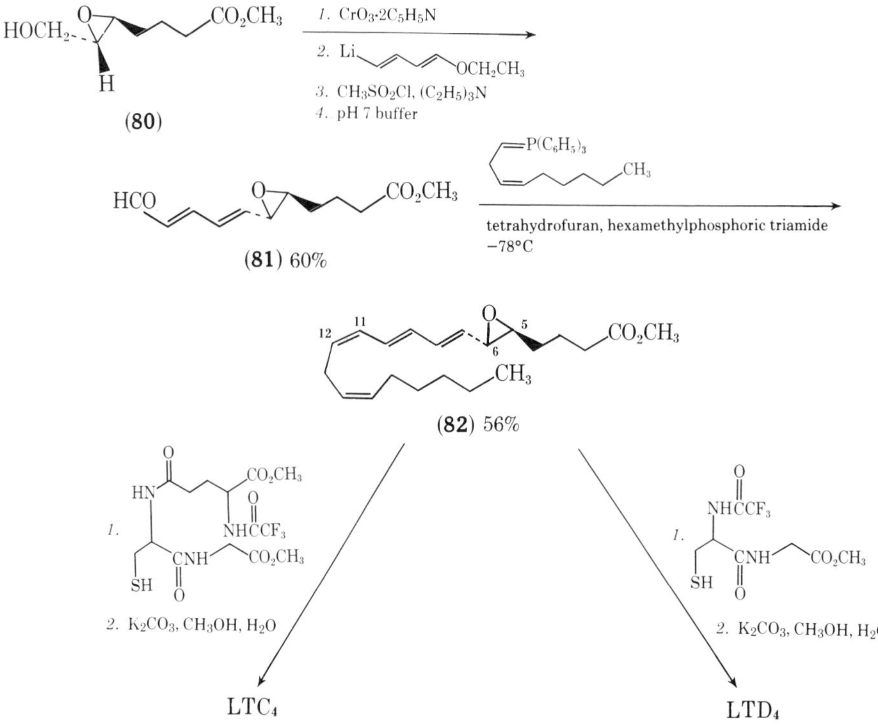

(**80**)

1. $CrO_3{\cdot}2C_5H_5N$
2. Li–CH=CH–CH=CH–OCH_2CH_3
3. CH_3SO_2Cl, $(C_2H_5)_3N$
4. pH 7 buffer

(**81**) 60%

tetrahydrofuran, hexamethylphosphoric triamide
−78°C

(**82**) 56%

1. (protected glutathione)
2. K_2CO_3, CH_3OH, H_2O

LTC_4

1. (protected cysteinylglycine)
2. K_2CO_3, CH_3OH, H_2O

LTD_4

This aldehyde is chain-extended to dienal (**81**), which is subsequently condensed with the indicated phosphonium ylid to afford the LTA_4 methyl ester (**82**). An important aspect of this Wittig condensation is that the product (**82**) stereoselectively forms with a *cis*-11,12-double bond. Reaction of (**82**) with a protected form of glutathione results in exclusive epoxide ring opening at *C*-6. Following base hydrolysis of the protecting groups, LTC_4 is isolated as the single principal product of the reaction. In similar fashion, LTD_4 is obtained from the condensation of (**82**) with the indicated protected form of cysteinylglycine (283).

The importance of this stereospecific total synthesis to the structural elucidation of the peptide-containing leukotrienes cannot be overemphasized. Although several modifications of Corey's original synthesis have been described, the steps outlined from dienal (**81**) to LTC_4 or LTD_4 still make up the most practical synthesis of these materials (134,284–285).

The total synthesis of the potent leukocyte chemoattractant LTB_4 was also pioneered by the Corey group, and the key steps in its formation are shown below.

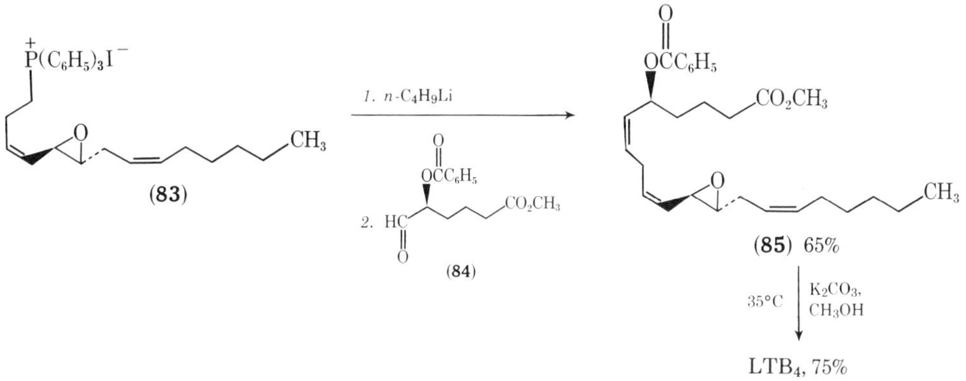

1. n-C_4H_9Li
2. (84)

(83) (84) (85) 65%

35°C K_2CO_3, CH_3OH

LTB_4, 75%

Wittig condensation of the ylid derived from phosphonium salt (**83**) with aldehyde (**84**) derived from 2-deoxy-D-ribose stereoselectively affords diene epoxide (**85**). Treatment of this intermediate with base effects protecting-group removal, double-bond migration, and epoxide ring opening to afford, ultimately and stereospecifically, LTB_4. Other syntheses of LTB_4 and its isomers have been reported (134,286–287).

Partial Syntheses. ***Prostaglandins from Plexaura homomalla.*** The discovery that (15R)-PGA_2 [*23602-72-4*] and its 15-acetate, methyl ester (**86**) were present in relatively large quantities (0.2–1.3% of the dry weight) in air-dried specimens of the gorgonian *Plexaura homomalla* stimulated research aimed at the utilization of these natural products in the synthesis of PGE_2 and $PGF_{2\alpha}$ (23–24,98). The conversion of (**86**) to $PGF_{2\alpha}$ is an example of this type of process. Base-catalyzed epoxidation of (**86**) produces the isomeric 10α,11α- and 10β,11β-epoxides (**87**) which, without further purification, are reduced to (15R)-PGE_2 [*37785-89-0*] (**88**). The C-9 ketone group of the trimethylsilyl derivative of (**88**) is then reduced with sodium borohydride. Following protecting-group removal, the allylic alcohol group at C-15 is selectively oxidized with 2,3-dichloro-5,6-dicyanobenzoquinone (DDQ) to yield 15-dehydro-$PGF_{2\alpha}$ [*35850-13-6*] (**89**). The tris(trimethylsilyl ether) derivative of this latter compound is finally reduced with zinc borohydride, the trimethylsilyl groups are removed with acid, and $PGF_{2\alpha}$ is isolated.

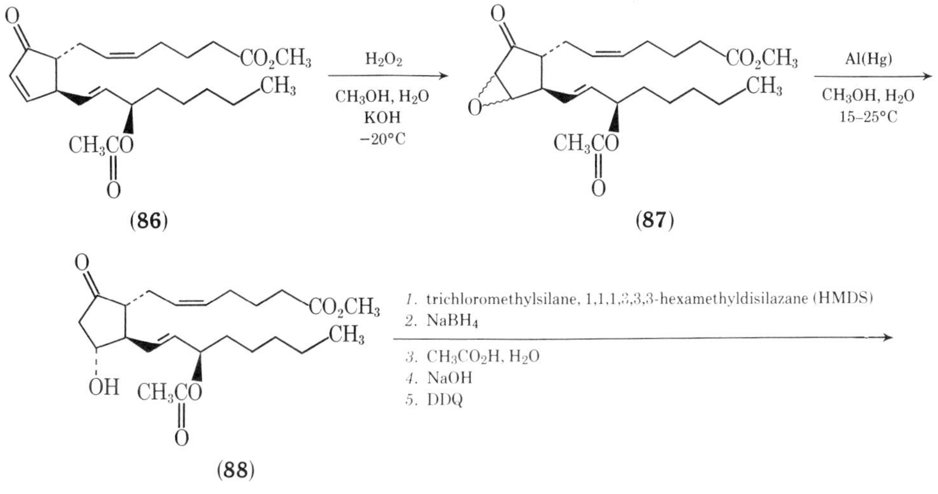

H_2O_2 / CH_3OH, H_2O, KOH, −20°C; Al(Hg) / CH_3OH, H_2O, 15–25°C

(86) (87)

1. trichloromethylsilane, 1,1,1,3,3,3-hexamethyldisilazane (HMDS)
2. $NaBH_4$
3. CH_3CO_2H, H_2O
4. NaOH
5. DDQ

(88)

(89) 38%

1. trichloromethylsilane, HMDS
2. $Zn(BH_4)_2$
3. CH_3CO_2H

$PGF_{2\alpha}$; 57%

(90)

1. H_2O_2, KOH
2. Al(Hg)
3. esterase

PGE_2

It was determined that some forms of *Plexaura homomalla* contain esterified derivatives of (15*S*)-PGA_2 [*13345-50-1*] (98). Since these derivatives have the natural 15*S* configuration at *C*-15, conversion to PGE_2 and $PGF_{2\alpha}$ is simplified. For example, PGE_2 is synthesized from the (15*S*)-PGA_2 derivative (**90**) in three short steps (98,288).

With the advent of efficient total syntheses, the use of coral-derived intermediates for the large-scale synthesis of prostaglandins is apparently no longer extensively employed.

Prostaglandin Interconversions. A number of synthetic procedures exist for the chemical interconversion of different prostaglandin families, and these procedures have been reviewed (289–291). Interconversions among the two-series prostaglandins, ie, those derived from arachidonic acid, are outlined below (38,43–46,56–58,98,107,197,260,292–296).

PGA_2, PGC_2, PGE_2, PGB_2, TXB_2, PGI_2, PGG_2, $PGF_{2\alpha}$, PGH_2, PGD_2

Prostaglandin E_2 can be selectively reduced to $PGF_{2\alpha}$ which can, in turn, be selectively reoxidized to PGE_2. As described earlier, $PGF_{2\alpha}$ can be converted into thromboxane B_2 (TXB_2), PGI_2, or PGD_2. Prostaglandin E_2 can be dehydrated to PGA_2, and the latter substance can be selectively isomerized to PGC_2. Alternatively, PGE_2, PGA_2, and PGC_2 can be directly isomerized to PGB_2 with aqueous base. Finally, PGA_2 can be converted to PGE_2. Further detail concerning these and other interconversions is given in refs. 289–291.

Modified Eicosanoids

As is evident from the biological activities listed in Table 3, the naturally occurring eicosanoids derived from arachidonic acid, as well as those derived from related unsaturated fatty acids, display a wide variety of biological activities. Some of these biological activities are desirable, eg, inhibition of blood-platelet aggregation and bronchodilation, whereas others are less desirable or undesirable, eg, aggregation of blood platelets, bronchoconstriction, and gastrointestinal smooth-muscle stimulation. The selectivity of action of the naturally occurring eicosanoids has been enhanced through the syntheses of many analogues of these substances. Other factors that have encouraged the syntheses of eicosanoid analogues include increasing the potency for a given biological response; prolongation of the duration of action of the short-lived, naturally occurring eicosanoids; enhancement of chemical stability; and identifying orally active analogues and those having simplified structures and maintaining desirable biological profiles.

Analogues of the Classical Prostaglandins. Table 4 lists 20 analogues which are either marketed or at advanced stages of development for the indicated utilities (59,135,137,296–298). One of the main objectives during the initial stages of many prostaglandin analogue research programs was prolonging the duration of biological action of the naturally occurring substances. Since the metabolic oxidation effected by *C*-15 prostaglandin dehydrogenase (15-PGDH) is very rapid, numerous analogues that contain alkyl substituents at or near *C*-15 have been synthesized. Table 4 structures (**91**), (**93**)–(**94**), (**97**), (**101**)–(**103**), (**105**), (**108**), and (**110**) exemplify this type of modification; these analogues are resistant or less susceptible to the action of the dehydrogenase. Susceptibility to degradation caused by other metabolic enzymes is also lessened by other molecular modifications. For example, the *C*-2, *C*-3 double bond in structures (**93**), (**102**), and (**110**) is believed to significantly retard β-oxidation and the incorporation of ω-substituents in structures (**92**), (**95**)–(**96**), (**98**)–(**100**), and (**107**) is believed to prevent ω- and ($\omega - 1$)-oxidation. Bovilene [*69381-94-8*] is a long-acting synthetic prostaglandin that regulates the timing of a cow's reproductive cycle and has been approved by the FDA for marketing in the United States. It is known generally as fenprostalene and is marketed abroad as Synchrocept B (299) (see Veterinary drugs).

Endoperoxide Analogues. The types of endoperoxide analogues that have been synthesized and biologically evaluated are summarized in refs. 133 and 300. In general, these analogues were designed to be more chemically stable than the naturally occurring substances PGG_2 and PGH_2, which have labile 2,3-dioxabicyclo[2.2.1]heptane ring systems. Notable among the many analogues that have been synthesized are those shown in Figure 4. Azo analogue (**111**) and the epoxymethano analogues (**113**) and (**114**) appear to mimic the biological actions of the naturally occurring endoperoxides, whereas the 15-deoxy analogues (**112**), (**115**), and (**116**) inhibit PGH_2-induced human platelet aggregation (301–303).

Thromboxane A_2 Analogues. Chemical stability has also been one of the principal goals in the synthesis of thromboxane A_2 analogues; the structures of some of these compounds are given in Figure 5 (133,304). Compound (**118**) inhibits PGH_2-induced blood-platelet aggregation; however, it mimics TXA_2 in its ability to constrict vascular tissue (305). Related compound (**117**) has also been synthesized in racemic form (306).

Table 4. Selected Analogues of Classical Prostaglandins

Compound	CAS Registry No.	Compound no., generic name (trade name)	Structure	Manufacturers	Present or potential utilities
15-methyl-$PGF_{2\alpha}$, tromethamine salt	*[58551-69-2]*	carboprost (Prostin 15M)	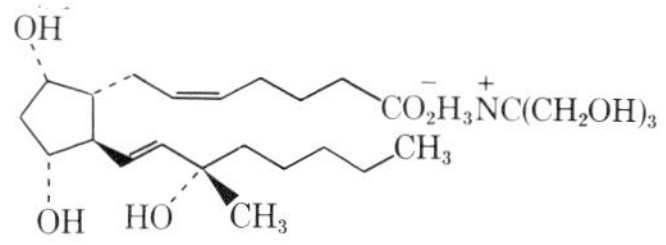 (91)	Upjohn, Kalamazoo, Mich.	termination of pregnancies, induction of menstruation, cervical dilation, refractory postpartal hemorrhage
16-phenoxy-17,18,19,20-tetranor-PGE_2, *N*-(methanesulfonyl)amide	*[54348-10-6]*	sulprostone (Sulglandin)	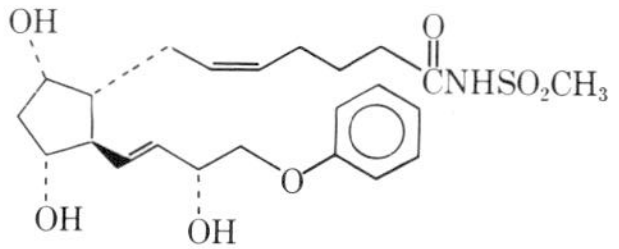 (92)	Schering, Berlin, FRG; Pfizer, New York, NY	termination of pregnancies, termination of postpartal hemorrhage, cervical dilation
(2*E*)-2,3-didehydro-16,16-dimethyl-PGE_1, methyl ester	*[64318-79-2]*	ONO-802; gemeprost	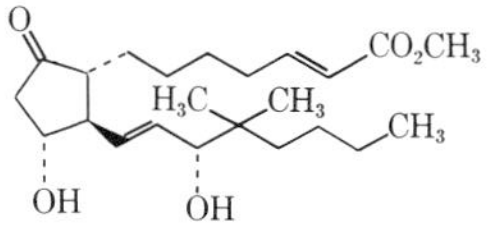 (93)	Ono, Osaka, Japan	termination of pregnancies, cervical dilation
9-deoxo-9-methylene-16,16-dimethyl-PGE_2	*[61263-35-2]*	meteneprost	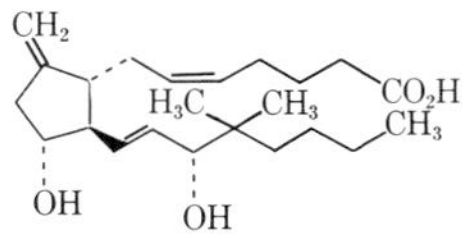 (94)	Upjohn	induction of menstruation, cervical dilation
16-[*m*-(trifluoromethyl)phenoxy]-17,18,19,20-tetranor-$PGF_{2\alpha}$	*[59685-93-7]*	ICI 81008; fluprostenol (Equimate)	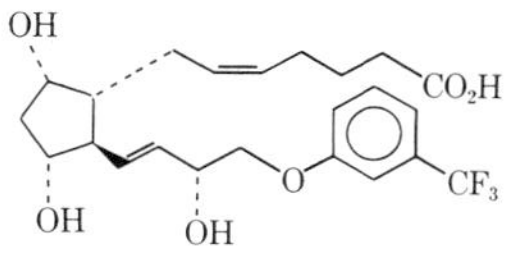 (95)	Imperial Chemical Industries (ICI), Cheshire, UK	synchronization of estrus and treatment of infertility in horses

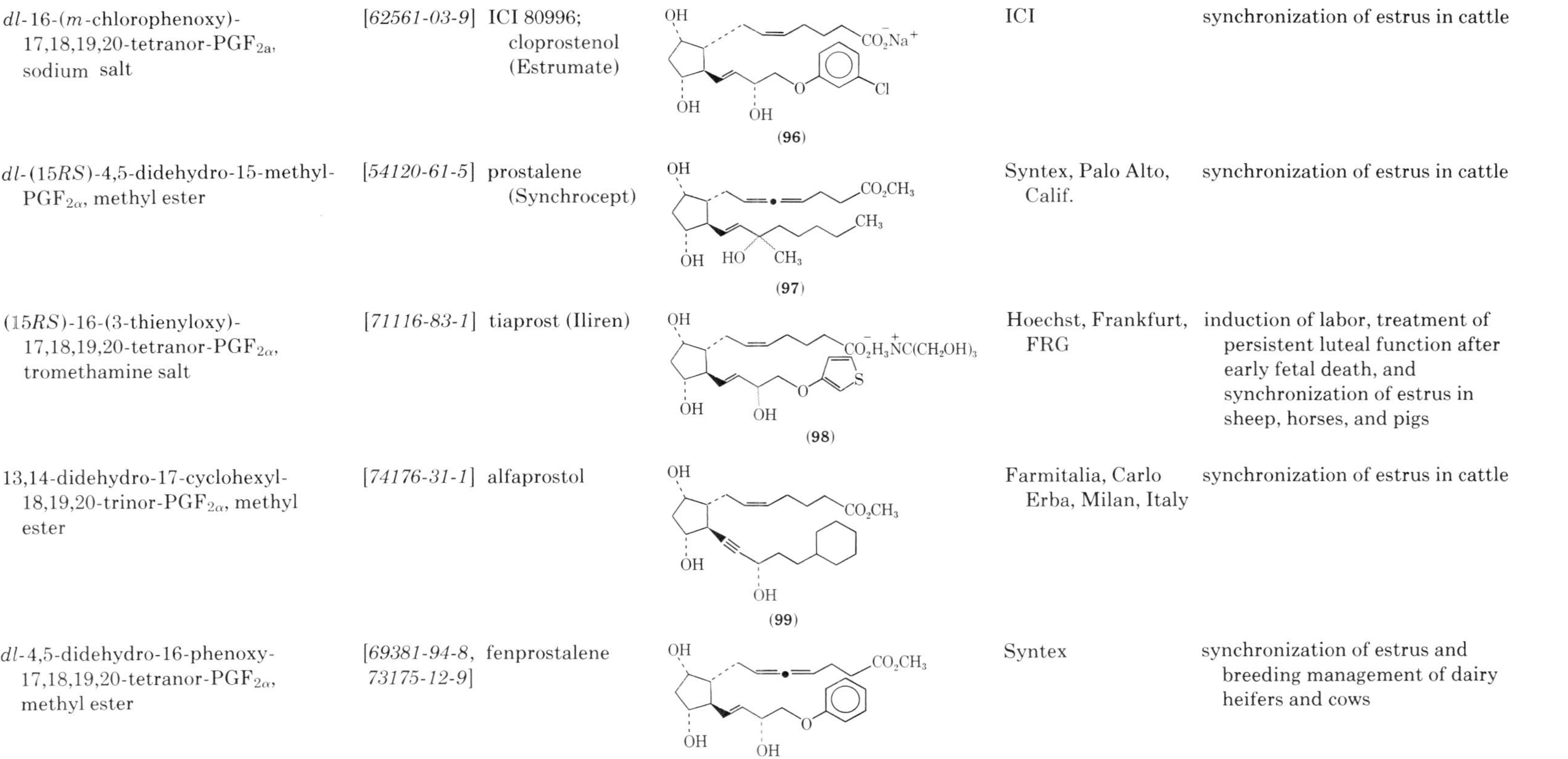

Chemical name	CAS Registry No.	Name	Structure	Manufacturer	Use
dl-16-(*m*-chlorophenoxy)-17,18,19,20-tetranor-PGF_{2a}, sodium salt	[*62561-03-9*]	ICI 80996; cloprostenol (Estrumate)	(**96**)	ICI	synchronization of estrus in cattle
dl-(15*RS*)-4,5-didehydro-15-methyl-$PGF_{2\alpha}$, methyl ester	[*54120-61-5*]	prostalene (Synchrocept)	(**97**)	Syntex, Palo Alto, Calif.	synchronization of estrus in cattle
(15*RS*)-16-(3-thienyloxy)-17,18,19,20-tetranor-$PGF_{2\alpha}$, tromethamine salt	[*71116-83-1*]	tiaprost (Iliren)	(**98**)	Hoechst, Frankfurt, FRG	induction of labor, treatment of persistent luteal function after early fetal death, and synchronization of estrus in sheep, horses, and pigs
13,14-didehydro-17-cyclohexyl-18,19,20-trinor-$PGF_{2\alpha}$, methyl ester	[*74176-31-1*]	alfaprostol	(**99**)	Farmitalia, Carlo Erba, Milan, Italy	synchronization of estrus in cattle
dl-4,5-didehydro-16-phenoxy-17,18,19,20-tetranor-$PGF_{2\alpha}$, methyl ester	[*69381-94-8*, *73175-12-9*]	fenprostalene	(**100**)	Syntex	synchronization of estrus and breeding management of dairy heifers and cows

Table 4. (*continued*)

Compound	CAS Registry No.	Compound no., generic name (trade name)	Structure	Manufacturers	Present or potential utilities
15-dehydro-16-phenoxy-17,18,19,20-tetranor-$PGF_{2\alpha}$, ethylene ketal, tromethamine salt	[*59619-82-8*]	ZK 71677	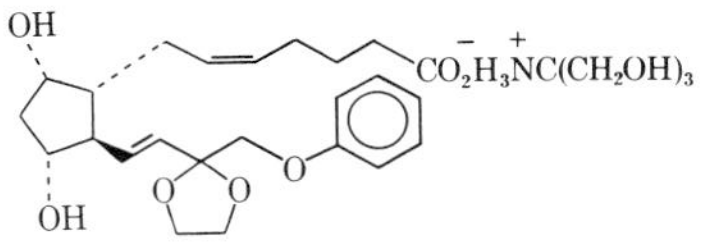 (**101**)	Schering	synchronization of estrus in cattle, sheep, and pigs
(2*E*,17*S*)-2,3-didehydro-17,20-dimethyl-PGE_1	[*74397-12-9*]	OP-1206	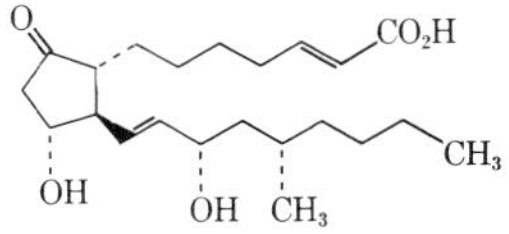 (**102**)	Ono	treatment of peripheral circulating disorders and ischemic heart disease
(15*R*)-15-methyl-PGE_2	[*55028-70-1*]	arbaprostil (Arbacet)	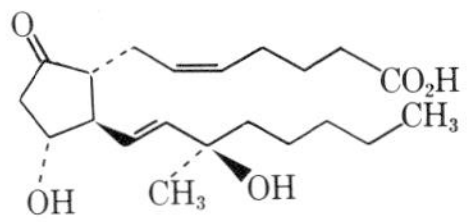 (**103**)	Upjohn	treatment of peptic ulcers, protection of gastric mucosa
dl-(16*RS*)-15-deoxy-16-hydroxy-16-methyl-PGE_1, methyl ester	[*59122-46-2*]	misoprostol	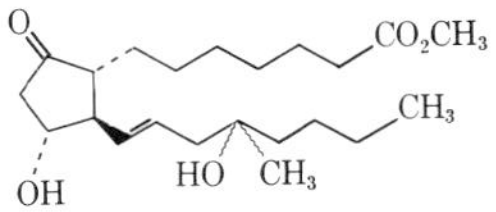 (**104**)	Searle, Chicago, Ill.	treatment of peptic ulcers, protection of gastric mucosa
(11*R*)-11-deoxy-11,16,16-trimethyl-PGE_2	[*69900-72-7*]	RO 21-6937	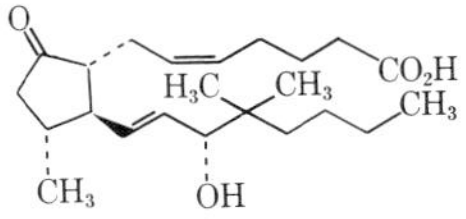 (**105**)	Hoffmann-La-Roche, Nutley, N.J.	treatment of peptic ulcers, protection of gastric musoca

dl-(16*RS*)-2-decarboxy-2-hydroxy-acetyl-15-deoxy-16-hydroxy-16-methyl-PGE_1	[*72079-25-5*]	CL-115,574	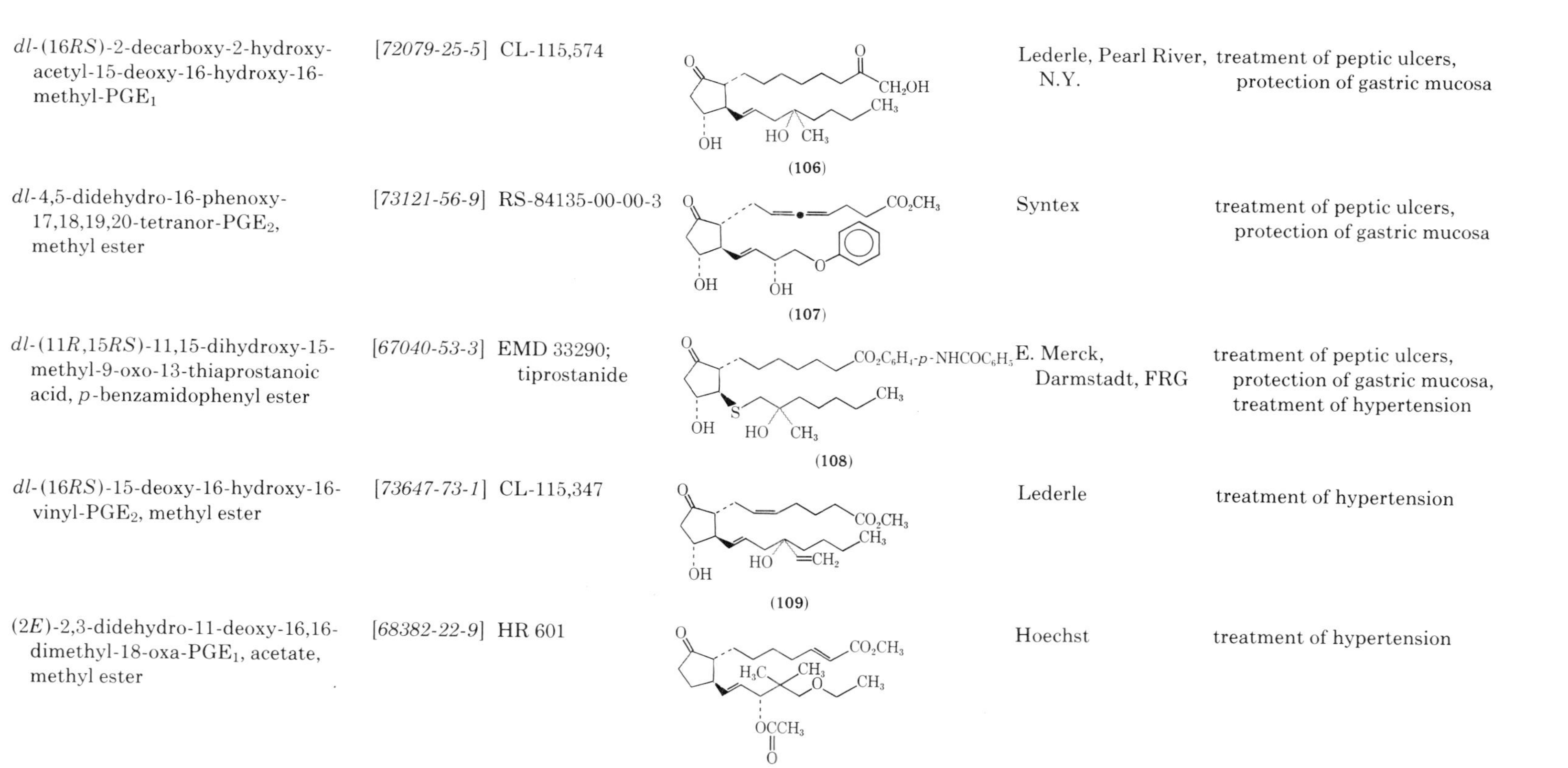 (**106**)	Lederle, Pearl River, N.Y.	treatment of peptic ulcers, protection of gastric mucosa
dl-4,5-didehydro-16-phenoxy-17,18,19,20-tetranor-PGE_2, methyl ester	[*73121-56-9*]	RS-84135-00-00-3	(**107**)	Syntex	treatment of peptic ulcers, protection of gastric mucosa
dl-(11*R*,15*RS*)-11,15-dihydroxy-15-methyl-9-oxo-13-thiaprostanoic acid, *p*-benzamidophenyl ester	[*67040-53-3*]	EMD 33290; tiprostanide	(**108**)	E. Merck, Darmstadt, FRG	treatment of peptic ulcers, protection of gastric mucosa, treatment of hypertension
dl-(16*RS*)-15-deoxy-16-hydroxy-16-vinyl-PGE_2, methyl ester	[*73647-73-1*]	CL-115,347	(**109**)	Lederle	treatment of hypertension
(2*E*)-2,3-didehydro-11-deoxy-16,16-dimethyl-18-oxa-PGE_1, acetate, methyl ester	[*68382-22-9*]	HR 601	(**110**)	Hoechst	treatment of hypertension

Table 5. Selected Analogues of Prostaglandin I_2 (Prostacyclin)

Compound	CAS Registry No.	Compound no., generic name (trade name)	Structure	Manufacturers	Present and potential utilities
(16*RS*)-6*a*-carba-16-methyl-18,18,19,19-tetradehydro-PGI_2	[*73873-87-7*]	ZK 36,374; ciloprost	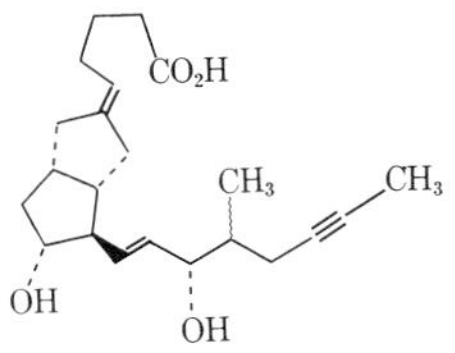 CO_2H, CH_3, CH_3, OH, OH	Schering, Berlin, FRG	extracorporeal circulation, treatment of coronary or vascular occlusions
(16*RS*)-5-cyano-16-methyl-PGI_2	[*71097-83-1*]	ZK 34,798; nileprost	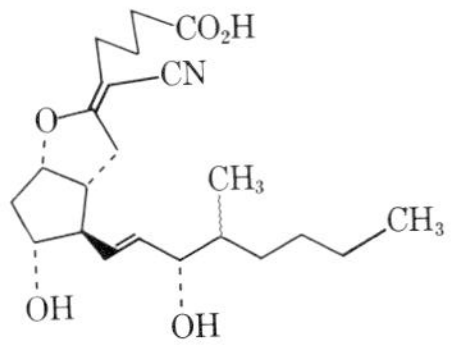 CO_2H, CN, O, CH_3, CH_3, OH, OH	Schering	treatment of peptic ulcers, protection of gastric mucosa, bronchodilation
dl-9-deoxy-9α,6-nitrilo-5-thia-$PGF_{1\alpha}$, methyl ester	[*80225-28-1*]	HOE-892	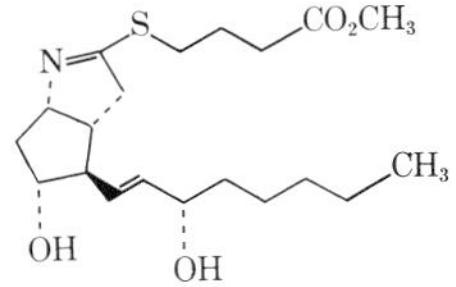 N, S, CO_2CH_3, CH_3, OH, OH	Hoechst, Frankfurt, FRG	treatment of coronary or vascular occlusions

6a-carba-15-cyclopentyl-16,17,18,19,20-pentanor-PGI_2	[*73996-62-0*] OP-41483	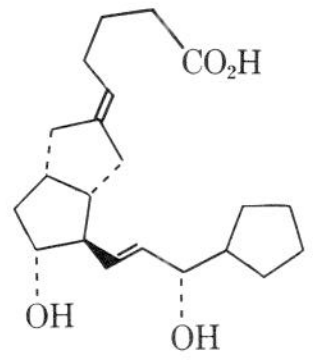 CO_2H; OH; OH	Ono, Osaka, Japan	extracorporeal circulation, treatment of coronary or vascular occlusions
7-oxo-PGI_2	[*79821-50-4*]	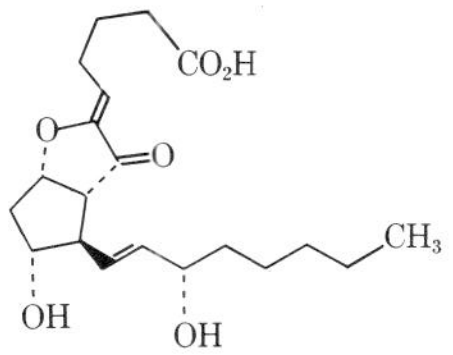 CO_2H; O; CH_3; OH; OH	Chinoin, Budapest, Hungary	extracorporeal circulation, treatment of coronary or vascular occlusions
(17*RS*)-6a-carba-20-isopropylidene-17-methyl-PGI_2	[*80951-69-5*] R 59,274	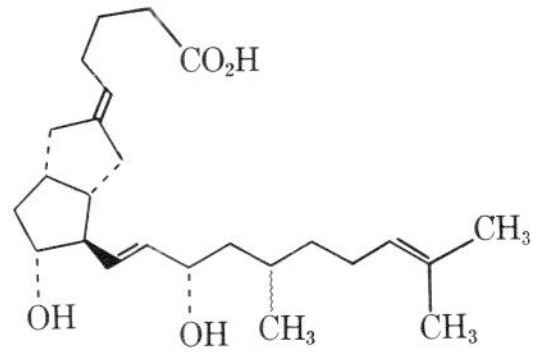 CO_2H; CH_3; OH; OH; CH_3; CH_3	Sankyo, Tokyo, Japan	extracorporeal circulation, treatment of coronary or vascular occlusions

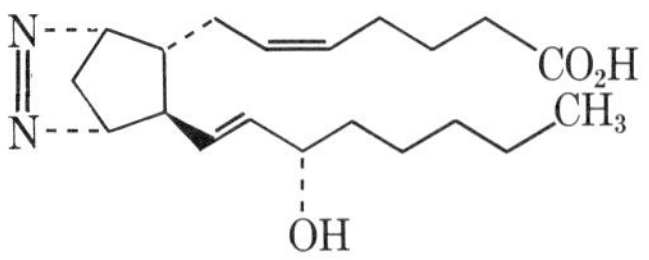

[*57712-08-0*]
(**111**)

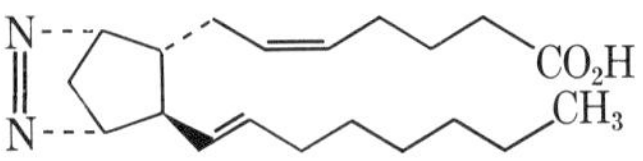

[*64192-56-9*]
(**112**)

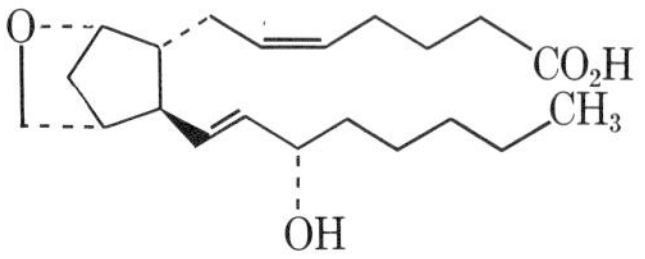

[*56985-32-1*]
(**113**)

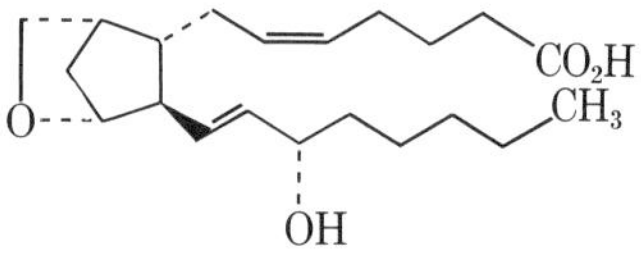

[*56985-40-1*]
(**114**)

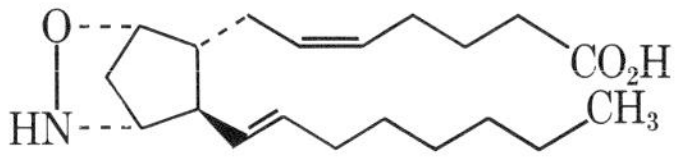

[*66464-51-5*]
(**115**)

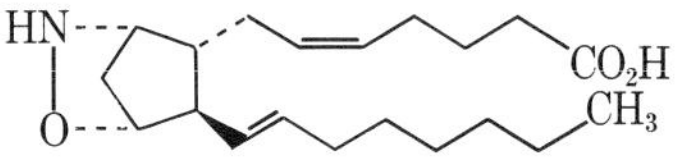

[*66464-50-4*]
(**116**)

Figure 4. Chemically stable endoperoxide analogues.

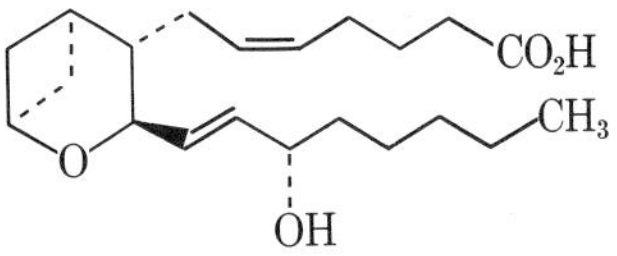

[*74243-73-5*]
(**117**)

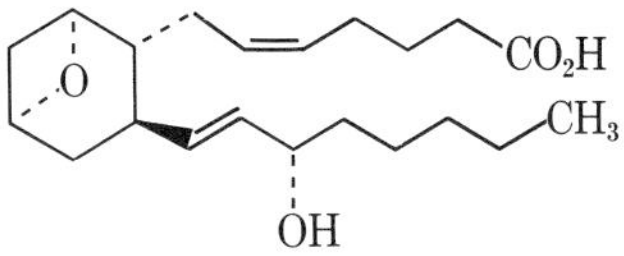

[*74271-62-8*]
(**118**)

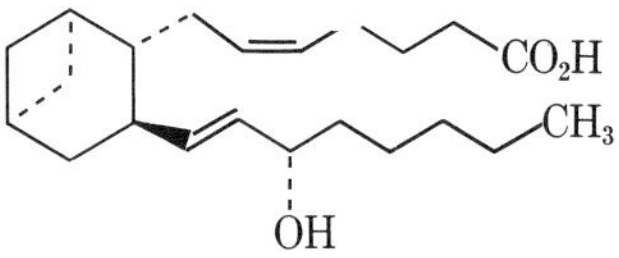

[*73543-47-2*]
(**119**)

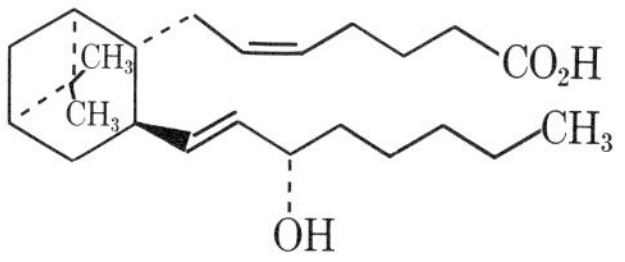

[*71154-83-1*]
(**120**)

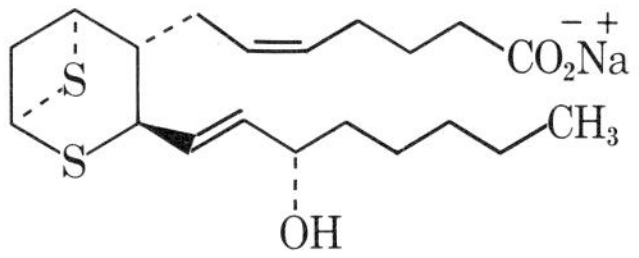

[*78685-32-2*]
(**121**)

Figure 5. Chemically stable analogues of thromboxane A_2.

The totally carbon-substituted analogues (**119**) and (**120**) have been prepared and their biological properties are reviewed in refs. 307–308. Finally, the dithia-TXA_2

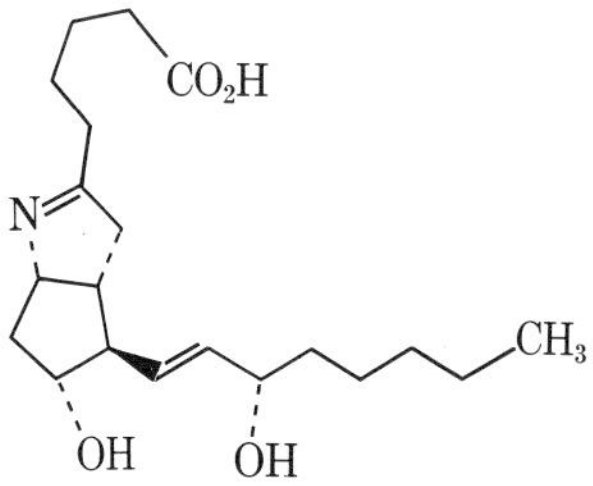

[*67496-60-0*]
(**122**)

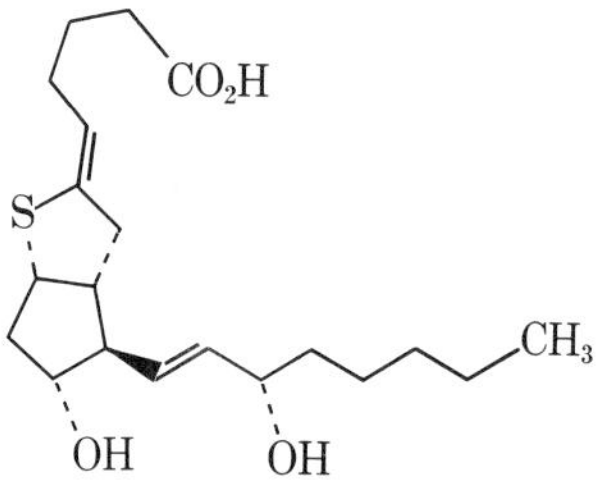

[*64868-63-9*]
(**123**)

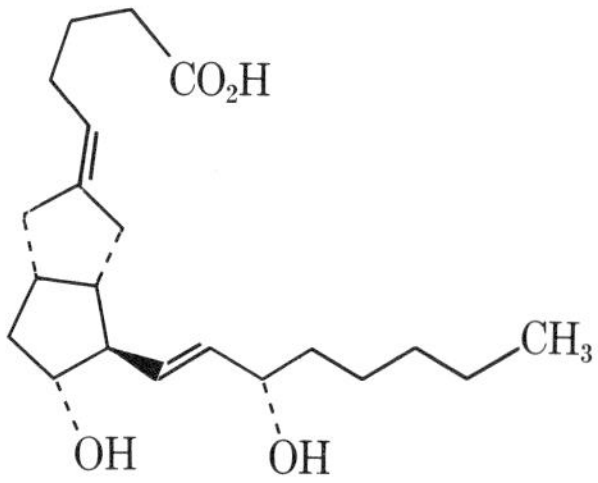

[*69552-46-1*]
(**124**)

Figure 6. Chemically stable analogues of prostaglandin I_2 (prostacyclin).

analogue (**121**) has been synthesized in racemic form, and it exhibits biological properties that are very similar to those of TXA_2 (309).

Prostaglandin I_2 (Prostacyclin) Analogues. The potent ability of prostaglandin I_2 to inhibit the aggregation of human platelets caused by various stimuli suggests its therapeutic potential. However, PGI_2 is quite unstable and has a half-life of ca 2 min under physiological conditions. As is the case with the endoperoxides, PGG_2 and PGH_2, and TXA_2, the syntheses of stable PGI_2 analogues has been pursued (133,310–311). Notable among the chemically stable ones initially synthesized are the nitrilo analogue (**122**), the thia analogue (**123**), and the carbon analogue (**124**) shown in Figure 6 (312–314). All of these effectively inhibit adenosine-diphosphate-induced aggregation of human blood platelets *in vitro*. Six analogues of PGI_2 which, as of early 1983, are in advanced stages of biological testing or development are listed in Table 5.

BIBLIOGRAPHY

1. N. A. Nelson, *J. Med. Chem.* **17,** 911 (1974).
2. R. A. Johnson, D. R. Morton, and N. A. Nelson, *Prostaglandins* **15,** 737 (1978).
3. B. Samuelsson, M. Hamberg, L. J. Roberts, J. A. Oates, and N. A. Nelson, *Prostaglandins* **16,** 857 (1978).
4. B. Samuelsson and S. Hammarstrom, *Prostaglandins* **19,** 645 (1980).
5. J. H. Fletcher, O. C. Dermer, and R. B. Fox, *Nomenclature of Organic Compounds*, American Chemical Society Advances in Chemistry Series, Vol. 126, Washington, D.C., 1974; *Selection of Index Names for Chemical Substances, Chem. Abstracts* **82,** Index Guide, American Chemical Society, Washington, D.C., 1982.
6. *Chem. Brit.*, 847 (Dec. 1982).
7. U. S. von Euler, *Klin. Wochenschr.* **14,** 1182 (1935).
8. S. Bergstrom and J. Sjovall, *Acta Chem. Scand.* **14,** 1693, 1701, 1706 (1960).
9. S. Bergstrom, L. Krabisch, B. Samuelsson, and J. Sjovall, *Acta Chem. Scand.* **16,** 969 (1962).
10. S. Bergstrom, F. Dressler, R. Ryhage, B. Samuelsson, and J. Sjovall, *Ark. Kemi* **19,** 563 (1962).
11. B. Samuelsson, *J. Am. Chem. Soc.* **85,** 1878 (1963).
12. S. Bergstrom, F. Dressler, L. Krabisch, R. Ryhage, and J. Sjovall, *Ark. Kemi* **20,** 63 (1962).
13. B. Samuelsson, *Biochim. Biophys. Acta* **84,** 707 (1964).
14. S. Abrahamsson, S. Bergstrom, and B. Samuelsson, *Proc. Chem. Soc. (London)*, 332 (1962).
15. S. Abrahamsson, *Acta Crystallogr.* **16,** 409 (1963).
16. D. H. Nugteren, D. A. van Dorp, S. Bergstrom, M. Hamberg, and B. Samuelsson, *Nature* **212,** 38 (1966).
17. D. A. van Dorp, R. K. Beerthuis, D. H. Nugteren, and H. von Heman, *Biochim. Biophys. Acta* **90,** 204 (1964).
18. S. Bergstrom, H. Danielson, and B. Samuelsson, *Biochim. Biophys. Acta* **90,** 207 (1964).
19. U.S. Pat. 3,296,091 (Feb. 19, 1964), P. F. Beal, G. S. Fonken, and J. E. Pike (to The Upjohn Co.).
20. P. F. Beal, J. C. Babcock, and F. H. Lincoln, *J. Am. Chem. Soc.* **88,** 3131 (1966).
21. G. Just and C. Simonovitch, *Tetrahedron Lett.*, 2093 (1967).
22. E. J. Corey, N. H. Anderson, R. M. Carlson, J. Paust, E. Vedejs, I. Vlattas, and R. E. K. Winter, *J. Am. Chem. Soc.* **90,** 3245 (1968).
23. A. Weinheimer and R. Spraggins, *Tetrahedron Lett.*, 5185 (1969).
24. M. Berte, *Prostaglandins Obtained from the Gorgonian Plexaura homomalla*, World Intellectual Property Organization Publication No. 710(E), 1981; J. L. Theodor in P. Crabbe, ed., *Prostaglandin Research*, Academic Press, Inc., New York, 1977, pp. 47–63.
25. E. J. Corey, I. Vlattas, and K. Harding, *J. Am. Chem. Soc.* **91,** 535 (1969).
26. E. J. Corey, R. Noyori, and T. K. Schaaf, *J. Am. Chem. Soc.* **92,** 2586 (1970).
27. S. M. M. Karim and G. M. Filshie, *Lancet* **1,** 157 (1970).
28. U. Roth-Brandel, M. Bygdeman, N. Wiqvist, and S. Bergstrom, *Lancet* **i,** 190 (1970).
29. J. R. Vane, *Nature* **231,** 232 (1971).
30. J. B. Smith and A. L. Willis, *Nature* **231,** 235 (1971).
31. M. Hamberg and B. Samuelsson, *Proc. Natl. Acad. Sci. U.S.A.* **70,** 899 (1973).

32. D. H. Nugteren and E. Hazelhof, *Biochim. Biophys. Acta* **326,** 448 (1973).
33. M. Hamberg, J. Svensson, T. Wakabayashi, and B. Samuelsson, *Proc. Natl. Acad. Sci. U.S.A.* **71,** 345 (1974).
34. M. Hamberg and B. Samuelsson, *Proc. Natl. Acad. Sci. U.S.A.* **71,** 3400 (1974).
35. M. Hamberg, J. Svensson, and B. Samuelsson, *Proc. Natl. Acad. Sci. U.S.A.* **72,** 2994 (1975).
36. N. A. Nelson and R. W. Jackson, *Tetrahedron Lett.*, 3275 (1976).
37. R. C. Kelly, I. Schletter, and S. J. Stein, *Tetrahedron Lett.*, 3279 (1976).
38. W. P. Schneider and R. A. Morge, *Tetrahedron Lett.*, 3283 (1976).
39. C. Pace-Asciak, *J. Am. Chem. Soc.* **98,** 2348 (1976).
40. S. Moncada, R. J. Gryglewski, S. Bunting, and J. R. Vane, *Nature* **263,** 663 (1976).
41. S. Moncada and J. R. Vane, *J. Med. Chem.* **23,** 591 (1980).
42. R. A. Johnson, D. R. Morton, J. H. Kinner, R. R. Gorman, J. C. McGuire, F. F. Sun, N. Whittaker, S. Bunting, J. Salmon, S. Moncada, and J. R. Vane, *Prostaglandins* **12,** 915 (1976).
43. E. J. Corey, G. E. Keck, and I. Szekely, *J. Am. Chem. Soc.* **99,** 2006 (1977).
44. R. A. Johnson, F. H. Lincoln, J. L. Thompson, E. G. Nidy, S. A. Mizsak, and U. Axen, *J. Am. Chem. Soc.* **99,** 4182 (1977).
45. K. C. Nicolaou, W. E. Barnette, G. P. Gasic, R. L. Magolda, and W. J. Sipio, *J. Chem. Soc. Chem. Commun.*, 630 (1977).
46. N. Whittaker, *Tetrahedron Lett.*, 2805 (1977).
47. I. Tomokozi, G. Galambos, V. Simonidesz, and G. Kovacs, *Tetrahedron Lett.*, 2627 (1977).
48. B. Samuelsson, P. Borgeat, S. Hammarstrom, and R. C. Murphy in B. Samuelsson, P. W. Ramwell, and R. Paoletti, eds., *Advances in Prostaglandin and Thromboxane Research*, Vol. 6, Raven Press, New York, 1980, pp. 1–18.
49. R. C. Murphy, S. Hammarstrom, and B. Samuelsson, *Proc. Natl. Acad. Sci. U.S.A.* **76,** 4275 (1979).
50. E. J. Corey, D. A. Clark, G. Goto, A. Marfat, C. Mioskowski, B. Samuelsson, and S. Hammarstrom, *J. Am. Chem. Soc.* **102,** 1436 (1980); for manuscript correction, see E. J. Corey, D. A. Clark, G. Goto, A. Marfat, C. Mioskowski, B. Samuelsson, and S. Hammarstrom, *J. Am. Chem. Soc.* **102,** 3663 (1980).
51. H. R. Morris, G. W. Taylor, P. J. Piper, M. N. Sanhonn, and J. R. Tippins, *Prostaglandins* **19,** 185 (1980).
52. L. Orning, S. Hammarstrom, and B. Samuelsson, *Proc. Natl. Acad. Sci. U.S.A.* **77,** 2014 (1980).
53. M. K. Bach, J. R. Brashler, S. Hammarstrom, and B. Samuelsson, *Biochem. Biophys. Res. Commun.* **93,** 1121 (1980).
54. E. J. Corey, A. Marfat, G. Goto, and F. Brion, *J. Am. Chem. Soc.* **102,** 7984 (1980).
55. R. A. Lewis, E. J. Goetzl, J. M. Drazen, N. A. Soter, K. F. Austen, and E. J. Corey, *J. Exp. Med.* **154,** 1243 (1981).
56. N. A. Porter, J. D. Byers, K. M. Holden, and D. B. Menzel, *J. Am. Chem. Soc.* **101,** 4319 (1979).
57. N. A. Porter, J. D. Byers, A. E. Ali, and T. E. Eling, *J. Am. Chem. Soc.* **102,** 1183 (1980).
58. R. A. Johnson, E. G. Nidy, L. Baczynskyj, and R. R. Gorman, *J. Am. Chem. Soc.* **99,** 7738 (1977).
59. N. A. Nelson, R. C. Kelly, and R. A. Johnson, *Chem. Eng. News* **60**(33), 30 (1982).
60. S. Bergstrom in R. T. Holman, ed., *Progress in Lipid Research*, Vol. 20, Pergamon Press, Ltd., New York, 1982, pp. 7–12; E. W. Horton, *Trends in Pharmacological Sciences* **1,** 59 (1979); D. I. Weisblat in E. M. Southern, ed., *The Prostaglandins: Clinical Applications in Human Reproduction*, Futura Publishing Company, Inc., Mount Kisco, N.Y., 1972, pp. 9–13; S. Bergstrom in E. M. Southern, ed., *The Prostaglandins: Clinical Applications in Human Reproduction*, Futura Publishing Company, Inc., Mount Kisco, N.Y., 1972, pp. 5–8; U. S. von Euler, *Ann. N.Y. Acad. Sci.* **180,** 6 (1971); S. Bergstrom in S. Bergstrom and B. Samuelsson, eds., *Prostaglandins*, Interscience Publishers, a division of John Wiley & Sons, Inc., New York, 1967, pp. 21–30; S. Bergstrom, *Science* **157,** 382 (1967); U. S. von Euler and R. Eliasson, *Prostaglandins*, Academic Press, Inc., New York, 1967; J. C. Colberg, *Prostaglandins Isolation and Synthesis*, Noyes Data Corporation, Park Ridge, N.J., 1973.
61. B. Samuelsson in *The Harvey Lectures, 1979–1980*, Academic Press, Inc., New York, 1981, pp. 1–40.
62. R. R. Gorman in H. V. Rickenberg, ed., *International Review of Biochemistry*, Vol. 20, University Park Press, Baltimore, Md., 1978, pp. 81–107.
63. K. H. Gibson, *Chem. Soc. Rev.* **4,** 489 (1978).
64. B. Samuelsson, M. Goldyne, E. Granstrom, M. Hamberg, S. Hammarstrom, and C. Malmsten, *Annu. Rev. Biochem.* **47,** 997 (1978).

65. B. Samuelsson, G. Folco, E. Granstrom, H. Kindahl, and C. Malmsten in F. Coceani and P. M. Olley, eds., *Advances in Prostaglandin and Thromboxane Research*, Vol 4, Raven Press, New York, 1978, pp. 1–25.
66. S. Hammarstrom, *Arch. Biochem. Biophys.* **214,** 431 (1982).
67. A. J. Marcus, *J. Lipid Res.* **19,** 793 (1978).
68. A. J. Marcus, H. L. Ullman, and L. B. Safier, *J. Lipid Res.* **10,** 108 (1969).
69. T. K. Bills and M. J. Silver, *Fed. Proc. Fed. Am. Soc. Exp. Biol.* **34,** 322 (1975).
70. E. J. Christ and D. H. Nugteren, *Biochim. Biophys. Acta* **218,** 296 (1970).
71. P. Cohen and A. Derksen, *Br. J. Haematol.* **17,** 359 (1969).
72. R. Eliasson, *Acta Physiol. Scand. Suppl.* **46,** 1 (1959).
73. C. Pace-Asciak and L. S. Wolfe, *Biochim. Biophys. Acta* **152,** 184 (1968).
74. R. F. Irvine, *Biochem. J.* **204,** 3 (1982).
75. B. Haye, S. Champion, and C. Jacquemin in B. Samuelsson and R. Paoletti, eds., *Advances in Prostaglandin and Thromboxane Research*, Vol. 1, Raven Press, New York, 1976, pp. 29–34.
76. M. Hamberg and B. Samuelsson, *Biochem. Biophys. Res. Commun.* **61,** 942 (1974); L. J. Roberts, R. A. Lewis, J. A. Lawson, B. J. Sweetman, K. F. Austen, and J. A. Oates, *Prostaglandins* **15,** 717 (1978); W. C. Hubbard, A. Hough, J. T. Watson, and J. A. Oates, *Prostaglandins* **15,** 721 (1978); W. C. Hubbard, A. J. Hough, A. R. Brash, J. T. Watson, and J. A. Oates, *Prostaglandins* **20,** 431 (1980).
77. P. Borgeat and B. Samuelsson, *Proc. Natl. Acad. Sci. U.S.A.* **76,** 2148 (1979).
78. P. Borgeat, M. Hamberg, and B. Samuelsson, *J. Biol. Chem.* **251,** 7816 (1976); for manuscript correction, see P. Borgeat, M. Hamberg, and B. Samuelsson, *J. Biol. Chem.* **252,** 8772 (1977).
79. N. A. Porter, R. A. Wolf, W. R. Pagels, and L. J. Marnett, *Biochem. Biophys. Res. Commun.* **92,** 349 (1980).
80. B. Samuelsson in B. Samuelsson and R. Paoletti, eds., *Advances in Prostaglandin, Thromboxane and Leukotriene Research*, Vol. 9, Raven Press, New York, 1982, pp. 1–17.
81. M. W. Greaves and W. McDonald-Gibson, *Br. Med. J.* **2,** 83 (1972).
82. G. P. Lewis and P. J. Piper, *Nature* **254,** 308 (1975).
83. F. Kantrowitz, D. R. Robinson, M. B. McGuire, and L. Levine, *Nature* **258,** 737 (1975).
84. A. H. Tashjian, E. F. Voelkel, J. McDonough, and L. Levine, *Nature* **258,** 739 (1975).
85. R. J. Gryglewski, B. Panczenko, R. Korbut, L. Grodzinska, and A. Ocetkiewicz, *Prostaglandins* **10,** 343 (1975).
86. M. W. Greaves, W. P. Kingston, and K. Petty, *Br. J. Pharmacol.* **53,** 470 (1975).
87. R. J. Flower, R. Gryglewski, H. K. Cedro, and J. R. Vane, *Nature* **238,** 104 (1972).
88. S. L. Hong and L. Levine, *Proc. Natl. Acad. Sci. U.S.A.* **73,** 1730 (1976).
89. G. J. Blackwell, R. Carnuccio, M. DiRosa, R. J. Flower, L. Parente, and P. Persico, *Nature* **287,** 147 (1980).
90. F. Hirata, E. Schiffmann, K. Venkatasubramanian, D. Salomon, and J. Axelrod, *Proc. Natl. Acad. Sci. U.S.A.* **77,** 2533 (1980).
91. E. Granstrom, U. Diczfalusy, M. Hamberg, G. Hansson, C. Malmsten, and B. Samuelsson in J. A. Oates, ed., *Advances in Prostaglandin, Thromboxane and Leukotriene Research*, Vol. 10, Raven Press, New York, 1982, pp. 15–58.
92. P. J. Piper, ed., *SRS-A and Leukotrienes*, Research Studies Press, a division of John Wiley & Sons, Inc., New York, 1981.
93. M. K. Bach, J. R. Brashler, H. W. Smith, F. A. Fitzpatrick, F. F. Sun, and J. C. McGuire, *Prostaglandins* **23,** 759 (1982).
94. L. J. Roberts, A. R. Brash, and J. A. Oates in ref. 91, pp. 211–225, and references cited therein.
95. F. F. Sun, B. M. Taylor, J. C. McGuire, and P. Y. K. Wong, *Kidney Int.* **19,** 760 (1981).
96. E. W. Horton, *Prostaglandins*, Springer-Verlag, New York, 1972, pp. 67–86.
97. C. W. Lee, R. A. Lewis, E. J. Corey, A. Barton, H. Oh, A. I. Tauber, and K. F. Austen, *Proc. Natl. Acad. Sci. U.S.A.* **79,** 4166 (1982).
98. W. P. Schneider, G. L. Bundy, F. H. Lincoln, E. G. Daniels, and J. E. Pike, *J. Am. Chem. Soc.* **99,** 1222 (1977).
99. Ref. 24, pp. 315–319.
100. W. P. Schneider, *J. Chem. Soc. Chem. Commun.*, 304 (1969).
101. E. J. Corey and C. R. Cyr, *Tetrahedron Lett.*, 1761 (1974).
102. R. C. Kelly, I. Schletter and R. L. Jones, *Prostaglandins* **4,** 653 (1973).
103. E. E. Nishizawa, W. L. Miller, R. R. Gorman, G. L. Bundy, J. Svensson, and M. Hamberg, *Prostaglandins* **9,** 109 (1975).

104. G. L. Bundy, The Upjohn Company, unpublished results, 1984.
105. E. J. Corey, T. K. Schaaf, W. Huber, U. Koelliker, and N. M. Weinshenker, *J. Am. Chem. Soc.* **92,** 397 (1970).
106. C. J. Sih, J. B. Heather, R. Sood, P. Price, G. Perruzzotti, L. F. H. Su Lee, and S. S. Lee, *J. Am. Chem. Soc.* **97,** 865 (1975).
107. J. E. Pike, F. H. Lincoln, and W. P. Schneider, *J. Org. Chem.* **34,** 3552 (1969).
108. R. A. Johnson, F. H. Lincoln, E. G. Nidy, W. P. Schneider, J. L. Thompson, and U. Axen, *J. Am. Chem. Soc.* **100,** 7690 (1978).
109. S. Hanessian and P. Lavalle, *Can. J. Chem.* **55,** 562 (1977).
110. I. Ernest, A. J. Main, and R. Menasse, *Tetrahedron Lett.* **23,** 167 (1982).
111. G. Kito, H. Okuda, S. Ohkawa, S. Terao, and K. Kikuchi, *Life Sci.* **29,** 1325 (1981).
112. P. W. Ramwell, J. E. Shaw, G. B. Clarke, M. F. Grostic, D. G. Kaiser, and J. E. Pike, *Prog. Chem. Fats Other Lipids* **9,** 231 (1968).
113. C. Hensby in ref. 24, pp. 89–120.
114. P. Crabbe, *Tetrahedron* **30,** 1979 (1974).
115. W. L. Duax and J. W. Edmonds, *Prostaglandins* **3,** 201 (1973).
116. J. W. Edmonds and W. L. Duax, *Prostaglandins* **5,** 275 (1974).
117. G. Kotovych, G. H. M. Aarts, T. T. Nakashima, and G. Bigam, *Can. J. Chem.* **58,** 974 (1980).
118. G. Kotovych and G. H. M. Aarts, *Org. Magn. Reson.* **18,** 77 (1982).
119. C. Chachaty, Z. Wolkowski, F. Piriou, and G. Lukacs, *J. Chem. Soc. Chem. Commun.*, 951 (1973).
120. G. F. Cooper and J. Fried, *Proc. Natl. Acad. Sci. U.S.A.* **70,** 1579 (1973); G. F. Cooper and J. Fried, *Proceedings of the First International Conference on Stable Isotopes in Chemistry, Biology and Medicine*, 1973, pp. 72–83.
121. S. A. Mizsak and G. Slomp, *Prostaglandins* **10,** 807 (1975).
122. J. D. Henderson, G. L. Longenecker, and B. Brown, *Chem. Eng. News*, 2 (Aug. 15, 1983).
123. E. Granstrom and H. Kindahl in J. C. Frolich, ed., *Advances in Prostaglandin and Thromboxane Research*, Vol. 5, Raven Press, New York, 1978, pp. 119–210.
124. E. Granstrom, *Prostaglandins* **15,** 3 (1978).
125. K. Green, M. Hamberg, B. Samuelsson, M. Smigel, and J. C. Frolich in ref. 122, pp. 39–118.
126. N. Anderson, C. H. Wilson, B. De, S. Tynan, J. Watkins, J. Callis, M. Giannelli, L. Harker, S. Hanson, and T. Eggerman in R. J. Hegyeli, ed., *Atherosclerosis Reviews*, Vol. 8, Raven Press, New York, 1981, pp. 1–38.
127. F. A. Fitzpatrick and M. A. Wynalda in K. K. Wu and E. C. Rossi, eds., *Prostaglandins in Clinical Medicine*, Year Book Medical Publishers, Chicago, Ill., 1982, pp. 35–47.
128. B. Samuelsson and R. Paoletti, eds., *Advances in Prostaglandin, Thromboxane and Leukotriene Research*, Vol. 9, Raven Press, New York, 1982.
129. N. H. Andersen and P. W. Ramwell, *Arch. Intern. Med.* **133,** 30 (1974).
130. M. Shodell, *Science* **83,** 79 (1983).
131. *Prevention*, 31.
132. S. M. Roberts and F. Scheinmann, eds., *New Synthetic Routes to Prostaglandins and Thromboxanes*, Academic Press, Inc., New York, 1982.
133. K. C. Nicolaou, G. P. Gasic, and W. E. Barnette, *Angew. Chem. Int. Ed. Engl.* **17,** 293 (1978).
134. D. A. Clark and A. Marfat in H. J. Hess, ed., *Annual Reports in Medicinal Chemistry*, Vol. 17, Academic Press, Inc., New York, 1982, pp. 291–300.
135. P. R. Marsham in F. D. Gunstone, ed., *Aliphatic and Related Natural Product Chemistry*, Vol. 1, The Chemical Society, London, 1979, pp. 170–235.
136. M. P. L. Caton and K. Crowshaw in G. P. Ellis and G. B. West, eds., *Progress in Medicinal Chemistry*, Vol. 15, Elsevier/North-Holland, Inc., New York, 1978, pp. 357–423.
137. M. P. L. Caton, *Tetrahedron* **35,** 2705 (1979).
138. E. J. Corey, T. Ravindranathan, and S. Terashima, *J. Am. Chem. Soc.* **93,** 4326 (1971).
139. E. J. Corey and P. L. Fuchs, *J. Am. Chem. Soc.* **94,** 4014 (1972).
140. N. M. Weinshenker, *Prostaglandins* **3,** 219 (1973).
141. S. Ranganathan, D. Ranganathan, and A. K. Mehrotra, *J. Am. Chem. Soc.* **96,** 5261 (1974).
142. S. Ranganathan, D. Ranganathan, and A. K. Mehrotra, *Tetrahedron Lett.*, 1215 (1975).
143. B. M. Trost and Y. Tamaru, *J. Am. Chem. Soc.* **97,** 3528 (1975).
144. E. J. Corey and H. E. Ensley, *J. Am. Chem. Soc.* **97,** 6908 (1975).
145. P. A. Bartlett, F. R. Green, and T. R. Webb, *Tetrahedron Lett.*, 331 (1977).
146. H. E. Ensley, C. A. Parnell, and E. J. Corey, *J. Org. Chem.* **43,** 1610 (1978).

147. J. S. Bindra and R. Bindra, *Prostaglandin Synthesis*, Academic Press, Inc., New York, 1977, pp. 187–245.
148. N. Inukai, H. Iwamoto, I. Yanagisawa, N. Nagano, T. Tamura, Y. Ishii, and M. Murakami, *Chem. Pharm. Bull.* **24,** 2566 (1976).
149. N. M. Weinshenker, G. A. Crosby, and J. Y. Wong, *J. Org. Chem.* **40,** 1966 (1975).
150. E. J. Corey and J. W. Suggs, *J. Org. Chem.* **40,** 2554 (1975).
151. M. Vandewalle, V. Sipido, and H. DeWilde, *Bull. Soc. Chim. Belg.* **79,** 403 (1970).
152. E. J. Corey, Z. Arnold, and J. Hutton, *Tetrahedron Lett.*, 307 (1970).
153. E. J. Corey and T. Ravindranathan, *Tetrahedron Lett.*, 4753 (1971).
154. D. Brewster, M. Myers, J. Ormerod, M. E. Spinner, S. Turner, and A. C. B. Smith, *J. Chem. Soc. Chem. Commun.*, 1235 (1972).
155. G. Jones, R. A. Raphael, and S. Wright, *J. Chem. Soc. Chem. Commun.*, 609 (1972).
156. F. Kienzle, G. W. Holland, J. L. Jernow, S. Kwok, and P. Rosen, *J. Org. Chem.* **38,** 3440 (1973).
157. E. J. Corey and B. B. Snider, *Tetrahedron Lett.*, 3091 (1973).
158. D. Brewster, M. Myers, J. Ormerod, P. Otter, A. C. B. Smith, M. E. Spinner, and S. Turner, *J. Chem. Soc. Perkin Trans. 1*, 2796 (1973).
159. R. B. Woodward, J. Gosteli, I. Ernest, R. J. Friary, G. Nestler, H. Raman, R. Sitrin, Ch. Suter, and J. K. Whitesell, *J. Am. Chem. Soc.* **95,** 6853 (1973).
160. J. S. Bindra, A. Grodski, and T. K. Schaaf, *J. Am. Chem. Soc.* **95,** 7522 (1973).
161. E. J. Corey and C. U. Kim, *J. Org. Chem.* **38,** 1233 (1973).
162. J. Van Hooland, P. De Clercq, and M. Vanderwalle, *Tetrahedron Lett.*, 4343 (1974).
163. P. De Clercq and M. Vandewalle, *Bull. Soc. Chim. Belg.* **83,** 305 (1974).
164. P. De Clercq, D. Van Haver, D. Tavernier, and M. Vandewalle, *Tetrahedron* **30,** 55 (1974).
165. G. Jones, R. A. Raphael, and S. Wright, *J. Chem. Soc. Perkin Trans. 1*, 1676 (1974).
166. E. J. Corey and B. B. Snider, *J. Org. Chem.* **39,** 256 (1974).
167. E. D. Brown, R. Clarkson, T. J. Leeney, and G. E. Robinson, *J. Chem. Soc. Chem. Commun.*, 642 (1974).
168. E. J. Corey, K. C. Nicolaou, and D. J. Beames, *Tetrahedron Lett.*, 2439 (1974).
169. R. Peel and J. K. Sutherland, *J. Chem. Soc. Chem. Commun.*, 151 (1974).
170. R. Coen, P. De Clercq, D. Van Haver, and M. Vandewalle, *Bull. Soc. Chim. Belg.* **84,** 203 (1975).
171. W. Van Brussel, J. Van Hooland, P. De Clercq, and M. Vandewalle, *Bull. Soc. Chim. Belg.* **84,** 813 (1975).
172. M. Samson, P. De Clercq, and M. Vandewalle, *Tetrahedron* **31,** 1233 (1975).
173. H. Shimomura, J. Katsube, and M. Matsui, *Agric. Biol. Chem.* **39,** 657 (1975).
174. A. Fischli, M. Klaus, H. Mayer, P. Schonholzer, and R. Ruegg, *Helv. Chem. Acta* **58,** 564 (1975).
175. G. A. Crosby, N. M. Weinshenker, and H.-S. Uh, *J. Am. Chem. Soc.* **97,** 2232 (1975).
176. S. Ranganathan, D. Ranganathan, and A. K. Mehrotra, *Tetrahedron Lett.*, 1215 (1975).
177. P. De Clercq, M. De Smet, K. Legein, F. Vanhulle, and M. Vandewalle, *Bull. Soc. Chim. Belg.* **85,** 503 (1976).
178. P. De Clercq, R. Coen, E. Van Hoff, and M. Vandewalle, *Tetrahedron* **32,** 2747 (1976).
179. I. Tomoskozi, L. Gruber, G. Kovacs, I. Szekely, and V. Simonidesz, *Tetrahedron Lett.*, 4639 (1976).
180. K. G. Paul, F. Johnson, and D. Favara, *J. Am. Chem. Soc.* **98,** 1285 (1976).
181. I. Ernest, *Angew. Chem. Int. Ed. Engl.* **15,** 207 (1976).
182. S. Takano, N. Kubodera, and K. Ogasawara, *J. Org. Chem.* **42,** 786 (1977).
183. E. D. Brown, R. Clarkson, T. J. Leeney, and G. E. Robinson, *J. Chem. Soc. Perkin Trans. 1*, 1507 (1978).
184. M. Naruto, K. Ohno, and N. Naruse, *Chem. Lett.*, 1419 (1978).
185. L. A. Paquette, G. D. Crouse, and A. K. Sharma, *J. Am. Chem. Soc.* **102,** 3972 (1980).
186. S. Goldstein, P. Vannes, C. Honge, A. M. Frisque-Hesbain, C. Wiaux-Zamar, and L. Ghosez, *J. Am. Chem. Soc.* **103,** 4616 (1981).
187. L. A. Paquette and G. D. Crouse, *Tetrahedron* **37**(Suppl. 1), 281 (1981).
188. I. Fleming and B.-W. Au-Yeung, *Tetrahedron* **37**(Suppl. 1), 13 (1981).
189. E. J. Corey, K. B. Becker, and R. K. Varma, *J. Am. Chem. Soc.* **94,** 8616 (1972).
190. J. Bowler, K. B. Mallion, and R. A. Raphael, *Synth. Commun.* **4,** 211 (1974).
191. E. J. Corey, K. C. Nicolaou, M. Shibasaki, Y. Machida, and C. S. Shiner, *Tetrahedron Lett.*, 3183 (1975).
192. J. Hutton, M. Senior, and N. C. A. Wright, *Synth. Commun.* **9,** 799 (1979).

193. S. Iguchi, H. Nakai, M. Hayashi, and H. Yamamoto, *J. Org. Chem.* **44,** 1363 (1979).
194. S. Iguchi, H. Nakai, M. Hayashi, H. Yamamoto, and K. Maruoka, *Bull. Chem. Soc. Jpn.* **54,** 3033 (1981).
195. A. L. Gemal and J.-L. Luche, *J. Am. Chem. Soc.* **103,** 5454 (1981).
196. F. Johnson, K. G. Paul, D. Favara, R. Ciabatti, and U. Guzzi, *J. Am. Chem. Soc.* **104,** 2190 (1982).
197. E. J. Corey and R. K. Varma, *J. Am. Chem. Soc.* **93,** 7319 (1971).
198. F. H. Lincoln, W. P. Schneider, and J. E. Pike, *J. Org. Chem.* **38,** 951 (1973).
199. T. K. Schaaf and E. J. Corey, *J. Org. Chem.* **37,** 2921 (1972).
200. C. Doria, P. Gaio, and C. Gandolfi, *Tetrahedron Lett.*, 4307 (1972).
201. K. B. Mallion and E. R. H. Walker, *Synth. Commun.* **5,** 221 (1975).
202. A. E. Green, M. A. Teixeira, E. Barreiro, A. Cruz, and P. Crabbe, *J. Org. Chem.* **47,** 2553 (1982).
203. E. J. Corey, H. Shirahama, H. Yamamoto, S. Terashima, A. Venkateswarlu, and T. K. Schaaf, *J. Am. Chem. Soc.* **93,** 1490 (1971).
204. R. C. Kelly, V. Van Rheenen, I. Schletter, and M. D. Pillai, *J. Am. Chem. Soc.* **95,** 2746 (1973), and references cited therein.
205. D. R. White, *Tetrahedron Lett.*, 1753 (1976).
206. R. F. Newton and S. M. Roberts, *Tetrahedron* **36,** 2163 (1980), and references cited therein.
207. E. D. Brown and T. J. Lilley, *J. Chem. Soc. Chem. Commun.*, 39 (1975).
208. K. Sakai and T. Kobori, *Tetrahedron Lett.* **22,** 115 (1981).
209. K. Kondo, T. Umemoto, K. Yako, and D. Tunemoto, *Tetrahedron Lett.*, 3927 (1978).
210. K. Sakai, T. Kobori, and T. Fujisawa, *Tetrahedron Lett.* **22,** 115 (1981).
211. J. Fried, C. H. Lin, J. C. Sih, P. Dalven, and G. F. Cooper, *J. Am. Chem. Soc.* **94,** 4342 (1972).
212. J. Fried, J. C. Sih, C. H. Lin, and P. Dalven, *J. Am. Chem. Soc.* **94,** 4343 (1972).
213. J. Fried and J. C. Sih, *Tetrahedron Lett.*, 3899 (1973); J. J. Partridge, N. K. Chadha, and M. R. Uskokovic, *J. Am. Chem. Soc.* **95,** 7171 (1973).
214. Ref. 147, pp. 99–145.
215. C. J. Sih, R. G. Salomon, P. Price, R. Sood, and G. Peruzzotti, *J. Am. Chem. Soc.* **97,** 857 (1975), and references cited therein.
216. C. J. Sih, J. B. Heather, R. Sood, P. Price, G. Peruzzotti, L. F. Hsu Lee, and S. S. Lee, *J. Am. Chem. Soc.* **97,** 865 (1975).
217. F. S. Alvarez, D. Wren, and A. Prince, *J. Am. Chem. Soc.* **94,** 7823 (1972).
218. A. F. Kluge, K. G. Untch, and J. H. Fried, *J. Am. Chem. Soc.* **94,** 7827 (1972), and references cited therein.
219. K. F. Bernady and M. J. Weiss, *Tetrahedron Lett.*, 4083 (1972).
220. M. B. Floyd and M. J. Weiss, *Prostaglandins* **3,** 921 (1973).
221. R. Pappo and P. Collins, *Tetrahedron Lett.*, 2627 (1972).
222. W. P. Schneider, G. L. Bundy, F. H. Lincoln, E. G. Daniels, and J. E. Pike, *J. Am. Chem. Soc.* **99,** 1222 (1977).
223. Y. Yura and J. Ide, *Chem. Pharm. Bull.* **17,** 408 (1969).
224. J. Katsube and M. Matsui, *Agric. Biol. Chem.* **33,** 1078 (1969).
225. R. Pappo, P. Collins, and C. Jung, *Ann. N.Y. Acad. Sci.* **180,** 64 (1971).
226. S. Yamada, M. Kitamoto, and S. Terashima, *Tetrahedron Lett.*, 3165 (1976).
227. M. Kitamoto, K. Kameo, S. Terashima, and S. Yamada, *Chem. Pharm. Bull.* **25,** 1273 (1977).
228. C. S. Subramaniam, P. J. Thomas, V. R. Mamdapur, and M. S. Chadha, *Tetrahedron Lett.*, 495 (1978).
229. A. Citterio and E. Vismara, *Synthesis*, 751 (1980).
230. R. Pappo, P. Collins, and C. Jung, *Tetrahedron Lett.*, 943 (1973).
231. S. Kurozumi, T. Toru, and S. Ishimoto, *Tetrahedron Lett.*, 4959 (1973).
232. L. Gruber, I. Tomoskozi, E. Major, and G. Kovacs, *Tetrahedron Lett.*, 3729 (1974).
233. M. B. Floyd, *Synth. Commun.* **4,** 317 (1974).
234. G. Stork, C. Kowalski, and G. Garcia, *J. Am. Chem. Soc.* **97,** 3258 (1975).
235. L. Gruber, E. Major, and I. Tomoskozi, *Acta Chim. Sci. Hung.* **87,** 183 (1975).
236. R. A. Ellison, E. R. Lukenbach, and C. Chiu, *Tetrahedron Lett.*, 499 (1975).
237. M. Kobayashi, S. Kurozumi, T. Toru, and S. Ishimoto, *Chem. Lett.*, 1341 (1976).
238. G. K. Cooper and L. J. Dolby, *Tetrahedron Lett.*, 4675 (1976).
239. G. Piancatelli and A. Scettri, *Tetrahedron Lett.*, 1131 (1977).
240. G. Piancatelli and A. Scettri, *Synthesis*, 116 (1977).
241. T. Shono, H. Hamaguchi, and K. Aoki, *Chem. Lett.*, 1053 (1977).

242. F. Naf and R. Decorzant, *Helv. Chem. Acta* **61,** 2524 (1978).
243. M. B. Floyd, *J. Org. Chem.* **43,** 1641 (1978).
244. T.-J. Lee, *Tetrahedron Lett.*, 2297 (1979).
245. M. Gill and R. W. Rickards, *J. Chem. Soc. Chem. Commun.*, 121 (1979).
246. M. Gill, H. P. Bainton, and R. W. Rickards, *Tetrahedron Lett.* **22,** 1437 (1981).
247. M. Gill and R. W. Rickards, *J. Chem. Soc. Perkin Trans. 1*, 599 (1981).
248. M. Gill and R. W. Rickards, *Aust. J. Chem.* **34,** 2587 (1981).
249. *Ibid.*, 1063 (1981).
250. R. J. Ferrier and P. Prasit, *J. Chem. Soc. Chem. Commun.*, 983 (1981).
251. G. Stork and T. Takahashi, *J. Am. Chem. Soc.* **99,** 1275 (1977).
252. G. Stork and M. Isobe, *J. Am. Chem. Soc.* **97,** 4745, 6260 (1975); G. Stork and G. Kraus, *J. Am. Chem. Soc.* **98,** 6747 (1976).
253. M. Suzuki, T. Kawagishi, T. Suzuki, and R. Noyori, *Tetrahedron Lett.* **23,** 4057 (1982).
254. J. Schwartz and Y. Hayasi, *Tetrahedron Lett.* **21,** 1497 (1980).
255. J. Schwartz, M. J. Loots, and H. Kosugi, *J. Am. Chem. Soc.* **102,** 1333 (1980).
256. R. E. Donaldson and P. L. Fuchs, *J. Am. Chem. Soc.* **103,** 2108 (1981).
257. G. Stork, T. Takahashi, I. Kawamoto, and T. Suzuki, *J. Am. Chem. Soc.* **100,** 8272 (1978).
258. M. Hayashi and T. Tanouchi, *J. Org. Chem.* **38,** 2115 (1973).
259. E. F. Jenny, P. Schaublin, H. Fritz, and H. Fuhrer, *Tetrahedron Lett.*, 2235 (1974).
260. E. E. Nishizawa, W. L. Miller, R. R. Gorman, G. L. Bundy, J. Svensson, and M. Hamberg, *Prostaglandins* **9,** 109 (1975).
261. D. P. Reynolds, R. F. Newton, and S. M. Roberts, *J. Chem. Soc. Chem. Commun.*, 1150 (1979).
262. T. W. Hart, D. A. Metcalfe, and F. Scheinmann, *J. Chem. Soc. Chem. Commun.*, 156 (1979).
263. A. G. Cameron and A. T. Hewson, *Tetrahedron Lett.* **23,** 561 (1982).
264. Y. Konishi, H. Wakatsuka, and M. Hayashi, *Chem. Lett.*, 377 (1980).
265. Ref. 147, pp. 291–330.
266. J.-B. Wiel and F. Rouessac, *J. Chem. Soc. Chem. Commun.*, 446 (1976).
267. D. F. Taber, *J. Am. Chem. Soc.* **99,** 3513 (1977).
268. J. Martel, A. Bladefont, C. Marie, M. Vivat, E. Toromanoff, and J. Buendia, *Bull. Soc. Chim. France*, Part 2, 131 (1978); J. Buendia, M. Vivat, E. Toromanoff, and J. Martel, *Bull. Soc. Chim. France*, 140 (1978).
269. A. X. Mitra, *Diss. Abstr. Int. B* **38,** 4816 (1978).
270. L. A. Paquette, G. D. Crouse, and A. K. Sharma, *J. Am. Chem. Soc.* **102,** 3972 (1980).
271. M. A. W. Finch, S. M. Roberts, G. T. Woolley, and R. F. Newton, *J. Chem. Soc. Perkin Trans. 1*, 1725 (1981).
272. R. R. Gorman, F. F. Sun, O. V. Miller, and R. A. Johnson, *Prostaglandins* **13,** 1043 (1977).
273. R. F. Newton, S. M. Roberts, B. J. Wakefield, and G. T. Woolley, *J. Chem. Soc. Chem. Commun.*, 922 (1981).
274. R. A. Johnson, F. H. Lincoln, E. G. Nidy, W. P. Schneider, J. L. Thompson, and U. F. Axen, *J. Am. Chem. Soc.* **100,** 7690 (1978).
275. U.S. Pat. 4,338,325 (July 6, 1982), R. A. Johnson, F. H. Lincoln, and J. E. Pike (to The Upjohn Co.).
276. M. A. W. Finch, S. M. Roberts, and R. F. Newton, *J. Chem. Soc. Chem. Commun.*, 589 (1980).
277. H. Ohrui and S. Emoto, *Agric. Biol. Chem.* **41,** 1773 (1977).
278. E. J. Corey, M. Shibasaki, J. Knolle, and T. Sugahara, *Tetrahedron Lett.*, 785 (1977).
279. E. J. Corey, M. Shibasaki, and J. Knolle, *Tetrahedron Lett.*, 1625 (1977).
280. O. Hernandez, *Tetrahedron Lett.*, 219 (1978).
281. A. G. Kelly and J. S. Roberts, *J. Chem. Soc. Chem. Commun.*, 228 (1980).
282. R. R. Schmidt and W. Abele, *Angew. Chem. (Suppl.)*, 616 (1982).
283. E. J. Corey, D. A. Clark, A. Marfat, and G. Goto, *Tetrahedron Lett.* **21,** 3143 (1980).
284. J. C. Buck, F. Ellis, and P. C. North, *Tetrahedron Lett.* **23,** 4161 (1982).
285. B. Samuelsson, *Science* **220,** 568 (1983).
286. Y. Guindon, R. Zamboni, C.-K. Lau, and J. Rokach, *Tetrahedron Lett.* **23,** 739 (1982).
287. R. Zamboni and J. Rokach, *Tetrahedron Lett.* **23,** 2631 (1982).
288. W. P. Schneider, G. L. Bundy, and F. H. Lincoln, *J. Chem. Soc. Chem. Commun.*, 254 (1973).
289. Ref. 24, pp. 124–128.
290. Ref. 147, pp. 337–348.
291. Ref. 131, pp. 49–50.

292. N. A. Porter, J. D. Byers, R. C. Mebane, D. W. Gilmore, and J. R. Nixon, *J. Org. Chem.* **43,** 2088 (1978).
293. E. W. Yankee, C. H. Lin, and J. Fried, *J. Chem. Soc. Chem. Commun.*, 1120 (1972).
294. E. J. Corey and H. E. Ensley, *J. Org. Chem.* **38,** 3187 (1973).
295. E. J. Corey and C. R. Cyr, *Tetrahedron Lett.*, 1761 (1974).
296. Ref. 24, pp. 223–313.
297. Ref. 147, pp. 373–510.
298. Ref. 132, pp. 50–54; 97–102; 119–131.
299. *Chem. Week*, 40 (Feb. 23, 1983).
300. Ref. 132, pp. 159–180.
301. E. J. Corey, K. C. Nicolaou, Y. Machida, C. L. Malmsten, and B. Samuelsson, *Proc. Natl. Acad. Sci. U.S.A.* **72,** 3355 (1975).
302. G. L. Bundy, *Tetrahedron Lett.*, 1957 (1975).
303. R. R. Gorman, G. L. Bundy, D. C. Peterson, F. F. Sun, O. V. Miller, and F. A. Fitzpatrick, *Proc. Natl. Acad. Sci. U.S.A.* **74,** 4007 (1977); G. L. Bundy and D. C. Peterson, *Tetrahedron Lett.*, 41 (1978).
304. Ref. 132, pp. 181–190.
305. K. M. Maxey and G. L. Bundy, *Tetrahedron Lett.*, 445 (1980).
306. E. J. Corey, J. W. Ponder, and P. Ulrich, *Tetrahedron Lett.*, 137 (1980).
307. K. C. Nicolaou, R. L. Magolda, and D. A. Claremon, *J. Am. Chem. Soc.* **102,** 1404 (1980); S. Ohuchida, N. Hamanaka, and M. Hayashi, *Tetrahedron Lett.*, 3661 (1979).
308. K. C. Nicolaou, R. L. Magolda, J. B. Smith, D. Aharony, E. F. Smith, and A. M. Lefer, *Proc. Natl. Acad. Sci. U.S.A.* **76,** 2566 (1979); M. F. Ansell, M. P. L. Caton, M. N. Palfreyman, and K. A. J. Stuttle, *Tetrahedron Lett.*, 4497 (1979).
309. S. Ohuchida, N. Hamanaka, and M. Hayashi, *J. Am. Chem. Soc.* **103,** 4597 (1981).
310. Ref. 132, pp. 191–241.
311. W. Bartmann and G. Beck, *Angew. Chem. Int. Ed. Engl.* **21,** 751 (1982).
312. G. L. Bundy and J. M. Baldwin, *Tetrahedron Lett.*, 1371 (1978).
313. K. C. Nicolaou, W. E. Barnette, G. P. Gasic and R. L. Magolda, *J. Am. Chem. Soc.* **103,** 3472 (1981) and references cited therein.
314. Ref. 132, pp. 221–227.

General References

References 24, 132, 134, 147, and 291 are also general references.

A. Mitra, *The Synthesis of Prostaglandins*, John Wiley & Sons, Inc., New York, 1977.

Advances in Prostaglandin, Thromboxane and Leukotriene Research, Vols. 1–11, Raven Press, New York.

R. M. Sparks, ed., *Prostaglandin Abstracts*, Vol. 1, IFI/Plenum, New York, 1974; R. A. Shelita, ed., *Prostaglandin Abstracts*, Vol. 2, IFI/Plenum, New York, 1975.

The Prostaglandins Bibliography, Medical Documentation Service, College of Physicians of Philadelphia, Philadelphia, Pa.

S. Moncada, R. J. Flower, and J. R. Vane in A. G. Gilman, L. S. Goodman, and A. Gilman, eds., *Goodman and Gilman's The Pharmacological Basis of Therapeutics*, 6th ed., MacMillan, New York, 1980, pp. 668–681.

DOUGLAS R. MORTON, JR.
The Upjohn Company

PUMPS

Pumps are required in nearly every process plant, from the largest petroleum-refining unit or chemical complex to the very smallest laboratory prototype operation. This article presents an overall view of pumping equipment of all types, together with basic pumping principles and enough guidance so that preliminary pump application and selection can be made by the process or project engineer in almost any situation. The designing of pumps is not discussed here; refs. 1–3 provide some information about pump design (see also Vacuum technology).

Basic Pumping Principles

Pumps are divided into two fundamental classes, dynamic and positive-displacement (Table 1). The dynamic type, most commonly encountered as the centrifugal pump, operates by increasing the velocity of the fluid as it passes through a rotating impeller, after which the velocity energy is converted to head or pressure as the fluid exits through a stationary volute or diffuser casing. Other dynamic types include axial-flow (propeller) pumps, regenerative turbine, partial-emission, pitot-tube pumps, plus other specialty types. Positive-displacement types include pumps in which a chamber is filled with liquid and, by the action of reciprocating pistons or plungers or the rotary motion of gears, screws, lobes, vanes, or similar mechanisms, pressure is imparted to the liquid. Diaphragm pumps, hydraulically or air-operated, can be considered a variety of the positive-displacement type.

Pump Operating Conditions. Before a pump selection can be made, the duty conditions must be known or developed, including type of liquid, density or specific gravity, temperature, viscosity, flow, inlet and outlet pressures, presence of solids, and corrosive or erosive material in the liquid. The engineer for a typical process installation is usually required to make an accurate estimate of the pumping system, starting with the source of the liquid (tank, vessel, pipeline, basin, etc) through the planned system layout to the terminal point. Piping sizes must be determined, based on suitable flowing velocities for the fluid or for the fluid mixture in the case of slurries. Pressure drops through the entire piping system must be accurately estimated, including all valves, fittings, process equipment such as heat exchangers, fired heaters, or boilers, and any losses through orifices or control valves. Differences in pressures between the suction and discharge locations, together with changes in elevation between the two (static head), complete the various elements of the system and make it possible to accurately estimate or design a complete pumping system. Most process plant designers utilize some form of preprinted pump calculation worksheet to aid in the collection of this information. This helps assure that, for a given system, all possible pumping situations, alternative routes, alternative pressures or temperatures, varying flow rates, varying fluid properties, etc, are considered. The most severe or limiting case is then chosen as the pump rated condition. A sample of such a calculation sheet is shown in Figure 1.

With the estimate of flow and total head or pressure required in hand, the engineer is ready to make a preliminary selection of pump type and size for the particular pumping system. Useful formulas for preliminary sizing are shown in Table 2.

Another helpful parameter for selecting the best type of pump is given by the

Table 1. Classification of Pump Types

Dynamic types	Dynamic types (cont'd.)
centrifugal, horizontal	regenerative turbine
chemical process (ANSI)	pitot-tube pump
refinery process (API)	disk pump (boundary layer)
horizontal-split, single-stage	elbow pump (propeller)
general service	vertical propeller pump
pipeline	partial-emission
close-coupled	*Positive-displacement types*
self-priming	reciprocating
canned-rotor	crank-driven
slurry	duplex
multistage-split-case	triplex
multistage-barrel-type	multiplex
centrifugal, vertical	direct-acting
turbine	controlled volume
closed-coupled	plunger
deep-well	diaphragm
submersible	hydraulically or mechanically actuated
can-type	tubular
volute	rotary
wet-pit	gear, internal or external gears
dry-pit	screw, two or three screws
special process	progressing cavity
slurry	vane
in-line	lobe
chemical (ANSI)	tubular
refinery (API)	cam and piston
high speed	

dimensionless term specific speed, defined as:

$$N_s = \frac{N\sqrt{Q}}{H^{3/4}}$$

where N = rotating speed, Q = flow, and H = head. When flow is given in units of dm^3/s, head in m, and speed in rpm, pump specific speeds for conventional dynamic-type pumps are ca 300–10,000 [with flow in gal/min (GPM) and head in ft, the range is ca 500–20,000]. If the calculated value is below ca 300 (or 500 in parenthetical units), a positive-displacement pump probably is required. Note that for higher head applications, two or more stages may be required; then the value for specific speed is based on head per stage or, by trial and error, the probable number of stages to achieve reasonable specific speed can be estimated by the specifying engineer. Figure 2 shows a comparison of impeller profiles for pumps of varying specific speeds.

Net positive suction head (NPSH) must be made *available* in any pumping system, large or small. This is the net amount of head or pressure available at the inlet of the pump to overcome inlet losses and to allow liquid to flow into the pump. Available NPSH is calculated as follows:

$$\text{NPSH}_A\ (\text{m}) = (P_s - P_{vp} - P_{frict} + P_{static}) \times \frac{102}{\rho}$$

where pressures are in kPa. Or,

$$\text{NPSH}_A\ (\text{ft}) = (P_s - P_{vp} - P_{frict} + P_{static}) \times \frac{2.31}{\text{sp gr}}$$

PUMP NO.	P-101			
SERVICE	Fract. bottoms	VISCOSITY at PT, mm^2/s	0.33	
FLUID	Hydrocarbon	NORMAL at PT, dm^3/s	82.5	(1308 GPM)
PUMPING TEMP, °C	350	OVERCAPACITY, %	10	
DENSITY, kg/m^3	530	RATED FLOW at PT, dm^3/s	91	(1442 GPM)

flow sketch

C−101

1.5 m

9 m

P−101

E−101

C−103

17.2 m

NPSH CALCULATION

VESSEL, kPa		800	(116 psi)			
LIQUID HEAD, kPa	8.3 m =	43	(6.2 psi)			
SUBTOTAL, kPa		843	(122 psi)			
PIPING LOSS, kPa		4	(0.6 psi)			
SUCTION, kPa		839	(121 psi)			
VAPOR PRESS, kPa		800	(116 psi)			
NPSH AVAIL., kPa		39	(5.7 psi)			
NPSH AVAIL., m		7.54				

DIFFERENTIAL HEAD CALCULATION

CASE	A via C-101		B via C-103		C	
CONDITION	NORMAL	RATED	NORMAL	RATED	NORMAL	RATED
PIPING LOSS, kPa		71.5	(10.4 psi)	81.5	(11.8 psi)	
ORIFICE ΔP, kPa		41.4	(6 psi)	41.4	(6 psi)	
EXCHANGER ΔP, kPa						
E-101		70	(10 psi)	70	(10 psi)	
OVERDESIGN FACTOR						
(10%, 50 min)		50		50		
Σ DYNAMIC ΔP, kPa	TOTAL	232.9	(33.8 psi)	242.9	(35.2 psi)	
CONTROL VALVE ΔP, kPa		90	(13 psi)	893	(130 psi)	
STATIC HEAD, kPa	9.5 m =	49	(7.7 psi)	1	(0.15 psi)	two-phase flow
TERM. PRESS, kPa		800	(116 psi)	35	(5 psi)	
DISCHARGE, kPa		1172	(170 psi)	1172	(170 psi)	
SUCTION, kPa		839	(122 psi)	839	(122 psi)	
NET DIFF. PRESS, kPa		333	(48 psi)	333		
OVERPLUS						
RATED DIFF. HEAD, kPa		333	(48 psi)	333	(48 psi)	
RATED DIFF. HEAD, m		64.1	(210 ft)	64.1		

Figure 1. Pump head calculation.

Table 2. Useful Pump Formulas

For determination of pump power requirements	
$\text{kW} = \dfrac{\text{dm}^3/\text{s}^a \times \text{head}^b\ (\text{m}) \times \text{density}^e\ (\text{kg/m}^3)}{102{,}000 \times \text{efficiency (decimal)}}$	$\left(\text{BHP}^c = \dfrac{\text{GPM} \times \text{head (ft)} \times \text{sp gr}}{3960 \times \text{efficiency (decimal)}}\right)$
$\text{kW} = \dfrac{\text{dm}^3/\text{s} \times \text{differential pressure (kPa)}}{1000 \times \text{efficiency (decimal)}}$	$\left(\text{BHP}^c = \dfrac{\text{GPM} \times \text{differential pressure (psi)}}{1715 \times \text{efficiency (decimal)}}\right)$
For conversion of pressure to head	
atmospheric pressure (sea level) = 101.3 kPa = 10.34 m water	(14.7 psi = 34 ft water)
$\text{head} = \dfrac{\text{kPa} \times 0.102\ \text{m/kPa}^d \times 1000}{\rho^e}$	$\left(\text{head, ft} = \dfrac{\text{psi} \times 2.31\ \text{ft/psi}^f}{\text{sp gr}}\right)$

[a] dm³/s or L/s = flow (GPM = flow, gal/min).
[b] Head, m = total pumping head.
[c] BHP = brake horsepower.
[d] 1034 m/101.3 kPa = 0.102.
[e] ρ = density (kg/m³).
[f] 34 ft/14.7 psi = 2.31.

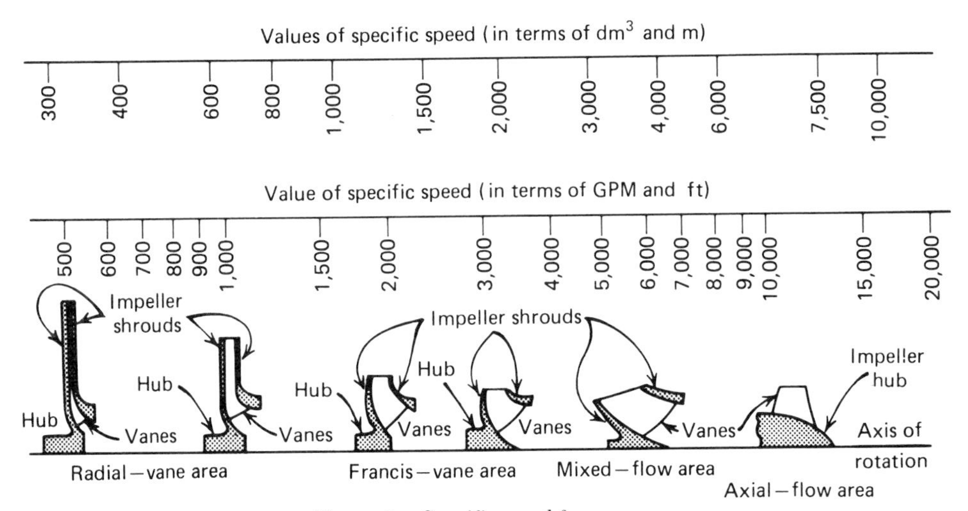

Figure 2. Specific speed for pumps.

for pressures in psi. For dynamic pumps, NPSH is usually expressed in m (or ft) of head. For positive-displacement pumps, which actually develop pressure, not head, it is customary to express NPSH in pressure units, kPa (psi).

The pump, in turn, *requires* a certain amount of NPSH ($NPSH_R$), which is a function of the pump design, the operating speed, and, for dynamic pumps, the flow rate. Thus, a particular centrifugal pump will have an $NPSH_R$ that varies throughout the flow range of the pump at a given speed and can be tested or calculated at any other speed. $NPSH_A$ must exceed $NPSH_R$ to avoid cavitation. When the two values are equal, the liquid vaporizes as it flows into the pump because of entrance losses in the impeller eye. This causes rapid formation of vapor bubbles, which collapse soon after forming, with such force as to cause severe internal damage to most pumps. Most pump installations are not designed to tolerate this condition; the customary rule on simple systems is to provide at least 0.5 m greater $NPSH_A$ than the pump requires at the rated condition. On large, high energy pumps such as pipeline pumps or boiler-feedwater pumps, or on other specialized services where service conditions may change, this margin should be increased significantly, to make as much as twice or even three times more $NPSH_A$ than the pump requires. This helps assure cavitation-free operation of the pump for long periods.

An example of the $NPSH_A$ calculation is also shown in Figure 1. A curve that may help the process or project engineer gain an idea of reasonable NPSH to be made available for ordinary process-pump selections is shown in Figure 3. This curve is based on the concept of suction specific speed, a number without dimensions, similar to specific speed, but with $NPSH_R$ substituted for pump head:

$$N_{ss} = \frac{N\sqrt{Q}}{NPSH_R^{3/4}}$$

In SI units, this value should be ca 4200–8000 (for non-SI units, ca 7000–13,000). These are the limits shown in each shaded area on Figure 3. For double-suction pumps, suction specific speed must be figured based on one half the total pump capacity since the double-suction impeller is actually two identical single-suction impellers back-to-back and the entrance to each side of the double-suction impeller is designed for one half of the total flow.

System head exists for every pumping system as the friction values in piping and other elements in the system vary with changing flows. It is often desirable on more complex pumping systems to calculate data that lead to the system-head curve, which can be overlaid on the proposed pump curve (Fig. 4). From this comparison, the rated condition and all alternative operating conditions for a given pump can be adequately estimated. This estimation is especially helpful in considering how control valves must function to absorb the differences in pump output and system requirement at lower-than-rated flow.

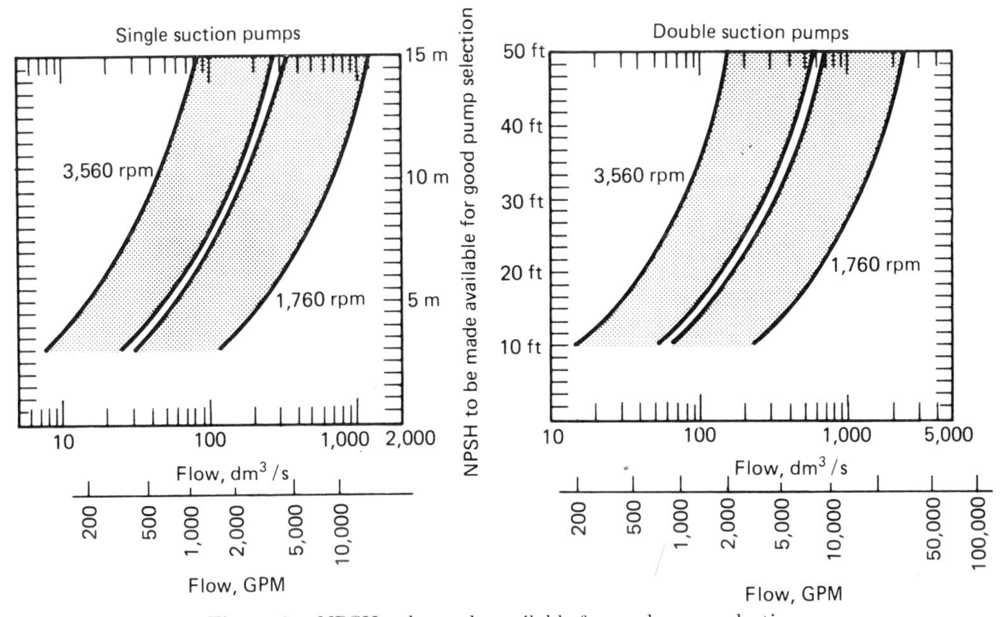

Figure 3. NPSH to be made available for good pump selection.

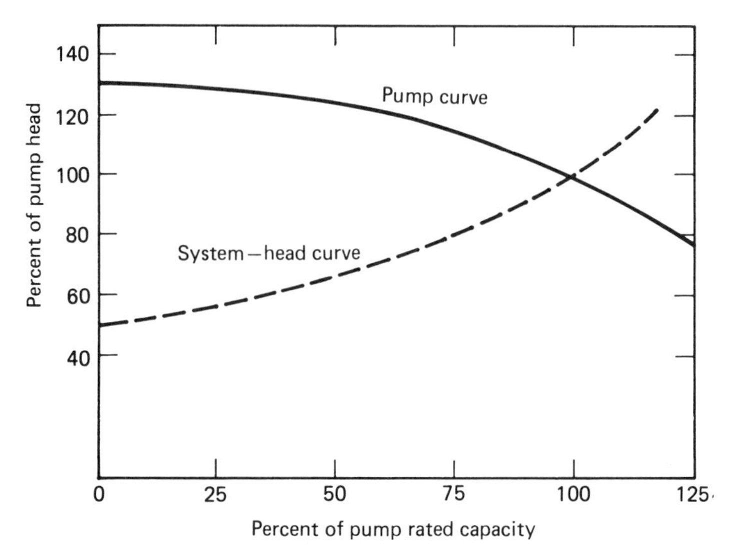

Figure 4. System-head curve.

Pump Application and Selection

Centrifugal Pump Application. Efficiency and power are calculated both for the rated point and for any and all other operating points along the pump curve. Each centrifugal pump impeller and casing can be optimized for only one maximum efficiency. Changes in flow rates to an impeller of fixed dimensions at constant speed affect certain of the velocity components in the impeller design, resulting in a change in head developed and a reduction in efficiency. Centrifugal pump performance for standard, predesigned pumps is given in pump manufacturers' curve books, widely available to the user engineer, enabling a preliminary selection to be made if desired. The curves are usually for water performance, taken from data of pump tests in the factory. Actual power may be quite different on the user's specific liquid: for example, half as much power is required for liquid half the weight of water, twice as much for liquid twice as heavy. See Table 2 for pump power formulas. Efficiencies of centrifugal pumps at the best operating point are ca 30–40% on very small, low specific speed pumps, and as much as >90% on the very largest pumps. Average values of 65–75% can be expected in most process-pump applications.

Impeller diameter for each pump application must be individually determined. To permit use of standard pump designs, rather than a custom-designed pump for every application, the impeller is trimmed to a diameter shown by previous test data for the same pump to be adequate for the required rated conditions. Often such trimming is held to the next highest fraction (eg, 5 mm). It is quite common to cut an impeller a little on the plus side (slightly oversized) to accommodate minor differences between the planned and actual operating conditions. It is not difficult to trim impellers after installation at the jobsite if a given pump is producing too much head.

For each impeller, there is a minimum diameter which is normally shown on the pump manufacturer's curve. Below the minimum, the hydraulic performance is so far from optimum that a different pump or a different impeller for the same casing should be used. If performance does not meet expectations with a given impeller of fixed dimensions, underfiling the blades at the exit or polishing the flow passages may sometimes improve performance.

The shape of performance curves varies with the blade angle, impeller geometry, number of blades, and specific speed. Figure 5 illustrates several types of pump head-capacity curves. The constantly rising curve is best for process applications, presenting no problems in control as the flow is reduced. The drooping curve is not desirable for operation below its stable range. Note that there are two capacities at which this type impeller produces exactly the same head. Such an impeller hunts between these points and process control may be difficult. The hooked curve or knee-shaped curve is more common in larger vertical turbine-type pumps with mixed-flow impellers of higher specific speed. Pumps with this type of impeller should not, as a rule, be operated at flow rates in or near the knee. The flat curve (radial bladed impeller) and the very steep curve (propeller pump, regenerative turbine pump, and others) are also shown to indicate the varieties that exist. The engineer should look carefully at the particular pump curve throughout its entire range to determine that it will be satisfactory in the application at any normal, expected operating point.

Variations in speed and diameter are in accordance with the affinity laws for changes in performance, as follows:

$$\frac{N_1}{N_2} = \frac{Q_1}{Q_2} = \sqrt{\frac{H_1}{\mathrm{H}_2}} \quad \text{and} \quad \frac{D_1}{D_2} = \frac{Q_1}{Q_2} = \sqrt{\frac{H_1}{H_2}}$$

with subscripts 1 and 2 indicating performance points at alternative speeds or impeller diameters. Variation in pump speed is an excellent way to match varying pumping-system needs with a given pump. Variable-frequency electric motors, variable-speed mechanical drives, hydraulic couplings, steam and gas-turbine drivers, and diesel-engine drivers all provide variable speeds for pumps.

Parallel pumping occurs when two or more pumps operate in parallel in the same system. If the pumps are identical, they should share the duty equally. Because of

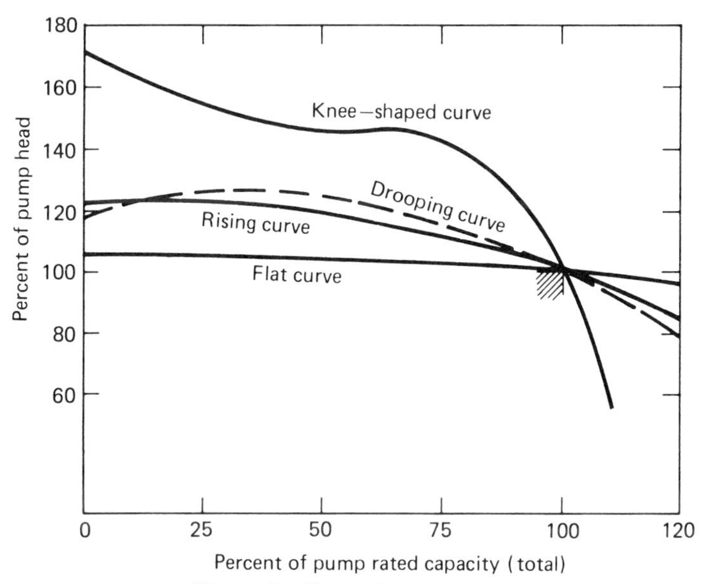

Figure 5. Types of pump curves.

manufacturing tolerances, however, two supposedly identical pumps may not prove to be completely the same. In such a case, one pump may assume more of the load than the other. Two pumps with somewhat different head-capacity characteristics also operate in an unequal manner when pumping in parallel. Two similar pumps having different types of drivers (such as one motor drive and one steam-turbine drive), should operate satisfactorily in parallel if both are at identical speeds. However, if the variable-speed pump is allowed to increase in speed, it will take a larger portion of the load or even cause the slower pump to operate at shutoff. Figure 6 gives an example of two pump curves in parallel where the pumps are dissimilar.

Pumps for viscous liquids and slurries perform very differently than they would on clear water. The Hydraulic Institute Standards (4) provide an industry-accepted method of derating a pump for a viscous liquid, based on data accumulated by experience. Reference 4 gives details on how to use this chart. Other methods may be offered by pump designers, especially on pumps for larger flows. The amount that a large pump must be derated because of viscous liquid may be significantly different than shown by the Hydraulic Institute method as the additional friction loss of the viscous fluid flowing through the pump has less effect on pump efficiency than in a small pump.

For slurries, the water performance of a pump must likewise be derated, based on the characteristics of the liquid–solid mixture to be pumped. Viscosity may be one of these considerations, but, in addition, the water performance is derated because the solids cause additional friction of the slurry flowing through the pump. Velocity of flow through the pump is most important as it affects wear in the pump and also as it relates to the ability of the solids to stay in suspension as they flow through the pump, not settling in the bottom part of the casing or becoming unevenly distributed inside. References 5–7 refer to the subject of slurry pumping, a rather specialized aspect of pumps.

Types of Dynamic Pumps. Most dynamic pumps are centrifugal, with either horizontal or vertical shafts. As sizes (specific speeds) increase, pumps are classified

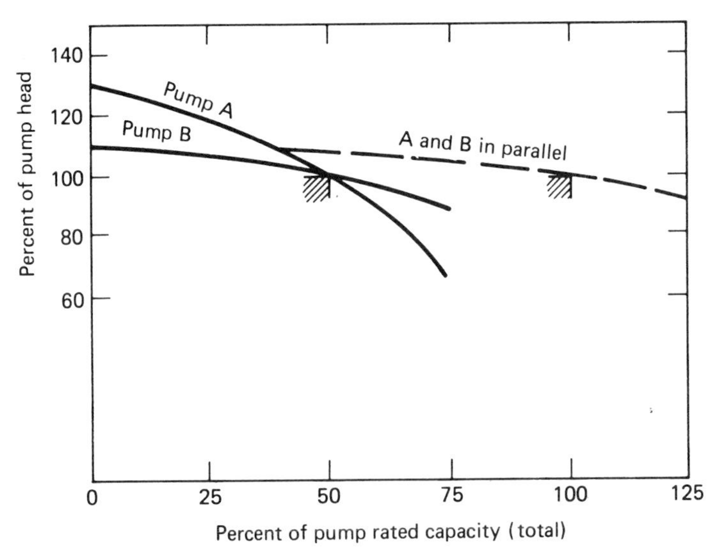

Figure 6. Pumps in parallel.

as mixed-flow and finally as axial-flow or propeller types. In very small sizes, there are several specialty types, the partial-emission pump, the regenerative turbine pump, the pitot-tube pump, and others. A few of the most widely used types are described briefly here.

The chemical process pump, or so-called ANSI type (Fig. 7), is a single-stage, overhung bearing bracket design, developed by most U.S. pump manufacturers in conjunction with the principal U.S. chemical companies and is now covered by ANSI Standard B73.1-1977 (8). External mounting dimensions, suction and discharge flange sizes, and orientations are standardized so that, size for size, a pump from one manufacturer is interchangeable with any other make in regard to its outer dimensions. The pump is designed to allow the pumping element and bearing bracket to be removed for maintenance without disconnecting the pump suction and discharge flanges. Flow sizes range from ca 1 dm^3/s (ca 15 GPM) with 25.4-mm discharge opening, to >200 dm^3/s (>3000 GPM) with 203.2-mm discharge. Some larger sizes are also available, but these are beyond the range given in the ANSI B73.1 Standard, although the basic design idea is the same. Standard materials of construction include ductile iron, carbon steel, 316 and other stainless steels, plastic, ceramic-coated, Teflon-lined, and others.

The refinery process pump, sometimes called the API-610 type (9), was developed for applications where higher temperature liquids must be pumped, such as in oil-refinery applications. In addition, these pumps, which were designed before the ANSI type, have the back pull-out feature so that pump rotor and bearing bracket can be removed without disconnecting suction and discharge flanges. The pump casing is supported at the horizontal centerline, so that approximately equal expansion of the casing takes place when pumping hot liquids. The casing is split at right angles to the

Figure 7. ANSI chemical-type pump. Courtesy of Duriron, Inc.

drive shaft and is constructed with a joint using a confined, metallic gasket, making the pump less susceptible to leakage, a feature especially important when handling flammable or toxic liquids. Larger flow sizes use double-suction impellers, which double the capacity for a given diameter and speed and provide inherent hydraulic balance so that there is a very minimum net hydraulic thrust load to bearings. Figures 8(**a**)

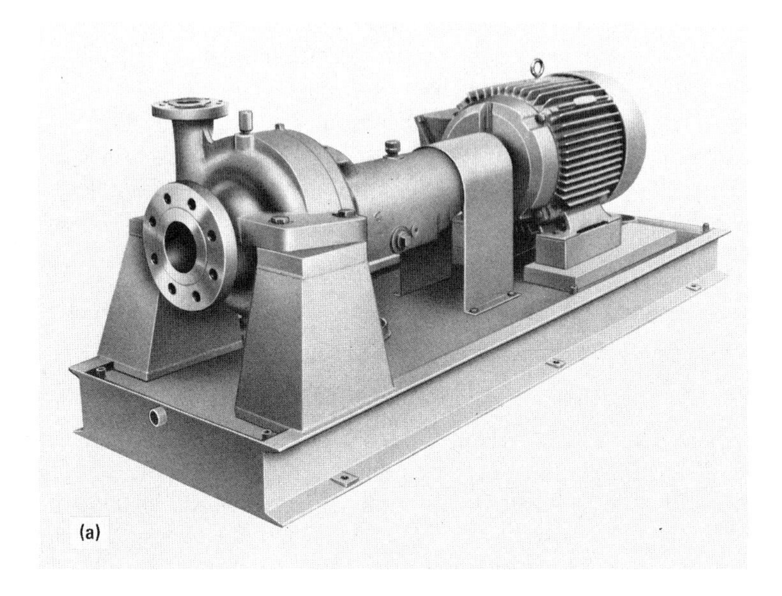

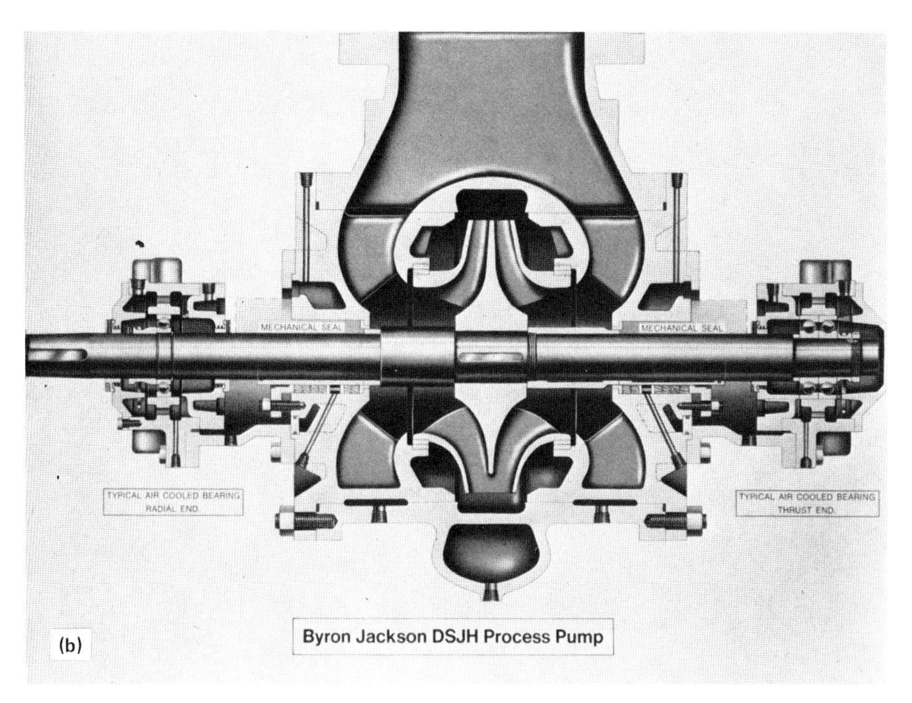

Figure 8. Photographs (**a**) (courtesy of Goulds Pumps) and sectional (**b**) (courtesy of Byron-Jackson Pumps, a division of Borg-Warner Corp.) view of API-610-type pump.

and 8(**b**) illustrate the basic types. Heavier (higher pressure) casing designs are likewise available, as are casing materials of steel or alloy steels, suitable for the most difficult and dangerous pumping jobs.

There are many other types of horizontal pumps. The horizontal-split-casing type, using a flat, unconfined casing gasket, a single-stage double-suction-type impeller, and between-bearing design (Fig. 9) has long been the standard for use on cooling water, circulating water, solution circulating, fire fighting, and in certain modified higher pressure versions, used as a pipeline pump for long distance pumping of hydrocarbons or other liquids. The latter types may operate at speeds of 3600 rpm or higher, developing heads up to 300 m in a single stage. For such pipeline applications, ample NPSH must be made available (see Fig. 4).

Close-coupled single-stage pumps are widely used for general service applications. This type pump is mounted directly on an electric-motor frame and shaft, using the bearings of the motor for support. No coupling is required; alignment problems are minimized. However, close-coupled pumps are not generally desired for hazardous or toxic fluids because they pose a greater danger of fire in event of mechanical seal leakage, as the seal or stuffing box is closer to the electric-motor driver.

Self-priming pumps are made in a limited range of sizes and find application in the process industries that require a horizontal pump to pull a suction lift. In such a case, it may not be practical or feasible to prime a conventional centrifugal pump manually or automatically each time it starts; the self-priming pump has a built-in

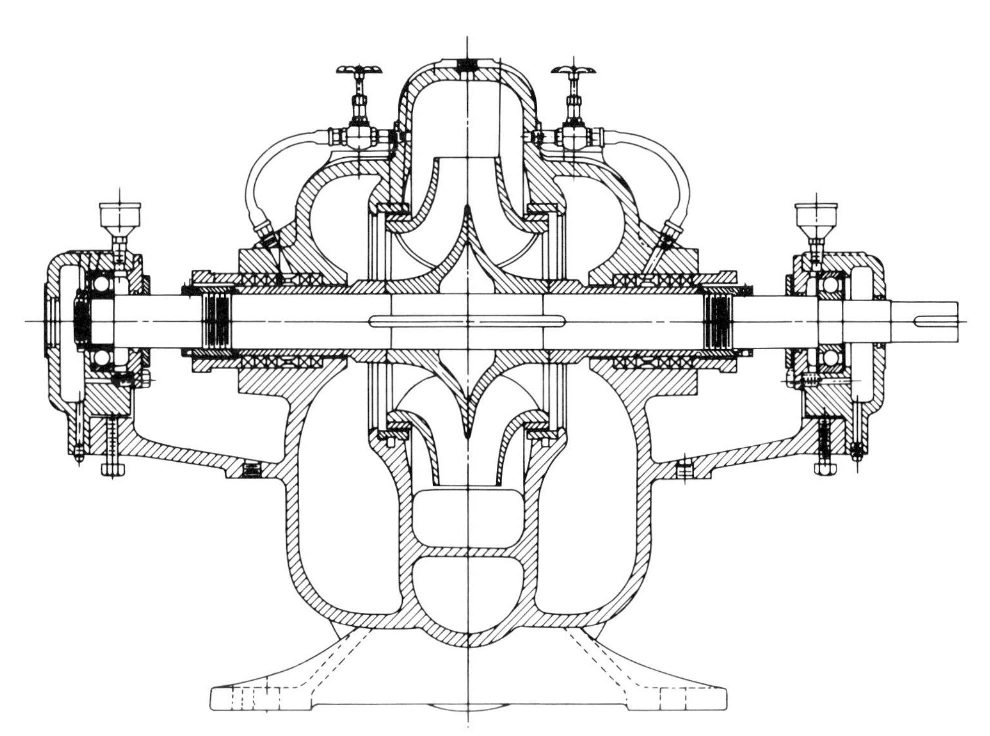

Figure 9. Sectional view of horizontal-split-case pump. Courtesy of Worthington Pumps, a division of McGraw-Edison Corp.

chamber that fills with liquid from the previous pumping cycle and is ready to start automatically the next time.

Canned-rotor pumps are used where no leakage of pumped fluid can be tolerated. Sizes range up to ca 100 dm^3/s (ca 1500 GPM), heads to ca 150 m in smaller sizes. A small portion of the pumped fluid recirculates through the motor bearings, with a thin stainless-steel shell separating the motor rotor from the stator and protecting the motor windings from contact with the pumped fluid. Because of this feature, overall efficiency for this type pump is lower than that for conventional types, but sometimes the canned-rotor pump is the best or only solution to pumping a very dangerous liquid.

Slurry pumps are generally single-stage, overhung bearing designs. The pumps are heavier construction than pumps for clear liquids, as the slurry services are generally more severe. They normally run at much lower speeds, often using v-belt or variable speed drives to suit the exact duty conditions or to make it possible to change speeds as wear increases. Replaceable wear plates of hard material such as high silicon iron resist wear on certain slurries; other applications require rubber-lined casings and rubber-coated impellers. Slurry pumping is a specialized subject and the reader is referred to specialized publications for additional material.

Horizontal multistage pumps are used when a single impeller cannot produce the required head. From two to as many as 14 or more impellers may be mounted on a single shaft to form the rotor for a pump for certain high head applications. Casings are split either on the horizontal or vertical axis. Diffuser or multiple volute construction is available for either type casing. There is no industry-wide agreement on where to switch from a horizontal-split casing to a vertical or "barrel" casing. Horizontally split casings have flat, unconfined gaskets, which are more difficult to maintain to avoid leakage at higher pressures. Many horizontal-split units are successfully being used, however, in pressures of 10–14 MPa (1500–2000 psi) and even higher, depending on the service and the location of installation. The barrel type (Fig. 10) can be applied on pressures higher than these or on lower pressures in process locations, or when the fluid is hot, or the density (specific gravity) low. This ensures that there will be no leakage from the casing joint on what is probably a dangerous, hot, and possibly flammable fluid.

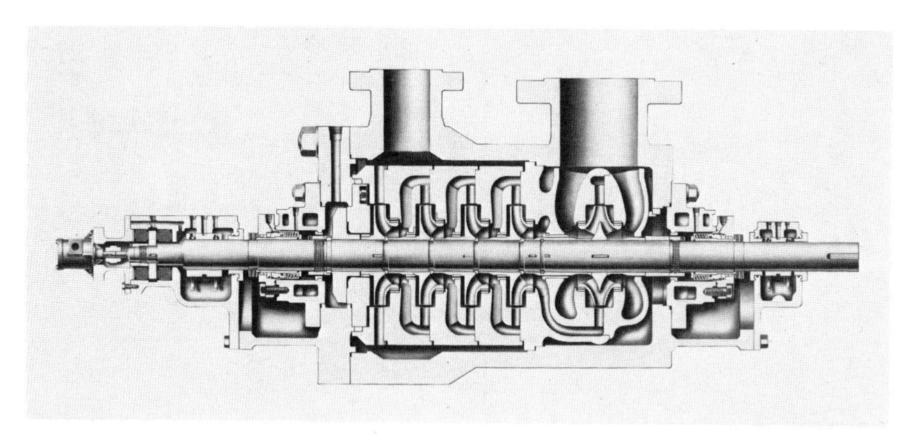

Figure 10. Sectional view of barrel-type multistage pump. Courtesy of Ingersoll-Rand Co.

Bearings and seals in centrifugal pumps are always very important parts to be considered by the engineer in preparing the pump specification. Most horizontal pumps operating at speeds 3600 rpm and below have ball bearings both for radial and thrust loads. Larger pumps, such as the multistage barrel type or large single-stage pipeline types, have sleeve-type journal bearings, sometimes tilt-pad types, often Kingsbury-type thrust bearings, or occasionally ball thrust bearings. API Standard 610 (9) shows criteria for determining whether ball bearings are adequate, based on the DN factor of the pump being less than 300,000 where D is the diameter of the bearing inner race (mm) and N is the rotative speed (rpm). Journal bearings usually require forced lubrication systems with lube-oil-system components mounted either on the pump baseplate or in a console not unlike that used for a process-type centrifugal compressor. In specifying a large pump with journal bearings, it is important to make clear what sort of lubrication-system components are desired.

Mechanical shaft seals are used almost exclusively in modern process-pump applications. Standard pump designs of the API-610 or ANSI types have been planned to accommodate seals. Low pressure chemical pumps and general service pumps generally use the unbalanced seal [Fig. 11(**a**)], but most process pumps have been designed with stepped-shafts and can accommodate the balanced seal, which is preferable as the pressures increase. Pusher seals are the current standard for all but the most difficult applications. Tandem seals [Fig. 11(**b**)] or double seals [Fig. 11(**c**)] are becoming mandatory in some applications where no leakage, however minor, can be tolerated. Bellows seals [Fig. 11(**d**)] can be applied at much higher temperatures, to at least 400°C, and are normally specified for pumping temperatures above 200°C. Most seals require some form of flushing to the seal faces to prevent excessive build-up of solid material and to provide cooling. API Standard 610 provides standard seal-flushing plans and auxiliary plans, which will aid the specifying engineer in using standardized, pre-engineered systems wherever possible. Mechanical seals are generally not used when slurry is being pumped because of the large amount of solid material likely to get to the seal faces. In most slurry pumps, stuffing boxes containing readily replaceable packing rings are found to be the best.

Cooling water is usually supplied for bearing brackets and stuffing-box jackets when pumping temperature is expected to be 150°C or higher. For temperatures above 300°C, the pump-support pedestals may also be water-cooled. Cooling-water piping systems are also outlined in API-610, which is a useful guide to choosing what may be correct for a given application. It is customary to have the pump-seal piping and cooling-water piping furnished with a common inlet and outlet connection, complete with any accessories such as filters, coolers, cyclone separators, sight-flow indicators, etc.

Vertical in-line pumps are arranged to be mounted directly in the piping system with no additional elbows or changes in direction of flow (Fig. 12). They are made in single-stage models, with the impeller mounted on a vertical shaft that is either an extension of a vertical motor shaft or, in some models, on a separate shaft connected through a rigid coupling to a vertical motor or vertical steam-turbine driver. Although this type of pump saves space, the vertical motor or steam turbine is generally somewhat more expensive than its horizontal counterpart. ANSI Standard B73.2-1975 (8) covers the vertical chemical type. The API-610-type liquid end is also available as a vertical in-line pump.

High speed versions of the vertical in-line type also exist, operating at speeds up

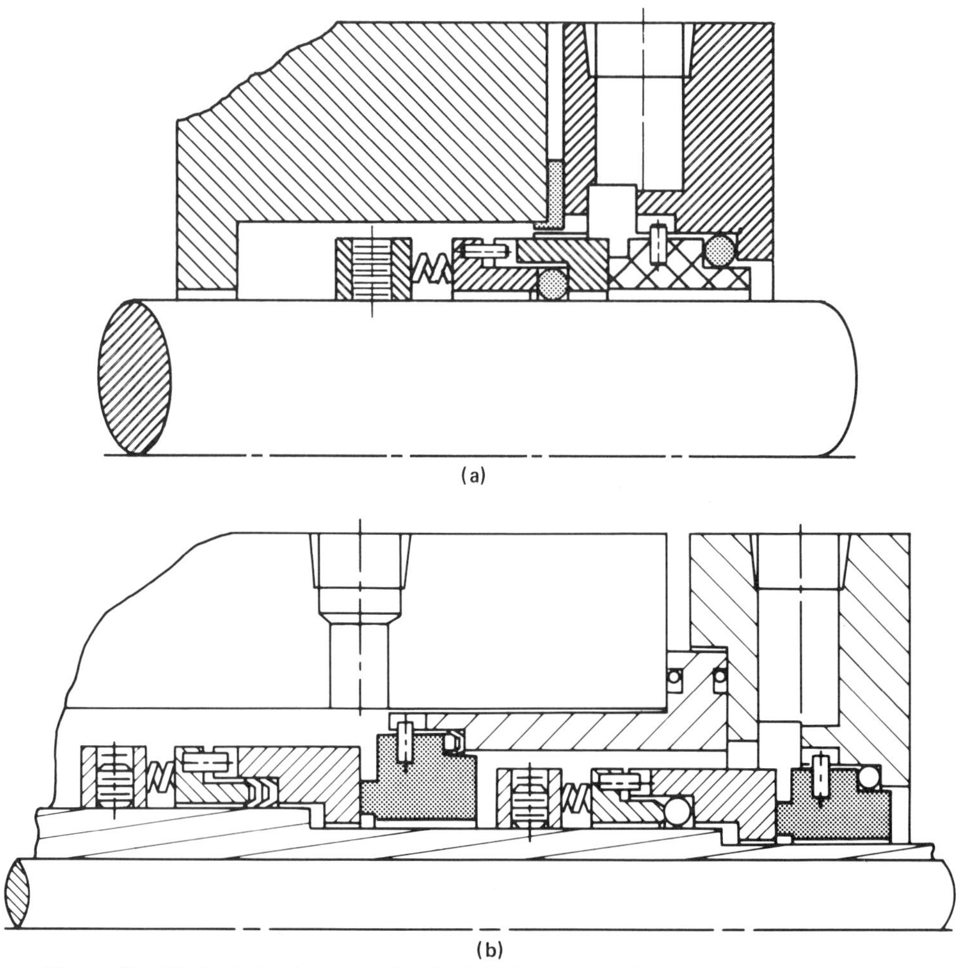

Figure 11. Mechanical seal types. (**a**) Inside "RO" dura seal, (**b**) double tandem seal, (**c**) double "RO" arrangement. Courtesy of Durametallic Corp. (**d**) Bellows type mechanical seal. Courtesy of EG&G Sealol.

to 20,000 rpm, producing very high heads in a single stage. In small sizes, this is correctly called a partial-emission pump, where the pumped fluid exits through a partial opening in the diffuser casing. In larger sizes, such pumps have a full diffuser. A built-in gear speed increaser is used to increase the pump speed from the normal 60 Hz electric-motor input speed. Specific speed can be used to determine whether a pump of this type is feasible for a given duty. For example, assume a pump is required for 12 dm^3/s (ca 190 GPM) flow at 1200 m head. Assuming a rotative speed of 20,000 rpm, calculate

$$N_s = \frac{20{,}000\sqrt{12}}{1200^{3/4}} = 340 \text{ (or 554 in non-SI units)}$$

which by reference to Fig. 2 is seen to be a reasonable specific speed for a small centrifugal-type pump.

The vertical-can pump is a vertical-shaft, multistage pump which can be arranged

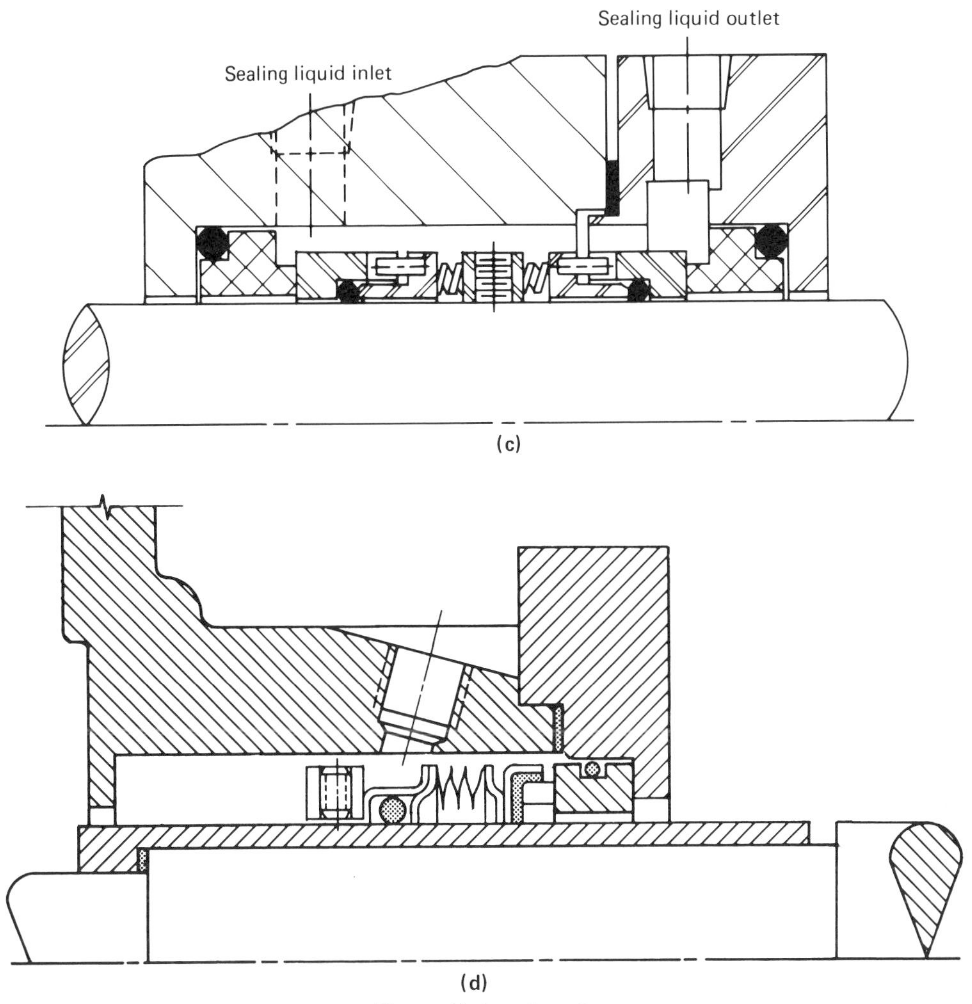

Figure 11. *(continued)*

to produce pumping heads exceeding 1200 m by operating at conventional 60 Hz speeds and utilizing as many stages as necessary to produce the head. A common application for this pump is found in low NPSH systems, where it is difficult or impossible to increase the $NPSH_A$ to a reasonable value. The solution provided by the vertical-can pump is to put the pump below grade level, in a can or barrel. The pump impellers are located at whatever distance below grade may be required to provide adequate NPSH for successful pumping. Figure 13 shows details of this pump type.

Other vertical pumps include the single-stage vertical volute type, used from the smallest size wet-pit sump pump applications up to the very largest circulating pumps, wet- or dry-pit types. Most large pumps in water-supply pumping (sizes to >6000 dm^3/s or >100,000 GPM) are vertical-shaft volute type (Fig. 14). Radial-flow, mixed-flow, or axial-flow impellers can be applied with these larger sizes. Figure 14 shows a mixed-flow impeller; the flow does not turn a complete 90° relative to the shaft, but may turn 30°, 45°, or 60° as it passes through the impeller. When the flow does not change direction, the pump is termed axial-flow, or propeller, type. A typical selection

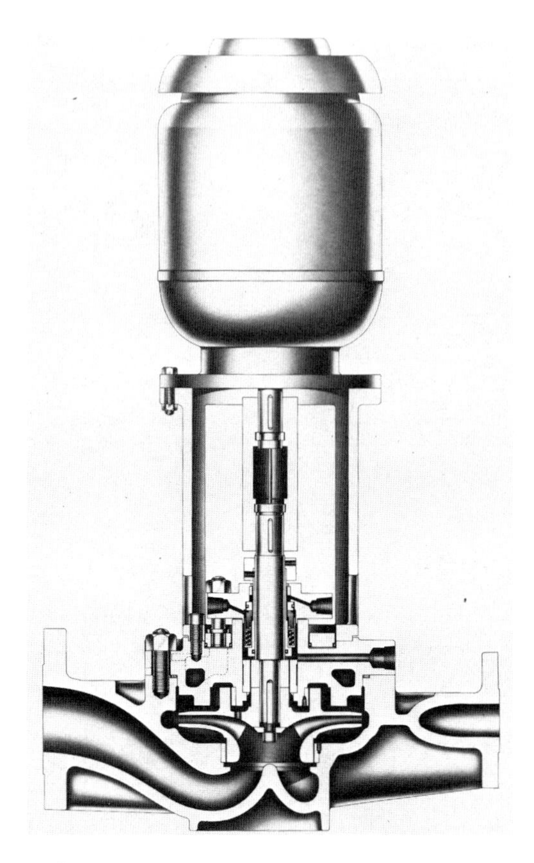

Figure 12. Sectional view of vertical in-line-type pump. Courtesy of Ingersoll-Rand Co.

for a mixed-flow impeller with N_s = 2900 (ca 4750 non-SI units) might run at 320 rpm, pumping 6300 dm^3/s (ca 100,000 GPM) at 18 m head. The overall efficiency of a large pump such as this would be close to 90%.

Vertical-turbine pumps are used in cooling-tower basins, river and lake water pumping, vertical firewater service, vertical process services, and for almost any application when pumping water from a well. For deep-well pumps, the pumping element is either submerged in the water well and connected to the driver on the surface by means of long vertical lineshafts, or it has a submersible motor, connected directly to the pumping element submerged in the well with the pump.

Variations of vertical pumps for the chemical- and petroleum-processing industries include cantilever shaft pumps, where no bearings come in contact with pumped fluid and steam-jacketed pumps (Fig. 15), conventional volute types with the addition of steam jacketing and other special details for pumping molten liquids at the required higher temperatures. Also available are vertical slurry pumps, rubber-lined or hard-metal construction, with special slow-speed drivers or speed-reduction methods, to pump solid–liquid mixtures from pits, tanks, flotation cells, etc.

The inertia pump is an oscillating type of dynamic pump (see Fig. 16). Its simplicity of design and favorable pumping characteristics are used in the chemical-processing and food industries.

Other dynamic pumps, mentioned earlier (regenerative turbine, pitot-tube, disk

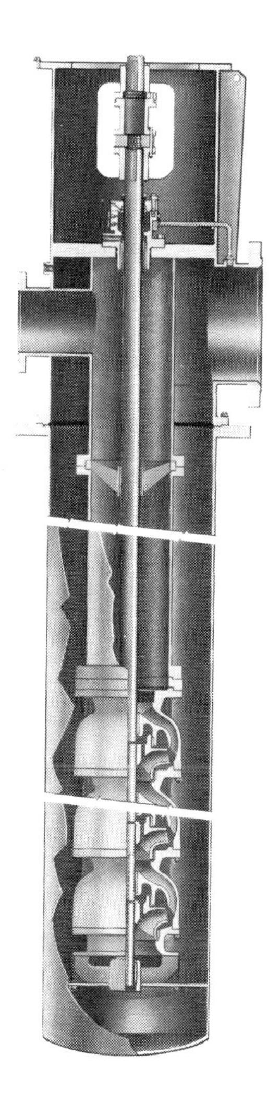

Figure 13. Sectional view of vertical can-type pump. Courtesy of Bingham-Willamette Co.

pump, etc) also have specialized areas of application and must be studied individually.

Positive-Displacement Pumps. Positive-displacement pumps represent a completely separate area of pumping equipment (Table 1). This class of pumps is divided into three general types: reciprocating, controlled-volume (including diaphragm pumps), and rotary pumps. Their basic pumping formulas are quite like those for dynamic pumps, with the important distinction that positive-displacement pumps produce pressure, not head, and are essentially fixed-volume machines if operated at a constant speed and stroke. The capacity can be changed by changing the speed. The pump volumetric efficiency changes, resulting in a small change in flow when the pump is operated at lower or higher pressures than the original rated values or when fluids of different viscosities are pumped at fixed speed. Normally, pump performance curves for positive-displacement pumps are not requested or given, or they may show a given pump performance at varying speed only.

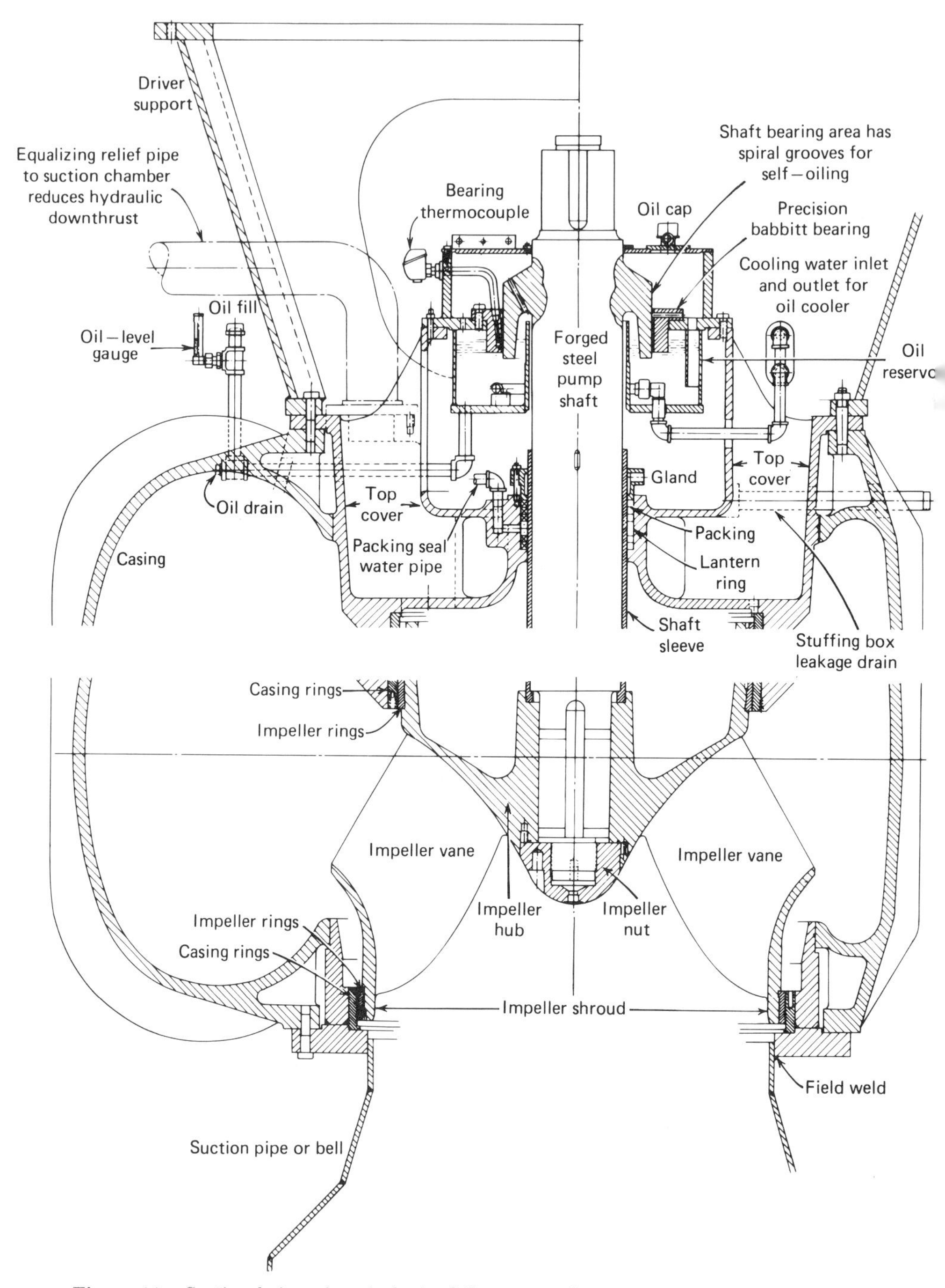

Figure 14. Sectional view of vertical mixed-flow pump. Courtesy of Allis-Chalmers Co.

Pulsating flow is inherent in reciprocating pumps as each piston or plunger alternates in one complete revolution from zero flow to maximum, then back to zero. Most reciprocating pumps have two or more cylinders, pistons, or plungers, arranged with the crankshaft in such a way that the pulsating flow is cancelled to some extent by the alternate strokes of the multiple plungers or pistons. It is usually a requirement of a process-pump system using a multicylinder reciprocating pump that pulsation dampers, snubbers, or surge chambers be applied before and after the pump to assure adequate filling of the pump on each cycle and to reduce pulsating flow downstream of the pump discharge. Reciprocating pumps are generally used for low volume, high pressure applications (below the lowest specific speed range of centrifugals), or in some larger flow applications where the pump efficiency may be notably higher than that of a small centrifugal. They have other special applications such as slurries; mud pumps in oil-field drilling; gathering pumps in oil fields where a wide variety of pump fluids must be handled; viscous liquids when the efficiency of the reciprocating type may be higher than a centrifugal; and numerous other applications.

The $NPSH_A$ for reciprocating pumps is calculated much as for dynamic pumps. In addition, the acceleration head in a given system must be considered to assure that the planned pumping system has adequate head to fill the pump on every stroke. Some useful information on this subject is provided in refs. 10–12.

Reciprocating pumps are either direct acting or crank driven. Direct-acting pumps have air, steam, or gas cylinders at one end of the pump, connected by piston rods and valve linkage to liquid cylinders at the other end. The drive end is actuated by the driving medium directly through the piston rods to the liquid end, with the reciprocating motion controlled by the valve gear. Operation at variable speed is inherent in this type; many such pumps perform very useful services in process applications where varying speeds, varying pumping rates, or varying fluids must be accommodated.

The crank-driven type has a crankshaft operating at low-to-medium speed (60–500 rpm) with connecting rods to pistons or plungers arranged either horizontally or vertically. Most have either two pistons, or three, five, or seven plungers operating from the same crankshaft to gain smoother flow and higher capacity. Figure 17 shows a typical crank-driven pump, illustrating how an electric-motor or steam-turbine driver is connected through a speed-reducing gear to obtain the desired slow speed operation.

Controlled-volume pumps, also known as metering or proportioning pumps, are applied when a precise amount of fluid must be pumped (see Flow measurement). These are basically a variation of the reciprocating type, with the added feature of adjustable stroke length ranging from zero to maximum. Thus, the pumping rate can be set manually or controlled automatically for any flow within the range, and the rate altered reproducibly by setting the stroke to predetermined values. The accuracy of this type pump is extremely good, making it mandatory for such services as chemical injection, boiler-water or cooling-water-treatment system injection, etc. Figure 18 shows a sectional view of a typical controlled-volume pump. Diaphragm-type liquid ends are widely used for this type pump, which has a single- or double-disk diaphragm separating the pumped fluid from a mechanical or hydraulic mechanism that actuates the diaphragm. Tubular diaphragms are also used on certain smaller, low pressure units. There is no contact between the pumped fluid and the drive-end hydraulic fluid or the atmosphere (during the life of the diaphragm), and thus truly leak-proof

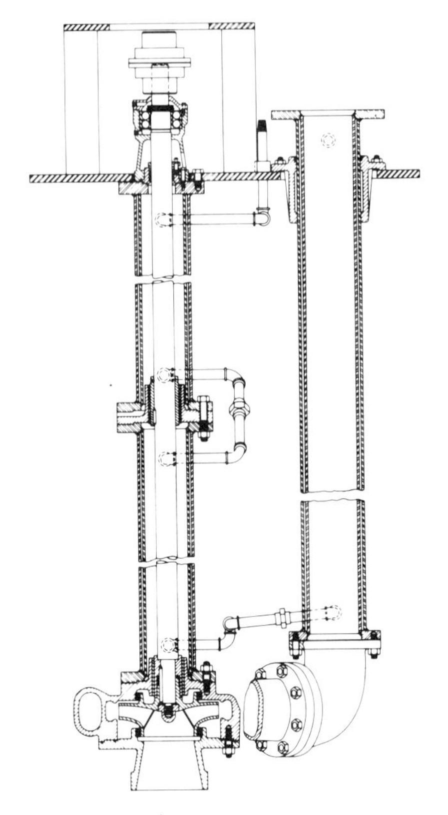

Figure 15. Sectional view of vertical steam-jacketed pump. Courtesy of Charles S. Lewis & Co., Inc.

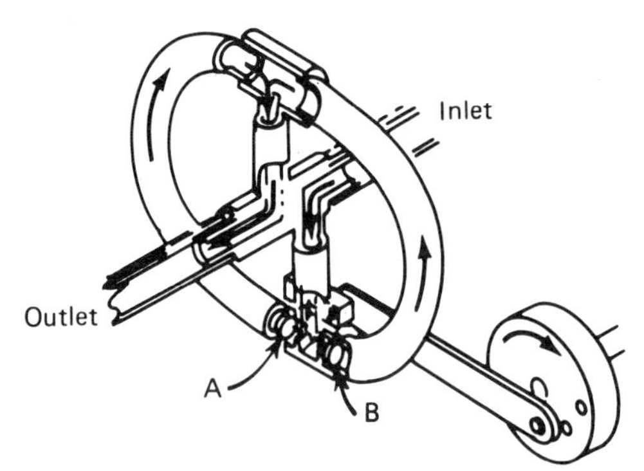

Figure 16. Inertia pump. Courtesy of Horst Dynamics, Inc.

pumping is achieved. Because by definition the requirement for precise metering of pumped fluids is usually for small-volume flows, such pumps are all relatively small in size, but they perform a unique pumping function that must not be overlooked. Pumps with two or more plungers or diaphragms attached to the same drive mechanism, or two or more individual pumps coupled together to the same drive motor, are often used to save space or to provide a central pumping system.

A rotary pump is defined in Hydraulic Institute Standards (4) as "a positive displacement pump, consisting of a chamber containing gears, cams, screws, vanes,

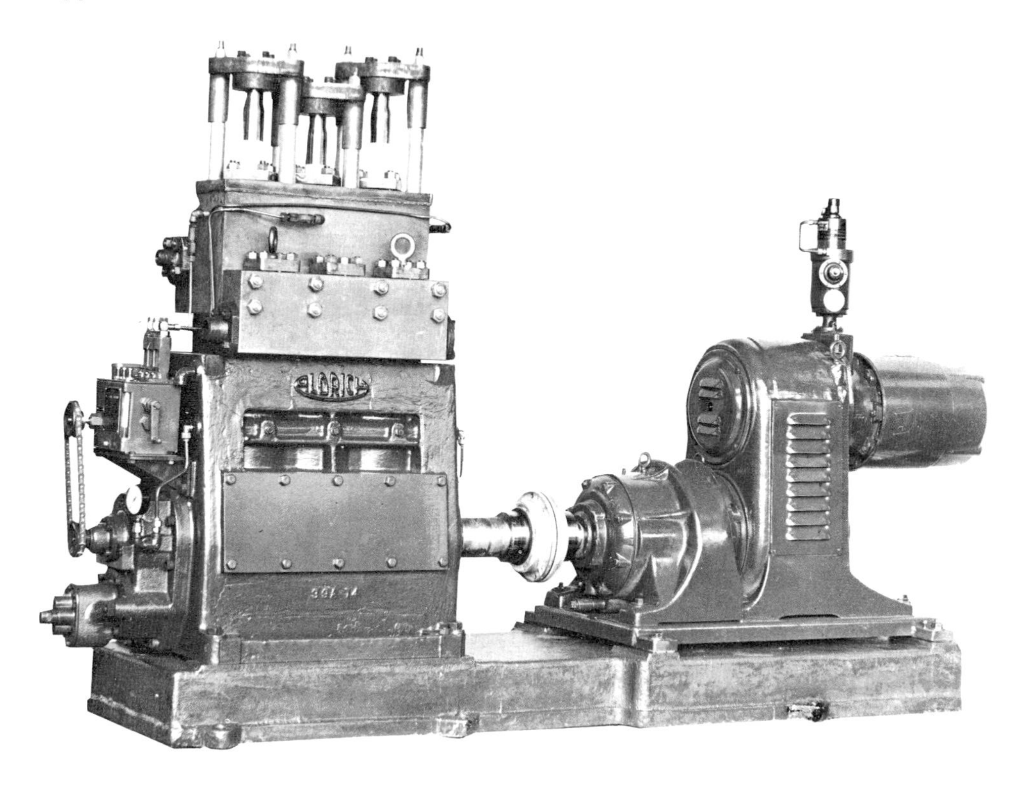

Figure 17. Triplex reciprocating pump. Courtesy of Ingersoll-Rand Co.

plungers or similar elements actuated by relative rotation of the drive shaft and casing, and which has no separate inlet and outlet valves." Many types exist, and for more details the reader is referred to this publication. Figures 19(**a**) and 19(**b**) illustrate two types.

The rotary pump has been found to be particularly useful in many applications, including the following: pumping viscous liquids or liquids of varying viscosity; pumping non-Newtonian fluids; pumping liquid with entrained air or gas; low flow pumping; certain high pressure situations; and pumping in systems requiring suction lift.

Comparison of a rotary pump with a centrifugal pump for pumping a viscous liquid shows why the rotary is often a better choice. This is especially true when the liquid viscosity is higher than about 100 mPa·s (= cP) or when it is variable because of changes in the pumping temperature or the pumped liquids. It may be desirable to make a preliminary selection of both pump types to decide which is best. Using the Hydraulic Institute method for rating a centrifugal pump on viscous liquid, then referring to manufacturers' published data for rotary pump performance, the specifying engineer can make a quick comparison that may help determine the best choice (Table 3). As the viscosity increases, the rotary pump becomes more attractive, to the extreme where it may be the only type that can be used on certain very viscous liquid applications.

Non-Newtonian fluids are also best handled by rotary pumps. The objective is to keep the velocity of flow as nearly constant as possible while the fluid flows through the piping, into the pump, and onward. The principle of operation of dynamic pumps,

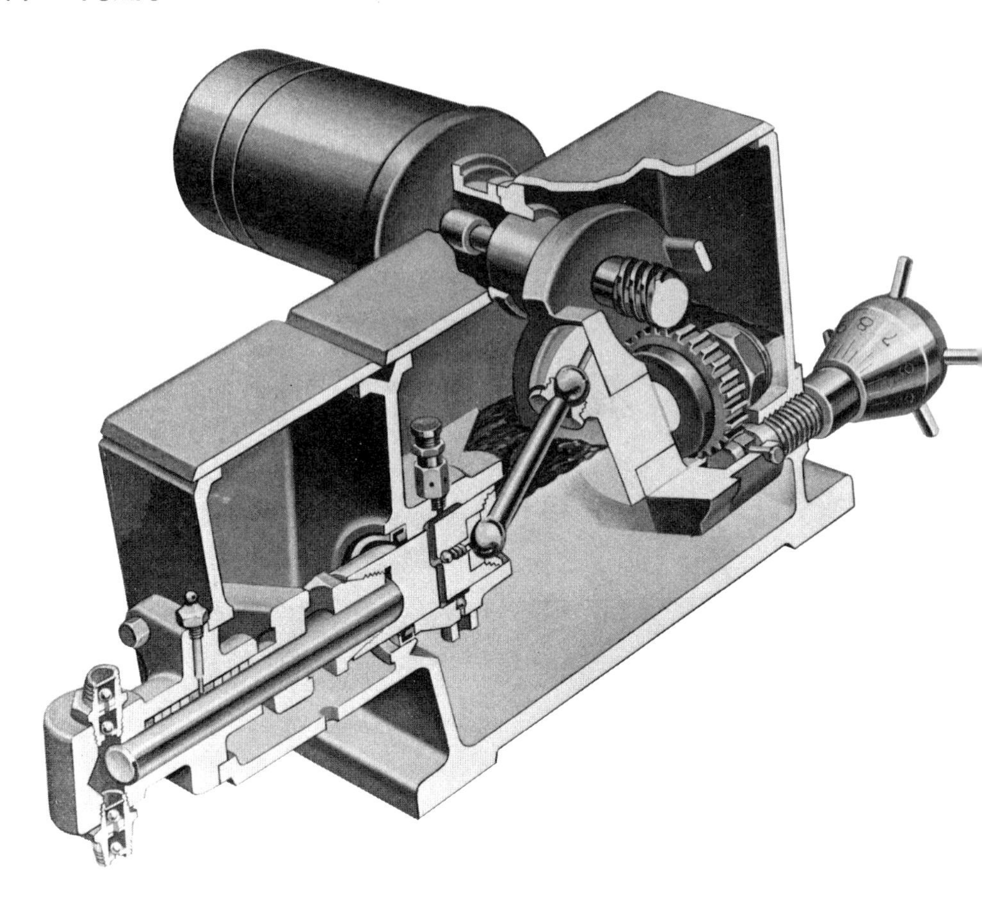

Figure 18. Sectional view of controlled-volume-type pump. Courtesy of Milton Roy Co.

to impart energy by means of increasing the velocity, is contradictory to this objective. Some rotary pumps, on the other hand, are specifically suited to this constant velocity concept; they are the best types to use for these special liquids.

The rotary screw pump (or sometimes the gear pump) has become the standard type for use on lube-oil- and seal-oil-pumping systems associated with large turbomachinery such as compressors and steam and gas turbines. It continues to produce a relatively steady rate of flow even in event of a pressure rise due to system fouling or of varying oil temperature and viscosity. Most rotary pumps are comparatively small, ranging from about 0.03 dm^3/s (ca 0.5 GPM) up to 250 dm^3/s (ca 4000 GPM). Efficiencies of 40–80% depend on size, viscosity, speed, and differential pressure. Although rotary pumps serve limited applications, they should not be overlooked in the many places where they are satisfactory.

Drivers for Pumps. The types of drivers used for pumps include all varieties of electric motors; both single-stage and multistage steam turbines; gas turbines and diesel or gas engines in certain specialized applications such as pipeline service, or in remote locations where gas or liquid fuels are more readily available than electric power. Diesel and gas engines are used as firewater pump drivers when the main,

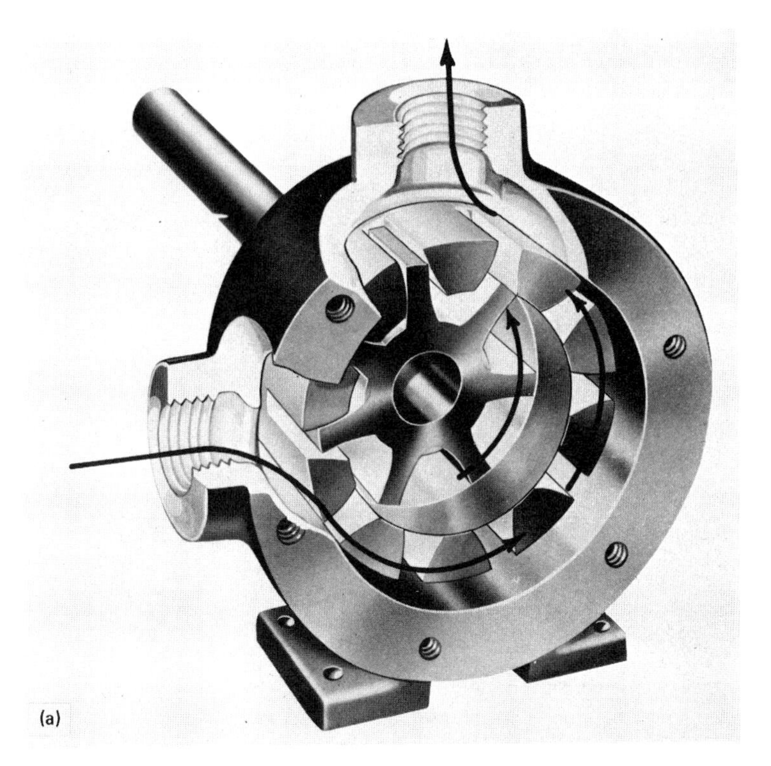

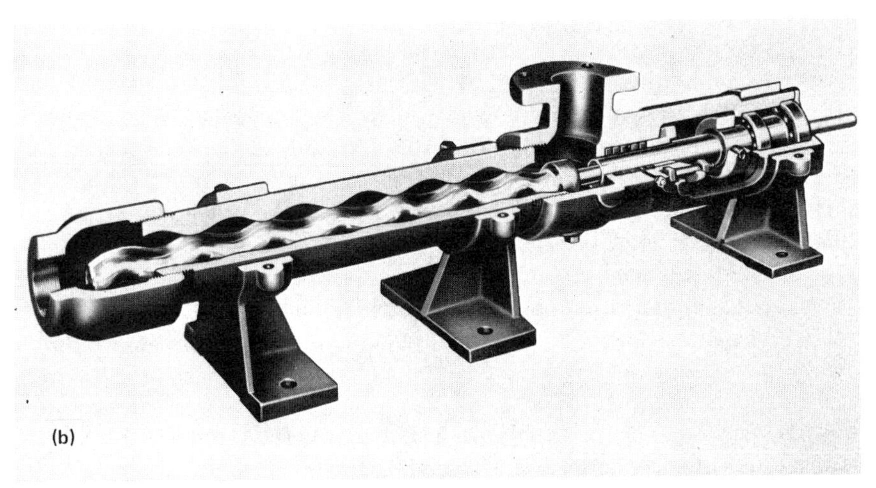

Figure 19. Sectional views of rotary pumps. (**a**) Courtesy of Viking Pump Division, Houdaille Industries, Inc. (**b**) Courtesy of Robbins & Myers, Inc.

electric-motor-driven pump has been forced out of service by the fire. In process-pump applications where uninterrupted pumping service is required and where steam is available directly at the process site, a steam-turbine-driven main pump or spare pump is often installed to take over during electric-power failure. In choosing the type of driver for a given application, careful evaluation must be made of the relative capital cost of the driver versus the cost of energy, electricity, steam, or gas. Occasionally, the economics of operating costs can be disregarded, however, in favor of the uninterrupted service requirement for process or safety reasons.

Table 3. Pump Evaluation for Viscous Liquid

Service and selection	Charge solution	Centrifugal	Rotary
liquid			
pumping temperature, °C	30		
viscosity at pumping temperature, mPa·s (= cP)	110		
density at pumping temperature, kg/m^3	950 (0.95 sp gr)		
capacity, dm^3/s	35 (555 GPM)		
differential pressure, kPa	600 (87 psi)		
pump differential head, m	64.4		
$NPSH_A$, m	5		
pump size			
API-610 type		$3 \times 4 \times 8\frac{1}{2}$	
gear type, mm			150
pump speed, rpm		3550	350
efficiency, %			
water		76	
viscous		63	80
efficiency correction[a]		0.83	
power required, kW		33.3	26
end-of-curve power, kW		40	
power at relief valve setting, kW			30
probable motor size, kW		40	30
$NPSH_R$, m		4.3	2.5

[a] Ref. 4.

The optimum driver arrangement is to match the driver speed to the pump speed, so that the two can be directly coupled. When this is not possible, as in the case of some gas-turbine, steam-turbine, and diesel-engine drives, or even in event of slow-speed or variable-speed pumps requiring motor drivers, then the use of speed-reducer or speed-increaser gears is common. In small sizes, especially for slurry pumps and positive-displacement pumps, use of V-belt or chain drives may be desirable. Consideration may be given to variable-speed motors, variable-frequency motors, hydraulic drives, eddy-current drives, etc, when variable-speed pumping is required.

Driver sizing for centrifugal pumps is usually done in one of two ways. Reference to a typical centrifugal-pump performance curve shows that the power requirement for a given pump changes in relation to the flow, with the peak power most often reached somewhat beyond the point of best efficiency. The most dependable and conservative method of sizing the driver is to examine the pump curve and choose a driver equal to or slightly larger than this maximum power at any point on the curve. This will provide enough capacity in the event of pumping conditions beyond those originally planned, plus extra power in event of changes in fluid properties such as viscosity or density (specific gravity). The other common method for driver sizing, as set forth in API-610 Standard, is to apply a percentage factor (10–25%) above the power required at the pump-rated conditions, and size the driver accordingly. This often saves gross oversizing of drivers and also provides more efficient driver operation as the driver then operates nearer its full-load point. Sizing the driver in this way is entirely satisfactory for many process-pumping systems as demonstrated by the system-head curve study (Fig. 4), which shows the maximum pumping point in a given system and the corresponding power requirement. No single rule can be given; each pumping system should be analyzed carefully to determine how much oversizing, if any, need be used.

For positive-displacement pumps, driver sizing is different. Since these pumps have fixed displacement and their output pressure does not vary with flow to any appreciable extent, there is no end-of-curve power to be ascertained. Instead, it is necessary to size the driver for the maximum pressure the system is designed to take. Often this can be an alternative or upset condition different from the pump rated condition. Most often, it is the setting of the relief valve downstream of the pump that limits the maximum pressure in the system. Such power requirements should be studied carefully, and the pump driver sized to handle the required power at the highest pressure, including full relief-valve accumulation. For pumps on viscous liquids, the maximum expected viscosity at the highest system pressure or relief-valve setting should be used to determine driver size.

As a minimum rule for sizing any type of driver (electric, steam, gas, or diesel) for any type of pump, the driver rating at the site conditions should not be less than 110% of the pump duty condition point on the heaviest (or most viscous) liquid to be pumped.

Economic Aspects of Pumping

Estimate of Pump Costs. Costs of pumps can be estimated in several ways, with varying degrees of accuracy. Generally, the most accurate price takes the longest time to obtain. A preliminary rough-order-of-magnitude (ROM) pricing method has been offered repeatedly in trade publications. The pumping requirements of flow and head are first obtained as discussed earlier. A factor equal to flow multiplied by head is calculated, and the type pump desired is decided upon (ANSI type, API-610 type, etc). Then, using the factor with the proper chart, an estimate of pump and driver cost can be obtained. In these charts, the costs are shown as a range rather than as a single line. Many variables influence the price of a pump, such as materials of construction, type of mechanical seal, cooling arrangements, types of tests, couplings, motor enclosure, and voltage, making it virtually impossible to present simple charts that cover all the possibilities. When estimating from Figures 20–24, the engineer should use the upper region of the shaded portion for a pump with more complex metallurgy or more accessories. To be conservative, if details like these are not yet known, use the upper lines throughout. The lower lines give closer costs for lower classes of metallurgy or for pumps with fewer accessories.

Use any preliminary estimating curves with care. Not every type of pump may be included—other curves could be constructed as the individual user feels necessary. Other pricing methods have been offered, including several computer programs in which the price list and pump data, size by size, are stored in the computer memory. When the individual data for each pump required are entered, the programs quickly select and price pumps of whatever size and type has been stored in the memory. At least one of these programs is available as public information (13). Pump prices have continued to increase during the last decade.

It must be understood that these estimates (Figs. 20–24) are the prices of the pump and driver fob manufacturers' works. They do not include freight to the installation site, nor any costs associated with installation, such as cost of foundation under the pump baseplate, piping or electrical connections, instrumentation or controls for the

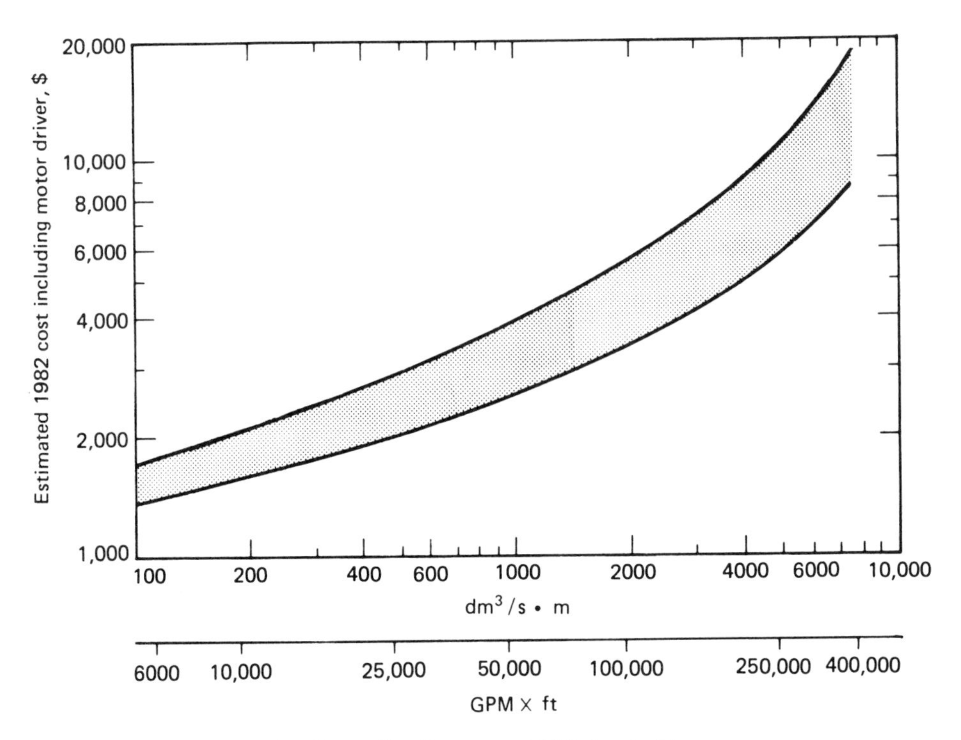

Figure 20. Cost-estimating curve, ANSI chemical-type pump.

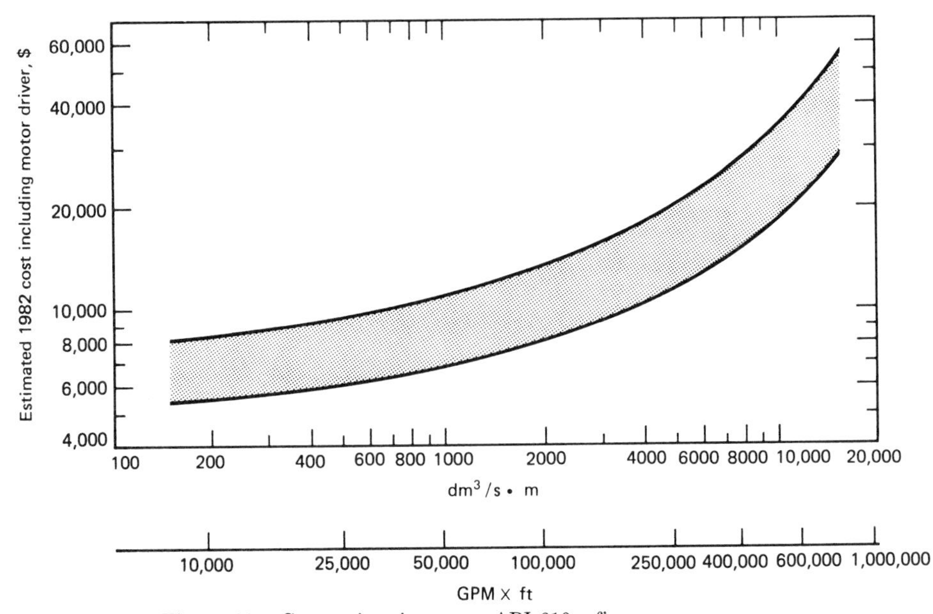

Figure 21. Cost-estimating curve, API-610 refinery process pump.

pump or the motor, motor starter, switchgear, etc. Experience gives some idea of how much the machinery cost must be increased to arrive at the total installed cost; this figure, too, varies with individual owner's special requirements, location of job site,

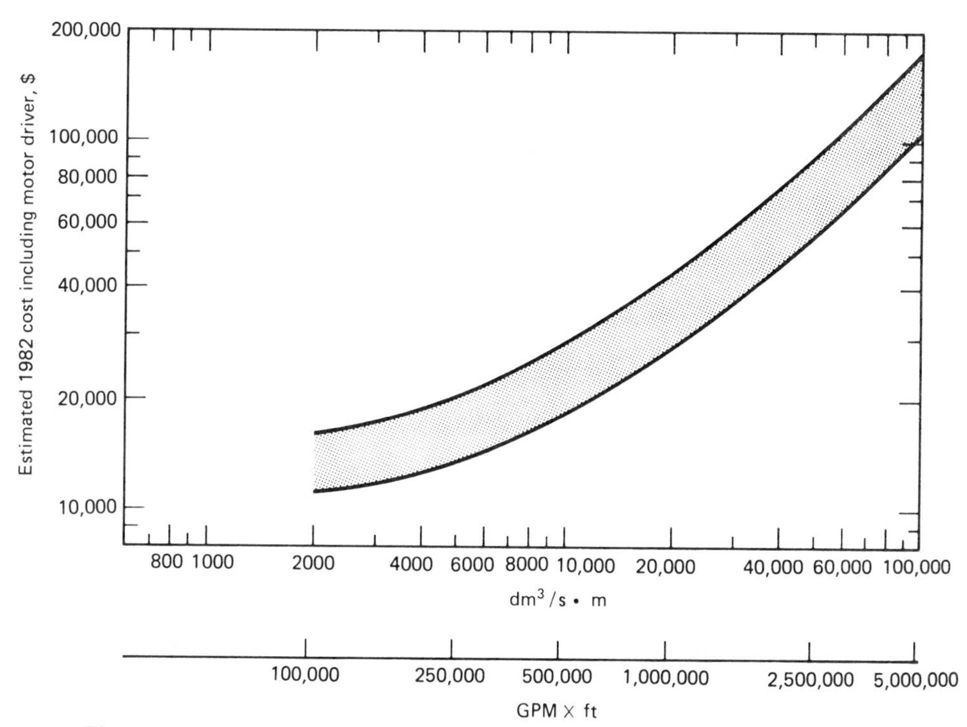

Figure 22. Cost-estimating curve, horizontal-split-case single-stage centrifugal pump.

labor costs at site, etc. Another ROM factor can be applied: the actual installed cost of the pump complete with its driver, baseplate, and coupling will be between two and three times the cost of the pump and driver.

A good estimate of pump costs can be achieved by actually selecting a pump make and model, referring to an up-to-date price list for that type pump, adding all the accessories and then adding the driver price from a similar up-to-date driver price list, and totaling these figures to obtain the net cost. Although some large user or contractor firms may maintain such price lists for in-house estimating, it is most common to obtain this level of pricing accuracy by contacting a pump manufacturer. The most accurate estimating method is the same as just described, except that formal written estimates are requested of one or more manufacturers. This obviously gives the most accurate results, but it usually also requires the longest time.

Evaluation of Pumping System Costs. Once the actual cost of a pump with driver is known, it is quite common to make an economic evaluation between two pumps which are offered for the same pumping requirement, to determine which is the more economical from an overall operating and installation cost. For example, the fractionator bottoms pump, rated at 38 dm^3/s (602 GPM) flow, 106 m head, and considering various manufacturers' printed performance curves for a pump to meet such conditions, might have efficiencies of 67–71%. Assume a cost for electric power was given at 3¢/kWh and that the pump is assumed to operate continuously for one year (ca 8500 h). Table 4 shows a simple evaluation for three pumps. It must be realized that fairly accurate cost estimates are needed to make this kind of comparison meaningful.

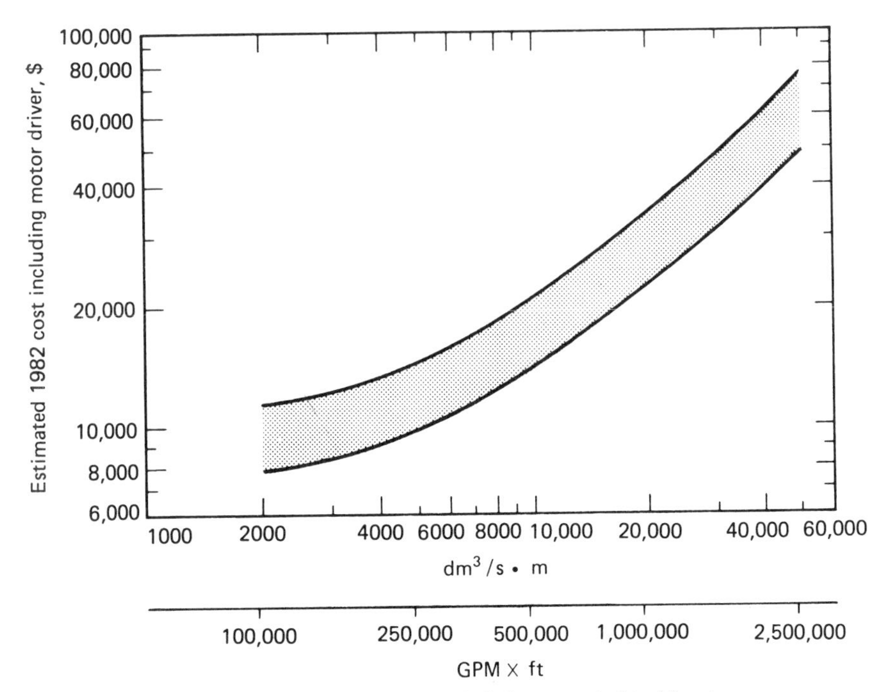

Figure 23. Cost-estimating curve, vertical short-coupled turbine-type pump.

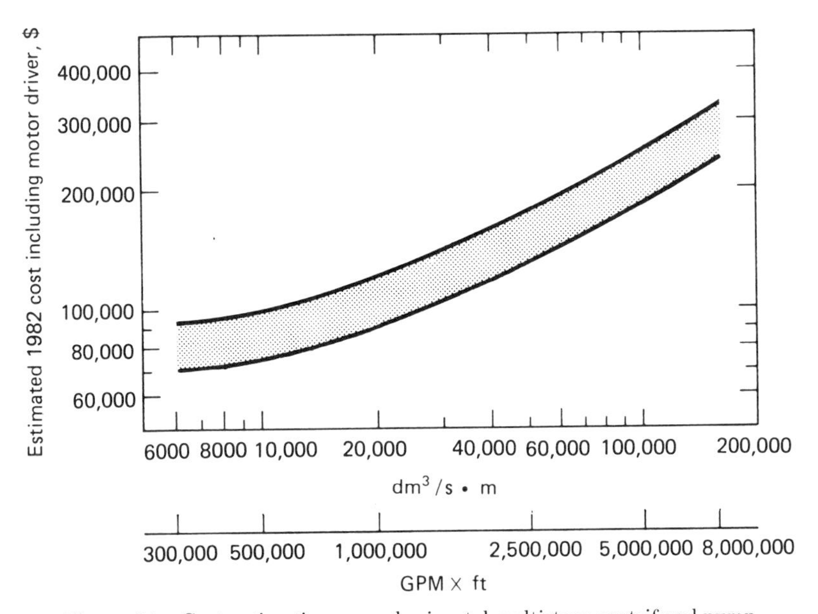

Figure 24. Cost-estimating curve, horizontal multistage centrifugal pump.

Evaluation of Pump Performance. ***Performance Testing.*** Testing of pumps before shipment from the manufacturer's works, or after installation in the field, or both, will prove whether the pump is doing the job it was designed for. All pumps undergo certain tests before shipment, and current practice in the process industries often

Table 4. Economic Evaluation of Pumps

Specifications and costs	Manufacturer		
	X[a]	Y[a]	Z[a]
pump model and size	4 × 6 × 10½	4 × 6 × 13¼	3 × 4 × 10½
number of stages	1	1	1
speed, rpm	3550	3550	3550
efficiency, %	67	68	71
power at rated point, kW	37.1	36.6	35
power at end of curve, kW	47	45	39
$NPSH_R$, m	3	4	4.6
impeller diameter, mm	235	260	236
cost, pump with motor driver, $	16,100	17,290	16,300
power evaluation, $	+535	+408	0
power cost basis (3¢/kWh, 8500 h/yr), $	9460	9333	8925

[a] Data for all manufacturers:

capacity, dm^3/s (GPM)	38 (600)
total head, m	106
temperature, °C	200
$NPSH_A$, m	5
density at temperature, kg/m^3	630
specific gravity	0.63
viscosity at temperature, mPa·s (= cP)	110

requires witnessing of such tests by the purchaser or his representative. On certain small, noncritical items, the witnessing of tests may be waived and certification obtained from the manufacturer that the pumps have passed.

Hydrostatic tests of pressure containing parts such as casings or cylinders should always be performed, at pressures of 150–200% of the pump design maximum allowable working pressure. On standard designs, with casings or cylinders taken from stock, all parts from a group of castings are tested to such a predetermined design value. On special designs, larger units, and custom-engineered items, it may be necessary to specify the test pressure clearly, as the manufacturer may interpret the 150–200% requirement to apply to the maximum working pressure required for this application. Thus, a nominal 4 MPa (ca 600 psi) casing could be applied for a 2 MPa service, and hydrostatically tested at 3 MPa, obviously not as revealing as a test performed at 6 MPa (ie, 150% of design).

For multistage pumps, especially barrel-type and sometimes also on horizontally split-type and vertical can-type, it is customary to perform the hydrostatic test of the casing in sections. This is done by inserting a blank disk into the casing at a certain point so that the hydrostatic test of the suction side of the pump can be performed at lower-than-discharge pressure. This is generally considered to be an acceptable test procedure. Most high pressure pumping systems are equipped with valves around each pump to block off an idle pump against system discharge pressure. Similarly, most multistage pumps have either opposed impellers or balance drums that allow stuffing-box pressure to be significantly lower than discharge pressure. Thus, the pressure on the suction side of the pump is controlled by system and pump design to a lower pressure and the sectional pressure testing is a safe procedure.

Mechanical running tests are performed on even the smallest pumps to be used for process duty to assure that adequate assembly has been performed and that the pump rotates freely with the clearances and fits which have been built. Certain types of small pumps are tested in groups, rather than each being set up on the test stand for an individual test.

The adequacy of such a mechanical running test should be evaluated in each case in the light of severity of service, importance to the process, etc. Normally, for pumps in process service or those used as auxiliary equipment to other process machinery, the mechanical running test is not considered to be sufficient.

A performance test is therefore specified for most process-type pumps to demonstrate the performance of each pump in the shop and prove its ability to meet the rated conditions. The preferred performance tests for centrifugal pumps follow ASME Power Test Code 8.2 and, for positive-displacement types, Power Test Code 7.1 (14). Hydraulic Institute Standards (4) also provide valuable information on the proper methods of conducting pump tests. Specifically, these standards require shop tests to be made under controlled conditions: suction conditions must be ideal, water is delivered to the pump suction from a still-well through ideal piping to eliminate prerotation, vortexing, and uneven flow, instrumentation is located to ensure that readings are not influenced by elbows or other pipe fittings that might disturb the flow pattern, and calibrated instruments are used and are sized for the rated pump flow.

Five or more performance points are usually required in a centrifugal-pump test, thus enabling an accurate test curve to be constructed. The test liquid is usually water, for safety reasons. (Rotary pumps are usually tested on oil.) Water performance is easily transformed to performance on hydrocarbons or other liquids based upon the relation of head to pressure and to power. Although the manufacturer makes a performance test without the purchaser being represented, and certifies that such a test was conducted and the results transmitted, the preferred way to test any pump in critical process duty is for the manufacturer to perform the necessary preparations, then allow the purchaser to witness the test. Witnessed performance tests are available for standard extra charges on most pumps. Test procedures (including codes and references to be followed) should be clearly specified in the original technical specification and purchase documents to avoid later misunderstanding.

A witnessed performance test using the actual job driver may not be feasible at the pump manufacturer's test facility. This is especially true on large motor-driven pumps, where the power, voltage, or other electrical characteristics of actual pump motor may not be compatible with the test-stand capabilities. Also, steam-turbine or gas-turbine drivers may be unsuited to the pump manufacturer's test stand and may require special tests to be conducted elsewhere.

An NPSH test may be needed in certain pumping applications, to prove the $NPSH_R$ for the particular pump by means of a suppression test. A vacuum is created on the suction side of the pump system while it is operating on water until the vapor pressure of the water is low enough to cause $NPSH_A$ to approach $NPSH_R$. Again, Hydraulic Institute Standards are specific about how this is to be done and how the results are to be interpreted. It is recommended that a shop-conducted NPSH suppression test be performed on any pump where the proposed or quoted $NPSH_R$ is ca 0.5 m of the calculated $NPSH_A$. On certain systems, such as multistage pumps for boiler-feed service, this limit may be raised, for example, to $NPSH_A - NPSH_R$ equal to ≥ 3 m. If a suppression test is specified, it is normal practice to witness the test.

In summary, it is usually good engineering practice to specify tests for every pump in process duty, however small or simple the pump may be. As the pumps increase in size and complexity, this rule becomes even more important. Finally, if tests are specified, it is good practice for the purchaser to witness them.

If centrifugal pumps miss their rated condition point as a result of oversizing of the impeller by the manufacturer and produce excess head at rated capacity, or if the pump is undersized by a small amount and the manufacturer wishes to polish the impeller vanes to produce a slightly higher head or efficiency, the pump should be retested after alteration. Similarly, if a pump is opened after a test for any other corrections, such as bearing fits, seal leakage, shaft polishing, adjustment of clearances, etc, a retest is a legitimate request to be made of the manufacturer.

More reference material on pump testing is found in the *Pump Handbook* (15).

Field Testing. Field evaluation of existing pumps must often be made, as it is not always possible to duplicate in the field the same ideal conditions that existed on the manufacturer's test stand. Piping arrangements for the actual installation may not provide sufficient straight length ahead of or after the pump for accurate measurement of flow. Instruments may not have been located at ideal points from a test standpoint, and readings may have been influenced by valves, elbows, or other fittings that disturb the flow pattern. The available instruments may not have the same degree of accuracy as those used on the test stand, or may not have been properly calibrated.

On the other hand, the field test in the actual pumping system is the final step in determining whether the pump does its job. Accordingly, if field-testing is to be performed, all of the above must be taken into consideration and field results compared with shop test results in light of the different piping and instrumentation arrangements. The shop test, if properly performed, should remain as the standard by which the pump is judged to be acceptable.

When pumps have been in service for some time, it may be necessary to evaluate performance on actual duty conditions to determine whether maintenance is required. Some pumping systems are designed with a great deal of overcapacity or surplus head built in, so that the pumps may run for years before becoming hydraulically unsuitable. It should be possible from field operating records to watch pump performance and evaluate when a pump is in need of maintenance or replacement. The pump head at a given capacity and speed is a fixed value when the pump is new. Field conditions may cause the pumped liquid to vary in temperature, viscosity, or specific gravity, or the suction or discharge pressure to differ from original specifications. These items must be carefully checked before concluding that a given pump is not doing its job.

Pump Revisions. Revisions to existing pumps may be needed for a number of reasons, including deterioration of pump performance after a reasonable length of operation, changed duty conditions in the given pumping system, or relocation of the pump to another site or another service.

Among the revisions readily made in the field maintenance shop are replacement of pump impeller or wearing rings to restore original performance or to meet new performance requirements. When a larger impeller is being installed in a given pump, the power requirements and thrust capabilities of the pump should be rechecked. If the driver must be increased in size or in speed (in the case of turbine drivers), additional system changes may be needed such as baseplate or piping changes, turbine revisions regarding steam flow or shaft limitations, etc.

Replacement of entire rotors in multistage pumps, other than to restore original rated performance, should first be reviewed carefully with the manufacturer. These pumps are considerably more complex and all aspects of their design need attention before an extensive change is made. Destaging of a multistage pump, that is, removing one or more stages to reduce total head, is a realistic procedure but likewise requires careful checking with the designer regarding its effect on thrust, shaft deflection, critical speed, hydraulic performance of the destaged pump, etc.

In more complex revisions, it may be desirable to return the pump to the manufacturer for rebuilding and possibly for retesting. The possible requirements of future revisions point clearly to the need for keeping records of original manufacturer's certified test curves, agreed-upon field-test performance data and retirement criteria, and accurate field-testing and measurement procedures for later use as pump performance deteriorates or process needs change.

Replacement of seals, bearings, sleeves, shafts, etc, should be considered as routine maintenance and not a revision to an existing unit.

Nomenclature

D = diameter, mm (in.)
H = head, m (ft)
N = rotative speed, rpm
N_s = specific speed, no dimensions
N_{ss} = suction specific speed, no dimensions
$NPSH_A$ = net positive suction head available, m (ft)
$NPSH_R$ = net positive suction head required, m (ft)
P_s = pressure, suction, kPa (psi)
P_{vp} = pressure, vapor, kPa (psi)
P_{frict} = pressure loss due to friction, kPa (psi)
P_{static} = pressure, static, kPa (psi)
Q = flow, dm^3/s (or L/s), (gal/min, GPM)
sp gr = specific gravity of fluid
ρ = density of fluid, kg/m^3
μ = viscosity (dynamic) of fluid, mPa·s (= cP)

BIBLIOGRAPHY

"Pressure Technique (Pumps)" in *ECT* 1st ed., Vol. 11, pp. 123–126, by L. H. Garnar, Worthington Corp.; "Pumps and Compressors (Pumps)" in *ECT* 2nd ed., Vol. 16, pp. 728–741, by R. Kobberger, Worthington Corp.; "Pressure Technique (Compressors)" in *ECT* 1st ed., Vol. 11, pp. 115–123, by E. L. Case and J. Charls, Jr., Worthington Corp.; "Pumps and Compressors (Compressors)" in *ECT* 2nd ed., Vol. 16, pp. 741–762, by J. J. Julian and J. F. Hendricks, Worthington Corp.

1. I. J. Karassik, W. C. Krutzch, W. H. Fraser, and J. P. Messina, *Pump Handbook*, McGraw-Hill Book Co., New York, 1976.
2. A. J. Stepanoff, *Centrifugal and Axial Flow Pumps*, John Wiley & Sons, Inc., New York, 1957.
3. S. L. Dixon, *Fluid Mechanics, Thermodynamics of Turbomachinery*, Pergamon Press, New York, 1966.
4. *Hydraulic Institute Standards*, 14th ed., Hydraulic Institute, Cleveland, Ohio, 1983.
5. R. E. McElvain and I. Cave, *Transportation of Tailings*, Warman Equipment International, Ltd., Madison, Wisc., Nov. 1972.
6. K. E. Wick and R. E. Lintner, *Pumping Slurries with Rubber-Lined Pumps*, Denver Equipment Division, Joy Manufacturing Co., Denver, Colo., Nov. 1973.
7. F. H. Love, *The Black Mesa Story*, Pipeline Engineering International, Nov. 1969.
8. Standards B73.1-1977, B73.2-1975, American National Standards Institute, New York.
9. *Centrifugal Pumps for General Refinery Services, API Standard 610*, 6th ed., American Petroleum Institute, Washington, D.C., Jan. 1981.

10. T. L. Henshaw, *Chem. Eng.*, (Sept. 21, 1981).
11. J. M. Bristol, *Chem. Eng.*, (Sept. 21, 1981).
12. *NPSH Handbook*, Bulletin 100.02, Milton Roy Co., 1981.
13. *Cost Oriented Systems Technique*, ICARUS Corporation, Rockville, Md.
14. *PTC 8.2 Power Test Code for Centrifugal Pumps* and *PTC 7.1 Power Test Code for Displacement Pumps*, American Society of Mechanical Engineers.
15. Ref. 1, Chapt. 14.

General References

Cameron Hydraulic Data, 15th ed., Ingersoll-Rand Co., Woodcliff Lake, N.J., 1977.

Flow of Fluids Through Valves, Fittings and Pipe, Technical Paper 410, Crane Co., Chicago, Ill., 1974.

W. H. Fraser in *Materials of Construction of Fluid Machinery and Their Relation to Design and Performance*, American Society of Mechanical Engineers, 1981.

T. Hicks, *Pump Selection and Application*, McGraw-Hill Book Co., New York, 1971.

F. A. Holland and F. S. Chapman, *Pumping of Liquids*, Reinhold Publishing Co., New York, 1966.

I. J. Karassik and R. Carter, *Centrifugal Pumps—Selection, Operation and Maintenance*, McGraw-Hill Book Co., New York, 1960.

R. Kern, *Chem. Eng.*, (April 28, 1975).

J. E. Miller, *paper presented at the American Society of Mechanical Engineers Petroleum Engineering Conference*, New Orleans, La., Sept. 18–21, 1966.

I. Taylor, *The Most Persistent Pump—Application Problem for Petroleum and Power Engineers*, ASME Paper 77-Pet-5, American Society of Mechanical Engineers.

RICHARD NEERKEN
The Ralph M. Parsons Company

R

REPELLENTS

Repellents are materials that affect insects and other organisms and disrupt their natural behavior. For bloodseeking insects, the desired result is to interrupt hostseeking and especially to interrupt biting of humans and animals to prevent the spread of disease. As pesticides become less effective owing to rapidly growing resistance in some locations, the expense and undesirability of using pesticides in the usual manner will undoubtedly increase (see also Insect-control technology).

In 1965, during the WHO Malaria Eradication Project, there were only 17 cases of malaria reported in the entire country of Sri Lanka (Ceylon), and the demise of the disease was confidently predicted by WHO. With the subsequent loss of inexpensive, persistent pesticides and the rise of broad-spectrum resistance to all pesticides, the number of cases per year increased sharply; 4×10^6 cases were reported in 1979. This phenomenon is by no means an isolated situation and is becoming possible, if not likely, worldwide. In addition, resistance of the malaria parasite, *Plasmodium falciparum*, to prophylactic drugs has been documented in many locations in Southeast Asia, India, and recently on the mainland of East Africa. This is particularly serious since the number of cases of malaria is estimated at 100×10^6 cases per year, resulting in as many as one million (10^6) deaths. The WHO has retreated from its projected goal of eradication and now hopes only to achieve control except in selected locations (1). On the U.S. mainland, there are locations like the San Joaquin Valley in California where heavy use of agricultural pesticides has caused all mosquitoes breeding there to become highly resistant, and spraying for control of adult mosquitoes is futile. There are periodic epidemics of mosquito-borne diseases on the U.S. mainland, such as Western Equine Encephalitis (41 cases, 1977), Eastern Equine Encephalitis (3 cases, 1977),

St. Louis Encephalitis (132 cases, 1977; 1300 cases, 1978), and California Encephalitis (139 cases, 1978) (2). Dengue fever, currently endemic in the islands of the Caribbean, is vectored by *Aedes aegypti*, a mosquito present in large numbers in the southeastern United States (3).

It has been suggested that the U.S. military, during operations in Southeast Asia during the 1960s and 1970s, evacuated more personnel because of vector-borne diseases, primarily malaria, than for combat injuries.

Chemical repellents have been in common use for many years and include such legendary materials as smudge fires and bear grease, from the days before window screens. It is very possible that animal fats could be efficaceous (see Biting Midge Repellents).

Types of Repellents

A critical review of techniques for the evaluation of insect repellents describes many test methods of evaluating repellents, including the following which are most often used (4).

Repellents in Cloth. Candidate repellents were applied to a knit cotton stocking or cloth patch at 33 g/m^2 cloth, usually as a 1% solution in acetone. Two hours later, the stocking or cloth patch was placed over an untreated nylon stocking on the arm of a subject, the hand was covered, and the arm exposed to 1500 female mosquitoes for one minute. If fewer than five bites were counted, the test was repeated at 24 h, then weekly until failure which is, by definition, five bites per minute. The standard mosquitoes used were *Aedes aegypti*, *Anopheles quadrimaculatus*, or *An. albimanus*. Candidate repellents in cloth tests are in the following classes: class 1, effective 0 d; 2, 1–5 d; 3, 6–10 d; 4, 11–21 d; or 5, >21 d. There is no standard repellent for *An. albimanus*.

Repellents on Skin. The candidate chemical is dissolved in ethanol and spread over one forearm of the human subject, as *N*,*N*-diethyl-*m*-toluamide [*134-62-3*] (DEET) (**1**) is similarly applied to the other forearm. Each arm is then exposed to 1500 avid female mosquitoes for 3 min at 30 min intervals. Effectiveness is based on complete protection, ie, the time until the first confirmed bite (one bite followed by another within 30 min).

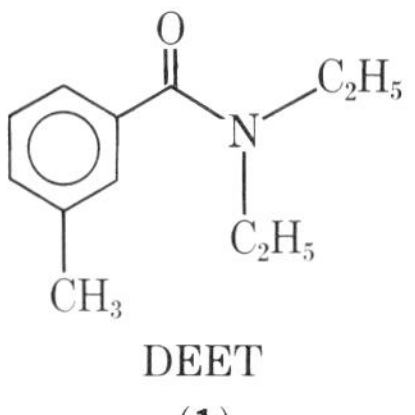

O, N, C_2H_5, C_2H_5, CH_3

DEET

(**1**)

Space-Borne Repellents. Air is drawn over a human arm through a 9.5-cm disk of cotton netting with 0.64-cm holes that has been treated with the candidate repellent; the air is then drawn into an olfactometer cage containing 125 avid female *Ae. aegypti*. The number of days the repellent prevented >10% of the mosquitoes from passing through the netting constituted "effectiveness." Test A used 3.89 g/g netting, and Test B used 1 g/g netting.

Skin-Patch-Tested Repellents. Small areas of human forearms are marked and treated with small amounts of repellent on a unit area basis to assure that the treatment rate is always the same between subjects (5). The patches are tested at 0 and 4 h against small numbers of mosquitoes (ca 15). This method does not consider creep, movement of repellent across the skin surface, or the interaction between two chemicals owing to such lateral movement of chemical. This method is commonly used with repellents evaluated at the Letterman Army Institute of Research in San Francisco, Calif.

Repellents Not Using Human Bait (No Attractant). A treated strip of fabric and a control strip are lowered into a container of crawling arthropods such as ticks, fleas, and mites. After a predetermined time, the strips are lifted and the animals remaining are counted and the percentage repellency determined.

Inanimate Attractant. Machines have been constructed by several groups to measure the intrinsic (initial) repellency of a compound when it is added to a warm, moist air stream to overcome the attractiveness of the airstream to mosquitoes. Such machines remove the factor of human odor in attempts to simplify the measurement of repellency. The use of the term intrinsic repellency implies that all chemicals have such a property at high enough dosage, but there is no standardized procedure to measure this property, which renders it difficult to use and interpret.

Animal Attractants. Numerous methods have involved the use of animals as attractants, then evaluating repellents as skin treatments or attached cloth treatments, often against crawling arthropods such as fleas, ticks, and mites. Animals such as gerbils, guinea pigs, camels, mice, shaved rabbits, and hairless dogs have been used, particularly when the toxicity was unknown.

Arthropod Repellents

Disease-bearing mosquitoes that are day- or night-biters include *Aedes* species that often transmit viruses to humans. Of 435 arboviruses (arthropodborne viruses) found in insects by 1980, 100 cause diseases in humans. The diseases include yellow fever, dengue haemorrhagic fever, and many forms of encephalitis: Japanese B, Venezuelan, West Nile, Chikungunya, Ross River, and Rift Valley fever. *Culex* species are vectors of several forms of encephalitis in the U.S.: Western Equine, Eastern Equine, St. Louis, and California encephalitis. Other insect species that bite humans and are vectors of diseases (scientific name; disease) include biting midges (*Ceratopogonidae*; bluetongue virus), blackflies (*Simulium*; river blindness), sandflies (*Phlebotamines*; leishmaniasis), tsetse flies (*Glossina*; sleeping sickness), tabanids or horseflies (*Tabanus*), clegs (*Haemotopota*), deerflies (*Chrysops*), and stableflies (*Stomoxys*). Crawling arthropods include cattle ticks (*Rhipicephalus*; heartwater fever, blackwater fever), soft ticks (*Dermacentor*, *Amblyomma*; rocky mountain fever, spotted fever), mites or chigger mites (*Trombidiidae*; scrub typhus), bedbugs (*Cimex*), and fleas (*Pulex*; plague) (6).

Mosquito Repellents, Clothing Impregnants. Clothing impregnants may be applied to socks, head nets, gloves, or conventional apparel by dipping into emulsions, manual application of liquid, or surface spraying with aerosols (qv). Repellents may be applied to cotton cloth using technical material dissolved in acetone, although this solvent may be unavailable or incompatible with some synthetic-fabric blends. Repellents may be applied to clothing on a large scale using 10% of a nonionic emulsifier and 90% technical material at 5 wt % of the cloth. Repellent jackets made of 0.64 cm cotton–

nylon mesh with drawstring hoods and front closures may be repeatedly treated with repellent such as DEET. There are several advantages: the repellent is applied to the mesh, not directly on skin, and repellents other than DEET may be used, such as repellents more effective than DEET against biting flies or other mosquitoes. The jacket may be retreated as necessary simply by addition of a measured quantity of concentrated repellent (such as 75 wt % DEET) to the jacket while in its storage pouch. An effective skin repellent such as DEET is lost much more slowly from the mesh jacket than from skin. This jacket has been added to the army supply system and is meant to be retreated with standard military-issue 75 wt % DEET. The DEET-treated jacket was very effective against mosquitoes in northern and southern states, blackflies in Maine, biting midges in Florida, and sandflies in Panama. The jacket was effective against heavy pressures of very small insects that could have passed through the mesh. Also, the jacket's wide spaces allow passage of air and moisture, making it more comfortable to wear than long-sleeved, heavily repellent-treated clothing or than skin-applied repellents, particularly in warm, humid climates with heavy insect populations (7).

The standard clothing impregnant adapted by the U.S. Armed Services, M-1960, was intended for large-scale utilization in military laundries for protection from mosquitoes, fleas, ticks, and chigger mites (8). It contains equal parts of benzyl benzoate [*120-51-4*], *N*-butylacetanilide [*91-49-6*], 2-ethyl-2-butyl-1,3-propanediol [*115-84-4*], and a third of a part of Tween 80 (see Emulsions). It has disadvantages, including odor, and is less effective as a mosquito repellent than DEET, although benzyl benzoate ensures that it is a good tick and mite repellent. The M-1960 formulation has been supplemented with aerosol formulation DEET for personal protection and clothing treatment or 75 wt % DEET:25 wt % ethanol in plastic squeeze bottles especially for skin treatment. A formulation problem with aerosols is that freon propellants have often been replaced with compressed hydrocarbon gases; such potential flammability is undesirable. Clothing treatments onto 100% cotton military uniforms using DEET have been shown to resist laundering better than treatments onto polyester–cotton blends, whereas 100% polyester retains repellents poorly (9).

Standard Mosquito Repellents. Dimethyl phthalate [*131-11-3*] (DMP) is a clear oil insoluble in water, soluble in organic solvents, and synthesized from phthalic acid esterified with methanol. DMP is a good repellent, used as a standard against *Ae. aegypti*, and is effective for 11–22 d on cloth. The famous repellent Rutgers 6-12 [*94-96-2*], 2-ethyl-1,3-hexanediol is a clear oil that is insoluble in water and miscible with organic solvents including ethanol. Screened during World War II, it is exceptionally effective against *Ae. aegypti*, lasting 196 d on cloth.

DEET has proved to be the most outstanding all-purpose individual repellent yet developed (10). Since its initial report as a promising repellent in 1954, DEET is considered the best all-around repellent with generally acceptable characteristics despite a continuing search for a superior chemical. The mode of action of successful repellents is unproven. DEET is the principal or only component in most of the effective mosquito repellents on the market (11), as can be seen from the listing in Table 1 of some recent entries to the commercial market (12). Since the introduction into the United States of Muskol (100% DEET) in 1978, a number of companies have begun marketing 100% DEET. Outdoor magazines reported that 100% DEET worked, was perhaps more desirable, and had to be applied less often than more dilute formulations. *Consumer Reports* advised campers to demand pure DEET repellent. Sales of DEET

Table 1. Mosquito Repellents Containing DEET

Company	New product date	Form	Brand name	Composition
Mowatt Enterprises, Brewer, Maine	1980	liquid	Ben's 100	100% DEET
S. C. Johnson & Son, Inc., Racine, Wisc.	1956	liquid	Off!	50% DEET
		liquid, aerosol, lotion, pump-spray	Deep Woods Off!	25–75% DEET
	1983	liquid	Maximum Strength Deep Woods Off!	100% DEET
Muskol Ltd., Schering-Plough, Inc., Kenilworth, N.J.	1956	liquid	Muskol	100% DEET
Bayer A.G., Miles Laboratories	1978 USA	alcohol solution	Cutter	75% DEET
Sterling Drug Co., d-Con Co.		solution		2-ethyl-1,3-hexanediol [*94-96-2*] plus DEET
Wisconsin Pharmacal Co., Jackson, Wisc.		liquid	Repel 100	100% DEET
		lotion	Repel Lotion	52% DEET
Ole Time Woodsman Inc., Worcester, Mass.	1981	liquid	Jungle Plus	100% DEET
	1960	liquid	Liquid Fly Dope	pine tar camphor [*76-22-2*] citronellal [*106-23-0*] DEET

repellents expanded rapidly to the present and currently are estimated at \$(50–75) $\times$ 10^6/yr in the United States alone (12–13). DEET costs ca \$6.90/L (\$26/gal) from a chemical manufacturer. A 30-cm^3 (one ounce) bottle that retails for \$3–4 and lasts the typical consumer a year contains \$0.20 of chemical.

Experimental Mosquito Repellents. A comparison of nine commercial repellents was made against *Ae. aegypti* using the skin-patch test. At zero time (measuring intrinsic repellency), Stabilene [*9003-13-8*] and MGK Repellent 326 [*136-45-8, 3737-22-2*] were significantly inferior to DEET, dibutyl phthalate [*131-11-3*], Indalone [*532-34-3*] (**2**), dimethyl phthalate [*84-74-2*], MGK Repellent 11 [*126-15-8*] (**3**), 2-ethyl-1,3-hexanediol [*94-96-2*], and Citronyl [*39785-81-4*] (**4**). Efficacy ranking by 4-h ED_{50} (50% biting rate on the treatment) was Indalone, Citronyl, 2-ethyl-1,3-hexanediol, and DEET, as the others had become ineffective. The relative superiority of DEET in comparison to other standard repellents was discussed (14).

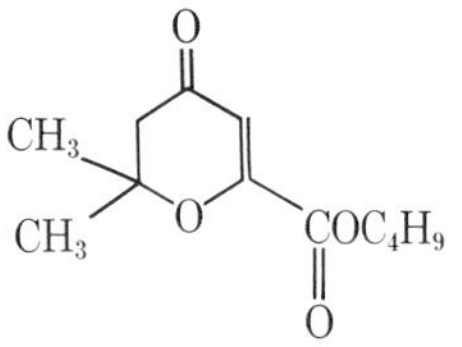

Indalone
(**2**)

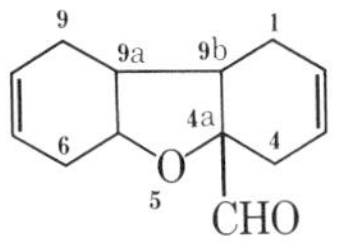

MGK Repellent II
(**3**)

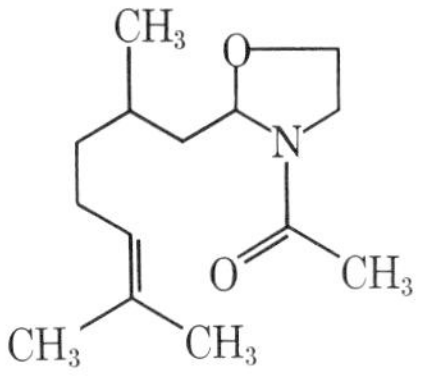

Citronyl
(**4**)

Seventy-one *N*,*N*-dimethylcarboxamides were tested on cloth against three species of mosquitoes; *Ae. aegypti*, *An. quadrimaculatus*, and *An. albimanus*. The amides were prepared via the acid chloride by treatment of the appropriate carboxylic acid with thionyl chloride, which was then added to a cold mixture of ether–water containing excess *N*,*N*-dimethylamine plus sodium hydroxide. Extraction and vacuum distillation gave materials that were used as distilled, with 95–98% purity by gc. Nine of thirty-eight *N*,*N*-dimethylbenzamides, three of seventeen *N*,*N*-dimethylphenylacetamides, two *N*,*N*-dimethyl-ω-phenylcarboxamides, and three *N*,*N*-dimethylalkanamides were class 5 repellents against all three species of mosquitoes. Typically, *N*,*N*-dimethylundecanamide [*6225-09-8*] provided 71-, 132-, and 59-d protection against *Ae. aegypti*, *An. quadrimaculatus*, and *An. albimanus*, respectively (15–16). Several of these materials were tested and found to be effective against blackflies in Maine.

Biting Midge Repellents. The genus *Culicoides* is found in fresh water, salt water, and tidewater environments in the southeastern United States and Caribbean, where they may be properly called biting midges, or more commonly sand flies, sand fleas, sand gnats, punkies, no-see-ums, or flying teeth. Because of their small size, they can pass easily through ordinary mosquito screens. Commonly used repellents are of varying effectiveness, depending upon the species involved (17).

In paired field tests, topically applied candidate-carboxamide repellents were compared to DEET at Parris Island, S.C., against *C. hollensis* and on the Gulf coast of Florida against *C. mississippiensis*. Repellency was determined three ways, ie, biting rates, length of protection, and coefficient of protection. DEET averaged 61 times greater protection than the untreated check, and all others averaged 130 times greater protection at Parris Island. In Florida, DEET averaged 28 times greater protection than an untreated check, whereas all other materials averaged 190 times greater protection. DEET was considered an effective repellent when applied to exposed skin as a 25 wt % formulation, but four novel alicyclic piperidines (carboxamides) were more effective than DEET: *N*,*N*-dipropylamino-3-cyclohexenecarboxamide [*68571-08-4*] (**5**), bp 93°C (60 Pa or 0.45 mm Hg); 2-methyl-1-piperidyl-3-cyclohexenecarboxamide [*69462-43-7*] (**6**), bp 110°C (103 Pa or 0.77 mg Hg); 1-piperidyl-3-cyclohexenecarboxamide [*52736-58-0*] (**7**), bp 108°C (60 Pa or 0.45 mm Hg), bp 114°C (27 Pa or 0.2 mm Hg); and 1-(hexahydro-1*H*-azepinyl) 2-methylcyclohexenecarboxamide [*67013-96-1*] (**8**), bp 115°C (120 Pa or 0.9 mm Hg). The first two were substantially more effective against *Culicoides*.

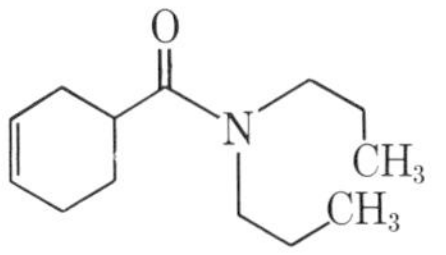

O, N, CH_3, CH_3

(**5**)

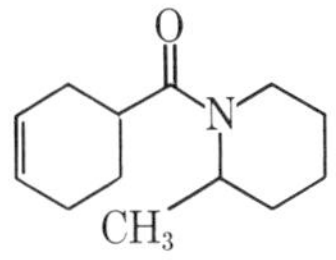

O, N, CH_3

(**6**)

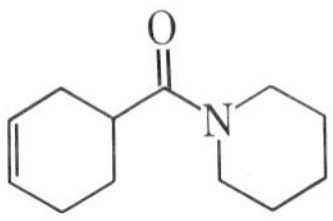

O, N

(**7**)

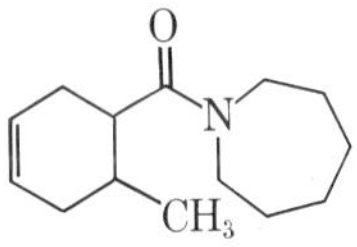

O, N, CH_3

(**8**)

They were typically prepared by addition of the appropriate alicyclic carbonyl chloride to an ether solution of the amine at 0°C; the solutions were warmed to room temperature overnight and then extracted, dried, and vacuum-distilled. The crude carboxamide was distilled under high vacuum. Among 117 compounds synthesized, structures found to be especially active included the 1-piperidyl- (**7**) and 1-[hexahydro-1*H*-azepinyl] (**8**) derivatives where the cyclohexenyl ring could be monounsaturated or have methyl branches such as in the 2-position (16,18).

Because of anecdotal information from South Carolina that baby oil was an effective repellent for biting midges, experiments were carried out using several commercial mineral-oil preparations, scented and unscented. The oils were shown to interfere with biting behavior and did therefore have a repellent effect, as when applied thickly to skin, the small biting midges would tend to become thickly coated and drown in the oil (19).

Phlebotomine Sandfly Repellents. The development of personal protection measures to control the attack of phlebotomine sand flies has been a minor area of research when compared with that done with mosquitoes and some other biting *Diptera*, largely because these species are geographically confined to what have been relatively underdeveloped tropical and subtropical areas of the world. With expanding world population, the development of these areas for agriculture and subsequent habitation is economically important to many countries. In addition, some of these regions are of strategic value for military installations and training grounds (20).

Only a few tests of repellents for Old World phlebotomine flies have been documented, none recently. Dimethyl phthalate and pyrethrum cream were partially protective on skin against *Phlebotomus papatasi* for 6 h (21). Tests with laboratory-reared *P. papatasi* showed that the duration of complete protection (no bites) provided by *N*,*N*-diethyl-*m*-toluamide (DEET), *o*-ethoxy-*N*,*N*-diethylbenzamide, *o*-chloro-*N*,*N*-diethylbenzamide, or *N*-butyryl-1,2,3,4-tetrahydroquinoline averaged at least 4 h, but perspiration contributed to a high rate of repellent loss (22). Investigations with repellent-treated netting indicated that DEET-treated bed nets gave complete protection throughout the night (23).

Phlebotomine sand flies are bloodsucking insects found primarily in the tropical and subtropical regions of the world. They are vectors of the various forms of leishmaniasis, bartonellosis, and numerous arboviruses, including the medically important sandfly-fever group. In Panama, phlebotomine sand flies are responsible for the transmission of *Leishmania braziliensis panamensis*, the causative agent of cutaneous leishmaniasis, a human disease among people living in close association with the forests. Cutaneous leishmaniasis and arboviruses affect not only Panamanians but U.S. military personnel who regularly train in jungle areas. A recent study showed that 1.6% of Army personnel who took part in the jungle warfare training course during a 3-wk period contracted leishmaniasis (24).

Personal protection methods were evaluated against phlebotomine sand flies in the bush in Panama. Arms were treated with 2 mL of 12 wt % max of the following repellents in paired tests with DEET, which was applied to the other forearm in the same manner: hexahydro-1-[(2-methylcyclohexyl)carbonyl]-1*H*-azepine [*52736-62-6*] (**9**); 1-(3-cyclohexen-1-ylcarbonyl)piperidine [*52736-58-0*] (**7**); *p*-isopropyl-*N*,*N*-dimethylbenzamide [*6955-06-2*] (**10**); DEET (**1**); and *N*,*N*-dipropylcyclohexanecarboxamide [*67013-94-9*] (**11**). Skin applications of the five selected repellents including DEET provided a mean coefficient of protection (CP) of 99.2% against the attack of at least three species of *Lutzomyia*. All of these repellents tested at the highest dosage gave good protection from the bites of at least three species of phlebotomine sand flies, two of which are important vectors of leishmaniasis. However, the azepine treatment gave complete protection and warrants further study (19).

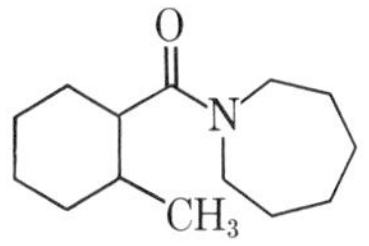

(9) 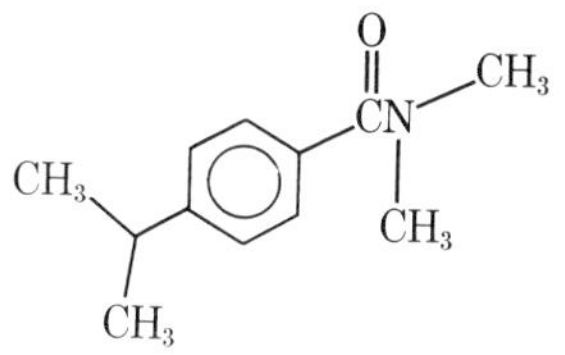 **(10)** 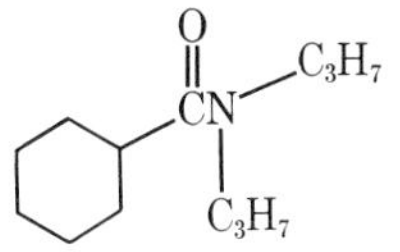 **(11)**

DEET-treated net jackets also provided good protection, but it was concluded that an additional application of repellent to the unprotected face was necessary for maximum protection. Permethrin (**12**) [*52645-53-1*]-treated clothing did not provide the protection expected. Sand fly behavior and resistance to quick knockdown were responsible for the numbers of bites recorded, and therefore, maximum protection from bites would require application of DEET or another suitable repellent to the exposed skin when wearing permethrin-treated clothing (19).

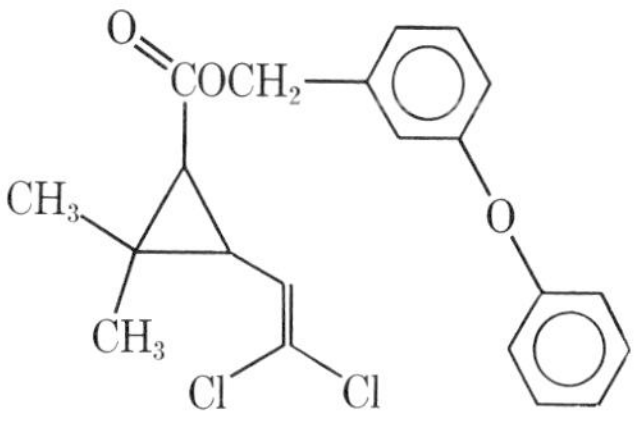

(12)

Table 2 compares the effectiveness of eight repellents to DEET.

Tick and Chigger Repellents. Prevention of tick attachment is possible by mechanically preventing access of the ticks to bare skin bordering or beneath the clothing, such as zippered long pants with cuffs tucked into boot or sock tops and a long-sleeved shirt worn inside the pants (26).

Repellents are best impregnated into clothing, on the skin on the wrists under the sleeves, on and above the socks, and around the neck on the exposed skin and under the collar. DEET (Off, Cutters, Repel, and a number of other products) can also be purchased as a 50% liquid and mixed with isopropyl alcohol (59 mL DEET and 1 L 2-propanol) to give sufficient material to impregnate clothing with a 5% solution and give 80–98% protection depending on tick density.

Permethrin has been available since 1982 as a clothing treatment in 10 states under EPA label 24C in an aerosol formulation called Permanone Tick Repellent (Fairfield American Corporation, Medina, N.Y.). A topical Permethrin treatment of clothing has shown good effectiveness against crawling insects when applied as a water-based formulation of 0.5–4% Permethrin. It gives extremely effective protection against ticks. Permethrin-treated cloth is practically odorless; a single treatment of ca 20-s spray (ca 0.2 g/m^2) adsorbs to the outer surface of clothing, does not contact

Table 2. Repellent Compounds with Their Relative Effectiveness[a] Compared to DEET[b]

Priority	Structure no.	Structure	Name	CAS Reg. No.	*Aedes aegypti*	*Aedes taenio-rhynchus*	*Anopheles quadri-maculatus*	*Stomoxys calcitrans*	*Chrysops atlanticus*	Black flies	*Culicoides* (sand flies)
1	**(7)**		1-(3-cyclohexen-1-ylcarbonyl)piperidine	*[52736-58-0]*	<	>	=	>	>	>	>
2	**(11)**		*N*,*N*-dipropylcyclohexanecarboxamide	*[67013-94-9]*	<	>	<	>	>	<	>
3	**(9)**		hexahydro-1-[(2-methylcyclohexyl)-carbonyl]-1*H*-azepine	*[52736-62-6]*	<	=	<	>	>	=	>
4	**(13)**		1-(3-cyclohexen-1-ylcarbonyl)hexahydro-1*H*-azepine	*[52736-59-1]*	<	<	=	>	=	>	
5	**(10)**		*p*-isopropyl-*N*,*N*-dimethylbenzamide	*[6955-06-2]*	<	=	=	=	=	<	>
6	**(14)**		1-[(6-methyl-3-cyclohexen-1-yl)-carbonyl]pyrrolidine	*[67013-95-0]*	<	=	<	=	=	=	
7	**(15)**		1-(cyclohexylcarbonyl)hexahydro-1*H*-azepine	*[68571-09-5]*	<	=	<	<	=	=	
8	**(16)**		1-[(2-methylcyclohexyl)carbonyl]-pyrrolidine	*[52736-60-4]*	=	<	=	=		=	

[a] >, <, and = indicate statistically significant differences of greater than, less than, and equal to, respectively.
[b] Ref. 25.

skin, and is long lasting (27). Labels have been issued in Kentucky, Tennessee, Alabama, South Carolina, Missouri, Arkansas, Oklahoma, and Virginia.

Permethrin as a clothing treatment acts more as a toxicant than a repellent, and though ticks may crawl on the clothing, the visit is only temporary and usually fatal within a few minutes. The lethal barrier provided has been shown to give 100% protection in tests in Oklahoma, Kentucky, and Florida (28). The materials listed above are effective repellents against chigger mites, also called chiggers or redbugs.

Cockroach Repellents

A recent, well-documented source for general information on cockroach control includes a chapter on repellents and another on toxicants (29). Transport of goods and materials provides universal opportunity for cockroach infestation in commodities, including shipments of bananas, laundry, dry cleaning, paper bags, corrugated cardboard boxes, empty beer and soft-drink bottles, cases in recycling locations, and trucks used for transporting these commodities. Although repellents may never completely stop the development of infestations, they may be helpful in preventing transport of cockroaches into uninfested areas. Some logical uses of repellents are on cardboard cartons for food and soft drinks, on beer crates, and in coin-operated vending machines, all of which provide excellent shelter and food (30).

Ten of the best cockroach repellents were chosen from over 200 compounds tested against *Periplaneta americana*. A good repellent could be used either alone or in conjunction with an insecticide as a residual treatment in business establishments or homes. Such effective, long-term repellents could become more useful in the future if the only toxicants available are short-term biodegradable materials. This is especially problematic when retreatment is expensive and rapidly becomes ineffective. Also, the cockroach's opportunistic nature of feeding and shelter-finding permits survival and flourishing when most but not all sites are treated. Similarly, the use of slow-acting toxicants such as borax and boric acid is not effective unless insects can be confined to treated surfaces. Obviously, this describes a laboratory test and is not applicable to the real world in which cockroaches may quickly leave an effectively treated area and fully recover from the sublethal effects.

Tabutrex (**3**) [*141-03-7*], R-11, R-55 (**17**) [*23885-27-0*], and R-874 (**8**) [*3547-33-9*] have been available for use by industry as cockroach repellents for ca 20 yr. Dibutyl succinate or Tabutrex (Glenn Chemical Co.) is formulated as an emulsion concentrate (20%) and an oil spray (2%). The oral LD_{50} (rat) is 8000 mg/kg. Treated surfaces remain 100% repellent to *Blattella germanica* for three weeks. In laboratory tests, cockroaches were repelled from wooden beverage crates for 15 wk (31).

Hexahydrodibenzofurancarboxaldehyde–butadienefurfural copolymer (**3**) (MGK R-11) (Phillips Petroleum Co.) is a pale yellow liquid with a fruity odor, miscible with many organic solvents, and compatible with most insecticides. A typical formulation contains 0.075% pyrethrins, 0.15% piperonyl butoxide, and 1% R-11. For treating the inside of cartons, R-11 is applied as a 1% emulsion incorporating 2% of the synergist MGK 264. On beer cartons, R-11 gave >80% repellency for 2 mo, reducing to 60% at 6 mo. MGK R-11 is now used in pet sprays and in repellents for personal use. Of four materials evaluated for their odor, this repellent was the most pleasant (32). The acute oral LD_{50} (rat) is 2500 mg/kg; the dermal LD_{50} is >2000 mg/kg.

tert-Butyl *N*,*N*-dimethyldithiocarbamate (**17**) or MGK R-55 (McLaughlin Gormley King Co.) is a rodent and insect repellent. It repels *B. germanica* from treated cartons for 90 d (at 2%) and for 63 d (at 1%). It is more odorous and toxic than MGK R-11 and MGK R-874.

$$(CH_3)_2N\overset{\overset{S}{\|}}{C}SC(CH_3)_3$$

(17)

2-Hydroxyethyl *n*-octyl sulfide (**18**) or MGK R-874 (Phillips Petroleum Co.) is a light amber liquid with a mild mercaptanlike odor, slightly soluble in water but miscible with most organic solvents. It is used with MGK 264, a pyrethrins synergist. Formulations commercially available are an emulsion concentrate diluted with water and applied at 1–5% by automatic spraying equipment and an oil solution used at one gram of active material per square meter. R-874 tested against German cockroaches is marginally more effective than R-55 and lasts twice as long as R-11. Toxicity is low; the acute oral LD_{50} (rat) is 8530 mg/kg; dermal LD_{50} is 13,590 mg/kg. Use of this repellent near food should not create a health risk (29).

$$HOCH_2CH_2S(CH_2)_7CH_3$$

(18)

A listing of compounds evaluated in the laboratory as cockroach repellents is available that summarized 872 synthetic compounds of 901 bioassayed from 1953 to 1974 (31). Fencholic acid [*512-77-6*] (**19**) (3-isopropyl-1-methylcyclopentanecarboxylic acid) was used as a standard repellent in tests conducted by placing twenty cockroaches in a glass crystallizing dish without food and water and offering them a choice of two cardboard shelters, one of which was treated with 1 ml or 2 ml of a 1% solution of the candidate in acetone. Counts were made daily for 7 d. There is currently no work going on in this area in USDA laboratories.

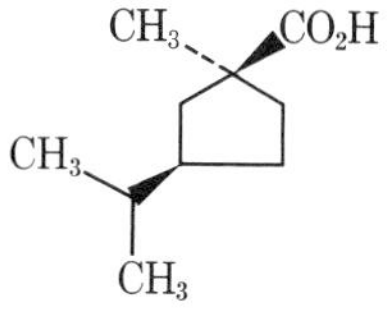

CH_3 CO_2H CH_3 CH_3

(19)

Another problem lies in the overlap of repellent–toxicant definition, in that many toxicants are known to have repellent effects (32). Pyrethrins often are used on ships to flush cockroaches from harborages during a treatment with another less activating toxicant.

In a survey of the components making up commercially available formulations of insecticides for cockroach control in the United States (toxicants, synergists, solvents, flushing agents, and emulsifiers), 121 different materials were examined (33). Tests showed that pyrethrins (considered repellents for some years), MGK 264 [*113-48-4*] (**20**), and Triton X100 [*9002-93-1*] (**21**) (an emulsifier) were noticeably repellent to both German and American cockroaches.

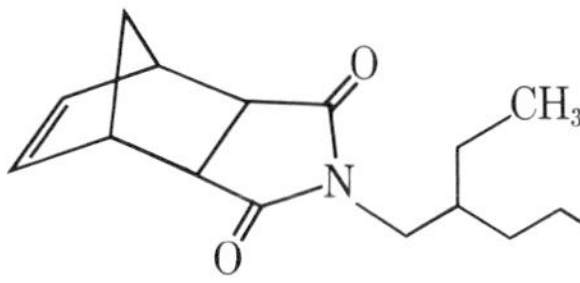

O N O CH_3 CH_3

(20)

CH_3 CH_3 CH_3 CH_3 CH_3 $-O\!\!+\!CH_2CHO\!+_{\!n}H$

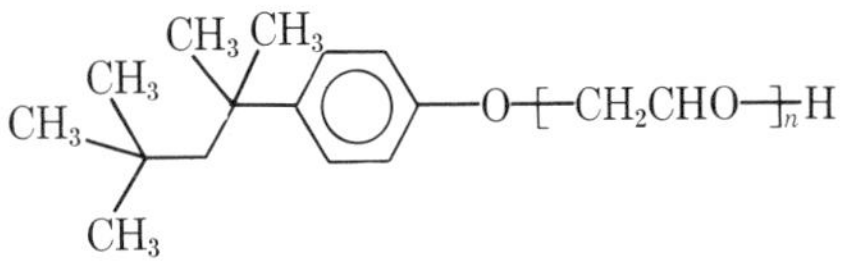

(21)

The list of repellent materials also included a number of surfactants (wetting agents) and deodorants, but in no case were solvents implicated (34). In laboratory studies for repellency, some formulations containing 0.5% organophosphates did not function as repellents, but diazinon [*333-41-5*] (**22**) (0.5%), propoxur [*114-26-1*] (**23**) (1%), synergised pyrethrins (1%), some synthetic pyrethroids, and bendiocarb [*22781-23-3*] (**24**) (1%) were repellent for a week or more (35). In an extensive test program of many insecticides, avoidance of treated surfaces was observed more frequently with diazinon than with any of the other materials (36). Diazinon (**22**) is the pesticide of choice for modern cockroach control.

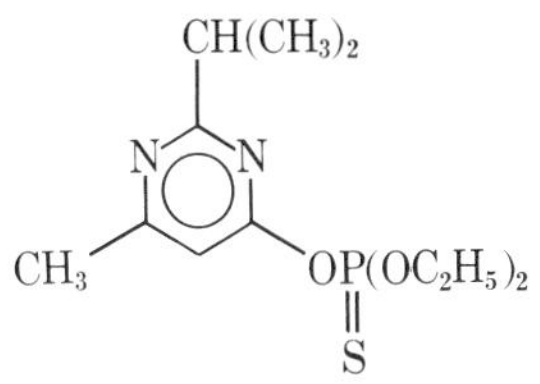

(**22**) 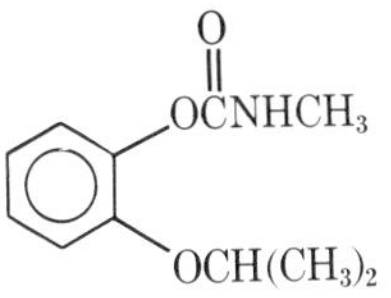 (**23**) 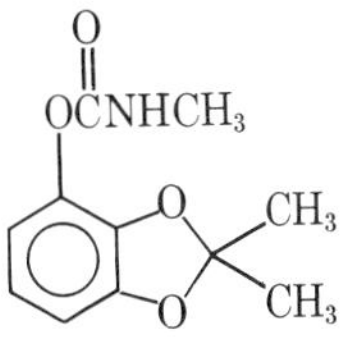 (**24**)

Sixty-two novel experimental carboxamides of 1,2,3,6-tetrahydropyridine were tested as repellents of German cockroaches, and five provided 100% repellency for 17 d in a stringent test (37).

Bird Repellents

Blackbirds, starlings, and sparrows are North American birds that cause serious damage to growing crops, costing at least $40 × 10^6/yr. Repellent devices such as propane cannons, scarecrows, metallic pinwheels, and recorded distress calls give temporary results at best, and when the birds become accustomed to the devices, the devices lose their effect.

According to a 1978 FAD report, sorghum is a vital staple food crop occupying 44 × 10^6 ha (10.9 × 10^7 acres), mostly in developing countries, including India, southern Asia, Latin America, the Sahelian zone of Africa, the Near East, and the Middle East. The most important sorghum pests are grain-eating birds such as the red-billed weaver bird in Africa. In many of these areas, control measures must be taken or most of the crop will be lost, but chemical repellents are expensive, are difficult to obtain, require special application equipment, and therefore would be an unlikely consideration. For these areas, it seems practical to breed the ability to resist bird depredation into the genetic composition of the plants, and much effort has been directed to this end in sorghum since 1960. High content of tannins is the characteristic most often associated with bird resistance in sorghum because these polyphenolics (tannins) produce astringency and thereby repellency. Unfortunately, the palatability, digestibility, and nutritional quality of foods are reduced in tannin-loaded food products (38).

Hydrolyzable tannins are present in small quantities in sorghum, and condensed tannins are responsible for the bulk of coagulation of proteins of the saliva and mucous membranes resulting in the astringent taste response. Polyphenolic condensed tannins or proanthocyanidins are a series of complex condensed 4-ketoflavan-3-ol [*577-85-5*] (**25**) and flavan-3,4-diol [*5023-02-9*] (**26**) molecules of 500–3000 mol wt (39). The subject of polyphenolic tannins has been reviewed recently (40).

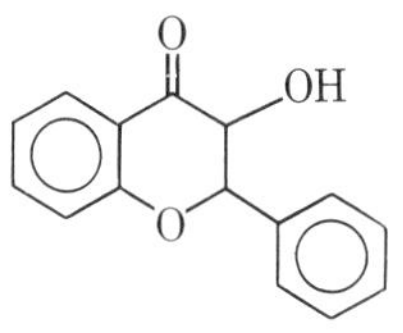

(25) 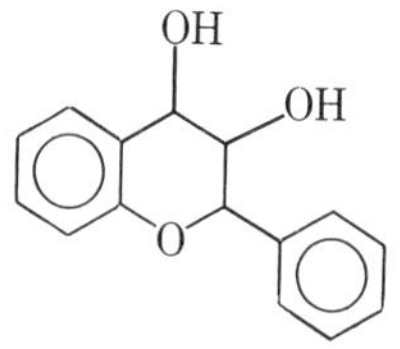 **(26)**

Some bird repellents are composed of viscous, sticky materials that birds dislike having on their feet (13). These compounds, Tanglefoot, Roost-No-More, and Tack-Trap are often based on incompletely polymerized isobutylene and thinned with aromatic solvents. They should be formulated to have the proper blend of tackiness and viscosity for the weather, method of application, and pest species. They are applied to leave sticky residues on perching locations in buildings and roosts in trees. Since these materials do not have an obnoxious odor, the birds must land on the treatment and learn its location in order to avoid it; ie, there are no long-range cues for conditioning.

Intoxicating chemicals are those that are not necessarily toxic (see Poisons, economic) but operate as primary repellents or secondary repellents (eg, emetics causing sickness or distress). Primary bird repellents are those whose mode of action is having a bad taste; immediate rejection of food is the desired result. These include condensed tannins obtained from sorghum and other cereal grains that have astringent properties, causing puckering of mucous membranes. However, they are effective only if other foods are available and are not effective in times of food shortages, since large flocks of migrating birds would be forced to feed or starve. Bird repellents were discussed in a recent review (41).

Aldicarb [*116-06-3*] (**27**) Temik, is classified as an insecticide, acaricide, and nematicide, and it has been used as a bird repellent. Its efficacy in Israel as a bird repellent was reported (42), as was treatment of sugar-beet seed to establish conditioned aversion in birds (43). The effect of high level bird-repellent treatments of Aldicarb (**27**) has been studied on the germination and development of sugar beets (44).

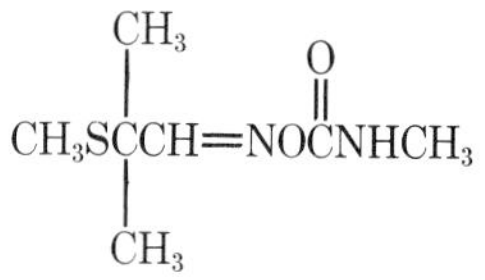

$$CH_3SC(CH_3)_2CH{=}NOC(=O)NHCH_3$$

(27)

Avitrol [*504-24-5*] (**28**), 4-aminopyridine, mp 155–158°C, bp 273°C, has repellent–toxicant properties for birds and is classed as a severe poison and irritant. This secondary bird repellent acts as a soporific when ingested from a broadcast bait, causing uncoordinated flight and distress calls and escape responses in nearby birds (45). Suspected contamination of drinking water with 4-aminopyridine was reported in toxicosis of Brahman cattle (46) and horses (47).

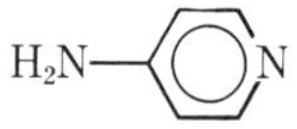

(28)

Methiocarb [*2032-65-7*] (**29**) Mesurol (Chemagro Bay) is classed as an insecticide and acaricide and is used as a slug and snail bait. It is registered for use as a bird repellent on corn seed and ripening cherries. Its uses on field and horticultural crops for bird repellency as an emetic were reviewed (48). Its efficacy in reducing bird damage in bait crops of sugar beets (49) and in treatments of sweet corn (50) has been studied. Methiocarb has been applied to wine grapes in Australia (51), Ohio (52), California, and Oregon (53), and to blueberries in New Zealand (54). Residues were reported in wine (qv), and its effect on the composition and flavor of the bottled wine was reported (55). Its efficacy in ripening sorghum in Canada and Senegal were reported (56–57), as were its residues and its sulfoxide and sulfone metabolites during efficacy studies against starlings in cherry orchards (58). Sorghum hybrids were treated with methiocarb, and grain yield and predation were studied (59). The conditioning response acquired is effective against skylarks (60) and in red-winged blackbirds persists in the laboratory up to 16 wk (38).

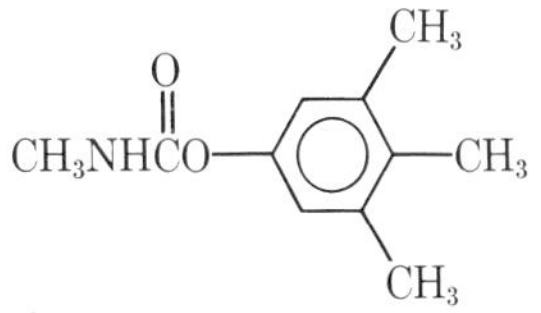

(29)

Anthrahydroquinones were patented as bird repellents (61), and anthraquinone [*84-65-1*] (**30**) Morkit, is used widely in Europe as a spray to protect growing crops and as a wood dressing. The synthetic pyrethroid decamethrin [*52918-63-5*] (**31**) was evaluated (62), as were other materials including bendiocarb (**24**) (63) and 20,25-diazocholesterol dihydrochloride [*1249-84-9*] (**32**) (Ornitrol), a steroid that inhibits embryo development when adsorbed or ingested as a seed treatment of bait corn (43,64).

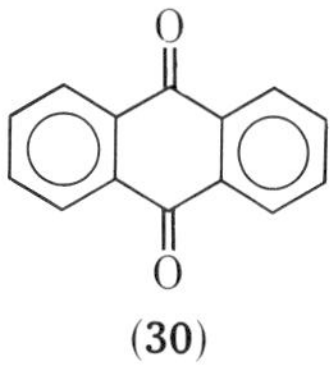

(30)

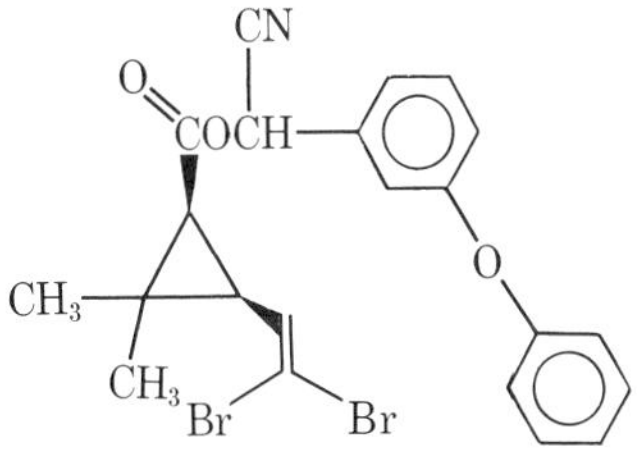

(31)

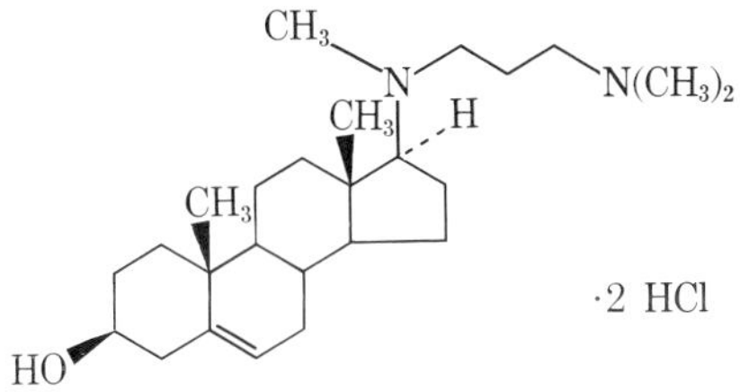

(32)

Deer, Rabbit, and Rodent Repellents

The concept of employing a nonlethal repellent to control wildlife depredation on crops arose early in agricultural history and has been pursued vigorously ever since. The present popularity of this idea reflects the realization that other methods (ie, reproductive inhibition, killing) are frequently impractical or socially unacceptable. Repellent is defined as a compound or combination of compounds that, when added to a food source, acts through the taste system to produce a marked decrease in the utilization of that food by the target species. The action can be primary, where the animal reacts to the taste of the repellent alone, or secondary, where the animal uses the taste of the repellent as a cue to later adverse effects. Nonlethal repellents to protect crops from vertibrate pests, together with some considerations for their use and development were reviewed recently, and several previous reviews were cited (38). It appeared that no consistently effective chemical repellent has been developed for vertibrate pests. Most recent attempts to find repellents failed to recognize that the purpose of a chemical repellent is to cause a hungry animal to stop feeding on the most readily accessible, abundant, and palatable food. The pest animal is expected to leave the area or make a change in food habits. The feeding activity of deer has become an increasingly important problem in the Pacific Northwest of the United States, especially where black-tailed deer and Roosevelt elk cause browse damage to Douglas fir seedlings. Some of the most promising early biological repellents were fractions of putrefying salmon, and some compounds were identified from volatiles. Later evaluations were conducted using deer and a multiple-choice, preference-testing apparatus and employing tetramethylthiuram disulfide [*137-26-8*] (**33**), TMTD or Thiram, a fungicide, as a standard repellent for competitive tests with repellent-treated food (65).

A pellet containing selenium embedded in the soil near a seedling causes the sapling to have "bad breath." Any wild deer that might be tempted to nibble the young shoots, thus damaging the valuable crop, are repelled by the odor of dimethyl selenide which is exuded from the seedling (66).

A fermented-egg product (FEP), patented as an attractive bait for synanthropic flies, has been shown to be attractive to coyotes and repellent to deer (67). This somewhat variable manufactured item was analyzed by gc and gc-ms in an attempt to prepare a necessarily complex synthetic mixture from the important volatile compounds in FEP that would be useful as a deer repellent. Volatile aliphatic fatty acids of 1–8 carbons were found, with 4, 6, 7, and 8 carbon acids most prevalent, and eight aliphatic amines were found, with trimethyl, isoamyl, amyl, and β-phenethylamine most prevalent. The relative concentrations were 77% fatty acids, 13% bases, and 10% (primarily) neutrals composed of at least 54 volatiles such as ethyl esters, dimethyl disulfide, and 2-mercaptoethanol. Since fatty acids and amines are found in anal glands of canid and felid predators, there are rational reasons to suggest that deer may be repelled by these materials. Synthetic formulations were evaluated to find a replacement for a patented fermented-egg-protein product (FEP) that attracts coyotes and repels deer. With the aid of a human odor panel, a systematic study resulted in a good match that duplicated the natural product in repelling deer and attracting coyotes. Ten aliphatic acids (C_2 to C_8), four amines (pentyl, hexyl, heptyl, and trimethyl), dimethyl disulfide, 2-mercaptoethanol, and 54 more volatiles (C_1 to C_5 esters of C_2 to C_8 acids) were tested as synthetic fermented egg (SFE) (68) in approximately the same proportions that were present in FEP. Repellency to deer was slightly better with 0.00009% SFE than with 0.004% FEP when tested in competition with 0.01% TMTD. Repellency of SFE was increased with higher treatments in the field, although weathering was a problem that caused decreased efficacy, suggesting trials of controlled-release formulations. Fourteen repellents were tested against white-tail deer in Pennsylvania in choice tests when treated onto shelled corn (69).

Treatment	*Summed total of ranks*
Meatmeal	26.5 A
BGR	27.5 A
Feather meal	39.0 A
Hinder/Repel	50.5 A
Hot sauce	51.5 A
Chew-not	55.5 A
Chaperone	57.0 A
42-S	57.5 A
Spotrete-F	61.5 A

The numbers above are the percentages of "browse" (eaten trees). The letter A represents top-rated repellents by Friedman's multiple comparative analysis.

Later, dogwood seedlings were treated with candidates and exposed to browsing deer. Hinder and BGR were nearly unbrowsed on the first two of four days, and only BGR was browsed less than controls and other candidates on the last two days.

Hinder or Repel is registered under the Federal Insecticide, Fungicide, and Rodenticide Act (FIFRA) Section 24C to repel deer and rabbits from fruit trees, vines, vegetables, field crops, forage and grain crops, ornamentals, nursery stock, and noncrop areas. It is best applied before damage occurs as an aqueous spray using conventional equipment or by painting and is claimed to last 3–4 wk, perhaps as long as 8 wk. Hinder contains 15% ammonium soaps of higher fatty acids (1.5% ammonia and 13% mixed rosin and fatty acids) and 85% inert ingredients (69). The material is sold in the western U.S. as Hinder and in the eastern U.S. as Repel or Sticker-spreader 268. Chaperone is the only material presently approved by EPA in Florida as a repellent for deer, mice, and rabbits. About ten materials were registered in Georgia in 1982, usually containing 4–22% Thiram (**33**) (see Table 3 for a non-comprehensive list) (70).

Area repellents are materials with a bad odor that are intended to keep animals away from a broad area. They include lion scent (lion or tiger manure), blood meal, tankage (putrefied slaughterhouse waste), bone tar oil, rags soaked in kerosene or creosote, and human hair (73). Few, if any, controlled tests have been run on these materials.

Health And Safety Factors

Toxicological testing has been carried out on many of the older, widely-used materials. Few of the newer compounds have been submitted for extensive toxicological testing because of cost, the problems of registration with the FDA, and the perceived necessity to make a profit with every new product. As a result of FDA regulations, many of the materials submitted as cloth repellents since 1970 have been tested at the USDA Agriculture Research Service, Insects Affecting Man and Animals Research Laboratory (Gainesville, Fla.). Successful compounds, after further testing, are then submitted

Table 3. Repellents for Deer, Rabbits, and Rodents

Repellent name	Active ingredient	Target organism	Manufacturer
Registered in Georgia			
Chaperone Rabbit and Deer Repellent	7% Thiram	deer, rabbits, meadow mice	Sudbury Laboratories, Inc., Sudbury, Mass.
Deer Ban	10% Thiram	deer	Animal Repellents, Inc., Griffin, Ga.
Deer-Away (MGK, BGR)	37% putrescent whole egg solids	deer, elk	Deer-Away, Minneapolis, Minn.
Magic Circle Deer Repellent	93.75% bone tar oil	deer	State College Laboratories, Reading, Pa.
Magic Circle Rabbit Repellent	20% Thiram	rabbits, meadow mice, deer	State College Laboratories, Reading, Pa.
Rabbit Rid	10% Thiram	rabbits	Animal Repellents, Inc., Griffin, Ga.
Science Rabbit and Deer Repellent	20% Thiram	deer, meadow mice, rabbits	Science Products Co., Inc., Chicago, Ill.
Z.I.P.	22% zinc dimethyldithiocarbamate–cyclohexylamine complex	deer, rabbits	Morton Chemical Co., Chicago, Ill.
TAT GO	22% Thiram	rabbits, deer, meadow mice	D. E. Linch Division, Walco-Linch Corp., Clifton, N.J.
Chew-not	20% Thiram	rabbits, deer, mice	Nott Manufacturing, Pleasant Valley, N.Y.
Gustafson 42-S	42% Thiram	rabbits, deer	Gustafson, Inc., Dallas, Texas
Hinder/Repel	15% ammonium soaps of higher fatty acids	deer, rabbits	T. H. Agriculture & Nutrition Co., Inc., Brea, Calif.
Registered in New York			
Bonide Rabbit and Deer Repellent	11% Thiram, 11% acrylic	rabbits, deer, field mice	Bonide Chemical Co., Inc., Yorksville, N.Y.
Hot Sauce Animal Repellent 2	2.43% Capsaicin (*trans*-8-methyl *N*-vanillyl-6-nonenamide)	rabbits, deer, meadow mice, pine mice	Miller Chemical & Fertilizer Corp., Hanover, Pa.
Registered in California			
R-55	21.9% *tert*-butyl *N*,*N*-dimethyldithiocarbamate, 3.8% related compounds	pocket gophers, other burrowing rodents	Philips Petroleum Co., Bartlesville, Okla.

to the U.S. Army Environmental Health Agency for extensive toxicological testing. Compounds are tested as skin repellents after passing the four standard toxicological tests: rabbit eye irritation, rabbit skin dermal, rat inhalation, and rat acute ingestion. All of the above, plus FDA regulations in the U.S. that have been adopted to reduce hazards, have drastically reduced the number of candidate chemicals submitted to the Gainesville laboratory for general screening since about 1975 and virtually eliminated chemicals submitted as candidate repellents. Since that time, only twelve materials have been submitted. More than half of these were submitted by one company, having been derived from optically-active isomers of a known repellent, citronellal (**33**). The racemic repellent was shown to be somewhat active as a repellent before World War II.

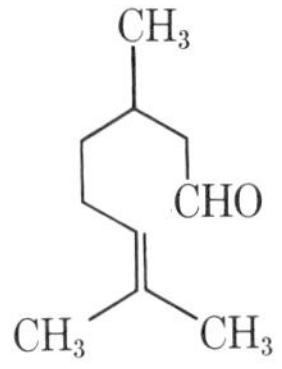

CH_3
CHO
CH_3 CH_3

(**34**)

Caveats

Dog repellents that are available commercially at present have been generally unsuccessful in laboratory tests. For example, lithium chloride treatments were usually rejected immediately with no ingestion, and bone-oil treatments that contained up to 0.1% of the active ingredient were still consumed (74). Oleoresin capsicum [*8023-77-6*] (essence of red pepper) did have an extended effect on coyotes, although the deer repellents mentioned above were attractive to coyotes (74). A capsicum-base aerosol repellent was described as potentially harmful (75).

Numerous articles in the popular press have stated that heavy consumption of Vitamin B_1 (thiamine) will stop attacks of biting and stinging insects on the thiamine-loaded human (see Vitamins, thiamine). This was investigated during World War II and recently at Gainesville. There is no scientific evidence that thiamine has any effect whatsoever on the attraction of *Ae. aegypti* to humans in olfactometer tests, whether taken internally to excess or applied externally (76).

Effects of repelling or disrupting ultrasonic devices on selected rodent species (77) have been extended by some producers of such devices to include repelling of cockroaches, mosquitoes, and other insects (see Ultrasonics, low energy). There is replicated scientific evidence that shows no effect of several sonic and ultrasonic frequencies (1,000–60,000 Hz) on German cockroaches in choice boxes, since the cockroaches were neither killed nor repelled (78).

Experiments with human arms in olfactometers showed no effect on the attraction of *Ae. aegypti* when sonic devices were used. Mosquito attraction was statistically the same whether or not any of several makes of small portable sonic devices (600–1000 Hz) reputed to repel mosquitoes were activated (78). Warnings were sent in the spring of 1983 to some distributors of ultrasonic pest-control devices that "statements that pertain to the efficacy of the product have not been substantiated and when used in connection with the product could be in violation of the FIFRA" (25).

An appealing idea for increasing the length of life of a repellent treatment is to chemically bond its molecule to dermophilic (skin-loving) compounds that then bind to the skin. Compounds containing 1,3-dihydroxyacetone and pendant repellent molecules were intensively investigated from 1966 to 1972 (79). Amino acid analogues of 2-ethyl-1,3-hexanediol were later prepared for testing (80), but results were not outstanding.

BIBLIOGRAPHY

1. *World Health*, 10 (Apr. 1982); E. A. Smith, *Mosq. News* **42,** 510 (1982).
2. P. Bres, *World Health*, 25 (Apr. 1982).
3. *Biology and Control of Aedes aegypti, Vector Topics No. 4*, U.S. Public Health Service, Center for Disease Control (CDC), Atlanta, Ga., 1979; *Dengue Surveillance Survey No. 9*, CDC, Atlanta, Ga., Feb. 1983.
4. C. E. Schreck, *Ann. Rev. Entomol.* **22,** 101 (1977).
5. F. E. Kellog and co-workers, *Can. Entomol.* **100,** 763 (1968).
6. A. Smith, *World Health*, 11 (Apr. 1982).
7. C. E. Schreck and co-workers, *Mosq. News* **37,** 455 (1977).
8. W. V. King, *Chemicals Evaluated as Insecticides and Repellents at Orlando, Fla.*, USDA, ARS, 1954.
9. C. E. Schreck and co-workers, *J. Econ. Entomol.* **73,** 451 (1980); C. E. Schreck and co-workers, *Soap Cosmet. Chem. Special.*, 36 (Sept. 1982).
10. D. E. Weidhaas, *Proc. Symp. Univ. Alberta*, 109 (June 1972).
11. *Consumer Rep.*, 304 (June 1982).
12. *Wall Street J.*, 1 (May 26, 1983).
13. *Pest Control, Retail Producers Guide*, Mar. 1983, p. 20.
14. M. D. Buescher and co-workers, *Mosq. News* **42,** 428 (1982).
15. T. P. McGovern, C. E. Schreck, and J. Jackson, *Mosq. News* **40,** 394 (1980).
16. U.S. Pat. 4,291,041 (Sept. 22, 1981), U.S. Pat. 4,356,180 (Oct. 26, 1982), and U.S. Pat. 4,298,612 (Nov. 3, 1981), T. P. McGovern and C. E. Schreck (to USDA).
17. C. E. Schreck and co-workers, *J. Med. Entomol.* **16,** 524 (1979).
18. C. E. Schreck and co-workers, *Am. J. Trop. Med. Hyg.* **31,** 1046 (1982).
19. C. E. Schreck and D. L. Kline, *Mosq. News* **41,** 7 (1981).
20. R. H. Grothaus and co-workers, *Mosq. News* **36,** 11 (1976).
21. A. B. Sabin and co-workers, *J. Am. Med. Assoc.* **125,** 693 (1944).
22. M. L. Schmidt and J. R. Schmidt, *J. Med. Entomol.* **6,** 79 (1969).
23. V. M. Safyanova, *Med. Parazitol. Parazit. Bolezni* **35,** 549 (1963).
24. E. T. Takafugi and co-workers, *Am. J. Trop. Med. Hyg.* **29,** 516 (1980).
25. *Pest. Toxic Chem. News* **11,** 22 (1983); *Science* **204,** 484 (1979).
26. *Outdoor Life*, 21 (June 1982).
27. C. E. Schreck and co-workers, *J. Econ. Entomol.* **75,** 1059 (1982).
28. C. E. Schreck and co-workers, *J. Med. Entomol.* **19,** 143 (1982).
29. P. B. Cornwell, *The Cockroach*, Vol. II, Associated Business Programmes, Ltd., London, 1976, pp. 157–190.
30. L. D. Goodhue and G. L. Tissol, *J. Econ. Entomol.* **45,** 133 (1952).
31. *Pest Control* **25,** 22 (1957).
32. *Laboratory Evaluations of Compounds as Repellents to Cockroaches, 1953–1974, Production Research Report No. 64*, Agricultural Research Service, USDA, Washington, D.C., Oct. 1976.
33. B. J. Smittle and co-workers, *Pest Control* **36,** 9 (1968).
34. *N.P.C.A. Tech. Release No. 15-69*, National Pest Control Association, Vienna, Va., 1969.
35. G. S. Burden, *Pest Control* **43,** 16 (1975).
36. J. M. Grayson, *Pest Control* **44,** 30 (1976).
37. T. P. McGovern and G. S. Burden, *J. Med. Entomol.* (in press); T. P. McGovern and co-workers, *J. Med. Entomol.* **12,** 387 (1975).
38. J. G. Rogers, Jr., *ACS Symp. Ser.* **67,** 150 (1978).
39. R. W. Bullard and co-workers, *J. Agric. Food Chem.* **28,** 1006 (1980).
40. R. W. Bullard and D. J. Elias, *Proc. Inst. Food Technol.*, 43 (June 1979).
41. E. N. Wright, ed., British Crop Prot. Counc. 23, *Bird Problems in Agric.*, British Crop Protection Council 23, BCPC publications, Croydon, UK, 1980, p. 164.
42. L. Benjamini, *Phytoparasitica* **9,** 89 (1981).
43. *Ibid.*, 3 (1981).
44. *Ibid.*, 11 (1981).
45. J. F. Besser, *Proceedings of the 7th Vertebrate Pest Control Conference*, Monterey, Calif., 1976, p. 11.
46. S. S. Nicholson and C. J. Prejean, *J. Am. Vet. Med. Assoc.* **173,** 1277 (1981).

47. G. A. Van Gelder in P. W. Pratt, ed., *Equine Medicine and Surgery*, 3rd ed., American Veterinary Publications, Santa Barbara, Calif., 1982, p. 197.
48. F. T. Crase and R. W. Dehaven in ref. 45, p. 46.
49. *Phytoparasitica* **8,** 151 (1980).
50. P. P. Woronecki and co-workers, *J. Wildlife Manage.* **35,** 693 (1981).
51. P. T. Bailey and G. Smith, *Aust. J. Exp. Agric. Anim. Husb.* **19,** 247 (1979).
52. *Proc. Bird Cont. Semin.* **8,** 59 (1982).
53. R. L. Hothem and co-workers, *Am. J. Enol. Vitic.* **32,** 150 (1981).
54. *Proc. N. Z. Weed Pest Control Conf.* **33,** 125 (1980).
55. A. C. Noble, *Am. J. Enol. Vitic.* **31,** 98 (1980).
56. R. R. Duncan, *Can. J. Plant Sci.* **60,** 1129 (1980).
57. G. Gras and co-workers, *Bull. Environ. Contam. Toxicol.* **26,** 393 (1981).
58. *Phytoparasitica* **8,** 95 (1979).
59. *Agron. J.* **73,** 290 (1981).
60. *Chemtech*, 710 (1982).
61. Jpn. Kokai Tokyo, Koho, 8183408 (July 8, 1981).
62. *Poult. Sci.* **60,** 1149 (1981).
63. *Res. Discl.* **211,** 420 (1981).
64. J. L. Guarino, *Proceedings of the 5th Vertebrate Pest Control Conference*, Davis, Calif., 1972, p. 108.
65. D. L. Campbell and R. W. Bullard, *Proceedings of the 5th Vertebrate Pest Conference*, Fresno, Calif., 1972.
66. *Chem. Br.* **19**(6), 469 (1983).
67. U.S. Pat. 3,846,557 (Nov. 5, 1974), M. S. Mulla and Y.-S. Hwang (to 3M Company).
68. R. W. Bullard and co-workers, *J. Agric. Food Chem.* **26,** 155 (1978).
69. W. Palmer, Penn State University, *Deer-Away Technical Report*, International Reforestation Suppliers, Eugene, Ore., 1980.
70. J. Jackson, Dept. of Extension, Forest Research, University of Georgia, Athens, Ga., 1983.
71. *Extension Publication 18*, No. 11, Dept. of Natural Resources, N.Y.S.C. Agriculture and Life Sciences, Cornell University, Ithaca, N.Y., 1980.
72. *Supplement No. 120*, Extension Wildlife and Sea Grant, University of California, Davis, Calif., Oct. 1979.
73. *Extension Information Bull. No. 146*, Cornell University, Ithaca, N.Y., 1978.
74. R. Teranishi, private communication, USDA Western Regional Laboratory, Albany, Colo., 1983.
75. *Vet. Human Toxicol.* **22,** 18 (1980).
76. C. E. Schreck, private communication, USDA Laboratory, Gainesville, Fla., 1983.
77. A. V. Scalingi, *Pest Control* **48,** 26 (1980).
78. C. E. Schreck, G. S. Burden, and J. C. Webb, *Envirn. Sci. Health*, in press.
79. R. P. Quintana and co-workers, *J. Econ. Entomol.* **65,** 66 (1972).
80. R. P. Quintana and co-workers, *J. Med. Chem.* **15,** 1073 (1972).

D. A. CARLSON
University of Florida

S

SEPARATIONS, LOW ENERGY

The separation and purification of chemicals and fuels consume large quantities of energy. Until recently, particularly in the United States, energy has been inexpensive and the optimal process design was usually inefficient in terms of energy consumption. The Arab oil embargo in 1973 caused a rapid increase in energy costs; thus, there was and continues to be an emphasis on energy efficiency in new process design and on the need to review existing plant operations in terms of whether retrofitting would improve energy efficiency.

For industrial separation of a liquid mixture into its components, distillation is generally the preferred, lowest-cost technique. Energy expenditures are, however, often far in excess of theoretical requirements (1). For complete separation of a liquid mixture into its components, the minimum work ($W_{\min}$), of separation is given by:

$$W_{\min,\mathrm{T}} = -RT \sum_j x_{j\mathrm{F}} \ln (\gamma_j x_{j\mathrm{F}}) \tag{1}$$

where γ_j is the activity coefficient for component j in the feed and $x_{j\mathrm{F}}$ is the mole fraction of component j in the feed. The minimum work for the separation of a mixture of 54 mol % ethylbenzene and 46 mol % styrene monomer at 110°C is

$$W_{\min} = -8.3144 \times 383.1 \times (0.54 \ln 0.54 + 0.46 \ln 0.46)$$

$$= 2198 \text{ J/mol } (525.3 \text{ cal/mol}) \text{ of feed} = 43.7 \text{ J/g } (10.4 \text{ cal/g}) \text{ of styrene monomer}$$

This minimum work of separation is only 1.5% of the energy required [2853 J/g (681.9

cal/g) of styrene] for the separation as performed in a well-designed tower (2). However, a comparison between the work of separation and the thermal energy requirements of the separation by distillation may be misleading. Since the energy input to the reboiler is required at a fairly low temperature, ie, 110°C with an 11°C delta between the steam and the reboiler temperature, the steam which heats the reboiler can be generated at high temperature and pressure and can be expanded through turbines to extract work before it is used for reboiler heating. For plants that practice this type of cogeneration, a more valid comparison is between the minimum work of separation and the thermodynamic availability (or exergy or lost work) requirements for the distillation separation. At a dead-state, ie, surroundings, temperature of 40°C for the ethylbenzene/styrene monomer separation, the availability requirements for steam-heating the reboiler are

$$\Delta B = \Delta H(T - T_0)/T_0 = 2853(383.1 - 313.1)/383.1$$
$$= 521 \text{ J/g } (125 \text{ cal/g}) \text{ of styrene monomer}$$

The minimum work of separation is still only 8.5% of this availability requirement. Modifications to conventional distillation, which can increase its energy efficiency, and alternative separations processes, which are energy-conservative, imply considerable energy-saving potential.

Energy Management

In both the chemical and petroleum-refining industries, steam is generated for a plant in a central boiler-plant facility at elevated pressures and this steam is piped around the plant at a few different, typically two or three, pressure levels. Steam may also be generated in waste-heat boilers within the plant where excess process heat is available, and this steam is also fed into the plant steam-piping system. In large petroleum refineries and chemical-plant complexes, the highest pressure at which steam is circulated is typically 4.1–4.8 MPa (600–700 psi) as the flanges for higher pressure are prohibitively expensive. In smaller facilities, the highest pressure usually does not exceed 2.8 MPa (400 psi). Other pressures at which steam is piped around the plant are typically 1 MPa (145 psi) and 0.34 MPa (50 psi). In petroleum refineries, hot oil heated in direct-fired heaters is also circulated as a heating medium for reboilers. For the bulk of the reboiling requirements, low pressure [0.34–1 MPa (50–ca 150 psi)] steam is all that is needed, and reboilers are heated by steam or hot oil piped to the reboiler from the circulating plant line. An opportunity exists for extraction of energy from the steam before it is used in reboilers by decreasing its pressure from the maximum at which it is generated in the steam-boiler plant to the pressure at which it is utilized in the reboiler in the steam turbine. This is practiced extensively throughout these industries for both mechanical and electrical power generation.

Economic Considerations

The only certain approach to determining whether fuel is conserved by a particular option is to evaluate its impact on plant steam and fuel balances. For plants practicing no cogeneration, preliminary evaluation of alternatives should be based on the respective energy requirements of the alternatives. For plants generating steam at high pressure and practicing cogeneration extensively in the sense of extracting mechanical energy from steam before using it as a heat source, preliminary evaluation of alternatives should be based on their respective availability requirements.

Energy Evaluation of Plants Practicing Cogeneration Extensively. Since many separations processes involve compression and refrigeration, there should be a basis for comparing energy costs of these processes with those requiring heat input at above-ambient temperatures. A useful approach for accomplishing this is given in ref. 3. The value of energy is assumed to be proportional to its availability, ie, to the amount of work which can be extracted from it so as to reduce it to dead-state conditions:

$$V = (\Delta B/\Delta H) \times Cp \times K \tag{2}$$

where ΔB is the availability of the stream or energy source relative to dead-state or ambient conditions, ΔH is the change in enthalpy in going from process conditions to the dead state, Cp is the cost of power, and K is a factor depending upon the type of energy source being considered.

$$\Delta B = \Delta H - T_0 \Delta S \tag{3}$$

where ΔS is the change in entropy in going from process conditions to the dead state and T_0 is the dead-state absolute temperature.

For refrigeration energy, K is calculated according to the formula:

$$K = F_c/(E_M \times E_F) \tag{4}$$

where E_F is the fluid efficiency factor, assumed to be 0.85; E_M is the mechanical efficiency of the turbine or compressor, assumed to be 0.7; and F_c is a capital factor, with a suggested value of 1.5, which takes into account the correlation between the equipment required to produce the refrigeration (ref) and the work requirement.

For a latent heat source, equation 2 reduces to

$$V_{ref} = Cp \frac{(T_r - T)}{T} \frac{1}{E_M} \times \frac{1}{E_F} \times F_c = 2.521\ C_P \frac{(T_r - T)}{T} \tag{5}$$

For compression energy, K may be considered equal to 1.47; this value is derived assuming a compressor efficiency of 75% and an additional 10% availability loss resulting from the compression being nonisothermal and the heat of compression being dissipated to cooling water. This availability loss increases with larger ratios of specific heats since the compression deviates from isothermal with an increase in the ratio. For thermal energy higher in temperature than dead-state conditions, K is equal to 1.0. However, it has been suggested that a factor of less than 1 may be appropriate, since a mechanical device must be used to extract work from the fluid or source (3).

Based on the preceding approach in evaluating a separation process, Figure 1 and Table 1 present an analysis of an ethylene–ethane splitter in an ethylene plant. It is assumed that the desired ethylene product is a liquid at the temperature and pressure at which it is produced by the separation. The column is therefore credited with the value of the energy required to compress, cool, and condense the ethylene, since this requires refrigeration. The column is debited with the value of the compression and refrigeration energy associated with the feed to the column and with the value of the energy associated with condenser refrigeration and reboiler heating. Since the C_2-splitter reboiler requires refrigeration, the column is credited with the value of the reboiler refrigeration energy. In the analysis, expansion turbines are assumed 85% efficient and heat exchangers are assumed to require an 11°C temperature differential driving force.

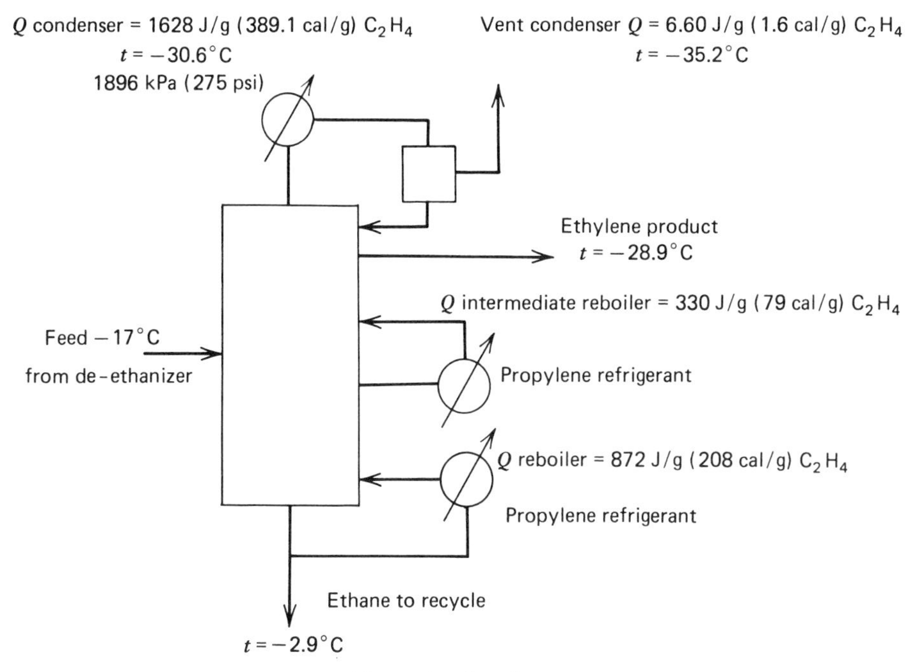

Figure 1. C_2 splitter.

Table 1. Energy Used in C_2 Splitter[a]

	Debit to column $/kg $C_2H_4 \times 10^6/C_P$	Credit to column $/kg $C_2H_4 \times 10^6/C_P$
condenser	1130.3	
vent condenser	5.1	
reboiler		334.4
interreboiler		177.9
feed compression	420.4	
feed refrigeration	200.8	
ethylene product compression[b]		338.4
ethylene product refrigeration[b]		266.3
ethane recycle refrigeration[c]		30.0
ethane recycle expansion[d]		68.6
Total	*1756.6*	*1215.6*

[a] Net value of energy usage input required for C_2 splitter: 541.0 $C_P/10^6$ $/kg C_2H_4.
[b] Assumes ethylene product is required as a liquid at product conditions, ie, at 1896 kPa (275 psi) and −30.6°C.
[c] Assumes ΔT for heat transfer = 11°C.
[d] Assumes an expander efficiency of 85%. If an expansion turbine is not used, this credit does not apply.

Compromise Between Capital Investment and Energy Requirements. For the comparison of alternative processes, it is necessary to consider capital investment requirements in addition to operating costs (see Distillation). For typical industrial payback and depreciation criteria, the following rule of thumb has been developed: if more than three dollars must be invested to save one dollar per year in energy costs, no net economic benefit accrues (4). For preliminary screening of alternatives, this rule may be useful.

Distillation versus Alternative Separation Processes

Distillation has many advantages compared to other separation processes (5), such as: the phases are fluids with relatively large density differences and large interfacial tensions (except near the critical point) which make countercurrent flow, staged contacting, and gravitational separation of the phases readily accomplished; the low liquid- and vapor-phase viscosities lead to high diffusivities and efficient mass transfer; small increments in boiling temperature can produce vapor-pressure differences sufficient to cause vapor to flow through the equipment without need for blowers; throughputs are high, stage heights are low, and the equipment is reliable and without moving parts (except for external liquid pumps); and no mass-separation agents are used and additional separations associated with these agents are thereby avoided.

However, there are cases where distillation is not as well suited for the desired separation as alternative techniques. There are also cases where separation by distillation is essentially infeasible and alternative techniques must be used. As energy costs increase, separation processes requiring less energy than distillation have broader applications. Distillation may not be suitable if: relative volatilities are low; classes of compounds with a broad overlapping range of boiling points are to be separated, eg, aromatics from nonaromatics; an impurity is to be separated from a more volatile liquid; a less-volatile organic compound is to be recovered from a water solution, eg, acetic acid from water; distillation would require extremes of temperature or pressure; heat-sensitive materials that might be degraded by distillation are to be recovered; and an azeotrope is present in the feed components. Table 2 lists a number of industrially important separations carried out by processes other than conventional distillation.

Modifications to Conventional Distillation

More Efficient Separation. ***Control Considerations.*** A basic source of energy waste in distillation processes is inadequate control over product purity. Because of the high penalty associated with failure to meet product specifications, columns are usually overdesigned to provide products that are purer than actually needed. Such columns must be operated with too large a reflux ratio, which results in increased internal flow rates. The large flow rates put extra heat loads on the condenser and the reboiler. In addition, there are capital cost penalties associated with a larger column diameter and added heat-exchanger area.

Incorporating improved control in the column reduces the extent of the overdesign. For new column construction, this results in lower capital and operating costs. If the controls are added to an existing column, the reduced reflux still results in lower operating costs through energy savings.

Factors which favor improved controls include high annual production volume, low relative volatility, and high distillate-to-product-rate ratio. A number of control strategies have been developed to reduce distillation requirements, including material-balance control, feedforward control, floating-pressure control, and combined reflux and distillate control.

In material-balance control (see Fig. 2), product purity is controlled by direct manipulation of the product flow rate rather than by direct manipulation of the re-

Table 2. Industrial Separations Based on Methods Other Than Conventional Distillation

Bulk separations	Separation process used
low relative volatility separations	
ethylene/ethane	external refrigeration heat pumping, intermediate reboiling, pasteurization
propylene/propane	vapor recompression
butenes/butadiene	extractive distillation, absorption, extraction
ethylbenzene/styrene monomer	tray retrofit, eg, Linde slotted sieves or packing
p-xylene/C_8-aromatics	displacement regeneration adsorption with zeolites; crystallization
pentenes/isoprene	extractive distillation
overlapping boiling ranges	
aromatics/aliphatics	extraction, extractive distillation
normal paraffins/isoparaffins, aromatics (C_4s–C_{25}s)	pressure-swing adsorption with zeolites
asphaltenes/heavy hydrocarbons (residua)	extraction with propane, supercritical-fluid extraction with pentane
azeotrope breaking	
ethanol/water	azeotropic distillation, thermal swing adsorption
recovery of lower volatility organics from water	
acetic acid/water	extraction, azeotropic, distillation
broad-cut fractionation	
crude distillation	sidedraws, intermediate condensers, heat integration
gas separations	
acid gases (CO_2, H_2S, SO_2)/light gases	absorption involving basic solutions, physical solvents; adsorption; extractive distillation
oxygenated organics/light gases, eg, ethanol/ethylene	absorption involving water
nitrogen/oxygen	heat cascading, heat integration, advanced heat exchange, pressure swing adsorption
hydrogen upgrading	external refrigeration heat pumping, heat integration, pressure-swing adsorption, membranes
separations involving thermally sensitive, high boiling materials	
penicillin/water	extraction, crystallization
caffeine/coffee	extraction involving methylene chloride; supercritical-fluid extraction with CO_2
hydrometallurgical processing	
rare earths, Group IVA and Group VA metals, etc	extraction
impurity removal	
organics/water	adsorption, extraction, stripping if the organics are more volatile than water
organics/gases	adsorption, absorption, condensation
water/gases	adsorption, absorption
water/organic liquids	adsorption, distillation if water is more volatile than the organics

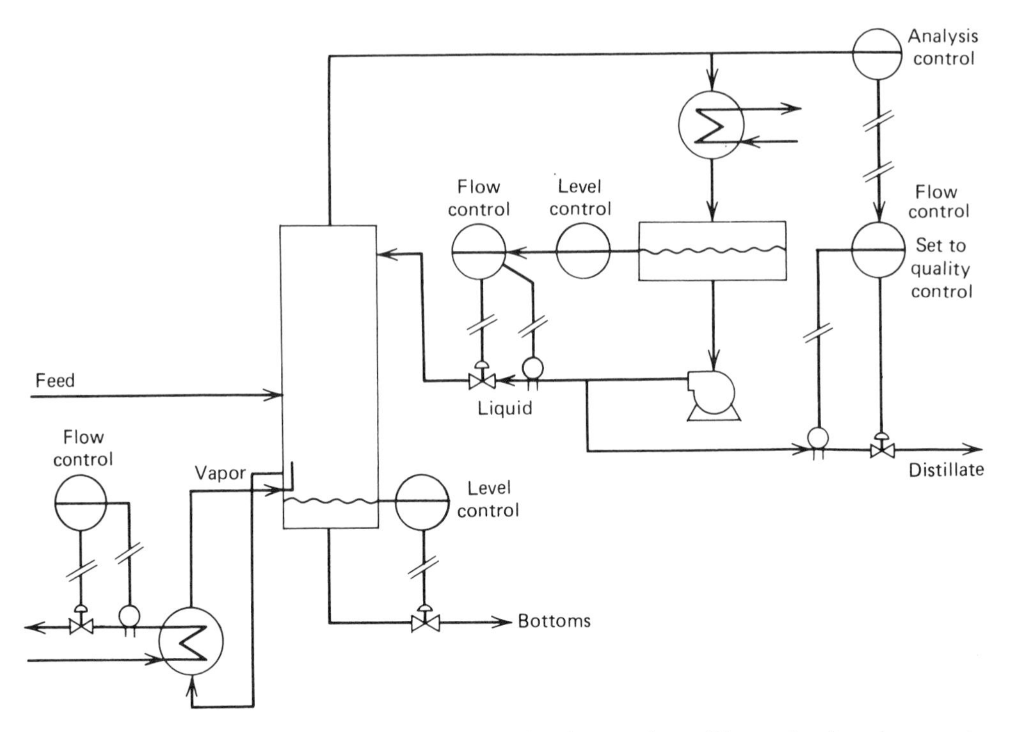

Figure 2. Direct material-balance control of a distillation column. The overhead product rate is held fixed and the reflux rate is adjusted so that product specifications are met (6).

boiler heat input or of the reflux rate (6–7). Generally, the smaller product flow is controlled.

Feedforward controls are normally used to account for flow variations by, for example, establishing a ratio of reboiler steam to feed flow rate (6,8). Feedforward is frequently combined with product-quality feedback control. Feedforward control to account for feed composition variation is costly and less likely to be cost-effective (6).

Floating pressure control has been proposed for use with air-cooled condensers where day–night and seasonal variations in ambient temperature significantly alter the cooling capacity of the condenser. Column pressure floats, ie, it decreases as the cooling capacity of the condenser increases.

In combined reflux and distillate control, there are the combined features of material-balance and reflux-boilup control systems for composition control; for example, a recently developed control scheme is shown in Figure 3 (9–10). The pressure-controller output represents total liquid flow leaving the condenser, which represents vapor flow leaving the column in the steady state. This output adjusts both distillate and reflux flows without changing their ratio. The composition analyzer adjusts the distillate-to-vapor ratio. A change in the feed rate appears at the base of the column as a change in level, and both bottoms rate and boilup are changed so as to maintain a constant composition.

Advantages and disadvantages of these control strategies are listed in Table 3. A common deterrent to the implementation of improved column control is the requirement for trained operators and maintenance personnel.

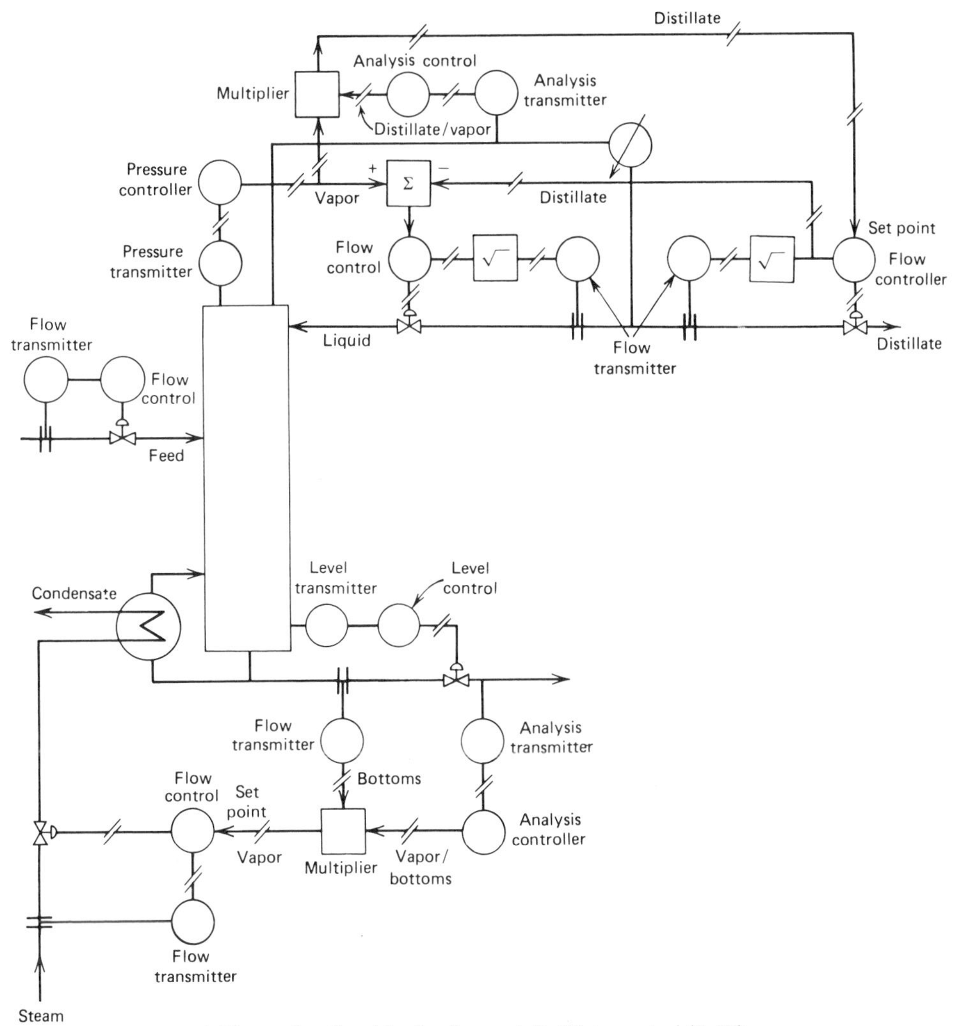

Figure 3. Combined reflux and distillate control (9–10).

Tray Retrofit. A retrofit with additional or with more efficient trays is energy-conserving by virtue of the decreased column reflux requirements, whereas in retrofits with lower pressure-drop internals, energy is conserved because there is decreased product degradation or because there are lower vacuum-system energy requirements. In some separations, eg, heavy and light waters, lower pressure-drop internals allow higher relative volatilities in the columns (11). Although the two types of retrofit are considered separately, in many cases more efficient internals also have lower pressure drops and both types of retrofits should be considered.

Retrofits with additional or more efficient trays. Most distillation columns have fewer trays and, therefore, operate at higher reflux ratios than if they were designed with energy costs in mind. Increasing the number of theoretical stages in the tower reduces the reflux and column energy requirements. Factors that increase the desirability of retrofit with additional or more efficient trays are fewer trays, low efficiency trays, high ratio of actual-to-minimum reflux, low relative volatility, high separation factor, and high column pressure.

Table 3. Improved Control Opportunities

Control scheme	Advantages	Disadvantages
material-balance control	reduces the sensitivity of the column to disturbances in the energy balance; can be adjusted to achieve a desired product purity; enables precise control over close separations; effective with high reflux-ratio columns; control of both top and bottom product purities to reduce reflux is apt to be economically attractive where required product purity is low and relative volatilities are low	directly affects external material balance; introduces delay in imposing external material balance on internal material balance; may require addition of feedforward control where feed flow rate, composition fluctuations, or both, are appreciable
feedforward control	reduces sensitivity of product quality to feed disturbances; ie, where flow variations exceed 10% or composition variations exceed 2%[a]	introduces additional control loops and more complex control; requires dynamic compensation
floating-pressure control	advantageous where relative volatility increases with decrease in pressure/temperature	requires composition analyzers and more complex controls, particularly to handle sudden fluctuations in condenser cooling capacity caused by sudden weather fluctuations
combined-reflux and distillate control	combines fast response with stability; simpler than feedforward control; does not require dynamic compensation	

[a] Ref. 8.

Imperial Chemical Industries, Inc., has developed a centrifugal distillation device, which is claimed to have a lower installed cost than conventional distillation devices (12). Suggested applications include the separation of heat-sensitive fine chemicals and off-shore gas-treatment facilities. Centrifugal devices have been in use for many years in liquid–liquid extraction but have not been applied to distillation, in part because of concern about possible increased maintenance costs and downtime.

Texas McCan has developed a column with offset halftrays and claims improved efficiency (13). The device has not yet been demonstrated in large-scale applications.

Retrofits with lower pressure-drop internals. A distillation column operating at subatmospheric pressures requires either a mechanical vacuum pump or a steam-fed jet ejector system to maintain column pressure. The energy requirements of these vacuum systems increase greatly with decreasing pressure at the column top. When a vacuum tower is retrofitted with lower pressure-drop internals, the decrease in column pressure drop can be applied toward decreasing the pressure at the reboiler or feed heater or towards increasing the pressure at the top of the column. Commercially available column internals for low pressure-drop applications are compared in refs. 1 and 14.

Several factors must be considered when deciding how best to utilize the decrease in column pressure drop brought about by the retrofit. If product degradation is a problem at a specific reboiler temperature, decreasing the pressure at the reboiler reduces the reboiler temperature, which should decrease the amount of product degradation. Even when product degradation is not a problem, decreasing the reboiler temperature reduces the requirements for high temperature energy, which is more valuable than lower temperature energy. If a furnace or other type of feed heater is being used with steam stripping, as in a refinery vacuum tower, lowering the reboiler pressure increases the fraction of the feed that is vaporized or reduces the steam-stripping requirements. Alternatively, the reduced pressure drop can be applied toward decreasing the energy input to the vacuum system, ie, mechanical energy for a vacuum pump or motive steam for a jet ejector, by increasing the pressure at the top of the column.

Separations Sequencing. Multicomponent separations generally require a sequence or train of distillation columns. The main problem in designing the most efficient train is the number of possible sequences, particularly if several components are being separated and if more than one separation technique is being considered. Algorithmic and evolutionary techniques have been studied, but heuristic guidelines seem to be the most useful (see Separation systems synthesis).

A composition node design method (CNDM) has been used to evaluate the least expensive surfaces for eight distillation systems used to separate ternary feeds (15). Comparisons with equivalent equilibrium-stage models indicate that CNDM results in small but acceptable overdesign, even when applied to complex, thermally coupled tower configurations. Design heuristics for an ideal, three-component mixture ABC, where A is the lowest boiling component and C is the highest, have been developed from these studies (see Guidelines).

Additional Column Draw to Eliminate a Column. In the fractionation of a multicomponent mixture, the components in the feed distribute through the column according to their boiling points; the extent of segregation of the individual components increases with increase in reflux and with the number of plates in the column. At total reflux, individual components concentrate where the temperature and pressure are such that their liquid- and vapor-phase mole fractions are equal (16). At finite reflux, these concentration peaks are less pronounced, but this segregation tendency is the basis for the use of side-stream withdrawals.

In many multicomponent separations, the withdrawal of side streams reduces the number of columns and the energy required. Side-stream removal can be advantageous where a mixture is to be separated into three or more products that need not be very pure or recovered in high yield, a stream contains both heavy impurities and a small quantity of relatively volatile light impurities, and a mixture to be separated contains an impurity which tends to concentrate at an intermediate location in a fractionating column.

Pasteurization. If the feed contains a small concentration of impurities that are more volatile than the desired distillate product, it often is possible to separate them from the distillate product in the same columns where the heavy impurities are removed. The distillate product is taken as a side cut and the impurity as the top product. Several trays are placed above the drawn product side stream to concentrate the impurity, and this top section is called the pasteurizing section. The light impurities are removed as a concentrated small stream from the top of the column and the overhead product is removed as a liquid side stream.

Since the impurity stream is small compared to the side-cut-product stream, the liquid and vapor flows in the pasteurization section are usually almost equal and correspond to nearly total reflux. If the impurity stream is withdrawn as a gas, the reflux of the impurity can be much less than the reflux of the side-stream product. The impurity concentration x_{iD} in the side-stream product is constrained by the inequality:

$$x_{iD} \geq \frac{FZ_{iF}}{KV} \qquad (6)$$

where x_{iD} = mole fraction of light impurity in the overhead side-stream product, Z_{iF} = mole fraction light impurity in the feed, K = constant in the light impurity vapor–liquid equilibrium expression $y = Kx$, F = molal feed rate, and V = molal vapor rate. From this relationship, it seems likely that the impurity concentration in the product decreases with increasing reflux.

The pasteurization technique is best suited for removal of impurities that are more volatile than the overhead product. However, if the impurities are so volatile as to be in the vapor phase, they may simply be removed by venting the condenser. Thus, there is an intermediate volatility for impurities most efficiently removed by pasteurization. Pasteurization is used industrially in the purification of ethylene, ethanol, methanol, propylene oxide, propylene glycol, acrylonitrile, and phenol, among other chemicals.

Intermediate products. Columns with side-stream product withdrawals are frequently used in petroleum refining where broad boiling-range products are the rule for all but the lowest molecular weight hydrocarbons (see Petroleum, refinery processes). Above the feed, side-stream products are removed usually as liquids and sufficiently high in the column that heavy components are fractionated to the desired degree. Feed components lighter than those wanted in the side-stream product must pass through the side-stream draw-off tray to reach the point in the column where they are to be removed; the side stream, therefore, contains some fraction of these components. The withdrawal of the side stream as a liquid reduces the concentration of these light components in the side stream since they concentrate in the vapor. A high reflux also reduces the concentration of undesired light and heavy components as it increases the column fractionating power; however, it also requires more energy. When necessary, side-stream products can be stripped of any light impurities in a side column with steam or by reboiling (Fig. 4). The vapors containing the light impurities are then returned to the main column at a location appropriate to their concentration. Atmospheric and vacuum distillation towers, catalytic cracking fractionators, and alkylation-plant depropanizers and deisobutanizers are examples of refinery towers involving extensive side-stream withdrawals (17). Side-stream product withdrawal may also be advantageous in the fractionation of C_8 aromatics (18). Side-stream products can also be withdrawn below the feed, but in this case it is advantageous to remove the product as a vapor to minimize the concentration of heavy impurities which must pass through the draw-off tray, since these components concentrate in the liquid phase. Vapor products, however, remove significant enthalpy from the tower and, therefore, are less often used.

Intermediate impurity removal. In certain multicomponent fractionations, feed impurities have intermediate volatilities and, therefore, accumulate and concentrate at intermediate column locations. In such cases, the impurities can be removed by withdrawal of a side stream into a separate column in which the impurity is removed. The purified side stream is then returned to the main column (19).

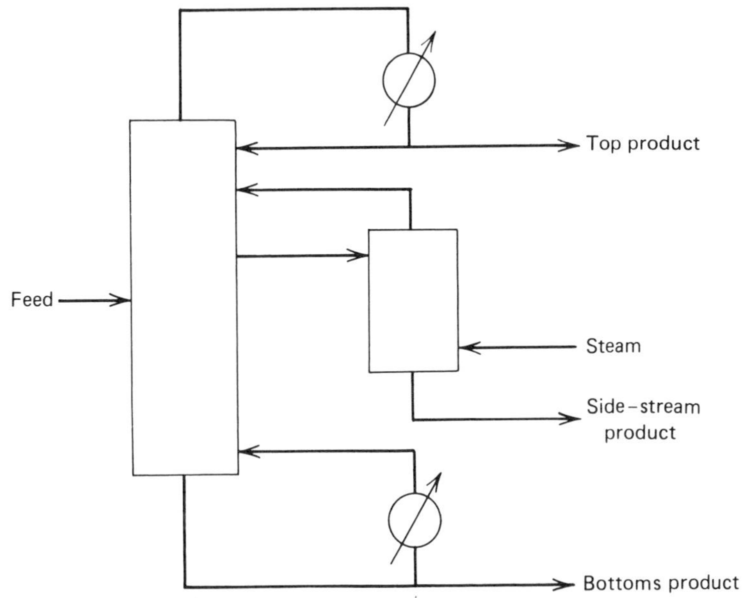

Figure 4. Side-stream product removal.

More Efficient Heat Utilization. ***Heat-Exchanger Selection.*** Efficient heat-transfer surfaces are desirable to minimize condenser and reboiler temperature-difference driving forces and area requirements (see Heat-exchange technology). Use of heat-pump, vapor-compression, reboiler-flashing, and heat-cascading techniques requires increased energy and, consequently, increased heat-exchanger temperature differences; thus, high heat-transfer coefficients are particularly important. The use of enhanced heat-transfer surfaces to reduce driving-force requirements in distillation has been applied to petroleum light ends and natural-gas processing plants (20). Other applications to distillation will take place, particularly in response to the increased use of energy-conserving options.

Enhanced heat-transfer surfaces. Enhanced heat-transfer surfaces of two principal types are available (21). One is Linde's porous boiling surfaces (high flux tubes), and involves application of a thin, metallic, microporous surface to either the inside or outside of heat-transfer tubing. The coating enhances boiling and condensing heat-transfer coefficients by a factor of 10–30 over those for a bare tube. High flux tubing is generally limited to clean-service applications. The other type of enhanced heat-transfer surface is based on the use of fins or flutes to extend the heat-transfer surface and, where the fins or flutes are discontinuous, to interrupt the boundary layer and reduce its thickness and resistance to heat transfer.

By use of high flux tubing and fluting in its air-separation plants, Linde can limit the ΔT in its air column condenser–reboiler, in which the high pressure column condenser supplies heat to the low pressure column reboiler, to 1.5–2°C. In olefins plants, these techniques have been applied to ethylene–ethane and propylene–propane fractionators, and particularly in the latter case when vapor recompression is used. An example of the savings possible for a propylene–propane splitter with high flux tubing is reported in a study conducted for a 6.8×10^4 metric tons per year propylene plant (20). High flux tubing resulted in a ΔT of 5.6°C in the heat exchanger of the reboiler–condenser, as opposed to 20°C ΔT in a conventional vapor-recompression scheme. Reduction of the compressor power requirement by 45% was possible; at $5.50/t high pressure steam, this reduction provides an annual savings of $360,000. Compared to a conventional C_3-splitter design, in which no vapor recompression is used, the high flux tubing with vapor recompression saves 79% of the steam requirement. However, the steam for conventional design is low pressure, whereas high pressure steam is required for driving the turbine in the vapor-recompression scheme. Nonetheless, if low pressure steam is valued at $3.85/t, yearly savings of $1,035,000 are possible. Other vapor-recompression applications involving improved heat-transfer surfaces include isobutylene–2-butene, and isobutane–butane fractionation.

Compact heat exchangers. Plate-and-frame and spiral-plate exchangers are examples of compact heat exchangers, which are being used increasingly in heat exchange between two process streams where close-approach temperature differences are important. Spiral-plate exchangers are more resistant to solids deposition and fouling than shell-and-tube exchangers, permit close-approach temperatures because of cocurrent flow, are well-suited for handling viscous fluids, and are fairly easy to clean (21). Plate-and-frame exchangers are compact and offer high heat-transfer rates with low pressure drop. They are also suited for handling viscous materials and are also easy to clean but do have a large gasketed area.

Air-cooled exchangers. Air-cooled and evaporative condensers are more expensive and bulkier than water-cooled exchangers but are used increasingly because they discharge directly to the atmosphere, whereas cooling water involves intermediate exchange, ie, exchangers plus a cooling tower; are less subject to fouling; may require lower utility costs; and require no water and are less apt to generate environmental discharge problems (21–24). However, the performance of water-cooled exchangers is less affected by environmental changes, and they are more economical for cooling to near-ambient temperatures.

Feed/Product Heat Exchange. Single-column heat integration implies the possibility of exchanging heat between the feed and product streams of a column to reduce distillation energy requirements. If the condenser temperature is above ambient, the exchange is between bottoms product and feed to preheat and/or partially vaporize the feed. For columns with refrigerated condensers, the object of heat integration is reduction of the condenser load, and the exchange of interest is between the top product and the feed so as to subcool the feed. Partial feed vaporization through preheating saves energy by reducing the reboiler heat requirements, but it increases the condenser load and vapor traffic in the rectifying section of the column and may limit column throughput. Similarly, subcooling the feed by exchange with a refrigerated condenser decreases the condenser load but increases the reboiler load and the liquid traffic in the stripping section of the column. Although feed/product heat exchange can save energy, it does alter the vapor–liquid traffic in the column and may be incompatible with the desired column operation because it reduces column throughput or leads to heat-exchanger limitations. Feed preheat may be uneconomical where cooling is expensive because of water shortages, high ambient air temperatures, or both (25).

In chemical and light hydrocarbon plants, there are frequently limited opportunities for single-column heat integration, because the bottoms product from one column is the feed to the next. If the bottoms product is at the appropriate temperature for feeding into the next column, it is unavailable for heat exchange. Single-column heat exchange is of principal importance in the crude preheat train for the atmospheric topping tower in a petroleum refinery; it can also be essential in cryogenics plants, although these require trains of towers.

The same general rules that apply in determining where single-column heat exchange should be implemented apply in the evaluation of heat exchange with other process, distillate, or bottoms streams.

Heat Pumping. Heat pumping results in reduced distillation energy requirements as heat is pumped from the condenser to the reboiler. According to the Second Law of Thermodynamics, in order for heat transfer from a low temperature to a high temperature reservoir, there must be an input of work. In a heat-pumping scheme, a compressor is used to supply the work. The working fluid is boiled at the condenser and heat is removed from the column. The vaporized working fluid is then compressed and passed to the reboiler where it condenses, supplying heat to the column. Finally, the working fluid expands and returns to the condenser. The amount of work energy added at the compressor is substantially less than the amount of heat energy which would have to be added to the reboiler for columns with above-ambient condenser temperatures or less than the refrigeration requirements for a column with a refrigerated condenser (see Refrigeration).

External refrigeration (Fig. 5), vapor recompression, and reboiler flashing are three variations of heat pumping. A separate refrigerant is the working fluid in external refrigeration, whereas the column overhead and bottoms, respectively, are the working fluids in vapor recompression and reboiler flashing. These variations of heat pumping are restricted to separations involving fairly narrow temperature ranges. If the temperature range is too great, it would be impossible to obtain a refrigerant, either external or column fluid, which could be compressed sufficiently to span the reboiler-to-condenser temperature range, or the work of compression would be too large for the process to be economical.

An attractive feature of heat pumping is that it represents a relatively simple change in the conventional column design; only the heat exchangers are altered sig-

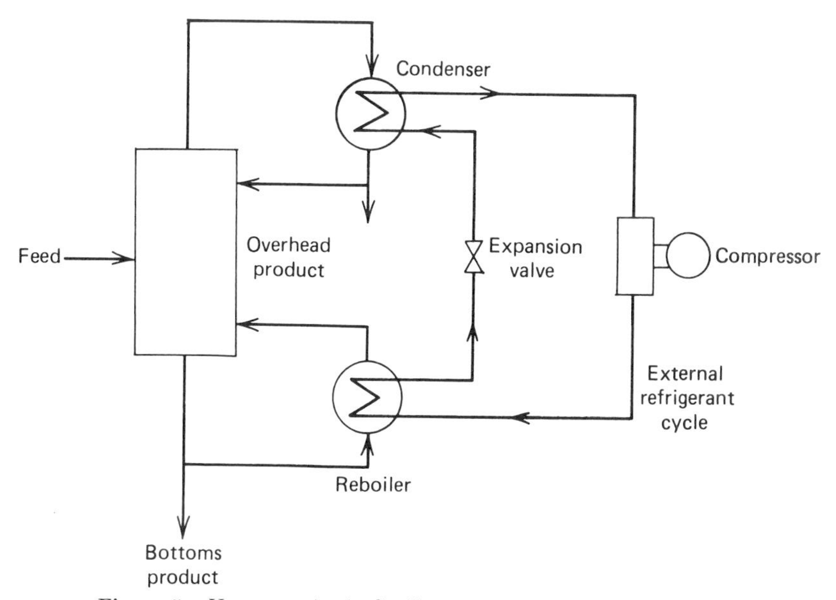

Figure 5. Heat pumping in distillation with an external refrigerant.

nificantly. However, even though this technique can substantially cut energy use by the column, the form of energy used, ie, compressor work, is more expensive than the form it replaces, ie, reboiler or condenser heat. Thus, for heat pumping to be economical, the ratio of the energy usage of the modified design to the energy usage of the conventional design must be less than the ratio of the cost of steam or refrigeration energy to the cost of work energy. The application of heat pumping to distillation columns in existing units and those being constructed is reviewed in ref. 26.

Intermediate Heat Exchange. Intermediate heat exchange (Fig. 6) is considered primarily where the intermediate exchange can be based on a less-expensive energy source than that required by the condenser or reboiler, eg, hot water instead of steam for reboiling or cooling water instead of refrigeration for condensing; or where systems are highly nonideal and the intermediate exchange permits the operating lines to approach the equilibrium curve more closely, ie, permits the use of lower reflux, over a significant portion of the fractionation. Since the temperature difference between the intermediate exchanger and the reboiler or condenser is less than the overall column temperature difference, its use can make heat pumping more attractive.

Intermediate condensing of vapor is widely used in petroleum refining by means of pump-around or pump-back loops, in which liquid is pumped from one tray location, through a heat exchanger where heat is added or removed external to the column, and then back to the column at the same location (pump-back) or a few trays removed from that location (pump-around). Crude distillation and catalytic cracking–fractionator towers incorporate such intermediate condensing loops.

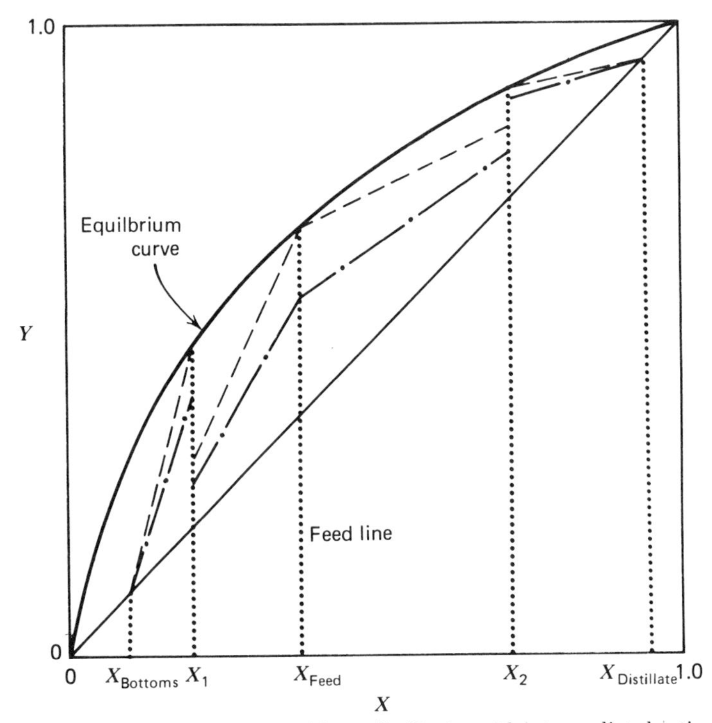

Figure 6. McCabe-Thiele diagram for a binary distillation with intermediate heating at X_1 and intermediate cooling at X_2. Because of the intermediate heat exchanges, the reflux ratios for the upper and lower operating lines (dot/dash lines) are actually lower than the minimum reflux ratio (dashed lines) for the case without intermediate heat exchange.

Heat Cascading (Multiple-Effect). Another method for more completely utilizing the energy supplied to a distillation process is to operate columns at different pressures and use the latent heat in the overhead product from a high pressure column to supply reboil energy to a lower pressure column. Such a two-effect design is conventionally used in the fractionation of air to produce nitrogen and oxygen, but it otherwise has not had much application in industrial separations.

Distillation with a Mass-Separating Agent. ***Extractive Distillation.*** In extractive distillation, a solvent is introduced near the top of a distillation column to increase the relative volatility of the materials being separated. The solvent is less volatile than the materials being separated and flows down to the bottom of the column, where it is removed with the extracted or less volatile material. A few trays are provided above the tray to which the solvent is introduced to fractionate the solvent from the overhead product. The solvent and extract mixture in the bottoms product is separated in a separate column by distillation or by other means, such as a water wash. The solvent is then cooled and recycled to the extractive-distillation column (Fig. 7).

Bringing about the desired increase in relative volatility usually requires large solvent flow: from one to ten times the molar rate of feed to the column. The improvement in relative volatility results entirely from changes in activity coefficients in the liquid phase, because the solvent is concentrated there and vapor-phase behavior deviates little from the ideal except as critical conditions are approached. In conventional extractive distillation, the solvent and its concentration are chosen so as to ensure that only one liquid phase exists within the distillation column.

Since extractive distillation depends for its effectiveness on selective enhancement of the nonideal solution behavior of one of the components or classes of components

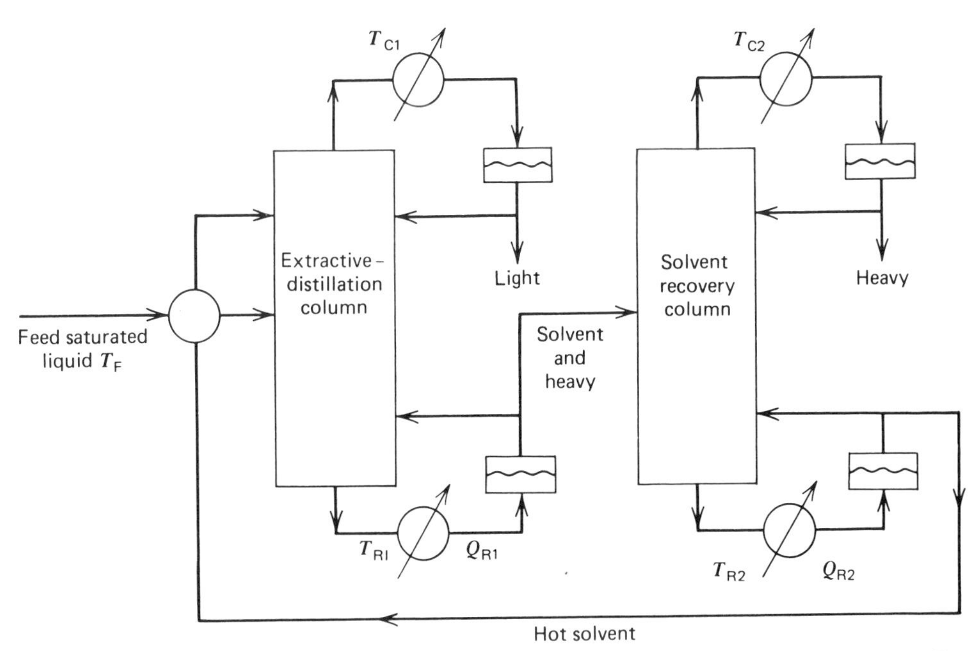

Figure 7. Typical extractive-distillation process scheme (18). T_{C1}, Temperature condenser 1; T_{C2}, temperature condenser 2; T_{R1}, temperature reboiler 1; T_{R2}, temperature reboiler 2; Q_{R1}, heat flux, reboiler 1; Q_{R2}, heat flux, reboiler 2; T_F, temperature feed.

to be separated, the materials to be separated must not form an ideal solution and must differ sufficiently in chemical type for this enhancement to be feasible. Extractive distillation is well-suited for separating materials that differ in polarity or hydrogen-bonding tendency and is used industrially for the separation of aromatics from paraffins, as in the extraction of benzene–toluene–xylene and C_8 aromatics from reformate, frequently in conjunction with liquid–liquid extraction; the separation of olefins from paraffins; the separation of diolefins from monoolefins, eg, butadiene from butenes; and the fractionation of oxygenated hydrocarbons. The solvents used in extractive distillation usually are polar or hydrogen-bonding and selectively decrease the volatility of the more polar or higher hydrogen-bonding-potential material relative to the less-polar or lower hydrogen-bonding-potential material.

Desired solvent properties include high selectivity. This should result more from a decrease in the activity coefficient of the extracted material, for which the solvent has affinity, than from an increase in the activity coefficient of the material which goes overhead. A high activity coefficient for the more volatile material, when it is dissolved in the solvent, is apt to increase tower height requirements (27). Another desired solvent property is high solubility of the feed material in the solvent. The critical factor is usually the solubility of the less-polar or lower hydrogen-bonding-potential material. This material generally is less soluble and predominates at the top of the column where the temperatures, and, therefore, the solubilities are apt to be lowest. The solvent should not be very volatile to minimize the number of trays above the solvent feed point required to fractionate the solvent from the overhead vapor or knockout trays and to simplify the separation of the solvent from the extract in the solvent-recovery tower. The volatility, however, must not be so low as to make the reboiler temperature in the solvent-recovery tower high, since the solvent must be cooled before feeding to the extractive-distillation tower. The energy lost in heating and cooling the recirculating solvent can be a main factor determining the energy required by the process. The solvent can, however, have negligible volatility if it can be regenerated by an alternative means, eg, steam stripping or liquid–liquid extraction, as with a water wash. In addition, the solvent should be relatively inexpensive.

Energy considerations. Extractive distillation increases the relative volatility of components to be separated and, therefore, reduces the reflux and energy requirements for the separation of the light and heavy key components. Offsetting these savings is the energy required to separate the solvent from the extracted material and that lost in heating and cooling the solvent as it cycles between the extractive-distillation and solvent-recovery columns. Much of this energy lost in solvent recirculation may be economically recovered by using the solvent to vaporize the feed or to reboil the contents of the extractive-distillation column (28).

Azeotropic Distillation. In azeotropic distillation (Fig. 8), a solvent is added, generally at the feed, to a distillation column to increase the relative volatility of the materials being separated. The solvent volatility is similar to that of the materials being separated. It passes overhead with the distillate product, normally forming a minimum-boiling, frequently homogeneous azeotrope.

In azeotropic distillation, the solvent is introduced as feed or occasionally with the reflux, is taken overhead, and is largely confined to the rectification or upper section of the distillation column; whereas, in extractive distillation, the solvent is added near the top, is removed with the bottoms, and is present in high concentration throughout the entire column, with the exception of a few solvent knockout trays at the top of the

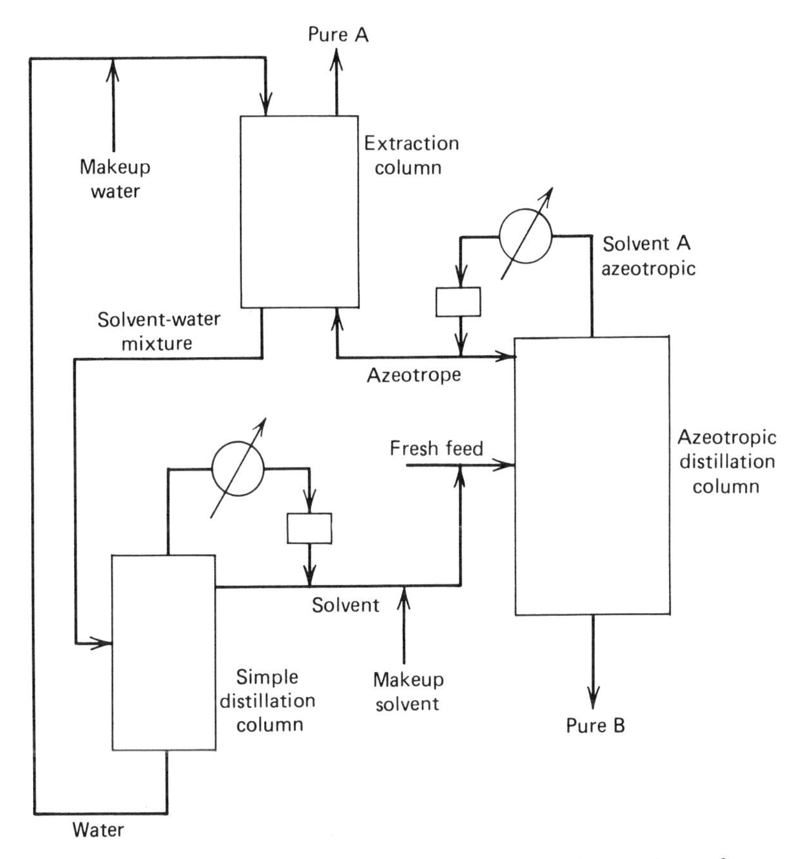

Figure 8. Typical scheme for azeotropic distillation. (The extraction step is not always needed.)

column. Although the added solvent should be miscible with the column liquid at the temperature and concentration in the column, two liquid phases frequently form, the one solvent-rich and the other distillate-product-rich, upon condensation with the overhead and subsequent cooling. This simplifies solvent recovery.

Azeotropic distillation is used principally in overhead distillation of a material that is present in the feed in low concentration and that is difficult to separate by straight distillation. In such a case, usually only a small concentration of azeotropic agent is needed. The additional energy requirements for vaporizing the agent are small, and recovery of the agent is simplified. Azeotropic distillation is used commercially in dehydration, eg, of acetic acid and ethyl alcohol; in the breaking of azeotropes, particularly where the azeotropic agent is immiscible with the overhead and can be separated by decantation; and in separations in which either the agent or the overhead product is water-soluble and the two can be separated by washing with water (29–30).

An azeotropic agent should preferably form a minimum-boiling azeotrope with one of the components to be separated; this allows for the ready separation of the azeotropic agent. As in extractive distillation, the agent amplifies nonideal solution behavior in the liquid phase: it increases the relative volatility of the feed components most dissimilar to it, eg, a nonpolar agent increases the volatility of the polar agent relative to the nonpolar components in the feed. The agent should be of high enough volatility to pass overhead, but low enough to have a sufficient liquid-phase concentration to exert the desired influence on the liquid-phase activity coefficients. The agent should be minimally soluble with the compound being driven overhead. However, it should be soluble enough in the column liquid to ensure complete miscibility and solution throughout the column so as to influence the material being driven overhead and to avoid the downcomer and flooding problems characteristic of two liquid phases. The agent should be easily recoverable from the overhead product as, for example, in systems where the agent forms a second phase with the condensed distillate product and can be separated by decantation or in systems where the agent or the distillate product is preferentially water-soluble and can be separated by water washing. The agent should have a low heat of vaporization, resulting in minimum vaporization energy requirements and should be inexpensive, nontoxic, and noncorrosive.

Nondistillation Techniques. ***Gas Absorption.*** In gas absorption, a gas containing a mixture of components contacts a solvent which extracts one or more of the compounds to be separated. The absorbed components are then desorbed by either pressure reduction or temperature increase, or both, and the solvent is recycled to the absorber. Absorption and stripping are normally carried out in a countercurrent manner in trayed or packed towers (see Absorption).

Absorption is widely used for acid-gas removal from natural gas, synthesis gas, refinery gas, stack-gas scrubbing, and acid-gas scrubbing associated with synfuels synthesis. It is also used for scrubbing acid gases from gas streams in the manufacture of chemicals, eg, sulfuric, nitric, hydrochloric, and hydrocyanic acids, silica gel, vinyl chloride, fluorocarbons, etc; scrubbing organic chemicals from vent gases in chemical manufacture; and in water scrubbing of reactor effluent gases to recover ethylene oxide, ethanol, acetaldehyde, methanol, formaldehyde, acetic acid, and acrylonitrile in the manufacture of these chemicals. Absorption is still used for light-ends recovery in refinery operations, but it has largely been displaced in this application by cryogenic processing.

Absorption is particularly applicable if a small amount of a high boiling component or an acid, eg, CO_2, SO_2, H_2S, etc, is to be removed from a gas stream. Absorption is also useful if a gas stream is far removed from its bubble point and the products are not required as liquids or as high pressure gases, a solvent is available with a high specificity for a component to be removed, or the gas stream is available at moderate-to-high pressure.

Solvent selection. Solvents may be either physical or chemical absorbents or a combination of the two. Physical absorbents are best suited for bulk removal of components having a high partial pressure in the feed stream, since the capacity of a physical solvent increases nearly proportionally with the partial pressure. Chemical absorbents are best suited if complete removal of a component is required or if the partial pressure of the component in the feed stream is low. Because of the high pressures used for acid-gas removal with physical solvents, equipment costs can increase significantly with equipment volume requirements and, thereby, place a premium on the use of efficient solvents and contacting equipment. With the increase in energy costs, use of chemical absorbents with low energy regeneration requirements has increased.

Desirable solvent properties include high selectivity for the components to be absorbed; high capacity; low volatility or high water solubility so as to be readily recoverable from the exhaust gases; low viscosity so diffusivities and mass-transfer rates are high; and low cost, toxicity, and corrosivity.

Energy requirements. If the absorption process regenerates the solvent by pressure reduction, energy inputs are required to pump the solvent from the low pressure stripper to the high pressure absorber, to compress the solute to the product pressure, to compress the feed where necessary to the absorber pressure, and to compress the solute to the absorber pressure if some of it is used to reboil the absorber. If solvent regeneration is by temperature increase, energy is required to heat the solute-rich solvent stream leaving the absorber to the temperature of the stripper, to reboil the stripper, and to circulate the solvent. Much of the energy required for solvent heating is conventionally supplied by countercurrent heat exchange of the solute-rich solvent against the regenerated solvent leaving the stripper and returning to the absorber. In both cases, energy requirements can be decreased by reducing the volume of the solvent that is to be recirculated. This volume can be reduced by keeping the absorber as close to isothermal conditions as possible by use of internal cooling or pumparounds to remove heat, if necessary; using enough stages in the absorber and stripper; using a selective, high capacity solvent; and using high gas pressures. If physical absorbents are used, the activity coefficient of the solute in the solvent necessarily is greater than one and the volume of the solvent to be circulated is proportional to this activity coefficient. Choosing a solvent whose solubility parameter is close to that for the solute is one means of ensuring that the solute activity coefficient is close to one.

Liquid–Liquid Extraction. In liquid–liquid extraction, a liquid containing a mixture of components contacts a partially miscible solvent, which extracts one or more of the components to be separated (see Extraction, liquid–liquid extraction). The process potentially is more selective than distillation because both phases are nonideal, which makes it possible to amplify differences in the materials being separated by proper choice of extractant.

Liquid–liquid extraction is usually carried out in the countercurrent manner with solvent recovery from the raffinate, solvent separation from the extracted material, and recirculation of the solvent (Fig. 9). The solvent may be either lower or higher boiling than the extracted material, but it generally has a boiling point and activity coefficient in the raffinate such as to permit the solvent to be stripped overhead from the raffinate. Otherwise, all of the raffinate would have to be distilled overhead to separate it from the solvent.

Despite the potential of liquid–liquid extraction for energy conservation relative to distillation, it is not widely used because of the additional complexities and costs it introduces into the process and process equipment. Key uses include: separation of aromatics from paraffins, where the overlap of boiling points of the materials to be separated is too broad to permit efficient extractive distillation; recovery of less volatile solutes, eg, acetic acid, from wastewater; and specialty petroleum processes, eg, desulfurization, deasphalting, and lube-oil refining, eg, dewaxing. Relative to extractive distillation, liquid–liquid extraction usually is economically attractive if more than one component is to be separated from a mixture, whereas extractive distillation is more economical if a single pure component is to be recovered (31).

Solvent selection. Because of the dominant effects of solvent selectivity and extractant solubility in the solvent on equipment size and process energy requirements, solvent selection plays a key role in the economics of liquid–liquid extraction. Desired solvent properties include high selectivity; high solubility of the extractant in the solvent; low cost; appropriate volatility, ie, low enough so as not to require special

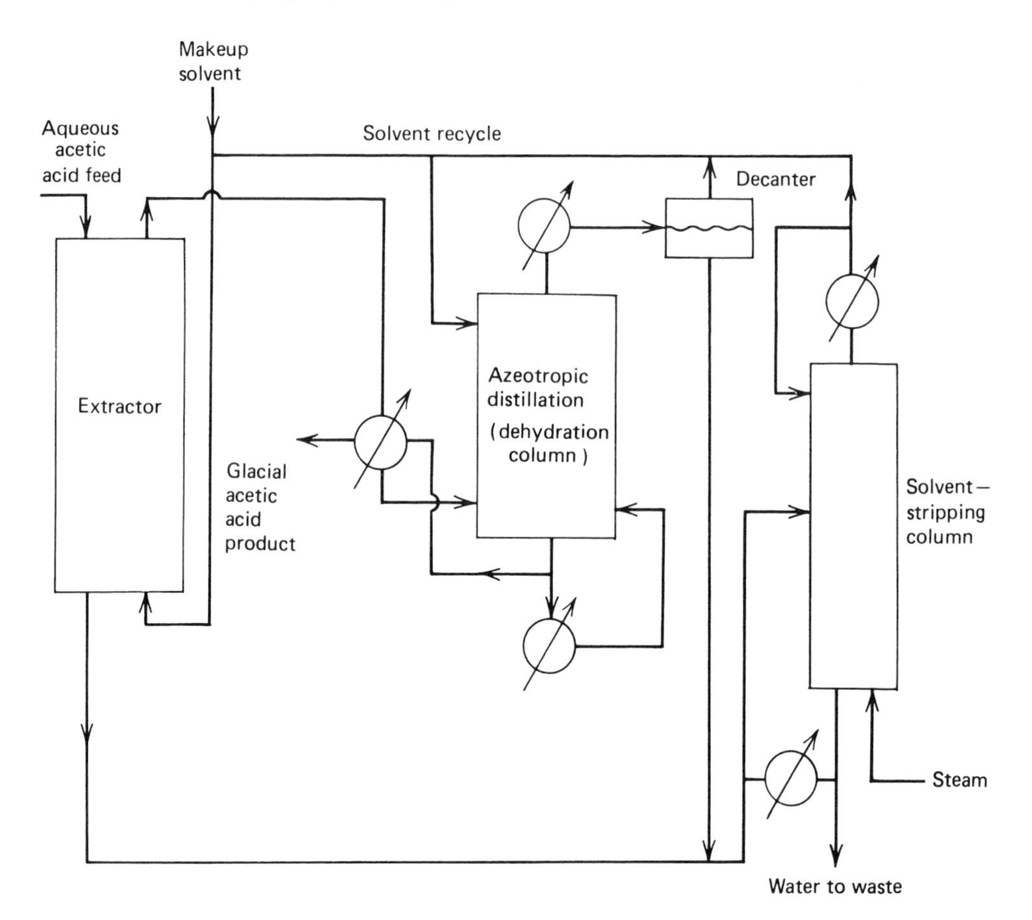

Figure 9. Solvent extraction with low boiling solvent. (The azeotropic distillation step is not always needed.)

handling but high enough to simplify recovery of the solvent from extracted material; a high enough density difference in the contacting phases to promote rapid phase separation; appropriate interfacial tension between the contacting phases during extraction, ie, it should be low enough to promote mass transfer but high enough to promote rapid phase separation; low heat of vaporization to reduce energy requirements for solvent recovery by distillation; chemical stability, nontoxicity, and noncorrosivity; low viscosity for facile pumping; and low solubility in the raffinate phase to reduce the cost of solvent recovery from the raffinate.

The current state of the art of solvent selection for liquid–liquid extraction is reviewed in ref. 32. For molecules that interact through dispersion forces only and that are not associated in solution, the selectivity may be related to the solubility parameter (33). For such solvents, there tends to be an inverse relationship between solvent selectivity and capacity. Since high selectivity and capacity cannot be simultaneously obtained with these solvents, commercial interest has focused on more polar and more hydrogen-bonding solvents. Approximate methods for predicting activity coefficients and selectivities, where these forces are present, are reviewed in refs. 32–34.

The highest selectivities and capacities result from solvents that react reversibly to form complexes with the material to be extracted. Combinations of solvents usually satisfy the many solvent-selection criteria; for example, one solvent may react with or complex the material to be extracted and the other solvent may function as a phase former and extractant for the complex and ensure good phase separability by controlling phase-density differences and interfacial tension. Such systems, however, make handling and processing more complex and make more difficult the determination of appropriate solvent compositions and their selectivities and capacities.

Energy requirements. As with extractive and azeotropic distillations, liquid–liquid extraction has the potential for reducing energy requirements compared to straight distillation by increasing the separation factor and reducing reflux requirements. Because the solvent must be removed from both the extract and the raffinate and usually by distillation, these separations must be simpler and require less energy than the distillation of the original feed to justify the use of extraction. Energy requirements for extraction are proportional to solvent recirculation rates and are determined by the ease of solvent recovery from the raffinate and extract phases and by the difference in operating temperatures between the extractor and the solvent-recovery columns.

Supercritical-Fluid Extraction. In supercritical-fluid extraction (SFE), a liquid or comminuted solid contacts a supercritical fluid at temperatures and pressures near the critical point and, thereby, permits the extraction of one or more components to be separated (see Supercritical fluids). The extraction is analogous to liquid extraction with the exception that the pressures and supercritical fluid's diffusivity are high and the fluid's density and viscosity are low, enabling rapid extraction and phase separation. Recovery of the extracted material is accomplished generally by lowering the pressure, which decreases the fluid's density and markedly lowers its solubility; raising the fluid's temperature at constant pressure is another possibility. The cycle is completed by recompressing the gas to extraction conditions. The process is based on the significant changes in the solvent power of supercritical fluids, which occur with changes in temperature and pressure near the critical point.

The process is particularly well-suited for the separation of high boiling or heat-sensitive materials which are difficult to separate by distillation or liquid extraction. Low polarity materials tend to be preferentially extracted. The food industry has been one of the first to commercialize the process, because fluids, eg, CO_2, that pose no health hazards and leave no residue, can be used for extraction (35). In the FRG, caffeine is extracted commercially from coffee (qv) with supercritical CO_2; more recent developments are processes for the extraction of hops, spices, and tobacco. The pharmaceutical industry will likely find SFE useful. Supercritical-fluid extraction has also been examined extensively as a means of upgrading heavy hydrocarbons. Kerr-McGee Refining Corporation has developed the residuum oil supercritical extraction (ROSE) process in which supercritical pentane is used to deasphalt atmospheric and vacuum-distillation residua (36). The process is in limited commercial use. Kerr-McGee and the UK National Coal Board have also done considerable work on the SFE of coal (qv) and of coal liquids (37).

Fluid selection. Studies to date have concentrated on readily available, low cost gases with moderate critical temperatures, eg, lower molecular weight hydrocarbons, CO_2, and water. Theoretical considerations indicate that, for maximum solvent power, the fluid should have a critical temperature close to the extraction temperature. However, theoretical prediction of solute solubilities in supercritical fluids is limited by lack of understanding of the dense-fluid state and of the highly asymmetric mixtures involved in SFE. A key advantage of supercritical-fluid extraction is the broad range of gases that are superior solvents at supercritical pressures and at slightly above critical temperatures.

Energy requirements. A SFE process involving a compression ratio of three is roughly equivalent in energy requirements to a typical separation by distillation (38). Energy savings can become larger if separation by distillation is difficult or if it is possible to operate a SFE process at low pressure ratios. Since SFE usually requires more expensive process equipment because of the high pressures involved, substantial energy savings or other process advantages must be present for it to be an attractive alternative.

Adsorption. In adsorption, a gas or liquid mixture of components contacts a microporous solid, which selectively adsorbs certain of the components (see Adsorptive separation). Desorption can be effected by raising the temperature, decreasing the pressure, or displacing the adsorbed material with another material. The process is normally carried out in fixed beds. In a batch operation, there are two or more beds and the material to be separated is fed into one bed where adsorption occurs while the second bed is being regenerated (39–41). Desorbent flows are normally in the vapor phase. In a continuous operation, countercurrent flow of solid and fluid is generated by periodic shifting of the column locations to which the feed and desorbent materials are fed (42–44). Feed and desorbent flows are normally liquid phase.

Regenerative adsorption processes may be categorized as follows (41,45–46). Thermal swing adsorption is a process in which desorption is accomplished by raising the adsorbent temperature, generally by heating fluid prior to its passing through the bed but occasionally by direct heating of the adsorbent. Following desorption, the bed is cooled to an appropriate temperature for adsorption; pressure-swing adsorption (PSA) is a process in which adsorption takes place at a relatively high pressure (see Fig. 10). The nonadsorbed product in the bed and particle voids is purged from the bed by reduction of the pressure to an intermediate level with the resultant desorbed material accomplishing the purge. The adsorbate is recovered and the bed is regenerated by reduction of the pressure to a low level. The purge material is usually compressed and recycled to the feed; purge-gas stripping is a process in which desorption results from the stripping action of a purge gas that is not adsorbed (see Fig. 11). The purge gas lowers the partial pressure of the adsorbed material. In heatless adsorption, regeneration is accomplished by purge-gas stripping at lower pressures with the cycling from adsorption to desorption frequent enough so that the heat of adsorption compensates for the endothermic heat of desorption. Displacement regeneration is a process in which desorption results from displacement of the adsorbed material by a second adsorbable fluid (see Fig. 12). Parametric pumping is a process in which the fluid being separated flows alternately back and forth through the adsorbent while the bed temperature or other variable affecting absorptivity is cycled between two levels. By appropriate coupling of the flow and temperature cycling, it is possible to produce both an effluent containing little adsorbate and an effluent containing highly concentrated adsorbate.

Adsorption is used as an alternative to distillation where highly polar or high molecular weight trace impurities must be removed, molecular isomers must be separated, or molecules of different polarities or hydrogen-bond character need separation. Typical applications include the removal of H_2O, H_2S, organic sulfides, and CO_2 from fluids, eg, in the drying and purifying of petroleum and petrochemical products, natural

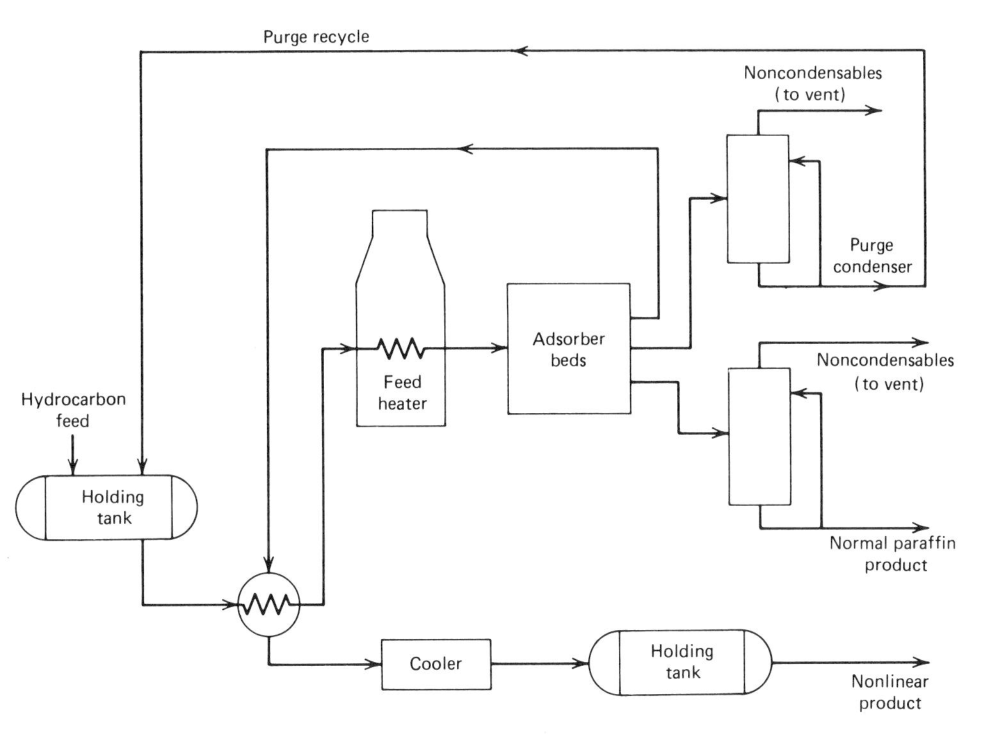

Figure 10. Pressure-swing adsorption.

gas, and air prior to liquefaction and separation (41,45,47). Adsorption of impurities is particularly useful where ultrahigh purities are required. Bulk separation applications include the separation of normal paraffins from petroleum distillates in both normal paraffin production and in the catalytic isomerization of C_5/C_6 gasoline components, in the purification of hydrogen, in the separation of *p*-xylene from *o*- and *m*-xylene and other C_8 hydrocarbons, in the separation of olefins from paraffins, and in the production of 80–90% oxygen from air from small oxygen plants that produce tons of oxygen per day (4,42,44,48–61).

A new process combining pressure-swing adsorption and parametric pumping has been commercialized (62). The pressure and flow direction at the inlet of an adsorption bed are cycled rapidly enough that the position of the adsorption front within the bed stays relatively fixed. The nonadsorbed material flows continuously from the outlet of the bed as a relatively pure stream, and the adsorbed material is purged from the inlet during the low pressure cycles. This process is being used for oxygen production but has many other potential applications.

Pressure-swing adsorption involving product gas for desorption only is useful where high product recovery is not needed. Purge-gas or displacement regeneration must be used to obtain high recovery.

Process selection. For drying and purification applications usually involving contaminant concentrations less than 5 mol %, thermal-swing adsorption is generally preferred (4). Regeneration is relatively simple and requires purge-gas heating or bed heating; a purge gas is usually required. Desorption results in low residual loading changes and large adsorbent loading changes during the cycle. Unfortunately, the regeneration cycle in thermal-swing adsorption is apt to be inefficient and to take a

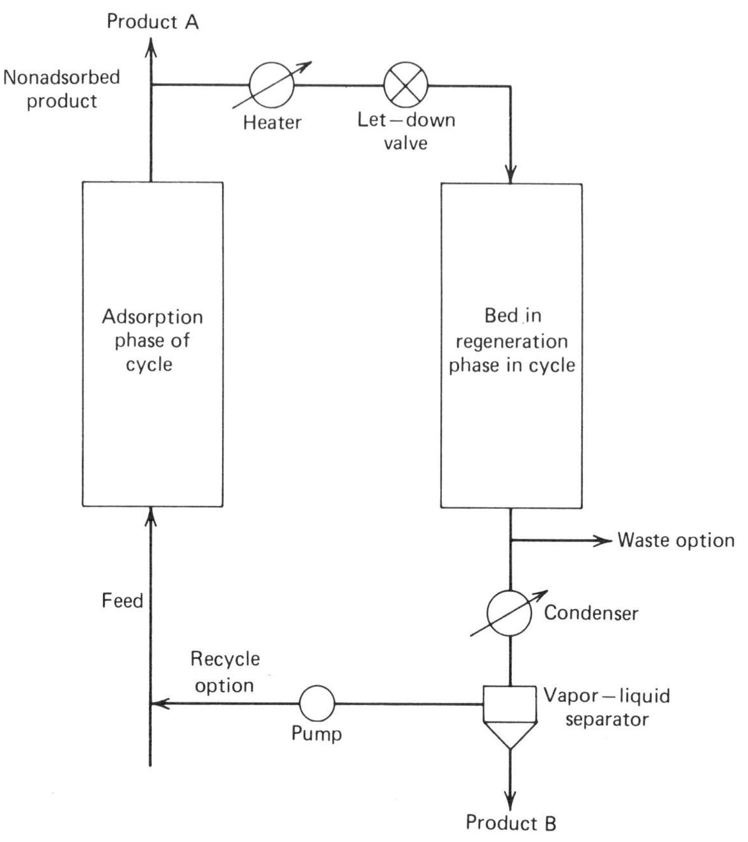

Figure 11. Regeneration with product gas purge, thermal and pressure-swing options, and recycle and waste options.

long time, ie, 8–24 h (41). Low impurity concentrations lead to infrequent regeneration requirements. Thermal-swing adsorption is generally not suited for higher feed concentrations of the sorbable component because of the bed heat capacity and long regeneration time. A nonisothermal adsorption process makes thermal-swing adsorption look promising, however, for dehydration of streams with water contents up to 20 wt % (63).

For bulk separations where the adsorbable component feed concentration is ca 7–50 wt %, pressure-swing, purge-gas stripping, and displacement are the preferred regeneration techniques (64). Pressure-swing regeneration has the advantages of enabling direct recovery of a high purity product and permitting short cycle time operation, and it is particularly attractive if the adsorbate loading changes significantly with pressure at high pressure levels. Thus, it is best suited for low molecular weight, less polar molecules, eg, low molecular weight hydrocarbons and gases such as hydrogen, helium, argon, air, and natural gas. Pressure-swing regeneration can be made economical, however, even for the separation of C_{10} and higher normal paraffins if operation is at high temperatures and pressures and desorption is at low pressures into a vacuum or compressor through a condenser (52,65).

In purge-gas stripping, the gas used to strip may be a slipstream of the nonadsorbed product leaving an adsorber bed, or it may be a nonadsorbed gas separate from

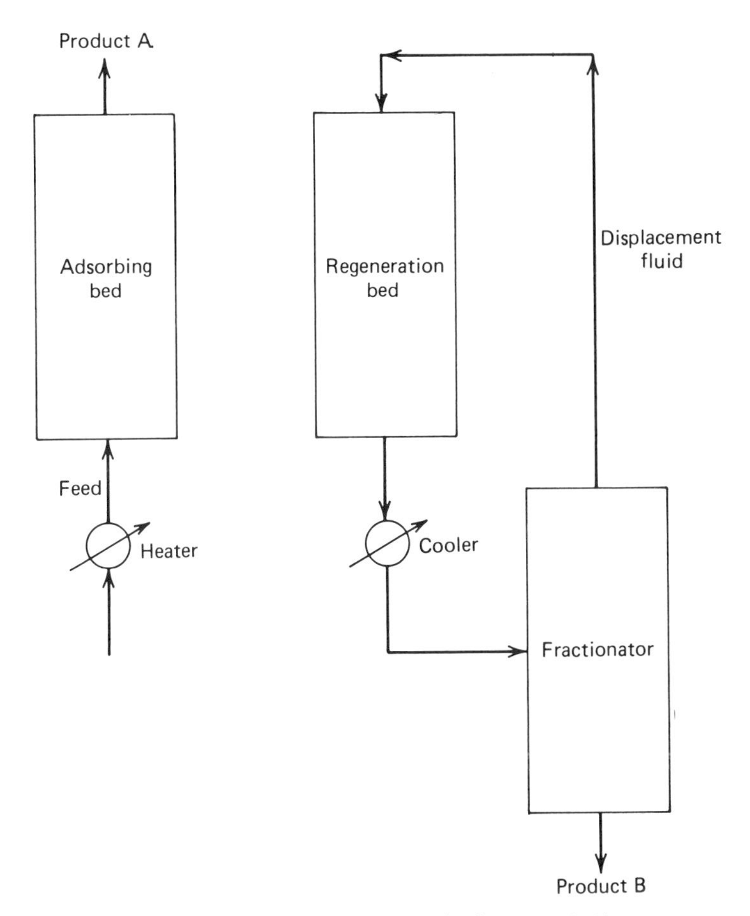

Figure 12. Regeneration with displacement fluid.

the process fluid. Purge-gas stripping is similar to pressure-swing adsorption and regeneration, in that the principal function of the purge gas is desorption by lowering the partial pressure of the adsorbate. The purge gas can also be used to desorb by bringing heat into the bed, but the bed is normally operated adiabatically with moderate changes in pressure. With a short enough cycle time, heatless adsorption conditions can be attained. The purity of the regeneration gas can be important, since it may affect the purity of the desorbed product. Purge-gas stripping is particularly simple if the effluent can be discarded.

If the pressure-swing adsorption requires vacuum regeneration, purge-gas stripping may be preferable, as the vacuum complicates adsorbate recovery and may result in air leaks with fires or catalyst poisoning. On the other hand, purge-gas stripping complicates adsorbate product recovery.

Energy requirements. For temperature-swing adsorption, the energy requirements per kilogram of adsorbate are determined by the bed-heating requirements:

$$Q_{\text{TSA}} = [(f/g)C_p\Delta T + \Delta H]/\eta \qquad (7)$$

where f = the ratio of the heat capacity of the adsorbent bed plus vessel to that of the bed alone and g = the ratio of the adsorbent-bed weight to the weight of adsorbate absorbed per cycle. Since the bed depth must be at least two to three times the length of the mass-transfer zone, this ratio is greater than that determined by the delta loading between equilibrium with the inlet and the loading of the regenerated bed. ΔH = the heat of desorption of the adsorbate, C_p = the specific heat of the adsorbent, ΔT = the difference between the adsorption and regeneration temperatures, and η = the ratio of the actual heat requirement to the minimum heat requirement. For fixed-bed adsorption of organic vapors (eg, toluene) from air, energy requirements are frequently ca 12,100 kJ/kg (2900 kcal/kg) of organic vapors. If the storage vessel is not cycled in temperature, as with fluid-bed carbon-adsorption systems, energy requirements can be nearly an order of magnitude lower.

For pressure-swing adsorption, whether conventional or parametric pumping in nature, the energy requirements are associated with adiabatic compression of the feed from the feed pressure to the supply pressure to the bed inlet. If the gas is assumed to compress ideally:

$$\frac{W}{\text{kg product}} = \frac{k}{(k-1)} \frac{RT_1\left[\left(\frac{P_2}{P_1}\right)^{k-1/k} - 1\right]}{f_{na} \cdot x_{naf} \cdot \text{MW}_{na}} \tag{8}$$

where W = work input required in kilojoules; $k = C_p/C_v$; R = gas constant = 8.3144 J/(mol·K); T_1 = inlet temperature, K; P_1 = feed pressure; P_2 = supply pressure; f_{na} = fractional recovery of the nonadsorbed product $\leq 1 - P$ desorption/P adsorption; x_{naf} = mole fraction of the nonadsorbed product in the feed; and MW_{na} = molecular weight of the nonadsorbed product.

From data reported in ref. 62 for pressure-swing adsorption (PSA) purification of oxygen, the energy requirement for PSA production of 90 wt % oxygen is estimated as (with k = 1.41):

$$\frac{W}{\text{kg } O_2} = \frac{1.41}{0.41} \times \frac{8.3144 \times 298 \left[\left(\frac{446}{101}\right)^{0.41/1.41} - 1\right]}{0.4 \times 0.21 \times 32} = 1712 \text{ kJ/kg } (409.2 \text{ kcal/kg}) \text{ } O_2$$

This is somewhat greater than the 1349 kJ/kg (322.4 kcal/kg) O_2 required for production of oxygen cryogenically, and the oxygen produced is not as pure as that produced cryogenically; however, these factors are compromises for the simpler nature of the plant in small-volume applications. For hydrogen purification, PSA is claimed to be competitive with cryogenic purification, even for large-volume applications.

Membrane Gas Separations. In gas separation by membrane processing, pressurized gas is introduced into a membrane module. Those components with higher permeabilities, ie, with higher diffusivity–solubility products, in the membrane, are preferentially transported across it. In this manner, separation is effected and the permeate is collected in a port connected to the lower pressure, downstream side of the membrane and the higher pressure residue is collected in a port connected to the upstream side of the membrane. The process is normally used to generate a single-stage separation. Staging, ie, using several modules in series with intramembrane compressors, is generally not economically favorable (66). A fairly new concept is the use of a continuous membrane column to effect a clean separation by circulating the feed gas through the column continuously and bleeding only a small fraction of the most permeable and least permeable gases at the top and bottom of the column, respectively (67).

Membrane processes are used in hydrogen purification and recovery in the petrochemical industry and in ammonia production and, on a limited scale, to separate CO_2 and/or H_2S from sour natural-gas streams (67–71). Membranes are also used in electrodialysis (qv), ultrafiltration (qv), dialysis (qv), microporous filtration, reverse osmosis (qv), and pervaporation (see Filtration; Colloids).

Membrane selection. The most useful membrane in gas separations is the Monsanto Prism system, which is a composite hollow-fiber membrane system made of polysulfone hollow fibers coated with thin, nonselective, highly permeable silicone rubber (68,72). This outer film fills pores in the polysulfone skin and, thereby, eliminates leaks. Dried cellulose acetate membranes are also used in spiral-wound module form for acid-gas separations (69–71).

Energy requirements. The energy requirements of this process, when it is used for hydrogen purification, are substantially those involved in the recompression of the permeate. For removal of acid gases from sour natural gas, the energy requirement is that associated with the pressure drop in the membrane modules of the gas which does not penetrate the membrane and is minor. If the permeate is CO_2 and is to be recompressed for oil-feed injection for secondary oil recovery, the energy requirement is that of permeate recompression.

The prime deterrent to wider use of membrane processing in gas separations is the difficulty of combining high permeability with high selectivity in a membrane. Selectivities possible through existing commercial membranes are indicated in Table 4 where several gas permeabilities for cellulose acetate and polysulfone membranes are listed. Permeation data for cellulose acetate flat-sheet membrane suggests an effective skin thickness of ca 0.07 μm, whereas Monsanto patent data indicate an effective skin thickness for polysulfone hollow fibers slightly $>0.2\ \mu$m (70,73).

Crystallization. In fractional crystallization, a component is separated from a liquid mixture by freeze crystallization involving either indirect heat exchange, direct-contact heat exchange, or vacuum to remove heat by vaporization (see Crystallization). The crystals are separated from and then washed of the mother liquor (the

Table 4. Polymer Gas Permeabilities [a]

Gas	Permeabilities, $(m^3 \cdot m)/(m^2 \cdot s \cdot GPa)$ [b]	
	Cellulose acetate	Polysulfone
NH_3		4
H_2O	75	
H_2	9	9.8
He	11	3.8
N_2O		6.2
H_2S	7.5	2.3
CO_2	4.5	5.2
O_2	0.75	0.83
CO	0.23	0.24
Ar		0.34
CH_4	0.15	0.19
N_2	0.14	0.14
C_2H_6	0.075	
C_2H_4		0.17

[a] Refs. 70, 73.
[b] To convert GPa to bar, multiply by 10^4.

larger the crystals, the less surface area to be washed), and then melted to yield product. The process can be staged but is seldom carried beyond a few stages because of cost and complexity. The most advanced continuous, multistage, countercurrent, commercially available crystallizers give separations equivalent to about five theoretical stages (74).

Crystallization is particularly applicable to a solid–liquid phase diagram of the eutectic type rather than the solid–solution type. Eutectics, however, also limit the degree of achievable separation. With the eutectic diagram, there is a temperature-composition region where the temperature is above the temperature of the eutectic point and the solution composition is richer in the component to be separated than the eutectic. Cooling the solution separates the desired component. This technique is used commercially to recover *p*-xylene from mixtures containing *m*-xylene (Fig. 13) (see also Xylenes and ethylbenzene). Distillation is impractical for this separation because of the very low relative volatilities. Fractional crystallization is feasible since the melting points of the two isomers differ markedly. Displacement regeneration adsorption also is effective for this separation and yields more *p*-xylene than does fractional distillation; therefore, it has been displacing fractional crystallization in this application. Crystallization is widely used industrially for the purification and recovery of inorganic salts and of those organic chemicals and pharmaceuticals that are crystalline solids at room temperature. Crystallization was used for dewaxing lubricating oils but has been displaced by solvent dewaxing in part because of the difficulty of separating the wax crystals from the cold and viscous oil. Similarly, this technique is not used in making fruit-juice concentrate as separation of the ice crystals from the juice concentrate is difficult. As in distillation, mass-addition agents can be used with crystallization to increase its selectivity and efficiency but at the penalty of increased process complexity. Miscible nonsolvents can be added to a solution causing precipitation of a solute. This enables high recovery of solute, frequently at high purities and at room temperature operation but requires a solvent-recovery step. Liquid isomers, eg, xylenes and cresols, can be separated by addition of another component to cause a clathrate or adduct to precipitate. The technique is used in the petroleum industry for manufacturing low pour point oils by urea dewaxing. Eutectics can be separated by adding a third component in extractive crystallization.

Energy requirements. Since heats of fusion are usually small relative to heats of vaporization, energy requirements for crystallization tend to be small. Crystallization that must be done at low temperatures may result in high unit energy costs. In a plant practicing cogeneration, the cost of refrigeration to −12°C is about four times the cost of 690 kPa (100 psi) steam.

Reverse Osmosis. In reverse osmosis, an applied pressure is used to reverse the normal osmotic flow of water across a semipermeable membrane. Although it is, in principle, applicable to other solvents, it has almost exclusively been applied to purification of water from salts and other low molecular weight solutes. High pressures are required to overcome osmotic pressure for even modest solute concentrations, and pumping power requirements and membrane capital costs dictate single-pass operations for most applications. High solute rejections are, therefore, required and can be achieved with a number of membrane materials, including cellulose acetate, aromatic polyamide, and, more recently, thin-film composites made by interfacial polymerization.

The principal application of reverse osmosis is to water desalination. Single-pass

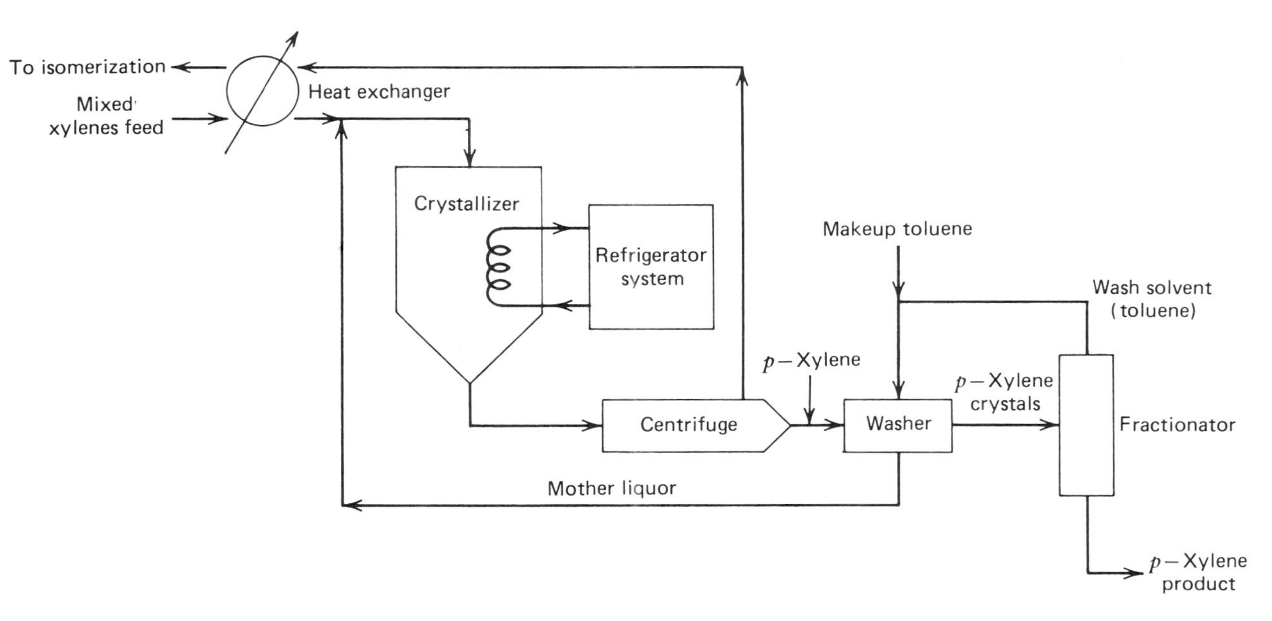

Figure 13. *p*-Xylene manufacture by crystallization.

desalination of seawater is possible and large plants are operating or are scheduled to be built for brackish-water and seawater desalination (66). Other applications range from pollution control to food processing (qv). In these latter applications, tubular or plate-and-frame modules are used because of their greater resistance to fouling, whereas in the desalination applications hollow-fiber and spiral modules predominate. Applications include metals recovery from plating waste; milk, whey, coffee, and maple syrup concentration, and treatment of pulp (qv) and paper (qv) effluents.

Energy requirements. The work required in reverse osmosis is the volumetric flow rate multiplied by the pressure drop across the membrane. Operating pressures for industrial applications are typically 2.8–4.1 MPa (400–600 psi) (75).

Toxicology. Some of the solvents used in separations are hazardous materials that pose health and safety problems, and efforts are being made to find suitable replacements. Benzene, for example, is still used as an azeotropic agent for the dehydration of ethanol and 2-propanol. Methylene chloride is still used for decaffeination of coffee, and chlorinated hydrocarbons in general are widely used as solvents, eg, developers in circuit-board manufacture. An arsenic catalyst is used in one of the acid-gas absorption processes (Giammarco-Vetrocoke). Where such hazardous materials are used, environmental considerations and regulations have led to process changes to reduce emissions and the risk associated with their use.

Guidelines for the Selection of Energy-Conserving Options

For some of the options, criteria for a simple payout time of less than a year are given. These payout times include no provisions for taxes and do not discount cash flows. The payout criteria are stated in terms of process and design variables. For other options, process parameters favoring the use of the options are indicated. These guidelines are intended as a first qualitative test of the merits of an alternate or modified separations process. A detailed analysis is a prerequisite to the determination of the quantitative benefits resulting from such a process change. A comparison of distillation with several other distillation techniques has been reported for the recovery of dimethylformamide from aqueous solution (76).

Control Retrofit. ***To Reduce Reflux.*** It is assumed that the retrofit enables a 20% reduction in the logarithm of the separation factor and that the Gilliland correlation (77) applies. The retrofit is based on representative tower designs, and the use of material balance or combined reflux and distillate control.

The simple payout time for such a retrofit is less than one year if (1):

$$(N\epsilon + 1) \ln \frac{\alpha}{[Q_c \sqrt{k(k-1)} \ln S]} < 0.3 \frac{C_E}{\$2.50} \cdot \frac{\$25{,}000}{\text{cost of retrofit}} \tag{9}$$

Equations 9 and 10 are theoretically based but include a number of simplifying assumptions. The empirical constant 0.3 in these equations is derived from a number of approximate Stanford Research Institute (SRI) tower designs made in the early-to-mid-1970s. These criteria should, therefore, be viewed as qualitative.

To Increase Throughput. The same as above is assumed for the use of a control retrofit to reduce flux.

The payout time for increased throughput is less than one year if (1):

$$(N\epsilon + 1) \ln \frac{\alpha}{[Q_c \sqrt{k(k+1)} \ln S]} < 0.3 \frac{1}{\$2.50}$$

$$\cdot \frac{\$25{,}000}{\text{cost of retrofit}} \text{ PPR} \left[\frac{C_E Q^*}{Q_R} + \frac{Z}{(Q_R - Q_S)}\right] \quad (10)$$

Feedforward Control Retrofit. This assumes that the retrofit enables a 10% energy savings.

The payout time for feed-forward control retrofit is less than one year if (1,7):

$$Q_C > 6 \text{ GJ/h } (\$2.50/C_E)(\text{cost of retrofit}/\$12{,}500) \quad (11)$$

and

$$(\Delta F/F) > 0.1; \text{ or } \Delta X_F/X_F > 0.02 \quad (12)$$

Floating-Pressure Retrofit. This assumes that the retrofit provides a 10% energy savings and the condenser is air-cooled.

The payout time for floating-pressure retrofit is less than one year if (1,6):

$$Q_C > 6 \text{ GJ/h } (\$2.50/C_E)(\text{cost of retrofit}/\$12{,}500) \quad (11)$$

and

$$\partial\alpha/\partial \ln P \le -0.04 \quad (13)$$

Tray Retrofit. ***With Additional or More Efficient Trays to Reduce Reflux.*** The retrofit increases the number of theoretical trays by 20% and the Gilliland correlation to apply; the retrofit is based on representative tower designs.

The simple payout time is less than one year if (1):

$$(N^2\epsilon^2) \ln \frac{\alpha}{[\ln S\sqrt{Pk(k-1)}]} < 150 \frac{C_E}{\$2.50} \cdot \frac{\$1100}{\text{cost of retrofit per m}^2} \quad (14)$$

Equations 14 and 15 are theoretical but include a number of simplifying assumptions, eg, the minimum reflux is large with respect to one and the ratio of ln S to ln α is large with respect to one. The constant 150 in these equations is derived from a number of approximate SRI tower designs. These criteria should, therefore, be considered qualitative.

With Additional or More Efficient Trays to Increase Throughput. The assumption is the same as for the tray retrofit to reduce reflux.

The simple payout time is less than one year if (1):

$$N^2\epsilon^2 \ln \alpha/[\ln S\sqrt{Pk(k-1)}] < \frac{150}{\$2.50}\left(\frac{\$1100}{\text{cost of retrofit per m}^2}\right) \times \text{PPR}\left[\frac{C_E Q^*}{Q_R} + \frac{Z}{(Q_R - Q_S)}\right] \quad (15)$$

With Lower Pressure-Drop Trays to Reduce Product Degradation. The simple payout for such a retrofit is the cost of the retrofit divided by the value of the incremental product. The retrofit can also be used to increase throughput by maintaining a constant pressure drop or to reduce jet ejector steam requirements by maintaining a constant reboiler pressure. In this latter case, with a tray retrofit of $1100/m^2 ($102/ft^2), the simple payout time is less than one year if (1):

$$\ln \frac{P_{SAR}}{P_{SBR}} \ge 0.110 \frac{Nd^2(1.158 \ln P_D - \ln P_{SBR} - 0.545)}{(Q_{MS})(C_E)} \quad (16)$$

Separations Sequencing and Additional Column Draw. Heuristic separation sequence selection factors include the following guidelines (32,78): the most difficult separations should be left until last in a sequence; products should be recovered one at a time, in order of decreasing volatility, in the column overheads; preferred sequences are those that divide the feed to each column equally between the distillate and bottoms flows on a molal basis; and very pure products should be separated last in a sequence.

More specific heuristics for sequencing and side draws for an ideal, three-component mixture ABC, where A is the lowest boiling component and C is the highest, are given by the ease-of-separation-index equations in Table 5 (15).

These heuristics are more likely to be applicable to refinery separations than to chemical separations since, in the former, relative volatilities are likely to be constant and high purities are less frequently required. The first and sixth heuristics in Table 5, although favored for the indicated feed compositions in the preceding analysis, are infrequently used industrially.

Pasteurization is favored if the impurity to be taken overhead is <1% of the feed and <10% of the top product, if the volatility of the impurity relative to that of the light key component is greater than three, and if the following inequalities are true for the desired purities of the overhead impurity X_A and the top product X_D (1):

$$X_D < 10(X_A - 0.9); \quad X_A > 0.1(X_D + 9) \tag{17}$$

Intermediate product removal is favored if a split of one key component between two products is desired and if the overall column temperature decrease exceeds 56°C.

Table 5. Ease-of-Separation-Index Equations[a]

Equations	Feed composition	Heuristic
$ESI = \alpha_{A/B}\alpha_{C/B} < 1.6$	B = 40–80 mol %, A ≃ C	separate feed into AB and BC in first column; remove A,C and C as products from second column
	B ≥ 50 mol %, C < 5 mol %	use single column with B as a side draw below feed
	B ≥ 50 mol %, A < 5 mol %	use single column with B as a side draw above the feed (similar to pasteurization)
	B < 15 mol %, A ≃ C	extract B as distillate product from vapor side draw below feed
	other	remove A or C first (whichever is more plentiful) and then separate other two
ESI ≥ 1.6 (B/C separation is more difficult than A/B separation)	C ≥ 50 mol %	remove C first, then separate A and B
	B > 50 mol %, 5 mol % < C < 20 mol %	separate feed into AB and BC in first column and produce A,B and C in second column
	B > 50 mol %, C < 5 mol %	remove B as a side-stream product below feed
	B > 50 mol %, A < 5 mol %	remove B as a side-stream product above feed
	other	extract B as distillate product from vapor side draw below feed

[a] Ref. 15.

Intermediate impurity side draw is favored if a column is operated at high reflux, eg, if the relative volatility of light to heavy key is less than 1.50, and if a buildup of an intermediate impurity that is less than 1% of the feed is likely.

More Efficient Heat Utilization. ***Heat Pumping.*** Conditions favoring heat pumping are listed in Table 6.

High Heat Flux Tubing and Augmented Heat-Transfer Surfaces. Advanced heat-exchange technology primarily is useful with heat pumping. It is favored, therefore, by the same conditions that favor heat pumping, heat cascading, or both.

Intermediate Heat Exchange. Conditions favoring intermediate heat exchange are listed in Table 7. Intermediate condensers should be considered for columns with one or more side draws, and intermediate reboilers should be considered if the top product is an azeotrope and if excess heat is available at a temperature intermediate between the reboiler and feed temperatures.

Heat Cascading. Conditions favoring heat cascading are listed in Table 8.

Feed/Product Heat Exchange. It is assumed that heat exchanger cost = \$3.92 C_{NSE} (area, m^2)$^{0.65}$ (Nelson steel exchanger cost index); cost of energy = \$2.50/GJ (\$0.60/Gcal); and all heat exchanged represents savings in energy.

The simple payout time for feed/product heat exchange is less than one year if (1):

Table 6. Conditions Favoring Heat Pumping[a]

Condition	External refrigeration	Vapor recompression	Reboiler flashing
overall column ΔT, °C	<36	<36	<36
condenser temperature, °C	<10	<38	<38
reboiler temperature, °C	>150		
column pressure, kPa[b]		<700	>700
$\partial\alpha/\partial \ln P$		<−0.04	<−0.04
compressor efficiency, ϵ_K		$\simeq\lambda_{LK}/C_{PLK}T_R$	$\simeq\lambda_{HK}/C_{PHK}T_R$

[a] Ref. 26.
[b] To convert kPa to psi, multiply by 0.145.

Table 7. Conditions Favoring Intermediate Heat Exchange

Condition	Condensers	Reboilers
reboiler temperature, °C		>150
condenser temperature, °C	<38	
feed composition	$X_{HK} < 0.3$	$X_{LK} < 0.3$

Table 8. Conditions Favoring Heat Cascading

Condition	Separation column	Thermal coupling	Reboiler–condenser coupling
overall column ΔT, °C	56	56	
reboiler heat load, GJ/h[a]	20	40	$Q_{R_i} \simeq Q_{C_j}$
reboiler temperature, °C			$T_{R_i} < T_{C_j} - 10$

[a] To convert J to cal, divide by 4.184.

$$\left[\frac{1}{F_H C_{P_H}} + \frac{1}{F_C C_{P_C}}\right]^{0.35} \Big/ (T_{H_{in}} - T_{C_{in}}) \leq 0.025\,(C_E/\$2.50)(480.5/C_{NSE}) \qquad (18)$$

Distillation with a Mass-Separating Agent. ***Extractive Distillation.*** Extractive distillation must show substantial energy savings relative to conventional distillation to be justified. It has been suggested that, if the relative volatility for a separation by conventional distillation is 1.1, it must be increased to at least 1.5 by extractive distillation to justify its use (79). Extractive distillation is less energy-intensive compared to conventional distillation if the following inequality is satisfied (1):

$$\left[1.2\left(\frac{\alpha_1}{\alpha_1 - 1} + \frac{M + 1 - X_{FLK}}{\alpha_2 - 1} - X_{FLK}\right) + \frac{MC_{P_S}(T_{BS} - T_{BL})}{\lambda_F}\right] \Big/ \left(1.2\left(\frac{1}{\alpha_0 - 1}\right) + X_{FLK}\right) < 1 \qquad (19)$$

This equation is based on the assumption that the feed to the extractive-distillation column is vaporized by heat exchange, with the solvent leaving the solvent recovery column reboiler. More energy is conserved if the solvent can be used to reboil the extractive-distillation column.

Azeotropic Distillation. Determination of the energy savings relative to conventional distillation is based on the following assumptions: no additional energy is required to separate the azeotroping agent from the products; the latent heats of vaporization for the azeotroping agent (λ_A) and for the light key component (λ_L) are equal; and the ratio of actual to minimum reflux is 1.2 for the conventional and azeotropic columns. It follows that the energy saved is, as shown in equation 20 (1):

$$Q_{saved} = 1.2\, F \cdot \lambda_{LK} \left(\frac{1}{\alpha_0 - 1} - \frac{M_A + 1}{\alpha_A - 1}\right) \qquad (20)$$

where F = feed rate, M_A = azeotropic agent recirculation rate divided by column reflux, k = ratio of actual to minimum reflux, and α_A, α_0 = relative volatility of key components in the presence or absence, respectively, of the azeotropic agent. For example, if an azeotroping agent could increase the relative volatility of the key components in a distillation column from 1.10 to 1.20, the azeotropic distillation would require less energy if M_A were less than 4; however, it would consume more energy if M_A were greater than 4.

Nondistillation Separations. ***Absorption.*** Distillation of gases frequently requires refrigeration, ie, it must be done under cryogenic conditions. As a result, an approach similar to that outlined in Table 1 for a C_2 splitter should be used to compare the value of the energy usage rather than the energy usage itself because of the higher cost associated with refrigeration energy.

Liquid–Liquid Extraction. If the relative volatility for a separation by conventional distillation is 1.1, it must be increased to at least 2.0 by liquid–liquid extraction to justify the latter's use (79). It has been shown that extraction requires less energy than distillation if the following two criteria apply (74):

The solvent is less volatile than the products being separated:

$$R_D > (1 + R_E + (1 + R_E')B/D_0)(T_S/T_{SE})(T_{SE} - T_0)/(T_S - T_0) - 1 \qquad (21)$$

Usually, $T_{SE} > T_S$. However, if $T_{SE} = T_S$, equation 21 becomes

$$R_D > R_E + (1 + R_E')B/D_0 \qquad (22)$$

If steam distillation is impractical for recovery of the extract and raffinate from the solvent, the reboiler temperatures for these distillations corresponds to the solvent boiling point at the pressures of the distillations. As the solvent boiling point increases, the relative volatility for these distillations increases and the reflux requirements (R_E and R'_E) decrease. The reboiler temperatures, however, also increase, increasing the sensible heat input into the solvent. Much of this sensible energy input can be recovered by heat exchange against the solvent–extract and solvent–raffinate streams, but the greater the heat exchanged, the larger and more costly the heat exchanger required. For these reasons, a solvent of intermediate volatility often proves desirable.

The solvent is more volatile than the products being separated:

$$R_D > ((1 + R_E)N_S/D_0 + (1 + R'_E)N'_S/B)(\lambda_{VS}/\lambda_V) - 1 \qquad (23)$$

It is generally desirable that the solvent is condensible by water cooling. Since the solvent is often selected to have a boiling point and activity coefficient in the raffinate which permits it to be stripped overhead from the raffinate, thereby obviating the removal of the entire raffinate overhead, a third case to consider is that in which the solvent is less volatile than the extract but can be stripped overhead from the raffinate. If the solubility of the solvent in the raffinate is X_{SRaf} and if the raffinate is assumed to correspond to the distillate bottoms, then the criterion, in this case, for extraction to require less energy than distillation is:

$$R_D > (1 + R_E)(T_{SE} - T_0)(T_S/T_{SE}) + (B/D_0)(X_{SRaf}/(1 - X_{SRaf})(1 + R'_E) \times (T'_{SE} - T_0)/(T_S - T_0)(T_S/T'_{SE})(\lambda_{VS}/\lambda_V) - 1 \qquad (24)$$

Since the solvent concentration in the raffinate is small, it is reasonable to assume that $T'_{SE} = T_S$. Also, if the mutual solubility of the solvent and raffinate is small and the operating reflux for this separation is assumed to be 1.2 times the minimum reflux, then $1 + R'_E$ may be approximated by $1 + 1.2/(P^\circ_{Sol}/P^\circ_{Raf} - X_{SRaf})$.

Equation 24 then simplifies to

$$R_D > (1 + R_E)(T_{SE} - T_0)(T_S/T_{SE}) + (B/D_0)(X_{SRaf}/(1 - X_{SRaf}) \times (1 + 1.2/(P^\circ_{Sol}/P^\circ_{Raf} - X_{SRaf})(\lambda_{VS}/\lambda_V) - 1 \qquad (25)$$

Supercritical-Fluid Extraction. Because of the process complexity of supercritical-fluid extraction relative to conventional distillation, it must offer substantial energy savings to justify its economic use.

Thermal-Swing Adsorption. Thermal-swing adsorption requires less energy than distillation if the following inequality is true (74):

$$R_D > (N_{ads}/D_0)(Q_{TSA}/\lambda_V)(T_S/T_{RG})(T_{RG} - T_0)/(T_S - T_0) - 1 \qquad (26)$$

Pressure-Swing Adsorption, Including Purge-Gas Regeneration. The energy requirements for pressure-swing adsorption are given by equation 8, which may be compared with the work requirements for separation by distillation to determine if pressure-swing adsorption offers an energy advantage relative to distillation. However, the energy requirements for separation of gases by distillation should take into account the higher cost of refrigeration energy. In addition, the comparison varies significantly with the conditions under which the gas to be separated is supplied and the purities and conditions at which the products are desired.

Crystallization. Melt crystallization has a lower energy requirement for separation than distillation if the following applies (73):

$$R_D > (R_C + 1)(\lambda_F/\lambda_V)(B_C/D_0)(T_S/T_{SC})(T_{SC} - T_0)/(T_S - T_0) + (1/\epsilon_P\epsilon_R)(T_S/T_0)(T_0 - T_{RC})/(T_S - T_0) - 1 \quad (27)$$

Reverse Osmosis. Spiral-wound modules, exclusive of pressure vessel, are sold in large quantities for less than ca \$32/m^2 (\$3/ft^2), whereas hollow-fiber modules are sold for less than \$11/m^2 (\$1/ft^2), complete with pressure vessel. Membrane life is typically 2–3 yr or longer. Capital costs for tubular and plate-and-frame modules are higher and, for fouling applications, membrane life is lower. Commercially useful membranes exhibit salt rejections of at least 95% for brackish-water treatment and 98.5% for seawater desalination and they provide water fluxes of at least 0.4 m/d (10 gal/d·ft^2 membrane area) at normal-use pressures (75).

Reverse osmosis (qv) is suited for moderate concentration of dilute solutions of low molecular weight solutes. Higher solute concentrations lead to excessive osmotic pressures.

Nomenclature

AC = analysis (quality) controller
AT = analysis transmitter
B = bottoms product molar flow rate
C_E = cost of energy, \$/GJ
C_P = molar heat capacity (J/mol, °C)
C_{PC} = heat capacity, cold fluid, kJ/(kg·K)
C_{PH} = heat capacity, hot fluid, kJ/(kg·K)
D = distillate flow rate
d = column diameter, m
D_0 = top product molar flow rate
F = flow rate (mol/h)
FC = flow controller
F_C = flow rate, cold fluid, kg/s
F_H = flow rate, hot fluid, kg/s
FT = flow transmitter
k = ratio of actual to minimum reflux
L = reflux flow rate
LC = level controller
M = solvent recirculation rate (mol solvent/mol feed)
N = number of trays in the column
N_{ads} = average rate of adsorption, mol/h
N_S = molal flow rate of solvent in distillation of solvent from extract
N_S' = molal flow rate of solvent in distillation of solvent from raffinate
PPR = annual plant production rate, kg/yr
Q^* = specific reboiler heat requirement, kJ/kg of product
Q = heat duty, GJ/h
Q_c = condenser heat load
Q_R = reboiler heat duty
Q_S = reduction in reboiler heat duty which would have resulted from retrofit if throughput had been held constant
P = column pressure, kPa
P° = vapor pressure
PC = pressure controller
PT = pressure transmitter

R = reflux ratio
R_D = external reflux ratio for conventional distillation (reflux rate divided by overhead product rate)
R_E = reflux ratio for distillation of extract from solvent
R'_E = reflux ratio for distillation of raffinate from solvent
S = separation factor $[(X_{D_{LK}})(X_{B_{LK}})/(X_{D_{HK}})(X_{B_{LK}})]$
SP = set point
T = column temperature, K
$T_{C_{in}}$ = temperature cold fluid entering exchanger, K
$T_{H_{in}}$ = temperature hot fluid entering exchanger, K
V = vapor flow rate
X_A = overhead impurity
X_D = top product
X_F = mole fraction of a feed component
X_{Raf} = mole fraction of raffinate
X_{SRaf} = mole fraction of solvent in raffinate
Z = additional profit ($/kg of additional production
α = relative volatility
α, α_0 = relative volatility for conventional distillation
α_1 = relative volatility in extractive distillation column
α_2 = relative volatility in extractive distillation solvent recovery column
ϵ = Murphree plate efficiency
ϵ_K = compressor efficiency
ϵ_P = power cycle efficiency (relative to a Carnot cycle)
ϵ_R = refrigeration efficiency
ΔF = fluctuation in flow rate
ΔX_F = fluctuation in mol fraction of a feed component
λ = latent heat of vaporization, J/mol
λ_F = latent heat of fusion, J/mol

Subscripts

A = impurity, azeotroping agent
ads = adsorption
BL = boiling point of the light key at the condenser pressure of the extractive-distillation column
BS = boiling point of the solvent at the reboiler pressure of the solvent-recovery column
C = condenser, cold stream, crystallization
D = steam jet ejector discharge, top product
F = feed
H = hot stream
HK = heavy key
i = ith column
j = jth column
LK = light key
MS = jet ejector motive steam
0 = cooling water temperature
R = reboiler
Raf = raffinate
RC = heat sink to the crystallizer
RG = supply temperature of the regeneration gas
S = solvent, steam
SAR = atop column after retrofit
SBR = atop column before retrofit
SC = heat source to the crystallizer
SE = heat source of the solvent–extraction distillation column
SE′ = heat source of the solvent–raffinate distillation column
Sol = solvent

TSA = thermal-swing adsorption
V = extract and raffinate
VS = solvent

BIBLIOGRAPHY

1. T. W. Mix, J. S. Dweck, M. Weinberg, and R. C. Armstrong, *Energy Conservation in Distillation*, NTIS Report No. DOE/CS/40259, National Technical Information Service, Washington, D.C., 1981.
2. J. C. Frank, G. R. Geyer, and H. Kehde, *Chem. Eng. Prog.* **65,** 79 (1969).
3. D. Steinmeyer in *Process Design for Energy Conservation*, AIChE Today Series, American Institute of Chemical Engineers, New York, 1975.
4. G. E. Keller II in E. P. Gyftopoulos, ed., *Industrial Energy-Conservation, Manual 9*, The MIT Press, Cambridge, Mass., 1982, p. 9.
5. P. L. T. Brian, *Staged Cascades in Chemical Processing*, Prentice-Hall, Inc., Englewood Cliffs, N.J., 1972, p. 103.
6. F. G. Shinskey, *Distillation Control for Productivity and Energy Conservation*, McGraw-Hill Book Co., Inc., New York, 1977, p. 163.
7. A. E. Nisenfeld, *Chem. Eng.* **76,** 169 (Oct. 1969).
8. O. Rademaker, J. E. Rijnsdorp and A. Maarleveld, *Dynamics and Control of Continuous Distillation Units*, Elsevier Scientific Publishing Co., Amsterdam, 1975.
9. C. J. Ryskamp, *Hydrocarbon Process.* **60**(6), 51 (1980).
10. F. G. Shinskey, *paper to be presented at the Pacific Area Chemical Engineering Conference*, Seoul, Republic of Korea, May 11, 1983.
11. F. E. Rush, E. I. du Pont de Nemours & Co., Inc., personal communication, 1983.
12. H. Short, *Chem. Eng.* **90**(4), 23 (1983); *Chem. Eng. News*, 26 (March 7, 1983).
13. *Chem. Week*, 30 (Oct. 19, 1983).
14. P. G. Nygren and G. K. S. Connolly, *Chem. Eng. Prog.* **67,** 49 (March 1971).
15. D. W. Tedder and D. F. Rudd, *AIChE J.* **24,** 303 (March 1978).
16. C. S. Robinson and E. R. Gilliland, *Elements of Fractional Distillation*, 4th ed., McGraw-Hill Book Co., Inc., New York, 1950, p. 335.
17. U.S. Pat. 3,371,032 (Feb. 27, 1968), P. A. Witt and J. E. Gantt (to Universal Oil Products Co.).
18. U.S. Pat. 3,522,153 (July 28, 1970), N. B. King (to The Badger Co., Inc.).
19. J. H. Bojnowski in R. B. McBride, ed., *Energy Conservation Opportunities*, Union Carbide Corp., Chemicals and Plastics Division, S. Charleston, W.Va., 1975.
20. C. W. Wolf, D. W. Werle, and E. G. Ragi, *Oil Gas J.*, 8 (Sept. 1, 1975).
21. P. E. Minton in ref. 18.
22. R. W. Maze, *Oil Gas J.*, 74 (Nov. 18, 1974).
23. R. W. Maze, *Oil Gas J.*, 125 (Nov. 25, 1974).
24. R. W. Maze, *Chem. Eng.*, 106 (Jan. 6, 1975).
25. K. N. Watkins, *Petroleum Refinery Distillation*, Gulf Publishing Co., Houston, Texas, 1973, p. 103.
26. H. R. Null, *Chem. Eng. Prog.*, 58 (July 1976).
27. R. Kumar, J. M. Prausnitz, and C. J. King, *Extractive and Azeotropic Distillation, Advances in Chemistry Series 115*, American Chemical Society, Washington, D.C., 1972.
28. T. W. Mix and J. S. Dweck in E. P. Gyftopoulos, ed., *Industrial Energy-Conservation*, Manual 13, The MIT Press, Cambridge, Mass., 1982.
29. J. Coates, *Chem. Eng.*, 121 (May 16, 1960).
30. J. A. Gerster, *Chem. Eng. Prog.* **65,** 42 (Sept. 1969).
31. *Chem. Eng. (London)*, 359 (Sept. 1972).
32. C. J. King, *Separation Processes*, 2nd ed., McGraw-Hill Book Co., Inc., New York, 1980.
33. R. F. Weimer and J. M. Prausnitz, *Hydrocarbon Process.* **44**(9), 237 (1965).
34. J. L. Humphrey and M. Van Winkle, *Ind. Eng. Chem. Process Des. Dev.* **7,** 581 (Oct. 1968).
35. K. Zosel, *Angew. Chem. Int. Ed. Engl.* **17,** 702 (1978).
36. J. A. Gearhart and L. Garwin, *Hydrocarbon Process.*, 125 (1976).
37. D. F. Williams, *Chem. Eng. Sci.* **36,** 1769 (1981).
38. C. A. Irani and E. W. Funk in *Recent Developments in Separation Science*, Vol. III, CRC Press, West Palm Beach, Fla., 1977, Part A, p. 171.
39. G. M. Lukchis, *Chem. Eng.*, 111 (June 11, 1973).

40. G. M. Lukchis, *Chem. Eng.*, 83 (July 9, 1973).
41. *Ibid.*, (Aug. 6, 1973).
42. D. B. Broughton, *Chem. Eng. Prog.* **64,** 60 (Aug. 1968).
43. D. P. Thornton, Jr., *Hydrocarbon Process.* **49,** 151 (Nov. 1970).
44. S. Otani, *Chem. Eng.*, 106 (Sept. 17, 1973).
45. H. L. Brooking and D. C. Walton, *Chem. Eng. (London)*, 13 (Jan. 1972).
46. R. A. Anderson and H. J. Springett, *Natural Gas Processing and Utilization Conference*, *Symp. Ser. No. 44*, Institute of Chemical Engineers, Rugby, UK, 1976, pp. 1-10–1-17.
47. J. J. Collins, *Chem. Eng. Prog.* **64,** 66 (Aug. 1968).
48. N. R. Iammartino, *Chem. Eng.*, 62 (April 28, 1975).
49. M. H. Hainsselin, M. F. Symoniak, and G. R. Cann, *paper presented at the National Petroleum Refiners Association 73rd Annual Meeting*, San Antonio, Texas, March 23–25, 1975.
50. M. F. Symoniak, R. A. Reber, and R. M. Victory, *paper presented at the American Petroleum Institute, Division of Refining Meeting*, Philadelphia, Pa., May 1973.
51. M. F. Symoniak, R. A. Reber, and R. M. Victory, *Hydrocarbon Process.* **52,** 101 (May 1973).
52. M. F. Symoniak and A. C. Frost, *Oil Gas J.*, 76 (March 15, 1971).
53. F. Schmeling and H. Heneka, *Oil Gas J.*, 1 (1969).
54. R. C. Ewing, *Oil Gas J.*, 1 (April 13, 1970).
55. G. C. Roy, J. W. Myers, and D. L. Ripley, *Hydrocarbon Process.* **53,** 141 (Jan. 1974).
56. K. J. Doshi, C. H. Katira, and H. A. Stewart, *AIChE Symp. Ser.* **67,** 19 (1971).
57. H. A. Stewart and J. L. Heck, *Chem. Eng. Prog.* **65,** 78 (Sept. 1969).
58. J. L. Wagner and H. A. Stewart, *paper presented at the Novel Separation Systems Symposium, 3rd Joint Meeting, I.I.Q.P.R. and American Institute of Chemical Engineers*, San Juan, Puerto Rico, May 17–20, 1970.
59. H. J. Bieser and G. R. Winter, *Oil Gas J.*, 74 (Aug. 11, 1974).
60. R. S. Atkins, *Hydrocarbon Process.* **49,** 127(Nov. 1970).
61. C. V. Berger, *Hydrocarbon Process.* **52,** 173 (Sept. 1973).
62. G. E. Keller II and R. L. Jones in W. H. Frank, ed., *Adsorption Separation of Gas Streams*, Symposium Series #135, American Chemical Society, Washington, D.C., 1980, pp. 275–286.
63. D. R. Garg and J. P. Ausikaitis, *paper presented at the AIChE Annual Meeting*, Los Angeles, Calif., Nov. 14–19, 1982.
64. M. F. Symoniak, Union Carbide Corporation, personal communication, 1976.
65. J. Grebbel, *Oil Gas J.*, 85 (Aug. 14, 1975).
66. H. K. Lonsdale, *J. Membr. Science* **10,** 81 (1982).
67. S. T. Hwang, K. H. Yuen, and J. M. Thorman, *Sep. Sci. Technol.* **15,** 1069 (1980).
68. W. A. Bollinger, D. L. MacLean, and R. S. Narayan, *Chem. Eng. Prog.* **78,** 27 (Oct. 1982).
69. W. J. Schell and C. D. Houston, *Chem. Eng. Prog.* **78,** 33 (Oct. 1982).
70. W. H. Mazur and M. C. Chan, *Chem. Eng. Prog.* **78,** 38 (Oct. 1982).
71. A. B. Coady and J. A. Davis, *Chem. Eng. Prog.* **78,** 44 (Oct. 1982).
72. J. M. S. Henis and M. K. Tripodi, *J. Membr. Sci.* **8,** 233 (1981).
73. U.S. Pat. 4,230,463 (Oct. 28, 1980), J. M. S. Henis and M. K. Tripodi (to Monsanto Company).
74. H. R. Null, *Chem. Eng. Prog.* **76,** 42 (1980).
75. J. E. Cadotte, R. S. King, R. J. Majerle, and R. J. Petersen, *J. Macromol. Sci. Chem.* **A15,** 727 (1981).
76. F. E. Rush, *Chem. Eng. Prog.* **76,** 44 (1980).
77. Ref. 16, pp. 348–349.
78. D. F. Rudd, J. E. Hendry, and J. D. Seader, *AIChE J.* **19,** 1 (1973).
79. M. Souders, *Chem. Eng. Prog.* **60,** 75 (Feb. 1964).

TOM MIX
Merix Corporation

SIZE SEPARATION

Size separation is the parceling of particulate material on the basis of size. Size separation is an important industrial unit operation which produces coarser and finer streams from a feed stream on a continuous basis in a single stage. Multiple stages of size-separation devices can be arranged to produce multiple streams of differing degrees of fineness. Multiple-staging (multiple stages combined with blending) is also practiced to produce two streams, the same as a single-stage separation.

The various types of size separation devices fall into two general categories: those that separate by forces of fluid dynamics; and those involving the probability of passing through an aperture. Fluid-dynamic size separation takes advantage of the differences in rates of travel, of particles in a fluid, due to differences in particle size. Probability size separation is based on the repeated presentations of particles to uniformly sized apertures in screens.

A method of presenting particle size data is the cumulative fraction (or percent) finer than (or coarser than) some size. However, the precise definition of particle size is very difficult (1). It is usually defined on the basis of a sieve analysis, or the ability of a particle to pass through a standardized square aperture. Obviously, even if a standard method of defining size is employed (nonsieving techniques can be corrected to an equivalent sieve size by utilizing a proper shape factor), the partitioning achieved by employing fluid-dynamic differences must also be attributed to other particle properties such as specific gravity or shape. Size separation implies that these properties are relatively constant among the particles, and that the size is the basis of the partitioning. If, on the other hand, there are differences among several properties that influence the separation, then the interpretation of the separation is not possible because it is sorting (see Gravity concentration). In order to interpret the size separation, the particles must be grouped into subpopulations of common specific gravity or shape.

Screening devices are used to make coarser separations; fine products having 95% passing ca 100 mm–50-μm size. Dry-screening devices have a lower recommended size of ca 500 μm. Wet-screening devices that produce 95% passing ca 500–50-μm size are continually being improved.

Fluid-dynamic separating devices are used to make fine separations; fine products having 95% passing ca 1000–1-μm size. However, different devices have different separating ranges. For example, hydraulic-settling classifiers produce fine streams in the range of 95% passing ca 1000–100-μm size. Hydraulic-cyclone classifiers, or hydrocyclones, produce fine streams in the range of 95% passing ca 500–5-μm size.

Generally, as the product size becomes finer, the capacity of the separating device decreases. Thus, there are devices that can be fed hundreds of metric tons per hour (MTPH) and produce 95% passing 50 μm, but a device which produces 95% passing 5 μm may have a capacity of ca 1 MTPH.

Evaluation of Size Separations

Before the separating devices are examined, it is helpful to understand how they are evaluated. The mass-flow relationships of a size separation are usually expressed on a relative basis rather than an absolute basis and have been assigned unique names.

For example, the ratio of the product stream to the feed stream is known as the yield. The ratio of the coarse stream to the fine stream is known as the circulation ratio, and the ratio of the feed stream to the fine stream is known as the circulation load, which is equal to 1 + circulation ratio. The yield is the same as the inverse of the circulation load.

Recovery. These ratios are not good indicators of the quality of the separation. Recovery (the ratio of the amount of material less than some size in the product stream to the amount of that material in the feed stream) is used as a measure of the quality of the separation. The amount of material less than some size in a stream is the product of the cumulative fraction less than the size and the mass flow rate of that stream. However, the recovery can be calculated as the cumulative percent less than the size in the product stream times the yield divided by the cumulative percent less than the size in the feed stream.

Recovery has been used as an expression of the efficiency of the separation, if the calculation is performed using the size at which the separation was made (often termed the cut size). However, the definition of the cut size is not always a simple one. In a perfect separation, the cut size is the size at which the cumulative fraction less than in the feed stream is equal to the yield (2). If this cut size definition is used, then the recovery efficiency is simply equal to the cumulative fraction less than the cut size in the product stream. However, a definition of separation efficiency only accepts recovery as an expression of separation efficiency if the quantity of material greater than the cut size in the product stream equals zero (3). Square-aperture screening is probably the only size separation process which could meet this condition.

Separation Efficiency. Classification efficiency is defined as a corrected recovery efficiency, ie, the recovery efficiency minus the ratio of the amount of material greater than the cut size in the product stream to the amount of material greater than the cut size in the feed. This correction is calculated as the cumulative percent greater than the cut size in the product stream times the yield divided by the cumulative percent greater than the cut size in the feed stream (3). Classification efficiency is the same as separation efficiency. The feed size distribution should be matched to the desired product size distribution; the feed should be at least 50% less than the cut size (4). The cut size is the 95% passing size in the product stream. The yield is probably $\frac{1}{3}$ for this rule of thumb. Therefore, it follows that the recovery efficiency would be:

$$\frac{95\%}{50\%} \times \frac{1}{3} = 63\tfrac{1}{3}\%$$

The classification efficiency would be:

$$\left[63\tfrac{1}{3}\% - \frac{5\%}{50\%} \times \frac{1}{3}\right] = 60\%$$

Quantitative Efficiency. Quantitative efficiency is the ratio of the sum of the amount of material less than the cut size in the product stream plus the amount of material greater than the cut size in the other stream to the feed rate (5). This calculation is made as the cumulative percent less than the cut size in the product stream times the yield plus the cumulative percent greater than the cut size in the other stream times the quantity (1 − yield). This expression is not a true efficiency (3). If the definition for the cut size is the size at which the cumulative percent less than in the product stream is equal to the cumulative percent greater than in the other stream

(6), the quantitative efficiency reduces to simply the cumulative percent less than the cut size in the product stream. The screening efficiency equation (7) for rectangular-hole screening is the same as the quantitative efficiency equation (5) for classification, and hence is also not a true efficiency. Most efficiency equations for size separation reduce to either recovery, classification efficiency, or quantitative efficiency.

Fractional Recovery. A major disadvantage of the separation efficiency is that it is a dependent evaluation criterion; by definition it depends upon the feed properties and hence cannot be used to compare different operating situations unless the same feed is used. Therefore, the AIChE selected an independent criterion, based upon the fractional recovery of material to the coarse stream, to evaluate size separations (8). The fractional recovery is the ratio of the amount of material in a narrow size fraction in the coarse stream, to the amount of material in that same narrow size fraction in the feed stream. The methodology plots the fractional recovery values for many different narrow size fractions against the upper size of the size interval and then uses the resulting curve to characterize the separation. Although the narrow size intervals can be a differential quantity (graphical differentiation of the cumulative fraction less than curves), an easier and equally valid approach is to expand the size interval from the differential quantity (a point value) to a geometric quantity (an interval value) using a ratio of $\sqrt{2}$ in establishing the range of the size interval (9).

The resulting curve is characteristic of the separation and can therefore be used to evaluate the separation. Consider the size-separation data given in Table 1; the nonsieve entries have been interpolated from actual size-distribution data. Interval percent values for the three streams are calculated by subtracting the cumulative percentage less than the lower size of the interval, from the cumulative percent less than the upper size of the interval, except for the sink size interval, as shown in Table 2. Now the circulation ratio can be determined from material balances as:

$$\frac{\text{cum \% < in the fine stream} - \text{cum \% < in the feed stream}}{\text{cum \% < in the feed stream} - \text{cum \% < in the coarse stream}}$$

Statistically, the size to be used for the circulation ratio calculation is the largest size for which the fine-material-percentage value is greater than the feed-interval material and the feed-material-percentage value is greater than the coarse interval material.

Table 1. Cumulative Size Distributions of Streams from a Size Separator, %

Sieve (mesh)	Size, μm	Feed stream	Coarse stream	Fine stream
12	1700	100.0	100.0	100.0
	1180	99.9	99.85	100.0
20	850	99.75	99.65	100.0
	600	98.5	98.0	100.0
40	425	94.0	92.0	100.0
	300	82.5	76.6	99.8
70	212	66.5	55.5	98.9
	150	50.0	34.75	95.0
140	106	35.0	18.8	82.8
	75	24.0	10.7	63.2
270	53	16.0	6.7	43.5
	38	10.5	4.25	28.9

Table 2. Interval Size Distributions of Streams from a Size Separator, %

Top size, μm of interval	Feed stream	Coarse stream	Fine stream
1180	0.15	0.2	0
850	1.25	1.65	0
600	4.5	6.0	0
425	11.5	15.4	0.2
300	16.0	21.1	0.9
212	16.5	20.75	3.9
150	15.0	15.95	12.2
106	11.0	8.1	19.6
75	8.0	4.0	19.7
53	5.5	2.45	14.6
38	10.5	4.25	28.9

If the former condition is met but the latter is not, then the calculation is an approximation, but is still recommended. For the data presented in Tables 1 and 2, the size is 106 μm (140 mesh), giving a circulation ratio of:

$$\frac{82.8 - 35.0}{35.0 - 18.8} = 2.95$$

The fractional recovery values for each size interval are calculated as:

$$\frac{\text{interval coarse value (circulation ratio)}}{\text{interval coarse value (circulation ratio)} + \text{interval fine value}}$$

The denominator represents the reconstituting of the feed stream in order to ensure a material balance. The values are given in Table 3. Plotting these values gives the curve presented in Figure 1.

The first thing that is obvious in Figure 1 is that the curve does not go to zero; there is fine material reporting to the coarse stream. This is typical of many size separation devices and can be characterized as an apparent by-passing of the feed material to

Table 3. Interval Fractional Recovery and Corrected Fractional Recovery Values

Top size, μm of interval	Fractional recovery	Corrected fractional recovery
1180	1.00	1.00
850	1.00	1.00
600	1.00	1.00
425	0.996	0.993
300	0.986	0.980
212	0.940	0.914
150	0.794	0.706
106	0.549	0.356
75	0.375	0.107
53	0.330	0.043
38	0.303	

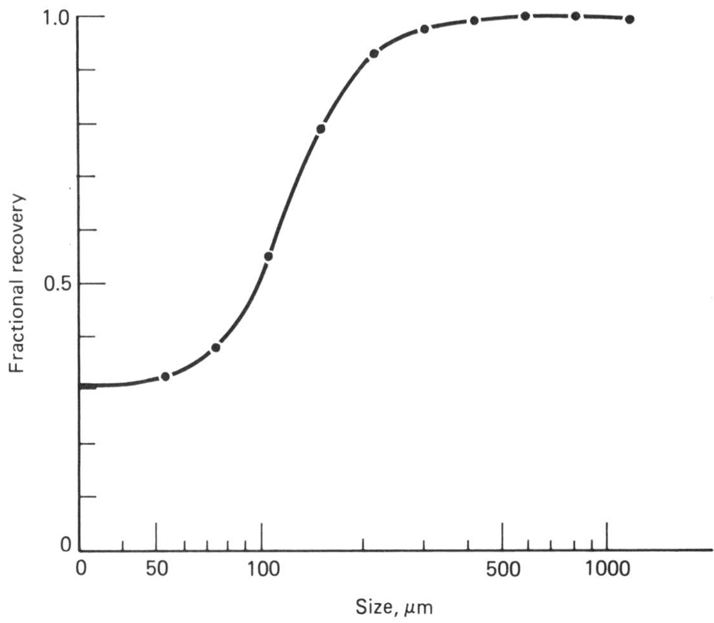

Figure 1. Fractional recovery curve.

the coarse stream. The same phenomenon can occur at the other end of the curve, but this is not typical since it represents apparent by-passing of the feed material to the fine stream. This occurrence is viewed as a device malfunction and therefore must be corrected. The apparent by passing of feed material to the coarse stream must be estimated and the fractional recovery values corrected in order to continue the evaluation. The actual values are corrected as:

$$\frac{\text{fractional recovery value} - \text{apparent bypass fraction}}{1 - \text{apparent bypass fraction}}$$

By assuming an apparent bypass value of 0.3 the values in Table 3 can be corrected as shown. These corrected values are plotted in Figure 2. From the corrected plot, two parameters can be determined that characterize the separation: the equiprobable size and the sharpness index. The equiprobable size, called by many, the cut size, is the size for which the corrected value is 50% (122.5 μm for the example in Figure 2). It is quite often assigned the symbol, d_{50}. The sharpness index is the ratio of the size for which the correct value is 25% to the size for which the corrected value is 75% (95 μm/158 μm or 0.6 for the example in Fig. 2). This ratio varies between 0 (terrible) and 1 (perfect).

These three parameters—the apparent bypass, the cut size, and the sharpness index—are used to evaluate the separation. These can be estimated, if a functional form such as the logistic function (see eq. 6) is assumed for the corrected curve, by employing nonlinear parameters estimation schemes. The process can be reversed and the separation predicted if these three parameters are known (see Prediction of Size Separation).

Separation Inefficiency. The apparent bypass value does not affect the size consist of the fine stream. It does affect the size consist of the coarse stream. It also increases the circulation ratio, and hence decreases the yield. The maximum recovery efficiency

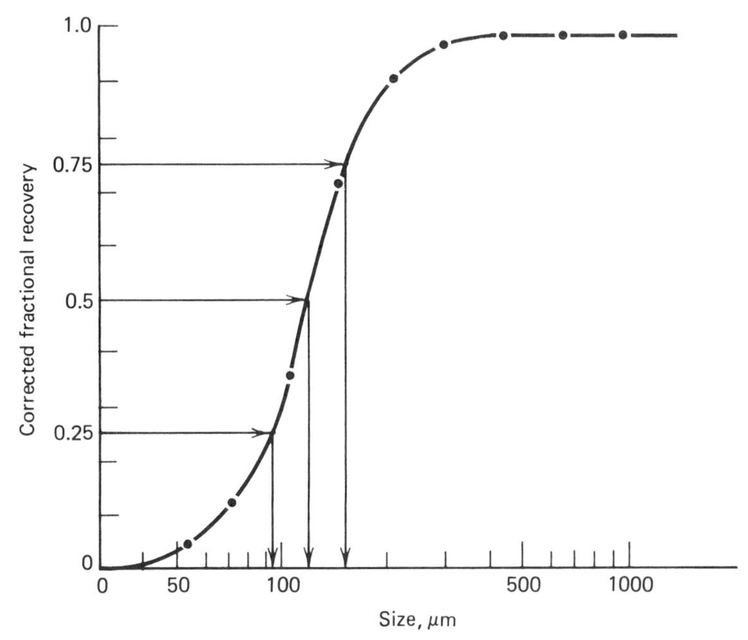

Figure 2. Corrected fractional recovery curve.

under perfect conditions (sharpness index equal to 1) is equal to one minus the apparent bypass value. Thus, if the apparent bypass value is 0.3, the maximum recovery efficiency is 70%. The recovery efficiency continues to decrease with decreasing sharpness-index values.

The 95% passing size of the fine stream is proportional to the d_{50} value. However, the proportionality constant varies with the sharpness-index.

Sharpness index	0.3	0.4	0.5	0.6	0.7	0.8	0.9
Proportionality constant	6.45	4.2	2.5	1.65	1.3	1.1	1.0

Thus, in order to produce a fine stream analyzing 95% passing 250 μm with a classifier whose sharpness index is 0.6, it must have a d_{50} value of 150 μm. However, a device with a sharpness index of 0.4 requires a d_{50} value of 60 μm.

Screening

Screening is a process whereby particles are presented in an appropriate manner to a series of apertures that allow undersize particles to pass through into the underflow and oversize particles to be retained in the overflow. In addition to sizing, screens are used for desliming (the removal of very small particles from much larger ones by draining and/or rinsing), dewatering (qv) (the removal of water from particles by draining) and prewetting (the addition of water to the particles by spraying).

By deriving a simple theory which views the screening process as a rate process and then examining the simple model, some observation can be made concerning the rate at which particles pass through the apertures. The rate depends upon: the open area of the screen, the total area of the screen, the quantity of material on the screen, the presentation per aperture·second, the density and shape of the material, and the fraction of oversize and near-size particles on the screen. The last factor makes use of the three classes of particles, ie, large, near-size, and small, defined by the relation of their size to the aperture size. (Near-size particles are those whose sizes are <1.5 times and >0.5 times the aperture size.) Hence, a small particle for one size screen may be a near-size particle for a smaller size screen. The large particles never pass through the screen, whereas the small particles pass through the screen very rapidly. It is the near-size particles that control the process since these are the particles that, if properly aligned, can pass through the aperture but not through in other alignments. In order to achieve quick, accurate screening results: the number of presentations per opening·second must be large (if the material being screened exhibits integrity); the amount of oversize material on the screen must be reduced (which can be achieved by stacking decks of screens); and the amount of material on the screen must be reduced as the aperture openings are reduced. By analyzing the potential of near-size particles to lodge in the aperture, thereby blocking the screen, it is also obvious that the amount of material on the screen must be reduced as the screen-aperture opening is reduced. The tendency to block a screen can be expected to differ with type of material.

Screening devices can have stationary or moving-screen decks. Moving screen decks can be designed as rotating cylinders (trommels) or vibrating surfaces. By far, the most popular screen-deck design for sizing is the inclined vibrating screen (Fig. 3). Vibration is produced on inclined screen decks by circular motion in a vertical plane of 3–12.5 mm amplitude at 700–1000 cycles per minute; newer high speed screening designs operate at 3600 cycles per minute. The vibration lifts the material, producing stratification. Stratification places the larger particles on top, and the smaller particles on the bottom, next to the apertures. With the screen deck on an incline, the material cascades down the slope, introducing the probability that the particle will either pass through the opening or over the screen surface. For horizontal screen decks, the motion must be capable of conveying the material without the assistance of gravity. Straight-line motion at an angle of approximately 45° to the horizontal produces a lifting component for stratification and a conveying component for probability of separation as the material passes across the horizontal screen surface.

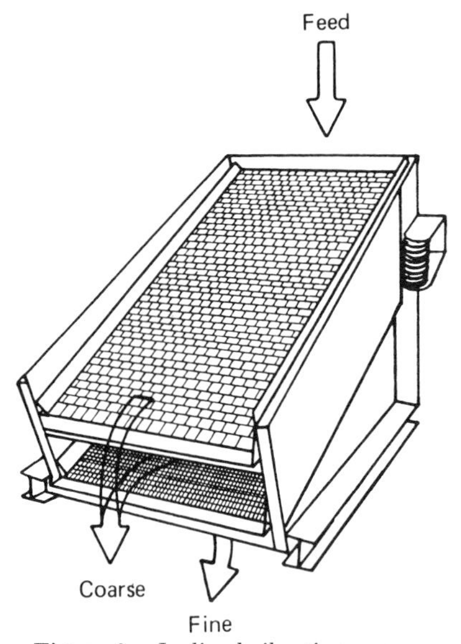

Figure 3. Inclined vibrating screen.

The speed, direction of rotation, and slope affect the binding of the screen with near-size particles, but the stroke is the predominant factor affecting the ejection of blinding particles. Generally, the stroke is reduced as the screen size decreases, and the speed increases.

The orientation of large particles on top of the bed prevents the bouncing around of the small particles which would reduce their exposure to the aperture. This also helps push near-size particles through the apertures, reducing blinding. Efficient sizing (the ability to remove undersize material from a given feed) increases with feed rate, passing through a maximum, then decreases, since the bed thickness increases with feed rate. There is an optimal bed thickness on the screen deck for efficient continuous sizing. Consequently, for a given rate of feed, the screen width is selected to control the depth of the bed at the discharge end; the feed end bed depth can be greater, particularly if the feed contains a large percentage of fines since the very fine particles are removed readily in the first few meters. Hence, the screen width sets the capacity, and the screen length establishes efficiency. The desired bed depth at the discharge of the screen is 3 (for 800 kg/m^3 material) to 4 (for 1600 kg/m^3 material) times the screen opening. It is calculated by dividing the volume of material being fed per unit time per unit width by the material travel velocity, which increases with an increase in the angle of inclination. Good design practice is that the minimum length to width ratio for screen decks should be 2:1.

The formula for selecting screen-deck size determines the area of screening surface needed to remove the undersize material from the feed before it is discharged from the end of the screen surface. This is the net-screening area, equal to the width times the length of the screen, less the deck parts (hardware) that reduce the available opening for screening such as clamp bars, hold-down bars, etc, and less the loss of available screening area owing to particles passing from one deck to another on a double- or triple-deck screen arrangement. If the required area exceeds the largest screen deck available (2.4 m × 7.4 m), then multiple units are used to provide the equivalent area.

The sizing procedure starts by dividing the quantity of material by the capacity of one square unit of the screen for a standard gradation of the type of material. The quantity of material can be the feed, the amount of overflow material or the amount of underflow material; the capacity factor must match the particular quantity selected. It should include any recirculated or surge material. The capacity factor is for a standard aperture geometry, percent open area, efficiency, bulk density, and dry screening operation. For example, the capacity factor [metric tons per hour (MTPH) passing through 1 m^2 of screen cloth] for different materials and aperture openings is (10):

Square opening	100 mm	75 mm	50 mm	40 mm	25 mm	20 mm	15 mm	6.5 mm	3 mm
Sand and gravel	69	55	44	38	32	29	25	14.5	9
Crushed stone	50	45	36	32	26	24	18.5	12	6
Coal	43	33	26	23	19.5	17	14	7	1.5

The estimated screen area is modified (by dividing the appropriate factors) for deviation from the standard conditions as described below.

Undersize (percent of material less than ½ of the screen opening in size) in the feed is expected to be 40%. As the percentage of fines in the feed increases, the deck area required is reduced. If the percentage of fines is less, than the screening surface required increases. The undersize factors are (10):

100%	90%	80%	70%	60%	50%	40%	30%	20%	10%
2.0	1.9	1.8	1.6	1.4	1.2	1	0.8	0.65	0.55

Oversize (percent of material greater than the screen opening in size) in the feed is expected to be 25%. As the percentage of oversize in the feed increases, the screening surface required increases. The oversize factors are (10):

95%	90%	80%	70%	60%	50%	40%	30%	20%	10%
0.4	0.55	0.7	0.8	0.85	0.9	0.95	1	1.05	1.10

Efficiency (the ability to remove undersize material from a given feed) is expected to be "commercially perfect" or 95%. As the efficiency desired is decreased (it can't be increased) the screening surface required decreases. The efficiency factors are (10):

95%	90%	85%	80%	75%	70%	60%
1	1.1	1.25	1.4	1.5	1.7	2.1

Wet screening (spraying the material on the screen deck with water in order to remove and screen the finer-size particles) requires greater screening surface for screen sizes <50 mm (this trend reverses <3 mm openings). The number of sprays required is determined by the total water necessary for good screening [10–25 L/min (LPM) per cubic meter per hour of feed, ie, L·h/(min·m^3), dimensionless] and by covering the screen with a curtain of water from side-to-side. The sprays are adjusted so that the water strikes the material at an angle of 15° and are staggered so that adjacent sprays do not strike each other. Water pressure to the sprays is 200–400 kPa (29–58 psi) (normally 300 kPa or ca 45 psi). In practice, spray water is also added in the feed chute immediately before the screen. The wet screening factors are (10):

Square opening	25 mm	20 mm	15 mm	6.5 mm	3 mm
	1.1	1.2	1.3	2.25	2.5

Percent open area of the screen is a function of the screen opening. A standard screen, hence open area, is used when determining the capacity factor. As the open area of the screen increases, the required screening surface decreases. Screen blinding is the equivalent of reducing the open area, hence increasing the required screening surface.

Aperture geometry (the shape of the screen openings) decreases the required screening surface for rectangular openings, and increases it for circular openings. However, any decrease may be offset by an increase owing to a lessening in the percent open area of the rectangular slot screen deck. For wire screening with the same small dimension and wire thickness, the percent open area increases when going from square to rectangular (2:1) to parallel rod geometry. Open area must be balanced against the wire diameter that gives adequate lift. Screen decks can be woven wire screens (including piano wire), perforated stock or cast. Materials used range from ferrous and nonferrous alloys to rubber and plastic. Screen decks made of rubber or plastic supposedly reduce blinding because they flex. Unique woven wire geometry supposedly reduces blinding. The addition of bouncing balls underneath the deck can also be used to reduce blinding. Air sprays similar in concept to water sprays are also used to reduce blinding.

Data for dry screening on a 20 mm square-aperture vibrating screen (11) indicate that the screen is relatively efficient, giving an apparent bypass value of 0.5%, sharpness index of 0.8, and a cut size of 20 mm. However, data for smaller apertures (½ mm) (12) indicate that the cut size does not reach the opening size (it is smaller) and that good sharpness indexes are only achieved after long screening times (long screen lengths). One manufacturer (Allis-Chalmers Corp., Appleton, Wisc.) does not recommend dry screening with apertures smaller than ½ mm.

Recently, there has been increased interest in screening slurries with small aperture (100-μm) vibrating screens. Mixed results have been reported (13). The ratio of cut size to aperture size decreases from 0.75 to 0.5 with increasing pulp density. The apparent by pass value increased with increasing feed rate and pulp density. However, other results (14) indicate that the type of screen cloth can affect these results, when going from a square to an elongated rectangular aperture.

There is a family of stationary screens known as cross-flow screens, used mainly for dewatering or desliming, that have been applied to slurry sizing. The original member was the sieve bend. These screens are characterized by a slotted deck, usually made of stainless-steel profile wire and set at an angle. The slurry flow is at right angles to the slots and depending upon the angle of inclination, the cut size value is ½–⅔ of the slot. Hence, it is possible to size at ½ mm using a 1 mm slot, which should reduce blinding problems. However, as the slot width is reduced, the apparent bypass value increases, approaching 100% for a slot width of 40 μm. Apparently this can be substantially reduced (apparent bypasses approaching 5%) by vibrating the screen deck (15).

Other approaches to reducing the blinding effect have included the development of so-called probability screens. Screen blinding led to the commercialization of a simple principle. The projected area of a rectangular aperture, normal to the horizontal, decreases as the angle of inclination of the screen deck containing the aperture increases. Thus, the probability of passage for a particle falling normal to the horizontal, which would normally pass through the aperture, decreases with an increasing angle of inclination. The probability of a particle lodging in the opening is minimal.

The commercial sizer (Fig. 4) consists of five superimposed screen decks each with a slightly steeper slope than the one above. The screen apertures are no less than twice the desired separation size. Two out-of-balance motors transmit a linear vibration to each deck and the resulting action fluidizes the feed so the particles are presented essentially normal to the horizontal. This design can achieve separation sizes of 50 mm down to 100 μm at high capacity with minimal blinding compared to conventional screens.

A recent screening design (Fig. 5), developed to handle feed materials with extensive damp, clayey fines content, is the rotating probability screen (Ropro). It was developed by the National Coal Board, UK, (16) after rejecting electrically heated decks, piano wires, loose rod decks, air sprays, bouncing balls, centrifugal screening, and flip-flop decks made from polymethane which stretched, relaxed, and fluidized the material to be screened.

The "screen" is created by fitting stainless steel rods to a vertical rotating shaft. The rods, radiating from the central hub, create a horizontal circular surface. The apertures between the radial elements of the screen progressively enlarge and the rods have no supporting members. These apertures are larger than the feed.

If the deck is stationary, all the feed material passes through the apertures; if the deck is rotated at high speeds, none of the feed material passes through the apertures. Thus, by controlling the speed of rotation, the screen size is regulated. Hence, the variable rotation speed creates a variable-aperture screen. Although no blinding is

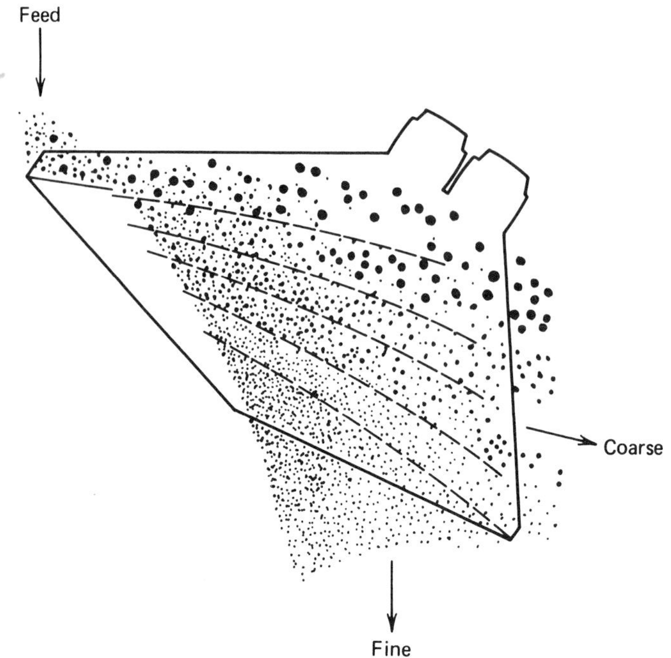

Figure 4. Hi-probe sizer.

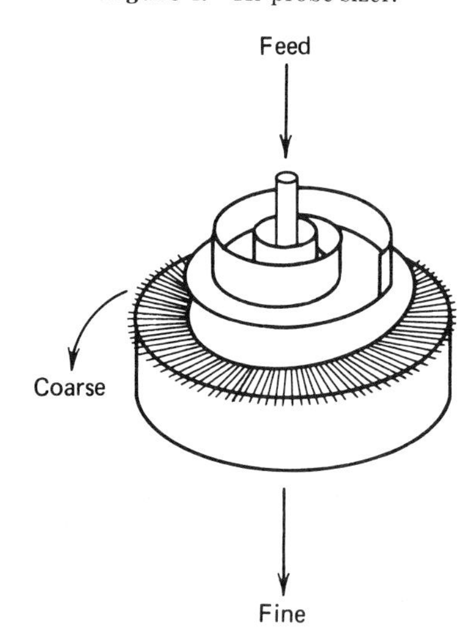

Figure 5. Rotating probability screen (Ropro).

claimed for this design, no screening efficiency data (the amount of fine material that stays with the coarse stream) are given (16). An upper deck, designed to scalp the feed to the lower deck, and a rotating table product-collecting system are recent modifications.

The approximate cost of many of the size-separation devices follows the form:

$$\$ = a(X)^b \tag{1}$$

where X is a suitable parameter and a and b are appropriate coefficients (17). The cost, in U.S. dollars, can be updated by cost-index-ratio adjustment since the dollars are for a Marshall and Swift mining and a milling industry index of 800 (17).

For vibrating screens, the suitable parameter is the screen deck area multiplied by the length in m^3. The appropriate coefficients are

Vibrating screens

X, m^3	a	b
single deck 0.5–45	7673.15	0.4069
double deck 0.5–45	8734.80	0.4256
triple deck 1.5–45	8999.92	0.4908

Cost includes drive- and feed-boxes, and excludes motor, starter, and screen cloth.

Hydraulic Classification

The settling-pool group of classifiers (Fig. 6) consists of a rectangular tank with an inclined floor [25 (min)–35 (max) cm/m, ie, ca 14–20°] that creates a pool. Feed slurry is introduced at the side of the tank and the overflow of fine particles and water exits through an overflow weir and box arrangement. The coarse particles settle to the bottom and are discharged over the upper edge of the tank after being dragged up the inclined floor by some mechanism, such as intermittently operating rakes, continuously operating drag conveyor, or continuously operating spirals.

The length of time the particles stay in the pool determines the distance they settle in the pool. Thus, the feed entrance must be located so that the velocity of the pulp toward the weir, together with the distance, allows sufficient time for the fine particles to be carried out over the weir as the coarse particles settle out below. It is assumed that if a particle settles more than twice the overflow weir crest, it will settle out. The coarse particles are agitated and washed as they are conveyed up the inclined floor, thus reducing the amount of undersize particles carried out in the coarse stream. The pulp velocity (hence time in the pool) can be varied by altering the weir geometry, the total volume flow, or the solids' concentration. All are basically changing the mean residence time (the ratio of the mass of holdup material to the feed rate) by varying the mass of holdup material or the feed rate. However, some changes may force the

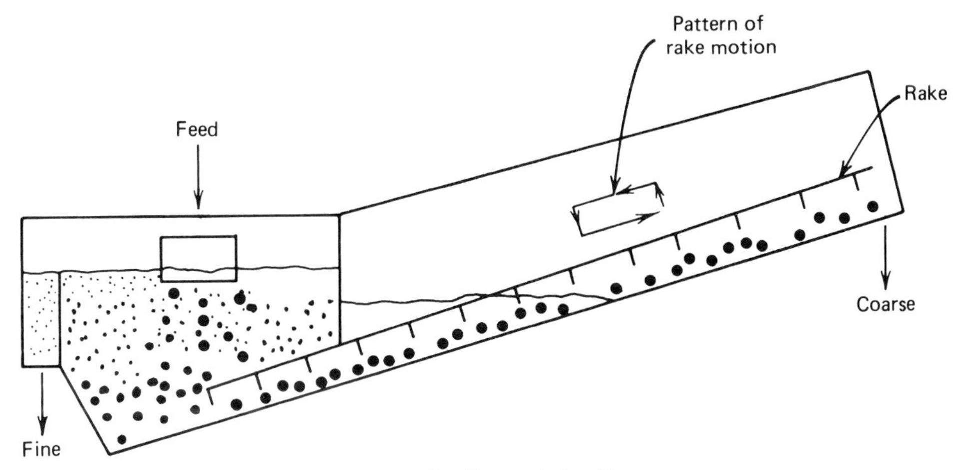

Figure 6. Settling-pool classifier.

pseudo-free settling conditions more towards hindered settling conditions, thus giving an unexpected result.

The separation mechanism of a settling pool should be very straightforward. Consider a cylinder of slurry made up of particles that can settle independently of one another under the influence of gravity (pseudo-free settling). If a cylinder is shaken initially so that all the particles are uniformly distributed throughout the suspension, all the particles immediately start to settle at rates determined by their size, shape, density, and other factors. In this exemplification, the upper half of the cylinder contains particles given to the overflow whereas the lower half contains particles given to the underflow. There is a certain size of particles that just settles from the top to the level at which the overflow and underflow are partitioned. The underflow contains: all particles having settling rates greater than the certain size; particles having settling rates less than the certain size because they were near the level between the overflow and underflow; and particles of all settling rates which were originally distributed into the underflow (so-called void filling material which represents apparent bypassing for this type of classifier).

If the movement of particles in a continuous classifier is viewed as simple-settling while moving uniformly across the pool (plug flow), then the same type of separation should be achieved. Such a model has been termed pulp-partitioning (18); the fractional recovery values exhibit an apparent bypass value and a relatively high sharpness index. If, on the other hand, there is mixing of the particles caused by turbulence, then the appropriate model is termed pulp-tapping (18) and the sharpness index is much lower and the d_{50} value is higher. Reality appears to lie between these two models.

The results for a DSF Dorr Classifier (1.8 m × 7 m), operating at 19 strokes per minute with a weir depth of 100 cm and a slope 19.4 cm/m, are given in reference 19. The set of fractional-recovery corrected values, when plotted, produce characteristic parameters of cut size equal to 240 μm, a sharpness index of 0.50 and an apparent bypass of 26%. However, the data are not sufficiently complete for accurate analysis. The corrected values are the same as settling factors. The settling factors have been estimated by assuming that the probability of particles in a size-distribution reporting to the underflow is equal to the ratio of their settling velocities. The settling velocity for particles <75 μm (−200 mesh) is assumed to be proportional to the square of their size; the settling velocity for particles >75 μm (+200 mesh) is assumed to be proportional to the size raised to the 3/2 power.

As the settling type of classifiers were used to produce finer size separations, the pool volume was increased. Today, a modern settling-type classifier can be operated to produce d_{50} values of 150–22.5 μm (20). However, the percent solids in the fine stream must be reduced from 25 to 7% correspondingly, while the coarse stream percent water increases from 15 to 25%; this increases the apparent bypass percentage ("void-filling" material). The ratio of the maximum fines stream flow rate (MTPH, solids) for the largest d_{50} value to that for the smallest d_{50} value is 157.5.

The maximum coarse stream flow rate (MTPH, solids) is set for a fixed size classifier. The variation in the maximum coarse stream flow rate is 1.6–150. The ratio of the maximum fine stream flow rate for the largest device to that for the smallest one is 27.5. The ratio of the maximum fine-stream flow rate to the maximum coarse-stream flow rate, for a 150 μm d_{50} value for the smallest device, is 5. Thus, the maximum fine-stream flow rate (MTPH, solids) for the largest device (maximum coarse-stream flow rate of 150 MTPH solids) operating at a 150 μm d_{50} value is 1.6·27.5·5 or

220 MTPH solids, and the maximum fine-stream flow rate when operating at a 22.5 μm d_{50} value is 220/157.5 or 1.4 MTPH, solids.

For spiral classifiers, the suitable parameter for costing is the spiral diameter, in cm (17) (see eq. 1 and vibrating screens under Screening). The appropriate coefficients are

Spiral classifiers

	X, cm	a	b
Simplex	60–125	385.49	1.008
	125–215	3.39	1.986
Duplex	60–130	489.32	1.043
	130–200	13.78	1.776

Cost includes motor.

Increasing the gravitational force, by developing centrifugal force, decreases the settling time of smaller particles. A centrifugal force can be imposed on the particles by rotating the slurry of particles (see Centrifugal separation). The classifying hydrocyclone (Fig. 7) is designed to rotate the slurry of particles by introducing it tan-

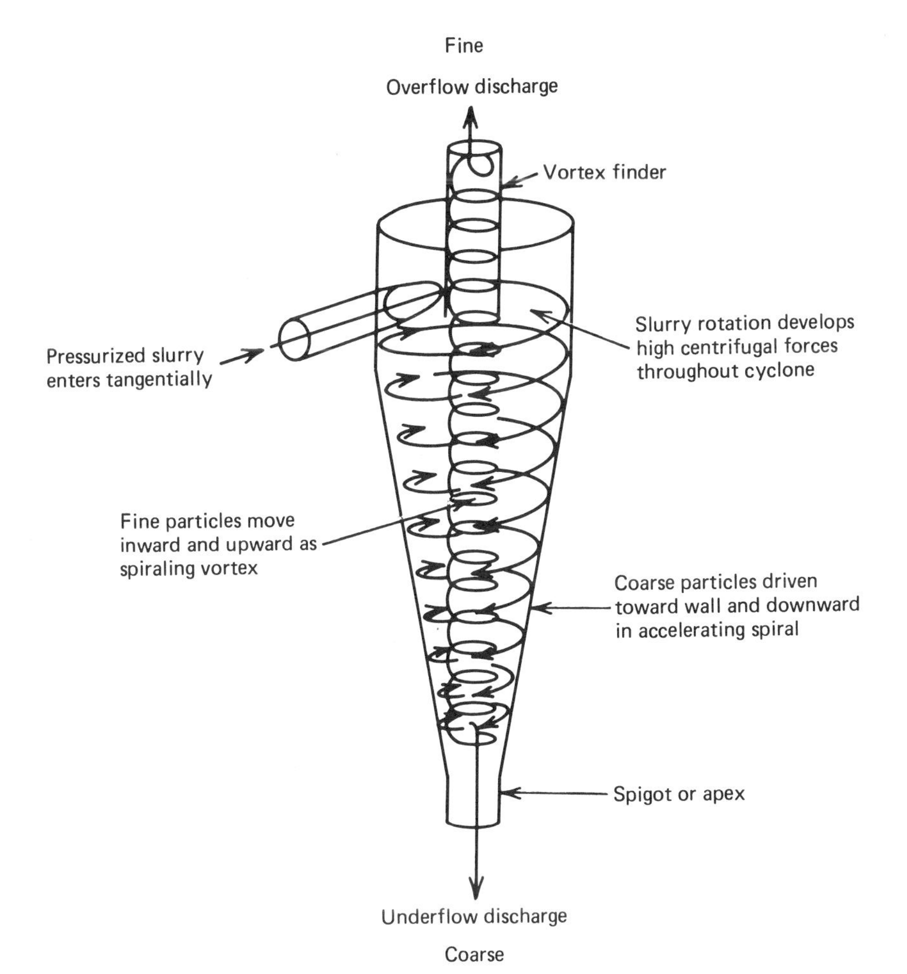

Figure 7. Hydrocyclone classifier.

gentially into a cylinder. A second, smaller diameter (20–40% of the chamber diameter) pipe (vortex finder) is placed through the exact center of the solid top of the cylinder, extending substantially below the feed inlet but above the bottom of the cylinder. The slurry that exits through the vortex finder is termed the overflow, which contains the finer particles. Attached to the bottom of the cylinder is an inverted truncated cone section whose included angle is typically 15–17° (range 10–30°). The slurry that exits through the cone section (apex) is termed the underflow, which contains the coarser particles.

As the fluid rotates in the hydrocyclone, forming an air core in the center, there are three fluid velocities of interest: the tangential velocity, the radial velocity, and the axial velocity (21). The tangential velocity increases (it is zero at the cyclone wall, increasing immediately) moving from the outer wall to the air core, reaching a maximum and then decreasing (there is a locus of constant tangential velocity). The radial velocity immediately increases from zero at the cyclone wall, but then, moving from the outer wall to the air core, decreases to zero. In the cone section, the axial velocity immediately drops from zero at the cyclone wall (a positive value indicates vertical movement) and then moving from the outer wall to the air core increases to positive values. Thus, the fluid motion here is down the wall of the cyclone to the apex and up the air core through the vortex finder. In the cylindrical section, the axial velocity goes negative again, approaching the vortex-finder wall. The fluid flow is then down the inner cyclone wall and the outer vortex-finder wall. There is a locus of zero axial velocity.

The diameter of the air core varies with the feed volumetric flow rate. If the rate is too low, there is no air core and all of the pulp leaves the cyclone as underflow; if the rate is too high, the air core expands closing off the apex and forcing all of the pulp to leave the cyclone as overflow. Consequently there is a minimum and maximum volumetric feed rate. Since the pressure drop is proportional to the square of the volumetric feed rate, the minimum and maximum rates can be monitored by the pressure drop. The ratio of the maximum pressure drop to the minimum pressure drop should be less than 4, meaning the maximum to minimum volumetric feed rate should be less than 2.

A particle entering the cyclone finds a point where its velocity, with respect to the fluid, is equal to the radial velocity and hence is at rest with respect to radial movement to or from the wall. Because the radial velocity changes with radius, the particle spirals down the cone section following a path defined by the particle velocity with respect to the fluid equal to the radial velocity. This path is the equilibrium orbit for all particles of this mass and shape. Larger particles move toward the wall to find their equilibrium orbit, unless it reaches the wall. Smaller particles move toward the air core to find their equilibrium orbit. Thus, a particle will pass out in the overflow if its equilibrium orbit is to the interior of the locus of zero axial velocity. Particles whose equilibrium orbit is toward the outside of the locus will pass out in the underflow. The particles whose equilibrium orbit lies on the locus define the equiprobable or d_{50} size. As the concentration of particles increase, this simple model (22) breaks down.

For a properly designed and operated cyclone, the sharpness index is constant, typically 0.6. The d_{50} and apparent bypass are a function of the cyclone geometry, the volumetric feed rate, the material specific gravity, the feed solids concentration and the pulp rheology. A "standard" cyclone geometry: inlet area = 0.05 (cylinder diameter)^2, vortex finder diameter = 0.35 (cylinder diameter), 0.35 (cylinder diameter) > apex diameter > 0.1 (cylinder diameter), and 20° > included angle > 10° is given in reference 23 in which,

$$d_{50},\ \mu\text{m} = 16.9\,\frac{(\text{cylinder dia, cm})^{0.66}}{(\text{pressure drop, kPa})^{0.28}(\rho_s - \rho_l)^{0.5}[1\text{–}1.9\ (\text{vol frac. solids})]^{1.43}} \qquad (2)$$

Here, ρ_s is the specific gravity of the solid and ρ_l is the specific gravity of the liquid. (This expression has no correction for pulp rheology.) From this expression it can be seen that the d_{50} value decreases with a decrease in the chamber diameter and an increase in the pressure drop. Hence, in order to make fine separations, small diameter cyclones operating at a high pressure drop (200–350 kPa or 29–51 psi) should be used. It can also be seen that the d_{50} value decreases with increasing solid specific gravity, but increases with increasing solids concentration.

The apparent bypass can be estimated by assuming it is approximately equal to the water split which is the percent of water in the feed that reports to the underflow. Calculation of the water split requires an estimate of the pulp split, the ratio of volumetric underflow rate to the volumetric feed rate. From the graphs in reference 23, the pulp split can be estimated as:

$$\text{pulp split} = \frac{3.67\,K_o\left[\dfrac{\text{apex dia}}{\text{cylinder dia}}\right]^2}{\sqrt{(\text{pressure drop, kPa})}} \qquad (3)$$

where K_o is 10–25. Assuming an operating pressure range of 35 to 100 kPa (ca 5–15 psi) gives:

$$K_o = 0.23\ (\text{pressure drop, kPa}) + 2.0 \qquad (4)$$

The water split is equal to the pulp split multiplied by the ratio of:

$$\frac{100 - \%\ \text{solids by volume in the underflow}}{100 - \%\ \text{solids by volume in the feed}}$$

The solids concentration of the underflow must be monitored in order to prevent a roping condition from developing (24). A spiraling solid underflow stream is referred to as a rope discharge. The underflow stream should appear as a hollow cone spray with a 20–30° included angle. Typically, the maximum solids concentration of the feed stream is 30 vol %, the overflow (fine) stream is 15 vol % and the underflow (coarse stream) is 55 vol %. In order to prevent roping, the underflow stream solids concentrate by volume must be less than 0.5385 times the (overflow vol % solids) plus 49.31.

Since the d_{50} value decreases with cyclone diameter (approximation: d_{50}, μm = $5\sqrt{\text{cylinder dia, cm}}$) and since the capacity also decreases with cyclone diameter (approximation: capacity, L/s = 0.022 [cylinder dia, cm]^2), high classification capacity at low d_{50} values can only be achieved by feeding many cyclones of the same diameter. One manufacturer (Krebs Engineers, Menlo Park, Calif.) offers cyclones in diameters of 2.5, 10, 15, 25, 38, 51, 66, 76, and 127 cm. Thus the manifolding of cyclones for parallel processing is very important since each cyclone must be fed at the same pressure drop and with the same amount of solids and size analysis. Radial manifolding is usually recommended in order to achieve uniform distribution of the feed; in-line manifolding usually gives poor distribution.

For hydrocyclones, the suitable parameter for costing is the cyclone dia, in cm (17) (see eq. 1 and vibrating screen under Screening). The appropriate coefficients are

	Wet cyclones	
X, cm	a	b
2.5–35	171.79	0.7582
35–130	16.01	1.430

Cost includes fittings and combination urethane–ceramic liners.

Pneumatic Classification

Pneumatic classification can conveniently be partitioned into classes: coarse (fine products above 95% less than 100 μm); intermediate (fine products between 95% less than 100 μm and 30 μm); and fine (fine products below 95% less than 30 μm). Pneumatic classification, like hydraulic classification, balances the force of gravity with drag forces (counter flow) in order to bring about a separation. The simplest example of counterflow classification would be an expansion classifier, depicted in Figure 8. The solids, dispersed and conveyed in a gas stream within a conduit, are introduced into an enlarged section (reducing the gas velocity) where large particles drop out but smaller particles stay with the gas as it exits through the outlet conduit (increasing the gas velocity).

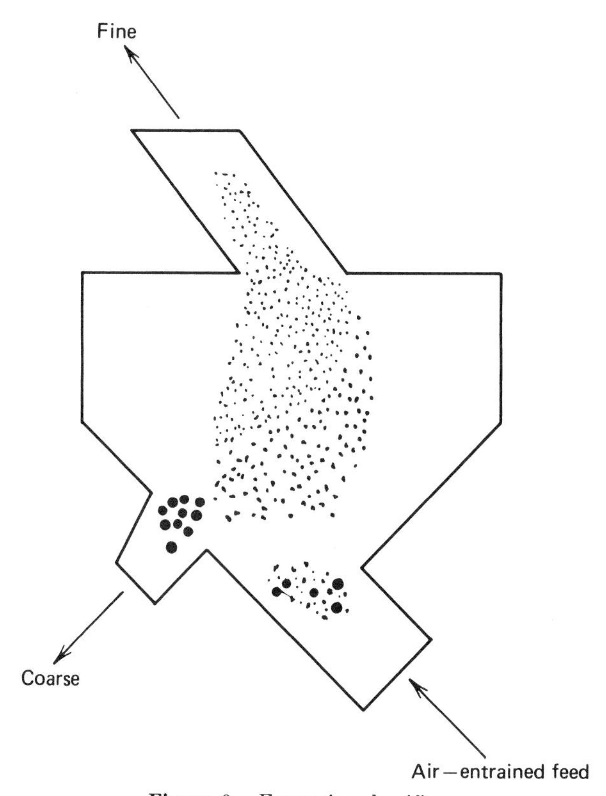

Figure 8. Expansion classifier.

The counterflow principle can be extended to movement in a centrifugal field, improving the selectivity of the classification process and the fineness of the product. The vane classifier, depicted in Figure 9, would be an example of a centrifugal version of a counter flow classifier. The solids to be classified enter the outer cone, dispersed and entrained in a gas stream. A cyclonic flow pattern is imparted on the feed stream before it passes through the adjustable vanes into the inner cone. As the vanes are closed down, an increase in the centrifugal motion causes more of the larger particles to strike the inner wall of the inner cone and drop out. The finer particles remain in the gas stream, exiting through the centrally located exit conduit. For a pilot scale

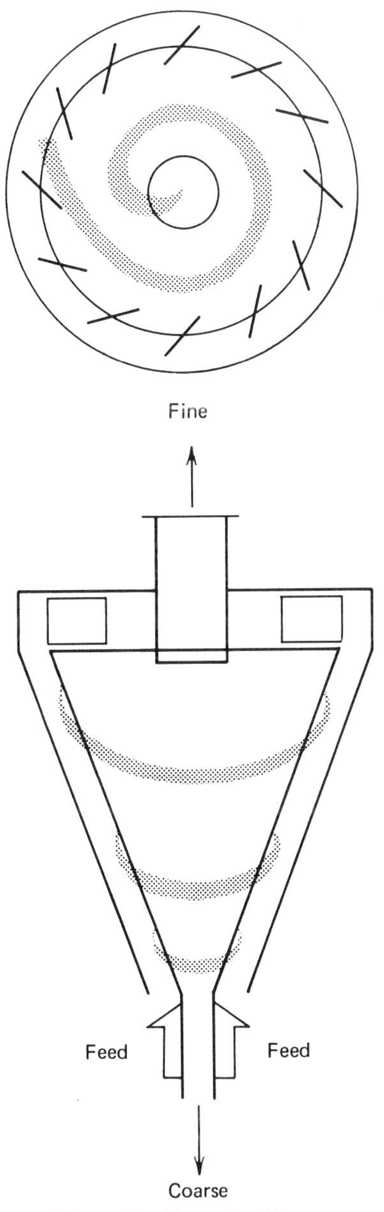

Figure 9. Vane classifier.

twin-cone classifier, the sharpness index was constant for a set vane position (25). As the vanes were adjusted from 100 to 50 to 25% open, the sharpness index increased from 0.325 to 0.5 but decreased to 0.3 at the 25% vane setting (see Table 4). Increasing the air rate, thereby also increasing the feed rate, caused a decrease in the apparent bypass, but an increase in the d_{50} value for each vane setting.

The fixed vane centrifugal counterflow principle is termed a free vortex and has been extended to fine classification (Apline Mikroplex Spiral) where d_{50} values of 2.5–60 μm are claimed at sharpness values of 0.8. Apparently there is no perceptible bypass. An alternative design is the forced vortex design in which the vanes are rotated. For example, the Donaldson Acucut, d_{50} values from ½–50 μm are claimed at sharpness indices of 0.8. Devices with capacities up to ca 4 MTPH are manufactured. The advantage claimed for the forced vortex over the free vortex is less sensitivity to solids loading.

Another pneumatic classification design is to balance inertial forces with drag forces (transverse flow) in order to bring about a separation. An example of this type of design is the classifier depicted in Figure 10. The feed, dispersed and entrained in a gas stream, enters the classifier from the top. The gas stream exits, after changing its direction of flow by 135°, and takes with it the finest particles. The larger particles continue down the bottom of the classifier where they exit. However, before they exit, a secondary air stream transverses the particle stream to remove any fine particles not removed in the primary classification step. These particles are swept back to the feed and given another chance to exit with the other fine particles. This secondary classification step is typical of a two-stage classification arrangement in which the coarse stream from the first stage is reclassified and the fine stream from the second stage is mixed back into the feed stream of the first stage.

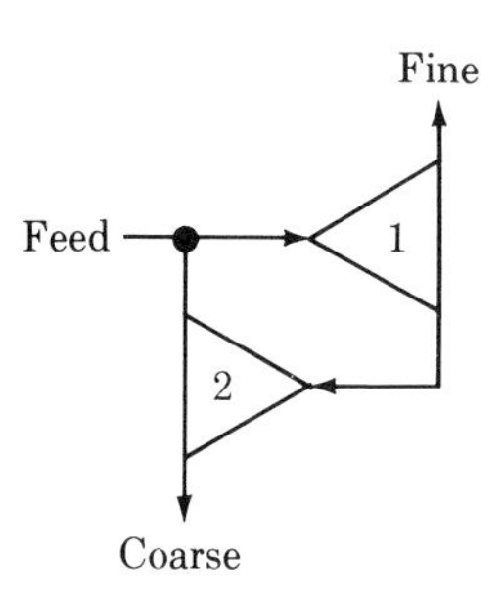

Classifiers of this design appear to exhibit no apparent bypass and seem to have average sharpness indexes (0.5–0.6). Classifiers of this design are used for coarse and inter-

Table 4. Characteristic Classification Parameters for a 0.75-m dia Twin Cone Classifier

Vane setting	100%	50%	25%
Sharpness index	0.325	0.5	0.3
Cut size, d_{50}, μm			
1700 m^3/h (1000 ft^3/min)	120	56	22
3400 m^3/h (2000 ft^3/min)	164	105	82
5100 m^3/h (3000 ft^3/min)	197	151	178
Apparent bypass			
1700 m^3/h (1000 ft^3/min)	0.05	0.35	0.55
3400 m^3/h (2000 ft^3/min)	0.05	0.20	0.30
5100 m^3/h (3000 ft^3/min)	0.05	0.05	0.05

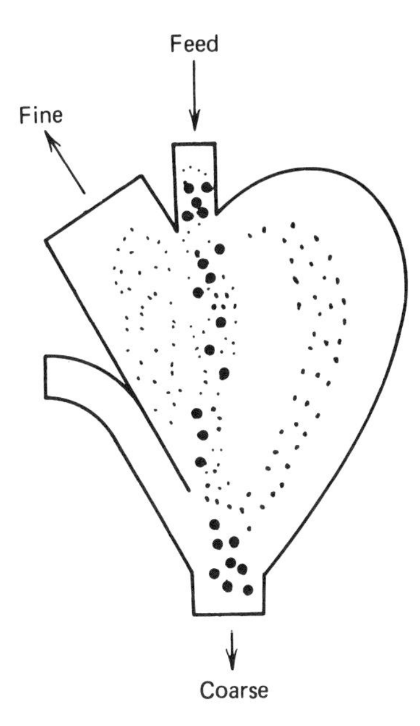

Figure 10. Inertial classifier.

mediate commercial classification. Since they have a set capacity, scale up is achieved by mounting several units radially.

Another classifier design incorporates both the gravitational and centrifugal counterflow principle. The tank-through flow classifier design is depicted in Figure 11. The feed, dispersed and entrained in the gas stream, enters the classifier where the larger particles drop out and exit out the bottom. The first particles are conveyed to the top where they encounter a variable-speed rotor, having essentially radial vanes. As a particle is carried into the rotor by the drag force exerted by the gas flow, its discharge is opposed by the centrifugal force caused by the rotational speed. Only the finest particles pass through the rotor and exit with the gas stream through the fine stream conduit. Typically, the exiting coarse particles are reclassified with an air stream whose velocity is adjustable. Another common industrial design, depicted in Figure 12, enters the feed and gas separately, dispersing and entraining the solids in the tank. Data for a pilot device (26) of the type depicted in Figure 12 are given in Table 5. The sharpness index was a constant 0.6.

There are several responses common to classifiers employing rotors. It is well known that increasing the feed rate, without any other changes, reduces the d_{50} value, the efficiency of the separation, and the percentage of the feed discharged as the fine fraction. The data in Table 5 demonstrate a consistent pattern, in all cases, of the d_{50} value decreasing to a minimum with increasing feed rate. The efficiency of the classification is reduced in all cases with increasing feed rate because the apparent bypass value increases. In all cases, the percentage of the feed discharged as the fine fraction (yield) also decreases with increasing feed rate.

For a given material, the effect of increased air flow on the d_{50} value is:

$$d_{50} \text{ is proportional to } (\text{m}^3/\text{min})^n[(\text{ft}^3/\text{min})^n] \quad 1 < n < 2$$

Assuming the air flow rate to be proportional to fan speed, then the data for a rotor rpm of 1000 gives $n = 1.2$.

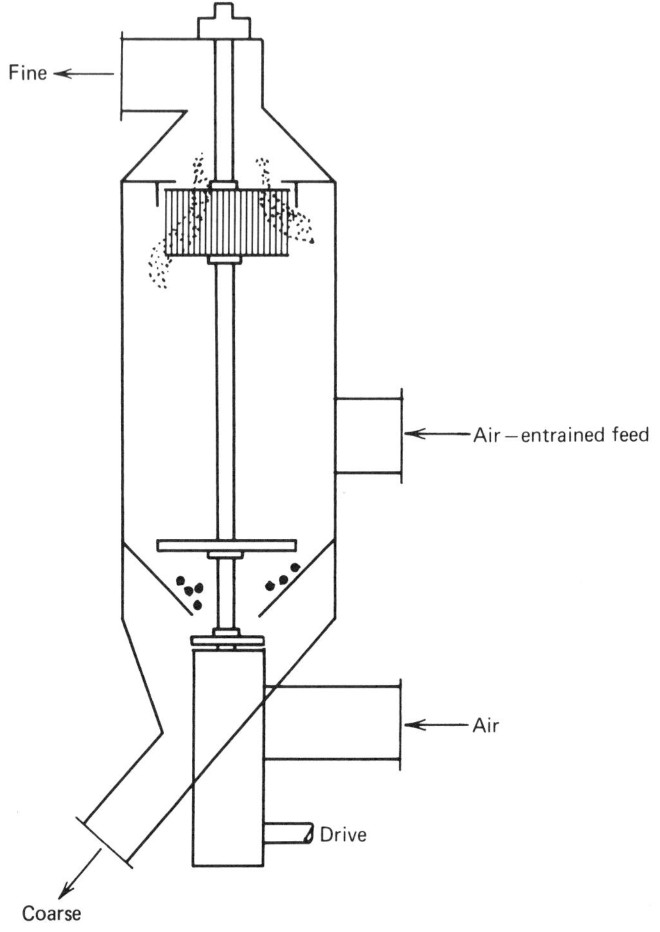

Figure 11. Tank through-flow classifier.

For a given material, the d_{50} value is inversely proportional to the rotor rpm. Although this pattern is not exact, the trend is present. For example, consider the 800 and 1400 rotor rpm data:

$$1.71 = \left(\frac{60}{35}\right) \simeq \left(\frac{1400}{800}\right) = 1.75$$

It is quite common in this design to separate the fine particles from the gas stream and recirculate the gas back to the classifier. Such an arrangement is a special two-stage arrangement.

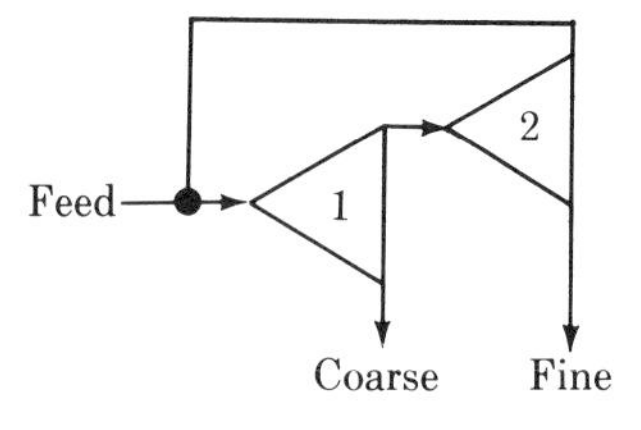

If any particles are recirculated with the gas stream (inefficient solid–gas separation), then there is a finite probability that they will exit with the coarse stream, leading to

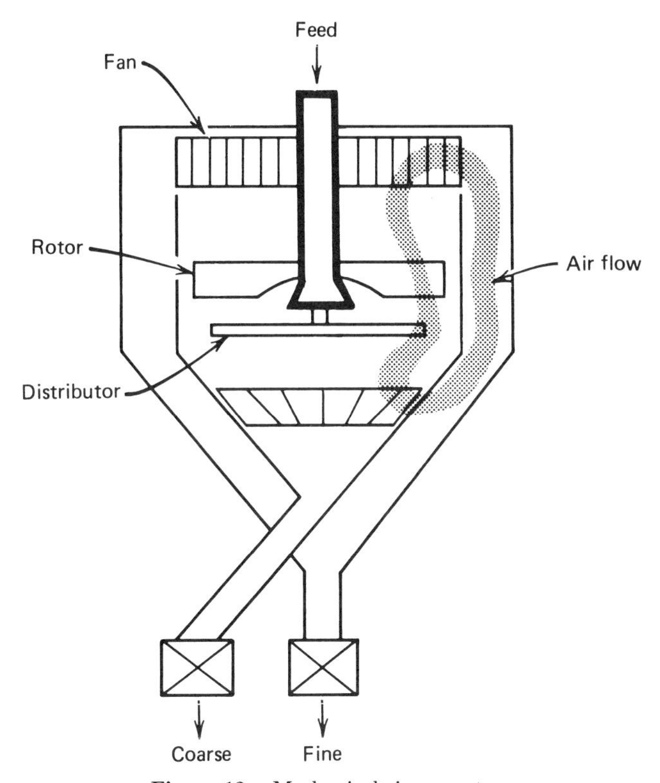

Figure 12. Mechanical air separator.

inefficient classification. This is particularly true if the recirculated gas is introduced into a secondary stage that is reclassifying the coarse particles before exiting. These coarse and intermediate classification devices can be sized to handle hundreds of metric tons per hour. For example, an 8 m-dia unit can handle 900 MTPH for a d_{50} value of 45 μm. Such a unit can be operated down to a d_{50} value of 22.5 μm, but at a reduced feed rate.

Tank flow through designs of the type shown in Figure 13 are used for fine classification. The device is supposed to achieve d_{50} values from 3 to 75 μm at sharpness indices of 0.75. However, capacity (9 MTPH and less) is not as great.

The concept of two stage classification in order to improve the classification has been very popular in both hydraulic (27) and pneumatic (28) classifier installations. There are four common arrangements, two for reclassifying the fine stream of the first stage of classification, and two for reclassifying the coarse stream of the first stage of classification. The latter is more commonly practiced, ie,

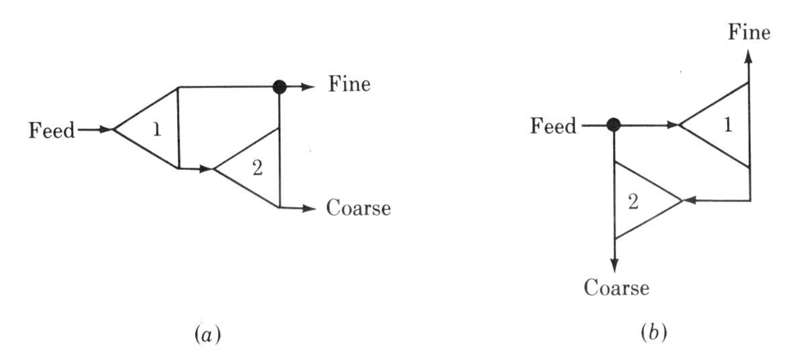

Table 5. Characteristic Classification Parameters for a 1.2-m dia Mechanical Air Separator

Fan, rpm	Rotor, rpm	Relative feed rate	d_{50}, μm	Apparent bypass	Yield
1400	600	1	62.5	0.05	0.57
1400	600	4	40	0.50	0.20
1400	600	7	40	0.75	0.11
1400	1000	1	32.5	0.10	0.33
1400	1000	4	30	0.75	0.09
1400	1000	7	30	0.80	0.07
1850	800	1	60	0.05	0.55
1850	800	4	40	0.30	0.29
1850	800	7	40	0.40	0.23
1850	1000	1	45	0.05	0.46
1850	1000	4	35	0.30	0.25
1850	1000	7	35	0.50	0.18
1850	1400	1	35	0.05	0.40
1850	1400	4	30	0.40	0.19
1850	1400	7	30	0.60	0.13

If both stages have the same d_{50} value (100 μm), sharpness indexes (0.6), and apparent bypasses (say 30%), then the overall parameters would be

	(*a*)	(*b*)
d_{50}	113	101
sharpness index	0.62	0.64
apparent bypass	9%	11%

Thus, classification efficiency is improved by increasing the sharpness index, and decreasing the apparent bypass.

If a classifier is to be utilized for dedusting (the production of a fine particle-free coarse stream) then it should not have an apparent bypass value.

The pneumatic classification system should be designed to handle hazardous dust (29). A hazardous dust is one which, when finely divided and suspended in air, will burn and/or is sufficiently toxic to be injurious to personnel health (see Air pollution control methods). At the least, almost any dust can be irritating to personnel because of inhalation or skin or eye contact.

Most solids when finely divided, suspended in air in the proper concentration and ignited, can produce violent explosions. The only "safe" materials are those which are already fully oxidized or hydrated.

Explosions are caused by the rapid generation of heat and/or gases, causing pressures to rise more rapidly than they can be dissipated or relieved. Dust explosions usually occur in pairs; the first involves dust already in suspension which jars dust from beams, ledges, etc, creating a second cloud to which the explosion propagates, resulting in a secondary explosion.

Before a process is designed in which dusty materials are to be handled, the flammability characteristics of the dust must be determined. Dust clouds have been ignited by open flames, electric sparks, hot filments of light bulbs, friction sparks, hot surfaces, static parks, spontaneous heating, welding and cutting torches, and other common sources of heat for ignition. Dust-ignition temperatures, for the most part,

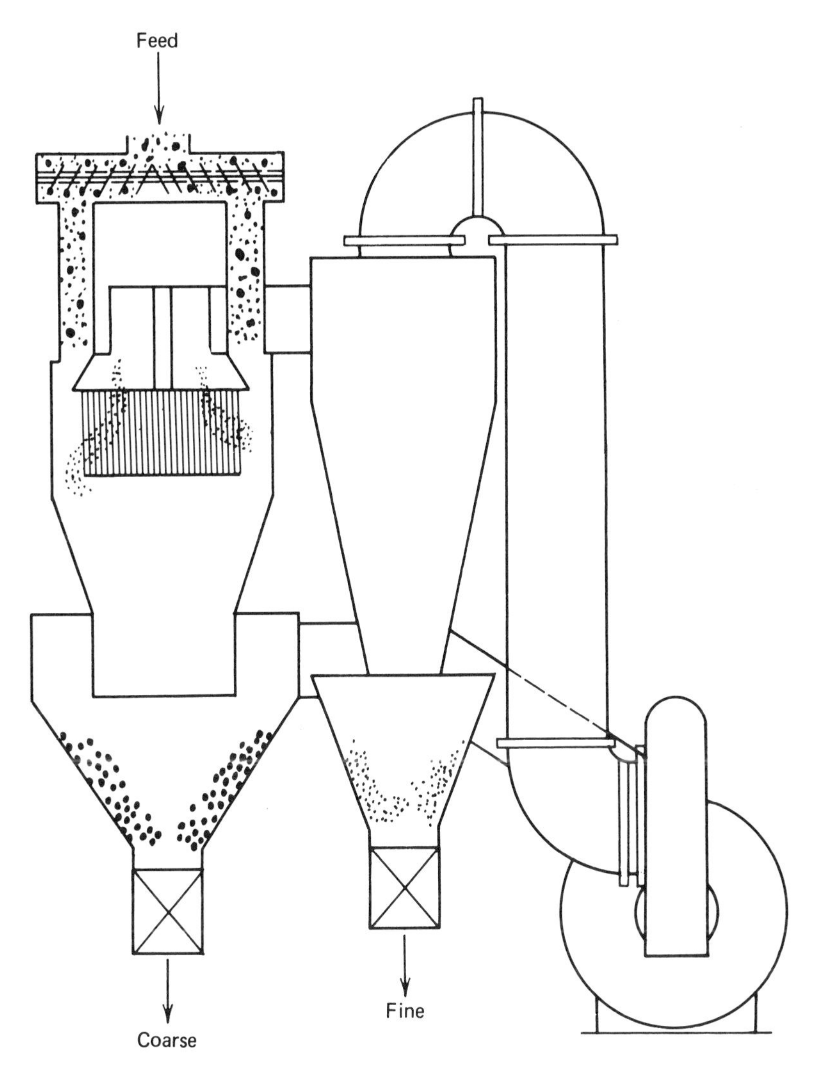

Figure 13. Recirculation-flow classifier.

are 300–600°C. A large majority of the reported minimum spark energy for ignition of dust was less than 100 mJ. These ignition energies are much lower than the energies that are found in stray electric currents and static electricity discharges. The finer the dust the more the concern with static electricity.

Dust explosions can be minimized if any leg of the basic fire triangle (source of ignition, oxygen, and fuel) is eliminated. Although elimination of the source of ignition and oxygen (lower to less than 10 wt %) are possible, the elimination of the "fuel" is not. Instead, it must be controlled by utilizing air-to-solid ratios throughout the system to keep dust concentrations below a minimum level or above a maximum level. Most organic materials, including plastics, have a minimum explosive concentration of 15 to 300 g/m^3, and that of metals can approach 500 g/m^3.

The equipment in which the dust is handled or stored should be designed to contain the pressure resulting from an internal explosion. Most dusts show maximum pressures of ca 345–735 kPa (50–107 psi); however, the role of pressure rise ranges from ca 700 to 7000 kPa/s (1–100 psi/s). Equipment-containment design can be coupled with explosive-venting design for the equipment and the building.

Sometimes the conveying velocities, typically 20–25 m/s, can be designed to exceed the flame-propagation rate as determined by tests. If the flame-propagation rate exceeds the conveying velocities, keeping in mind that velocities vary throughout the system, consideration would be given to the isolation of parts of the system by "choking"—the use of rotary values, screen conveyors, bins, etc.

Proper ventilation and housekeeping minimizes secondary explosions. Dust collectors of the dry type should be located outside the building, and provided with conduction bags and adequate explosion venting to a safe location.

Prediction of Size Separation

The separation can be predicted if the appropriate fractional recovery values can be estimated. The fractional recovery values, $r(x_i)$, are given by

$$r(x_i) = (1 - a - b)c(x_i; \kappa, d_{50}) + a \tag{5}$$

where a is the apparent bypass fraction of the feed to the coarse stream; b is the apparent bypass fraction of the feed to the fine stream; and $c(x_i; \kappa, d_{50})$ is the corrected curve function with characteristic parameter d_{50}, the equiprobable size, κ, the sharpness index; and x_i is the upper size of ith size interval. If the logistic function is used as the corrected curve function, then

$$c(x_i; \kappa, d_{50}) = \frac{1}{1 + (x_i/d_{50})^{\ln[9]/\ln[\kappa]}} \tag{6}$$

The prediction methodology is demonstrated in Table 6. The mass of feed material (column 3) is simply taken as the percent material in each size fraction. The fractional-recovery values (column 4) were calculated by setting $\kappa = 0.6$, $d_{50} = 122.5$ and $a = 0.3$ in the logistic function for this example. The mass of coarse material (column 5) is calculated by multiplying column 3 by column 4. The size distribution of the coarse stream (column 6) is calculated by dividing each entry in column 5 by the sum of column 5 (these values are then multiplied by 100%). The mass of fine

Table 6. Predicting Size Separation Using Fractional-Recovery Values

Relative	Feed		Fractional	Coarse		Fine	
Size	Cumulative	Mass, %	recovery	Mass, %	Interval	Mass, %	Interval
1.000	99.9	0.15	1.00	0.15	0.2	0	0.0
0.707	99.75	1.25	1.00	1.25	1.7	0	0.0
0.500	98.5	4.5	1.00	4.5	6.0	0	0.0
0.353	94.0	11.5	0.996	11.4	15.3	0.1	0.4
0.250	82.5	16.0	0.986	15.8	21.2	0.2	0.8
0.177	66.5	16.5	0.940	15.5	20.8	1.0	3.9
0.125	50.0	15.0	0.794	11.9	16.0	3.1	12.2
0.088	35.0	11.0	0.549	6.0	8.0	5.0	19.7
0.063	24.0	8.0	0.375	3.0	4.0	5.0	19.7
0.044	16.0	5.5	0.330	1.8	2.4	3.7	14.6
0.031	10.5	10.5	0.300	3.2	4.3	7.3	28.7
		100.0		74.5		25.4	

material (column 7) is calculated by subtracting column 5 from column 3. The fine stream size distribution (column 8) is calculated by dividing each entry in column 7 by the sum of column 7 (again multiplied by 100%). The product yield is the sum of column 7 divided by the sum of column 3. Relative sizes (interval top size divided by the maximum size) are used because enough intervals must be created in order to insure that the sink (final) interval fractional recovery value is equal to the apparent bypass fraction (feed to coarse). Otherwise, the prediction is in error.

BIBLIOGRAPHY

"Size Separation" in *ECT* 1st ed., Vol. 12, pp. 520–523, by W. A. Lutz, F. L. Bosqui, and A. D. Camp, The Dorr Company; "Size Separation" in *ECT* 2nd ed., Vol. 18, pp. 366–399, by F. L. Bosqui, Consultant.

1. T. Allen, *Particle Size Measurement*, 3rd ed., Chapman & Hall, New York, 1981.
2. H. Rumpf and K. Leschonski, "Principles and Recent Methods of Air Classification," *Selected Translated Paper*, Vol. 1, Universität Karlsruhe, FRG, 1972.
3. N. F. Schultz, *Trans. Soc. Min. Eng.* **247,** 81 (1970).
4. R. T. Hukki and H. Eklund, *Trans. Soc. Min. Eng.* **232,** 265 (1965).
5. D. A. Dahlstrom, PhD thesis, Northwestern University, 1949.
6. F. C. Bond, *Rock Prod.* **96,** 120 (Dec. 1967).
7. J. W. Leonard, *Trans. Soc. Min. Eng.* **256,** 185 (1974).
8. *Particle Size Classifiers, A Guide To Performance Evaluation, AIChE Equipment Testing Procedure*, Equipment Testing Procedure Committee, AIChE, 1980.
9. *Testing Sieves and Their Uses, Handbook 53*, W. S. Tyler, Inc., 1981.
10. *Rock Talk Manual*, K1078, Kennedy Van Saun Corp., Danville, Pa., 1978.
11. R. J. Batterham, K. R. Weller, T. E. Norgate, and C. J. Birkett, *Particle Technology*, Amsterdam, 1980.
12. T. Brereton and K. R. Dymott, *Proceedings 10th IMPC*, IMM 1974, pp. 181–194.
13. J. Slechta, I. J. Taggart, B. A. Firth, and E. Gallagher, "An Evaluation of a Novel Vibrating Fine Screen, 1981, unpublished report.
14. R. S. C. Rogers, *Powder Technol.* **31**(1), 135 (1982).
15. W. Benson and C. Burgess, *Proceedings of the 5th ICPC*, Dept. of the Interior, Washington, D.C., 1966, pp. 297–311.
16. D. Jenkinson, *Min. Congr. J.* **68,** 29 (July 1982).
17. A. L. Mular, "Mining and Mineral Processing Equipment Cost and Preliminary Capital Cost Estimations," *Special Vol. 25*, CIM, Montreal, 1982.
18. R. E. Riethman and B. M. Brunnel in Mular and Bhappu, eds., *Mineral Processing Plant Design*, 2nd ed., AIME, New York, 1980, Chapt. 16.
19. E. J. Roberts and E. B. Fitch, *Trans. Soc. Min. Eng.* 1113 (1956).
20. H. Schubert and T. Neesse, *Proceedings of the 10th IMPC*, IMM, 213 (1974).
21. D. F. Kelsall, *Trans. Inst. Chem. Eng.* **30,** 87 (London), (1952).
22. D. Bradley, *The Hydrocyclone*, Pergamon Press, New York, 1965.
23. R. A. Arterburn in Mular and Jergensen, eds., *Design and Installation of Comminution Circuits*, AIME, New York, 1982, Chapt. 32.
24. A. L. Mular and N. A. Jull in Mular and Bhappu, eds., *Mineral Processing Plant Design*, AIME, New York, 1982, Chapt. 17.
25. Committee on Comminution and Energy Consumption, *National Academy of Sciences Report No. NMAB*, 1981.
26. L. G. Austin and P. T. Luckie, *ZKG* **29,** 452 (Oct. 1976).
27. R. S. C. Rogers, A. M. Hukki, C. J. Steiner, and R. A. Arterburn, *Preprint 81-125*, SME-AIME, New York, 1981.
28. R. T. Hukki and T. Airaksinen in Somasundaran, ed., *Proc. International Symposium of Fine Particle Processing*, AIME, New York, 1980, Chapt. 11.
29. H. Boyd, Hugh Boyd & Associates, Tulsa, Oklahoma, 1981, private communication.

PETER LUCKIE
The Pennsylvania State University

SUPERCRITICAL FLUIDS

In the late 1970s, a renewed interest in supercritical-fluid extraction (SFE) intensified due to increased scrutiny by the U.S. government of the hazards of conventional industrial solvents; more stringent pollution-control laws; the search for a wider base of petrochemical feedstocks; and a change in process economics caused by rapidly increasing energy costs. The number of recent review papers (1–7), with 30 papers presented at the 1981 AIChE annual meeting, is indicative of the magnitude of this activity (8). A variety of SFE processes are being developed despite previous hindrances resulting from the lack of understanding of the complex phase behavior, and the requirements of pressures of 5 to >30 MPa (725–4350 psi).

Supercritical-fluid extraction is a hybrid unit operation utilizing the advantages of both distillation and liquid extraction. It has the benefit that slight changes in the temperature and pressure in the critical region cause extremely large changes in the solvent density and thus its dissolving power. In comparison with conventional processes, SFE offers considerable flexibility for an extractive separation using the variables of pressure, temperature, choice of solvent, and additives called entrainers. Heavy nonvolatile substances dissolve in supercritical fluids (dense gases, compressed gases, supercritical gases, high pressure gases), typically 2–7 orders of magnitude in excess of the amount based on the ideal gas law. This is due to the high density of the fluid, which can approach that of a liquid. Thus, SFE offers both high solubility extraction based on the enhancement of vapor pressure and nearly complete solvent–extract separation which is accomplished by reducing the solvent density to the gaseous state.

Development of SFE

The phenomenon of solubility enhancement in dense gases was discovered in the late 1870s when the effects of pressure on the solubility for the system potassium iodide–ethanol were observed (9). The next development was the discovery of the effects of supercritical water in geological processes (4), and of methane in the formation and migration of petroleum (10). In 1943, the first proposal for a practical application of SFE was submitted describing a method for deasphalting petroleum oils as reviewed in ref. 4. A pressure-controlled separation for deasphalting petroleum residues was described (11); however, a less-energy-intensive temperature-controlled separation was eventually developed (12). Supercritical methane was used for fractionation of crude oil, extraction of lanolin from wool grease, and ozocerite wax from ores (11). At the Max Planck Institute in 1962, solubility enhancement in the reaction of triethylaluminum with high pressure ethylene was recorded (13).

Physical Properties of Supercritical Solutions

The density of a fluid is extremely sensitive to pressure and temperature near the critical point ($P_r = 1$, $T_r = 1$) as shown for pure carbon dioxide in Figure 1. Consider the simple case of the solubility of a solid in this fluid. At ambient pressure, the density is 0.002 g/cm^3; thus, the solubility of a solid in the gas is low and is given by the vapor pressure over the total pressure. At the critical point, the density is 0.468 g/cm^3 which

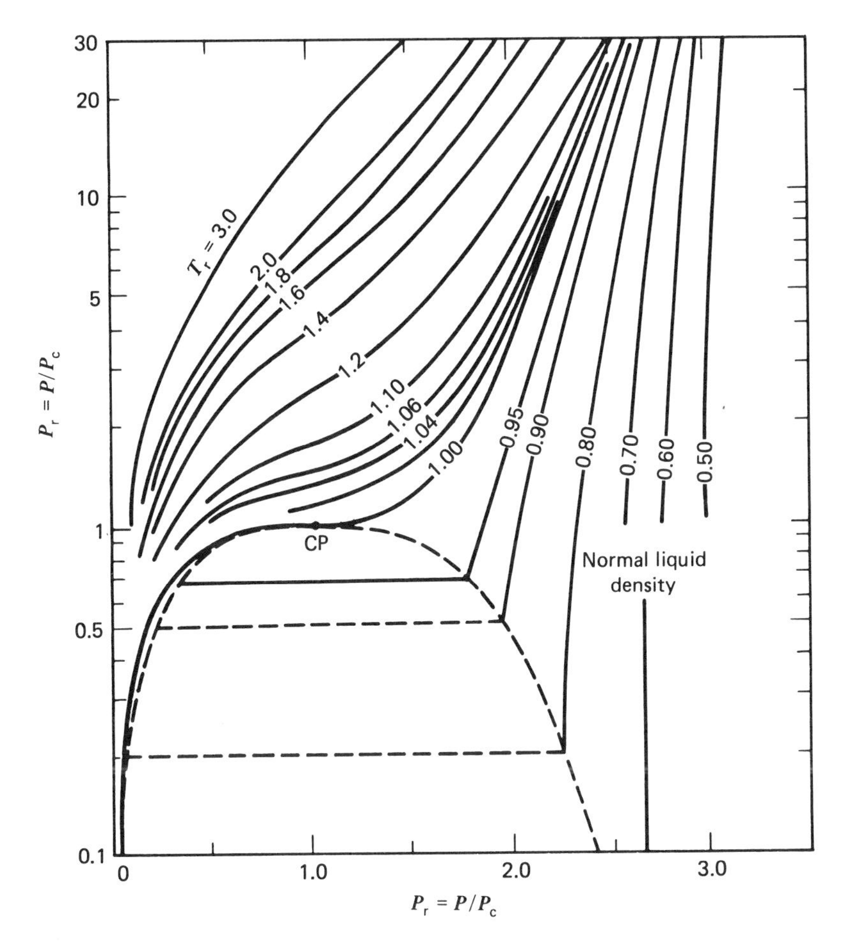

Figure 1. Pressure versus density isotherms for pure carbon dioxide (T_c = 304 K, P_c = 7.38 MPa, ρ_c = 0.468 g/cm^3). To convert MPa to psi, multiply by 145.

is nearly comparable to that of organic liquids at ambient pressure leading to similar solubility.

The effects of temperature and pressure on the solubility of naphthalene in compressed ethylene are shown in Figure 2 (14–15). The solubility versus temperature isobars have positive slopes in the high pressure region, and the opposite for low pressures. This is due to the two competing effects of temperature on the vapor pressure of the solute and the density of the fluid. In Figure 1, the density is more sensitive to temperature near the critical pressure than at higher pressures. At these near-critical pressures, this density decrease with increasing temperatures dominates the effect on vapor pressure leading to negatively sloped isobars. At the higher pressures, the dominant temperature effect is on the vapor pressure giving positively sloped isobars.

Based on Figure 2, one can envision two basic types of SFE consisting of an extraction stage and a separation stage. In the pressure-controlled type of extraction, the compressed solvent dissolves the solute in an extraction vessel, then the solution is expanded in the separation stage to precipitate the extract. Finally, the solvent is

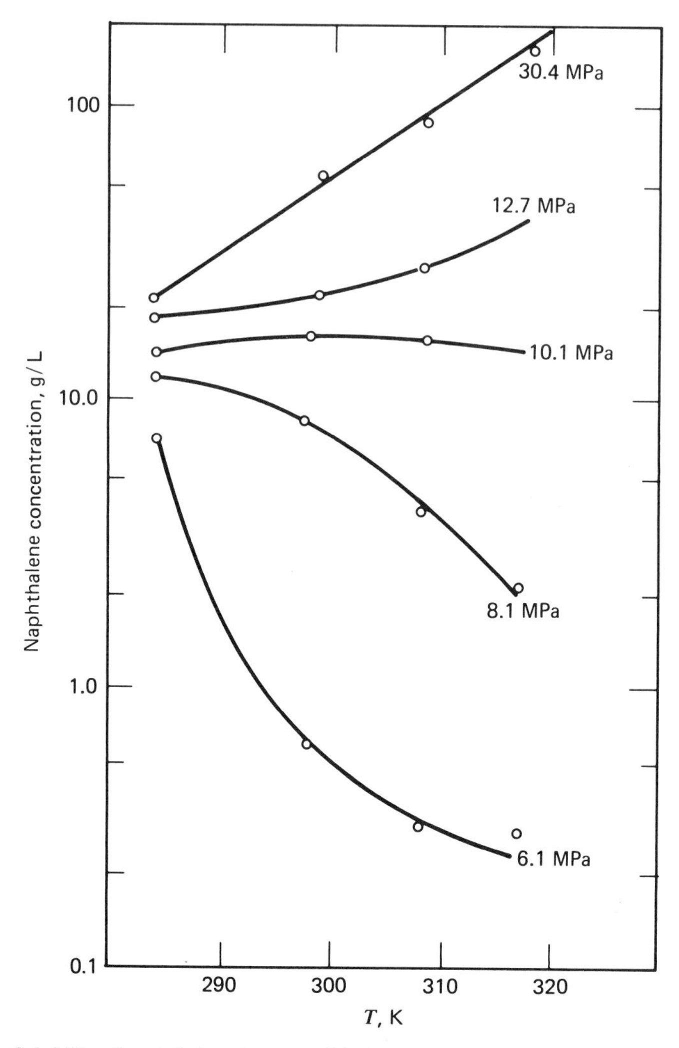

Figure 2. Solubility of naphthalene in supercritical ethylene (T_c = 282 K, P_c = 5.0 MPa) as a function of temperature and pressure. To convert MPa to psi, multiply by 145.

recompressed for recycle. The temperature-controlled type involves the same kind of extraction stage, except that the extract is precipitated by heating the solution to lower the solvent density. The density is then increased for recycle by isobaric cooling. This type is highly energy efficient as the heat is transferred directly between the heating and cooling stages, and the nearly isobaric conditions minimize compression energy.

In an SFE process, it is desirable to limit the reduced temperature of the solvent to ca 0.8–1.5 as the density, and thus the dissolving power is low at high reduced temperatures. For example, the reduced density of carbon dioxide is less than 0.2 at $T_r = 3$ and $P_r = 1$ (see Fig. 1). This is demonstrated in Figure 3 by comparing the solubility of phenanthrene in a series of supercritical gases at a constant temperature of 313 K and pressure of 40 MPa (5800 psi) (16). Whereas the solubility is negligible for the gases with critical temperatures considerably below 313 K, it is substantially

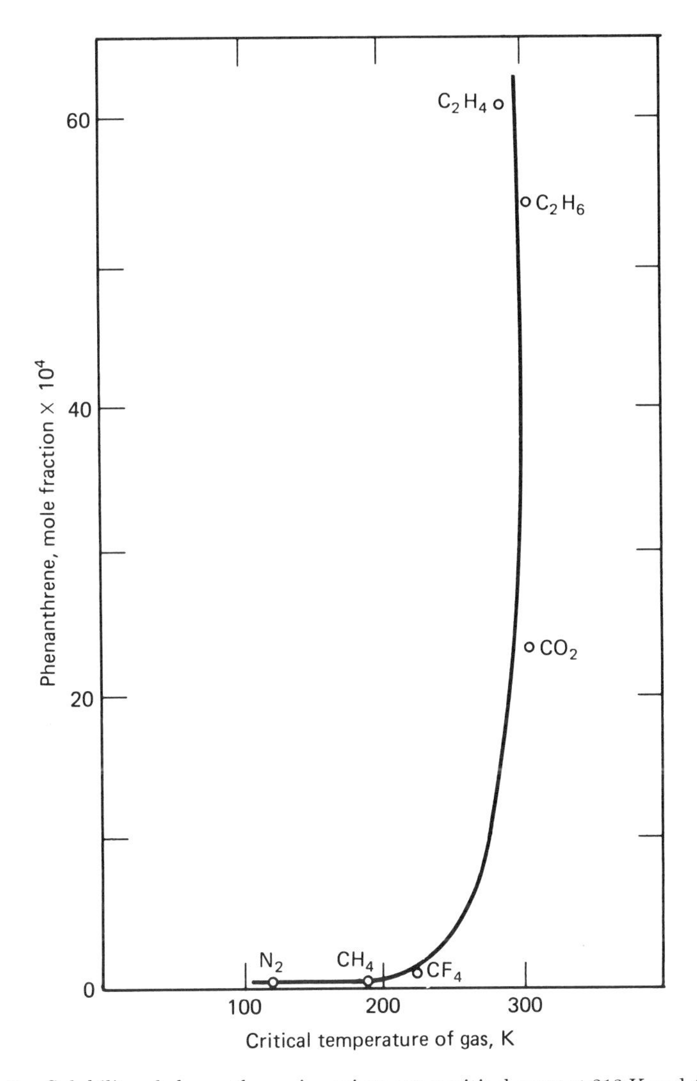

Figure 3. Solubility of phenanthrene in various supercritical gases at 313 K and 40 MPa.

higher for the others. This example is oversimplified for instructive purposes; better correlations would result using the density as the independent variable.

In addition to the density, the viscosity and diffusivity for typical supercritical fluids are intermediate between those of a liquid and a gas as shown in Table 1. Although a supercritical fluid has a density approaching that of a liquid for high solvent

Table 1. Physical Properties of a Typical Gas, Supercritical Fluid, and Liquid

Property	Gas	Supercritical fluid	Liquid
density, g/cm^3	10^{-3}	0.3	1
viscosity, mPa·s (= cP)	10^{-2}	0.1	1
diffusion coefficient, cm^2/s	0.1	10^{-3}	5×10^{-6}

capacity, the diffusivity is orders of magnitude greater giving improved mass-transfer rates. In addition, the lower viscosity also provides advantages such as enhanced solid settling rates during precipitation. For these reasons, supercritical solvents are superior to liquids for penetrating the micropores of a solid structure such as coal (see Table 2) (3). Further detail for the prediction of the diffusivity, viscosity, Schmidt number, and mass-transfer coefficient is given in ref. 7.

Advantages of SFE

The advantages of SFE compared to distillation and liquid extraction are described below.

In SFE, there are many options available for achieving and controlling the desired

Table 2. Solubility Data for Supercritical Solid–Fluid Equilibria Systems at Near-Critical Conditions[a]

Supercritical fluid–solid equilibria systems	Ref.
H_2 (T_c = 33.2 K)	
N_2	*b*
O_2	*b*
CO	*b*
Ne (T_c = 44.4 K)	
CH_4	*b*
N_2 (T_c = 126 K)	
CO_2	*b*
CH_4 (T_c = 191 K)	
neopentane	*b*
C_2H_4 (T_c = 282 K)	
naphthalene	17
phenanthrene	17
anthracene	17
pyrene	18
2,3-dimethylnaphthalene (DMN)	*b*
2,6-DMN	*b*
hexachloroethane	*b*
p-chloroiodobenzene	*b*
benzoic acid	*b*
hexamethylbenzene	18
fluorene	18
CHF_3 (T_c = 298 K)	
anthracene	*b*
papaverine	*b*
thebaine	*b*
noscapine	*b*
morphine	*b*
CO_2 (T_c = 304 K)	
naphthalene	*b*
phenanthrene	18
anthracene	18
pyrene	18
2,3-DMN	*b*
2,6-DMN	*b*
hexachloroethane	*b*
p-chlorophenol	*b*
benzoic acid	*b*
hexamethylbenzene	18
fluorene	18
triphenylmethane	18
1-octadecanol	*b*
stearic acid	*b*
diphenylamine	*b*
phenol	*b*
2,4-dichlorophenol	*b*
biphenyl	*b*
codeine	*b*
thebaine	*b*
noscapine	*b*
papaverine	*b*
morphine	*b*
C_2H_6 (T_c = 305 K)	
naphthalene	18
phenanthrene	18
anthracene	18
triphenylmethane	18
n-C_3H_8 (T_c = 370 K)	
anthracene	*b*
NH_3 (T_c = 406 K)	
anthracene	*b*
H_2O (T_c = 647 K)	
quartz	*b*
glucose	*b*
Na_2CO_3	*b*

[a] At least part of the range $0.9 < T_r^{solvent} < 1.15$ was investigated for these systems.

[b] References for binary solid–supercritical fluid equilibria data are available in Ref. 7.

selectivity, which is extremely sensitive to variations in pressure, temperature, and choice of solvent.

The extract is virtually free of residual solvent. The solvent can be recovered with minimal losses by either isobaric heating or isothermal decompression to the gaseous state.

Supercritical fluids can be used to vaporize thermally labile nonvolatile substances at moderate temperatures, at which they are nondistillable. At these conditions, the selectivity may be too low for liquid extraction.

Nontoxic, nonhazardous supercritical carbon dioxide can be used in the food and pharmaceutical industries without contaminating the product. Liquid extraction leaves residual toxic organic solvent in the product even after energy-intensive distillation or vacuum processing stages.

Although the density and thus the solvent capacity of a supercritical fluid is nearly comparable to that of a liquid, the lower viscosity and higher diffusivity provide advantages in transport rates. For example, the settling rate is higher for precipitates (19), and the mass-transfer rate is improved for solvent diffusion through solid phases.

By reducing the density of the solvent over a continuum, the extract may be fractionated into numerous components, even if they have similar volatilities.

Additional components, usually intermediate in volatility between the solute and supercritical solvent, can be used for further manipulation of the phase behavior, eg, volatility enhancement of the solute.

The ability of the supercritical fluid to vaporize nonvolatile compounds at moderate temperatures reduces the energy requirements compared to distillation. This is particularly evident for temperature-controlled extraction, where recompression work is minimal (1–7).

Supercritical-Fluid Phase Behavior

Classification of Phase Boundaries for Binary Systems. These classifications are usually based on P–T projections of mixture critical curves and three-phase equilibria lines (4,7,20–25). The experimental data are usually obtained by the simple synthetic method in which the pressure and temperature of a homogeneous solution of known concentration are manipulated to precipitate a visually observed phase. Unfortunately, this method does not usually give the composition of this phase, and the complementary pressure–composition (P–x) diagrams are not obtained. Six classes of binary P–T diagrams are identified, each one located along a continuum, which traverses the degree of dissimilarity in the intermolecular forces (21) (see Fig. 4).

Class 1. Class 1 binary P–T diagrams include, for example, CO_2–n-hexane and CO_2–benzene. This is the simplest case in which the P–T projection of the three-dimensional pressure–temperature–composition (P–T–x) diagram consists of a vapor-pressure curve for each pure component, and a critical line.

Class 2. Class 2 binary P–T diagrams include, for example, CO_2–octane, CO_2–dodecane, and CO_2–2-hexanol. As the mutual solubility of the components decreases, an upper-critical solution-temperature- (UCST) versus-pressure line appears that depicts *ll* immiscibility. This line starts at the *llg* three-phase line. At point M, notice that the heavy component can be precipitated by a small temperature increase, small

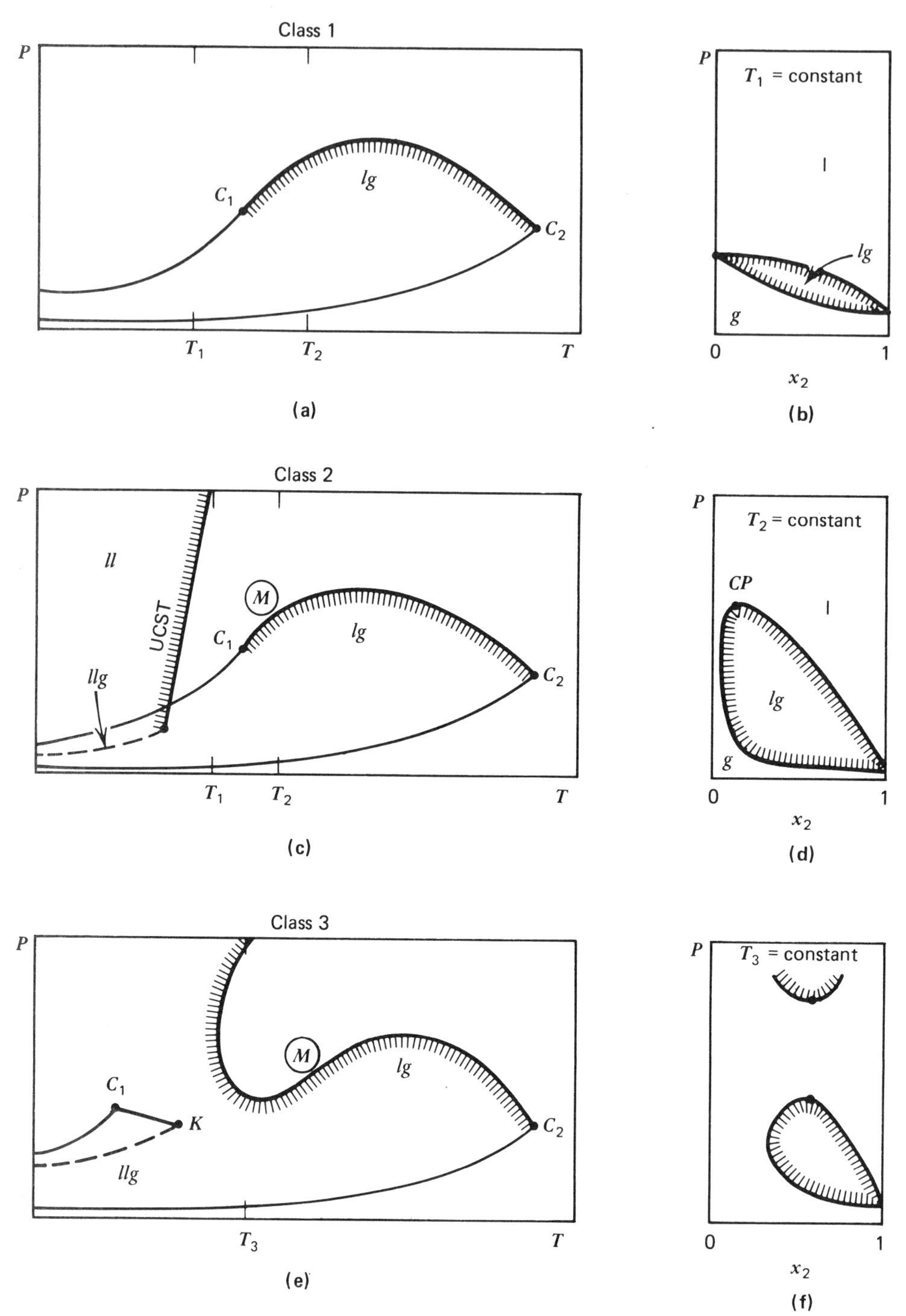

Figure 4. P–T diagrams and selected P–x isotherms depicting critical curves and phase boundaries for binary systems (— = pure component vapor-pressure curve, — = critical curve, - - - = three-phase line).

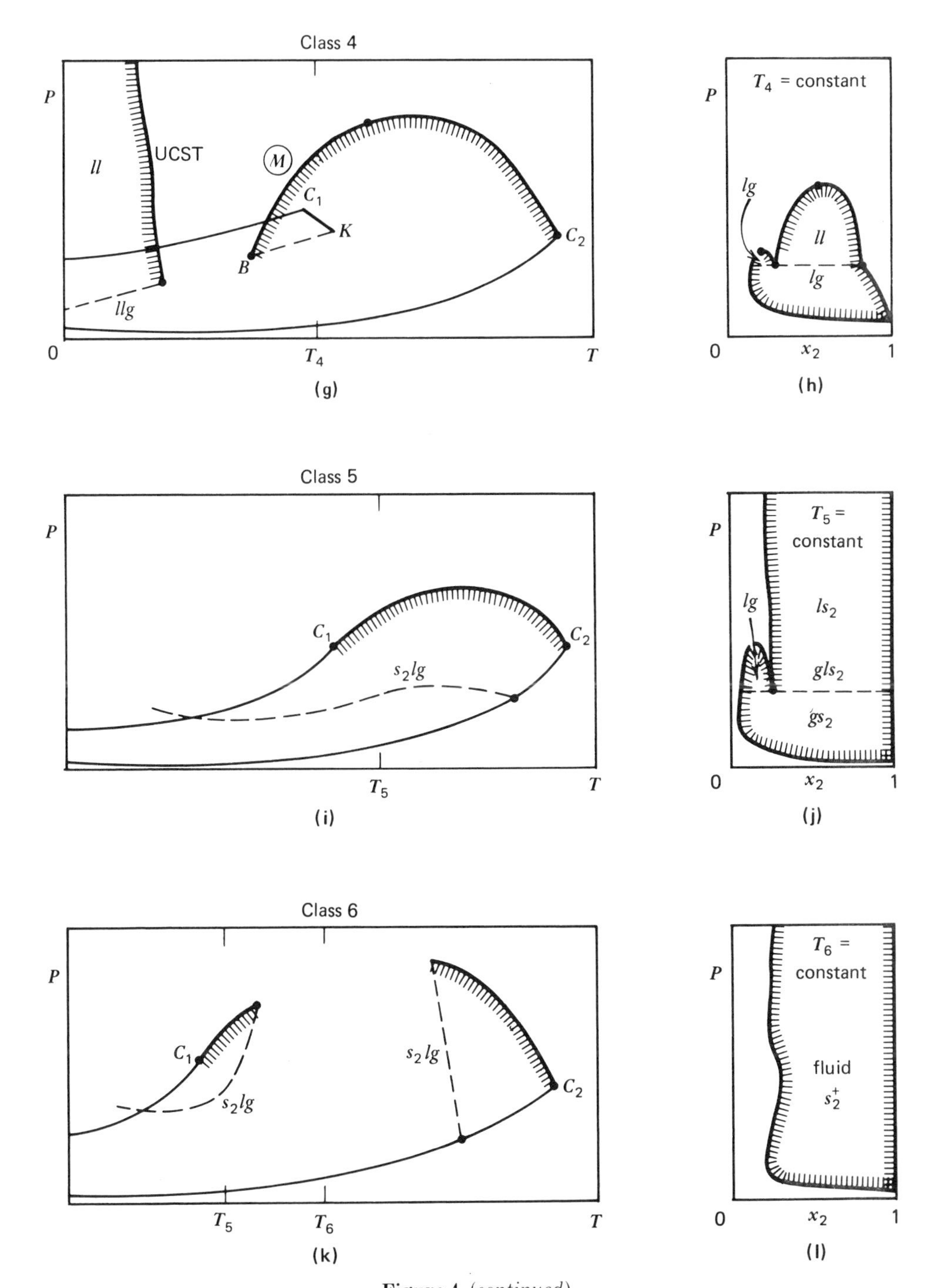

Figure 4. *(continued)*

pressure decrease, large temperature decrease, or extremely pronounced pressure increase. The first two are more feasible for SFE.

Class 3. Class 3 binary P–T diagrams include, for example, CO_2–hexadecane, CO_2–squalane, and CO_2–water. As the dissimilarities in the forces and the volatility of the components increase, the *ll* line shifts to higher temperatures. Here, it interrupts the critical curve. An additional critical curve connects C_1 and the critical end point K, the terminal point on the *llg* line. At even lower mutual miscibility, the critical curve originating at C_2 does not display a pressure maximum or minimum (not shown), thus yielding two types of gas–gas equilibria (21).

Class 4. Class 4 binary P–T diagrams include, for example, CO_2–nitrobenzene and ethane–squalane. The only new phenomenon in this case is the presence of two distinct *llg* lines. The first is like the one in class 2, whereas the second is similar to class 3, except that it merges with the critical curve. Along BK, two liquid phases are in equilibrium with the gas, and at K the liquid phase rich in the gas merges with the gas phase. This class has been subdivided to include an additional one without the *ll* line (eg, ethane–ethanol) (7).

Class 5. Class 5 binary P–T diagrams include, for example, ethane–1,4-dichlorobenzene. In these last two classes, the triple-point temperature of the solid is higher than C_1. In class 5, the s_2lg line is at lower pressures than the critical curve, with a phase boundary line J originating on the s_2lg line.

Class 6. Class 6 binary P–T diagrams include, for example, CO_2–naphthalene and CO_2–diphenylamine. The s_2lg line extends to pressures high enough to cut the critical curve at the upper (U) and lower (L) critical end points. At T_5, Figure 4(**j**) is applicable, whereas a temperature between points U and L leads to the P–x diagram in Figure 4(**l**), which does not include an s_2lg line. The composition is extremely sensitive to temperature and pressure in the vicinity of U or L (7). Therefore, solubilization and precipitation of the extract require only a small amount of energy.

Temperature-controlled and pressure-controlled SFE may be accomplished in all of these classes. For example, a reduction in pressure at constant temperature from point M leads to phase separation, eg, see Figure 4(**c**). An increase in temperature at constant pressure accomplishes phase separation in classes 1–4 and in some regions of classes 5 and 6. It is not always necessary to form a new phase for separation in SFE. For example, the solubility of naphthalene in ethylene is extremely sensitive to pressure and temperature in the two phase fluid–solid region as shown in Figure 2 [not as well seen in the curvature of Fig. 4(**l**)].

Binary Systems. Methane with the n-paraffins through pentane gives class-1 behavior, whereas it gives class 4 with n-hexane, class 6 with n-C_7–C_{10} paraffins and class 3 with the branched C_{30} paraffin squalane (2). With the n-paraffins, ethane does not become immiscible until C_{19}, and propane until C_{38}. For hydrocarbon solvents of six or more carbon atoms, immiscibility is only found for polymer solutes (20). Ethylene exhibits behavior that is similar to ethane although it is a somewhat better solvent for hydrocarbons. Despite the differences in chemical structure, the solvent capacity of carbon dioxide is comparable to ethylene for hydrocarbon solids (18). Unlike ethane, CO_2 is miscible with the C_2–C_4 alcohols. Further detail for the classes of behavior of binary systems can be found in tabulated form (4).

The miscibility behavior is influenced primarily by the volatility, size, and polarity rather than the structure, aromaticity, or degree of saturation. For example, tetralin, decalin, and the C_{13}–C_{16} alkanes all exhibit class 3 behavior with CO_2, whereas cyclohexane, benzene, *n*-hexane, and 1-hexene show class 1. The maximum pressure on the critical curve is also related to the volatility differences between the following solutes and CO_2: *n*-butane (8 MPa or 1160 psi), *n*-octane (16 MPa or 2320 psi), *n*-hexadecane (30 MPa or 4350 psi), and squalane (40 MPa or 5800 psi). The characteristics of polar solvents such as water and ammonia are discussed in ref. 23.

The solubility of solid hydrocarbons in solvents such as ethylene, ethane, and carbon dioxide has been measured recently, primarily using dynamic experimental techniques (17–18,26–29) to develop a pressure–temperature–composition (P–T–x) data base for modeling SFE. Table 2 lists systems in which solid–fluid equilibria have been measured at near-critical conditions of interest in SFE. Unless indicated otherwise, the references and temperature and pressure ranges can be found in Tables 2 and 3 of ref. 7. Correspondingly, Table 3 lists liquid–fluid equilibria systems. Additional systems are provided in the extensive reviews of Paulaitis and co-workers (7), and of Randall (6). The latter also provides descriptions of systems measured specifically for the development of practical applications, and reviews supercritical-fluid chromatography (SFC) (30–32). Compared with conventional gas chromatography, SFC offers two advantages: lower separation temperatures for thermally labile substances and the alternative of pressure programming to achieve fractionation instead of temperature programming. Although SFC is not as effective as high pressure liquid chromatography (HPLC) for analytical purposes, it may be superior for preparative applications due to the ease of reprecipitation from the mobile phase.

Ternary Systems. A near-supercritical solvent can be used to modify the mutual miscibility of a binary mixture. Small changes in the temperature or pressure of the mixture may have a pronounced effect on the area of the immiscible region, as shown on a ternary diagram (33–35). Furthermore, the addition of another component or

Table 3. Solubility Data for Supercritical Fluid–Liquid Equilibria Systems at Near-Critical Conditions[a,b]

Liquid–fluid equilibria system	Liquid–fluid equilibria system
C_2H_4 (T_c = 282 K)	ethanol
1-methylnaphthalene	1-butanol
eicosane	water
squalane	C_2H_6 (T_c = 305 K)
hexadecane	*n*-pentane
bicyclohexyl	hexadecane
ethanol	eicosane
methanol	squalane
diphenylamine	1-methylnaphthalene
CO_2 (T_c = 304 K)	bicyclohexyl
n-pentane	diethyl ether
n-decane	ethanol
i-octane	methanol
toluene	1-butanol
carbon tetrachloride	water
diethyl ether	diphenylamine
methanol	

[a] At least part of the range $0.9 < T_r^{solvent} < 1.15$ was investigated for these systems.
[b] References for supercritical fluid–binary liquid equilibria data are available in ref. 7.

entrainer to a mixture already consisting of a supercritical component can modify extraction selectivities markedly. The entrainer can effect the system by amplifying the volatility of the solutes, by densifying the solvent, or by dilating a liquid condensed phase, thus increasing its solvent capacity (2). It can be used to reduce, as well as increase, miscibility.

Ternary systems near the critical point of the supercritical solvent may display regions of three-phase *llg* behavior, which are analogous to the binary case, as shown for acetone–water–ethylene in Figure 5 (33–34). Ethylene, when dissolved in many water–organic mixtures at elevated pressures, decreases the mutual solubility of the liquid components. This can be observed as an introduction of a region of immiscibility or an increase in the size of an existing immiscibility gap. In Figure 5, the selectivity is 71 for the acetone solubility in the ethylene-rich versus the water-rich phase. Twenty-six organic compounds consisting of acetone, aldehydes, alcohols, and others were also studied, each with the ethylene–water system. This was the first reference to the use of supercritical fluids as extraction media. For the system ethylene–methyl ethyl ketone–water, small loadings of ethylene below its critical pressure greatly reduced the mutual miscibility yielding a selectivity of 705. This salting-out effect can be used to dehydrate many types of organic compounds, with numerous companies interested in ethanol–water separation discussed below.

The role of supercritical water in the separation of aliphatic from aromatic hydrocarbons has been reviewed (2,36). The class 3 critical curve for the water–benzene system exhibits a temperature minimum at significantly lower temperatures than

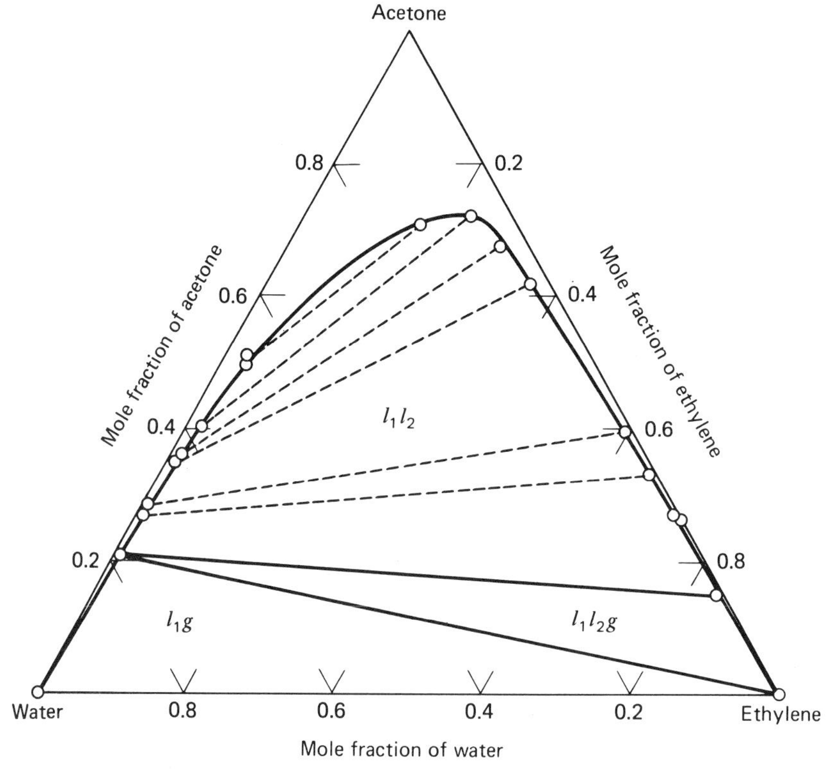

Figure 5. Ternary phase diagram for ethylene–water–acetone at 288 K and 4.93 MPa (715 psi).

water–n-heptane. This basis for an extractive separation was also observed for a branched chain aliphatic, thus demonstrating the general applicability of the phenomenon. The mutual solubilities of hydrocarbons along the three-phase *llg* equilibrium line at conditions near the critical point of CO_2, methane, or ethylene have been investigated systematically (37). For the system methane–decane–dotriacontane, the dissolved gaseous methane dilated the liquid solvent decane, thus enhancing the solubilities of dotriacontane particularly near its melting point. This phenomenon has important applications in mobilizing large molecular components in tertiary oil recovery.

The combination of a supercritical as well as a liquid solvent provides flexibility in utilizing the advantages of both types of solvents. The class 3 supercritical CO_2–decalin system can be changed to class 1 (miscible) by the addition of benzene (2). The addition of acetone as an entrainer to the quasibinary supercritical CO_2–glycerides system leads to an increase of glycerides in the gas phase, with a decrease in the separation factor between glycerides (38). The entrainer improved the economics of the process by allowing for a lower extraction pressure. In a system of two solids in equilibrium with a supercritical fluid, the solubility of a given solid may be significantly enhanced when the second solid also dissolves in the fluid (39). These benefits resulting from supercritical solvents with or without entrainers should lead to a wide variety of new separation processes.

Theory of Supercritical-Fluid Phase Behavior

Quantitative correlation, and to a lesser degree prediction of supercritical-solution phase behavior, is beginning to emerge, based on recent advancements made in high pressure solution thermodynamics using computers (40–42). There are two important problems in developing these thermodynamic models: these mixtures are often highly asymmetric, in terms of the size and energy differences between the components reflected, for instance, in the critical properties; and the isothermal compressibility of the fluid is ca 1000 times that of a triple-point liquid. Dissolution of a solid in the highly compressible fluid leads to strong attraction that causes the mixture to contract with very large negative values of the partial molar volume of the solute. Most of this discussion is therefore focused on the simplest case of solid–fluid equilibria, in which the solubility of the fluid in the solid phase is negligible.

Thermodynamic models for SFE are based on the principle that the fugacities of component i, f_i are equal for all phases at equilibrium at constant temperature and pressure. In the supercritical-fluid phase, the fugacity can be represented by an expression for an expanded liquid (L) or a dense gas (V):

$$f_i^{\mathrm{L}} = f_i^{0\mathrm{L}}(P^0)\gamma_i(P^0,x_i)x_i \exp \int_{P^0}^{P} \frac{\bar{v}_i(x_i,P)dP}{\mathrm{RT}} \quad (1)$$

$$f_i^{\mathrm{V}} = \phi_i y_i P \quad (2)$$

where x_i is the expanded liquid phase mole fraction, γ_i the activity coefficient for deviations from Raoult's law, the superscript ^{0} implies reference, $\bar{v}_i$ in the Poynting correction is the partial molar volume, and y_i is the vapor phase mole fraction. The fugacity coefficient ϕ_i which characterizes nonideal behavior relative to the ideal gas can be calculated from an equation of state with the associated mixing rules.

The following expanded liquid models have been used: the Poynting correction has been absorbed into a pressure-dependent activity coefficient, which was calculated using regular solution theory (RST) (30,43–44); and the activity coefficient was treated as a constant and $\bar{v}_i$ was calculated from the Redlich-Kwong equation of state (45). In the dense gas treatments, the crucial property ϕ_i has been obtained using second virial coefficients (1,46), cubic equations of state (17,28,47–48), and statistical mechanical perturbation theory (18,49). Given ϕ_i, the calculation of y_i is straightforward for solid–fluid equilibria. The solubility of a solid, component 2, in the fluid based on equation 2 is given by

$$y_2 = \frac{P_2^s}{P} \left\{ \frac{\phi_2^s \exp\left[\frac{v_2^s (P - P_2^s)}{RT}\right]}{\phi_2} \right\} \tag{3}$$

where P_2^s is the vapor pressure with the superscript s denoting saturation. The enhancement factor E, shown in braces in equation 3, is defined as the real solubility divided by the ideal solubility. This equation assumes that the solid phase is pure. Enhancement factors of 10^4 are common for supercritical systems, whereas the squalane–carbon dioxide system gives an enhancement factor greater than 10^{10}. The solubility data can be reduced to a simple form by plotting the logarithm of the enhancement factor versus density which results in a fairly linear form (17).

The original attempts to model SFE (46) utilized the virial equation truncated at the second coefficient to obtain the fugacity coefficient; however, this equation of state is applicable only to ¼ to ½ of the critical density. The gross deviations in the predictions of the solubility-versus-pressure isotherms for naphthalene–carbon dioxide using the ideal gas and truncated virial equations are illustrated in ref. 7. The extension of the virial equation is limited by lack of knowledge about higher order coefficients. Regular solution theory was used to model the fugacity of an expanded liquid by modifying equation 1. The new pressure-dependent activity coefficient was calculated from RST, and implicitly included the Poynting correction. The solubility parameters used in RST, which were now a function of pressure, could be obtained from P–V–T–enthalpy data, developed originally for supercritical-fluid chromatography (30), or an equation of state. Regular solution theory assumes a value of zero for the excess volume and entropy of mixing, which is unrealistic for supercritical fluids at conditions where the compressibility is high. For example, the partial molar volume of the solute reaches negative infinity in the limit of the solvent critical point (50). However, modified RST has been successful (44), particularly at higher pressures away from the highly compressible region, and other solution theories such as nonrandom two liquid (NRTL) or universal quasi-chemical (UNIQUAC) can probably be applied. The fugacity and activity coefficients for systems of interest in mineralogy and geology are calculated in ref. 51.

Most recent investigators have chosen the dense gas approach utilizing two or three parameter cubic equations of state that are suitable for compressible mixtures (17,28,47–48). These models are used to calculate the fugacity coefficient of the solute, which is typically on the order of 10^{-4}, and is the primary contributor to the enhancement factor. The logarithms of ϕ_2 are nearly linear in the most important variable, the unlike-pair-attraction parameter; therefore, an error of 1% in this parameter gives a 30% error in ϕ_2 and thus y_2. This error magnification in taking the exponential of log ϕ_2 is much less important at lower pressures where ϕ_2 approaches unity. For these reasons, the more important concern in these models is not which of the better cubic equations of state is used but, instead, how the size parameters and particularly the attraction parameters are obtained. The unlike-pair-attraction parameters are usually fit from the experimental solubility data, and display a weak temperature dependence. The agreement is poor between the fitted parameters and the ones predicted using critical properties supplemented with the commonly used deviation parameter k_{ij} (40). This is not surprising since it is nearly impossible to choose a corresponding-states theory based on critical properties that can describe these asymmetric mixtures, in which the constituents have vastly different critical properties. This difficulty has been circumvented by successfully cross-correlating the unlike-pair-attraction parameter with the heat of vaporization of the pure solute. Another benefit of this approach is that heat-of-vaporization data are more available than the critical properties for solids (17–18).

In the near-critical region, ie, at reduced pressures (P/P_c) less than ca 1.3 and temperatures (T/T_c) below ca 1.2, the fluid is highly compressible and the cubic equations of state along with conventional random mixing rules predict a lower-than-actual solubility (18). The attractive forces lead to large negative values for the partial molar volume of the solute and the mixture becomes highly nonrandom. Several investigators used statistical mechanical perturbation theory to model this region (18,49).

Although models for solid–fluid equilibria are presently giving valuable information about supercritical solutions primarily at dilute conditions, there are numerous additional problems that require attention: polar and hydrogen-bonded species, liquid condensed phases containing dissolved fluids, highly concentrated supercritical solutions in which solute–solute interactions are important, and multicomponent systems. Models which calculate mixture critical curves for multicomponent mixtures would also be useful for SFE. This calculation uses a lattice-gas treatment (52), whereas others have used a Peng-Robinson equation (7) and other cubic equations of state (53). The significant growth in the theoretical understanding of these solid–fluid systems in the last decade provides the basis for answering these further problems necessary to benefit the development of SFE.

Economic Aspects

The consensus of the review papers (2,4,7) is that SFE processes must be considered case by case to determine the economic feasibility compared with distillation and liquid extraction. Generally, these processes involve higher capital costs for high pressure equipment which may be more than offset by reduced utility costs. Both costs have been subject to hyperinflation in recent years, which complicates prediction of the process economics. The SFE process can often use small pressure or temperature changes for solvent recovery to avoid the massive latent-heat requirements needed for the alternative processes. Specifically, the temperature-controlled extraction mentioned above is highly efficient as the heat is transferred directly between the heating and cooling stages, at nearly isobaric conditions with minimal compression energy. For a hypothetical pressure-controlled separation using propane, the energy requirements, which can often dictate the feasibility of SFE, were favorable to distillation if the compression ratio was less than three (2). In some smaller-scale applications, particularly in the food industry, SFE with carbon dioxide (an inexpensive solvent) does not require expensive processing to remove residual toxic solvent, as needed in liquid extraction. Furthermore, SFE may offer extremely narrow fractions in specialty chemicals refining, which are virtually unobtainable using conventional techniques.

As many of the developing commercial applications are in a competitive and infantile stage, few examples of the process economics are available. In the temperature-controlled Residuum oil supercritical extraction (ROSE), discussed in the section on petroleum applications, the utility costs are reduced by 60% compared with conventional processing, as well as the capital costs (12). In biotechnology, the economics are influenced strongly by energy costs for concentration of the dilute products of fermentation. A DOE report on ethanol–water separation cited an energy advantage of about 3.5 MJ/kg (ca 1500 Btu/lb) of ethanol for the carbon dioxide SFE process relative to distillation (7). The economics for the separation of glycerides with carbon dioxide as the solvent and acetone as the entrainer are evaluated in ref. 38. The entrainer which enhanced the volatility of the glycerides allowed a pressure reduction of 13.5 MPa (ca 1960 psi) to 8 MPa (1160 psi) for reduced capital costs and a savings of 30% in operating costs. To limit the costs for the pressure vessels, a tall, slim column was recommended which necessitates a relatively low reflux ratio. Therefore, SFE, less energy-intensive and more discriminating for nonvolatile substances, will be economically advantageous in cases in which these factors compensate for the generally higher capital costs for high pressure equipment.

Environmental Considerations

It is precisely the safety features of carbon dioxide that make it an attractive solvent for the food and pharmaceutical industries. It is environmentally nonhazardous, nontoxic, nonflammable, noncorrosive, and inexpensively available in high purity. In the food industry, it is not regarded as a foreign substance or additive, and is an inert that does not react with food constituents, unlike organic solvents such as hexane or methylene chloride (54). In general, SFE does not leave residues in the product, which can be a problem in conventional processes (see Extraction, liquid–solid; Extraction, liquid–liquid; Solvents, industrial). The hazards of solvent flammability, solvent content in the work atmosphere, and corrosiveness are described in this encyclopedia for a number of compounds which are commonly used as supercritical solvents [see, for example, Ammonia; Water; Methanol; Ethanol; Propyl alcohols; Hydrocarbons C_1–C_6; Ethylene; Benzene; Toluene; Fluorine compounds, organic; Nitrogen (nitrous oxide); Fluorine compounds, inorganic (sulfur hexafluoride)]. Halogenated solvents are being severely regulated (see Industrial hygiene and toxicology; Toxicology). Other environmental considerations include liquid effluents and gaseous emissions (see Wastes, industrial).

As the ASME Code Section VIII, Division 1 is limited to 20.7 MPa (3000 psi), most of the designs for SFE are restricted below this pressure, if possible, to avoid further regulations. At these pressures, the stored energy is much larger for highly compressible supercritical fluids compared to incompressible liquids. The requirements for the high pressure design of process equipment including corrosion effects have been discussed elsewhere in this encyclopedia (see High pressure technology).

Applications of SFE

Food and Pharmaceutical Applications. Nontoxic carbon dioxide, which is relatively inexpensive, nonflammable, noncorrosive, and natural, can be used to extract thermally labile food components at near-ambient temperatures. The food product is not contaminated with residual solvent, as in the case of conventional toxic liquid solvents such as methylene chloride. Not only can these supercritical solvents remove oils, or substances such as caffeine from food substrates, but they can also fractionate mixtures such as glycerides and vegetable oils into numerous components. Compared with liquid carbon dioxide, the solvent capacity of supercritical carbon dioxide is superior, with an extraction rate at least 2.5 times higher due to a higher diffusivity, and a wider range of possible operating temperatures.

A process for the removal of caffeine from coffee beans using supercritical carbon dioxide was patented in the United States in 1974, and a commercial plant went on stream in the FRG in 1978. The prolific patent activity for decaffeination in 1981 that resulted in six U.S. patents, including General Foods and Nestlé, is discussed in ref. 7. Three supercritical CO_2 decaffeination processes that use various techniques for solvent recovery from the caffeine-containing effluent of the extraction stage are described in ref. 13. One process uses water to wash the caffeine from this mixture, and another uses activated carbon to adsorb the caffeine. The third and most simple method involves charging a vessel containing the beans and activated carbon to supercritical conditions to cause the caffeine to migrate to the fluid phase where it is subsequently adsorbed by the carbon.

Carbon dioxide is used to extract α-acids from hops, which are used to give beer (qv) its characteristic bitter taste (54–55). Although the yields are similar to those using toxic methylene chloride, the extract's color, composition, and texture are more controllable. Other flavors and fragrances that are discussed include lilac, lemon peel, black pepper, almonds, nutmeg, and ground chilies (see also Flavors and spices). In another application using carbon dioxide, the denicotinization of tobacco was complicated by the high solubility of the aroma-rich substance (54). The process must be staged to dearomitize the tobacco, then to denicotize the dearomitized tobacco, and finally to homogeneously impregnate the aroma into the denicotinized tobacco. The nicotine reduction of 94.7% easily meets the stipulated level. Unlike processes using liquid organic solvents, this one does not cause the leaves to acquire a rubbery texture.

The USDA has been testing supercritical carbon dioxide for extraction of oils, particularly triglycerides, from soybean flake and corn germ (7) (see Vegetable oils; Soybeans and other oilseeds). This process provides an oil low in phosphorus compounds, unlike conventional hexane extraction, which additionally requires vacuum distillation to remove these compounds. Numerous patents resulting from work at the Max Planck Institute include deodorization of soybean, palm, and peanut oils, with simultaneous removal of free fatty acids using CO_2; oil extraction from soybean flakes, corn, and bones using propane, ethane, CO_2, or N_2O; and partial hydrogenation and deodorization of fats or oils with CO/H_2 (6). Air Products and Chemicals obtained U.S. licensing rights to the Max Planck Institute patents, which may lead to some commercial processes. Critical Fluid Systems (Cambridge, Massachusetts) (7,56) is developing the use of supercritical fluids to produce low oil-content snack foods, such as defatted potato chips, with improved nutritional value and shelf life.

The use of supercritical fluids for the extraction of flavors and fragrances in natural products is discussed in ref. 57. A microextractor coupled with a thin-layer chromatography plate has been used to remove compounds including limolene, menthol, caffeine, and triglycerides from plant materials such as caraway fruits, peppermint leaves, camomile flowers, raw coffee, sunflower seeds, and sesame seeds (58–59); the alkaloids (qv) that had relatively polar groups such as hydroxyls were extracted more successfully with nitrous oxide than carbon dioxide. Supercritical ethane is used to separate cod-liver oil into 50 fractions, some having molecular weights as high as triglycerides, which are inseparable using conventional distillation (13). In pharmaceutical applications, supercritical fluids can potentially extract drugs from plants, without the side effects of chemical decomposition that can result from extraction with organic liquid solvents (60).

Petroleum Applications. The energy costs for distillation or liquid solvent extraction have increased tenfold during the last decade; therefore, the less-energy-intensive SFE can be attractive if the capital costs for high pressure equipment are not prohibitive. Kerr-McGee developed a ROSE process in the 1950s. However, it was not commercialized until the late 1970s, when crude oil and utility costs were no longer inexpensive (12). Several feedstocks, including atmospheric and vacuum-distillation residua, were converted to cat-cracker feed and lubricating-oil stocks using supercritical pentane. After mixing the residuum and pentane (Fig. 6), the insoluble asphaltenes were recovered containing the undesirable heavy metals and carbon residues. The resin and oil fractions were sequentially precipitated from the solvent by stepwise isobaric temperature increases that reduced the solvent density. Compared with conventional propane deasphalting which requires large amounts of latent heat, this temperature-controlled extraction reduces capital and energy costs as it efficiently exchanges heat, as explained above. The higher process temperature gave favorable flow conditions for the viscous asphaltenes. The extraction yields were higher with less carbon residue and heavy metal contamination, which poison hydrotreating catalysts.

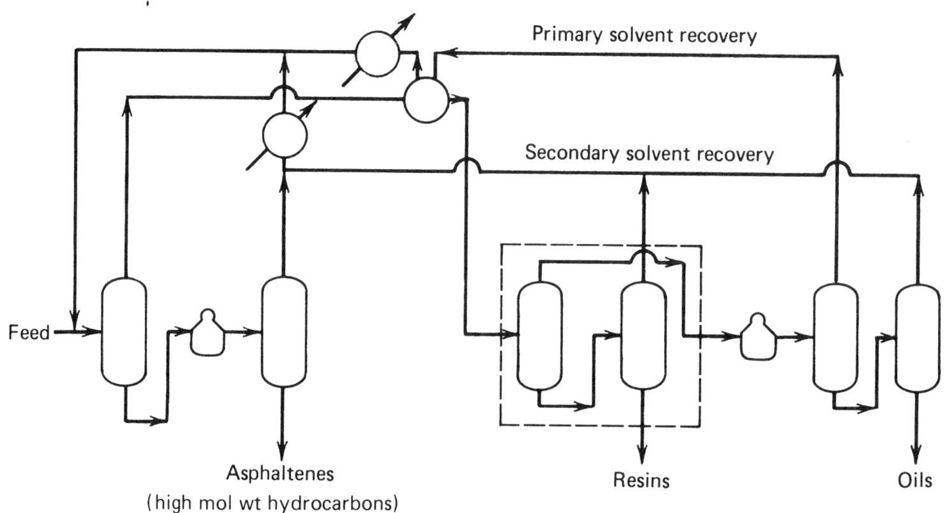

Figure 6. Residuum oil supercritical-extraction (ROSE) process using pentane with reduced energy requirements.

Several types of dense gas injections have been tested for tertiary oil recovery from petroleum reservoirs (2). The propane slug and enriched gas-drive processes achieve miscibility at moderate pressures (7–25 MPa or 1000–3600 psi), but are not economically feasible for most reservoirs as they use expensive hydrocarbons. Although carbon dioxide flooding requires higher pressures (>28 MPa or >4060 psi) to achieve miscibility, it is less expensive and is in the field-development stages (4).

Supercritical-fluid extraction can be used to fractionate low vapor-pressure oils, which would be difficult using either high vacuum or molecular distillation, since the impurities have about the same volatility as the primary homologues. Furthermore, the solubility of these species is sufficiently high in liquid solvents that it would not give desirable selectivities. In contrast, the solvent power in SFE can be carefully controlled using pressure and temperature to fractionate synthetic or natural oils such as silicones, fluoroethers, α-olefins, and petroleum resins. The Phasex Corporation fractionated fluoroethers and silicone oil, which contained high molecular weight components, by using stagewise reductions in pressure (7). Air Products and Chemicals may market a crankcase oil re-refining process (7). *n*-Paraffins are separated from wax distillate using supercritical fluids and molecular sieves (61). In these oil fractionations, SFE appears to be markedly more effective in achieving separation compared to the relatively unsuccessful alternatives.

Synthetic Fuels Applications. Supercritical fluids are utilized in coal processing in the critical solvent deashing (CSD) process of Kerr-McGee used in a 6 t/d solvent refined coal demonstration plant, and the coal-SFE process of the UK's National Coal Board (see Coal). The primary purpose of CSD is to remove the ash and insolubles in the effluent from the dissolution stage to provide a clean feed for the upgrading stage (62). In CSD, subcritical toluene, or a similar aromatic solvent, contacts the dissolved coal at a ratio of 2.5:1 at ca 560 K and 5.5 MPa (ca 800 psi). Although most of the organic material passes through the overflow, a small amount precipitates with the ash. The solids settle rapidly in this low viscosity solution even at solvent upflow velocities greater than 25 mm/s (90 m/h). The first stage overflow is heated to recover the solvent for recycle, with the cleaned coal withdrawn as a molten fluid. In addition to providing high settling rates and a low ash content in the cleaned coal of 0.1 wt %, this process offers enormous potential benefits that are currently unproven. Instead of using the second stage for solvent recovery only, it could potentially be replaced by a series of fractionators to obtain specific chemical components for selective solvent recycle to the dissolver, or for feed to various types of downstream processing units.

In the coal-SFE process, a mixture of bicyclic aromatic/naphthenic hydrocarbons is used not only to dissolve coal but to stabilize the intermediate products from the thermal decomposition (4,63). The rate of extraction is determined by the coal decomposition rate, the rate of transfer of the volatile materials to the supercritical phase, and the repolymerization rate of the extractable material to insoluble species. The high diffusivity of the fluid provides favorable mass-transfer rates (3). The products that have a molecular weight of 400 are extracted readily, whereas those of 1000 require higher pressures and longer times. This process selectively extracts the lighter hydrogen-rich material from coal giving a maximum yield of 50%, with the remaining coal used to produce hydrogen for downstream upgrading and to produce power and steam. The porous char has excellent characteristics for gasification, and the high quality product can be used as a chemical feedstock. Again, as in CSD, the solvent recovery and solids removal are favorable. This process was tested at a scale of 5 kg/h, with the design for a 1-t/h pilot plant completed in 1981. Solvents for SFE of coal have been systematically studied (64–65), and the chemical nature of supercritical extracts is described in ref. 66.

Supercritical-fluid extraction has been proposed to derive alternative sources of fuel from tar sands (qv) (4,67–68), lignite (qv) (69), wood (qv) (69–70), and oil shale (qv) (4), although these processes are far from commercialization. The Standard Oil Company of Indiana used supercritical water with a density above 0.2 g/cm^3 to recover the bitumen from tar sand, and Williams used supercritical tetrahydrofuran [583 K, 10 MPa (1450 psi)] or toluene [668 K, 10 MPa] to remove 95% of the organic matter from Athabasca tar sands (4). Compared to oil-shale retorting at 870 K, which extracted 71% of the kerogen, supercritical toluene at 713 K and 10 MPa removed 88% (4), whereas similar treatment with nitrogen removed only about 17% (71). MODAR, Inc., claims supercritical water will be used eventually for liquefaction and gasification of forest products and fossil fuels because it exhibits an affinity for, and an ability to react with, organic compounds (72). This is caused by the loss of hydrogen bonding in water in the critical region [647 K, 22 MPa (3200 psi)]. With a glucose–water system as a model, supercritical water limits charring during gasification by preventing organic molecules from clustering and undergoing condensation–polymerization reactions (73).

Other Applications. The remaining applications involve the separation of chemicals from aqueous solutions, specialty chemical separations, environmental control, crystal growth, solids settling, and polymer processing (7,73–79). Supercritical solvents, especially carbon dioxide, may benefit the development of large scale biotechnology by reducing the large energy costs for concentrating the dilute products of fermentation. A number of investigators (7,74–75), along with Critical Fluid Systems and the DOE, are exploring ethanol–water separation to determine the maximum achievable concentration, energy requirements, and process economics. Concentration of a 10% ethanol to 80% (w/w) ethanol solution requires 3.5 MJ/kg less energy using supercritical CO_2 compared to distillation (7).

The distribution coefficient of the organic in the carbon dioxide phase over the water phase is the primary factor influencing the economics. Since this coefficient is only 0.1 for the carbon dioxide–water–ethanol system, a large solvent/feed ratio is needed requiring significant recovery of the process energy to be economical. The envisioned process is based on a pressure-controlled extraction (7), with a minimal operating pressure and a minimal pressure reduction to keep capital costs and recycle energy low. The carbon dioxide, which is recovered from the alcohol-concentrated extract by distillation, is recycled and recompressed, while the energy supplied to the vapor is used as a reboiler heat source. In contrast, the distribution coefficients are on the order of ten or more for higher alcohols, esters, and aromatics, which may lead to future processes. A patent has been issued for the removal of organic acids from dilute aqueous solutions of salts of organic acids using supercritical fluids (76).

Critical Fluid Systems concluded that a supercritical carbon dioxide extraction was advantageous economically for the regeneration of activated carbon, used for the adsorption of pesticides, or other pollutants in wastewater treatment (77–78). The conventional thermal-regeneration process that uses a reforming atmosphere of combustion gases and steam at ca 1200 K is capital and energy intensive, and emits corrosive gases. Owing to the difficulties of feeding solids to a high pressure vessel, the supercritical regeneration would most likely be done in multiple vessels in parallel using a pressure-controlled extraction. One difficulty is that irreversibly adsorbed compounds, which are not soluble in the carbon dioxide, may build up on the carbon. However, this process does not suffer from the disadvantage of adsorbate degradation owing to cycles of heat-related expansion and contraction.

In another environmental application, MODAR, Inc., has developed supercritical wastewater-treatment systems for hazardous or toxic organics in various scales up to maximum of 95 m^3/d (ca 25,000 gal/d) of organic wastes in up to 1900 m^3 (500,000 gal) H_2O/d (71–72). At supercritical conditions, water is an excellent solvent for the oxidation of complex organic substances. This process can treat a broad spectrum of hazardous wastes with 99.99% destruction efficiencies for organic chlorides, sulfides, and phosphates. The energy requirements are self-generated, with markedly improved energy efficiencies compared to high temperature incineration.

Although a number of companies are known to be developing supercritical fractionation and processing methods for polymers, the specific applications are proprietary. Three given in ref. 7 are extraction of residual solvents and volatiles, extraction of unreacted monomers and low molecular weight oligomers, and fractionation of polymers. In another application, supercritical-fluid nucleation offers the possibility of controlling particle size and distribution without limitations of temperature and solvent impurity (72). This is attractive for heat labile dyes, fine chemicals, pharmaceuticals, and to prepare process feeds that require a specific particle size. Finally, supercritical water at ca 670 K and 100–200 MPa (14,500–29,000 psi) has been proposed as a solvent for hydrothermal breeding of synthetic quartz crystals (4). A vast majority of these applications utilize separations based on the phase-equilibria behavior, and a whole new field could follow using supercritical fluids to control chemical reactions. A conference entitled *Supercritical Fluids: Their Chemistry and Application* discussed this concept, along with a wide variety of aspects of SFE such as physical properties, theoretical models, entrainers, and energy applications (80).

BIBLIOGRAPHY

1. P. M. F. Paul and W. S. Wise, *The Principles of Gas Extraction*, Mills and Boon, London, 1971.
2. C. A. Irani and E. W. Funk, *Recent Developments in Separation Science*, Vol. III, CRC Press, West Palm Beach, Fla., 1977, Part A, p. 171.
3. N. Gangoli and G. Thodos, *Ind. Eng. Chem. Prod. Res. Dev.* **16,** 208 (1977).
4. D. F. Williams, *Chem. Eng. Sci.* **36,** 1769 (1981).
5. G. Brunner and S. Peter, *Chem. Ing. Tech.* **53,** 529 (1981).
6. L. G. Randall, *Sep. Sci. Technol.* **17,** 1 (1982).
7. M. E. Paulaitis, V. J. Krukonis, R. T. Kurnik, and R. C. Reid, *Rev. Chem. Eng.* **1,** 179 (Apr.–Jan. 1983).
8. M. E. Paulaitis, J. M. L. Penninger, P. Davidson, and R. D. Gray eds., *Chemical Engineering at Supercritical-Fluid Conditions*, Ann Arbor Science Publishers, Ann Arbor, Mich., 1983.
9. J. B. Hannay and J. Hogarth, *Proc. R. Soc. London Ser. A* **29,** 324 (1879).
10. D. L. Katz and F. Kurata, *Ind. Eng. Chem.* **32,** 817 (1940).
11. T. P. Zhuze, *Petroleum (London)* **23,** 298 (1960).
12. J. A. Gearhart and L. Garwin, *Oil Gas J.* **74**(24), 63 (1976).
13. K. Zosel, *Angew. Chem. Int. Ed. Engl.* **17,** 702 (1978).
14. Y. V. Tsekhanskaya, M. B. Iomtev, and E. V. Mushkina, *Russ. J. Phys. Chem.* **38,** 1173 (1964).
15. G. A. M. Diepen and F. E. C. Scheffer, *J. Phys. Chem.* **57,** 575 (1953); *J. Am. Chem. Soc.* **70,** 4081 (1948).
16. J. Eisenbeiss, Final Report, Contract No. DA-18-108-AMC-244(A), Southwest Research Institute, San Antonio, Texas, 1964.
17. K. P. Johnston and C. A. Eckert, *AIChE J.* **27,** 773 (1981).

18. K. P. Johnston, D. H. Ziger, and C. A. Eckert, *Ind. Eng. Chem. Fundam.* **21,** 191 (1982).
19. G. Brunner, D. Stuetzer, and S. Peter in ref. 8, p. 435.
20. J. S. Rowlinson, *Liquids and Liquid Mixtures*, 3rd ed., Butterworths, London, 1982.
21. G. M. Schneider, *Angew. Chem. Int. Ed. Engl.* **17,** 716 (1978).
22. G. M. Schneider in B. Vodar and P. Marteau, eds., *High-Pressure Science and Technology, 7th AIRAPT Conference*, Vol. 2, 1980, p. 685.
23. E. U. Franck, *Pure Appl. Chem.* **53,** 1401 (1981).
24. W. B. Street in ref. 8, p. 3.
25. E. J. Shimshick, *Chemtech*, 374 (June 1983).
26. R. A. van Leer and M. E. Paulaitis, *J. Chem. Eng. Data* **25,** 257 (1980).
27. M. E. McHugh and M. E. Paulaitis, *J. Chem. Eng. Data* **25,** 326 (1980).
28. R. T. Kurnik, S. J. Holla, and R. C. Reid, *J. Chem. Eng. Data* **26,** 47 (1981).
29. M. B. King in ref. 8, p. 31.
30. J. C. Giddings, M. N. Meyers, L. McLaren, and R. A. Keller, *Science* **162,** 67 (1968).
31. T. H. Gouw and R. E. Jentoft, *J. Chromatogr.* **68,** 303 (1972).
32. E. Klesper, *Angew. Chem. Int. Ed. Engl.* **17,** 738 (1978).
33. D. Todd and J. Elgin, *AIChE J.* **1,** 20 (1955).
34. J. C. Elgin and J. J. Weinstock, *J. Chem. Eng. Data* **4,** 3 (1959).
35. S. R. M. Ellis, *Br. Chem. Eng.* **16,** 358 (1971).
36. C. A. Irani and D. J. McHugh, *High-Pressure Science and Technology, 6th AIRAPT Conference*, Vol. 1, 1979, p. 600.
37. N. C. Huie, K. D. Luks, and J. P. Kohn, *J. Chem. Eng. Data* **18,** 311 (1973).
38. S. Peter and G. Brunner, *Angew. Chem. Int. Ed. Engl.* **17,** 746 (1978).
39. R. T. Kurnik and R. C. Reid, *Fluid Phase Equilibria* **8,** 93 (1982).
40. J. M. Prausnitz, *Molecular Thermodynamics of Fluid Phase Equilibria*, Prentice-Hall, Englewood Cliffs, N.J., 1969.
41. R. C. Reid, J. M. Prausnitz, and T. K. Sherwood, *The Properties of Gases and Liquids*, 3rd ed., McGraw-Hill Book Co., New York, 1977.
42. J. M. Prausnitz and P. L. Chueh, *Computer Calculations for High Pressure Vapor-Liquid Equilibria*, Prentice-Hall, Englewood Cliffs, N.J., 1968.
43. J. M. Prausnitz, *ACS Symp. Ser.* **60,** 11 (1977); *NBS Technical Note 316*, National Bureau of Standards, Washington, D.C., 1965.
44. A. Vetere, *Chem. Eng. Sci.* **34,** 1393 (1979).
45. M. E. Mackay and M. E. Paulaitis, *Ind. Eng. Chem. Fundam.* **18,** 149 (1979).
46. A. H. Ewald, W. B. Jepson, and J. S. Rowlinson, *Discuss. Faraday Soc.* **15,** 238 (1953).
47. G. Brunner and H. Hederer in ref. 36, p. 527.
48. S. Peter and H. Wenzel, *EFCE Publ. Ser., Phase Equil. Fluid Prop. Chem. Ind.* **11,** 355 (1980).
49. K. S. Shing, D. Jonah, V. Venkatasubramanian, and K. E. Gubbins in ref. 8, p. 221.
50. M. E. Paulaitis, K. P. Johnston, and C. A. Eckert, *J. Phys. Chem.* **85,** 1770 (1981).
51. J. R. Holloway, *NATO Adv. Study Inst. Ser. Ser. C* **C30,** 161 (1977).
52. L. A. Kleintjens and R. Koningsveld, *J. Electrochem. Soc.* **127,** 2352 (1980).
53. U. Deiters and G. M. Schneider, *Ber. Bunsenges. Phys. Chem.* **80,** 1316 (1976).
54. P. Hubert and O. G. Vitzthum, *Angew. Chem. Int. Ed. Engl.* **17,** 710 (1978).
55. H. Brogle, *Chem. Ind. (London)*, 385 (June 19, 1982).
56. *Chem. Eng. News*, 52 (Nov. 15, 1982).
57. A. B. Caragay, *Perfum. Flavor.* **6,** 43 (1981).
58. E. Stahl, W. Schilz, E. Schutz, and E. Willing, *Angew. Chem. Int. Ed. Engl.* **17,** 731 (1978).
59. E. Stahl, E. Schuetz, and H. K. Mangold, *J. Agric. Food Chem.* **28,** 1153 (1980).
60. P. M. Kohn, P. R. Savage, and S. McQueen, *Chem. Eng. (N.Y.)* **86**(6), 41 (1979).
61. P. Barton and D. F. Hajnik, *ACS Symp. Ser.* **135,** 221 (1980).
62. R. M. Adams, A. H. Knebel, and D. E. Rhodes, *Chem. Eng. Prog.* **75,** 44 (1979).
63. R. R. Maddocks, J. Gibson, and D. F. Williams, *Chem. Eng. Prog.* **75,** 49 (1979).
64. J. E. Blessing and D. S. Ross, *ACS Symp. Ser.* **71,** 171 (1978).
65. N. P. Vasilakos, *Fourth Quarterly Report on Solvent and Chemical-Reaction Effects in Supercritical Extraction of Coal*, DOE No. DE-FG22-81PC40801, U.S. Department of Energy, Washington, D.C., Sept. 1982.
66. T. G. Martin and D. F. Williams, *Philos. Trans. R. Soc. London Ser. A* **300,** 183 (1981).
67. W. Eisenbach, K. Niemann, and P. Gottsch in ref. 8, p. 419.

68. T. R. Bott, *Chem. Ind. (London)*, 228 (March 15, 1982).
69. A. Olcay, T. Tugrul, A. Calimli, and R. Ceylan in ref. 8, p. 409.
70. U.S. Pat. 4,308,200 (June 29, 1981), H. A. Fremont (to Champion Int. Corp.)
71. *Chem. Eng. News*, 46 (Sept. 12, 1983).
72. W. Worthy, *Chem. Eng. News*, 16 (Aug. 3, 1981).
73. U.S. Pat. 4,338,119 (June 7, 1982); Belg. Pat. 888,696 (Aug. 26, 1981), M. Modell (to Modar, Inc.).
74. M. A. McHugh, M. W. Mallett, and J. P. Kohn in ref. 8, p. 113.
75. M. S. Kuk and J. C. Montagna in ref. 8, p. 101.
76. U.S. Pat. 4,250,331 (Feb. 10, 1981), E. J. Shimshick (to Du Pont).
77. R. P. De Filippi, V. J. Krukonis, R. J. Robey, and M. Modell, *Supercritical Fluid Regeneration of Activated Carbon for Adsorption of Pesticides*, EPA-600/2-80 054, National Technical Information Service, Washington, D.C., March 1980.
78. J. C. Fetzer, J. A. Graham, R. F. Arrendale, M. S. Klee, and L. B. Rogers, *Sep. Sci. Technol.* **16**, 97 (1981).
79. *Chem. Week*, 33 (Jan. 5, 1983).
80. *Supercritical Fluids: Their Chemistry and Application*, Meeting of the Royal Society of Chemistry, Faraday Division, Cambridge, UK, Sept. 13–15, 1982, see *Fluid Phase Equilib.* **10**, 135 (1983).

KEITH JOHNSTON
University of Texas, Austin

T

TOXICOLOGY

The noxious effects of certain substances have been appreciated since the time of the ancient Greeks. However, it was not until the Middle Ages that certain principles of toxicology became formulated as a result of the thoughts of Philippus Aureolus Theophrastus Bombastus von Hohenheim-Paracelsus (1493–1541). Among a variety of other achievements, he embodied the basis for contemporary appreciation of dose–response relationships in his often paraphrased dictum: "Only the dose makes a poison."

Subsequently, further concepts of toxicology came from the work and writings of Bonaventura Orfila (1787–1853), a Spanish physician. Among his most important accomplishments were the delineation of the discipline of toxicology, attempts to correlate chemical nature and toxic effect, and the establishment of bases for the specialization of forensic toxicology (see Forensic chemistry). The twentieth century has seen significant development of the discipline of toxicology, and its establishment as a professional activity. This has been, at least in part, due to a phenomenal increase in the number of synthetic industrial chemicals and pharmaceutical preparations. Relying particularly on developments in the chemical, physical, and general biological sciences, toxicologists have made substantial advances in defining the nature and mechanisms of toxic injury, in determining factors which may influence the expression of a toxic effect, in developing methodologies, and in obtaining information on the toxicity of a multitude of discrete or combined substances. The historical development of toxicology has been reviewed (1).

There are about as many definitions of toxicology as there exist textbooks on the subject. Although they differ in detail, central to all good definitions is the concept that toxicology is concerned with the potential of chemicals, or mixtures of them, to produce harmful effects in living organisms. Toxicology is a study of the interactions between chemicals and biological systems in order to quantitatively determine the potential for such chemicals to produce injury which results in adverse health effects in intact living organisms, and to investigate the nature, incidence, mechanism of production, and reversibility of such adverse effects.

In the context of the above definition, adverse health effects are taken to mean those which are detrimental either to the survival or to the normal functioning of the individual. This definition is intended to highlight the following points with respect to phenomena investigated in toxicology.

Materials that produce harmful effects must come into close structural or functional relationship with the tissue or organ they may affect. As a result, they can physically or chemically interact with particular biological components in order to effect the toxic response.

Investigations are carried out using a variety of biological systems, including observations on exposed whole animals (*in vivo* studies) or on appropriately treated isolated tissues and cells, homogenates of tissues, or cultured lower organisms (*in vitro* studies).

If possible, there should be measurement of the toxic effect in order to quantitatively relate the observations made to the degree of exposure (exposure dose). Ideally, there is a need to quantitatively determine the toxic response to several differing exposure doses, in order to determine the relationship, if any, between exposure dose and the nature and magnitude of any effect. Such dose–response relationship studies are of considerable value in determining whether an effect is causally related to the exposure material, in assessing the possible practical (in-use) relevance of the exposure conditions, and to allow the most reasonable estimates of hazard.

A prime consideration in toxicology, and a principle reason for the need of dose–response relationship information, is the potential for a material to produce harmful effects. Thus, different materials may produce similar toxic effects, but the exposure doses to produce a just-detectable (threshold) effect, or a given degree of injury, may vary significantly among the materials. Hence, the magnitude of the injury produced will vary for similar exposure doses to different materials. Also, with a particular material that is known to cause a variety of toxic effects, the individual effects may be produced only by differing exposure conditions. In assessing the relevance of any induced toxic effects to practical situations, and in comparing and contrasting different materials, potency is clearly a central issue. Toxicity describes the nature of harmful effects produced with respect to the conditions necessary for their induction; ie, the toxicity of a material is its potential to produce biological injury.

Although studies are carried out by *in vitro* and *in vivo* procedures, it is the primary aim of toxicology to determine the potential for harmful effects in the intact living organism, usually with an ultimate goal of assessing the significance of the findings with respect to humans.

As an ideal, any series of toxicological investigations should attempt to define the following: the nature of the harmful effects, ie, the basic injury produced; the incidence and severity of the effects as functions of the exposure dose; the mechanisms by which the effects are produced, ie, the fundamental biological interactions and consequent biochemical and biophysical aberrations which are responsible for the initiation and maintenance of the toxic responses; the detection of the effects, ie, the development of methodologies for the specific recognition and quantitation of the toxic effects; and whether there is reversibility of the toxic injury. This may involve a determination of whether spontaneous resolution of injury occurs after cessation of exposure (ie, healing), or if it is possible to induce reversibility of toxic injury by antidotal or other measures (ie, treatment).

Toxicity, the potential to produce harmful effects, is to be clearly differentiated from hazard, which is the likelihood that a particular material will exhibit its known toxicity under specific conditions of use.

Classification of Toxic Effects

The diagram shown in Figure 1 gives a basis for the classification of toxic effects according to site and degree of exposure. In order to cause tissue injury, a substance must come into contact with an exposed body surface; this may be skin, eye, or the lining membranes of the respiratory and alimentary tracts. Toxic effects may be produced where a material comes into contact with a body surface, these being referred to as local effects. However, material may be absorbed from a contaminated site and be disseminated by the circulatory system to various body organs and tissues. As a consequence, toxic injury may be produced in tissues and organs remote from the site of primary contamination; these are referred to as systemic effects. As discussed below, systemic effects may be produced by the parent material that is absorbed, or by conversion products following absorption. They may be restricted to one organ or tissue system, or affect multiple organs and tissues. Many materials may cause both local and systemic toxicity.

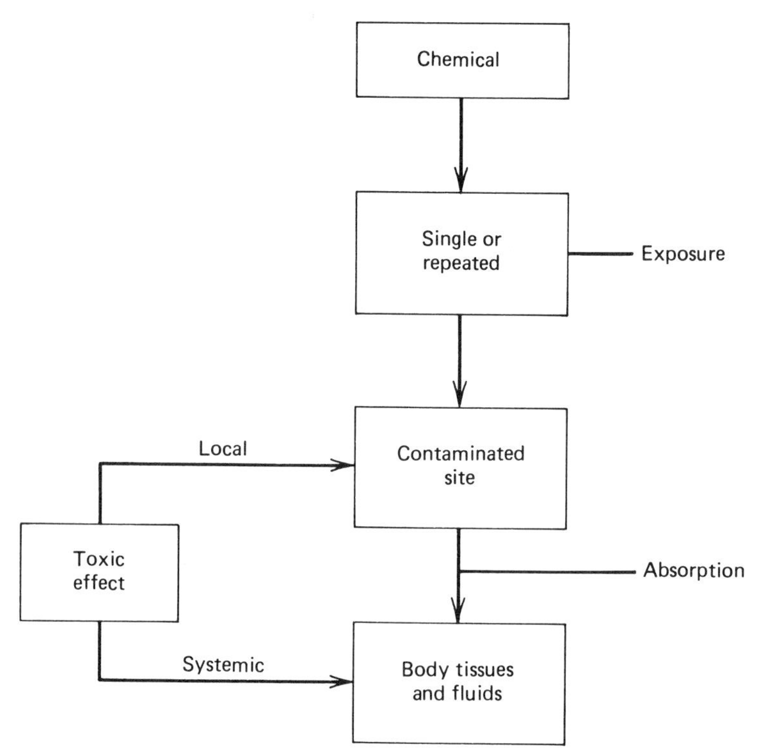

Figure 1. Schematic representation to show the basis for classification of toxic effects into local and systemic by single or repeated exposures.

The nature of a toxic effect, and the probability of its occurring, is often related to the number of exposures. The classification of toxic effects, and descriptions of toxicology tests, may be qualified by the number of exposures which elicit toxic effects. The following terms are convenient in this respect.

Acute exposures involve a single exposure to the test chemical in order to determine if this is effective in producing immediate, delayed, or persistent effects.

Short-term repeated exposures involve consecutive daily exposures to the test chemical, which are continued over a period of a few days to a few weeks, but usually not more than 5% of the lifespan of the animal. These test conditions are sometimes referred to as subacute, but this is a misleading term which should be avoided in order to prevent confusion between single and repetitive exposure toxicity.

Subchronic exposures involve consecutive daily exposures to the test material for a period amounting to usually no more than 10–15% of the lifespan of the test species.

Chronic exposures involve consecutive daily exposure to the test material over the lifespan of the test species, or a great portion of it.

The above terminology is useful to classify toxic effects with respect to their development as a function of the number of exposures. For example, acute toxic effects are those resulting from a sufficiently large single challenge to the test chemical, and developing within a short time of that exposure; in contrast, chronic toxic effects are those resulting from repetitive exposures over a significant proportion of the lifespan.

It is important to remember that some materials of low acute toxicity may have a significant potential for producing harmful effects by repeated exposure, and *vice versa*. This stresses the need for a complete overview of the toxicity of a chemical by acute and repeated exposure in the process of hazard evaluation.

The following additional descriptive terms are also useful for the classification of toxic effects.

Latent effects occur either only after there has been a significant period free of toxic signs following exposure, or those occurring after resolution of acutely toxic effects which appeared immediately following exposure. They are also referred to as delayed-onset effects.

Persistent effects do not resolve, and may even become more severe after removal from the source of exposure. They can occur as a consequence of acute or repeated-exposure conditions. Thus, the use of the term persistent should be clearly differentiated from the implication of the use of the description of an effect as chronic. It should be noted, however, that some chronic effects may be persistent; an example is malignant neoplasia.

Cumulative effects are those where there is progressive injury and worsening of the toxic effect as a result of repeated-exposure conditions. Each exposure produces a further increment of injury adding to that already existing. Many materials known to induce a particular type of toxic effect by acute exposure can also elicit the same effect by a cumulative procedure from repetitive exposure to a dose less than that causing threshold injury by acute exposure.

Transient effects are those where there is repair of toxic physical injury or the reversal of induced biochemical aberrations.

Some examples of toxic effects produced by different chemicals, and classified according to the preceding guidelines, are shown in Table 1.

Table 1. Examples of Differing Types of Toxic Effect Classified According to Time Scale for Development and Site Affected

Time scale	Site	Effect	Chemical	Ref.
acute	local	lung damage	hydrogen chloride	2
	systemic	hemolysis	arsine	3
	mixed	lung damage methemoglobinemia	oxides of nitrogen	4
short-term	local	sensitization	ethylenediamine	5
	systemic	peripheral neuropathy	methyl-*n*-butyl ketone	6
	mixed	respiratory irritation kidney injury	pyridine	7
chronic	local	bronchitis	sulfur dioxide	8
	systemic	liver angiosarcoma	vinyl chloride	9
	mixed	emphysema kidney damage	cadmium	10
latent	local	pulmonary edema	phosgene	11
	systemic	neuropathy	organophosphates	12
		lung fibrosis	paraquat	13

Depending on the circumstances of exposure, any given material may produce more than one type of toxic effect. Therefore, when describing toxicity for a particular material, it is necessary to define the following: whether the effect is local, systemic, or mixed; the nature of the injury; the organs and tissues affected; and the conditions of exposure, including route of exposure, number of exposures, and magnitude of exposure.

The Nature of Toxic Effects

The biological response to chemical insult may take numerous forms, depending on the physicochemical properties of the material and the conditions of exposure. Listed below are some of the more significant and frequently encountered types of injury or toxic response; they may be defined in terms of tissue pathology, altered or aberrant biochemical processes, or extreme physiological responses.

Inflammation. This describes the local and immediate biological response to tissue injury (14). There is increased blood flow, leak of blood plasma into the tissues, and migration of particular blood cells to the affected area; these have protective functions. A process of repair follows. Depending on the duration of the inflammatory response, and the type of cells in the affected tissue, inflammation may be described as acute or chronic. Acute inflammation is rapid in onset with early and complete healing of the injured area, and produced by locally irritant chemicals. In chronic inflammation, there is persistence of the aggravating agent, such as insoluble particles, or continual repetitive exposure to an irritant material. A characteristic of chronic inflammation is that tissue destruction and the inflammatory process continue at the same time that healing processes are in operation. This may cause the development of excessive amounts of fibrous tissue (scar tissue), which may be sufficient to impair organ or tissue function; eg, in the lung there may be chronic progressive fibrotic disease.

Degeneration. This is a generic description for a variety of abnormal changes, visible on microscopy, that occur in tissue cells as a response to toxic injury. Acutely induced degenerative changes may be reversible, but repetitive exposure can cause progression of the degenerative changes resulting in cell malfunction and, ultimately, cell death.

Necrosis. This term is used to describe the circumscribed death of tissue, and may be a consequence of many pathological processes induced by chemical injury.

Immune-Mediated Hypersensitivity Reaction. The immune system, as one of its primary functions, protects against invasion by foreign biological and other materials of potential harm. Such materials (antigens) stimulate various immune mechanisms in the host which cause functional elimination of the antigenic material. In some instances, there is an excess biological reactivity to the antigen, and a state of hypersensitivity develops (15). In the context of toxicology, the most important of such immune-mediated hypersensitivity reactions occur in skin and lungs. In skin, and following an appropriate period for the induction of immune defense mechanisms, the hypersensitivity reaction is recognized as an exaggerated inflammatory response at the site of application of the material; such materials are causes of allergic contact dermatitis (16). There is now an increasing awareness of the potential for immune-mediated hypersensitivity reactions by the inhalation of antigenic materials. Inhaling such materials results in the induction of a state of immunity against the antigen, which exhibits itself as a hypersensitivity reaction affecting the respiratory tract, and is clinically recognized as asthma (17–18). A classical cause of immune-mediated hypersensitivity reaction effecting the respiratory tract caused by industrial chemicals is toluene diisocyanate (19).

Immunosuppression. Since a primary function of the immune system is protection against pathogenic foreign materials, any substance capable of producing a suppression of immune function will have a deleterious effect on such protective mechanisms, including defense against infective agents. Also, in view of the possible role of immunologic factors in neoplastic cell growth rate and other aspects of carcinogenesis, the influence of immunosuppressive chemicals in tumorigenesis is always a factor to be considered.

Neoplasia. Neoplasms are abnormal masses of cells in which growth control and divisional mechanism are impaired, resulting in aberrant proliferation and growth. Neoplasms are basically classified as benign or malignant. Benign neoplasms grow locally and without erosion of surrounding tissues. Adverse effects produced by benign neoplasms are due either to mechanical compressive effects, or to the liberation of biologically active materials from the tumor cells. Malignant neoplasms (cancers) may erosively invade surrounding tissues and become disseminated throughout the body, setting up secondary deposits of malignant-cell proliferation (metastasis). Induction of neoplasia is referred to as tumorigenesis or oncogenesis; the term carcinogenesis is used to describe the development of malignant neoplasms. The mechanisms of carcinogenesis, as yet incompletely understood, have been reviewed exhaustively (20–24).

Mutagenesis. Chemically induced mutagenesis involves an interaction between the causative agent and cellular constituents, which produces deoxyribonucleic acid (DNA) damage that is heritable. Chemically induced mutations can occur in somatic or germ cells, and may be reflected in altered structure or function of the cell. DNA damage can be classified broadly into that which can be visualized by light microscopic examination of the chromosomes (cytogenetics), and that which occurs at a strictly molecular level and is not visible by microscopy. The former may be visible as breaks, loss of chromosomal material, or rearrangement of segments of the chromosomes; this is frequently referred to as clastogenesis. DNA damage which is restricted to focal molecular lesions is often specifically referred to as mutagenesis. The implications for harmful effects from mutagenic events are multiple. If mutagenesis occurs in rapidly proliferating tissue, there may be abnormalities in the differentiation and proliferation of cells; should this occur in the embryo, a teratogenic effect may result. However, a variety of mechanisms may be involved in teratogenesis from differing materials, and a material which is devoid of mutagenic potential cannot necessarily be regarded as being devoid of teratogenic potential.

It is generally conceded where genotoxic carcinogens are concerned that an early irreversible stage in the complex process of carcinogenesis is likely to involve a mutagenic event. This has resulted in the use of mutagenicity testing procedures as screening methods for the detection of potentially carcinogenic materials. With appropriate test procedures, there is usually a reasonable correlation between the mutagenic potential of a material and its tumorigenicity as demonstrated in conventional chronic toxicity studies. Further, and of current concern, is the possibility that chemically induced mutations in germ cells could result in heritable alteration of cellular function, some of which might be deleterious to health or survival. Numerous excellent texts on mutagenesis are available (25–27).

Enzyme Inhibition. Some materials produce toxic effects by inhibition of biologically vital enzyme systems, leading to an impairment of normal biochemical pathways. The toxic organophosphates, for example, inhibit the cholinesterase group of enzymes. An important factor in their acute toxicity is the inhibition of acetylocholinesterase at neuromuscular junctions, resulting in an accumulation of the neurotransmitter material acetylcholine and causing muscle paralysis (28) (see Neuroregulators).

Biochemical Uncoupling. The energy liberated by normal biochemical processes is stored in high energy phosphate molecules, eg, adenosine triphosphate. Uncoupling agents, such as dinitrophenol, interfere with the synthesis of these high energy phosphate molecules, resulting in the continual excess liberation of energy as heat.

Lethal Synthesis. This is a process in which the toxic substance has a close structural similarity to normal substrates in biochemical reactions. As a result, the material may be incorporated into the biochemical pathway and metabolized to an abnormal and toxic product. A classical example is fluoroacetic acid, which is accepted in the place of acetic acid in the Krebs tricarboxylic acid cycle. The result is formation of fluorocitric acid, which is an inhibitor of aconitase and thus blocks energy production in the citric acid cycle.

Teratogenesis. Teratogenic effects are those resulting in the development of a structural or functional abnormality in the fetus or embryo. Depending on the nature of the material, teratogenic effects may be produced by a variety of mechanisms; these include mutagenesis, induction of chromosomal aberrations, interference with nucleic acid and protein synthesis, substrate deficiencies, and enzyme inhibition. With respect to the induction of structural abnormalities in development, the most critical time for exposure is during the early stage of gestation when the greatest degree of cell differentiation and definitive organ formation are occurring. However, there is increasing interest and concern about the effects of exposure to foreign chemicals in the later stages of gestation, which may induce functional, including behavioral, abnormalities. A number of excellent reviews on teratogenesis are available (29–32).

Sensory Irritation. Although not strictly a toxic effect, peripheral sensory irritation is important in many occupational health considerations. Materials described as peripheral sensory irritants are capable of interacting with sensory nerve receptors in body surfaces, producing local discomfort and related reflex effects. For example, with the eye there is pain, excess lachrymation, and involuntary closure of the eyelids (blepharospasm); inhaled sensory-irritant materials cause respiratory tract discomfort, increased secretions, and cough. Although these effects may be regarded as protective since they warn of exposure to a potentially harmful material, they are also distracting and thus likely to predispose to accidents. For this reason, information on sensory-irritant effects may be used extensively in assessing the suitability of exposure guidelines for workplace environments. A number of excellent review papers dealing with sensory irritation are available (33–34).

Factors Influencing Toxicity

During the design, conduct, and evaluation of toxicology studies, there is a constant need to be aware of the numerous factors that may influence the nature, severity, and probability of induction of toxic injury. Some of the more important are listed below.

Number of Exposures. Some toxic effects are produced in response to a single exposure of sufficient magnitude, while others require multiple exposures for their development (Table 1).

Magnitude of Exposure. As discussed in detail later, the magnitude of the exposure will influence both the likelihood of an effect being produced and its severity.

Species Tested. In addition to the variation in susceptibility to chemically induced toxicity among members within a given population, there may be marked differences between species with respect to the relative potency of a given material to produce toxic injury. These species differences may reflect variations in physiological and biochemical systems, differences in distribution and metabolism, and differences in uptake and excretory capacity.

Route of Exposure. As discussed below, the route of uptake may have a significant influence on the metabolism and distribution of a material. Differences in route of exposure may influence the amount of material absorbed, and its subsequent fate. These differences may be reflected in variation in the nature and magnitude of the toxic effect.

Time of Dosing. The time of day, or day of the year, may influence the toxic response. These changes reflect diurnal and seasonal variations in biochemical and physiological profiles, which may influence toxicity through a variety of mechanisms.

Formulation. The formulation of a material may have a significant influence on its potential to cause toxic injury. For example, solvents may facilitate or retard the penetration and absorption of a chemical, resulting in enhancement or suppression of a toxic response, respectively.

Impurities. The presence of impurities may modify the toxic response, particularly if they have high toxicity.

The above are given as but a few examples of the factors which may influence the expression toxicity. The subject has been reviewed in detail (35–36).

Routes of Exposure

In order to induce a toxic effect, local or systemic, the causative material must first come into contact with an exposed body surface—these are the routes of exposure. In normal circumstances, and depending on the nature of the material, the practical routes of exposure are by swallowing, inhalation, and skin and eye contact. In addition, and for therapeutic purposes, it may be necessary to consider intramuscular, intravenous, and subcutaneous injections as routes of administration.

Swallowing. If sufficiently irritant or caustic, a swallowed material may cause local effects on the mouth, pharynx, esophagus, and stomach. Additionally, carcinogenic materials may induce tumor formation in the alimentory tract. Also, the gastrointestinal tract is an important route by which toxic materials are absorbed. The sites of absorption and factors regulating absorption have been reviewed (37–38).

Skin. The skin may become contaminated accidentally or, in some cases, materials may be deliberately applied. Skin is a principal route of exposure in the industrial environment. Local effects that are produced include acute or chronic inflammation, allergic reactions, and neoplasia. The skin may also act as a significant route for the absorption of systemically toxic materials. Factors influencing the amount of material absorbed include the site of contamination, integrity of the skin, temperature, formulation of the material, and physicochemical characteristics, including charge, molecular weight, and hydrophilic and lipophilic characteristics. Determinants of percutaneous absorption and toxicity have been reviewed (31–34,37–42).

Inhalation. The potential for adverse effects from materials dispersed in the atmosphere depends on a variety of factors including physical state, concentration, and time and frequency of exposure. Gases and vapors reach the alveoli. However, the solubility in water of a gas or vapor influences the depth of its penetration into the respiratory tract. Thus, the differences in solubility of chlorine and hydrogen chloride influence the depth of penetration or location and the irritant action of the two gases. The distribution of particles and fibers is determined by their size. In general, particles of mass median aerodynamic diameter greater than 50 μm do not enter the respiratory system; those greater than 10 μm are deposited in the upper respiratory tract; those in the range of 2–10 μm are deposited progressively in the trachea, bronchi, and bronchioles; and only particles of $\leq$(1–2) μm reach the alveoli. It follows that larger respirable particles are more likely to cause local reactions in the upper airway than in the gas-exchanging tissues. The potential for alveolar involvement is greater with small-diameter particles. Factors governing the deposition of particles in the lung have been reviewed extensively (43–46). The aerodynamic behavior of fibers is such that those having diameters >3 μm are unlikely to penetrate into the lung. In general, fibers having a diameter of $\leq$3 μm and length not greater than 200 μm gain access to the lung. Fibers longer than 10 μm may not be readily removed by the normal pulmonary clearance mechanisms. Several studies indicate that maximum biological activity is associated with fibers less than 1.5 μm dia and more than 8 μm in length (47–50).

The likelihood that materials will produce local effects in the respiratory tract depends on their physical and chemical properties, solubility, reactivity with fluid-lining layers of the respiratory tract, reactivity with local tissue components, and (with particulates) the site of deposition. Depending on the nature of the material, and the conditions of the exposure, the types of local response produced include acute inflammation and damage, chronic inflammation, immune-mediated hypersensitivity reactions, and neoplasia.

The degree to which inhaled gases, vapors, and particulates are absorbed, and hence their potential to produce systemic toxicity, depends on their solubility in tissue fluids, any metabolism by lung tissue, diffusion rates, and equilibrium state.

Eye. Adverse effects may be produced by splashes of liquids or solids, and by materials dispersed in the atmosphere. The eye is particularly sensitive to peripheral sensory irritants in the atmosphere. Toxic effects that may be induced include transient acute inflammation, persistent damage, and, occasionally, sensitivity reactions. Toxicologically significant amounts of material may be absorbed by the periocular blood vessels in cases of splash contamination of the eye with materials of high acute toxicity (51).

Multiple Exposures

Although toxicology testing is often performed with only a single material or a material in a relatively inert solvent, in most practical situations there is simultaneous exposure to multiple chemicals and thus a potential for complex biological interactions. The following descriptive terms are useful to classify such effects.

Independent is an effect in which each material exerts its own effect irrespective of the presence of another.

Additive effects involve materials producing similar toxic effects where the magnitude of the response is numerically equal to the sum of the effect produced by each individual material.

Antagonism is applied to a situation where two chemicals, given together, interfere with each other's action or where one interferes with the action of the other. The result will usually be a decrease in toxic injury. A special case of antagonism is in studies on antidotal action.

Potentiation is applied to a condition where one material, of relatively low toxicity, enhances the expression of toxicity by another chemical. The result may be a larger response or more severe injury than that produced by the toxic chemical alone. A particular example is an enhancement of the absorption of a material of known toxicity by a surface-active material.

Synergism is applied to a situation where the effect of two chemicals, given together, is significantly greater than that expected from considerations on the toxicity of each material alone. This differs from potentiation in that both materials contribute to the toxic injury, and the net effect is always greater than additive.

Exposure to combinations of chemicals does not always necessarily produce clearly distinguishable interactions. Each situation must be considered in detail with due regard to all the factors required to be analyzed in the process of hazard evaluation (52–53).

Fate of Absorbed Chemicals Relative to Toxicity

The induction of systemic toxicity may involve a variety of complex interrelationships between the absorbed parent material, any conversion products, and their concentration and distribution in body tissues and fluids. The general pathway that a material may follow after its absorption is shown schematically in Figure 2.

Materials may be absorbed by a variety of mechanisms. Depending on the nature from the material and the site of absorption, there may be passive diffusion, filtration

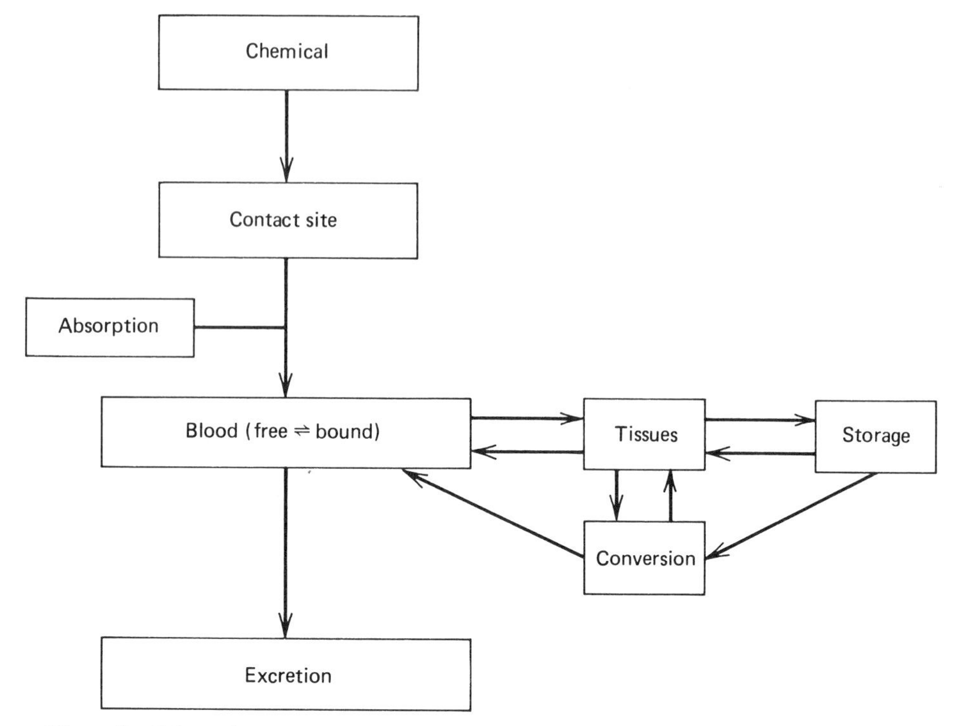

Figure 2. Schematic representation of the possible fate of a chemical absorbed from a primary contact site.

processes, facilitated diffusion, active transport and the formation of microvesicles for the cell membrane (pinocytosis) (54). Following absorption, materials are transported in the circulation either free or bound to constituents such as plasma proteins or blood cells. The degree of binding of the absorbed material may influence the availability of the material to tissue, or limit its elimination from the body (excretion). After passing from plasma to tissues, materials may have a variety of effects and fates including no effect on the tissue, production of injury, biochemical conversion (metabolized or biotransformed), or excretion (eg, from liver and kidney).

The metabolism of a material may result in the formation of a transformation product of lower intrinsic toxicity than the parent molecule; ie, a process of detoxification has occurred. In other cases, the end result is a metabolite, or metabolites, of intrinsically greater toxicity than the parent molecule; ie, metabolic activation has occurred. Some examples of detoxification and metabolic-activation processes are given in Table 2.

The kidney is an important organ for the excretion of toxic materials and their metabolites, and measurement of these substances in urine may provide a convenient basis for monitoring the exposure of an individual to the parent compound in his or her immediate environment. The liver has as one of its functions the metabolism of foreign compounds; some pathways result in detoxification and others in metabolic activation. Also, the liver may serve as a route of elimination of toxic materials by excretion in bile. In addition to the liver (bile) and kidney (urine) as routes of excretion, the lung may act as a route of elimination for volatile compounds. The excretion of materials in sweat, hair, and nails is usually insignificant.

Table 2. Examples of Metabolic Detoxification and Metabolic Activation of Chemicals by Biological Systems

Chemical	Transformation	Conversion	Ref.
cyanide, CN^-	detoxification	enzyme conversion to less acutely toxic thiocyanate	55
benzoic acid, $C_6H_5CO_2H$	detoxification	conjugation with glycine to produce less-toxic hippuric acid	56
isoniazid, 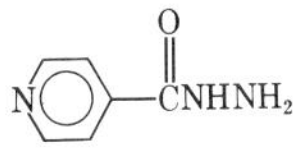 [pyridine ring]–$\overset{O}{\overset{\|}{C}}NHNH_2$	detoxification	*N*-acetylation to less-toxic acetyl derivative	57
parathion, 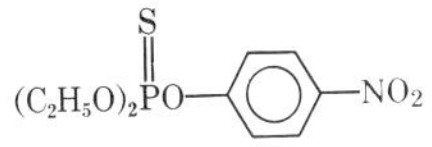 $(C_2H_5O)_2\overset{S}{\overset{\|}{P}}O$–[benzene ring]–$NO_2$	activation	converted by oxidative desulfuration to paraoxon, a potent cholinesterase inhibitor	58
carbon tetrachloride, CCl_4	activation	microsomal enzyme-mediated metabolic activation to hepatotoxic trichloromethyl radical	59
2-acetylaminofluorene, 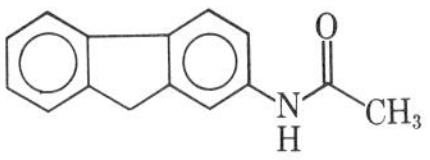 [fluorene]–$NH\overset{O}{\overset{\|}{C}}CH_3$	activation	*N*-hydroxylation to the more potent carcinogen *N*-hydroxyacetylaminofluorene	60

Parent substances and metabolites may be stored in tissues, such as fat, from which they continue to be released following cessation of exposure to the parent material. In this way, potentially toxic levels of a material or metabolite may be maintained in the body. However, the relationship between uptake and release, and the quantitative aspects of partitioning, may be complex and vary between different materials. For example, volatile lipophilic materials are generally more rapidly cleared than nonvolatile substances, and the half-lives may differ by orders of magnitude. This is exemplified by comparing halothane and DDT (see Anesthetics; Insect-control technology).

Both the metabolism of a material and its potential to cause toxic injury may vary with the route of exposure, although, as indicated below, the magnitude of the dose and duration of dosing may influence this relationship. For example, materials that are metabolically activated by the liver are likely to exhibit a comparatively greater degree of toxicity when given perorally than when absorbed in the lung or across the skin. This is largely related to the anatomical routes of transport. Thus, the greatest proportion of material absorbed from the gastrointestinal tract passes via the portal vein directly to the liver. In contrast, materials absorbed as a result of respiratory exposure or skin contact initially pass to the lung and then into the systemic circulation, with only a small fraction of the cardiac output being delivered to the liver through the hepatic artery (Fig. 3). By similar reasoning, materials that are detoxified by the liver may be significantly less toxic by swallowing than by either inhalation or penetration across the skin. An example of the influence of route on toxicity is presented in Table 3. When assessing the relevance of metabolism in acute toxicity testing, and particularly when comparing toxicity by different routes of exposure, both the magnitude of the dose and the time period over which it is given must be considered. For example, when a single large dose (a bolus) of a metabolically activated material is given by gavage, it may be almost completely metabolized, resulting in the rapid de-

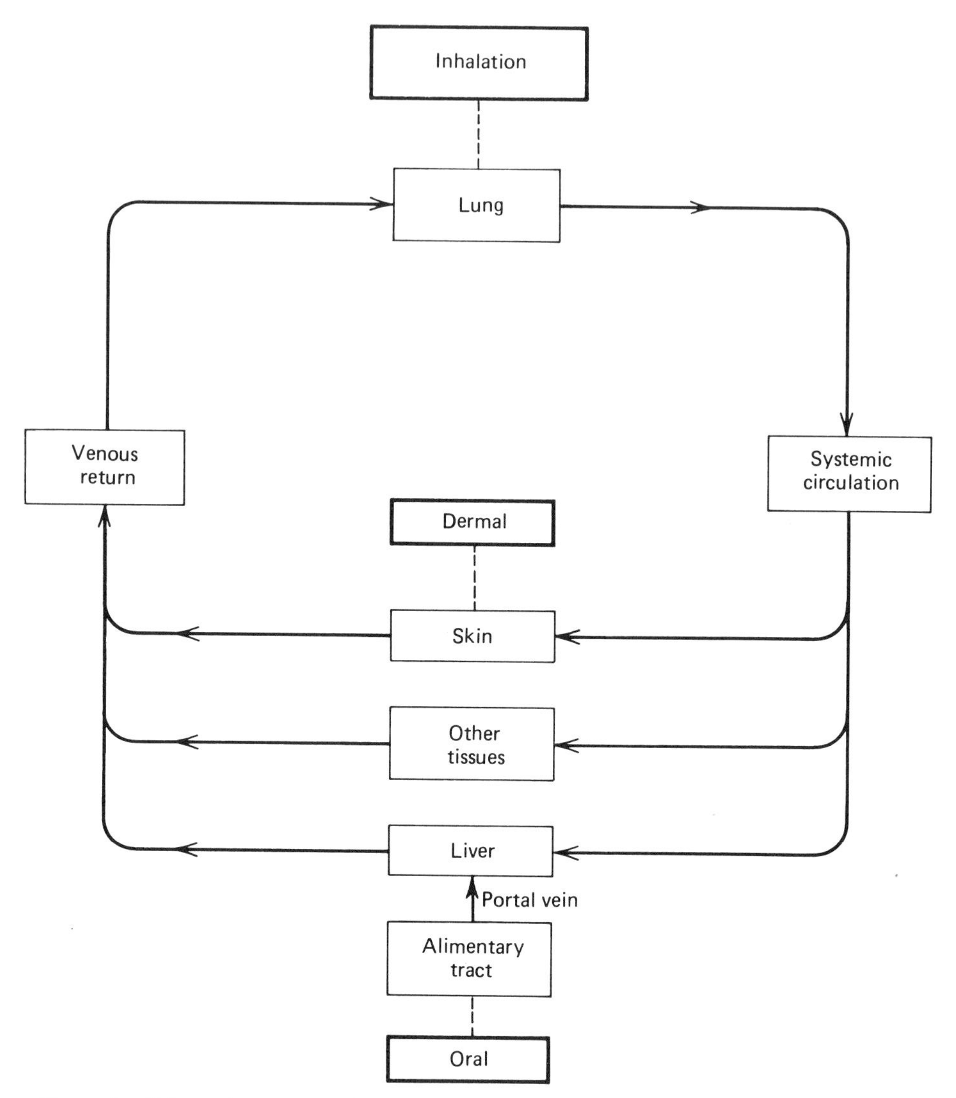

Figure 3. Schematic representation to show the anatomical basis for differences in the quantitative supply of absorbed material to the liver. By swallowing (oral route), the main fraction of the absorbed dose is transported directly to the liver. Following inhalation or dermal exposure, the material passes to the pulmonary circulation and thence to the systemic circulation, from which only a proportion passes to the liver. This discrepancy in the amount of absorbed material passing to the liver may account for differences in toxicity of a material by inhalation and skin contact compared with its toxicity by swallowing, if metabolism of the material in the liver is significant in its detoxification or metabolic activation.

velopment of acute toxic injury. When the same material is given orally at a slower rate, eg, by continuous inclusion in the diet, then there is a slow and continual absorption and metabolism of the material, and in these circumstances the rate of generation of the toxic species may approach that which occurs from the continuous absorption resulting from persistent exposure to an atmosphere of the material. The influence of dose magnitude–time relationships also apply to the interpretation of results with materials detoxified by the liver. With such materials, slow continuous peroral administration of the material results in slow titration to the liver and a high

Table 3. Example of the Influence of Route on the Acute Toxicity of Potassium Cyanide to the Rabbit (Female)

Route	LD_{50} (mg/kg) with 95% confidence limits
intravenous	1.89 (1.66–2.13)
intraperitoneal	3.99 (3.40–4.60)
oral	5.82 (5.50–6.31)
percutaneous	22.3 (20.4–24.0)

proportion of the material being detoxified. In this instance, the anticipated differential toxic effect between the oral and inhalation routes of exposure occurs. However, if a bolus of the material is introduced into the stomach, then the endogenous hepatic detoxification mechanisms may be exceeded, and unmetabolized material may enter the systemic circulation and initiate toxic injury.

Principal factors that determine the likelihood of toxic effects being produced and their severity include the rates at which the causative substances (parent molecule, toxic metabolite, or both) reach the tissues and the absolute amounts of materials to which the tissues are exposed. These determine the dose of material received at the target tissues.

With respect to environmental-exposure conditions, the probability of adverse effects occurring depends on many factors; the more important of these include magnitude, duration, and number of exposures. These conditions determine the amount of material to which an organism is exposed (the environmental-exposure dose), and thus relates to the amount of material available for absorption at the contact site (the absorbed dose). The absorbed dose is an important factor that determines the amount of material available for distribution to body tissues, the amount of toxic metabolites formed, and thus the likelihood of inducing a toxic effect. Opposed to the influence of absorption and buildup of metabolites is the elimination of these materials from the body. Thus, for any given environmental-exposure situation, the probability of inducing a toxic effect depends primarily on the dynamic equilibrium between the rate of absorption of the environmental chemical and the rate of excretion of the absorbed material and its metabolites. This relationship clearly determines the residence time for materials and metabolites in various body tissues and fluids, their fluctuation in concentration with time, the potential for storage, and availability for binding with macromolecules and structural cellular components.

The amount of material in contact with the absorbing surface is a principal determinant of the absorbed dose. In general, the higher the concentration, the greater the absorbed dose. However, when mechanisms other than simple diffusion across a concentration gradient are operating, a simple proportionate relationship between concentration and absorbed dose may not exist. In such cases, a rate-limiting factor could result in proportionately smaller increases in absorbed dose for incremental increases in concentration at the contact site. Also, and particularly when an active transport mechanism is involved in the absorptive process, there may be saturation and a ceiling value for absorption.

It is important to appreciate that the magnitude of the absorbed dose, the relative amounts of biotransformation product, and the distribution and elimination of metabolites and parent compound seen with a single exposure, may be modified by repeated exposures. For example, repeated exposure may enhance mechanisms responsible for biotransformation of the absorbed material, and thus modify the relative proportions of the metabolites and parent molecule, and thus the retention pattern of these materials. Clearly, this could influence the likelihood for target organ toxicity. Additionally, and particularly when there is a slow excretion rate, repeated exposures may increase the possibility for progressive loading of tissues and body fluids, and hence the potential for cumulative toxicity.

It is clear from the above considerations that the absorbed dose, and the distribution, excretion, and relative amounts of the absorbed material and its metabolites may be quantitatively different for acute and repeated exposures. This modifies the potential for the absorbed material to produce adverse effects by a given route of exposure.

Dose–Response Relationships and Their Toxicological Significance

The importance of determining a relationship between the magnitude of the exposure and the frequency of occurrence of a toxic effect is considered in detail below.

An observation which is fundamental to the interpretation of toxicology information is the variation in susceptibility to potentially harmful chemicals of individual members within a given population. Thus, if a group of animals of stated species and strains are exposed to a particular material by a given route of exposure, then as the exposure dose is increased so does the proportion of animals exhibiting a toxic effect. However, the biological variability with respect to individual susceptibility of animals to toxic materials is not a simple linear relationship. The most typical response is represented by a sigmoid curve (Fig. 4), indicating that a very small proportion of the exposed population is more susceptible (hypersensitive) and a few more resistant (hyposensitive) to the chemical. The majority of the population, however, responds over a defined exposure-dose range around an average. This variability also exists among human populations, and the magnitude of the average response and the existence of a hypersensitive group clearly influences judgments on the hazards of materials studied.

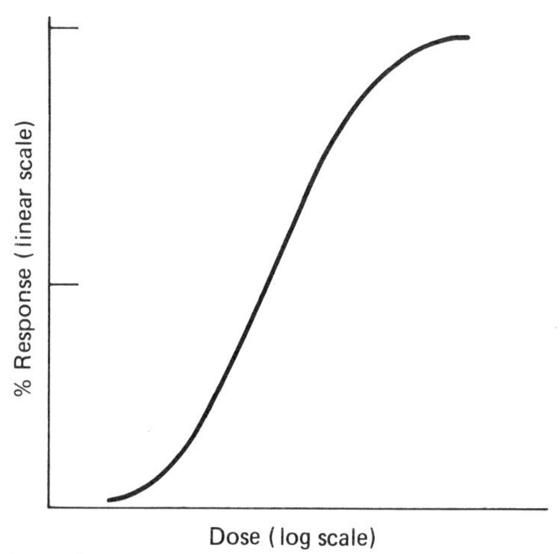

Figure 4. Typical sigmoid curve for the response of a biological system to chemical injury.

In addition to the effect of biological variability in group response for a given exposure dose, the magnitude of the dose for any given individual also determines the severity of the toxic injury. In general, the considerations for dose–response relationship with respect to both the proportion of a population responding and the severity of the response are similar for local and systemic effects. However, if metabolic activation is a factor in toxicity, then a saturation level may be reached.

Dose–response relationships are useful for many purposes; in particular, the following: if a positive dose–response relationship exists, then this is good evidence that exposure to the material under test is causally related to the response; the quantitative information obtained gives an indication of the spread of sensitivity of the population at risk, and hence influences hazard evaluation; the data may allow assessments of no-effects and minimum-effects doses, and hence may be valuable in respect of assessing hazard; and by appropriate considerations of the dose–response data, it is possible to make quantitative comparisons and contrasts between materials or between species.

Considerable caution is necessary in making quantitative comparisons between different materials, even when considering the same toxic end point. This can be conveniently illustrated using, as an example, death in response to a single exposure, ie, acute lethal toxicity. Studies to determine acute lethal toxicity by a particular route are usually conducted as described below.

Several groups of animals of a particular species and strain are given different doses of the test material; members within the same group receive similar doses. The animals are subsequently observed, usually for two weeks, and the number of mortalities noted. In most cases, the dose–mortality curve has a typical sigmoid form. This may be conveniently converted to a linear form by log–probit plot (Fig. 5). A frequently used numerical means to allow comparison of the lethal potential of different materials is to quote a particular level of mortality. Since the largest proportion of deaths is distributed around the 50% mortality level, this forms a convenient reference point, and is referred to as the median lethal dose $_{50}$ (LD_{50}). The LD_{50} may, therefore, be defined as the dose, calculated from the dose–mortality data, which will cause the death of half of the population exposed. In view of the multiplicity of factors that influence

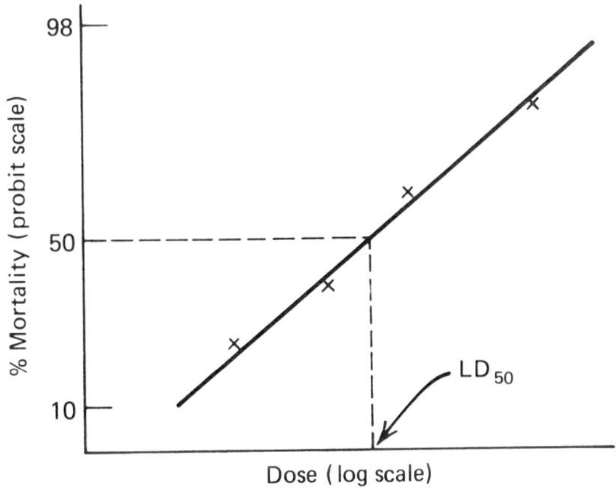

Figure 5. Dose–response regression line for mortality data (represented by X) expressed by log–probit plot.

toxicity, the LD_{50} is valid only for the specific conditions of the test. It is, of course, possible to compare other levels of lethality, such as the LD_{10} and the LD_{90}; the difference between these limits will give an indication of the range of doses causing lethal toxicity to the majority of the population studied. They also give an indication of the relationship of hypersensitive and hyposensitive groups to the average response.

The LD_{50} is calculated from data obtained by using small groups of animals and usually for only a few dose levels. Therefore, there is an uncertainty factor associated with the calculation. This can be defined by determining the 95% confidence limits for the particular levels of mortality of interest (Fig. 6). The 95% confidence limits give the dose range for which there is only a 5% chance that the LD_{50} will be outside.

In defining acute level toxicity for the purposes of comparing different materials, the LD_{50} itself is not sufficient; but the LD_{50} and the 95% confidence limits should be quoted as a minimum. For example, and as demonstrated in Figure 7, two materials (A and B) with different LD_{50} values, but overlapping 95% confidence limits, are to be considered not statistically significantly different with respect to mortality at the

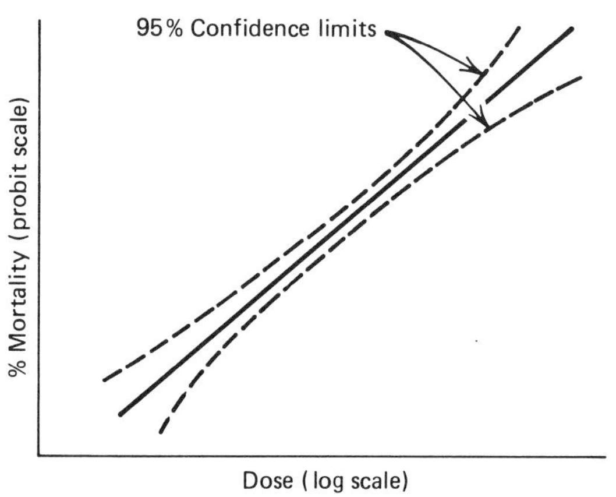

Figure 6. Typical 95% confidence limits for dose–mortality regression data.

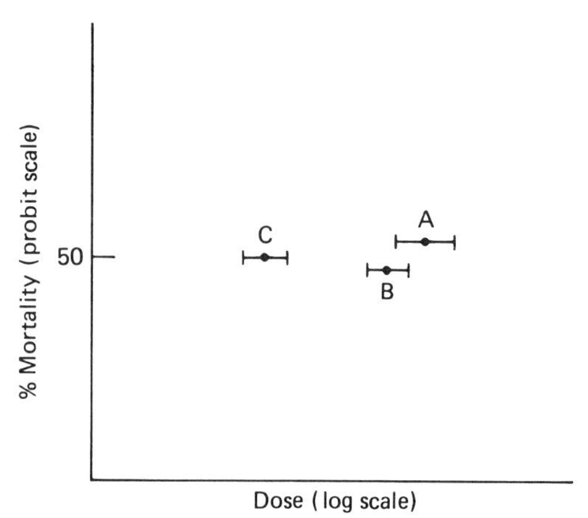

Figure 7. Comparison of three materials on the basis of LD_{50} and 95% confidence internal data. Materials A and B are not statistically significantly different. Material C, however, has 95% confidence limits at the LD_{50} level which do not overlap those of A or B; it is statistically significantly more lethally toxic than either.

50% level; this is based on the fact that there is a statistical probability that the LD_{50} of one material could lie in the 95% confidence limits of the other, and *vice versa*. Conversely, when there is no overlap in 95% confidence limits, as shown with material C, it may be concluded that the LD_{50} values are statistically significantly different.

A more complete comparison of the lethal potential of two or more materials requires that attention be paid to the slopes of the dose–response regression lines. For example, two materials having statistically significantly similar LD_{50} values (based on 95% confidence-limit comparisons) and parallel dose–response regression lines also have statistically similar LD_{10} and LD_{90} values (Fig. 8). In contrast, materials having similar LD_{50} values (based on overlapping 95% confidence limits), but differing slopes for the regression lines, may have widely differing and separated 95% confidence limits at the LD_{10} and LD_{90} levels (Fig. 9). Such a significant difference in lethal toxicity at the hypersensitive and hyposensitive regions may markedly influence considerations with respect to the relative hazards of the two materials. For example, as shown in Figure 9, with the material having the steeper slope it is clear that once a lethal dose is reached, only a small increase in dose is necessary to affect the whole population. Thus, this material may present significant problems with respect to acute overexposure situations. In contrast, the material having a shallow dose–response regression line may present problems with respect to the development of toxic effects in a small hypersensitive group.

When making comparisons of lethal toxicity, it must be remembered that different mechanisms may be involved with different materials, and these need to be taken into account. Also, comparisons of acute toxicity should take note of differences in time to death, since marked differences in times between dosing and death may influence hazard evaluation procedures and their implications. In a few instances, it may be possible to calculate two LD_{50} values for mortality; one based on early death due to one mechanism, and a second based on delayed deaths due to a different mechanism (61).

Although acute lethal toxicity has been used as an example, the principles dis-

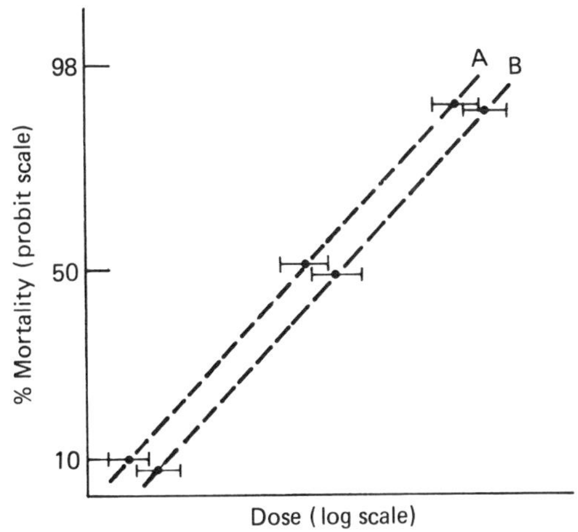

Figure 8. The two materials, A and B, have overlapping 95% confidence limits at the LD_{50} level. Since the slopes of the dose–mortality regression lines for both materials are similar, there is no statistically significant difference in mortality at the LD_{10} and LD_{90} levels. Both materials may be assumed to be lethally equitoxic over a wide range of doses, under the specific conditions of the test.

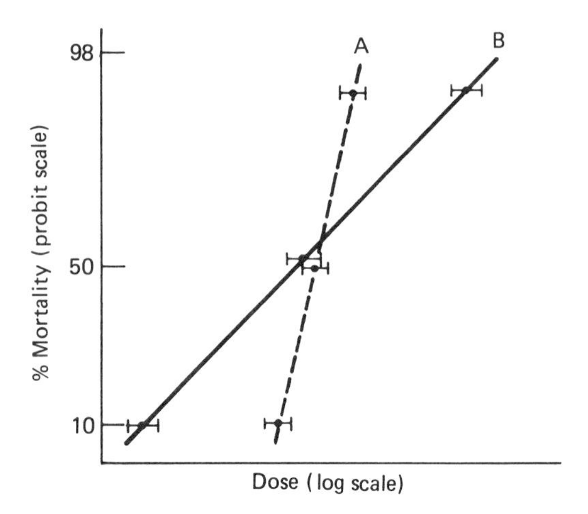

Figure 9. Two materials, A and B, have statistically similar LD_{50} values but, because of differences in the slopes of the dose–mortality regression lines, there are significant differences in mortality at the LD_{10} and LD_{90} levels. Material A is likely to present problems with acute overexposure to large numbers of individuals in an exposed population when lethal levels are reached. With Material B, because of the shallow slope, problems may be encountered at low doses with hypersensitive individuals in the population.

cussed apply in general to other forms of toxicity capable of being quantitated in terms of dose–response relationships.

Testing Procedures

For descriptive purposes, toxicology testing procedures can be conveniently subdivided into general and specific forms. General toxicology studies are those in which animals are exposed to a test material under appropriate conditions and then examined for all types of toxic effects that the monitoring procedures employed allow. Specific toxicological studies are those in which exposed animals are monitored specifically for a defined toxic end point or effect.

There are many guidelines that need to be followed and which are common to all types of toxicity testing, the most important of which are as follows:

There should be sufficiently large numbers of animals to allow a quantitative determination of the average response and the range of responses, including the demonstration of hypersensitive populations. When objective procedures are undertaken, these should be sufficient to allow valid statistical comparison to be made between treated and control groups.

Sufficient numbers of control animals should be employed. The use of such controls allows a determination of normal values for features monitored in the study, background incidence of pathology in the population studied, detection of the onset of adverse conditions (eg, infection) which are unrelated to, and detrimental to, the conduct of the study, and deviation of monitored features between controls and exposed animals, which may indicate a treatment-related effect.

Vehicle control animals may be necessary to allow an assessment of the possible contribution of the vehicle to any effects observed in exposed animals.

Exposure should be by the practical route. Other conditions, such as number and magnitude of exposures, should include at least one level representative of the practical situation; monitoring should be appropriate to the needs for conducting the study; and when practically and economically possible, pharmacokinetic observations should be undertaken in order to better define the relationship of dose to metabolic thresholds.

Factors in the design and conduct of studies have been published (62–64).

General Toxicology Studies. For reasons inherent in both the toxicity assessment procedure and the design of studies, it is usual to proceed in sequence from acute to the various stages of multiple-exposure studies. Acute studies give information on the type of toxic injury produced by a single exposure, including the effects of massive overexposure. The fact that a particular type of toxic injury is not produced by an acute exposure does not necessarily imply the absence of potential for that type of injury by the chemical, since multiple exposures may be necessary to induce the effect. However, effects produced by acute or relatively short-term repeated exposure may also be produced by longer-term repeated exposures, and at lower concentrations. Hence, in addition to giving information on potential for toxicity, the acute and short-term repeated studies are used to give guidance on exposure conditions to be followed in longer-term repeated-exposure studies. The type of monitoring to be employed will depend on a variety of considerations, including the chemistry of the material, its known or suspected toxicology, the degree of exposure, and the reason for conducting the test. In general, since the multiple-exposure studies are more likely to produce the widest spectrum of toxic effects, it is usual to employ the most extensive monitoring in these studies.

The types of monitoring employed to assess the functional status of the living animal and for the detection of injury in dead animals may include the following.

General Observations. Animals are inspected at frequent intervals in order to discover any departure from normal appearance and function, the presence of abnormal patterns of behavior, and any other differences from the control animals. Simple observation of the animals may give information of considerable importance in assessing potential for toxicity and giving preliminary guidance on the nature of any injury.

Body Weight. The detection of a decrease in the rate of gain in body weight, in comparison with controls, may be one of the earliest indications of the onset of toxic effects, particularly if it follows a dose–response relationship. Rarely, a relative increase in body weight may indicate a metabolic or endocrinological effect.

Food and Water Consumption. Measurement of changes in food and water consumption may indicate a toxic potential, and can give guidance on the reason for abnormal body weight gains.

Hematology. The functional status of blood and of the blood-forming tissues can be assessed by tests which include red and white blood cell counts, platelet counts, clotting time, coagulation tests, and examination of bone marrow. Such tests, in addition to detecting abnormalities, may also allow differentiation between primary and secondary effects on blood and blood-forming tissues (65).

Chemical Pathology. Also referred to as clinical chemistry, this monitoring procedure involves the measurement of the concentration of certain materials in the blood, or of certain enzyme activities in serum or plasma. A variety of methods exist that allow (to variable degrees of specificity) the definition of particular organ or tissue injury, the nature of the injurious process, and the severity of the effect (66).

Urinalysis. Urine is collected at various times and examined with respect to its volume, specific gravity, and the presence of abnormal constituents. The results may indicate kidney damage or suggest tissue injury at other sites (67).

Pathology. Animals are examined macroscopically at autopsy following death during the study or at planned sacrifice. This may show features apparent to the naked eye which are abnormal and suggestive of tissue damage (gross pathology). Sections of tissue examined under the light microscope allows a detailed evaluation of the interrelationships and structural integrity, or otherwise, of cells and intercellular materials. In this way, normality of tissue may be confirmed, or a specific pathological diagnosis attached to induced or coincidental tissue injury (histopathology) (68).

Organ Weight Determinations. Measurement of the weight of organs removed at autopsy is an integral part of most toxicology studies. This information may provide an indication of changes in these organs, although they have to be carefully related to the state of hydration and nutrition of the animal.

Special Investigations. Existing information on toxicology, or suspicions based on the chemistry of the material, may indicate the possibility for a particular type of organ toxicity. In such cases, it may be appropriate to incorporate certain special investigational procedures into the general toxicology study. For example, if there is evidence or suspicion that a material is capable of inducing progressive lung dysfunction, it is of importance that there be periodic monitoring of animals by respiratory function testing in any subchronic or chronic inhalation study with the material.

Types of Studies. Studies may be carried out by single exposure or by repeated exposure over variable periods of time. The design of any one study, including the monitoring procedures, is determined by a large number of factors including the nature of the test material, route of exposure, known or suspected toxicity, practical use of the material, and the reason for conducting the study.

Acute Toxicity Studies. These studies should provide the following information: the nature of any local or systemic adverse effects occurring as a consequence of a single exposure to the test material; an indication of the exposure conditions producing the adverse effects, in particular, information on dose–response relationships, including minimum and no-effects exposure levels; and data of use in the design of short-term repeated-exposure studies.

Acute toxicity studies are often dominated by consideration of lethality, including calculation of the median lethal dose. By routes other than inhalation, this is expressed as the LD_{50} with 95% confidence limits. For inhalation experiments, it is convenient to calculate the atmospheric concentration of test material producing a 50% mortality over a specified period of time, usually 4 h; ie, the 4-h LC_{50}. It is desirable to know the nature, time to onset, dose–related severity, and reversibility of sublethal toxic effects.

Short-Term Repeated Studies. These studies should give information about the potential for cumulative toxicity and allow the detection of toxicity, other than neoplasia, not detected in acute studies. Studies are generally carried out by exposing animals by an appropriate route for 5 d/wk for 1–4 wk. As a minimum, the conduct of these tests should include observations for signs of toxicity; measurement of body weight, food intake, and water consumption; autopsy; and gross pathology. Other monitoring requirements are dictated by the reasons for conducting the test.

Subchronic Studies. Although short-term repeated-exposure studies provide valuable information about toxicity over this time span, they may not be relevant for assessment of hazard over a longer time period. For example, the minimum and no-effects levels determined by short-term exposure may be significantly lower if exposure to the test material is extended over several months. Also, certain toxic effects may have a latency which does not allow their expression or detection over a short-term repeated-exposure period; for example, kidney dysfunction or disturbances of the blood-forming tissues may not become apparent until subchronic exposure studies are undertaken.

Typically, subchronic inhalation studies involve exposing the animals for 6 h/d, 5 d/wk for about 3 months. For feeding studies, the material is frequently included in the diet, providing unpalatability is not a problem. As with the shorter-term studies, several dose levels are used, together with a control group. Because of the potential for a wide spectrum of effects and the cost of conducting the basic test, a significant amount of relevant monitoring is employed in order to detect the nature, onset, progression, and severity of any toxic effects. Ideally, a small proportion of animals should be kept for several weeks after the end of the exposure period in order to determine the reversibility of any induced toxic effects. Subchronic exposure conditions usually detect all potential long-term repeated-exposure toxicity, except for neoplasia.

Chronic Toxicity Studies. With the exception of tumorigenesis, most types of repeated-exposure toxicity are detected by subchronic exposure conditions. Therefore, chronic exposure conditions are usually conducted for the following reasons: if there is a need to investigate the tumorigenic potential of a material; if it is necessary to determine a no-effects or threshold level of toxicity for lifetime exposure to a material; and if there is reason to suspect that particular forms of toxicity are exhibited only under chronic exposure conditions.

For the above reasons, chronic exposure studies are frequently designed in such a way that it is possible to combine observations for tumorigenesis and non-neoplastic tissue injury. Chronic studies are usually extensively monitored. It is common practice to sacrifice animals at intervals during the study in order to detect the onset of any tissue injury. For two-year exposure studies, it is most meaningful to have interim sacrifices at 12 and 18 months.

Guidelines are available for conducting acute or repeated-exposure studies by inhalation (69–71), application to the skin (72–73), or perorally (74–75).

Specific Toxicology Studies. Many procedures, both *in vivo* and *in vitro*, are available to detect specific organ toxicity or quantitatively monitor for particular end points or effects. Although many of these studies are directed at measuring a particular toxic effect for hazard-evaluation purposes, some are employed as screening or short-term tests to determine the potential of a material to induce chronic toxic effects or those with a long latency period. In this context, screening means an experimental approach that allows the rapid and cost-effective prediction of the likelihood that a material exerts a particular type of adverse biological activity. Such approaches should be based on studies showing the method gives a high degree of correlation with conventional and credible methods for detecting the particular toxic end point. Some of the most commonly employed special toxicology methods and approaches are listed below.

Primary Irritancy Studies. These studies are employed to determine the potential of materials to cause local inflammatory effects in exposed body surfaces, notably skin and eye, following acute or short-term repeated exposure. In general, the approach involves applying the test material to the surface of the skin or eye, and observing for signs of inflammation, their duration, and resolution. Reviews have been written about the conduct of primary eye irritation (51,76–77) and primary skin irritation studies (78–79).

Studies for Immune-Mediated Hypersensitivity. Allergenic materials may produce hypersensitivity reactions by skin contact or by inhalation. In conventional tests for determining allergenic potential by skin contact, the basic approach involves repeatedly applying test material to the skin, or under the skin, in order to induce a state of hypersensitivity. After a latent period, the skin is challenged with test material to determine if an exaggerated local response, typical of delayed hypersensitivity contact dermatitis, has been produced. Details of the test procedures are available (80–82).

Experimental methods for determining the potential of materials to produce hypersensitivity reactions by inhalation use procedures to detect hyper-reactivity of the airways as demonstrated by marked changes in resistance to air flow, and the detection of antibodies in blood serum (83).

Neurological and Behavioral Toxicology. Observations on animals in general toxicology studies may indicate a potential for injury to the nervous system, particularly if there are abnormalities of movement, gait, and reaction to the environment. When it is known or suspected that a material may produce structural or functional damage to the nervous system, there should be incorporation of special methods into general toxicology studies in order to determine the nature and extent of any neurological injury. This may include the use of simple observational test batteries in order to better assess the clinical status of the animal (84–85), more detailed examination of the potentially affected areas of the nervous system by light and electron microscopy (86–87), and the use of selective biochemical procedures (88).

However, in order to precisely define the nature of a neurotoxic process, its mechanism of production, and the quantitative determinants for the effect, it may be necessary to conduct specific studies. These may involve the use of electrophysiological, pharmacological, tissue-culture, and metabolism techniques (89–92). Special observational methods are available for behavioral studies (93–94).

Sensory Irritation. In view of the ability of peripheral sensory irritants to produce local discomfort with related reflex effects, they may cause harassment and be a predisposing factor in accidents. Methods for quantitatively assessing the potential of materials to produce sensory irritation by inhalation (33,95) and contamination of the eye (96–97) are available. Techniques for assessing peripheral sensory irritant potential by inhalation are based on measuring a reflexly induced decrease in breathing rate. Those for assessing eye irritation are based on blepharospasm (involuntary closure of eyelids) in animals, and blepharospasm and discomfort in human subjects. The information so obtained may be useful in considerations on assigning suitable exposure guidelines for materials in the workplace atmosphere (98).

Teratology. At present, most studies that are conducted to determine the teratogenicity of materials are aimed primarily at assessing structural defects of development. Basically, these studies involve administering the test material to the pregnant animal during the period of maximum organogenesis; for rats, it is usual to expose on days 6–15, and for rabbits on days 6–18. The day before anticipated normal parturition, fetuses are delivered by Caesarian section, to prevent the cannabilization of any deformed fetuses by the mother. Resorbed and dead fetuses are counted. The viable fetuses are sexed, weighed, measured (crown–rump length); some are used for examination of the integrity of the skeleton; and the remainder are dissected to determine the presence of any soft-tissue abnormalities. Additionally, observations are made for pathology in the maternal reproductive system. To allow for dose–response considerations, several exposure levels are used. Control groups should include animals that are untreated and others given the vehicle alone. The design and conduct of conventional teratology studies for the detection of structural abnormalities of development have been extensively reviewed (99–101).

Recently, there has been increasing interest in the development of test methods to assess the possible adverse functional effects of exposing the fetus, both early and late in gestation (102–103).

Reproductive Toxicology Tests. In contrast to teratology studies, which are aimed at assessing adverse effects on the developing fetus, reproductive studies cover a much wider spectrum of developmental biology. They are designed to assess the potential for adverse effects on gonads, fertility, gestation, fetuses, lactation, and general reproductive performance. Exposure to the chemical may be over one or several generations. Because of the necessarily comparatively low doses used over these long-term studies, there may be insufficient sensitivity to detect most potentially teratogenic materials in conventional multigeneration reproduction studies. Tests for reproductive toxicity have been reviewed (104–106).

Metabolism and Pharmacokinetics. Since the potential for systemic toxicity of a material may be highly dependent on its distribution, residence time, and bioconversion, studies on metabolism and pharmacokinetics can be of fundamental importance with respect to interpretation of the significance of conventional toxicology studies, determination of mechanisms of toxicity, relation of environmental exposure conditions to target organ toxicity, selection of dosages, and the design of further toxicology studies (see Pharmacodynamics).

Metabolism is concerned with a determination of the biotransformation of the parent material, the sites at which this occurs, and the mechanism of the biotransformation.

Pharmacokinetic studies are designed to quantitatively measure the rate of uptake and metabolism of a material and determine the absorbed dose; to determine the distribution of absorbed material and its metabolites among body fluids and tissues, and their rate of accumulation and efflux from the tissues and body fluids; to determine the routes and relative rates of excretion of test material and metabolites; and to determine the potential for binding to macromolecular and cellular structures.

Pharmacokinetic studies should allow an assessment of the relationship between the environmental-exposure conditions and the absorbed dose, and how these influence the doses of test material and metabolites received by various body tissues and fluids, and the potential for storage. Numerous texts are available on the design and conduct of metabolism and pharmacokinetic studies (107–109).

Mutagenicity. Studies to determine the potential for materials to produce mutagenic events may be conducted *in vitro* and *in vivo.* The most widely used test system for mutagenic potential has been the Ames procedure (110). This is based on the ability of mutagenic chemicals to cause certain bacteria to regain their ability to grow in media deficient in an essential amino acid. Other tests of the *in vitro* type make use of various end points, indicating a mutagenic event, in mammalian cells grown in culture. *In vivo* studies involve the exposure of animals to the test chemical after which cells are removed, usually from blood or bone marrow, and examined for chromosomal abnormalities or for focal mutagenic events using a biochemical or morphological marker. The tests for assessing mutagenic potential of chemicals have been extensively reviewed (111–114).

A positive result in a mutagenicity test system is not *per se* a directly useable end point in toxicity evaluation. There is general agreement that materials exhibiting a mutagenic potential, particularly by an *in vivo* approach, need to be reviewed with particular respect to their possible genetic, teratogenic, and direct carcinogenic activity. The relationship between mutagenicity tests, both *in vivo* and *in vitro*, and the ability of a chemical to produce genetically transmitted adverse effects in the progeny of exposed individuals is unclear and the subject of much debate and research. A variety of mechanisms may be concerned in the induction of teratogenic effects by differing materials, of which one is mutagenicity. Thus, a material that exhibits a mutagenic potential in appropriate tests should be suspect of being teratogenic, and appropriate laboratory studies may be required. However, it is of the utmost importance to be aware that many other mechanisms for teratogenesis exist, and a material devoid of mutagenic potential may not necessarily be devoid of teratogenic potential. Currently, perhaps the most common application of mutagenicity studies is to assess the carcinogenic potential of materials. Thus, there is a strong positive correlation between the mutagenicity of materials and their ability to cause a neoplastic response in appropriately conducted chronic toxicity studies. Also, there is a considerable body of evidence to indicate that an early critical and irreversible stage in carcinogenesis is the induction of a mutagenic event in the affected cell line. For these reasons, materials which have been shown to be mutagens are suspect of being chemical carcinogens and this may necessitate, and assign a priority for, chronic exposure studies. However, this facet of mutagenicity testing requires considerable caution in its application. There is a need to look at the effect of metabolic activation, both *in vivo* and *in vitro*; the nature of the end point indicating a mutagenic potential; the potency of a material (or metabolite) in inducing events characteristic of chemical carcinogenesis; the possible influence of route on *in vivo* tests; and to determine that any response was not due to the presence of contaminants.

Review of Toxicology Studies

The review and interpretation of toxicology studies is a professional matter, requiring experience in both the laboratory conduct of such studies and the practice of applied toxicology. Although all studies should be reviewed on a case-by-case basis, there are some general considerations to be kept in mind during the review process, described below.

The reviewer should establish that the laboratory reporting the study has the necessary scientific credibility, experience, and expertise in the area investigated. It should be confirmed that adequate quality-control facilities are in place, and good laboratory practices and procedures followed.

The objectives of the study should be precisely stated, and the work presented in a clear and coherent matter with all detail necessary to allow the reviewer to make his or her own assessment of the study. It should be confirmed that the overall design of the protocol satisfies the needs of the objectives of the study.

The material tested should be specified, including nature, relative proportions of any impurities, and stability over the test period. All details of the conduct of the study should be presented. It must be established that the methods employed for exposing and monitoring the animals are appropriate and sufficiently specific for the end points or effects planned to be studied.

Attention should be paid to the sufficiency of the study with respect to determining significance and assessing hazard; for example, whether the number of control and test animals is sufficient to allow detection of biological variability in response and for comparative statistical procedures.

There should be sufficient dose–response information to allow decisions on causal relationships and relevance.

The results of the study should allow decisions on whether injury is a direct result of toxicity or secondary to other events. In addition to confirming causal relationship between exposure to the test material and development of injury, the study should be reviewed in order to assess whether information is available to determine if the effect is due to parent material or metabolite.

In evaluating numerical information, it is important to remember that, although an effect may be statistically significant, this does not necessarily imply that the effect is of adverse biological significance. Conversely, a change or trend which is determined not to be statistically significant may be of biological consequence. Quantitative information, particularly when this involves dose–response considerations, should be reviewed against the background of the study as a whole and the perspective of normal biological variations.

Hazard Evaluation Procedures and the Role of Toxicology

Hazard is the likelihood that the known toxicity of a material will be exhibited under specific conditions of use. It follows that the toxicity of a material, ie, its potential to produce injury, is but one of many considerations to be taken into account in assessment procedures with respect to defining hazard. The following are equally important factors that need to be considered: physicochemical properties of the material; use pattern of the material and characteristics of the environment where the material is handled; source of exposure, normal and accidental; control measures used to regulate exposure; the duration, magnitude, and frequency of exposure; route of exposure and physical nature of exposure conditions (eg, gas, aerosol, liquid); population exposed and variability in exposure conditions; and experience with exposed human populations.

Considerations on the above information allow the exposure conditions to be defined and reviewed in the light of the known toxicity of the material being examined.

Relevance of Toxicology in Hazard Evaluation

Ideally, available information on the toxicology of a material should allow the following to be determined as part of a hazard evaluation procedure: nature of potential adverse effects; relevance of the conditions of the toxicology studies to the practical in-use situation; the average response, range of responses, the presence of a hypersensitive group, and an indication of minimal or no-effects levels; identification of factors likely to modify the toxic response; effects of acute gross overexposure (ie, accident situations); effects of repeated exposures; recognition of adverse effects; assistance in the definition of allowable and nonallowable exposure conditions; assistance in the definition of monitoring requirements; guidance on the need for personal and collective protection measures; guidance on first-aid, antidotal, and medical support needs; relevance of toxicity to coincidental disease; and definition of "at risk" individuals (eg, pregnant and fertile females; genetically susceptible individuals).

Information of the above type can only be obtained from carefully designed studies. In some instances, it may be economically impossible to conduct a complete spectrum of toxicology testing and, in such circumstances, it is necessary to carefully consider the most appropriate investigational approaches and their order of conduct for a hazard evaluation of a particular material under its anticipated conditions of use. The relevance and credibility of a toxicological study can be no better than its study design and conduct allow. It is of the utmost importance that meticulous detail be given in the planning of a study and preparing the protocol for that study (71).

The cost of toxicology studies varies, for example, from a few thousand dollars for simple acute toxicity tests to perhaps a million (10^6) dollars for an extensively monitored chronic inhalation study. However, the precise costs for particular studies vary with the design and content of the protocol and the laboratory conducting the study. Those needing toxicology testing of their materials should arrange for a careful and critical independent opinion on the nature of the testing required, and then obtain estimates for the timing and costing of these from several laboratories. Advice should be sought about the credibility of these laboratories and their abilities to conduct particular studies. Sponsors should arrange for independent audit of the conduct and reporting of their studies.

BIBLIOGRAPHY

"Industrial Hygiene and Toxicology" in *ECT* 1st ed., Vol. 7, pp. 847–870, by C. H. Hine, University of California, and L. Lewis, Industrial and Hygiene Associates; "Industrial Toxicology" in *ECT* 2nd ed., Vol. 11, pp. 595–610, by D. W. Fassett, Eastman Kodak Company; "Industrial Hygiene and Toxicology" in *ECT* 3rd ed., Vol. 13, pp. 253–277, by G. D. Clayton, Clayton Environmental Consultants, Inc.

1. L. J. Casarett and M. C. Bruce in J. Doull, C. D. Klaassen, and M. O. Amdur, eds., *Casarett and Doull's Toxicology*, Macmillan Publishing Co., Inc., New York, 1980, Chapt. 1.
2. *Medical and Biological Effects of Environmental Pollutants, Chlorine and Hydrogen Chloride*, National Academy of Sciences, Washington, D.C., 1976.
3. M. Sittig, *Hazardous and Toxic Effects of Industrial Chemicals*, Noyes Data Corporation, N.J., 1979, p. 39.
4. R. Morley and S. J. Silk, *Ann. Occup. Hyg.* **13,** 101 (1970).
5. R. L. Baer, D. L. Ramsey, and E. Biondi, *Arch. Dermatol.* **108,** 74 (1973).
6. *Occupational Exposure to Ketones*, Publication No. 78-173, Department of Health, Education, and Welfare, National Institute for Occupational Safety and Health, Center for Disease Control, Washington, D.C., 1978.
7. E. Browning, *Toxicity and Metabolism of Industrial Solvents*, Elsevier Publishing Co., Amsterdam, 1965, p. 304.
8. C. Zenz, ed., *Developments in Occupational Medicine*, Year Book Medical Publishers, Inc., Chicago, Ill., 1980, p. 403.
9. C. Maltoni, G. Lefemine, A. Ciliberti, G. Cotti, and D. Carretti, *Environ. Health Persp.* **41,** 3 (1981).
10. *Ann. Occup. Hygiene* **20,** 215 (1977).

11. *Occupational Exposure to Phosgene*, Publication No. 76-137, U.S. Department of Health, Education, and Welfare, National Institute for Occupational Safety and Health, Center for Disease Control, Washington, D.C., 1976.
12. C. S. Davis and R. J. Richardson in P. S. Spencer and H. H. Schaumburg, eds., *Experimental and Clinical Neurotoxicology*, Williams and Wilkins, Baltimore, Md., 1980, Chapt. 36.
13. K. Fletcher in B. Ballantyne, ed., *Forensic Toxicology*, John Wright & Sons, Ltd., Bristol, UK, 1974, p. 86.
14. S. L. Robbins, and R. S. Cotran, *Pathologic Basis of Disease*, W. B. Saunders Co., Philadelphia, Pa., 1979, Chapt. 3.
15. J. B. Walter and M. S. Israel, *General Pathology*, Churchill-Livingstone, Edinburgh, UK, 1979, Chapt. 14.
16. R. Pittelkow in C. Zenz, ed., *Occupational Medicine*, Year Book Medical Publishers, Inc., Chicago, Ill., 1975, Chapt. 12.
17. W. R. Parkes, *Occupational Lung Disorders*, Butterworths, London, 1982, Chapt. 12.
18. A. Seaton in W. K. C. Morgan and A. Seaton, eds., *Occupational Lung Diseases*, W. B. Saunders Co., Philadelphia, Pa., 1975, Chapt. 12.
19. *Occupational Exposure to Diisocyanates*, Publication No. 78-215, Department of Health, Education, and Welfare, National Institute for Occupational Safety and Health, Center for Disease Control, Washington, D.C., Sept. 1978.
20. R. E. Kouri, *Genetic Differences in Chemical Carcinogenesis*, CRC Press Inc., West Palm Beach, Fla., 1980.
21. M. Sorsa in H. Vainio, M. Sorra, and K. Hemminki, eds., *Occupational Cancer and Carcinogenesis*, Hemisphere Publishing Corporation, Washington, D.C., 1979, p. 57.
22. J. H. Weisburger and G. M. Williams in ref. 1, Chapt. 6.
23. S. H. Yuspa in D. Schottenfeld and J. F. Fraumeni, eds., *Cancer Epidemiology and Prevention*, W. B. Saunders Co., Philadelphia, Pa., 1982, Chapt. 3.
24. *Science* **218,** 975 (Dec. 3, 1982).
25. D. Brusick, *Principles of Genetic Toxicology*, Plenum Press, New York, 1980.
26. W. G. Flamm and M. A. Mehlman, eds., *Mutagenesis*, Hemisphere Publishing Corporation, Washington, D.C., 1978.
27. V. W. Mayer and W. G. Flamm in A. L. Reeves, ed., *Toxicology: Principles and Practice*, Vol. 1, John Wiley & Sons, Inc., New York, 1981.
28. J. M. Arena, *Poisoning*, Charles C Thomas, Springfield, Ill., 1976, Chapt. 2.
29. R. H. Schwarz and S. J. Yaffe, eds., *Drug and Chemical Risks to the Fetus and Newborn*, Alan R. Liss, Inc., New York, 1980.
30. M. R. Juchau, ed., *The Biochemical Basis of Chemical Teratogenesis*, Elsevier/North Holland, New York, 1981.
31. C. A. Kimmel and J. Beulke-Sam, eds., *Developmental Toxicology*, Raven Press, New York, 1981.
32. K. S. Khera, *Fund. Appl. Tox.* **1,** 13 (1981).
33. Y. Alarie, L. Kane, and C. Barrow in ref. 27, Chapt. 3.
34. Y. Alarie in B. K. J. Leong, ed., *Inhalation Toxicology and Technology*, Ann Arbor Science Publishers, Inc., Ann Arbor, Mich., 1981, p. 207.
35. G. Zbinden and M. Flury-Roversi, *Arch. Toxicol.* **47,** 77 (1981).
36. J. Doull in ref. 1, Chapt. 5.
37. C. D. Klaassen in ref. 1, Chapt. 3.
38. R. S. Chhabra, *Environ. Health Persp.* **33,** 61 (1979).
39. T. A. Loomis in V. A. Drill and P. Lazar, eds., *Current Concepts in Cutaneous Toxicity*, Academic Press, New York, 1980, p. 153.
40. D. J. Birmingham in V. A. Drill and P. Lazar, eds., *Cutaneous Toxicity*, Academic Press, New York, 1977, p. 53.
41. P. H. Dugard in F. N. Marzulli and H. I. Maibach, eds., *Dermatotoxicology and Pharmacology*, Hemisphere Publishing Corp., Washington, D.C., 1977, Chapt. 22.
42. M. G. Bird, *Ann. Occup. Hyg.* **24,** 235 (1981).
43. C. N. Davies in B. Ballantyne, ed., *Respiratory Protection*, Year Book Medical Publishers, Inc., Chicago, Ill., 1981, Chapt. 4.
44. *Airborne Particles*, Committee on Medical and Biological Effects of Environmental Pollutants, National Research Council, University Book Press, Baltimore, Md., 1979.
45. W. K. C. Morgan in ref. 18, Chapt. 3.

46. W. R. Parkes in ref. 17, Chapt. 3.
47. I. M. Asher and P. P. McGrath, eds., *Symposium on Electron Microscopy of Microfibers, Proceedings of the First FDA Office of Science Symposium*, Stock No. 017-012-00244-7, Superintendent of Documents, U.S. Government Printing Office, Washington, D.C., 1976.
48. M. S. Stanton, *J. Natl. Cancer Inst.* **48,** 797 (1972).
49. M. S. Stanton, *J. Natl. Cancer Inst.* **67,** 965 (1981).
50. J. S. Harington, *J. Natl. Cancer Inst.* **67,** 977 (1981).
51. B. Ballantyne in B. Ballantyne, ed., *Current Approaches in Toxicology*, John Wright and Sons, Ltd., Briston, UK, 1977, Chapt. 12.
52. K. J. Freundt, *Occup. Health Safety*, 10 (Aug. 1982).
53. E. J. Ariens, A. M. Simonis, and J. Offermeir, *Introduction to General Toxicology*, Academic Press, New York, 1976, Chapt. 7.
54. F. E. Guthrie in E. Hodgson and F. E. Guthrie, eds., *Introduction to Biochemical Toxicology*, Elsevier Publishing Co., New York, 1980, Chapt. 2.
55. W. B. Jakoby, R. D. Sekura, E. S. Lyon, C. J. Marcus, and J.-L. Wang in W. B. Jakoby, ed., *Enzymatic Basis of Detoxification*, Vol. 2, Academic Press, New York, 1980, Chapt. 11.
56. R. G. Killenberg and L. T. Webster, Jr., in ref. 55, Chapt. 8.
57. W. W. Weber and I. B. Glowinski in ref. 55, Chapt. 9.
58. A. de Bruin, *Biochemical Toxicology of Environmental Agents*, Elsevier Publishing Co., Amsterdam, 1976, p. 185.
59. R. O. Recknagel and L. A. Glende, *Crit. Rev. Toxicol.* **2,** 265 (1973).
60. J. A. Timbrell, *Principles of Biochemical Toxicology*, Taylor and Francis, Ltd., London, 1982, Chapt. 7.
61. R. P. Maickel and D. P. McFadden, *Res. Commun. Chem. Pathol. Pharmacol.* **26,** 75 (1979).
62. C. L. Galli, S. D. Murphy and R. Paoletti, eds., *The Principles and Methods in Modern Toxicology*, Elsevier/North Holland, Amsterdam, 1980.
63. E. J. Gralla, ed., *Scientific Considerations in Monitoring and Evaluating Toxicological Research*, Hemisphere Publishing Corp., Washington, D.C., 1981.
64. *Chem. Eng. News* **61**(4), 42 (1983).
65. S. R. M. Bushby in G. E. Paget, ed., *Methods in Toxicology*, Blackwell Scientific Publications, Oxford, UK, 1970, Chapt. 13.
66. A. E. Street in ref. 65, Chapt. 12.
67. D. T. Plummer in J. W. Gorrod, ed., *Testing for Toxicity*, Taylor and Francis, London, 1981, Chapt. 12.
68. *Principles and Methods for Evaluating the Toxicity of Chemicals, Part 1, Environmental Health Criteria No. 6*, World Health Organization, Geneva, 1978, Chapt. 5.
69. D. Poynter in ref. 51, Chapt. 8.
70. B. Ballantyne in B. Ballantyne, ed., *Respiratory Protection*, Chapman and Hall, London, 1981, Chapt. 5.
71. H. N. MacFarland in W. J. Hayes, ed., *Essays in Toxicology*, Vol. 7, 1976, p. 121.
72. B. P. McNamara in M. A. Mehlman, R. E. Shapiro, and H. Blumenthal, *New Concepts in Safety Evaluations*, Vol. 1, Hemisphere Publishing Corp., Washington, D.C., 1976, Part 1, Chapt. 4.
73. R. P. Giovacchini in ref. 40, p. 31.
74. P. S. Elias in ref. 62, p. 169.
75. *Principles and Procedures for Evaluating the Toxicity of Household Substances*, National Academy of Sciences, Washington, D.C., 1977, Chapt. 2.
76. R. Heywood in ref. 67, Chapt. 17.
77. T. O. McDonald and J. A. Shadduck in ref. 41, Chapt. 4.
78. A. M. McCreesh and M. Steinberg in ref. 41, Chapt. 5.
79. A. B. Lansdown, *J. Soc. Cosmet. Chem.* **23,** 739 (1972).
80. W. E. Parish in ref. 67, Chapt. 20.
81. F. Marzulli and H. C. Maguire, Jr., *Fd. Chem. Toxic.* **20,** 67 (1982).
82. G. Klecak in ref. 41, Chapt. 9.
83. M. H. Karol in ref. 34, p. 233.
84. S. C. Gad, *J. Toxicol. Environ. Health* **9,** 691 (1982).
85. H. A. Tilson, P. A. Cabe, and T. A. Burne in ref. 12, Chapt. 51.
86. S. Norton, *Environ. Health Persp.* **26,** 21 (1978).
87. P. S. Spencer, M. C. Bischoff, and H. H. Schaumburg in ref. 12, Chapt. 50.

88. T. Damstra and S. C. Bondy in ref. 12, Chapt. 56.
89. A. J. Dewar in ref. 67, Chapt. 15.
90. C. L. Mitchell, ed., *Nervous System Toxicology*, Raven Press, New York, 1982.
91. L. Manzo, ed., *Advances in Neurotoxicology*, Pergamon Press, Oxford, UK, 1980.
92. K. Presad and A. Vernadakis, eds., *Mechanisms of Action of Neurotoxic Substances*, Raven Press, New York, 1982.
93. I. Geller, W. C. Stebbins, and M. J. Wayner, eds., *Test Methods for Definition of Effects of Toxic Substances on Behavior and Neuromotor Function, Neurobehavioral Toxicology*, Vol. 1, Suppl. 1, 1979.
94. B. Weiss and V. G. Laties, eds., *Behavioural Toxicology*, Plenum Press, New York, 1975.
95. B. Ballantyne, M. F. Gazzard, and D. W. Swanston in ref. 51, Chapt. 11.
96. B. Ballantyne and D. W. Swanston, *Acta Pharmacol. Toxicol.* **35**, 412 (1974).
97. E. J. Owens and C. L. Punte, *Am. Ind. Hyg. Assoc. J.* **24**, 262 (1963).
98. *TLVs for Chemical Substances in Workroom Air Adopted by ACGIH for 1981*, Publications Office, American Conference of Governmental Industrial Hygienists, Cincinnati, Ohio.
99. T. F. X. Collins and E. V. Collins in ref. 72.
100. J. L. Schardein, *Drugs as Teratogens*, CRC Press, Cleveland, Ohio, 1976.
101. D. Neubert, H.-J. Merker, and T. E. Kwasigroch, eds., *Methods in Prenatal Toxicology*, Georg Thieme Publishers, Stuggart, FRG, 1977.
102. H. B. Pace, W. M. Davis, and L. A. Borgen, *Ann. N.Y. Acad. Sci.* **191**, 123 (1971).
103. I. Coyle, M. J. Wagner, and G. Singer, *Pharmacol. Biochem. Behav.* **4**, 191 (1976).
104. K. S. Rau and B. A. Schwetz in L. Breslow, ed., *Annual Review of Public Health*, Vol. 3, 1982, p. 1.
105. J. P. Griffin in ref. 51, Chapt. 4.
106. I. C. Munro in ref. 62, p. 125.
107. "*Chemobiokinetics and Metabolism,*" *Principles and Methods for Evaluating the Toxicity of Chemicals*, Part 1, Environmental Health Criteria No. 6, World Health Organization, Geneva, 1978, Chapt. 4.
108. D. B. Tuey in E. Hodgson and F. E. Guthrie, eds., *Introduction to Biochemical Toxicology*, Elsevier Publishing Co., New York, 1980, Chapt. 3.
109. P. J. Gehring, P. G. Watenabe, and G. E. Blau in ref. 72, Chapt. 8.
110. J. McCann and B. N. Ames in ref. 26, Chapt. 5.
111. D. Brusick, *Principles of Genetic Toxicology*, Plenum Press, New York, 1980.
112. F. Vogel and G. Rohrborn, eds., *Chemical Mutagenesis in Mammals and Man*, Springer-Verlag, Berlin, 1970.
113. B. J. Kilbey, M. Legator, W. Nicols, and C. Ramel, eds., *Handbook of Mutagenicity Test Procedures*, Elsevier Scientific Publishing Company, Amsterdam, 1977.
114. A. W. Hsie, J. P. O'Neill, and V. K. McElheney, eds., *Mammalian Cell Mutagenesis: The Maturation of Test Systems*, Banbury Report No. 2, Cold Spring Harbor Laboratory, Cold Spring Harbor, New York, 1979.

General References

References 1, 3, 60, 67, 68, and 108 are also general references.

T. A. Loomis, *Essentials of Toxicology*, Lea and Febiger, Philadelphia, Pa., 1974.

V. K. Brown, *Acute Toxicity in Theory and Practice*, John Wiley & Sons, Inc., Chichester, UK, 1980.

E. Boyland in R. S. F. Schilling, ed., *Occupational Health Practice*, Butterworths, London, 1981, Chapt. 24.

M. A. Cooke in A. W. Gardner, ed., *Current Approaches to Occupational Medicine*, John Wright and Sons, Ltd., Bristol, UK, 1979, Chapt. 9.

C. R. Richmond, P. J. Walsh, and E. C. Copenhaver, eds., *Health Risk Analysis*, The Franklin Institute Press, Philadelphia, Pa., 1980.

N. H. Proctor and J. P. Hughes, *Chemical Hazards of the Workplace*, J. B. Lippincott Co., Philadelphia, Pa., 1978.

Registry of the Toxic Effects of Chemical Substances, Publication No. 81-116, Vols. 1 and 2, National Institute for Occupational Safety and Health, Superintendent of Documents, U.S. Government Printing Office, Washington, D.C., 1980.

G. D. Clayton and F. E. Clayton, eds., *Patty's Industrial Hygiene and Toxicology*, Vols. 2A, 2B, and 2C, John Wiley & Sons, Inc., New York, 1981.

W. M. Grant, *Toxicology of the Eye*, Charles C Thomas, Springfield, Ill., 1974.
E. Cronin, *Contact Dermatitis*, Churchill Livingstone, Edinburgh, UK, 1980.
L. Fishbein, *Potential Industrial Carcinogens and Mutagens*, Elsevier Scientific Publishing Co., Amsterdam, 1979.
Toxicology Abstracts, Cambridge Scientific Abstracts, Bethesda, Md. (monthly publication).
Industrial Hygiene Digest, Industrial Health Foundation, Pittsburgh, Pa. (monthly publication).
V. M. Traina, *Medicinal Research Reviews* **3**(1), 43 (1983).

BRYAN BALLANTYNE
Union Carbide Corporation